HANDBOOK OF SPACE ENGINEERING, ARCHAEOLOGY, AND HERITAGE

ADVANCES IN ENGINEERING

A SERIES OF REFERENCE BOOKS, MONOGRAPHS, AND TEXTBOOKS

Series Editor

Haym Benaroya

Department of Mechanical and Aerospace Engineering
Rutgers University

Published Titles:

Handbook of Space Engineering, Archaeology and Heritage, *Ann Garrison Darrin and Beth Laura O'Leary*

Spatial Variations of Seismic Ground Motions: Modeling and Engineering Applications, *Aspasia Zerva*

Fundamentals of Rail Vehicle Dynamics: Guidance and Stability, *A. H. Wickens*

Advances in Nonlinear Dynamics in China: Theory and Applications, *Wenhu Huang*

Virtual Testing of Mechanical Systems: Theories and Techniques, *Ole Ivar Sivertsen*

Nonlinear Random Vibration: Analytical Techniques and Applications, *Cho W. S. To*

Handbook of Vehicle-Road Interaction, *David Cebon*

Nonlinear Dynamics of Compliant Offshore Structures, *Patrick Bar-Avi and Haym Benaroya*

Upcoming Titles:

Lunar Settlements, *Haym Benaroya*

HANDBOOK OF SPACE ENGINEERING, ARCHAEOLOGY, AND HERITAGE

EDITED BY
ANN GARRISON DARRIN
BETH LAURA O'LEARY

CRC Press is an imprint of the
Taylor & Francis Group, an **informa** business

CRC Press
Taylor & Francis Group
6000 Broken Sound Parkway NW, Suite 300
Boca Raton, FL 33487-2742

Printed in the United States of America on acid-free paper
10 9 8 7 6 5 4 3 2 1

International Standard Book Number: 978-1-4200-8431-3 (Hardback)

Library of Congress Cataloging-in-Publication Data

Handbook of space engineering, archaeology, and heritage / editors, Ann Garrison Darrin and Beth Laura O'Leary.
p. cm. -- (Advances in engineering)
"A CRC title."
Includes bibliographical references and index.
ISBN 978-1-4200-8431-3 (hard back : alk. paper)
1. Space archaeology. 2. Space vehicles. 3. Space debris. 4. Space environment. 5. Aerospace engineering--History. 6. Cultural property. I. Darrin, Ann Garrison. II. O'Leary, Beth Laura. III. Title. IV. Series.

TL788.6.H36 2009
629.4--dc22 2009015980

Visit the Taylor & Francis Web site at
http://www.taylorandfrancis.com

and the CRC Press Web site at
http://www.crcpress.com

Contents

SECTION I All Sky Survey

SECTION VIII Space Policy and Preservation

SECTION IX The Future and Space Archaeology

SECTION X The Mind and the Cosmos

Preface

Since 2003, the terms *space archaeologists* and *space heritage* have been used regularly at the World Archaeological Congress in its annual meetings, at the Society for American Archaeology, and at the International Committee on Monuments and Sites in Australia.

The definition of space archaeology is *the archaeological study of material culture related to space exploration that is found on Earth and in outer space, that is, exoatmospheric material that is clearly the result of human behavior.* This includes all material culture in the aerospace and aeronautical realms that relates to the historical development and support of exoatmospheric activities. Space archaeology is a subdiscipline within archaeology, which also comprises other specialized fields, such as maritime or underwater archaeology and cultural resource/heritage management.

This book is an ambitious project to set a strong foundation in the social and physical sciences for future studies. Our space heritage is vast, recent, and ongoing, and one work can only begin to do the field justice. It has also been a challenge to join anthropologists, historians, physicists, and engineers in the same work, but the reward was worth it. The authors have brought together a strong interdisciplinary approach grounded in the sciences and hope you will find it informative, thought provoking, educational, and enjoyable.

Ann Garrison Darrin
The Johns Hopkins University Applied Physics Laboratory
Laurel, Maryland

Beth Laura O'Leary
New Mexico State University
Las Cruces, New Mexico

Acknowledgments

The editors cannot express enough love and gratitude for their respective spouses, Armond Darrin and John Hyndman, for their enduring support and patience. Ann also thanks her (grown) kids. Beth thanks her friends, who always listened. Ann has been fortunate to have the support of her institution, The Johns Hopkins University Applied Physics Laboratory. Beth acknowledges the support of New Mexico State University. Also, the editors thank their colleagues who diligently read and discussed ideas, many at critical points in the evolution of the book.

This book, like so many good ideas, started over shared drinks among friends. Ann Darrin's friendship with Hanna Szczepanowska began through Hanna's son, Rafal. When the concept of the book was in its infancy, Ann knew she needed the social sciences lead and contacted Beth O'Leary, an archaeologist who had led the Lunar Legacy Project. Beth and Ann began a fast long-distance relationship that blossomed into a friendship over the book. They met in person for the first time well into the first draft at Beth's colloquium in May 2008 at The Johns Hopkins University Applied Physics Laboratory. It was a joy to work with Jonathan Plant of Taylor & Francis; the editors' special thanks go out to him. Patricia M. Prettyman was one of the key individuals in the completion of the book. Beth, Ann, and Trisha cannot count the many hours invested but truly enjoy their ongoing friendship. Working with 43 contributors is like nailing Jell-O to a tree!

This book is dedicated to Rafal and to all of the most wonderful contributors.

It is all about relationships.

Editors

ANN GARRISON DARRIN

Ann Garrison Darrin has worked at The Johns Hopkins University Applied Physics Laboratory for more than 10 years. Ann Darrin is a member of the principal staff and is the manager in the Milton S. Eisenhower Research Center for the Aerospace and Materials Sciences group. Prior to joining the laboratory, Ann worked at NASA Goddard Space Center in aerospace engineering and was the Division Chief for Assurance Technologies. She is the author of numerous papers and an author and editor of the book *MEMS and Microstructures for Aerospace Applications*. As a technologist, Ann holds numerous patents and has participated in several exciting, albeit small, technology "firsts" in space. Ann is the founder and cochair of the MEMS Alliance Mid-Atlantic and holds degrees from the Pennsylvania State University and the University of Maryland University College.

BETH LAURA O'LEARY

Beth Laura O'Leary has taught anthropology at New Mexico State University since 1991. She has worked in cultural resource management for the federal government, private firms, and universities. She is currently vice chairperson of the Cultural Properties Review Committee, a governor-appointed policymaking board on historic preservation for the state of New Mexico. Beth is a recognized expert in the emerging field of space archaeology and heritage. In 1999, she received a grant from the New Mexico Space Grant Consortium to document the archaeological assemblage at the *Apollo 11* lunar landing site and to investigate ways to manage and preserve it for the future. She has cochaired several international symposia on space heritage in the United States, Australia, Canada, and Ireland and is a member of the World Archaeological Congress Space Heritage Task Force. Beth has a BA from Mount Holyoke College and a PhD in anthropology from the University of New Mexico.

Contributors

Robert Barclay
Conservation Consultant
Ottawa, Ontario, Canada

Janet L. Barth
NASA Goddard Space Flight Center
Greenbelt, Maryland

Paul J. Biermann
The Johns Hopkins University Applied Physics Laboratory
Laurel, Maryland

Bradley G. Boone
The Johns Hopkins University Applied Physics Laboratory
Laurel, Maryland

Randall C. Brooks
Canada Science and Technology
Ottawa, Ontario, Canada

John B. Campbell
James Cook University
Queensland, Australia

P. J. Capelotti
Pennsylvania State-Abington
Abington, Pennsylvania

Daniel E. Clemens
The Johns Hopkins University Applied Physics Laboratory
Laurel, Maryland

Ann Garrison Darrin
The Johns Hopkins University Applied Physics Laboratory
Laurel, Maryland

William Doleman (retired)
New Mexico State Historic Preservation
Sante Fe, New Mexico

Michelle M. Donegan
The Johns Hopkins University Applied Physics Laboratory
Laurel, Maryland

Stephen E. Doyle
International Institute of Space Law
Shingle Springs, California

Jerold Emhoff
The Johns Hopkins University Applied Physics Laboratory
Laurel, Maryland

R. Michael Furrow
The Johns Hopkins University Applied Physics Laboratory
Laurel, Maryland

Roger Gerke
New Mexico State University
Las Cruces, New Mexico

Robert Gold
The Johns Hopkins University Applied Physics Laboratory
Laurel, Maryland

Alice Gorman
Flinders University
Adelaide, South Australia

Jeremy Hansen
The Johns Hopkins University Applied Physics Laboratory
Laurel, Maryland

William S. Heaps
NASA Goddard Space Flight Center
Greenbelt, Maryland

Barbara Leary
The Johns Hopkins University Applied Physics Laboratory
Laurel, Maryland

Ralph D. Lorenz
The Johns Hopkins University Applied Physics Laboratory
Laurel, Maryland

Theresa McCracken
Cartoonist
Waldport, Oregon

Ralph L. McNutt, Jr.
The Johns Hopkins University Applied Physics Laboratory
Laurel, Maryland

Douglas Mehoke
The Johns Hopkins University Applied Physics Laboratory
Laurel, Maryland

Thomas S. Mehoke
The Johns Hopkins University Applied Physics Laboratory
Laurel, Maryland

Anne Millbrook
Historian and Author
Bozeman, Montana

Susan Milbrath
Florida Museum of Natural History
Gainesville, Florida

Guy Murphy
Mars Society Australia
Victoria, Australia

Beth Laura O'Leary
New Mexico State University
Las Cruces, New Mexico

Greg Orndorff
The Johns Hopkins University Applied Physics Laboratory
Laurel, Maryland

Robert Osiander
The Johns Hopkins University Applied Physics Laboratory
Laurel, Maryland

Paul Ostdiek
The Johns Hopkins University Applied Physics Laboratory
Laurel, Maryland

Larry J. Paxton
The Johns Hopkins University Applied Physics Laboratory
Laurel, Maryland

Patricia M. Prettyman
The Johns Hopkins University Applied Physics Laboratory
Laurel, Maryland

Cheryl L. B. Reed
The Johns Hopkins University Applied Physics Laboratory
Laurel, Maryland

Aaron Q. Rogers
The Johns Hopkins University Applied Physics Laboratory
Laurel, Maryland

Jennifer L. Sample
The Johns Hopkins University Applied Physics Laboratory
Laurel, Maryland

Dirk H. R. Spennemann
Charles Sturt University
Albury, Australia

Patrick A. Stadter
The Johns Hopkins University
Applied Physics Laboratory
Laurel, Maryland

Edward Staski
New Mexico State University
Las Cruces, New Mexico

Philip J. Stooke
University of Western Ontario
London, Ontario, Canada

Richard Sturdevant
Air Force Space Command
Peterson Air Force Base
Colorado Springs, Colorado

Theodore S. Swanson
NASA Goddard Space Flight Center
Greenbelt, Maryland

Hanna Szczepanowska
National Air and Space Museum
Smithsonian Institute
Washington, DC

Section I

All Sky Survey

Archaeologist meets Aerospace Engineer:
Match made in Heaven?

1 Introduction

Ann Garrison Darrin and Beth Laura O'Leary

CONTENTS

"Physical or Social Science?"

Developing a handbook of space engineering, archaeology, and heritage from an interdisciplinary approach is a fascinating and complex journey in interpretation, juxtaposition, and parallelism. Unique and varying perspectives on our exploration of space are brought together in one conjoined document. This introduction discusses the parallel universes of the editors and authors as engineers, archaeologists, scientists, historians, conservationists, anthropologists, and heritage/cultural resource managers as they have come together to meld their unique insights into one compilation about space. Early in the discussion it became clearly evident that even the usage of terms and expressions varied greatly among this diverse community as we all jointly explored the same topic areas. The *Handbook* is divided into ten sections, each drawn along a unique theme. The goal was to let the perspectives of the contributors contrast and complement each other with the hope that the "hard" and "soft" sciences would merge in an intra- and interdisciplinary approach. The variations and similarities in viewpoints from the engineering/physical scientist versus the archaeologist/social scientist perspective are discussed.

One of the general areas of agreement is that the material record of space is important. For many of the engineers, the design and production of items destined for space exploration was a long labor and represents major parts of their careers. The artifacts may represent a test of theories created by the physicists. The intellectual challenges are manifest in what they helped to build and launch. For the archaeologist, those items represent a way of understanding and valuing the Space Age. Their significance is judged within the historic context of their creation. For example, the location of one human footprint on the Moon in situ is an archaeological feature that defines one of the ultimate events of humankind and is a part of the physical and temporal record of all the engineering research and development that made it possible. To the preservationist, important artifacts need to be conserved in museums. To the cultural resource and heritage managers, legal ways need to be found to protect and preserve artifacts, or at the very least the information about the artifacts that can answer important questions about history. Both groups agree on the significance and value of documenting space exploration.

WHAT IS SPACE ARCHAEOLOGY?

In the next few chapters there is more detail on the field of space archaeology. The subdiscipline is so new that there is still discussion on what it should be called. There are various usages of the terms *aerospace archaeology* and *exoarchaeology*, as well as facets or areas of study within space archaeology called *space heritage* and the cultural landscape of space, or *spacescape*. The main area of research, whatever the term, is clearly within the field of archaeology. It is reliant on archaeological theory and method, especially within the framework of behavioral archaeology. Traditionally, archaeology is part of the larger discipline of anthropology. Anthropology is a wide, inclusive discipline that embraces the study of humankind in all places and at all times. As defined by Staski (Chapter 3 of this volume) and others, archaeology is the study of relationships between material culture and human behavior. There are no temporal limits on legitimate archaeological research: the immediate present and recent past can be studied, as well as the distant past. Archaeology has no spatial limits; it can be done anywhere on earth, as well as off-earth. This handbook takes the broadest definition that space archaeology is the archaeological study of material culture relevant to space exploration that is found on earth and in outer space (i.e., exoatmospheric material) and that is clearly the result of human behavior. This definition expands to include all material culture in the aerospace and aeronautical realms that relate to the development and support of exoatmospheric activities: everything from launch pads to satellites to landers on Mars.

This *Handbook* has a section on archaeoastronomy, or the study of the material culture of the traditional astronomical practices and related beliefs of cultures (e.g., Stonehenge in Great Britain proved to be a prehistoric astronomical observatory with both cultural and social meanings). Because so much of human cultural behavior is based on belief (some would even argue that scientific method is a cultural belief system), this *Handbook* includes cultural anthropology. There is a discussion of ethnoastronomy, which is the study of human belief systems as they relate to cosmologies or the study of traditional astronomical beliefs and practices among indigenous cultures (Milbrath, Chapter 10 of this volume).

The *Handbook* also moves into speculative realms. One of the most intriguing studies (Doleman, Chapter 46 of this volume) is an archaeological investigation of the purported Roswell UFO crash site. The most speculative material in the handbook is exoarchaeology, which concerns itself with the role archaeologists might play in understanding the material culture of extraterrestrials or other intelligent species, once the physical evidence is scientifically verifiable (Campbell, Chapter 38 of this volume). This is followed by a discussion of the realm of science fiction.

INTERDISCIPLINARY APPROACH TO SPACE ARCHAEOLOGY

Given the broad scope of the discipline of archaeology, there is a need for an intra- and interdisciplinary approach to research. Traditionally, archaeology draws upon many fields besides anthropology: history, art history, linguistics, classics, ethnology, ethnography, geography, geology, geoarchaeology, physics, information sciences, chemistry, statistics, palynology, paleoecology, paleontology, zooarchaeology, preservation, and ethnobotany. With space archaeology we have included aerospace disciplines, including thermal, mechanical, and electrical systems; propulsion; and chemical, material, and aerospace engineering. Physicists participating in this work included those individuals with a research emphasis in near-Earth and deep space, along with the traditional experimentalists.

CULTURAL LANDSCAPES VERSUS OBJECT PERSPECTIVES

The cultural landscape of space, or spacescape, includes three principal categories as reflected in Gorman (Chapter 16 of this volume). First, the spacescape is a designed and intentionally created landscape; second, it is organically evolving from human interactions both past and ongoing; third, it is an associative cultural landscape that reflects the relationship of humans to an area often beyond the material evidence, in narrative, the arts, religions, and ceremonial behavior.

From an engineering perspective there are systems of systems rather than spacescapes. Systems of systems group like objects and activities by technology or by mission type. To the engineer, a material object destined for space, be it a probe, orbiter, station, or voyager, is considered a system. The system is composed of subsystems. If it is large and complex, engineers refer to systems of systems. The systems-of-systems view dominates when one considers constellations of communication, weather, earth observing (i.e., spying), and other complex activities. It is important to understand this building-block perspective because the engineer will see an object (subsystem) and be prone to look at the specific set of interests at a particular time. Why is the command and data handling system not sending a specific command? Why is the secondary battery not charging at the predicted rate? These object views are predicated on the question "Is the object acting in the manner or with the value as designed and predicted, perhaps through a model?"

AN OVERVIEW OF THE NAVY NAVIGATION SATELLITE SYSTEM

Transit, pictured on the cover,[1] had its inception just days after the launch of *Sputnik* on October 4, 1957. Choosing an image of the material record of space was difficult, but the editors decided on Transit as one of the most important early satellites at the beginning of the Space Age. Two scientists of The Johns Hopkins University Applied Physics Laboratory (JHUAPL)—George Weiffenbach and William Guier—were able to determine Sputnik's orbit by analyzing the Doppler shift of its radio signals during a single pass. Frank McClure, then chairman of JHUAPL's Research Center, went a step further by suggesting that if the satellite's position was known and predictable, the Doppler shift could be used to locate a receiver on Earth; in other words, one could navigate by satellite.[2] The Transit system was the precursor to today's global positioning systems. The plucky little Transits represent a microcosm of Space Age development.

With early funding from the Advanced Research Projects Agency, the development of the system began at JHUAPL in 1958, and the U.S. Navy assumed responsibility for the system in 1960. By the end of 1964 JHUAPL had designed, built, and launched fifteen navigation satellites and eight related research satellites and had established a worldwide network of tracking stations. The first prototype satellite was launched in September 1959, and the system entered Navy service in 1964. The system provided passive, accurate, reliable, all-weather global navigation for Navy submarines and surface ships. In the summer of 1967, Vice President Hubert Humphrey announced that the system was available to commercial ships and aircraft of all nations. The Transit constellation consisted of two types of spacecraft, designated Oscar (left) and Nova (right). The final constellation consisted of six satellites (all Oscars) in a polar orbit with a nominal 600-nautical-mile altitude, three ground control stations, and receivers (i.e., the system's users). Of the six satellites, three Oscars provided navigation service, whereas three other Oscars were stored-in-orbit or in situ spares. Today, the legacy of Transit may be seen in the application of many of its innovations and discoveries to scientific and engineering endeavors and to products that benefit humankind in at least three areas: space technology systems, science systems, and medical innovations.

Space Technology Applications

Transit added to great advances in the knowledge of Earth's gravitational field, the first measurements of low-energy protons in the polar regions and the discovery of auroral field-aligned currents, and it detected and measured the artificial radiation belt created by the Starfish Prime high-altitude nuclear test in 1962. This early work led to programs for NASA and other agencies.

Historic Artifact

Transit technology spurred the development a rechargeable cardiac pacemaker, programmable implantable medication system, and automatic implantable defibrillators.

The last Transit satellite launch was in August 1988. At the end of 1996, Transit, the Navy Navigation Satellite System, was retired after more than 32 years of continuous, successful service to the U.S. Navy. Some of the satellites are still in use and transmitting under the Navy Ionospheric Monitoring System (NIMS). Dozens of the satellites have deorbited, but dozens more are still silently orbiting, some for more than 45 years. As space debris, they help to create a landscape of true significance to our terrestrial-based lives.

Transit as material culture is part of the larger cultural landscape of space. It has significance in the history of space exploration and development. It can be studied and evaluated for its contribution to the evolution of science in the twentieth and twenty-first centuries, as well as for its heritage value. Transit is an important part of humankind's technological heritage, and its current status as debris should be questioned when considering what should be preserved for future generations. Transit is a great place to begin to look at space engineering, archaeology, and heritage. Through Transit we see great innovations that have affected every space system since the early 1960s; we can see our human presence in space through our materials, respective innovations, and the Cold War rapid military investment that spun off civilian applications. A discussion of Transit also yields an understanding of the space engineering and archaeological terminology. Words such as *constellations*, *in situ*, and *heritage* have different meanings depending on the usage.

TOWARD A COMMON TERMINOLOGY

A building-block perspective dominates for engineers and contrasts with the archaeological and scientific perspectives. Without offending the scientific contributors to this work, it is important to note their very wide-screen perspective, with a broad view of spatial and temporal constraints. An artifact, to the engineer, may be the smallest object that can be often be examined for why predicted and actual behavior differ. A ding on the space shuttle's windshield, for example, would be an artifact indicating that the shield had encountered a small piece of debris, perhaps a paint flake.

For the space archaeologist, an artifact is any portable object used, modified, or made by humans.[3] It does not have to be considered complete or whole for it to have meaning. Artifacts in the archaeological record are rarely complete, and most sites and features at sites are not pristine and perfectly preserved but exhibit complex patterning in terms of their location, reuse, and changes from environmental factors (i.e., erosion, etc.). For example, a primary flake and the stone core from which it was struck are the both by-products of behavior. The flake is from when the stone tool was being made, and the core is an artifact that has evidence of the stage of lithic reduction that was reached before it was discarded. To use an example from space archaeology, the discarded space boots of Neil Armstrong at the *Apollo 11* lunar landing site are artifacts. The footprints made by the space boots represent features at the site where the astronauts landed and took off from another celestial body for the first time in human history. The trails that the astronauts created are part of the material record of space. To use the artifacts mentioned above in the engineer's perspective, the shuttle windshield in its dinged condition and the debris (i.e., the paint flake) that caused it are both artifacts. An archaeologist seeks to describe the material type and condition of these artifacts and their patterning at a site in reference to both the human behavior that created them and the natural processes that may have had an effect(s) on them. It is the relationship between the artifact or feature and human behavior that interests an archaeologist. The archaeologist is looking for the spatial and temporal patterning in the material record in order to understand the behavior that created it and to further evaluate its significance.

The different perspectives of the hard and soft sciences when viewing humankind's material record in space is further noted in the terminology section below. Again, the subsystem approach of the engineer contrasts with the holistic approach of the archaeologists and is reflected in word usage. Often a word will have synchronous meanings, such as *environment,* but other terms, such as *heritage* and *artifact*, will have totally asynchronous usages. Some terminology differences are not as obvious and are influenced by perspective. When viewing the environment, the engineer or scientist will again resort to a subset viewpoint and speak of the radiation environment, the thermal environment, the launch environment and so forth. The view of environment as an aggregate of all of these elements is seen when we say "space environment," which then would include conditions and influences.

The archaeologist sees an aggregate environment containing the surroundings, conditions, or influences that include the interactions of humans with the archaeological record. Environments change from the past to the present. An archaeologist may seek to reconstruct grasslands near playas in the Pleistocene where a prehistoric site was found by looking at the pollen, soils, and topography. The grasslands may now be gone, but through an examination of pollen (palynology) retrieved at a site, a clear picture of the past climate and vegetation can be drawn. Here there are some interesting threads. Evidence of human interactions in the past does not always wait dormant in space, just as it does not on earth. Debris fields caught in Earth's magnetic field are evolving trails of what happened before and continues to happen as societies contribute to a material presence in space. Sites in space are multicomponent in an archaeological sense in that they represent multiple layers of use. When we look at the material record it is stagnant, as the past and present are both elements of the spacescape.

The term *heritage* is used by the space engineer or scientist relative to hardware. Heritage hardware is a space engineering concept driven by the constant fear of failure and risk, or more politely put, risk mitigation. Heritage hardware uses known designs that have been built, tested, and flown before and are proven space hardy. In order to reduce the amount of risk, the cardinal rule of space hardware is "Test as it will be used." Although this principle seems obvious, violations of it open the doors to risk and, in many cases, mission failures. "Testing as it will be used" is difficult and costly. On Earth, we cannot pull the hard vacuum of space and must approximate it. For example, the difficulty in emulating a reduced gravity environment or cosmic radiation makes Earth-based testing inherently inadequate. Complex systems are especially costly and time-consuming to test. To mitigate risk, the program manager turns to heritage hardware. This is the dilemma: the desire to fly only that which has flown before means that when one inserts new technology or upgrades systems, flying even relatively current technology systems is often perceived as too much risk.

Heritage for the archaeologist is a complex philosophical idea. It is used most commonly outside the U.S. Heritage belongs to a group of people, a nation, a population, or an individual. Its most basic definition is that which is or may be inherited. It refers to the artifacts, structures, features, and sites that exist in any region or for any part of history that has a material referent and also those that do not. In the United States, cultural resources are usually a synonym for heritage. There is said to be heritage management or cultural resource management. Because it is thought to be able to be managed, it has to be evaluated for its significance and appropriate actions taken to preserve it. One example of a heritage scenario is Antarctica, which is managed by treaty and agreements of several nations. Are the ruins of a research station or remote huts in Antarctica from ca. 1912 places to be preserved? What are the impacts to them in terms of weather, temperature, and tourism? What are the costs of investigating them, their condition, and their conservation needs? How does one weigh the competing priorities to preserve them or treat them as waste? The cultural heritage of space must be decided upon for the future or it stands to be lost through decay, destruction, or neglect.

TOWARD COMMON TERMINOLOGIES AND TAXONOMIES

The following table contains some of the most relevant terms used in this book. Appendix A includes a more complete glossary and guide to the use of terms.

	Engineering/Physical Science: Common Aerospace Usage	**Social Sciences: Archaeological Context Usage**
Artifact	Material evidence of an event. Any feature that is not naturally present but is a product of an extrinsic agent, or method; (chemistry) a substance or structure not naturally present in the matter being observed but formed by artificial means, as during preparation of a microscope slide.	Any portable object used, modified, or made by humans; e.g., stone projectile points, ceramics, or solar panels.

(*Continued*)

	Engineering/Physical Science: Common Aerospace Usage	Social Sciences: Archaeological Context Usage
Assemblage	Not a term used. Closest definition alignment might be *subsystem* or *system* depending on context.	A group of artifacts recurring together at a particular time and place and representing the sum of human activities.
Constellation	A group of stars that make a shape, often named after mythological characters, people, animals, and things. Most often used with weather, navigation or communication satellite constellations.	A group of stars that make a shape, often named in human cultures after mythological characters, people, animals, and narratives.
Decay	To decline from a sound or prosperous condition; **orbital decay** is the process of prolonged reduction in the height of a satellite's orbit. "Decayed" means an item has fully deorbited.	To decline from a sound condition, to fall into ruin or to undergo decomposition. Also used in radiocarbon dating, which is an absolute dating method that measures the decay of the radioactive isotope of carbon (C^{14}) in organic material.
Feature	Not a term used regularly in aerospace except related to geology or geography.	A nonportable artifact, such as hearths, architectural elements, and footprints.
Heritage	Used in *heritage hardware*: hardware that has been demonstrated spaceworthy.	The material and nonmaterial record inherited from the past, usually synonymous with cultural resources in the United States; implies significance and that it can be managed.
Impact	Physical site of any object that has touched another object.	Something that can affect the qualities of a site, artifact, or features, usually adversely.
In Situ	In place; in situ processing, e.g., data processing on board a spacecraft.	In the original place. Usually concerns artifacts that are nonportable, e.g., a hearth or soil stains.
Preservation	The term *passivation* is widely used here. To coat (a semiconductor, e.g.) with an oxide layer to protect against contamination and increase electrical stability.	(Also called historic preservation.) Includes the identification, evaluation, recordation, documentation, curation, acquisition, protection, management, rehabilitation, restoration, stabilization, maintenance, research, interpretation, conservation, and education and training regarding the foregoing activities or any combination of the above. Also the act or process of applying measures necessary to sustain the existing form, integrity and materials of a prehistoric or historic property.[4]
Legacy	A term used to indicate space materials that have a known flight history.	Material culture that is handed down through generations. Usually material of value (e.g., the lunar legacy of the first human landing site on the Moon).

OVERVIEW AND LIST OF CONTRIBUTORS

The following provides a guide to the organization and those involved in writing chapters for the *Handbook*.

Part I is called "All Sky Survey" because it provides the basics and foundation for further discussion relative to space engineering, archaeology, and heritage. It is a view of the sky from the perspective of archaeologists and physical scientists. This first part is a trilogy:

- Introduction to the discipline of archaeology and how space archaeology and heritage have evolved;
- Introduction to space basics, a grounding in what one should know about space; and
- Introduction to space flight, how to get to and stay in space.

Dr. Edward Staski, professor of anthropology and former director of the University Museum at New Mexico State University, provides a basic definition of archaeology as the study of the relationships between material culture and human behavior, focusing on the intriguing idea that there are no temporal limits on archaeology. The very recent past can be studied, as well as the ancient past, and because there are no spatial limitations, it can be done on Earth and in outer space. He introduces the term *space archaeology* as it is used in this book. Coeditor and chapter contributor **Dr. Beth Laura O'Leary** explains how the fledging field of space archaeology and heritage is evolving, who the players are, and various areas of expertise in management and preservation. Dr. O'Leary is a professor of anthropology at New Mexico State University (NMSU) and led the NASA-funded project to recognize the 1969 *Apollo 11* Moon landing site as a significant historic property under federal preservation law. This introductory area leads to an overview of the basics of space. **Dr. Jerold Emhoff** and **Dr. Michelle M. Donegan**, both of the JHUAPL, an aerospace engineer and an astronomer, respectively, provide a basic understanding of the space environment. The last part of the trilogy provides an understanding of the elements of space flight and support. **Mr. R. Michael Furrow**, a Ground Operations Specialist at JHUAPL, supplies an overview of the vital ground operations segment. **Dr. Daniel E. Clemens**, an aerospace engineer with a background in both electric and solid rocket propulsion, and **Ms. Barbara Leary**, a propulsion engineer who supports major programs such as the U.S. Standard Missile Program, explain the launch segment. To complete the basics, a cursory overview of space vehicles and payloads is provided by **Ms. Ann Garrison Darrin** and **Mr. Thomas S. Mehoke**, both of the Research Center at JHUAPL.

Part II is called "Cultural Astronomy" because this perspective, although tangential to other chapters of this book, is necessary because it acknowledges the importance and meaning of the sky in world cultures. Indigenous peoples were interested in the stars, the sun, and the moon long before there was a Space Age. **Dr. Susan Milbrath** authors the chapter that provides an overview of ethnoastronomy and archaeoastronomy. Dr. Milbrath is the curator of Latin American art and archaeology at the Florida Museum of Natural History, University of Florida. She is an expert on the Mesoamerican worldview links between astronomy and seasonal ceremonies.

Part III is called "Introduction to the Space Age" because it sets the Space Age in a historic context. Leading off this section is the award winning author of a book on aviation history, **Dr. Anne Millbrooke**. An author, educator, aviator, and expert in the area of the history of science and technology, Dr. Millbrooke provides a historic overview for the Space Age. **Dr. Robert Osiander**, a principal staff member and winner of Inventor of the Year at JHUAPL, developed the pre-Sputnik chapter. A physicist who trained at Technische Universität München, his chapter includes many of the contributions Germany and German scientists made to the pre-Space Age. **Mr. Greg Orndorff**, of the National Security Space area at JHUAPL, and **Dr. Rick Sturdevant**, a historian for the U.S. Air Force, team to provide and overview of the Cold War period. **Dr. Patrick Stadter**, JHUAPL, an electrical engineer and space communication expert, discusses the nonmilitary developments in space, with an emphasis on commercial applications. In contrast, **Dr. Ralph L. McNutt, Jr.**, supplies the civilian or scientific perspective to the maturing Space Age. Dr. McNutt, JHUAPL, is the project scientist for the MESSENGER mission, coinvestigator on NASA's New Horizons (Pluto–Kuiper Belt) mission, a team member of the Cassini Ion Neutral Mass Spectrometer investigation, and a science team member of two Voyager investigations.

Part IV is titled "The Landscape of Space" because it addresses the variety and complexity of the material resources of space, both on and off the earth, as part of the cultural landscape of humans. **Dr. Alice Gorman** of Flinders University and expert in space "junk" provides a thorough overview of the concepts of the cultural landscape (or spacescape). Dr. Gorman is an honorary research fellow at the University of New England, Australia, and cochair of the World Archaeological Congress Space Heritage Task Force. Orbital objects still functioning and those nonfunctioning (debris) are discussed by **Dr. Daniel E. Clemens**, followed by **Dr. Robert Osiander** and **Dr. Paul Ostdiek**. Dr. Ostdiek, a member of the principal staff at JHUAPL, brings both a civilian and a military space technology development background to an overview of orbital debris. **Dr. Alice Gorman** then provides a discussion of orbital debris from the cultural heritage perspective. **Dr. Robert Gold** reviews spacecraft and objects left on planetary surfaces. Dr. Gold, JHUAPL, was the Science Payload Manager for the Near Earth Asteroid Rendezvous (NEAR) mission that was the first to orbit and land on an asteroid. Asteroid 4955 Gold was named for his work on the NEAR mission. To round out the "Landscape of Space" section, **Dr. P.J. Capelotti** provides a catalog of manned exploration of the Moon. Dr. Capelotti is a senior lecturer in anthropology and American studies at Penn State University Abington College in Abington, Pennsylvania. He is an expert in the area of the archaeology of exploration and the author of *Sea Drift: Rafting Adventures in the Wake of Kon-Tiki*, *By Airship to the North Pole: An Archaeology of Human Exploration*, and a forthcoming textbook, *The Exploring Animal: An Introduction to Archaeology from Seafarers to Spacefarers*.

Part V of the book brings together diverse topics under "Space Forensics and Mystery Solving." This part is about what is unknown in the world of space engineering and problem solving when it comes to malfunctions and lost or failed missions.

Ms. Ann Garrison Darrin and **Ms. Patricia Prettyman**, both of JHUAPL, look at the potential of future studies to provide answers to space hardware mysteries. **Dr. Dirk H.R. Spennemann** and **Dr. Guy Murphy** discuss failed Mars mission landings sites. Dr. Spennemann is a professor in cultural heritage management at Charles Sturt University in Albury, Australia. His main research interests are in the area of future studies, focusing on heritage by examining issues such as the conceptual understanding of emergent heritage(s) and the recognition of heritage sites and objects of future heritage value such as space and robotic culture. **Dr. Philip J. Stooke**, a professor at the University of Western Ontario, provides a discussion of lost spacecraft. Dr. Stooke has compiled a comprehensive lunar geography, *The International Atlas of Lunar Exploration*, and has developed cartographic methods and data sets for nonspherical worlds. He has been involved with the National Space Society and has written for *Ad Astra*. **Dr. Ralph D. Lorenz**, JHUAPL, author of several books ranging from *Titan Unveiled* and *Space Systems Failures* to *Spinning Flight: Dynamics of Frisbees, Boomerangs, Samaras and Skipping Stones*, takes a fresh look at and discusses space hardware that is not in space because of design, accident or economic or sociopolitical reasons.

Part VI is titled "Environmental Effects and the Material Record"; in this section, an interdisciplinary approach is used to provide an overview of the impacts of the space environment on materials and systems. **Mr. Roger Gerke**, a graduate student in Anthropology at NMSU, teams with **Dr. Edward Staski** to lead off the chapter. Individual disciplines are covered by their respective specialists:

- **Dr. Jennifer L. Sample**, JHUAPL Research Center section supervisor, nanotechnologist, and material scientist
- **Ms. Janet L. Barth**, NASA, Goddard Space Flight Center, an expert in the radiation environment; Ms. Barth is a NASA Goddard senior fellow
- **Mr. Douglas Mehoke**, Space Department, JHUAPL, thermal engineer of the space department and a member of the principal professional staff
- **Dr. William S. Heaps**, NASA Goddard Space Flight Center (GSFC), an atmospheric physicist and light detection and ranging (LIDAR) expert
- **Mr. Theodore S. Swanson**, NASA, GSFC chief technologist of mechanical engineering; Mr. Swanson is a NASA Goddard senior fellow
- **Ms. Ann Garrison Darrin**, JHUAPL, former NASA electrical parts engineer and former division chief of Assurance Technologies
- **Dr. Michelle M. Donegan**, JHUAPL, Research Center, surfaces of objects specialist
- **Mr. Paul J. Biermann**, JHUAPL, a materials engineer with emphasis in test and evaluation who has participated in the conservation of space flight hardware

Part VII recognizes the importance of preserving significant space artifacts in several case studies. It also includes how to evaluate significance of objects and structures and looks at an archaeological investigation done at a putative UFO crash site. **Ms. Hanna Szczepanowska** leads off with her work on the conservation of space artifacts at the Smithsonian Institution, including the now declassified *Corona* spy satellite components. **Dr. Robert Barclay**, conservation consultant, and Dr. Randall Brooks of the Canada Science and Technology Museum, Ottawa, Canada, provide a fascinating discussion of what becomes a heritage item and what is discarded. They also discuss in situ preservation of space craft. **Dr. William Doleman** of the New Mexico Laboratory of Anthropology demonstrates investigation approaches to a terrestrial site in Roswell, New Mexico, where a UFO purportedly landed. His archaeological work, including geophysical investigations. The archaeological investigation was paid for by the SciFi Channel.

Part VIII is about space policy and preservation, with an overview of the international space community and space law. **Steve Doyle, Esq.**, introduces the legal framework. Mr. Doyle is a retired attorney who specialized in Space Law and Policy. **Dr. Beth Laura O'Leary** draws heavily on her experiences in the Lunar Legacy Project for her chapter "One Giant Leap: Preserving Cultural Resources on the Moon." Following that chapter, **Dr. Dirk H.R. Spennemann** provides a second chapter to the book, which discusses how to manage crash analysis and heritage concerns and the ethics of conflicting values about these issues. The final chapter in this section is authored by **Aaron Q. Rogers**, JHUAPL Space Department systems engineer, and **Ms. Ann Garrison Darrin**, who discuss mission planning for a space future. They propose changes to the Phase B development cycle to include concerns of preservation.

Part IX looks at the future and space archaeology and moves us forward; this section is led by **Dr. Beth Laura O'Leary** in her chapter "Planning for the Future Preservation of Space." **Dr. Ralph L. McNutt, Jr.**, contributes Space Exploration: The Next 100 Years. A bit of a futurist, Dr. McNutt has authored and contributed to numerous mission studies, including an interstellar probe study. **Dr. P.J. Capelotti** provides a chapter that proposes how to do archaeological field research in space.

Part X is called "The Mind in the Cosmos"; in it, we have included thoughts on some avenues for exoarchaeology in the solar system and how science fiction writers inspired and enhanced our thirst for going to space. We include the work of **Dr. John B. Campbell**, James Cook University, cochair of the World Archaeological Congress Space Heritage Task Force, who ventures into a search for extraterrestrial or other intelligent species. This is followed by various analyses of science fiction and its historical development. **Mr. Thomas S. Mehoke**, JHUAPL Research Center, discusses technology and material culture in science fiction. **Dr. Larry J. Paxton**, JHUAPL Space Department scientist with interests in the atmospheres of Mars, Venus, and Earth, takes us deep into the Drake equation and Fermi's paradox. **Dr. Bradley G. Boone**, JHUAPL Research Center, provides a thorough coverage of the historiography of science fiction.

CONCLUDING REMARKS

The editors apologize in advance if topics, missions, space hardware, or organizations have not been included in this volume. The body of knowledge is too large for one book. We feel the *Handbook of Space Engineering, Archaeology, and Heritage* launches this exciting field.

NOTES AND REFERENCES

1. Courtesy of the Johns Hopkins University Applied Physics Laboratory. Oscar spare for museum display.
2. Guier, W.H., and Weiffenbach, G.C. 1998. The Legacy of Transit. 1998. *The Johns Hopkins University Applied Physics Laboratory Technical Digest* 19(1): 14.
3. Renfrew, C., and A.P. Bahn. 2007. *Archaeology Essentials: Theories, Methods and Practices*. London: Thames & Hudson.
4. King, T.F. 2004. *Cultural Resource Laws & Practice*. 2nd ed. Walnut Creek, CA: Altamira Press.

2 Archaeology: The Basics

Edward Staski

CONTENTS

INTRODUCTION

Since the 1970s, a growing number of American archaeologists have defined *archaeology* as that subdiscipline of anthropology that studies the relationships between material culture and human behavior.[1] *Anthropology*, in turn, can be defined as that social science concerned with the totality of the human experience, in all times and all places. This is a mainstream definition of anthropology in the United States, and while the definition of archaeology offered here would seem to logically follow from it, the absence of temporal limitations makes it one of at least two competing characterizations of the discipline. Some archaeologists see their work as necessarily restricted to investigations of the past.

Others see archaeology as capable of studying the past but not necessarily compelled to do so. This latter position is the one taken in this chapter. Indeed, archaeologists can study both the past and the present and make important, substantive contributions to knowledge that are unique and unavailable from other disciplines. Put in blunt terms, archaeologists can study the past and the present and in both cases contribute knowledge that other disciplines cannot. This fact is particularly important (albeit contestable) when it comes to studying the present, because the present can be studied by practitioners of so many different scholarly disciplines.

In what follows an additional contestable and undoubtedly minority position among archaeologists is put forward: space archaeology—archaeological investigations of exoatmospheric material culture—is also a legitimate, scholarly branch of the discipline. At least, it has the potential of becoming such a legitimate, scholarly endeavor, but potentiality in lieu of actuality pertains only because it is such a new and emerging specialization. It will become mainstream practice. Making this claim is akin to sticking one's professional neck out. Many traditional archaeologists will object (even cringe) when they read this, but the argument is nevertheless sound. One stands on firm ground, so to speak, when taking this position on space archaeology.

SOME DEFINITIONS

The term *space archaeology* means different things to different people. It is useful to offer a few brief definitions from the perspective of archaeology.

First, space archaeology sometimes refers to the archaeological study of space; not outer space, but one of the three dimensions used by archaeologists to conceptualize and order patterns of material culture and patterns of behavior (the other two dimensions are time and form). This type of space archaeology has been fundamental to archaeology since its inception as a scholarly discipline, and some sort of spatial analysis, at one scale or another and to one degree or another, has been essential to all archaeological research. There have been degrees of emphasis nevertheless, with some archaeologists giving greater weight to the spatial dimension than others.[2]

Landscape archaeology, a still-emerging methodological and theoretical specialization that studies how people interact with their social and natural environments, has recently renewed archaeological interest in the spatial element.[3] The concept of the archaeological or cultural landscape, a geographic expanse that reflects and is the result of interacting environmental and cultural forces, has become the basic unit of study among landscape archaeologists. With the advent of geographical information systems and their productive application to archaeological concerns, this emphasis on space has flourished in recent years.[4]

Second, space archaeology is a term sometimes used to describe a particular type of remote sensing that involves the analysis of satellite or high-altitude aircraft imagery so that archaeologists can better observe the distribution of material culture across landscapes on the earth (mostly features and structures across relatively large areas or regions, and just as often below as on the surface;[5] see discussion below). It might be better to call this type of remote sensing strategy archaeology *from* space. Regardless, it is a fruitful approach to discovering and identifying sites, mapping known sites in remarkable detail, and retrieving data from inaccessible and dangerous places. Using satellite technology has also helped archaeologists in the field simply to communicate with one another and their colleagues elsewhere. More importantly, it has resulted in some significant discoveries and reinterpretations of settlement patterns and settlement systems. Some of the better known research has focused on the sites of Angkor (a World Heritage Site, and location of the temple Angkor Wat, it is in the Kingdom of Cambodia, formally Kampuchea)[6] and Ubar (popularly called "the lost city of Ubar" and branded "the Atlantis of the sands" by Lawrence of Arabia, it is in remote southern Oman, on the Arabian Peninsula).[7] Additional productive space archaeology of this kind has focused on a number of long-distance roads and trails that cannot be easily observed on the ground or even at low altitudes, including a number of Roman roads in the Near East, the Silk Road in China, the road network radiating out of Chaco Canyon in New Mexico, and others.[8]

Space archaeology is also another term for archaeoastronomy, the archaeological study of traditional astronomical practices and related belief systems in all cultures and societies as revealed by patterns of material culture (again, mostly features and structures).[9]

Archaeoastronomy should not be conflated with the numerous outlandish claims that have been made regarding how astonishingly sophisticated astronomical knowledge is (or was) within certain preindustrial societies. Such claims include the extraordinary assertion that the Dogon of Mali (Africa) had knowledge that the star Sirius is a binary, without the aid of adequate technology to observe this fact and giving support to the notion that these and other people have been visited and educated by extraterrestrials.[10] Another such claim is that Stonehenge and other megalithic monuments throughout Great Britain and Europe were constructed by people thousands of years ago to predict and mitigate the results of comet and asteroid impacts on the Earth. According to this view, these individual monuments together formed a sophisticated (and seemingly efficacious) network of astronomical observatories capable of accomplishing these extraordinary feats.[11] These and many other similar assertions—just search the Internet for Chichen Itza, Easter Island, the Great Pyramids of Giza, Machu Picchu, the Nazca Lines, the Sphinx, Teotihuacan, and other such well-known sites for examples—have been thoroughly debunked and have nothing to do with authentic archaeoastronomy.

Indeed, archaeoastronomy is serious scholarship, providing insight into how people of various cultures have conceptualized the cosmos and their place within it. Archaeoastronomy has also revealed how these various cosmologies and world views have related to more practical, technological endeavors such as subsistence practices. Knowledge of the spatial arrangements and motions of various objects in the sky allows people to know the seasonal cycle of the year and the related climatic/environmental patterns to expect, thus improving subsistence-related choices.

A fourth definition of space archaeology is more speculative than real. It concerns the perceived future role archaeologists might play in studying the material culture of extraterrestrials, living or not, once such material is discovered and verified as genuine. This type of space archaeology is also known as *xenoarchaeology* and sometimes *exoarchaeology*. And a fifth definition of space archaeology is truly on the fringe and unacceptable within rational, scholarly circles. It concerns the archaeological study of material culture not on Earth already claimed to be of extraterrestrial origin. Investigation of the well-known Cydonian mesa known as the "Face on Mars" is a good example of this sort of thing.[12] The archaeological claims have been thoroughly debunked.

A sixth definition of space archaeology is the one used in this chapter. It is the archaeological study of material culture found in outer space, that is, exoatmospheric material that is clearly the result of human behavior. Some might want to expand this definition and include all material culture in the aerospace and aeronautical realms that relates to the development and support of exoatmospheric activities. There is no problem with defining space archaeology in this wider sense. Among the archaeologists, physicists, astronomers, and engineers who have an interest in space archaeology, there is at present no single set of parameters that delimits the definition to everyone's satisfaction. Generally speaking, and all things being equal, it might be better in the long run to adopt the broader characterization, but because of the conceptual argument made about archaeology in this chapter emphasis here is on the narrower definition: the archaeological study of those materials *out there*. Regardless, this type of space archaeology, whether narrowly or broadly conceived, can be serious scholarship. It must be, if it is ever to be accepted into the discipline's mainstream.

WHAT IS ARCHAEOLOGY?

Freeing archaeology of temporal limitations and focusing on material culture/human behavior relationships has made possible four distinct archaeological strategies that underscore the holistic nature of archaeological research. These four strategies emerge when the temporal foci of investigative settings (i.e., what you study) and research goals (i.e., what you want to learn about) are considered. This epistemological model of archaeology is illustrated by the following matrix:[13]

	Understand the past	**Understand the present**
Study the past	Traditional archaeology	Archaeological ethnography
Study the present	Ethnoarchaeology	Modern material culture studies

Source: Adapted from Reid, J.J., M.B. Schiffer, and W.L. Rathje, *American Anthropologist* 77, 864–869 1975.

The four strategies, traditional archaeology, ethnoarchaeology, archaeological ethnography, and modern material culture studies, enhance the scientific rigor of the discipline by either generating testable hypotheses (past-past and present-present strategies: traditional archaeology and modern material culture studies) or providing data that allow those hypotheses to be tested (past-present and present-past strategies: archaeological ethnography and ethnoarchaeology).

Following the chart's structure, *traditional archaeology* can be defined as studying the past in order to better understand the past. This is the definition of archaeology that everyone recognizes and that most professional archaeologists practice. It is an appropriate and valuable strategy for increasing our knowledge of the human experience, and for that vast period of time before writing—prehistory—it is the only strategy available. About 95%–99% of human existence (the percentage used depends on how you define *human*) can be understood only by doing traditional archaeology.

Ethnoarchaeology is defined as the study of the present in order to better understand the past.[14] It entails archaeologists doing traditional ethnography from an archaeological perspective, spending extended periods of time with groups of living people in order to observe directly *both* their patterns of behavior and relatable patterns of material culture. The goal is to improve inferences regarding patterns of behavior that can no longer be observed directly because they are reflected in patterns of material culture in the archaeological record. Reaching this goal requires the proper application of analogy.

Archaeologists have been doing ethnoarchaeology for more than 100 years, since the earliest days of the discipline. It lost popularity during the mid-twentieth century, only to become enormously popular again with the advent of *new archaeology* and the explicitly scientific approach that practitioners of this movement stipulated.[15] It remains quite a popular and fruitful archaeological strategy to this day.

Archaeological ethnography is the study of the past in order to better understand the present. It is an explicitly applied form of archaeology that underscores the relevance that certain archaeologically obtained information can have when it comes to addressing current situations and mitigating current problems (as are modern material culture studies, described below). As such, it is a rare variant of archaeological practice. Examples include creating models for present-day sustainable agriculture from prehistoric and traditional examples,[16] reconstructing human impacts on the environment from archaeological data in order to better account for current ecological conditions,[17] and including indigenous histories and traditions along with archaeological data to construct contemporary interpretations of the past for a range of stakeholders.[18]

Some archaeologists use the term archaeological ethnography to mean ethnoarchaeology as defined in this chapter.[19] Others use it to describe their explicitly holistic approach to anthropological research.[20] Rathje's use of the expression[21] is most acceptable, though his own research on refuse management should be placed in the modern material culture category (see below).

Finally, *modern material culture studies* involve the study of the present in order to better understand the present.[22] The past is not involved at all. More than any other type of archaeology, modern material culture studies call attention to the fact that archaeology need not be the study of the past exclusively.

Best known of all the modern material culture studies is the long-running Projet du Garbage—the Garbage Project—originated in 1973 by archaeologist William Rathje at the University of Arizona and elsewhere.[23] Rathje recognized that a large part of the archaeological record consists of people's refuse. He also knew that patterns of refuse offer archaeologists great insight into patterns of consumption and use. More, or at least different, information about such behaviors can be gleaned from refuse, which is nonreactive, than from interview and questionnaire responses, which are largely shaped by attitudes and context and not necessarily true reflections of how people act. With these facts in mind, Rathje initiated the Garbage Project with a simple assumption: we can learn a lot about living people by studying them in the same way we study dead people, as archaeologists, interested in relating patterns of material culture to patterns of behavior. We cannot learn as much about living people, or at least we cannot learn the same things, in any other way.

Researchers have used data from the Garbage Project to measure rates of household alcohol consumption in Tucson, Arizona, for example, and to compare these data with ethnic, economic, and other social and cultural categories of people.[24] For decades, sociologists and other social scientists had observed an apparent correlation between particular ethnic identities and patterns of alcohol use and abuse, yet their data had come from interviews, questionnaires, and other reactive descriptions and displays of alcohol use. These were all reflections of *attitudes*, not necessarily *behaviors*, and it was not surprising that such attitudinal patterns should correlate with patterns of ethnic identity. The archaeological data from the Garbage Project, however, showed no such correlation between actual behavior and ethnic identity.

Amounts of alcohol actually consumed, as reflected in the archaeological record, did not seem related to ethnic identity at all.

Further consideration of this particular research is unimportant in this chapter. What is relevant to this discussion is simply that something was learned about the present by doing archaeology of the present, something that could not have been learned otherwise. Clearly, then, there is nothing in the conceptualization of archaeology used here or in these definitions of archaeological strategies that make it necessarily the study of the past. Archaeologists *can* study the past, because patterns of material culture are often preserved well enough in the archaeological record that past patterns of human behavior can be inferred from them with confidence. Archaeologists do not *have to* study the past, however. So long as the relationships between patterns of material culture and patterns of human behavior are being studied, the study is archaeology.

The reason most archaeologists disagree with this view involves their conviction that we can reach a much more complete and in-depth understanding of the present by applying the methods of other social sciences (e.g., cultural anthropology, sociology, or economics). But as illustrated by the research project described previously, these alternative approaches, while contributing significant knowledge and understanding of the human experience, seldom focus on material culture and behavior in the way archaeology is designed to do. Additionally, practitioners of these other social sciences seldom consider change over time, and even more rarely change over long periods of time, as having had influence in shaping the present. The outcome is that time and process are often ignored as explanatory factors when it comes to accounting for present-day conditions. This conclusion is wrong-headed, and archaeologists know it. Long-term changes are often critical when trying to understand the broad patterns and conditions of human experience that can be observed in the present day.

SPACE ARCHAEOLOGY

Despite these facts, getting professional archaeologists to agree to eliminate the temporal boundaries on the discipline has been difficult because of the fact that the past (as a temporal unit) seems so remarkably different and separate from the present. During most of the history of archaeology, eliminating any spatial boundaries ("all times and all places") has not been so problematic, precisely because such spatial limitations did not exist. All human activity, and all the material remains of all human activity, have been restricted to one planet and thus have been considered part of one mega-assemblage, so to speak. That has been the case at least until recently. Since the advent of space exploration, however, and the creation of an exoatmospheric archaeological record, the situation has changed dramatically. Now, we have a small assemblage of materials that, for no other reason than their place of deposition, are apparently considered beyond the pale of legitimate archaeological inquiry. This situation persists, even as the historic significance of and need to preserve space-related objects and artifacts *on Earth* are becoming more widely recognized and appreciated.[25] Currently, there is resistance to expanding *both* the temporal and spatial dimensions of archaeology.

Viewing the present as a legitimate temporal category for archaeological study is analogous to viewing the exoatmospheric archaeological record as a legitimate spatial arena for archaeological investigation. Both are minority views. Both are resisted by the mainstream archaeological community. Both are nevertheless correct views in that both result in a better strategy for doing archaeology by putting emphasis on material-behavioral relationships rather than on temporal-spatial parameters.

Furthermore, doing space archaeology will result in knowledge unobtainable by other methods and unavailable from other data sources, just as doing archaeology in the present results in information and knowledge unobtainable otherwise. The current inventory and site map of materials left on the Moon at Tranquility Base by the *Apollo 11* astronauts, for example, generated by archaeologists who understand the significance of archaeological formation processes such as discard, loss, and abandonment, and appreciate that not every behavior is written down for perpetuity, are undoubtedly more exhaustive and accurate than any inventory or map based on available and authorized documentary records.[26] And this list was generated without benefit of a return visit to the site. Imagine what might be learned if archaeologists could actually survey this and the other five areas of human visitation on the Moon. What might be learned that cannot be known in any other way?

It should now be clear that material culture in the present, operating in ongoing cultural and social systems, is not meaningfully or significantly isolated from material culture from the past, currently in the archaeological record. Such material does not exist in an historical or cultural vacuum. Analogously, exoatmospheric artifacts do not exist in an historical or cultural vacuum either (although in this case many of them exist in an actual, physical vacuum or near-vacuum). Indeed, these objects can be viewed as being part of a much larger assemblage of materials that, until a certain time and stage in technological development, were confined to the Earth but that afterwards could enter the archaeological record somewhere else. This larger assemblage consists of all objects—artifacts, features, and structures—associated with aerospace technology.

The assemblage is enormous, and a good number of these objects are in fact recognized as historically significant. Many are listed on the National Register of Historic Places. Thus, Robert Goddard's rocket launching site in Auburn, Massachusetts, is listed, as is his house in Roswell, New Mexico. So are numerous National Historic Landmarks and other significant sites at Cape Canaveral in Florida (the Cape Canaveral Air Force Station, the Headquarters building, and Launch Complex 39); at the Lyndon B. Johnson Manned Space Flight Center in Houston, Texas (the Apollo Mission Control Center and the Space Environment Simulation Laboratory); at White Sands Missile Range in New Mexico (the V-2 launch site); in Huntsville, Alabama (the neutral buoyancy space simulator, the propulsion and structural test facility, the Redstone test stand, and a Saturn V launch vehicle and dynamic test stand); at Vandenberg Air Force Base in Lompoc, California (Space Launch Complex 10); at the Jet Propulsion Laboratory in Pasadena, California (the Space Flight Operations Facility); and elsewhere.

At a known time in the past, on October 4, 1957, when the Soviet Union successfully launched *Sputnik 1*,[27] a sample of objects in this aerospace assemblage started leaving Earth. At later times, some came back to Earth, while others remained in space. Still others found themselves "deposited" on the Moon, Mars, and other extraterrestrial bodies. A number of these archaeological sites created somewhere other than Earth have clear historical and cultural significance, at least as much significance as their counterparts on this planet that are listed on the National Register. Which is more significant, the Lunar Landing Research Facility at the Langley Research Center in Hampton, Virginia, a National Historic Landmark, or the six actual lunar module descent stages that took men to the Moon and enjoy no historic designation or recognition simply because they were left there? Which is a more important place, Launch Complex 39A, from which *Apollo 11* was launched on July 16, 1969, and which since 2000 enjoys National Register status as part of the John F. Kennedy Space Center Multiple Property Submission, or Tranquility Base, its destination on the Moon, which because of its location does not?

METHODOLOGICAL CONSIDERATIONS

Guided by explicit research designs, specifying research questions to be addressed, expected observations to be made in the empirical record, and project-specific methods to be followed, archaeologists face the task of observing and gathering appropriate material and generating relevant data. The material consists of the actual artifacts, features, structural remains, and other physical remnants that exist in the archaeological record. The data are the various measurements that archaeologists make from the material in order to describe, categorize, analyze, and interpret that record. Just which materials are observed and gathered and just which data are generated are determined by the research questions set forth in the research design and the strategies to address them put forward by the archaeologist doing the work.

Until quite recently, *excavation*—the controlled removal of physical remains from beneath the ground surface, designed to ensure the recording of spatial and formal information before it is destroyed—was the primary way archaeologists gathered materials and generated data. Over the past several decades, however, excavation has become increasingly rare. *Archaeological survey*—the systematic observation of the ground surface and the recording of relevant landscape features and material culture upon it—is today far and away the more common strategy. There are several reasons why there has been this shift in strategies; the reasons have intriguing connections to what appears will be the proper methodology for space archaeology.

First, excavation takes a great deal of time, when done correctly. Recovering materials from the subsurface without damaging them and ensuring that as few data as possible are lost requires slow, meticulous work. Second, excavation is usually very expensive, primarily because of the time factor but also because it demands special equipment and is very labor intensive.

Because this strategy is slow and costly, it follows that it allows only a small amount of the subsurface to be explored, and herein lies the third reason excavation has become so rare. Many of the most interesting and significant research questions being asked by archaeologists these days—Why did agriculture and animal domestication arise independently in several places around the world at about the same time? Why did cities and complex societies do the same, shortly thereafter? and so forth—require data from across regions, not from individual sites. It is impossible to excavate a region, of course, or even an adequate portion of a region to properly address these and many other questions. Excavation is not the answer.

The fourth reason that excavation is so rare today is perhaps the most important. Excavation results in the destruction of the archaeological record. Responsible archaeologists try to obtain all the materials and data they can, so that they can address their research questions in the most thorough manner possible. They often try to gather additional materials for future data generation too, to be conducted by themselves or by others who come after them. Ideally, all potential material and data should be preserved in one form or another, because if they are not the process of excavation will destroy them. But this ideal can never be achieved. No matter how comprehensive the research design, and no matter how meticulous the researcher (and no matter how much time and money are spent), some material and data will be lost forever.

Enter archaeological survey, a relatively quick and inexpensive field strategy that allows material to be gathered and data to be generated from across regions while not destroying much (if any) of the archaeological record.

Archaeological surveys can be accomplished in several ways. The methods chosen are determined by the type of material and data being sought, as directed by the research design, and by the nature of the environment in which the material and related data can be found. Usually a combination of approaches is used so that a broad range of data will be made available. The most common strategy involves direct inspection of the ground surface. Archaeologists walk over the landscape in a prescribed, systematic fashion and record what they see and where they see it. The resulting data are often used to narrow the scope of the survey while increasing survey resolution; that is, archaeologists will walk across a smaller sample area closer to one another so that they can observe surface features more carefully. These resulting data in turn might be used to repeat the process, with even a smaller, sample area surveyed with even greater intensity. Some surveys will include the excavation of a limited number of test units at this time as well, so that the subsurface can be sampled in areas of perceived high investigative potential. Archaeological survey is a multistage process, with area covered decreasing and resolution increasing as each stage is reached.

Archaeologists are not limited to walking across the landscape to get an initial sense of the archaeological record across great territories. In many instances, and especially when the territories to be covered are vast and environmental conditions are hostile, survey archaeologists turn to various remote sensing technologies in order to get the data they require.[28] Aerial photographs and satellite images, for example, are often used to reveal large-scale patterns in the archaeological record across vast areas that would be difficult to cover on the surface. The patterns, clearly visible from afar, are often invisible to surface-bound observers.

Reliance on remote sensing technologies of various sorts is becoming more commonplace in archaeology as survey replaces excavation as the primary method of data collection and as remote sensing technologies become more sophisticated and user-friendly. Remote sensing in archaeology consists of a wide range of strategies of varying sophistication and expense. Increasing numbers of archaeologists conduct remote sensing in one sense or another. Their activities range from conducting proton magnetometer, electrical resistivity, and ground-penetrating radar surveys on the surface of sites to locate anomalies underground, to searching archives for early-twentieth-century low-altitude aerial photographs to learn about historical landscapes, to relying on LANDSAT and other satellite imagery to gain large-scale views of entire regions and the archaeological resources they contain. All of these remote sensing practices were originally developed to do something other than archaeological research, yet these and other like data-gathering strategies have become mainstream, accessible, and useful to all archaeologists. Their greatest appeal is that they can be used to understand the nature and extent of the subsurface archaeological record without impacting its integrity.

This growth in the legitimacy and relevancy of remote sensing in archaeology in general is important for space archaeology in particular, especially when it comes to the potential future acceptance of space archaeology as a legitimate endeavor worthy of serious attention and praise, for not only will we increasingly rely on space technology for remote sensing capabilities in our studies of archaeological resources on Earth, we will also quite certainly rely on the same technology in our attempts to conduct comparable studies of archaeological resources in space. Indeed, because space is such a hostile and exotic environment for people, space archaeology will very likely depend upon sophisticated remote sensing capabilities more than any other archaeology has done.

This discussion brings to mind one final consideration regarding the future development of space archaeology. It derives from the fact that robots in space will be doing a lot more than simply remotely sensing exoatmospheric archaeological resources. These craft will have more important tasks to perform than conducting archaeological research, tasks for which they were designed. As with aerial and space photography today, their archaeological remote sensing capabilities, and specifically the application of these capabilities for the benefit of archaeology, will be afterthoughts.

Many of the robots' primary tasks will involve the design, manufacture, and manipulation of other objects. It is intriguing, then, to consider that the robots doing this work will be not only archaeological resources in their own right but creators of other archaeological resources. They will be the stuff *and the creators of the stuff* we archaeologists will want to preserve and study.[29] What will happen when these robots assume significant levels of autonomy and are no longer directly or completely dependent on human commands for their actions? How will we evaluate the significance of an archaeological record fashioned by a mechanical device sophisticated enough to have made a series of decisions on its own, the result of which will have been the formation of this archaeological record? Will we still be dealing with material *culture* at that point? Will our definition of archaeology have to be expanded even more than is proposed in this chapter? Answers to these questions will be proposed by other scholars, at another time.

CONCLUSION

The definition of archaeology—the study of the relationships between patterns of material culture and patterns of human behavior—sets no temporal limits on the practice of archaeology. Neither does this definition set any spatial limits on the discipline. Archaeology can be performed anywhere on the earth, and anywhere off the earth. Space archaeology—the study of material culture found in outer space and its relationship to human behavior—is a serious, legitimate pursuit that has recently been recognized as important and necessary. The amount of exoatmospheric material culture has increased dramatically in recent years. Many exoatmospheric objects are historic and significant, and many of them face the increasing prospect of being damaged or destroyed by various adverse impacts. It is imperative that we treat them as valuable objects worthy of legitimate archaeological inquiry. The various chapters in this book illustrate these facts and will facilitate the inevitable increase in the recognition of their veracity among both archaeologists and nonarchaeologists alike.

NOTES AND REFERENCES

1. Rathje, W.L., and M.B. Schiffer. 1980. *Archaeology*, vii. New York: Harcourt Brace Jovanovich. Staski, E., and J. Marks. 1992. *Evolutionary Anthropology*. Fort Worth, TX: Harcourt Brace Jovanovich.
2. Hodder, I.R., and C. Orton. 1979. *Spatial Analysis in Archaeology*. New York: Cambridge University Press. Robertson, E.C., J.D. Seibert, D.C. Fernandez, and M.U. Zender, eds. 2006. *Space and Spatial Analysis in Archaeology*. Albuquerque, NM: University of New Mexico Press.
3. Kelso, W.M., and R. Most, eds. 1990. *Earth Patterns: Essays in Landscape Archaeology*. Charlottesville, VA: University Press of Virginia.
4. Allen, K., S.W. Green, and E.B. Zubrow, eds. 1990. *Interpreting Space: GIS and Archaeology*. London: Taylor and Francis. Wheatley, D., and M. Gillings. 2002. *Spatial Technology and Archaeology: The Archaeological Applications of GIS*. London: Taylor and Francis.
5. Ebert, J.I. 1984. Remote Sensing Applications in Archaeology. In *Advances in Archaeological Method and Theory*, ed. M.B. Schiffer, 293–362. Vol. 7. New York: Academic Press.
6. Higham, C. 2004. *The Civilization of Angkor*. Berkeley, CA: University of California Press.
7. Clapp, N. 1999. *The Road to Ubar: Finding the Atlantis of the Sands*. Boston: Mariner Books/Houghton Mifflin.
8. Obenauf, M.S. 1991. Photointerpretation of Chacoan Roads. In *Ancient Road Networks and Settlement Hierarchies in the New World*, ed. C.D. Trombold, 34–41. Cambridge, UK: Cambridge University Press. Sever, T.L., and D.W. Wagner. 1991. Analysis of Prehistoric Roadways in Chaco Canyon Using Remotely Sensed Digital Data. Ibid., 42–52. Sheets, P., and T.L. Sever. 1991. Prehistoric Footpaths in Costa Rica: Transportation and Communication in a Tropical Rainforest. Ibid., 53–65.
9. Aveni, A.F. 1989. *World Archaeoastronomy*. New York: Cambridge University Press. Aveni, A.F. 2001. *Skywatchers*. Austin, TX: University of Texas Press. Ruggles, C., and N. Saunders, eds. 1993. *Astronomies and Cultures*. Boulder, CO: University Press of Colorado.
10. Temple, R. 1975. *The Sirius Mystery*. New York: St. Martin's Press.
11. Knight, C., and R. Lomas. 2001. *Uriel's Machine: Uncovering the Secrets of Stonehenge, Noah's Flood and the Dawn of Civilization*. Beverly, MA: Fair Winds Press.
12. Hoagland, R. 1996. *The Monuments of Mars: A City on the Edge of Forever*. Berkeley, CA: North Atlantic Books.
13. Reid, J.J., M.B. Schiffer, and W.L. Rathje. 1975. Behavioral Archaeology: Four Strategies. *American Anthropologist* 77(4): 864–869.
14. See classic statements in (1) Binford, L R. 1978. *Nunamiut Ethnoarchaeology*. New York: Academic Press; (2) Cramer, C., ed. 1979. *Ethnoarchaeology: Implications of Ethnography for Archaeology*. New York: Columbia University Press; (3) Gould, Richard A., ed. 1978. *Explorations in Ethnoarchaeology*. Albuquerque, NM: University of New Mexico Press; (4) Gould, Richard A., ed. 1980. *Living Archaeology*. New York:

Cambridge University Press; (5) Hodder, I.R. 1982. *Symbols in Action: Ethnoarchaeological Studies of Material Culture*. New York: Cambridge University Press; (6) Longacre, W.A., ed. 1991. *Ceramic Ethnoarchaeology*. Tucson, AZ: University of Arizona Press; (7) Yellen, J.E. 1977. *Archaeological Approaches to the Present*. New York: Academic Press.
15. Binford, L.R. 1962. Archaeology as Anthropology. *American Antiquity* 28(2): 217–225.
16. Erickson, C. 1992. Applied Archaeology and Rural Development: Archaeology's Potential Contribution to the Future. *Journal of the Steward Anthropological Society* 20(1–2): 1–16. Erickson, C. 2003. Agricultural Landscapes as World Heritage: Raised Field Agriculture in Bolivia and Peru. In *Managing Change: Sustainable Approaches to the Conservation of the Built Environment*, ed. J.M. Teutonico and F. Matero, 181–204. Oxford, United Kingdom: Getty Conservation Institute in collaboration with US/ICOMOS, Oxford University Press.
17. Briggs, J.M., K.A. Spielmann, H. Schaafsma, K.W. Kintigh, M. Kruse, K. Morehouse, and K. Schollmeyer. 2006. Why Ecology Needs Archaeologists and Archaeology Needs Ecologists. *Frontiers in Ecology and the Environment* 4(4): 180–188. Rick, T.C., and J.M. Erlandson. 2003. Archeology, Ancient Human Impacts on the Environment, and Cultural Resource Management on Channel Islands National Park, California. *CRM: The Journal of Heritage Stewardship* 1(1): 86–89.
18. Shackel, P.A. 2004. *Places in Mind: Archaeology as Applied Anthropology*. London: Routledge. Little, B.J. 2002. *Public Benefits of Archaeology*. Gainesville, FL: University Press of Florida.
19. Watson, P.J. 1979. *Archaeological Ethnography in Western Iran*. Tucson, AZ: University of Arizona Press.
20. Meskell, L. 2005. Archaeological Ethnography: Conversations Around Kruger Park. *Archaeologies* 1(1): 81–100.
21. Rathje's use of the expression. Rathje, W.L. 1978. Archaeological Ethnography … Because Sometimes It Is Better to Give than to Receive. In *Explorations in Ethnoarchaeology*, ed. R.A. Gould, 49–75. Albuquerque, NM: University of New Mexico Press.
22. Gould, R.A., and M.B. Schiffer, eds. 1981. *Modern Material Culture: The Archaeology of Us*. New York: Academic Press. Rathje, W.L. 1979. Modern Material Culture Studies. In *Advances in Archaeological Method and Theory*, ed. M.B. Schiffer, 1–37. Vol. 2. New York: Academic Press. Rathje, W.L., and C. Murphy. 2001 *Rubbish! The Archaeology of Garbage*. Tucson, AZ: University of Arizona Press. Buchli, V., and G. Lucas, eds. 2001. *Archaeologies of the Contemporary Past*. New York: Routledge.
23. Rathje and Murphy, 2001 (note 22).
24. Staski, E. 1983. Alcohol Consumption among Irish-Americans and Jewish-Americans: Contributions from Archaeology. PhD diss., University of Arizona.
25. Flanagan, J. 2007. Sputnik at 50: The Legacy of the Space Race. *Common Ground: Preserving Our Nation's Heritage* 12(4): 24–37.
26. See O'Leary, B.L. 2006. The Cultural Heritage of Space, the Moon, and Other Celestial Bodies. *Antiquity* 80: 307. http://antiquity.ac.uk/ProjGall/oleary/index.html; O'Leary, B.L. 2007. Historic Preservation at the Edge: Archaeology on the Moon, in *Space and on Other Celestial Bodies. Keynote Address at the ICOMOS 2007 Conference on Extreme Heritage*, Cairns, Australia; and Staski, E., and Gerke, R. 2009. Chapter 26 in this volume.
27. Flanagan 2007 (note 25).
28. Ebert 1984 (note 5). Johnson, J.K. 2006. *Remote Sensing in Archaeology: An Explicitly North American Perspective*. Tuscaloosa, AL: University of Alabama Press.
29. A more in-depth consideration of this issue is found in Spennemann, D.H.R. 2007. On the Cultural Heritage of Robots. *International Journal of Heritage Studies* 13(1): 4–21.

3 Evolution of Space Archaeology and Heritage

Beth Laura O'Leary

CONTENTS

An Archaeologist's Perspective.

INTRODUCTION

Space archaeology begins not with research by a team of archaeologists, but with two astronauts.[1,2,3] As part of the *Apollo 12* mission on November 19, 1969, astronauts Charles "Pete" Conrad and Alan Bean were instructed to observe and record information about *Surveyor 3,* an unmanned probe that had soft-landed on the Moon's Ocean of Storms 2 years and 7 months earlier. The *Apollo 12* astronauts had manually landed their spacecraft about 180 m from where the *Surveyor 3* was located.[4] Bean and Conrad conducted the first "archaeological" studies on the Moon (Figure 3.1). Although their mandate was to look at the general effects of space and the conditions of the Moon in particular on a spacecraft, their investigation was archaeological in nature. Their documentation of the *Surveyor 3* included photographs of the impressions made by the *Surveyor*'s footpads. They noted the condition of the spacecraft and took samples of its remote sampling arm and pieces of tubing, and they also removed the probe's television camera.[5] They bagged and labeled the artifacts and returned them to Earth for analyses of any changes. A biological analysis brought exciting news that the bacterium *Streptococcus mitis* was present on the returned camera, although it was later determined that it had been put there on Earth before *Surveyor*'s flight and had somehow survived more than two and a half years on the Moon's surface. Capelotti notes that this was the first example of

FIGURE 3.1 Alan Bean and *Surveyor 3* on the surface of the Moon. (Courtesy of NASA.)

> extraterrestrial archaeology and—perhaps more significant for the history of the discipline—formational archaeology, the study of environmental and cultural forces upon the life history of human artifacts in space.[6]

The foundation of the archaeological inquiry into space flight began with archaeologists who were interested in aircraft wrecks and who proposed that they could contain significant data.[7] According to Capelotti (Chapter 45 of this volume) it was Ben Finney, who explored the technology of the Polynesians settling the Pacific Ocean, who first suggested in 1993 that it might be worthwhile to think about space sites created by both the United States and Soviet Union. Nonterrestrial archaeology began with underwater or marine archaeology in the 1930s, with a scientific investigation by Carl Ekman of a sunken sixteenth-century warship near Kalmar, Sweden and in 1963 in the United States with the first American conference on underwater archaeology. As a subdiscipline, underwater archaeology supports the idea that archaeological studies can be undertaken on not just terrestrial environments, but anywhere humans have left material culture. Fischer's[8] description of the these early efforts by underwater archaeologists in many ways parallels what is now the field of space archaeology and heritage.

The field of remote sensing from space was also embraced by archaeologists as a tool for identifying archaeological sites that were hard to detect by normal surficial pedestrian surveys. Interestingly, remote sensing technology such as LANDSAT, which has allowed archaeologists to look from space to earth for archaeological resources, has rarely been used in the other direction to help archaeologists look for sites in space and on other celestial bodies. Future lunar orbiters and telescopes will have the capability of looking at the first lunar landing site at a scale archaeologists customarily use to map and evaluate sites for potential significance. In 1984, the U.S.

National Park Service commissioned a "Man in Space" Historic Landmark Theme Study that inventoried sites in the United States that epitomized the space program in order to evaluate and include them as properties in the National Register of Historic Places (National Register) and/or as National Historic Landmarks.[9,10,11] None of the sites in the study were outside the United States, and no sites in space or on other celestial bodies were considered, although the space sites are arguably the critical component of their significance.[12] The eight sites that were associated with the Apollo program in the United States on the National Register or as National Historic Landmarks include the Neutral Buoyancy Space Simulator, the Propulsion and Structural Test Facility, and Saturn V Dynamic Test Stand at the George C. Marshall Space Flight Center in Huntsville, Alabama; the Space Flight Operations Facility and Twenty-Five Foot Space Simulator at the Jet Propulsion Laboratory in Pasadena, California; the Pioneer Deep Space Station at Fort Irwin, California; the Unitary Plan Wind Tunnel at Moffet Field, California; and the Central Instrumentation Facility at Kennedy Space Center, Florida.

EXOARCHAEOLOGY

In a provocative article in 1999, William Rathje looked at the amazing amount of orbiting debris as an archaeological resource.[13] Rathje is well known for the archaeological subdiscipline of "garbology." He was one of the first to look at modern material culture—specifically refuse in Tucson, Arizona, in the 1970s—as a legitimate archaeological study. A large part of the archaeological record consists of discarded items (e.g., stone chips or plastic bottles), and all periods of discard can provide information about patterns of consumption and use (Staski, Chapter 2 of this volume). Rathje may have been the first archaeologist to coin the term *exoarchaeology*, or "the study of artifacts in outer space," and to call those who study it "exoarchaeologists."[14]

SPACE JUNK VERSUS ARTIFACTS

What is for archaeologists a precious artifact can be considered by others as junk. NASA has been tracking bits and pieces of objects in space (a.k.a. "space junk") and reports on an orbiting cloud of space junk left behind by all those nations who have launched objects into space. Johnson writes that the Earth is surrounded by flotsam that "resembles angry bees around a beehive, seeming to move randomly in all directions."[15] According to Rathje, by 1989 there were about ten thousand detectable resident space objects, only 5% of which were functioning spacecraft.[16] In 1997 the International Space Station missed a collision with the Progress supply freighter. Significant amounts of space junk are from approximately 150 satellites that have blown or fallen apart and left fragments over 10 centimeters (4 inches) to be tracked from Earth. The garbage in space includes objects jettisoned (much of it in garbage bags) from the Soviet *Mir* space station. Artifacts in space have fallen intentionally or unintentionally back to Earth. The 150-ton *Skylab* crashed in the Australian outback in 1979. The *Mir* space station itself collapsed in a fiery heap into the Pacific Ocean off Australia in 2001. At the time of Rathje's article, NASA estimated there were another four hundred thousand space artifacts too small to detect, millions of flakes of paint, and detritus such as urine and fecal matter found on a recovered satellite after 6 years in orbit (Gorman, Chapter 19 of this volume).[17] Johnson in 1999, in his seminal article about the early phase of the exploration of the Moon (1966–1976), looked at human debris in and from lunar orbit, which includes twenty-nine manned and robotic missions that placed more than forty objects into lunar orbit.[18] When all landed or crashed spacecraft and related debris are included, there are more than eighty sites on the Moon comprising more than 100 metric tons of human debris on the lunar surface.[19] In truth, NASA has its own brand of exoarchaeologists trying to keep track of all this material culture because of its potentially devastating effects on current and future missions.

Space is similar to all other frontiers, such as the Camino Real, a historic trail dating from 1588 that started in Chihuahua, Mexico, and ended on the frontier around Santa Fe, New Mexico, where there are patterns of discarded items. Rathje calls frontiers of all kinds "junk magnets."[20] Today, in part, space archaeology concerns itself with the discard practices of what were once a few nations during the Cold War (1946–1989) to the current global practices of many nations. It can be argued that as a legitimate subdiscipline of archaeology, investigations to see how and why any group of humans discard their material culture can be focused on space and other celestial bodies. This might include, in Rathje's estimation, the *Pioneer 11* spacecraft, which has left our solar system.[21]

MUSEUM CURATION AND PRIVATE COLLECTIONS

Special museums such as the Smithsonian Air and Space Museum have collected many of the firsts in space such as the *Apollo 11* capsule. They also have a large range of moveable items of space heritage in their collections. Many of the smaller objects on the open market vary from items owned by astronauts and those bits and pieces of items "flown in space" that can be purchased on the Internet or at auction. A lunar core sample from the Soviet's *Luna 16*, which touched down on the Moon in 1970, collected the sample, and flew it back to Earth, sold at Sotheby's in 1993 for more than $400,000. At the same auction, the title to but not possession of the *Luna 21* and *Lunokhod 2* (both Soviet robotic spacecraft are still resting on the Moon), was sold for $68,500.[22] There continues to be a popular collecting mania for items that have been in space. It has been argued that it is only a question of time before historic archaeological objects may be retrieved from the Moon.

LUNAR LEGACY PROJECT

In many instances, this chapter will presage the rest of book written from the archaeological, museum, and historical perspectives, as many of the authors in this volume were the facilitators of the birth of space heritage. My own research into space archaeology and heritage began when a graduate student, Ralph Gibson, asked me during a Cultural Resource Management seminar at New Mexico State University (NMSU) in 1999 whether U.S. federal preservation law applied on the Moon. The answer to that question is still being investigated and debated. In what may be the earliest instance of funded space archaeological research, the New Mexico Space

Grant Consortium granted monies (1999–2001) to two NMSU students (Gibson and John Versluis), myself, and Dr. Jon Hunner (NMSU Department of Public History) to investigate this and other issues. What is called the Lunar Legacy Project focused on one site, the *Apollo 11* landing site at Tranquility Base on the Moon, as the most obvious site worthy of preservation. We created the Lunar Legacy Web site, which discusses the project.[23] It is one of the earliest investigations of Tranquility Base solely as a historical archaeological site. What was initially thought to be a simple matter of retrieving archived records, maps, and photos to describe the archaeological assemblage was not so simple. Of course, it would be imperative to revisit the site as archaeologists to accurately record and map the site, but NASA did not choose to fund the Lunar Legacy project to that level. One of the outcomes of the project was the completion of an MA thesis in anthropology by Ralph Gibson that examines the historic context of the *Apollo 11* lunar landing site and its significance in the Cold War, describes the site, and discusses the need for its preservation.[24] Working with Dr. Leslie Brown, an astrophysicist at Connecticut College, Gibson and I did a poster presentation on space heritage at the American Astronomical Society Meetings jointly with the American Association of Physics Teachers in San Diego in 2001.[25] The poster was met with interest and discussion among the astronomers, astrophysicists, and educators attending the conference. The Lunar Legacy project and the *Apollo 11* lunar site are described as a historic preservation case study (O'Leary, Chapter 40 of this volume). Interestingly, one of the other graduate students in the same Cultural Resource Management seminar in 1999, Michael Stowe, wrote an article on the Lunar Legacy project for *Discovering Archaeology*.[26] This set off public interest with many popular articles and radio and television interviews.[27,28] What fascinated the public were the ideas that archaeology could be applicable to outer space and that a recent site on the Moon could be thought of historic, although it was less than 50 years old. Many of the space events were experienced by millions of people, if only vicariously through various media such as radio and television. What also fired the public imagination was that first Moon landing site, with its unique assemblage of the earliest evidence of humans on another celestial body, was in a gray area of protective laws that made the Apollo 11 Tranquility Base site vulnerable to future destruction or to unregulated tourism, which Spennemann later called "extreme cultural tourism."[29] The Web site and publicity created by the Lunar Legacy Project allowed many archaeologists with similar interests in the archaeology of outer space to find each other.

SPACE HERITAGE

It seems that the beginning of the twenty-first century inspired the discussion of the space sites and objects as a collection of historic archaeological properties at both professional archaeological and preservation conferences. One of the earliest meetings of what now may be called "space archaeologists" or "space heritage archaeologists" (the terms are still in flux today, along with the lesser used "exoarchaeologists") occurred in 2003 at the Fifth World Archaeological Congress (WAC-5) in Washington, D.C. The theme under which this session occurred was titled "The Heavens Above: Archaeoastronomy, Space Heritage and SETI." It included sessions on the established field of archaeoastronomy, which since 1960 has looked at how premodern peoples understood the workings of the solar system scientifically and in their own cultural context.[30,31] The work of archaeoastronomers has focused on diverse international sites such as Stonehenge in the United Kingdom; Chaco Canyon, New Mexico; and Kaua'i, Hawaii. These prehistoric sites have physical evidence of sophisticated astronomical knowledge about the solstices and the movement of celestial bodies. While archaeoastronomists study the astronomical practices of past cultures, the sister field of ethnoastronomy is the study of cultural narratives about the skies and perceptions of the cosmos in indigenous astronomies (Milbrath, Chapter 10 of this volume).[32] Also in the conference was Doug Vakoch from the Search for Extraterrestrial Intelligence Institute (SETI). SETI is exploring anthropological approaches to interstellar messages and communication. His continuing work is on how anthropology and archaeology can contribute to scientific practice by examination the theoretical presuppositions of scientists and suggesting alternative perspectives.[33] Archaeology can provide analogies for understanding extraterrestrial civilizations through material culture. A Space Archaeology session organized at the WAC-5 under this theme by John Campbell, James Cook University, Australia, and Beth O'Leary, NMSU, was called "Space Heritage and the Potential for Exoarchaeology in the Solar System: National and International Perspectives." This symposium focused on three main areas: definitions of space archaeology, description and evaluation of the significance of classes of artifacts in space and the landscape that is created by the exploration of space, and the logistics and strategies for the development of space heritage legislation, regulation, and policy. The idea of cultural resources on other celestial bodies besides the Moon, such as Venus, the asteroid Eros, and those further out in the Solar System, was also discussed (Campbell, Chapter 46 of this volume).[34]

WORLD ARCHAEOLOGICAL CONGRESS SPACE HERITAGE TASK FORCE

The area of archaeology that focuses on cultural resources and how they are managed and treated legally is generally referred to as cultural resource management (CRM) in the United States, while in much of the English-speaking world CRM is subsumed under the rubric of cultural heritage or heritage management. The WAC-5 addressed cultural heritages issues in space. The first outcome of the conference was a resolution by the WAC-5 for the global archaeological community to recognize that

> the material culture and places associated with space exploration are significant at individual, local, organizational, national and international levels. As space industries and eventual space colonization develop in the 21st century, it is necessary to consider what and how elements of this cultural heritage should be preserved for the benefit of present and future generations.[35]

The second outcome was the creation of a World Archaeological Congress Space Heritage Task Force, whose goals in 2003 were ambitious and far reaching.[36] Both the WAC-5 resolution and the creation of the task force represent the first attempt to plan for the incorporation of space heritage into systems of protection and preservation. The steps required to undertake this goal include identifying themes related to space exploration, addressing significance including "exceptional" significance, providing criteria for evaluating importance at different levels, and exploring the existing avenues for preservation (e.g., the World Heritage Convention) on national and international levels. This would include cooperation with many kinds of partners such as space agencies, aerospace industries, engineers, astronautical and astronomical associations, etc., in the appropriate management of the cultural heritage of space.

Since 2003, space archaeology as a developing discipline has become more inclusive of different approaches and concerns about its identity and its goals. In 2004, the WAC Space Heritage Task Force met again in Montreal, in conjunction with the symposium organized for the Society for American Archaeology (SAA) by O'Leary and Campbell titled "The Outer Limits: Theory and Method in Space Heritage." The SAA is an international organization that since 1934 has been dedicated to research, interpretation, and protection of archaeological heritage and to stimulating interest and research in American archaeology and aiding in the conservation of its resources.[37] The SAA symposium codified that space heritage was an appropriate term for sites and artifacts associated with the exploration of space that humans created and are creating on and off-world such as on the Moon and Mars. It focused on the question of the cultural context of space and how to best document and preserve space heritage artifacts, features, and sites as historic properties. Besides codifying the legal issues, it broadened the idea of space history configured as the "Space Race" between the United States and USSR and the claimed universality of human drive to conquer outer space.[38] Gorman has argued for an investigation of the economic, social, and political inequalities between spacefaring and nonspacefaring nations, between men and women, and between nations and their colonies.[39] As part of her work she looked at the cultural significance of Woomera in Australia, a rocket testing range that displaced and threatened the culture of aboriginal peoples living in the area. Gorman has also focused on objects in outer space and in Earth orbit.[40,41] She carries Rathje's earlier work one step further. One of her major contributions to the study of space archaeology is the idea that space is not "empty" either physically or ideologically.[42] Beginning with the launch of *Sputnik 1* in 1957, Earth orbit has accumulated more than eight to ten thousand trackable objects consisting of satellites, launch vehicle upper stages, mission-related debris, and space junk. Gorman argues that orbital space is now an organically evolving cultural landscape of the Space Age.[43] Both the materials in it and their location in this landscape, or spacescape, may have significance in social, historical, scientific, and aesthetic terms according to the internationally recognized guidelines of the Burra Charter.[44] The increasing amount of debris constitutes a threat for the successful development and delivery of space services and development. Gorman argues that space agencies will have to consider the necessity of removing material from orbital spaces, as they do now from locations on Earth.[45] Proposals include destruction using ground-based lasers, missile technology, electrodynamic tethers, and intervention missions. In the longer term, some orbital material may be the subject of commercial salvage operations. If the owners of orbital material choose, they could move objects into a space graveyard that could serve as a destination for space tourists. Not every object poses the same risk to ongoing and future space operations. There should be way of assessing the risk presented by different debris size classes, so that considerable leeway is created for preserving significant orbital objects such as *Vanguard 1* (the oldest human object in space), Australia's *FedSat* scientific satellite, and other important satellites. Gorman's ongoing investigations discuss the heritage value of orbital material and suggest avenues for managing the archaeological record of human endeavors beyond the atmosphere.[46]

CULTURAL LANDSCAPE OF SPACE

Gorman is also one of the first archaeologists to look at space as part of a cultural landscape (albeit at a much larger scale than any such landscapes on Earth). The idea of landscapes as the focus of archaeological research is a recent one. King claims its origins lie with John Brinkerhoff Jackson and his magazine *Landscape* in the 1951 and geographers such as Yi-Fu Tuan.[47] It has now become part of the archaeological vocabulary.[48,49,50,51,52] Using this approach, sites are viewed as more than discrete locations on a map; they are considered as part of a larger context imbedded with other sites, the natural environment, and those places that are associative with cultural, religious, or artistic values that may or may not have material culture associated with them. For example, Mount Taylor is an 11,301-foot mountain in New Mexico that is named Tsoodził in Navajo, thought of as one of the topographic features that serves to mark directions in the Navajo sacred geography.[53] It is also an important traditional mountain to many Puebloan peoples. In 2008, Mount Taylor was listed by the New Mexico Cultural Properties Review Committee as a Traditional Cultural Property on the State Register of Cultural Properties.[54]

The World Heritage Convention defines cultural landscape as "the combined works of nature and man."[55] Space can be seen as a landscape that is the ultimate wilderness, where, as of yet, humans have not met any indigenous inhabitants. Space, with its celestial bodies, is a very powerful associative landscape for all cultures in their belief systems, creation narratives, mythologies, and traditional environmental knowledge. In reality, space and its celestial bodies belong to all cultures in the world.[56] Gorman adopts a three-tiered approach for places associated with space: designed space landscapes (e.g., launch facilities), organic landscapes in orbit and on celestial bodies (e.g., satellites and landers), and an area beyond the solar system where spacecraft such as *Voyager 1* have ventured.[57]

The concept of landscape is useful because it is so inclusive of material culture and its nonmaterial components. What Gorman desires is inclusion in the spacescape of those outside spacefaring nations and their attendant scientists and engineers, such as people who were part of space history but whose participation was either forced (i.e., slave labor at Nazi factories at Peenemünde, Germany) or whose traditional lands were co-opted without their permission, as was the case with the Kokatha and Pitjantjatjara people for the Woomera Rocket Range in Australia.[58] To assess significance in a holistic way as the World Heritage Convention and the idea of landscape advocates, the complexity of cultural values of all participants needs to be addressed, especially when their cultural values are in conflict. The heritage site in Peenemünde, Germany, is where Wernher von Braun directed work on the V2 rocket and is touted as an important place where the Space Age started. Von Braun, of course, is recognized as one of the fathers of the Space Age, but more recently has been called to task for his part in the deaths of workers at his rocket facilities.[59] Peenemünde is also part of a landscape that includes the Mittelbau complex, where those from concentration camps were made to work on rockets. This Nazi complex accounts for the deaths of almost half of the 60,000 prisoners who worked there as slaves.[60] Gorman states that "Wernher von Braun and Peenemünde illustrate some of the conflicting meanings … in understanding the significance of space heritage."[61] The goal of preservation for future generations is to assess the cultural value and significance of space heritage that should address designed elements on Earth, objects in orbit, and intangible elements, such as the associations with people who participated or were impacted in the history of space exploration. A discussion of the plan for the future preservation of the space landscape is given in Chapter 43 of this volume.

If the landscape approach is pursued then its components can be quite broad in terms of facilities and objects that are associated with space exploration. In fact, the idea encompasses places, structures, features, and objects that may have been used in the history of space exploration and reused for other purposes not associated with space. Spennemann, a professor of cultural heritage management at Charles Sturt University, Australia, advocates a look at the naval heritage of the U.S. space program and how now-obsolete naval vessels were called upon to track and recover space capsules and astronauts from the world's oceans.[62] Lamenting that many of these vessels were critical to the success of the U.S. space programs, Spennemann points out that their heritage value has seldom been addressed. Currently, most of vessels involved have either been sunk or scrapped. The United States Navy Ship (USNS) *Vanguard*, which Spennemann feels is the most significant of the tracking ships that recovered all of the crewed Apollo missions, was still around at the time he wrote the article; however, it is currently marked for disposal by the U.S. Department of Transportation Marine Administration.[63,64] When the elements of space heritage are viewed in their entirety, the number and kinds of technological buildings, structures, objects, and artifacts rank in the millions, and a way to access significance and prioritize preservation is a necessity, especially when it concerns preservation and curation of these resources (Barclay and Brooks, Chapter 37 of this volume; Szczepanowska, Chapters 35 and 36 of this volume).

PRESERVATION AND CURATION OF SPACE HERITAGE

The curation of space heritage was also addressed at the 2004 SAA symposium. In that symposium, two conservators from Canada presented their work. Robert Barclay and Randall Brooks have written a fascinating paper about space heritage from a museum curator's point of view, arguing that the fiery demise of the *Mir* space station was a metaphor for how space heritage is in danger of disappearing (Barclay and Brooks, Chapter 37 of this volume).[65] The *Mir* was the first permanent manned presence in Earth's orbit and is now lost. Museums may have replicas, drawings, and the few items that have been returned to Earth from space, but they are not the same as the real thing in place. The Canada Science and Technology Museum in Ottawa has the actual command module from the *Apollo 7*, with its scorch marks from reentry, along with evidence of its subsequent dismantling by NASA scientists, but many other important historical space objects are lost. Museum professionals argue that representative examples of all space technologies need to considered for preservation. Preservation could also occur not only in space museums like theirs, but actually in space. One strategy for preservation is for objects that because of their size or fragility can best be maintained in space. Barclay and Brooks argue that parking significant objects in situ in space might be the answer.[66] Several ideas have been put forward as to how to park important space objects like the Hubble telescope someplace in space, because chronologically it now sits on the knife edge between "transient" and "durable"—roughly translated, between space junk to be discarded and a historic property to be preserved.[67] If space tourism is going to be part of the future, then the Hubble telescope after about 2015 (its projected useful life span) may be a tourist destination. The apt analogy that is used for preserving space objects is the raising of the *Vasa,* a seventeenth-century vessel, from the bottom of Stockholm harbor. This was not a viable objective until the 1960s. Its restoration and exhibition allows visitors to experience the culture, politics, and trade of Europe at that time. At the SAA symposium, Brooks[68] looked at how to develop criteria for scientific and technological collections concerning space, and Barclay[69] looked at the challenges of the deterioration of objects in space compared with those in terrestrial museums and the possible avenues of research in caring for them.

ARCHAEOLOGY OF THE INACCESSIBLE

Space archaeology seemed a natural choice in a symposium at the University of Sheffield, Sheffield, United Kingdom, in 2005 titled "The Archaeology of the Inaccessible," organized by the Theoretical Archaeology Group (TAG). TAG was founded as a national body in 1979 with the aim of promoting debate and discussion of issues in theoretical archaeology. The symposium presented archaeological projects that have sought to retrieve and analyze the stories of past humanity in places that humans cannot safely go. Besides the presentation on the archaeology of outer space by O'Leary,[70] the presentations included the site of recent prison in Northern Ireland by McAtackney,[71] the drowned landscape in the southern North Sea by Fitch et al.,[72] and a minaret in war-torn Afghanistan by Thomas and Gascoigne.[73] Aitchison argued that it is it the very untouchability of such sites has made those archaeological remains all the more valuable a resource, and the methods developed for accessing that data can be considered to define the very cutting edge of capabilities.[74] It may have been natural processes or the subsequent action of humans that have rendered these places unreachable, but there are also places that required such a great technological investment for humanity to reach them in the first instance that they have simply never been returned to.[75] Martin Carver, editor of *Antiquity,* one of the foremost journals in archaeology, published an editorial on the conference and heritage on the nearest celestial body.[76] The same issue of *Antiquity* also included an article on space heritage by O'Leary.[77]

Working in a parallel track with other researchers, Spennemann has published a series of articles in journals concerned with heritage management and space policy.[78,79,80,81,82] In his earlier papers he discussed the terrestrial and extraterrestrial heritage of the Apollo program as it faced renewed interest in the twenty-first century and a potential return to the Moon. Recognizing that the literature on the topic is small, he presents an argument that the Apollo program comprises a variety of physical sites and artifacts: those on Earth associated with its development, artifacts that went into space and returned to Earth, artifacts still in space, and the lunar samples brought back from the Moon.[83] The lunar samples (i.e., samples of the regolith and rocks), he argues, are part of "natural heritage," although they remain part of the archaeological record because they were collected by astronauts who in many cases left "borrow" areas, which are human features on the lunar landscape just like the footprints. Spennemann and others, most notably O'Leary[84] and Gorman and O'Leary,[85] have argued that the *Apollo 11* lunar landing site "will always remain as the site where humans first set foot onto another celestial body."[86] In Spennemann's estimation it is the "ultimate heritage site, both in terms of significance to humanity as a whole, but also in terms of heritage preservation as a single site."[87] The first lunar landing site on the Moon can boast almost complete, exact preservation from the time it was created to the present, which sites on Earth cannot (Staski and Gerke, Chapter 26 of this volume).

Cultural resource management for anything on the Moon or other planets is not within anyone's national boundaries. Most cultural resource management law rests on the physical location of the site or object being under the sovereignty and jurisdiction of the nation-state that owns it. Spennemann[88] compares the preservation jurisdiction jungle in space to the German warship *Bismark,* discovered in 1989. Its legal owner is still the German government, but that government can only request that that shipwreck be respected as a graveyard for its crew; the request is not legally binding on those who wish to visit the site. In space or on other celestial bodies there is not a real sense of trespass. Some nations, such as the United States, have relatively encompassing historic preservation legislation, while other nations have their own set of criteria for evaluating historic properties and differing applications of heritage law. What many would consider important sites in the space program, such as the radiotelescope at Parkes, Australia, that relayed much of the television coverage of the first walk on the Moon to the world, is not listed on Australia's National Register or State Register.[89] The Honeysuckle Creek facility in Australia, equally important for its communication role in the Apollo 11 mission in 1969, exists now only as a concrete pad with a few interpretative signs.[90]

SPACE TOURISM

If space tourism happens in the future, how will lunar heritage be managed? Tourists on the Moon can obliterate the very footprints and integrity of the first astronaut tracks. As Spennemann[91,92] and Gorman and O'Leary[93] discuss, there is an ethical question of how to allow human access now or in the future to any historic site on the Moon that has the potential to destroy it. In all cultures on earth, unethical individuals collect souvenirs as mementos of their visits to archaeological sites, despite laws protecting the sites and punishing this when it happens. In the face of mass tourism or in the absence of law and lack of respect for all antiquities due to political, social, or economic conditions, it becomes even harder to protect even recent and very remote antiquities. The trend toward "extreme cultural tourism" and technological advances have allowed increased visits to places like Mount Everest, deep underwater sites of shipwrecks such as the *Titanic*, and Antarctica.[94] This has caused real impacts to both natural and cultural resources and has forced the international community to try to find solutions and set up rules for tourism.

Space tourism began with nonastronauts such as Christa McAuliffe, the teacher who died in the *Challenger* crash in 1986. Other tourists who paid for their flight include Dennis Tito in 2001, an American who paid millions to be trained in Russia and went onboard the International Space Station. For the Russians this kind of tourism was an important source of income. It was initially condemned by NASA, but an agreement was reached and rules delineated so that three other space tourists, including most recently, in 2006, a woman, Anousheh Ansari, have gone to the space station and returned. The issues of ownership, visitation, and how to visit lunar sites without greatly impacting the qualities that evidence their significance are unresolved. Taking only pictures and *not* leaving any footprints may be the ultimate solution because humanity cannot afford to lose sites "that are associated with the greatest adventure embarked on so far."[95]

Spennemann and Murphy[96] have also looked at cultural heritage of Mars and the human traces left on the planet beginning with the crash landing of the USSR's *Mars 2* in 1971. The red planet is one of the few celestial bodies besides the Moon that has human artifacts on it. In the broadest sense, an archaeological study of Mars in a cultural landscape would include successful landings and their attendant material culture on the planet's surface, as well as failed landings. There were 37 launch attempts, and the majority of them were unsuccessful, especially in the early days when the technology was first being tried and there was not much information about the conditions on Mars. The physical conditions of the Martian surface include dust storms and high winds, great changes in temperature, and high levels of ultraviolet light, all factors that would affect the material and locational integrity of sites and artifacts. Surprisingly, the *Viking 1* lander, which remains on Mars, is considered part of the Smithsonian Air and Space Museum. Importantly, on Mars, Spennemann and Murphy[97] identify a series of what they broadly call "artifacts" that include technological (e.g., heat shields), environmental (e.g., rover tracks), and cultural, which may not have a physical manifestation (e.g., a named landscape topographic feature). The exact location of many of these artifacts is unknown. Again, they stress the need for future management strategies.

ROBOTIC HERITAGE

What lies, in part, at the edge of theory and method of archaeology and cultural resource management is the recent work of Spennemann[98] on the cultural heritage of robots and actual archaeological investigation at the purported UFO 1947 crash site in Roswell, New Mexico (Doleman, Chapter 38 of this volume). Spennemann's[99] article is meant as a discussion of the future of human heritage and what is termed emergent heritage—technologies that are still in use and form an ongoing technological investment. He is in agreement with Staski (Chapter 2 of this volume) that archaeology can investigate the present, and therefore, protective measures can also be applied to contemporary technology. Spennemann[100] goes further still to discuss not only the need to preserve robotic heritage (i.e., the material correlates of robots) but the heritage created by robots in the form of their artificial intelligence. Earlier, Capelotti in 1996[101] had written that this "bring[s] us into a dicey area of inquiry, if for no other reason than its seemingly obvious implication that tool[s] of human ingenuity are increasingly less our servant and more our master." Spennemann[102] wants to expand the discourse on heritage in a cultural landscape and the artifacts and places created and modified by robots and rovers. Lunar heritage is rife with all kinds of robotics even in the earliest stages, with the Luna robotics sent by the Soviets and the American *Surveyor* to the robotics used today on the surface of Mars. Capelotti writes about an archaeological record "created by the actions of machines that in some cases operate on the other side of the system from the hand and mind that presumably control them."[103] Although far from mainstream archaeology, both Spennemann[104] and Capelotti[105] argue for finding ways to value and preserve a new type of heritage that is part human and part machine. In the future, if there is a legal standing for the artificial intelligence and rights of robots, then their heritage must be recognized. Earth's robotic heritage exists both on the planet and in outer space.

ROSWELL UFO CRASH SITE

Also within the realm of space archaeology but on its outskirts is the work done by William Doleman. While much speculation and science fiction and many Hollywood movies have dealt with aliens coming to Earth, few archaeological studies have been done to try to locate them. In 2002, the University of New Mexico's Office of Contract Archaeology, with funding from the Sci Fi Channel, did some testing at the putative Roswell, New Mexico, UFO crash site. Doleman as principal investigator tested the site as an archaeologist while using geophysical studies. It was a location where in 1947 a UFO had been "sighted" and where a rancher claimed it left physical marks in the soil and a debris field. A summary of that project is included in this volume (Doleman, Chapter 38 of this volume).

While it is true that the early cadre of science fiction writers such as H.G. Wells may have inspired humans to go into space, no real proof exists that we have ever been visited or influenced by those coming from outer space. Space archaeology and heritage does not embrace in any way the nonscientific theories and methods of those like Erich von Däniken, who essentially wants to "prove" that ancient peoples did not have the intellectual abilities to create architectural monuments at a large and complex scale.[106] While insulting the actual intelligence of *Homo sapiens* in the past, who had the same brain capacity as contemporary ones, he uses the concept of "unexplained" mysteries as "proof" that humans could not have created the marvelous feats of the engineering that have scientifically been dated to the past. In his way of thinking, humans must have had to have had help from more intelligent aliens from space. Also, he compares prehistoric symbols and images to how astronauts look. For example, in his scheme the ruler of the ancient Maya engraved on a tomb becomes an example of an ancient astronaut, a total misuse of the concept of analogy.

EXTRATERRESTRIAL INTELLIGENCE

Others have looked for signs of extraterrestrial intelligence. The astronomer Arkhipov[107] has looked at the *Clementine* lunar orbiter images as a kind archaeological reconnaissance on the Moon, but there are no defined ruins in his studies that would convince an archaeologist. Archaeology is not a quasi-science that is dependent upon beliefs. Even if many people in our contemporary world believe in UFOs or aliens landing on Earth, there are no material correlates. Those who believe that the Moon landing was a hoax created in a television studio could be dissuaded by sending an archaeologist back to the original site where Armstrong and Aldrin walked, because the artifacts they left in 1969 would still be there.

We may speculate, like Drake, that there is a way to estimate the number of technologically advanced civilizations that might exist in our Galaxy. He conceived a means, called the Drake equation, to mathematically estimate the number of worlds that might harbor beings with technology sufficient to communicate across the vast gulfs of interstellar space.[108] Other civilizations may exist in the cosmos, but so far we have no material evidence of any.

Two of the most recent conferences on space heritage happened in 2007 and 2008. A conference titled "Extreme Heritage" was organized by the Australian International Council on Monuments and Sites (ICOMOS) in 2007.[109] It followed a theme similar to that of the TAG meetings and took place in Cairns, Australia, at James Cook University. ICOMOS is primarily concerned with the philosophy, terminology, methodology, and techniques of conservation. It is a nongovernment professional organization formed in 1965. It is closely linked to UNESCO and serves as an advisory to the World Heritage Convention. The ICOMOS Australia conference theme covered a wide spectrum of natural disasters, military conflicts, and sites in remote locations. It reiterated the challenges of managing cultural heritage in the changing and volatile modern world. It included the topic of climate change and heritage.[110] One of the invited keynote addresses, "Historic Preservation at the Edge," by O'Leary,[111] was given on the 39th anniversary of the *Apollo 11* lunar landing. A symposium titled "The Heritage of Off-World Landscapes," organized by Alice Gorman and John Campbell, included their work and a presentation by Dirk Spennemann[112] on failed missions on Mars as forensic investigation scenes. Spennemann explored the management and ethics of conflicting values between those doing crash analyses of space vehicles, including such tragedies as the *Challenger*, and heritage conservation. Brent Biddington,[113] a nonarchaeologist specializing in Australia's future in space, presented a fascinating paper on how humans will value, store, and care for data, whether in space or on Earth and how to capture the impact of cyberspace on some of the more extreme aspects of human behavior in the distant future. One of the best outcomes was a discussion with John Hurd (ICOMOS International Advisory Committee), who suggested the formation of an ICOMOS International Scientific Committee on Space Heritage.[114] ICOMOS has scientific international committees on various cultural heritage themes and issues. Members consist of internationally renowned expert specialists in each subject. The international scientific committees include ICOMOS technical groups, which undertake research, develop conservation theory, guidelines, and charters; the committees also foster training for better heritage conservation, promote international exchange of scientific information, and carry out common projects.[115]

Another recent conference was held in Dublin, Ireland, in June 2008 at the Sixth World Archaeological Congress, where a series of papers on space heritage was given in a symposium called "Nostalgia for Infinity: Exploring the Archaeology of the Final Frontier." WAC-6 continues the work of the original WAC-5 Space Heritage symposium and the task force.

CONCLUSION

The future of space archaeology, like that of any emerging science, cannot be predicted; the field waits to be developed by a dedicated cadre of varied professionals in the discipline. Its early history has focused on several critical areas: defining it as a legitimate subdiscipline of archaeology, providing its historic context during the Cold War and in the contemporary period, addressing the quality and quantity of the archaeological record, how to document this record, and lastly, beginning to think about how and why humanity needs to preserve it for future generations.

REFERENCES

1. Rathje, W. 1999. An Archaeology of Space Garbage. *Discovering Archaeology* October: 108–112.
2. Capelotti, P.J. 2004. Space: The Final (Archaeological) Frontier. *Archaeology* 57(2): 48–55.
3. Capelotti, P.J. 1996. A Conceptual Model for Aerospace Archaeology: A Case Study from the Wellman Site, Virgohamna, Dansköya, Svalbard. PhD diss., Rutgers University, University Microfilms, 9633681.
4. NASA. 1972. Analysis of Surveyor 3 Material and Photographs Returned by Apollo 12. NASA SP-28. Washington, DC: U.S. Government Printing Office.
5. Ibid.
6. Capelotti 2004 (note 2).
7. Gould, R.A., ed. 1983. *Shipwreck Archaeology*. Albuquerque, NM: University of New Mexico Press.
8. Fischer, G. 1993. The History of the ACUA. Keynote Retrospective at the 1993 Conference on Historical and Underwater Archaeology, Kansas City, Kansas. http://www.acuaonline.org/ACUA_Hist.pdf.
9. Butowsky, H.A. 1984. *Man in Space: A National Historic Landmark Theme Study*. Washington, DC: U.S. National Park Service, U.S. Department of Interior.
10. Butowsky, H.A. 1986. Man in Space: These Are the Voyages of. *CRM* 9(2): 5–7.
11. Butowsky, H.A. 1987. The Man in Space: The Voyage Continues. *CRM* 10(6): 8–11.
12. O'Leary, B. 2008. Historic Preservation in Space: Archaeology on the Moon. Paper presented at the Society for Historical Archaeology, Albuquerque, NM.
13. Rathje 1999 (note 1).
14. Rathje 1999 (note 1).
15. Johnson, N. 1998. Monitoring and Controlling Debris in Space. *Scientific American* 279(2): 62–67.
16. Rathje 1999 (note 1).
17. Rathje 1999 (note 1).
18. Johnson, N.L. 1999. Man-Made Debris in and from Lunar Orbit. *American Institute of Aeronautics and Astronautics*. IAA-99-IAA.7.1.03.
19. Ibid.
20. Rathje 1999 (note 1).
21. Rathje 1999 (note 1).
22. Kluger, J. 1994. The Bloc on the Block—Auction of Soviet Space Memorabilia. *Discover*, April 1. http://discovermagazine.com/1994/apr/theblocontheblоc363.
23. Lunar Legacy Web site. http://spacegrant.nmsu.edu/lunarlegacies.
24. Gibson, R. 2001. Lunar Archaeology: The Application of Federal Historic Preservation Law to the Site Where Humans First Set Foot upon the Moon. MA thesis, New Mexico State University.

25. O'Leary, B., R. Gibson, J. Versluis, and L. Brown. 2001. Preserving a Lunar Legacy. Poster presented at the joint meetings of the American Astronomical Society and the American Association of Physics Teachers, San Diego, CA.
26. Stowe, M. 2000. One Small Step for a Man, One Giant Job for Archaeologists. *Discovering Archaeology.* October: 11.
27. Dye, L. 2000. Group Seeks Protection for Lunar Remains. ABC News.com, accessed on the Lunar Legacy Web site. http://spacegrant.nmsu.edu/lunarlegacies.
28. Lunar Legacy Web site (note 23).
29. Spennemann, D. 2007. Extreme Cultural Tourism From Antarctica to the Moon. *Annals of Tourism Research* 34(4): 898–918.
30. Ruggles, C. 2003. Archaeo-Astronomy: Conflicting Perspectives. Paper presented at the Fifth World Archaeological Congress, Washington, DC.
31. Carlson, J. 2003. Archaeo-Astronomy in the Americas: A Perspective for the New Millenium. Paper presented at the Fifth World Archaeological Congress, Washington, DC.
32. Chamberlain, V.D., J.B. Carlson, and M.J. Young, eds. 2005. *Songs from the Sky: Indigenous Astronomical and Cosmological Traditions of the World.* College Park, MD: Ocarina Books, Ltd.
33. Vakoch, D. 2003. Anthropology, Archaeology and Interstellar Communication Science and the Knowledge of Distant Worlds. Symposium at the Fifth World Archaeological Congress, Washington, DC. Bogner Regis, UK: Center for Archaeoastronomy.
34. Campbell, J.B. 2003. Assessing and Managing Human Space Heritage in the Solar System: The Current State of Play and Some Proposals. Paper presented at the Fifth World Archaeological Congress. http://godot.university.sa.edu.au/wac/paper.php?paper=150.
35. World Archaeological Congress Space Heritage Task Force. http://ehlt.flinders/au/wac/site/active_space.php.
36. Ibid.
37. Society for American Archaeology. http://www.saa.org.
38. O'Leary, B. 2004. Federal Law and the Archaeological Sites on the Moon. Paper presented at the Society for American Archaeology, Montreal, Canada.
39. Gorman, A. 2003. The Cultural Landscape of Space. Paper presented at the Fifth World Archaeological Congress, Washington, DC. Bogner Regis, UK: Center for Archaeoastronomy.
40. Gorman, A. 2005. The Cultural Landscape of Interplanetary Space. *Journal of Social Anthropology* 5(1): 85–107.
41. Gorman, A. 2007. Leaving the Cradle of Earth: The Heritage of Low Earth Orbit, 1957–1963. Paper presented at the Extreme Heritage Conference ICOMOS Australia, Cairns, Australia.
42. Gorman, A., and B. O'Leary. 2007. An Ideological Vacuum: The Cold War in Outer Space. In *A Fearsome Heritage: Diverse Legacies of the Cold War*, eds. J. Schofield and W. Cocroft, 73–92. Walnut Creek, CA: Left Coast Press.
43. Gorman 2007 (note 41).
44. ICOMOS Australia 1999. Burra Charter. http://www.icomos.org/australia.
45. Gorman 2005 (note 40).
46. Gorman 2007 (note 41).
47. King, T. 2003. *Places That Count.* Walnut Creek, CA: Altamira Press.
48. Cleere, H. 1995. Cultural Landscapes as World Heritage. *Conservation and Management of Archaeological Sites* 1: 63–69.
49. Groth, P., and T.W. Bressi, eds. 1997. *Understanding Ordinary Landscapes.* New Haven, CT: Yale University Press.
50. National Register of Historic Places. n.d. 30. *Guidelines for Evaluating and Documenting Rural Historic Landscapes.* Washington, DC: National Park Service.
51. National Register of Historic Places. 1990. 38. *Guidelines for Evaluating and Documenting Traditional Cultural Properties.* Washington, DC: National Park Service.

52. Hirsh, E., and M. O'Hanlon, ed. 1995. *The Anthropology of Landscape: Perspectives on Place and Space.* Oxford, United Kingdom: Clarendon Press.
53. Schwarz, M.T. 1997. *Molded in the Image of Changing Woman: Navajo Views of the Human Body and Personhood.* Tucson, AZ: University of Arizona Press.
54. Clark, H. 2008. Mount Taylor Granted an Emergency Listing as Cultural Property. *Albuquerque Journal*, February 23, 2008.
55. World Heritage Convention. 1998. Operational Guidelines for the Implementation of the World Heritage Convention. Section 35-4. http://whc.unesco.org/archive/opguide08-en.pdf.
56. O'Leary, B. 2006. The Cultural Heritage of Space, the Moon and Other Celestial Bodies. *Antiquity*, 80(307). http://antiquity.ac.uk/ProjGall/oleary/index.html.
57. Gorman 2005 (note 40).
58. Ibid.
59. Brzezinski, M. 2007. *Red Moon Rising: Sputnik and the Hidden Rivalries That Ignited the Space Age.* New York: Henry Holt and Company.
60. Neufield, M.J. 1996. *The Rocket and the Reich: Peenemünde and the Coming of the Ballistic Missile Era.* Cambridge, MA: Harvard University Press.
61. Gorman 2005 (note 40).
62. Spennemann, D. 2005. The Naval Heritage of the US Space Programme: A Case of Losses. *Journal of Maritime Research,* October. http://www.jmr.nmm.ac.uk/spennemann.
63. Ibid.
64. U.S. Department of Maritime Administration. 2008. National Defense Reserve Fleet Inventory for Month Ending May 31, 2008. Division of Sealift Operation (MAR-612) on June 3, 2008. http://www.marad.dot.gov/Offices/Ship/Current_Inventory.pdf.
65. Barclay, R., and R. Brooks. 2002. In Situ Preservation of Historic Spacecraft. *Journal of the British Interplanetary Society* 55(5–6): 173–181.
66. Ibid.
67. Ibid.
68. Brooks, R. 2004. Cultural Heritage in Space. Paper presented at the Society for American Archaeology, Montreal, Canada.
69. Barclay, R. 2004. The Conservation of Historic Spacecraft beyond Earth. Paper presented at the Society for American Archaeology, Montreal, Canada.
70. O'Leary, B. 2005. The Ultimate Inaccessibility? Archaeology on the Moon, in Space and on Other Celestial Bodies. Paper presented at the Theoretical Archaeology Group, University of Sheffield, Sheffield, United Kingdom.
71. McAtackney, L. 2005. Issues Surrounding Accessing the Archaeology of the Recent Past: Long Kesh/Maze Prison Site, Northern Ireland. Paper presented at the Theoretical Archaeology Group, University of Sheffield, Sheffield, United Kingdom.
72. Fitch, S., V. Gaffney, and K. Thomson. 2005. A River Runs through It: Reconstructing the Drowned Landscapes of the Southern North Sea. Paper presented at the Theoretical Archaeology Group, University of Sheffield, Sheffield, United Kingdom.
73. Thomas, D., and A. Gascoigne. 2005. The Minaret of Jam Archaeological Project (Afghanistan). Paper presented at the Theoretical Archaeology Group, University of Sheffield, Sheffield, United Kingdom.
74. Aitchison, K. 2005. The Archaeology of the Inaccessible. Symposium at the Theoretical Archaeology Group, University of Sheffield, Sheffield, United Kingdom.
75. Ibid.
76. Carver, M. 2006. Editorial. *Antiquity* 80: 307.
77. O'Leary 2006 (note 56).
78. Spennemann, D. 2004. The Ethics of Treading on Neil Armstrong's Footprints. *Space Policy* 20: 279–290.
79. Spennemann 2005 (note 62).

80. Spennemann, D. 2006. Out of This World: Issues of Managing Tourism and Humanity's Heritage on the Moon. *International Journal of Heritage Studies* 12(4): 356–371.
81. Spennemann, D., and G. Murphy. 2007. Technological Heritage on Mars: Towards a Future of Terrestrial Artifacts on the Martian Surface. *Journal of the British Interplanetary Society* 60: 42–53.
82. Spennemann, D. 2007. On the Cultural Heritage of Robots. *International Journal of Heritage Studies* 13(1): 4–21.
83. Spennemann, D. 2004 (note 78).
84. O'Leary 2006 (note 56).
85. Gorman and O'Leary 2007 (note 42).
86. Spennemann, D. 2004 (note 78).
87. Spennemann, D. 2004 (note 78).
88. Ibid.
89. Ibid.
90. Honeysuckle Tracking Station Web site. Canberra, Australia. http://www.waymarking.com/waymarks/WM1KQV.
91. Spennemann, D. 2004 (note 78).
92. Spennemann and Murphy 2007 (note 81).
93. Gorman and O'Leary 2007 (note 42).
94. Spennemann 2007 (note 29).
95. Ibid.
96. Spennemann and Murphy 2007 (note 81).
97. Spennemann and Murphy 2007 (note 81).
98. Spennemann 2007 (note 82).
99. Spennemann 2007 (note 82).
100. Spennemann 2007 (note 82).
101. Capelotti 1996 (note 3).
102. Spennemann 2007 (note 82).
103. Capelotti 1996 (note 3).
104. Spennemann 2007 (note 82).
105. Capelotti 1996 (note 3).
106. Von Däniken, E. 1968. *Chariots of the Gods?* New York: Putnam Books.
107. Arkhipov, A. 1999. Preliminary Search for Ruin-Like Formations on the Moon. *Meta Research Bulletin* 8(4): 49–54.
108. Drake, F. Drake's Equation. http://www.setileague.org/general/drake.htm.
109. ICOMOS Australia. 2007. http://www.icomos.org/australia.
110. Ibid.
111. O'Leary, B. 2007. Historic Preservation at the Edge: Archaeology on the Moon, in Space and on Other Celestial Bodies. Paper presented at the Extreme Heritage Conference ICOMOS Australia, Cairns, Australia.
112. Spennemann, D.H.R., and G. Murphy. 2007. Lost in Space: Failed Mars Mission Landing Sites as Forensic Investigation Scenes. Paper presented at the Extreme Heritage Conference ICOMOS Australia, Cairns, Australia.
113. Biddington, B. 2007. Reconstructing the Reality of Remote and Complex Systems in the Future. Paper presented at the Extreme Heritage Conference ICOMOS Australia, Cairns, Australia.
114. Hurd, J. Personal communication, July 22, 2008.
115. ICOMOS International Scientific Committee. http://www.international.icomos.org/isc_eng.htm.

4 Space Basics: The Solar System

Jerold Emhoff

CONTENTS

INTRODUCTION

The solar system is extremely important to space archaeology, since all human-made objects but two are somewhere inside it. The two exceptions are *Voyager 1* and *Voyager 2*, which are outside of the solar system completely.[1] The *New Horizons* spacecraft will most likely also leave the solar system within the next 20 years (as will its third-stage booster rocket and two other pieces of debris).[2]

Besides these, all other objects made by humans exist in the environment of our solar system. This consists of the sun, eight planets and their 166 known moons, three dwarf planets and their 4 known moons, an asteroid belt, the Kuiper Belt, comets, and other, smaller objects. Figure 4.1 shows the orbits of all the planets, as well as Pluto, in a scale drawing. Even though the sun is enormous compared with the planets, it cannot be seen in this figure because it is 100 times smaller than the distance between itself and the Earth. This gives an indication of just how vast the distances are in space, even inside our own solar system. The sizes of the planets vary widely as well, as Figure 4.2 shows.

Each of these bodies has its own environment and properties that can affect human-made objects orbiting or landing on them. This chapter will give an overview of the characteristics of these celestial bodies. Further information about the planets and other objects in our solar system can be found in the appropriate texts.[3,4]

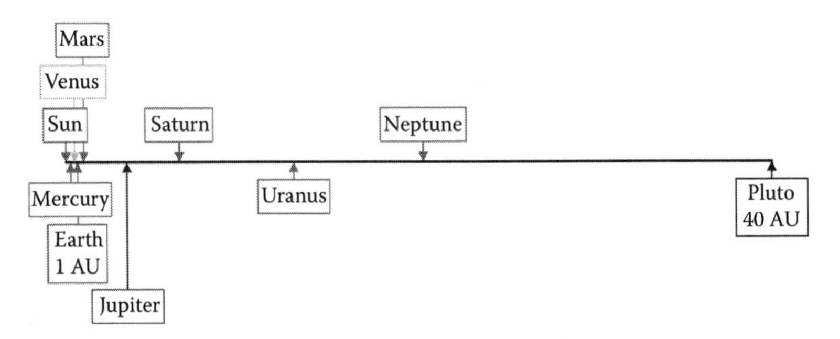

FIGURE 4.1 A scale diagram of the planetary orbits. On this scale, human eyes cannot even see the shape of the sun, which is 100 times smaller than the distance between it and the Earth.

FIGURE 4.2 Approximately scale picture of the planets and the sun. All eight planets are shown in order, starting from Mercury at the top left and ending with Neptune at the lower right.

SUN

Our sun, called Sol, is a medium-sized star with a diameter of about 1.39×10^6 km, 109 times that of the Earth. The sun is composed mainly of hydrogen and helium, with trace amounts of various other elements. Its temperature depends on the location: the core is estimated to be at about 1.5×10^7 K, the surface is about 5,778 K, and the corona is about 5×10^6 K. Its mass is approximately 2×10^{30} kg, more than 330,000 times more massive than the Earth, accounting for 99.8% of the mass in the entire solar system. Figure 4.3 shows an image of the sun's 60,000°C to 80,000°C layer.

The sun follows a 11-year cycle of activity, with high activity and many sunspots at solar maximum, and low activity and few sunspots at solar minimum. Coronal mass ejections (CMEs) are much more common during solar maximum as well, and are more powerful. A CME is an explosion of plasma that erupts from the sun's surface and is accelerated away by the strong magnetic fields of sunspots. The plasma is highly ionized, and when it passes over a spacecraft it can cause damage and failures due to the resulting electromagnetic pulse.

SOLAR WIND

The sun is continuously emitting a stream of plasma called the solar wind. This plasma permeates the solar system and to some extent defines the boundaries of the system. Like the sun, the solar wind is composed mainly of ionized hydrogen and

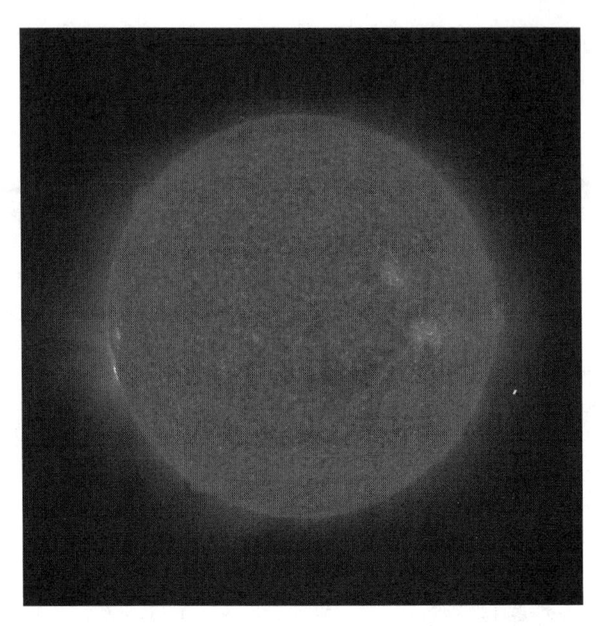

FIGURE 4.3 An image of the sun taken by the ultraviolet imager onboard the STEREO spacecraft. The image shows the layer of the sun with temperatures between 60,000°C and 80,000°C. (Courtesy of NASA.)

helium, with trace elements such as highly ionized oxygen. It is a neutral plasma, so the ions are accompanied by electrons in approximately the same density. The solar wind is accelerated as it leaves the sun, reaching velocities of 500–800 km/s.

The solar wind interacts with the atmospheres and magnetic fields of the planets as well as the interstellar medium. The Earth has a strong magnetic field, which protects it from the bulk of the solar wind radiation, but solar wind particles can follow the magnetic field lines to the North and South poles, creating the aurora borealis. The solar wind forms a bow shock with the interstellar medium at the edge of the solar system, where the solar wind stops and the interstellar medium begins.

EARTH

The Earth, pictured in Figure 4.4, is the third planet from the sun, has a diameter of about 13,000 km, a mass of just under 6×10^{24} kg, and a magnetic field strength between 30 and 60 microtesla. It has a slightly elliptical (eccentricity of 0.0167) orbit 1.5×10^{8} km from the center of the sun. This distance is defined as 1 astronomical unit (AU). The primary atmosphere of the planet reaches about 100 km above the surface, only 1.5% of the Earth's radius. The atmosphere and magnetic field form a bubble inside the solar wind, with a bow shock occurring about 13 Earth radii from the planet.

The magnetic field also results in regions of trapped radiation called the Van Allen radiation belts. These are tori of ionized particles that follow the magnetic field lines of the Earth. Since the field lines go into the surface at the magnetic poles,

FIGURE 4.4 Picture of the Earth taken during the *Apollo 17* mission. (Courtesy of NASA.)

the radiation approaches the surface as well, but is stopped by the dense atmosphere. Even so, the ionized particles can cause phenomena such as the aurora borealis, especially when excited by a coronal mass ejection.

Objects in orbit about the Earth can experience damage due to the atmosphere or radiation from the Van Allen belts or from the solar wind and CMEs. There are also many pieces of debris in orbit about the Earth that bombard anything in orbit. The debris may take the form of a micrometeoroid, a cast-off part from a satellite, or even something as small as a paint flake. Because of the high velocities in orbit, even extremely small pieces of debris can cause severe damage to a spacecraft.

Moon

Earth's only moon orbits the planet in an elliptical orbit with a semimajor axis of about 384,000 km and eccentricity of 0.0549. This results in a closest approach to the Earth's center (perigee) of about 363,000 km and a furthest distance (apogee) of about 406,000 km. It completes an orbit every 27.3 days. The Moon, pictured in Figure 4.5, is approximately 3,474 km in diameter, 0.273 times the size of the Earth. Its mass is 7.35×10^{22} kg, 0.0123 times the mass of the Earth. The Moon has no

FIGURE 4.5 A picture of the Moon taken by the *Galileo* spacecraft. (Courtesy of NASA, the Jet Propulsion Laboratory, and the U.S. Geological Survey.)

atmosphere to speak of, and the surface temperature depends greatly on whether or not it is facing the sun. During the lunar day, the temperature is 107°C on average, and at night it averages –153°C. Gravity on the Moon is 16% that on Earth, and it has a magnetic field much weaker than the Earth's.

The Moon rotates synchronously with the Earth, meaning that the same side of the Moon always faces the Earth. This has led the nonvisible side of the Moon to be called the "dark side," but this is in fact a misnomer as that side is struck by direct sunlight every day.

The surface of the Moon is covered in a fine dust called regolith. This dust can be very detrimental to human-made objects, as well as human beings, because of its small size, jaggedness, and tendency to acquire static charge. The surface is also a harsh environment because of radiation from the sun. Without an atmosphere or magnetic field to block radiation, any object on the surface receives the full force of any coronal mass ejection.

PLANETS

Our solar system has eight planets and three dwarf planets. Pluto was once considered a planet, but it has been downgraded to dwarf planet status. The planets, their moons, and their properties are detailed here.

Mercury

Mercury is the closest planet in the solar system to the sun, and as such it is one of the most difficult to study. It has an elliptical orbit with a semimajor axis of 0.387 AU and eccentricity of 0.2, completing an orbit around the sun once every 88 Earth days. It has a diameter of 4,878 km, 0.383 times smaller than the Earth. Mercury has a mass of 3.3×10^{23} kg, 0.055 times that of Earth, and a magnetic field approximately 100 times smaller than Earth's. The planet rotates much more slowly than the Earth, taking 59 Earth days to complete a single rotation. A picture of Mercury is shown in Figure 4.6.

Mercury has no significant atmosphere, with the gas at the surface coming from the solar wind, as well as decomposition of elements in the surface. The surface temperature ranges from 427°C to −183°C, depending on the location relative to the sun. This is an extremely harsh environment because of the large temperature swings as well as the closeness to the sun's radiation. Mercury has no moon.

Mercury has been visited in the past by the *Mariner 10* spacecraft, which took magnetic field data as well as images of roughly half of the surface. The *MESSENGER* spacecraft performed the first of three flybys in January 2008 and the second in October 2008, with the third to occur in September 2009. Following that, *MESSENGER* will enter into an elliptical orbit about the planet in 2011. The probe has instruments for imaging, spectroscopy, and magnetic field measurements.

FIGURE 4.6 Image of Mercury taken by the *MESSENGER* spacecraft. (Courtesy of NASA, the Johns Hopkins University Applied Physics Laboratory, and the Carnegie Institute of Washington.)

Venus

Venus is the Earth's inside neighbor, notable for its highly corrosive atmosphere and high surface temperatures. Venus has a slightly elliptical orbit about the sun, with a semimajor axis of 0.723 AU and eccentricity of 0.0068. It takes 224.7 Earth days to orbit the sun and completes a rotation only once every 243 Earth days, the slowest planetary rotation in the solar system. It has a diameter of 0.95 that of Earth and a mass 0.815 times that of Earth, a very weak magnetic field, and no moon.

Venus has a significant atmosphere composed primarily of carbon dioxide, with highly corrosive clouds of sulfur dioxide. These clouds can be seen in Figure 4.7, where Venus' upper atmosphere is shown in ultraviolet light. The atmosphere is very dense, especially at the surface, where the pressure is about 92 times that at the surface of Earth. It is also a hot planet, with surface temperatures of more than 460°C, because of an extremely strong greenhouse effect.

The combination of the sulfur clouds, high temperatures and pressures, and lack of a magnetic field give Venus one of the most difficult environments in the solar system for exploration by human-made objects. The sulfur corrodes any objects moving

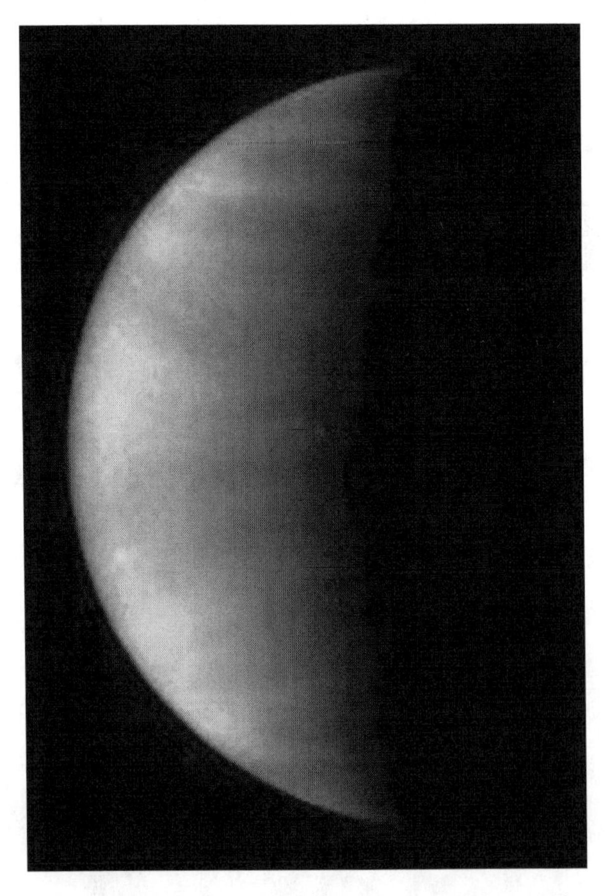

FIGURE 4.7 The cloud tops of Venus as imaged by the Hubble Space Telescope in ultraviolet light. (Courtesy of NASA and the Jet Propulsion Laboratory.)

through the atmosphere, radiation from the sun is able to reach the surface without a magnetic field to stop it, and the temperature and pressure make it difficult to operate electronics and machinery. Even so, many human-made objects have explored Venus to date. The U.S. *Mariner 2* spacecraft orbited the planet briefly in 1962, measuring temperatures and the magnetic field. *Mariner 5* performed a flyby at a lower altitude, making additional atmospheric measurements in 1967. *Mariner 10* imaged the planet on its way to Mercury in 1974. The American *Pioneer Venus Orbiter* and *Pioneer Venus Multiprobe* visited the planet in 1978.

The Soviet Venera spacecraft were primarily designed for entry into the atmosphere of Venus, and in some cases direct landing or impact on the surface. *Venera 3* impacted the surface in 1966 but did not transmit any data, while *Venera 4* parachuted through the atmosphere in 1967, one day before *Mariner 5* reached the planet, making several measurements, including atmospheric temperature and composition. *Venera 5* and *Venera 6* both reached the planet in 1967, only 5 days apart. The Venera spacecraft missions continued up to *Venera 16* in 1983, performing various missions on the surface and in the atmosphere. Notably, *Venera 9* and *Venera 10* sent back images of the surface in 1975. The last visit to Venus was in 1985, by the Soviet Vega probes.

Mars

Mars is the outer neighbor of the Earth, orbiting in an elliptical orbit with a semimajor axis of 1.524 AU and eccentricity of 0.0933. Pictured in Figure 4.8, it has a diameter of 6,792 km, 0.533 times that of Earth, and a mass of only 0.107 times Earth's. It orbits the sun once every 687 days and rotates at almost the same rate as Earth, having 24 hours and 39 minutes in a day. The planet has a tilt close to that of the Earth, so it also experiences seasons, although at a slower rate. Mars has no magnetic field, and it has two small moons called Phobos and Deimos.

The atmosphere of Mars is composed almost entirely of carbon dioxide, at a pressure that varies depending on the altitude, but is generally 1% that of Earth. The temperature on the surface is –46°C on average. During the Martian winter, ice caps composed of carbon dioxide form, and water also exists on the planet in the form of ice. Some evidence of small amounts of liquid water has been seen, although the liquid itself has not. The surface is very dusty, with regular dust storms occurring as well. This can cause problems for operations on the surface, although it is speculated that the high winds of the storms have actually lengthened the service lives of the *Spirit* and *Opportunity* rovers by cleaning their solar arrays.

Phobos and Deimos

Phobos is the larger and closer of Mars' two moons. Its orbit has a semimajor axis of about 9,377 km and has a diameter of roughly 22 km. Deimos is even smaller, with an approximate diameter of 12 km, and has an orbital radius of 23,460 km. Both moons share characteristics with asteroids in terms of density and albedo.

Exploration

Mars has been visited by several spacecraft, including three ground rovers, although many of the missions failed for various reasons. By the end of 2006, thirty-seven

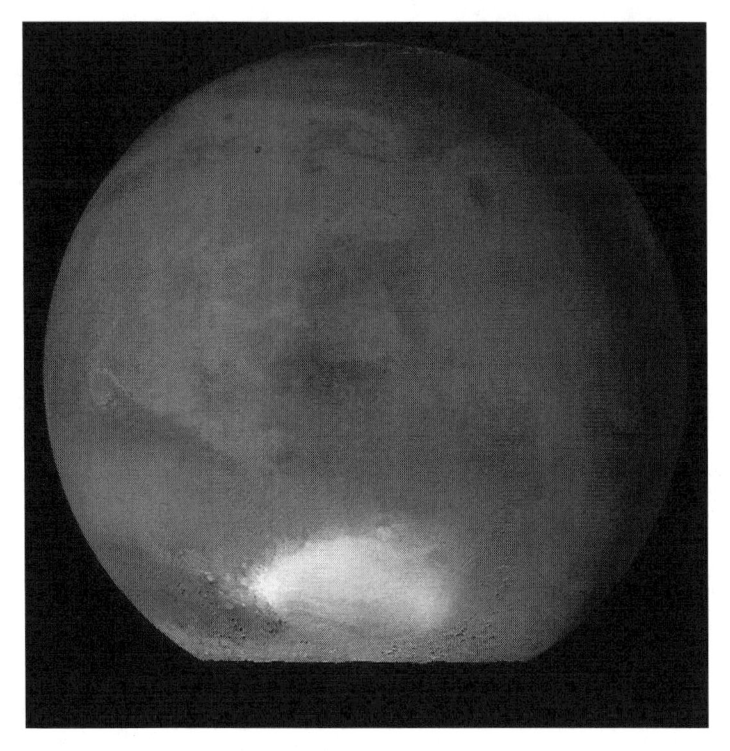

FIGURE 4.8 A composite image of Mars created from images collected by the Mars Global Surveyor. In this image, Mars is experiencing a northern summer and southern winter. (Courtesy of NASA, the Jet Propulsion Laboratory, and Malin Space Science Systems.)

launches had been attempted for Mars probes, of which only eighteen reached the planet, and many of those failed to communicate after reaching the surface.

The first flyby was in 1964 by the U.S. *Mariner 4* spacecraft, which acquired the first close-up images of the surface. The Soviet *Mars 2* and *Mars 3* probes landed on the surface in 1971, but contact was lost with both after impact. The U.S. *Viking 1* and *Viking 2* spacecraft had landers that both successfully reached the surface in 1976, where they operated for 6 and 3 years, respectively, returning color images of the surface during that period. Following this, the Soviet *Phobos 1* and *2* spacecraft were launched in 1988, although contact with *Phobos 1* was lost before it reached the planet, and *Phobos 2* failed before it could release landers to the surface of Phobos.

More recently, the U.S. *Mars Global Surveyor* orbited the planet successfully for 10 years from 1996 to 2006 before failing. Shortly after the *Surveyor* was launched in 1996, the U.S. *Mars Pathfinder*, which consisted of a lander containing a ground rover, was launched. The rover lasted for 3 months on the surface, more than three times longer than initially planned. Currently, two other rovers, *Spirit* and *Opportunity*, are present on the surface. They landed on the surface in 2004 and continue to perform new missions far beyond their initial expected life. Mars has also been targeted by NASA and the European Space Agency (ESA) for future exploration by humans, possibly sometime in the 2030s or 2040s.

Jupiter

Jupiter, shown in Figure 4.9, is the largest planet in the solar system, with a diameter of 143,000 km, 11 times that of Earth. Jupiter is a gas giant, meaning that it is composed mostly of gases rather than solid material, although solid material most likely exists in its center. The mass of Jupiter is 1.9×10^{27} kg, 318 times the mass of Earth, and 2.5 times more massive than all other planets in the solar system combined. Its orbit has a semimajor axis of 5.20 AU, with an eccentricity of 0.0488. It completes a single orbit every 11.9 years and rotates once every 10 hours, the fastest rotation in the solar system. However, different latitudes of the surface rotate at different velocities because of its gaseous nature.

The planet is composed mostly of hydrogen (90%) and helium (10%), along with smaller amounts of methane, ammonia, and other trace molecules. The atmosphere is covered with clouds of ammonia crystals that forms bands, as well as storms such as the Great Red Spot. Jupiter's magnetic field is about 14 times stronger than that of the Earth, resulting in a significant magnetosphere. The planet also has light dust rings due to material that has been ejected from some of its moons. The temperature depends heavily on the altitude, with extremely high temperatures expected in the core and low temperatures at higher altitudes.

Jupiter has 63 known moons, although 47 of these are smaller than 10 km in diameter. The four largest moons, called the Galilean moons, are Io, Europa, Ganymede, and Callisto.

FIGURE 4.9 Simulated globe picture of Jupiter, composed of four images taken by the *Cassini* spacecraft. (Courtesy of NASA, the Jet Propulsion Laboratory, and the University of Arizona.)

Io

Io is the innermost of the large moons, with an average orbital radius of 421,000 km and a diameter of 3,640 km. Io is very active volcanically because of its proximity to Jupiter. The variation of the planet's gravitational field causes friction in Io's core, resulting in heating and geological activity.

Europa

Europa orbits Jupiter at an altitude of 671,000 km and has a diameter of 3,138 km. It is believed that a layer of liquid water exists below the surface of Europa, heated by gravitational friction from Jupiter.

Ganymede

Ganymede is the largest of Jupiter's moons and the largest in the solar system. It has a diameter of 5,268 km, about 0.413 times that of Earth and larger than that of Mercury. It has an orbital radius of 1,070,000 km, with only a slight amount of eccentricity. Ganymede is believed to be composed mostly of silicate rock and water ice. A layer of liquid water is theorized to exist approximately 200 km below the surface, between ice layers. It has its own magnetosphere as well, although it does not approach the magnitude of Jupiter's surrounding field. The average surface temperature on Ganymede is 110 K (–163°C).

Callisto

Callisto has an orbital radius of 1,883,000 km, with low eccentricity. Its diameter is 4,820 km, just below that of Mercury, although its mass is lower by a factor of three. It is believed to be composed of equal amounts of rocks and ices, with water, carbon dioxide, silica, and organics being detected on its surface. Because of its distance from Jupiter and the other Galilean moons, it does not experience the geothermal heating that affects the others.

Exploration

Jupiter and its moons have been visited by several spacecraft, mostly during fly-bys or gravity-assist maneuvers. *Pioneer 10* and *Pioneer 11* performed fly-bys in 1973 and 1974, respectively, performing imaging of Jupiter and several of its moons. They also encountered strong radiation fields near the planet. *Voyager 1* and *Voyager 2* flew by Jupiter on their way to other planets farther out in the solar system. The *Ulysses* probe performed a gravity-assist maneuver around Jupiter in 1992, and its orbit passed the planet again in 2004. The *Cassini* and *New Horizons* spacecraft also flew by during gravity-assist maneuvers in 2000 and 2007, acquiring high-resolution images of the planet and its moons. The only probe that has orbited Jupiter to date is the *Galileo* spacecraft, which was in orbit for seven years starting in 1995. It was deliberately crashed into Jupiter in 2003 to avoid impacting on Europa and contaminating possibly existing life on that moon.

Saturn

Saturn is the sixth planet from the sun and the second largest after Jupiter. Its orbit has a semimajor axis of 9.58 AU, with an eccentricity of 0.0557. Like Jupiter, it is a gas giant, with a diameter of 120,540 km (9.45 Earths) and a mass of 5.68×10^{26} kg (95 times that of the Earth). It orbits the sun once every 30 years, but the rotation rate of the planet is unknown because the surface elements rotate at different speeds. Saturn has an intrinsic magnetic dipole that is slightly weaker than Earth's.

Saturn's composition is similar to that of Jupiter, with about 93% molecular hydrogen and 7% helium, along with ammonia, methane, ethane, acetylene, and other molecules. The upper clouds of the atmosphere are composed mainly of ammonia crystals, while clouds at lower altitudes are composed of water or ammonium hydrosulfide. Saturn is most famous for its rings of ice, which can be seen clearly in Figure 4.10. These rings extend from 6,630 to 120,700 km above the surface and are only 20 m in thickness on average. They are composed primarily of water ice, with some amorphous carbon. Pieces of the ice can be as small as a dust particle and as large as several meters in diameter. Probe data suggest that a very thin oxygen atmosphere exists near the rings, because of ultraviolet radiation from the sun interacting with the water ice.

Saturn has a large number of moons, although the exact number is difficult to determine as the difference between a moon and a piece of ring ice is not always clear. To date, sixty moons have been identified, with seven of these large enough to be of note: Titan, Mimas, Enceladus, Tethys, Dione, Rhea, and Iapetus.

Titan

Titan is by far the largest of the moons of Saturn, and is second in moon mass only to Ganymede. It orbits Saturn with a semimajor axis of 1,221,000 km and an eccentricity of 0.0288. Titan's diameter is 5,152 km (0.404 Earths), and its mass is 1.35×10^{23} kg, about 80% more massive than Earth's Moon. Titan is unique in that it is the only known moon with its own significant atmosphere. The atmosphere is composed almost entirely of molecular nitrogen.

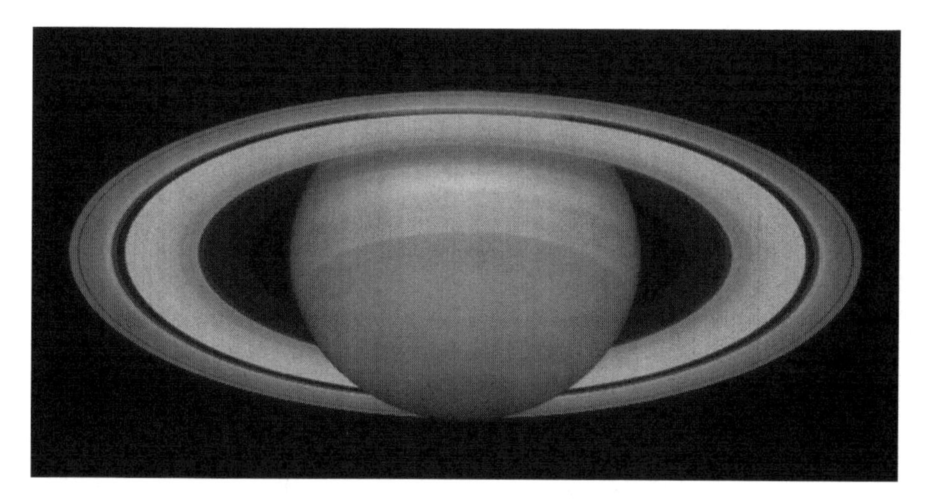

FIGURE 4.10 Picture of Saturn taken by the Hubble Space Telescope. This image was taken while Saturn was at its highest inclination, corresponding to the winter solstice in the Northern Hemisphere. [Courtesy of NASA and the Hubble Heritage Team (Space Telescope Science Institute/Association of Universities for Research in Astronomy).]

Mimas

Mimas has an orbital radius of about 185,000 km and a diameter of 400 km. It is most likely composed of water ice with a small amount of rock. It has a distinctive impact crater 160 km across named Herschel, after the moon's discoverer.

Enceladus

Enceladus' orbital radius is 238,000 km, and its diameter is 500 km. It is the sixth-largest moon of Saturn, and it is believed to have a higher percentage of silicates and iron in its composition than Saturn's other moons. Enceladus is active geologically, with its surface reforming constantly and plumes of liquid water jetting from beneath the surface. The *Cassini* spacecraft flew through one of these plumes in an attempt to determine their composition. The spacecraft's instruments found the presence of water and carbon dioxide, as well as various hydrocarbons.

Tethys, Dione, Rhea, and Iapetus

Tethys orbits Saturn with a radius of 295,000 km and has a diameter of 2,160 km. It is composed almost entirely of water ice. Its most distinctive feature is the Ithaca Chasm, a large valley 100 km wide and 3–5 km deep that runs for 2,000 km along the surface of the moon. Dione has a diameter of 2,200 km and an orbital radius of 378,000 km. It is composed primarily of water ice like most of the other moons, but it has a higher density, indicating a denser core, possibly of silica rock.

Rhea orbits with a radius of 527,000 km and has a diameter of about 1,500 km, the second-largest of Saturn's moons. It is most likely composed of 75% water ice and 25% silica rock. Iapetus is unique in that it has a large difference in brightness from one hemisphere to the other, so much so that its discoverer, Giovanni Cassini, was able to see it only during one half of its orbit around Saturn. It has an orbital radius of 3,560,000 km and a diameter of 3,000 km. It is composed mostly of water ice, by 80%, with the rest being rocky materials.

Exploration

Only a few spacecraft have visited Saturn. The first was a fly-by by *Pioneer 11* in 1979, which acquired some low-resolution images. Both *Voyager 1* and *Voyager 2* flew by the planet in 1980 and 1981, obtaining high-resolution images of the planet and the larger moons, as well as radar estimates of densities and temperatures. The primary Saturn probe to date is the *Cassini* orbiter. It entered into orbit in 2004, and has been returning high-resolution images and data since then. It also released the *Huygens* probe to the surface of Titan. *Cassini*'s primary mission will extend to 2008, but additional missions will probably be added until the spacecraft discontinues operation.

URANUS

Uranus is the seventh planet from the sun and the third largest in our solar system. Its orbit has a semimajor axis of 19.23 AU, with an eccentricity of 0.044. Uranus has a diameter of 51,118 km (four times that of Earth) and a mass of 8.68×10^{25} kg (14.54 times that of Earth). It orbits the sun once every 84 years, and rotates once every 17.25 hours. However, it has an axial tilt of 98°, so it rotates in a rolling motion relative to the sun, as opposed to the spinning-top motion of most other objects in the solar system. This means that one side of the planet receives sunlight continuously for part of the year, while the other side is in the dark. Then these sides reverse in the second half of the orbit. Figure 4.11 displays a composite image of the sunlit hemisphere of Uranus taken by *Voyager 2*.

Uranus is similar in composition to Saturn and Jupiter, although it has a larger amount of ice in its atmosphere because of its lower temperatures. Its upper atmosphere is about 83% hydrogen and 15% helium. Its core is small compared with the other planets, and it has very low internal heating. Uranus has a system of rings surrounding it, with thirteen rings currently known. They are composed mostly of dark particulate matter, making imaging difficult. The planet also has a strange magnetic field that is tilted at a large angle from the rotation axis, leading to a highly asymmetric field around the planet.

Uranus has twenty-seven known satellites, with the most significant being Miranda, Ariel, Umbriel, Titania, and Oberon. The moons are the least massive of any gas giant in our solar system. They are composed mostly of ice and rock, with roughly equal amounts of each.

Voyager 2 is the only human-made object to visit Uranus to date. It acquired images, magnetic field data, and data on the structure and composition of the atmosphere.

FIGURE 4.11 A composite image of Uranus assembled from images taken by *Voyager 2*'s narrow-angle camera. (Courtesy of NASA and the Jet Propulsion Laboratory.)

Neptune

Neptune is the eighth planet from the sun in our solar system and is now considered the planet farthest from the sun. That distinction was formerly given to Pluto before its status as a planet was revoked. Neptune has an elliptical orbit with a semimajor axis of 30.1 AU and an eccentricity of 0.011. Its diameter is 49,528 km, about four times that of Earth, and has a mass of 1.02×10^{26} kg, about 17 times that of Earth. It is nearly the same in size as and slightly more dense than Uranus. It orbits the sun once every 165 years and rotates once every 16 hours.

Neptune's atmosphere is about 80% hydrogen and 19% helium, with about 1% methane and other trace molecules. Other than the larger presence of methane, its composition is very similar to that of Uranus, although the higher density indicates that more rock is present. Neptune has higher internal heat than Uranus as well, leading to strong weather patterns. A large spot similar to the Great Red Spot of Jupiter has been observed in the planet's atmosphere, and it can be seen in Figure 4.12. Neptune has a magnetic field that, like that of Uranus, is tilted at a large angle from its axis of rotation.

Triton

Neptune has 13 known moons, of which Triton is the only one of significance. Triton is 2,700 km in diameter and has an orbital radius about the planet of 355,000 km. Triton

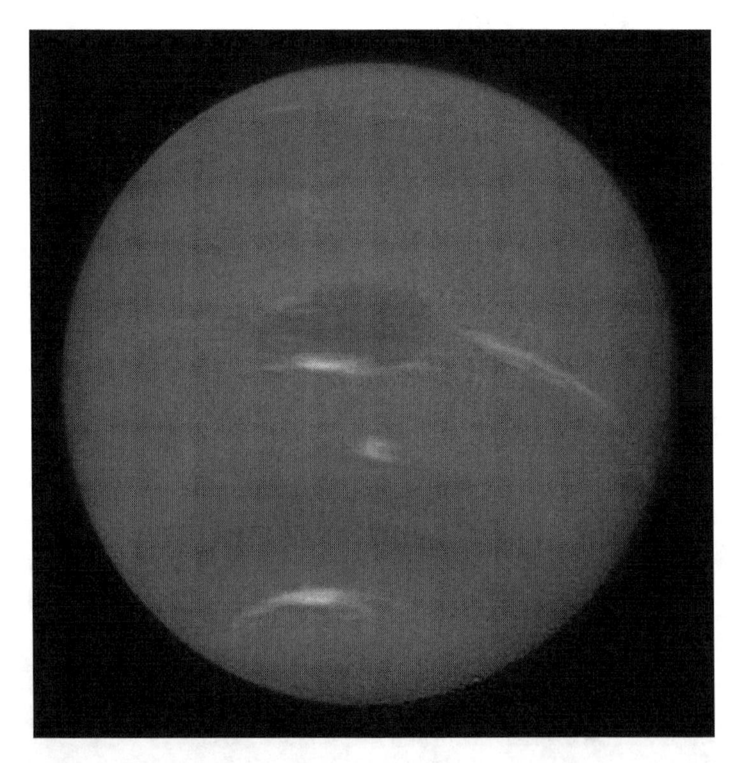

FIGURE 4.12 Image of Neptune taken by *Voyager 2*, showing the Great Dark Spot in the center. (Courtesy of NASA and the Jet Propulsion Laboratory.)

is the only moon in the solar system that orbits its planet in the opposite direction of the planet's rotation (known as a retrograde orbit). The moon's orbital axis is also on a large angle to the orbital plane of Neptune, such that its orbital axis points toward the sun, much like Uranus. Triton has also been observed to be geologically active.

Neptune has only been visited by one human-made object: the *Voyager 2* spacecraft made its closest approach to the planet in 1989. *Voyager* imaged the planet as well as Triton and acquired data on the atmosphere and magnetic field.

DWARF PLANETS

A dwarf planet is defined as an object orbiting the sun that is rounded by its own gravity, but has not cleared the neighborhood around its orbit of smaller objects. So far, three objects have been given this classification: Ceres, Pluto, and Eris.

Ceres

Ceres orbits the sun in the region of the asteroid belt between Mars and Jupiter. It has an elliptical orbit with a semimajor axis of 2.77 AU and an eccentricity of 0.080. It has a diameter of 960 km, representing by far the largest object in the asteroid belt, in both size and mass. It is probably composed mostly of water ice and hydrated minerals. No spacecraft have visited Ceres to date, although the U.S. Dawn mission has launched and plans to visit the dwarf planet in 2015.

Pluto

Pluto lies in the Kuiper Belt, nominally beyond the orbit of Neptune. It was considered to be a planet until 2006, when it was reclassified as a dwarf planet. It has an extremely elliptical orbit with a semimajor axis of 39.5 AU and an eccentricity of 0.25. This means that its closest approach to the sun is 29.7 AU, and its farthest point away is 49.3 AU. This brings it closer to the sun than Neptune during its perihelion. Its orbit is also highly inclined out of the ecliptic compared with the other planets in the solar system. Pluto is about 2,400 km in diameter, and its composition is similar to that of other Kuiper Belt objects, being a mix of ice and rock.

Pluto has three known moons: Charon, Nix, and Hydra. Charon is only half the diameter of Pluto, although it is lighter by a factor of ten. Nix and Hydra have only recently been discovered, and not much is currently known about them.

Pluto has not yet been visited by any human-made objects. However, the New Horizons mission launched in 2006 and successfully performed a gravity-assist maneuver around Jupiter in 2007 to set its course toward Pluto. It will reach the planet in 2015.

Eris

Eris is the largest of the three dwarf planets, orbiting the sun in a region beyond the Kuiper Belt known as the scattered disc. Its elliptical orbit has a semimajor axis of 68 AU and an eccentricity of 0.44, so its closest approach of 38 AU is much nearer the sun than its aphelion of 98 AU. It has a diameter of approximately 2,600 km and was first discovered in 2003. One moon, named Dysnomia, has been found to date.

ASTEROID BELT

The asteroid belt is defined roughly as the region of space between Mars and Jupiter. There are many orbiting bodies in this region of space, but the largest are Ceres, Vesta, Pallas, and Hygeia. Each of these has a diameter of more than 400 km, and together they represent more than half of all the mass in the asteroid belt, although only Ceres has enough size to obtain a rounded shape and the classification dwarf planet. More than two hundred of the known asteroids are over 100 km in diameter, and some data have shown that there are between 700,000 and 1.7 million asteroids more than 1 km in diameter.

The asteroids in this belt are mainly of three types of composition. The first is carbonaceous or C-type, which have large amounts of carbon in their material. These are the most common type of asteroid. Next are the silicate asteroids, or S-type, which are made up mostly of silicate rock and some metal, but little carbon. S-type asteroids are found closer to the sun in general, within 2.5 AU, and account for about 17% of all asteroids. The last common type is the metal-rich M-type, which are composed of an iron-nickel mix. These make up about 10% of the asteroids in the belt.

Ten different spacecraft have passed through the asteroid belt to date, and none have experienced a collision. Although there are a large number of objects in the belt, the amount of space they occupy is so enormous that collisions are very unlikely. The odds of collision for a probe are estimated at one in a billion. None of the spacecraft have performed any missions on the asteroids other than fly-bys, but the Dawn mission plans to explore Vesta and Ceres in 2012 and 2015.

KUIPER BELT AND BEYOND

The Kuiper Belt is an expanse of space around the sun defined as the region from Neptune's orbit of 30 AU to about 55 AU from the sun. It is similar to the asteroid belt in that it contains many bodies, although it encompasses a much larger part of space and the total mass of its members is much larger than that of the asteroid belt. The number of known Kuiper Belt objects is over 1,000 to date, with more than 70,000 objects larger than 100 km in diameter theorized to exist there. These objects are composed mainly of ices of various types: water, methane, and ammonia. The primary member of the Kuiper Belt is the dwarf planet Pluto. The planet Neptune has a large effect on the orbits of Kuiper Belt objects because of its gravity.

SCATTERED DISC

The region at the outer edge of the Kuiper Belt and encompassing the part of space beyond that is called the scattered disc. Very little is known about this region of space, and the objects that exist in it are theorized to have been scattered by the gravity of the outer planets, mainly that of Neptune. The condition for membership is that an object's perihelion must be greater than 35 AU.

OTHER OBJECTS

There are several other objects in the solar system that are not included in the above descriptions. These include comets, trojans, and centaurs.

COMETS

Comets are objects that orbit the sun and, when close enough, exhibit small atmospheres and tails due to heating. Some comets are in permanent orbit, while others may only pass through the solar system once. The number of known comets is over three thousand, which includes several hundred one-time visitors. Comets range in size from 50 km in diameter down to a half kilometer and are usually composed of rock, water ice, dust, and frozen gas.

The comet Wild-2, shown in Figure 4.13, was visited in 2004 by the *Stardust* spacecraft, which captured material from the comet's "atmosphere" and took close-up images of the nucleus. The spacecraft returned the captured material to Earth afterward.

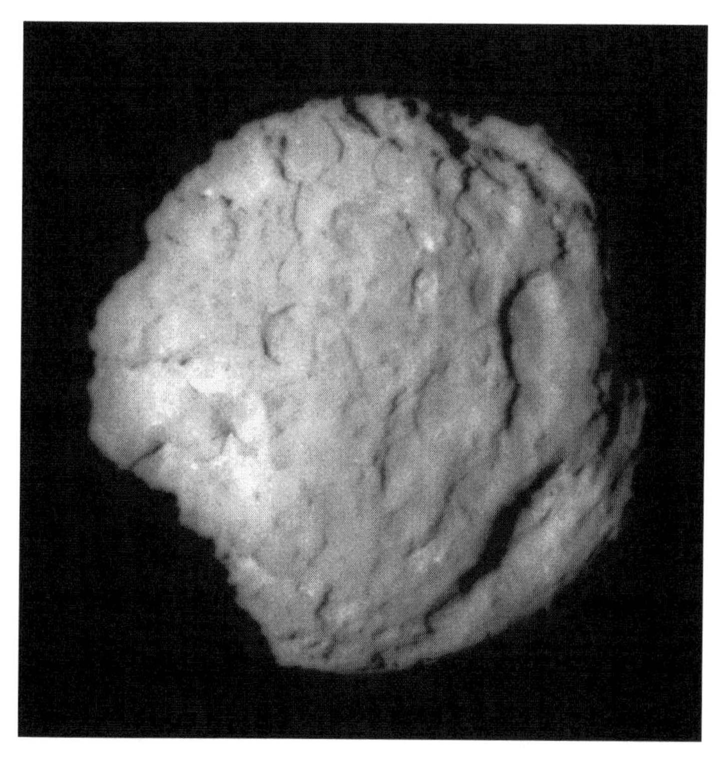

FIGURE 4.13 A picture of the comet Wild-2, taken by a camera on the *Stardust* spacecraft. *Stardust* passed through the coma of the comet, collecting samples of the material that is constantly shedding from the surface. (Courtesy of NASA and the Jet Propulsion Laboratory-Caltech.)

Trojans

Trojans are asteroids that have the same orbit as a planet, typically orbiting near the L_4 and L_5 Lagrange points for the planet. They exist around many of the planets, including Mars, Jupiter, and Neptune.

Centaurs

Centaurs are a class of asteroid that tends to cross the orbits of the larger planets, often in highly elliptical orbits. Centaurs are believed to have generally unstable orbits on timescales of 10^6–10^7 years.

CONCLUSION

Our solar system is a big place, and the planets, few though they may be, show an enormous amount of diversity. That makes human missions to the planets all the more attractive. For the purposes of space archaeology, it is important to understand the environment and properties of the planets and other objects in the solar system, but also the reasons for our visits to them.

REFERENCES

1. Cooper, K., and E. Stone. 2007. "The Voyagers at 30—The Man On a Thirty Year Mission." *Astronomy Now* 21(8): 22–25.
2. Guo, Y., and R.W. Farquhar. 2002. New Horizons Mission Design for the Pluto-Kuiper Belt Mission. *Proceedings of the AIAA/AAS Astrodynamics Specialist Conference and Exhibit*, August 5–8, 2002. AIAA Paper number 2002-4722.
3. McFadden, L.-A., P. Weissman, and T. Johnson. 2006. *Encyclopedia of the Solar System*. 2nd ed. San Diego, CA: Academic Press.
4. de Pater, I., and J.J. Lissauer. 2001. *Planetary Sciences*. Cambridge, United Kingdom: Cambridge University Press.

5 Space Basics: Orbital Mechanics

Jerold Emhoff

CONTENTS

INTRODUCTION

Most human-made objects in space are in orbit around something, whether it be the Earth, the Moon, the sun, or another planet. Very few objects can be said to not be in orbit, namely *Voyager 1* and *Voyager 2*. These objects have achieved escape velocity from the sun and so will not return. The location and motion of every other human-created object in space is governed by orbital mechanics: the combination of gravitational fields, velocity, and atmospheric drag that determines where an object is at any given time and where it will be in the future.

This chapter will begin by giving a brief history of orbital mechanics, followed by descriptions of some of the important orbits and concepts. Details on some of the mathematics and physics involved in orbital mechanics are presented later in the chapter. For a more in-depth and complete explanation of orbital mechanics, a text focused on the subject should be consulted.[1,2]

KEPLER AND NEWTON

Orbital mechanics began with the work of Johannes Kepler,[3] who lived from 1571 to 1630. Kepler was a German mathematician and astronomer who studied the earlier works of Tycho Brahe. Brahe had created very precise tables of the motion of the planets, and Kepler was able to determine from these tables that the planets follow three laws in their motions:

1. The orbit of every planet is an ellipse, with the sun at one of the foci.
2. A line joining a planet and the sun sweeps out equal areas during equal intervals of time as the planet travels along its orbit.
3. The squares of the orbital periods of planets are directly proportional to the cubes of the semimajor axes of their orbits.

The first law simply states that the orbits of the planets are elliptical (not circular, as thought by previous astronomers). The second law indicates that the orbiting object speeds up as it moves closer to the central body and slows down as it moves farther away while following its elliptical orbit. The third law says that for a given central body, the time it takes to complete one orbit (the period) and the semimajor axis of the orbit are related as follows.

$$\frac{P^2}{a^3} = k \tag{5.1}$$

Here k is a constant that depends on the central body being orbited, P is the period of the orbit, and a is the semimajor axis of the orbit.

These laws were the first to accurately describe the motion of the planets about the sun, and they hold true for all pairs of orbiting objects, not just for the sun-planet pairs. However, Kepler's laws do not describe *why* the planets follow these paths. They merely describe the result of some underlying physical law, a law discovered by Isaac Newton.

Isaac Newton is often credited with the discovery of gravity, but in fact what he actually discovered was the law that governs the gravitational force.[4] Gravity was known of before Newton, but he described how the force changes with distance and the masses of the objects involved. The equation found by Newton for the gravitational force between two objects is

$$F_G = \frac{GMm}{r^2}, \tag{5.2}$$

where G is the gravitational constant, M is the mass of one object, m is the mass of the other, and r is the distance between them. In words, the equation states that the gravitational force between two objects increases as the mass of either object increases, but the force decreases more rapidly as the objects move farther apart. So, if you decreased the mass of one object by one-fourth, the effect would be the same as doubling the distance between the objects. When applied to the planets and the sun, this equation and Newton's three laws of motion verify each of Kepler's three laws.

Together, Kepler's laws and Newton's law of gravitation form the cornerstone of orbital mechanics. These laws accurately describe the motion of all of the planets, as well as most moons, although these orbits can be complicated by other moons. However, there is an additional factor for some objects, especially those in orbit near a planet's surface: atmospheric drag.

ATMOSPHERIC DRAG

When an object is in orbit close to a planet and its atmosphere, it will encounter drag from that atmosphere. Although space is often considered to be a complete vacuum, with absolutely nothing in it, this is not true. At all points in space, there is some level of matter, whether in the form of a thinned planetary atmosphere or particles of hydrogen and helium from the solar wind. While the solar wind is not dense enough to be significant, planetary atmospheres become denser the closer an object is to the surface.

The atmospheric density determines how high an orbit has to be in order to for it to be reasonably stable. In other words, if an orbit is too low, it requires a constant thrust in order to counteract the drag and maintain the orbit. The higher the orbit is, the lower the drag and the less work that has to be done to stay in orbit. If an orbit is high enough, the drag is so small that it has no effect at all.

One of the first spacecraft ever put into orbit is *Vanguard 1*,[5] which was launched in 1958. *Vanguard* is in an elliptical orbit with an altitude of about 2,300 km, and it will stay in orbit for about another 240 years. It will eventually deorbit because of atmospheric drag, but with its high orbit, the drag is very low, so it will take a very long time. *Vanguard*'s original mission was actually to aid in the determination of atmospheric drag, since scientists at the time had only estimates of the drag in orbit. *Vanguard* showed that those estimates were actually much too low, both initially and later, when calculations were made about when it would deorbit. Original calculations estimated it would take over 2,000 years to fall, but later estimates showed that seasonal fluctuations, as well as solar activity, would reduce the lifetime of the satellite to about 240 years.

The physics of drag on objects in orbit is basically the same as on Earth. Drag is caused by particles bouncing off a surface, transferring momentum, and slowing the object down. Just like on the ground, a more aerodynamic shape decreases the drag by reducing the amount of momentum transferred by particles impacting. Even so, drag is not usually a major factor in spacecraft design.

SPECIFIC ORBITAL CHARACTERISTICS

Most of the satellite orbits around the Earth fall into a few specific categories. The most common are low Earth orbit (LEO), medium (or middle) Earth orbit (MEO), and geosynchronous Earth orbit (GEO). Figure 5.1 shows the locations of these three orbits with respect to the Earth. There are also sun-synchronous, graveyard, and Molniya orbits, each with special characteristics.

LOW EARTH ORBIT

Low Earth orbit is usually defined as the region of space from 160 to 2,000 km from the Earth's surface. This is the most active orbit for human-made objects because it is the easiest to reach. All of the space stations—*Skylab*, *Mir*, and the *International Space Station*—have been in LEO. All launch vehicles are capable of servicing this orbit, although the size of the vehicle determines the size of the launched satellite.

All launches must go through LEO, even if that is not the final destination, so in addition to the live satellites launched into LEO, there is a large amount of space debris such as rocket stages and other parts in the orbit range. This makes LEO the most cluttered orbit and the most dangerous in terms of possible collisions or impacts with debris.

The change in velocity required to reach LEO is approximately 10 km/s. This figure includes compensation for atmospheric drag when launching. Atmospheric drag is also significant once in LEO because of the closeness to the Earth. The amount of velocity lost per year while in orbit depends on the aerodynamic characteristics of the spacecraft. The value can vary from 1 km/s at the lowest altitudes down to millimeters per second at higher altitudes. Higher altitudes are more sensitive to the solar cycle as well, as the Earth's atmospheric drag is smaller and counts for less than the solar wind drag.

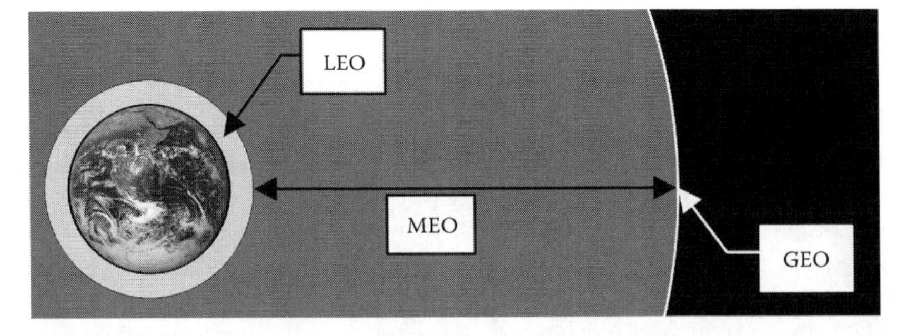

FIGURE 5.1 Locations for three of the orbital classes. Low Earth orbit is between 160 and 2,000 km away from the Earth, medium Earth orbit reaches from 2,000 to 35,700 km, and geosynchronous orbit is 35,780 km from the surface.

Medium Earth Orbit

Medium Earth orbit is used almost exclusively for navigation satellites such as the global positioning satellites or Galileo constellations. The orbital altitude of MEO is the region between LEO and GEO, 2,000 to 35,700 km, although the navigation constellations are around the 20,000-km point. This orbit is generally undesirable because it requires much more effort to reach than LEO, without the geostationary advantages of GEO. The Van Allen radiation belts also cross this orbit, so satellites in this region can encounter large amounts of radiation.

Geosynchronous Earth Orbit

Geosynchronous orbits have an average altitude of about 35,780 km above the Earth's surface. Objects at this altitude take 1 day to complete a single cycle, the same as the Earth. Any satellite in this type of orbit visits the same spot on Earth at the same time every day, so it is a useful orbit for communications, weather, or observation. However, this orbit requires a much larger amount of energy to reach, so it is much less populated than LEO. An additional velocity of about 4 km/s is required to move from LEO into GEO, a 40% increase over the velocity needed to just reach LEO.

A special case of a geosynchronous orbit is the geostationary orbit. Geostationary orbits have zero inclination; in other words, they orbit directly above the equator. This means that they are always over the same spot on Earth, at all times. These orbits were first postulated as being useful for communications satellites by the science fiction writer Arthur C. Clarke in 1945.[6]

To reach a geosynchronous orbit, a special temporary orbit called a geosynchronous transfer orbit (GTO) is usually used. These orbits transfer a satellite from a LEO orbit with a finite inclination to a GEO orbit with zero inclination, positioned in the correct spot over the Earth. A spacecraft may leave behind rocket staging parts in GTO while the satellite continues on to GEO.

Graveyard Orbit

When a satellite in geosynchronous orbit nears the end of its life and is almost out of propellant, an attempt may be made to place it into a graveyard orbit. These orbits are slightly higher in altitude than a geosynchronous orbit, by a few hundred kilometers. When a spacecraft is failing, there is a desire to make sure it will not cause issues for other, active satellites. Ideally, it would be deorbited and allowed to burn up on reentry, removing a potential piece of space debris. However, spacecraft in geosynchronous orbits are so far from the Earth that it would take a significant amount of velocity change in order to reach the atmosphere: about 1,500 m/s. Instead, the spacecraft is sped up slightly to raise it into a graveyard orbit, requiring only around 10 m/s of velocity change. Once in the orbit, there is very little drag from the Earth's atmosphere, so only the solar wind has any effect at all. This means that the dead spacecraft will stay in the graveyard orbit for a very long time, several hundred years at least. The United Nations,[7] U.S. government,[8] and the U.S. Federal Communications Commission have regulations requiring geosynchronous satellites to be put into a graveyard orbit at the end of life.

Sun-Synchronous Orbits

Sun-synchronous orbit (SSO) is a subset of LEO, where the spacecraft is over the same spot on Earth at the same time each day. This is accomplished by tuning the orbit's inclination and altitude to provide the specific characteristics desired. The Earth under these orbits will always have the same lighting by the sun, so they can be useful for ground observation. The orbits can also be tuned such that they sit on the divide between night and day, so the solar arrays are always in contact with the sun, or so the satellite can always observe the sun in a scientific mission. The altitude of SSOs is typically between 600 and 800 km.

Molniya Orbits

A Molniya orbit is a highly elliptical, highly inclined orbit that is used primarily for communications and spy satellites. Objects in these orbits spend most of the time in their orbits over a single region of Earth. A typical orbit has a 12-hour period and spends 8 hours of that time over the useful portion of the planet, so for continuous coverage only three satellites are needed. Communications satellites are able to send and receive signals from that part of the planet, and spy satellites have a long period of observation. Figure 5.2 shows a representative Molniya orbit, where the spacecraft spends most of its time observing North America. Molniya orbits were first used by the Soviet Union in 1964–1965 for the Molniya communications satellites. The Soviet Union also pioneered the use of these orbits for spy satellites, which allowed a nearly constant view of the continental United States, along with U.S. missile sites. The United States also used these types of orbit to observe the Soviet Union and Russia for the same reason.

Lagrange Points

The Lagrange points are a set of points in space where the gravitational force is balanced between two large objects.[9] These points were discovered in 1772 by the mathematician Joseph-Louis Lagrange, and they were verified in the 1900s by the discovery of the Trojan asteroids in Jupiter's orbit by astronomer Max Wolf. The most commonly used set of points are the Sun-Earth points, although there are Lagrange points for the Earth-Moon system, as well as for all other planets and the sun or their moons. Some of these points are theoretically stable, meaning that the gravitational forces will actively maintain an object's position at that point. Other points are unstable, and if the object leaves the Lagrange point, gravity will pull it farther away. However, in a real case such as our solar system, there are many effects that keep the stable points from being strictly stable, such as the solar wind and the gravitational pull of other planets. Five Lagrange points, diagrammed in Figure 5.3, exist for every two-body system; these points are denoted L1 through L5. The L1 point lies on a line between the two objects; for the Sun-Earth system, this is about 1.5 million kilometers from the Earth. At this point, the gravitational pull toward the sun is equal to the pull toward the Earth plus the effect of centripetal force. This is the most popular point for human missions, with four NASA missions to date making use of the point,

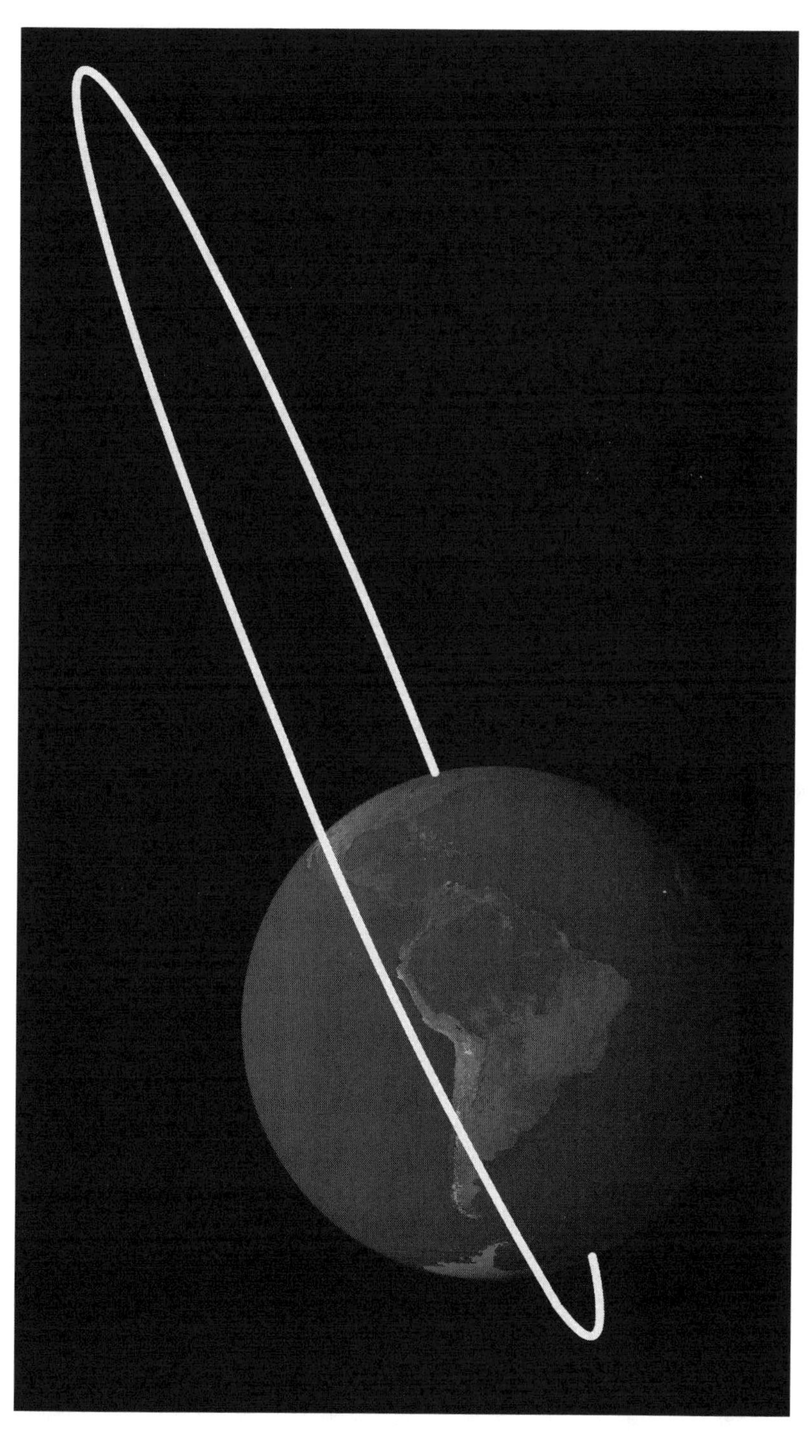

FIGURE 5.2 A representative Molniya orbit, with a long viewing time of North America.

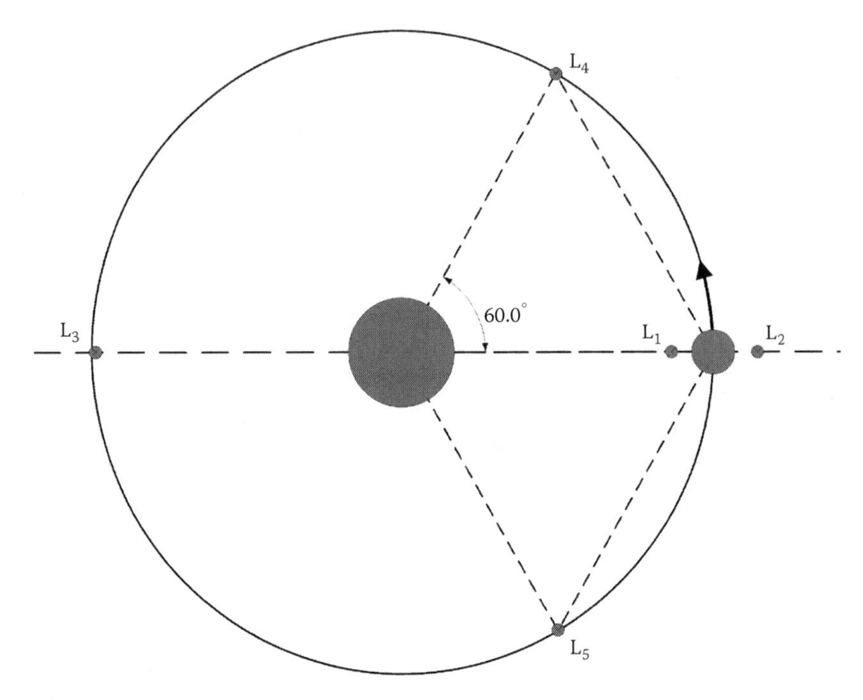

FIGURE 5.3 Diagram of the Lagrange point locations in a system with one very massive body and one much smaller, as in the Sun-Earth system.

including the active Solar and Heliospheric Observatory and Advanced Composition Explorer. Both of these spacecraft make use of the unobstructed view of the sun to perform science related to the sun and solar wind. In the Earth-Moon system, the L1 point allows easy change of orbit with very little effort, which may be very useful for future missions. This point is theoretically stable in some instances, but in reality a special type of orbit around the point called a Lissajous orbit is required for near-perfect stability.[10]

L2 is also on the line between the two objects, but is instead on the outside of the second, smaller object (Earth, in the case of the Sun-Earth system). At this point, the centrifugal force is equal to the sum of the gravitational force of the Earth and the sun. For the Sun-Earth system, this point is also about 1.5 million kilometers away from the Earth. This orbit is currently home to the Wilkinson Microwave Anisotropy Probe, and future missions such as the James Webb Space Telescope (JWST) are planned that will make use of this point. In the case of the JWST, the L2 point has the advantage that the sun is blocked by the Earth, so the temperature is extremely low. This, along with a large sun-shield, blocks external infrared radiation that would normally interfere with the telescope's sensors. This point is also nominally unstable, but it can be stabilized with a Lissajous orbit.

The L3 point is similar to the L2 point, except it is on the other side of the larger mass (the sun in the Sun-Earth system). Again, the centripetal force balances the sum of the gravitational attraction of both the sun and the Earth. This point actually lies slightly inside the Earth's orbit, directly across the sun from the Earth's current position. This point is highly unstable because of the gravitational pull of other planets.

L_4 and L_5 lie slightly outside the path of the smaller body's orbit, ahead of and behind it by 60° along the arc. At these points, the gravitational and centripetal forces balance in a complicated way related to the ratios of the masses of the objects in the system. These points are stable in reality for the Sun-Earth system and are home to the Trojan asteroids in Jupiter's orbit.

ORBITAL MECHANICS PRIMER

In this section, a few concepts related to orbital mechanics are described in more detail, starting with the simple physics of circular orbits, the geometry of ellipses, and then discussion of some orbital maneuvers.

CIRCULAR ORBITS

Objects in orbit have to balance the force of gravity with the centripetal force—the force of an object's momentum. Any object moving in a straight line wants to continue moving in that direction, resisting changes. This is stated by Newton's first law: "Every body perseveres in its state of being at rest or of moving uniformly straight forward, except insofar as it is compelled to change its state by force impressed."[11] Objects in orbit are continually changing direction because of the gravitational force and continually resisting that change because of momentum. This balance is described by the following equation:

$$\frac{GMm}{r^2} = \frac{mu^2}{r}, \tag{5.3}$$

where G is the gravitational constant, M is the mass of the object being orbited, m is the mass of the object in orbit, r is the distance from the center of the orbited object to the center of the orbiter, and u is the velocity of the orbiter. The gravitational side of this equation states that the force decreases as the square of the distance from the source of the gravity. For Earth orbit, this is the distance from the center of the Earth, and M is the mass of the Earth. The other half of the equation represents the required centripetal force to hold the object in orbit, which depends on the distance to the center of the rotation. It also depends on the velocity of the object along its path, and the mass of the object. Since this mass, as well as the distance, appears on both sides of the equation, it can be cancelled out, revealing that the mass of the object does not matter to the balance of forces. Rearranging gives the following:

$$GM = u^2 r. \tag{5.4}$$

This equation shows that there is a balance between the orbital velocity and the height of the orbit, and this balance is a constant for a given central body being orbited. This means that to achieve a given orbit altitude r, the object must be moving with velocity u. To have a higher orbit, the velocity has to increase. In Earth orbit, GM is called μ, and it has a value of 398,600 km^3/s^2. This gives the following equation for orbital velocity in Earth orbit:

$$u = \frac{631.3481}{\sqrt{r}}, \tag{5.5}$$

where u is in km/s and r is in km.

Elliptical Orbits

Circular orbits are useful for understanding the basics of orbits, but orbits are often elliptical, not circular. In fact, a circular orbit is simply a special type of elliptical orbit. All of the planetary orbits in our solar system are elliptical, and many spacecraft have elliptical orbits as well.

Basics of Ellipse Geometry

An ellipse is defined as a curve where the sum of the distance between two points is a constant. These two points are called the foci of the ellipse. The major axis is the distance between the two farthest ends of the ellipse, and the minor axis is the distance between the two nearest ends. The semimajor and semiminor axes, defined as a and b, respectively, are half the length of the major and minor axes. The diagram in Figure 5.4 shows an ellipse and its corresponding properties.

An ellipse's shape is defined by its eccentricity e:

$$e = 1 - \frac{r_p}{a} = \frac{r_a}{a} - = \sqrt{1 - \frac{b^2}{a^2}}. \tag{5.6}$$

Here, r_p is the radius at periapsis, or the closest distance between the ellipse and one of its foci. r_a is the radius at apoapsis, which is the largest distance between the ellipse and one of its foci.

Note that the terms periapsis and apoapsis take different forms in orbital mechanics depending on the central body. For Earth orbit, the terms become perigee and apogee (the closest distance to the Earth and the farthest distance from it), and for solar orbit, the terms are perihelion and aphelion.

Orbital Maneuvers

Objects in orbit are not necessarily fixed in their path, so long as they have sufficient propulsion to change their velocity in the appropriate way.

Hohmann Transfers

To change the altitude of an orbit, a Hohmann transfer is often used. The transfer is a special ellipse where the beginning point is tangent to the current orbit, and

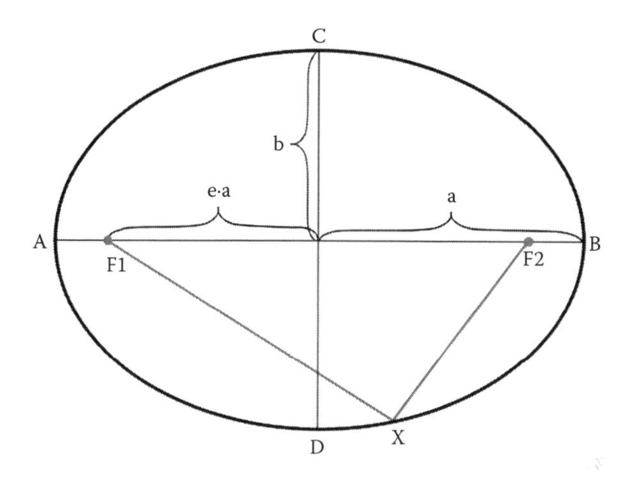

FIGURE 5.4 Diagram of an ellipse. The sum of F1 and F2 is a constant, and the length a is called the semimajor axis.

the ending point is tangent to the final orbit, tracing out half of the transfer ellipse. Hohmann transfers are the most fuel-efficient way to move between two circular orbits. Two fuel burns are required: the first enters into the elliptical Hohmann orbit, and the second adds enough velocity to circularize at the desired final orbit. In both cases, the velocity change is in the direction of travel, tangent to the current orbit.

One-Tangent Transfers

If a faster orbit change is needed than can be obtained from a Hohmann transfer, a one-tangent transfer can be used instead. In this case, the initial change in path is tangent to the current orbit, but the intersection with the new orbit is not tangential. So the initial applied velocity is in the direction of travel, but the second is perpendicular to the desired orbit.

Orbital Plane Change

The plane of the orbit can be changed by applying thrust perpendicular to the current orbit. The inclination of the plane is most often changed for objects in Earth orbit. The inclination of an Earth orbit is the angle of the plane of the orbit, referenced against the equatorial plane, so any satellite whose orbit is directly above the equator has an orbit inclination of zero degrees. For solar orbit, all of the planets roughly fall inside the same plane, called the ecliptic. Inclination in solar orbit is referenced against the ecliptic plane.

Gravity Assist

For some desired orbits, a large change in velocity is required. This can require an enormous amount of propellant, or it may not be possible at all. An alternative is to use a gravitational-assist maneuver. In such a maneuver, the spacecraft is sent close to a massive object, such as Jupiter, such that the gravitational field of the object changes the direction of the spacecraft. In Jupiter assists, the spacecraft velocity relative to the sun can be increased enough for the spacecraft to leave the solar system entirely, or the velocity can be decreased in order to achieve a solar orbit. For example, *Voyager 1*, the human-made object farthest from Earth, used both Jupiter and Saturn gravity assists to achieve its solar system escape velocity.

Deviations from Theory

The preceding explanations and descriptions generally describe orbital motion in a simplified way. For the most part, they give an excellent degree of accuracy; however, there are small deviations that should be kept in mind. For Earth orbit, the planet is not perfectly spherical, and the gravitational field is not perfectly constant. These two points can cause slight perturbations to occur in orbits. Also, the Sun and Moon have some gravitational effect on objects in Earth orbit.

Orbits also cannot be maintained indefinitely in many cases. Space is not a perfect vacuum, so some amount of drag is always acting on any spacecraft. For orbits close to the Earth, this drag is significant because of the atmosphere, while for orbits around the sun the drag is small and is due to the solar wind. The solar wind drag also varies depending on the solar cycle—during solar maximum, drag is much higher than at solar minimum.

CONCLUSION

This chapter was intended to give a reader not necessarily knowledgeable in physics and space a brief introduction into the basic processes and nomenclature used in orbital mechanics. The orbital descriptions give will allow the reader to have a basic understanding of what those orbits mean when encountered elsewhere in the book. The descriptions of the basic orbital mechanics physics will also impart a sense of what behavior to expect from orbiting objects.

REFERENCES

1. Bate, R.R., D.D. Mueller, and J.E. White. 1971. *Fundamentals of Astrodynamics*. New York: Dover Publications.
2. Prussing, J.E., and B.A. Conway. 1993. *Orbital Mechanics*. Oxford, United Kingdom: Oxford University Press.
3. Caspar, M. 1959. *Kepler*. London: Abelard-Schuman.
4. Newton, I. 1999. *The Principia: Mathematical Principles of Natural Philosophy*. Trans. I. Bernard Cohen and Anne Whitman, assisted by Julia Budenz. Berkeley, CA: University of California Press.
5. Green, C.M., and M. Lomask. 1971. *Vanguard—A History*. Washington: Smithsonian Institution Press.
6. Clarke, A.C. 1945. Extra-Terrestrial Relays—Can Rocket Stations Give Worldwide Radio Coverage? *Wireless World* 51: 305–308.
7. Committee on the Peaceful Uses of Outer Space, United Nations General Assembly. 1996. Steps Taken by Space Agencies for Reducing the Growth or Damage Potential of Space Debris A/AC.105/663.

8. U.S. Government. 2001. U.S. Government Orbital Debris Mitigation Standard Practices. Washington, DC: White House Office of Science and Technology Policy.
9. Tyson, N.D. 2007. *Death by Black Hole*. New York: W.W. Norton.
10. Gomez, G., J. Masdemont, and C. Simo. 1998. Quasihalo Orbits Associated With Libration Points. *Journal of the Astronautical Sciences* 46(2): 135–176.
11. Newton 1999 (note 4).

6 Space Basics: Getting to and Staying in Space

Michelle M. Donegan

CONTENTS

INTRODUCTION

On March 11, 2008, the space shuttle *Endeavour* launched from Cape Canaveral, Florida, as mission STS-123. The mission lasted nearly 16 days, covered over nearly 10M km, and carried a crew of seven astronauts. The shuttle's external tank, which was discarded before reaching orbit, contained supercold liquid hydrogen and liquid oxygen totaling about a half-million gallons. All this propellant was burned in just 8.5 minutes, delivering *Endeavour* to an altitude of 341 km.

Just this small sample of a single space mission invites a host of questions. For example, why did the launch need so much propellant? At what point did the shuttle enter "space," and who gets to decide that? What is it like to be in space? Why isn't getting to space a routine affair?

DIFFERENT DEFINITIONS OF SPACE

Where does space begin? This is a question that does not have a single, agreed-upon answer. If one defines space as the end of Earth's atmosphere, there is no sharp beginning to space—the atmosphere's density decreases smoothly with increasing altitude. NASA defines *aeronautical and space activities* as "research into, and the solution of, problems of flight within and outside the Earth's atmosphere."[1] This rather nonspecific definition is in the National Aeronautics and Space Act, which

FIGURE 6.1 The International Space Station, orbiting Earth at an altitude of 390 km. (Courtesy of NASA.)

established NASA in 1958, and depends on where one considers the atmosphere to end. A more specific definition, used by the Ansari X PRIZE,*[2] is 100 km, which is also the threshold set by the World Air Sports Federation.[3] This number is known as the Kármán line and forms the boundary between "aeronautics" and "astronautics." The Kármán line is named for the Hungarian-American engineer and physicist Theodore von Kármán, whose career in aeronautics and astronautics included advances in the characterization of supersonic and hypersonic airflow. At approximately the Kármán line, the atmosphere becomes too thin to provide enough aerodynamic lift to support a vehicle's weight, so that a velocity faster than orbital velocity (and therefore significant thrust) is required in order to maintain that altitude. The trajectories of items below the Kármán line will degrade because of atmospheric drag and eventually crash. That same 100-km figure is used by NASA to determine who is and is not an astronaut, while the U.S. Air Force uses 80 km as their threshold for awarding astronaut wings.

For comparison, the International Space Station (Figure 6.1) orbits at 390 km above the Earth's surface, while the space shuttle (Figure 6.2) flies orbits in the 300–400-km range. The inner Van Allen radiation belt† begins at around 700 km.

* The Ansari X PRIZE offered a cash prize of US$10M for launching a crewed, reusable spacecraft twice within a 2-week span. The prize was won by the SpaceShipOne spaceplane in October 2004.

† The Van Allen radiation belts are two regions of plasma that form an inner and outer belt; the charged particles are trapped by Earth's magnetic field.

FIGURE 6.2 The space shuttle *Atlantis*, lifting off from Cape Canaveral, Florida, on its way to Earth orbit, February 7, 2008. (Courtesy of NASA.)

BASIC PHYSICS OF ROCKETS

In order to reach space, a rocket requires a minimum Δv—that is, that its velocity be changed by the application of thrust. Conventional rockets achieve a change in velocity (acceleration) by expelling propellant in the opposite direction, preferably at high speed. Conservation of momentum leads to the rocket equation, which is as follows:

$$\Delta v = v_e \ln\left(\frac{M_i}{M_f}\right), \tag{6.1}$$

where Δv is the change in the rocket's velocity, v_e is the effective exhaust velocity, M_i is the initial mass before the rocket is fired, and M_f is the mass of the rocket after the propellant has been expelled. Large values of Δv can be achieved by expelling mass at high speed, using up a lot of propellant, or both.

A rocket launched from Earth would need to reach a certain velocity, known as the escape velocity, to entirely escape the Earth's gravity. That is, the projectile's kinetic energy at that speed is exactly equal to its potential energy at its starting position. The escape velocity is defined as follows:

$$v_{esc} = \sqrt{\frac{2GM_E}{r_E}} \tag{6.2}$$

where v_{esc} is the escape velocity, $G = 6.67 \times 10^{-11}$ m^3/(kg·s^2) is the gravitational constant, M_E is the mass of the Earth, and r_E is the radius of the Earth. This works out to be 11.2 km/s (approximately Mach 34) for a starting point on the surface of the Earth. Orbital velocities, on the other hand, are lower than the escape velocity, and the Δv needed to achieve a certain orbit includes the desired orbital velocity, as well as the losses expected from sources such as atmospheric drag.

These two equations underscore the difficulty level of launching spacecraft from Earth into space. A very large value of Δv is required, so that the rocket must carry a large amount of propellant. Propellant is generally the single biggest item in the mass budget for a rocket and represents a large part of the launch costs. For example, the space shuttle weighs approximately 2M kilograms at liftoff; at landing, its mass is only a little over 100,000 kg.[4] The orbiter itself is thus only a small minority of the total liftoff mass. The shuttle travels only as far as low Earth orbit because, by design, it only has enough propellant to get there.

The shuttle, like all Earth-orbiting vehicles to date, uses multistage rockets to achieve orbit. It casts off rocket hardware during its ascent, including the solid rocket boosters, to reduce the mass that is accelerated in subsequent stages. A launch to Earth orbit using a single stage has never been accomplished, primarily because of the very large amount of propellant mass that would be needed to put a significant payload into orbit. As the amount of propellant mass increases, the fraction of the total mass that is made up of propellant grows to be prohibitively high, and even more mass is needed to hold that much propellant. Therefore, only multistage rockets have been feasible.

LAUNCH WINDOWS

New Horizons, NASA's Pluto–Kuiper Belt mission, launched on January 19, 2006, eight days after the first launch attempt. Additional launch opportunities, known as launch windows, were available after January 19. However, February 2, 2006, was the last such window that would also allow the spacecraft to follow a trajectory that included a Jupiter gravity assist, such that it will reach its destination years earlier than without the flyby. Most missions do not have such a restriction, but all launches do need to deal with the problem of launch windows to some extent.

Launch windows are a consequence of the Δv required to perform a maneuver that ensures the spacecraft reaches a certain destination in space at a given time. The rocket equation governs the dependence of Δv on the available propellant. For many Earth orbits, the timing of launch is relatively unimportant and so the launch window is very flexible. If the mission is interplanetary or includes a rendezvous with another satellite, space station, or other craft, then the timing becomes more critical.

SPACE AND SPACEPLANES

Anyone who has ever watched a science fiction movie has likely seen some portrayals of space that are physically implausible or that defy the realities of physics. For example, a spaceship bursts into "warp drive," accompanied by a loud noise—but there is no noise in space. Sound consists of vibrations traveling through a medium, such as air. Since space is essentially a vacuum, even the most powerful engines would be silent.

Dreams of the spacefaring future also often include a true spaceplane—a spacecraft that takes off and lands like a conventional airplane. The space shuttle (United States) and *Buran* (Soviet Union) were conceived in this vein and do in fact land like airplanes, but their takeoffs remain those of rockets. They do, however, achieve orbital spaceflights. On the other hand, recent commercial efforts have produced spaceplanes that do take off as well as land by using aerodynamic lift, but they reach only as far as suborbital altitudes. For example, two flights of NASA's X-15 aircraft reached altitudes above 100 km in 1963.[5] *SpaceShipOne* achieved suborbital spaceflight in 2004,[6] although it was first carried to high altitude by an airplane and dropped, with a rocket then applying the thrust needed to achieve spaceflight. One difficulty in building a true orbital spaceplane lies in harnessing sufficient thrust to propel a spacecraft to the higher orbital altitudes, though both NASA and commercial efforts are under way to overcome this problem. Additionally, as discussed earlier in this chapter, capability of reaching orbit using a single stage is very difficult to achieve because of the amount of propellant required.

BEING HUMAN IN SPACE

For human beings, space travel is no simple affair, and it is not for the squeamish. The acceleration experienced by an astronaut when the space shuttle launches is actually relatively mild[7] at $3g$; the Apollo astronauts, however, experienced as much as $9g$ at launch. Similar stresses are present during landing. Once in space, astronauts in Earth orbit do not experience the force of gravity. This is not, however, because there is no gravity in space. In fact, the strength of Earth's gravity at the orbital altitude of the shuttle and the International Space Station is still about 83% of its magnitude at the Earth's surface. The sensation of weightlessness occurs because the astronauts are in a state of free-fall, with no acceleration occurring except that provided by gravity.

Life aboard the space shuttle or on a space station comes with its own challenges. The astronauts are, of course, entirely dependent upon their craft's life support system,[7] called the Environmental Control and Life Support System by NASA. The following life support parameters are those used aboard the space shuttle.[8]

- The cabin of the spacecraft must maintain suitable interior temperature, which may require cooling or heating at various points in its orbit. The cabin air temperature on the space shuttle is kept between 18°C and 27°C.
- Humidity is also controlled and is maintained in the range of 30%–75%.
- An appropriate mixture of oxygen and nitrogen must be provided (80% nitrogen, 20% oxygen), the cabin air pressure must be kept within a relatively narrow range (total pressure of 101 kPa), and the air must be filtered and recycled.
- Food and water have to be carried aboard the spacecraft. The space shuttle includes four tanks for drinkable water, with a total capacity of 300 kg.

- Waste needs to be recycled or else disposed of in a safe and hygienic way. Human waste is processed by an integrated waste collection system. The liquid components are sent to the shuttle's 75-kg wastewater storage tank, and fuel cell power plants recycle them back into potable water. Wastewater can also be dumped overboard if necessary. Solid components are simply stored. This system is integrated with the toilet itself.
- Astronauts must wear spacesuits when outside the protection of their spacecraft. Without the pressurization of the suit, people would die of suffocation as their lungs emptied; the pressure differential caused by the near-vacuum of space would cause the air to leave their lungs. In such a low-pressure environment, their blood would also boil. Both these effects would take place after less than a minute of exposure.

The challenges to long-term human exploration are myriad. The current record for time spent continuously in space is held by Russian cosmonaut Valeriy Polyakov, who spent more than 14 months on the *Mir* space station from 1994 to 1995. Such long periods of time spent in space can have adverse effects on the human body, known as deconditioning. The long-term effects of microgravity, for example, include reduced muscle strength and endurance, decreased bone density, and loss of aerobic and cardiovascular capacity. For example, *Mir* cosmonauts generally had to be carried from their landing vehicle, and those who had completed longer stays in space required rehabilitation to regain their strength.[9] Many of these effects are reversible, though bone density decrease is not. Some effects, such as reduction in muscle strength, can be mitigated by exercise while in space, and space missions include exercise facilities and scheduled physical activity for that very reason. Radiation exposure is also a concern, even though the Earth's magnetic field provides some protection and spacecraft designers include shielding for the human occupants. Astronauts are already exposed to cumulative doses higher than they would receive on Earth; a long-term mission, such as to Mars, could have serious health consequences for space travelers, including cataracts, infertility, and an increased risk of developing cancer.[9] Although this does not preclude space exploration, much research will be required to alleviate the effects of space travel so that the risks do not overwhelm the rewards. Finally, advanced life support systems would be required, including thorough recycling of air and water.

CONCLUSION

This chapter has introduced many of the basics of human space flight. Although a single, clear demarcation of where space begins is necessarily artificial, since the atmosphere fades smoothly in density with increasing altitude, the most commonly used definitions have been introduced. The basic physics of rockets has also been presented, including the key parameter Δv, which governs the fuel-versus-payload tradeoff inherent in spaceflight and influences the choices of launch windows. The challenges involved in building a viable spaceplane have been discussed, along with efforts to date. Finally, the implications for the human body of space travel have been considered, particularly with regard to long-term exploration.

REFERENCES

1. U.S. Government. National Aeronautics and Space Act. 1958. http://www.nasa.gov/offices/ogc/about/space_act1.html.
2. X PRIZE Foundation. Ansari X PRIZE. 2008. http://www.xprize.org/x-prizes/ansari-x-prize.
3. FAI. 2004. FAI Astronautic Records Commission—100 km Altitude Boundary for Astronautics. http://www.fai.org/astronautics/100km.asp.
4. NASA. 2005. Space Shuttle Basics. http://www.spaceflight.nasa.gov/shuttle/reference/basics/index.html.
5. Jenkins, D.R. 2000. *Hypersonics Before the Shuttle: A Concise History of the X-15 Research Airplane.* SP-2000–4518. NASA.
6. Scaled Composites. 2007. SpaceShipOne Makes History: First Private Manned Mission to Space. http://www.scaled.com/projects/tierone/062104-2.htm.
7. Sellers, J.J., Astore, W.J., Giffen, R.B., and Larson, W.J. 2005. *Understanding Space: An Introduction to Astronautics.* Space Technology Series, ed. Douglas H. Kirkpatrick. 3rd ed. New York: The McGraw-Hill Companies, Inc.
8. NASA. 1988. Shuttle Reference Manual. http://spaceflight.nasa.gov/shuttle/reference/shutref/index.html.
9. Harrison, A.A. 2001. *Spacefaring: The Human Dimension.* Berkeley, CA: University of California Press.

7 Ground Segment: Ground Systems and Operations

R. Michael Furrow

CONTENTS

INTRODUCTION

The ground segment of a space system encompasses the terrestrial assets used to interface with the remote elements of the system, capture and process the science and operational data generated by the system, and command and control the system so that it can meet its design goals. The ground systems and networks used to interface with space vehicles are thus critical to effectively operating those systems and achieving their scientific and operational goals. This chapter details the characteristics and development of ground systems, including current ground systems such as the Deep Space Network, Tracking and Data Relay Satellite System, and the Universal Space Network. It provides a method of understanding the manner and means by which remote space systems are controlled and should provide insight into how the limitations of space communications affect system development, relative to purely terrestrial systems that are easily accessible.

A spacecraft's ground segment is not nearly as glamorous as a spacecraft streaking into the heavens on a tongue of flame and smoke, but without it, the spacecraft is nothing more than an expensive firework.

GROUND SEGMENT COMPONENTS

As illustrated in Figure 7.1, most modern spacecraft ground systems consist of six functional components: operations center, ground-to-spacecraft communications, data operations center, spacecraft simulators, support teams, and ground

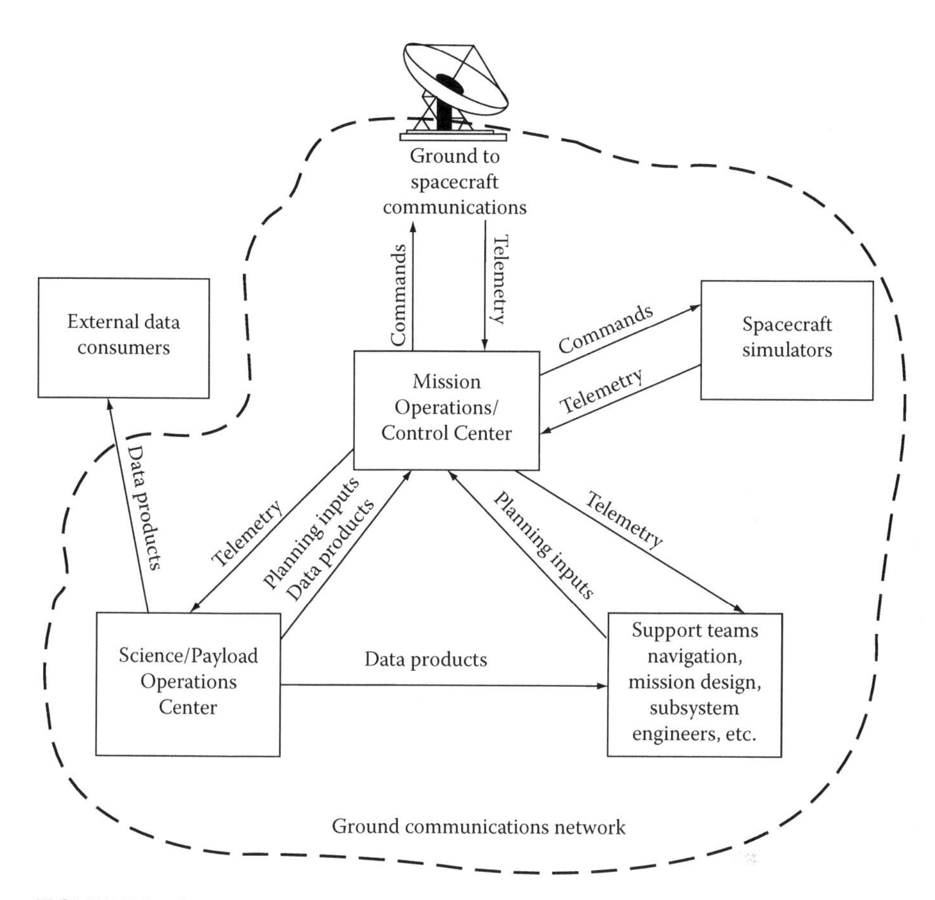

FIGURE 7.1 Ground system components.

communications network. These components can vary from mission to mission and are sometimes named or arranged differently, but in general, all ground systems have them.

Operations Center

The operations center is the heart of the ground system.[1] It goes by various monikers depending on the mission, but they all have the same basic function. Some common names for operations centers are mission operations center (MOC), spacecraft operations center, mission control center (MCC), launch control center, launch operations center, and space flight operations facility. The operations center provides the facilities, hardware, software, procedures, and personnel required to monitor and control a spacecraft. Most missions have two operations centers: one that controls the launch vehicle, and one that controls the spacecraft. Launch operations centers are normally located near the launch site and are active from slightly before launch until spacecraft separation. Spacecraft operations centers are typically geographically removed from the launch site and are active from well before launch through the end of the mission. An operations center can be extremely complex and is typically implemented with a mixture of computer and communications hardware and software. It is home to the operations team, which consists of a variety of personnel, including flight controllers, operations analysts, the operations manager, the mission manager, and spacecraft subsystem engineers.[2] Since the operations center is the focal point for all spacecraft operations, it usually has direct connections to all the other components of the ground system. Utilizing information from these other components, the operations center performs the following functions.

Mission Planning

Mission planning is the process of developing the command sequences that will be executed on the spacecraft. Good planning is essential to a mission's success. Utilizing input from various support teams, such as navigation, instruments, and mission design, operations analysts generate and test all command sequences that will be sent to the spacecraft. These sequences generally undergo multiple reviews before finally being transmitted.

Command Processing

Command processing takes human-readable commands, validates them, and converts them into a format that can be sent to the ground station for transmission to the spacecraft. This process can differ from mission to mission, depending on the standards and protocols that are implemented. In a common scenario, commands are first compared against a command dictionary to ensure that they are valid and have the proper format. Then they are converted to binary, packetized, framed, encoded, and possibly encrypted. A connection is then established to a ground station, and the processed commands are forwarded for transmission.

Missions handle verification of command receipt by the spacecraft in different ways. Some missions simply monitor telemetry to detect dropped commands and then manually resend them. Other missions implement protocols that require the commands to be buffered until spacecraft receipt has been verified via telemetry feedback. When receipt is confirmed, the commands are discarded from the buffer. If, for some reason, the spacecraft does not receive the command after a period of time, the command is automatically retransmitted to the spacecraft. The decision to choose one form of command receipt verification over the other is sometimes based on the distances involved. In the case of Earth-orbiting missions, where the communications times or round-trip light times are small and the verification of command receipt is received in seconds, command buffering and automatic retransmission makes a lot of sense. For interplanetary missions, where round-trip light times can be measured in hours, this method might not be feasible.

Telemetry Processing

Telemetry processing takes raw spacecraft telemetry received from the ground stations and converts it into a human-readable format. This is basically the inverse of the commanding process and also can vary depending on the standards being used. Usually, a connection is established to the ground station, and the operations center receives the raw telemetry. The raw telemetry is decrypted, decoded, and deframed, and individual telemetry points are extracted from the packets. These telemetry points are then converted to human-readable form (decommutated) and displayed to the user. Processed telemetry may also be fed to other operation center components for archiving or to support automatic command verification and retransmission.

Flight Software Uplink Processing

Flight software uplink processing takes a flight software binary image delivered from the flight software development team and converts it into a form that can be ingested by the commanding system. This process can vary greatly, depending on the flight hardware and software as well as the standards being used. If the spacecraft has an onboard file system, the process can be as simple as combining a load command and the software image into a single file and passing it to the command system. On the opposite end of the spectrum, the image might need to be broken into many small pieces, with multiple commands for each piece, before being sent to the command system. Once the new flight software image is transmitted to the spacecraft, it must be verified that nothing was corrupted during transmission. The most rigorous way to accomplish this is to dump the image from the spacecraft back to the ground and compare it with what was originally uplinked. This method is very reliable but can take a lot of time if the image or round-trip light times are large. Other methods involve computing a checksum or cyclic redundancy check for the image on board and dumping it to the ground for comparison. This method requires less time but is not quite as reliable.

Spacecraft Health and Safety Monitoring

This involves monitoring housekeeping telemetry to ensure that the spacecraft is operating as expected. On many missions, the telemetry system will automatically generate alarms to the operators if telemetry values are outside the expected nominal values. Long-term trending of certain telemetry points can also alert operators that corrective action needs to be taken to avoid potential problems in the future. Some deep space missions with long cruise phases rely on beacon tones instead of telemetry to report a gross spacecraft status. On preset days, for a period of time, the spacecraft transmits a tone that can represent green, red, or various levels of red. If the operations center detects a green tone, all is well with the spacecraft; otherwise, there is a problem and some action must be taken. Beacon operations are less expensive than normal telemetry operations because they are shorter and require fewer personnel.

Telemetry Archiving

Telemetry archiving is the process of storing telemetry in a format that can be retrieved later. In general, the archiving system receives raw and processed telemetry from the telemetry system and saves it to some type of long-term storage device. The archive system must also support data retrieval. Many missions have some form of archive server that users can access to request telemetry data. Archive servers typically have filtering capabilities that allow users to request data for specific time ranges or from specific spacecraft subsystems or instruments.

FIGURE 7.2 Dr. Goddard observes launch site. (Courtesy of NASA.)

Ground System Monitoring

Ground system monitoring is the ground equivalent of spacecraft health and safety monitoring. Many missions have a dedicated console or application that monitors the ground system status. It queries for and receives status from ground system components, consolidates it, and displays it to the operators. Some systems will also e-mail or page operators when a component needs attention. Network traffic volume, disk utilization, process status, CPU utilization, and device connectivity are examples of resources that are typically monitored.

Planning and development of the operations center starts at mission inception. Because of its many functions and the large number of interfaces it supports, the operations center's development and support team can be fairly large and is an integral part of the mission team. Because of its criticality, much of the operations center's software must be developed with the same degree of rigor and testing as flight software. A command that is improperly formatted on the ground can be just as detrimental to the spacecraft as one that is improperly executed by the onboard software. If spacecraft telemetry is not interpreted or displayed properly by the ground software, it could cause the operations team to take an action that could endanger the spacecraft.

The operations center is a critical resource for many missions long before the spacecraft is launched. During spacecraft integration and testing, it is often used to send commands and receive telemetry for spacecraft checkout. This allows the spacecraft to be tested as it will be flown. The mission operations team starts utilizing the

FIGURE 7.3 Gemini Mission Control, 1965. (Courtesy of NASA.)

operations center many months before launch for activities such as running mission simulations and anomaly resolution exercises.

Operations centers have been around in one form or another since the beginnings of spaceflight, from the very simple to the immensely complex. Figure 7.2 shows Dr. Robert H. Goddard at the control center of one of his early launch sites. Notice the hard-wired controls that allowed him to fire, release, or stop firing the rocket housed in the launch tower. Figure 7.3 shows a view of the MCC in Houston, Texas, during the *Gemini 5* flight of 1965. The screen at the front of the MCC was used to track the progress of the spacecraft. Figure 7.4 shows the MOC for NASA's New Horizons mission to Pluto, operated by the Johns Hopkins University Applied Physics Laboratory, during the spacecraft's closest approach to Jupiter on January 28, 2007.

In the past, launch control centers could be hazardous places to work because of their close proximity to the launch vehicle. Launch vehicle explosions have killed or injured numerous people since the beginning of spaceflight. Today, this hazard has been greatly reduced. Thanks to advances in electronics and communications, it is no longer necessary to locate launch control centers adjacent to launch pads. Figure 7.5 shows the launch control center for Launch Complex 39 at NASA's Kennedy Spaceflight Center, which is located 5.5 km from launch pad 39A.

Ground-to-Spacecraft Communications

The ground-to-spacecraft communications component of the ground segment is mission operations' sole link with the spacecraft. Whether the mission is in low Earth orbit or deep space, they all rely on ground stations to track and communicate

FIGURE 7.4 Mission Operations Center for NASA's New Horizons mission to Pluto operated by the Johns Hopkins University Applied Physics Laboratory during Jupiter closest approach. (Courtesy of the Johns Hopkins University Applied Physics Laboratory.)

with the spacecraft. Ground stations vary greatly, depending on the type of missions they support. Deep space missions require very large antennas with sensitive receivers and powerful transmitters capable of sending signals to and receiving signals from the edge of the solar system and beyond.[3] Earth-orbiting missions require smaller antennas, but they must be able to track spacecraft that move rapidly across the sky. Some low Earth-orbiting missions have antenna view periods that are only a few minutes long. Geostationary missions, which typically maintain continuous communications with the ground, can utilize limited motion or fixed antennas.

Some missions maintain their own ground-to-spacecraft communications component, while others rely on government or commercial ground stations. For deep space missions, the primary communications providers are NASA's Deep Space Network (DSN) and the European Space Agency's (ESA's) European Space Tracking (ESTRACK) network.[4,5] The DSN consists of three ground station complexes with a variety of antennas, ranging from 26 to 70 m in diameter.[6] The stations are located in Canberra, Australia; Madrid, Spain; and Goldstone, California. Since these locations are separated by roughly one-third of Earth's diameter, almost any point in the sky is visible to at least one of the stations 24 hours a day. ESA's ESTRACK network has two deep space stations, each with a 35-m antenna, located in New Norcia, Australia, and Cebreros, Spain.[7] Figure 7.6 shows the 70-m antenna at DSN's station in Canberra, Australia. Considerably more ground-to-spacecraft communications options are available for Earth-orbiting missions. There are a variety of government funded ground station networks, such as NASA's Tracking and Data Relay Satellite

FIGURE 7.5 Launch Control Center next to the Vehicle Assembly Building at NASA's Kennedy Spaceflight Center Launch Complex 39. (Courtesy of NASA.)

System (TDRSS), the Indian Space Research Organization's Telemetry Tracking and Command Network, NASA's Ground Network, ESA's ESTRACK network, the U.S. Air Force Satellite Control Network, and Japan's National Space Development Agency's (NASDA) ground network.[8,9,10,11] NASA's TDRSS is unique in that it can transmit data to and receive data from client spacecraft over 100% of their orbit via a single ground station complex. TDRSS utilizes a constellation of seven relay satellites in geosynchronous orbit, which are distributed to provide global coverage. As illustrated in Figure 7.7, data is uplinked from the White Sands Ground Complex to the relay satellites and then forwarded to the client spacecraft and vice versa. Many commercial ground station networks, such as the Universal Space Networks (USN), provide support for Earth-orbiting missions. Some missions maintain their own dedicated ground stations. It is not unusual for a mission to utilize the services of multiple ground station providers during the life of a mission. NASA's Thermosphere Ionosphere Mesosphere Energetics and Dynamics mission, operated by the Johns Hopkins University Applied Physics Laboratory, uses TDRSS, USN, and its own ground station.[12]

The number of ground stations has exploded from the handful of Soviet and U.S. stations built in the late 1950s to the thousands that are operational today. Some of the early sites, such as DSN's first antenna, Deep Space Station (DSS) 11 (the Pioneer site), still exist. Although deactivated in 1981, it was designated a National Historic Landmark in 1985.[13] Other sites have not been as fortunate. As shown in Figure 7.8, the original site of DSS-46 at Honeysuckle Creek, which received pictures beamed back from *Apollo 11*, is marked only by a spire. DSS-46's 26-m antenna (shown in Figure 7.9) was moved to DSN's Canberra station and upgraded in 1981.

FIGURE 7.6 NASA's 70-m DSN Antenna at Canberra, Australia. (Courtesy of NASA.)

Data Operations Center

Most missions, with the exception of communications and navigation satellites, have some form of data operations center. For civilian space missions, they are sometimes known as science operations centers or payload operations centers (POCs). For military or intelligence missions, they have various names depending on the organization. However, all data operations centers have the same basic functions: provide planning input to the operations center for instrument/payload observations, receive telemetry from the spacecraft via the operations center, and generate products for data consumers both internal and external to the mission.[14] Data operations centers are also usually responsible for the long-term storage of both raw telemetry and refined data products. Data products can vary, depending on the spacecraft's instrument/payload suite. They can be anything from images to particle counts to spectra. Most modern data operations centers have some form of automated data pipeline, where raw telemetry is fed into the front of the pipeline, and finished data products

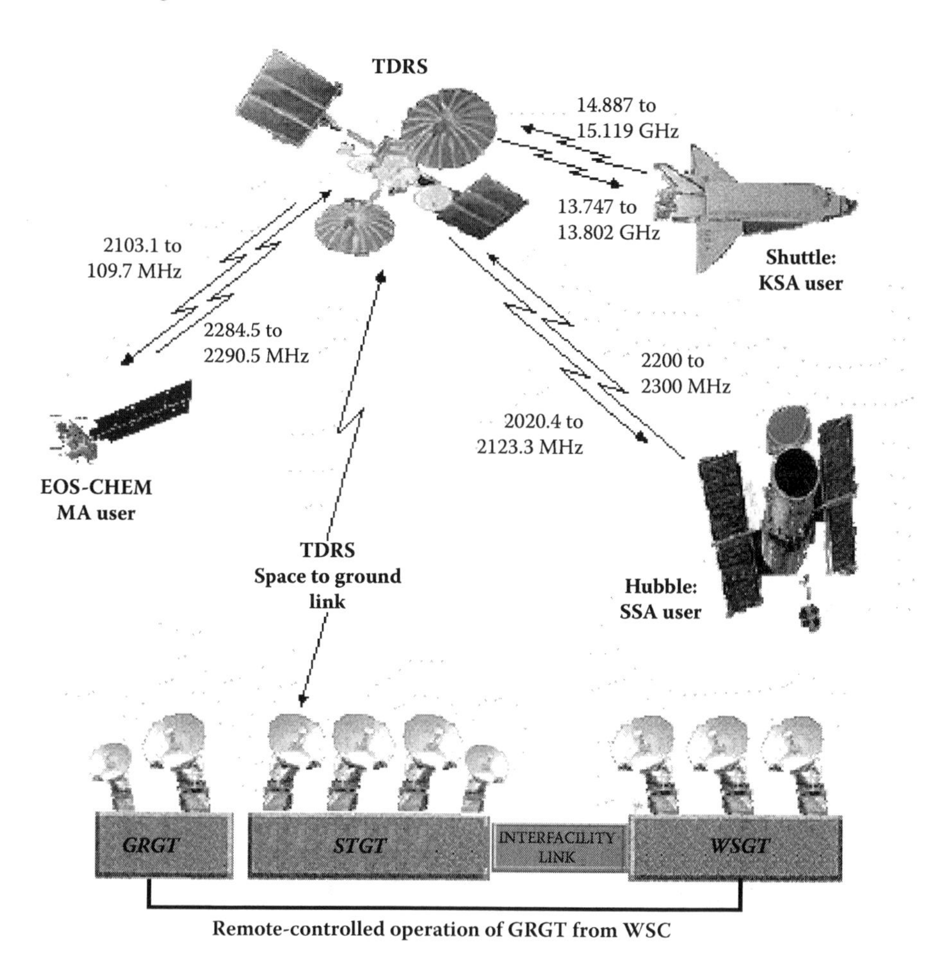

FIGURE 7.7 NASA's TDRSS uses relay satellites to communicate with Earth-orbiting spacecraft. (Courtesy of NASA.)

come out at the end, with various intermediate products in between.[15] These products are then cataloged, archived, and made available to the data consumers.

Data operations center personnel play a major role in spacecraft planning. They work closely with mission operations personnel to plan instrument/payload command sequences. For orbital and flyby missions, instrument/payload operations are often tightly coupled to spacecraft maneuvers and resource availability. For instance, the spacecraft might need to be slewed to a particular attitude before a camera can take a picture of some area of interest. Observational campaigns are sometimes planned months in advance and require many iterations and lots of simulator time to perfect. On other missions, such as some types of observatories, resources are not as constrained and the instruments are not nearly as dependent on spacecraft maneuvers. The spacecraft on these missions typically point at one position (such as the center of the sun), and the instrument teams are free to control their

FIGURE 7.8 The steel spire that marks the spot where the DSS-46 antenna used to stand. (Courtesy of Mick Stanic.)

instruments without regard to the rest of the spacecraft. For this type of decoupled operation, instrument commands can be sent directly from the POC to spacecraft via the operations center. The operations center receives the instrument commands from the POC and forwards them to the spacecraft with only minimal validation, such as origination source confirmation.[16] This is sometimes called bent-pipe instrument commanding.

FIGURE 7.9 NASA's DSN DSS-46 26-m antenna, originally from the Honeysuckle Creek Station. (Courtesy of NASA.)

Spacecraft Simulators

Spacecraft simulators are an extremely important component of the ground segment. On many missions, command sequences are not sent to the spacecraft unless they have first been vetted on a spacecraft simulator.[17] Spacecraft simulators are also used prelaunch for testing and training exercises, such as mission simulations. There are

FIGURE 7.10 Hardware simulator for NASA's *MESSENGER* spacecraft. (Courtesy of the Johns Hopkins University Applied Physics Laboratory.)

two types of spacecraft simulators: faster than real-time software-based simulators and real-time hardware-based simulators.

Spacecraft software-based simulators are software applications that model a spacecraft's functionality with varying degrees of precision. They generally ingest spacecraft and instrument command sequences in some form and output reports that describe the state the spacecraft would be in if the command sequence had actually been executed on board. Software simulators normally run at faster than real time. This allows the operations team to determine the result of a command sequence in just a few minutes or hours, instead of the days, weeks, or months of real time necessary to execute it on the spacecraft. Depending on their fidelity, software simulators are sometimes not sufficient to completely validate operational command sequences. If this is the case, the command sequences are run through hardware-based spacecraft simulators. Hardware-based spacecraft simulators are constructed from operational flight hardware or engineering models and execute actual flight software. Figure 7.10 is a photo of a hardware-based simulator for NASA's *MESSENGER* spacecraft. Hardware-based simulators typically have a higher fidelity than software-based simulators. A command sequence executed on a hardware simulator should produce the same effect as if it were executed on the actual spacecraft. Unlike software-based simulators, hardware-based simulators typically execute commands in real time. In addition to validating critical command sequences, hardware-based simulators are used to check out new flight software before it is loaded to the spacecraft. In many cases, flight software development teams have a dedicated hardware simulator for software testing.

Spacecraft simulators often play a key role in anomaly diagnosis and resolution. The conditions that were present on the spacecraft when an anomaly occurred can be recreated on a simulator in an attempt to reproduce the situation on the ground for further study. The *Apollo 13* command module simulator was invaluable for developing and testing a startup sequence that enabled the crew to return safely to Earth.

Support Teams

Ground segment support teams are a major contributor to the success of any mission. These support teams consist of all the various groups and organizations that are required to provide mission operations the information they need to successfully operate the spacecraft. Common support teams include navigation, mission design, ground system engineers, network engineers, system administrators, and spacecraft subsystem engineers (such as guidance and control, command and data handling, propulsion, thermal, autonomy, and radio frequency, to name just a few). These teams provide support as needed and often staff consoles in the operations center during critical events. For instance, the navigation team might provide the operations team with orbit determination information derived from operations center provided telemetry products and data operations center provided optical navigation images. Interface control documents (ICDs) are sometimes developed to describe how the various teams interact with the operations center and each other. ICDs define in detail the services each team provides and how the services are delivered. ICDs describe things like archive files, navigation products, time correlation files, and planning files.

Ground Communications Network

The ground communications network is the almost invisible infrastructure that allows the ground segment to operate as a unified whole. This network consists of all the mission's telecommunications and information technology equipment and can be extremely complex.[18] As illustrated by Figure 7.11, even a simplified ground network diagram can be fairly complicated. The ground communications network consists of a collection of routers, switches, multifunction gateways, firewalls, dedicated leased lines, Ethernet cables, fiber optics, Internet connections, Web servers, and voice communications equipment, as well as a host of supporting software and configuration files. Ground network planning and design starts at the very beginning of a mission. The ground network engineers work closely with the rest of the spacecraft team and network engineers from supporting organizations to develop a robust set of network requirements and implement them to meet the mission's ground communications needs.

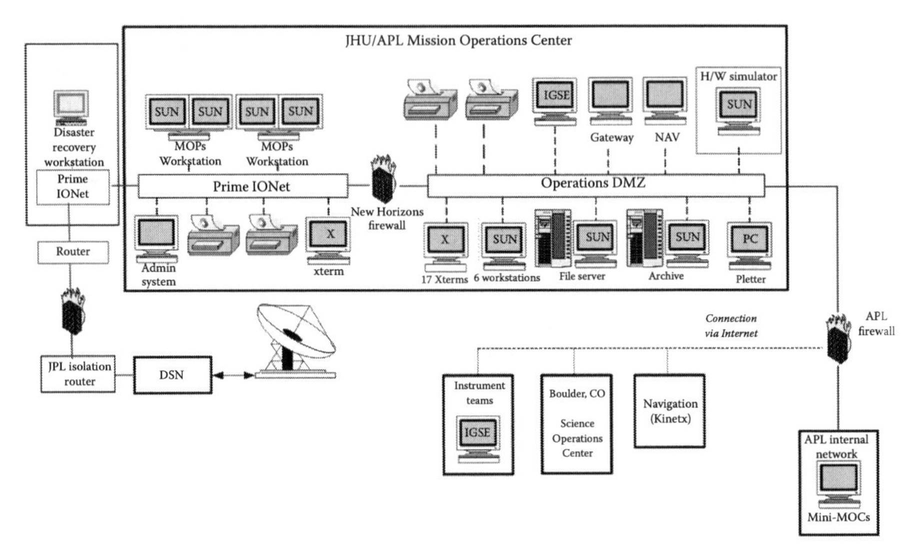

FIGURE 7.11 Simplified postlaunch ground network configuration for NASA's New Horizons mission, operated by the Johns University Applied Physics Laboratory. (Courtesy of Johns Hopkins University Applied Physics Laboratory.)

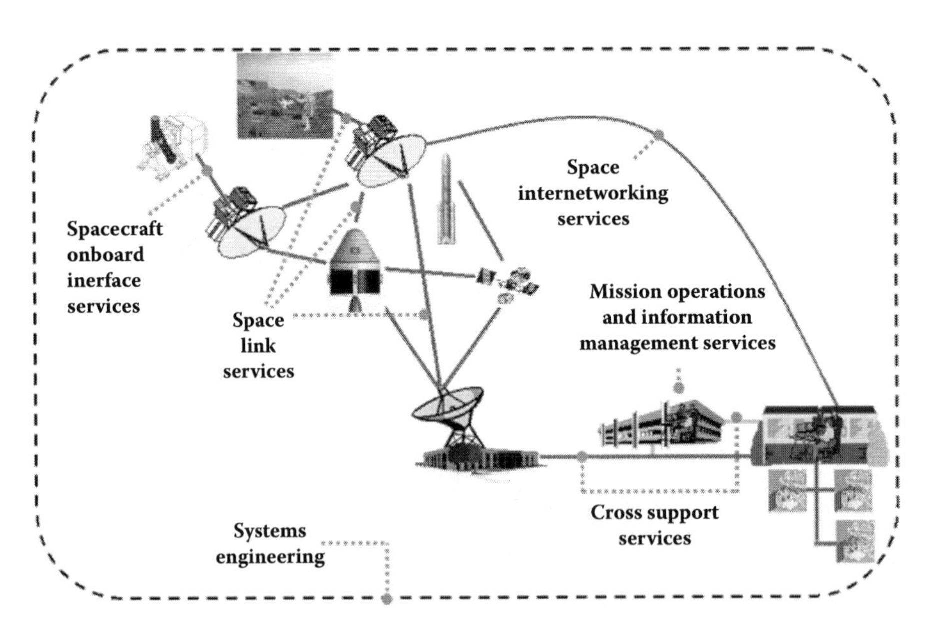

FIGURE 7.12 CCSDS interface standards. (From CCSDS.)

STANDARDS

Communications standards play a key role in facilitating mission interoperability. Ideally, through adherence to standards, missions can share such vital resources as ground stations. The Consultative Committee for Space Data Systems (CCSDS) is one of the major standards organizations for the space industry. It was formed in 1982 by several major space agencies, including NASA and ESA.[19] CCSDS was chartered to study the problems of cross-support and to develop standardized solutions to the challenges of exchanging space mission data. As illustrated by Figure 7.12, CCSDS develops standards for space link services, space Internet working services, spacecraft onboard interface services, cross-support services, mission operations and information management services, and system engineering services. Of primary importance to the ground segment are the space link services standards and the cross-support services standards. Among other things, the space link services standards address how commands and telemetry are formatted for transmission to and from the spacecraft. The space link extension (SLE) services standards in the cross-support services area describe how ground stations and operations centers exchange data. Since ESA's ESTRACK network and NASA's DSN currently support CCSDS SLE for both commands and telemetry, each agency is able to provide ground station support for the other agency's missions.[20]

CONCLUSION

Spacecraft ground systems are vital to the success of any space mission. They provide the resources that allow Earthbound operators and scientists to control spacecraft and receive the data that are the ultimate goal of any mission. Ground systems are extremely complex, and their development and maintenance require as much or more effort than that required for the spacecraft they support. They have evolved over the years, from Robert Goddard's primitive launch shack to a world wide network of control centers, antenna complexes, and data centers. The next time you are amazed by a picture from some distant spacecraft or reach your destination safely with the help of your GPS navigation system, remember the humble ground system behind the scenes that helps make it all possible.

REFERENCES

1. Miau, J., and R. Holdaway. 2000. *Reducing the Cost of Spacecraft Ground Systems and Operations. Space Technology Proceedings*. Dordrecht, The Netherlands: Kluwer Academic Publishers.
2. Wheaton, M.J. 2006. An Overview of Ground System Operations. The Aerospace Corporation. http://www.aero.org/publications/crosslink/spring2006/01.html.
3. Hall, R.C. 1977. *Lunar Impact: A History of Project Ranger*. NASA History Series. Washington, DC: Scientific and Technical Information Office, National Aeronautics and Space Administration.
4. Imbriale, W.A. 2003. *Large Antennas of the Deep Space Network*. Deep-Space Communications and Navigation Series. Hoboken, NJ: Wiley-Interscience.

5. Svedhem, H., D.V. Titov, D. McCoy, J.-P. Lebreton, S. Barabash, J.-L. Bertaux, P. Drossart, et al. 2007. Venus Express—The First European Mission to Venus. *Planetary and Space Science* 55(12): 1636–1652.
6. Jet Propulsion Laboratory. 2006. DSN: Antennas. http://deepspace.jpl.nasa.gov/dsn/antennas/index.html.
7. Jet Propulsion Laboratory. 2008. Deep Space Network Home Page. http://deepspace.jpl.nasa.gov/dsn.
8. Lewis, D. 1993. Indian Tracking and Data Acquisition Facilities for Low Earth and Geosynchronous Satellites. *IEEE Aerospace and Electronic Systems Magazine* 8(2): 19–28.
9. Teles, J., M.V. Samii, and C.E. Doll. 1995. Overview of TDRSS. *Advances in Space Research* 16(12): 67–76.
10. Air Force Satellite Control Network. http://www.fas.org/spp/military/program/nssrm/categories/sfssoafs.htm.
11. TDRSS Overview. http://msp.gsfc.nasa.gov/tdrss/oview.html.
12. Rodberg, E.H., W.P. Knopf, P.M. Lafferty, and S.R. Nylund. 2003. TIMED Ground System and Mission Operations. *Johns Hopkins APL Technical Digest (Applied Physics Laboratory)* 24(2): 209–220.
13. Mudgway, D.J. 2001. *Uplink-Downlink: A History of the Deep Space Network 1957–1997*. Washington, DC: National Aeronautics and Space Administration Office of External Relations.
14. Winters, H.L., D.L. Domingue, H.C., Choo, T.H., Espiritu, R., Hash, C., Malaret, E., Mick, A.A., Skura, J.P., and Steele, J. 2007. The MESSENGER Science Operations Center. *Space Science Reviews* 131(1–4): 601–623.
15. Meade, P.E. 2005. New Horizons Data Management and Archiving Plan. http://pds-smallbodies.astro.umd.edu/missions/newhorizons/new_horizons_pdmp.pdf.
16. Eichstedt, J., and W.T. Thompson. 2007. STEREO Ground Segment, Science Operations, and Data Archive. http://www.springerlink.com/content/e8j8331x422p0824/fulltext.pdf.
17. Holdridge, M.E., and A.B. Calloway. 2007. Launch and Early Operation of the MESSENGER Mission. *Space Science Reviews* 131(1–4): 573–600.
18. Griffith, G., M. Miguel, and J. Handiboe. 2007. Common Ground System Architecture for Multiple Missions. Paper presented at the Seventh International Symposium on Reducing the Costs of Spacecraft Ground Systems and Operations, Moscow, Russia.
19. Consultative Committee for Space Data Systems. 2003. Procedures Manual for the Consultative Committee for Space Data Systems. http://public.ccsds.org/publications/archive/A00x0y9.pdf.
20. Alexander, C., S. Gulkis, M. Frerking, M. Janssen, D. Holmes, J. Burch, A. Stern, et al. 2005. The U.S. Rosetta Project: NASA's Contribution to the International Rosetta Mission. Paper presented at the 2005 IEEE Aerospace Conference, Big Sky, MT.

8 Launch Segment: Launch Vehicles and Spacelift

Daniel E. Clemens and Barbara Leary

CONTENTS

INTRODUCTION

A fundamental and essential driver in the development and deployment of all space systems is the means of spacelift used to launch systems into space. It is this launch segment that defines the reach of the space system, constrains its mass and volume, and thus greatly impacts its implementation. Like many of humankind's greatest achievements, space launch was born in an atmosphere of conflict, spawned by machines of war. But while some rockets carried the means for humankind's potential destruction, others carried the means by which humanity could extend its footprint beyond the Earth and reach out into the solar system. This chapter will detail the development and implementation of the launch vehicle systems providing the performance and evolution of those systems and thus the basis for the commensurate evolution of the space vehicles launched by those systems. Also discussed are trends in launch vehicle operations, which are driven by a consistent need for affordable and reliable access to space. Finally, a discussion of launch vehicle basics is included.

LAUNCH VEHICLE EVOLUTION

Aside from the necessary preparation and training, all space missions truly begin with the liftoff of the launch vehicle. The key function of a launch vehicle is to lift a payload from the surface of the Earth and successfully place it into orbit around the Earth or on a trajectory for transfer elsewhere in the solar system. The only type of propulsion system that is currently capable of performing these functions is a rocket. Jet engines are incapable of reaching orbital velocities. In addition, they utilize atmospheric air for propellant and, thus, cannot properly function in or near the vacuum of space. By contrast, a rocket is defined as a propulsion system that carries all of its propellant onboard. Referring to the state of the propellant used, chemical rockets are classified as either solid or liquid rockets.

PIONEERS OF MODERN ROCKETRY

The first serious technical discussion of a liquid rocket engine has been credited to a Russian, Konstantin Tsiolkovsky (1857–1935), a deaf, self-taught high school teacher of physics and mathematics, who began contemplating rockets and spaceflight in about 1883.[1,2,3,4] In 1903, he published his first work, titled "Exploitation of Cosmic Expanse via Reactive Equipment." In this paper, and many more that followed, he discussed the utilization of rockets for investigating the upper atmosphere, as well as space. He is credited as being the first to give the equation for Earth escape velocity and was the first to derive the relationship between the velocity achieved by a rocket-propelled vehicle and the mass of propellant expelled. Today, this equation is sometimes called the Tsiolkovsky equation, but it is more widely known simply as the rocket equation.

Tsiolkovsky conceptualized many ideas, including multistage rockets, rudders for steering, gyroscopic stabilization, weightlessness, artificial satellites, and space stations. He analyzed the energy content in several propellants and determined that liquid propellants were preferable to solid propellants for spaceflight because of their higher available energy density (energy per unit mass). He also stated the need for cooling of the combustion chamber and recommended using the fuel flow as coolant. He studied the heat release of many liquid propellant combinations using liquid oxygen (LOx) as the oxidizer and several fuels, including alcohol, various hydrocarbons, and liquid hydrogen (LH_2). He then correctly concluded that LOx and LH_2 was the best propellant combination for spaceflight. Although Tsiolkovsky was the first to write seriously about these concepts, he never built or tested any rocket engines. That would be left to an American, Robert Goddard.

Robert Goddard (1882–1945) was a physics professor at Clark University in Massachusetts who developed the first liquid rocket engines in the world.[5,6,7,8] His early work included studies of solid rocket motors, but he eventually gave that up because he realized that liquid propellants offered higher performance. A patent from 1914 belonging to Goddard shows the first realistic design for a pump-fed liquid rocket engine, and another 1914 patent shows a two-stage vehicle. His 1919 paper "A Method of Reaching Extreme Altitudes," now considered one of the great pioneering works on rocket science, brought him attention; however, much of it was negative, as he was often ridiculed by an uninformed press. He designed, built, and tested engine components, and he operated engine assemblies with various propellant combinations using LOx as the oxidizer and ether, gasoline, alcohol, and kerosene as fuels.

FIGURE 8.1 Dr. Robert H. Goddard and a LOx-gasoline rocket in the frame from which it was fired on March 16, 1926, at Auburn, Massachusetts. (Courtesy of NASA.)

The world's first flight of a liquid rocket engine was on March 16, 1926, in Auburn, Massachusetts. This rocket is shown in Figure 8.1 mounted in its pyramidal launch frame. The thrust chamber, consisting of the combustion chamber and the nozzle, can be seen at the top, connected by feed lines to the propellant tanks at the bottom. This vehicle flew to an altitude of 12.5 m with a range of 56 m in about 2.5 s. Goddard launched rockets in Auburn, Massachusetts, from 1926 to 1930, but he moved to Roswell, New Mexico, to avoid scrutiny and work in relative seclusion until 1941. He conducted hundreds of component tests and static firings, and out of about 50 flight tests, 31 were successful.

Goddard pioneered many technologies, among which were methods for cooling, ignition, propellant feed, high-strength lightweight tanks, baffles to suppress propellant sloshing, valves and flow control, and thrust vector control. Interestingly, however, he had a relatively minor impact on the development of liquid rocket engines in the United States because of his reluctance to disclose his designs and the results of his testing. Only a small portion his total body of work was published during his lifetime. His collected works were published in 1970, 25 years after his death, but by then, rocket technology had matured without the full benefit of his pioneering work. In 1945, he saw a captured German V-2 and found that many parts of the design were strikingly similar to his own. The timeline of the evolution of rocket technology is likely to have been different if Goddard had been more open with his work and had more financial support from a government that recognized its importance.

Hermann Oberth (1894–1989), a Romanian-born German, is regarded as one of the three great pioneers of rocketry, along with Goddard and Tsiolkovsky.[9,10,11] As a young boy, he was inspired by Jules Verne's *From Earth to the Moon*. Oberth studied physics, and in 1922, he presented a doctoral dissertation on rocket-powered flight. It described the mathematical theories of rocketry and conceptual designs for practical rockets, but it was rejected for being too fantastic. He published it anyway in 1923 with the title "The Rocket into Planetary Space." In 1929, he published an expanded work, titled "Ways to Spaceflight," which also described electric rockets. His writings inspired the emergence of several rocket clubs in Germany, most notably the Society for Spaceflight, in which he became an active member and mentor. Though he was a scientist at heart and not a particularly good engineer, he fired his first liquid rocket engine in 1929 with the help of students from the Technical University of Berlin, who were also members of the Society for Spaceflight. One of his most important assistants was Wernher von Braun.

WORLD WAR II AND THE BEGINNING OF THE COLD WAR

Wernher von Braun (1912–1977) is widely regarded as the preeminent rocket scientist of the twentieth century. He began his studies with spaceflight his primary passion, influenced by and working beside Hermann Oberth. When the Nazis came to power in Germany, rocket-based weapon systems became a national priority. A military research grant was awarded to von Braun to fund his doctoral work on liquid rocket engines, and he became the central in the rocket research program of the German Army. Von Braun was the lead designer of the Aggregate series of ballistic missiles. The most successful, and most notorious, of these was the A-4, better known as the V-2.[12,13,14,15]

The Germans gained proficiency in the areas of large liquid rocket engines, supersonic aerodynamics, and guidance and control, which led to the successful development of the V-2, or Vengeance Weapon 2. The V-2, shown in Figure 8.2, was the first large-scale production liquid propellant ballistic missile and the first human-made object to achieve suborbital spaceflight. Operated using LOx and alcohol propellants to produce about 15 kN of thrust, it was used by the Germans against Allied targets in World War II, especially London and Antwerp.

Active government rocket research programs existed in the United States and USSR through the 1930s and early 1940s, although their technologies were not as advanced as that of Germany.[16,17] At the end of World War II, the United States and USSR competed to retrieve as many of the V-2s and German rocket engineers as possible. In order to avoid capture by the Soviets, von Braun and many of his staff surrendered to the Americans and were brought to the United States through Operation Paperclip. There they began work for the U.S. Army on the Hermes missile program,

FIGURE 8.2 A German V-2 on its transport trailer. The V-2 is about 46 ft. long and 5 ft. in diameter (UK government; public domain).

an attempt to copy and then expand the capabilities of the V-2.[18] The Soviets had a similar program employing many of the German rocket engineers who did not go with the Americans.[19] The first of their developments was the R-1.

Von Braun and his team moved to the Redstone Arsenal in Huntsville, Alabama, to work for what would become the Army Ballistic Missile Agency. It was there that the Redstone missile was developed under von Braun's leadership.[20] Redstone was a medium-range ballistic missile that used ethyl alcohol and LOx as propellants and produced approximately 330 kN of thrust at takeoff.[21] It was first flown in 1953 and was used for the first live nuclear missile tests by the United States. The Jupiter intermediate-range ballistic missile was an outgrowth of the Redstone that burned LOx and rocket propellant 1 (RP-1; kerosene) to produce about 670 kN of thrust at takeoff.[22]

In the USSR, development of the R-7 Semyorka began in the early 1950s under the leadership of Sergei Korolev* (1907–1966), regarded by many as the father of the Russian space program.[23,24] The R-7 was a 34-m-long, 3-m-diameter two-stage rocket using LOx and kerosene to produce approximately 3.9 MN of thrust at takeoff. It used four strap-on liquid booster engines and one core liquid engine constituting the first stage. The first-stage boosters were dropped, and the core continued to burn for the second stage. When it was first successfully tested in August of 1957, the R-7 became the world's first intercontinental ballistic missile (ICBM).

At the same time, the U.S. Air Force was working to develop the Atlas and Titan ICBMs.[25,26] Atlas was America's first ICBM, and it was first tested in 1957. The Atlas used balloon-type propellant tanks made of thin stainless steel with very little rigid support structure; the pressure in the tanks provided the necessary structural rigidity for the rocket. This allowed a lightweight design. At this time, there was some doubt

* Trained and employed as an aircraft designer, Korolev was accused of subversion during Stalin's Great Purge in 1938 and imprisoned for 6 years, which included several months spent in a Siberian gulag. Eventually, he became a rocket engineer and was appointed to lead the Soviet space program. However, his identity and position were kept secret, and his role in the success of the Soviet space program was not publically acknowledged until after his death.

that rocket engine ignition could be successfully achieved in space, leading to the "stage-and-a-half" design of the Atlas, which had all three of its engines ignited on the ground. Two of the engines, but not the accompanying propellant tanks, were jettisoned after burnout, and the third remained operational for an additional period of time. The Atlas burned LOx and RP-1 to produce approximately 1.6 MN of thrust at takeoff. The Titan missile was developed in parallel to Atlas in case of delays. Titan was the first multistage ICBM built by the United States. It was a two-stage rocket using LOx and RP-1 and first flew in 1958. Titan produced approximately 1.3 MN of thrust at takeoff and had a longer range than Atlas because of its stage design.

SPACE AGE

In 1955, the United States announced its intention to place a scientific satellite into orbit in late 1957 as part of the International Geophysical Year program.[27,28,29,30,31] The three possible launch vehicles were the Air Force's Atlas, the Army's Redstone, and the Navy's proposed Vanguard design. President Eisenhower wanted the first U.S. satellite in space to be launched by a civilian-designed vehicle because the laws regarding the legality of spacecraft flyover of another country had not been established and he did not want the adversarial image of military space programs to hamper the discussions. Even though the Redstone launch vehicle could be ready first, the Navy's Vanguard rocket was selected for this task because it was designed at the Naval Research Laboratory, which was seen more as a scientific organization than a military establishment. The Vanguard launch vehicle was designed specifically for space launch, whereas the others were modified ballistic missiles. The Vanguard was a three-stage rocket designed to deliver approximately 10 kg of payload into orbit. The first stage burned LOx and RP-1 to produce about 135 kN of thrust.

Things changed on October 4, 1957, when the Soviets launched *Sputnik 1* aboard a modified R-7 ICBM and became the first nation in the world to put a satellite into orbit. The modified R-7 was called the Sputnik launch vehicle, and it was capable of delivering an impressive 1,300 kg of payload to low Earth orbit (LEO). *Sputnik 2*, carrying a dog, was launched the following month. The Soviets clearly had a substantial lead over the Americans in terms of thrust and, thus, payload capacity.

The Americans attempted a response to the launch of *Sputnik 1* by launching their own small satellite aboard a Vanguard rocket in December 1957. However, the launch failed when a first-stage malfunction caused the rocket to lose thrust, fall back into the launch pad, and explode. This is shown in Figure 8.3. The major disappointment caused the United States to allow von Braun and the Army to proceed with their plan to launch a satellite aboard a Jupiter-C, which was a modified Redstone. The Jupiter-C was a three-stage suborbital vehicle with a 330-kN liquid propellant first stage and solid-propellant second and third stages.[32] A fourth stage was added to the Jupiter-C, and it was renamed Juno 1.

Juno 1 had a payload capacity of approximately 10 kg and, in February of 1958, launched the first U.S. satellite into orbit, *Explorer 1*, which enabled the discovery of the Van Allen radiation belts. Many believe that the Americans would have beaten the Soviets into space if the Army had originally been given permission to

FIGURE 8.3 December 1957 launch failure of the Vanguard rocket intended to place the first U.S. satellite into orbit. (Courtesy of NASA.)

proceed with its proposal; however, the Soviets still would have led in payload capacity. In March of 1958, the *Vanguard 1* satellite was successfully launched aboard a Vanguard rocket and is now the oldest human artifact in space.

In the USSR, the R-7 ICBM continued to evolve and would eventually become one of the most flown, most reliable launch vehicles in the world.[33,34,35,36] A small third stage was added to the missile, and the new vehicle was called the Vostok. It had an LEO payload capacity of almost 5,000 kg and was used to launch the Luna

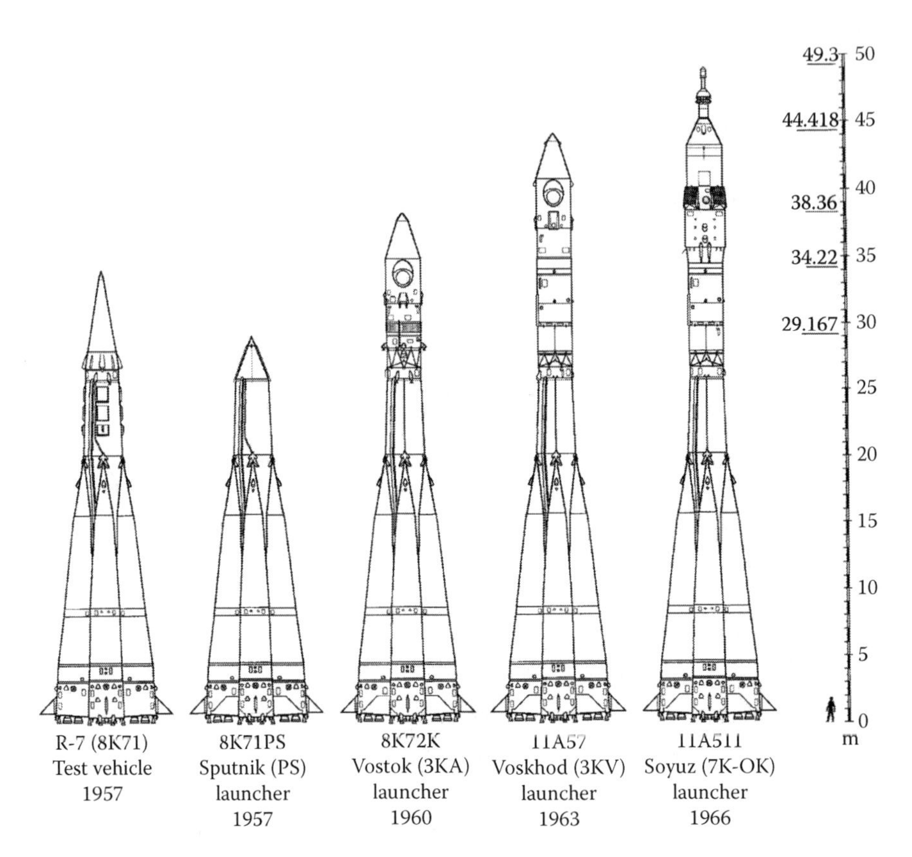

FIGURE 8.4 Evolution of the R-7 ICBM. The R-7 spawned the Sputnik, Vostok, Voskhod, Soyuz, and Molniya space launch vehicles, becoming one of the most flown, most reliable launch vehicles in the world and claiming several historical firsts. (Courtesy of NASA.)

spacecraft to the Moon in 1959. *Luna 1* was the first spacecraft to leave Earth orbit, *Luna 2* was the first to impact another planetary body, and *Luna 3* captured the first images of the far side of the Moon. The Vostok launch vehicle was designed for the human spaceflight program and it was used to launch Yuri Gagarin on April 12, 1961, making him the first human to fly in space and the first to orbit the Earth. Vostok flew from 1959 to 1991.

A larger third stage was added to the R-7, producing the Voskhod, which had an LEO payload capacity of almost 6,000 kg. The Voskhod was human-rated, capable of launching interplanetary spacecraft, and flew from 1963 to 1976. The Soyuz launch vehicle is another human-rated R-7 derivative with an improved third stage. The Soyuz was first flown in 1965 and continues to fly today, serving as the launch vehicle for Russia's human spaceflight program. Soyuz is the world's longest operating launch vehicle. The Molniya is still another R-7 derivative. It had a fourth stage and flew from 1961 to 2007, delivering spacecraft to interplanetary and high-energy earth orbits. The R-7 allowed to Soviet Union to claim several historical firsts, and its evolution is depicted in Figure 8.4. This is by far the most frequently flown and most reliable line of launch vehicles in the world, with nearly 1,700 launches by the end of 2007 and a reliability rate of about 95%.

In the United States, Redstone and Atlas launch vehicles were modified to launch humans into space for the single-member crewed Mercury program.[37,38] In May of 1961, a Mercury-Redstone rocket launched the *Freedom 7* spacecraft with astronaut Alan Shepard aboard for the first U.S. suborbital human spaceflight. In February of 1962, John Glenn became the first American astronaut to orbit the Earth in the *Friendship 7* spacecraft launched aboard a Mercury-Atlas launch vehicle. These launches are shown in Figure 8.5. In addition to several crewed Mercury missions, the modified Atlas launch vehicles were used with the Agena and Centaur upper stages for interplanetary missions to Mercury, Venus, and Mars. The Centaur, featuring the RL-10 engine that burned LOx and LH_2, was the world's first high-energy upper stage.

FIGURE 8.5 (a) Launch of *Freedom 7* on a Mercury-Redstone rocket with Alan Shepard aboard for the first American manned suborbital space flight. (b) Launch of *Friendship 7* on a Mercury-Atlas rocket with John Glenn aboard. (Courtesy of NASA.)

FIGURE 8.6 *Atlas II* (a) and *Atlas V* (b) on the launch pad. The strap-on solid rocket boosters are visible at the aft end of the launch vehicle. (Courtesy of NASA.)

Commercial development of Atlas began in the late 1980s. This new commercial family had four main members: Atlas I, II, III, and V.[39] Atlas II and Atlas V are shown in Figure 8.6. The payload capacity was continually enhanced through booster and upper stage engine upgrades for higher performance, the addition of strap-on solid rocket boosters for higher takeoff thrust, and stage lengthening for increased propellant capacity. The latest member of the family is the Atlas V, first launched in 2002, whose most powerful variant is capable of delivering 25,000 kg to LEO.

The Titan family evolved over four decades providing launch services to the U.S. government.[40] The Titan II, first flown in 1964, generated 1.9 MN of thrust at takeoff using Aerozine 50 [a blend of hydrazine, N_2H_4, and unsymmetrical dimethyl hydrazine (UDMH)] and nitrogen tetroxide (N_2O_4) propellants and was capable of delivering 3,600 kg to LEO. That propellant combination is storable and hypergolic, meaning that the fuel and oxidizer ignite on contact. Titan II rockets were used to launch the two-member crewed Gemini missions. Titan IIIC was first flown in 1965 and had a 13,000-kg LEO payload capacity. It utilized strap-on solid rocket boosters for additional thrust and launched several deep-space probes, including the Viking, Voyager, and Helios spacecraft, using a Centaur upper stage. The last of the Titan family, Titan IV, came into service following the loss of the space shuttle *Challenger*. It had a takeoff thrust of 15 MN and a 21,000-kg LEO payload capacity. The Titan

FIGURE 8.7 Titan II (a) and Titan IV (b) launch vehicles. The Titan family was retired in 2005. (Courtesy of NASA.)

launch vehicle was expensive to operate and was retired in 2005 after the development of the Atlas V and Delta IV heavy-lift launch vehicles. Titan II and Titan IV are shown in Figure 8.7.

In 1957, Wernher von Braun began the design of the Saturn launch vehicles.[41] He suggested the need for a heavy-lift launch vehicle with a thrust of approximately 6.7 MN capable of delivering 9,000–18,000 kg to LEO. The Advanced Research Projects Agency approved the program in 1958, and development began. In 1960, however, the Department of Defense decided it had no immediate application for such a large vehicle and turned the Saturn program, along with von Braun's team, over to NASA. At the time, NASA was already working on a large booster called the Nova. In 1961, President Kennedy called on NASA to put a man on the Moon by the end of the decade. Both the Nova and Saturn launch vehicles were considered, but the decision was made to proceed with Saturn largely because of the fact that much of the supporting infrastructure (factories, transport system, etc.) for Saturn was already in place.

The *Saturn I*, which first flew in 1961, was a two-stage rocket capable of delivering 9,000 kg to LEO. The eight-engine first stage consisted of eight Redstone propellant

FIGURE 8.8 *Saturn V* second stage being hoisted off its transport barge onto a test stand. (Courtesy of NASA.)

tanks clustered around one Jupiter tank. The H-1 engines burned LOx and RP-1 to produce 6.7 MN of thrust at takeoff. The second and third stages used RL-10 engines burning LH_2 and LOx. The Saturn IB, first flown in 1966, was an upgraded version of the Saturn I using the new J-2 engine on the second stage, which consumed LOx and LH_2 propellants.

The next member of the Saturn family was the mighty Saturn V. The Saturn V was a 111-m long, 10-m diameter, three-stage launch vehicle capable delivering 118,000 kg to LEO or 47,000 kg to lunar orbit. The first stage used five F-1 engines burning LOx and RP-1 to produce 34 MN of thrust at takeoff. The second stage used five J-2 engines, and the third stage used a single J-2. The Saturn V was so large that its first two stages had to be transported by barge from their factories in Louisiana and California to Florida for assembly. Figure 8.8 shows a Saturn V second stage being hoisted off its transport barge onto a test stand. As of 2008, the Saturn V remains the most powerful launch vehicle ever put into production, enabling the three-member crewed Apollo lunar missions and *Skylab*, America's first space station, which was actually a modified Saturn V third stage. NASA launched thirteen of these rockets

FIGURE 8.9 Launch of *Apollo 11*, the first lunar landing mission, carrying astronauts Neil Armstrong, Buzz Aldrin, and Michael Collins. (Courtesy of NASA.)

between 1967 and 1973 for Apollo and *Skylab*, with a 100% success rate. Figure 8.9 shows the launch of *Apollo 11*, the first lunar landing mission, and Figure 8.10 shows Wernher von Braun standing behind the first stage of a Saturn V.

The Apollo lunar module (LM) ascent stage could also be considered a launch vehicle. It is the only vehicle to have launched humans from the surface of another celestial body. Figure 8.11 shows a diagram of the LM. The LM ascent stage consisted of the crew cabin with its equipment for radar, communications, guidance, navigation, and control, the ascent engine, and fuel tanks. It carried as its payload two astronauts and lunar geologic samples to be returned to the Earth. The ascent engine burned N_2O_4 and Aerozine 50 to produce 15.6 kN of thrust. This thrust level is sufficient for launch in the low lunar gravity. Reliability was clearly very important, and the choice of hypergolic propellants was made in order to minimize the chances of an engine misfire. The ascent stage successfully lifted off from the lunar surface six times, with a 100% reliability rate. All of the ascent stages from the missions that actually landed were either deliberately crashed onto the lunar surface or were left in lunar orbit and eventually crashed on their own; however, the ascent stage from *Apollo 10* remains in solar orbit.

FIGURE 8.10 Dr. Wernher von Braun standing behind the first stage of a Saturn V. Four of the five F-1 engines are visible in the picture. (Courtesy of NASA.)

In 1959, the USSR began development of a large heavy-lift launch vehicle, comparable in power to the Saturn V, under the direction of Sergei Korolev.[42,43] Korolev proposed the N-1 launch vehicle, a five-stage rocket capable of delivering 75,000 kg to LEO, which had the potential to enable a crewed Soviet lunar landing. The N-1, shown in Figure 8.12, was a very complex rocket powered by a thirty-engine first stage, an eight-engine second stage, a four-engine third stage, and two single-engine upper stages. Korolev died during the development in 1966, but the work continued.

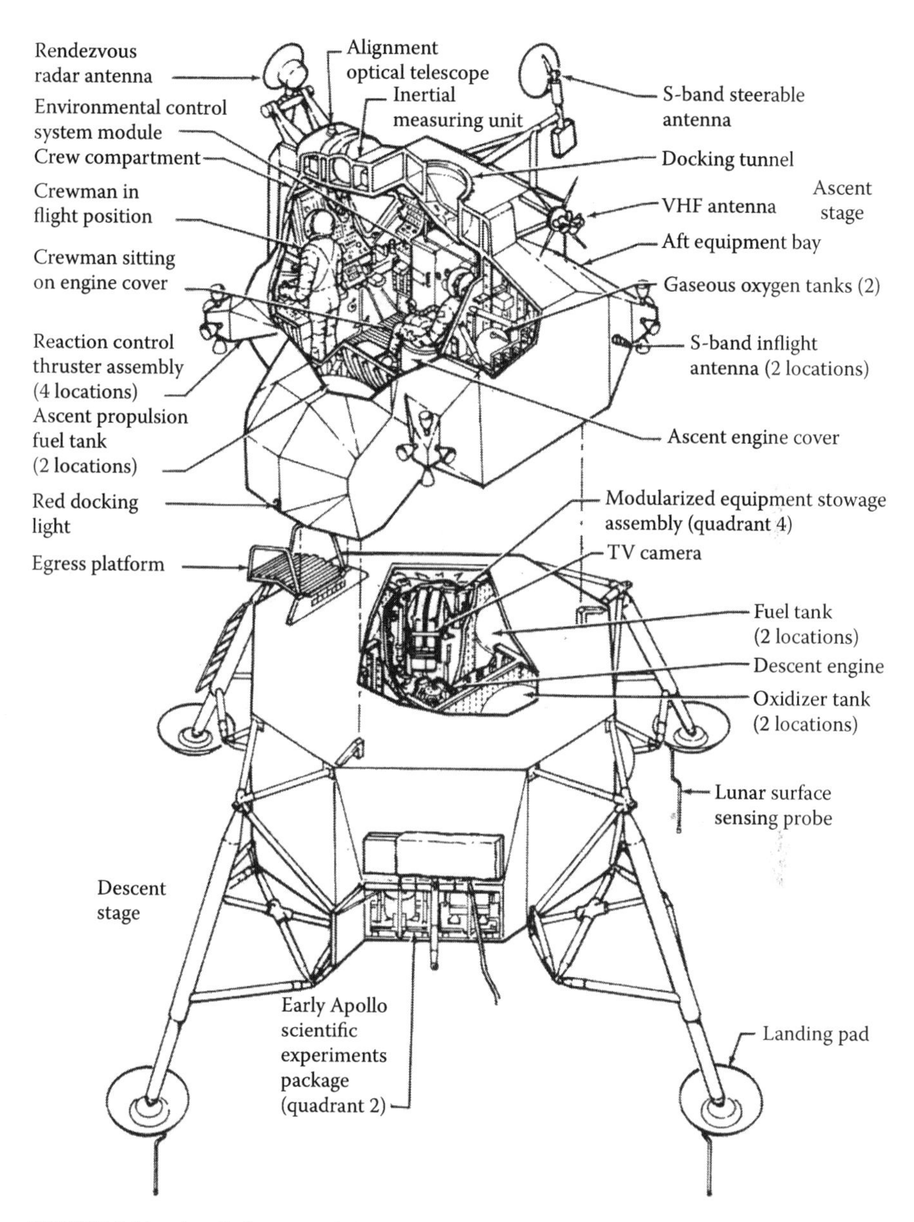

FIGURE 8.11 Apollo lunar module. (Courtesy of NASA.)

From the beginning, the effort was underfunded, and the first launch attempt, in 1969, ended in catastrophic failure. Three more launch attempts over the next three years also ended in complete failure, and the program was eventually canceled before the Soviets could attempt any crewed lunar missions with the vehicle.

The next significant evolution in human space launch was the development of the U.S. space shuttle following the cancellation of the Apollo program.[44] The

FIGURE 8.12 Two Soviet N-1 rockets on the launch pad in July 1969. (Courtesy of NASA.)

space shuttle was the first spacecraft designed to be reusable, and its maiden flight occurred in 1981. It generates a total thrust of 30 MN at takeoff using three liquid main engines and two solid rocket boosters. Approximately two minutes after takeoff, the solid rocket boosters are jettisoned. Later they are recovered and refurbished for use on subsequent missions. The reusable main engines continue to burn LH_2 and

FIGURE 8.13 Scaled comparison of the Saturn V, space shuttle, and the Ares launch vehicles that NASA plans to use for future crewed spaceflight missions. (Courtesy of NASA.)

LOx supplied by the large external propellant tank, which is also jettisoned prior to orbital insertion and burns up in the atmosphere. To return to Earth, the orbiter glides unpowered. It is capable of delivering a crew (typically seven members) and approximately 24,000 kg of payload to LEO. The shuttle has been used to carry large payloads to orbit, service spacecraft already in orbit, conduct on-orbit science missions, and facilitate crew rotation of the International Space Station. In addition to launching and deploying several large classified payloads, the shuttle has been used to launch many of the world's most productive scientific spacecraft, such as the Hubble Space Telescope. Out of over one hundred missions flown, two catastrophic losses have occurred: *Challenger* exploded during ascent in 1986, and *Columbia* disintegrated during reentry in 2003. Figure 8.13 shows a comparison of the Saturn V, space shuttle, and the Ares launch vehicles that NASA plans to use for future crewed spaceflight missions. Figure 8.14 shows the shuttle being transported on the back of a Boeing 747.

In the midst of the American–Soviet space race during the 1960s, other countries slowly began to develop space launch capabilities, typically built on the foundation of a domestic ballistic missile program. Table 8.1 shows a timeline of first orbital launches for spacefaring nations with independent launch capability. Some of these vehicle families are still operational today and are highly prominent.

In April of 1970, China became the fifth nation to independently launch a spacecraft using the Long March 1 launch vehicle.[45] The Long March 1 was a three-stage rocket burning N_2O_4 and UDMH propellants with a 1-MN takeoff thrust and a 300-kg LEO payload capacity. This launch vehicle was related to the Dongfeng missile

TABLE 8.1
Timeline of First Orbital Launches for Nations with Independent Launch Capability[21]

Nation/Region	Year	Launch Vehicle	Takeoff Thrust, kN	Payload Capacity
USSR	1957	Sputnik (R-7 Semyorka)	3,900	500 kg to LEO
United States	1958	Juno (Jupiter-C)	330	10 kg to LEO
France	1965	Diamant	290	85 kg to LEO
Japan	1970	Lambda	970	26 kg to LEO
China	1970	Long March	1,020	300 kg to LEO
United Kingdom	1971	Black Arrow	220	110 kg to LEO
Europe	1979	Ariane	2,450	1850 kg to GTO
India	1980	Satellite Launch Vehicle	420	40 kg to LEO
Israel	1988	Shavit	410	160 kg to LEO

GTO, geostationary transfer orbit; LEO, low Earth orbit.

designed by Hsue-shen Tsien† (1911–), regarded as the father of modern Chinese rocketry.[46] The Long March continued to evolve, and in 2003, a Long March 2 successfully launched the *Shenzhou 5* spacecraft, carrying China's first astronaut into orbit and making China the third nation to launch humans into space.

A consortium of European countries, led by France, has developed the Ariane launch vehicle.[47,48] The first successful flight of the Ariane 1 was in 1979. Ariane 1 generated almost 2.5 MN of thrust at takeoff and had a GTO payload capacity of 1,850 kg. Today, the latest member of the Ariane family is *Ariane 5*, which has a liftoff thrust of 15.4 MN and a 10,000-kg GTO payload capacity. The evolution of the Ariane launch vehicle family is shown in Figure 8.15.

TRENDS IN LAUNCH VEHICLE OPERATIONS

Figure 8.16 shows the total number of successful launches each decade for the United States, USSR/Russia, and other countries.[49] It can be seen that overall, the number of successful space launches worldwide peaked in the 1970s, with about 1,163 launches that decade, and about 74% of those were of Russian origin. In total, since 1957, the Russians have been responsible for approximately 62% of successful launches, while the United States and other countries have contributed approximately 30% and 8%, respectively.

These trends are broken down further in Figure 8.17, which shows the number of successful launches each year for the United States, USSR/Russia, and other

† The Chinese-born Tsien, a world renowned fluid dynamicist who was the protégé of Theodore von Karman, worked for the U.S. government as an important contributor to the early ballistic missile program of the World War II era and was one of the founding members of the Jet Propulsion Laboratory at the California Institute of Technology. However, during the Red Scare of the 1950s, he was accused of having Communist ties. Although his colleagues defended him and he was never formally charged with any crime, the ordeal led to his repatriation to China, where his work would become the foundation of the Chinese rocket program.

FIGURE 8.14 Space shuttle being transported on the back of a Boeing 747. (Courtesy of NASA.)

countries. The rapid increase in the number of successful space launches during the space race of the 1950s and 1960s is apparent. Even though the United States led the Russians in the number of successful launches in the late 1950s and early 1960s, the Russians had a substantial lead in payload capacity. After the development of the Saturn V and the American victory in the Moon race, the number of U.S. launches declined greatly, while the number of Soviet launches continued to rise through the 1980s. When the Soviet Union collapsed and the Cold War came to a close, the number of Russian launches decreased sharply and continued to fall through the 1990s. U.S. launches increased during that time, in the wake of the space shuttle *Challenger* disaster, as the United States developed new vehicles with heavy-lift capabilities to provide an alternative to the shuttle. Other nations had lagged far behind, but since the mid-1960s, they have steadily increased their capabilities, and they now rival both the United States and Russia for total launches. The European Ariane launch vehicle now captures about half of the worldwide commercial launch market.

The launch is typically considered to be the most dangerous phase of spaceflight because of the tremendous amount of energy that must be generated and released.[50] Controlling this energy is a serious challenge. Figure 8.18 shows the launch success rate each decade for the United States, Russia, and other countries. The reliability of United States and Russian launch vehicles increased greatly through the 1950s and 1960s. By the 1970s, the technology was fairly mature, and since then, the success rate has remained fairly flat at roughly 95% overall (crewed and noncrewed).

Crewed launch vehicles have been more reliable than noncrewed vehicles. In the history of U.S. human spaceflight, the first and only loss of crew during launch occurred when the shuttle *Challenger* exploded shortly after liftoff. The other human-rated launch vehicles of the Mercury, Gemini, and Apollo missions, all intended to

FIGURE 8.15 Evolution of the Ariane launch vehicle family (European Space Agency, Ducros; public domain).

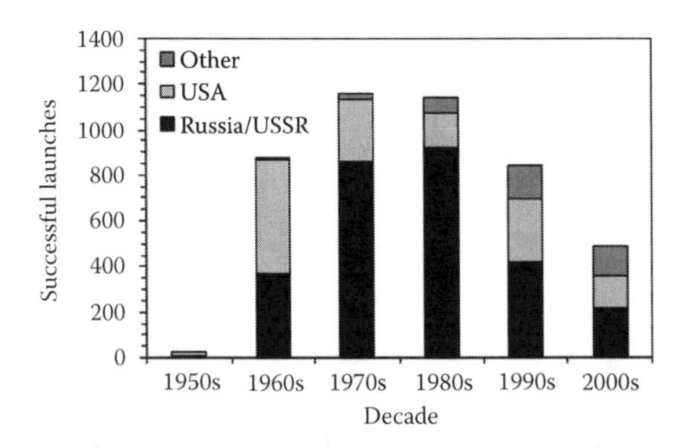

FIGURE 8.16 Total number of successful launches worldwide each decade beginning in October 1957 and ending in December 2007.

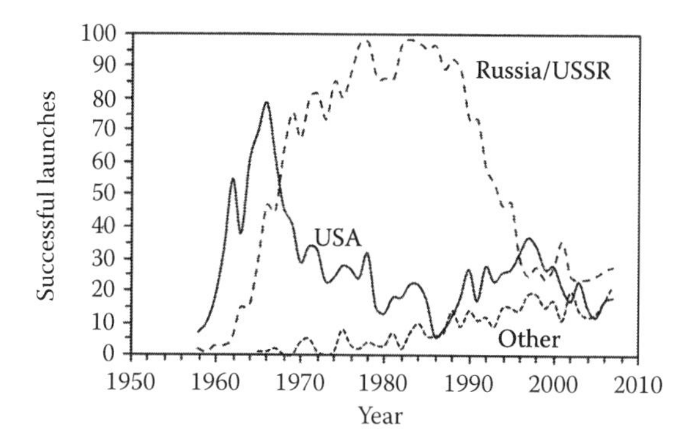

FIGURE 8.17 Number of successful launches for various countries each year beginning in October 1957 and ending in December 2007.

enable the United States to land humans on the Moon, flew with 100% reliability. In fact, out of about 250 total crewed space launches worldwide (United States, Russia, and China) to date, *Challenger* has been the only known failure.

The launch technology of other countries lagged behind the United States and Russia, and accordingly, so did their reliability. However, by the 1980s, their success rate had approached that of the superpowers. Without major human launch programs over this period to scrutinize, the reliability of the noncrewed launch vehicles has been only slightly lower than that of the United States and Russia.

LAUNCH VEHICLE BASICS

The main components of a launch vehicle are the propulsion system, the guidance, navigation, and control (GN&C) system, and the payload fairing (for noncrewed flight). The propulsion system, which includes the propellant, propellant tanks, and engines, makes up the majority of the vehicle and provides the thrust necessary to

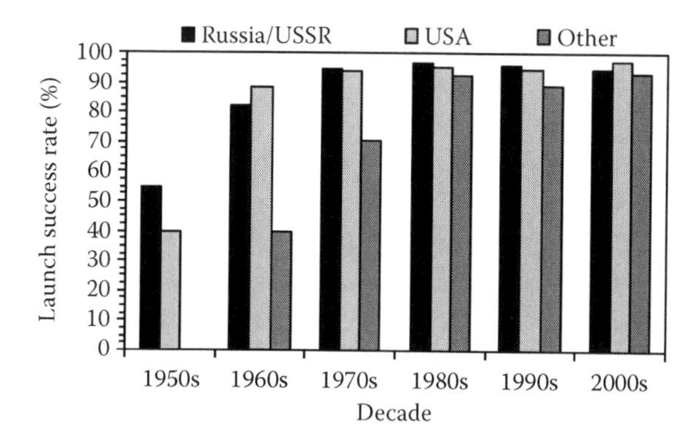

FIGURE 8.18 Launch success rate for various countries each decade beginning in October 1957 and ending in December 2007.

lift it into space. The GN&C system directs the vehicle on the proper trajectory to reach the desired orbit, while the payload fairing houses the spacecraft and protects it from the harsh launch environment. Considerations for launch vehicle selection include payload envelope and mass, desired orbit, launch site and national origin, payload/vehicle interface, cost and reliability, availability, and launch environment.

Two key parameters of propulsion system performance evaluation are thrust, τ, and specific impulse, I_{sp}.[51] Thrust is the reaction force generated on a rocket by an expelled propellant, and specific impulse is defined as the ratio of thrust to weight flow of the expelled propellant, given by

$$I_{sp} = \tau / \dot{m}g, \tag{8.1}$$

where $\dot{m}$ is the propellant mass flow rate and g is the acceleration due to gravity at the surface of the Earth. Specific impulse is an indicator of how much energy can be extracted from the resources onboard and how well that energy can be converted to exhaust kinetic energy. Propellants with higher energy density offer higher specific impulse and consume less propellant for a given propulsive maneuver.

The change in spacecraft velocity, Δv, is an important parameter because to reach a particular orbit, a certain Δv is required. The relationship that determines the amount of Δv gained by a rocket of a given mass that expels a given amount of propellant assuming continuous mass expulsion is known as the rocket equation. If M_i and M_f are the initial and final rocket masses, respectively, before and after a burn, then the rocket equation states

$$\Delta v = I_{sp} g \ln(M_i / M_f). \tag{8.2}$$

Since the difference between the initial and final masses is equal to the propellant mass, M_p, this can be rearranged to yield the fraction of the initial mass that is comprised of propellant such that

$$M_p / M_i = 1 - \exp(-\Delta v_{eff} / I_{sp} g). \tag{8.3}$$

Here, the effective Δv, Δv_{eff}, is the sum of the spacecraft Δv necessary to reach a desired orbit, $\Delta v_{s/c}$, and the Δv necessary to overcome gravitational and atmospheric drag losses, Δv_{grav} and Δv_{drag}, such that

$$\Delta v_{eff} = \Delta v_{s/c} + \Delta v_{drag} + \Delta v_{grav}. \tag{8.4}$$

Figure 8.19 shows the propellant fraction necessary to achieve a certain Δv as a function of specific impulse, assuming a one-stage burn. Gravity and drag losses are typically 1.5–2 km/s for launch vehicles, and the minimum orbital velocity in LEO is about 7.6 km/s, so the minimum effective Δv necessary to reach orbit from the surface of the Earth is about 9.1 km/s. Since most launch vehicles have I_{sp} in the range of 300–450 s, it can be seen that high propellant fractions are necessary.

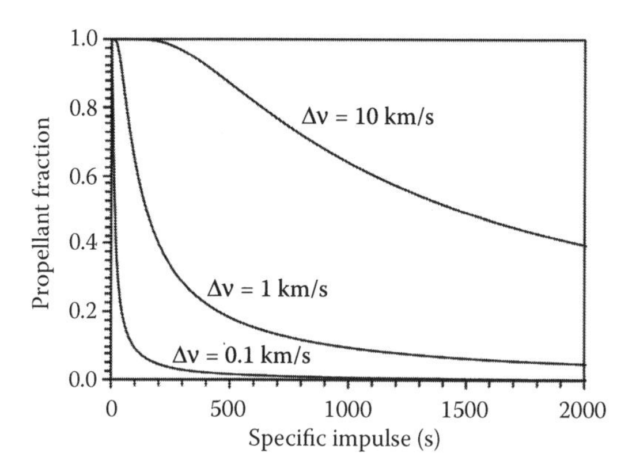

FIGURE 8.19 Propellant fraction, M_p/M_i, necessary to obtain a certain Δv as a function of specific impulse (I_{sp}), assuming a one-stage burn.

The propellant mass of a launch vehicle and the volume it occupies is very high, and accordingly, a significant amount of the vehicle structure is comprised of propellant tank mass. As the propellant is expelled, that tank mass becomes dead weight that must still be lifted. Sending a significant payload into orbit on a single-stage rocket is nearly impossible. However, a rocket can be segmented into multiple stages, each with its own propulsion system, structure, and propellant. After propellant from a given stage is consumed, the structure and engine can be dropped, thereby lightening the load for the subsequent stage. As the number of stages increases, the delivered Δv also increases; however, the Δv increase diminishes as the number of stages increases, so most launch vehicles have no more than three or four stages. Strap-on boosters are frequently added to the first stage of launch vehicles to augment takeoff thrust, depending on the mass and final destination of the payload.

The thrust power generated by a heavy-lift launch vehicle is greater than the equivalent electrical power output of a typical nuclear power plant. Onboard electrical power supplies are incapable of producing this amount of electricity, so electric propulsion systems cannot be used for launch vehicles. Nuclear thermal propulsion concepts, where thermal energy derived from a nuclear power source is added to a propellant gas, were theorized and tested in the 1950s and 1960s.[52] However, they were deemed infeasible for various reasons, one of which being that they were open-loop systems that produced large amounts of radioactive exhaust. The only type of rocket propulsion system feasible for launch is a chemical rocket.

The two main types of chemical propulsion systems are liquid rockets and solid rockets, referring to the type of propellant employed.[53,54] Liquid rockets are much more complex than solid rockets because of the need for pumps and plumbing; solid rockets do not require this. Compared with solids, liquid rockets generally have higher reliability and higher specific impulse, can be throttled, can be stopped and started multiple times, and can be operated for longer durations. However, storage of liquids can be more difficult, especially for cryogenic propellants such as LH_2 and LOx. Most orbital launch vehicles use a liquid rocket engine for the primary propulsion

TABLE 8.2
Approximate Sea-Level I_{sp}, Advantages, and Disadvantages of Common Liquid Propellant Combinations Used for Launch Vehicles

Fuel	Oxidizer	I_{sp}, s	Advantages	Disadvantages
LH_2	LOx	400	• Highest I_{sp}	• Cryogenic • Large LH_2 tank because of low density
RP-1	LOx	300	• RP-1 is storable • Higher density than LH_2	• Lower I_{sp} than LH_2
N_2H_4/MMH/UDMH	N_2O_4	280	• Storable	• Toxicity • Corrosivity

I_{sp}, specific impulse; LH_2, liquid hydrogen; LOx, liquid oxygen; MMH, monomethyl hydrazine; RP-1, rocket propellant 1; UDMH, unsymmetrical dimethyl hydrazine.

system and solid rocket motors as boosters for thrust augmentation. Solids are also frequently employed in upper stages.

The most common liquid propellant combinations are LOx/LH_2, LOx/RP-1, and N_2O_4/hydrazine, or blends of hydrazine, monomethyl hydrazine, and UDMH. Table 8.2 shows the approximate I_{sp}, along with some advantages and disadvantages of these propellant combinations. The most common solid propellant combination for launch vehicles is a mixture of ammonium perchlorate, hydroxyl-terminated polybutadiene, and aluminum, with a sea-level I_{sp} of about 265 s.

For noncrewed missions, the spacecraft is housed in the payload fairing during flight. The fairing protects the spacecraft from the harsh launch environment, in which the payload would otherwise experience high thermal and aerodynamic loads. In addition, the spacecraft must survive the severe vibrational environment of launch. The fairing is jettisoned late during ascent after the thermal and aerodynamic loads have been reduced to acceptable levels. The volume and dimensions of the payload are constrained by the usable size of the fairing; thus, the spacecraft must be designed to meet these requirements. A range of fairing sizes exists for the various launch vehicles.

The GN&C system is responsible for controlling the vehicle from its initial position on the ground to its destination at altitude. Navigation refers to the determination of the vehicle position, velocity, and attitude. Navigation is typically performed by an inertial measurement unit, which is a system of accelerometers and gyros used to plot the vehicle velocity vector given very accurate and precise knowledge of initial position and velocity. Guidance refers to the process of steering the vehicle to the target position in the most efficient path, and control refers to the execution of the guidance commands.

CONCLUSION

Rocket propulsion systems and launch vehicles were the primary technologies that enabled the Space Age. Over the past 50 years, thousands of spacecraft have been launched, and most have been successfully delivered to orbit. Worldwide, hundreds of rockets have been conceived, designed, built, and tested to various degrees for that purpose. Performance and reliability have increased greatly over that time, but the launch is still typically the most dangerous phase of spaceflight. As such, it is also, arguably, the most exhilarating.

REFERENCES

1. Sutton, G.P. 2003. History of Liquid Propellant Rocket Engines in Russia, Formerly the Soviet Union. *Journal of Propulsion and Power* 19(6): 1008–1037.
2. Mikhailov, V.P. 1993. The Contributions of K.E. Tsiolkovsky and Other Native Scientists to Technology of Rocket Launching. In *History of Rocketry and Astronautics*, ed. T.D. Crouch and A.M. Spencer, 65–72. American Astronautical Society History Series, vol. 14. San Diego, CA: Univelt.
3. Kosmodemjansky, A.A. 1994. Konstantin Eduardovich Tsiolkovsky and the Present Times. In *History of Rocketry and Astronautics*, ed. R.D. Launius, 165–173. American Astronautical Society History Series, vol. 11. San Diego, CA: Univelt.
4. Rauschenbach, B.V. 1993. The Development of Space Flight Theory by Soviet Scientists. In *History of Rocketry and Astronautics*, ed. L.H. Cornett, 141–146. American Astronautical Society History Series, vol. 15. San Diego, CA: Univelt.
5. Sutton, G.P. 2003. History of Liquid Propellant Rocket Engines in the United States. *Journal of Propulsion and Power* 19(6): 978–1007.
6. Goddard, R.H. 1970. *The Papers of Robert H. Goddard,* ed. E.C. Goddard and G.E. Pendray. Vols. 1–3. New York: McGraw-Hill.
7. Goddard, R.H. 1948. *Rocket Development*, ed. E.C. Goddard and G.E. Pendray. New York: Prentice-Hall.
8. Pendray, G.E. 1964. Pioneer Rocket Development in the United States. In *The History of Rocket Technology*, ed. E.M. Emme, 19–28. Detroit, MI: Wayne State University Press.
9. Elder, J. 1997. The Experience of Hermann Oberth. In *History of Rocketry and Astronautics*, ed. J.D. Hunley, 277–318. American Astronautical Society History Series, vol. 20. San Diego, CA: Univelt.
10. Roth-Oberth, E., and R. Layritz. 2001. The Personality of the Rocket Pioneer Professor Hermann Oberth. In *History of Rocketry and Astronautics*, ed. D.C. Elder and C. Rothmund, 203–208. American Astronautical Society History Series, vol. 23. San Diego, CA: Univelt.
11. Rohrwild, K. 1998. The History of the UFA Rocket. *History of Rocketry and Astronautics*, ed. P. Jung, 3–26. American Astronautical Society History Series, vol. 22. San Diego, CA: Univelt.
12. Hermann, R. 1994. The Supersonic Wind Tunnel Installations at Peenemünde and Kochel and Their Contributions to the Aerodynamics of Rocket-Powered Vehicles. In Launius (note 3), 39–56.
13. Dannenberg, K., and E. Struhlinger. 1998. Rocket Center Peenemünde: Personal Memories. In Jung (note 11), 27–39.
14. Dornberger, W.R. 1964. The German V-2. In Emme 1964 (note 8), 29–45.
15. Debus, K.H. 1989. From A-4 to Explorer 1: A Memoir. In *History of Rocketry and Astronautics*, ed. K.R. Lattu, 215–262. American Astronautical Society History Series, vol. 8. San Diego, CA: Univelt.
16. Malina, F.J. 1964. Origins and First Decade of the Jet Propulsion Laboratory. In Emme 1964 (note 8), 46–66.
17. Prishchepa, V.I. 1989. History of Development of First Space Rocket Engines in the USSR. *History of Rocketry and Astronautics*, ed. F.I. Ordway, 89–104. American Astronautical Society History Series, vol. 9. San Diego, CA: Univelt.

18. London, J.R. 1993. Brennschluss Over the Desert: V-2 Operations at White Sands Missile Range, 1946–1952. In Cornett (note 4), 335–367.
19. Harford, J. 2001. What the Russians Learned from German V-2 Technology. In Elder and Rothmund (note 10), 401–424.
20. Von Braun, W. 1964. The Redstone, Jupiter, and Juno. In Emme 1964 (note 8), 107–121.
21. Isakowitz, S.J. 1991. *International Reference Guide to Space Launch Systems*, 2nd ed., Washington, DC: American Institute of Aeronautics and Astronautics.
22. Braun, J.H. 1997. Development of the Jupiter Propulsion System. In Hunley (note 9), 133–144.
23. Rauschenbach, B.V. 1989. S.P. Korolev and Soviet Rocket Technology. In Ordway (note 17), 283–290.
24. Mishin, V.P. 1997. The Role of Academician Sergei P. Korolev in the Development of Space Rocket Vehicles for Lunar Exploration with the Help of Manned Spaceships. In Hunley (note 9), 247–255.
25. Martin, R.E. 1995. The Atlas and Centaur Steel Balloon Tanks: A Legacy of Karel Bossart. *History of Rocketry and Astronautics*, ed. J. Becklake, 285–317. American Astronautical Society History Series, vol. 17. San Diego, CA: Univelt.
26. Perry, R.L. 1964. The Atlas, Thor, Titan, and Minuteman. In Emme 1964 (note 8), 107–121.
27. Pickering, W.H. 1993. The Beginning of the U.S. Space Program: A Memoir. In Cornett (note 4), 211–221.
28. Hall, R.C. 1964. Early U.S. Satellite Proposals. In Emme 1964 (note 8), 67–93.
29. Friedman, H. 1993. International Geophysical Year to International Space Year. In Cornett (note 4), 285–291.
30. Hagen, J.P. 1964. The Viking and the Vanguard. In Emme 1964 (note 8), 122–141.
31. Hall, R.C. 1997. The Origins of U.S. Space Policy: Eisenhower, Open Skies, and Freedom of Space. *History of Rocketry and Astronautics*, ed. P. Jung, 75–105. American Astronautical Society History Series, vol. 21. San Diego, CA: Univelt.
32. Isakowitz 1991 (note 21).
33. Isakowitz 1991 (note 21).
34. Tokaty, G.A. 1964. Soviet Rocket Technology. In Emme 1964 (note 8), 271–284.
35. Konyukhov, S.N., and V.A. Pashchenko. 2001. The History of Space Launch Vehicle Development. In Elder and Rothmund (note 10), 451–458.
36. Rauschenbach, B.V. 1997. From the Development History of the Vostok Spacecraft. In Hunley (note 9), 151–158.
37. Bland, W.M. 1964. Project Mercury. In Emme 1964 (note 8), 212–240.
38. Sharpe, M.R., and B.B. Burkhalter. 1995. Mercury Redstone: The First American Man-Rated Space Launch Vehicle. In Becklake (note 25), 341–388.
39. Isakowitz, S.J., J.B. Hopkins, and J.P. Hopkins. 2004. *International Reference Guide to Space Launch Systems*. 4th ed. Reston, VA: American Institute of Aeronautics and Astronautics.
40. Isakowitz et al. 2004 (note 39).
41. Isakowitz 1991 (note 21).
42. Isakowitz 1991 (note 21).
43. Mishin 1997 (note 24).
44. Isakowitz 1991 (note 21).
45. Isakowitz et al. 2004 (note 39).
46. Thomas, S. 1997. Theodore von Karman's Caltech Students. In Jung (note 31), 3–39. American Astronautical Society History Series, vol. 21. San Diego, CA: Univelt.
47. Isakowitz et al. 2004 (note 39).
48. Russo, A. 2001. Launching Europe into Space: The Origins of the Ariane Rocket. In Elder and Rothmund (note 10), 35–49.

49. McDowell, J. Jonathan's Space Homepage. http://www.planet4589.org/space.
50. Sackheim, R.L. 2003. Overview of United States Space Propulsion Technology and Associated Space Transportation Systems. *Journal of Propulsion and Power* 22(6): 1310–1333.
51. Sutton, G.P., and O. Biblarz. 2001. *Rocket Propulsion Elements*, 7th ed. New York: John Wiley & Sons.
52. Dewar, J.A. 1991. Project Rover: The United States Nuclear Rocket Program. *History of Rocketry and Astronautics*, ed. J.L. Sloop, 109–124. American Astronautical Society History Series, vol. 12. San Diego, CA: Univelt.
53. Sackheim, R.L., and S. Zafran. 1999. Space Propulsion Systems. *Space Mission Analysis and Design*, 3rd ed., ed. by J.R. Wertz and W.J. Larson, 685–718. Torrance, CA: Microcosm.
54. Loftus, J.P., and C. Teixeira. 1999. Launch Systems. *Space Mission Analysis and Design*, 3rd Ed., ed. J.R. Wertz and W.J. Larson, 719–744. Torrance, CA: Microcosm.

9 Space Segment: Space Vehicles and Payloads

Ann Garrison Darrin and Thomas S. Mehoke

CONTENTS

INTRODUCTION

The raw data necessary to support archaeological and heritage research comes from the artifacts of the systems designed, built, and operated by engineers on Earth. Basic understanding of the functional decomposition of these systems, the design principles for the components contained within, and the construction of these devices is essential to establishing the framework for study in this area. This chapter will present general system decomposition for space vehicles and the sensor payloads that they host across the spectrum of space applications. This includes an overview of both robotic and crewed systems in the near-Earth environment (e.g., terrestrial weather, navigation, terrestrial Earth sensing, communications, space shuttle, and space station) and robotic systems in the deep space environment (e.g., planetary exploration, space weather, remote sensing, and lunar landing). This review of the space segment includes a discussion of physical characteristics, construction, function, and material properties of deployed systems that will provide a basis for future artifact identification and cataloging. From the earliest times, humankind has looked to the sky and anthropomorphic behavior to all objects seen above. It is a fitting place to start our discussion of all human-created objects above in terms of the same personification. Along these lines, it is easy and entertaining to equate the fundamental elements of all spacecraft to the human body model. At some level all spacecraft are robots, either semiautonomous or autonomous. A spacecraft has a bus (skeleton and skin), a power system (heart), a control system (brain), the ability to sense (eyes and ears), and the ability to transmit (voice). Layers of sophistication may be added relative to controls.

SYSTEMS FOR SPACECRAFT

In the human body model, the brain is the site of reason and intelligence, which includes such components as cognition, perception, attention, memory, and emotion. The brain is also responsible for control of posture and movements. It makes possible cognitive, motor, and other forms of learning. The brain can perform a variety of functions automatically, without the need for conscious awareness, such as coordination of sensory systems (e.g., sensory gating and multisensory integration), walking, and homeostatic body functions, such as blood pressure, fluid balance, and body temperature. The cerebellum controls balance and movement; without it, movements would not be coordinated. This model holds true but does not, perhaps, evoke much emotion, although it is not uncommon to hear engineers and scientists describe the performance of spacecraft in spirited terms, and nomenclature of many satellites and instruments uses common proper nouns: *Lewis and Clark*, *LISA*, *Thelma*, *Minerva*, and *Minuteman*.

Our discussion of the human body model will start at the brain, with cognition, perception, attention, and memory, more commonly known as the command and data handling system (C&DH). The brain of the spacecraft receives commands from the communications subsystem, interprets the inputs by validating and decoding the commands, and distributes the commands to the appropriate spacecraft subsystems and components. The C&DH monitors the health and well-being of the spacecraft; these data are referred to as "housekeeping data" by the other spacecraft subsystems and components. This monitoring is done autonomously, just as the brain coordinates sensory data. The memory of the spacecraft is held in this "brain" through packaging the data for storage on a solid-state recorder or transmission to the ground via the communications subsystem.

The heavy "thought processing" for control and movement in the cerebellum is extended to the guidance, navigation, and control system (GNC) and the attitude control system. *Guidance* refers to the calculation of the commands (usually done by the C&DH subsystem) needed to steer the spacecraft where it is desired to be, *navigation* means determining a spacecraft's orbital elements or position, and *control* means adjusting the path of the spacecraft to meet mission requirements. Continuing with our human body model, the need for balance and orientation is provided by an attitude control system. The attitude control subsystem consists of sensors and actuators, together with controlling algorithms. The attitude control subsystem permits proper pointing for the science objective, sun pointing for power to the solar arrays, and Earth pointing for communications. On some missions, the GNC and attitude control are combined into one subsystem of the spacecraft.

Leaving the brain and moving to the heart, we look at the power systems. Spacecraft need an electrical power generation and distribution subsystem for powering the various spacecraft subsystems. For spacecraft near the sun, solar panels are frequently used to generate electrical power. Spacecraft designed to operate in more distant locations, such as Jupiter, might employ a radioisotope thermoelectric generator (RTG) to generate electrical power. Electrical power is sent through power conditioning equipment before it passes through a power distribution unit over an electrical bus to other spacecraft components. Batteries are typically connected to the bus via a battery charge regulator, and the batteries are used to provide electrical power during periods when primary power is not available, for example, when a low Earth orbit (LEO) spacecraft is eclipsed by the Earth.

Moving to the skeleton and the skin, we look at the structures and the thermal control system, which is monitored by the C&DH, as the human body's homeostatic system is monitored by the brain. Spacecraft must be engineered to withstand launch loads imparted by the launch vehicle, and they must have a point of attachment for all the other subsystems. Depending on the mission profile, the structural subsystem might need to withstand loads imparted by entry into the atmosphere and landing on the surface of another planetary body. Thermal-controlled spacecraft must be engineered to withstand transit through the Earth's atmosphere and the space environment. They must operate in a vacuum with temperatures potentially ranging across hundreds of degrees Celsius, as well as (if subject to reentry) in the presence of plasmas. This environment places material requirements such that high-melting-temperature, low-density materials [such as beryllium, carbon composite (C-C), or tungsten], or ablative C-C composites are needed. Depending on the mission profile, spacecraft may also need to operate on the surface of another planetary body. The thermal control subsystem can be passive, dependent on the selection of materials with specific radiative properties, or active, making use of electrical heaters and certain actuators, such as louvers, to control temperature ranges of equipments within specific ranges.

Mobility for the launch vehicle is discussed in Chapter 8 of this *Handbook* as propulsion. In orbit, mobility is provided through the propulsion subsystem, depending upon whether or not the mission profile calls for propulsion. The *Swift* spacecraft is an example of a spacecraft that does not have a propulsion subsystem. Typically, though, LEO spacecraft [e.g., *Terra* (EOS AM-1)] include a propulsion subsystem for altitude adjustments (called drag make-up maneuvers) and inclination adjustment maneuvers. A propulsion system is also needed for spacecraft that perform momentum management maneuvers. Components of a conventional propulsion subsystem include fuel, tankage, valves, pipes, and thrusters. The thermal control subsystem interfaces with the propulsion subsystem by monitoring the temperature of those components and by preheating tanks and thrusters in preparation for a spacecraft maneuver. Thus, the metaphor of the human body is an apt one for spacecraft and its systems.

INTRODUCTION TO PAYLOADS

In addition to the various components discussed thus far that keep a spacecraft operational, every spacecraft is also customized with a payload, which provides customization for that spacecraft's mission purpose. This payload is dependent upon the mission of the spacecraft and is typically regarded as the part of the spacecraft that "pays the bills." Typical payloads could include scientific instruments (e.g., cameras, telescopes, or particle detectors), cargo, or a crew.

Scientific payloads usually fall into two categories: imagers and detectors. Imagers can be either short range or long range and are designed to take pictures of a region of interest (e.g., planetary bodies or stars) at a particular wavelength of light (e.g., visible, infrared, or ultraviolet). These devices thus consist of many optical components, such as lenses, and detectors, such as a charge-coupled device imager, to receive and convert the light into a usable image. One specific type of imager that allows some ability to detect the composition of what it sees is an infrared spectrometer, which can provide specific information as to the atoms and functional groups that make up whatever object or molecule emitted that infrared radiation. Other regions on the electromagnetic spectrum, such as gamma rays or x-rays, can also be used in a similar manner for elemental detection from a distance. These devices often use more exotic materials (e.g., hyperpure crystals of germanium in gamma ray detectors), as traditional optical materials are not efficient or do not have the ability to sense at these wavelengths.

Along with this means of detection through electromagnetic radiation analysis, spacecraft payloads often have detectors, as well, that measure particles directly, such as ion or neutral mass spectrometers. These particle spectrometers contain a sensor that reacts when it encounters the particle of interest, by fluorescing or producing an electric signal, an amplifier that increases the signal to detectable levels, and the electronics necessary to count and quantify these signals. One of the more common neutron detectors is the scintillation counter, which uses a fluorescent means of sensing.

Finally, one of the more specialized types of payload is human cargo and the facilities necessary to ensure their survival and allow them to complete their specific mission. The most important system for human survival in outer space is the life support system. This system must protect occupants from the radiation and meteorites present in the space environment, as well as supplying them with a pressurized atmosphere of oxygen, nitrogen, carbon dioxide, and other gases normally present in Earth's atmosphere and providing a supply of clean water and food. As water and food cannot be manufactured in space, they must be brought from Earth at the onset of the mission. A biological wastewater processor allows water and oxygen to be recycled from human waste, prolonging the spacecraft's supply of these essential materials, as shown in Figure 9.1.

SPACE MISSION TYPES

Depending on the purpose or mission, spacecraft will be designed as dependent more on sensing or on receiving/transmitting/observing, and if mobility is required, as for rovers, there may be some ability to navigate through wheels or tractor drives (limbs). Artificial satellites are classified according to their mission. There are six main types of artificial satellites (orbiters), all of which are sensing-dependent: [1]

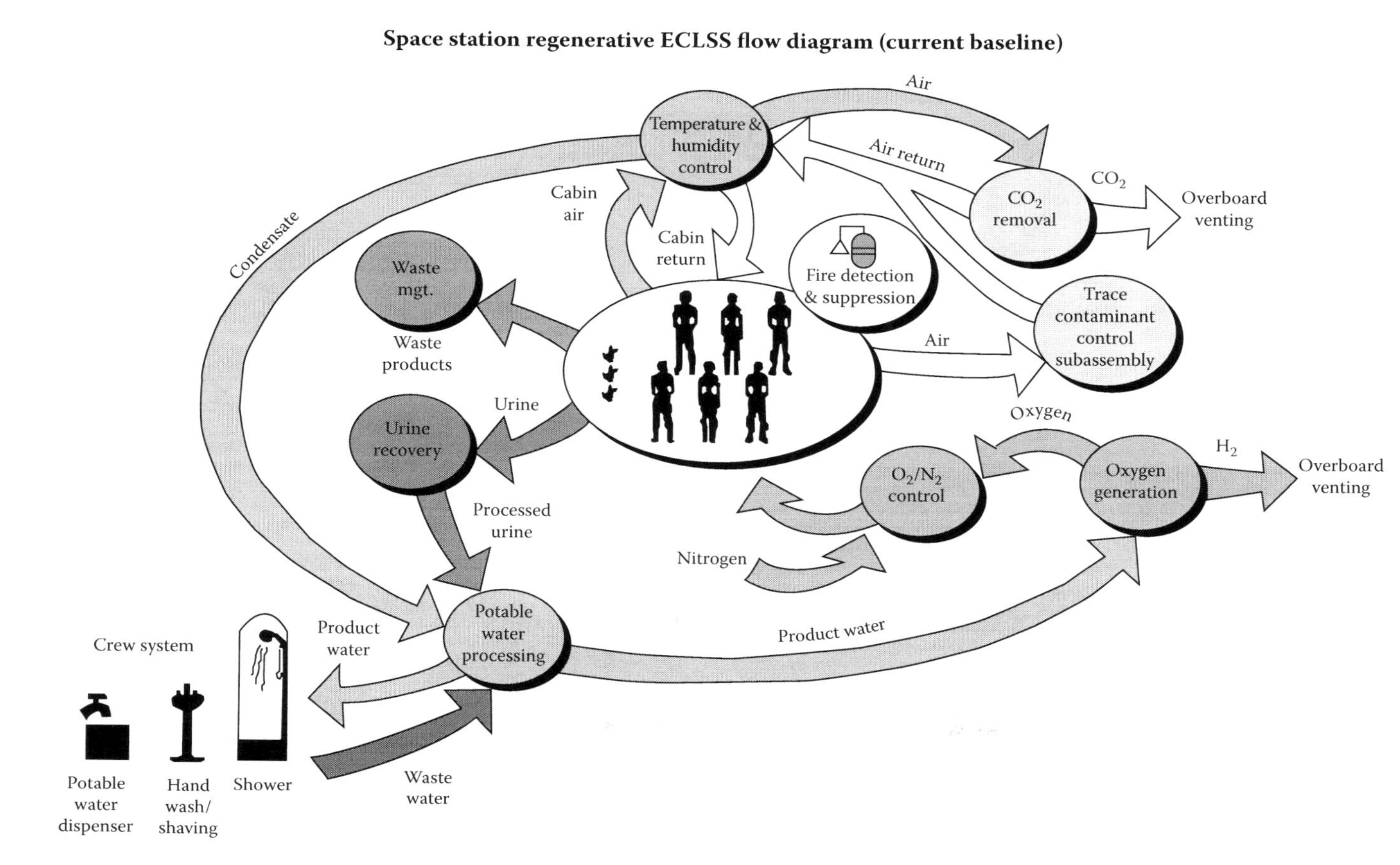

FIGURE 9.1 Spacecraft life support system. (Courtesy of NASA.)

1. Scientific research
2. Weather
3. Communications
4. Navigation
5. Earth observing
6. Military

Each of these major categories is discussed in the following section, along with illustrations.[2]

- Scientific research satellites gather data for scientific analysis. These satellites are usually designed to perform one of three kinds of missions. (1) Some gather information about the composition and effects of the space near Earth. They may be placed in any of various orbits, depending on the type of measurements they are to make. (2) Other satellites record changes in Earth and its atmosphere. Many of them travel in sun-synchronous, polar orbits. (3) Still others observe planets, stars, and other distant objects. Most of these satellites operate in low-altitude orbits. Scientific research satellites also orbit other planets, the Moon, and the sun. Various science missions' satellites and spacecraft are shown in Figure 9.2.

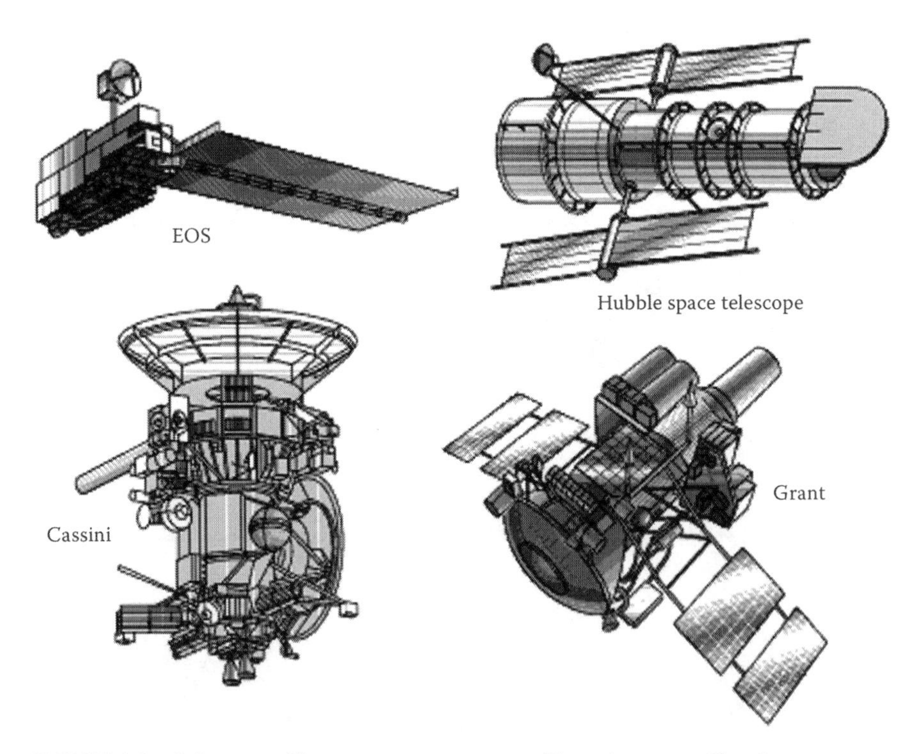

FIGURE 9.2 Science satellites—preceptor systems. (From Aerospace Corp.)

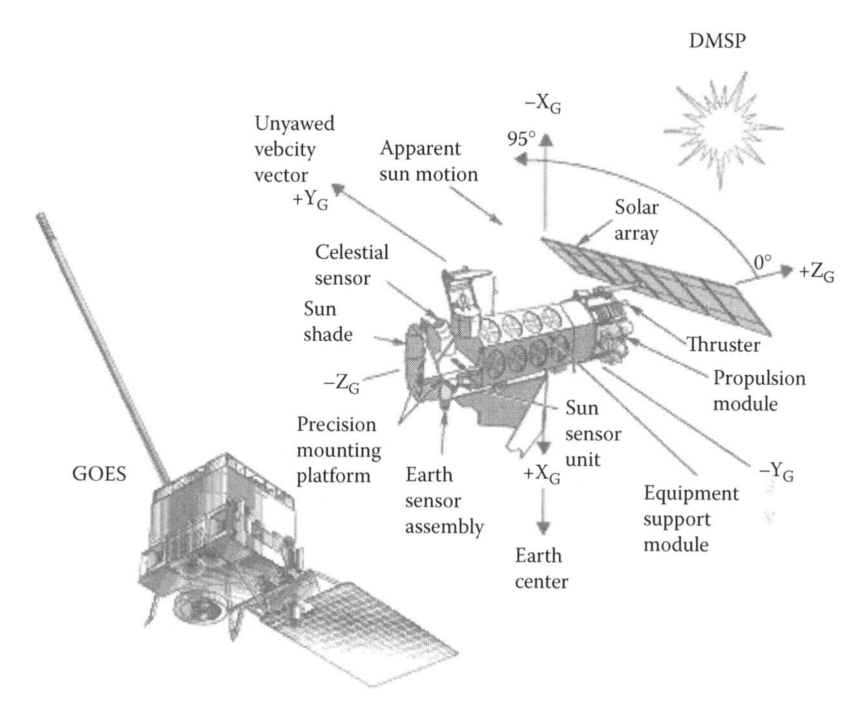

FIGURE 9.3 Weather satellites—perceiving (multiple sensors—Big Eyes). (From Aerospace Corp.)

- Weather satellites observe the atmospheric conditions over large areas. A variety of weather satellites are shown in Figure 9.3. Many weather satellites travel in a sun-synchronous, polar orbit, from which they make close, detailed observations of weather over the entire Earth. Their instruments measure cloud cover, temperature, air pressure, precipitation, and the chemical composition of the atmosphere. Because these satellites always observe Earth at the same local time of day, scientists can easily compare weather data collected under constant sunlight conditions. The network of weather satellites in these orbits also functions as a search-and-rescue system. They are equipped to detect distress signals from all commercial, and many private, planes and ships. Other weather satellites are placed in high-altitude, geosynchronous orbits. From these orbits, they can always observe weather activity over nearly half the surface of Earth at the same time. These satellites photograph changing cloud formations. They also produce infrared images, which show the amount of heat coming from Earth and the clouds.
- Communications satellites serve as relay stations, receiving radio signals from one location and transmitting them to another as shown in Figure 9.4. A communications satellite can relay several television programs or many thousands of telephone calls at once. Communications satellites are usually

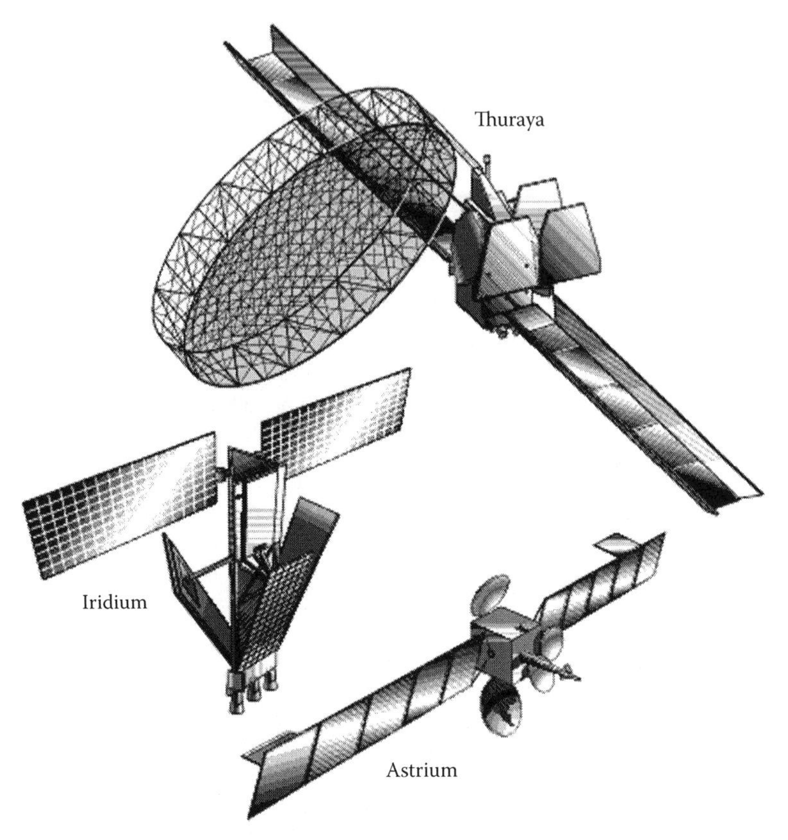

FIGURE 9.4 Communication satellites (multiple antennas—Big Ears). (From Aerospace Corp. With permission.)

put in a high-altitude, geosynchronous orbit over a ground station. Often, a group of low-orbit communications satellites are arranged in a network or constellation, working together by relaying information to each other and to users on the ground.

- Navigation satellites enable operators of aircraft, ships, and land vehicles anywhere on Earth to determine their locations with great accuracy. The global positioning system satellites are shown in Figure 9.5. Navigation satellites operate in networks, and signals from a network can reach receivers anywhere on Earth. The receiver calculates its distance from at least three satellites whose signals it has received. It uses this information to determine its location.
- Earth-observing satellites are used to map and monitor our planet's resources and ever-changing chemical life cycles. Figure 9.6 demonstrates several types of Earth-observing satellites. They follow sun-synchronous, polar orbits. Under constant, consistent illumination from the sun, they take pictures in different colors of visible light and nonvisible radiation.

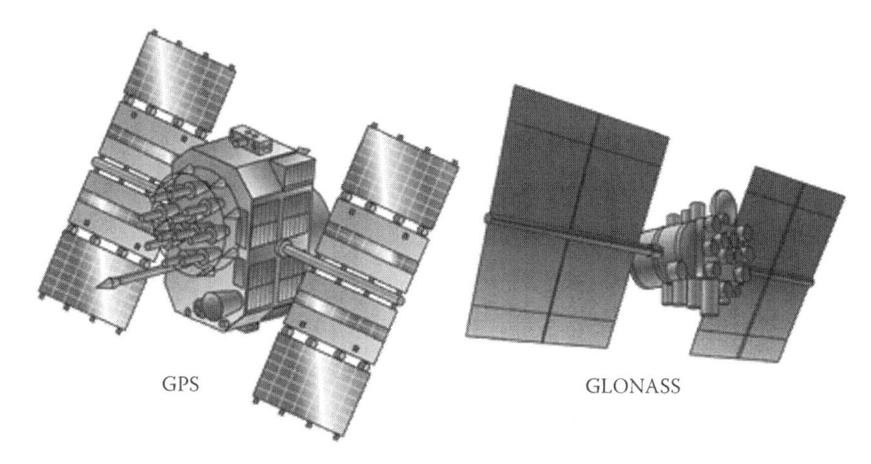

FIGURE 9.5 Navigational satellites. (From Aerospace Corp.)

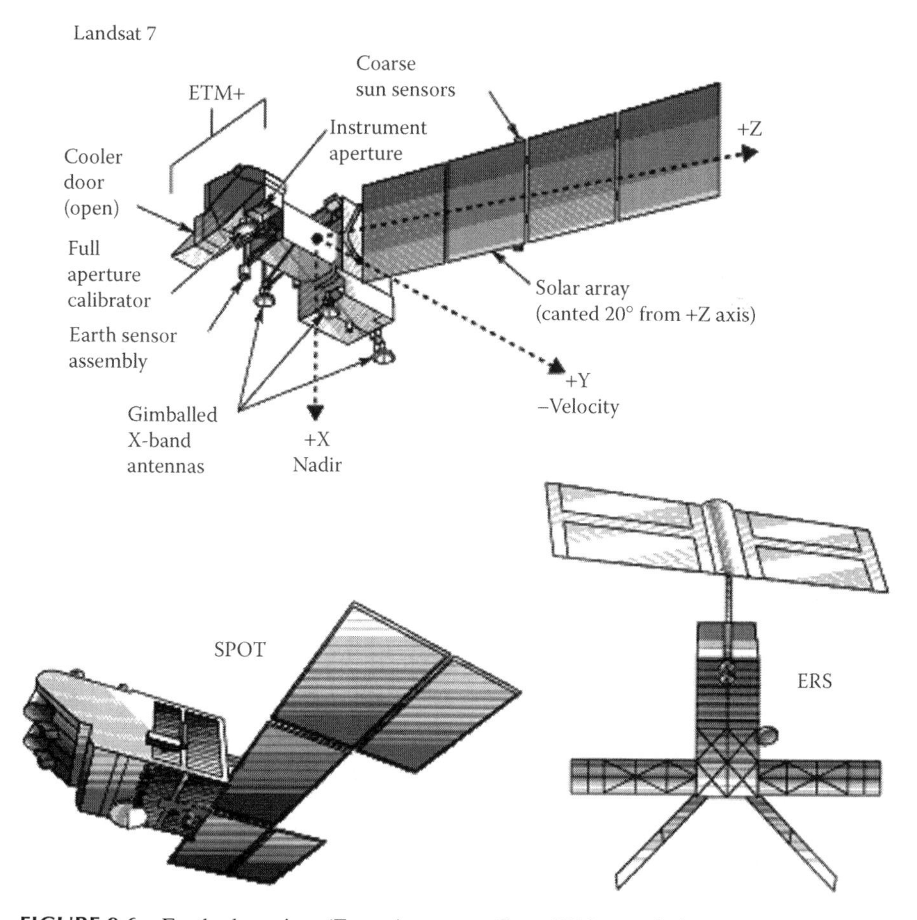

FIGURE 9.6 Earth observing. (From Aerospace Corp. With permission.)

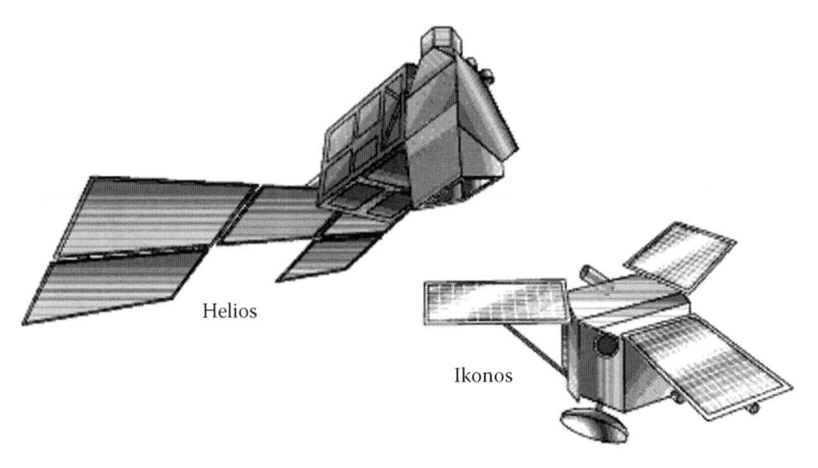

FIGURE 9.7 Surveillance satellites—Big Eyes (cameras). (From Aerospace Corp. With permission.)

Earth-observing satellites are used to locate mineral deposits, to determine the location and size of freshwater supplies, to identify sources of pollution and study its effects, and to detect the spread of disease in crops and forests.

- Military satellites include weather, communications, navigation, and Earth-observing satellites used for military purposes. Some military satellites—often called "spy satellites"—can detect the launch of missiles, the course of ships at sea, and the movement of military equipment on the ground. Several surveillance satellites are shown in Figure 9.7.

PROBES

The most classic example of a spacecraft, space probes are typically designed to study a specific planetary body by flying by it, orbiting it, crashing into it, or landing on it. Probes require the ability to precisely change their trajectories, accomplished by advanced propulsion systems. Probes also contain a source of power, usually consisting of either solar panels or a radioactive isotope, such as in an RTG. Finally, probes contain a payload of sensing equipment to monitor and measure certain aspects of the planetary body of interest, and they may contain robotic rovers for delivery to the surface of the planetary body.

Flyby missions require the least amount of navigation, as the probe just needs to get near the object and then can fly off in any direction. Famous flyby missions include both the *Voyager 1* and *Voyager 2* probes, which flew by many of the planets in our solar system before heading out toward interstellar space; having reached solar escape velocities, they are now interstellar probes and are still transmitting data back to Earth. An impact mission requires more careful calculation to precisely guide the probe to the surface of the body, but no secondary features are needed to slow the probe or guide it safely to a landing on the surface. Similarly, an orbiter mission requires more calculation in order to get the probe into a stable orbit around the planetary body. However, once the probe is in orbit, its trajectory is relatively stable, and it can remain in orbit for many years. A lander mission is perhaps the trickiest of all the mission types, as the probe must be guided to the surface of the planetary body intact but then also often carries a payload such as a robotic rover that is then released onto the surface of the body to explore and gather data about the surface. Examples of such rover-containing landers are the ones that carried the *Spirit* and *Opportunity* rovers to the surface of Mars.

ROVERS

Rovers are vehicles designed for space exploration—in particular, designed to traverse the surface of a planetary body. Rovers are often designed to be robotically controlled (either autonomously or remotely) but have also been designed to allow human drivers or passengers. Because of this central purpose revolving around transportation, a central component in the design of a rover is understandably its means of locomotion. This locomotion is often in the form of a four-wheeled or multiwheeled car-like vehicle, but it can sometimes be more complex and exotic.

The first roving vehicle to land on any celestial body was the *Lunokhod 1*, shown in Figure 9.8. The eight-wheeled rover landed on the Moon on November 17, 1970, and was remotely controlled from the Earth for a span of 322 Earth days, during

FIGURE 9.8 *Lunokhod 1*. (Courtesy of NASA.)

which it analyzed lunar soil, measured cosmic rays and x-rays, and took pictures of the lunar surface. Its body contained equipment to collect data about lunar soil (extendable devices that impacted the soil to test its density), spectrometers and detectors to measure x-rays and cosmic rays, four television cameras, and one x-ray telescope. The vehicle was powered by solar panels during the day and kept warm at night by a Polonium-210 heat source. The location of the *Lunokhod 1* on the lunar surface is unknown at this time.

The first manned rover for spacecraft exploration was deployed on July 31, 1971, during the *Apollo 15* mission to the Moon. The subsequent *Apollo 16* and *17* missions (the three Apollo J-class missions) also brought their own lunar roving vehicles (LRVs), as shown in Figure 9.9. These rovers allowed astronauts to extend their exploration range from feet to miles; however, the range of these rover trips was still limited to the distance their air supplies would last in the event the rover broke down and they were forced to walk back to the lander. The LRVs were all four-wheeled and could seat two astronauts in extravehicular activity (EVA) suits for extended EVA at up to 8 mi/h. Structurally, the rover was constructed primarily of an aluminum base, and for its primary function of locomotion, the lunar rover was equipped with four wheels (each with its own electric drive motor), front and rear steering motors, and two 36-volt silver-zinc potassium hydroxide batteries for power. In addition to these components, the rover contained a color television camera with a high-gain antenna capable of broadcasting live footage back to Earth, as well as film for black-and-white pictures.

FIGURE 9.9 Lunar roving vehicle. (Courtesy of NASA.)

STATIONS

The most expansive type of space construction undertaken thus far in our history of sending spacecraft into space is certainly that of the space station. These stations have the life support capabilities to house human occupants for an extended period of time, as well as docking ports to allow other spacecraft to dock with them, allowing the replenishing of consumable food and water or other necessary items. In contrast with other spacecraft, space stations do not require propulsion once in orbit, and they consist primarily of modules for living and research, as well as massive solar arrays to supply these modules with the necessary power. The primary mission of these space stations has been to conduct experiments on biology and fluid physics in space, taking advantage of the station's unique environment, namely the weightlessness that comes from being in constant free fall. Space stations have also provided extensive data on the long-term effects that living in space has on the human body. This research is invaluable in determining the necessary specifications for life support and other systems to allow even longer manned missions in space, such as those that would be required for colonization of the Moon or other planetary bodies.

Whereas early space stations were constructed completely on Earth and then launched into space (often requiring a massive lift vehicle), the more recent *Mir* and the International Space Station (ISS) were modular in design, allowing the addition of new modules to be attached to the core module and assembled in orbit. The earliest space stations were the Salyut stations, *Salyut 1* being launched in 1971, and followed by six additional Salyut stations spanning the years up through 1986. Two other Soviet stations launched during this time, the *DOS-2* in 1972 (failed during launch) and the *Cosmos 557* (reentered after 11 days), did not last long enough to house a cosmonaut crew. The American-built *Skylab* was also launched and was operational during that time (1973–1974). Once in orbit, the space stations were manned by a crew of two or three astronauts, who would be replaced by a new crew after several months. Space stations have the important requirement for life support systems that can remain fully operational for years, often requiring astronauts to perform repairs on the outside of the station via tethered spacewalks. These early stations were all designed and launched in one piece without the plan or ability to add on new sections, with the *Salyut 7* serving as a proof-of-concept for the transition from the monolithic station design to that of a modular one.

These early, monolithic stations were followed by the launching of a more modular design, as seen in *Mir* and the ISS. This new design of space station permitted the later addition of modules, providing more mission flexibility as well as expandability for the potential to accommodate an increased crew size in the future. *Mir,* launched in February 1986, was the first truly modular space station put into orbit, ending up with a total of seven modules, and was the first long-term, continuously lived-in structure in space. The modules are largely cylindrical in shape, with one or two ports on their ends. These modules are either connected directly to another module or are joined by a spherical node with five ports at the end of the core module. Each module has a specific purpose, serving the crew as living quarters, life support system, or research laboratories. *Mir*'s core module was the site of the main station living quarters and five docking ports. The *Kvant* and *Kvant-2* modules both housed

FIGURE 9.10 *Mir* space station—modules. (Courtesy of NASA.)

the station's life support systems, including waste recycler. The *Kristall*, *Spektr*, and *Priroda* modules were all research stations, consisting largely of various sensing instruments to look at Earth or the stars. Finally, a docking module was added to make it easier for U.S. space shuttles to dock with *Mir*. *Mir* could dock with both a manned U.S. space shuttle and a manned Russian *Soyuz-TM*; it also had a docking port for unmanned *Progress-M* cargo spacecraft. Crewed by 28 crews over its 15-year life span, the *Mir* still holds the record for the longest continuously manned space vehicle, at 8 days short of 10 years, the streak having ended when the station was de-orbited in 2001. The *Mir* space station is shown in Figure 9.10. The ISS, launched in 1998, has now been continuously inhabited since November 2, 2000, by 17 crews to date.

Some of the new issues that have arisen in the use of these long-lifetime space stations are those of maintenance and long-term survivability of the inhabitants. As space stations cannot return to Earth for repairs when damaged, the crew must often perform potentially risky EVAs to repair damage caused by extended exposure to the radiation and micrometeorites in the space environment. Proposed future forms of space stations often include a means of generating an artificial gravity, usually by creating a circular station, such as a rotating wheel or sphere, which provides a centrifugal force on anything on the interior surface of the outer edge of the wheel or sphere as the station rotates. This form of space station has been theorized by both scientists and science fiction writers from the beginning of the twentieth century, but it was particularly popularized in the 1968 film *2001: A Space Odyssey.*

LUNAR OR PLANET-BASED HABITAT

The research on prolonged human living in space will eventually lead to the construction of a lunar or planet-based habitat. As with the space station, these habitats will require expansive life support systems to maintain an environment suitable for human life. As opposed to orbiting space stations, however, these stationary habitats may be able to take advantage of in situ resource utilization to create materials such as water or raw ore that would otherwise need to be brought from Earth. Habitats would also most likely need to be constructed near the poles of the planetary body so that their solar panels would have access to direct sunlight as much as possible, as power would have to be stored or obtained from another source for use during extended periods of darkness. The effects of microgravity on the human body would also have to be researched in more detail so that the habitat would be equipped to counter any negative effects such an environment would have on its occupants.

SOLAR SAILS

Solar sails are a proposed form of spacecraft propulsion using large mirrors to reflect light and produce a minuscule, yet useful, acceleration. These sails would need to be either folded before launch and then deployed once in space, or constructed entirely in space should the materials be too delicate to survive the folding and launch process. This sail would reflect light emitted by either the sun or an Earth-based laser, using the pressure that arises from the physical bouncing of the photons off the sail surface. While this pressure is very low ($\sim 10^{-5}$ Pa), the vacuum of space is at a much lower pressure ($\sim 10^{-10}$ Pa) so the pressure of the reflecting photons on the solar sail could produce a positive force without requiring any fuel. This technology could theoretically be employed in slowing a spacecraft to reach its final destination, or more easily to make fine adjustments in orbit without expending any fuel.

INFLATABLES

Inflatables are an attractive and potentially advantageous means of creating a large volume of easily constructible living area in space. Essentially, the inflatables would be made of Kevlar and Mylar woven around an air bladder, which can be used for structural support once filled with an atmosphere. The interior of the air bladder, if filled with a nitrogen-oxygen atmosphere, could be used to maintain an environment capable of sustaining human life in space. The outer covering of the inflatable would need to shield inhabitants from radiation and be resistant to puncturing by micrometeorites. Inflatables are an attractive design because they can exist as a much more compact shape for launch, and then once in space be deployed to a more useful size. The *Genesis 1,* developed by the private firm Bigelow Aerospace and launched in 2006, was the first inflatable habitat to be sent into orbit and is serving as a proof-of-concept to determine the long-term viability of the inflatable materials in the space environment. A proposed inflatable module for the ISS, Transhab, was pursued by NASA for a time but plans were stopped because of increasing costs of the ISS and a House of Representatives resolution banning NASA from further research into inflatable habitat modules for the ISS.

BALLOONS (TITAN AND MARS)

A balloon could be used instead of a wheeled rover in exploration of planetary bodies that have some type of atmosphere. A balloon-suspended payload could fly over the surface of a planet and take measurements, avoiding the numerous obstacles rovers must avoid on the land. So far, the only planetary balloon missions have been the Venus Vega missions, both launched in December 1984 to study the Venusian cloud system. Each spacecraft carried a 5-kg payload to be suspended from a helium-filled balloon that was 11 ft. in diameter when deployed. In addition to a helium-filled design, the Montgolfiere balloon design can be used, in which the local atmosphere is heated and used to generate a buoyant force, lifting the balloon. This design is being considered for balloon missions to Mars and Titan, where they would be able to soft-drop payloads in a much more controlled manner than parachutes and could also make measurements of the surface from various altitudes. Balloons would be especially useful on Titan, where the sum of radiant and convective heat transfer coefficients is nine times less than on Earth, and as buoyancy is inversely related to temperature, the total required power to lift an equivalent mass would be about one hundred times less on Titan than on Earth.[3]

TECHNOLOGY DEVELOPMENTS

Significant technology developments have enabled generations of spacecraft to evolve. These technology developments are in components and technologies and in techniques and processes. Key developments in components and technologies include the dramatic performance improvements in solar cells and the development of three-axis stabilized spacecraft for guidance control and on-orbit stabilization.[4] Ultraprecise oscillators[5] for precision clocking and revolutionary advances in communication[6,7,8,9,10] have advanced military and civilian exploration and commercial space.

The development of three-axis stabilized spacecraft is an excellent reflection upon technology advances. Early spacecraft relied on spin stabilization, where the entire spacecraft rotates around its own vertical axis, spinning like a top. This keeps the spacecraft's orientation in space under control. Spin stabilization is the simplest way to keep the spacecraft pointed in a certain direction. These gyroscope-like spinners of early spacecraft most often had a cylindrical shape and rotate at one revolution every second.

A disadvantage to this type of stabilization is that the satellite cannot use large solar arrays to obtain power from the sun. Thus, it requires large amounts of battery power. Another disadvantage of spin stabilization is that the instruments or antennas also must perform "despin" maneuvers so that antennas or optical instruments point at their desired targets. Spin stabilization was used for NASA's *Pioneer 10* and *11* spacecraft, the *Lunar Prospector*, and the *Galileo Jupiter* orbiter.[11] In the mid-1970s, several satellites were built using three-axis stabilization. They were more complex than the spinners, but they provided more despun surface to mount antennas and made it possible to deploy very large solar arrays. The greater the mass and power, the greater the advantage of three-axis stabilization appears to be. Perhaps the surest indication of the success of this form of stabilization was the switch of Hughes, closely identified with spinning satellites, to this form of stabilization in the early 1990s.[12] The National Oceanic and Atmospheric Administration launched *GOES-8*, the first three-axis body-stabilized geostationary environmental satellite in April 1994.[13] With three-axis stabilization, satellites have small spinning reaction wheels or momentum wheels, which rotate so as to keep the satellite in the proper orientation. Additionally, some spacecraft may also use small propulsion-system thrusters to correct spacecraft positions. *Voyagers 1* and *2* stay in position using three-axis stabilization. The shape of three-axis stabilization systems will vary greatly and is driven by the mission type. Large, deployable antennas and solar arrays may provide for more powerful systems. The three-axis stabilized control system revolutionized satellite systems.

CONCLUSION

Among the more than 8,700 objects launched into space, there is a wide range of missions and styles. Spacecraft are driven by the purpose, and certainly, form follows function. The incredibly high costs associated with launch means that nothing is extraneous. Every gram of the spacecraft has a contribution. Although emblems, decals, flags, and ashes have been flown in space, especially with the crewed flights, for public relations purposes, it is rare to see "extras." A few notables are electronic media carrying names or other time capsule material, but again these activities are driven by public relations concerns and not by the mission itself. From the small, round, early spinners to large structures of tens of meters, spacecraft reflect both technology evolution and the mission drivers. Constellations of weather, communication, navigation, and Earth-observing satellites have made our world a more global community. The *Voyager* craft have extended our reach to new realms, while the sheer magnitude of the Cold War's footprint in space reminds us of the world we live in.

REFERENCES

1. Oberright, J.E. 2004. Satellite, Artificial. World Book Online Reference Center. World Book, Inc. http://www.worldbookonline.com/wb/Article?id=ar492220.
2. Gilmore, D.G. 2002. *Spacecraft Thermal Control*. El Segundo, CA: The Aerospace Press.
3. Jones, J., et al. 2005. Montgolfiere Balloon Missions for Mars and Titan. Third International Planetary Probe Workshop, Attiki, Greece, June 27, 2005. Pasadena, CA: Jet Propulsion Laboratory, National Aeronautics and Space Administration. http://trs-new.jpl.nasa.gov/dspace/handle/2014/38360.
4. Herther, J.C., and J.S. Coolbaugh. 2006. Genesis of Three-Axis Spacecraft Guidance, Control, and On-Orbit Stabilization. *Journal of Guidance, Control, and Dynamics* 0731-5090 29(6): 1247–1270.

5. Norton, J.R., J.M. Cloeren, and P.G. Sulzer. 1996. Brief History of the Development of Ultra-Precise Oscillators for Ground and Space Applications. In *Proceedings of the Annual IEEE International Frequency Control Symposium*, 47–57.
6. Brady, M. 2002. The Geostationary Orbit and Satellite Communications: Concepts Older than Commonly Supposed. *IEEE Transactions on Aerospace and Electronic Systems* 38(4): 1408–1409.
7. Marsten, R.B. 1997. Dream Come True: Satellite Broadcasting. *IEEE Transactions on Aerospace and Electronic Systems* 33(1): 361–381.
8. Pritchard, W.L. 1984. History and Future of Commercial Satellite Communications. *IEEE Communications Magazine* 22(5): 22–37.
9. O'Neal, D.M. *Commercial Communications Satellites History and Highlights*. Los Angeles: Hughes Space and Communications.
10. Posner, E.C., and R. Stevens. 1984. Deep Space Communication Past, Present, and Future. *IEEE Communications Magazine* 22(5): 8–21.
11. U.S. Centennial of Flight Commission. Spin and Three-Axis Stabilization. http://www.centennialofflight.gov/essay/Dictionary/STABILIZATION/DI172.htm.
12. Whalen, G. Communications Satellites: Making the Global Village Possible. http://www.hq.nasa.gov/office/pao/History/satcomhistory.html.
13. Davis, G. 2007. History of the NOAA Satellite Program. NOAA Satellite and Information Service, USA. *Journal of Applied Remote Sensing* 1, 012504.

Section II

The Sky: A Cultural Perspective

"I hate changing to Daylight Saving Time."

Chapter 10 Archaeoastronomy, Ethnoastronomy, and Cultural Astronomy

Susan Milbrath

10 Archaeoastronomy, Ethnoastronomy, and Cultural Astronomy

Susan Milbrath

CONTENTS

INTRODUCTION

Ethnoastronomy and archaeoastronomy are closely related, but each emphasizes different sources of information and requires different skills in collecting data. Cultural anthropologists have led the way in developing research on ethnoastronomy, whereas the most prominent researchers in archaeoastronomy are astronomers, archaeologists, and art historians. Archaeoastronomy focuses on what can be learned about ancient astronomy through study of archaeological sites. Architectural orientations and cosmological symbols are sometimes studied in tandem with historical accounts that record the religious practices of local indigenous populations. Formalized studies in archaeoastronomy developed relatively recently, in the 1960s, when astronomers and archaeologists began to systematically investigate astronomical orientations at archaeological sites. Ethnoastronomy became a focus of study about a decade later, when ethnographers began to gather integrated data on traditional astronomical beliefs and practices among indigenous cultures. Even though anthropologists have recorded information on native calendars and cosmological beliefs for over a century, indigenous astronomy has often been marginalized in broader studies of cultural practices. Remaining rooted in the framework of cultural anthropology, ethnoastronomy now focuses attention on traditional indigenous astronomy.

There is a chronological distinction between archaeoastronomy and ethnoastronomy, but the two are so closely linked that the term *cultural astronomy* is gaining currency in reference to the integrated fields of study. Cultural astronomy is by no means a fully developed discipline with a defined methodology and theoretical framework; instead it is an interdisciplinary study involving people trained in a wide variety of fields. Cultural astronomy, as described by Von Del Chamberlain and Jane Young, reflects three underlying themes that link astronomy to humankind throughout history: 1) people seek to explain what they observe because they cannot separate their concept of self from their concept of the universe; 2) people observe the heavens to pace the events in their lives, most notably food-related activities, but also events related to the human life cycle; 3) people participate in the celestial cycles by conducting rituals founded on the belief that human activities can ensure that the natural cycles will repeat in proper sequence. These universals emerge in all studies of cultural astronomy, because our shared view of the celestial dome integrates our perceptions of the cosmos.[1]

Cultural astronomy also embraces research on the history of astronomy, a field that most often depends on written texts. Since the data in this chapter are largely drawn from New World studies, discussion of written texts will be limited to ethnohistorical accounts written shortly after the European conquest and the few indigenous texts predating the European conquest. Preconquest indigenous writing is known only from Mesoamerica, an area spanning from Mexico south to Honduras. The most advanced form of writing comes from the Maya area, where logosyllabic texts use both logographs and syllables. Recent advances in decipherment have revolutionized the study of texts recorded on monuments and in painted books known as codices.

Ethnoastronomy, archaeoastronomy, and the history of science are all focal points in the study of cultural astronomy. For the purposes of this chapter, ethnoastronomy and archaeoastronomy are treated separately because integrating information from both fields is sometimes not possible, owing to the different forms of data, different methods of data collection, and the cultural and chronological distinctions involved. Although astronomical texts can be considered apart from the archaeological sites themselves and studied in the context of the history of astronomy, here Preconquest written texts are incorporated in the section on archaeoastronomy because they provide information about which astronomical phenomena are likely to be recorded at archaeological sites. Ethnohistorical accounts, often useful in studies of the history of astronomy, are included in discussions of both archaeoastronomy and ethnoastronomy to provide a bridge between past and present.

HISTORY OF ASTRONOMY AND ARCHAEOASTRONOMY IN THE OLD WORLD

The basic research on archaeoastronomy in literate civilizations has often been overshadowed by the abundant written records. Because our scholarship has an established attachment to evidence derived from the written word, archaeoastronomy receives less appreciation than matters based on written texts, which are a mainstay in the history of astronomy.[2] This certainly is the case in China, where the archaeological sites themselves are rarely a focus for studies of astronomy, largely because there is so much information in written texts. Study of Chinese texts in the context of the history of science has culminated in impressive multivolume works, such as Joseph Needham's *Science and Civilization in Ancient China* (1954–2004). Texts also dominate research on the history of astronomy in classical civilizations. The history of Greek astronomy is well known because Greek astronomical texts have been integral to the development of Western Astronomy. Archaeoastronomy in the classical world has received little attention, despite an early study of Greek temple orientations in 1892 by Francis C. Penrose in the journal *Nature* and clear evidence in classical texts that the Romans used celestial alignments to lay out their city streets.[3]

Research on Egyptian archaeoastronomy first developed in the nineteenth century, less than 50 years after the first decipherment of ancient texts. As early as 1867, astronomers began studying the Egyptian pyramids in terms of possible astronomical alignments.[4] Following up on a study by Henrik Nissen in 1885 (*Rheinisches Museum* XL), Sir Norman Lockyer published a detailed investigation of astronomical alignments in Egyptian temples (*The Dawn of Astronomy*, 1894). Some of these early studies pushed the limits of credulity, and there was a subsequent decline in research focusing on Egyptian sites.[5] Interest was revived by an American astronomer, Gerald D. Hawkins, who documented astronomical alignments at a number of sites, including lunar and solar orientations at Karnak in Thebes.[6]

Renewed interest in research focusing on the astronomy of cultures known only from the archaeological record was sparked by research on Neolithic Stonehenge in the 1960s. A summer solstice alignment at Stonehenge had been proposed as early as 1740 by the antiquarian William Stukeley, but the solar alignments were not systematically studied until Lockyer surveyed Stonehenge in 1901.[7] By the early twentieth century, astronomical interpretations of megalithic sites had fallen into disfavor. Under the influence of Darwinian evolutionism, scholars rejected the notion that primitive cultures could have possessed sophisticated astronomical knowledge.[8] The subject of megalithic astronomy was dramatically reopened in a 1963 article by Hawkins. He employed a computer to calculate solar and lunar horizon positions in 1500 BC and correlated them with sighting lines evident at Stonehenge. Around the same time, C.A. Newham, an amateur archaeologist, published a similar study involving a site survey that more accurately reflected the true alignments.[9] Despite analytical errors, Hawkins published a popular book demonstrating that megalithic stones were aligned to mark astronomical observations at least as far back as 2500 BC.[10] There followed a series of articles by astronomer Fred Hoyle, who corroborated certain aspects of the research Hawkins had published but questioned the mechanism of eclipse prediction.[11] In a broader study undertaken in 1973 to survey many different megalithic sites, Alexander Thom, an engineer, confirmed the importance of solar and lunar alignments and demonstrated that Stonehenge had counterparts among other megalithic sites.[12] More recently, the orientations to the summer solstice sunrise and the winter solstice sunset have been incorporated in a broader interpretation of Stonehenge as the largest Neolithic cemetery in England.[13] Studies at Stonehenge helped establish that research on archaeoastronomy must be firmly grounded in the context of local landscape and the built environment of archaeological sites, requiring interdisciplinary studies involving both archaeologists and astronomers.

HISTORY OF ASTRONOMY AND ARCHAEOASTRONOMY IN THE NEW WORLD

The earliest studies of astronomy in the New World were based exclusively on written texts, following broader research developments in the history of the astronomy. Ernst Förstemann, a librarian in Dresden, was the first scholar to decipher specific astronomical texts in 1880, when he began to work on the Dresden Codex, a painted book (screen-fold codex) from Yucatan, Mexico. In 1906, he analyzed the calendrical content of the Venus tables and lunar tables in his commentary on the Dresden Codex.[14] By 1924, Astronomer R.W. Willson had identified a Mars table on pages 43–45 of the Dresden Codex.[15] A major advance in our knowledge of Maya astronomy was made in a 1930 publication by John Teeple, a chemical engineer who published a comprehensive study of astronomical cycles in Maya calendar texts in monumental art and codices.[16] These early studies resulted from a new understanding of the dates and time periods recorded in the texts, without benefit of the advances in decipherment of noncalendric Maya texts that have revolutionized the field.

Although the Maya captured the limelight early on in the studies of New World astronomy, codices from other areas of Mesoamerica also attracted an early interest. These Pre-Columbian painted books record numerous dates and complex visual imagery, although they lack fully developed hieroglyphic writing. The first breakthrough in our understanding of the astronomical content of these codices was made by Eduard Seler, who documented Venus periods in calendar records represented in the Codex Borgia and related ritual manuscripts.[17] Seler's work on astronomical imagery in the Borgia Group codices remains influential today, even though there have been many advances in our understanding of astronomical events portrayed.

Another early researcher, Zelia Nuttall, was the first scholar to call attention to the importance of the solar zenith in the New World, when the sun passes directly overhead at noon in the tropics.[18] She also was the first to identify images of observatories in ancient Mixtec codices. These symbols are now recognized as Mixtec place signs set in the context of narrative events, but they could designate cities where observatories were located. The dynasties and political events represented in Mixtec codices have been the subject of much study over the course of the twentieth century, but astronomical events have received little attention.

During the first half of the twentieth century, archaeologists turned to the sites themselves as a source of information on Maya astronomy, conducting the first studies to measure astronomical orientations. Oliver J. Ricketson, an American archaeologist, documented astronomical alignments at Chichen Itza as early as 1928, and by 1937 he had identified solstice and equinox orientations in the "E-Group" buildings at Uaxactun.[19]

Scattered publications alluded to the importance of astronomy in ancient Mesoamerica, but formalized study of archaeoastronomy in the New World did not develop until the early 1970s, when Anthony Aveni began publishing papers on astronomical alignments in Mesoamerica and organized the first of many conferences to be documented in an edited volume: *Archaeoastronomy in Pre-Columbian America* (1975). Two articles in that volume represent seminal works synthesizing information from an entire culture area. Archaeologist Michael Coe focuses on Mesoamerican records of indigenous astronomy preserved in historical accounts, Pre-Columbian codices, and contemporary ethnographic accounts.[20] A similar approach is seen in a study of Pueblo astronomy by anthropologist Florence Ellis, who combines a number of different sources of information preserved in art and historical accounts of the Southwest. [21]

Research on New World archaeoastronomy greatly expanded shortly after Aveni published a 1980 book exploring the role astronomy played in Mesoamerica: *Skywatchers of Ancient Mexico* (revised in 2001), a work that also incorporates information from other areas of North America and South America.[22] Another landmark volume in the field of cultural astronomy edited by Aveni, *Archaeoastronomy in the New World* (1982), includes significant data drawn from iconography, ethnography, and ethnohistory, as well as study of Mayan texts. The codices and colonial period texts related to astronomy are the focus of Aveni's edited volume (*The Sky and Mayan Literature*, 1992), which helped to advance our understanding of astronomical texts in the codices.[23]

A key publication in the study of ethnoastronomy is a 2005 volume based on an international conference held in Washington, D.C., in 1983 and edited by Von Del Chamberlain, John Carlson, and Jane Young.[24] *Songs from the Sky* includes studies from across the globe, but here I focus on articles detailing the ethnoastronomy of New World cultures. In the next two sections I also integrate information on indigenous ethnoastronomy from a variety of other sources, including information incorporated into "Star Gods of the Ancient Americas," a traveling exhibition that opened in 1982 at the American Museum of Natural History.[25] Some studies featured in the following sections are closely linked to astronomical calendars, focusing on the more pragmatic needs of agricultural societies. Others are integrated with mythology, positing larger concepts of cosmology and first origins.

ETHNOASTRONOMY IN THE UNITED STATES AND CANADA

Reenacting skylore is often part of performance events, as in the case of the Northwest Coast of North America, spanning from the First Nations area of Vancouver, Canada, north to the Tlingit area of Alaska. A desire to draw back the light is an important part of the winter ceremonials enacted in the Raven cycle, which dramatizes the theft of daylight and the retrieval of light from a dark box.[26] Darkness is a place of magic and the world of the dead, relating to night itself but also to the seasons. Raven, who appears before sunlight is brought to the world, may embody the planet Venus bringing the light of dawn.[27] The Raven carrying the sun up to the sky is a well-known image in Northwest Coast art. This imagery may have a more specific astronomical

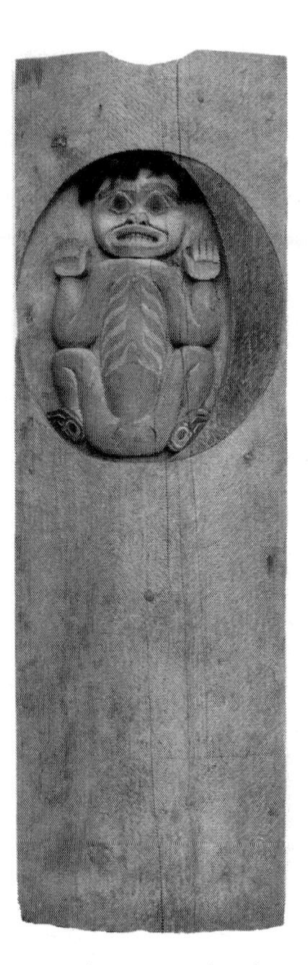

FIGURE 10.1 Northwest Coast moon housepost showing emaciated figure representing the waxing moon. (From the National Museum of the American Indian, NMAI 17/8013. With permission.)

meaning when we consider that Venus, as the morning star, rises just ahead of the sun, metaphorically carrying the sun up into the sky.

Astronomical themes in Northwest Coast art deserve further study, for there are a great variety of intriguing images. For example, house panels from the nineteenth century Tlingit community of Yakutat in Alaska represent the four phases of the moon, each with a different form of lunar imagery (Figure 10.1).[28] The first-quarter moon is an emaciated figure, showing that the moon is not yet full, a visual metaphor immortalized in art.[29] A focus on lunar imagery probably developed because this coastal community depended on maritime activities and the tides controlled by the moon's position and phases.

Skylore is central to creation cosmology among the Anishinaubaeg (Anishinabe), also known as Ojibway or Chippewa, who live north and south of United States-Canada border, as far east as Quebec and as far west as Alberta.[30] The sun is the father of the Anishinaubaeg, and the moon was put in the sky to remind people of their genesis and to honor women. After the great flood, Sky-Woman gave birth to the Anishinaubaeg, and they reached land when an island emerged on the back of a turtle. The cosmos records their past, present, and future, and their moral course in life is derived from the skies. Ojibwa shamans from the Great Lakes area recount that a renowned shaman had predicted the 1910 passage of Halley's comet. The apparition of such a great comet allowed the shaman to enter a hole in the sky to gain access to the spirit world located above the night sky.[31] Visual images of the 1910 comet appear in historical records painted by tribes in the Great Plains, an area extending from Texas north to Alberta. These "Winter Counts" also include other astronomical events, such as eclipses and the great 1833 meteor shower.[32]

Records of astronomical events among the more sedentary Plains groups living to the east on the prairies, prior to displacement by European-American settlers, take a different form, more often encoded in narratives associated with ritual bundles. For example, one of the Pawnee meteor bundles contained a grinding stone that was said to have come from sky when a fireball was observed by the villagers.[33] Starlore is extensive among the Pawnee and includes a number of important concepts. The Milky Way was variously referenced as a stream, an animal's path, or a path for souls of the dead among the Pawnee of Kansas and Nebraska.[34] They recounted that long ago in time the stars came down to earth and set up their camps according to their places in the sky, an example later followed by the people when they established their villages.[35] The creator designated the North Star as the chief of all the gods that were placed in the heavens.[36] The northern sky is also important among the Lakota, who trace the seven Sioux tribes to the "seven council fires," representing the seven stars in the Big Dipper, and they identify another important icon, the Thunderbird, as stars in Draco and Ursa Minor.[37]

Among the nomadic tribes living on the High Plains, prophetic visions inspired decoration of ritual and religious objects with images of the sun, moon, morning star, and Big Dipper.[38] Even though most Plains calendars were regulated by lunar months, the sun was the most important cosmological entity, sometimes being equated with the Creator God.[39] Solar and lunar imagery predominates in the Sun Dance, which was timed by the full moon nearest the summer buffalo hunt.[40] Among the Crow, the lodge itself is the sun's lodge and the Sun Dance dolls represent Moon-woman.[41] These dolls, traditionally attached to the cedar center post, had screech owl plumes and sometimes lunar crescents.[42]

Starlore of the High Plains makes a compelling connection between their people, the sacred landscape, and the heavens. A Kiowa story tells about seven sisters who rose up to sky to become a constellation (the Big Dipper) at the Devil's Tower, a sacred site located in the Black Hills of northeastern Wyoming.[43] Ronald Goodman notes that "the stories of the Lakota oral tradition are sacred—not literature, not myth, not folklore." [44] The Lakota relate the sacred landscape of the Black Hills to the dome of heaven. The bright winter stars of the Race Track form a star map that reflects the pattern of the Black Hills. Traditionally, an annual ceremonial journey through the Black Hills mirrored the path of the sun through these important constellations. This direct connection between the history of Plains people and their concept of the divine makes it clear why the Black Hills are so sacred to their traditions.

Although many of the most important features of Plains astronomy are encoded in the actual landscape, the tipi is an important astronomical image. Indeed, the sky itself was visualized as an all-encompassing tipi among traditional nomadic Plains tribes.[45] The rich imagery of Blackfoot astronomy is evident in painted designs recorded on early tipis and altars from the area bordering Montana and Canada. For example, Clark Wissler notes that the Blackfoot Water-Monster Tipi was decorated with stars representing the Big Dipper ("seven stars"), the Pleiades ("bunched stars"), a Maltese cross depicting the morning star, and the Water-Monster, a feathered serpent "who came from the sun." [46]

The circular earth lodge was a form of observatory among nineteenth century agricultural societies of the eastern Plains, especially the Pawnee. In contrast to the mobile Plains tipis, the fixed location of the earth lodge provided an ideal setting to make repeated astronomical observations.[47] The Skidi Pawnee calculated the year by watching the stars, and they named certain months for constellations visible during those months.[48] They began the ceremonial year when the Real Snake constellation (stars in Scorpio) was visible for the longest period of time. The four main support posts in the lodge symbolized four stars marking intercardinal directions, serving as cosmic pillars holding up the four quarters of heaven. The smoke hole overhead represented the Circle of Chiefs (Corona Borealis), which passes nearly overhead in the Skidi Pawnee area. The smoke hole served as an ideal portal for observing the changing seasonal position of sun and stars. The east-facing entry could also have played a similar role, and in some lodges light from the rising sun at the equinox illuminated an altar in the interior dedicated to the evening star. Two entry posts were marked with stars representing twin morning stars. According to Skidi Pawnee traditions, the morning star was Mars, rather than Venus.[49] Mars was apparently honored in an infamous Pawnee astronomical ceremony involving the sacrifice of a young Ogala Sioux woman to honor the morning star, a rite last performed in 1838.[50]

In the Southwest (Arizona, New Mexico, and the southernmost parts of Utah and Colorado), the major cultural groups are the Navajos, Apaches, and the Pueblos, which include the Eastern Pueblos living along the Rio Grande and the Western Pueblos, referring collectively to the Zuni, Acoma, Hopi, and Tewa community of Hano on Hopi's First Mesa. Coordinating solar and lunar time was a major consideration for Pueblo calendar specialists and remains of great interest today. The full moon nearest the winter solstice is celebrated as the turning point of the year.[51] Although the Pueblo ceremonial calendar is regulated mainly by observations of the sun, most Pueblos begin their months with the appearance of the first crescent. The winter solstice is the major seasonal ceremony among traditional Pueblos, a time of year when solar and lunar cycles are realigned.[52] Matilda Cox Stevenson described and photographed a Zuni sun shrine in New Mexico where the sun priest would seat himself to observe the alignment of shadows cast by the rising sun on specific monolithic features.[53] Prayer sticks were planted at the new moon and full moon, and this practice apparently continues today among the Zuni.[54] The idealized Zuni winter solstice ceremony occurs on the full moon in December, to be followed by a New Year fire marking the start of the ritual year.[55]

Early accounts of the Pueblos note that a religious specialist (Flute Chief) associated with Mother Earth watched the sun and moon during the agricultural season, from winter solstice to summer solstice, while observations from the summer solstice to the winter solstice were monitored by a "sun watcher" (Sun Chief) linked with Father Sun.[56] Working with these religious specialists were representatives of the Sky or Star God and the elder of the Twin War Gods. The War Twins may be linked with the morning and evening stars.[57]

In the 1880s and 1890s, Alexander Stephen and Walter Fewkes recorded a Hopi horizon calendar in Arizona that marked time by the changing position of the sun, and they documented certain ceremonies that were timed by the position of important star groups. [58]Among the Hopi of Walpi, observations focused on the sun's movement along the horizon, with the agricultural events (January–July) timed by changes in the sun's movement along the eastern horizon and winter ceremonies timed by the sun's position at sunset over the San Francisco Peaks to the west.[59] Masked kachina dances are performed from the winter solstice to the summer solstice, and in mid-July the kachinas return to their mountain homes.[60]

Solar imagery is prominent in origin tales and artistic images among the Pueblos. The Zuni say that people were brought up from the underworld to provide the sun with worshippers.[61] Both the Zuni and Hopi depict the sun disk as a blue human-like face. In the Flute Dance, performed 2 months after the summer solstice, Hopi dancers wear this sun symbol, representing the sun itself as Tawa.[62] Tawa also appears as a Hopi kachina, one of several kachinas linked with the sun.[63] Ahulani, the winter solstice (Soyal) kachina, represents a child.[64] This suggests that he may be a symbol of the sun reborn on the winter solstice.[65] At the winter solstice ceremony, Ahulani appears as a dramatic transformation of an eagle impersonating the sun.[66] The eagle is symbolic of the sun throughout the Southwest.[67] The macaw, a favorite image among the Pueblos, is also closely linked to solar imagery. The Zuni say that Sun Father holds up macaw features to make daylight.[68] The Ahul or Sun Kachina, traditionally decorated with macaw feathers, appears in the Hopi winter solstice ceremony.[69] The macaw is associated with the southeast, the direction of the winter solstice sunrise in Hopi cosmology.[70]

In ceremonial buildings known as kivas, Pueblo Indians traditionally observed the appearance of constellations in the entrance overhead to time rituals, measuring the passage of time by observing the Pleiades, the Big Dipper, and Orion.[71] Zuni carvings and paintings often depict the Galaxy, Orion, the Pleiades, Ursa Major, Ursa Minor, and North Star; also prominent are images of Venus as the evening star, a warrior of the Sun Father, and Venus as the morning star representing the warrior who comes before the sun.[72] Among the Hopi, the moon and morning star are depicted on the Chief Kiva walls, and certain parts of the winter solstice ceremony are timed by the position of Orion and the Pleiades.[73] Ceramics excavated in a Tewa ruin show the morning star, the Pleiades, and Orion on bowls for offering corn meal to the sun.[74] The Hopi Mastop kachina, decorated with star groups representing the Big Dipper and the Pleiades, appears in the winter solstice ceremony, a time of year that these constellations are especially prominent.[75] Elaborate astronomical images abound on Zuni altars and ritual attire.[76] For example, the Zuni Corn Maiden's headdress, depicting a blue-faced sun and a cut-out crescent moon with twin stars framed by black and white bands representing the Milky Way, is worn in ceremonies following the winter solstice (Figure 10.2).[77] Zuni legends recount that a ball game was played for sunlight at the time of creation, and thereafter a god of the Milky Way saved the Corn Maidens and restored fertility on earth.[78]

FIGURE 10.2 Zuni Corn Maiden's headdress showing the sun, moon, twin stars, and the galaxy. (From the National Museum of the American Indian, NMAI 10/8739. With permission.)

Astronomical practices among the Pueblos and Navajos have some shared features, but there are inherent differences in the cosmological systems.[79] The Pueblos, an agrarian population living in clustered settlements, emphasize the sun and moon as the paramount celestial beings, whereas Father Sky and Mother Earth (Changing Woman) are most important among the Navajos living on a large reservation spanning four state borders in what is called the "Four Corners" area. The Navajos also place great emphasis on observation of the stars, which is apparently linked to their origins in a more nomadic tradition and their current residential patterns, living as ranchers in isolated houses.

Among the Navajos, Sun is the ultimate source of power and light, and his union with Changing Woman engendered the Warrior Twins, Monster Slayer and Child of Water, who made an epic journey to visit their Sun-Father and acquire the power to slay monsters and restore order to the world.[80] Moon is a weaker brother of Sun, and together they divide day and night. Sun controls the seasons and the day, while Moon controls the night, the months, and vegetation.[81] Although the sun's movement along the horizon is responsible for the seasons, specific constellations are seen as seasonal markers. The orderly nature of the stars was established by Black God. He started by placing the Pleiades on his brow and was able to place the most important Navajo star groups in the sky before Coyote scattered the remaining stars.[82] The Black God sand painting ceremony in the Nightway chant, a winter ceremony that lasts for nine nights, provides detailed information on stars encoded in ceremonial events.[83]

Navajo star ceilings in Canyon de Chelly in northeastern Arizona apparently date to the eighteenth century, when the Navajos first adopted Pueblo star motifs.[84] These star ceilings developed from Pueblo IV star ceilings, indicating that the Navajos were influenced by the Tewa Pueblos.[85] No constellations seem to be represented, and the purpose of the paintings may be to provide protection and beneficial effects at sacred sites. The stars may have been intended to prevent rock falls by "holding up" the ceilings, not unlike the Pueblo tradition of painting feathers on rock ceilings for that purpose.

The Mescalero Apaches of New Mexico are sun watchers who track the solstices, but they also carefully watch the moon and keep a detailed lunar count.[86] They observe that when the "horns" of the moon point up there will be no rain, an overlap with the Pueblos of the Southwest.[87] The commensuration between the solar and lunar calendars is not a problem because the Apaches keep separate counts, coordinating the two by starting the lunar year at the first visible crescent after the summer solstice.[88]

ETHNOASTRONOMY IN LATIN AMERICA

Moving to the south, to the area of Mesoamerica, the most detailed information about ethnoastronomy comes from highland Guatemala, where researchers themselves have become shamans or religious practitioners.[89] This has led to a more sophisticated understanding of astronomy in specific cultural contexts. Such studies reveal some unique astronomical concepts in individual communities in the Maya region, an area spanning from Yucatan, Mexico, south to just beyond the border of Honduras. Some astronomical concepts seem to be shared by different communities. The Milky Way is visualized as a serpent or a road and is sometimes considered to be the realm of the dead, and there is a widespread belief that the souls of the dead are transformed into stars.[90] Accounts of other cosmological concepts seem quite diverse, even within the same language group.[91] Explanations of where the sun goes at night show considerable variation. For example, the Chol in Chiapas, Mexico, say the sun goes into his house during the night, but other Maya groups visualize the sun climbing down a tree or passing through a cave in the underworld, traveling to the underworld in a gourd, or carried by dwarfs through the underworld.[92] A number of complex astronomical images have resulted by the merger of Maya concepts with Catholicism. The sun is widely linked to Christ and the moon to the Virgin Mary. The Virgin often takes on multiple aspects, as among the people of Santiago Atitlan, Guatemala, who say that the twelve Marias represent the lunar months, joined in intercalary years by a monstrous female called Francisca Batz'bal representing the thirteenth month.[93]

Indigenous Maya farmers generally predict planting time and seasonal change by watching the sun, moon, and certain stars.[94] The Pleiades are still observed to time the planting at beginning of the rainy season among a wide range of Maya populations.[95] A number of Maya groups believe that the lunar phases affect rainfall, but often other astronomical observations are involved. For example, among the Chorti, predictions about rainfall in the agricultural cycle are most often based on observations of the moon, Pleiades, and the Milky Way and occasionally on the luminosity of Venus.[96]

Astronomical observations are just one of many indicators of seasonal cycles. The turtle stars among the Yucatec Maya, variously identified as Gemini or neighboring stars in Orion, may be linked to the season that sea turtles are laying eggs.[97] Gemini and Orion are very prominent in the month when turtles lay their eggs, and this natural cycle was noted for the month of Mac in the eighteenth-century Chilam Balam of Chumayel. The Quiche Maya watch for the movements of migrating hawks as a seasonal sign, and they also recognize a hawk constellation in the sky, coordinating the hawk migration with the position of the constellation.[98]

Among some Tzotzil communities in Chiapas, four gods hold up the heavens at the four corners, corresponding to the solstice positions of the sun on the horizon.[99] Among the Yucatec Maya of Yalcoba, the four corners of the sky are described as being equivalent to the solstice extremes.[100] The contemporary Maya often link the sun to Christ, but some early accounts provide a better understanding of Pre-Columbian Maya view of the sun. One early source links the macaw to the noon sun, and people who were ill made offerings to the sun at noon to ask for healing, a practice that continues today among the Yucatec Maya.[101] Maya priests of Yalcoba make petitions to the noon sun, when they believe that there is a sky opening that provides a conduit directly up to the sun.[102]

Similar patterns of continuity in astronomical concepts are seen in comparing ethnographic accounts with terminology in sixteenth-century Yucatec dictionaries and chronicles.[103] Friar Landa's chronicle, written in 1566, tells us that the Maya told time at night by observing the Pleiades ("rattlesnake's rattle"), Gemini ("fire drill"), and Venus ("lucero"). Venus and the Pleiades bear these names even today.[104] Other interesting survivals of colonial period concepts include references to Venus as the "wasp star," the moon goddess as a patroness of childbirth, the identification of the Milky Way with a serpent.[105]

Beyond the Maya area, detailed studies of ethnoastronomy in Mesoamerica are scarce, but one outstanding resource is Frank K. Lipp's study of Mixe astronomy in southern Oaxaca, Mexico.[106] He includes data on stars, solar observations, the moon, and Venus, as well as various calendars in use in the area. A Sacred Round of 260 days and numbered yearbearers, reminiscent of Aztec and Mixtec calendars, integrates with an agricultural calendar of the 365 days. This calendar has a short "month" of 5 days at the end of a cycle of paired 20-day months. There is no formal intercalation, but every 4 years, 1 day is "lowered" into the 5-day month.[107] There are also a little-known cycle of 845 days involving two different almanacs in the Sacred Round, a ritual count that completes its cycle in 3,380 days (13 × 260 days), and another 3,180-day cycle that may relate to Venus.[108] Much remains to be learned about astronomy among the various groups living in Oaxaca. For example, the Mixtec link animal cycles and meteorology with astronomical events, noting a relationship between the seasonal emergence of ants, the onset of the rains, and the planet Venus.[109] Similar constructs have been recorded among the Barasana of the Amazon, but here the emergence of leaf-cutter ants at the beginning of the rainy season is linked with the zenith of Betelgeuse, which is known as the Ant star.[110]

The astronomical concepts of various lowland groups in South America are well documented among the Warao in the Orinoco delta of Venezuela, the Yekuana of Colombia, the Bororo of southwestern Brazil, and numerous groups in the Amazon.[111] One shared feature among communities located near the tropic of Capricorn is that the vertical axis of the sun overhead is a powerful cosmological concept. For example, Yekuana residences are patterned on the structure of the tropical sky, where the December solstice sun passes overhead at zenith. The transition between the dry season and rainy season at the time of the solar zenith establishes a seasonal opposition that is mirrored in significant star groups.

Among the Caribs living along the coasts of Guyana, Surinam, and French Guiana, important principles of astronomy can be extracted from mythology. Using the structuralist framework of Claude Levi-Strauss, Edmundo Magaña deduces that the opposition of Pleiades and Scorpius is incorporated in the agricultural cycle and observations of the solstices, but the Caribs also linked the solstices to a triad formed by the Pleiades, Orion, and Canis Major. Apparently all the stars used for navigation have declinations that fall within the solstice extremes.[112]

Studying the Arawak, one of the most widely distributed South American language groups, Fabiola Jara notes that the best documented Arawak concepts are in the Guianas.[113] Here Orion's dawn rise heralds the dry season, the preferred season for fishing, and the dawn rise of the Pleiades announces the New Year and the beginning of the agricultural season. The dawn rise of the Southern Cross, visualized as a curassow, coincides with that bird's mating season. The ordering of the night sky conveys zoological information, and the asterisms are used as mnemonic devices. The dawn rise of the capybara constellation (Aires), announcing the time for planting maize in the dry season, is also the season that an animal is easy prey when it feeds on savanna grasses. To the Arawak, Scorpius is a water boa that rises at dawn to announce the December rains, and Antares is variously referred to as the eye of the snake or prey in its belly.[114] By contrast, among other language groups living south of the Amazon, the celestial snake is an anaconda, but there are different interpretations about which stars form the snake.

Research on ethnoastronomy among the Cariçara, representing only 220 inhabitants living on a small island off the coast of São Paulo, Brazil, focuses on their perspective of natural phenomena of time, space, and place, and their concepts of earth-sky relations.[115] Meteorites and "special stones" found on the island are related to their notion that falling stones create thunder. The Cariçara explain that the earth oscillating horizontally around the local vertical axis (zenith-nadir) results in the changing solar positions and the solstice extremes. The solstice extremes are marked by watching the sun's movement along the horizon in relation to neighboring islands.

In a broad study of ethnoastronomy in Peru, Peter Roe sees shared cosmological concepts as part of a metacosmology that spread from the lowlands to highland Peru prior to 1000 BC. Focusing on Shipibo concepts from the lowlands, he notes that a primary form of opposition is evident in accounts of the sun and moon, which travel at opposite ends of a celestial canoe.[116] The canoe voyages of the sun and moon refer to an east-west path related to the equinoxes, following the course of a celestial river, while Cayman's Canoe (identified as the Pleiades, Hyades, and Orion), traveling across the current of the river, is linked with the solstices. The celestial river in turn relates to seasonal cycles, for the annual flooding disperses the fish and concentrates the animals. On the other hand, the dry season confines the fish to the river, making them easier to catch, while animals are dispersed and harder to hunt. Because the Shipibo travel exclusively by canoe in the Amazonian lowlands, references to the Milky Way as a path are essentially the same as Andean images of the Milky Way as a river.[117]

Among the Amahuaca, who live in the rain forest of the eastern Andes, on the border of Peru and Brazil, certain asterisms are associated with sunny weather and the season of gardening, while others bring the rainy season.[118] Names for planets other than Venus are only rarely recorded by ethnographers, so it is noteworthy that the Amahuaca have several names for planets, including Jupiter, known as the "large star" and "wife of the moon." A number of celestial observations resonate in other areas of the New World cultures, such as the notion that the when the moon's horns point up the weather is dry and sunny.

Andean ethnoastronomy in Peru is well documented in studies of the Quechua by Gary Urton.[119] In the area of Cuzco, similar stars and constellations are recognized from one Quechua community to the next, including star-to-star groups and dark cloud constellations that are formed by interstellar dust amidst dense stars in the Milky Way. At least 16 of the stars and constellations recorded among the Quechua today bear the same names they had among the Incas in the sixteenth century. The Quechua view the heavens in term of animal imagery and metaphorically relate the behavior of specific animals to constellations identified with those animals. Urton also recorded names for Venus ("dawn of the earth/time star") and the outer planets, which are called "dawn star" or "zenith star." The planets are observed in the morning to determine when to plant the annual crops. The calendar is a social construction, integrating agricultural tasks with religious festivals. Among the celestial representations in calendar periods, only the Pleiades (the "storehouse") and Orion (the "plow") are directly associated with agriculture. Observations of their heliacal rise and set are coordinated with different points in the agricultural cycle.

In the Andes of Chile, the southern limits in our survey of ethnoastronomy, the Mapuche and Aymara emphasize a day-night and east-west opposition in their cosmology.[120] The east and rising sun are associated with waterfalls and springs, energy and vitality, whereas the west and setting sun are linked to the sea, darkness, and death. The principal indicators of time during the day are the sun and the dual apparitions of Venus, while time at night is reckoned by observations of Orion's Belt, the Southern Cross, the Pleiades, the Milky Way, and other star groups.

The rich astronomical traditions recorded in areas of the New World are not only interesting in terms of ethnoastronomy, but they also inform us about concepts that may be traced back through time in cultures that are known from ethnohistorical and archaeological records. Evidence of continuity provides a more nuanced picture of cultural astronomy in areas such as the Andes, Mesoamerica, and the Southwest.

ARCHAEOASTRONOMY IN THE UNITED STATES AND CANADA

Moving into the realm of past civilizations, study of archaeoastronomy inevitably involves contested evidence. Unlike ethnoastronomy, where the traditions

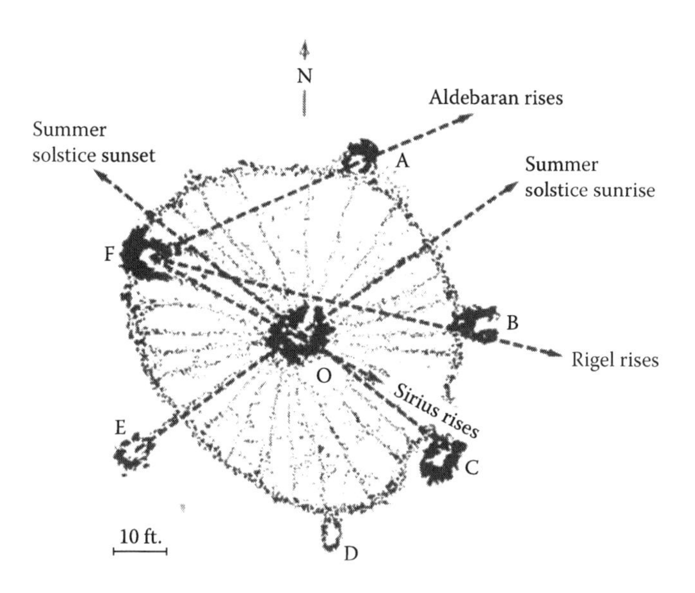

FIGURE 10.3 Prehistoric astronomical alignments of Big Horn Medicine Wheel in the Great Plains. (From Aveni, A.F., *Skywatchers*, University of Texas Press, Austin, TX, 2001. With permission.)

of indigenous people are a principal source of information, studies of archaeoastronomy are interpretations of the past based on evidence from a wide variety of sources, including measurements of alignments at archaeological sites and analyses of archaeological data and ancient art. Information from ethnohistorical and ethnographic accounts also has been incorporated in the discussion, when there is a degree of cultural continuity between the past and present.

We begin in the northern latitudes with the Great Plains, where prehistoric stone circles called Medicine Wheels or Council Circles are found scattered from Colorado north to Canada. Studying astronomical alignments dating circa AD 1400–1700 at Big Horn Medicine Wheel in Wyoming, John Eddy found solar, lunar, and stellar alignments that seem to be related to intervals of 28 days (Figure 10.3).[121] The accuracy of the stellar alignments has been questioned because the horizon position of stars in northern latitudes is difficult to estimate, but Eddy maintains that these alignments are still visible today.[122] David Vogt studied all known examples of Medicine Wheels and concluded that their wide distribution "weakens the observatory theory, since, if the main function of Medicine Wheels was calendrical, there would be no need for so many calendrical 'instruments.'"[123] He notes that when many sites are compared, the spoke orientations seem to be random. He hypothesizes that Medicine Wheels were probably burial and commemorative structures, rather than calendrical constructions. Nonetheless, it is possible that Big Horn functioned as a true observatory, while other structures had different functions and were only loosely modeled on this sacred structure. A similar pattern is seen in the case of the Maya E-Groups, which are discussed under "Archaeoastronomy in Latin America."

There has been relatively little archaeoastronomical field study conducted in the southeastern United States. There may be very early evidence (circa 1000 BC) for an interest in astronomy at Poverty Point in northeastern Louisiana, which has an avenue aligned to the winter solstice and several other stellar alignments.[124] Aveni points out that the predominant orientation is to the solstice in a sample of over 30 sites surveyed in the Mississippi valley. In southern Illinois, a very large earthwork at Cahokia, a Mississippian site that flourished around AD 1000, has been interpreted as astronomically significant.[125] A key structure was identified as the "American Woodhenge" by Warren Wittrey, no doubt inspired by studies of Stonehenge.[126] This circular configuration of holes once held posts that marked the horizon positions of sunrises and sunsets on significant seasonal dates when viewed from a central post set slightly off center.[127]

A particularly rich area for the study of archaeoastronomy is to be found in the Southwest, where Anasazi sites, such as Hovenweep, Chaco Canyon, and Mesa Verde, date to the twelfth century or earlier. A tower-shaped structure at Hovenweep in southern Utah has been identified as an observatory or sun room.[128] Two ports are aligned so that the sun enters on the solstices, and a set of doorways is aligned to the equinox, defining a complete solar calendar. In New Mexico, solar alignments are evident at Pueblo Bonito in Chaco Canyon, where a large D-shaped complex of rooms has two corner windows oriented to the winter solstice sunrise, the turning point of the year among Pueblo people today.[129] Another group of buildings in Chaco Canyon may have lunar standstill alignments, along with cardinal and solsticial orientations.[130] Recent explorations of an Anasazi site at Wupatki National Monument in Arizona, a wall with three portals facing the flat eastern horizon, seems to show alignments to solar dates anticipating agricultural events and associated ceremonies that are still of importance among the Hopi today.[131]

Ray Williamson, who began measuring astronomical alignments in Chaco Canyon as early as 1972, has proposed a number of significant alignments for Anasazi sites in the Southwest.[132] Williamson's interpretation of Casa Rinconada, a Great Kiva in Chaco Canyon with a series of intriguing niches and a "window," remains controversial. The current effect of light from the summer solstice sunrise entering an opening on the northeast side and falling on one of the niches is "partly an accident of reconstruction and partly a consequence of the lack of a complete reconstruction," according to Michael Zelik.[133] Furthermore, some of the niches may have been covered over at the time the kiva was in use, and one of the roof posts most probably blocked the path of the sunlight. Zelik adds that that is no record that kiva structures were used to observe the sun among the historic Pueblos. Given the importance of starwatching in kivas among historical Pueblos, stellar alignments at Anasazi sites have not been adequately explored, even though some alignments to the Pole Star have been documented.[134]

ARCHAEOASTRONOMY IN LATIN AMERICA

Moving south into Mesoamerica, the study of building orientations and hieroglyphic texts has greatly enhanced our understanding of Maya astronomy. Glyphic studies indicate that the Maya of the Late Classic period (AD 600–800) generally timed warfare based on astronomical observations.[135] An important study of the Temple of the Sun at Palenque in Chiapas, Mexico, has revealed multiple solar alignments in the Late Classic Maya period.[136] The main panel, representing the "jaguar war god," is illuminated by light from the rising sun on the solar nadir in November, marking the onset of the dry season and the season of warfare.[137] Six months later on the solar zenith, coinciding with the beginning of the rainy season in May, a beam of light at sunrise falls precisely on the southeastern corner of the sanctuary, while some weeks later, the June solstice sunrise lights the left interior corner of the main temple. Other solar orientations seem to mark the December solstice and the equinoxes, making the structure a fully functioning "Temple of the Sun." Palenque, with its intriguing orientations and a wealth of texts recorded on monuments, provides us detailed information on the role of astronomy in Maya dynastic history. Glyphic texts indicate that rituals were timed by certain astronomical events, and the life histories of rulers at Palenque were even adjusted to coincide with certain key events.[138]

Recent research on recorded dates found on monumental carvings has been enhanced by modern computer programs that recreate the sky at different locations on specific dates. Computer programs that serve as mini-planetariums abound, but they must be used with caution lest we impose our scientific precision on past civilizations. Astronomical tables, such as Bryant Tuckerman's 1964 American Philosophical Society monograph, *Planetary, Lunar, and Solar Positions, AD 2 to 1649*, are still very useful because it is often easier to search for spans of dates in such tables.

Although recorded dates are of considerable interest in studies of Maya astronomy, there is an equally rich resource in the study of alignments in architectural sites. The archetype for Maya solar observatories is the E-Group at Uaxactun in the Peten of Guatemala. Aveni's comparative study of E-Group constructions at other sites suggests that Uaxactun was a true observatory, serving as a model for other constructions.[139] Here the observer positioned on a pyramid in Group E viewed the changing horizon position of the sun rising over three range buildings that mark the sun's solstice extremes and the equinoxes. Numerous other architectural groups in the Maya area seem to copy this format, but the alignments do not accurately mark the equinoxes or solstices, indicating that these were not precise copies of the original E-Group construction. One recent study suggests that orientations of E-Groups were more generally to the 18° ecliptical band, where the sun, moon, planets, and zodiacal constellations are located.[140]

The solar cycle was a major focus of interest in the astronomical orientations of Maya structures in Mesoamerica.[141] Orientations to the solstice extremes predominate from the Preclassic through the Early Classic in the Maya lowlands. Some orientations in the Terminal Classic period (AD 800–1050) in the northern lowlands of Yucatan at Puuc sites such as Uxmal seem keyed to observations centered around dates anticipating the solar zenith. The importance of the solar zenith seems to be clear in alignments that mark dates anticipating zenith sunrise or sunset events by 20 or 40 days, alignments that are fairly common in the Maya world to the south in

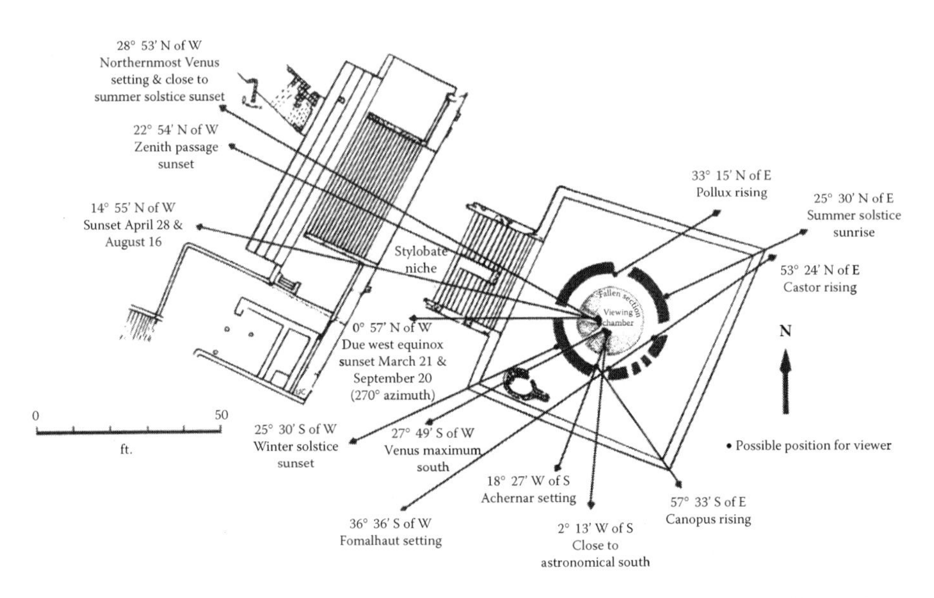

FIGURE 10.4 Maya astronomical alignments of the Caracol Observatory, Chichen Itza. (Modified from Milbrath, S., *Star Gods of the Maya*, University of Texas Press, Austin, TX, 1999. With permission.)

Peten.[142] Towers constructed in the Chenes region of Campeche, Mexico, may have been used to mark observations of the zenith passages linked along a meridian line running north to Uxmal.[143]

The Castillo at Chichen Itza in Yucatan is a radial pyramid that encodes a number of calendric principles and astronomical alignments. The temple faces the zenith sunset in late May, allowing a beam of light to penetrate the inner temple at sunset around the beginning of the rainy season.[144] The setting of the Pleiades viewed through the doorway in April may have announced the upcoming zenith. The play of light and shadow on the Castillo creates a dramatic effect on the equinox, when a pattern of rattlesnake markings appears on the serpent-headed balustrade.[145]

Chichen Itza's Caracol not only looks like an observatory, it clearly functioned in that capacity.[146] The Caracol has windows and doors aligned to observe the setting sun at the equinox and solar zenith, and the corners of the main platform mark the solstice extremes (Figure 10.4).[147] A stairway niche has walls with two different orientations and a red column on the south side of the niche and another column on the north side that is painted black. Seated or standing in front of the black column, the observer would see the solar zenith sunset (May 24 and July 20) aligned with the north wall (22°54' north of west; Figure 10.4). Positioned in front of the red column, the viewer would see the south wall has an orientation to the northern extreme of Venus, parallel to one of angles seen in the western window of the tower (28°53' north of west; Figure 10.4).[148] These paired alignments indicate that the two events were linked together in a longer cycle involving Venus, which reaches its horizon extremes once every 8 years. These observations were coordinated with the seasonal cycle over the course of 8 years, the length of the Maya Venus Almanac (and also the Greek octaeteris cycle).[149]

Because of the thick walls of the Caracol tower, multiple orientations are evident. An alignment from the inner left corner of Window 1 to the outer right corner, marking the northern extreme of Venus, repeats the stairway niche orientation. The other side of the window has an angled view to the equinox sunset (Figure 10.4). The midline of Window 1 is centered on the setting sun on April 30 and August 13, dividing the year into segments of 105 and 260 days, intervals that are repeated in orientations seen at a number of other Mesoamerican sites.[150] The window also frames the Pleiades a few degrees over the western horizon at dusk in late April, marking its last visibility in the west prior to the onset of the rainy season in May.[151]

Chichen Itza's Upper Temple of the Jaguars may also be aligned for observing the Pleiades hovering over the western horizon at the onset of the rainy season, but the solar alignment seems to be paramount.[152] The 286° alignment (16° north of west) marks the sun's horizon position on April 29 and August 13, dates that are 105 days apart.[153] Here the alignment reflects the horizon position of the sun on a date about a month before the onset of the rainy season. The orientation is intended to mark the division of the year into 105- and 260-day periods, reflecting the integration of the ritual calendar cycle in the solar year.[154] The alignments marking 105- and 260-day intervals, quite widespread in Mesoamerica, seem to have been used to track divisions of the 365-day solar year in relation to the 260-day calendar, a unique ritual calendar found only in Mesoamerica. The pattern extends also to Classic period Teotihuacan in Central Mexico and to the Postclassic Aztec capital of Tenochtitlan, where observations of the sun along the horizon as it passed over Cerro Tlamacas marked 260- and 105-day intervals on April 30 and August 13.[155]

Mayapan, located to the west of Chichen Itza, but at virtually the same latitude, has a Postclassic Round Temple that seems to copy the earlier Caracol at Chichen Itza.[156] An alignment perpendicular to the base of Mayapan's Round Temple (286°02' azimuth) is virtually the same as the midline of a window in the Caracol (286°06' azimuth), a solar orientation to the horizon position of the sun on April 30 and August 13, dividing the year into 260- and 105-day periods. Niches inside the temple repeat solsticial alignments seen in the Caracol's main platform and the equinox orientation of one of the windows.[157] Alignments to Venus horizon extremes and horizon positions of the Pleiades are not apparent; however, orientations to the brightest stars in the Southern Cross, Taurus, and Scorpius have been proposed.[158]

Although solar orientations seem predominate in the Maya area, the alignments to the extreme horizon positions of Venus seen at Chichen Itza, near the northern limit of the Maya area, are repeated at Copan in Honduras, located at the southern limit of the Maya area. Copan Temple 22, constructed in the eighth century, has a window aligned to the northern extreme of Venus at the beginning of the rainy season.[159] The northerly extreme of Venus visible in the western window corresponded to the beginning of the rainy season and the southerly extreme to the end of the rainy season, delimiting the agricultural season from May to November.

In Central Mexico, to the west of the Maya area, solar and stellar orientations seem to be paramount, and alignments to Venus horizon extremes are not well documented. Teotihuacan, which served as the Classic period capital of Central Mexico until it was abandoned around AD 550, has a principal east-west alignment that links the horizon position of the Pleiades with intervals that divide the year into segments of 105 and 260 days. A 1967 study of the orientations at Teotihuacan by James Dow noted that east-west axis is aligned to the Pleiades setting. At the time the site was constructed, this star group rose at dawn in late May on the solar zenith (May 18 or 19), a date that marked the beginning of the rainy season.[160] The Pleiades orientation has been questioned, because the Pleiades are not very bright and consequently the star group disappears (extinction) several degrees above the horizon.[161] Nonetheless, the pecked cross on Cerro Colorado accurately marks the setting position of the Pleiades when viewed from a similar cross in the center of the city.[162] Furthermore, the last visibility of the constellation in the west coincided with a solar date in April that could have been used to coordinate solar and stellar time. Today, as in the past, the sun sets along the principal east-west axis at the site in late April and mid-August, dividing the year into 105- and 260-day segments, and the April date falls 20 days before the first solar zenith and 40 days after the vernal equinox.[163]

Further study of the solar alignments at Teotihuacan suggests that the east-west axis of the Pyramid of the Sun and the Ciudadela align to sunrises and sunsets on two different sets of dates, in accord with an observational calendar composed of intervals that included multiples of 20 days, essentially dividing the year into four equal segments.[164] The 15.5° (285.5° azimuth) orientations mark the quarter days, with sunrises on February 11 and October 29 being separated by an interval of 260 days, and sunsets on April 30 and August 13 by an interval of 105 days. The April and August dates are also marked by light-and-shadow effects in the so-called astronomical caves 1 and 2 of Teotihuacan. Similar alignments are seen at Xochicalco, located to the south of Teotihuacan but still in the central highlands area. Here individual buildings have orientations that mirror the two main astronomical alignments found at Teotihuacan, with the central Acropolis marking the sunrises on February 12 and October 30 and the sunsets on May 1 and August 14. The sunrise dates in February and October mark an interval of 260 days that precisely overlaps with the agricultural season.

In one of the Teotihuacan caves, a stela and altar are positioned beneath the zenith tube so that the monuments are lit by a beam of light entering at noon on the two annual solar zenith dates (May 19 and July 25).[165] Other modifications of the cave interior allow light from the noon sun at the solstices and equinoxes to illuminate specific areas of the chamber. An observatory cave at Xochicalco also has a tube or shaft that allows light from the zenith sun at noon to fall on the center of the cave floor.[166] The noon sun traces a longer trajectory because a beam of light first falls on the cave floor in late April at noon and is last seen in the cave in mid-August, marking a 105-day segment of the year and effectively dividing the year into the same intervals seen in alignments elsewhere in Mesoamerica.[167]

Ivan Šprajc's recent studies demonstrate that in Mesoamerica, structures are most often oriented to solar positions that mark intervals that are multiples of 13 and 20 days, and the most frequently recurrent dates mark crucial moments in the agricultural cycle.[168] If one structure faces a solar position on a given day, another structure at the same site often will face the sun's position 13 or 20 days later, codifying intervals basic to the Mesoamerican ritual calendar. It seems that many of these sites were selected because the observer could see the sun move over prominent mountain peaks at intervals of 13 or 20 days, which are subdivisions of the 260-day ritual cycle (13 × 20 days). Šprajc points out that in the 260-day ritual calendar, dates that are separated by multiples of 13 days had the same trecena numeral, whereas those separated by 20 days had the same day sign.[169]

Beyond astronomical alignments, there is a wealth of additional data on Mesoamerican astronomy, including dated images in monumental art, astronomical tables, and historical annals in the codices. Historical dates abound in calendar records on Maya monuments, and these can be studied in relation to astronomical events and ongoing decipherment of noncalendric texts.[170] Aztec annals include records of a number of astronomical events.[171] Mixtec historical codices have barely been explored in terms of the representation of astronomical events. Celestial images appear to be represented in the context of historical events. For example, a flaming orb with a dart on Codex Nuttall page 43 appears with the date 12 Reed, corresponding to AD 1063, a date associated with the birth of 8 Deer Jaguar Claw.[172] The disk resembles a comet in the Aztec Florentine Codex, but the dart is a feature seen on another orb representing an image of a shooting star.[173] Perhaps 8 Deer's birth was somehow linked with a bright meteor shower or the passage of a comet, such as Halley's comet, visible in AD 1066.

Recent advances in decipherment have aided the study of astronomical texts in the three surviving Maya codices, dated to the Postclassic period (AD 1100–1550), named for the cities where they are now located (Dresden, Paris, and Madrid). These codices remain a major focus of study for Maya astronomy, and research continues on tables recording events involving Venus, Mars, the moon, eclipses, and seasonal cycles reflected in rituals and astronomical events.[174]

In the Maya area, ethnographic analogy allows ancient practices to be understood through the survival of core elements in the agricultural cycles of indigenous cultures, including a fixed agricultural calendar running from February through October that is also codified in site orientations throughout Mesoamerica.[175] The Madrid Codex (pages 11–18) depicts a 260-day ritual calendar that relates to a 260-day fixed agricultural calendar running from February through October. A feathered snake on this sequence of pages may symbolize Venus with the Pleiades, represented as the rattlesnake's rattle (tzab), a name still preserved today in Yucatan (Figure 10.5).[176] Eclipse signs, depicted by winged panels, may allude to subdivisions of the 260-period used to record eclipse intervals. The rattlesnake representing the Pleiades appears in the Paris Codex as one of 13 constellations, each suspended from eclipse signs that are spaced at intervals close to the eclipse half-year.[177]

In Central Mexico, the ritual calendar of 260 days in the codices is also linked with astronomical imagery. Dates in the 260-day calendar (the tonalpohuallli) may

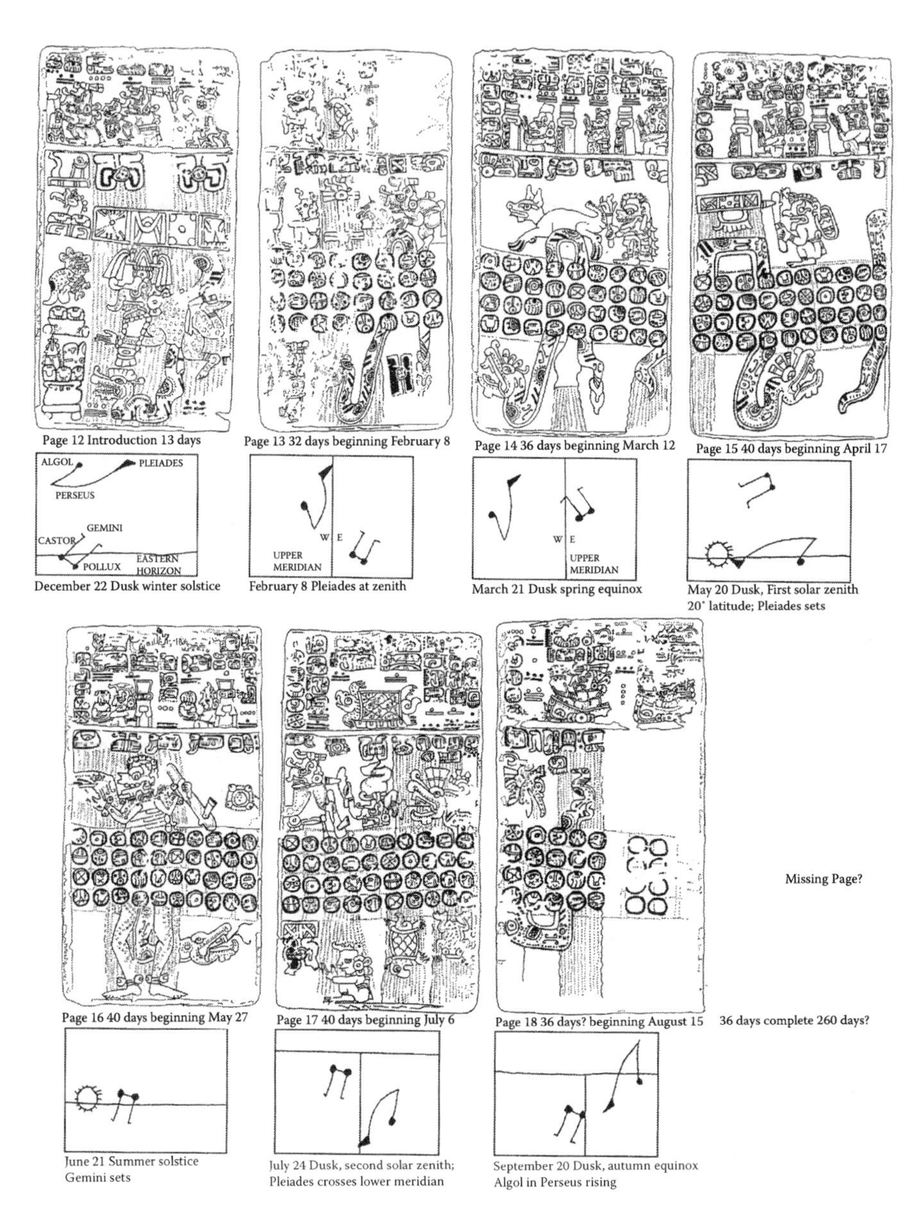

FIGURE 10.5 Maya seasonal star calendar in the Madrid Codex, pages 11–18 (Modified from Milbrath, S., *Star Gods of the Maya*, University of Texas Press, Austin, TX, 1999. With permission.)

be used to record an eclipse interval on pages 39–40 of the Codex Borgia.[178] Codex Borgia pages 27–28 record astronomical events with dates from the yearbearer cycle of 52 years and dates in the 260-day calendar.[179] The yearbearer cycle on pages 49–52 represents astronomical events and seasonal activities linked with a cycle of 52 years in the fifteenth century.[180] In another section of the Codex Borgia (pages 29–46), Venus was observed in relation to periods recorded in a 365-day festival calendar that began around the winter solstice.[181] Such evidence of "real-time" events recorded in the Codex Borgia has led to a much better understanding of the astronomical content in this manuscript.

Study of Mesoamerican archaeoastronomy is aided by study of historical records written by the Spanish friars and native informants in the early decades after the conquest. For example, a sixteenth-century chronicler's account says that the Aztec emperor Moctezuma wanted the principal temple in the capital of the Aztec empire aligned so that the sun would mark the time of the annual equinox festival.[182] Excavations at Tenochtitlan, the Aztec capital until the Spanish conquest in 1521, have confirmed an alignment to the equinox sunrise in the recently excavated twin temples on the Templo Mayor pyramid.

The solar cult was highly developed among the Aztecs, and the sun god had many different names, being known variously as "the valiant warrior," "the turquoise prince," and "he who goes forth shining" (Tonatiuh); this deity embodied the sun disk itself. Imagery of the sun was also linked to various birds, including the blue heron, the macaw, the roseatte spoonbill, the eagle, and the hummingbird.[183] These images may encode different diurnal or seasonal aspects of the sun, such as the solar god Huitzilopochtli ("the hummingbird of the south"), who was reborn every year in late November.[184] In a famous Aztec migration legend, Huitzilopochtli, the brother of the moon goddess (Coyolxauhqui), decapitated his disgruntled sister in a dramatic conflict, not only representing the conquest of populations that worshipped the moon but also symbolizing the sun conquering the moon during lunar eclipse events.[185] A male aspect of the moon, known as Tecuciztecatl, was more cowardly than the sun in the legend of the creation of the sun.[186] The Aztecs also visualized a rabbit on the face of the moon, and they depicted stars as "eyes of the night."

To forecast the future, Aztec priests watched the sky for omens, most notably comets and eclipses.[187] As part of his duties as king of the Aztecs, Moctezuma was advised that he must rise at midnight to observe the constellations that marked the four parts of heaven: the yohualitqui mamalhuatztli ("fire drill," Hyades or Orion's Belt and Sword), the citlaltlachtli ("star ball court," a northern constellation), the tianquiztli ("marketplace," the Pleiades), and colotlixayac ("scorpion," stars in Scorpius), and near dawn he was to observe the morning star and the xonecuilli, a constellation of uncertain identity.[188] The most important planet was Venus, which was visualized as a god called Quetzalcoatl ("feathered serpent"), but the planet had multiple personalities relating to its patterns of visibility. The first rise of Venus as the morning star was an especially ominous event.[189]

The Codex Telleriano-Remensis tells us that the Pleiades ruled the year except in the month of Tecuiluitontli.[190] The most important ceremony involving the Pleiades took place in the month of Panquetzaliztli, when the Aztecs performed New Fire Ceremony every 52 years. This Calendar Round ceremony was timed by the midnight zenith of the Pleiades in November, exactly 6 months from the solar zenith. This observation was probably also coordinated with the full moon around the solar nadir, which the Maya today calculate by observing the full moon pass overhead at zenith.[191] The Aztecs feared cataclysmic earthquake and the destruction of the sun by an eclipse when the Pleiades reached midnight zenith on the day 4 ollin ("earthquake"), a date derived from their 260-day ritual calendar.[192] The sun's position in

Scorpius coincided with the solar nadir at the beginning of the dry season, just as the sun in conjunction with the Pleiades represented the seasonal opposite, the solar zenith at the beginning of the rainy season. Among the ancient Maya, the Pleiades and Scorpius also symbolized different seasons, and the sun's position crossing the Milky Way marked the beginning of the rainy season in May and the onset of the dry season in November.[193]

In the Peruvian Andes, the Pleiades and the tail of Scorpius represented seasonal opposites among the Incas, a concept that survives among their modern Quechua descendants. Indeed, many of the Inca asterisms are known today among the Quechua.[194] The Pleiades and Scorpius were probably considered the harbingers of the solstices and seasonal change, but in the Andes the seasons are reversed relative to Mesoamerica. The Inca site of Machu Picchu seems to confirm observations of this seasonal opposition of these asterisms.[195] The semicircular wall of the Torreon has a window that was used to observe the Pleiades rising just prior to the June solstice sun at the onset of the dry season. A second window is oriented to observe the rise of Scorpio's tail at the beginning of the rainy season just prior to the December solstice. In Cuzco, the Inca temple known as the Coricancha ("the Golden Enclosure"), a building of semicircular shape used as a form of observatory, was aligned to observe the rise point of the Pleiades, repeating an orientation seen at Machu Picchu (Figure 10.6).[196] When the Pleiades rose in November at dusk there was a pilgrimage to a shrine called moon Granary (quillay collca) in the month of their longest visibility.[197] Thunder, the weather god of Inca times, produced rain by dipping into the waters of the Milky Way, and his seven eyes were the Pleiades.[198] Today the Thunder God is known as Santiago, who rides a white horse across the sky.[199]

A sixteenth-century chronicler recounts that the Incas calculated the dates for sowing and harvest by observing the changing position of the sun on the horizon and the rays of sunlight entering a window.[200] On the horizon surrounding Cuzco, Inca pillars marked the position of the rising and setting sun at the solstices and other important dates in the agricultural calendar.[201] The Coricancha or a nearby marker (the ushnu) in the plaza may also have been the observation point for a series of ceque lines and pillars. Seasonal transitions at the solar zenith and nadir were probably marked by pillars that were observed from the Coricancha, known as the Temple of the Sun.[202] Although the pillars were apparently removed by the Spaniards, similar solar markers have been found overlooking the Temple of the Sun in Lake Titicaca, and among the Quechua today horizon observations also determine planting dates.[203] Tom Zuidema has reconstructed an Inca sidereal lunar calendar, using a system of directions encoded in the ceque lines recorded in ethnohistorical sources.[204] Although there is little doubt that such observations were important to the Incas, alternate observation points in the Cuzco valley have been proposed, based on markers for the ceque lines found in archaeological context.[205]

On the north Coast of Peru, the chroniclers tell that observations of the Pleiades were important in the Moche calendar, and such observations are confirmed by alignments to horizon positions of the Pleiades at sites dating prior to the Inca conquest.[206] A Pleiades alignment in Peru is evident at Chavin de Huantar, a highland site of the Early Horizon that dates back to the first millennium BC.[207] Cerro Sechin,

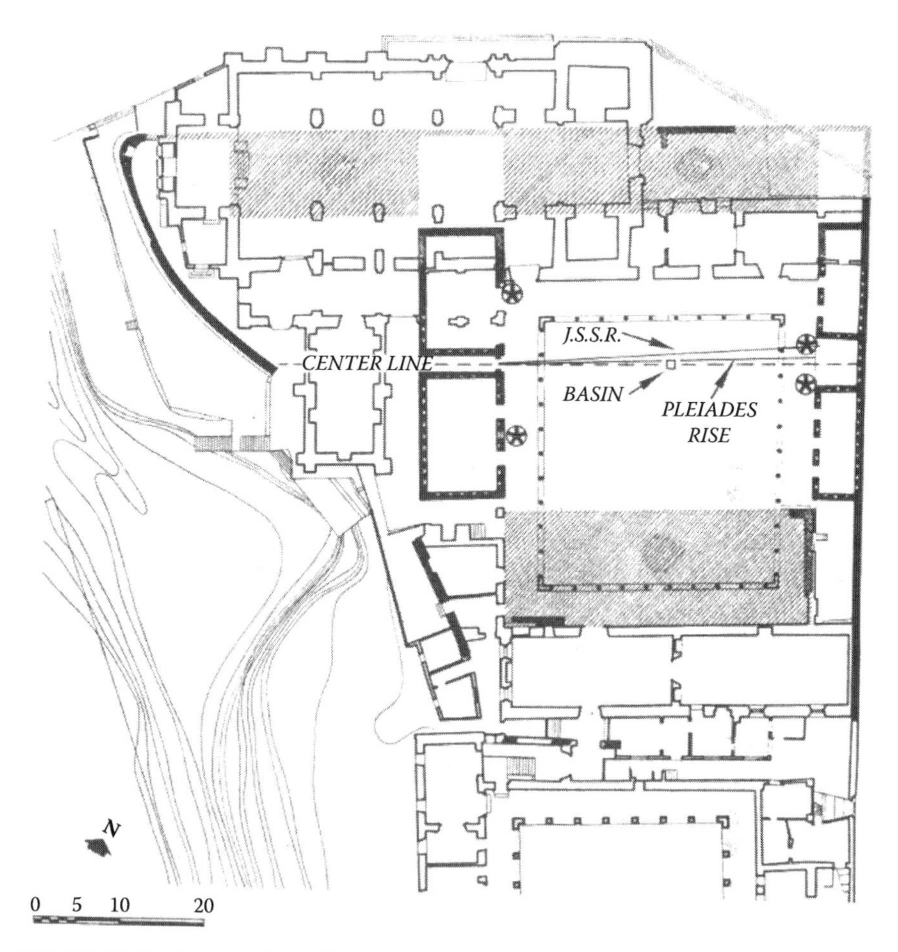

FIGURE 10.6 Astronomical alignments in the Inca of the Coricancha temple in Cuzco, Peru. (From Zuidema, R.T., Catachillay, in *Ethnoastronomy and Archaeoastronomy in the American Tropics*, ed. A.F. Aveni and G. Urton, 203–230, New York Academy of Sciences, New York, 1982. With permission.)

an earlier site on the north coast, exhibits orientations to the rising sun on the zenith passage date. There are also alignments to the equinox sunset at Cerro Sechin and the December solstice at Chavin de Huantar, as well as possible alignments to the lunar extremes.[208]

The evidence from Peru indicates a long tradition of astronomical observations and alignments that are closely linked with the built environments of monumental architecture and the landscape itself, where mountainous peaks marked the changing seasonal position of the sun. A horizon calendar based on the changing seasonal position of the sun is just one of a number of astronomical features that Peruvian sites share with other archaeological sites to the north, recalling the horizon calendars of Mesoamerica and the Southwest.

CONCLUSION

The New World has rich resources for the study of indigenous astronomy, but we can only skim the surface in this chapter, leaving aside certain important astronomical alignments known from the archaeological record, such as Alta Vista in the Chalchihuites cultural sphere of northern Mexico, and ethnographic cultures with well-known astronomical traditions, such as the Chumash of California. Even though this chapter is by no means comprehensive, it does clarify that cultural astronomy in certain geographical areas can be studied as a continuum. The ideal situation is when we can trace astronomical concepts preserved in ancient art and architecture through the epoch of first European contact, the period of many important chronicles, up through the ethnographic present. In the Southwest, where Pueblo populations reside in the area they had occupied prior to European contact, preservation of language has proved to be key in understanding certain astronomical concepts, and important information is also preserved in their traditional rituals and agricultural practices. Among the Maya and Quechua, continuity in language and geographical location has also preserved many indigenous astronomical constructs and rituals.

With the dome of heaven as a space we all share, it is not surprising that many concepts appear repeatedly in our records of ethnoastronomy and archaeoastronomy. There are many overlaps in patterns in cultural astronomy in different areas, extending far into the past. The celestial events themselves establish some uniform patterns, and keying seasonal events to changes in the skies remains an integral part of our relationship with the cosmos. Some of these shared concepts clearly serve pragmatic goals, as in predictions of seasonal change and cycles in nature, but others seem linked with some innate bond with the celestial realm. Religion certainly is one way of formulating this relationship. Space travel is another. We travel in spaceships, while the North American Indians tell us they have traveled through space in shamanic visions, and voyages after death to the stars or Milky Way are a mainstay of New World starlore. Research on archaeoastronomy and ethnoastronomy helps trace the cultural landscape of space back through time, establishing a link with our contemporary fascination with space so evident in modern technological achievements in space exploration.

NOTES AND REFERENCES

1. Chamberlain, Von Del, and M. Jane Young. 2005. Introduction. In *Songs from the Sky: Indigenous Astronomical and Cosmological Traditions of the World,* ed. V.D. Chamberlain, J.B. Carlson, and M.J. Young, xi–xiv. College Park, MD: Center for Archaeoastronomy.
2. Aveni, Anthony F. 2008. Personal communication.
3. Aveni, Anthony F. 2001. *Skywatchers. Revised and Updated Version of Skywatchers of Ancient Mexico*. Austin, TX: University of Texas Press (see specifically p. 260).
4. Iwaniszewski, Stanislaw. 1987. Arqueoastronomía y ciencia. *Antropología y Técnica* 2: 119–137.
5. Krupp, Edwin C. 1997. Archaeoastronomy. In *History of Astronomy: An Encyclopedia*, ed. J. Lankford, 21–30. New York: Garland Publishing.
6. Hawkins, Gerald D. 1973. *Beyond Stonehenge.* New York: Harper & Row.

7. Krupp, Edwin C. 1978. The Stonehenge Chronicles. In *In Search of Ancient Astrronomies,* ed. E.C. Krupp, 81–132. New York: Doubleday & Co. (see specifically p. 85).
8. Mitchell, John. 1977. *Secrets of the Stones: The Story of Astro-archaeology*, 45–47. New York: Penguin Books.
9. Krupp 1978 (note 7).
10. Hawkins, Gerald D. 1965. *Stonehenge Decoded.* Garden City, NY: Doubleday & Co.
11. Krupp 1978 (note 7), 105–109.
12. Thom, Alexander, and Archibald Stevenson Thom. 1978. Rings and Menhirs: Geometry and Astronomy in the Neolithic Age. In Krupp 1978 (note 7), 39–801.
13. Alexander, Caroline. 2008. If the Stones Could Speak: Searching for the Meaning of Stonehenge. *National Geographic* 213(6): 34–59.
14. Förstemann, E. 1906. Commentary of the Maya Manuscript in the Royal Public Library at Dresden. *Papers of the Peabody Museum of Archaeology and Ethnology, Harvard University*, vol. 4, no. 2. Cambridge, MA: The Peabody Museum.
15. Willson, Robert W. 1924. Astronomical Notes in the Maya Codices. *Papers of the Peabody Museum of American Archaeology and Ethnology, Harvard University,* vol. 6, no 3. Cambridge, MA: The Peabody Museum.
16. Teeple, John E. 1930. *Maya Astronomy.* Carnegie Institution of Washington, pub. 403. Contributions to American Archaeology, no. 2. Washington, DC: Carnegie Institution of Washington.
17. Seler, Eduard. 1904. Venus Period in the Picture Writings of the Borgian Codex Group. *Bureau of American Ethnology Bulletin* 28, 353–392. Washington, DC: Smithsonian Institution.
18. Nuttall, Zelia. 1901. The Fundamental Principles of Old and New World Civilizations. *Papers of the Peabody Museum of American Archaeology and Ethnology* 2: 1–602.
19. Aveni, Anthony F., Anne S. Dowd, and Benjamin Vining. 2003. Maya Calendar Reform? Evidence from Orientations of Specialized Architectural Assemblages. *Latin Americna Antiquity* 14(2): 159–178. Rickertson, Oliver G. 1928. Notes on two Maya Astronomic Observatories. *American Anthropologist* 18: 434–444.
20. Coe, Michael D. 1975. Native Astronomy in Mesoamerica. In *Archaeoastronomy in Pre-Columbian America,* ed. A.F. Aveni, 3–31. Austin, TX: University of Texas Press.
21. Ellis, Florence H. 1975. A Thousand Years of the Pueblo Sun-Moon-Star Calendar. In Aveni 1975 (note 21), 59–87.
22. Aveni, Anthony F. 1980. *Skywatchers of Ancient Mexico.* Austin, TX: University of Texas Press.
23. Aveni takes a broader perspective incorporating Old and New World astronomy in his most recent books, such as *Conversing with the Planets: How Science and Myth Invented the Cosmos* (1992), New York: Times Books; *Ancient Astronomers* (1993), Washington, D.C.: Smithsonian Books; *Stairways to the Stars: Skywatching in Three Ancient Cultures* (1997), New York: John Wiley and Sons; and *People and the Sky: Our Ancestors and the Cosmos* (2008), New York: Thames and Hudson. Other books featuring cross-cultural studies of ancient astronomy worldwide include E.C. Krupp's *Echoes of the Ancient Skies: The Astronomy of Lost Civilizations* (1983), New York: Harper and Row; *Beyond the Blue Horizon: Myths and Legends of the Sun, Moon, Stars, and Planets* (1991), New York: HarperCollins; and *Skywatchers, Shamans, and Kings* (1997), New York: John Wiley and Sons. In addition, beginning in the 1970s a wide range of articles on cultural astronomy worldwide have been published in *Archaeoastronomy:* a supplement to *The Journal for the History of Astronomy* and in *Archaeoastronomy: The Journal of Astronomy and Culture* and *Archaeoastronomy* (formerly *The Bulletin of the Center for Archaeoastronomy*).
24. Chamberlain et al. 2005 (note 1).
25. Milbrath, Susan. 1982. *Star Gods of the Ancient Americas: Exhibition Script.* New York: Museum of the American Indian (now the Smithsonian National Museum of the American Indian).

26. Vogt, David. 2005. Raven's Universe. In Chamberlain et al. 2005 (note 1), 38–48.
27. Ibid., 45.
28. National Museum of the American Indian, NMAI 17/8013.
29. Milbrath 1982 (note 25), interpretative text for NMAI 17/8013.
30. Johnston, Basil H. 2005. Sky Tales from the Anishinaubaeg. In Chamberlain et al. 2005 (note 1), 147–150.
31. Conway, Thor. 1985. Halley's Comet Legends Among the Great Lakes Ojibwa Indians. *Archaeoastronomy: The Journal for the Center for Archaeoastronomy* VIII: 98–105.
32. Hollabaugh, Mark. 2005. Celestial Imagery in Lakota Culture. In *Current Studies in Archaeoastronomy,* ed. J.W. Fountain and R.M. Sinclair, 243–260. Durham, NC: Carolina Academic Press.
33. Chamberlain, Von Del. 1982. *When Stars Came Down To Earth: Cosmology of the Skidi Pawnee Indians of North America*, 244, 248. Los Altos, CA: Ballena Press. Murie, James R. 1981, Ceremonies of the Pawnee. Part 1: The Skiri, 199–200. In *Smithsonian Contributions to Anthropology* 27, ed. D.R. Parks. Washington, DC: Smithsonian Institution.
34. Chamberlain 1982 (note 33), 112.
35. Ibid., 123.
36. Ibid., 212.
37. Goodman, Ronald. 1990. *Lakota Star Knowledge: Studies in Lakota Stellar Theology.* Rosebud Sioux Reservation, SD: Sinte Gleska College (see specifically p. 45).
38. Note 25.
39. Simms, Stephen C. 1903. Traditions of the Crows. *Field Columbian Museum Publications 85. Anthropological Series* 2(6). Chicago: Field Columbian Museum (see specifically p. 300).
40. Highwater, Jamake. 1977. *Ritual of the Wind: North American Indian Ceremonies, Music, and Dances.* New York: Francis P. Harper (see specifically p. 64).
41. Lowie, Robert. H. 1915. The Sun Dance of the Crow Indians. *American Museum of Natural History Anthropological Papers* 16: 1–50. New York: America Museum of Natural History (see specifically p. 14).
42. Milbrath 1982 (note 25), interpretative text for NMAI 12/6426. Wildeschut, William. 1975. Crow Indian Medicine Bundles, ed. J.C. Ewers. *Contributions from the Museum of the American Indian Heye Foundation* XVII. New York: Museum of the American Indian Heye Foundation (see specifically p. 33).
43. Momaday, N. Scott. 2005. The Seven Sisters. In Chamberlain et al. 2005 (note 1), 32–37.
44. Goodman, Ronald. 2005. Lakota Star Knowledge. In Chamberlain et al. 2005 (note 1), 140–146 (quote p. 142). See also Goodman 1990 (note 37).
45. Kehoe, Alice B. 2005. Ethnoastronomy of the North American Plains. In Chamberlain et al. 2005 (note 1), 127–139.
46. Wissler, Clark. 1912. Ceremonial Bundles of the Blackfoot Indians. *American Museum of Natural History Anthropological Papers* VIII, Pt. 2: 65–289 (quote p. 237).
47. Chamberlain, Von Del. 1982. The Skidi Pawnee Earth Lodge as an Observatory. In *Archaeoastronomy in the New World,* ed. A.F. Aveni, 183–194. Cambridge, United Kingdom: Cambridge University Press.
48. Weltfish, Gene. 1965. *The Lost Universe.* New York: Basic Books (see specifically p. 360).
49. Murie 1981 (note 33).
50. Williamson, Ray A. 1984. In *Living the Sky: The Cosmos of the American Indian.* Boston: Houghton Mifflin Co. (see specifically pp. 218–220).
51. Ellis 1975 (note 21). Bunzel, Ruth L. 1932. Introduction to Zuñi Ceremonialism. *Bureau of American Ethnology Annual Report* 47, 467–544 (see specifically pp. 512, 534). Zelik, Michael. 1988. Astronomy and Ritual: The Rhythm of the Sacred Calendar

of the U.S. Southwest. In *New Directions in American Archaeoastronomy,* ed. A.F. Aveni, 183–198. Oxford, United Kingdom: BAR International Series 454 (see specifically pp.187, 191).
52. Note 21.
53. Stevenson, Matilda C. 1905. The Zuñi Indians. *Twenty-Third Annual Report of the Bureau of America Ethnology*. Washington, DC: Smithsonian Institution (see specifically pp. 117–118).
54. Zelik, Michael. 1986. Ethnoastronomy of the Historic Pueblos: Moon Watching. *Archaeoastronomy* 10 (Supplement to *Journal for the History of Astronomy* 17): S1–S22 (see specifically p. S4).
55. Zelik 1988 (note 51) (see specifically p. 191). Bunzel 1932 (note 51) (see specifically p. 512). Bunzel, Ruth L. 1932. Zuñi Ritual Poetry. *Bureau of American Ethnology Annual Report* 47, 611–836 (see specifically pp. 637–638).
56. Ellis 1975 (note 21), 72–74. Stephen, Alexander M. 1936. *Hopi Journal of Alexander M. Stephen*, ed. E.C. Parsons. *Columbia University Contributions to Anthropology* 23. New York: Columbia University (see specifically p. 4).
57. Zelik 1988 (note 51), 190. Young, M. Jane. 2005. Astronomy in Pueblo and Navajo World Views. In Chamberlain et al. 2005 (note 1), 49–64.
58. Stephen 1936 (note 56), 51. McCluskey, Stephen C. 1982. Historical Astronomy: The Hopi Example. In Aveni 1982 (note 47), 31–57.
59. Ellis 1975 (note 21), 64. McCluskey 1982 (note 58), Figs. 3, 5.
60. Titiev, Mischa. 1944. Old Oraibi: A study of the Hopi Indians of Third Mesa. *Papers of the Peabody Museum of American Archaeology and Ethnology* 22, no. 1. Cambridge, MA: Harvard University (see specifically p. 129).
61. Bunzel, Ruth L. 1932. Zuñi Origin Myths. *Bureau of American Ethnology Annual Report* 47, 545–610 (see specifically pp. 584–585).
62. Titiev 1944 (note 60) (see specifically p. 149).
63. Colton, Harold S. 1959. *Hopi Kachina Dolls*. Albuquerque, NM: University of New Mexico Press (see specifically p. 54). Tyler, Hamilton A. 1964. *Pueblo Gods and Myths*. Norman, OK: University of Oklahoma Press (see specifically p. 151).
64. Waters, Frank. 1969. *Book of the Hopi*. New York: Ballantine Books (see specifically p. 189).
65. Milbrath 1982 (note 25), interpretative text for NMAI 18/6259.
66. Fewkes, Jesse W. 1903. Hopi Kachinas. *Bureau of American Ethnology. Twenty-first Annual Report, 1899–1900.* Washington, DC: Smithsonian Institution (see specifically p. 122).
67. Ellis 1975 (note 21) (see specifically p. 81).
68. Curtis, Edward S. 1926. *North American Indian*, vol. 17. New York: Johnson Reprint Co. (see specifically p. 104).
69. Stephen 1936 (note 56), 66. Wright, Barton. 1977. *Hopi Kachinas*. Flagstaff, AZ: Northland Publishing (see specifically p. 36).
70. Milbrath 1982 (note 25), interpretative text for NMAI 18/6832.
71. Ellis 1975 (note 21) (see specifically pp. 74, 85). Stevenson 1905 (note 53) (see specifically p. 432). Cushing, Frank H. 1896. Outlines of Zuñi Creation Myths. In *Thirteenth Annual Report of the Bureau of American Ethnology*, 325–447. Washington, DC: Smithsonian Institution (see specifically pp. 392, 432).
72. Stevenson 1905 (note 53), 23–27.
73. Stephen 1936 (note 56) (see specifically pp. 39, 47, 51, 812, 906, 969, 977, Fig. 143).
74. Ellis 1975 (note 21) (see specifically p. 74).
75. Titiev 1944 (note 60) (see specifically pp. 110, 143). Colton 1959 (note 63) (see specifically p. 21).
76. Stevenson 1905 (note 53).

77. Milbrath 1982 (note 25), interpretative text for NMAI 10/8739, Collection of the National Museum of the American Indian. Stevenson 1905 (note 53), pl. 28. Bunzel, Ruth L. 1932. Zuñi Katchinas. *Bureau of American Ethnology Annual Report* 47, 837–1086 (see specifically p. 919).
78. Bunzel 1932 (note 61) (see specifically pp. 597–599).
79. Young 2005 (note 57).
80. Sandner, Donald. 1979. *Navaho Symbols of Healing*. New York: Harcourt Brace Jovanovich (see specifically pp. 74–75).
81. Griffin-Pierce, Trudy. 1992. *Earth Is My Mother, Sky Is My Father: Space, Time, and Astronomy in Navajo Sandpainting*. Albuquerque, NM: University of New Mexico Press (see specifically p. 75).
82. Haile, Bernard. 1977. *Starlore among the Navaho*. Santa Fe, NM: William Gannon (see specifically pp. 1–4).
83. Griffin-Pierce 1992 (note 81). Griffin-Pierce, Trudy. 2005. Black God: God of Fire, God of Starlight. In Chamberlain et al. 2005 (note 1), 73–79.
84. Chamberlain, Von Del, and Polly Schaafsma. 2005. Origin and Meaning of Navajo Star Ceilings. In Chamberlain et al. 2005 (note 1), 80–98.
85. Ibid. (see specifically p. 88).
86. Farrer, Claire R. 2005. When is a Month? The Moon and Mescaleros. In Chamberlain et al. 2005 (note 1), 123–126.
87. Ellis 1975 (note 21).
88. Farrer 2005 (note 86) (see specifically p. 126).
89. Tarn, Nathaniel, and Martin Prechtel. 1986. Constant Inconstancy: The Feminine Principle in Atiteco Mythology. In *Symbol and Meaning beyond the Closed Community: Essays in Mesoamerican Ideas*, ed. G.H. Gossen, 173 184. Studies in Culture and Society, vol. 1. Albany, NY: Institute for Mesoamerican Studies, State University of New York at Albany. Tedlock, Barbara. 1992. *Time and the Highland Maya*, rev. ed. Albuquerque, NM: University of New Mexico Press.
90. Milbrath, Susan. 1999. *Star Gods of the Maya: Astronomy in Art, Folklore, and Calendars*. Austin, TX: University of Texas Press (see specifically pp. 40–41).
91. Lamb, Weldon. 2005. Tzotzil Maya Cosmology. In Chamberlain et al. 2005 (note 1), 163–172.
92. Milbrath 1999 (note 90) (see specifically pp. 21–23). Lamb 2005 (note 91).
93. Tarn and Prechtel 1986 (note 89).
94. Milbrath 1999 (note 90) (see specifically pp. 13–15, 30–31).
95. Ibid., 38.
96. Girard, Raphael. 1949. *Los chortis ante el problema maya*, vol. 2. Mexico City: Colección Cultura Precolumbiana. Girard, Raphael. 1962. *Los maya eternos*. Mexico City: Libro Mex Editores.
97. Milbrath 1999 (note 90) (see specifically pp. 39, 59, 266–267).
98. Tedlock, Barbara. 1985. Hawks, Meteorology and Astronomy in Quiché-Maya Agriculture. *Archaeoastronomy: The Journal of the Center for Archaeoastronomy* VII: 80–89.
99. Lamb 2005 (note 91) (see specifically p. 168).
100. Sosa, John. R. 1989. Cosmological, Symbolic, and Cultural Complexity among the Contemporary Maya of Yucatan. In *World Archaeoastronomy: Selected Papers from the Second Oxford International Conference on Archaeoastronomy*, ed. A.F. Aveni, 130–142. Cambridge, United Kingdom: Cambridge University Press (see specifically p. 132).
101. Milbrath 1999 (note 90) (see specifically p. 94).
102. Sosa 1989 (note 100) (see specifically pp. 139–140).

103. Milbrath 1999 (note 90). Lamb, Weldon. 1981. Star Lore in the Yucatec Maya Dictionaries. In *Archaeoastronomy in the Americas,* ed. R.A. Williamson, 233–248. Los Altos, CA: Ballena Press.
104. Tozzer, Alfred M. 1941. Landa's Relación de las Cosas de Yucatán. *Papers of the Peabody Museum of Archaeology and Ethnology, Harvard University*, vol. 18. Cambridge, MA: The Peabody Museum (see specifically p. 258).
105. Milbrath 1999 (note 90) (see specifically pp. 34, 40, 141, 282).
106. Lipp, Frank J. 2005. Mixe Calendrics, Ritual and Astronomy. In Chamberlain et al. 2005 (note 1), 173–179.
107. Ibid. (see specifically p. 174).
108. Ibid. (see specifically p. 176).
109. Katz, Esther. 1994. Meteorolgía popular Mixteca: tradiciones indígenas y europeas. In *Time and Astronomy at the Meeting of Two Worlds*, ed. S. Iwaniszewski, A. Lebeuf, A. Wiercinski, and Mariusz S. Ziókowski, 105–122. Warsaw: Universidad de Varsoiva Centro de Estudios Latinamericansos.
110. Hugh-Jones, Stephen. 1982. The Pleiades and Scorpius in Barasana Cosmology. In *Ethnoastronomy and Archaeoastronomy in the American Tropics*, ed. A.F. Aveni and G. Urton, 183–202. Annals of the New York Academy of Sciences, vol. 385. New York: New York Academy of Sciences (see specifically p. 193).
111. Aveni 2001 (note 3) (see specifically p. 321). Fabian, Stephen M. 1992. *Space-time of the Bororo of Brazil.* Gainesville, FL: University of Florida Press. Reichel-Dolmatoff, Gerardo. 1971. *Amazonian Cosmos*. Chicago: University of Chicago Press. Reichel-Dolmatoff, Gerardo. 1982. Astronomical models of social behvior among some Indians of Colombia. In Aveni and Urton 1982 (note 110), 165–182. Wilbert, Johannes. 1973. Eschatology in a participatory universe: Destinies of the soul among the Warao Indians of Venezuela. In *Death and the Afterlife in Pre-Columbian America,* ed. E.P. Benson, 163–189. Washington, DC: Dumbarton Oaks, Research Library and Collections. Wilbert, Johannes. 1981. Warao Cosmology and the Yekuana Round House Symbolism. *Journal of Latin American Lore* 7(1): 37–72.
112. Magaña, Edmundo. 2005. Tropical Tribal Astronomy: Ethnohistorical and Ethnographic Notes. In Chamberlain et al. 2005 (note 1), 244–263.
113. Jara, Fabiola. 2005. Arawak Constellations: A Bibliographic Survey. In Chamberlain et al. 2005 (note 1), 264–280.
114. Ibid., 271.
115. Campos, Marcio D'Olne. 2005. Búzios Island: Knowledge and Belief among a Fishing and Agricultural Community on the Coast of the Sate of São Paulo. In Chamberlain et al. 2005 (note 1), 236–243.
116. Roe, Peter G. 2005. Mythic Substitution and the Stars: Aspects of Shipibo and Quechua Ethnoastronomy Compared. In Chamberlain et al. 2005 (note 1), 193–227 (see specifically pp. 194, 196).
117. Ibid. (see specifically pp. 207–208, 218–219).
118. Woodside, Joseph Holt. 2005. Amahuaca Astronomy and Star Lore. In Chamberlain et al. 2005 (note 1), 228–235.
119. Urton, Gary. 1981. Astronomy and Calendrics on the Coast of Peru. In Aveni and Urton 1982 (note 110), 231–248. Urton, Gary. 2005. Constructions of the Ritual-Agricultural Calendar in Pacariqtambo, Peru. In Chamberlain et al. 2005 (note 1), 180–192.
120. Grebe Vicuña, María Ester. 1994. Concepción del tiempo en las culturas indígenas sur-andinas. In Iwaniszewski et al. 1994 (note 109), 297–314.
121. Aveni 2001 (note 3) (see specifically Fig. 115). Eddy, John A. 1978. Archaeoastronomy of North America: Cliffs, Mounds, and Medicine Wheels. In Krupp 1978 (note 7), 133–164.

122. Eddy, John A. 1975. Reply to Jonathan Reyman. 1975. Big Horn Medicine Wheel: Why Was it Built? *Science* 188: 278–279. Schaefer, Bradley E. 1986. Atmospheric Extinction Effects on Stellar Alignments. *Archaeoastronomy* 10 (Supplement to the *Journal for the History of Astronomy* 17): S32–S42.
123. Vogt, David. 1993. Medicine Wheel Astronomy. In *Astronomies and Cultures*, ed. C.L. Ruggles and N.J. Saunders, 162–201 (quote p. 183). Boulder CO: University Press of Colorado.
124. Aveni 2001 (note 3) (see specifically pp. 307–308).
125. Eddy 1978 (note 121).
126. Wittry, Warren L. 1964. An American Woodhenge. *Cranbrook Institute of Science Newsletter* 33(9): 102–1907.
127. Aveni 2001 (note 3) (see specifically pp. 304–305).
128. Williamson, Ray A. 1981. North America's Multiplicity of Astronomies. In Williamson 1981 (note 103), 61–80.
129. Reyman, Jonathan E. 1976. Astronomy, Architecture and Adaptation at Pueblo Bonito. *Science* 193(4257): 957–962.
130. Sofaer, Anna. 1994. Chacoan Architecture: A Solar–Lunar Geometry. In Iwaniszewski et al. 1994 (note 109), 265–278.
131. Bates, Bryan C. 2005. A Cultural Interpretation of an Astronomical Calendar (Site #WS 833) at Wupatki National Monument. In *Current Studies in Archaeoastronomy,* ed. J.W. Fountain and R.M. Sinclair, 133–150. Durham, NC: Carolina Academic Press.
132. Williamson 1984 (note 50). Williamson 1981 (note 128). Williamson, Ray A. 1982. Casa Rinconada: A Twelfth Century Anasazi Kiva. In Aveni 1982 (note 47), 205–219.
133. Zelik, Michael. 1984. Summer Solstice at Casa Rinconada: Calendar, Hierophany or Nothing? *Archaeoastronomy: The Bulletin of the Center for Archaeoastronomy* VII(1–4): 76–81 (quote p. 78).
134. Aveni 2001 (note 3), 306–307.
135. Aveni, Anthony F., and Lorren D. Hotaling. 1994. Monumental Inscriptions and the Observational Basis of Maya Planetary Astronomy. *Archaeoastronomy* (Supplement to *Journal for the History of Astronomy*) 19: S21–S54. Lounsbury, Floyd. 1982. Astronomical Knowledge and Its Uses at Bonampak. In Aveni 1982 (note 47), 143–168. Justeson, John. S. 1989. The Ancient Maya Ethnoastronomy: An Overview of Hieroglyphic Sources. In Aveni 1989 (note 100), 76–129.
136. Mendez, Alonzo, Edwin L. Barnhart, Christopher Powell, and Carol Karasik. 2005. Astronomical Observations from the Temple of the Sun. *Archaeoastronomy: Journal for Astronomy in Culture* XIX: 44–73.
137. On Maya seasonal warfare: Milbrath 1999 (note 90) (see specifically p. 196).
138. Lounsbury, Floyd. 1989. A Palenque King and the Planet Jupiter. In Aveni 1989 (note 100), 246–259.
139. Aveni et al. 2003 (note 19).
140. Aylesworth, Grant R. 2004. Astronomical Interpretations of Ancient Maya E-Group Architectural Complexes. *Archaeoastronomy: The Journal of Astronomy in Culture* XVII: 34–66.
141. Aveni 2001 (note 3) (see specifically pp. 246–250, Fig. 86). Aveni, Anthony F., and Horst Hartung. 1986. Maya City Planning and the Calendar. In *Transactions of the American Philosophical Society,* vol. 76, no. 7, 1–84. Philadelphia: The American Philosophical Society (see specifically p. 17). Aveni, Anthony F., and Horst Hartung. 1991. Archaeoastronomy and the Puuc Sites. In *Arqueoastronomía y etnoastronomía en Mesoamérica*, ed. J. Broda, S. Iwaniszewski, and L. Maupomé, 65–96. Mexico City: Universidad Nacional Autónoma de México. Aveni, Anthony F., and Horst Hartung. 2000. Water, Mountain, and Sky: The Evolution of Site Orientations in Southeast

Mesoamerica. In *Precious Greenstone Precious Feathers/In Chalchihuitl in Quetzalli*, ed. E. Quiñones Keber, 55–68. Culver City, CA: Labyrinthos.
142. Aveni 2001 (note 3) (see specifically p. 249).
143. Tichy, Franz. 1994. Four Towers on the Meridian of Uxmal in the Archeological Region of Chenes/Campeche, Mexico. Astronomical Instruments for the Observation of the Solar Zenith Passages. In Iwaniszewski et al. 1994 (note 109), 279–290.
144. Milbrath 1999 (note 90) (see specifically pp. 66, 263, Fig. 3.1a).
145. Aveni 2001 (note 3) (see specifically pp. 298–300).
146. Aveni, Anthony F., Sharon L. Gibbs, and Horst Hartung. 1975. The Caracol Tower at Chichen Itza: An Ancient Astronomical Observatory? *Science* 188(4192): 977–985.
147. Milbrath 1999 (note 90).
148. Aveni 2001 (note 3) (see specifically Fig. 100).
149. Milbrath 1999 (note 90) (see specifically pp. 51, 158, 292).
150. Aveni, Anthony F., Susan Milbrath, and Carlos Peraza Lope. 2004. Chichén Itzá's legacy in the Astronomically Oriented Architecture of Mayapán. *RES: Anthropology and Aesthetics* 45: 123–143 (see specifically p. 137, Table 3).
151. Aveni 2001 (note 3) (see specifically Fig. 103a).
152. Milbrath 1999 (note 90) (see specifically p. 263).
153. Milbrath 1999 (note 90) (see specifically Fig. 3.1a). Galindo Trejo, Jesús. 1994. *Arqueoastronomía en la América Antigua*. Madrid: Editorial Equipo Sirius (see specifically p. 127).
154. Aveni 2001 (note 3). Galindo Trejo 1994 (note 153).
155. Šprajc, Ivan. 2000a. Astronomical Alignments at the Templo Mayor of Tenochtitlan, Mexico *Archaeoastronomy* 25 (Supplement to the *Journal for the History of Astronomy* 31): S11–S40. Šprajc, Ivan. 2000b. Astronomical Alignments at Teotihuacan, Mexico. *Latin American Antiquity* 11(4): 403–415.
156. Aveni et al. 2004 (note 150).
157. Aveni et al. 2004 (note 150) (see specifically Table 3).
158. Galindo Trejo, Jesús. 2007. Un análisis arqueoastronómico del edificio circular Q152 de Mayapán. *Estudios de Cultura Maya* 29: 63–81 (see specifically Table 2).
159. Aveni, Anthony F., Michael Closs, and Horst Hartung. 1993. An Appraisal of Baudez' Appraisal of Archaeoastronomy at Copan and Elsewhere. *Indiana* 13: 87–95. Šprajc, Ivan. 1987–1988. Venus and Temple 22 at Copán: Revisited. *Archaeoastronomy: The Bulletin for the Center for Archaeoastronomy* 10: 88–97.
160. Dow, James A. 1967. Astronomical Orientations at Teotihuacan: A Case Study in Astroarchaeology. *American Antiquity* 32: 326–334.
161. Chiu, Bella C., and Phillip Morrison. 1980. Astronomical Origin of the Offset Street Grid at Teotihuacan. *Archaeoastronomy* 2 (Supplement to *Journal for the History of Astronomy* 17): S55–S64.
162. Aveni 2001 (note 3) (see specifically p. 227).
163. Aveni 2001 (note 3) (see specifically p. 228).
164. Šprajc 2000 (note 159) (see specifically pp. 408–409).
165. Galindo Trejo 1994 (note 153) (see specifically pp. 147–148).
166. Aveni, Anthony F., and Horst Hartung. 1981. The Observation of the Sun at the Time of Passage through the Zenith in Mesoamerica. *Archaeoastronomy* 3 (Supplement to the *Journal for the History of Astronomy* 12): S1–S16. Galindo Trejo 1994 (note 153) (see specifically p. 146).
167. Galindo Trejo 1994 (note 153) (see specifically pp. 144–146).
168. Šprajc 2000a, 2000b (note 155). Šprajc, Ivan. 2001. *Orientaciones astronómicas en la arquitectura prehispánica del centro de México*. México, D.F. Instituto Nacional de Antropología e Historia. Šprajc, Ivan. 2004. Astronomical Alignments in Río Bec Architecture. *Archaeoastronomy: The Journal of Astronomy in Culture* XVII: 98–107.

169. Šprajc, 2000a (note 155).
170. Aveni 2001 (note 3). Aveni, Anthony F., and Edward Calnek. 1999. Astronomical Considerations in the Aztec Expression of History: Eclipse Data. *Ancient Mesoamerica* 10(1): 87–98.
171. Aveni 2001 (note 3) (see specifically Fig. 10).
172. Byland, Bruce E., and John M.D. Pohl. 1994. *In the Realm of 8 Deer: The Archaeology of Mixtec Codices*. Norman, OK: University of Oklahoma Press.
173. Milbrath 1999 (note 90). Aveni and Hotaling 1994 (note 135). Lounsbury 1982 (note 135). Lounsbury 1989 (note 138). Justeson 1989 (note 135). Schele, Linda, and David Freidel. 1990. *A Forest of Kings: The Untold Story of the Ancient Maya*. New York: William Morrow.
174. Milbrath 1999 (note 90). Aveni, Anthony F. 1992. The Moon and the Venus Table: An Example of the Commensuration in the Maya Calendar. In *The Sky in Mayan Literature*, ed. A.F. Aveni, 87–101. Oxford, United Kingdom: Oxford University Press. Bricker, Harvey M., and Victoria R. Bricker. 2007. When Was the Dresden Codex Venus Table Efficacious? In *Skywatching in the Ancient World: New Perspectives on Cultural Astronomy*, ed. C. Ruggles and G. Urton, 95–121. Boulder, CO: University Press of Colorado. Bricker, Victoria R., and Harvey M. Bricker. 1986. The Mars Table in the Dresden Codex. *Research and Reflections in Archaeology and History: Essays in Honor of Doris Stone*, ed. E. Wyllys Andrews V, 51–79. Middle American Research Institute, pub. 47. New Orleans: Tulane University. Bricker, Victoria R., and Harvey M. Bricker. 1989. Astronomical References in the Table on Pages 61–69 of the Dresden Codex. In Aveni 1989 (note 100), 232–245. Bricker, Victoria R., and Harvey M. Bricker. 1992. Zodiacal References in the Maya Codices. In *The Sky and Mayan Literature*, ed. A.F. Aveni, pp. 148–183. Oxford: Oxford University Press. Lounsbury, Floyd. 1978. Maya Numeration, Computation, and Calendrical Astronomy. *Dictionary of Scientific Biography*, vol. 15, suppl. 1, ed. C. Coulston-Gillispie, 757–818. New York: Charles Scribner's Sons. Paxton, Merideth. 2000. The Alamanc on Pages 10b-11b of the Madrid Codex and the Burner Cycle of the Maya. In Quiñones Keber 2000 (note 141), 83–100. Thompson, John Eric S. 1972. *A Commentary on the Dresden Codex: A Maya Hieroglyphic Book*. Philadelphia: The American Philosophical Society. Thompson, John Eric S. 1974. Maya Astronomy. *Philosophical Transactions of the Royal Society of London* 276: 83–98. Vail, Gabrielle. 2004. A Reinterpretation of *Tzol'kin* Almanacs in the Madrid Codex. In *The Codex Madrid: New Approaches to Understanding an Ancient Maya Manuscript*, ed. G. Vail and A. Aveni, 215–254. Boulder, CO: University Press of Colorado. Vail, Gabrielle and Victoria Bricker. 2004. *Haab* Dates in the Madrid Codex. In Vail and Aveni 2004 (note 174), 171–214.
175. Milbrath 1999 (note 90) (see specifically pp. 15, 59). Šprajc 2000a, 2000b (note 155). Galindo Trejo 2007 (note 158).
176. Milbrath 1999 (note 90) (see specifically pp. 259–261).
177. Milbrath 1999 (note 90) (see specifically p. 257, Fig. 7.2a). Bricker and Bricker 1992 (note 174).
178. Milbrath, Susan. 2007. Astronomical Cycles in the Imagery of Codex Borgia, 29–46. In Ruggles and Urton 2007 (note 174), 157–209.
179. Aveni, Anthony F. 1999. Astronomy in the Mexican Codex Borgia. *Archaeoastronomy* 24 (Supplement to the *Journal of the History of Astronomy* 30): S1–S20. Bricker, Victoria R. 2001. A Method for Dating Venus Almanacs in the Borgia Codex. *Archaeoastronomy*, 26 (Supplement to the *Journal of the History of Astronomy* 32): S21–S43.
180. Hernández, Christine. 2004. "Yearbearer Pages" and Their Connection to Planting Almanacs in the Borgia Codex. In Vail and Aveni 2004 (note 174), 321–366.
181. Milbrath 2007 (note 178).

182. Aveni 2001 (note 3), 236–238. Aveni, Anthony F., Edward Calnek, and Horst Hartung. 1988. Myth, Environment, and the Orientation of the Templo Mayor of Tenochtitlan. *American Antiquity* 53: 287–309.
183. Milbrath 1982 (note 25). González Torres, Yólotl. 1975. *El culto a los astros entre los mexicas*. Mexico City: SEP/SETENTAS (see specifically p. 217). Sahagún, Fray Bernardino de. 1950–1982. *Florentine Codex: General History of the Things of New Spain*. 2nd ed. of Book 1 (1981) and Book 2 (1970). 12 vols. Trans. A.J.O. Anderson and C.E. Dibble. Salt Lake City, UT: School of America Research and University of Utah Press. Townsend, Richard F. 1979. *State and Cosmos in the Art of Tenochititlan*. 20. Washington, DC: Dumbarton Oaks (see specifically p. 36).
184. Milbrath, Susan. 1980. Star Gods and Astronomy of the Aztecs. In *La antropología americanista en la actualidad, homenaje a Raphael Girard*, 289–304. Mexico City: Editores Mexicanos Unidos.
185. Milbrath, Susan. 1997. Decapitated Lunar Goddesses in Aztec Art, Myth, and Ritual. *Ancient Mesoamerica* 8(2): 185–206. Sahagún 1950–1982 (note 183) (see specifically III: 1–5).
186. Milbrath, Susan. 1995. Gender and Roles of Lunar Deities in Postclassic Central Mexico and Their Correlations with the Maya Area. *Estudios de Cultural Nahuatl* 25: 45–93. Sahagún 1950–1982 (note 183) (see specifically VII: 4–7).
187. Aveni 2001 (note 3) (see specifically pp. 16–19, Fig. 5).
188. Ibid. (see specifically pp. 35–36).
189. Seler 1904 (note 17).
190. Quiñones Keber, Eloise. 1995. *Codex Telleriano-Remensis: Ritual, Divination, and History in a Pictorial Aztec Manuscript*. Austin, TX: University of Texas Press (see specifically p. 252).
191. Milbrath 1999 (note 90), 14, 292. Milbrath 1980 (note 184). Krupp, Edwin C. 1982. The "Binding of the Years," The Pleiades, and the Nadir sun. *Archaeoastronomy: The Bulletin for the Center for Archaeoastronomy* 5(1): 9–13. Sahagún 1950–1982 (note 183) (see specifically V: 143).
192. Milbrath 1980 (note 184). Milbrath 1997 (note 185), 202. Sahagún 1950–1982 (note 183) (see specifically II: 147, VII: 2, 38).
193. Milbrath 1999 (note 90) (see specifically pp. 281–282).
194. Aveni 2001 (note 3), 310–311. Urton, Gary. 1981. *At the Crossroads of the Earth and Sky: An Andean Cosmology*. Austin, TX: University of Texas Press (see specifically pp. 132–133, Table 7).
195. Dearborn, David S., and Ray E. White. 1982. Archaeoastronomy at Machu Picchu. In Aveni and Urton 1982 (note 110), 249–260. Dearborn and White. 1983. The "Torreon" of Machu Picchu as an Observatory. *Archaeoastronomy* 5 (Supplement to *the Journal of the History of Astronomy* 14): S37–S49. Dearborn and White. 1989. Inca Observatories: Their Relation to the Calendar and Ritual. In Aveni 1989 (note 100), 462–469. Urton 1981 (note 195) (see specifically p. 71).
196. Dearborn and White 1989 (note 193), 466. Zuidema, R. Tom. 1982. Catachillay. In Aveni and Urton 1982 (note 110), 203–230 (see specifically pp. 212–214).
197. Milbrath 1982 (note 25). Molina, Christobal de. 1959. *Fábulas y Ritos de los Incas* (1573). Buenos Aires: Editorial S.R.L. (see specifically pp. 66, 72–73).
198. Aveni 2001 (note 3) (see specifically p. 310, Fig. 11d). Rowe, John. 1946. Inca Culture at the time of the Spanish Conquest. In *Handbook of South American Indians*, ed. J.H. Steward, 183–330. *Bureau of American Ethnology Bulletin* 143, vol. 2. Washington, DC: Smithsonian Institution (see specifically pp. 294–295).
199. Miskin, Bernard. 1940. Cosmological Ideas among the Indians of the Southern Andes. *Journal of American Folklore* 53: 225–41 (see specifically p. 239).
200. Dearborn and White 1989 (note 195).

201. Aveni 2001 (note 3) (see specifically pp. 310–321).
202. Aveni 2001 (note 3). Zuidema, R. Tom. 1977. The Inca Calendar. In *Native American Astronomy*, ed. A. Aveni, 219–259. Austin, TX: University of Texas Press.
203. Aveni 2001 (note 3) (see specifically p. 318). Dearborn, David S., Matthew T. Seddon, and Brian S. Bauer. 1998. The Sanctuary of Titicaca: Where the Sun Returns to the Earth. *Latin American Antiquity* 9(3): 240–258. Urton 1981 (note 195).
204. Zuidema 1982. The Sidereal Lunar Calendar of the Incas. In Aveni 1982 (note 47), 59–107.
205. Bauer, Brian A. and David Dearborn. 1995. *Astronomy and Empire in the Ancient Andes*. Austin, TX: University of Texas Press (see specifically pp. 97–100).
206. Urton, Gary. 1982. Astronomy and Calendrics on the Coast of Peru. In Aveni and Urton 1982 (note 110), 231–247. Urton, Gary, and Anthony F. Aveni. 1983. Archaeoastronomical Fieldwork on the Coast of Peru. *In Calendars in Mesoamerica and Peru: Native American Computations of Time,* ed. A.F. Aveni and G. Brotherston, 221–234. Oxford, United Kingdom: BAR International Series 174 (see specifically p. 225).
207. Romano, Giuliano. 1994. Orientaciones astronomicas en Chavín de Huantar y Cerro Sechin. In Iwaniszewski et al. 1994 (note 109), 335–343. Urton and Aveni 1983 (note 206).
208. Romano 1994 (note 207).

Section III

Introduction to the Space Age

The Cold War and the Space Race:
The "Sputnik Nudge."

11 History of the Space Age

Anne Millbrooke

CONTENTS

INTRODUCTION

In 1947 Robert A. Heinlein published *Rocket Ship Galileo*, a "science fiction adventure," about four American astronauts who flew to the Moon, where they found and fought Nazis. Theirs was by no means the first fictional space trip. Jules Verne's *From the Earth to the Moon* (1865) and H.G. Wells' *The First Men in the Moon* (1901) had long been classics. Heinlein's book became the inspiration for the 1950 movie *Destination Moon*, set firmly in the postwar Cold War: the race to the Moon was a military arms race. In the movie, the fictional General Thayer explains that "there is absolutely no way to stop an attack from outer space" so "the first country that can use the Moon for the launching of missiles will control the Earth."[1] In fact, the Cold War defined much of the early Space Age from the World War II and postwar work on missiles and rockets to satellites and other space vehicles—in the United States and in the Soviet Union. Only as the Soviet Union and Cold War crumbled did other countries develop spaceflight programs.

The Space Age as a historical era coincided with the Cold War, and that is no coincidence. The atomic bomb dropped on Hiroshima in 1945, the successful launch and orbits of the Soviet satellite *Sputnik* in 1957, Soviet cosmonaut Yuri Gagarin becoming the first person in space in 1961, American astronauts landing and walking on the Moon in 1969, the various space stations and space transportation systems developed thereafter, and other Space Age events happened in the context of the Cold War rivalry between the United States and the Soviet Union, allies of these two superpowers, and the unaligned (Third World) nations whose allegiance both sides sought (Figure 11.1). The Space Age, defined by space travel, continued after the explosion of the space shuttle *Columbia* in 1986, the fall of the Berlin Wall in 1989,

The Huntsville Times

Man Enters Space

'So Close, Yet So Far,' Sighs Cape

U.S. Had Hoped For Own Launch

Soviet Officer Orbits Globe In 5-Ton Ship

Maximum Height Reached Reported As 188 Miles

Hobbs Admits 1944 Slaying

Praise Is Heaped On Major Gagarin

'Wacker' Stands By Story

Reds Deny Spacemen Have Died

To Keep Up, U.S.A. Must Run Like Hell'

Reds Win Running Lead In Race To Control Space

FIGURE 11.1 Yuri Gagarin on the launch pad before becoming the first person in space, April 12, 1961. (Courtesy of NASA.)

and the collapse of the Soviet Union in 1991, but it continued without the direction and funding of the military imperative of the Cold War space rivalry and without the popular and political support of earlier times. There were, however, more nations seeking international stature and scientific and technical development through space programs.

A popular "age" is a period generally recognized by the public. For the Space Age, the initial public awareness came in 1957 with the Soviet launch of the satellite *Sputnik*. That event marks the transition of spaceflight from science fiction to science fact, albeit not the end of science fiction by any means, as amply illustrated by the subsequent popularity of the three *Star Wars* movies, the later three prequel *Star Wars* movies, the several *Star Trek* television series, and *Star Trek* movies.[2]

SPUTNIK

The Soviet Union launched the first *Sputnik* satellite in early October 1957. The next day the Soviet newspaper *Pravda* reported, "As a result of very intensive work by scientific research institutes and design bureaus the first artificial satellite in the world has been created. On October 4, 1957, this first satellite was successfully launched in the USSR. According to preliminary data, the carrier rocket has imparted to the satellite the required orbital velocity of about 8,000 meters per second."[3] It was the carrier rocket, an R-1 military rocket, as much as the satellite that caused much of the concern in the First World, the Free World, the West. The fact that the Union of Soviet Socialist Republics—a member of the Second World, the Communist World, the East—had a rocket that powerful and had achieved a spaceflight before the West suggested a Soviet lead in the scientific and technological arms race. Indeed, there was fear that the Soviets would develop the capability to attack the United States from outer space, or at the least that the powerful Soviet rockets might be used to launch intercontinental missiles.

In 1957 the Soviets did more than launch the world's first artificial satellite. On November 3, 1957, they successfully launched their second satellite, *Sputnik 2*, with a dog named Laika on board. Then, on December 6, the United States failed to launch the Navy's Vanguard rocket with a satellite on board. Finally, on January 31, 1958, an Army team, led by the German immigrant Wernher von Braun, succeeded in launching the first American satellite, *Explorer 1*, on a Jupiter-C rocket. Although the early satellite launches of both the United States and the Soviet Union were part of formal, published programs of the International Geophysical Year of 1957–1958, the actual success of *Sputnik* brought what had been quiet military rocket programs to the public attention as a space race. The *Explorer* carried scientific instruments that allowed the discovery of radiation belts named after the scientist James A. Van Allen. As Van Allen later recalled, "We were treated like heroes, rescuing the honor of the United States in this great Cold War with Russia by having a successful satellite."[4]

Just as the Atomic Age had inspired "Atomic" cafes, "Atomic" cocktails, and even bikini swimwear (too hot to handle, like the Bikini nuclear test site in the Marshall Islands), the Space Race acquired a public language of its own. *Flopnik*, for example, became the derisive term applied to the United States' first and failed launch of a rocket with a satellite on board. *Spookniks* became a term for the spying possibilities of Soviet satellites. The Atomic Age and the Space Age merged at the thought that one or the other superpower might deliver an atomic bomb from a space-launch-capable rocket, a spacecraft, or a space base.

Among the consequences of *Sputnik* was the development of satellite communications, both commercial and defense systems.[5] A significant consequence was the establishment in 1958 of two space agencies in the United States, one civilian and one military. The National Aeronautics and Space Administration (NASA) and the Advanced Research Projects Agency became part of the growing Cold War military-industrial complex. NASA, for example, acquired not only the programs of the National Advisory Committee for Aeronautics but also the Navy's Vanguard rocket program and the Army's Explorer rocket program. According to a NASA chief historian (Roger D. Lanius), "First, NASA's projects were clearly cold war propaganda weapons that national leaders wanted to use to sway world opinion about the relative merits of democracy versus the communism of the Soviet Union.... Second, NASA's civilian effort served as an excellent smoke screen for the DOD's military space activities, especially for reconnaissance missions."[6] The Soviets used their space programs similarly.

SPACEFLIGHT

The romance of flight, whether of birds, balloons, or airplanes, extended with spacecraft into outer space, but spaceflight remained experimental through many projects, including the Soviet one-person Vostok (East), two-person Voshkod (Rise), and lunar Soyuz (Union) programs, and the American one-person Mercury, two-person Gemini, and lunar Apollo programs (Figure 11.2). Using the one-person capsules, both nations acquired valuable experience: one astronaut at a time orbiting the Earth and integrating man and machine, the machine being not only the spacecraft but also the spacesuit and mission equipment. As spacecraft got large enough to carry two people, the nations experimented with an extravehicular activity that quickly became known as a spacewalk. Early in 1965 the Russian astronaut Alexei Leonov had made the first spacewalk in history, tethered to a Voskhod capsule. Good publicity followed for the Soviet Union. Three months later American astronaut Edward H. White accomplished the first American spacewalk, tethered to a Gemini capsule, but also using a Hand-Held Maneuvering Unit. Similar good publicity followed for the United States. Such was the propaganda value of the space programs (Figure 11.3).

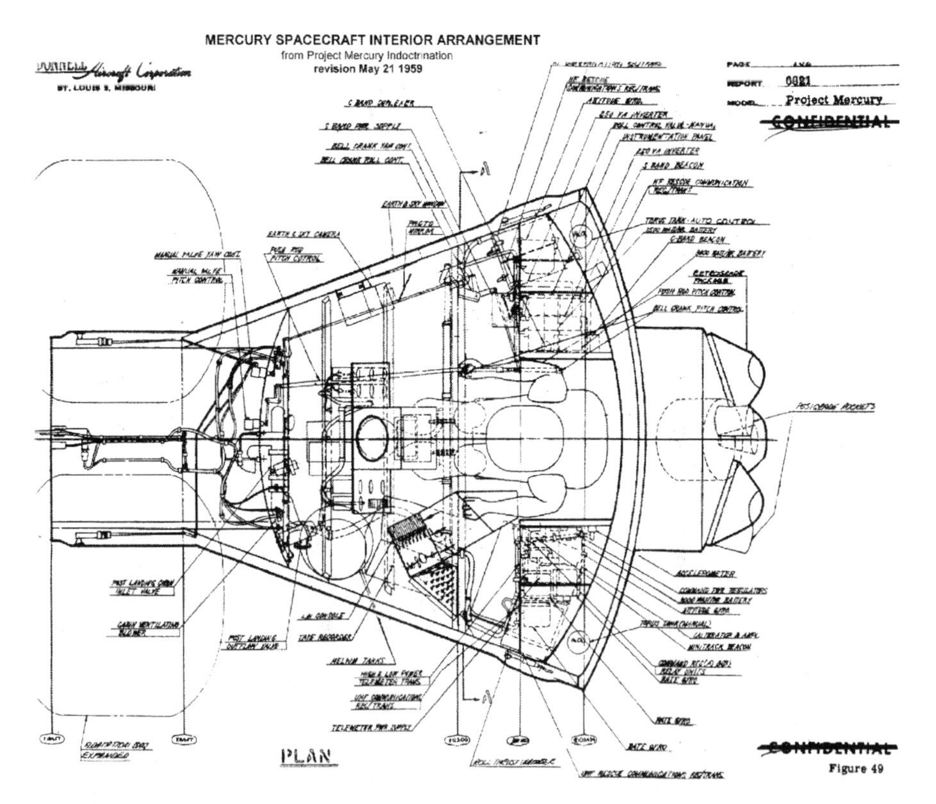

FIGURE 11.2 Top view of the interior arrangement of the *Mercury* spacecraft, with seating for one. (Courtesy of NASA.)

FIGURE 11.3 During the *Gemini 4* mission, astronaut Edward White performed the United States' first extravehicular activity, or spacewalk. (Courtesy of NASA.)

Initially, the Soviets seemed to be ahead in the Space Race: first satellite (*Sputnik*, October 4, 1957), first animal in space (the dog Laika, November 3, 1957), the first man in space (Yuri Gagarin, April 12, 1961). In May 1961, President John F. Kennedy acknowledged the Soviet lead, attributed it to their big rockets, and predicted it would continue for a while. But he had a plan to overtake the Soviets and achieve leadership in space. He announced, "Finally, if we are to win the battle that is now going on around the world between freedom and tyranny … Now it is time to take longer strides—time for this nation to take a clearly leading role in space achievement, which in many ways may hold the key to our future on earth."[7] He continued, "We go into space because whatever mankind must undertake, free men must fully share." Kennedy concluded, "I believe that this nation should commit itself to achieving the goal, before this decade is out, of landing a man on the moon and returning him safely to the Earth."

But, as Kennedy predicted, the Soviets continued to capture headlines with their space achievements:

- First woman in space—Valentina Tereshkova, June 16, 1963
- First spacewalk—Alexei Leonov, March 18, 1965
- First soft landing on the Moon—the probe *Luna 9*, February 3, 1966
- First automatic docking in space—unmanned *Kosmos 186* and *Kosmos 188*, October 30, 1967
- First docking in space of two manned spacecraft—*Soyuz 4 and Soyuz 5,* January 14–15, 1969

Whether in control of the ground station or of an astronaut on board, early spacecraft were being flown by test astronauts in the tradition of test pilots. There were accidents through the years, bad accidents in the early lunar tests. Three American astronauts—Virgil "Gus" Grissom, Edward H. White, and Roger Bruce Chaffee—died in a fire during a launch pad test of the *Apollo 1* spacecraft (January 27, 1967). Soon thereafter the Soviets reached a dubious first: first person to die in spaceflight, when the crash of the first *Soyuz* spacecraft killed Vladimir Komarov (April 24, 1967). These accidents reminded the public of the dangers undertaken in the race to reach the Moon. The Apollo accident review board concluded, "Those organizations responsible for the planning, conduct, and safety of this test failed to identify it as being hazardous."[8] That prompted a redesign of the Apollo capsule and the Apollo safety program. *Apollo 7* finally achieved the first manned Apollo flight in October 1968, when it orbited the Earth 163 times. *Apollo 8* orbited the Moon. These and other flights tested equipments, systems, and astronauts in preparation for a lunar landing.

Despite the strong Soviet push under Sergei Pavlovich Korolev, the United States won the race to place a man on the Moon: Neil A. Armstrong, July 20, 1969. The successful *Apollo 11* mission involved a Saturn V rocket as the launch vehicle, the *Columbia* command module that orbited the Moon, and the *Eagle* landing module. Five more Apollo missions reached the Moon: *Apollo 12, 14, 15, 16*, and *17*. Thus, between July 1969 and April 1972, twelve American astronauts walked on the Moon. Russian astronauts on the Moon, zero. The Americans had achieved unquestioned leadership in the Space Race.

In addition to manned spaceflight, NASA and the Soviet space program sponsored unmanned programs that sent probes to explore to the Moon, planets, and solar system.[9] But popular attention focused mostly on the Space Age manned flights. The first Soviet Salyut (Salute) space station entered orbit in April 1971. In all, the Soviets launched nine Salyut modules during the 11-year program; six Salyuts were research stations and three were military reconnaissance stations. The American *Skylab 1, 2,* and *3* orbited the Earth in 1973 and 1974. These space stations enabled astronauts of the two countries to spend extended lengths of time in space and to conduct experiments while there, and the stations had enough space for weightless floating inside the spacecraft. That experience expanded with the Soviet *Mir* (Peace) space stations and the reusable American space shuttles (Figure 11.4).

The Soviets designed reusable space transportation vehicles, selected the Buran (Snowstorm), and even successfully launched an unmanned Buran, but they canceled the program. In contrast, the United States produced and operated a fleet of reusable vehicles: *Columbia, Challenger, Enterprise, Discovery,* and *Atlantis*. The *Columbia* was the first to reach orbit, on April 12, 1981. On return, the shuttle became a lifting-reentry vehicle, and astronauts Robert L. Crippen and John W. Young landed the shuttle like an airplane. The *Columbia* made its next flight in November and thereby became the first spacecraft to be reused. Space shuttles thereafter flew missions to recover satellites and to move people and equipment, even spacecraft. In 1983, on space shuttle flights, Sally Ride became the first American woman in space, and Guion Bluford became the first black astronaut.

Inside these spacecraft, as space writer Harry L. Shipman noted, "Velcro takes the place of gravity."[10] The invention of Velcro (the brand name for a hook-and-loop fastener) derived from the observation that cockleburrs' hooks stuck in the hair loops of the inventor's dog. That inventor, George de Mestral, received a patent in 1955—in time

FIGURE 11.4 Launched from the Baikonur Cosmodrome in Kazakhstan, this Russian Soyuz space capsule is leaving the International Space Station in October 2001. On board are the Soyuz taxi crew, Commander Victor Afanasyev and Flight Engineer Konstantin Kozeev, and a French flight engineer named Claudie Haignere, who represented the European Space Agency. (Courtesy of NASA.)

for Space Age applications, literally in space. Astronaut Joseph P. Allen traveled into space on the space shuttles *Columbia* and *Discovery*. He described the experience:

> During the first few days in space, the act of simply moving from *here* to *there* looks so easy, yet is challenging. The veteran of zero gravity moves effortlessly and with total control, pushing off from one location and arriving at his destination across the flight deck, his body in the proper position to insert his feet into Velcro toe loops and to grasp simultaneously the convenient handhold, all without missing a beat in his tight work schedule. In contrast, the rookies sail across the same path, usually too fast, trying to suppress the instinct to glide headfirst and with vague swimming motions. They stop by bumping into the far wall in precisely the wrong position to reach either the toe loops or the handholds.[11]

The reusable space shuttle carried manned maneuvering units (MMUs) on three flights, and six astronauts wore the backpack units, à la Buck Rogers, on a total of nine sorties for a total of 10 hours and 22 minutes in 1984. Bruce McCandless became the first person to fly free, untethered in space, when he wore an MMU in February. He described the first step off the shuttle, 150 nautical miles above Earth, as "a heck of a big leap."[12] The MMUs were to enable astronauts to perform services outside the shuttle, specifically to recover satellites for repair, but the 1984 sorties were the only time the units were used. The robotic arm replaced the maneuvering unit.

The space shuttle program suffered two major disasters. The *Challenger* exploded after liftoff on January 28, 1986. The crew of five men and two women died. According to the Presidential Commission that investigated the accident: "The specific failure was the destruction of the seals that are intended to prevent hot gases from leaking through the joint during the propellant burn of the rocket motor."[13] The *Challenger* had been starting its tenth mission. The first of the space shuttles to enter service in 1981, the *Columbia* went on the 114th shuttle mission in 2003. It reentered Earth's atmosphere on February 1st. The spacecraft literally fell apart, and the seven astronauts on board died. This time, a piece of insulating foam had broken off the external propellant tank on takeoff and knocked some reinforced carbon insulation off the wing, which then overheated and failed on reentry. The *Columbia* Accident Board concluded that NASA's culture and history contributed to the accident, because the "agency [was] trying to do too much with too little."[14] The board continued, "The recognition of human spaceflight as a developmental activity requires a shift in focus from operations and meeting schedules to a concern with the risk involved." This brought to many minds the conclusion of the early Apollo accident board. Spaceflight was still developmental.

BEYOND THE SUPERPOWERS

Although Europe was caught between the superpowers during the Cold War, the nations of Europe tried at times to establish independence from the Cold War rivals. To develop their own technologies, economies, space programs, the European nations cooperated on various programs. The Commission Préparatoire Européenne pour la Recherche Spatiale started the work toward a European satellite program and a European launch vehicle. Established in 1962, the European Space Research Organization (ESRO) established and operated a satellite program that used American launch vehicles and the Vandenberg launch site in California. Established in 1964, the European Launcher Development Corporation (ELDO) used the British Blue Streak vehicle for the first stage, French technology for the second stage, and German technology for the third, while Italy developed a satellite and the Netherlands and Belgium worked on tracking and telemetry. Australia joined ELDO when the European organization selected Woomera for the launch test site. The four tests there failed to launch a satellite. ELDO moved launch tests to Kourou, French Guiana, in South America.

The two European collaborations—ESRO and ELDO—merged in 1973. The resulting European Space Agency continued the work of developing and operating a European space program. The European Space Agency selected its first astronauts in 1977. Riding Russian and American spacecraft, European astronauts have visited *Skylab*, *Mir*, and the International Space Station. Why does Europe continue to develop its space program? According to the European Space Agency, "Today's space systems are the key to the understanding and management of the World, to the provision of goods and services in the global marketplace, and to regional and global security and peacekeeping."[15]

The International Space Station is a cooperative program involving the United States, Russia, the European Space Agency, Japan, and Canada. A Russian Proton rocket and an American space shuttle carried the first two modules into space in 1988, and in 2000 Soyuz spacecraft carried the first crew to reside at the station. As a matter of fact, with the aging of the space shuttle fleet and the groundings after the *Challenger* and *Columbia* accidents, the venerable Soyuz spacecraft have become the workhorses, transporting crews, supplies, and equipment to the International Space Station. The Russian Federal Space Agency (Roskosmos) literally sells flights on the Soyuz; the price in January 2006 was just over US$20M.[16] The Soyuz line has been in service since 1968; the first flight in 1967 ended in the astronaut's death. The current model rendezvouses and docks at the International Space Station. Shedding the orbital and service modules for reentry, Soyuz is not reused, as only the reentry capsule parachutes to Earth.[17]

Since the 1950s, China has had a space program, albeit a small one initially. As a young communist country, China received space technology and technical assistance from the Soviet Union until 1960. During the 1960s, China focused more on developing nuclear weapons than on spacecraft for a while; China detonated its first nuclear bomb in 1964 and first hydrogen bomb in 1967. Yet trade opened with the West: West Germany provided a communication package for a Chinese satellite, and the United States provided radiation-hardened electronic chips for Chinese meteorological satellites. In 1966 China decided to pursue human spaceflight, but for years, launching satellites remained the main space operation. The Chinese government again approved a human spaceflight program in 1992, using the country's three satellite launch centers for tests. An unmanned Chinese spacecraft orbited the Earth in November 1999. The Chinese launched a spacecraft carrying animals into orbit in January 2001 and launched an unmanned spacecraft into orbit in March 2002 and another in December 2002. China achieved manned spaceflight on October 15–16, 2003, when Yang Liwei rode the *Shenzhou 5* around the Earth 14 times. Among the motivations behind the Chinese space program and the Chinese astronaut Yang was national prestige. As Yang said as he boarded the spacecraft, "I will not disappoint the motherland."[18] He did not disappoint the motherland. China sent two astronauts into space in 2003. A spaceflight and spacewalk took place in September 2008.

India, South Africa, and Brazil are among the other nations developing space capabilities.

LITERATURE OF THE SPACE AGE

The American space program, at least the civilian side of it, is well documented, especially by NASA. The NASA History Series includes a four-volume set called *NASA Historical Data Book*, which covers NASA resources, programs, and projects, 1958–1978.[19] The NASA History Series also includes a seven-volume set entitled *Exploring the Unknown: Selected Documents in the History of the U.S. Civil Space Program.*[20] The first director of NASA, T. Keith Glennan, recounts *Birth of NASA.*[21] *"Before This Decade Is Out ..."* is a collection of personal accounts of the Apollo program.[22] The NASA History Series also covers the history of various NASA facilities (such as the Johnson, Marshall, and Stennis space centers, the Langley Research

Center, and the Dryden Flight Research Center) and space programs (such as the Mercury, Apollo, Skylab, and space shuttle programs).[23]

Many books published independent of NASA also cover the U.S. space program. Autobiographies present the stories of different participants in the space program. Homer H. Hickam Jr.'s *Rocket Boys* tells of boys inspired by *Sputnik* to build their own rockets.[24] Alan Shepard and Deke Slayton wrote *Moon Shot*, about the race to the Moon from the astronauts' perspective.[25] Similarly, biographies, such as Ernst Stuhlinger and Frederick I. Ordway's two-volume *Wernher von Braun, Crusader for Space,* provide a wealth of information.[26] On the fifth anniversary of the Aerospace Medical Association, Eloise Engle and Arnold Lott wrote an early history of biomedical research, *Man in Flight.*[27] More specific to the space program is John A. Pitts' *The Human Factor: Biomedicine in the Manned Space Program to 1980*, in the NASA History Series.[28] Another example of a book focused on an aspect of the space program is Lillian D. Kozioski's *U.S. Space Gear.*[29] Dennis R. Jenkins provides a thorough overview of the space shuttle through the first one hundred missions.[30] There is also some literature about the military side of the space program, such as the Industrial College's *National Security Management: National Aerospace Programs.*[31]

Some books cover Soviet as well as American space programs. An excellent volume covering the Space Race in the Cold War context is Walter A. McDougall's ... *The Heavens and the Earth: A Political History of the Space Age.*[32] Philip Baker has written *The Story of Manned Space Stations.*[33] Much of the Soviet space story remains to be told, but a good introduction is James Harford's *Korolev: How One Man Masterminded the Soviet Drive to Beat America to the Moon.*[34] These are just some examples of vast and growing body of literature on the Space Age.

CONCLUSION

No one has visited the Moon since 1972. The space shuttle is scheduled for retirement in 2010. Its successor will not be ready until at least 2015. Soyuz spacecraft have been carrying most astronauts to and from the International Space Station, but the Soyuz is an old technology too, older than the space shuttle. Spaceplanes remain a popular idea for a twenty-first-century space vehicle: for example, NASA's orbital space plane is intended to replace the space shuttle, and Scaled Composites' commercially designed *SpaceShipOne* won the Ansari X Prize in 2004. Russia has looked at the Klipper (Clipper) winged crew vehicle concept. The European Space Agency launched a new automated transfer vehicle, a cargo hauler, called *Jules Verne* in March 2008; it went to the International Space Station. Reentry and a fiery destruction over the Pacific Ocean occurred in late September 2008. Currently unmanned, the one-time-use spacecraft might be further developed for crew capability. As recently as September 2008, NASA Administrator Michael Griffen reminded Congress of the importance of priority and status in what is now an international space environment: "A Chinese landing on the moon prior to our own return will create a stark perception that the U.S. lags behind not only Russia, but also China, in space."[35] Planetary probes still try to answer the question, "Are we alone?" But national pride, scientific research, and technological development aside, the Cold War rivalry that fueled and funded the historic Space Age is over.

NOTES AND REFERENCES

1. *Destination Moon*. 1950. Directed by George Pal. Written by Robert A. Heinlein, et al. Wade Williams 50th Anniversary Edition, DVD, 2000. Universal Studios.
2. *Star Wars Episode IV: A New Hope*. 1977. Directed and written by George Lucas. TCF/Lucasfilm. *Star Wars Episode V: The Empire Strikes Back*. 1980. Directed by Irvin Kershner with others. TCF/Lucasfilm. *Star Wars Episode VI: Return of the Jedi*. 1983. Directed by Richard Marquand. Written by Lawrence Kasdan and George Lucas. TCF/Lucasfilm, 1983. The Star Wars prequels were *Star Wars Episode I: The Phantom Menace* (1999), *Star Wars Episode II: Attack of the Clones* (2002), and *Star Wars Episode III: Revenge of the Sith* (2005). Regarding *Star Trek*, there were several television series: the original *Star Trek*, produced in 1966–1969; *The Animated Series* in 1973–1974; *The Next Generation*, 1987–1994; *Deep Space Nine*, 1993–1999; *Voyager*, 1995–2001; and *Enterprise*, 2001–2005. The Star Trek films were *Star Trek: The Motion Picture* (1979), *Star Trek II: The Wrath of Khan* (1983), *Star Trek III: The Search for Spock* (1984), *Star Trek IV: The Voyage Home* (1986), *Star Trek V: The Final Frontier* (1989), *Star Trek VI: The Undiscovered Country* (1991), *Star Trek: Generations* (1994), *Star Trek: First Contact* (1996), *Star Trek: Insurrection* (1998), and *Star Trek: Nemesis* (2002).
3. *Pravda*, October 5, 1957, as translated and quoted in *Exploring the Unknown, Selected Documents in the History of the U.S. Civil Space Program.* Vol. 1, *Organizing Exploration*. John M. Logsdon, Linda J. Lear, Jannelle Warren Findley, Ray A. Williamson, and Dwayne A. Day, eds. NASA History Series. 1995. Washington, DC: National Aeronautics and Space Administration.
4. Van Allen, James A., as quoted in Paul A. Hanle. 1982. The Beeping Ball That Started a Dash into Outer Space. *Smithsonian*, October: 148–162. See also Sullivan Walter. *Assault on the Unknown, the International Geophysical Year*. New York: McGraw-Hill Book Company, 1961.
5. Butrica, Andrew J. *Beyond the Ionosphere, Fifty Years of Satellite Communications*. NASA History Series, NASA SP-4217. Washington, DC: NASA History Office, 1997.
6. Launis, Roger D. *NASA: A History of the U.S. Civil Space Program*. Malabar, FL: Krieger Publishing Company, 1994 (see specifically pp. 34–35).
7. Kennedy, John F. Special Message to the Congress on Urgent National Needs, May 25, 1961. in *Public Papers of the Presidents of the United States: John F. Kennedy ... 1961*, 396–406. Washington, DC: United States Government Printing Office, 1962.
8. Apollo 204 Review Board. Final Report, April 1967. In *NASA Historical Reference Collection*. NASA History Division, Washington, DC. http://history.nasa.gov/Apollo204/content.html.
9. Nicks, Oran W. *Far Travelers, the Exploring Machines*. NASA SP-480. Washington, DC: NASA Scientific and Technical Information Branch, 1985. For specific case studies, see (1) Viking Lader Imaging Team. *The Martian Landscape*. NASA SP-425. Washington, DC: NASA Scientific and Technical Information Office, 1978. (2) Caidin, Martin, and Jay Barbree, with Susan Wright. *Destination Mars in Art, Myth, and Science*. New York: Penguin, 1997.
10. Shipman, Harry L. *Humans in Space, 21st Century Frontiers*. New York: Plenum Press, 1989 (see specifically p. 97).
11. Allen, Joseph P., with Russell Martin. *Entering Space: An Astronaut's Odyssey*, rev. ed. New York: Stewart, Tabori and Chang, 1985 (see specifically p. 75).
12. McCandless, Bruce, II. Quoted in Craig Covault. Astronauts Evaluate Maneuvering Backpack. *Aviation Week & Space Technology,* February 13, 1984 (see specifically p. 16).
13. Presidential Commission on the Space Shuttle Challenger Accident. *Report of the Presidential Commission on the Space Shuttle Challenger Accident.* 1986. NASA History Office. http://history.nasa.gov/rogersrep/51lcover.htm.

14. Columbia Accident Investigation Board. *Report Volume 1*. 2003. NASA. http://www.nasa.gov/columbia/home/CAIB_Vol1.html.
15. European Space Agency. http://www.esa.int.
16. Berger, Brian. NASA Strikes Deal for Soyuz Flights. Space.com (CNN.com). http://www.cnn.com/2006/TECH/space/01/06/nasa.soyuz.flights/index.html.
17. A good source on the Soyuz is the RussianSpaceWeb.com site, at http://www.russianspaceweb.com/soyuz_flight.html.
18. BBC News. Profile: China's First Spaceman. October 15, 2003. http://news.bbc.co.uk/2/hi/asia-pacific/3192844.stm.
19. Van Nimmen, Jane, Leonard C. Bruno, Robert L. Rosholt, Linda Neuman Ezell, Ihor Gawdiak, and Helen Fedor, eds. *NASA Historical Data Book*. NASA History Series, vol. I, NASA SP-4012. Washington, DC: NASA History Office, 1988–1994.
20. Logsdon, John M., Russell J. Acker, Dwayne A. Day, Jannelle Warren Finley, Stephen J. Garber, Roger D. Launius, Linda J. Lear, David H. Onkst, Ray A. Williamson, et al., eds. *Exploring the Unknown: Selected Documents in the History of the U.S. Civil Space Program*. NASA History Series, vols. I–VII, NASA SP-4407. Washington, DC: NASA History Office, 1995–2008.
21. Glennan, T. Keith. *The Birth of NASA: The Diary of T. Keith Glennan*. NASA History Series, NASA SP-4105. Washington, DC: NASA History Office, 1993.
22. Swanson, Glen E., ed. *"Before This Decade Is Out ...": Personal Reflections on the Apollo Program*. NASA History Series, NASA SP-4223. Washington, DC: NASA History Office, 1999.
23. Dethloff, Henry C. *Suddenly, Tomorrow Came ...: A History of the Johnson Space Center*. NASA History Series, NASA SP-4307. Houston, TX: NASA Lyndon B. Johnson Space Center, 1993. Dunar, Andrew Je, and Stephen P. Waring. *Power to Explore: A History of the Marshall Space Flight Center, 1960–1990*. NASA History Series, NASA SP-4313. Washington, DC: NASA History Office, 1999. Herring, Mack R. *Way Station to Space: A History of the John C. Stennis Space Center*. NASA History Series, NASA SP-4310. Washington, DC: NASA History Office, 1997. Hallion, Richard P. *On the Frontier: Flight Research at Dryden, 1946–1981*. NASA History Series, NASA SP-4303. Washington, DC: NASA Scientific and Technical Information Branch, 1984. Swenson, Loyd S., Jr., James M. Grimwood, and Charles C. Alexander. *This New Ocean: A History of Project Mercury*. NASA History Series, NASA SP-4201. Washington, DC: NASA History Office, 1998. Compton, William David. *Where No Man Has Gone Before, a History of Apollo Lunar Exploration Missions*. NASA History Series, NASA SP-4214. Washington, DC: NASA Scientific and Technical Information Division, 1989. Compton, W. David, and Charles D. Bensen. *Living and Working in Space: A History of Skylab*. NASA History Series, NASA SP-4208. Washington, DC: NASA Scientific and Technical Branch, 1983. Heppenheimer, T.A. *The Space Shuttle Decision, NASA's Search for a Reusable Space Vehicle*. NASA History Series, NASA SP-4221. Washington, DC: NASA History Office, 1999.
24. Hickam, Homer H., Jr. *Rocket Boys: A Memoir*. New York: Delacorte Press, 1998.
25. Shepard, Alan, and Deke Slayton, with Jay Barbree and Doward Benedict. *Moon Shot: The Inside Story of America's Race to the Moon*. Atlanta, GA: Turner Publishing Company, 1994.
26. Stuhlinger, Ernst, and Frederick I. Ordway III. *Wernher Von Braun, Crusader for Space, 2 vols., A Biographical Memoir* and *An Illustrated Memoir*. Malabar, FL: Krieger Publishing Company, 1994.
27. Engle, Eloise, and Arnold S. Lott. *Man in Flight, Biomedical Achievements in Aerospace*. Annapolis, MD: Leeward Publications, 1979.
28. Pitts, John A. *The Human Factor: Biomedicine in the Manned Space Program to 1980*. NASA History Series, NASA SP-4213. Washington, DC: NASA Scientific and Technical Information Branch, 1985.

29. Kozioski, Lillian D. *U.S. Space Gear: Outfitting the Astronaut*. Washington, DC: Smithsonian Institution Press, 1994.
30. Jenkins, Dennis R. *Space Shuttle: The History of Developing the National Space Transportation System; The First 100 Missions*. 3rd ed. Stillwater, MN: Voyageur Press, 2001.
31. Foster, Robert A., Robert T. Herres, Bernard S. Morgan, and Franics W. Penney. *National Security Management: National Aerospace Programs*. Washington, DC: Industrial College of the Armed Forces, 1972.
32. McDougall, Walter A. ... *The Heavens and the Earth: A Political History of the Space Age*. New York: Basic Books, 1985.
33. Baker, Philip. *The Story of Manned Space Stations: An Introduction*, New York: Springer, 2007.
34. Harford, James. *Korolev: How One Man Masterminded the Soviet Drive to Beat America to the Moon*. New York: John Wiley & Sons, 1997.
35. Kaufman, Marc. NASA's Star Is Fading, Its Chief Says. *Washington Post*. September 14, 2008, A3.

12 From *Vengeance 2* to *Sputnik 1*: The Beginnings

Robert Osiander

CONTENTS

INTRODUCTION

It is hard to say what to consider as the major ingredients of the pre-Space Age—the development of rockets, the planetary sciences, or the writings and fantasy of the science fiction authors of the time.

Rockets have been around for a long time. They were invented by the Chinese and used for wars, against the Mongols circa AD 1223,[1] as well as for entertainment, as recorded in AD 1264, in a celebration in honor of Emperor Lizong.[2] It was this perspective of creating a long-range weapon that has driven generations to develop rockets and rocket technology and ultimately led to *Sputnik 1*'s signal in 1957.

The work of Copernicus, Kepler, Galileo, and others in astronomy and planetary sciences changed the way we would look at the world, and their works moved the Moon closer, from the heavens dominated by the gods into the empty space around the Earth, just above the sky. With their works, humankind moved closer to the dream of exoatmospheric exploration.

Many dreams were initiated in the 1800s with a hoax article by Richard Adams Locke in the *New York Sun* about life on Mars.[3] Later in the century, an Italian astronomer, Giovanni Virginio Schiaparelli, discovered dark areas on Mars connected by lines he called *canali,* which translates to "channels" or "grooves," but it also invites the more enchanting translation of "canals." Percival Lowell decided that these could only be water-carrying canals and devoted his life to an observatory in Arizona and the possibility to life on Mars.[4,5] Among those of the next generation fascinated by Lowell's writings was Robert Hutchings Goddard. Goddard would later become known as the "father of American rocketry."

A number of French writers speculated about travel to the Moon and the planets in and after 1865, including Achille Eurad, Alexandre Dumas, Henri de Parville, and Jules Verne. Verne's *From the Earth to the Moon* was a well-researched science fiction story describing many technologies and observations. This work would inspire future rocket scientists, such as Goddard, Konstantin Tsiolkovsky, and Hermann Oberth. It is interesting that in his story, Jules Verne launched his rocket to the Moon from Florida, foretelling the later selection of the Kennedy Space Flight Center for the actual launch of the first (U.S.) rocket to the Moon for a moonwalk. At the Pan American Exposition in Buffalo, New York, in 1901, the Aerial Navigation Company offered a tour to the Moon in the spaceship *Luna*, an extravaganza organized by Frederick Thompson.[6]

ROCKET SCIENTISTS

The transition from science fiction to rocket science can easily be pinned to three names: Konstantin Tsiolkovsky, Robert Goddard, and Hermann Oberth. Driven by the dream created by Jules Verne and others, they used physics and mathematics to get there. In 1883, Konstantin Edvardovich Tsiolkovsky, a Russian, wrote *Free Space*, about the effects of zero gravity. In 1903, after hundreds of monologues and fiction stories, he published *Exploring Space with Reactive Devices,* filled with calculations and narratives about sending a spacecraft into space. This work included the definition of thrust, the design of liquid propellants made up with an oxidizer, multistage rockets, and escape velocities.

At the same time, Robert Hutchings Goddard, a professor of physics at Clark University in Worcester, Massachusetts, who as a teenager was thrilled by H.G. Wells' *The War of the Worlds*,[7] did research on ways to reach altitudes beyond the limit of balloons. Between 1914 and 1919, he received 70 patents for rockets and rocket apparatuses, including such fundamental patents on the design of the nozzle combustion chamber that allows the introduction of liquid fuel into the chamber and the design of a multistage rocket for high-altitude flight. He published his research in *A Method of Reaching Extreme Altitudes* in the Smithsonian Collection in 1919. It was a description of how to build a two-stage solid-propellant rocket and included a discussion of the feasibility of reaching the Moon with a rocket.[8] Despite the public ridicule following the publication of this research, Goddard launched the world's first liquid-propelled rocket on March 16, 1926, in Auburn, Massachusetts, from his outdoor laboratory, an open field on the Asa Ward farm. The rocket reached an altitude of 41 ft. and flew to the ground 184 ft. from the launching frame after 2.5 seconds. For the final launch at Auburn, on July 17, 1929, an 11-foot rocket carried an aneroid barometer, thermometer and camera which all operated successfully and were recovered. This launch received enough attention, especially from Charles A. Lindbergh, to obtain funding from the Guggenheim Foundation and the Smithsonian to move his laboratory to Roswell, New Mexico[9] (Figure 12.1).

FIGURE 12.1 Aviator Charles Lindbergh took this picture of Robert H. Goddard's rocket as he peered down the launching tower on September 23, 1935, in Roswell, New Mexico. (Courtesy of NASA.)

Currently, the Goddard Rocket Launching Site in Auburn is a National Historic Landmark and is indicated with markers located at the 9th Fairway of the Pakachoag Golf Course.

On December 30, 1930, one of Goddard's rockets reached an altitude of 2,000 ft. and a speed of 500 mi/h, and in 1934, he launched the first rocket equipped with a gyroscope to a height of 4,800 ft., a horizontal distance of 13,000 ft.[10] Unfortunately, Goddard's research was not fully recognized, and his work not seriously studied by American scientists until years later. Some of Goddard's rockets—a replica of the 1926 rocket and an original from 1941—can be seen in the Air and Space Museum in Washington, D.C. The Goddard Laboratories in Roswell have been recreated in a wing of the Roswell Museum and Art Center, and the Goddard House in Roswell is on both the State Register of Cultural Properties and the National Register of Historic Places.

In 1908 Germany, a 14-year-old schoolboy, Hermann Julius Oberth, fascinated by the books of Jules Verne, constructed a model rocket and conceived the first multistage model rocket. After World War I, he went back to study physics in Munich, Germany. In 1922. his PhD thesis *Die Rakete zu den Planetenräume* (*By Rocket into Planetary Space*) was rejected, and he published the visionary work privately. This book, and the expanded 429-page version entitled *Wege zur Raumschiffahrt* (*Ways to Spaceflight*), published in 1929, are probably the most influential books on the future of space flight.

PRE–WORLD WAR II ROCKETRY

Following the writings of Goddard, Oberth, and others, a number of rocket research organizations were created. In part, the members were enthusiasts of space flight, and with some government support had a goal to develop guided missiles. In Germany, the *Verein für Raumschiffahrt* (VfR) (the Spaceflight Society), an amateur rocket group, was founded 1927 by Johannes Winkler. Its member list included the who's who in rocket engineering in Germany: Max Valier, Willy Ley, and Walter Neubert. They were later joined by Klaus Riedel, Rudolf Nebel, Wernher von Braun, Hermann Oberth, Walter Hohmann, Kurt Heinisch, Eugen Sänger, Rolf Engel, and up to five hundred other members who produced a periodical called *Die Rakete* (*The Rocket*). Hohmann's book, *Die Erreichbarkeit der Himmelskorper* (*The Attainability of Celestial Bodies*) was published in 1925 and was so technically advanced that it was consulted years later by the National Aeronautics and Space

FIGURE 12.2 Rocket ship from *Frau im Mond*, directed by Fritz Lang. (From Mark Wade, Astronautix.)

Administration (NASA). The VfR promoted space travel with public relations, motivating actions such as the (failed) demonstration of a rocket for Fritz Lang's science fiction movie *Frau im Mond* (Figure 12.2). Valier, in collaboration with the German automobile manufacturer Fritz von Opel, tested the first rocket-powered automobile in 1928 in addition to gliders, train cars, and snow sleds. They also conducted rocket experiments, and opened the first *Raketenflugplatz* (rocket airfield) in Berlin-Tegel (Figure 12.3).[11] Mirak, a "Minimum Rocket" with liquid propellant, was based on Oberth's *Kegelduese* (conical rocket motor) and conceived and realized by Rudolf Nebel and Klaus Riedel. This rocket flew more than a hundred times in 1931–1932 at the *Raketenflugplatz* and convinced the German Army of the practicality of the rocket as a weapon of war. With the increasing influence of Nazism in German society, German rocket development was taken over by the Army and the VfR was dissolved.

The American Rocket Society was founded in 1930 as the *American Interplanetary Society* (AIS) by G. Edward Pendray, David Lasser, Laurence Manning, and others. Parallel to Goddard's efforts, they performed work in the testing and design requirements of liquid-fueled rockets and successfully launched multiple rockets up to 382 ft. in altitude and a distance of 1,338 ft. Their work was discontinued in World

RAKETENFLUG

Aufruf!

Seit Jahrzehnten arbeitet die deutsche Wissenschaft und Technik an dem Raketenproblem. Endlich sind wir so weit, daß greifbare Erfolge vorhanden sind. Zur Weiterführung und zum Ausbau der Errungenschaften fehlt uns, die wir uns mit den kleinsten Mitteln bisher geholfen haben, das Geld. Das Ausland hat, in dem Bestreben uns unsere bisherigen Erfolge zu entreißen, ungeheure Anstrengungen gemacht. Dies zu

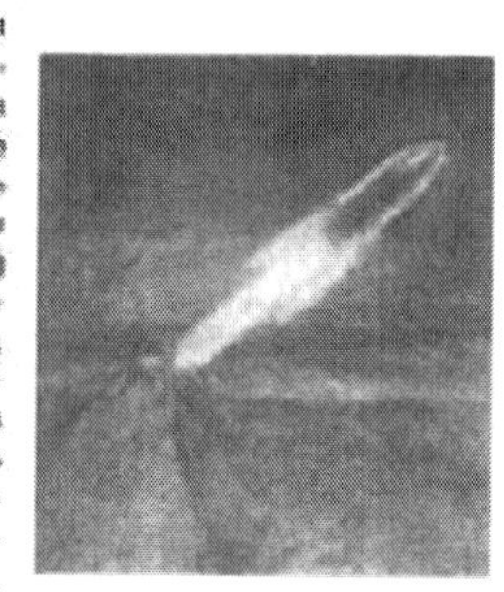

verhindern, muß jedem Deutschen am Herzen liegen. Möge jeder nach seinen Verhältnissen hierzu einen Beitrag geben, damit uns die Früchte jahrzehntelanger mühevoller Arbeit nicht entgehen. Deutschland wird durch die Lösung des Raketenproblems

Raketenflugplatz Berlin

Verein für Raumschiffahrt E. V.

Leiter: Dipl.-Ing. Rudolf Nebel, Reinickendorf-West, Tegeler Weg

FIGURE 12.3 Flyer "Rocketflight" calling for the help of all Germans to support the "Rocket-Port" in Berlin. (From Mark Wade, Astronautix.)

War II, and the AIS was absorbed into the American Institute of Aeronautics and Astronautics in 1963.[12]

The British Interplanetary Society was founded in 1933 by P.E. Cleator. It is the oldest still-existing organization in the world that exclusively promotes astronautics and space exploration. It did not perform any experiments, since the Explosives Act of 1875 prevented any private testing of liquid-fuel rockets in the United Kingdom. In the late 1930s, the group devised a project of landing people on the Moon by a multistage rocket.

In Russia, rockets had been considered as a way to defend the revolution, and the Communist Party established a Laboratory for Development of the Invention of Engineer N.I. Tikhomirov (rocket-propelled weapons) in Moscow. In 1928, this facility became the Gas Dynamics Laboratory (GDL), a laboratory for research in rocketry led by Nikolai Tikhomirov.[13] In 1929, Valentin Glushko, one of the pioneers of the Soviet rocket program, joined the GDL to investigate electrical and liquid-fueled rocket engines. Between 1930 and 1934, the GDL tested a large number of liquid-fueled rockets. At the same time (1931), Sergei Korolev (also spelled Korolyov), a Soviet space pioneer, and Fridrikh A. Tsander, in support of the Society for the Advancement of Defense, Aviation and Chemical Technology, founded the Group for Investigation of Reactive Movement (GIRD). First in Leningrad and Moscow and later other cities, GIRD then became Jet Propulsion Research Institute (RNII). In 1933, the GDL became a Leningrad branch of RNII, and in 1934, Glushko and the personnel of the GDL relocated to Moscow, to continue research on the liquid-propellant engines. In 1936, the RNII was renamed NII-3.[14] In the same year, the work on ballistic and cruise missiles, which employ liquid-fuel propulsion, was consolidated in a single department led by Korolev. One of the Russian rocket designs emerging from these tests, GIRD-X, which weighed 65 pounds, was 8.5 ft. long and 6 in. wide and reached a maximum altitude of 3 miles during a test on November 25, 1933 (Figure 12.4). Another of the Russian rockets, *Aviavnito*, weighed 213 pounds, was 10 ft. long and 1 ft. wide and reached an altitude of 3.5 miles in 1936.[15]

By 1937, Stalin's "cleansing of the Bolshevik elite" brought long prison terms to Glushko and Korolev. Andrei Kostikov became the director of what was now NII-3, tasked to finalize the development of unguided short-range missiles, which became known during the World War II as Katyusha rockets.

WORLD WAR II

Among the scientific achievements of World War II, the German development of the V-2 rocket probably is on par with the American nuclear bomb both in importance and in terms of investment by their respective countries. The efforts of the German army in the missile program overshadowed any other such efforts in Russia, Britain, Japan, or the United States, although every nation had a major rocket program, in most cases focusing on short-range missiles.

Just before the German advance in October 1941, the NII-3 was evacuated to the Ural region.[16] The former director, Korolev, was sentenced to 8 years in the Gulag, and he was assigned to a *Sharaska*, which was effectively a slave-labor camp for

FIGURE 12.4 GIRD-X rocket. This 65-pound rocket reached a maximum altitude of 3 miles during a test on November 25, 1933. (From Mark Wade, Astronautix.)

scientists and engineers to work on projects assigned by the communist party leadership. Under Glushko, he worked on rocket plane boosters and various rocket designs until 1944.[17] When information about a German secret weapon—the A-4 ballistic missile—reached Moscow in July 1944, the investigation of German technology became a primary task of the NII-3.

The United States had a number of military rocket programs. Supported by the Navy at Annapolis, Maryland, Midshipman Robert C. Truax began experimenting with liquid-fueled rocket engines in 1936 and developed several small experimental rocket engines that burned a combination of compressed air and gasoline. In 1945, the U.S. Navy had 12,000 factories working on a rocket production. The most successful and famous of these is probably the bazooka missile launcher, developed after 1940, which fired a tank-piercing 3.5-pound missile, 21 in. long, with destructive power up to 750 m.

By 1940, Britain had developed a similar 3-in. rocket, which was used successfully against German airplanes. By 1942, these missiles were developed into air-to-surface missiles nearly 6 ft. in length and capable of traveling nearly 1,000 mi/h.

They were used operationally in the Royal Navy, primarily against submarines. Toward the end of the war, the Stooge, a radio-guided missile with a range of 8–9 miles and a top speed of 500 mi/h, was developed to carry a 220-pound warhead against airplanes.[18]

In Germany, the VfR had been forced to disband in the winter of 1933–1934, and private rocket experiments ceased at the *Raketenflugplatz* facility in January 1934. Wernher von Braun went to work officially for the German Army at Versuchsstelle Kummersdorf-West, a static testing site for ballistic missile weapons. In his memoirs, von Braun states "Our feelings towards the Army resembled those of early aviation pioneers, who, in most countries, tried to milk the military purse for the own ends and who felt little scruples as to the possible future use of their brainchild."[19,20] Under the direction of Captain Walter Dornberger, the Kummersdorf team, with Wernher von Braun, Major von Richthofen, and Ernst Heinkel, designed and built the A-1 (Aggregate-1) rocket. Powered by a combination of liquid oxygen and alcohol and with a gyroscope to provide stability during flight, it could develop a thrust of about 660 pounds. Its successor, the A-2, employed separate alcohol and liquid oxygen tanks, and in December 1934, two A-2 rockets, nicknamed Max and Moritz (after the twins in the German version of the *Katzenjammer Kids* cartoon strip), were launched from the North Sea island of Borkum and reached an altitude of about 1,700 m (6,500 ft).[21]

In April 1937, all of the German rocket testing was relocated to a top-secret base, the *Heeresversuchsstelle Peenemünde* (Army Experimental Station, Peenemünde) on the Baltic Coast. The German Air Ministry, the army, and the German government spent more than 70 million dollars on the construction of Peenemünde, a dream facility for the fabrication of long-range missiles. Von Braun assembled a team of first-rate engineers, designers, and administrators, including Walter Thiel, Rudolf Herrmann, Herman Steuding, and others. As their first task, they developed the A-3 rocket, a 1,650-pound, 21-foot-long rocket, which burned a combination of liquid oxygen and alcohol. Its propulsion system worked very well, and great progress was made on the guidance and control systems, but it became what von Braun would later call "a successful failure."[22]

By 1938, Germany had begun invading huge portions of Eastern Europe, and Adolf Hitler began recognizing the need for an effective ballistic missile weapon. The German Ordnance Department requested that the Peenemünde team develop a ballistic weapon with a range of 150–200 miles that could carry a 1-ton explosive warhead and could be transported on existing railways, which required size compatibility with tunnels and bends. These criteria led directly to the development of the A-4 rocket (Figure 12.5). The Peenemünde team had, often in secret collaboration with academia and industry, made a number of breakthroughs in key technologies, especially the motor and the guidance system, and on October 3, 1942, an A-4 rocket lifted off and disappeared into the sky over the Baltic Sea to come down 190 km (120 miles) away as the first man-made object to touch the edge of space at an altitude of about 80 km (50 miles).[23]

Photographs taken by a reconnaissance mission over Usedom, Germany, in May 1943 (Figure 12.6), as well as prisoner of war interrogations and intelligence information, had begun to expose the German secret weapon. In August 17, 1943,

FIGURE 12.5 Period views of one of the A-4/V-2 launch sites at Peenemünde. (Courtesy of NASA.)

the Allies ran operation Hydra, a three-hundred-plane raid against Peenemünde, destroying many of the facilities and killing an estimated 735 people. Within a few days, Heinrich Himmler orchestrated the construction of an underground factory to built V-2s (Vengeance 2), as the A-4 was called now, with slave labor. The factory, called Mittelwerk, was in a lonely valley near Nordhausen, Germany, in the Harz Mountains. Inmates from the concentration camp Buchenwald, mostly Russian, Polish, French, and Jewish, were moved to the Mittelwerk facility and a nearby camp called Dora, both of which became infamous for their terrible living and working conditions. More than 30,000 prisoners had died building the V-2 by the end of the war, and 5,400 people were killed by the more than 3,000 V-2s launched against Antwerp, London, Liege, and Paris.[24,25] The last V-2 was launched from Peenemünde on February 14, 1945,[26] and the Mittelwerk camp was liberated in April 11, 1945, after the German SS had moved many of the prisoners.[27]

As is so often the case in history, the palaces and monuments we admire, such as the pyramids in Egypt or South America, are based on the exploitation of prisoners, slaves, and peasants. It is our duty to give those workers credit and to honor them by preserving the places of their suffering in our minds as much as we honor the engineers and their achievements.

FIGURE 12.6 1943 Royal Air Force reconnaissance photograph of V-2 rockets at Peenemünde Test Stand VII. (Courtesy of NASA.)

DIVIDING UP THE GERMAN ROCKETS

When in 1945 the Soviet Army advanced through the Baltic states, von Braun and his colleagues met secretly and decided to move to a region expected to see U.S. invasion. Starting in February, most of the personnel were moved to Nordhausen, and the facilities in Peenemünde were blown up to make them useless for the Soviet Army. Some of the buildings have now been restored, and Peenemünde can be visited. When Hitler ordered all research materials to be destroyed, the rocketeers ignored the order and instead moved 14 tons of documents to a mine shaft in Dornten, Germany, 50 miles northwest of Nordhausen, and sealed it off. Five hundred members of the rocket team were moved to Oberammergau in Bavaria; Nordhausen was entered on April 11, 1945, and the U.S. Army recovered parts for about a hundred complete V-2 rockets. Project HERMES moved these parts, as well as the documents from the mine, to the White Sands Proving Ground in the United States just before the area was to be turned over to the Red Army.[28,29] On May 2, five days before the collapse of the third Reich, von Braun and the others at Oberammergau surrendered themselves to the U.S. Army (Figure 12.7) and were moved to Garmisch for interrogation. Most scientists wanted to go to the United States to continue their research on rockets, von Braun was asked to write a report on the future of rocketry, and he used this opportunity to document his vision of rockets serving military as well as scientific and civilian use. In July 1945, Operation Overcast was put into effect to safeguard the most valuable members of Peenemünde's rocket engineers. Around 350 of these men, with their families, were brought to the United States. In September, the first seven rocket specialists, including von Braun, entered the United States as "wards of

FIGURE 12.7 U.S. soldiers, Walter Dornberger, Herbert Axter, Wernher von Braun, Hans Lindenberg, and Bernhard Tessmann (partially cropped) after the scientists surrendered to the Allies in 1945. (Courtesy of the National Archives.)

the Army," to circumvent immigration requirements.[30] In March 1946, the operation was renamed Paperclip, and it assembled about 1,600 scientists, engineers, researchers, and managers from the German rocket factory into the United States.[31] Many of their personnel files were laundered to eliminate the connections with any of the SS or Nazi organizations. The political and technical gains overruled the moral dilemma, and only later would some come forward to expose this hypocrisy: for example, Alden Whitmen in his 1971 *Obituary Book*,[32] and songwriter Tom Lehrer who in the 1960s penned these lyrics:[33]

> Don't say that he's hypocritical,
> Say rather that he's apolitical.
> "Once the rockets are up, who cares where they come down?
> That's not my department," says Wernher von Braun.
> Some have harsh words for this man of renown,
> But some think our attitude
> Should be one of gratitude,
> Like the widows and cripples in old London town,
> Who owe their large pensions to Wernher von Braun.

The second White Russian Army had taken Peenemünde on May 5, 1945. By then, Sergei Korolev had been commissioned into the Red Army with a rank of colonel and had been flown to Germany along with other experts to recover V-2 rocket technology. Of course, they had to take whatever was left from the U.S. raids, and most attempts to lure some of the rocket scientists to their side failed. Russia got about 150 scientists and engineers. Most of them, with the exception of Helmut Groettrup, were involved only in production and had not been in von Braun's close circle. They worked with the Russians in Germany, and later, after 1946, in the USSR. A new institute was created for this purpose, the NII-88, in the suburbs of Moscow, and Branch 1 of the NII-88 was set up on Gorodomlya Island for the German engineers. Korolev became the chief designer of long-range missiles and began to reproduce V-2 rockets, designated R-1, from the blueprints of the disassembled V-2s. A total of eleven rockets were launched, but because the members of the German team were cut out of the Russian technology, their morale was affected, and in 1950 the Ministry of Defense stopped any work related to long-range rockets by the German team and repatriated the German engineers and their families.

For the British, all that was left of the Nazi rocket program after the United States and Russia had taken their share were eight V-2 rockets. In a project called Backfire, these were launched under the eyes of Glushko, Theodor von Karman, William H. Pickering (California Institute of Technology), and other military and scientific observers. Walter Dornberger, one of the team leaders, was taken to England to stand trial for the German attacks on civilians and was released for lack of a case—killing civilians in England was equal to the English killing civilians in Germany. He later joined the other German scientists in the United States.

FROM WORLD WAR II TO *SPUTNIK*

In the United States, von Braun and his coworkers assembled the V-2 rockets and launched approximately sixty-seven of them between 1946 and September 1952, when the program ended. The Navy was building a guided missile itself in its Rocket Sonde Research Section for the investigation of the physical phenomena in and the properties of the upper atmosphere. It had the Army's agreement that the empty warheads on the V-2s could carry scientific experiments. A panel for V-2 upper atmosphere research was formed at Princeton University in February 1946, with representatives from the Army Air Forces, the Naval Research Laboratory (NRL), Harvard University, Princeton University, the National Bureau of Standards, and the Johns Hopkins University Applied Physics Laboratory (JHUAPL). One of the fifty-eight scientists, James van Allen from JHUAPL, was interested in cosmic rays, as well as the structure and dynamics of the upper atmosphere. He also suggested taking pictures of Earth from high altitude. Many of the terms and technologies used today were first introduced there, such as a radio connection from space to Earth (*downlink*) or from Earth to space (*uplink*) and the instrument readings, which were referred to as *telemetry*.

One of the V-2s launched in the United States exceeded an altitude of 100 miles, sending ultraviolet data on the Sun. Some of the V-2s carried mice, and four of them

FIGURE 12.8 Launch of Bumper, July 24, 1950. (Courtesy of NASA.)

carried monkeys, anticipating manned space flight. The Bumper-WAC (Figure 12.8) combined the V-2 with a WAC Corporal missile, which was designed and built in a cooperative effort between Douglas Aircraft and the Guggenheim Aeronautical Laboratory and in association with the Jet Propulsion Laboratory at California Institute of Technology. This two-stage rocket reached a record height of 244 miles in February 1949.[34]

The U.S. Army's major motivation was to develop long-range missiles for long-range artillery. When the Navy suggested a joint satellite committee in 1946, the Army not only rejected the idea but tried to claim a primary responsibility for any military satellite vehicle. The Army contracted with the Douglas Aircraft Company at El Segundo, California, for a feasibility study called project RAND, launching objects into orbits. A result of this study was a proposal for building reconnaissance satellites and missions. In the meantime, the Air Force had been formally generated in 1947, and it also considered missiles and satellites an extension of its charter. In 1948 the RAND Corporation was spun off from Douglas as an independent non-profit think tank contracted to the Air Force as a representative of all military forces satellite requirements. This helped to ensure the Air Force's dominance in space, as well as a 1949 Department of Defense board ruling that space was to be the Air Force's exclusive domain.[35]

Other companies started to build their own rockets, in part contracted by the Navy and the Air Force. Bell Aerospace built a rocket plane that flew Chuck Yeager,

FIGURE 12.9 A Viking rocket on its launch pad, ca. 1949. (Courtesy of NASA.)

who was the first human to exceed the speed of sound, in 1947, in an X-1 experimental plane designed to reach to the edge of space. Von Karman's company, Aerojet General, developed the Aerobee in 1946, a sounding rocket that could reach around 230 km. It provided constant telemetry and was recovered by a parachute. Between 1947 and 1985, a total of 1,037 Aerobees were launched. Each military branch had its own missile developments: Nike-Ajax, Nike-Hercules, and Nike-Zeus as well as the Jet Propulsion Laboratory (JPL) missiles—Private, Corporal, and Sergeant for the Army and Hound Dog, Mace, Matador, Navaho, and Skybolt for the Air Force. The Navy decided to build its own American rocket, and NRL contracted with the Glenn L. Martin Company to initially design and built ten and later another four high-altitude research rockets called Viking (Figure 12.9); the first were launched May 3, 1949. The major contributions of these rockets were their high-altitude photographs, such as the *Viking 11* images taken from 158 miles above the Earth.[36]

Meanwhile, in 1950, von Braun and his German engineers had moved to the Redstone Arsenal in Huntsville, Alabama, designing an improved V-2 called Redstone, first launched in 1953, as the Army's workhorse. This was a medium-range ballistic missile, used for the first live missile nuclear tests and developed into Jupiter, which would later carry the first U.S. satellites and astronauts into space.

In Russia, the group under Korolev at NII-88 began working on more advanced designs in 1947. The R-2 doubled the range of the V-2, and the later model R-3

FIGURE 12.10 A Russian R-5 rocket. (From Mark Wade, Astronautix. With permission.)

achieved a range of 3,000 km. The project was canceled in 1952 since Glushko could not get the engines to develop the required thrust. Work began on the R-5, which had a 1,200-km range and completed a successful first flight by 1953 (Figure 12.10). The first true intercontinental ballistic missile was the 7 *Semyorka* (code-named SS-6 Sapwood by the North Atlantic Treaty Organization).[37] This two-stage rocket had a payload of 5.4 tons, sufficient to carry a Soviet nuclear bomb a distance of 7,000 km. The R-7 successfully launched on August, 1957, sending a dummy payload to the Kamchatka Peninsula.

With the Cold War gaining speed, the U.S. administration was convinced by its Security Council to build small scientific satellites in order to develop reconnaissance satellites.[34,38] On July 29, 1955, the White House announced that the United

States intended to launch into Earth orbit a scientific satellite during the International Geophysical Year (IGY). Such a project had been suggested in 1950 by James van Allen (JHUAPL) and his colleagues, under them L.V. Berkner (Navy Bureau of Aeronautics Engineering Division), and S. Fred Singer (University of Maryland) as a part of the third International Polar Year to study Earth's oceans, atmospheres, and outer space.[39]

As an entry for the IGY, the three armed forces agreed on Orbiter, using the Army's Redstone booster, a Navy satellite, and the Air force tracking facilities. In a surprise move, the Navy Research Laboratory suggested its own entry, an advanced Viking, as the launcher for what it called the Vanguard mission. There are different accounts why U.S. President Dwight D. Eisenhower opted for the Navy's Vanguard mission. Using civilian sounding rockets was deemed more appropriate for a scientific mission and would leave space above national borders. With the United States having a covert satellite program (WS117) to launch spy satellites on the U.S. Air Force Thor intermediate range ballistic missiles, there was a concern that the USSR would object to satellites overflying the Soviet Union. Another speculation was a lingering hostility toward the "Nazi" Germans.[40]

The Russian Academy of Sciences (without the approval of Soviet President Nikita Khrushchev) committed on January 30, 1956, to launch a satellite, later named *Sputnik*, within 2 years as part of the IGY program.[41,42] Korolev was still intrigued by the use of rockets for space travel, which met resistance in the military and among party members. In 1953, he first proposed the use of the R-7 design for launching a satellite into orbit. In 1957, during the International Geophysical Year, Korolev was finally able to win Khrushchev's support because of a successful launch of the R-7 in competition with the United States, suggesting that the USSR should try to be the first country to launch a satellite.

W. Burrows, author of *This New Ocean*, found in interviews with Sergei Kapitsa that the actual reason for launching *Sputnik* early was a typographical error. In a routine intelligence report about a U.S. conference in October that was to discuss launching a satellite, the "conference" was left off. Expecting the United States to launch in October, the actual development of *Sputnik* was performed in less than a month. *Sputnik* was a very simple design, a polished metal sphere with a transmitter, thermal measuring instruments, and batteries (Figure 12.11). Korolev personally managed the assembly. On October 4, 1957, the USSR launched a R-7 rocket that put *Sputnik 1* into orbit and woke up the United States. All the U.S. scientists at the Rockets and Satellites Conference in Washington, D.C.—at a reception in the USSR Embassy—could do was congratulate their Soviet colleagues.

Only a month later—November 3, 1957—*Sputnik 2* was launched. This satellite was filled with scientific instrumentation, such as solar radiation sensors, and a passenger: a terrier named Laika, whose life signs were transmitted back to Earth for an entire week, until an injection of poison killed her.[43]

In the public, *Sputnik* created both public excitement about space flight and the opinion that the United States lagged behind the USSR in space technology. As a result, the administration pushed scientific education in the National Defense Education Act on September 2, 1958, and accelerated the race to space with the launch of Vanguard and TV3 (Test Vehicle 3). The explosion of *Vanguard 3*, or

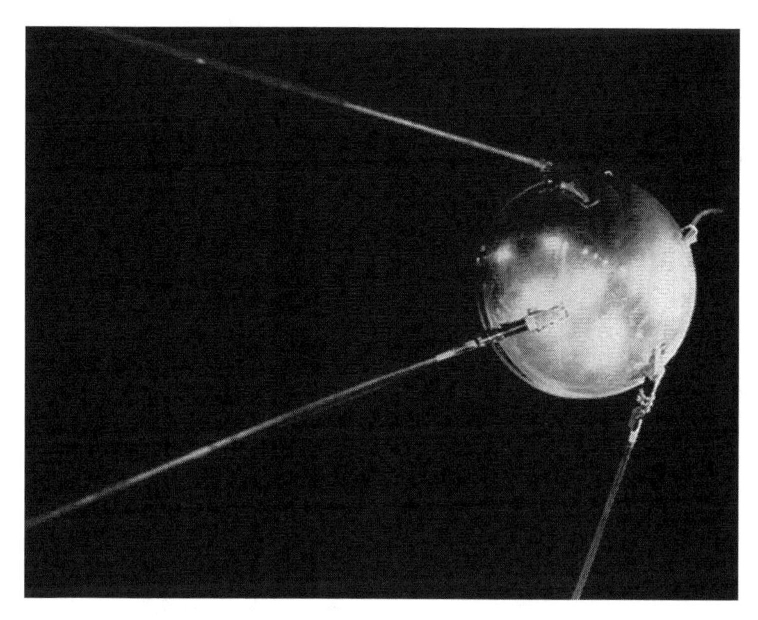

FIGURE 12.11 A replica of *Sputnik 1*, the first artificial satellite in the world to be put into outer space. (Courtesy of NASA-JPL.)

"Kaputnik" as it was called in the press, on December 6, 1957, at Cape Canaveral, Florida, was a severe blow to the image of the U.S. space program.

The Army Ballistic Missile Agency under von Braun proposed the Jupiter C intermediate-range ballistic missile, based on a Redstone booster, as an alternate launch vehicle. On January 31, 1958, *Explorer 1*, a JPL-developed satellite carrying cosmic ray and micrometeorite detectors, was launched successfully and steered JPL's future away from missiles into exploration (Figure 12.12).[44] For the next 20 years, the Explorer program would become one of the longest and most successful space programs. The Navy's satellite, TV-4, later called *Vanguard 1*, was launched successfully into orbit on March 17, 1958, and is presently the oldest remaining human-made object in space.

In their proposal to President Eisenhower on July 29, 1958, "A National Mission to Explore Outer Space,"[38] the members of the U.S. (IGY) satellite committee proposed a National Space Establishment. After brief competition between the different armed forces,[45] President Eisenhower signed the National Aeronautics and Space Act and established NASA, formed from employees and laboratories of the National Advisory Committee for Aeronautics, to conduct all non-military activity in space.

CONCLUSION

While we enjoy the discoveries of space exploration and achievements such as satellite television, global positioning systems, and so forth, it is an unfortunate reflection on our history that the infrastructure and the early advances were driven by militaristic exploitation and wartime conditions. Most of the dreamers of spaceflight could

FIGURE 12.12 Technicians lower *Explorer 1*, the first American satellite, onto the launch vehicle's fourth-stage motor (NASA-JPL).

achieve some of their goals only by developing rockets and missiles for the military. One of the biggest advances to space, the development of the A-4, the basis for most missiles and launch vehicles, was developed for a fascist regime and was paid for in part with the lives of prison workers in German concentration camps. It was renamed a "vengeance weapon" (thus the moniker V-2). One cannot separate the continuing arms race from humankind's desire to exploit exoatmospheric space. With the generation of NASA for civilian space exploration, a new era had begun, and for the years to come the race to space was paralleled by the efforts to deliver long-range missiles for military use.

REFERENCES

1. Hanson, C. The Mongol Siege of Xiangyang and Fan-ch'eng and the Song Military. http://www.deremilitari.org/resources/articles/hanson.htm.
2. Crosby, A.W. *Throwing Fire: Projectile Technology Through History*, 100–103. Cambridge, United Kingdom: Cambridge University Press, 2002.
3. Locke, R.A., and J.N. Nicollet. *The Moon Hoax; Or, A Discovery that the Moon Has a Vast Population of Human*. New York: W. Gowans, 1859.
4. Burrows, W.E. *This New Ocean*. New York: Random House, 1998.
5. Lowell, P. *Mars*. 1895. http://www.bibliomania.com/2/1/69/116/frameset.html.
6. Burrows 1998 (note 4).
7. Wells, H. G. *The War of The Worlds*. London: William Heinemann, 1898.
8. Goddard, R.H.: *A Method of Reaching Extreme Altitudes*. Smithsonian Miscellaneous Collection, vol 71 (2). Washington, DC: Smithsonian Institution, 1919.
9. Goddard, E.C., and G.E. Pendray, eds. *The Papers of Robert H. Goddard*. New York: McGraw-Hill, 1970.

10. National Park Service. The Goddard Rocket Launching Site. http://www.nps.gov/history/nr/travel/aviation/god.htm.
11. http://www.astronautix.com/sites/rakplatz.htm.
12. Crouch, T.D. *Rocketeers and Gentlemen Engineers: A History of the American Institute of Aeronautics and Astronautics...and What Came Before*. Library of Flight Series, ed. Washington, DC: Aldrin B. National Air and Space Museum, AIAA, 2006.
13. http://www.russianspaceweb.com/rockets.html.
14. http://www.russianspaceweb.com/rnii.html.
15. http://www.spaceline.org/history/3.html.
16. http://www.russianspaceweb.com/rnii.html.
17. Ibid.
18. http://www.spaceline.org/history/3.html.
19. von Braun, W. *Behind the Scenes*. W. Von Braun Papers at the Space and Rocket Center. Huntsville (see specifically pp. 9–10).
20. Neufeld, M.J. *The Rocket and the Reich*. New York: The Free Press, 1995 (see specifically p. 22).
21. Ibid. (see specifically p. 38).
22. Ibid. (see specifically p. 71).
23. Ibid. (see specifically p. 164).
24. Sellier, A. *A History of the Dora Camp*. Chicago: Dee, I. van R., in Association with the US Holocaust Memorial Museum, 2003.
25. http://www.astronautix.com/sites/peeuende.htm.
26. Ibid.
27. http://en.wikipedia.org/wiki/Mittelbau-Dora.
28. Burrows 1998 (note 4) (see specifically p. 112).
29. von Braun, W., and F. Ordway. *Space Travel: A History*. New York: Harper & Row, 1967.
30. Ordway, F., and M.R. Sharpe. *The Rocket Team*. Apogee Books Space Series 36. Wheaton, IL: CG Publishing, 1979.
31. Burrows 1998 (note 28).
32. Whitman, A. *The Obituary Book*. New York: Stein and Day, 1971.
33. http://www.guntheranderson.com/v/data/wernherv.htm.
34. White Sands—Bumper WAC Missile Launches. http://www.wsmr.army.mil/pao/FactSheets/bump.htm.
35. Bulkeley, R. *The Sputnik Crisis and Early United States Space Policy*. Bloomington, IN: Indiana University Press, 1991.
36. Baumann, R.C., and L. Winkler. *Rocket Research Report No. XVIII: Photography from the Viking 11 Rocket at Altitudes Ranging up to 158 Miles* (see specifically Picture 127). Washington, DC: Naval Research Laboratory, 1955.
37. http://en.wikipedia.org/wiki/R-7_Semyorka.
38. National Security Council. *NSC 5522: US Satellite Program*. Washington, DC: Author, 1955.
39. Burrows 1998 (note 4).
40. Ibid. (see specifically p. 171).
41. Tikhonravov, M. Report on an Artificial Satellite of the Earth (in Russian). In *Materialy po istorii kosmicheskogo korabl 'vostok'*, ed. B. V. Raushenbakh, 5–15. Moscow: Nauka, 1991.
42. Shternfeld, A. *Soviet Space Science*. New York: Basic Books, 1959.
43. Ibid.
44. Burrows, W.E. *Exploring Space*. New York: Random House, 1986.
45. Newell, H. *Beyond the Athmospheres*. NASA SP-4211. Washington, DC: NASA, 1980.

13 Space Race and the Cold War

Richard Sturdevant and Greg Orndorff

CONTENTS

Post Cold War Skills.

INTRODUCTION

From the late 1940s to the Soviet Union's collapse in 1991, that country and the United States engaged in a Cold War, politically and militarily. In terms of both national prestige and military security, space became a competitive arena. This chapter outlines, in a categorical and technological sense, the predominant national-security aspects of that competition and their evolution over the course of the Cold War.

Those with archaeological interest should note that creation occurs multiple times for space systems—first in the mind of the creator, second on paper, and finally instantiated in hardware. Products of the creation process provide the archaeologist with a multitude of artifacts associated with the end item, including contracts, program documentation (unclassified and classified), briefings, drawings, specifications, pictures, and renderings, to name a few. In addition to the spaceflight hardware itself, there may be duplicates in the form of test articles, engineering models, and even, possibly, a flight spare that was never used. The intent behind instantiating a new spacecraft or rocket can be discerned by discovering those documents used at the highest political levels in government. Fortunately for the researcher, governments like to document policies. One such resource for the archaeologist focused on U.S. space systems is National Security Council (NSC) documents. For example, in NSC 5814, dated June 20, 1958, President Dwight D. Eisenhower codified a preliminary statement of U.S. policy on outer space that acknowledged that space-based systems would be put into operation to contribute directly to national security. NSC documents tend be classified (and subsequently downgraded) when setting direction with regard to establishing space systems for national security purposes. When declassified, documents from the Central Intelligence Agency (CIA) would also contribute to understanding and preservation strategies. Artifacts related to the Space Race generally fall into one of three categories: launch vehicles, satellites, and terrestrial support elements.

LAUNCH VEHICLES

Given the demonstrated capabilities of the German V-2 ballistic missile near the end of World War II, the United States and the Soviet Union both initiated programs in the immediate postwar years to develop long-range ballistic missiles. The U.S. Army Air Forces contracted with Convair Corporation in 1946 for the MX-774, a prototype for what became the Atlas intercontinental ballistic missile (ICBM). At the same time, Wernher von Braun's team of German rocket engineers, having been brought to the United States under Operation Paperclip, worked at Fort Bliss, Texas and White Sands Proving Ground, New Mexico to improve on the V-2 design. By the mid-1950s, the technologies associated with production of thermonuclear weapons made it possible to produce smaller, lighter warheads, which prompted acceleration of missile-development efforts by von Braun's Army team, which had been directed to move to Redstone Arsenal, Alabama, under the command of General John Medaris in 1949 and by General Bernard Schriever's Air Force team at Western Development Division (WDD), which had been established in July 1954 at

Los Angeles, California. From those roots came nearly all Cold War–era American expendable, space-launch boosters.

The Army, relying largely on its traditional arsenal approach for missile development, produced the Redstone rocket, a direct descendant of the V-2. Contracts with Chrysler Corporation for the airframe, North American's Rocketdyne Division for the engine (a modified version of one developed for the Air Force's Navaho), and Ford Instrument Company for the inertial guidance system aided the Army's Redstone development program. The first Redstone launch took place from Cape Canaveral, Florida, on August 2, 1953. Its offspring, the Jupiter intermediate-range ballistic missile (IRBM) first launched successfully on May 31, 1957. Technical evolution of the Jupiter IRBM led to the Jupiter-C or Juno space launcher by late 1958 and the Mercury-Redstone that sent two American astronauts into suborbital space in the early 1960s.

Meanwhile, the Air Force's WDD became its Ballistic Missile Division in 1958. Instead of relying on the Army's arsenal concept, the Air Force used an almost exclusive contract approach to missile development, including a contract with Ramo-Wooldridge (later TRW) Corporation for systems integration and engineering on all three of its early ballistic missile programs—the Thor IRBM, plus the Atlas and Titan ICBMs.

The Thor program, initially directed by Navy Captain Robert Truax, who was assigned to the Air Force's WDD team, began in 1955. The first Thor, manufactured by Douglas Aircraft Company, launched successfully in September 1957. Soon thereafter, Thor became the main stage for launching satellites such as GRAB and Discoverer/Corona. Thor boosters, with thrust augmentation or a variety of upper stages, continued to launch satellites through the 1970s, and beginning in 1960, a modified Thor became the first stage of NASA's Delta space launcher, which continued to evolve through the remainder of the Cold War years.

Using one and one-half stages and a balloon-tank concept espoused by Convair engineer Karel Bossart, the Atlas family of space launchers originated with Project MX-774 in 1946 and the follow-on MX-1593 in 1951. Approval for full-scale Atlas ICBM development came in January 1955, and under the management of Colonel Maurice Cristadoro, Jr., the first Atlas A flight occurred in June 1957. An Atlas launched *SCORE*, the world's first communications satellite, in December 1958. During 1961–1963, Atlas boosters successfully launched four Mercury astronauts on orbital flights. Atlas space-launch vehicles continued to evolve in various configurations with Agena or Centaur upper stages throughout the Cold War.

Apprehensive about the chances for success of Atlas development, an undertaking more complex than the Manhattan Project to develop an atomic bomb, General Schriever opted for parallel procurement of a second ICBM, designated Titan, based on different technologies. The Glenn L. Martin Aircraft Company, prime contractor for the two-stage Titan, began work in May 1955, and the first Titan I ICBM test occurred in February 1959. In 1960, Titan program manager Colonel Albert "Red" Wetzel decided against making incremental changes to Titan I and chose, instead, a complete configuration change for Titan II. The first test launch of a Titan II ICBM occurred in March 1962, and all of NASA's Gemini astronauts went into orbit atop man-rated Titan II launch vehicles during 1965–1966. After the Titan II rockets were decommissioned as ICBMs in the late 1980s, the Air Force had Martin

Marietta Corporation refurbish fourteen to serve as space-launch vehicles for small to medium payloads. Other evolutionary design paths populated the Titan family of Cold War space boosters with Titan III variants from the mid-1960s into the 1980s and, following the space shuttle *Challenger* catastrophe in 1986, the Titan IV for lifting heavier payloads into higher orbits.

Military use of the space shuttle, a partially reusable, crewed transportation system, had been decided even before President Richard Nixon formally approved shuttle procurement in January 1972. In the early 1980s, the newly established Air Force Space Command even contemplated acquisition of a dedicated fleet of military shuttles, and construction of a military Shuttle Operations and Planning Complex (SOPC) began as part of the Consolidated Space Operations Center on the plains east of Colorado Springs, Colorado. The Air Force never received funding for a military shuttle fleet, however, and the *Challenger* explosion led to cancellation of the SOPC project. Nonetheless, the first military mission aboard the space shuttle took place from June 27 to July 4, 1982, with the first flight dedicated fully to military activities coming in January 1985 and the second in October 1985. Before the end of the Cold War, a total of eight space shuttle missions were dedicated to Department of Defense interests, and another, nondedicated mission in November 1991 launched an Inertial Upper Stage with a Defense Support Program early-warning satellite.

Soviet launch-vehicle development roughly duplicated the U.S. pattern, beginning with establishment of a Special Committee for Reactive Technology in the late 1940s to develop long-range ballistic missiles. Sergei Korolev's team successfully launched an R-7 Semyorka, the world's first ICBM, in August 1957 and used another R-7, labeled SL-1, to launch the world's first artificial satellite two months later. Subsequent Soviet space boosters (e.g., SL-3 Vostok, SL-4 Soyuz, and SL-6 Molniya) all descended directly from the SL-1, and many of those that did not (e.g., SL-8 Kosmos, SL-11 Tsyklon-M, and SL-14 Tsyklon) originated from other types of long-range ballistic missiles. Table 13.1 shows the spacelift timeline, with the shaded background indicating Soviet activity and the italicized lettering indicating nations other than the United States or USSR.

SATELLITES

During the Cold War, the United States and the Soviet Union developed military satellite systems for virtually all of the purposes identified in the first Project RAND report of May 1946: intelligence, early warning, nuclear detection, meteorology, communications, and navigation. Mikhail Tikhanravov initiated a similar feasibility study on the Soviet side in 1947. In the years immediately following World War II, most of the work was theoretical and focused on subsystems, as manifested in the various RAND "feedback" reports that culminated with the famous 1954 summary report authored by Lipp and Salter. Not until the late 1950s did the Army, Navy, and Air Force devote many resources to producing prototype systems, primarily for gathering electronic and photographic intelligence. A mixture of prototypes and first-generation operational satellite systems marked the 1960s, with fully operational military space systems for the full range of previously envisioned applications becoming available in the 1970s. Generally speaking, with the exception of ocean

TABLE 13.1
Timeline of Spacelift Events

U.S. events are not shaded. USSR events are shaded. Other nation's efforts are italicized.

Spacelift		Spacecraft		Supporting ground elements	
Event	**Year**	**Event**	**Year**	**Event**	**Year**
Atlas ICBM development starts	1951	*Sputnik 1* and *Sputnik 2* with dog named Laika	1957	Long Range Proving Ground at Cape Canaveral activated	1949
R-7 development approved	1953	First launches each of Explorer, Pioneer, and Vanguard series	1958	Cape Canaveral has its first attempted launch, a V-2	1950
Titan ICBM development starts	1955	Luna (lunar) series starts: first to orbit the Moon	1959	First Redstone launches from Cape Canaveral	1953
Thor and Jupiter IRBMs approved	1955	*Pioneer 4* becomes first sun orbiter	1959	Construction starts on Tyuratam test site (later the Baikonur Cosmodrome)	1955
Thor launched successfully	1957	Corona (reconnaissance) series starts	1960		
R-7 demonstrates ballistic trajectory and launches first artificial satellite	1957	*Tiros 1*, first weather satellite, launched	1960	Construction starts on Plesetsk ICBM center	1957
Jupiter IRBM first launch	1957	First man in space in *Vostok 1*	1960	Vandenberg AFB named and has first missile launch, a Thor IRBM	1958
First Atlas flight	1957	Venera (Venus) series starts	1961	Dedicated ground stations and control center for DMSP	1965
Titan I missile flies first test mission	1959	Ranger (lunar) series starts	1961	Start formation of special command and control network (KIK) of 13 measurement stations (NIPs)	1957
Atlas/Agena first orbital launch	1968	Mercury (manned) flights begin	1961	USAF activates 6594th Test Wing for TT&C	1959

(*Continued*)

TABLE 13.1 *(Continued)*

Spacelift		Spacecraft		Supporting ground elements	
Discover XIV launches first spy satellite	1960	*Telstar 1* beams the first live transatlantic telecast	1962		
France launches its first satellite, Arterix, on a Diamant A rocket	*1965*	Mariner (planetary) series starts	1962	Sea-based measurement stations deployed onboard Krasnodar, Ilichevsk, and Dolinsk	1960
Human-rated Titan II launches Gemini astronauts	1965	Argon (reconnaissance) series starts	1962	United States acquires Eniwetok and Kwajalein Atolls in Marshall Islands for instrumentation complexes to support launches from Vandenberg	1960
Soyuz rocket delivers the first Yantar spy satellite	1966	Marsnik (Mars) series starts	1962	U.S. completes initial TT&C infrastructure with operations center and remote tracking stations	1961
Tsyklon-2 rocket flies its first mission from Baikonur	1967	Lanyard (reconnaissance)	1963		
Proton supports translunar trajectory	1967	Cosmos (Earth orbiters) series starts, includes Polyot-1 (anti-satellite prototype) and Zenit-1 (first spy satellite)	1963	Three new NIPs added to support lunar program	1962
First Saturn flight for lunar mission	1968	Zond (lunar) series starts	1965		
Japan's Lambda 4 rocket launches Ohsumi, *a test satellite*	*1970*	First Defense Meteorological Satellite	1965	Deep space tracking facilities introduced in Ussuriisk and Evpatoria	1962
China launches Dong Fang Hong-1 on a Long March 1 rocket	1970	Gemini (manned) flights begin	1965	Air Force Satellite Control Network modified to support multiple missions, standardizes, and adds RTSs	1963

(Continued)

TABLE 13.1 *(Continued)*

Spacelift		Spacecraft		Supporting ground elements	
The United Kingdom launches its Prospero satellite on a Black Arrow rocket	*1971*	Surveyor (lunar) series starts	1966	Bezhitsa and Ristna command and control ships replace older Krasnodar and Llichevsk	1966
N1 fails fourth test launch in a row and dooms efforts to land a man on the Moon	1972	Lunar Orbiter (lunar) series starts	1966	Later to become the "Blue Cube" at Sunnyvale AFS, site preparation starts for MOL operations center	1966
A Saturn V launches *Skylab*	1973	Soyuz (ferry) flights begin	1967	Five brand-new ships join command and control fleet	1967
Energia-Buran system officially decreed	1976	*Apollo 8* orbits the Moon	1968	Later to become the Blue Cube at Sunnyvale Air Force Station (AFS), site preparation starts for a MOL operations center	1967
India launches its Rohini 1 satellite on SLV-3 rocket	1980	MOL (manned military space station) program canceled	1969	Eniwetok Atoll placed on caretaker status	1969
Space shuttle *Columbia* lifts off from Cape Canaveral	1981	Nuclear-powered Radar Ocean Reconnaissance Satellite launched on Cosmos 469	1971	Two more ships join fleet: *Academik Sergei Korolev* and *Cosmonaut Yuri Gagarin*	1970
President Reagan announces Strategic Defense Initiative, an antimissile defense shield	1983	*Salyut* (manned civil space station) mission commenced	1971	Four airborne NIPs introduced	1974
First of eight space shuttle missions dedicated to military activities	1985	Last Corona (#145) launched	1972	Four more ships enter service	1975
Last of 365 Agenas (upper stage for Thor, Atlas, and Titan boosters) launched	1987	*Skylab* (manned civil space station) launched	1973	The Air Force Western Test Range retires from service the last of its range ships, the *USNS Sunnyvale*	1975

(Continued)

TABLE 13.1 *(Continued)*

Spacelift		Spacecraft		Supporting ground elements	
First Energia superbooster is launched from Baikonur carrying a Polyus military payload	1987	*Almaz* (manned military space station) launched	1974		
Energia launches Buran space shuttle on its only flight, an unpiloted test	1988	Viking (Mars) series starts	1975	Construction begins on Space Launch Complex 6 at Vandenberg to support the shuttle	1979
		Pioneer (Venus)	1978	Command and control ship *Marshall Nedelin* joins fleet	1982
		Space Transportation System (STS) *Columbia* (winged transport) launched	1981	Construction for a new operations center in Colorado Springs, Colorado, starts	1983
		Vega-1 (Venus and comet flyby)	1984	Launch Complex 34, used to launch Saturns out of Cape Canaveral, is declared a national historic landmark	1984
		Phobos 1 and *Phobos 2* (Venus and comet flyby)	1988	Command and control ship *Marshall Krylov* joins fleet	1986
		Magellan (Venus orbiter)	1989	SLC-6 at Vandenberg ordered to be mothballed	1988
		Galileo (Jupiter orbiter and probe)	1989		
		First *Mir* (manned civil space station) element launched	1986		
		Buran (winged transport) flight	1988		

Presidents Gorbachev and Bush declare the Cold War officially over: 1989

ICBM, intercontinental ballistic missile; IRBM, intermediate-range ballistic missile; MOL, Manned Orbiting Laboratory; NIP, measurement station; RTS, remote tracking station; TT&C, telemetry, tracking, and command.

surveillance, the launch of Soviet satellites for different kinds of defense-related applications lagged behind U.S. programs by several years. The last decade of the Cold War witnessed the launch of more advanced satellites for the same purposes first stated in the late 1940s. New applications such as space-based radar or lasers remained largely theoretical.

Intelligence satellites began with the successful 1960 launches of the Navy's GRAB, which intercepted electronic signals from Soviet radars, and the Air Force's (mostly CIA-funded) Discoverer/Corona, which collected imagery for return to Earth via reentry capsules. Over the course of a quarter century, the Corona satellites evolved into Argon, Lanyard, and several other more capable imaging platforms that continued to employ reentry capsules for recovery of exposed photographic film. From the mid-1970s to the end of the Cold War, electro-optical and radar imaging satellites capable of transmitting data from orbit also entered the U.S. inventory of national-security assets.

On the Soviet side, *Zenit*, the first Soviet photoreconnaissance and electronic-intelligence (ELINT) satellite, went into orbit in April 1962. Several more generations of Zenit film-return satellites appeared throughout the remainder of the Cold War, along with Yantar electro-optical models in 1982 that allowed digital transmission of imagery to ground stations. For purposes of ocean reconnaissance for the Soviets, nuclear-powered Radar Ocean Reconnaissance Satellites began orbiting in the early 1970s, but technical and design problems along with safety concerns among other nations resulted in the program's cancellation in the late 1980s. At the same time, work on an ELINT Ocean Reconnaissance Satellite system led to on-orbit testing in October 1975 and an operational, four-satellite constellation by 1990. The first Tselina O satellite for ELINT launched in October 1967, but the complete Tselina O system did not achieve operational certification until March 1972.

Early-warning satellites grew out of the Air Force's Weapon System 117L (WS-117L) program, which had been broken down into Discoverer/Corona, Samos, and MIDAS in the late 1950s. MIDAS (Missile Defense Alarm System) prototype satellites tested the concept of detecting the infrared signature of long-range missiles launched from the Soviet Union or elsewhere. In 1970, the first Defense Support Program (DSP) satellite marked the beginning of an operational space-based early-warning system that endured into the twenty-first century. The primary DSP ground station was located at Buckley Air Force Base (AFB), Colorado. Oko or US-KS satellites for Soviet early warning became operational initially in 1976, with a full on-orbit constellation of nine satellites in highly elliptical orbits first achieved in 1980. The Soviet military began supplementing that HEO constellation with Oko satellites in geosynchronous orbit (GEO) during the period 1984–1987, before placing the first Prognoz or US-KMO into GEO in 1990.

For purposes of detecting nuclear explosions, the Air Force and scientists at the Los Alamos National Laboratory designed Vela satellites. The first satellites procured via a fixed-price incentive contract, the Velas monitored compliance with the Limited Test Ban Treaty that went into effect on October 10, 1963. Launched in three pairs in 1963–1965, with another three pairs of advanced models going up in 1967–1970, the last Vela continued to operate until it was turned off in 1984. In 1969–1970, the advanced Vela satellites first detected naturally occurring gamma-ray bursts from deep space. By the last years of the Cold War, nuclear-detection subsystems on global positioning system (GPS) and DSP satellites had replaced the dedicated Vela system.

For national-security purposes, meteorology satellites originated as a means of enabling operators of Corona photoreconnaissance satellites to avoid wasting valuable film footage if clouds obscured the Earth's surface. Having begun officially as Program II in August 1961 and designated Program 35 not long thereafter, the military weather satellites were envisioned only as a stopgap pending the ability of the civil TIROS program to meet the requirement. By 1962, when the first military weather satellites went into orbit, the interim program had achieved permanent status and yet another designation—Program 417. Its use expanded immediately to provide information to U.S. aircraft flights during the Cuban Missile Crisis, evacuation of civilians from the Congo, and aerial operations in Vietnam. Declassification of the system came with yet another name for it: the Defense Meteorological Satellite Program (DMSP). In 1994, it became the first operational military satellite system ever turned over to a civilian organization, the National Oceanic and Atmospheric Administration, when a presidential directive made it part of the National Polar-Orbiting Operational Environmental Satellite System.

Development of satellites for dedicated military communications initially became the Army's responsibility. *SCORE* (Signal Communication by Orbiting Relay Equipment) carried an Army Signal Corps payload to deliver President Eisenhower's message of peace to the world in December 1958. The Army also oversaw development of the world's first military communications satellite, the experimental *Courier 1-B*, which launched in October 1960. When Advent, a more ambitious Army project to field the world's first geosynchronous satellite, ran afoul of technological difficulties before launch of even a single payload, the Air Force gained approval for what became, in 1966, the world's first operational military communications satellite system—the Initial Defense Satellite Communications System (IDSCS). TRW Systems Group received an Air Force contract in 1969 to deliver a more advanced, geostationary model, and the Air Force launched the first pair of Defense Satellite Communications System (DSCS) II satellites in November 1971. The last operational DSCS II remained in service until October 1998, well after the end of the Cold War. Meanwhile, General Electric earned a contract in 1976 to produce an even more capable, jam-resistant DSCS III, the first of which launched in October 1982 with the last DSCS II. The DSCS III constellation continued to serve as the long-haul, high-capacity communications system for worldwide command and control of U.S. armed forces beyond the Cold War and into the twenty-first century.

On the basis of experience gained by using IDSCS in the late 1960s, North Atlantic Treaty Organization (NATO) asked the U.S. Air Force to contract in April 1968 with the Philco-Ford Corporation for a variant of the IDSCS satellite. The first NATO communications satellite orbited in 1970, and some later versions launched during 1976–1984 remained operational into the 1990s. Meanwhile, the United Kingdom had arranged with the U.S. Air Force in 1967 to procure from Philco-Ford a pair of IDSCS "augmentation" satellites for British military use. Dubbed Skynet satellites, newer models built by Marconi Space Systems went to space in 1974, and three even more advanced Skynet 4 models built by British Aerospace Dynamics began orbiting in 1988–1990.

Beginning in 1965 and continuing for a decade, the Massachusetts Institute of Technology Lincoln Laboratory designed a series of nine experimental satellites to improve U.S. military capabilities. Technologies that later went into U.S. military satellites (e.g., solid-state electronics, electronically de-spun antennas, onboard signal processing, cross-linking, and extremely high frequency reception and transmission) emerged from the Lincoln Experimental Satellite project.

The U.S. Navy's quest for space-based tactical communications led to Hughes Aircraft Company designing and building a single, geostationary Tactical Communications Satellite, which the Air Force launched in February 1969. This satellite aided recovery of NASA Apollo capsules by connecting aircrews with carriers and ground stations. Success with this experimental satellite prompted the Navy to acquire a Fleet Satellite Communications (FLTSATCOM) system, designed and manufactured by TRW Corporation, that became fully operational in January 1981 and remained so throughout the Cold War years. A portion of the Air Force Satellite Communications (AFSATCOM) system for presidential and national-security communications, including command and control of nuclear forces, also rode aboard FLTSATCOM. Two kinds of classified, polar-orbiting spacecraft also hosted AFSATCOM payloads: the Satellite Data System and "Package D."

During the 1970s, an elevated level of concern about the inability of existing communications satellites to survive in a nuclear war prompted the U.S. Air Force to undertake development of a system that could operate through all levels of conflict—a satellite with protection against the destructive effects of electromagnetic pulse. The service contracted with Martin Marietta Corporation in 1983 for development of hardened Milstar satellites, but the Cold War ended 2 years before the first of those satellites went into space.

Soviet experimentation with military communications satellites began with Strela store-dump platforms in August 1964, but operational versions of Strela did not appear until 1970. Meanwhile, in May 1967, the first Tsiklon model orbited to provide naval communications. Parus and Altair/Luch, in 1974 and 1985, respectively, began fulfilling that same function. Korund satellites came into use for general Soviet military communications in 1970, followed by Radugas: Gran in 1975 and Globus in 1989. For the Soviet military's data relay requirements, Geyzer/Potok satellites became available in May 1982.

Military navigation satellites began with the Navy's Transit system, largely developed by Johns Hopkins University Applied Physics Laboratory. The Transit series of spacecraft made many contributions and provides many perspectives. Following this first successful employment of space-based navigation aids benefiting military requirements, the Department of Defense created a joint program office, headed by Air Force Colonel Bradford Parkinson, for acquisition of the NAVSTAR GPS in 1972. Unlike Transit, which provided only latitude and longitude, GPS furnished precise positioning in three dimensions and timing to the billionths of a second. Drawing heavily on concepts and designs from Timation satellites that Roger Easton's team developed at the Naval Research Laboratory and to a lesser extent on those from the Air Force's Program 621B, which Ivan Getting oversaw at Aerospace Corporation, the first GPS Block 1 prototype satellites launched in 1978. A Block II operational constellation did not begin until February 1989. By the end of the Cold War period in 1991, the Air Force had launched only three-fourths of the Block II satellites required for a fully operational, 24-satellite constellation. Declaration of full operational capability for GPS would not occur until 1995.

For space-based navigation, the Soviet navy relied initially on Tsiklon satellites that also provided secure communications capability, but a satisfactory operational constellation did not evolve until Parus (Tsiklon B) satellites completely replaced the original Tsiklon models in 1978. By then, the Soviet Union had identified a need for its military forces to have a single, extremely accurate, space-based radio-navigation system, which a government decree in 1976 labeled the Global Navigation Satellite System (GLONASS). Although the first GLONASS satellite launched in October 1982, the Soviet Union collapsed and the Cold War ended before GLONASS achieved initial operational capability.

TERRESTRIAL SUPPORT TO LAUNCH AND SPACE SYSTEMS

Both the United States and the Soviet Union built facilities and networks for receiving telemetry from their satellites, tracking their orbits, and sending them operational commands. The telemetry, tracking, and commanding functions (generally referred to by the abbreviation TT&C) were required to ensure that satellites remained in their intended orbits and satisfactorily performed their respective missions. As military space systems multiplied and the number of active satellites mounted, both superpowers adjusted their TT&C capabilities and capacities to accommodate the growing workload. In addition to fixed sites around the world, both countries used aircraft and ships for TT&C functions.

The U.S. Air Force activated the 6594th Test Wing, its first unit for satellite operations, on April 4, 1959, to support the Corona reconnaissance program from an interim control center at Palo Alto, California, and stations at three operating locations: Edwards AFB, California; Chiniak Point on Kodiak Island, Alaska; and Annette Island, Alaska. In November 1959, the Air Force purchased approximately 11.5 acres of land with a newly constructed Building 171 in Sunnyvale, California, as the site for a permanent Satellite Test Annex. On March 1, 1961, the permanent control center in Building 171 (later designated Building 1001) became fully operational. By the end of that year, the Air Force had configured most of what would become known organizationally as the Air Force Satellite Control Facility (AFSCF) or network—the Satellite Test Center (STC) at Sunnyvale and remote tracking stations (RTSs) at Vandenberg AFB in California, New Boston in New Hampshire, Kaena Point in Hawaii, and Thule Air Base in Greenland. Over the next 40 years and beyond, the AFSCF provided TT&C support for Corona/Discoverer, many other kinds of U.S. intelligence-gathering satellites (so-called national systems), a host of different experimental or operational military satellites (e.g., DMSP, DSP, DSCS, GPS, and Milstar), and even some nonmilitary satellite missions.

Of course, the AFSCF underwent numerous changes as technology advanced and mission requirements evolved. Initially, no two RTSs had the same configuration, and each type of satellite had a unique TT&C system. Aerospace Corporation oversaw the first significant AFSCF upgrade—the Multiple Satellite Augmentation Program—in 1963 to increase data processing at the control center, simplify the RTSs, add additional RTS sites, and standardize equipment to enable support of more than one satellite at a time. A second modernization effort—the Advanced Data System project—in the late 1960s included new software, a new communications system, standardized transmission frequencies, and satellite beacons; a Space-Ground Link Subsystem integrated previously separate TT&C subsystems, which significantly reduced the time it took for an RTS to send and receive encrypted data from a satellite. To support the Manned Orbiting Laboratory (MOL), the Air Force began site preparation in October 1967 for an Advanced STC on 8.2 acres of land purchased from Lockheed Corporation. When the MOL program was cancelled in June 1969, however, the fate of the nearly completed $1.5M Advanced STC became questionable; a decision to continue construction of the structure, which became known as the "Blue Cube," led to its occupancy by AFSCF mission control centers in 1970. In January 1971, the Air Force changed the official designation of the Satellite Test Annex to Sunnyvale Air Force Station.

During the 1980s, numerous hardware and software upgrades continued to improve AFSCF capabilities. The Data Systems Modernization (DSM) program at the beginning of the decade centralized, at Sunnyvale, command and control of the entire AFSCF network and allowed more efficient use of existing resources. Under DSM, the RTSs essentially became relay stations. A multiyear Automated Remote Tracking Station project, begun in the late 1980s and ongoing at the end of the Cold War, increased the reliability and operational capacity of the RTSs, which allowed a 40% reduction in manning at those sites. Meanwhile, vulnerability of the facilities at Sunnyvale to terrorist attacks or earthquakes prompted groundbreaking in May 1983 for a new Consolidated Space Operations Center (CSOC) east of Colorado Springs, Colorado. A gradual transition of U.S. military satellite operations from Sunnyvale to the CSOC began in 1985. When the Cold War ended in 1991, both the CSOC and Sunnyvale remained operational.

The Soviet Union began design work in January 1956 on a military satellite system, which included a Command Measurement Complex (KIK) for TT&C. A formal decree dated September 3, 1956, authorized KIK development, which initially comprised seven primary stations across the country's vast territory and under control of the Strategic Rocket Forces. This system basically relied on existing equipment for the Soviet MRV-2M system for air defense and interceptor control, along with ICBM tracking systems. In December 1957, the original KIK network control center at Bolshevo moved to Moscow. The KIK performed TT&C not only for military spacecraft, but for all Soviet orbital, lunar, and interplanetary piloted or robotic space flights. This single system satellite TT&C stood in contrast to the two separate U.S. systems—the AFSCF for military satellites and NASA's for civil spacecraft.

During the 1960s, KIK capabilities expanded in several ways. Three ocean-going stations entered the inventory in 1960 onboard the *Krasnodar*, *Ilichevsk*, and *Dolinsk*; the *Beshitsa* and *Ristna* replaced the latter two ships in 1965–1966; and five new vessels—*Cosmonaut Vladimir Komarov*, *Kegostrov*, *Morzhovets*, *Nevel*, and *Borovichi*—joined the fleet in 1967. Two new ground stations to support planetary launches, along with five others—two for deep-space tracking and three for communication with interplanetary probes—entered service in 1962, and another five came on line within the next 8 years. Various TT&C systems also became operational (e.g., Zenit; Pluton and Saturn for sending and receiving data, respectively, on interplanetary missions; and Kub and Baza for geosynchronous satellites with large, programmable memories). The KIK operated nine different systems by 1966. Although Andrei Grigorevich Karass began work in 1962 on autonomous relay of data to the command center, much of the KIK communications during those early years traveled over telephone and telegraph links that used hand commutators and vacuum-tube equipment.

The 1970s witnessed significant improvement in KIK. Imposition of common control requirements in 1971 resulted in efforts to perfect automated guidance, which culminated with installation of the Central Automatic System (TsAK) in KIK centers beginning in 1979 and acceptance of TsAK for military satellite missions in 1981. New telemetry systems entered the inventory, along with modernized and better-integrated communication lines, control stations, processing equipment, and training programs for operators. Nonetheless, the different types of equipment at KIK tracking stations continued to proliferate from the early 1970s into the 1980s, causing Lieutenant General N.F. Shlykov to head a program to rationalize the equipment base. Another six ships joined the sea-based TT&C fleet—Academik Sergei Korolev and Cosmonaut Yuri Gagarin in 1970, plus Cosmonaut Pavel Belyaev, Cosmonaut Vladimir Volkov, Cosmonaut Georgi Dobrovolskiy, and Cosmonaut Viktor Patsaev in 1975. In addition, four airborne tracking stations joined to KIK in 1974 and became key to refining the orbital measurements needed to calculate corrective maneuvers. Finally, the Golitsyno-2 command center at Kraznoznamensk, northwest of Moscow, was upgraded to become the central hub for coordination for all the ground, sea, and airborne TT&C facilities.

The KIK continued its evolution during the last decade of the Cold War. Its stations received new equipment for microprocessor-controlled displays, plus systems with better navigation parameters. New tracking systems—Kub-Kontur and Taman-Baza—entered the network, along with the complementary KOS quantum-optical system. Increased power gave the KIK's improved Romashka radio telemetry system four times the range of earlier systems. Two improved tracking ships, *Marshal Nedelin* in 1982 and *Marshal Krilov* in 1986, entered the sea-based portion of the network. Authorization for full development of a Mobile Overland Command and Tracking System (PN KIP), code named Fazan, came on August 29, 1980, and system testing began within a few years. Before the end of 1988, the Soviet Union completed a major reorganization of its satellite TT&C infrastructure that established command points under the Rokot project and designated three main control-and-tracking centers—Yevpatoriya, Yeniseisk, and Ula-Ude—along with five trial centers—four at Baikonur and one at Plesetsk.

TERRESTRIAL SYSTEMS TO WATCH LAUNCH AND SPACE SYSTEMS

For national-security purposes, the emergence of ICBMs and Earth-orbiting satellites in the late 1950s compelled both the United States and the Soviet Union to develop and field systems for detecting and tracking rocket launches and human-made objects in space. Those systems included various kinds of radar equipment and optical sensors. In some cases, the radars performed double duty by watching for potentially hostile long-range, nuclear-tipped missiles and simultaneously tracking orbital objects. As the number of human-made objects in space—active payloads, dead satellites, spent boosters, and assorted other debris—grew from single digits to hundreds, then thousands, an up-to-date "space catalog" became increasingly important for determining whether specific objects posed a military threat, for avoiding collisions between objects, and for predicting when and where large pieces of debris in decaying orbits might impact Earth. After 1965, all the early-warning and space-surveillance data collected by sensors worldwide went to a processing and correlation center deep inside Cheyenne Mountain near Colorado Springs, Colorado.

The U.S. Air Force began planning for a Ballistic Missile Early Warning System (BMEWS) in 1955 and won approval in 1957 to begin its construction. Three locations—Thule Air Base in Greenland, Clear Air Station in Alaska, and Fylingdales Moor in the United Kingdom—had BMEWS radars that included General Electric's AN/FPS-50 stationary, parabolic reflectors 165 ft. high and 400 ft. wide for searching and RCA's AN/FPS-49 mechanical trackers with 80-ft.-diameter dish antennas that could rotate 360 degrees and swivel vertically 90 degrees. Starting with Thule in 1987, the Air Force modernized all the BMEWS sites with Raytheon's solid-state, phased-array radars. During the Cold War years, many other U.S. radar systems (e.g., Cobra Dane on Shemya in the Aleutian Island chain and the AN/FPS-17 detection-AN/FPS-79 tracker combination at Pirinclik, Turkey) also provided early warning of missile launches from the Soviet Union. In 1975, the Air Force requested proposals from industry for a solid-state phased array radar system to detect sea-launched ballistic missiles, and this led to four operational Pave Phased Array Warning System (PAWS) radar sites—Cape Cod, Massachusetts; Beale, California; Robins, Georgia; and Eldorado, Texas—in the 1980s that also supplied space surveillance data. On April 4, 1980, Cape Cod became the first Pave PAWS radar site to achieve initial operational capability, followed by Beale, Robins and, on May, 8 1987, Eldorado.

While many different systems, especially various types of radars, supplied space-surveillance data as contributing sensors during the Cold War, several dedicated U.S. space-surveillance systems existed. In preparation for tracking scientific satellites during the International Geophysical Year, which was scheduled for July 1957 to December 1958, Smithsonian Astrophysical Observatory director Fred Whipple and The Ohio State University professor J. Allen Hynek arranged for optical expert James Baker and mechanical specialist Joseph Nunn to design and build a tracking camera. In addition to those fielded by the Smithsonian, the U.S. Air Force ordered five Baker-Nunn cameras in 1959 for deployment at sites around the globe. Initially, Project Moonwatch volunteers detected satellites visually and provided sufficiently precise, preliminary orbital data for the Baker-Nunn camera crews to know where to point their instruments. Some of the Baker-Nunn cameras remained operational as dedicated military sensors into the early 1990s.

Skeptical about the ability of optical systems to track the first U.S. satellite, Milton Rosen, technical director for the Vanguard project, pressed for electronic detection and tracking of the satellite's radio beacon. Roger Easton from the Naval Research Laboratory implemented the Minitrack receiving system that used radio interferometry to triangulate signals transmitted from the satellite. When it became clear that the U.S. military required better capability for detecting and tracking "dark" or nonradiating satellites, Easton devised the Naval Space Surveillance system, which consisted of both transmitters and receivers stretched across the United States along the 33rd parallel, became fully operational in February 1959, and continued to operate into the twenty-first century.

Although a disastrous fire destroyed the AN/FPS-85 radar during acceptance testing at Eglin AFB, Florida, in January 1965, it became operational 4 years later after extensive reconstruction. The AN/FPS-85, the first phased array radar designed specifically for space surveillance, could track 200 known objects simultaneously, in contrast to earlier sensors that handled only one at a time. Furthermore, its latitudinal location allowed the AN/FPS-85 to pick up low-inclination objects, which greatly increased the overall capacity of the U.S. Space Surveillance Network. This radar underwent several upgrades to keep it operational throughout the Cold War period.

To improve optical surveillance and, ultimately, to replace its Baker-Nunn cameras, the U.S. Air Force fielded the Ground-Based Electro-Optical Deep Space Surveillance (GEODSS) system in the 1980s. After a prototype station at White Sands Missile Range demonstrated GEODSS potential in September 1975, the Air Force began actual operations at Stallion Range Center, White Sands Missile Range, New Mexico, in May 1982. Three GEODSS sites—Socorro, New Mexico; Taegu, South Korea; and Maui, Hawaii—achieved initial operational status in March 1983, and another, on the island of Diego Garcia in the Indian Ocean, became operational in January 1987.

Soviet space surveillance relied heavily on large, ground-based radars. In the 1960s, the core of that capability was the Dnestr-M/Dnepr "Hen House" radar system, which the Daryal-UM Large Phased Array Radar system began augmenting in the 1980s. Like the United States, the Soviet Union also received data from optical and electro-optical sensors throughout its realm. Construction began in the 1980s on the most advanced of the optical surveillance facilities, the Okno system in Nurek, Tajikistan, but its completion did not occur until 1999, well after the Soviet Union collapsed and the Cold War ended.

FIGHTING IN SPACE

Both the United States and the Soviet Union developed anti-ballistic missile (ABM) and antisatellite (ASAT) systems during the Cold War years. The need for ABM became apparent when Germany began launching V-2 missiles in the last months of World War II and, despite fluctuating levels of attention by both superpowers, persisted throughout the Cold War. On the U.S. side, the Army Air Forces began ABM feasibility studies in the late 1940s with the MX-794 Wizard and MX-795 Thumper projects. In September 1953, Soviet military leaders asked the Central Committee of the Communist Party for permission to study ABM feasibility, and their V-1000 Griffon successfully intercepted a ballistic missile in March 1961. During the 1960s, despite the questionable effectiveness of ABM systems, the U.S. Army's Nike-Zeus interceptor missile evolved to become the Spartan exoatmospheric interceptor, which was accompanied 2 years later by the smaller Sprint endoatmospheric interceptor. To counter the Spartan-Sprint combination, the Soviet military developed the medium-range Galosh and short-range Gazelle ABM missiles.

The superpowers decided to deploy ABM systems in the 1960s, beginning with the Soviet Union's construction in 1962–1963 of the A-35 system to protect Moscow. President Lyndon Johnson countered with Sentinel in 1967 to defend U.S cities. In 1969, however, President Richard Nixon reoriented the U.S. system to protect ICBM fields instead of cities, and he renamed it Safeguard. Completion of the single U.S. Safeguard complex permitted under the 1972 ABM Treaty occurred in October 1975, but that complex in northeastern North Dakota ceased operations after only 2 months. The Air Force, however, took over from the Army the AN-FPQ-16 Perimeter Acquisition Radar Attack Characterization System from that complex and continued operating it for purposes of early warning and space surveillance into the twenty-first century. Meanwhile, in 1978, the Soviet Union began work on an improved A-135 ABM complex around Moscow but failed to complete the project before the Cold War ended.

By the early 1980s, advances in several key technologies significantly improved the prospects for U.S. development of a truly reliable ABM system based on the kinetic-kill principle but supported by high-powered lasers on the ground, aboard aircraft, or in space. In March 1983, President Ronald Reagan announced a research-and-development program to take advantage of cutting-edge technologies. His administration formally established the Strategic Defense Initiative (SDI) in January 1984. A preponderance of the SDI effort focused on space-based systems for detection, tracking, and interception of ballistic missiles and their warheads. The Space-Based Infrared System (SBIRS), which ultimately would replace and improve upon the Defense Support Program's early-warning capability, would include sensors in geosynchronous, highly elliptical, and low Earth orbits. In 1990, the portion of SBIRS projected for low Earth orbits became known as Brilliant Eyes. Similarly, the space-based, kinetic-kill portion of SBIRS received the name Brilliant Pebbles. The SDI program, albeit restructured and renamed Global Protection Against Limited Strikes by President George H.W. Bush on the eve of the Soviet Union's collapse, continued beyond the twentieth century.

The Soviet Union challenged the United States during 1969–1983 with operational deployment of a Fractional Orbital Bombardment System, which could place payloads carrying nuclear-armed reentry vehicles into orbital trajectories. This could enable an attack on the United States from a southerly direction, where the fewest ground-based U.S. sensors existed for early warning. That vulnerability diminished substantially, however, with completion of an on-orbit DSP constellation, conversion of the AN/FPS-85 at Eglin AFB for use as an early-warning radar, and construction of the Pave PAWS radars at Robins AFB in Georgia and Eldorado AFS in Texas. In both the Strategic Arms Limitation Treaty II of 1979 and the Strategic Arms Reduction Treaty I of 1991, the signatories pledged not to "produce, test, or deploy" nuclear or other weapons of mass destruction in fractional orbits. If the use of ballistic missiles in the 1940s prompted an immediate quest for antimissile systems, the launch of Earth-orbiting spacecraft in the 1950s sparked U.S. and Soviet military interest in ASAT capabilities. The superpowers focused on two methods of intercepting and destroying a satellite: direct ascent, which involved firing an interceptor from the Earth's surface or an aircraft on a trajectory to converge with the orbiting satellite, and co-orbital, which involved placing the interceptor in orbit and maneuvering it to converge with the target satellite.

Both the U.S. and Soviet armed forces also undertook development of piloted space interceptors capable of performing ASAT, reconnaissance, surveillance, bombardment, and other missions. Combining the Hywards, Brass Bell, and Robo projects from the early 1950s, the U.S. Air Force planned for System 464L or Dyna-Soar near the end of 1957. The need for a more powerful booster to launch Dyna-Soar led to development of the Titan III. Because of its varied mission capabilities and highly publicized nature, Soviet leaders perceived Dyna-Soar as a direct military threat if it became operational; knowing this, President John F. Kennedy's administration decided to reduce some of the tension in Soviet-American relations by removing military subsystems from the Dyna-Soar program and emphasizing that the spacecraft's primary purpose was for orbital research. To reflect that programmatic shift, the Department of Defense changed Dyna-Soar's name to X-20 on June 26, 1962.

When Secretary of Defense Robert McNamara cancelled Dyna-Soar on December 10, 1963, he simultaneously authorized the Air Force's development of a military space station—the MOL. Designed primarily to improve upon timeliness of reconnaissance and surveillance over the film-returning capsules of the early spy satellites, MOL would allow military operators to zoom in on installations of interest in real time. In addition to military reconnaissance duties (still largely classified), the military occupants would conduct defense-related experimentation in Earth orbit. The Soviets launched and operated a manned military space station for Earth observing called Almaz. The Almaz configuration included a recoilless gun and missiles for self-defense.

The Air Force had begun planning for a Military Orbital Development System around the same time as the renaming of Dyna-Soar as X-20, and the service initially intended to use so-called Blue Gemini capsules as vehicles to ferry military astronauts to the orbiting station—basically a pressurized Titan IIIC upper-stage tank in subsynchronous, polar orbit. Although the Blue Gemini concept withered by January 1963 in the face of NASA suspicions that the Air Force was angling for a larger role in overall human spaceflight program, the MOL program that McNamara authorized later that year included use of modified Gemini capsules—Gemini Bs—to ferry military astronauts to and from the MOL. On November 3, 1966, NASA successfully completed the first and only test flight of a Gemini B, albeit an unmanned capsule. By the time the MOL program was cancelled on June 10, 1969, a total of seventeen military astronauts had been selected and were training for orbital missions beginning in 1972. The Air Force continued throughout the remainder of the Cold War to pursue the dream of military members orbiting Earth in military spacecraft of one kind or another, but that "military man in space" concept went no further than military members completing defense-related missions on NASA spacecraft.

Meanwhile, Soviet engineers had begun designing a piloted Raketoplan interceptor in 1960, and their launch of an unpiloted scale model in March 1963 marked the world's first test flight of a lifting reentry vehicle. Despite suspension of Raketoplan development by early 1965, almost simultaneous incorporation of the basic concept into a new program called Spiral kept the Soviet pursuit of an experimental piloted ASAT system going until September 1978.

Operational, direct-ascent ASAT weapons emerged in the 1960s when scientists discovered that high-altitude nuclear explosions could damage satellites' electronic components. In Program 505 "Mudflap" during 1963–1966, the U.S. Army kept a single DM-15S, a modified Nike Zeus missile equipped with a nuclear warhead, on alert at Kwajalein Atoll in the Pacific. Thereafter, during 1964–1970, the U.S. Air Force's Program 437 ASAT system maintained two nuclear-tipped Thor missiles in readiness on Johnston Island. Galosh ABM missiles, equipped with nuclear warheads and deployed around Moscow, served secondarily as direct-ascent ASAT weapons for the Soviet military.

Co-orbital ASAT systems received early attention from both sides during the Cold War. In 1956, the U.S. Air Force began exploring a co-orbital satellite inspector/interceptor called SAINT but cancelled the project 6 years later without launching even a prototype. Similarly, in 1959, Soviet designer Vladimir Chelomey conceived the Istrebitel Sputnikov (IS) for maneuvering close to a target in space and detonating a conventional warhead that would pepper it with shrapnel. On-orbit testing of IS satellites, begun in 1967–1971, was interrupted temporarily by the 1972 ABM Treaty and resumed during 1976–1982. In response to these Soviet actions, the U.S. Air Force renewed its quest for a non-nuclear ASAT capability in 1977, which led to Vought Corporation's development of an Air-Launched Miniature Vehicle (ALMV). The only full test of the ALMV direct-ascent ASAT system took place in September 1985, when an F-15 fighter aircraft released an ALMV that intercepted and destroyed *Solwind P78-1*. A self-imposed Soviet moratorium on further ASAT testing and congressional pressure on the U.S. side led to cancellation of the ALMV in 1988.

HUMAN CONSTRUCT

The ways in which the U.S. and Soviet militaries organized their respective space capabilities differed substantially, but the organizational structures changed markedly in both countries over the course of the Cold War. In the United States, the Air Force established WDD under its Air Research and Development Command in July 1954 to oversee accelerated development of the Atlas ICBM, and WDD gained responsibility for military satellite development in February 1956. When the first American ICBMs became operational in 1959, they came under the control of Strategic Air Command, but as space boosters, they remained under the control of WDD's successor organizations, which were assigned to Air Force Systems Command (AFSC) after its establishment in April 1961. Responsibility for acquisition and launch of all U.S. military satellites remained with WDD—renamed Air Force Ballistic Missile Division in 1957, Space Systems Division in 1961, Space and Missile Systems Organization in 1967, Space Division in 1979, Space Systems Division in 1989, and Space and Missile Systems Center in 1992. Although this organization and its parent command continued to perform TT&C for U.S. military satellites through the AFSCF, the parent command changed on October 1, 1987, from AFSC to Air Force Space Command (AFSPC), the latter having been established on September 1, 1982, as the first U.S. military command for space operations. AFSPC also gained from AFSC responsibility for military space launches on October 1, 1990. During the 1980s, the Army and Navy also created operational commands for space, and a joint United States Space Command, activated in September 1985, included all of the service commands as components.

Unlike the United States, where space-related responsibilities resided in all the independent services, the Soviet Union placed management of all space-related activities in a separate directorate under its Strategic Rocket Forces (RVSN), essentially a separate military service. In November 1964, that directorate became known as TSUKOS, for centralized creation of new space systems and resolution of operational issues; its designation changed to Main Space Systems Directorate (GUKOS) in March 1970 to reflect a reorganization and expansion of its role. On November 10, 1981, a decree removed GUKOS from under the RVSN and placed it directly under the General Staff, which essentially placed it on equal footing with the RVSN. The GUKOS, whose name changed in 1986 to the United Space Systems Directorate (UNKS), sought to synergistically maximize the capabilities of the various types of Soviet satellites and to coordinate space surveillance, early-warning, and ABM capabilities for enhanced support to air-defense and other Soviet military forces. Not long after the breakup of the Soviet Union, units of the UNKS underwent a restructuring into the Military Space Forces, which remained directly under the Russian General Staff.

CONCLUSION

In December 1989, Presidents Gorbachev and Bush declared the Cold War officially over. During the Cold War, space was largely thought of as part of the rarefied but terrifying domain of nuclear warfare. Once demonstrated operationally, military satellites were used principally to monitor nuclear missile facilities, provide early warning should they be fired, and maintain secure communications between commanders and nuclear-strike forces. The United States and Russia accepted that spy satellites provided a degree of mutual reassurance in nuclear arms control.

Over the course of the Cold War, both the Soviet and U.S. defense establishments oversaw the evolution of Earth-orbiting spacecraft from the status "definite possibility," as stated in Theodore von Karman's *Toward New Horizons* report of 1945, to an absolute necessity for the modern world. Conceptual studies done by both the Soviet and U.S. military in the late 1940s and early 1950s led, in the 1960s, to satellite prototypes or first-generation operational models for nearly every space-based application envisioned in those studies. By the end of the Cold War, space-based systems designed and intended primarily for strategic purposes were being employed at a rapidly accelerating pace, especially by U.S. military forces, to support theater or tactical operations. Numerous artifacts remain both terrestrially and in space to substantiate the period. Archaeological endeavors can only add to the gradual disclosure of Cold War secrets and increase insight into the earliest period of spaceflight, when two superpowers entered a technological sprint to gain an advantage on the world stage via the domain of space.

FURTHER READING

United States

Armacost, Michael H. *The Politics of Weapons Innovation: The Thor-Jupiter Controversy*. New York: Columbia University Press, 1969.

Arnold, David Christopher. *Spying from Space: Constructing America's Satellite Command and Control Systems*. College Station, TX: Texas A&M University Press, 2005.

Baucom, Donald R. *The Origins of SDI, 1944–1983*. Lawrence, KS: University Press of Kansas, 1992.

Beard, Edmund. *Developing the ICBM: A Study in Bureaucratic Politics*. New York: Columbia University Press, 1976.

Berhow, Mark A. *US Strategic and Defensive Missile Systems 1950–2004*. New York: Osprey Publishing, 2005.

Davies, Merton E., and William R. Harris. *RAND's Role in the Evolution of Balloon and Satellite Observation Systems and Related U.S. Space Technology*. R.3692-RC. Santa Monica, CA: The RAND Corporation, 1988.

Day, Dwayne A., John M. Logsdon, and Brian Latell, eds. *Eye in the Sky: The Story of the Corona Spy Satellites*. Washington, DC: Smithsonian Institution Press, 1999.

DeVorkin, David H. *Science with a Vengeance: How the Military Created the US Space Sciences after World War II*. New York: Springer-Verlag, 1992.

Dewar, James A. *To the End of the Solar System: The Story of the Nuclear Rocket*. Lexington, KY: University Press of Kentucky, 2003.

Divine, Robert A. *The Sputnik Challenge: Eisenhower's Response to the Soviet Satellite*. New York: Oxford University Press, 1993.

Erickson, Mark. *Into the Unknown Together: The DOD, NASA, and Early Spaceflight*. Maxwell AFB, AL: Air University Press, 2005.

Feyock, Stephanie, comp. *Presidential Decisions: National Security Council Documents*. Washington, DC: The George C. Marshall Institute, 2006.

Getting, Ivan A. *All in a Lifetime: Science in the Defense of Democracy*. New York: Vantage Press, 1989.

Hall, R. Cargill, comp. *Presidential Decisions: National Security Council Documents—Supplement*. Washington, DC: The George C. Marshall Institute, 2006.

Houchin, Roy. *US Hypersonic Research and Development: The Rise and Fall of Dyna-Soar, 1944–1963*. New York: Routledge, 2006.

Hunley, J.D. *Preludes to U.S. Space-Launch Vehicle Technology: Goddard Rockets to Minuteman III*. Gainesville, FL: University Press of Florida, 2008.

Hunley, J.D. *U.S. Space-Launch Vehicle Technology: Viking to Space Shuttle*. Gainesville, FL: University Press of Florida, 2008.

Lewis, Jonathan E. *Spy Capitalism: ITEK and the CIA*. New Haven, CT: Yale University Press, 2002.

Lindgren, David T. *Trust but Verify: Imagery Analysis in the Cold War*. Annapolis, MD: Naval Institute Press, 2000.

Lonnquest, John C., and David F. Winkler. *To Defend and Deter: The Legacy of the United States Cold War Missile Program*. Champaign, IL: U.S. Army Construction Engineering Research Laboratories, 1996.

McDougall, Walter A. ... *The Heavens and the Earth: A Political History of the Space Age*. 2nd ed. Baltimore, MD: Johns Hopkins University Press, 1997.

MacKenzie, Donald A. *Inventing Accuracy: A Historical Sociology of Nuclear Missile Guidance*. Cambridge, MA: MIT Press, 1990.

Neufeld, Jacob. *The Development of Ballistic Missiles in the United States Air Force, 1945–1960*. Washington, DC: Office of Air Force History, 1990.

Richelson, Jeffrey T. *America's Secret Eyes in Space: The U.S. Keyhole Spy Satellite Program.* New York: HarperBusiness, 1990.
Richelson, Jeffrey T. *America's Space Sentinels: DSP Satellites and National Security.* Lawrence, KS: University Press of Kansas, 1999.
Ruffner, Kevin C., ed. *Corona: America's First Satellite Program.* CIA Cold War Records Series. Washington, DC: History Staff, Center for the Study of Intelligence, CIA, 1995.
Spires, David N. *Beyond Horizons: A History of the Air Force in Space, 1947–2007.* 2nd ed. Peterson AFB, CO: Air Force Space Command, 2007.
Stares, Paul B. *The Militarization of Space: U.S. Policy, 1945–1984.* Ithaca, NY: Cornell University Press, 1985.
Stumpf, David K. *Titan II: A History of a Cold War Missile Program.* Fayetteville, AR: University of Arkansas Press, 2000.
Temple, L. Parker. *Shades of Gray: National Security and the Evolution of Space Reconnaissance.* Reston, VA: AIAA, 2005.
Walker, Chuck. *Atlas: The Ultimate Weapon by Those Who Built It.* Burlington, Ontario, Canada: Collector's Guide Publishing (Apogee Books), 2005.

USSR

Harvey, Brian. *Russia in Space: The Failed Frontier?* Chichester, United Kingdom: Springer-Praxis, 2001.
Gorin, Peter A. Black "Amber": Russian Yantar-Class Optical Reconnaissance Satellites. *Journal of the British Interplanetary Society* 51(8): 309–320, 1998.
Johnson, Nicholas L., and David M. Rodvold. *Europe and Asia in Space 1993–1994.* Colorado Springs, CO: Kaman Sciences Corporation, 1995.
Lantratov, Konstantin. Soyuz-Based Manned Reconnaissance Spacecraft. *Quest: The History of Spaceflight Quarterly* 6(1): 5–21, 1998.
Podvig, Pavel, ed. *Russian Strategic Nuclear Forces.* Cumberland, RI: MIT Press, 2001.
Siddiqi, Asif A. The Soviet Co-Orbital Anti-Satellite System: A Synopsis. *Journal of the British Interplanetary Society* 50(6): 225–240, 1997.
Siddiqi, Asif A. Staring at the Sea: The Soviet RORSAT and EORSAT Programmes. *Journal of the British Interplanetary Society* 52(11/12): 397–416, 1999.
Siddiqi, Asif A. *Challenge to Apollo: The Soviet Union and the Space Race, 1945–1974.* Washington, DC: NASA History Division, 2000.
Whitmore, Paul. Red Bear on the Prowl: Space-Related Strategic Defense in the Soviet Union, Part I. *Quest: The History of Spaceflight Quarterly* 9(4): 22–30, 2002.
Whitmore, Paul. Red Bear on the Prowl: Space-Related Strategic Defense in the Soviet Union, Part II. *Quest: The History of Spaceflight Quarterly* 10(1): 54–62, 2003.
Zaloga, Steven J. *The Kremlin's Nuclear Sword: The Rise and Fall of Russia's Strategic Nuclear Forces, 1945–2000.* Scranton, PA: Smithsonian Institution Press, 2002.

14 Origin and Developments in Commercial Space

Patrick A. Stadter

CONTENTS

INTRODUCTION

During the more than 60 years of commercial space development, the primary markets have been in communication, navigation, imagery, and spacelift. Interest in emerging markets, such as space tourism, has increased, though to date this market remains one that encompasses a low-volume, high-cost segment of the commercial market. This chapter focuses on each of these functional areas in turn, including the origin and development of the business enterprise that characterizes the commercial space aspects of each function. It should be noted that a more in-depth consideration of commercial space enterprises must account for the incidental and indirect market interactions that leverage space systems. For example, consider global financial transactions, such as credit cards, which rely on space systems for timely, broadly distributed transactions; although they are not a direct space system, the impact of space communications on financial markets illustrates the indirect contributions of commercial space to a critical aspect of the world economy. Another illustration of commercial space on broader markets is the manner in which space technology development can drive terrestrial technologies—either directly through dual-use technology, such as robust integrated circuit developments, or indirectly, through the development of a highly capable, technical, and experienced workforce.

ORIGINS AND ARTHUR C. CLARKE

Best known internationally for his science fiction writing, many credit Arthur C. Clarke as the father of the commercial space industry because of his prescient 1945 paper establishing the idea of communication relays in space. This idea was born of his view of the confluence of then-emerging technologies of rocketry, wireless communications, and radar. In an article titled "Extra-Terrestrial Relays" and published in the October 1945 edition of *Wireless World*, Clarke presented the idea of synchronous orbiting "rocket stations" that could provide worldwide radio frequency connectivity.[1] The geostationary orbit, honored as the Clarke Orbit, is described in the paper: "One orbit, with a radius of 42,000 km, has a period of exactly 24 hrs. A body in such an orbit, if its plane coincided with that of the earth's equator, would revolve with the earth and would thus be stationary above the same spot on the planet." Indeed, Clarke's article nearly remained unpublished, as it was considered too futuristic, and Clarke himself noted that his patent counsel suggested that there was no value in owning a patent on it because it was so far-fetched.

COMMUNICATIONS

Arthur C. Clarke's 1945 concept for a system of geosynchronous satellites, depicted in Figure 14.1, characterized a means to provide worldwide communication relay functionality for applications such as television and telephony. Clarke largely viewed this as a commercial effort, indicating its value as being "indispensable to a world society."[2] Early efforts that enabled commercial space communication systems were experimental government programs to advance space-borne communications technology and demonstrate proof of concept. These included Signal Communication by Orbiting Relay Equipment in 1958 by the Advanced Research Projects Agency,[3] Echo in 1960 and 1964 for the National Aeronautics and Space Administration (NASA),[4] and Courier in 1960 by the Army Signal Research and Development Laboratory.[5]

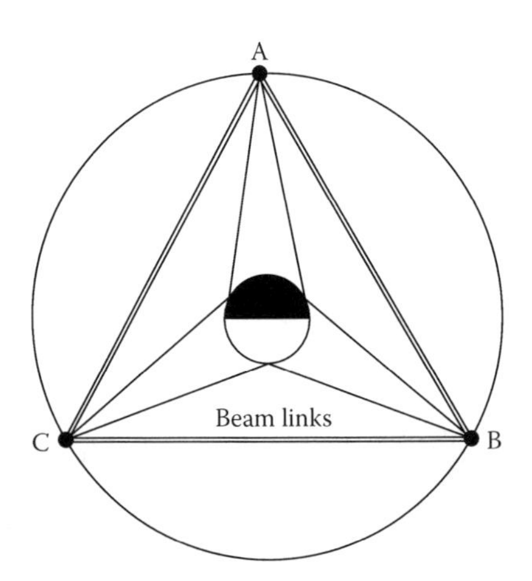

FIGURE 14.1 Geosynchronous communication system concept defined by Arthur C. Clarke.

Much of this experimentation led to the first operational space-based military communication service in 1967 in the form of the Initial Defense Communication Satellite Program, with 28 satellites placed into orbit; the system was declared operational in 1968 as military activity increased in Vietnam, and it was subsequently renamed the Initial Defense Satellite Communication System.[6] Commercial space communication service, however, was actually realized 2 years *earlier* than its military counterparts, in 1965 with *Intelsat I*. A timeline of the development of Intelsat and commercial space communications service during the 1960s provides an informative view of the technical and geopolitical influences that enabled the groundbreaking commercial space communication achievements.[7]

- December 20, 1961: The United Nations General Assembly adopts Resolution 1721, stating that global satellite communications should be made available on a nondiscriminatory basis.
- August 31, 1962: President John F. Kennedy signs the Communications Satellite Act, with the goal of establishing a satellite system in cooperation with other nations.
- August 20, 1964: The International Telecommunications Satellite Consortium (INTELSAT) is established on the basis of agreements signed by governments and operating entities.
- April 6, 1965: *Early Bird* (*Intelsat I*) is launched into synchronous orbit (Figure 14.2). This is the world's first commercial communications satellite, and "live via satellite" is born.

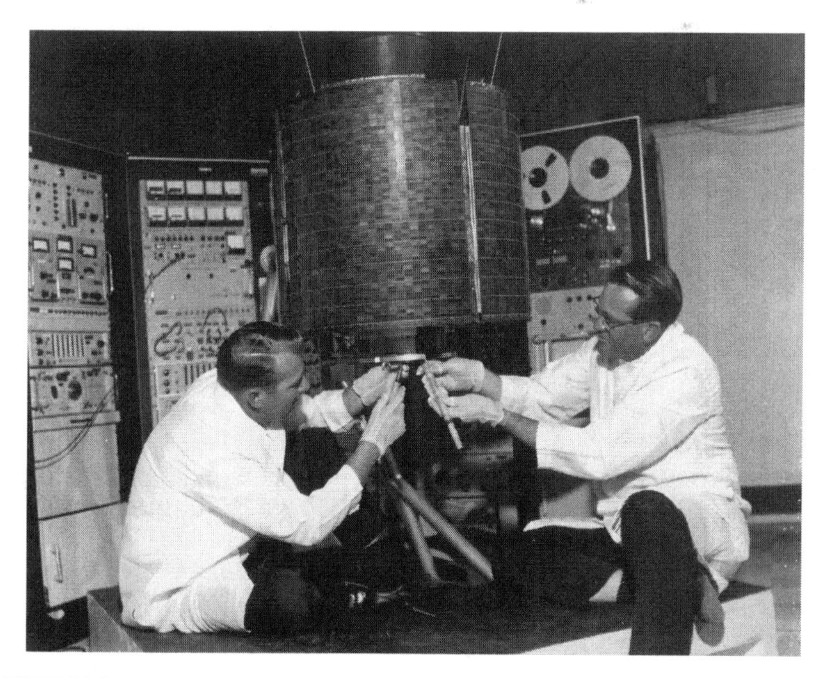

FIGURE 14.2 *Intelsat I*, also called *Early Bird*, was the world's first commercial communications satellite. Shown are technicians working on *Early Bird*. (Courtesy of NASA.)

- June 28, 1965: *Early Bird* begins providing television and voice services. Officials in the United States and Europe exchange greetings in a transatlantic ceremony introducing the new service.
- January 26, 1967: Commercial satellite service between the United States and Japan is established, with live television coverage in both countries.
- July 1, 1969: The world's first global satellite communications system is completed, with the *Intelsat III* satellite covering the Indian Ocean Region.
- July 20, 1969: Intelsat transmits television images of the Moon landing around the world—a record 500 million television viewers worldwide see Neil Armstrong's first steps on the Moon live via Intelsat.

NAVIGATION

In the autumn of 1957, in the wake of the Soviet Union's October 4 launch of its *Sputnik* satellite, scientists and engineers at The Johns Hopkins University Applied Physics Laboratory (JHUAPL) in Maryland adapted experiments in microwave spectroscopy and modified a shortwave receiver—creating one that could pick up very sensitive radio signals with a wire—to listen to the signals emanating from *Sputnik*. Upon processing the tape recordings with the aid of a wave analyzer, the results revealed a clear Doppler shift as a function of the motion of the satellite. While waiting for the satellite's next pass over the United States, the team realized that the slope of the Doppler shift could be used to ascertain the distance to *Sputnik*, and with a time estimate of *Sputnik*'s arrival over Washington, as broadcast by a Moscow shortwave radio station that had serendipitously been recorded, the satellite's orbit could be predicted with great accuracy.[8] For nearly 6 months, the team at JHUAPL perfected the techniques of orbit prediction by tracking the signals from *Sputnik II* and the first U.S. satellite, *Explorer I*, which was launched at the end of January 1958. It was near the conclusion of this period of research that one of the team's managers, Frank McClure, realized that if one can find a satellite from a listening station on Earth using the Doppler-shift data, then one can find the listening station on Earth from the orbit, and within a week he and a JHUAPL colleague had designed a navigation system on this principle.[9] The system would consist of four basic elements: (1) a satellite containing a highly precise crystal-driven clock or cycle counter, a frequency generator, and a dual-frequency radio transmitter to beam signals to Earth; (2) a network of tracking stations to measure the frequency of the received satellite signals; (3) an injection station, or communication channel, to permit ground engineers to insert the predicted orbital positions of the satellite (which they calculated using the previous day's tracking data) into the spacecraft's memory every 12 hours; and, of course, (4) the shipboard navigation set to receive and interpret the signals broadcast by the satellite. The proposed navigation system came to be known as Transit.[10] Developing the Transit system required a sequence of prototype spacecraft that ultimately proved the efficacy of the system implementation (Table 14.1) and build-out of the production system through a transition to industry.

The capability developed and implemented by the Transit system was the foundation for refining space-based navigation that the Department of Defense sought in 1973, when the follow-on global positioning system (GPS) was first conceived.[11]

TABLE 14.1
Experimental and Preproduction Transit Satellites

Mission Number	Launch Date	Status	Comments
1A	September 17, 1959	Failed to orbit	
1B	April 13, 1960		Navigation and tracking experiments
2A	June 22, 1960		Navigation and tracking experiments
3A	November 30, 1960	Failed to orbit	
3B	February 21, 1967		Navigation and tracking experiments
4A	June 29, 1967		Fleet navigation trials
4B	November 15, 1961		Fleet navigation trials
5A1	December 19, 1962		Failed 20 hours after launch
5A2	April 5, 1963		
5A3	June 16, 1963		Memory problems in orbit: never operational
5BN-1	June 28, 1963		Prototype with nuclear power source
5BN-2	December 5, 1963		Operational; satellite with nuclear power source
5BN-3	April 21, 1964	Failed to orbit	
5C1	June 4, 1964		Prototype with solar power only
O-4	June 24, 1965		Operational prototype
O-6	December 22, 1965		Operational prototype
O-8	March 25, 1966		Operational prototype
O-9	May 19, 1966		Operational prototype
O-10	August 18, 1966		Operational prototype
O-12	April 14, 1967		Operational prototype
O-13	May 18, 1967		Operational prototype
O-14	September 25, 1967		Operational prototype
TRIAD/ TIP-I	September 2, 1972		Transit improvement program (TIP) (failed 60 days after launch)
TIP-II	October 12, 1975		Never operational; solar panel problems
TIP-III	September 1, 1976		Never operational; solar panel problems
O-11/ TRANSAT	October 28, 1977		Operational prototype and beacon for range calibration

All of the experimental and prototype Transit satellites listed above were built for the Navy by The Johns Hopkins University Applied Physics Laboratory. Production Transit satellites were called Nova and were built by RCA Astro-Electronics under a Navy contract. The first Nova satellite was launched May 15, 1981.

The first GPS Navstar satellite was launched in 1978, and the full constellation of 24 was populated in 1993. While Transit leveraged Doppler-based navigation, the GPS system continuously broadcasts its own position and time. A GPS receiver receives this information from at least four satellites and computes the location of the receiver and the current time. While the U.S. GPS system dominates global navigation, other countries are considering or beginning to implement their own space-based navigation system, including the Russian Global Navigation Satellite System, the European Union's Galileo system, and the Chinese Compass/Beidou system concept.

The commercial aspect of space-based navigation is an interesting study in government-industry interaction. Specifically, the U.S. government (through the U.S. Navy and the Department of Defense) developed, implemented, and maintained the Transit system; similarly, the Department of Defense currently maintains the GPS system. The U.S. government provides signals to compute civilian navigation accuracies free of cost through GPS worldwide, and this dual civilian/government use was considered part of the design of the system from its inception. The ranges of applications that have evolved have been a strong impetus for the development of the commercial navigation system receiver market. Application breadth ranges from agriculture and surveying/mapping through trucking/shipping, aviation, archaeology, and emergency search-and-rescue support. These markets were slow to develop, remaining specialized and costly during the 1980s due mostly to the limited GPS coverage because the constellation was not fully populated until 1993.[12] As of 2006, the GPS market worldwide was more than US$20B.[13]

IMAGERY

The genesis of operational space-based imagery can be traced to the administration of U.S. President Dwight D. Eisenhower, which implemented a program to augment the U-2 aircraft reconnaissance program that surveiled the USSR.[14] Concern over U-2's vulnerabilities resulted in the establishment of what would become the National Reconnaissance Office in September 1961.[15] In fact, the Discoverer program announced in early 1958 had the stated goal of orbiting a series of scientific payloads, but it was actually a cover for the first U.S. photoreconnaissance satellite program, Corona. After a string of failures through 1960, the *Discoverer XIV* satellite was launched with the Corona photoreconnaissance camera-based payload and successfully performed its mission by returning imagery of approximately 1,650,000 square miles of the Soviet Union with a resolution on the order of 35 feet. The Corona program continued until 1972 and provided a strategic foundation to reconnaissance programs that would follow.[16] Corona launch sequence and discussion of conservation efforts relative to Corona can be found in Chapter 36 of this volume.

It was not until the 1980s that a series of events had an influence on the development of a commercial space imagery market,[17] including the following:

- Hybrid government/commercial satellite programs, such as NASA's Landsat program
- Development of purely commercial systems and successful launch of systems such as *OrbView-1/Microlab-1*, which was the first commercial weather satellite, and *UoSAT-5*, which was the first commercial microsatellite
- Planning and launch of satellites in the minisat category (100–500 kg) such as *TOMS/Earth Probe* and *OrbView-2/SeaStar*

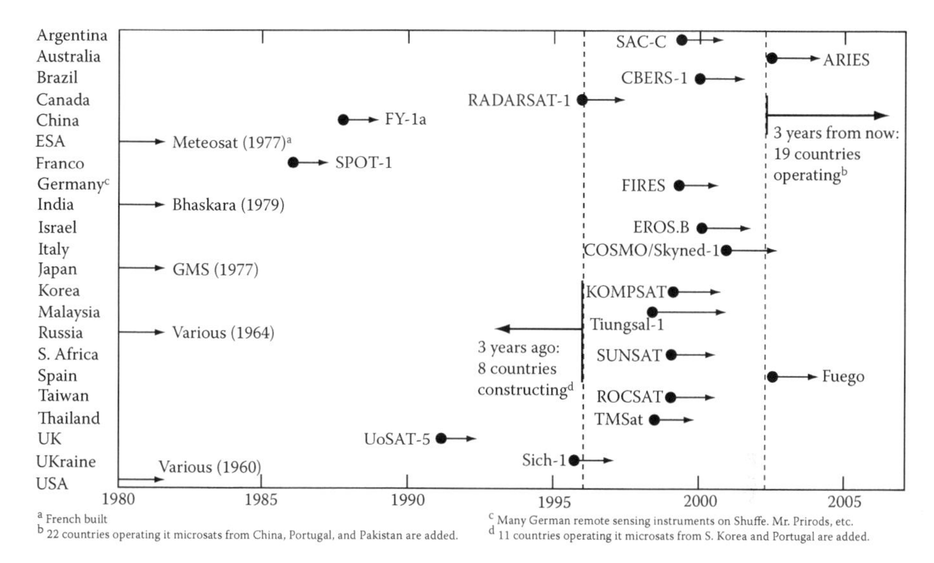

FIGURE 14.3 Worldwide commercial remote sensing system development showing initial system deployments by country, 1980–2007. (From Aerospace Corp.)

- Development of the microsat market (10–100 kg), including the commercial success of Surrey Satellite Technology Ltd. in the United Kingdom; and
- Development of government purchase of end-user data products, rather than satellite systems, from industry (e.g., NASA's approach to the *OrbView-2/SeaStar* ocean color program).

These and other developments helped develop the commercial imagery market on a worldwide scale. Analysis by The Aerospace Corporation[18] captures this development, and Figure 14.3 illustrates initial deployment of commercial imagery by country. In addition to the growing breadth of worldwide participation in commercial imagery, the quality and spectral content of imagery products have also seen significant growth. This growth is in part due to advances in imagery technology during the past several decades, but it is also due in part to improvements in processing and information extraction. The latter can be observed in specialized information that is realized by commercial data products on platforms with spectral content and resolution that was typically the domain of military and/or intelligence systems during the development of space-based imagery.

Electro-optical imagery can be categorized into classes as a function of the observation bands on the deployed system. The three general classes are as follows.[19]

- *Panchromatic*: characterized by a single, broad spectral band. Panchromatic imagery is often used for maximum spatial resolution because of the ability of the broad band to allow collection of more light relative to narrower bands.

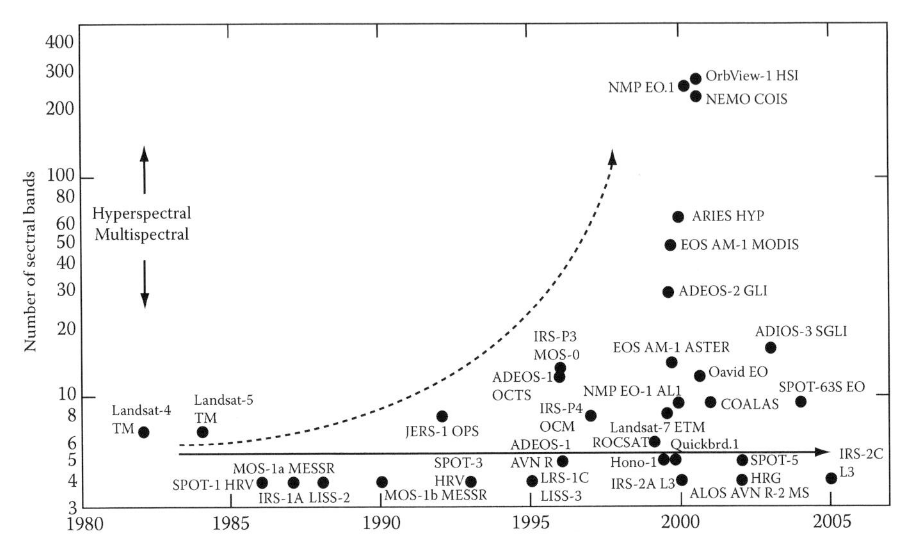

FIGURE 14.4 Trends in spectral bands for civil spaceborne visible/infrared imaging instruments, 1980–2007. (From Aerospace Corp.)

- *Multispectral*: characterized by up to tens of moderately wide spectral bands. Multispectral imagery provides additional information over panchromatic imagery well suited to specialized feature extraction, object discrimination, object classification, and object characterization based on spectral properties.
- *Hyperspectral*: characterized by systems utilizing one hundred or more narrow, contiguous spectral bands. Much like multispectral imagery, hyperspectral imagers are well suited to specialized target object characterization.

Trends in nonmilitary electro-optical systems, specifically related to the number of spectral bands characterized by the panchromatic, multispectral, and hyperspectral classifications, appear in Figure 14.4. As commercial imagers push strongly into the hyperspectral domain imagery products, the market forces that have developed include resource exploration (e.g., raw materials, petrochemicals), vegetation characterization, and farming.

SPACELIFT

A discussion of the development of commercial spacelift rightly begins with government efforts to develop launch vehicles for the purposes of lofting satellites into orbit during the Cold War between the United Sates and the Soviet Union. Two critical figures in the evolution of military rocketry into launch vehicles are Dr. Wernher von Braun and Dr. Robert H. Goddard. Wernher von Braun was responsible for the development of the German V-2 rocket, intended for combat by the Germans in World War II. It was after the war, in 1945, that he was recruited to the United States in the covert Operation Overcast.[20] After years of helping the United States develop its intercontinental ballistic missile program, Von Braun went on to develop the Saturn V launch vehicle for the NASA that was the spacelift foundation of the Apollo program. Also influential in the development of modern rocketry, Robert H. Goddard was the first person to develop and launch a liquid-fueled rocket, on March 16, 1926, and his development of liquid-fueled rocketry is a foundation on which modern launch vehicles are built. During the early development of launch vehicles, the focus remained principally on military and civilian scientific efforts, with the United States and the Soviet Union representing the primary developers. With the birth of the commercial space communications industry and passage in the United States of the Communications Satellite Act of 1962 under President Kennedy, the way was opened to private ownership of satellites. Launch capability, however, remained the domain of governments. The United States developed families of expendable launch vehicles, including the Atlas, Thor/Delta, and Titan, which still remain the basis for many current launch vehicles.[21] Other governments developed launch capability, including the Soviet Union, with the Proton and Cosmos launch vehicles. In an effort to reduce the cost of launch, the United States embarked on the development of the space shuttle (or Space Transportation System) in 1969; its first launch was achieved in 1981.[22] The shuttle was to be partially reusable, and a basis for its advertised reduction in launch costs was that it would serve as a means to launch commercial satellite payloads, resulting in regular, frequent flights. In fact, by policy until the 1986 loss of the *Challenger*, the space shuttle was the sole public-sector provider of launch services to the world market.[23] In the United States, two significant policy decisions affected the development of the commercial launch vehicle market: the Commercial Space Act (1984),[24] which allowed private enterprise to develop and operate expendable launch vehicles, and the Launch Services Purchase Act (1990),[25] which ordered NASA to purchase launch services commercially when viable for its primary payloads. Privatization of launch services even occurred in the former Soviet Union, when in 1994 Russia commenced selling part of its ownership of S.P. Korolev Rocket and Space Corporation Energia and became a minority owner in 1997. Immediately following the *Challenger* accident, the U.S. government pursued a policy to reinvigorate the expendable launch vehicle industry,[26] and market forces were sufficient to support the development of the current workhorses of the U.S. commercial launch industry, including the Delta and Atlas Evolved Expendable Launch Vehicles, as depicted in Figures 14.5 and 14.6. Private investment was also drawn to the launch vehicle industry, particularly in the late 1980s and beyond. At least nine companies were created to develop commercial launch services, among them Orbital Sciences Corporation, which in 1987 proposed the Pegasus air-launched vehicle. Other small launch vehicle endeavors have developed in the intervening years, including those that produced Microcosm's Sprite and the Space Exploration Technologies Corp. (SpaceX) Falcon (pursuing small, medium, and heavy lift services). In addition to these private endeavors, supported by government contracts, in 2006 NASA held a competition for Commercial Orbital Transportation Services (COTS), which serves to support development of commercial launch and transportation services to the International Space Station and as part of the Exploration Initiative created to develop a sustained human presence on the Moon and further exploration of Mars and beyond. The two initial COTS awards were given to Rocketplane Kistler (RpK)

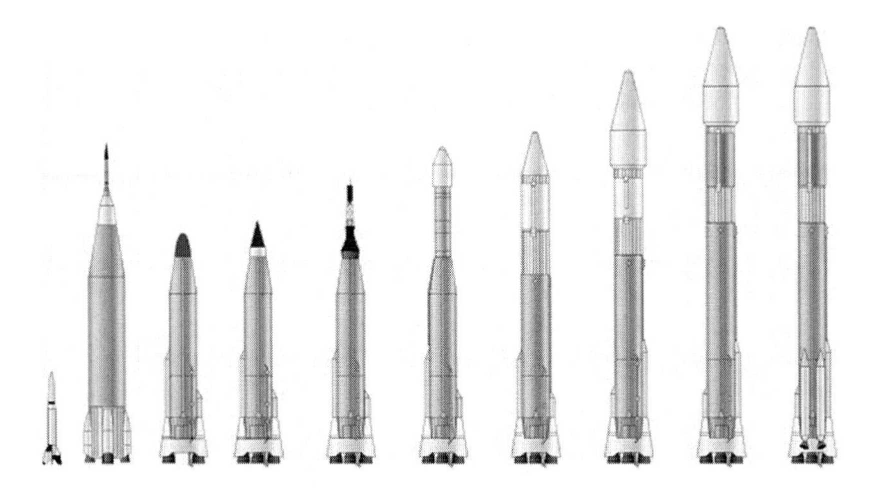

FIGURE 14.5 Atlas orbital launch vehicle. Country: United States. Status: retired 2004. (From Mark Wade, Astronautix.)

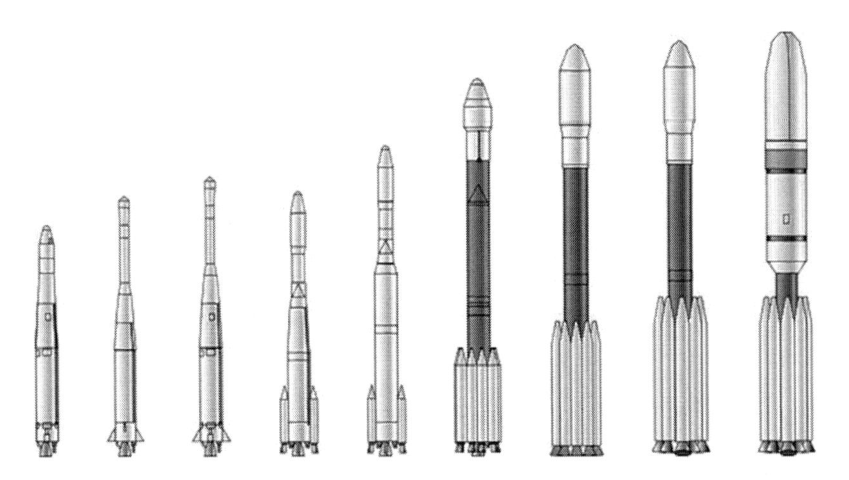

FIGURE 14.6 Delta orbital launch vehicle. Country: United States. Status: active. (Mark Wade, http://www.astronautix.com.)

and SpaceX, which subsequent recompeted, with Orbital Corp. winning a contract with the continued financial troubles of RpK.

CONCLUSION: CURRENT TRENDS

One of the more interesting trends in commercial space is that of space tourism. In part, the confluence of the worldwide dispersion of space system capability, the dissolution of the Soviet Union but continued international investment in the International Space Station (ISS), and the availability of capital with an invigorated entrepreneurial flavor in the period of the 1990s and 2000s has led to several pathfinders in private access to space. One of the most publicized was that of the self-financed personal space flight arranged by Dennis Tito, accepted by the Russian Space Agency for travel on a Soyuz to the ISS in 2001. Further efforts to invigorate the concept of space tourism include the Ansari X Prize, which offered US$10M to the first private team to build and launch a spacecraft capable of carrying three people to 100 km above the Earth's surface, twice within 2 weeks.[27] This prize was claimed on October 4, 2004, by Mojave Aerospace Ventures for the flight of *SpaceShipOne*, which was designed by Burt Rutan of Scaled Composites and funded by Virgin Galactic. These efforts have resulted in a nascent space tourism industry, currently dominated by Virgin Galatic's plan to begin regular space tourist flights based on the *SpaceShipOne* design; initial costs of a trip were targeted at US$200,000. Given time, reliability, and market forces, space tours may become available at affordable rates for the average private adventurer.

REFERENCES

1. *Wireless World* paper. October 1945. 305–308.
2. Ibid.
3. Brown, S.P., and G.F. Senn. 1960. Project SCORE. *Proceedings of the IRE* 48(4).
4. Special Issue on Project Echo. 1961. *Bell System Technical Journal* 40(4).
5. Senn, G.F., and P.W. Siglin. 1960. Courier Satellite Communication System. *IRE Transactions on Military Electronics* MIL-4(4).
6. Martin, D.H. 2001/2002. A History of U.S. Military Satellite Communication Systems. *Space Communications* 3(1). http://www.aero.org/publications/crosslink/winter2002/index.html.
7. http://www.intelsat.com/about-us/history/intelsat-1960s.asp.
8. http://history.navy.mil/books/space/Chapter1.htm
9. Klingaman, W.K. 1993. *APL, Fifty Years of Service to the Nation: A History of the Johns Hopkins University Applied Physics Laboratory.* Laurel, MD: The Laboratory.
10. Ibid.
    Federici, G. 1997. *From the Sea to the Stars: A History of U.S. Navy Space and Space-Related Activities*. http://www.history.navy.mil/books/space.
11. http://www.beyonddiscovery.org/content/view.page.asp?I=464.
12. Stansell, T. 2006. *Dual-Use GPS*. Institute of Navigation. http://www.rms-ion.org/Presentations/Dual-Use%20GPS.pdf.
13. Pellerin, C. 2006. United States Updates Global Positioning System Technology: New GPS Satellite Ushers in a Range of Future Improvements. http://www.america.gov/st/washfile-english/2006/February/20060203125928lcnirellep0.5061609.html.
14. Moorman, T.S., Jr. 1999. The Explosion of Commercial Space and the Implications for National Security. *Airpower Journal*. http://www.airpower.maxwell.af.mil/airchronicles/apj/apj99/spr99/moorman.html.
15. Haines, G.K. 1997. *The National Reconnaissance Office (NRO): Its Origins, Creation and Early Years.* Chantilly, VA: National Reconnaissance Office.
16. Guerriero, R.A. 2002. Space-Based Reconnaissance. *Army Space Journal.* http://www.armyspace.army.mil/spacejournal/toc.asp?IID=2.
17. Glackin, D.L., and G.R. Peltzer. Civil, Commercial, and International Remote Sensing Systems and Geoprocessing. http://www.aero.org/publications/glackin/glackin-1.html.
18. Ibid.

19. Ibid.
20. Ordway, F.I., III, and M.R. Sharpe. 1979. *The Rocket Team.* Apogee Books Space Series, no. 36. Burlington, Canada: Collector's Guide Publishing.
21. Launius, R.D., and D.R. Jenkins, eds. 2002. *To Reach the High Frontier: A History of U.S. Launch Vehicles*. Lexington, KY: University Press of Kentucky.
22. Heppenheimer, T.A. 1999. *The Space Shuttle Decision: NASA's Search for a Reusable Space Vehicle*. Washington, DC: National Aeronautics and Space Administration.
23. Congressional Budget Office. 1986. *Setting Space Transportation Policy for the 1990s*. Washington, DC: Congressional Budget Office.
24. http://www.reagan.utexas.edu/archives/speeches/1984/103084i.htm.
25. http://www.space-frontier.org/commercialspace/lspalaw.txt.
26. Launius and Jenkins (note 21).
27. http://www.xprize.org/x-prizes/ansari-x-prize.

15 Maturing Space Age

Ralph L. McNutt, Jr.

CONTENTS

The Space Age is approaching middle age by some accounts, having recently turned 50. Maturity suggests time for reflection (if not worse): initial ideals and great dreams are not always realized, but on the other hand, hard work has, hopefully, brought some major successes. The latter is certainly true of our push into space.

INTRODUCTION: BEGINNINGS AND OVERVIEW

EARTH ORBIT

After 50 years of effort, the good news is that we know what it takes to get into space and be successful there—that is also the bad news, because we have also learned that successes do not come cheaply and failures are seldom graceful. Born of Cold War competition, achievement in space continues, but not at the whirlwind pace of early days. The downside is that the continued successes are not noticed as often as they were once upon a time. Some statistics can provide insight into those accomplishments though. In the 50 years since the launch of the 83.6-kg *Sputnik 1* from Tyuratam on October 4, 1957, the human race has successfully launched some 8,458 items into low Earth orbit (LEO) or beyond, with a combined mass in excess of 28,900 metric tons,* more than the displacement of an Ohio-class nuclear missile submarine. The space infrastructure still has military satellites, but it has expanded to include commercial satellites, notably for communications, but also for weather monitoring and prediction and for survey, as well as scientific satellites that look back to our own planet to understand its dynamics and out to the stars and galaxies to understand theirs. In addition to these sophisticated robotic extensions of ourselves, the International Space Station (ISS) has provided humanity's first off-world settlement. The vast majority of these items are all in LEO, with shielding of human crews against radiation effects still provided by the magnetic field of the Earth, which forms a bubble, or *magnetosphere,* in the expanding supersonic atmosphere of the Sun, or solar wind, direct knowledge of which is entirely a product of the Space Age itself. The magnetosphere, enhancing the shielding provided by our planet's atmosphere from the ubiquitous flux of galactic cosmic rays in the solar system, provides this shielding just beyond geostationary orbit, at ~6 Earth radii (R_E ; 1 R_E = 6,378 km ~ 3,963 mi at the equator) to ~10 R_E in the sunward direction and further as one goes toward the magnetospheric flanks (Figure 15.1).

FIGURE 15.1 Schematic of the Earth's magnetosphere (right) formed by the dynamic pressure of the solar wind from the Sun (left) distorting the intrinsic magnetic field of the Earth. (Courtesy of NASA.)

* An analysis (McNutt, Jr., unpublished) shows that the first 50 years of satellite launches, beginning with *Sputnik 1* in 1957 and going through *Dawn* in 2007, includes some 8,458 items on 4,898 launches, of which 4,560 (93.1%) were successful. Of the total, there were 241 human-crewed orbital flights (plus 7 suborbital flights), including 118 flights of the space shuttle. Although the shuttle flights account for only 2.4% of the total launches, they account for about 13,105 of the 28,914 metric tons lofted to orbit (45.3%). (Note that all of this information has been compiled from a variety of unclassified sources, many of which incorporated "best guesses" for masses of military systems.)

To the Moon

Casting further away to the Moon and cislunar space (cislunar came from Latin and means "on this side of the moon" or "not beyond the moon"), out to some 60 R_E, the protection of the Earth's magnetic field is no longer assured. The Moon is considered a first step in moving back away from the Earth, and there is a new interest around the world in returning humans and more robotic probes to the Moon (probes now in orbit about the Moon or soon to be include *Chang'e 1*, China; *Kaguya*, Japan; *Chandrayaan-1*, India; and *Lunar Reconnaissance Orbiter*, United States[1]). From 1968 to 1974, only twenty-seven men (three each on *Apollo 8, 10, 11, 12, 13, 14, 15, 16,* and *17*), all American astronauts, have traveled so far from Earth, and of those, only twelve have actually walked on the lunar surface (two each from *Apollo 11, 12, 14, 15, 16,* and *17*).

Beyond the Moon

Expeditions even further afield, in particular to Mars, but not limited to Mars only, have been discussed as serious engineering objectives going as far back at the later 1940s (Wernher von Braun's *Das Marsprojeckt*).[2] Although a human mission to Mars was called for explicitly in the Bush Administration's Vision for Space Exploration (now the U.S. Space Policy), achieving this goal before 2040 is considered optimistic by most, unless significant new resources (money) are made available in the near future. Even less-challenging human missions beyond the Moon to the Sun-Earth Lagrange point downstream in the solar wind from the Earth (L_2) or to near-Earth asteroids as practice for Mars[3] remain ideas for later decades. In the meantime, the upstream Sun-Earth Lagrange point (L_1) has been used for a variety of current robotic spacecraft [the *Solar and Heliospheric Observatory* (*SOHO*) and the *Advanced Composition Explorer* (*ACE*)], as well as stopping points for previous missions (*Wind* and *Genesis*), and the L_2 point is the planned operational location for the James Webb Space Telescope, the successor to the Hubble Space Telescope, but operating in the infrared part of the spectrum. Both L_1 and L_2 are located ~1.5M km from the Earth along the Sun-Earth line. While these are points of gravitational equilibrium, they are unstable, and spacecraft must be actively controlled to remain in halo orbits about these balance points in space (Figure 15.2).

Interplanetary space has been the realm of robotic probes only, and will likely remain so for several decades to come, as is discussed in later chapters. Again, with origins in the heady days of the Cold War, both the United States and Soviet Union began with a competition first of robotic probes to the Moon, and then to Venus and Mars. The Soviet *Luna 1* was the first human-made object to achieve escape velocity from the Earth and orbit the Sun after missing the Moon by some 6,400 km in early January 1959.[4] The American 6.1-kg *Pioneer 4* (Figure 15.3) followed its Soviet counterpart 2 months later. The Soviet *Luna 3* obtained the first view of the far side of the Moon, the side that is not visible from Earth.

In 1961, the robotic assault on the planets began in earnest with the second Soviet Venera launch on February 12. Although the spacecraft failed prior to a Venus flyby at ~100,000 km, the spacecraft did achieve a heliospheric orbit, as did the earlier

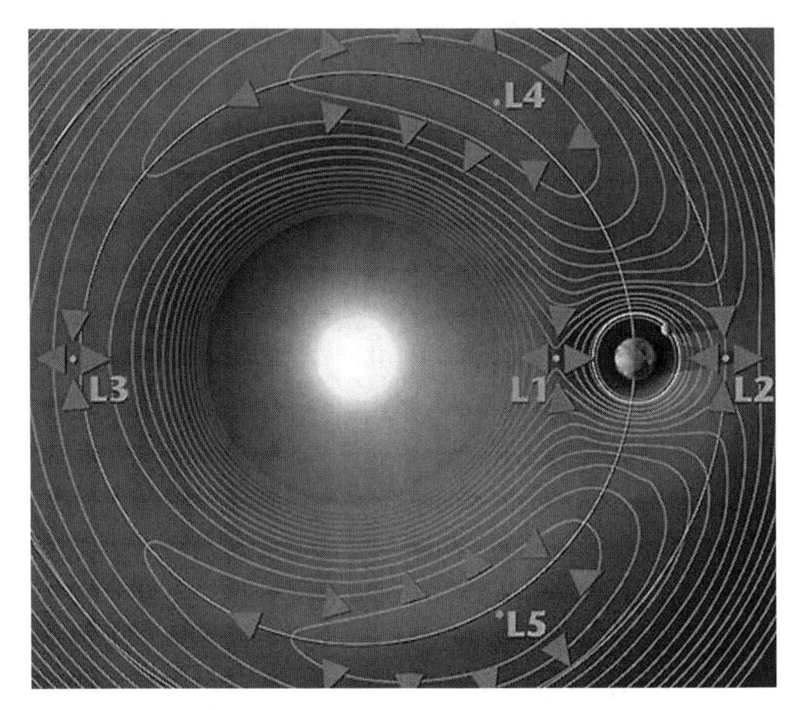

FIGURE 15.2 Schematic of the five Lagrange points in the Sun-Earth system (not to scale) in a frame of reference rotating with the Earth. Contours show equal gravitational potentials. Points L_1, L_2, and L_3 are at unstable points and L_4 and L_5 are at stable points for articles located there. (Courtesy of NASA.)

U.S. *Pioneer 5* probe. During this year, the Jet Propulsion Laboratory (JPL) began developing a series of deep-space Ranger spacecraft, outfitted with a variety of scientific experiments. During these years, both countries faced problems and failures with both spacecraft and launch vehicles. *Ranger 4* became the first U.S. probe to impact the Moon on April 26, 1962.

As Soviet efforts focused on system commonality for both Venus and Mars probes in 1962, JPL received approval for using the Ranger spacecraft as a starting point for designing a robotic spacecraft to function as a flyby mission to Venus. The first attempt, *Mariner 1*, was destroyed by the range safety officer 293 seconds after launch. A second unit was readied and was launched on an Atlas Agena B rocket on August 27, 1962 (Figure 15.4A and B). In the first successful mission to another planet, *Mariner 2* passed by Venus on December 14, 1962 at just under 35,000 km (Figure 15.5). *Mariner 2* showed that the planet possessed thick clouds and a surface temperature of at least 425°C. The 203.6-kg spacecraft carried six instruments, set a communications distance record of 87.4M km, and provided conclusive proof of the supersonic quasiconstant extension of the upper atmosphere of the Sun, the solar wind. A month prior to the Venus flyby of *Mariner 2*, the Soviet Union launched *Mars 1*, the first spacecraft to fly past Earth's closest outer planetary neighbor in the solar system.

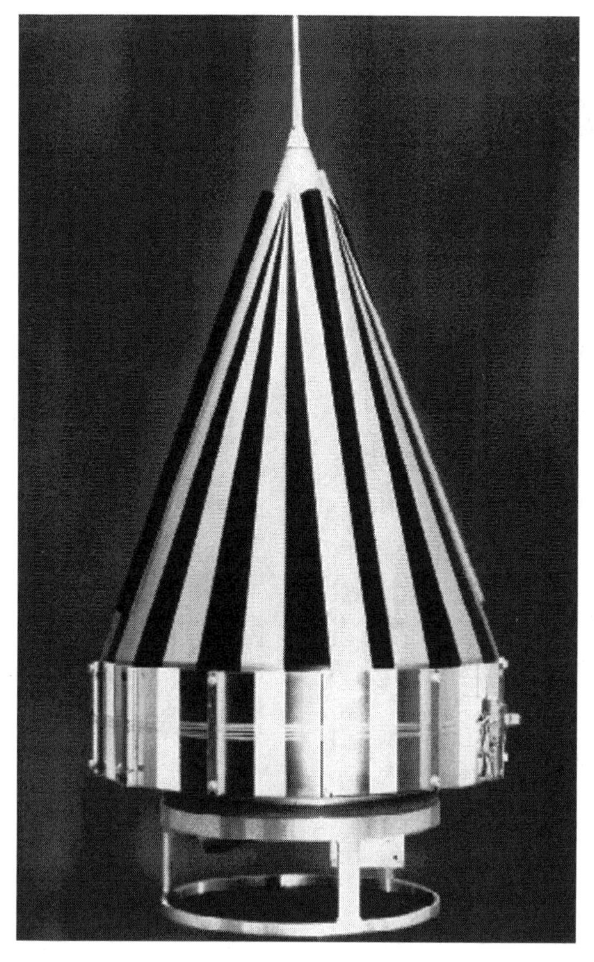

FIGURE 15.3 *Pioneer 4*. (Courtesy of NASA.)

PUTTING "SCIENCE" INTO "SPACE SCIENCE"

Roughly concurrent with the establishment of the National Aeronautics and Space Administration (NASA) in the United States, the National Academy of Sciences established the Space Studies Board to serve as a focus for space research. The board acted as a focus for the establishment of space science as a discipline (or discipline group) with an eye toward developing a methodology for selecting appropriate scientific investigations. This sort of thinking had already been inaugurated with the activities of the International Geophysical Year (IGY), with a significant focus on both Antarctica and space.[5]

An initial survey of space research was conducted as part of a summer study,[6] and a more detailed report was issued in 1966.[7] Various working groups were set up with a focus on various bodies in the solar system and an overall recommendation for "planetary exploration as the most rewarding scientific objective for the

FIGURE 15.4A *Mariner 2* being readied for launch. (Courtesy of JPL.)

1970–1985 period." Recommendations included a combination of scientific objectives and goals, as well as means for pursuing those goals, both programmatically and technically. Items included (1) study of the use of Saturn V launch vehicles to enable scientific exploration of the planets, (2) a dropsonde into Venus's atmosphere, (3) improvements in deep-space communications capabilities, including possible use of laser communications, (4) improvements in rocket pointing control, (5) use of an Orbiting Astronomical Observatory for planetary observations, and (6) NASA support of ground-based optical observatories.

The report also includes detailed scientific priorities for investigating Mars, the Moon, Venus, the outer planets, small solar system objects, and Mercury. A

FIGURE 15.4B *Mariner 2* in flight configuration (JPL).

technology chapter took up the question of deep-space information transfer, with a focus on communications, broken down into modulation of laser beams, return of high-resolution photographic film images, and improvements in radio communications over those that would accrue with the completion of the 210-ft. (64-m) dishes then under construction for what would become the Deep Space Network 70-m dishes. The suggested approach was to mass-produce dishes of between 85- and 120-ft. diameters and array these to increase the capabilities of ground receipt of data. This overall format and outline has been used many times over in scientific planning of the missions on the cutting edge—and these have remained the missions to the planets and beyond.

With the human exploration program focused on fulfilling President John F. Kennedy's challenge, hardware development accelerated with incremental steps taken through Mercury, Gemini, and Apollo. At the same time, weather and communications satellite development and implementation continued, but the peak of robotic development went into LEO and geosynchronous satellites for a variety of military applications, including launch detection, communications, photoreconnaissance, and signals intelligence—some of which are only now being fully declassified.

Planetary probe trajectories were scrutinized with a eye toward enabling robotic missions farther and farther from Earth.[8] At this time, it was realized that a planetary

FIGURE 15.5 A long scroll of data from Venus, seen in front of JPL's *Mariner 2* mission board. (Courtesy of NASA-JPL.)

alignment that occurs only ever 176 years was approaching—a means to provide for a "grand tour" of the solar system beyond the asteroid belt.[9]

The small Pioneer spacecraft were being launched into heliocentric orbits to monitor effects in interplanetary space for astronaut safety in Apollo (*Pioneer 5, 6, 7, 8,* and *9*), while the Mariner spacecraft, building upon the *Mariner 2* success, were targeting Venus (*Mariner 5*) and Mars (*Mariner 6, 7,* and *9*). Planetary probes in the Mars and Venera series were being launched by the Soviet Union with increasing success as well. With international meetings such as those of the Committee on Space Research held in a neutral country, scientific results could be shared by the two superpowers.

"I Was Strolling on the Moon One Day ..."

With *Apollo 11* hardly back from the Moon, scientific and engineering programs were in full swing, and scientific plans to benefit from the investment in hardware were made.[10] Exploration plans and recommendations for *Apollo 12* through *Apollo 20* focused on lunar age measurements, geochemistry and petrology, geophysics, and geology and geomorphology, all with plans for study of returned samples in appropriate laboratories on Earth. Approaches to instrumentation and investigation selection were described drawing upon experiences both with previous missions and, indirectly, upon experience gleaned in the Antarctic during the IGY.

While the era of manned lunar exploration was short—from the *Apollo 11* landing on July 20, 1969, through the departure of *Apollo 17* from the Taurus-Littrow lunar landing site on December 14, 1972—the missions did provide the largest cache of totally unprocessed material from beyond the Earth and enabled a revolution in our understanding of the Moon's origin and the chronology of the Earth-Moon system.[11] The 382 kg of "moon rocks" was a scientific treasure trove and certainly invigorated the debate for the scientific need for returning samples to Earth-based laboratories for analysis (Table 15.1).[12]

Changing Priorities

Set against the backdrop of a costly, unpopular war in Vietnam, and with a successful conclusion of the race to the Moon—the United States won—NASA's budget fell under the axe of the Bureau of the Budget (the precursor of the Office and Management and Budget), and follow-on missions to the Moon were no longer options. Not only were plans for *Apollo 20* scrapped, but also the flight hardware for *Apollo 18* and *Apollo 19* was mothballed, and the Saturn V would never fly again.[13] As the portions of NASA that had supported human flight began to look toward a new space transportation system that could enable cheaper transfer to orbit, the robotic drive to the planets—including the Moon as a footnote—continued.

Scientific panels met under the aegis of the National Research Council to provide independent advice to NASA—which was also read by the Congress—on an appropriate scientific strategy for exploring the solar system. The most recent work, a "Decadal Survey," was produced in 2002,[14] with a long genealogical heritage building upon how both human and robotic space exploration had progressed since Apollo (Figure 15.6).[15]

Congressional hearings in 1975[16] fed into a large study by NASA undertaken for the NASA Administrator that year. The final report[17] provided a comprehensive look (some would say a wish list) at what NASA was doing and supposedly could and should do for the final quarter of the twentieth century. While the hearings put forward some very expansive and ambitious goals (e.g., solar power satellites by Peter Glaser, a program of interstellar voyages and exploration by Robert Forward, and a full-up commercial space presence by Kraft Ehricke, to name but a few), the *Outlook for Space* report provided an overall summary of Earth-resource, in-space-observatory, space-plasma-physics, and planetary-science spacecraft, along with how all of these efforts would cross-reference across the agency.

TABLE 15.1
Human Lunar Explorations (NASA)

Mission number	Disposition	Number of lunar samples returned	Launch date	Landing date	Landing site	Landing site latitude	Landing site longitude	Extra vehicular activity time (hours)	Traverse (km)
7	Earth orbit checkout – first manned flight								
8	Lunar flyby								
9	Earth orbit checkout of Lunar Excursion Module (LEM), first manned LEM flight								
10	Lunar orbit								
11	Lunar landing	58	July 16, 1969	July 20, 1969	Mare Tranquilitatis	0.674° N	23.473° E	2.53	0.25
12	Lunar landing	69	November 14, 1969	November 19, 1969	Oceanus Procellarum	3.014° S	23.419° W	7.75	1.35
13	Lunar flyby, landing aborted								
14	Lunar landing	227	January 31, 1971	February 5, 1971	Fra Mauro	3.645° S	17.471° W	9.38	3.45
15	Lunar landing	370	July 26, 1971	July 30, 1971	Hadley Rille	26.132° N	3.634° E	19.13	27.9
16	Lunar landing	731	April 16, 1972	April 20, 1972	Descartes	8.973° S	15.499° E	20.23	27
17	Lunar landing	741	December 7, 1972	December 11, 1972	Taurus-Littrow	20.188° N	30.755° E	22.07	35
18	Canceled lunar landing								
19	Canceled lunar landing								
Totals		**2,196**						**81.1**	**95.0**

Source: Adapted from Heiken, G., D. Vaniman, and B.M. French, eds. 1991. *Lunar Sourcebook: A User's Guide to the Moon*. New York: Cambridge University Press. Also see http://nssdc.gsfc.nasa.gov/planetary/lunar/apolloland/html.

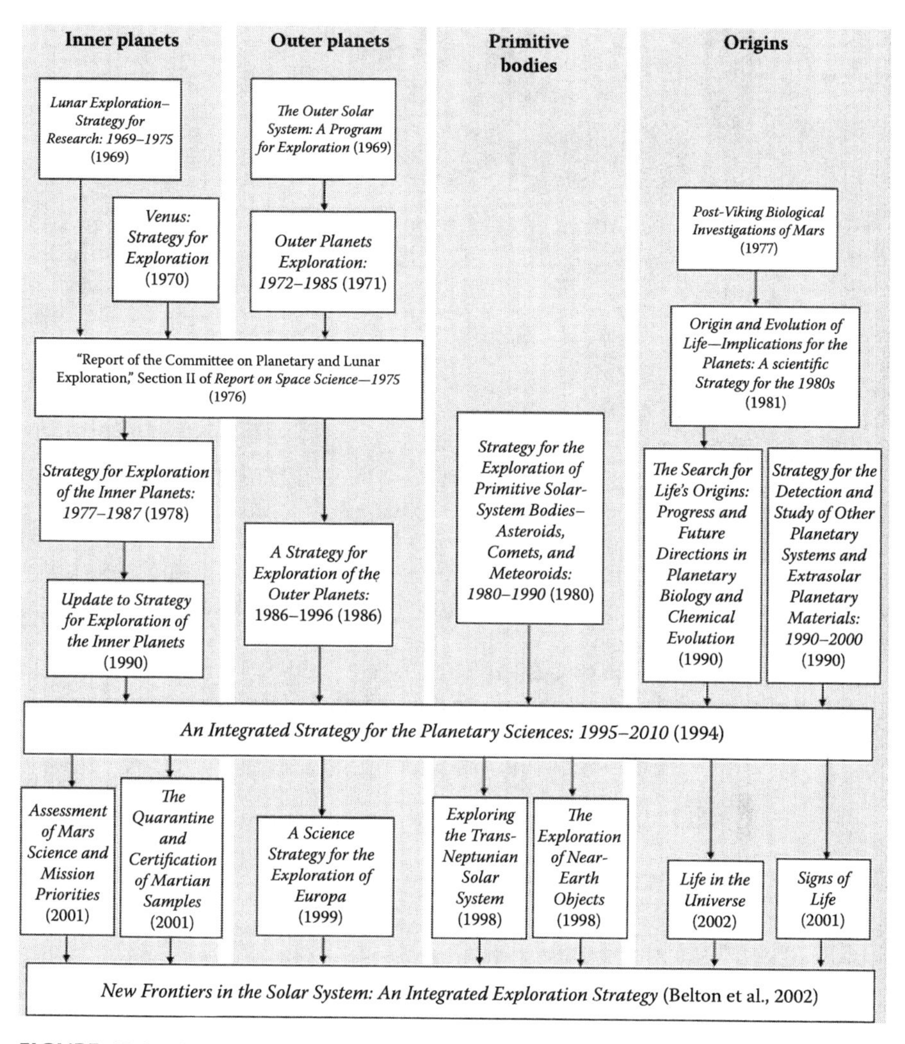

FIGURE 15.6 The "family tree" of Space Science Board advice to NASA regarding the exploration of the solar system. (Courtesy of NASA.)

By this time, the *Mariner 10* probe had reached and flown by Mercury three times, following the proof of the utility of gravity assists by using a boost from Venus. *Pioneer 10* and *Pioneer 11* had proven that spacecraft could successfully cross the asteroid belt, Mariner-Jupiter-Saturn probes (later renamed *Voyager 1* and *Voyager 2*) were being readied for a scaled-back Grand Tour mission, and *Pioneer 12* and *Pioneer 13* were being readied for resumption of U.S. exploration of Venus, a planet of Soviet exclusivity for some years, with the first image of the surface of another planet provided by *Venera 9* following its landing on October 22, 1975 (Figure 15.7). With all of the ongoing successes in exploration of the solar system by the Soviet Union and the United States, as well as the joint Apollo-Soyuz Test

FIGURE 15.7 The single image returned from *Venera 9* following landing on Venus. Transmission from the surface ceased 53 minutes after landing. Raw telemetry was 6-bit and about 115 × 512 pixels. The image shown follows a significant amount of processing of the raw image. (Courtesy of Don P. Mitchell, http://www.mentallandscape.com/Copyright.html.)

Project in Earth orbit by the two superpowers, the Moon receded as a human outpost goal during this phase of the Cold War.

Plans for manned Mars missions were shelved as well,[18] and with no identifiable need for nuclear power in space, continuing tight budgets for NASA, and the illness of sponsor Senator Anderson, the Nuclear Engine for Rocket Vehicle Application (NERVA) program[19] was terminated in 1972. Significant NASA funds were shifted to support the space shuttle program with a new start in fiscal year (FY) 1973. Nuclear thermal rocket development work in the U.S. came to a hard stop with the FY 1974 budget.

In 1975, the twin Viking probes left for Mars, entered Mars orbit a year later, and provided landers for the surface. Corrected for inflation, the Viking program was—and remains—the most expensive robotic program conducted by the United States. The failure to find life on the Red Planet after the large expenditure basically shut down Mars initiatives in the United States for the next 20 years.

THE "OUTLOOK" AND THE TRACK RECORD

In looking forward, the Outlook report provided a framework used implicitly for NASA Roadmaps and Academy Surveys in looking at planetary exploration for the next 25 years. Tables 15.2a and 15.2b illustrate both how the 1975 program fared and how plans evolved through various programmatic actions. Of the 29 missions proposed, with an estimated development cost of $15B (in 1975 U.S. dollars), the concepts evolved into a total of forty-seven missions, of which one had been lost and nineteen completed or in development by September 2003. By this time, many of these missions had been completed in the new competitive Discovery program, which was started in 1992.[20] Some of the missions were too expensive or had technological challenges that could not be accomplished under Discovery rules and are labeled as Discovery II—a provisional name that effectively became the New Frontiers line of competed missions.

The Discovery program began with a workshop at the San Juan Capistrano Research Institute in 1992. Although one hundred concepts for missions from the planetary community were initial submitted, twenty-seven were withdrawn by the submitters prior to discussion at the workshop (Table 15.3a–d). A brief depiction of science merit versus risk of implementation, as perceived for these missions at that time, is given in Table 15.4.

TABLE 15.2a

Disposition of *Outlook for Space* Planetary Missions: Part I

Program Supplements (1983); Augmentations (1986) (System Number)	Solar System Exploration Division Strategic Plan (1991)	SSES Recommendations (1994)	Space Science Enterprise Strategic Plan (1997)	Space Science Enterprise Strategic Plan (2000)	Implemented Mission	Type	Transport on	Illustrative Flight Dates	Implemented Launch Vehicle	Implemented Launch
	Mercury orbiter, candidate for FY 1999 FY 2003		Mercury Orbiter, post-2005	MESSENGER in development	MESSENGER	Discovery	Shuttle/ Tug + SEP	1985	Delta II 2925H	2004
							Shuttle/ Tug + SEP	1999		
Venus Atmospheric Probe, supplement mission	Discovery Venus Probe, candidate for FY 1999 FY 2003		Venus Laboratory post-2005			Rejected Discovery				
				Venus Surface Sample Return post-2007			Shuttle/ Tug	1994		
Mars Surface Probe, supplement mission	Mars Network (MESUR: Mars Environmental Survey), recommended for 1995	MESUR Network abandoned because of cost								

(Continued)

TABLE 15.2a (*Continued*)

Program Supplements (1983); Augmentations (1986) (System Number)	Solar System Exploration Division Strategic Plan (1991)	SSES Recommendations (1994)	Space Science Enterprise Strategic Plan (1997)	Space Science Enterprise Strategic Plan (2000)	Implemented Mission	Type	Transport on	Illustrative Flight Dates	Implemented Launch Vehicle	Implemented Launch
Mars Aeronomy Orbiter, supplement mission					Mars Climate Orbiter, lost near orbit insertion	Strategic			Delta II 7425	1998
						Rejected Discovery				
Mars Sample Return (~1996–1998), augmented program mission (1080)	Mars Sample Return, candidate for FY 1999 FY 2003		Mars Surveyor Program Sample Returns, First of three launched in 2005	Mars Exploration Program, Sample Return; Mars 2011			Shuttle/ Tug	1988		
Galileo Probe (approved mission) (1081)		Galileo in transit to Jupiter; Measure-Jupiter	Completed		Galileo Probe	Strategic	Titan IIIE/ Centaur	1980	Shuttle/PAM D with Galileo Orbiter	1989

Saturn Flyby/ Probe, supplement mission (1082)		Saturn Probe, complement Cassini					Shuttle/ IIUS	1984		
			Titan Organic Explorer post-2005	Titan Explorer post-2007			Shuttle/ Tug	1991		
Uranus Flyby/ Probe, supplement mission (1084)	Uranus Orbiter/Probe, candidate for FY 1999 FY 2003						Shuttle/ Tug	1984		
							Shuttle/ Tug	1997		
Comet Nucleus Sample Return, augmented program mission (1086)	Comet Nucleus Sample Return, candidate for FY 1999 FY 2003			Comet Nucleus Sample Return						
Comet Atomized Sample Return, supplement mission (1086)				Stardust in transit	Stardust	Discovery	Shuttle/ Tug + SEP	2000	Delta II	1999

(*Continued*)

TABLE 15.2a (*Continued*)

Program Supplements (1983); Augmentations (1986) (System Number)	Solar System Exploration Division Strategic Plan (1991)	SSES Recommen-dations (1994)	Space Science Enterprise Strategic Plan (1997)	Space Science Enterprise Strategic Plan (2000)	Implemented Mission	Type	Transport on	Illustrative Flight Dates	Implemented Launch Vehicle	Implemented Launch
Venus Radar Mapper (1988), core program mission (1087)		Extended Magellan mission completed 1994	Completed		Magellan	Strategic	Shuttle/ Tug	1989	Shuttle/IUS	1989
		Venus landers	Venus Laboratory post-2005				Shuttle/ US	1985		
				Lunar Giant-Basin Sample Return post-2007			Shuttle/ US	1988		
	Lunar Surface Science, candidate for FY 1999 FY 2003									
Mars Geoscience/ Climatology Orbiter (1990), core program mission (1090)		Mars Observer lost 1993, begin recovery of science ($574M, FY 1992$)	Mars Surveyor Program Orbiters	Mars Exploration Program, Orbiters; Mars Reconnaissance Orbiter 2005 in development	Mars Global Surveyor + Mars Odyssey	Strategic	Atlas Centaur	1981		

Galileo Orbiter (approved mission) (1091)		Galileo in transit to Jupiter	Galileo in operation in Jupiter orbit	Galileo Europa Mission (Galileo Extended Mission)	Galileo	Strategic	Shuttle/ US or Titan III/ Centaur	1985	Shuttle/ PAM D	1989
	Jupiter Grand Tour, candidate for FY 1999 FY 2003	Measure-Jupiter		Jupiter Polar Orbiter post-2007	INSIDE Jupiter, Phase II Discovery competition	Discovery				
			Europa Orbiter, derived from Galileo magnetometry results	Europa Orbiter						
			Europa Lander, follow Europa Orbiter post-2005	Europa Lander						
				Europa subsurface Explorer post-2007; follow Europa Lander						

(Continued)

TABLE 15.2a (*Continued*)

Program Supplements (1983); Augmentations (1986) (System Number)	Solar System Exploration Division Strategic Plan (1991)	SSES Recommen-dations (1994)	Space Science Enterprise Strategic Plan (1997)	Space Science Enterprise Strategic Plan (2000)	Implemented Mission	Type	Transport on	Illustrative Flight Dates	Implemented Launch Vehicle	Implemented Launch
			Io Volcanic Observer, derived from Galileo and Measure-Jupiter concept							
Titan Probe/ Radar Mapper (1988-1992), core program mission (1092)		Huygens in development	Huygens in transit with Cassini	Huygens in transit with Cassini	Huygens/ESA	Strategic	Shuttle/ Tug	2000	Titan IV SRMU/ Centaur with Cassini	1997
							Shuttle/ Tug	1995		

Multiple Mainbelt Asteroid Orbiter/Flyby, supplement mission (1094)	Mainbelt Asteroid Rendezvous, candidate for FY 1999 FY 2003			Multiple asteroid mission/ protoplanet explorer post-2007	Dawn, Phase II Discovery Competition	Discovery	Shuttle/ Tug	1987		
Earth-Approaching Asteroid Rendezvous, supplement mission (1094)	Discovery Near Earth Asteroid Rendezvous, recommended for 1994	NEAR (approved mission); first Discovery mission	NEAR in transit to 433 Eros	In completion	NEAR-Shoemaker	Discov p. ery			Delta II	1996
			Mars Surveyor Program landers	Mars Exploration Program, landers/rovers; Mars 2003 in development	Mars Polar Lander plus DS-2 penetrometers, all lost during landing	Strategic			Delta II 7425	1999
		Mars Pathfinder (approved mission); 2nd Discovery mission		Completed	Mars Pathfinder	Discovery	Shuttle/ US or Titan IIIE/ Centaur	1985	Delta II	1996

(*Continued*)

TABLE 15.2a (*Continued*)

Program Supplements (1983); Augmentations (1986) (System Number)	Solar System Exploration Division Strategic Plan (1991)	SSES Recommendations (1994)	Space Science Enterprise Strategic Plan (1997)	Space Science Enterprise Strategic Plan (2000)	Implemented Mission	Type	Transport on	Illustrative Flight Dates	Implemented Launch Vehicle	Implemented Launch
Saturn Orbiter, supplement mission (1096)		Cassini Orbiter in development	Cassini in transit with Huygens	Cassini in transit with Huygens	Cassini	Strategic	Shuttle/ Tug	1986	Titan IV SRMU/ Centaur with Huygens	1997
				Saturn Ring Observer post-2007						
	Pluto Flyby/ Neptune Orbiter, recommended for 1996			Neptune Orbiter post-2007			Shuttle/ Tug/SEP	2000		
		Pluto Fast Flyby	Pluto/Kuiper Express	Pluto-Kuiper Express	Pluto Kuiper Belt competition	Discovery II prototype				
Lunar Geoscience Orbiter, supplement mission (1104)	Lunar Observer, recommended for 1994		Completed		Clementine	Department of Defense/ NASA	Atlas/ Centaur	1980	Titan IIG	1994
			Completed		Lunar Prospector	Discovery			Athena II	1998

						Shuttle/ United States	1983		
						Shuttle/ United States	1988		
Voyager 1 and 2 flybys completed (1107)		Completed		Voyager 1 and Voyager 2	Strategic	Titan IIIE/ Centaur x 2	1977	Titan IIIE/ Centaur x 2	1977
Voyager 2 flyby completed (1108)		Completed		Voyager 2	Strategic	Titan IIIE/ Centaur	1979	Titan IIIE/ Centaur	1977
	Voyager 2 flyby completed	Completed		Voyager 2	Strategic	Titan IIIE/ Centaur	1992	Titan IIIE/ Centaur	1977
			CONTOUR in development	CONTOUR	Discovery	Titan IIIE/ Centaur	1980	Delta II	2002
			Deep Impact in development	Deep Impact	Discovery				
Comet Rendezvous/ Asteroid Flyby (1990-92), core program mission (1111)	Comet Rendezvous Asteroid Flyby cancelled 1992; Rosetta participation		Rosetta in development	Rosetta	ESA	Titan III/ Centaur or Shuttle	1981	Ariane 5	2003

(*Continued*)

TABLE 15.2a (*Continued*)

Program Supplements (1983); Augmentations (1986) (System Number)	Solar System Exploration Division Strategic Plan (1991)	SSES Recommendations (1994)	Space Science Enterprise Strategic Plan (1997)	Space Science Enterprise Strategic Plan (2000)	Implemented Mission	Type	Transport on	Illustrative Flight Dates	Implemented Launch Vehicle	Implemented Launch
	TOPS [Toward Other Planetary Systems (Orbital)], candidate for FY 1999 FY 2003	ASEPS, 1 (Astronomical Studies of Extrasolar Planetary Systems-Phase 1/ Earth-orbital mission)	Origins Program begun with notional mission set		(Moved from SSE to ASO)					
				Genesis in transit	Genesis	Discovery			Delta II	2001

TABLE 15.2b
Disposition of *Outlook for Space* Planetary Missions: Part II

System Number (1975)	Name (1975)	Projected Cost in Millions FY 1975$: 1976–1980	1981–1985	1986–1990	1991–2000	Total	True Cost in Millions $, Development Only: Total	True Cost Base Year for Fixed $: Total	Disposition (September 2003)	Remarks (October 2001)	Resolution Required to Enable (October 2001)	New Frontiers in the Solar System ("Decadal Survey") Recommended Missions 2003–2013	NASA Space Science "Vision Mission" Studies	Project Prometheus
1066	Mercury Orbiter with penetrometers		190	75		265	172	1999 Phases A–D	MESSENGER in development, launch in May 2004					
1078	Mercury Sample Return				2,000	2,000			Mercury Sample Return	Nuclear Thermal Rocket needed	Trip times and vehicle mass prohibitive without multistage nuclear thermal rocket engines			
									Venus Atmosphere Probe	Discovery candidate				
1079	Venus Surface Sample Return				2,300	2,300			Venus Surface Sample Return	Augmented Mars Sample Return (balloon ascent?)	Needs Mars Sample Return as a technology demonstration; thermal management and balloon ascent need significant ATD			

(*Continued*)

TABLE 15.2b (*Continued*)

System Number (1975)	Name (1975)	Projected Cost in Millions FY 1975$					True Cost in Millions $, Development Only	True Cost Base Year for Fixed $						
		1976–1980	1981–1985	1986–1990	1991–2000	Total	Total	Total	Disposition (September 2003)	Remarks (October 2001)	Resolution Required to Enable (October 2001)	New Frontiers in the Solar System ("Decadal Survey") Recommended Missions 2003–2013	NASA Space Science "Vision Mission" Studies	Project Prometheus
									MESUR Network	Mars Scout candidate		Mars Medium 2; Mars Long-Lived Lander Network		
									Mars Climate Orbiter	Lost at orbit injection				
									Mars Aeronomy Observer; SEC Roadmap	Discovery candidate		Mars Small 2; Mars Upper-Atmosphere Orbiter		
1080	Mars Surface Sample Return		100	1,540	60	1,700			Mars Sample Return	Mars Exploration Program, 2011 goal		Mars Large 1; Mars Sample Return		
1081	Jupiter Atmospheric Probes	100	55	50		205			Galileo Probe completed; impact Jupiter					
1082	Saturn Atmospheric Probes		150	20	5	175			Saturn Atmospheric Probes	Discovery 2 candidate				

1083	Titan Orbiter with Penetro-meter		200	100	300			Titan Explorer, Huygens followon	Nuclear Thermal Rocket and/or aerocapture	Planning requires successful completion of Huygens mission and advanced propulsion and/or aerocapture		Titan Explorer, Study Case 16
1084	Uranus Atmo-spheric Probe	200	35		235			Uranus Atmospheric Probe	Nuclear Thermal Rocket and/or aerocapture	"Reasonable" (<~15 years) flight times require Jupiter flyby		
1085	Asteroid Sample Return			650	650			Asteroid Sample Return	Discovery 2 candidate			
1086	Comet Sample Return							Comet Nucleus Sample Return	Not currently in development		Solar System Medium 5; Comet Surface Sample Return	
				700	700	190	1999- Development Cap True cost may be smaller	Stardust in transit				
1087	Venus Orbiter Imaging Radar with penetro-meters	10	205	10	225	585	1992	Magellan completed				
1088	Venus Lander	450	250		700			Venus Lander	Discovery 2 candidate		Solar System Medium 4; Venus In-Situ Explorer	

(Continued)

TABLE 15.2b (*Continued*)

System Number (1975)	Name (1975)	Projected Cost in Millions FY 1975$ 1976–1980	1981–1985	1986–1990	1991–2000	Total	True Cost in Millions $, Development Only: Total	True Cost Base Year for Fixed $: Total	Disposition (September 2003)	Remarks (October 2001)	Resolution Required to Enable (October 2001)	New Frontiers in the Solar System ("Decadal Survey") Recommended Missions 2003–2013	NASA Space Science "Vision Mission" Studies	Project Prometheus
1089	Lunar Sample Return (Highlands)		90	580		670			Lunar Giant-Basin Sample Return	Discovery 2 candidate		Solar System Medium 2; South Pole-Aitken Basin Sample Return		
									Lunar Surface Science	Discovery candidate				
1090	Mars Polar Orbiter with penetro-meters	100	70			170	MO (574) + MGS + Mars Odyssey + MRO 05 Development Cost	1992 for MO	Mars Exploration Program, Orbiters; Mars Reconnaissance Orbiter 2005 in development					
1091	Jupiter Orbiters, spinning/three-axis		135	45		180	1,401	1992	Galileo Europa Mission (Galileo Extended Mission), completed					
									INSIDE Jupiter, Phase II Discovery competition, not selected			Solar System Medium 3; Jupiter Polar Orbiter with Probes		Jupiter Icy Moons Orbiter

					Europa Orbiter, cancelled	Payload competition awaiting selection; price estimate over $1B		Solar System Large 1; Europa Geophysical Orbiter
					Europa Lander	Nuclear Thermal Rocket needed; extreme radiation environment; RTG needed for power; specialized, major ATD needed		
					Europa Subsurface Explorer			
					Io Orbiter			
1092	Titan Lander	700	700	ESA	Huygens in transit with Cassini			
1093	Uranus Orbiter	350	350		Uranus Orbiter	Nuclear Thermal Rocket and/or aerocapture	Decreased interest following Voyager 2 flyby of Neptune and Triton; 98 deg inclination restricts equatorial plane capture to equinox arrivals	

(Continued)

TABLE 15.2b (*Continued*)

System Number (1975)	Name (1975)	Projected Cost in Millions FY 1975$ 1976–1980	1981–1985	1986–1990	1991–2000	Total	True Cost in Millions $, Development Only Total	True Cost Base Year for Fixed $ Total	Disposition (September 2003)	Remarks (October 2001)	Resolution Required to Enable (October 2001)	New Frontiers in the Solar System ("Decadal Survey") Recommended Missions 2003–2013	NASA Space Science "Vision Mission" Studies	Project Prometheus
1094	Asteroid Rendezvous with penetrometer Plus Laser		60	160	40	260	105	1992	Dawn Discovery in Phase II competition, selected NEAR-Shoemaker completed					
1095	Mars Lander/ Rover						Mars Exploration Rovers 03		Mars Exploration Program, Landers/ Rovers; Mars 2003 Rovers in transit			Mars Medium 1; Mars Smart Lander	Search For Evidence of Past Life (Study Case 13); Exploration of Hydrothermal Habitats (Study Case 14); Search for Present Life (Study Case 15)	
		95	500	35		630	300	Upper limit from AP article at Pathfinder web site	Mars Pathfinder completed					

1096	Saturn Orbiter		200	100		300	1,457	1992	Cassini in transit with Huygens			Solar System Small 2; Cassini Extended	
									Saturn Ring Observer	Nuclear Thermal Rocket and/ or Nuclear Electric Propulsion	Multiple maneuvering capabilities seen as driver for NEP development		
1097	Neptune Orbiter				450	450			Neptune Orbiter	Nuclear Thermal Rocket and/or aerocapture	"Reasonable" (<~15 years) flight times require Jupiter flyby; capture requires nuclear stage or aerocapture ballute in unknown environment		Neptune Orbiter with Probes, Study Case 17
									New Horizons now in development			Solar System Medium 1; Kuiper-Belt Pluto Explorer; New Horizons selected for flight	
1104	Lunar Polar Orbiter					0	Clementine Cost		Clementine completed				
		85	15			100	LP Cost		Lunar Prospector completed				

(*Continued*)

TABLE 15.2b (*Continued*)

System Number (1975)	Name (1975)	Projected Cost in Millions FY 1975$ 1976–1980	1981–1985	1986–1990	1991–2000	Total	True Cost in Millions $, Development Only Total	True Cost Base Year for Fixed $ Total	Disposition (September 2003)	Remarks (October 2001)	Resolution Required to Enable (October 2001)	New Frontiers in the Solar System ("Decadal Survey") Recommended Missions 2003–2013	NASA Space Science "Vision Mission" Studies	Project Prometheus
1105	Lunar Orbiter with penetro-meters		180			180			Lunar Orbiter with penetrometers	Discovery 2 candidate				
1106	Lunar Rover Unmanned		250	300		550			Lunar Rover Unmanned	Discovery 2 candidate				
1107	Jupiter-Saturn Flyby	350				350	776	1992	Voyager planetary encounters completed					
1108	Uranus Flyby	165	15			180								
1109	Neptune Flyby			50	175	225								
1110	Comet Flyby/Fly-Through	90				90	CONTOUR Cost Deep Impact Cost		CONTOUR, mission lost Deep Impact in development					

1111	Comet Rendezvous	150	60			210	ESA	Rosetta waiting to launch
								Origins Program moved from Solar System Exploration
							Genesis Cost	Genesis in transit
	Total Cost in Millions of FY 1975$	1,135	2,730	3,645	7,540	15,050		

TABLE 15.3a Atmospheres Subgroup (14 Proposals)

Category	Subcategory	PI (Concept Reference Number)	PI's Institution	Number of Coinvestigators	Mission Name	Abbreviation	Description	Launch Vehicle	Implementation Risk	Science Merit
Terrestrial planets	Venus	Richard Goody (4)	Harvard	15	Venus Multiprobe Mission	VMPM	Fourteen small entry probes over one hemisphere of Venus		1	5
		Samuel Gulkis (12)	JPL	5	Venus Orbiter–Deep Atmosphere Temperature Sounder	DATS	Synoptic data on atmosphere from surface to 50 km altitude		2	4
		F. W. Taylor (16)	Oxford University	1	Venus Atmospheric Dynamics Imaging Radiometer	VADIR	Dynamics of the atmosphere of Venus from the surface to 90 km altitude		3	4
		Larry W. Esposito (17)	University of Colorado	7	Venus Composition Probe		"Free-flyer" probe enters Venus atmosphere in daylight after 4-month flight; measurements on parachute descent and then on down to the ground	Titan II or Delta II	2	5
		K. Baines (38)	JPL	10	Venus 4-D Discovery Mission		Dynamics, chemistry, and thermal structure of Venus atmosphere using NIMS, CCD camera, and thermal IR scanner; 45°, 33,400-km circular orbit		3	4
		Charles C. Counselman III (98)	MIT	2	Discovery Mission Concept to Investigate Venus' Rotation and Atmospheric Dynamics using Grounded and Floating Radio Beacons		Monitor rotation of the solid portion of Venus, circulation of the lower atmosphere, and atmosphere-surface coupling; release twelve radio beacons (six fall to surface and six remain aloft); do Earth-based observation of the beacons for 12 years		3	3

	James Arnold (99)	University of California at San Diego	13	University Cooperative Venus Mission		Study of the minor and trace molecule concentrations in the Venus atmosphere above cloud top and their variation with time, and study plasma composition and properties from ionosphere and up		3	2
Mars	Verner Suomi (3)	University of Wisconsin–Madison	4	Martian Climate Variability; a Microsat Approach		Radio occultation of Mars atmosphere with four microsats	Taurus XL/S	3	3
	Sanjay Limaye (49)	University of Wisconsin–Madison	13	Mars Operational Environmental Satellite	MOES	Weather system and diurnal behavior of Martian atmosphere and surface; two instruments; 25° inclination, 216-minute, 2,250-km circular orbit	Delta II	3	4
	John Langford (51)	Aurora Flight Sciences Corporation	0	Mars Atmospheric Aircraft Platforms		Small Mars aircraft for MESUR mission; visual imaging or other science		2	1
	Timothy Killeen (79)	University of Michigan	17	A Mars Upper Atmosphere Dynamics, Energetics and Evolution Mission	MUADEE	Exploration of Mars upper atmosphere and ionosphere with seven remote sensing and in situ instruments	Delta	2	5
	Daniel T. Lyons (80)	JPL	9	"The Litle Dipper" Mars Aeronomy, Gravity, and Radio Science		Neutral gas composition and density of the Martian atmosphere		3	2
	Jacques E. Blamont (92)	University of Paris	0	No name		Redistribution of responsibilities for Mars exploration, including a join U.S./Soviet technical team; does not meet Discovery Program requirements			
Giant Planets	Len Tyler (74)	Stanford University	5	Radio Science and Astronomy Mission: Giant Outer Planet Orbiters	RSAM	Radio science orbiter for any of the giant outer planets		2	4

TABLE 15.3b Dust, Fields, and Plasmas Subgroup (15 Proposals)

Category	Subcategory	Concept Reference Number	PI	PI's Institution	Number of Coinvestigators	Mission Name	Abbreviation	Description	Launch Vehicle	Implementation Risk	Science Merit
Cosmic dust		1	Friedrich Hörz	NASA-JSC	6	Cosmic Dust Collection Facility		Instrument facility on space station Freedom; determine composition and trajectories of cosmic dust particles	Space station	2	4.5
		22	J. Derral Mulholland	POD Associates, Inc.	9	Spatio-Temporal Monitoring of Space Debris		Fly a capacitor-type micrometeoroid impact detector as secondary payload on other Discovery spacecraft; not a stand-alone mission concept		2	2.5
		24	William Hayden Smith	Washington University	1	A SPACE Experiment	SPACE	Space Particle Analysis by Collisional Excitation (SPACE): infer composition of small particles in Earth orbit or other locations in space by observing emitted light from particle impact; instrument proposal only		3	2

Cometary dust	78	W. Merle Alexander	Baylor University	10	Comet Coma Sample Return	CCRS	Comet nucleus flyby near perihelion, closest approach <100 km with return trajectory to Earth; samples collected by four different means, impact parameters recorded, and plasma components measured. Sample propulsively captured into Earth orbit and retrieved by shuttle		2	3.5
Meteors, micrometeoroids	7	Thomas Wdowiak	University of Alabama–Birmingham	3	Ultraviolet Imaging Spectroscopy of Meteors		Analysis of middle to far ultraviolet spectral data of meteoric debris of cometary origin using the Quicksat spacecraft bus launched to an equatorial or polar orbit around Earth		2.5	3
Solar wind	9	Don Burnett	Caltech	9	Solar Wind Sample Return Mission		Fly outside Earth's magnetosphere, expose materials to the solar wind for 2 years, and return to Earth for analysis	"Piggy-back" launch possibility	3	3.5

(*Continued*)

TABLE 15.3b (*Continued*)

Category	Subcategory	Concept Reference Number	PI	PI's Institution	Number of Coinvestigators	Mission Name	Abbreviation	Description	Launch Vehicle	Implementation Risk	Science Merit
Planet fields, particles, plasmas, etc.	Mercury	60	Robert C. Ready	LANL	16	A Mercury Interior, Surface & Environment Mission Concept		Provide/develop fields and particle instruments to be carried on a Mercury orbiter(s) based upon Mercury Orbiter Science Working Team Concept in NASA TM-4255; instruments would include a magnetometer, ion mass spectrometer, electron reflectrometer, and neutron detector		4	3
	Venus	35	J. H. Waite, Jr.	Southwest Research Institute	10	A Planetary/ Heliospheric Reconnaissance of Dynamics: Ionosphere, Thermosphere, and Exosphere	APHRODITE	Exploration of Venus thermosphere, exosphere, and ionosphere; (1) characterize neutral wind systems in the upper atmosphere, (2) characterize the dynamics of plasma flow in the ionosphere and nearby solar wind; elliptical, polar orbit at Venus		3.5	5

	37	Chris Russell	UCLA	13	Venus Cloud Structure and Dynamics Lightning Observations Upper Atmospheric Loss Process Discovery	CLOUD	Study structure and dynamics of the Venus clouds using the nightside thermal infrared to backlight clouds from below, use lightning as a proxy for vertical convection to determine where strong vertical convection occurs in the clouds, evaluate the importance of lightning in the chemistry of the Venus atmosphere, and determine the accretion rate and loss of atmosphere of Venus	2	4
Moon	84	Donald Shemansky	University of Southern California	6	A Proposal for Atmospheric Exploration of the Moon		Measure content and morphology of the lunar atmosphere; determine source processes and utilize the Moon as a detector of small objects entering the inner solar system	3	4

(Continued)

TABLE 15.3b (*Continued*)

Category	Subcategory	Concept Reference Number	PI	PI's Institution	Number of Coinvestigators	Mission Name	Abbreviation	Description	Launch Vehicle	Implementation Risk	Science Merit
	Jupiter	2	Glenn Orton	JPL	13	Jupiter Polar Orbiter	JPO	Determine processes taking place in the magnetic field and charged particle environment that influence high latitude neutral atmosphere and ionosphere; dawn-dusk, polar, 90-day, elliptical orbit with initial 15-Jupiter radius (Rj) perijove; 18-month mission; small spinning spacecraft	Delta II	3.5	4
		13	Paul Feldman	Johns Hopkins University	7	Earth-Orbital UV Jovian Observer		One instrument—a spectrographic imaging telescope—to Earth-Sun L1 point; 9 months of a 1-year mission are dedicated to observation of the Jovian system		1.5	5

	95	James Warwick	Radiophysics, Inc.	5	Polar Orbiters for Giant Planet Exploration	JSO	Proposed Jupiter Skimming Orbiter (JSO) to be in a 1.01 RJ x 10 RJ polar orbit; carry instrumentation to measure electromagnetic, electrostatic, and magnetic close-in environment of Jupiter	3	3
Earth	39	Mark Hickman	NASA-LeRC	2	Magnetospheric Mapping and Current Collection in the Region from LEO to GEO		In-house, center project to fly a kilowatt-class solar electric propulsion vehicle with instrumentation to support plasma current collection and magnetospheric mapping from a highly inclined, low-altitude Earth orbit through the Van Allen radiation belts and plasma environment to a moderately inclined geosynchronous orbit	1.5	2

(*Continued*)

TABLE 15.3b (*Continued*)

Category	Subcategory	Concept Reference Number	PI	PI's Institution	Number of Coinvestigators	Mission Name	Abbreviation	Description	Launch Vehicle	Implementation Risk	Science Merit
Comas		93	Michael Mendillo	Boston University	1	Satellite for Imaging Planetary Alkaline Comas	SIPAC	Earth-orbiting mission to study tenuous extended atmosphere of Mercury, the Moon, and Jupiter; Pegasus launch vehicle is used to put a modified Ball QuickStar satellite into Earth orbit; single instrument is telescope optics with three CCD units	Pegasus	1.5	4

TABLE 15.3c Small Bodies Subgroup (23 Proposals)

Category	Subcategory	Concept Reference Number	PI	PI's Institution	Number of Coinvesti-gators	Mission Name	Abbreviation	Description	Launch Vehicle	Launch Date(s)	Implementation Risk	Science Merit
Comets	Nucleus	5	Marcia Neugebauer	JPL	5	A Comet Impact Mission	CIM	Cometary nucleus flyby mission near perihelion; impactor system detached prior to encounter impacts just preceding flyby; impactor provides enough kinetic impact energy to produce a large crater and ejecta that are observed by the trailing spacecraft and remotely from Earth			2	3
		18	Joseph Veverka	Cornell University	9	Comet Nucleus Tour	CONTOUR	Flyby of three comets (Encke, Tempel-1, d'Arrest) on a single 5-year mission launched in August 2003 by a Delta II (7925), employing multiple Earth gravity assists for retargeting purposes. Science focus is on nucleus structure, composition, and processes with data obtained from three instruments: imager, dust analyzer, and neutral/ion mass spectrometer	Delta 7925	TBD	1	4

(Continued)

TABLE 15.3c (*Continued*)

Category	Subcategory	Concept Reference Number	PI	PI's Institution	Number of Coinvestigators	Mission Name	Abbreviation	Description	Launch Vehicle	Launch Date(s)	Implementation Risk	Science Merit
		73	William Hayden Smith	Washington University	3	Comet Nucleus Observer	CNO	Proposal for an instrument to do spectral imaging and mapping of a comet nucleus and innermost coma during a rendezvous mission; no details of mission or spacecraft given			4	3
		76	William V. Boynton	University of Arizona	6	Comet Nucleus Penetrator		Penetrator deployed into the nucleus of a comet following rendezvous; at least three comet targets appear feasible: SW-3 is primary target for launch in 2001; penetrator similar to that on CRAF, but augmented by module for delivery; data relay direct to Earth			1	5

	88	John Kummer	Lockheed Palo Alto Research Laboratory	5	Solar System Exploration Cryogenic Telescope	SSECT	One-year mission to deploy a cryogenically cooled telescope and spectrometer in GEO to investigate a wide range of cometary phenomena and examine asteroids and small satellites; 881-kg spacecraft to be launched to orbit by a Delta 7925 launch vehicle in 2001 (other launch opportunities available)	Delta 7925	2	3
Coma	14	Ben Clark	Martin Marietta	4	Comet Coma Rendezvous Sample Return	CCR-SR	Cometary nucleus rendezvous at or near perihelion; collection of particulate and gas samples followed by direct return of samples aboard an entry vehicle with recovery on the Earth's surface; requires a foreign partner to provide Earth return system		3	3

(*Continued*)

TABLE 15.3c (*Continued*)

Category	Subcategory	Concept Reference Number	PI	PI's Institution	Number of Coinvestigators	Mission Name	Abbreviation	Description	Launch Vehicle	Launch Date(s)	Implementation Risk	Science Merit
		23	Glenn C. Carle	NASA ARC	8	Cometary Coma Chemical Composition	C4	Cometary nucleus rendezvous at or near perihelion followed by 100 days of scientific operations. At least four comet targets appear feasible with Temple 1 as primary target for a launch in 1999. Coma sampling by modified CIDEX and NGIMS. Spin-stabilized, solar-powered spacecraft			2	4
		46	Arden Albee	California Institute of Technology	9	Flyby Sample Return via SOCCER		Sample collection portion of the Japanese SOCCER project; baseline presumes an August 2000 launch to comet Finley with Earth return in August 2004; space shuttle is assumed to retrieve the payload			3	3

General	26	John Brandt	University of Colorado	12	Small Comet and Interplanetary Hydrogen Discovery Ultraviolet Solar System Observer	SCIH/UVSSO	Determine spatial density and physical properties of small comets and continue the role of IUE as a follow-on phase of the mission; three-axis stabilized spacecraft with ultraviolet imagers and a high-resolution telescope spectrograph launched into low Earth orbit in 1999 on a Pegasus (?) XL	Pegasus XL	1	3
	29	James Burch	Southwest Research Institute	14	Comet Activity Probe	CAP	Cometary nucleus rendezvous near perihelion with observations continuing to 3 astronomical units; at least four targets appear feasible with Temple (?); launch in 1999; imaging, dust detection, charged particle, and ?; spin-stabilized, solar-powered spacecraft		2	3

(*Continued*)

TABLE 15.3c (*Continued*)

Category	Subcategory	Concept Reference Number	PI	PI's Institution	Number of Coinvestigators	Mission Name	Abbreviation	Description	Launch Vehicle	Launch Date(s)	Implementation Risk	Science Merit
		40	Paul Weissman	JPL	10	SOCCER Pathfinder	SOCCER	U.S./Japan dual spacecraft Kopff comet flyby and coma sample return mission; U.S.-built and -launched spacecraft is-launched November 2001 to serve first as a navigational pathfinder for the Japanese spacecraft; Japanese spacecraft collects and returns samples to Earth; U.S. spacecraft is retargeted for a flyby of the asteroid Icarus in 2005			2	4
		90	S. Alan Stern	Southwest Research Institute	10	Chiron Discovery Flyby		Send the spare Pluto Flyby spacecraft to fly by the distant comet 2060 Chiron to address objectives relating to cometary science and Chiron's size, shape, polar obliquity, atmosphere, surface morphology, surface composition, internal structure, and surface activity			3	4

Asteroids	Near-Earth	6	Michael Belton	National Optical Astronomy Observatories	5	Small Missions to Asteroids and Comets	SMACS	SMACS involves separate launches of four small spacecraft on Pegasus XL boosters in the 1998–2000 time frame to a primitive object (2100 Ra-Shalom, a C-object); a highly evolved igneous object (1986 DA, a M-type); a moderately active cometary nucleus (P/Finley); and an extinct or dormant comet nucleus (3200 Phaethon, F-type)	Pegasus XL	1998-2000	1	4
		11	Robert Housley	Rockwell International	2	Asteroid Sample Return Mission		A "simple, unadorned" mission to rendezvous with an S-type or C-type asteroid, collect at least 1 kg of surface samples, and return them to Earth via aerocapture to low Earth orbit followed by entry and parachute descent to a nonwater landing site			3	3
		32	Daniel Britt	University of Arizona	22	Rendezvous with Earth Approaching Asteroids	REAAct	Four spacecraft launched in pairs 1 year apart; elliptical lunar parking orbit to await discovery of new objects and then sent to rendezvous; backup missions to known objects; instruments include CCD imager, IR point spectrometer, ? spectrometers that are landed on the asteroid surface			3	4

(Continued)

TABLE 15.3c (*Continued*)

Category	Subcategory	Concept Reference Number	PI	PI's Institution	Number of Coinvestigators	Mission Name	Abbreviation	Description	Launch Vehicle	Launch Date(s)	Implementation Risk	Science Merit
		77	Eugene Shoemaker	U. S. Geological Survey	13	Near Earth Asteroid Returned Samples	NEARS	Sample acquisition and return to Earth reentry/landing of a set of small samples from six different sites on the surface of a NEA target body; proposed to meet cost objectives via significant hardware heritage from NEAR spacecraft and GE reentry capsule			2	4
	Mainbelt	47	Joseph Veverka	Cornell University	11	Mainbelt Asteroid Exploration/Rendezvous	MASTER	One of two complementary alternative missions with identical payloads; launch in 2001 or 2003; rendezvous and then orbit the mainbelt asteroids Iris or Vesta; three-axis stabilized spacecraft utilizing solar power and bipropellant thrusters (a new class C design configuration "with significant subsystem heritage")			1	5

Pluto	54	B. Murray	Caltech	2	Pluto/Charon Flyby Mission	Fast flyby of Pluto and Charon as a reconnaissance mission with a battery-powered spacecraft; meet cost objectives with small staff and simplified spacecraft design; imaging based on Mars Observer camera, also includes a radio atmospheric occultation experiment		3	3
Phobos	75	Bradley Edwards	LANL	6	Prospector Mission	Geologic and geochemical composition of solar system objects "using advanced instrument capabilities"; Delta II 7925 sends a spacecraft to Phobos with a high-resolution x-ray fluorescence imager (elemental abundances) and visible /near-IR spectrometer (mineralogy)	Delta II 7925	3	3
	100	Thomas Duxbury	JPL	14	Joint Russian/U.S. Phobos Sample Return Mission	United States to supply remote and in situ instruments, sample return vehicle, and participants in mission planning and operations; primary goal is to collect and retrieve samples from Phobos and perform detailed studies of these samples on Earth to increase understanding of Phobos' composition, history, and evolution		3	5

(Continued)

TABLE 15.3c (*Continued*)

Category	Subcategory	Concept Reference Number	PI	PI's Institution	Number of Coinvestigators	Mission Name	Abbreviation	Description	Launch Vehicle	Launch Date(s)	Implementation Risk	Science Merit
	Io	85	William Smythe	JPL	10	Io Mapper		One year study of Io's volcanism using a single imaging instrument that improves on *Galileo's* spatial and spectral capability: a combined visual infrared camera and radiometer; proposal requires prior development of the Pluto spacecraft to meet the Discovery cost goal			3	3
	Instrument	20	David F. Blake	NASA ARC	3	Chemistry and Mineralogy using combined X-Ray Fluorescence and X-Ray Diffraction	CHEMIN	Writeup from J. Appleby file: The goal is to land an x-ray diffraction/x-ray fluorescence instrument on the surface of Mars (or other solid solar system body) to perform chemical and mineralogical analysis of surface material. X-ray diffraction analysis has never been performed on any previous space mission. This is not a complete mission proposal			4	3

61	William J. Borucki	NASA ARC	6	Frequency of Earth-sized Planets	FRESIP	1.2-m telescope in high Earth orbit; photometric survey of fields of 6,000 F, G, and K stars within a single FOV and 90–560 parsecs to detect transits of Earth-sized planets; confirm occurrence of three observed transits, so the mission is about 3 years in length	3	3

TABLE 15.3d Solid Bodies Subgroup (21 Proposals)

Category	Subcategory	Concept Reference Number	PI	PI's Institution	Number of Coinves-tigators	Mission Name	Abbreviation	Description	Launch Vehicle	Launch Date(s)	Implementation Risk	Science Merit
Terrestrial planets	Mercury	15	Paul D. Spudis	LPI	14	Mercury Polar Flyby		Send a spacecraft similar to Mariner 10 to Mercury on a flyby trajectory that is 2:1 resonant with Mercury in order to provide one or two subsequent returns; objective is to characterize and study Mercury's polar caps and complete imaging reconnaissance of the planet			2	4.5
		28	Faith Vilas	JSC	10	Inner Planet Spectrographic Imaging Telescope	IPSIT	Earth orbiting satellite to observe and study composition and distribution of Mercury's surface mineralogy and tenuous atmosphere; also, observations of other inner solar system objects (e.g. Venus and Mars, NEAs, comets) can be made when Mercury cannot be observed; the instrument is a 50-cm telescope with UV, visible, and infrared spectrographs; planned lifetime of the satellite is 5 years			1	3.5

34	Robert Nelson	JPL	19	Hermes Global Orbiter: A Mission to Mercury		Remote sensing observations of the planet's surface, its atmosphere, and its magnetosphere; payload consists of telescope system for passive and active photopolarimetry, UV spectrometer, and a magnetometer; after orbit insertion, the nominal mission lifetime is 1 year	3	4.5
52	Duane O. Muhleman	Caltech	1	Mercury Imaging and Radar Ranging Orbital Reconnaissance	MIRROR	Small spacecraft in orbit at Mercury to return the first global coverage of the entire surface and precisely locate and map the extent of the polar cap; the trajectory is E-VVMM-M with a lightweight production spacecraft that supports a Delta II launch; payload scaled down to two instruments and managed in a low-cost university mode at Caltech	2.5	4.5
53	Albert E. Metzger	JPL	13	Mercury Mapping Orbiter Mission		Mercury orbiter mission "utilizing a unique lightweight and low cost spacecraft"; payload of four instruments consisting of a UV/visible camera, GRS, XRFS, and a magnetometer; primary objective is planetary observations with solar, heliospheric, and celestial data obtained "only as instruments and mission lend themselves to that secondary objective"	3	4.5

(Continued)

TABLE 15.3d (*Continued*)

Category	Subcategory	Concept Reference Number	PI	PI's Institution	Number of Coinvestigators	Mission Name	Abbreviation	Description	Launch Vehicle	Launch Date(s)	Implementation Risk	Science Merit
		66	Bruce Bills	NASA GSFC	15	A Mercury Polar Orbiter Mission	Mallcu	Perform the first global survey of Mercury, characterizing surface geology, topography, and gravity and magnetic fields			3	4.5
		96	Stan Peale	UC Santa Barbara	8	Mercury Geophysics Mission		Determine whether Mercury has a molten core through gravity field measurements using both an orbiter and lander; 5-year mission to be launched on a Delta II	Delta II		3	3.5
	Venus	42	Michael Malin	MSSS	4	Venus Geophysical Network Pathfinder		Proof-of-concept, Venus hard lander measures and returns surface and geophysical data for 1 year; payload consists of seismometer, meteorology sensors, magnetometer, and surface imager; concept requires RTG-powered active refrigeration of pressure vessel that continas all electronics; sensor heads of several instruments are mounted outside of the dewar			4.5	5

	55	James Head, III	Brown University	15	Discovery Venera Surface-Atmosphere Geochemistry Experiments	SAGE	Venera-class lander "to a designated target of high scientific interest on Venus"; instruments to measure lower atmosphere constituents, surface geochemistry, mineralogy, and geology		2	4
	81	Ellen R. Stofan, R. Stephen Saunders	JPL	9	Venus Interior Structure Mission	VISM	Study of interior of Venus using seismometry; employ a PVO-type spacecraft with three probes, each with a seismometer; each lander and seismometer is capable of operating for more than 30 days on the Venus surface, transmitting data back to an orbiting platform for transmittal to Earth		4.5	5
Mars	83	David A. Paige	UCLA	16	The Mars Polar Pathfinder		"Subsurface exploration of the northern Martian polar cap by a modified MESUR Pathfinder lander system"; payload includes radar for subsurface layering to 5 km, thermal probe to measure ice quantities to 100 m, and subsurface camera deployed by auger to 50 cm; 2002 launch	2002 launch	2	5

(Continued)

TABLE 15.3d (*Continued*)

Category	Subcategory	Concept Reference Number	PI	PI's Institution	Number of Coinves-tigators	Mission Name	Abbreviation	Description	Launch Vehicle	Launch Date(s)	Implementation Risk	Science Merit
		86	Wallace T. Fowler	University of Texas	2	Mars Gravity Measurement/ Surface Penetrator Assembly Mission		Three optional Mars mission options; obtain high-precision gravity models, subsurface water, and elemental composition measurements; options are (1) gravity mapper and three penetrators, (2) two small low-orbit orbiters and two penetrometers, or (3) option (2) augmented by MO signals			2	3
Moon		43	Jeff Plescia	JPL	1	Lunar Interior Explorer Mission		Provide the same type of data provided by the Japanese Lunar A mission (lunar seismic, heat flow, and core structure) but at a more comprehensive/global level			2	3.5
		44	Jeff Plescia	JPL	9	Lunar Geophysical Explorer	LGE	Lunar orbiter mission to address the LEXSWG science measurement priorities not directly measured by Lunar Scouts I and II, including gravity, topography, remnant magnetics, heat flow, and lunar atmosphere; proposed spacecraft platform is similar to that proposed by Boeing for the Scouts (I and II)			1	4

58	Peter Bender	University of Colorado	13	Lunar Interior Structure		Place three microwave transponders on the front side of the lunar surface to improve dynamical studies of lunar rotation and tidal distortion by 2 orders of magnitude; 2-year investigation period; "should significantly improve our understanding of the interior structure of the moon providing important constraints on the formation and tidal evolution of the Earth-Moon system"		2	3.5
64	Lee Mason	LeRC	0	Combined Lander and Instrumented Rover	CLIR	Lunar rover, 14-day, near-side mission with an integrated walking lander/rover; "concept is simple and very lightweight, with a total payload mass within the capability of the OSC Taurus launch vehicle"; rover controlled semiautomatically with an advertised traversal range of 10 km during its 2-week primary mission	Taurus	2	0
65	Bruce Bills	NASA GSFC	12	Lunar Polar Orbiter	Koati	One-year lunar polar orbiter mission to obtain global topographic and gravity field maps of the moon supported by contextual global imaging; based on GSFC Lightsat spacecraft launched on a Taurus-class small expendable launch vehicle; mission operations conducted through a Wallops Island ground station	Taurus-class	1	4

(*Continued*)

TABLE 15.3d (*Continued*)

Category	Subcategory	Concept Reference Number	PI	PI's Institution	Number of Coinvestigators	Mission Name	Abbreviation	Description	Launch Vehicle	Launch Date(s)	Implementation Risk	Science Merit
		72	William Hayden Smith	Washington University	9	Lunar Ultraviolet Infrared Spectrometer		Spacecraft in a 100-km poolar orbit with an ultraviolet infrared spectrometer to accomplish the primary objective of obtaining "accurate, detailed global maps of geochemical and minealogical properties of lunar surface materials"			4	0
		87	Red Whittaker	Carnegie Mellon University	3	Lunar Lava Tube Explorer		"An integrated, self-sufficient lander/rover will traverse hundreds of kilometers, perform a variety of scientific experiments, map the surface and subsurface, and transmit high-definition images of the lunar landscape"			2	0
		94	David Scott	Scott Science and Technology, Inc.	5	Ulysses: A Return to the Hadley Apennine		"Primary objective is to prove the concept of conducting Apollo-type planetary exploration missions with low-cost, flexible, robust hardware and operations"; two microrovers to explore selected surface features in the vicinity of the Apollo 15 site			2	3

97	Elaine Hanson	Colorado Space Grant Consortium	9	Lunar Educator	Small, <200 kg, spinning spacecraft placed into lunar polar orbit; primary goals are to increase understanding of lunar polar regions and educate college students in realities of spacecraft design and operation; payload is an imager plus an ultrastable oscillator for radio science/gravity field determination	1	3

TABLE 15.4 Summary of Perceived Merit versus Risk for the 1992 Discovery Mission Concepts

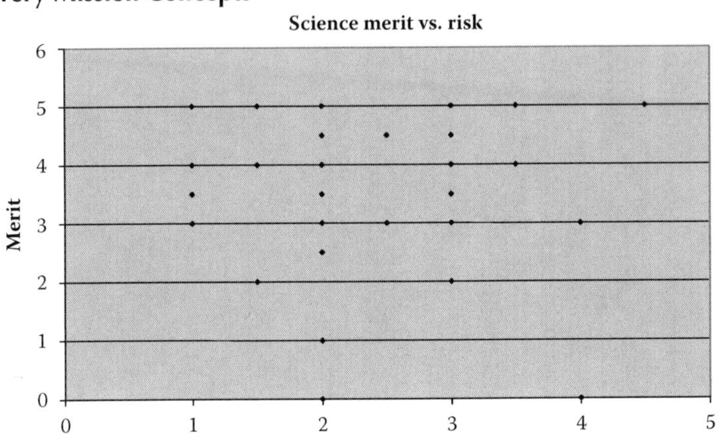

After twelve rounds of Discovery selections and two rounds of Mars Scout selections (Discovery sized missions to Mars), six missions have completed successfully (*NEAR*, *Mars Pathfinder*, *Lunar Prospector*, *Stardust*, *Genesis*, and *Deep Impact*), two have been lost (*CONTOUR* and *Mars Phoenix*), two are in progress (*MESSENGER* and *Dawn*), and two are currently under development (*Kepler* and *GRAIL*), with one mission selection pending (the next *Mars Scout*) and four "missions of opportunity" (*Aspera-3*, *M3*, *EPOXI*, and *Stardust-NExT*) in various stages.

CURRENT STATUS

PROJECTED LAUNCHES AND LAUNCH VEHICLE MARKET

With the demise of the Soviet Union and the concomitant end of the Cold War, the military need for satellites continued, but at a less frenetic pace that that of the previous three decades. Cost became more of an issue while launch vehicle providers were looking at a more expansive commercial market, primarily focused on global communications. By installing constellations of satellites in low to medium Earth orbit, global communications—that were also affordable—were perceived as a new market with significant growth, as well as profit margin potential. The Iridium system (originally named for the 77 orbital electrons of an iridium atom) was conceived as a breakthrough global communications system based upon a constellation of satellites in low Earth orbit. While the system and others like it are operating today, it had an early, rocky financial beginning, with the founding company going into bankruptcy 9 months after its start.[21] Since then, the launch vehicle market has been fairly flat, with fewer than about seventy launches per year, worldwide. In addition, a large fraction of these have been commercial payloads with an emphasis on communications.[22,23] Current projections are for just under 2,000 new payloads between now and 2027, including 730 civil payloads, of which 29% are for NASA.

The Goldin Years

In 1992, Daniel S. Goldin became the longest serving NASA Administrator, a position he held for almost 9 years. Early on he challenged the science community with pushing the limits of imagination, while also making robotic exploration "better, faster, cheaper" and backing human missions to the slot of just what was necessary and no more.[24]

The Vision and the Prometheus Excursion

With the accession of Sean O'Keefe to the post of NASA Administrator in December 2001, the new (at least perceived) emphasis within NASA switched to fiscal stewardship. O'Keefe had served previously under presidential appointments, most recently as the Deputy Director of the Office of Management and Budget. With continued overruns at NASA driven by work on the International Space Station, the shuttle, and robotic programs, a new move was afoot to control costs while still having a robust space program.

On January 14, 2004, the George W. Bush Administration unveiled new Space Exploration Vision,[25] with four programmatic points:

1. Implement a sustained and affordable human and robotic program to explore the solar system and beyond.
2. Extend human presence across the solar system, starting with a human return to the Moon by the year 2020, in preparation for human exploration of Mars and other destinations.
3. Develop the innovative technologies, knowledge, and infrastructures both to explore and to support decisions about the destinations for human exploration.
4. Promote international and commercial participation in exploration to further U.S. scientific, security, and economic interests.

The notion of returning to the Moon echoed the Space Exploration Initiative (SEI) that had been developed under the administration of President George H.W. Bush more than a decade earlier. Overseen by the NASA Administrator Thomas Paine, the SEI had encompassed many of the same goals but was dead on arrival and was quietly shelved, with its projected price tag of some $500B.[26] With the new Vision, more care was taken to seek a "sustainable" program while also obtaining high-level buy-in.[27] A system of "spiral development" such as that used in various Department of Defense (DoD) programs was brought to NASA as a conceptual framework for enabling Vision tasks. On the robotic front, O'Keefe made a major commitment to restarting and bringing to fruition the promise of nuclear electric power and propulsion in order to enable the next level of solar system travel.

Nuclear electric propulsion (NEP) has been viewed as a parallel technology to the nuclear thermal propulsion efforts under the NERVA program from the 1950s. Unlike NERVA, a direct outgrowth of Cold War efforts, NEP, using a nuclear reactor in space coupled with some type of electric propulsion, had been recognized as a potential *sine qua non* for human exploration of Mars from the 1950s,[28] because of the financial, weight, and logistical limitations of other forms of propulsion, such as chemical or ion. Fuel efficiency will be crucial in providing a manned mission to Mars the greatest probability of success in both arriving at the planet and returning to Earth. Following von Braun's early studies for Mars exploration with chemical propulsion, Ernst Stuhlinger had begun the advocacy and engineering studies for a more robust approach using NEP. Developments in the 1960s included mission design work showing a need for specific masses of less than ~30 kilogram per kilowatt electric (kg/kWe) specific mass for the power/plant propulsion system, development of the Kaufmann ion bombardment engine and Hall-effect engines, and design work under the Space Nuclear Applications Program (SNAP) for nuclear power in space. The only U.S. nuclear reactor flown and operated in space was *SNAPSHOT*, which included the SNAP-10 reactor with a power output of 533 W for 1,000 hours prior to a non-nuclear electronics system failure.[29] SNAP evolved to produce and perfect radioisotope power units for space applications. The United States continued a variety of study efforts on nuclear reactors, including the Timberwind and SP-100 programs, but without the hardware development and test levels performed under NERVA. The Soviet Union went on to develop several reactor-powered radar-ocean-reconnaissance satellites (RORSATs) using fast reactors and thermionic converters. Power levels combined with low orbital altitudes enabled all-weather tracking of U.S. deep-water naval deployments. The low altitudes would have led to control and lifetime issues for solar-array power systems; the thermionic converter approach, while robust, was limited by the amount of cesium used in the approach, limiting the eventual lifetime of the RORSATs. At the end of their useful life, the units were boosted into long-lived, safe orbits, where more than thirty of these Cold War remnants still reside. Failures, such as the impact of *Cosmos 954* into Canada, produced major environmental impacts on the ground; their use in space produced major impacts on orbiting gamma ray and x-ray telescopes due to positron injection into the Earth's magnetosphere.[30,31]

The new initiative in NEP, given the name Project Prometheus, was for reactor use in deep space, allowing propulsion to the outer reaches of the solar system.[32] Expertise from the naval nuclear reactor community was sought as a means of trying to break through previous technical barricades. Physics constraints, however, led to the same approach that had most recently been studied under SP-100: the use of a fast reactor and shadow shield in order to minimize the system mass (naval reactors and their commercial power station derivatives are thermal reactors; that is, moderators are used to thermalize the neutrons, and delayed neutrons are used to moderate the reactivity and control the output power). For power plants and large ships (e.g. nuclear-powered submarines and aircraft carriers), mass is not an engineering driver, while quite the opposite is true for spacecraft design. A fast reactor eliminates moderator mass; however, the radiation shadow shield remains massive in order to shield the electronics from fast neutrons and gamma rays and is typically as massive as the reactor itself. To minimize the shadow-shield mass and the flux, space-reactor designs have always evolved toward a reactor situated on a boom, located as far from the electronics—or crew—as possible (Figure 15.8). This approach had been studied in the early 1990s in a DoD project, testing the

FIGURE 15.8 Nuclear electric propulsion concept.

usability of the approach under the Nuclear Electric Propulsion Space Test Project (NEPSTP), based on a purchased Soviet TOPAZ-2 space reactor, a U.S. satellite bus built by the Applied Physics Laboratory, and various ion and Hall-effect engines. The effort was put forward by Pete Worden, then of the Strategic Defense Organization and more recently the director of NASA's Ames Research Center. The effort was abandoned soon after President Bill Clinton assumed office. The first Prometheus mission was to be a significant one: to explore the Jovian system with focus on Europa—a significant goal of the Decadal Survey—with an implementation dubbed the Jupiter Icy Moon Orbiter (JIMO). Elements of NASA and Department of Energy worked for 2.5 years on the concept, and a National Academy Study was carried out for NASA to identify all potential uses of this "new" technology.[33] A significant concern that developed early was that while the NEP system could enable a larger scientific payload to be transported to a distant destination with more power available there, transit times were not shorter than those obtainable with chemical propulsion and ballistic trajectories. Large velocity changes—up to tens of kilometers per second—were possible with Prometheus, but there were time penalties involved.

The JIMO study was a fairly massive undertaking, but after a 2.5-year study and the expenditure of $464M,[34] the project was dropped and quietly shelved. Run-out costs for JIMO, not including the launch vehicle, were estimated as $16.3B plus an additional $5.2B for launch vehicle(s).[35] While a robust estimate, it was simply out of scope for a robotic mission (and the Academy, at NASA's behest, was looking at multiple missions), the same problem that befell the original Voyager program to Mars in the 1960s and the original Grand Tour of 1970[36]—and both of those were projected to cost far less. Perhaps more worrying was the realization that a 3-year technology validation mission—not unlike what had been envisioned by Worden for NEPSTP some 13 years earlier[37]—would be required because of worries about guaranteed autonomous reactor operation for 10 years in deep space.

Technically, JIMO also brought home the problem with all of the space NEP technology studies to date: they were all underpowered. The same technological hurdle had been faced by attempts to construct airplanes in the late nineteenth and early twentieth centuries. Hiram Maxim's steam-powered airplane (Figure 15.9) of 1894 was a huge contraption, with the mass driven by that of the power plant. Successful powered flight relied upon the 200-lb, 12-hp custom gasoline engine developed and used by the Wright brothers on their 340-kg Wright Flyer.[38]

A selection of point designs[39] illustrates the problem: the mass penalty associated with thermal-to-electric power conversion inefficiency for NEP systems. A regression for these designs shows that the power system has an implied specific mass of ~70 kg/kWe, more than a factor of 2 of what is needed for these systems to provide the promised jump in propulsive capability. The problem is getting rid of waste heat in space. While naval vessels and power plants can use convection of an external medium (water) to carry away rejected heat—about 70%–75% of the power output

FIGURE 15.9 Hiram Maxim's steam-powered airplane. (From Desktop Aeronautics. With permission.)

by the reactor core—in space one must radiate the heat to deep space. This imposes high-temperature operation that further pushes all material limits. While gas-cooled thermal reactors with Brayton-cycle, turbo-mechanical generators can reach efficiencies as high as ~50%,[40] fast reactors tend to run at lower efficiencies and for high-temperature operation need active convective transfer of a working fluid throughout a large radiator-panel assembly. The latter, in turn, requires a relatively heavy support structure. When all of these items are combined, one obtains the ~70 kg/kWe figure of merit—somewhat larger than the ~65 kg/kWe combination as implemented on the DS-1 solar-electric-propulsion spacecraft.

Practical use of NEP systems (inherently heavy because of significant loss in mass efficiencies for reactors with thermal output of less than ~100 kWth), is in need of a paradigm shift to reduce significantly the mass associated with waste heat disposal. Both more efficient conversion and lower radiation system mass can, and will have to, play a part.

The Prometheus work did enable technical advances in ion engines and radioisotope power supplies, subsystem assemblies that played roles in the overall design. The study also provided a better indicator of the design and engineering issues at the system level, as well as input on development, cost, and risk issues, that must be addressed before a high-power (>100 kWe) NEP system can be developed for flight. The lack of progress with, and projected high costs of implementing, Prometheus has led to termination of almost all NEP technology development efforts by NASA.

DECADAL SURVEYS AND THE WAY FORWARD

Throughout the Space Age, there have been more scientific targets for observation and exploration—and difficult ones to reach—than there are funds to realize fully the desires of the scientific community. Great progress in the deep-space science arena continues to be made with robotic exploration. The Mars exploration program has seen the successful landing of *Mars Phoenix* and looks forward to the upcoming launch of the *Mars Science Laboratory*, even as cost growth remains an issue.[41] The Decadal Survey documents, produced as community consensus documents, continue to hold their ground in spite of new discoveries (e.g. the water vents on Enceladus) and new fiscal realities. The real question—and one that continuously does arise—is exactly what missions make sense for both the science community and other stakeholders.

In space physics (currently referred to as *heliophysics*), there have always been more mission concepts than resources for implementing them. The current "fleet" of spacecraft, including but not limited to *ACE*, the joint NASA/ESA *SOHO*, *Voyager 1* and *Voyager 2*, and the near-twin *Solar Terrestrial Relations Observatory* (*STEREO*) spacecraft continue to make key scientific observations. Nonetheless, important key goals remain just out of reach, requiring the implementation of a Solar Probe to explore the inner heliosphere and an Interstellar Probe to explore the space beyond.[42,43]

The "Decadal Survey" for planetary science[44] was used to identify missions that have subsequently been included as those "authorized" for proposals in NASA's New Frontiers line. While New Horizons is on its way to Pluto, and the Juno mission was selected for implementation, the other missions remain open for proposals in the future.[45]

The Present and Near-Term

The decades of the current Decadal Surveys for both Space Physics and Planetary Exploration will soon be up. New surveys will be conducted in the United States, but the current expectation is that these will be evolutionary studies rather than revolutionary ones. Further consolidation of commercial operations in LEO can be expected in the near term, including the possibility of "space tourism," but deep space will remain out of the commercial sphere, at least in the near term, with the possible exception of support in probe and launch vehicle provision for whichever governments continue to sponsor scientific and research missions.

In the immediate future of human space travel, NASA goals are completion of the ISS, development of the Orion Crew Exploration Vehicle and associated launch vehicles (*Ares I* and *Ares V*), and transitioning to these for transportation.[46] That transportation includes operations as needed to the ISS, as well as returning to the Moon by 2020.

Costs remain an issue. Previous large engineering developments, exclusive of wars per se, undertaken by the United States at the national level have included the building of the Panama Canal, development of the atomic bomb under the Manhattan Project, and the original Apollo project. While the canal development led to significant return on investment in commercial trade, that project, as well as the others, had its roots in national policy.

The Panama Canal officially opened on August 15, 1914, after a U.S. expenditure of $352M over a 10-year period; the price including French efforts beginning in 1880 was $639M, including 80,000 people and 30,000 deaths in total.[47] The Manhattan Project is estimated as having cost $1.9B in 1944 dollars,[48] and the official cost of the 11-year Apollo program was $25.4B in FY 1973 dollars.[49] The defense price deflator from the 2009 Federal Budget[50] yields a Manhattan project cost in estimated 2009 dollars of $21B, an Apollo cost of $110B, and a Panama Canal cost of $6.45B.[51] As a fraction of the gross domestic product (GDP) for the basis years, the numbers are 0.86% (Manhattan Project), 1.83% (Apollo), and 1.03% (Panama Canal). Note that expenditures were for ~6 years for the Manhattan Project, 11 years for Apollo, and 10 years for the Panama Canal. The FY 2009 NASA total budget request is ~$17B, and the projected U.S. GDP is $15,027B, making the (current) yearly NASA request ~0.11% of the U.S. GDP. Hence, a 10-year plan at this level is about what previous large projects cost as a fraction of the U.S. GDP, when all of those funds were focused on the project. However, the proposed NASA exploration budget for FY 2009 is $3.5B,[52] only about 0.02% of the GDP. Thus, the yearly government expenditure for human deep-space exploration is approximately one-fifth of these previous project expenditures on an annualized basis.

CONCLUSION

The combination of commercial, civil, and military programs from around the world has led to an Earth-orbital robotic infrastructure that continues to be maintained and used profitably. Humans have established a (so-far) permanent outpost in space with the ISS, but the enabling aspect of humans in space envisioned by

Clarke[53] has not played out. Access to space remains expensive. Most of the cost is due to infrastructure maintenance and does not scale well because of the (relatively) limited number of flights,[54] fewer than 5,000 launches in 50 years as compared with the 6,240 (suborbital) V-2 rockets manufactured by Germany during World War II and its 2,890 "successful" launches.[55] By way of comparison, in 2005 there were 82 air carriers in the United States operating 8,225 aircraft out of 575 certified airports, with almost 10 million departures (9,701,709 in 2004), 583,689 million passenger-miles, and 15,707 ton-miles of freight. This compares with 18 (worldwide) commercial space launches in 2005, none with human passengers. Air transportation overall constituted $50.0B in 2005, or 0.4% of the U.S. GDP for that year.[56]

Deep space (beyond Earth orbit, including cislunar space) remains the province of robotic scientific exploration. The Apollo missions to the vicinity of the Moon (*Apollo 8, 10*, and *13*) and its surface (*Apollo 11, 12, 14, 15, 16*, and *17*) between December 1968 and December 1972 are the only exceptions. Commercial (i.e., non-government-funded) missions to deep space continue to be discussed, but without firm foundation or concept at the end of the first 50 years of space exploration.

REFERENCES

1. NASA. 2008. Lunar and Planetary Science: The Moon. http://nssdc.gsfc.nasa.gov/planetary/planets/moonpage.html.
2. Von Braun, W. 1991. *The Mars Project*. Urbana, IL: University of Illinois Press (91 pp.).
3. Farquhar, R. W., D.W. Dunham, Y. Guo, and J.V. McAdams. 2004. Utilization of libration points for human exploration in the Sun–Earth–Moon system and beyond. *Acta Astronautica* 55 (3–9) August–November: 687–700. New Opportunities for Space. Selected Proceedings of the 54th International Astronautical Federation Congress.
4. Siddiqi, A.A. 2002. *A Chronology of Deep Space and Planetary Probes 1958–2000*. Monographs in Aerospace History, vol. 24. Washington, DC: National Aeronautics and Space Administration (243 pp.).
5. Stoneley, R. 1960. The International Geophysical Year. *Nature* 188(4750): 529–532.
6. National Academy of Sciences–National Research Council. 1962. *A Review of Space Research: The Report of the Summer Study Conducted under the Auspices of the National Academy of Sciences at the State University of Iowa, Iowa City, Iowa, June 17–August 10, 1962*. Washington, DC: National Academy of Sciences.
7. National Academy of Sciences–National Research Council. 1966. *Space Research: Directions for the Future, Report of a Study by the Space Science Board, Woods Hole, Massachusetts, 1965*. Washington, DC: National Academy of Sciences–National Research Council.
8. Niehoff, J.C. 1966. Gravity-Assisted Trajectories to Solar-System Targets. *Journal of Spacecraft and Rockets* 3(9): 1351–1356.
9. Heacock, R.L. 1980. The Voyager Spacecraft. *Proceedings of the Institution of Mechanical Engineers* 194: 211–224.
10. Space Studies Board. 1969. Lunar Exploration. In *Report of a Study by the Space Science Board*. Washington, DC: National Academy of Sciences (40 pp.).
11. Spudis, P.D. 2005. Lunar Science Overview. Electronic Presentations from the Space Resources Roundtable VII: Leag Conference On Lunar Exploration, October 25–28, 2005, League City, TX. http://www.lpi.usra.edu/meetings/leag2005/presentations/tues_am/05_spudis.pdf.

12. Heiken, G., D. Vaniman, and B.M. French, eds. 1991. *Lunar Sourcebook: A User's Guide to the Moon.* New York: Cambridge University Press.
13. Portree, D.S.F. 2001. *Humans to Mars: Fifty Years of Mission Planning, 1950–2000.* Monographs in Aerospace History, no. 21. NASA SP-2001-4521. Washington, DC: NASA History Division (see specifically 152 pp.).
14. National Research Council. 2003. *New Frontiers in the Solar System: An Integrated Exploration Strategy*. Washington, DC: The National Academies (232 pp.).
15. Space Studies Board. 2005. *Space Studies Board Annual Report 2005*. Washington, DC: National Research Council of the National Academies (see specifically p. 131).
16. Teague, O.E., K. Hechler, T.N. Downing, D. Fuqua, J.W. Symington, W. Flowers, R.A. Roe, M. McCormack, G.E. Brown, D. Milford, et al. 1975. Future Space Programs 1975. In *Subcommittee on Space Science and Applications of the Committee on Science and Technology*. Washington, DC: U.S. Government Printing Office (356 pp.).
17. Allen, J.P., J.L. Baker, T.C. Bannister, J. Billingham, A.B. Chambers, P.B. Culbertson, W.D. Erickson, J. Hamel, D.P. Hearth, J.O. Kerwin, W.C. Phinney, et al. 1976. Outlook for Space: Report to the NASA Administrator by the Outlook for Space Study Group. In *NASA SP*-386, D.P. Heath, ed. Washington, DC: Technical Information Office, NASA.
18. Portree 2001 (note 13).
19. Gunn, S. 1992. The Case for Nuclear Propulsion. *Threshold* 9: 2–11.
20. Krimigis, S.M., and J. Veverka. 1995. *Genesis of Discovery.* Journal of Astronomical Sciences 43: 345–347.
21. Wikipedia. Iridium (Satellite). http://en.wikipedia.org/wiki/Iridium_%28satellite%29.
22. Cáceres, M. 2008. Back-to-Back Growth Years for Space Launches. *Aerospace America.* March: 20–22.
23. Cáceres, M. 2008. A Look at the Next 20 Years. *Aerospace America*. January: 20–22.
24. Thompson, E., and J. Davis. 2004. Daniel Saul Goldin, NASA Administrator, April 1, 1992–November 17, 2001. http://history.nasa.gov/dan_goldin.html.
25. NASA. 2004. *The Vision for Space Exploration*. Washington, DC: NASA (32 pp.).
26. Dick, S.J. 2008. *Summary of Space Exploration Initiative*. http://history.nasa.gov/seisummary.htm.
27. Aldridge, E.C., Jr., C.S. Fiorina, M.P. Jackson, L.A. Leshin, L.L. Lyles, P.D. Spudis, N. deGrasse Tyson, R.S. Walker, and M.T. Zuber. 2004. *A Journey to Inspire, Innovate, and Discover: Report of the President's Commission on Implementation of United States Space Exploration Policy*. Washington, DC: Government Printing Office.
28. Portree 2001 (note 13)
29. Aftergood, S. 1989. Background on Space Nuclear Power. *Sci. Global Sec.* 1: 93–107.
30. Hones, E.W., and P.R. Higbie. 1989. Distribution and Detection of Positrons from an Orbiting Nuclear Reactor. *Science* 244(4903): 448–451.
31. Rieger, E., et al. 1989. Man-Made Transients Observed by the Gamma-Ray Spectrometer on the Solar Maximum Mission Satellite. *Science* 244(4903): 441–444.
32. Taylor, R. 2005. *Prometheus Project: Final Report*. Pasadena, CA: Jet Propulsion Laboratory (+215 pp.).
33. National Research Council. 2006. *Priorities in Space Science Enabled by Nuclear Power and Propulsion*. Washington, DC: The National Academies Press.
34. Taylor 2005 (note 32) (see specifically §13.3).
35. Taylor 2005 (note 32) (see specifically §13.2).
36. Heacock 1980 (note 9).
37. Herbert, G.A., and G.E. Cameron. 1993. NEPSTP—An International Testbed for Xenon Electric Propulsion. In *AIAA, AIDAA, DGLR, JSASS, 23rd International Electric Propulsion Conference,* paper IEPC-93-055. Seattle, WA: AIAA.
38. Sakrison, D. 2003. Steve and Jim Hay and the Wright 1903 Aircraft Engine. *Sport Aviation* 38–48.

39. Taylor 2005 (note 32) (see specifically §F.2).
40. LaBar, M.P. 2002. *The Gas Turbine—Modular Helium Reactor: A Promising Option for Near Term Deployment*. La Jolla, CA: General Atomics (8 pp.).
41. National Research Council. 2006. *Principal-Investigator-Led Missions in the Space Sciences*. Washington, DC: The National Academies Press (120 pp.).
42. National Research Council. 2003. *The Sun to the Earth—and Beyond: A Decadal Research Strategy in Solar and Space Physics*. Washington, DC: The National Academies Press (177 pp.).
43. Möbius, E. 2003. Where Do We Go with Solar and Heliospheric Physics? In *Solar Wind Ten: Proceedings of the Tenth International Solar Wind Conference*, AIP Conference Proceedings 679, 799–806. Pisa, Italy: American Institute of Physics.
44. National Research Council. 2003. *New Frontiers in the Solar System: An Integrated Exploration Strategy*. Washington, DC: The National Academies Press (232 pp.).
45. National Research Council. 2008. *Opening New Frontiers in Space: Choices for the Next New Frontiers Announcement of Opportunity*. Washington, DC: National Academies Press (58 pp.).
46. NASA. 2008. *Constellation: NASA's New Spacecraft: Ares and Orion*. http://www.nasa.gov/mission_pages/constellation/main/index.html.
47. McCullough, D. 1977. *The Path Between the Seas—The Creation of the Panama Canal, 1870–1914*. New York: Simon and Schuster.
48. Brookings. 2002. *U.S. Nuclear Weapons Cost Study Project*.
49. NASA. 1974. *1974 NASA Authorization, Hearings on H.R. 4567, 93/2, Part 2*, in *House, Subcommittee on Manned Space Flight of the Committee on Science and Astronautics*, Washington, DC: U.S. Government Printing Office (1271 pp.).
50. United States. 2009. *The Budget for Fiscal Year 2009, Historical Tables*. Washington, DC: U.S. Government Printing Office (194 pp.).
51. Johnston, L., and S.H. Williamson. 2003. *The Annual Real and Nominal GDP for the United States, 1789–Present*. Economic History Services.
52. United States. 2008. *The Budget for Fiscal Year 2009: National Aeronautics and Space Administration*. Washington, DC: U.S. Government Printing Office (see specifically pp. 1079–1090).
53. Clarke, A.C. 1946. The Challenge of the Spaceship. *J. Brit. Int. Soc.* 5: 66–81.
54. Sackheim, R.L. 2006. Overview of United States Space Transportation Technology and Associated Space Transportation Systems. *Journal of Power and Propulsion* 22(6): 1310–1333.
55. Walker, J. 1993. A Rocket a Day Keeps the High Costs Away. *Encyclopedia Astronautica*. http://www.astronautix.com/articles/arosaway.htm.
56. Department of Transportation. 2008. *National Transportation Statistics*.

Section IV

The Landscape of Space

Stairway to Heaven.

16 Cultural Landscape of Space

Alice Gorman

CONTENTS

INTRODUCTION

Cultural landscapes have been defined as "the combined works of nature and man [sic]" and illustrate

> the evolution of human society and settlement over time, under the influence of the physical constraints and/or opportunities presented by their natural environment and of successive social, economic and cultural forces, both external and internal.[1]

Since the development of the V2 rocket in Germany during World War II, places and objects associated with the exploration of space have proliferated. On Earth, there are launch sites, tracking stations, research and development facilities, and crash sites. In orbit there are satellites, spacecraft, and debris, and on the surface of celestial bodies there are probes, spacecraft and landing sites. Together, these sites and objects form a cultural landscape that represents the Space Age.

Instead of regarding space places and objects in a purely national context, a cultural landscape approach enables us to perceive broad patterns in the human interaction with outer space. These patterns can be chronological, geographical, or technological. It also allows the recognition that human technology and activities have been constrained by environmental factors and have impacted on the environment in their turn. Human material is not just inserted into an inert substrate: by entering space, humans have changed it from an empty vacuum into a cultural landscape, from merely space into a place. This has implications for how we study it and how we manage it as an environment.

CULTURAL LANDSCAPE APPROACHES

Since the early 1990s, cultural landscape approaches have become more prevalent in both archaeological studies and cultural heritage management.[2,3,4,5,6] In this view, a site is a relational entity, embedded in a landscape in which "off-site" areas may be just as informative.[7] Although archaeologists have always been interested in paleoenvironments, the idea that the landscape in which the site is situated is also the product of human activity has gained increasing currency.

Cultural landscapes conflate the traditional distinction between natural and cultural. It is no longer possible to say that a place or landscape is either/or: on the surface of the Earth, there are very few places that have not been shaped by both natural and cultural processes. Even the most remote or inhospitable places have been occupied by indigenous people in the past and in the present. "Wilderness" areas, assumed to have high value because they have been virtually untouched by human actions, are in fact often areas from which indigenous people have been removed by force, by the ravages of introduced diseases, or by rapid cultural transformations following European colonization.[8,9,10]

Even without human interaction, all landscapes can be seen as cultural. The geographer D.W. Meinig argued that "any landscape is composed of not only what lies before our eyes but what lies within our heads."[11] Similarly, what we understand to be "nature" is always mediated by culture (e.g., Merchant,[12] but see also Plumwood[13] for an argument about the autonomy of the nonhuman world). Cosgrove resolves this tension by acknowledging that

> landscapes have an unquestionably material presence, yet they have come into being only at the moment of their apprehension by an external observer, and thus have a complex poetics and politics.[14]

As well as capturing the interdependence of nature and culture, a landscape approach collapses the distinction between ancient and recent, old and new, past and present, and the values accorded to each. As Anschuetz et al.[15] have said, "Landscapes incorporate aspects of mythic, past and current histories concurrently: they have a quality of simultaneity." Contemporary structures and technologies are not sharply divided from those of the past: they are a continuum of human activity that takes a particular trajectory according to the historical, social, and environmental features of a place. Thus a twenty-first-century industrial structure can be seen not as an "eyesore," impacting on the visual values of a "natural" landscape, or on the heritage values of a nineteenth-century streetscape, but as a legitimate part of the landscape representing the most recent forces that have shaped it.

This opens a space for interrogating not only the dominant processes shaping a cultural landscape, but also those of the past and present. The landscape we see or experience is the accumulation of all past actions and processes: prehistoric occupation and land management practices; industrialization and pastoralism; and colonialism and globalization. In examining the cultural landscape, continuity of indigenous cultures can be acknowledged.[16]

In considering a cultural landscape, we can move away from seeing value in just the material culture itself, or only in monumental, historic, or aesthetic remains. The landscape is analyzed as an entirety in which the whole is more than the sum of its parts. The evidence of interactions between humans and their environment can be seen to have heritage value in itself.

DEFINING CULTURAL LANDSCAPES

A useful scheme to understand the diversity of cultural landscapes is captured in the United Nations Educational, Scientific and Cultural Organization's (UNESCO's) guidelines to the World Heritage Convention, which are used to assess the significance of cultural landscapes of outstanding universal value.[17] Three principal categories of cultural landscape are defined:

1. The designed or intentionally created landscape, such as gardens or parklands.
2. The organically evolved landscape. This results from an initial social, economic, administrative, and/or religious imperative and has developed its present form by association with and in response to its natural environment. Such landscapes reflect that process of evolution in their form and component features. They fall into two subcategories:
    - A relict (or fossil) landscape is one in which an evolutionary process came to an end at some time in the past, either abruptly or over a period. Its significant distinguishing features are, however, still visible in material form.
    - A continuing landscape is one that retains an active social role in contemporary society closely associated with the traditional way of life, and in which the evolutionary process is still in progress. At the same time it exhibits significant material evidence of its evolution over time.[18]
3. The associative cultural landscape, with religious, artistic, or cultural associations rather than evidence of material culture alone. This intersects with UNESCO's formulation of intangible heritage, manifested in the domains of oral traditions and expressions, performing arts, social practices, rituals and festive events, traditional crafts, and knowledge and practices concerning nature and the universe.[19]

The World Heritage Convention promotes an "integrated recognition of cultural and natural values,"[20] and any cultural landscape may have designed, organic, and associative aspects. Using these definitions as a starting point, can outer space be conceptualized as a cultural landscape?

SPACESCAPE

Cultural landscapes on Earth were created progressively as humans moved out of Africa to colonize virtually all environments on the globe. The move into space has been qualitatively different in its requirement for technological support—humans cannot survive unaided in the atmosphere-less microgravity environment—and the fact that space is not "land" as such. Space is a different place: separate from the Earth, inaccessible and distant.

To consider the spacescape both as a cultural landscape and as an archaeological record, it seems necessary to reconceptualize it as a place in a continuum with Earth. Archaeology is essentially Ptolemaic and geocentric: with Earth as a stationary center, artifacts and places in space cannot be assessed as related within the same landscape as space artifacts and places on Earth. Our land and landscapes are confined by the thin envelope of the atmosphere, where the pressure, temperature, and mixture of gases are just right. We define our world by where we can breathe and by what we can see.

Reconceptualizing the relationship between space and Earth can give us another perspective on the extent and nature of the cultural landscape of space. For a start, the end of the atmosphere may not provide a natural boundary protecting us from the emptiness of space. While terrestrial nations send material up into orbit and beyond, celestial material falls to Earth, contributing 40,000 tons to our mass every year.[21] Tides are created by the moon, solar cycles affect weather, and Earth's gravitation field is influenced by the movements of stars.[22] The Earth is, in fact, an open system in a dynamic interchange with the cosmos.

This is far from a new idea—many indigenous and premodern worldviews see the heavens and Earth as part of the same system.[23,24,25,26] The recognition that, in order to understand the Earth, one must investigate space was an integral part of the International Geophysical Year of 1957–1958. A decentering of the Earth has already been accomplished by the most influential photographs in the world—Earth seen from space, as taken by various Apollo missions.[27]

In the contemporary era of space exploration, the amount of traffic between Earth and low Earth orbit means that we can no longer regard space as remote.[28] Hays and Lutes, discussing a theory of spacepower, suggest that the next phase of the Space Age may be defined by "a shift from a geocentric perspective to a solar system perspective."[29]

To reconfigure Earth and space, we need new variables. One way to do this is to use gravity and energy in a dynamical system to describe the relationship between space places and artifacts throughout the solar system.[30,31,32] We can describe the location of artifacts according to the amount of energy they possess, following Lagrange,[33] and their movements over time in regard to stable and unstable equilibria (Figure 16.1). The Earth becomes an attractor, drawing objects to it over time as their energy decreases (the equivalent of taphonomy in archaeological science). Within a certain gravitational regime, described by the curvature of space-time, everything will end up at a point of low energy on the surface of the Earth. Where the curvature is less, a satellite flung into orbit with enough energy will fall toward Earth without ever reaching it. Other celestial bodies create gravity wells that are attractors for certain regions of the state-space, with the Lagrange points as unstable equilibria.

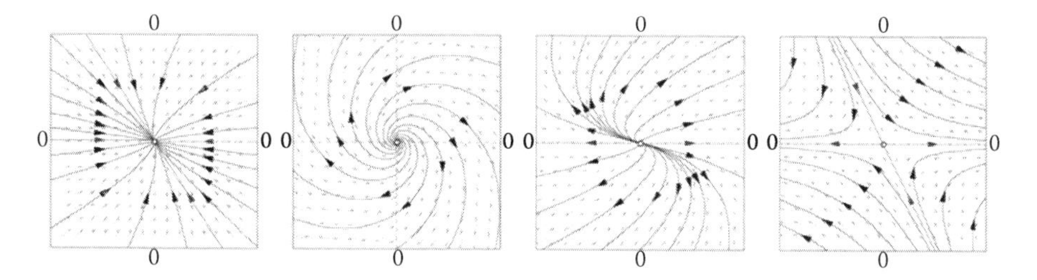

FIGURE 16.1 Linear dynamical systems. (Courtesy of Ethan Hein.)

Within this system, it is possible to describe the movements of artifacts over time in the same mathematical language, whether they are on Earth, in orbit, or on another celestial body. In this unified picture, the landscape of Earth is just a subset of the spacescape, with thresholds defined by gravity and energy.[34]

Our task now is to understand the cultural spacescape and its heritage implications. The places and objects related to space exploration could be considered as part of a spacescape encompassing the terrestrial surface and planetary surfaces, orbits, and interplanetary spaces within the solar system and the rest of the cosmos: what we see from Earth as the night sky.[35] Using the UNESCO categories, terrestrial spacescapes are often planned or intentionally designed; orbital and interplanetary spacescapes have evolved organically; and the night sky is predominantly an associative spacescape.

DESIGNED LANDSCAPES

The primary example of designed space landscapes are terrestrial launch sites. Their design has a number of features in common: prohibited access, enforced by barriers such as fences and entrance control features; location in sparsely populated areas; extensive systems of roads and rail to convey hardware to launch sites; and the actual launch pads themselves with their associated infrastructure. Early Cold War rocket ranges tended to be located in desert/steppe regions, as was the case in Algeria (Colomb-Béchar/Hammaguir), Australia (Woomera), the United States (White Sands), and the USSR (Kapustin Yar), all established in 1947. These were landscapes of defense, requiring "a hardening of space, a reinforcement of boundaries and distinctions, and the fortification of property, place or nation."[36]

In these landscapes, often configured as remote and empty places, the infrastructures of rocket launching were rapidly overlain on environments created by the activities of indigenous people (e.g., Aboriginal people in Australia and Puebloan and Athapaskan people in New Mexico) and later colonizers (English, French, Mexican, Russian). Those who lived and worked at these launch sites often conceptualized them as radical disjunctions between the Stone Age and the Space Age.[37,38] By considering them as cultural landscapes, we are encouraged to look instead at continuities and interactions rather than these disjunctions.

The Australian and British governments came to an agreement in 1946 to collaborate on the development of missiles at the Long Range Weapons Establishment

(LRWE) at Woomera in South Australia.[39] An area larger than the entire United Kingdom was set aside as "prohibited" land, and surveyors mapped out the village of Woomera near the rangehead area and pushed roads through the formerly pathless desert so that recovery crews could retrieve the wreckages of missiles and launch vehicles. There was no existing urban base around which to construct the rocket range: accommodations, water, electrical distributions systems, and roads all had to be constructed from scratch.[40] Woomera Village itself was designed according to the latest postwar town planning ideas,[41] and like the later space town of Kourou in French Guiana, the use of space in the village was aimed at creating social harmony. As the program grew from missiles to rocket tests and satellites, new launch areas and tracking stations were built across the desert, with more roads and infrastructure.

But this was no empty desert. Aboriginal people from many different groups lived across the entire vast area of the range (Figure 16.2). The launch path northwest crossed the Central Aborigines Reserve on the border with Western Australia. The area around Woomera Village and the rangehead was the traditional country of the Kokatha people.[42] The advent of the rocket range did not cause their disappearance, although it did prevent their access to ceremonial places at the appropriate times (Andrew Starkey, personal communication). While members of Woomera's Natural History Society were out collecting stone tools and geological samples from the desert, Kokatha people were trying to make a living on pastoral stations and in the urban center of Port Augusta.

The LRWE employed Native Patrol Officers to ensure that Aboriginal people living on the range were warned about impending launches and their dangers.[43] The increased presence of white Australians led to exchanges of goods and competition

FIGURE 16.2 Mural at Woomera Village. (Author's image.)

for scarce water: this was particularly noticeable at the Giles Meteorological Station, where both groups used the same water hole.[44] As the anthropologist A. P. Elkin had suggested, the presence of the range accelerated cultural changes associated with colonization.[45]

By looking at Woomera as a landscape, rather than just focusing on the material culture of the armaments/space program, we can see a more complex significance beyond that of the space technology. The places where indigenous people lived, and their movements through the landscape, were now constrained by the presence of the village, the rangehead, and the launch areas. Particular places were loci of cultural exchange, like the Giles Meteorological Station and the Gove tracking station, established in the Northern Territory for the Europa satellite program. Across the desert, rocket components now lay scattered among the vast numbers of stone tools that were testament to the long occupation of the land by Aboriginal people. In this landscape, there is a pattern common to other early Cold War launch sites and even later ones such as Kourou.[46] These landscapes, expressing the power and technological prowess of emerging spacefaring nations, are also landscapes of expropriation and exploitation[47] and are thus contested spaces, as continuing protests at Woomera and Kourou attest.[48,49,50]

Much of the original infrastructure of now-defunct space programs at Woomera has been dismantled, destroyed, or abandoned. Once the heralds of a spacefaring future, the launch structures are decaying like prehistoric monuments, becoming archaeological sites in their turn (Figure 16.3). Their cultural significance rests as

FIGURE 16.3 Space Age henge: the former Island Lagoon tracking station, Woomera. (Author's image.)

much in the story of local interactions with Aboriginal people on the range as it does in the history of the technology. Australia is no longer a spacefaring nation, but the Kokatha and Pitjantjatjara are still surviving in their own land.

ORGANICALLY EVOLVED LANDSCAPES

From terrestrial launch sites, based in nation-states, spacecraft are hurled into Earth's orbit, where the space hardware of many nations forms a cosmopolitan mixture in the global commons of space. Prior to the launch of *Sputnik 1* in 1957, orbital space was an associative landscape. Now, clouds of orbital debris have come to resemble a ring system like that surrounding Saturn.

Orbital space can be seen as an organically evolved landscape because the relationship between its elements has not been designed, even though the missions themselves were. It is not without structure, of course—particular orbital regimes are more used than others, and the orbital configurations of hardware within these orbits are partially determined by the location of the launch site and the purpose of the mission. Within each orbital regime, there are different types of space hardware. For example, low Earth orbit contains a higher proportion of space stations, upper rocket stages, and amateur satellites, such as the OSCAR series.[51] Medium Earth orbit contains navigation satellites, and geostationary orbit and the higher graveyard orbit are dominated by telecommunications satellites.

Taking a cultural landscape approach, we must look not only at the spatial relationships between the elements of human material culture, but also at the whole environment of Earth orbit. Spacecraft have not simply been added to an empty substrate but are impacted by, and impact in their turn, a complex system of high energy particles, plasmas, geomagnetic storms, solar cycles, meteor swarms, and upper atmospheric phenomena. The impacts of the space environment on satellite materials and functions have been extensively studied, but the effects of the orbital spacecraft population on the space environment, outside of the sphere of atmospheric impacts, have received less attention. Orbital debris is now part of this environment, and it is important to consider how it operates as a total system, rather than isolating human factors from the rest of the space environment. This may have ramifications for the environmental management in space.

This spacescape thus demonstrates the interaction of human material culture with the orbital environment, and this has cultural significance in its own right. Unlike terrestrial environments, however, we cannot read the landscape by identifying older surfaces or archaeological remains, or excavating the evidence of paleoenvironments that no longer survive. In its erasure of past and present, the orbital spacescape exemplifies the simultaneity of the cultural landscape.[52] Space-time has no provision for a "now" or the passing of time.[53] Essentially it is isotropic, the same in all directions.[54]

While it is true that the earliest generation of satellites were all launched into low Earth orbit, other orbital regimes were rapidly colonized: it took only 7 years from the first satellite for geostationary orbit to be attained. Satellites of all ages are in the same "stratigraphic" layers; this in itself is a feature of the spacescape determined by the state-space of the dynamical system. In this cultural landscape, the 1963 *Syncom*

3 may have as its neighbor the 2008 *Vinasat 1*, the first telecommunications satellite launched by Vietnam. And yet the material culture of orbit is amenable to chronological analysis.

The passage of time in orbit can be read on the bodies of the spacecraft as surfaces and components decay gradually (or catastrophically) through exposure to the space environment, and in the changing designs, materials, and technology as the goals of the space industry changed and the conditions of orbit were better understood. This is, however, no simple stimulus/response mechanism: like all human artifacts, these spacecraft also represent social perceptions, political aspirations, and ideologies: they have imprinted space with cultural meanings.[55]

ASSOCIATIVE LANDSCAPES

The associations of landscapes can be visual, kinetic, or factory, or acoustic, spiritual or cultural. While the actual physical components of a landscape may change little over time, their associations can alter or be radically transformed.[56] Indigenous cultural landscapes are the exemplar of associative landscapes, such as the Tongariro National Park in New Zealand and Uluru Kata-Tjuta National Park in Australia, both inscribed on the World Heritage List for their associative cultural values as well as for natural values.

Celestial phenomena have contributed to the creation of cultural landscapes on Earth, as "many traditional societies carefully track the movements of the celestial bodies in order to know how to act in the earth. ... The correlation of terrestrial and celestial phenomena enable[s] them to move in space and time."[57] Archaeoastronomers have studied the manifestation of celestial knowledge in terrestrial landscapes, ranging from representations of the night sky and calendrical systems in rock art, to the underlying logic of built structures, to sites specifically designed to align with the movements of celestial bodies. Despite this, the perception of the sky has rarely been incorporated into the assessment of cultural landscapes because of the traditional dichotomy drawn between the heavens and the Earth in Western systems of knowledge.[58]

On the ground, terrestrial space sites have associations driven by memory and experience, by moving and acting within a landscape. Of the orbital and planetary spacescapes, we usually have only one experience: of looking above, usually at night, when one can now see satellites circling in Earth orbit in addition to planets, moons, stars, and galaxies. The presence of human material culture in orbit has transformed the complexion of the night sky.

Although every person on Earth has a similar visual perspective of the night sky, its interpretation depends on culturally dependent worldviews.[59] The same constellations, such as the Pleiades, have formed part of diverse cosmologies across the world. The associations of this spacescape are under threat, both as light pollution obscures our ability to view the night sky and as "traditional perceptions and understandings of the sky are increasingly exposed, affected or replaced by cosmologies and astronomies derived from our modern Western culture."[60]

Of course, this has led to new understandings of the heavens that are evidence of changing human interactions with space, as greater numbers of indigenous people use satellite services such as telecommunications to subvert the impact of globalization.[61] Seeing the sky as a cultural landscape allows the contemporary dynamism of indigenous engagements with technology to be valued, rather than trapping them in an antiquated notion of authenticity that perpetuates the distinction between the Stone Age and the Space Age.[62]

In the post-1957 world, space has additional intangible elements. Places and objects in the spacescape are connected by telemetry (the information that flies unseen between spacecraft and ground stations) and the invisible routes taken by rockets and satellites in the atmosphere and in orbit, which can be described by equations rather than mapped. These are also part of the associative landscape.[63] Space hardware is the tangible, durable evidence of ephemeral digital information that fills the space between Earth orbit and the Earth's surface.

CONCLUSION: SPACESCAPES AS HERITAGE

In this chapter, I have considered how different kinds of places in space can be viewed as cultural landscapes according to the widely accepted definitions of the World Heritage Convention. These landscapes and spacescapes have heritage value because they capture the interaction of human and environmental processes at a particular time in the historical engagement with space. I have also attempted to reframe how we regard Earth and space by situating places and objects within a dynamical system. In effect, this brings Western conceptions of space closer to traditional indigenous views.

A cultural landscape approach emphasizes continuities rather than divisions between Earth and space, past and present, and technology and society. On the surface of the Earth, in orbit, and in the night sky, places and artifacts are connected by their physical, historical, and cultural relationships. Understanding these connections empowers a more inclusive assessment of their significance and provides a robust basis for managing the heritage values of a unique material culture.

REFERENCES

1. UNESCO Intergovernmental Committee for the Protection of the World Cultural and Natural Heritage. 2005. *Operational Guidelines for the Implementation of the World Heritage Convention*. Paris: World Heritage Centre (see specifically p. 83).
2. Cleere, H. 1995. Cultural Landscapes as World Heritage. *Conservation and Management of Archaeological Sites* 1: 63–68.
3. Gosden, C., and L. Head. 1994. Landscape—A Usefully Ambiguous Concept. *Archaeology in Oceania* 29: 113–116.
4. Hirsch, E., and M. O'Hanlon, eds. 1995. *The Anthropology of Landscape: Perspectives on Place and Space*. Oxford, United Kingdom: Clarendon Press.
5. Jacques, D. 1995. The Rise of Cultural Landscapes. *International Journal of Heritage Studies* 2: 91–101.
6. Knapp, A., and W. Ashmore. 1999. Archaeological Landscapes, Constructed, Conceptualized, Ideational. In *Archaeologies of Landscape: Contemporary Perspectives*, ed. W. Ashmore and A.B. Knapp, 1–30. Malden, MA: Blackwell Publishers.
7. Foley, R. 1981. *Off-site Archaeology and Human Adaptation in Eastern Africa: An Analysis of Regional Artifact Density in the Amboseli, Southern Kenya*. BAR International Series 97. Oxford: British Archaeological Research.

8. Cosgrove, D. 1985. Prospect, Perspective and the Evolution of Landscape Ideas. *Transactions of the British Institute of Geographers* 10: 52–53.
9. Denevan, W.M. 1992. The Pristine Myth: The Landscape of the Americas in 1492. *Annals of the Association of American Geographers* 82: 369–385.
10. Plumwood, V. 2006. The Concept of a Cultural Landscape: Nature, Culture and Agency in the Land. *Ethics and the Environment* 11(2): 120.
11. Meinig, D. 1979. The Beholding Eye: Ten Versions of the Same Scene. In *The Interpretation of Ordinary Landscapes: Geographical Essays*, ed. D. Meinig and J.B. Jackson, 33–48. New York: Oxford University Press. (see specifically p. 33).
12. Merchant, C. 2003. *Reinventing Eden: The Fate of Nature in Western Culture*. New York: Routledge.
13. Plumwood 2006 (note 10).
14. Cosgrove 1985 (note 8) (see specifically p. 50).
15. Anschuetz, K.F., R.A. Wilshusen, and C.L. Scheick. 2001. An Archaeology of Landscapes: Perspectives and Directions. *Journal of Archaeological Research* 9(2): 157–211. (see specifically p. 186).
16. Anschuetz et al. 2001 (note 15) (see specifically p. 159).
17. UNESCO 2005 (note 1) (see specifically pp. 83–84).
18. Ibid. (see specifically p. 84).
19. UNESCO. 2008. What Is Intangible Heritage? http://www.unesco.org/culture/ich/index.php?pg=00002.
20. Lennon, J., B. Egloff, A. Davey, and K. Taylor. 1999. *Conserving the Cultural Values of Natural Areas.* Discussion Paper for Australian Heritage Commission. Canberra, Australia (see specifically p. 11).
21. Clark, N. 2005. Ex-orbitant Globality. *Theory, Culture and Society* 22(5): 165–185.
22. Briggs, J., and F.D. Peat. 1989. *Turbulent Mirror: An Illustrated Guide to Chaos Theory and the Science of Wholeness*. New York: Harper and Row.
23. Chamberlain, V.D. 1982. *When the Stars Came Down to Earth: Cosmology of the Skidi Pawnee Indians*. College Park, MD: Ballena Press.
24. Gorman, A.C. 2007. *The Gravity of Archaeology*. Paper presented at the New Ground Australasian Archaeology Conference, September 21–26, University of Sydney.
25. Gorman, A.C. 2006. *From the Stone Age to the Space Age: High Technology and Indigenous Heritage.* Paper presented to the Australian Space Development Conference, July 19–21, Canberra, Australia.
26. Wright, P. 1975. Astrology and Science in Seventeenth-Century England. *Social Studies of Science* 5(4): 399–422.
27. Cosgrove, D. 1994. Contested Global Visions: One-World, Whole-Earth, and Apollo Space Photographs. *Annals of the Association of American Geographers* 84(2): 270–294.
28. MacDonald, F. 2007. Anti-Astropolitik—Outer Space and the Orbit of Geography. *Progress in Human Geography* 31(5): 592–615.
29. Hays, P.L., and C.D. Lutes. 2007. Towards a Theory of Spacepower. *Space Policy* 23: 206–209 (see specifically p. 207).
30. Gorman 2007 (note 24).
31. Katok, A., and B. Hasselblatt. 1995. *Introduction to the Modern Theory of Dynamical Systems*. New York: Cambridge University Press.
32. Palis, J., and W. de Melo. 1982. *Geometric Theory of Dynamical Systems: An Introduction.* New York: Springer-Verlag.
33. Thornes, J.B. 1983. Evolutionary Geomorphology. *Geography* 68: 225–235.
34. Gorman 2007 (note 24).
35. Gorman, A.C. 2005. The Cultural Landscape of Interplanetary Space. *Journal of Social Archaeology* 5(1): 85–107.

36. Gold, J., and G. Revill. 2000. Landscape, Defence and the Study of Conflict. In *Landscapes of Defence*, ed. J. Gold and G. Revill, 1–20. London: Prentice Hall (see specifically p. 15).
37. Gorman, A.C. 2005. *From the Stone Age to the Space Age: Interpreting the Significance of Space Exploration at Woomera.* Paper presented to the symposium Home on the Range: the Cold War, Space Exploration and Heritage, November 5, at Woomera, South Australia. Flinders University.
38. Gorman, A.C. 2009. The Archaeology of Space Exploration. In *Space Travel and Culture*, ed. M. Parker and D. Bell, 129–142. Oxford: Wiley-Blackwell Publishing:
39. Morton, P. 1989. *"Fire Across the Desert": Woomera and the Anglo-Australian Joint Programme, 1946–80.* Canberra, Australia: Australian Government Publishing Services.
40. Ibid.
41. Garnaut, C., P.-A. Johnson, and R. Freestone. 2002. The Design of Woomera Village for the Long Range Weapons Project. *Journal of the Historical Society of South Australia* 30: 5–23.
42. Gorman 2005 (note 35).
43. Morton 1989 (note 39).
44. Gorman, A.C. In press. Beyond the Space Race: The Significance of Space Sites in a New Global Context. In *Contemporary Archaeologies: Excavating Now*, ed. A. Piccini and C. Holthorf. Bern: Peter Lang.
45. Elkin, A.P. 1947. Aborigines Not Doomed by Rocket Project. (Letter) *Sydney Morning Herald*, June 20.
46. Gorman, A.C. 2007. La Terre et l'Espace: Rockets, Prisons, Protests and Heritage in Australia and French Guiana. *Archaeologies: Journal of the World Archaeological Congress* 3(2): 153–168.
47. Gold and Revill 2000 (note 36).
48. Gorman 2005 (note 35).
49. Redfield, P. 2002. The Half-Life of Empire in Outer Space. *Social Studies of Science* 32(5–6): 791–852.
50. Redfield, P. 2005. *Space in the Tropics: From Convicts to Rockets in French Guiana.* Berkeley, CA: University of California Press.
51. Baker, K., and D. Jansson. 1994. *Space Satellites from the World's Garage—The Story of AMSAT.* Paper presented to the National Aerospace and Electronics Conference, May 23–27, Dayton Ohio.
52. Anschuetz et al. 2001 (note 15) (see specifically pp. 157–211).
53. Fraser, J.T. 2005. Space-Time in the Study of Time: An Exercise in Critical Interdisciplinarity. *KronoScope* 5(2): 151.
54. Eddington, A. 1959. *Space, Time and Gravitation.* New York: Harper. 48.
55. Gorman, A.C., and B.L. O'Leary. 2007. An Ideological Vacuum: The Cold War in Space. In *A Fearsome Heritage: Diverse Legacies of the Cold War*, ed. J. Schofield and W. Cocroft, 73–92. Walnut Creek, CA: Left Coast Press.
56. Anschuetz et al. 2001 (note 15) (see specifically p. 165).
57. Australia ICOMOS. 1995. The Asia-Pacific Regional Workshop on Associative Cultural Landscapes. A Report to the World Heritage Committee. http://whc.unesco.org/archive/cullan95.htm.
58. Ibid. (see specifically p. 3).
59. Ibid. (see specifically Fig. 5).
60. Ibid. (see specifically p. 2).
61. Gorman (note 44).
62. Gorman 2006 (note 25).
63. Iwaniszewski, S. 2006. *Astronomy in Cultural Landscapes: New Challenges for World Heritage Issues.* Paper presented at the Forum UNESCO 10th International Seminar, Cultural Landscapes in the 21st Century, April 2005.

17 Orbital Artifacts in Space

Daniel E. Clemens

CONTENTS

INTRODUCTION

The artifacts left in orbit from our earliest days of spaceflight are not simply machines but are relics of our past that demonstrate our drive to explore and to conquer the unknown. Exploration of any kind is dangerous, and space exploration is certainly no exception; lives are sometimes lost. That is the nature of the world in which we live, but it is also the nature of the human spirit to persevere in the wake of devastating tragedy and to prevail over seemingly insurmountable challenges. These artifacts stand as a testament to the limitless capacity of the human mind and to the bravery and dedication of the men and women who have chosen to take up this burden in service to all of humanity.

Spaceflight has revolutionized global communications, surveillance, navigation, atmospheric and environmental sciences, meteorology, and, of course, astronomy and space sciences. This chapter will detail the types of objects found in the various orbits around Earth and elsewhere. Different orbits facilitate certain types of missions, and the type of mission determines the types of artifacts that are found in a particular region of space. Specific examples of these missions will be given, focusing on spacecraft currently flying in space as opposed to those that have reentered the Earth's atmosphere or landed on another planetary body. The chapter will begin with an overview of the spacescape and later examine the populations of the various orbits in more detail.

OVERVIEW OF THE SPACESCAPE

For the purposes of this discussion, space will be divided into the regions of low Earth orbit (LEO), geosynchronous (or geostationary) Earth orbit (GEO), medium Earth orbit (MEO), and other orbits. The majority of space missions are flown in LEO, which is defined approximately as the region of space around the Earth with an altitude of 160–2,000 km.[1] Mission types flown in this region of space include communications, Earth observation, and space science. All crewed orbital flights have been flown in LEO, including shuttle and space station missions, with the exception of the lunar missions. GEO is an Earth-synchronous orbit at an altitude of approximately 35,786 km and is used primarily for communications satellites.[2] MEO is defined as an orbit between LEO and GEO. Various types of missions are flown in this region of space, but one of the most common examples is the global positioning system navigation satellite constellation. Other orbits include different Earth orbits, such as Molniya orbits and highly elliptical orbits that cross through multiple regions of space, interplanetary (solar) orbits, and extrasolar orbits. Figure 17.1 shows an image of approximately 900 satellites currently orbiting the Earth that are cataloged in the NASA J-Track 3D spacecraft tracking tool.[3] This represents a fraction of the several thousand human-made objects that are currently in Earth orbit.

Humans began flying orbital space missions in October of 1957 when the USSR successfully placed *Sputnik 1* in orbit.[4] Since that time, thousands of spacecraft have been flown. Depending on its orbit, a spacecraft may fall out of orbit after its useful lifetime has ended and burn up (partially or completely) in the Earth's atmosphere. However, many spacecraft do not reenter or do so after a very long period of time. Space is speckled with thousands of human-made objects including "live" and "dead" spacecraft, the spent upper stages used to place them in orbit, and various pieces of orbital debris. The U.S. Strategic Command (USSTRATCOM) is tasked with tracking objects in space through the Space Surveillance Network (SSN).[5] The SSN is a global network comprising nearly thirty space surveillance facilities, including phased-array and conventional radars, and electro-optical telescopes. The facilities of the SSN are spread around the world, but most are located in North America. Together, these sensors perform 300,000–400,000 observations each day, tracking more than 15,000 human-made objects currently in Earth orbit that are generally 10

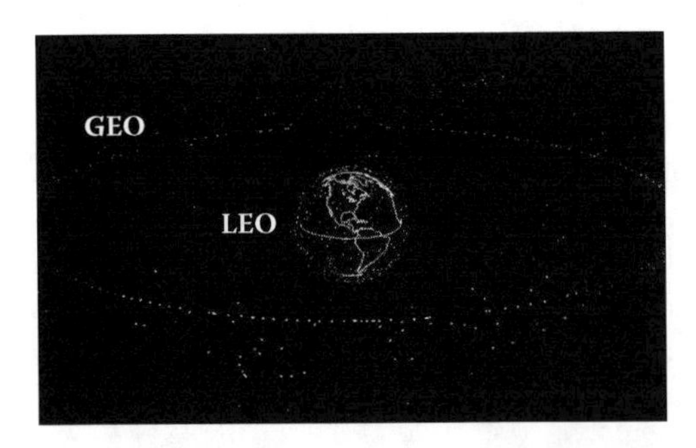

FIGURE 17.1 Image of approximately 900 Earth-orbiting satellites cataloged in the NASA J-Track 3D spacecraft tracking tool.

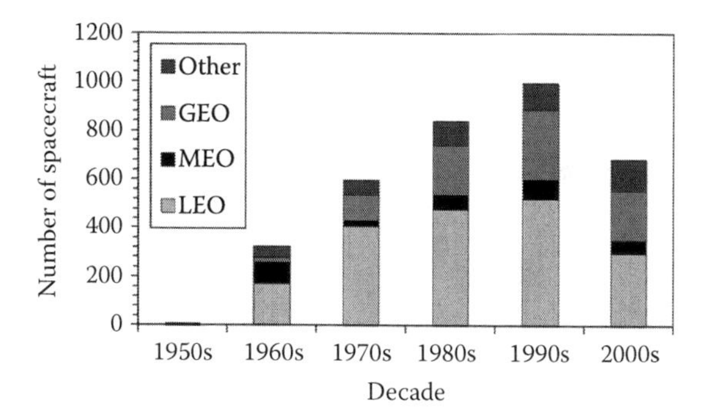

FIGURE 17.2 Approximate number of spacecraft currently found in various orbits around the Earth and elsewhere originating in each decade beginning in October 1957 and ending in December 2007. The data include upper stages but not orbital debris.

cm or larger. About 10% of those are functional spacecraft, 15% are launch vehicle components, and about 75% are fragments and inactive satellites. Some collision avoidance analyses, as well as atmospheric reentry assessments, are performed by USSTRATCOM. Since tracking began with *Sputnik*, more than 35,000 objects have been cataloged, more than 20,000 of which have reentered.

Figure 17.2 shows the approximate number of spacecraft originating in each decade, including only those currently found in orbits around the Earth and elsewhere, for the time span of October 1957 to December 2007.[6] The data include upper stages but not orbital debris. As expected, the highest number of spacecraft are found in LEO. It is also interesting to note the development of GEO utilization over time. Out of about 3,455 total spacecraft currently flying in Earth orbit and elsewhere (as of December 2007), approximately 54% reside in LEO, 9% in MEO, 24% in GEO, 11% in other Earth orbits, and 2% outside Earth orbit. The populations of these orbits will be examined in more detail in the upcoming sections. Figure 17.3[7] demonstrates the orbital path and proximity to the Earth of certain spacecraft that reside in various Earth orbits.

Table 17.1 shows what are thought to be the oldest satellites currently in orbit that were designed, built, and launched by each of the spacefaring nations with a demonstrated independent launch capability.[8,9] Many of these were among the nation's first satellites and were primarily intended to demonstrate the technology to safely deliver a functioning spacecraft to orbit.[10] *Prospero* is unique in that it was the first and only satellite designed, built, and launched by the United Kingdom. Following the delivery of *Prospero* to orbit, the British government decided to abandon its domestic space launch program on the basis that it would be cheaper to purchase launch services from the United States or the European Space Agency for future satellites.

ARTIFACTS IN LEO

The majority of space missions are flown to LEO. The types of missions flown there include space science, Earth observation, navigation, telecommunications, and all

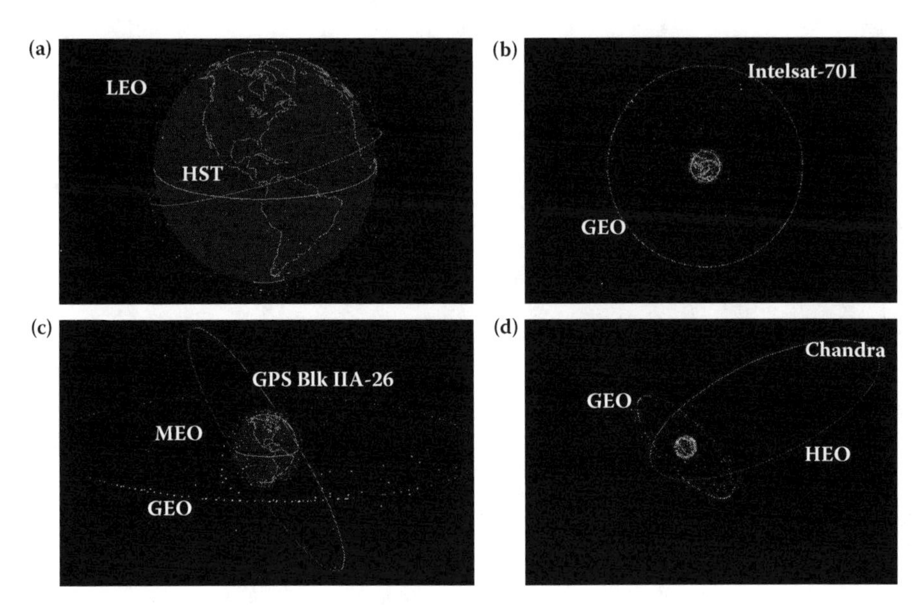

FIGURE 17.3 Orbital paths of certain spacecraft residing in various Earth orbits depicted using the NASA J-Track 3D spacecraft tracking tool: (a) Hubble Space Telescope (HST) in low Earth orbit; (b) *Intelsat-701* in geosynchronous Earth orbit; (c) *Navstar GPS Block IIA-26* in medium Earth orbit; (d) *Chandra X-Ray Observatory* in a highly elliptical orbit (HEO).

TABLE 17.1
Oldest Satellites Currently in Space (as of December 2007) That Were Designed, Built, and Launched by Each of the Spacefaring Nations with Demonstrated Independent Launch Capability

Nation	Satellite	Year	Orbit	Intended Purpose
United States	*Vanguard 1*	1958	MEO	Technology demonstration; oldest human artifact in space
USSR	*Luna 1*	1959	Solar	Lunar contact; first human artifact to leave Earth orbit
France	*Asterix**	1965	LEO	Technology demonstration
Japan	*Tansei 1*	1971	LEO	Technology demonstration
China	*Dong Fang Hong**	1970	LEO	Technology demonstration
United Kingdom	*Prospero**	1971	LEO	Technology demonstration
India	*IRS-P2*	1994	LEO	Earth sensing; resource mapping
Israel	*Ofeq 5*	2002	LEO	Reconnaissance

* Nation's first satellite.
LEO, low Earth orbit; MEO, medium Earth orbit.

crewed spaceflight (except the lunar missions). LEO is a region of space packed with a variety of operational spacecraft and, in addition, many nonoperational spacecraft and a great deal of orbital debris. Figure 17.4 shows the approximate number of spacecraft currently found in LEO originating in each decade from various countries.[11] Spacecraft in LEO and elsewhere played a key role in the Cold War and were used for intelligence, surveillance, reconnaissance, communications, navigation, and weather prediction. The Space Age began with the flight of the Soviet *Sputnik*, but the United States took the lead in spacecraft development and deployment during the 1960s. During the 1970s and 1980s, however, the Soviets launched far more satellites into orbit than the United States (see also Figures 8.16 and 8.17.). After the fall of the Soviet Union, the number of Russian spacecraft sent into orbit declined significantly, while the number of U.S. spacecraft increased, driven by the deployment of large numbers of commercial telecommunications satellites.[12] Other nations have increased their activity substantially over the past two decades. Out of about 1,877 spacecraft currently flying in LEO (as of December 2007), approximately 53% are of Russian/Soviet origin, 33% are from the United States, and 14% belong to other countries.

Naturally, space science missions were among the first endeavors of the Space Age. The only human-made object left in LEO from the 1950s is *Explorer 7*, which was built by the United States and launched in 1959.[13] It carried a payload of science instruments for measuring solar and cosmic radiation, micrometeorite impact, and the Earth's magnetic field. The solar radiation experiments failed because of the high energy flux of the Van Allen radiation belts, while the micrometeorite experiment failed because of technical problems. *Explorer 7* was very successful in helping to elucidate the interaction between solar activity, the Earth's magnetic field, and the Van Allen belts, and it explained the increase in auroral activity following solar flares.[14] *Explorer 7* also carried the first meteorological payload in the form of a radiometer, a device that measures the radiation balance of the Earth. It is the strong latitude dependence of the difference between incoming solar radiation and outgoing terrestrial infrared radiation that drives atmospheric circulation and the planet's weather patterns.[15]

The Hubble Space Telescope (HST), built by the United States and launched in 1990, was the first large optical telescope to operate successfully above the

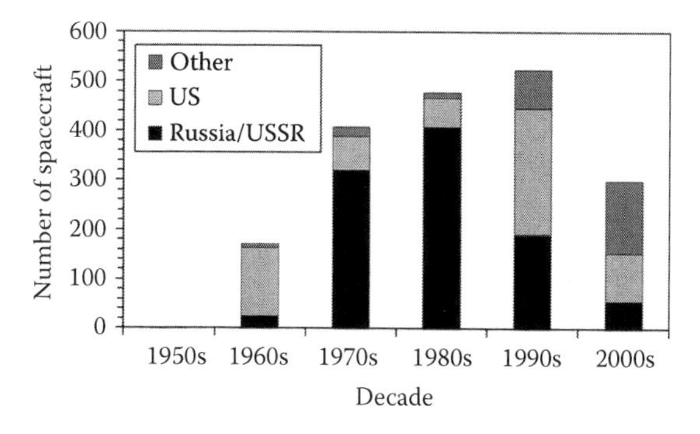

FIGURE 17.4 Approximate number of spacecraft from various countries currently found in low Earth orbit originating in each decade beginning in October 1957 and ending in December 2007. The data include upper stages but not orbital debris.

atmosphere.[16] HST was part of NASA's Great Observatories program, along with the Compton Gamma Ray Observatory, also in LEO, the *Chandra X-Ray Observatory*, in a highly elliptical Earth orbit, and the Spitzer Space Telescope, in solar orbit. These observatories were designed to complement each other by detecting different regions of the electromagnetic spectrum. Equipped with a 2.4-m-diameter primary mirror and a sophisticated set of gyros for high-precision pointing, HST could produce images from the near-ultraviolet to the near-infrared of objects far more distant and faint than any previously seen from the surface of the Earth.[17] HST was deployed in LEO (Figure 17.3a) by the shuttle and was designed to be periodically serviced by the shuttle for upgrades and repairs. Upon checkout in orbit, a spherical aberration was discovered in the primary mirror, which caused images to be out of focus. In addition, thermal cycling, caused by frequent shifts between the sunlit and dark portions of orbit, caused excessive flexing of the solar arrays, which led to pointing errors. Corrective actions were performed during the first servicing mission in 1993. Since then, HST has become not only one of the most prolific scientific instruments in history, producing data for thousands of peer-reviewed journal articles and countless astronomical discoveries, but has also captured the attention of the general public with its spectacular imagery.[18]

Since the beginning of the space program, satellites have been designed to observe the Earth. This was driven initially by the Cold War desire to locate enemy ballistic missile facilities and spy on other military installations and troop movements. For the United States, satellite-based observation platforms were preferable to high-altitude aerial reconnaissance aircraft such as the U-2 spyplanes, which were not only provocative to the USSR but also became vulnerable to Soviet ground-to-air missiles. Altogether, the United States and Russia/USSR, as well as many other countries, have developed and fielded hundreds of military surveillance spacecraft.[19]

Earth observation also includes atmospheric, environmental, and meteorological studies. *TIROS 1* (Television and InfraRed Observation Satellite), built by the United States and launched in 1960, was the first dedicated weather satellite.[20] It was designed to test the feasibility of producing and utilizing television cloud cover pictures from a spacecraft. The satellite was equipped with two 1.27-cm-diameter vidicon television cameras, one wide angle and one narrow angle, for taking Earth cloud cover pictures. Those pictures were either transmitted directly to a ground receiving station or were stored in an onboard tape recorder for playback later, depending on whether the satellite was within or beyond the communication range or line-of-sight of the station. The satellite performed normally for a little over 2 months until experiencing a power failure, but the mission was considered an unquestionable success. The pictures taken by *TIROS 1* showed weather formations and organized patterns that were previously unknown to meteorologists, who expected to see more chaos than order.[21] It was followed by nine more TIROS satellites (*TIROS 2–10*), launched by 1965, and *TIROS-N*,[22] launched in 1978, most of which are still in orbit. *TIROS 3* provided the first hurricane detection by a satellite, and *TIROS 9* provided the first photomosaic of the entire Earth.

Satellite navigation technology attracted the attention of the militaries early in the space program. During the flight of *Sputnik 1*, researchers at the Johns Hopkins University Applied Physics Laboratory discovered that they could determine the orbit of the spacecraft by listening to the Doppler shift of its transmitted signal, since they knew precisely the time and their position on the ground.[23] They deduced that the process could work in reverse—if they knew time and spacecraft position with high precision, they could determine where they were on the ground. Satellite navigation was born with the Transit series of spacecraft, several of which are still in LEO. Transit was used by the U.S. Navy for navigation of ships and submarines at sea anywhere, anytime, in any weather, which represented a great leap forward.[24] Transit was followed by Timation, which was the precursor to and testbed of the global positioning system. *Timation 1* was launched in 1967.

Communications satellites have been flown since the beginning of the space program, and several of the first are still in LEO. *Courier 1B*, built by the United States and launched in 1960, was the second active repeater communications satellite (after the U.S. *SCORE*, launched in 1958).[25] Passive reflector satellites were flown that simply bounced transmissions off their surfaces from one ground station to another. By contrast, Courier tape-recorded uplinked messages while flying over a transmitting ground station and downlinked them when in view of a receiving station. The command system failed after 3 weeks of operation.

In the 1990s, several large commercial communications constellations were developed by the United States. The Iridium constellation was one of the largest and was designed to provide access for personal telephone and data communications anywhere in the world.[26] Original plans called for a seventy-seven-satellite constellation, plus additional spares, but the program was scaled back to sixty-six, eleven in each of six polar orbits. Soon after becoming operational, very serious financial difficulties were encountered by the company because of a lack of a strong customer base among the general public, and the company declared bankruptcy.[27] The satellite phone could not be used indoors and was bulky compared with other mobile phones, which were becoming more compact. However, the phone is very useful at sea, in the air, and on the ground in remote areas where terrestrial cellular networks are unavailable. The satellites were purchased and are now operational. Other large commercial communications constellations, such as Orbcomm[28] (twenty-six satellites) and Globalstar[29] (forty-eight satellites), had similar financial problems.

The largest spacecraft in LEO is the International Space Station (ISS). First conceived in 1984, the ISS comprises multiple modules, built by the various international partners, which include the United States, Russia/USSR, Canada, Japan, and the European Space Agency.[30] The first two of these modules, the Russian Zarya and the U.S. Unity, were launched into orbit and attached in 1998. Since then, construction of the ISS has represented the largest construction project ever undertaken in space. The ISS is used to conduct basic and applied research in many different areas of biology, chemistry, and physics.[31] The U.S. segment has even been designated as a National Laboratory and is especially useful for researching issues related with the long-term human exploration and habitation of space.[32] Continuously occupied since 2000, the ISS is typically inhabited by a crew of three astronauts and has also been visited by the world's first space tourists, transported on the Russian Soyuz spacecraft.

During the late 1990s, a new spin was put on the concept of crewed spaceflight. In 1997, *Celestis 1*, a private commercial capsule, was deposited in space carrying some of the cremated remains of twenty-four people, including Gene Roddenberry, the creator of *Star Trek*, and Timothy Leary, the 1960s counterculture icon.[33] It was followed by *Celestis 2* in 1998, carrying another thirty "passengers," and *Celestis 3* in 1999, carrying thirty-six more.

ARTIFACTS IN GEO

The utility of GEO for communications was recognized as early as 1945, 12 years before the launch of *Sputnik 1*, when science-fiction writer and futurist Arthur C. Clarke first wrote about it in *Wireless World* magazine. The utilization of GEO was slowed initially by the low payload fractions of contemporary launch vehicles and debate over whether GEO was optimal for communications, given the increase in signal delay compared with lower-altitude orbits and terrestrial transmission stations.[34] Today, satellites in GEO are not only used for telecommunications but also Earth-observation missions, including surveillance and meteorology. Technology development and demonstration began in the 1960s and, in the 1970s, many countries began operating satellites in GEO. Since then, the number of spacecraft in GEO has steadily increased, as shown in Figure 17.5.[35] This trend was driven by the increase in demand from countries other than Russia and the United States. Russia preferred Molniya orbits because of the penalty paid for launching from high latitudes to GEO. Out of about 824 spacecraft currently flying in GEO (as of December 2007), approximately 34% are of U.S. origin, 18% are Russian/Soviet, and 48% belong to other countries. Many GEO satellites are moved into higher "graveyard" orbits near the end of their operational life.

The oldest spacecraft residing in GEO, and the first to be sent there, are the spin-stabilized Syncom satellites, built by the United States.[36] These spacecraft were intended to demonstrate the ability to place a satellite in the desired orbit, adjust the spacecraft attitude to properly direct the antenna beam, and keep it on its orbital station in communication with the ground. *Syncom 1*, launched in 1963, was designed to be the first test of a communications satellite in geosynchronous orbit. It nearly reached its intended destination, but communications were lost during the final second of its orbit-circularizing apogee motor burn.[37] It was determined at the time that

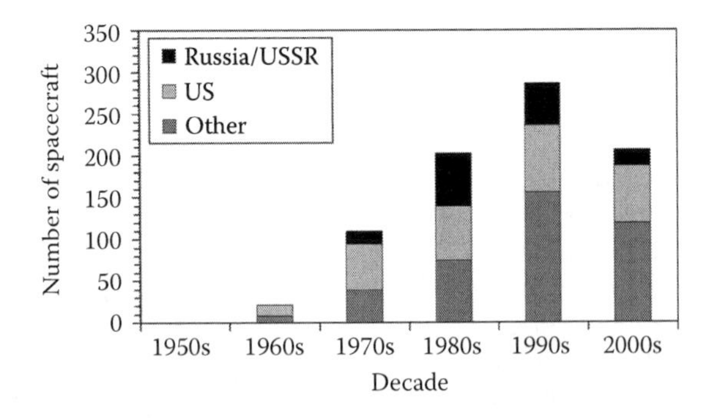

FIGURE 17.5 Approximate number of spacecraft originating in each decade from various countries currently found in geosynchronous Earth orbit beginning in October 1957 and ending in December 2007. The data include upper stages but not orbital debris.

a high-pressure nitrogen tank failure caused the loss of the spacecraft. *Syncom 2*, also launched in 1963, was the first satellite to function successfully in geosynchronous orbit, with an orbital inclination of 33°. *Syncom 3*, launched in 1964, was the first functional geostationary spacecraft.

Syncom 3 was followed by *INTELSAT 1*, better known as *Early Bird*, which was the first fully operational commercial communications satellite.[38] Launched in 1965, *Early Bird* was similar in design to Syncom but was capable of handling 240 transatlantic telephone circuits, compared with the single-circuit Syncom at comparable mass.[39] The International Telecommunications Satellite (INTELSAT) was an international consortium with, at one time, more than 130 member countries.[40] INTELSAT became a private company and now operates under the name Intelsat. This company controls a fleet of dozens of spacecraft in GEO (see Figure 17.3b) that provide telecommunications services to more than a hundred countries worldwide.

Symphonie 1 was built by France and Germany and launched in 1974. It was the first three-axis stabilized geostationary communications spacecraft.[41] Modern GEO communications spacecraft now share this characteristic.

ARTIFACTS IN MEO

MEO refers to orbits between LEO and GEO and is much less populated than either of those regions. Figure 17.6 shows the approximate number of spacecraft originating in each decade currently found in MEO.[42] Various types of missions have been flown in this region of space, including space science, reconnaissance, communications, and navigation. Out of about 305 spacecraft currently flying in MEO (as of December 2007), approximately 40% are of Russian origin, 53% are from the United States, and 7% belong to other countries.

The oldest spacecraft residing in MEO are the Vanguard satellites, built by the United States. *Vanguard 1*, a 1.5-kg, 16-cm spherical probe, was launched in 1958, making it the oldest human-made object in space.[43] It carried a radio transmitter and the first solar cells for power in space. It was originally expected to remain in orbit for 2,000 years, but orbital perturbations due to the gravitational fields of the

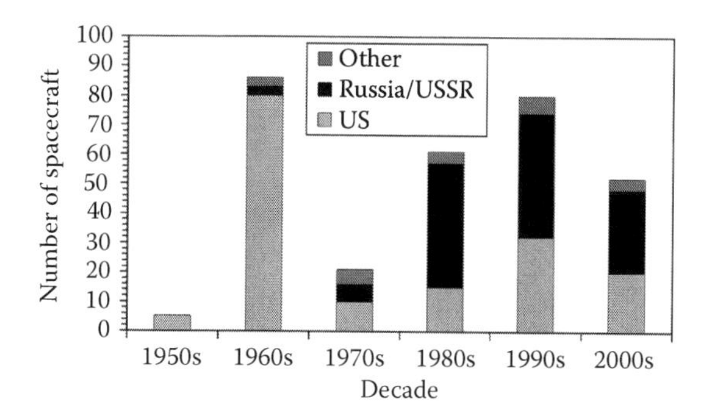

FIGURE 17.6 Approximate number of spacecraft originating in each decade from various countries currently found in medium Earth orbit beginning in October 1957 and ending in December 2007. The data include upper stages but not orbital debris.

Sun and Moon, as well as radiation pressure from sunlight, have caused the orbit to decay. Another cause of orbital decay, which was one of the most significant discoveries of *Vanguard 1*, is the expansion and contraction of the Earth's atmosphere with the solar cycle. Like any gas, the atmosphere expands when it is heated and contracts when it is cooled. As the atmosphere expands, it extends to higher and higher orbits, causing the satellites there to experience drag, which lowers their altitude. Ever since its discovery, this effect has been the primary factor for estimating the orbital lifetime of a spacecraft. Presently, *Vanguard 1* is expected to remain in orbit for about another 200 years.[44] Studies of *Vanguard 1*'s orbit also revealed two important discoveries about the nonspherical shape of the Earth and the effects of its magnetic field on the motion of a satellite. *Vanguard 2* and *Vanguard 3* were both launched in 1959. *Vanguard 2*, a 10.4-kg, 51-cm sphere, was intended to study cloud cover; however, excess fuel in the upper stage caused it to impact the satellite after separation. The impact caused the satellite to tumble, making data recovery very difficult. The 25.4-kg *Vanguard 3* carried a payload of instruments intended to measure solar radiation, the Earth's magnetosphere, and cosmic dust impacting the Earth. Its orbit kept it inside the Van Allen belts, preventing it from obtaining solar radiation information, although it did gather important data used to map the magnetosphere and the Van Allen belts.

The United States sent a relatively large number of spacecraft to MEO during the 1960s. The Orbiting Vehicle series of satellites, built by the U.S. Air Force, was used to conduct research on the space environment primarily for the purpose of improving future military spacecraft designs.[45] The spacecraft of the Initial Defense Communications Satellite Program constituted the first satellite constellation.[46] They provided the U.S. Department of Defense (DoD) with its first "geosynchronous" communications system and were used for reconnaissance and communications during the Vietnam War.[47] These satellites were placed in subsynchronous orbits and drifted about 30° each day. These were replaced by truly geosynchronous satellites in the 1970s when the United States, along with the rest of the world, began flying to GEO on a more frequent and consistent basis.

Azur, launched in 1969 by the United States, was the first flight of a German-built satellite. This is ironic since German rocket technology prior to the end of World War II was superior to that of both the United States and Russia, and it was this launch vehicle technology that was the primary enabler of the Space Age.[48] After the war, German rocket scientists were instrumental in the development of American and, to a lesser extent, Russian launch vehicle technologies. Azur performed research on the inner zones of the Van Allen belts.

One important group of spacecraft residing in MEO is the global positioning system (GPS). GPS is a satellite-based navigation system developed and operated by the U.S. DoD that provides highly accurate position and time information to military and civilian users worldwide.[49] The space segment of GPS consists of a minimum of twenty-four Navstar satellites, four equally spaced in each of six orbital planes separated by 60°, with an approximate altitude of 20,200 km and orbital period of 12 hours (Figure 17.3c). This spacing provides coverage such that, typically, at least six satellites are in line of sight from any point on Earth. By receiving a signal from at least four satellites simultaneously, position and velocity in three-dimensional space can be determined with a high degree of accuracy. The first GPS satellites were launched in 1978, and since then, upgrades have continually replaced older satellites on orbit. Block II GPS satellites, first launched in 1989, also carry a nuclear detonation detector onboard.[50]

Russia also developed a satellite navigation system called the Global Navigation Satellite System (GLONASS), which is similar to GPS.[51] The first of these satellites was launched in 1982. Many GLONASS satellites were launched throughout the 1980s and 1990s because of their relatively short lifespan.[52] Because of a lack of sufficient funding, the system began to degrade significantly in the mid-1990s. However, the system has been revived, and although it is not fully operational at the time of this writing, it is anticipated that it will be in the near future.

ARTIFACTS IN OTHER ORBITS

So far, the discussion of the spacescape has focused on the spacecraft residing in LEO, MEO, and GEO as they have been defined. Clearly, not all satellites, whether in Earth orbit or elsewhere, are confined to one of these specific regions. Some spacecraft in orbit around the Earth have highly elliptical orbits (HEOs) that carry them through multiple regions of space, and of course, many spacecraft reside outside Earth orbit altogether. The intended purposes of these spacecraft span the full range of mission possibilities, from communications and surveillance to interplanetary science and discovery.

Figure 17.7 shows the approximate number of spacecraft originating in each decade currently found in other orbits around the Earth and elsewhere.[53] Out of about 449 such spacecraft (as of December 2007), approximately 34% are of Russian/Soviet origin, 33% are from the United States, and 33% belong to other countries. In the early years of the Space Age, the United States and Russia/USSR sent several spacecraft on interplanetary voyages. The Cold War drove activity on both sides. For example, the Vela satellites in Earth orbit are a group of twelve spacecraft built by the United States that went into service between 1963 and 1970; they were equipped to

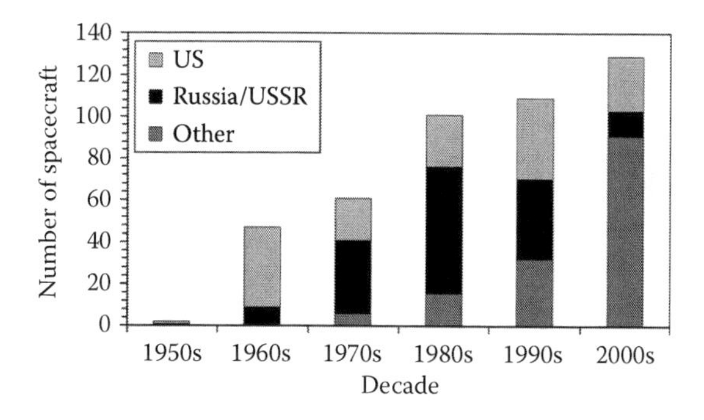

FIGURE 17.7 Approximate number of spacecraft originating in each decade from various countries, currently found in other orbits around the Earth and elsewhere beginning in October 1957 and ending in December 2007. The data include upper stages but not orbital debris.

detect nuclear detonations that would violate of the Atomic Test Ban Treaty of 1963.[54] During the 1970s and 1980s, the number of satellites of Russian origin increased greatly because of their extensive use of the highly elliptical Molniya orbits.[55] After the collapse of the Soviet Union, the number of Russian spacecraft declined. U.S. activity has been somewhat unsteady, as in other regions of space, while countries beside Russia and the United States have steadily increased their activity. Although a great many spacecraft have been sent outside Earth orbit, the majority reside in Earth orbit. Figure 17.8 shows only those spacecraft currently flying outside Earth orbit.[56] Out of approximately 84 such spacecraft (as of December 2007), about 54% are of American origin, 29% are Russian, and 18% belong to other countries.

One spacecraft in a HEO is the *Chandra X-Ray Observatory*. Built by the United States and launched in 1999, *Chandra* is part of NASA's Great Observatories program.[57] The primary science objectives were to determine the nature of various celestial objects, to determine the physical processes taking place in and between astronomical objects, and to study the history and evolution of the universe.[58] Observations are made of x-rays emanating from high-energy regions of the universe, such as supernova remnants, x-ray pulsars, black holes, neutron stars, and hot galactic clusters.[59] In the course of executing its mission, the orbital altitude of *Chandra* varies from approximately 10,000 to 140,000 km (Figure 17.3d).

Two satellites from the 1950s remain in space outside Earth orbit. *Luna 1*, built by the USSR and launched in 1959, was the first spacecraft to leave Earth orbit and now orbits the Sun.[60] Initially intended to hit the Moon, *Luna 1* missed by 6,000 km and eventually entered a heliocentric orbit. Though it did not accomplish its original objective, *Luna 1* discovered that the Moon has no magnetic field of its own. This led scientists toward the conclusion that the lunar core is inactive, unlike the Earth's active molten iron core. *Pioneer 4*, built by the United States, was also launched in 1959. *Pioneer 4* was the first American spacecraft to enter solar orbit. It performed a lunar flyby and on its way further mapped the Van Allen belts and the Earth's magnetic field.

Many objects remain in orbit from the 1960s. *Pioneer 5*, built by the United States and launched in 1960, is in solar orbit between Earth and Venus. It studied the Sun's

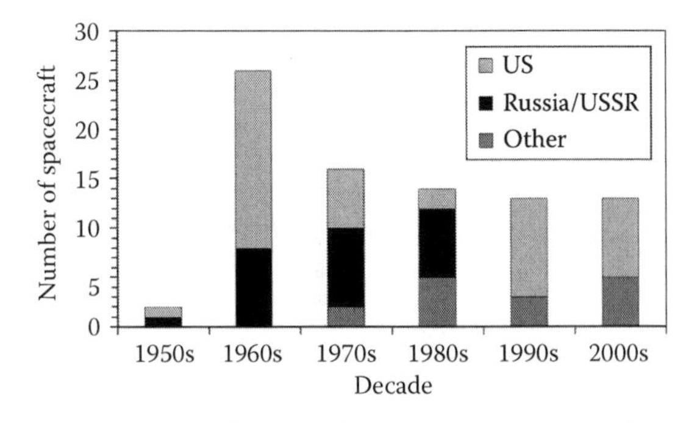

FIGURE 17.8 Approximate number of spacecraft originating in each decade from various countries currently found outside Earth orbit beginning in October 1957 and ending in December 2007. The data include upper stages but not orbital debris.

magnetic field, solar flares, and the nature of interplanetary particles, and in doing so, set the record (at the time) for the longest communications link at 36,200,000 km.[61] *Pioneer 6, 7, 8,* and *9,* put into service over the latter half of the decade, were a constellation of four satellites designed to study the solar wind. Intended for a half-year lifespan, they provided detailed information about solar activity for nearly two decades and were used to plan all crewed missions to the Moon.

Venera 1, built by the USSR and launched in 1961, was the first spacecraft sent to Venus.[62] Communication with the satellite was lost before it reached its target, but more valuable data were collected about the near-Earth space environment. In 1962, the USSR made the first attempt to send a spacecraft to the Red Planet with *Mars 1*, but it too suffered a loss of communication and failed. Communications loss continued to plague the USSR during subsequent attempts in 1964 to send probes to Venus and Mars; *Zond 1*, sent to Venus, and *Zond 2*, sent to Mars, were both failures. *Zond 3*, launched in 1965, performed a lunar flyby and sent back high-resolution pictures of the far side of the Moon.

Mariner 2, built by the United States and launched in 1962 to perform a flyby of Venus, was the first planetary probe to reach another planet and send back data.[63] It measured the temperature and rotation rate of the planet and showed that Venus has a hot surface, a slow retrograde rotation, and a day that lasts approximately 117 Earth days. The United States returned to Venus with *Mariner 5* in 1967, which returned a wealth of valuable and unexpected data pertaining to the nature of the atmosphere that contradicted contemporary theories. *Mariner 4*, built by the United States and launched in 1964, was the first spacecraft to reach Mars and return data successfully.

Snoopy, the Lunar Module (*LM-4*) ascent stage from the Apollo 10 mission in 1969, is in a heliocentric orbit. This spacecraft was used during the second Apollo mission to the Moon and was the first LM to travel there.[64] *Apollo 10* was a rehearsal for *Apollo 11*, performing all operations except the actual landing. Snoopy is the only intact LM ascent stage left in orbit out of those that were sent into space. The others either burned up in Earth's atmosphere or crashed into the lunar surface. Some were crashed deliberately to test and calibrate lunar seismometers.[65] The *Apollo 10* Command and Service Module, *Charlie Brown*, has been on display at the Science Museum in London.

In the 1970s, the United States launched several spacecraft that are now the most distant human-made objects from the Earth. *Pioneer 10*, launched in 1972, was the first to be sent to the outer solar system and was the first to reach and study Jupiter.[66] Following its flyby of Jupiter, it followed a solar escape trajectory. The spacecraft is heading in the direction of Aldebaran, a star in the constellation Taurus. Aldebaran is more than 68 light years from Earth, requiring about 2 million years to complete the trip. *Pioneer 11*, launched in 1973, was the second mission to Jupiter and the outer planets and the first to perform a flyby of Saturn. It too followed a solar escape trajectory. Because they were intended to leave the solar system and, hence, might eventually be encountered by an intelligent life-form, *Pioneer 10* and *11* each carry a plaque that shows a drawing of a man and a woman, the location of the Sun in the galaxy, and a depiction of our solar system indicating where the spacecraft originated.[67]

Launched in 1977, *Voyager 1* and *2* were a pair of spacecraft designed to explore the outer planets and the interplanetary environment.[68] The Voyager spacecraft were also designed to leave the solar system, and like *Pioneer 10* and *11*, they contain artifacts that indicate their origin, should they be encountered by another life-form. In this case, each carries a gold-plated copper disk, on which sounds and images are recorded. The package also includes a cartridge and needle, and an engraved message explaining the origin of the spacecraft and instructions for playing the disk. Assembled by a NASA committee chaired by Carl Sagan, the images and sounds were selected to display the diversity of life on the planet. The disk contents include 115 images, and the audio includes greetings in 55 languages, portions of 27 musical compositions, and 35 other natural and human-made sounds. An ultrapure source of uranium-238 is electroplated onto a 2-cm area of the recording's protective jacket. The uranium-238, with a half-life of 4.5 billion years, was included to facilitate the determination of the elapsed time since launch. In addition to having a more sophisticated human-identifying artifact package than *Pioneer 10* and *11*, the Voyager spacecraft are moving away from the solar system in the opposite direction. In 1998, the distance of *Pioneer 10* was surpassed by *Voyager 1*, which is now the most distant human-made object in space. Unlikely as it may be, it is fascinating to think about what it would be like for an intelligent life-form to encounter one these spacecraft far off in the future.

CONCLUSION

Spaceflight has truly revolutionized global communications, surveillance, navigation, and studies of the atmosphere, environment, weather, Earth sciences, and, of course, astronomy and space sciences. In five decades of spaceflight, humans have designed, built, and launched well over 6,000 spacecraft, more than half of which remain in space. Much like the Egyptian pyramids, China's Great Wall, and other "wonders of the world," these orbital artifacts collectively represent one of humankind's greatest engineering accomplishments. Worldwide, spaceflight is emblematic of the vigorous pursuit of profound scientific and humanitarian achievement that defines great societies. Countless numbers of people have watched with awe what has been accomplished in space, and as a result, countless more have been inspired to pursue science, engineering, and mathematics, leading to equally inspiring accomplishments here on the ground. Every time we think we have found the limit of our capacity to meet a challenge, all we have to do is look to the sky to be reminded that our potential has no limit. The future will be bright if coming generations can look to the achievements of the generations that came before and accept for themselves nothing less than greatness.

REFERENCES

1. Wertz, J.R. Orbit and Constellation Design. In *Space Mission Analysis and Design*, ed. J.R. Wertz and W.J. Larson, 159–202. 3rd ed. Torrance, CA: Microcosm, 1999.
2. Boden, D.G. Introduction to Astrodynamics. In Wertz and Larson 1999 (note 1), 131–158.
3. NASA. Satellite Tracking. http://science.nasa.gov/realtime.
4. Bilstein, R.E. *Orders of Magnitude: A History of the NACA and NASA, 1915–1990.* NASA SP-4406. Washington DC: NASA, 1989.

5. United States Strategic Command. USSTRATCOM Space Control and Space Surveillance Fact Sheet. http://www.stratcom.mil/resources-fact.html.
6. McDowell, J. Jonathan's Space Homepage. http://www.planet4589.org/space.
7. NASA (note 3).
8. McDowell (note 6).
9. Zimmerman, R. *The Chronological Encyclopedia of Discoveries in Space*. Westport, CT: Oryx Press, 2000.
10. Zimmerman 2000 (note 9).
11. McDowell (note 6).
12. NASA. National Space Science Data Center, Master Catalog. http://nssdc.gsfc.nasa.gov.
13. McDowell (note 6).
14. Zimmerman 2000 (note 9).
15. Williamson, M. *Spacecraft Technology: The Early Years*. London: Institution of Electrical Engineers, 2006.
16. Zimmerman 2000 (note 9).
17. NASA (note 12).
18. Doxsey, R. 15 Years of Hubble Space Telescope Science Operations. AIAA Paper 2006-5936, 2006.
19. NASA (note 12).
20. NASA (note 12).
21. Williamson 2006 (note 15).
22. Schnapf, A. TIROS-N—Operational Environmental Satellite of the 80's. *Journal of Spacecraft* 18(2), 1981.
23. Klingaman, W.K. *APL—Fifty Years of Service to the Nation*. Laurel, MD: Johns Hopkins University Applied Physics Laboratory, 1993.
24. Zimmerman 2000 (note 9).
25. Zimmerman 2000 (note 9).
26. Swan, P.A. A Revolution in Progress: IRIDIUM LEO Operations. AIAA Paper 97-3954, 1997.
27. Zimmerman 2000 (note 9).
28. Deckett, M. Orbcomm—A Description and Status of the LEO Satellite Mobile Data Communication System. AIAA Paper 94-1135, 1994.
29. Dietrich, F.J. The Globalstar Satellite Communication System: Design and Status. AIAA Paper 98-1213, 1998.
30. NASA (note 12).
31. Robinson, J.A., D.A. Thomas, and T.L. Thumm. NASA Utilization of the International Space Station and the Vision for Space Exploration. AIAA Paper 2007-139, 2007.
32. Uhran, M.L. Emergence of a US National Laboratory on the International Space Station. AIAA Paper 2008-797, 2008.
33. Zimmerman 2000 (note 9).
34. Williamson 2006 (note 15).
35. McDowell (note 6).
36. Williamson 2006 (note 15).
37. Bentley, R.M., and Owens, A.T. SYNCOM Satellite Program. *Journal of Spacecraft* 1(4): 395–399, 1964.
38. Zimmerman 2000 (note 9).
39. Williamson 2006 (note 15).
40. Wernek, S.B. The INTELSAT System: An Overview. AIAA Paper 94-4606, 1994.
41. Williamson 2006 (note 15).
42. McDowell (note 6).
43. Zimmerman 2000 (note 9).

44. NASA (note 12).
45. Zimmerman 2000 (note 9).
46. Williamson 2006 (note 15).
47. NASA (note 12).
48. Williamson 2006 (note 15).
49. United States Air Force—Los Angeles Air Force Base. Global Positioning System Fact Sheet. http://www.losangeles.af.mil/library/factsheets/index.asp.
50. NASA (note 12).
51. Zimmerman 2000 (note 9).
52. NASA (note 12).
53. McDowell (note 6).
54. Zimmerman 2000 (note 9).
55. Popescu, J. *Russian Space Exploration.* Oxon, United Kingdom: Gothard House Publications, 1979.
56. McDowell (note 6).
57. Tananbaum, H. Science Highlights from the Chandra X-Ray Observatory. AIAA Paper 2000-5130, 2000.
58. NASA (note 12).
59. Williamson 2006 (note 15).
60. Zimmerman 2000 (note 9).
61. Zimmerman 2000 (note 9).
62. Popescu 1979 (note 21).
63. Zimmerman 2000 (note 9).
64. NASA (note 12).
65. Williamson 2006 (note 15).
66. Nunamaker, R.R. Pioneer 10/11 Mission Results. AIAA Paper 75-196, 1975.
67. NASA (note 12).
68. Jones, C.P., and T.H. Risa. The Voyager Spacecraft System Design. AIAA Paper 81-0911, 1981.

18 Introduction to Space Debris

Robert Osiander and Paul Ostdiek

CONTENTS

Sources of Space Debris.

INTRODUCTION

The remains of a culture offer insight into its nature, and therefore the study of such debris can provide important evidence that affects how that culture is later understood. These remains are sometimes hidden in unknown places. They are rarely found neatly laid out, ready for analysis; mostly they are found broken and scattered. Studies of humankind's spacefaring cultures yields important artifacts in museums, and some documents in file cabinets are carefully laid away for study. Other studies might take one to the less-well-tended, abandoned facilities that birthed and nurtured success in space. Operating space systems are messy enterprises, and one cannot help but leave artifacts, or debris, scattered about. Orbital space debris consists of objects primarily orbiting Earth that are created by humans and no longer have a useful purpose. Examples are burnt rocket stages, rocket slag, defective satellites, screws, paint flakes, and explosion fragments. For our current society, they reflect the maturing space age, the globalization of space, the Cold War, and the end of the Cold War. Unlike many lost or ancient cultures, though, humankind's space culture is recent, and its trash-bin has been carefully watched over the years. The exact origin is known for more than 10,000 objects in the U.S. Strategic Command's catalogue, which was gathered by a number of ground-based radar facilities and telescopes, as well as by a space-based telescope.[1] Exotic lost examples are the outer glove lost by Edward White (*Gemini 4*) on America's first spacewalk, as well as tools and cameras lost by other astronauts. Most of those unusual objects have reentered the atmosphere of Earth within weeks, because of the low orbits where they were released and their small sizes, and are not major contributors to the space debris environment.

ORBITAL DEBRIS

One of the oldest pieces of "space junk" probably lies on the bottom of the Bosporus Strait in Turkey. In AD 1623, Larari Hasan Celebi, from Ottoman Turkey, launched himself into the air with a rocket filled with 54 pounds of gunpowder and glided back to ground with wings.[2] Two respected Turkish sources about the incident convinced Turkish and Norwegian scientists to suggest looking for the hardware on the bottom of the Bosporus.

Like this historic example, much of what we consider space junk or debris never actually makes it to orbit. Many pieces, some of which are unfortunate launch disasters or just regularly burned out stages of rockets, fall back to Earth. U.S. and European rocket launches are always directed over water, and this is where most of these pieces end up. Russian rockets, launched in the deserts of Kazakhstan, fall in this region, which is covered with space debris.

This chapter emphasizes those artifacts which actually make it into orbit. Once the satellite is in a stable orbit in low Earth orbit (LEO) or above, most of its parts, if it disintegrates, will actually stay there for a long time. That is true as long as the satellite or its components do not experience any change in speed, which could be caused by atmospheric drag for lower orbits, collision with other space debris or meteorites, or even via a propulsive maneuver. The lifetime of all orbital debris depends on their size and their altitude. In LEO, an object below 400 km will deorbit within a few months because of atmospheric drag and the gravitational attraction of Earth, Moon, and Sun, while above 600 km it might stay there for tens of years.

The first modern piece of exoatmospheric orbital debris stems from the first exoatmospheric launch. During the launch of *Sputnik 1* in 1957, the launch vehicle upper stage was discarded into LEO. Since then, the orbit has been filled with satellites (presently estimated to be about 3,600),[3] rocket bodies, fragmentations, and other human-made objects. During the first 30 years of spaceflight, few operators disposed of their spacecraft in a controlled way. In the past 20 years, this attitude has changed somewhat, and approaches have been developed to control debris via space policy.

The approximately 10,000 catalogued objects around Earth are distributed as follows.

- Operational spacecraft: 7%
- Old spacecraft: 22%
- Miscellaneous fragments: 41%
- Rocket bodies: 17%
- Mission-related objects: 13%

Figure 18.1 shows a computer-generated image of known objects in Earth orbit. The estimates for uncatalogued objects larger than 1 cm are somewhere between 50,000 and 600,000.[4]

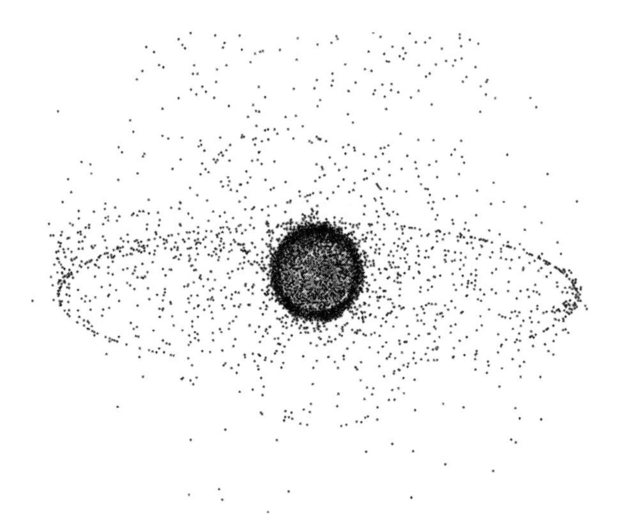

FIGURE 18.1 NASA's computer-generated image of known objects in Earth orbit. Most (95%) of the dots represent debris, not active spacecraft. Dots are not scaled to the size of Earth (http://orbitaldebris.jsc.nasa.gov).

OLD SPACECRAFT

Old spacecraft are in most cases intact structures that have completed their mission, or satellites that had a nondestructive malfunction that shortened their lifetime. Recent policies, described later, plan for deorbiting these craft or sending them up to a graveyard orbit, not only to remove them as the primary threat but also to reduce their chances of being hit and generating a fragmentation debris field.

The oldest piece of space hardware still in orbit is *Vanguard 1* (TV4). It was built by the Naval Research Laboratory and launched by the Navy on March 17, 1958, into a stable orbit with an apogee of 3,969 km (2,466 miles) and a perigee of 650 km (404 miles), where it should remain for at least another 200 years to as much as 1,000 years. Vanguard's predecessors, the first Soviet satellites, *Sputnik 1* and *Sputnik 2*, and the first U.S. satellite, *Explorer 1*, have since deorbited. The first attempt to launch a Vanguard satellite, *TV3*, shown in Figure 18.2, was a spectacular failure. This major setback for the United States is shown in Figure 18.3.

The Vanguard scientific mission demonstrated that the Earth is slightly pear-shaped, not perfectly round, and provided information on air density and temperature ranges in the upper atmospheres, and micrometeorite impact. Technically it validated that solar cells can be used for several years to power the spacecraft electronics. Vanguard became silent in 1964, but it still serves the science community by providing information about the effects of the Sun, Moon, and atmosphere on satellite orbits.

FRAGMENTATION DEBRIS

Fragments of spacecraft make up about 40% of the catalogued debris. The fragments can be generated in several ways. One possibility is a catastrophic destructive

FIGURE 18.2 TV3 satellite, identical to *Vanguard 1*. (Courtesy of NASA.)

FIGURE 18.3 TV3's Vanguard rocket exploded at the launch pad, but the nose cone and satellite survived. TV3 can now be seen in the Air and Space Museum. (Courtesy of NASA.)

event. This could be intentional, such as an antisatellite (ASAT) demonstration or the destruction of a nonfunctional spacecraft, or unintentional, such as a collision or accidental explosion. Deterioration of older spacecraft, mostly due to atomic oxygen and thermal cycling, could be another source, as well as parts of spacecraft that are detaching, such as thermal blankets, protective shields, and parts of solar panels.

The first on-orbit breakup event in space history was an accident. It is accredited to the Ablestar upper stage that put the U.S. Navy's *Transit-4A* navigation satellite and its companions, *INJUN* and *SolRad III* (also known as *GREB III*), safely into orbit. On June 29, 1961, a Thor-Ablestar rocket carried the three spacecraft when it was launched from Kennedy Space Center. The Ablestar upper stage executed two main propulsion burns and a small retro burn to deploy the three spacecraft.[5] Seventy-seven minutes after placing *Transit-4A* into its operational orbit, the Ablestar upper stage exploded, creating 298 trackable fragments. These fragments were still in orbit 40 years later.[6]

An event of even greater magnitude in debris creation occurred on February 19, 2007, when a Russian Briz-M booster rocket stage exploded in orbit over Australia. The booster had been launched on February 28, 2006, carrying an Arabsat-4A communication satellite but malfunctioned before it could use all of its fuel. Its debris field is at a lower altitude than the recent Chinese ASAT test, and much of its debris reentered the atmosphere in a relatively short time. As of February 21, 2007, more than 1,000 fragments had been identified[7].

The Chinese ASAT test, which occurred on January 11, 2007, is presently considered the largest space debris incident in history.[8] As of December 2007, it was estimated to have created more than 2,300 pieces of trackable debris (approximately golf ball size or larger), more than 35,000 pieces 1 cm or larger, and 1 million pieces 1 mm or larger. The debris event is more significant than previous ASAT tests in that the debris field is in a higher orbital plane, resulting in deorbit times of 35 years and greater. Figure 18.4 shows the orbit planes of some of the debris. In June 2007, NASA moved the Terra environmental spacecraft to prevent impacts from the ASAT test debris.[9]

Earlier, during the Cold War, the United States and the USSR were deliberately destroying decoy satellites with high-speed, bullet-like projectiles, which created thousands of fragments.[10] Many satellites have been deliberately destroyed, especially military satellites, to prevent them from falling into the hands of foreign powers. In 1967, a secret U.S. defense intelligence operation known as "Operation Moondust" was tasked to retrieve fallen space junk to collect information on foreign satellites.[11]

Even today, this might be one of the reasons to shoot down satellites. A recent debris event happened early in 2008, when the U.S. Navy used a missile interceptor to destroy a spy-satellite gone out of control that was rapidly descending to Earth, containing a large amount of frozen, toxic hydrazine propellant. Located at the core of the satellite propellant tank, it was protected from aeroheating and might have survived reentry until the surrounding structure melted away deep within the atmosphere. This example provides an all-encompassing entry into the problem of space debris. The classified satellite, speculated to be an experimental imagery satellite built by Lockheed Martin, was launched from Vandenberg Air Force Base in California in December 2006 aboard a Delta II rocket. Shortly after the satellite reached orbit, ground controllers lost the ability to control it, and from that time on the satellite was in a slowly decaying orbit, declining about 15–20 km per month to a circular orbit at about 275 km above the Earth. On February 26, 2008, a missile

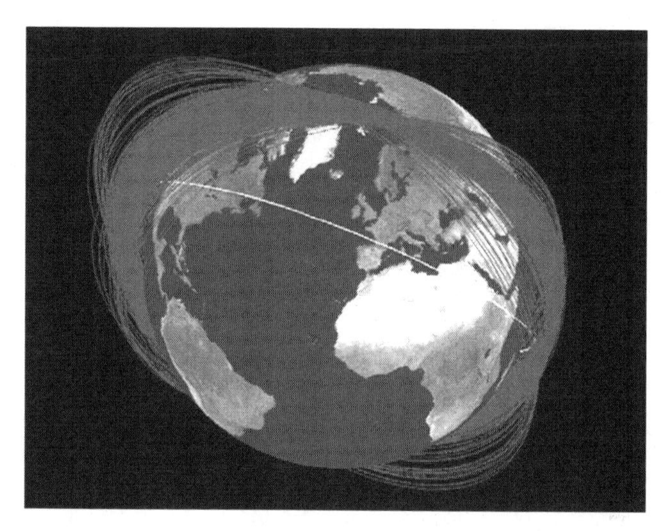

FIGURE 18.4 Known orbit planes of Fengyun-1C debris 1 month after its disintegration by the Chinese ASAT test (NASA Orbital Debris Program Office).

interceptor launched from a U.S. Navy warship struck the satellite and succeeded in destroying the propellant tank, according to information from the Pentagon, which had reached the conclusion based on studying remnants from the missile strike. This event was carefully planned to limit the impact of debris to orbiting spacecraft. Most of the fewer than 3,000 tracked pieces of debris from the strike were smaller than a football and were expected to reenter and burn up in the atmosphere within the following weeks; it was unlikely that any of them would remain intact to impact the ground. The contrast to the Chinese ASAT event mentioned above is striking. The Chinese intercept of the *Fengyun-1C* weather satellite occurred in high orbit, generating an expanding cloud of debris that will very slowly deorbit and some day will greatly limit the ability to deploy more satellites. One year after the intercept, the Space Surveillance Network observed that only 22 of approximately 2,600 officially catalogued fragments had deorbited.[12]

Rivaling the impact of the Chinese ASAT test is the very recent Cosmos-Iridium collision. On February 10, 2009 (while the authors were reviewing proofs of this book), a dead 16-year-old Russian *Cosmos 2251* communications spacecraft collided with an active *Iridium 33* communications satellite. The two spacecraft met at a closing velocity of 7 mi/s 790 km above Siberia. The debris cloud generated is not yet fully measured, but debris was potentially thrown into heavily used orbits. "Nearby" are the newly launched *NOAA N-Prime* weather satellite and the NASA A-Train series of Earth-observing science satellites. Further, NASA had planned to launch its *Orbiting Carbon Observatory* into this region of space about two weeks after the collision.[13]

ROCKETS AND ROCKET BODIES

Rocket bodies are accountable for 19% of the tracked orbital debris.[14] Figure 18.5 shows the objects for a Russian Proton launch that end up as debris, as well as their

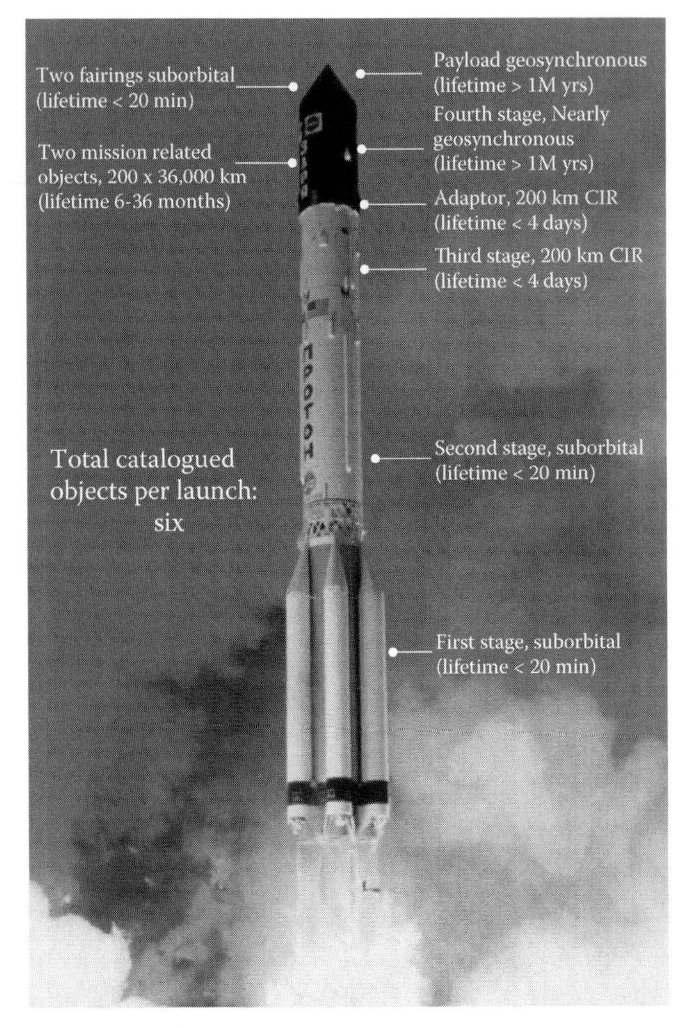

FIGURE 18.5 Space debris resulting from a Proton launch. Rocket bodies are accountable for 19% of the tracked orbital debris.

typical lifetime for a launch to geosynchronous orbit (GEO). Space debris from rocket launches is a driver for the future evolution of the space debris population, because of their large dimension and the potentially explosive residual propellant. In addition, for a typical launch vehicle, the space shuttle being probably the lone exception, many parts of the launcher become space debris.

One historical example is Asteroid J002E3, discovered on September 3, 2002 and now considered by many to be human-made. Further examination revealed the object was not a natural asteroid but instead the S-IVB third stage of the *Apollo 12* Saturn V rocket. Initially, the object was thought to be a rock orbiting the Earth. This was surprising, though, as Moon-Earth-Sun gravity perturbations would have long ago ejected it from Earth orbit. Then, University of Arizona astronomers found that the "rock" had a spectrum very similar to the white titanium dioxide paint used by NASA on the Saturn V rockets. Further analyses then indicated that the object had been orbiting the Sun for 31 years and had last been in the vicinity of the Earth in 1971. That timeline pointed to the *Apollo 14* mission, but NASA knew the whereabouts of all hardware used for this mission. Its third stage, from the *Apollo 14* mission, was deliberately crashed into the Moon for seismic studies. Attention then turned to the S-IVB third stage of *Apollo 12*. NASA had originally planned to direct the S-IVB into a solar orbit, but the propulsion event did not provide enough energy to escape the Earth-Moon system. Instead the stage ended up in a semistable orbit around the Earth. It passed the Moon on November 18, 1969, and eventually vanished until it was rediscovered in 2002.

In 1999, a large piece of space junk—a spent Russian rocket—was headed toward the International Space Station (ISS) and missed it by a mere 7 km (4 mi). The station's safety protocol called for a move of the space station, but it did not happen because of a misfiring of one of the thrusters. In 2008, the ISS was successfully moved into a higher orbit to escape from the path of a drifting Pegasus rocket.[15] NASA predicts two maneuvers a year for the ISS to evade space debris.[16]

MISSION-RELATED DEBRIS

Mission-related debris makes about 14% of the orbital debris. This includes items related to the functional operation of the satellite itself and comprises all other human-made items that are the result of spaceflight. Examples are explosive bolts, vehicle shrouds, and lids of covered telescopes and other fragile equipment. During operation, solid rocket motors can release condensed Al_2O_3 particles that are generally small in mass, but the flux rate is high. These particles can have diameters from 0.1 μm to 3 cm (1.2 in.). As result of the harsh space temperature environment, small pieces of paint detach and become space debris.

In 1962, as a result of a Department of Defense (DoD) project, now known as the Westford Needles, a large number of copper needles were intentionally released in an attempt to lay a radio-reflective ring around Earth. These dipoles would serve as an artificial scattering medium for radio signals in the centimeter band. This effort was criticized by astronomers because of optical and radio pollution. The first experiment did not work as a radio reflector. The second one was successful, but only in deploying, and the more than 300 million needles have contributed to the space debris problem. Most have deorbited, but still new needle populations are being discovered by radar and optical measurements from Earth. Another example is the droplets, from the Radar Ocean Reconnaissance SATellite (RORSAT), which is the western name given to the Soviet *Upravlyaemyj Sputnik Aktivnyj*. They were released coolant droplets (NaK, liquid metal) from the nuclear reactors used for powering the spacecraft, when the reactors were separated from the inactive spacecraft.

Mission-related debris are generally small in diameter. They are hard to detect with the current observation methods, which can trace space debris from a diameter of about 10 cm (4 in.) and larger, depending on the debris altitude. The amount of released debris can be quite large. For example, 200 pieces of mission-related space debris were linked to the Russian space station *Mir* during its first 8 years of operation. Most was dumped intentionally, but there are also examples of astronauts who lost items during extravehicular activity. Space debris even smaller—in the form of electrons—have been released in the 1960s as a result of nuclear testing above 200 km, such as the U.S. Starfish Prime detonation on July 9, 1962. Ten known satellites were lost because of radiation damage, some immediately after the explosion.[17] The Starfish Prime explosion injected enough fission spectrum electrons with energies up to 7 MeV to increase the fluxes in the inner Van Allen belt by at least a factor of 100. Effects were observed out to 5 Earth radii. The Starfish electrons that became trapped[18] and dominated the inner zone environment (~2.8 Earth radii at the equator) for 5 years and were detectable for up to 8 years in some regions. A significant effect was the electromagnetic pulse generated by the test. This caused power mains surges in Oahu, Hawaii, knocking out street lights, blowing fuses and circuit breakers, and triggering burglar alarms (and this was in the days before microelectronics).

IMPACT OF ORBITAL DEBRIS

A major source of concern is the possible impact of orbital debris into active space hardware. One example is the solar panels of the Hubble Space Telescope (HST). The solar arrays were supplied by the European Space Agency (ESA) and retrieved in March 2002 after more than 8 years in space. Thousands of impact craters were detected on the 41-m^2 solar arrays, the biggest about 8 mm in diameter. About 174 complete penetrations of the 0.7-mm-thick array were observed, with impact craters ranging in size from 3 μm to 8 mm. A chemical analysis of impact crater residue distinguishes between impacts due to naturally occurring micrometeoroids and those due to artificial space debris.

The deadly objects can be split into a few groups. Smaller debris of <1-cm diameter can be mitigated by protective shielding, such as Whipple Shield technology (employed by the ISS). Whipple Shields use a multilayer system similar to the armor employed on tanks and military vehicles to defeat armor-piercing rounds fired from a high-velocity cannon. The outer shields are penetrated, but the fragment's kinetic energy is reduced and the inner shields remain intact. Other debris, in the 1–10-cm range, are too small and numerous to be individually tracked but can cripple or kill any craft they hit. Risk assessments are based on a spacecraft's cross-sectional area, its orbital altitude and flight path, the assumed size of debris objects, and the geometry of a collision event and relative speed, among other factors.

Besides the impact of small particles, collision events happen about once per decade. According to H. Klinkrad,[19] "If you calculate the combined profile area of all satellites in orbit, you find that the average time between destructive collisions is about 10 years." Considering that even a single 10-cm debris collision event could wipe out a multimillion-Euro spacecraft or hit the (manned) ISS, a risk of even one impact per decade suddenly becomes very serious.

Examples of debris impacts include the HST's first servicing mission in 1993, which found a hole over 1 cm in diameter in a high-gain antenna. In 1996, France's *Cerise* military reconnaissance satellite was struck and severely damaged by a catalogued Ariane upper-stage explosion fragment; a 4.2-meter portion of *Cerise's* gravity gradient stabilization boom was torn off. As of 2001, space shuttle windows had

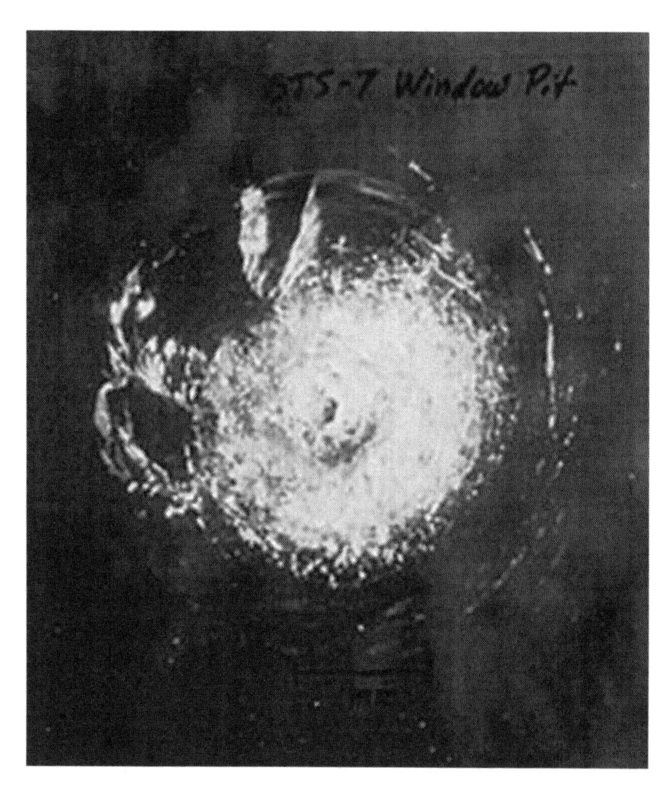

FIGURE 18.6 Orbital debris impact strike found on the space shuttle *Challenger* window on mission STS-007. (Courtesy of NASA.)

been replaced 80 times because of submillimeter object impacts. Figure 18.6 shows a debris impact strike in the space shuttle window on mission STS-007.

KESSLER SYNDROME

The orbital debris expert Donald Kessler has proposed a scenario wherein the amount of debris in low Earth orbit threatens the ability to launch new satellites into useful orbits. In his Kessler Syndrome, the volume of space debris in low Earth orbit is so high that objects in orbit are frequently struck by the debris, creating even more debris and a greater risk of further impacts. A domino effect or chain reaction develops when a collision creates shrapnel that, in turn, leads to more collisions and more shrapnel. The 2008 Disney-Pixar animated film *WALL-E* illustrates and exaggerates the Kessler Syndrome in a scene showing a huge starship literally plowing its way through a dense shroud of spacecraft and debris orbiting the Earth.

Because each launch puts debris into low Earth orbit and because each successful launch feeds a new object into this maelstrom, the Earth becomes entombed in an escalating cloud of debris. Taken to its limit, this debris might prevent humankind from using satellites for many generations. The Earth's atmosphere helps dampen this syndrome. In LEO, collisions between the debris and air molecules create a drag that helps deorbit and consume the debris, but it still takes months to clear these zones. Unfortunately, it takes millennia to clear the debris at higher altitudes.

DEBRIS MANAGEMENT

NASA leads the international community in debris management through the work of its Orbital Debris Program Office at the Johnson Space Center. This office works to improve understanding of the debris environment and ways to control its growth. This includes modeling and measurements of the environment, as well as efforts to build technical consensus on ways to protect space systems operating within that environment. Formally, NASA's work is divided into the categories of modeling, measurement, protection, mitigation, and reentry.

Orbital debris models are maintained to describe the current environment, as well as to project what the future may be like pending mitigation actions taken today. These models come in two forms: engineering and evolutionary.

DEBRIS MODELING

Engineering models, such as ORDEM2000, are used to understand the risk debris impact poses for spacecraft. The model describes LEO from 200 to 2,000 km in altitude and provides estimates of debris spatial density and flux. It is also used to benchmark ground-based measurements. ORDEM2000 incorporates in situ and ground-based observations of objects from 10 μm to 10 m in size.

Evolutionary models, such as EVOLVE and LEGEND, are designed to predict characteristics of future debris environments. EVOLVE is a one-dimensional simulation model that places launched objects or fragments of explosion/collisions into orbits and calculates how these orbits evolve over periods from a few decades to a few centuries. Its use typically begins by calculating the current environment based on the historical record of launches and breakups. The model's results are periodically compared with satellite catalogs and data from the Haystack radar.

NASA's new evolutionary model is called LEGEND. It is a full-scale, three-dimensional LEO to GEO environment debris model that covers orbits from 200 to 50,000 km in altitude. LEGEND provides data such as the number, type, size distribution, spatial density, velocity distribution, and flux of debris fragments as small as 1 mm. The model includes an historical simulation (1957 to present) used to validate techniques used for projecting data into the future. An important feature of the model is its ability to evaluate the probability of collisions in three dimensions.

DEBRIS MEASUREMENT

Models are only as good as the data from which they are formed and typically are valid only within the bounds defined by those data. To improve a model, or to expand its usefulness, new data are required. For this reason, there are active programs to measure Earth orbital debris. These measurements are conducted from the ground with radar and optical telescopes.

NASA has utilized a variety of radar systems for this purpose.[20] These include collaborative searches with Germany using the radar at the the *Forschungsgesellschaft für Angewandte Naturwissenschaften* (FGAN–Research Establishment for Applied Science) and the U.S. military using its radars. NASA has used the U.S. Army's radar system at the Kwajalein Atoll in the Pacific Ocean, the U.S. Air Force's FPS-85 radar at Eglin Air Force Base in Florida, and the Perimeter Acquisition Radar Characteristic System in North Dakota, also operated by the Air Force.[21] In addition, the Massachusetts Institute of Technology (MIT) Millstone radar and the MIT Lincoln Laboratory Haystack radar, located in Massachusetts, have been employed.

The Haystack radar is NASA's main source of measurement data for debris from 1 to 30 cm in size. The Haystack radar has been collecting orbital debris data for NASA since 1990. This radar stares into specific regions of space, statistically sampling the debris that flies through its field-of-view. These data provide information about the debris population size and location. The Haystack data indicate that more than 100,000 pieces of debris fragments as small as 1 cm orbit the Earth.

Radar and optical telescopes each see the debris differently. Each tool provides a different insight into the environment. Some debris elements are best seen with radar because they do not reflect sunlight well. Other objects reflect sunlight very well but do not interact well with radar. Optical telescopes can more easily detect debris objects at higher altitudes, such as GEO. NASA has previously used two optical telescopes for measuring orbital debris: a 3-m-diameter liquid mirror telescope, and a charge-coupled device (CCD) equipped with a 0.3-m Schmidt camera, which is referred to as the CCD Debris Telescope. Current optical measurements employ the University of Michigan's Michigan Orbital Debris Survey Telescope (MODEST), the Meter-Class Autonomous Telescope (MCAT), and the NASA Air Force Maui Optical and Supercomputing (AMOS) for the NASA AMOS Spectral Study (NASS) systems.[22]

MODEST watches most of the orbital slots assigned to the continental United States, from 25° W to 135° W. MCAT is a wide field-of-view, 1-m-aperture telescope. It is a joint NASA and U.S. Air Force resource deployed as part of the High Accuracy Network Orbit Determination System. During twilight hours, MCAT samples low-inclination orbits in a "track before detect" mode, and it performs a GEO search later in the night. MCAT relies upon the AMOS site, as does the NASS Spectral Study. NASS studies the spectral properties of debris fragments in order to understand their physical characteristics. These properties are determined by using low-resolution reflectance spectroscopy, the absorption features, and overall shape of the spectra.

The understanding gained from modeling and remote observation can guide the design of future space systems to survive orbital debris strikes. It is important to add knowledge gained from measurement of hypervelocity impacts upon the very materials those designs may use. The value of this activity comes when designers can safely reduce the mass of shielding material. Overdesigning the debris protection can result in unnecessary expense and loss of opportunity. These studies, conducted at NASA and DoD facilities, connect design to the risk the debris environment poses to the spacecraft. The studies also help understand the impact features found on spacecraft that have returned from space. This work is focused at the Hypervelocity Impact Technology Facility at NASA's Johnson Space Flight Center and is supported by other NASA and DoD facilities.[23]

DEBRIS MITIGATION

Perhaps the best way to protect future space systems is to control the growth of debris in orbit. The major spacefaring nations have made this a high priority. The goals are to prevent the creation of new debris and allow existing debris to deorbit. To do this, satellites need to be designed to operate in orbits with fewer debris. They also need to be able to maneuver to avoid collisions with large fragments and withstand impact from small debris. Finally, when their useful mission comes to an end, they need to be removed from orbit.

Toward these ends, the United States, Japan, France, Russia, and the ESA have established guidelines to mitigate orbital debris. NASA first issued a comprehensive set of orbital debris mitigation guidelines. Then, the U.S. government issued its Orbital Debris Mitigation Standard Practices, based on these guidelines. In 2002 a consensus was reached among the space agencies of ten countries, as well as the ESA, to adopt a set of guidelines put forward by the Inter-Agency Space Debris Coordination Committee (IADC). These guidelines were formally presented to the Scientific and Technical Subcommittee of the United Nations Committee on the Peaceful Uses of Outer Space in February 2003.

When the large Iridium commercial communications constellation was designed, special emphasis was placed on deorbiting dead spacecraft. Also, when NASA decided it was too dangerous to fly a servicing mission to HST using the shuttle, it was planned to deorbit the spacecraft because of the debris issue. NASA reversed that decision after strong outcry from the scientific community and general public, but both the Iridium and Hubble examples underscore how seriously debris mitigation is now taken.

DEBRIS REMOVAL

Formulated in 2002, the IADC Debris Mitigation Guidelines require spacecraft owners to protect the commercially valuable LEO and GEO zones. Requirements include limiting debris during normal operations, suppressing deliberate break-up of rockets or payloads, and properly disposing of spacecraft and upper stages by moving them to graveyard orbits or deorbiting them into the atmosphere.

The graveyard orbit is located about 400 km above GEO, well above the very sparse population of air molecules found in the low orbits. The graveyard is an orbit where spacecraft are intentionally placed at the end of their operational life to lower the probability of collisions with operational spacecraft and the generation of additional space debris. It is typically used for GEO satellites, when the change in velocity (Δv) required to perform a deorbit maneuver would be about 1,500 m/s, whereas reorbiting it to a graveyard orbit would only require about 11 m/s. Between 1996 and 2001, there were about 58 satellites in GEO that reached the end of their operational life; unfortunately, 38% of these were simply abandoned without following the international recommendation of a burial in the graveyard orbit.

Atmospheric drag is significant for any satellite low in LEO [400 km (or 240 miles) and less]. Drag reduces the spacecraft's orbital velocity and, thus, its altitude. Typically, smaller objects have less mass and therefore are decelerated faster by the weak drag of air molecules present in the lower orbits. A propulsive maneuver is required for drag makeup to maintain the desired orbit. Once all available propellant is exhausted, the spacecraft can no longer overcome the effects of drag. It will eventually reenter the atmosphere in a way similar to the U.S. spy satellite mentioned earlier, the Russian space station *Mir*, or the U.S. space station *Skylab*.

Deorbiting practices help reduce the number of dead spacecraft and spent rocket upper stages orbiting the Earth. These objects reenter Earth's atmosphere through either orbital decay (uncontrolled entry) or with a controlled entry. Orbital decay is achieved by firing engines to lower the orbit deeper into the atmosphere to increase atmospheric drag. Controlled entry normally requires using a large propulsion system to drive the spacecraft to enter the atmosphere at a steeper flight path angle. It will then enter at a precise latitude and longitude, typically into the Pacific Ocean, which is the preferred burial ground for space hardware.

During descent, the satellite encounters aerodynamic and aerothermal loads that cause the structure to disintegrate and burn. Weaker components, such as the solar array wings, break off first at altitudes of about 90–95 km. At altitudes between 84 and 72 km, aerodynamic forces typically exceed the allowable structural load limits.

After spacecraft break up, the resulting individual fragments will continue to fall and experience aeroheating. Some will survive to impact the Earth, but many spacecraft component materials, such as aluminum, which has a low melting point, burn away at a higher altitude. However, objects made of material with a high melting point (e.g., titanium, stainless steel, beryllium, carbon-carbon), will survive to much lower altitudes or even reach the Earth's surface. Some objects are contained inside a housing that protects the object from aeroheating for a while. This allows that object to fall nearer to the ground. Other objects are made of material of very high melting temperatures and so light, or aerodynamic, that they impact with a very low velocity.

In the example of *Mir*, a controlled deorbit maneuver, 27 tons of debris landed in the South Pacific, while *Skylab*, in an uncontrolled deorbit, delivered large pieces of junk into the outback of Australia.

TERRESTRIAL DEBRIS VERSUS SPACE DEBRIS

Terrestrial-based archaeologists analyze debris such as debitage, which is the waste material produced during lithic reduction in the production of chipped stone tools. The study of space debris has a direct correlation, but a significant difference. On Earth we see materials in situ as static. That is, relative to another object, materials at rest do not appear to be moving (although in archaeological sites, lithic debitage, depending on its size, may be moved through erosion and by human reuse). The one major exception might be in marine archaeology, where flotsam may well be more often in motion. Refuse and debris on Earth tend to be stratified and follow the law of superposition. The law of superposition is a key axiom based on observations of natural history that is a foundational principle of sedimentary stratigraphy and so of other geology-dependent natural sciences. It states that sedimentary layers are deposited in a time sequence, with the oldest on the bottom and the youngest on the top. Superposition in archaeology and especially in stratification, even in looking at debris, is slightly different, as the processes involved are somewhat different from geological processes. Human activities in the process impact or deform the layers. Some archaeological strata (often defined by their contexts or where they are found within the layers) are created by undercutting previous strata. An example would be that the silt backfill of an underground drain would form some time after the ground immediately above it. Other modifications, such as landfills, may modify the stratification. Thus, when humans are involved, there is a need for additional interpretation to correctly identify chronological sequences.

Since the early 1970s, the American archaeologist William Rathje has been studying debris (garbology), and studying material waste is a fundamental key to the subdiscipline of urban archaeology (Staski, Chapter 2 of this volume). Methodical, large-scale waste disposal is relatively recent in terrestrially based civilizations. In the first years of our space exploration and exploitation, there was no attempt to manage space debris. Even after those first years, a very global approach took half a century to develop. Our knowledge base on space debris (like space debris fields) is expanding rapidly. Just as modern civilization terrestrially copes with waste, so does the space community.

Space debris in orbit is dynamic, and as opposed to the law of superposition, will follow the laws of orbital mechanics. There is no stratification with a geospatial-temporal resolution. Terrestrially, we tend to see time as a linear force. The past is old, and "old" is relative. In space, time is the fourth dimension to define a location of an item.

This distinct contrast is noted in the use of the term *in situ*, which for the Earth-bound archaeologist indicates a fixed location. In orbit, in situ is where the object of interest is at the time of interest. This contrast of dynamic to static positioning yields on-orbit debris that does not follow the law of superposition. For planet-bound objects, it is more complex than directly following the laws of superposition, as here the method of arriving will significantly change the location in the stratified layers. One can envision the impactor probe versus a gentle lander. However, the law of superposition is more easily applied on planetary and other small bodies.

Another distinction from terrestrial-based study is the danger that most space debris poses to space assets. The same dangers are seen on Earth when areas such as landfills pose environmental risks, but the danger is perhaps not as great as that posed by space debris. This threat means that deorbiting may be the only solution to debris in lower LEO, no matter how culturally significant it may be.

CONCLUSION

This chapter discusses space debris, its characteristics and history. The heritage significance of debris is covered more thoroughly in the next chapter (Gorman, Chapter 19 of this volume). Today, it is hard to conceptualize a project such as the Westford needles, but military interests still trump scientific and environmental interests, as demonstrated in the 2007 Chinese ASAT demonstration.[24]

REFERENCES

1. Stokes, G.H., von Braun, C., Sridharan, R., Harrison, D. and Sharma, J. The Space-Based Visible Program. *Lincoln Laboratory Journal* 11, p. 205 (1998).
2. Winter, F.H. Who First Flew in a Rocket? *Journal of the British Interplanetary Society* 45: 276 (1992).
3. Proceedings of the Third European Conference on Space Debris, p 641, European Space Agency (2001).
4. According to the *ESA Meteoroid and Space Debris Terrestrial Environment Reference*, the MASTER-2005 Model. http://master-model.org/index.html.
5. Johnson, N. *History of On-Orbit Satellite Fragmentation*, 14th ed. NASA/TM-2008-214779. 30 (2008).
6. Klinkrad, H. *Space Debris*. London: Springer Praxis Books (2006).
7. Than, K. Rocket Explodes Over Australia, Showers Space with Debris. Space.com (February 21, 2007). http://www.space.com/news/070221_rocket_explodes.html.
8. http://www.centerforspace.com/asat/.
9. http://www.space.com/news/070706_sn_china_terra.html.
10. Challoner, J. *Space*. London: Channel-4 Books, 97 (2000).
11. Milne, A. *The Space Debris Crisis*, p. 49. Westport, CT: Praeger (2002).
12. NASA. Satellite Breakups During First Quarter of 2008. *Orbital Debris Quarterly News*. 12(1): 2 (2008)
13. Morring Jr., F., Butler, A., and Mecham, M. Orbital Collision Won't Be the Last. *Aviation Week & Space Technology*, February 15, 2009.
14. Belk, A., J.H. Robinson, M.B. Alexander, W.J. Cooke, and S.D. Pavelitz. *Meteoroids and Orbital Debris: Effects on Spacecraft*. NASA Reference Publication 1408. Huntsville, AL: NASA MSFC (1997).
15. Milne 2002 (note 11) (see specifically p. 76).
16. Rossi, A. and Valsecchi, G. B. Collision Risk Against Space Debris in Earth Orbits. *Celestial Mechanics and Dynamical Astronomy* 95: 345–356 (2006).
17. Vette, J.I. *The NASA/National Space Science Data Center Trapped Radiation Environment Model Program (1964–1991)*. NSSDC 91-29. Greenbelt, MD: NASA/Goddard Space Flight Center, National Space Science Data Center (1991).
18. Teague, M.J., and E.G. Stassinopoulos. *A Model of the Starfish Flux in the Inner Radiation Zone*. X-601-72-487. Greenbelt, MD: NASA/Goddard Space Flight Center (1972).
19. Klinkrad 2006 (note 6).
20. http://orbitaldebris.jsc.nasa.gov/measure/radar.html.
21. http://srmsc.org/par2010.html.
22. http://orbitaldebris.jsc.nasa.gov/measure/optical.html.
23. http://orbitaldebris.jsc.nasa.gov/protect/impacts.html.
24. Morring et al. 2009 (note 13).

19 Heritage of Earth Orbit: Orbital Debris—Its Mitigation and Cultural Heritage

Alice Gorman

CONTENTS

INTRODUCTION

Since the launch of *Sputnik 1* in 1957, human material culture in the form of satellites, launch vehicle upper stages, mission-related debris, and "space junk" has proliferated in Earth orbit. There are now significantly more than 10,000 trackable objects circling the Earth between low Earth orbit at around 200 km and the "graveyard orbit" around the geostationary ring at 35,000 km above the surface of the Earth.

Only a small portion of this material is operational spacecraft; the rest is classed as orbital debris. Space industry is now at the stage where collision with orbital debris is a serious threat for the continued provision of satellite-based services, such as navigation, telecommunications, meteorology, and earth observation.

This situation constitutes an environmental management problem for space industry. In the short term, measures to control the proliferation of debris have included changing mission design and operation practices as recommended the U.S. National Aeronautics and Space Administration (NASA), the European Space Agency (ESA), and the United Nations.[1,2,3] However, it is widely recognized that more active measures to remove debris from orbit will be required in the future. Proposals have included the destruction of debris using ground- and space-based lasers and intervention missions by specialized spacecraft.

While the necessity of some active management is accepted by all, the problem is slightly more complex than a consideration of the technical difficulties suggests. As discussed in Chapter 16, orbital space constitutes an organically evolved cultural landscape as defined by the World Heritage Convention.[4] Objects now classed as orbital debris may have social, historical, aesthetic, and scientific significance for nations, communities, groups, and individuals who will have an interest in decisions made about their long-term survival. It is not just the threat of collision that needs to be managed: proposals for orbital debris cleanup must also consider how to manage the cultural values of the orbital spacescape.

This does not mean that everything must be saved. In this chapter, I look at some of the issues that can help us make well-grounded decisions about what to preserve in situ and what to let go. There are two facets to this: an assessment of the risk posed to space operations by different debris classes, and the assessment of the significance of orbital objects according to the categories of the Burra Charter, Australia's primary heritage management document.[5] The Charter, while designed specifically for Australia, has been recognized as a simple and powerful set of heritage management guidelines that sets standards worthy of being emulated at an international level. To demonstrate how significance can be assessed as the basis for sound management decisions, I look at the oldest surviving human object in orbit: the *Vanguard 1* satellite.

ORBITAL DEBRIS

Orbital debris has been defined as any human-manufactured object in orbit that does not currently serve a useful purpose and is not anticipated to in the foreseeable future.[6] Approximately 4,200 launches have occurred since 1957, leaving more than 10,000 trackable objects larger than 10 cm in orbit (Figure 19.1).[7] Only 7% of these are operation spacecraft; 52% are decommissioned satellites, upper stages, and mission-related objects, and 41% are debris from the fragmentation of orbital objects.

Operational and decommissioned spacecraft include scientific and telecommunications satellites, weather and earth observation satellites, navigation and surveillance satellites, satellite constellations, and military satellites. Upper stages include the durable Agena, in use from the time of the Gemini program to the mid-1980s, and those of the Ariane family of rockets, first launched in 1979.

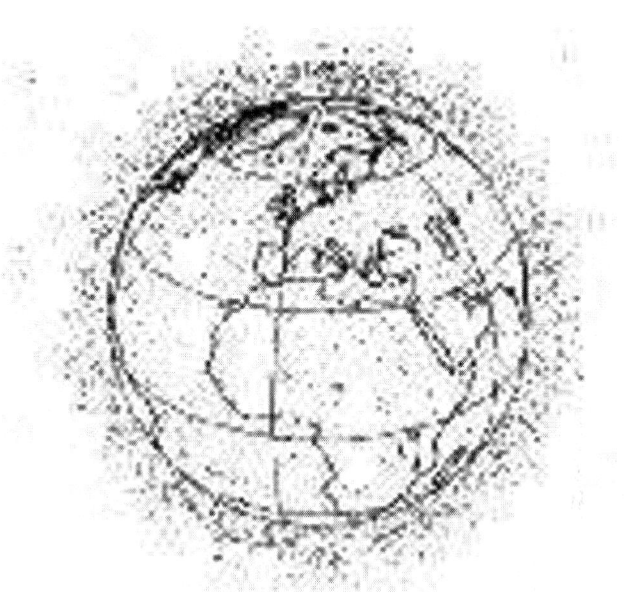

FIGURE 19.1 This computer-generated image shows the thousands of satellites, spent rocket stages, and breakup debris in low Earth orbit. (Courtesy of NASA.)

Mission-related debris derive from deployments and separations of spacecraft, which typically involve the release of items such as separation bolts, lens caps, flywheels, nuclear reactor cores, clamp bands, auxiliary motors, launch vehicle fairings, and adapter shrouds.[8] Solid rocket motors used to boost satellite orbits have contributed other objects, such as motor casings, aluminum oxide exhaust particles, nozzle slag, motor-liner residuals, solid-fuel fragments, and exhaust cone bits resulting from erosion during the burn, to the debris population.

Fragmentation debris are derived primarily from the explosion of satellites and launch vehicle upper stages, both of which tend to remain in orbit after the completion of their mission.[9] Explosion can occur when residual liquid fuel components accidentally mix, or when fuel or batteries become over-pressurized. There are also cases where spacecraft have been deliberately detonated, to prevent reentry and/or to conceal their presence or purpose. More than 124 breakups have been verified so far, and the rate of breakup increases each year. Another major source of debris is material degradation from a range of environmental effects, resulting in the production of particulates, such as flakes of paint and insulation. Figure 19.2 shows the "energy flash" when a projectile launched at speeds up to 27,000 km/h impacts a solid surface at the Hypervelocity Ballistic Range at NASA's Ames Research Center in Mountain View, California. This test is used to simulate what happens when a piece of orbital debris hits a spacecraft in orbit.

Debris is concentrated in the orbital configurations most commonly used in space operations, defined by altitude above the earth's surface, inclination, and eccentricity. The orbit employed depends on the purpose of the satellite and the location of the launch site. Most objects are in the nominally circular orbits, low Earth orbit (LEO)

FIGURE 19.2 Energy flash of an orbital debris hit. (Courtesy of NASA.)

or geosynchronous Earth orbit (GEO).[10] Medium Earth orbits are less widely used, to avoid the Van Allen radiation belts.

In low Earth orbit, aerodynamic drag acts as a "natural cleansing mechanism," causing objects to reenter the atmosphere and (mostly) burn up.[11] At 400 km or below in altitude, it may take only a few months for objects to reenter. However, above 600 km, objects can remain in stable orbits for a few decades up to thousands of years. Satellites in GEO are beyond the reach of atmospheric affects although still subject to the vagaries of the space environment.

Within these orbital regimes, there are areas of higher debris density. In low Earth orbit, debris builds up near polar inclinations from sun-synchronous satellites and at altitudes near 800, 1,000, and 1,500 km.[12,13] There are an estimated 70,000 pieces of debris about 2 cm in size at the 850–1,000 km altitude.[14] Objects in geosynchronous and Molniya orbits and constellations of navigation satellites cause another peak in density at 25,000 km. The highest peak is at 42,000 km, consisting of objects at or near the geostationary ring.[15] Within the GEO region, peaks of debris occur at the following:

- The equatorial inclination
- 28.5°, due to the latitude of the main U.S. launch site at the Kennedy Space Center
- 63°, from Molniya and GLONASS satellites

Modeling the debris environment is reliant on data collected from optical and radar tracking. Debris over 10 cm is tracked by U.S. Space Command (USSPACECOM), using twenty-five land-based radars and optical telescopes in the Space Surveillance Network. Over the former USSR's territory, debris is tracked by the Russian Space

Surveillance Centre,[16] ESA maintains the DISCOS database of space debris. Only debris above a certain size can be tracked in this way, and visibility depends on altitude: in GEO, an object must have a diameter of 1 m to be visible, while in LEO, radar can detect pieces as small as 5 mm. The population of debris below this size is estimated on the basis of impacts on returned spacecraft surfaces.[17]

MANAGING ORBITAL DEBRIS

Mitigation strategies for orbital debris can be broadly divided into three types: (1) design and operational solutions, (2) Earth-based removal programs, and (3) intervention missions.

Design and Operational Mitigation

This strategy is aimed at controlling the amount of new debris that enters the system by designing spacecraft and missions to minimize mission-related debris and the potential for fragmentation. Design solutions include using tethered lens caps and bolt catchers, shielding or augmenting components to withstand impact, and the use of operating voltages below arc thresholds.

Operational measures include postmission maneuvers to place the spacecraft within the range of aerodynamic drag or in a graveyard orbit and expelling remaining propellants and pressurants to prevent accidental explosion.[18] The NASA guidelines for limiting orbital debris recommend that an object should not remain in its mission orbit for more than 25 years.[19,20]

Earth-Based Removal Programs

Other proposals have examined the prospect of removing debris between 1 and 10 cm in diameter in LEO by ground-based lasers (e.g., NASA's Project Orion and Electro-Optical Systems).[21,22] The laser ablates particles from the surface of the debris, creating enough thrust to edge it into reentry.[23] Space-based laser removal has also been considered, for example, to move debris out of the path of the International Space Station, but is considered too costly in time and energy to be feasible at this stage.

Intervention Missions

Intervention missions include the use of specialized spacecraft that actively remove objects from orbit. A study undertaken by QinetiQ investigated scenarios for removing decommissioned satellites in GEO using a reorbiting spacecraft, concluding that this was plausible if not yet feasible.[24] For LEO, a Royal Melbourne Institute of Technology group has proposed the use of space-based electrodynamic tethers to capture and remove debris.

At this point in time, only design and operational mitigation is used on space missions. Before any active debris mitigation measures are implemented, the question of whether spacecraft currently classed as debris have any cultural heritage value needs to be addressed. What do we want to save for future generations? Should significant spacecraft be left in situ or removed to Earth for curation and display? If we accept that orbital space is a cultural landscape, the most appropriate management response is to leave spacecraft where they are. But this raises a critical question: can cultural heritage values be managed without compromising safety or service delivery?

ARCHAEOLOGY AND HERITAGE OF ORBITAL OBJECTS

The primary concern of archaeology is to understand role of material culture in human social and environmental engagement. Despite popular perceptions, archaeology is not necessarily concerned only with the ancient: there is much to be learnt from contemporary material culture, particularly as it brings with it the added dimensions of extensive documentation and personal memories. It is often in the disjunctions between these that the most interesting stories wait to be told.

Space objects fall into the field known as historical archaeology or postmedieval archaeology, which is concerned with the global expansion of industrialized European nations, the growth of capitalist economies, and interactions with indigenous people in the colonies. Within this broad field, there are subdisciplines, such as military archaeology and contemporary archaeology, which have useful theoretical perspectives.

Spacecraft can be regarded as archaeological artifacts, the material record of a particular phase in human social and technological development. They have research potential in terms of understanding human interaction with the space environment. But they also more than this. Material culture can be regarded as heritage: objects from the past that have meaning in the present and that are important to the identity and well-being of communities.

We know that people see the material culture of space exploration as important: for example, of all the Smithsonian institutions in Washington, D.C., the most popular is the National Air and Space Museum. The attraction is actually seeing the artifacts such as the Gemini capsule, spacesuits, and pieces of *Skylab*.[25] One can read books and documents, watch film footage, and view photographs, but nothing can convey the same information or meaning as the actual object itself. So we know that at least some people in some places value these objects. The question we need to address in managing heritage values during orbital debris mitigation is for whom are they significant, and why?

The significance of space material culture is often assumed to be self-evident. One of the most commonly cited rationales for space exploration, referred to in countless books, documentaries, and museum displays, is that space exploration is the natural outcome of an innate human urge to explore. Thus, space objects are perceived to have a globally understood meaning that appeals to our common human nature.[26] Space exploration is seen as the most recent manifestation of a fundamental curiosity that led humans out of Africa, across the seas from the Old World to the New World, and inevitably into space.

Another popular model for understanding the significance of space material culture is what I have called the Space Race model.[27] In this formulation, objects and places have significance for their contribution to the Cold War confrontation between the United States and the USSR. This model focuses on these two states, ignoring the achievements of other countries, such as France, Britain, China, Japan, and Australia, and the contributions made by "developing" nations who often hosted launch sites, ground stations, and so forth. The Space Race model emphasizes competitiveness rather than cooperation in space, with an implicit Darwinian overtone: spacefaring nations (mostly imperial, industrial, and white) demonstrate technological fitness by their success in space enterprises. The relationship of space exploration to inequalities between the developed and developing world is unexplored, and indeed unproblematic, in the Space Race scenario.[28]

Of course, these approaches do capture something meaningful about the significance of space material culture, but it is far from the whole story. To obtain a deeper and more inclusive understanding of heritage significance, I turn to the guidelines adopted by the Australian National Committee of the International Council on Monuments and Sites in the Burra Charter.[29] As well as providing a methodology for assessing significance, the Burra Charter sets out principles for preservation and conservation. Its main tenet is "Do as much as is necessary and as little as possible" in order to retain the cultural significance of a place or object.

The Burra Charter outlines four different categories of significance:

1. Aesthetic: considerations of form, scale, color, texture, material, smells and sound, and setting
2. Historic: association with historic figures, events, phases, or activities
3. Scientific: importance in terms of rarity, quality, representativeness, and the degree to which a place can contribute further substantial information
4. Social: the qualities for which a place has become a focus of spiritual, political, national, or other cultural sentiment to a majority or minority group

The Burra Charter also stresses that significance may be multivocal, and Article 6.3 states that the "co-existence of cultural values should be recognized, respected and encouraged, especially in cases where they conflict."

These kinds of significance are used successfully as the basis for museum collections of space artifacts around the world and have been used in nominating space sites on Earth for heritage listing. In the next section I want to apply them to the oldest artifact in Earth orbit: the *Vanguard 1* satellite.

VANGUARD 1

The *Vanguard 1* satellite, launched successfully on March 17, 1958, is now the oldest manufactured object in orbit (Figure 19.3). It is in a highly stable LEO orbit with every prospect of remaining there for perhaps another 600 years.

Unlike *Sputnik 1* and *Explorer 1*, *Vanguard* was launched using scientific sounding rockets rather than missile technology to avoid a military "taint." As the launch was part of the International Geophysical Year program, the Vanguard team recruited a network of volunteers across the world to carry out visual tracking in Project Moonwatch.[30] This community involvement played an important role in configuring the project as scientific, cooperative, and inclusive.

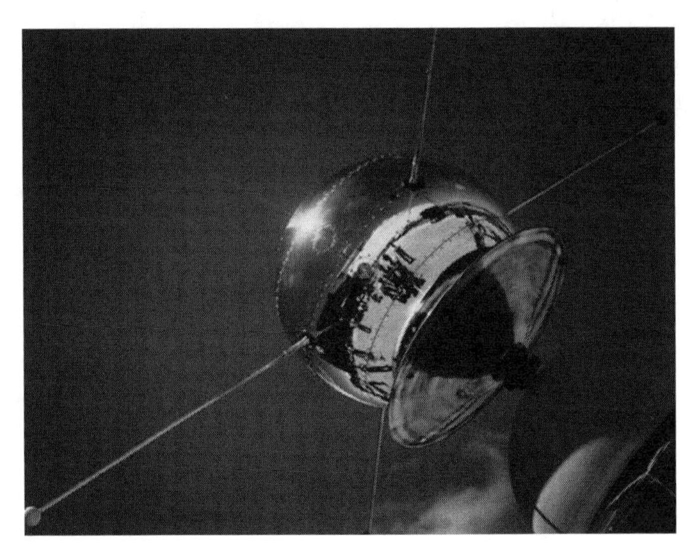

FIGURE 19.3 *Vanguard 1* satellite. (Courtesy of NASA.)

In terms of the Burra Charter's categories of significance, we might attempt an assessment along the following lines.

Aesthetic

In design, *Vanguard 1* is spherical, with antennae attached at right angles to the body, made of silver aluminum. It was renowned for its small size, being dubbed the "grapefruit satellite" by Khrushchev. The size and the design reflect the cost of placing material in orbit and perhaps also mutual influence with USSR designs—*Vanguard* and *Sputnik* are remarkably similar. Satellites are no longer manufactured to a spherical design, so this shape is indicative of an early phase, where satellites were seen not so much as an earth-circling spaceship as a miniature moon (bébé lune, or baby moon). The spherical shape continued to be used in the USSR for crewed shariks (descent capsules), but by 1960 this design was becoming rare.

Historic

Vanguard 1 is associated with the Cold War and the International Geophysical Year of 1957–1958. It was the third satellite to be successfully launched and the second U.S. satellite. It represents the first experimental phase of space exploration. Analysis of *Vanguard*'s orbital perturbations revealed that the Earth was "pear-shaped." *Vanguard* represents the conflicting motivations and rationales for space exploration in the critical period of the 1950s, when the United Nations also first moved to set up the principles of the Outer Space Treaty. Although it was designed as a peaceful scientific satellite, it was also an ideological weapon, a "visible display of technological prowess" aimed at maintaining the confidence of the free world and containing Communist expansion.[31,32] *Vanguard*'s design and mission reflect the competing models of cooperation and confrontation in space, at a time when there were no rules, laws, or guidelines to structure the human-orbital interaction.[33]

Scientific

There are three aspects to the scientific significance of *Vanguard 1*.

- Representativeness: *Vanguard 1* is the sole survivor of all satellites launched in 1957–1958, and one of the 23% of satellites remaining in orbit from 180 launches between 1957 and 1963 (the first successful geosynchronous launch, Table 19.1). As such, it is unique: there is no equivalent satellite if *Vanguard* should be destroyed.
- Relevance to space research: There is no other object that has been in space as long as *Vanguard 1*; hence it is the only artifact that can inform us of the long-term impact of the space environment on human materials.
- Relevance to archaeological research: Although there is extensive documentation about the history of *Vanguard 1*, this does not convey the same information as the object itself and its relationship to other artifacts in Earth orbit. A recording of the physical features of *Vanguard 1* (when this is possible) may reveal discrepancies between documentation and reality, and aspects of technological processes that are not longer in use.

Social

The satellite represents the expectations that the United States would be first in space, and its failure to be so caused nationwide doubt and panic. At the time, the community esteem in which *Vanguard* was held was very low, and it was the butt of jokes in both the United States and the USSR. Nevertheless, the staff of the Naval Research Laboratory responsible for the project and the international network of Project Moonwatch volunteers must have had considerable emotional investment in

TABLE 19.1
Satellites in Earth Orbit (1957–1963)

Year	Number of Launches	Number of Satellites Remaining	Country of Origin
1957	2	0	USSR
1958	8	1	United States
1959	13	3	United States
1960	23	7	United States
1961	38	9	United States
1962	61	10	Canada, United States
1963	59	14	United States

its success. With the passage of time, *Vanguard*'s social significance has changed. Later assessments have acknowledged as Project Vanguard as "the progenitor of all American space exploration today."[34] The satellite has Internet-based fan groups, including one that tracks its location in real time. It is now actually considered to be a "vanguard." It is esteemed at an international level as the oldest human artifact in space, at a national level as oldest U.S. artifact in space, and at the local level by the space-buff communities who follow its progress.

This cursory significance assessment demonstrates how the Burra Charter principles can be applied to orbital objects. Following the guidelines, significance assessment should then be used as the basis for management. So what then is the best management policy to preserve the cultural significance of this satellite? Options include destruction as part of an active debris mitigation program, removal to the safekeeping of a terrestrial museum as soon as practicable, or—simply nothing.

As is clear from the assessment of significance above, *Vanguard* does have high cultural significance, so destruction is not an appropriate option. It is also clear that part of *Vanguard*'s scientific and social significance is its presence in orbit. Two Burra Charter principles can be applied here: first, that we should do as much as is necessary and as little as possible, and second, that the setting of an object or place should be retained as part of its cultural significance. Removal to Earth would diminish the cultural significance of the satellite and should not be considered appropriate management unless this would prevent its destruction.

SURVEY OF EARLY SATELLITES IN ORBIT

Vanguard 1 is not the only satellite that may have heritage significance from the early years of space exploration. The following is a brief survey of other whole satellites still in low and medium Earth orbits from the period between the launch of *Sputnik 1* in 1957 to the launch of the first geosynchronous satellite, *Syncom 1*, in 1963.

Data come from a publicly available database of objects tracked by USSPACECOM. The information presented here focuses on satellites that had been launched intentionally into Earth orbit rather than toward the Moon, sun, or other planets. Spacecraft that have been lost or deliberately deorbited, landed, or decayed, as well as rocket bodies, mission-related debris, and fragmentation debris, have been excluded.

Of 180 satellites launched between 1957 and 1963, 41 remain in Earth orbit (23%). All except the Canadian *Alouette 1* originated from the United States (Figure 19.4; Table 19.1). They occupy low Earth orbit at both equatorial and polar inclinations, sun-synchronous orbits, and medium Earth orbit. The function of the satellites is not always clear-cut: many scientific and other missions were undertaken for military applications, and information about others is still classified, but even in this early period, the satellites cover the range of functions that are still predominant today, with a fairly even distribution among scientific, meteorological, navigation, communications, and defense-related missions.[35]

There are only a small number of satellites still in orbit from this early period, and each one could be argued to demonstrate an aspect of developing space technology. They include *Vanguard 1, 2,* and *3*; *Explorer 7*; *TIROS 1*, the first weather satellite; *Transit 4A* and *4B*, which carried the first nuclear power sources on a spacecraft;

FIGURE 19.4 *Alouette 1*. (Courtesy of NASA.)

Telstar 1, the first active telecommunications satellite (Figure 19.5); and the Westford needles and their release capsule.[36] In this range we can see technological trajectories: from nuclear power to solar power, from passive telecommunications to active, from spherical baby moons to more diverse designs, increasing size, and increasing height above Earth's surface.[37] The material is dominated by U.S. spacecraft: how might this be interpreted by archaeologists of the future? What can we learn about the early space programs by what is left in orbit, as opposed to the documentary record? We cannot anticipate future research directions, but one day orbital objects will tell their own stories, if they survive.

RISK ASSESSMENT

If significant spacecraft are left in orbit, does this merely contribute to the orbital debris problem? The next step is to assess the actual risk posed by heritage spacecraft to operational spacecraft. This involves a consideration of the damage caused by different size classes of debris, and the actual probability of collision.

Orbital debris can be divided into three size classes.

- Large: diameter greater than 10 cm. Large debris can be optically tracked.
- Medium: diameter between 1 mm and 10 cm. Tracking depends on size and altitude.
- Small: diameter less than 1 mm. This is the largest population of orbital debris, and these items cannot be tracked.

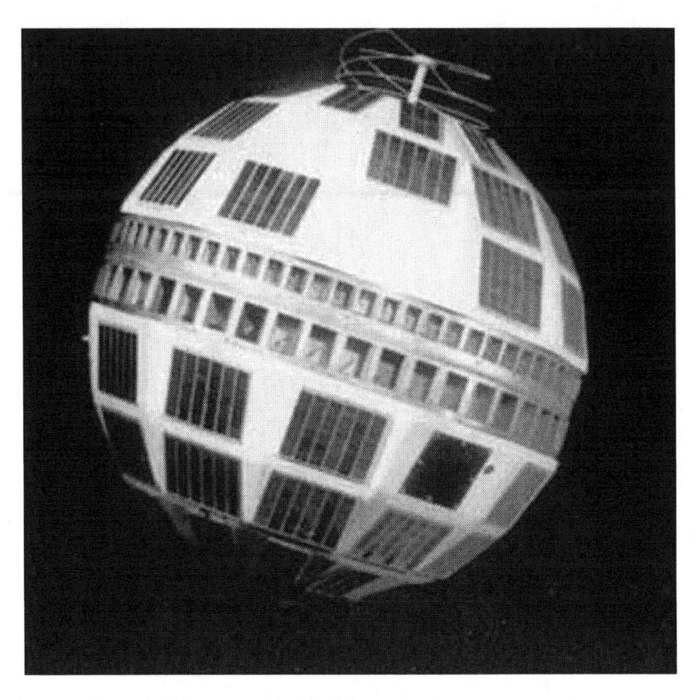

FIGURE 19.5 *Telstar 1.* (Courtesy of NASA.)

The results of collision with a piece of debris include mechanical damage, material degradation, and, occasionally, catastrophic breakup of operational spacecraft. Even tiny particles can cause significant damage because the impacts occur at hypervelocity (i.e., when the magnitude of the impact velocity is greater than the speed of sound in the impacted material).[38] In LEO, the average relative velocity of space debris at impact is 10 km/s (36,000 km/h).[39] Average relative velocities in GEO are much lower, about 200 m/s (720 km/h), but collisions at this speed can still cause significant damage. In terms of impact, a 10-cm fragment in geosynchronous orbit has roughly the same damage potential as a 1-cm fragment in LEO. A 1-cm geosynchronous fragment is roughly equivalent to a 1-mm low Earth orbit fragment.[40]

Collision with an object in the large size class (>10 cm) can cause fragmentation and breakup, a significant source of new orbital debris. Impact from the medium debris class, 1 mm to 10 cm, can cause significant damage and mission failure.[41] Penetration by a debris fragment 1 mm to 1 cm in size, through a critical component, can result in the loss of the spacecraft. Fragments greater than 1 cm can penetrate and damage most spacecraft.[42] Although objects in the small size class rarely cause catastrophic breakup, they can erode sensitive surfaces, such as payload optics.[43]

However, it is also necessary to consider the frequency with which such collisions occur for the different size classes. There is a direct relationship between the numbers of debris in each size class, the relative velocities in different orbital regimes, and the probability of impact.

TABLE 19.2
Mean Time between Impacts on a Satellite with a Cross-Section Area of 10 m² in Low Earth Orbit

Height of Circular Orbit (km)	Mean Time between Collisions (Years)		
	Objects 0.1–1.0 cm	Objects 1–10 cm	Objects >10 cm
500	10–100	3,500–7,000	15,000
1,000	3–30	700–1,400	20,000
1,500	7–70	1,000–2,000	30,000

Source: Courtesy of United Nations, *Technical Report on Space Debris*, United Nations, New York, 1999.

In LEO, the population of large debris is much lower than the medium and small classes, but the severity of impact is much greater when collision does occur because of the high relative velocity.[44] Despite this, collisions with large objects over 10 cm are rare, and there are only a few recorded breakups due to catastrophic collisions.[45] The density of the small debris class, however, is such that spacecraft in LEO experience continuous bombardment by very small particles.

Lower relative velocities and greater distances between objects in GEO significantly reduce the probability of collision.[46] Because of the increasing use of disposal orbits after mission completion, the rate of debris accumulation is slower than in LEO. Approaches between operational spacecraft and tracked objects can be predicted and evasive maneuvers undertaken to avoid collision. Because most objects in the geosynchronous ring move along similar orbits, objects in GEO are more likely to collide with a meteoroid than with debris.[47] However, untracked debris in GEO is not as well modeled as that in LEO.[48]

Table 19.2 illustrates the mean time between collisions with objects in the three size classes in different orbits. This table also demonstrates that the greatest risk of impact derives from the small debris size class in LEO. The larger the piece of debris and the higher the orbit, the less likely it is that a collision will occur. However, the medium debris class, 1 mm to 10 cm, is the most destructive. Medium debris are far more numerous than the large class, have a higher risk of collision, and can cause significant damage and mission failure.[49] Apart from indicating that any active orbital debris mitigation program should target this size class rather than large objects, this also suggests that costly intervention missions aimed at removing large decommissioned spacecraft will have minimal impact on the debris problem.

In the first instance, then, preserving orbital debris larger than 10 cm, which includes whole but defunct satellites, upper stages, and mission-related debris such as the famous glove lost by Edward White in 1965 (currently tracked by Electro-Optical Systems[50]) in their orbital locations can be done without compromising the safety and operation of crewed and uncrewed missions. If an object, such as

Vanguard 1, has been identified as having heritage value, then it can be excluded from any future debris mitigation projects that involve deorbiting. Potentially catastrophic approaches can be avoided by on-orbit maneuvers. As any active debris mitigation proposal must be designed to exclude operating spacecraft, it should simply be a matter of appropriate planning to avoid objects of cultural significance.

It is also possible that debris in other size classes may have cultural significance, particularly with regard to its representativeness. An initial reaction may assume that any significance will be extremely low, but this cannot be determined without a systematic investigation, which I will not attempt here.

CONCLUSIONS

Early conceptions of the Space Age imagined it as technological utopia, constructed of clean, metallic surfaces buffering the population from the disorder of a messy, organic past. In the contemporary world, it is acknowledged that continuity and connection to the past are vital in maintaining the well-being of communities, as the world becomes increasingly globalized. The destruction of cultural heritage has accelerated with the growth of population, development, and industrialization, and UNESCO, through the World Heritage Convention, recognizes that "that deterioration or disappearance of any item of the cultural or natural heritage constitutes a harmful impoverishment of the heritage of all the nations of the world."

There is also a growing interest in the archaeology and cultural heritage of the more recent past, covering events and phases such as the two world wars, the nuclear industry, and the Cold War.[51,52] Heritage authorities around the world are now protecting landscapes shaped by these events. It would be wrong to assume that simply because a place or object is "recent" we know all about it: rapid technology change and military and commercial secrecy may, in some cases, mean that we understand even less about a Cold War site than an Iron Age hill fort. Space places and objects are no exception.

To date, all considerations of the orbital debris problem have focused on the risk posed to satellite services and crewed missions. The potential for space debris mitigation to impact on cultural heritage values has not been examined. In this discussion, I have argued that orbital debris can have cultural heritage significance, and preserving significant orbital objects in the large size class in situ does not add to the risk posed by orbital debris to space missions. From this, it follows that the implementation of active debris mitigation strategies, such as deorbiting into the atmosphere or into graveyard orbits, should consider what impact this will have on the cultural landscape of orbital space and on the object as part of that landscape.

In the absence of legal instruments, cultural heritage in orbit could be protected by agreed guidelines. In 1999, an environmental symposium at the UNISPACE conference recommended that the concept of international environmental impact assessments be developed for all proposed space projects "that might interfere with scientific research or natural, cultural and ethical values of any nation."[53] Although cultural impacts were identified primarily as affecting the night sky as seen from Earth, this could apply equally to orbital debris.

Following terrestrial models such as those used in Australia, an environmental impact assessment for an orbital enterprise might include the following:

- Identification of objects of significance at international, national, and agency levels
- Identification and consultation with stakeholders (designers, scientists, government, industry, clients, and users of the service)
- Significance assessment, including aesthetic, historic, scientific, social, or spiritual value for past, present, or future generations[54]
- Identification of impacts on orbital heritage (e.g., damage, destruction, alteration of current orbit, increased risk of collisions)

Management options may include undertaking no active measures, monitoring of the position of significant objects, changing the orbit of the significant object to reduce the risk of damage, or redesigning the mission to avoid impacts on the significant object.

Before active debris mitigation strategies are implemented, there is an opportunity to assess the nature of the material record in orbit and ensure that objects of significant cultural heritage value are not lost. What would future generations of space tourists think if they found that *Vanguard 1* was destroyed needlessly through lack of forethought?

FURTHER READING

Belk, C.A., Robinson, J.H., Alexander, M.B., Cooke, W.J. and Pavelitz, S.D. 1997 *Meteoroids and Orbital Debris. Effects on Spacecraft.* NASA Reference Publication 1408. Huntsville, AL: NASA.

Campbell, J.W. 1996. *Project Orion: Orbital Debris Removal Using Ground-Based Sensors and Lasers.* NASA-TM-108522. Huntsville, AL: NASA.

Centre for Orbital Reentry and Debris Studies. http://www.aero.org/capabilities/cords.

Chapman, S. 1959. *IGY: Year of Discovery. The Story of the International Geophysical Year.* Ann Arbor, MI: University of Michigan Press.

Clark, P.S. 1994. Space Debris Incidents Involving Soviet/Russian Launches. *Journal of the British Interplanetary Society* 47(9): 379–391.

Cocroft, W., and Thomas, R. 2003. *Cold War: Building for Nuclear Confrontation 1946–1989.* London: English Heritage.

Crowther, R. 1994. The Trackable Debris Population in Low Earth Orbit. *Journal of the British Interplanetary Society* 47(4): 128–133.

European Space Agency. 2006. Robotic GEostationary orbit Restorer (ROGER). http://www.esa.int/TEC/Robotics/SEMTWLKKKSE_0.html.

European Space Operations Centre. 2003. Space Debris Spotlight. http://www.esa.int/SPECIALS/ESOC/SEMHDJXJD1E_0.html.

Gorman, A.C. 2005. The Cultural Landscape of Interplanetary Space. *Journal of Social Archaeology* 5(1): 85–107.

Gorman, A.C. 2007. Leaving the Cradle of Earth: The Heritage of Low Earth Orbit 1957–1963. Paper presented at the Australia ICOMOS Conference: Extreme Heritage, July 19–21, Cairns, Australia.

Gorman, A.C., and O'Leary, B.L. 2007. An Ideological Vacuum: The Cold War in Space. In *A Fearsome Heritage: Diverse Legacies of the Cold War,* ed. J. Schofield and W. Cocroft, 73–92. Walnut Creek CA: One World Archaeology, Left Coast Press.

Green, C.M., and Lomask, M. 1970. *Vanguard: A History.* NASA SP-4204. The NASA Historical Series. Washington DC.

Hypervelocity Impact Test Facility. http://www.wstf.nasa.gov/Hazard/Hyper/Default.htm.

ICOMOS Australia. 1999. Burra Charter. http://www.icomos.org/australia.

NASA. NASA Management Instruction 1700.8—Policy for Limiting Orbital Debris Generation. Washington, DC: Office of Safety and Mission Assurance.

NASA. NASA Safety Standard 1740.14—Guidelines and Assessment Procedures for Limiting Orbital Debris. Washington, DC: Office of Safety and Mission Assurance.

Osgood, K.A. 2000. Before Sputnik: National Security and the Formation of US Outer Space Policy. In *Reconsidering Sputnik: Forty Years since the Soviet Satellite*, ed. R.D. Launius, J.M. Logsdon, and R.W. Smith, 197–229. Amsterdam: Harwood Academic Publishers.

Smith, B. 2004. It's the Artifacts, Stupid! *The Mineralogical Record* 35(2): 106–107.

Sullivan, W. 1999. Report on the Special IAU/COSPAR/UN Environmental Symposium: Preserving the Astronomical Sky (International Astronomical Union Symposium 196). http://www.iau.org/IAU/Activities/environment/s196rep.html.

United Nations. 1999. *Technical Report on Space Debris*. New York: United Nations.

UNESCO Intergovernmental Committee for the Protection of the World Cultural and Natural Heritage. 2005. Operational Guidelines for the Implementation of the World Heritage Convention. World Heritage Centre. http://whc.unesco.org/archive/opguide05-en.pdf.

Woodford, J. 2004. A Blast from the Past. *Sydney Morning Herald*, July 10.

REFERENCES

1. NASA. NASA Management Instruction 1700.8—Policy for Limiting Orbital Debris Generation. Washington, DC: Office of Safety and Mission Assurance.
2. NASA. NASA Safety Standard 1740.14—Guidelines and Assessment Procedures for Limiting Orbital Debris. Washington, DC: Office of Safety and Mission Assurance.
3. United Nations. 1999. *Technical Report on Space Debris*. New York: United Nations.
4. UNESCO Intergovernmental Committee for the Protection of the World Cultural and Natural Heritage. 2005. Operational Guidelines for the Implementation of the World Heritage Convention. World Heritage Centre. http://whc.unesco.org/archive/opguide05-en.pdf.
5. ICOMOS Australia. 1999. Burra Charter http://www.icomos.org/australia/.
6. Crowther, R. 1994. The Trackable Debris Population in Low Earth Orbit. *Journal of the British Interplanetary Society* 47(4): 128–133.
7. European Space Operations Centre. 2003. Space Debris Spotlight. http://www.esa.int/SPECIALS/ESOC/SEMHDJXJD1E_0.html.
8. Belk, C.A., Robinson, J.H., Alexander, M.B., Cooke, W.J., and Pavelitz, S.D. *1997 Meteoroids and Orbital Debris. Effects on Spacecraft*. NASA Reference Publication 1408. Huntsville, AL: NASA.
9. Crowther 1994 (note 6).
10. Crowther 1994 (note 6).
11. Crowther 1994 (note 6).
12. Crowther 1994 (note 6).
13. Belk et al. (note 8).
14. Centre for Orbital Reentry and Debris Studies. http://www.aero.org/capabilities/cords/.
15. Crowther 1994 (note 6).
16. Clark, P.S. 1994. Space Debris Incidents Involving Soviet/Russian Launches. *Journal of the British Interplanetary Society* 47(9): 379–391.
17. Belk et al. (note 8).
18. Osgood, K.A. 2000. Before Sputnik: National Security and the Formation of US Outer Space Policy. In *Reconsidering Sputnik: Forty Years since the Soviet Satellite*, ed. R.D. Launius, J.M. Logsdon, and R.W. Smith, 197–229. Amsterdam: Harwood Academic Publishers.
19. NASA (note 1).
20. NASA (note 2).

21. Woodford, J. 2004. A Blast from the Past. *Sydney Morning Herald*, July 10.
22. Campbell, J.W. 1996. *Project Orion: Orbital Debris Removal Using Ground-Based Sensors and Lasers*. NASA-TM-108522. Huntsville, AL: NASA.
23. Ibid.
24. European Space Agency. 2006. RObotic GEostationary orbit Restorer (ROGER). http://www.esa.int/TEC/Robotics/SEMTWLKKKSE_0.html.
25. Smith, B. 2004. It's the Artifacts, Stupid! *The Mineralogical Record* 35(2): 106–107.
26. Gorman, A.C. 2005. The Cultural Landscape of Interplanetary Space. *Journal of Social Archaeology* 5(1): 85–107.
27. Ibid.
28. Ibid.
29. ICOMOS Australia 1999 (note 5).
30. Chapman, S. 1959. *IGY: Year of Discovery. The Story of the International Geophysical Year.* Ann Arbor, MI: University of Michigan Press.
31. Green, C.M., and Lomask, M. 1970. *Vanguard: A History.* NASA SP-4204. The NASA Historical Series. Washington DC.
32. Osgood 2000 (note 18). (see specifically p. 216).
33. Gorman, A.C., and O'Leary, B.L. 2007. An Ideological Vacuum: The Cold War in Space. In *A Fearsome Heritage: Diverse Legacies of the Cold War,* ed. J. Schofield and W. Cocroft, 73–92. Walnut Creek CA: One World Archaeology, Left Coast Press.
34. Ibid.
35. Gorman, A.C. 2007. Leaving the Cradle of Earth: The Heritage of Low Earth Orbit 957–1963. Paper presented at the Australia ICOMOS Conference: Extreme Heritage, July 19–21, Cairns, Australia.
36. Ibid.
37. Ibid.
38. Hypervelocity Impact Test Facility. http://www.wstf.nasa.gov/Hazard/Hyper/Default.htm.
39. Belk et al. (note 8).
40. Crowther 1994 (note 6).
41. Belk et al. (note 8).
42. Crowther 1994 (note 6) (see specifically Fig. 11).
43. Crowther 1994 (note 6).
44. United Nations 1999 (note 3).
45. Crowther 1994 (note 6).
46. United Nations 1999 (note 3).
47. Belk et al. (note 8).
48. United Nations 1999 (note 3).
49. Belk et al. (note 8).
50. Woodford 2004 (note 21).
51. Gorman and O'Leary 2007 (note 33).
52. Cocroft, W., and Thomas, R. 2003. *Cold War: Building for Nuclear Confrontation 1946–1989.* London: English Heritage.
53. Sullivan, W. 1999. Report on the Special IAU/COSPAR/UN Environmental Symposium: Preserving the Astronomical Sky (International Astronomical Union Symposium 196). http://www.iau.org/IAU/Activities/environment/s196rep.html.
54. UNESCO 2005 (note 4).

20 Spacecraft and Objects Left on Planetary Surfaces

Robert Gold

CONTENTS

INTRODUCTION

Humans have launched spacecraft that have flown past many celestial objects. Eight bodies have been orbited, and seven (Earth, the Moon, Venus, Mars, asteroids 433 Eros and 25143 Itokawa, and Saturn's moon Titan) have had landers. These landed missions have left their primary object, and often several secondary objects, on the surface for future rediscovery. These are important heritage objects; however, the physical state of these objects may also be the subject of future studies of "space weathering" because of the environments in which they reside.

In addition to the missions that have landed, two missions have collided with celestial bodies at high speeds. The Deep Impact mission sent a probe to collide with comet

Tempel 1 in order to create a plume for analysis by the primary spacecraft. The Galileo mission sent a probe into the Jovian atmosphere, and the Galileo spacecraft ended its long tour of the Jovian system by "crashing" into Jupiter as its final act. Because of the high energy release by the impacts of the Galileo spacecraft and the Deep Impact probe, there are probably no traces of these human-made objects to be found. However, objects that have soft-landed on other bodies and were left on these bodies remain outposts of humankind's exploration and are waiting to be found again.

PLANETARY ORBITERS AND LANDERS

Space objects that have been left on Earth, the Moon, and Mars are dealt with in other chapters. This chapter considers the small and unusual set of missions to asteroids, comets, moons, and Venus. Table 20.1 lists the first missions to orbit each planet or other type of celestial body. Only those spacecraft that have landed on celestial bodies have left artifacts for our future study. Table 20.2 lists the historic first landings on these bodies.

ASTEROID MISSIONS

NEAR EARTH ASTEROID RENDEZVOUS SPACECRAFT ON 433 EROS

The NASA Galileo mission was the first to pass close enough to an asteroid to resolve its shape as more than a point of light in a telescope. The images of 951 Gaspra and

TABLE 20.1
First Missions to Orbit Celestial Bodies

Date	Target	Mission	Nation
October 4, 1957	Earth	Sputnik 1	USSR
April 3, 1966	Moon	Luna 10	USSR
November 14, 1971	Mars	Mariner 9	United States
October 22, 1975	Venus	Venera 9	USSR
December 7, 1995	Jupiter	Galileo	United States
February 14, 2000	Asteroid Eros	NEAR	United States
July 1, 2004	Saturn	Cassini	United States

TABLE 20.2
First Missions to Land on Other Bodies

Date	Target	Mission	Nation
February 3, 1966	Moon	Luna 9	USSR
December 15, 1970	Venus	Venera 7	USSR
December 2, 1971	Mars	Mars 3	USSR
February 12, 2001	Asteroid Eros	NEAR	United States
January 14, 2005	Titan	Cassini-Huygens	ESA/United States

243 Ida showed them to be irregularly shaped bodies with a great deal of surface cratering. But the first really close observation of an asteroid came from the Near Earth Asteroid Rendezvous (NEAR) mission. NEAR was the first of NASA's Discovery line of planetary missions. The Discovery line was started in response to concerns that planetary missions had become so expensive that they could be flown only about once per decade. The Discovery-class missions are smaller missions with focused scientific goals that are financially capped to a small fraction of the cost of a flagship mission such as Cassini. NASA had envisioned a near-Earth asteroid rendezvous mission for a number of years. It was a contender for a future opportunity; however, it was judged a lesser priority than the flagship missions of Galileo to Jupiter and Cassini to Saturn. With the advent of the Discovery mission concept in 1992, the NEAR mission could be funded.

Development of the NEAR mission was assigned to the Johns Hopkins University Applied Physics Laboratory after a competition with the Jet Propulsion Laboratory for relevant mission concepts. NEAR was developed very rapidly, for an interplanetary spacecraft, and it was launched after a 27-month development period, on February 17, 1996. A mission to asteroid 433 Eros required a great deal of onboard fuel for the required deep-space maneuvers and orbit insertion. To keep the overall mass low, the designers kept the spacecraft relatively simple, with almost no moving parts. It weighed 787.8 kg at launch, and 320 kg of this was fuel (Figure 20.1).

NEAR imaged asteroid 253 Mathilde as it flew by on June 27, 1997, and imaged 433 Eros on December 23, 1998. However, because of a problem firing its main engine, NEAR passed by Eros and did not enter into orbit around it until February

FIGURE 20.1 Artist's conception of the NEAR spacecraft configuration in flight. The solar panels, high-gain antenna, large velocity-change thruster, and scientific instruments are all in fixed positions, with almost no moving parts. (Courtesy of Johns Hopkins University Applied Physics Laboratory.)

14, 2000. This was the first human-made object to orbit a highly nonspherical body. Eros is 33 × 13 × 13 km, about the size of Manhattan Island. NEAR used its five instruments to conduct a detailed study of Eros during the following year.

Staying close to such an oddly shaped body required many orbital corrections, and toward the end of the year in Eros orbit, the operations team had to confront the fact that the spacecraft was running out of fuel. The team, with NASA Headquarters approval, decided to make a "controlled descent" to the surface of Eros to get some high-resolution images of the surface. The phrase "controlled descent" was chosen because it sounded much better than "crash." The guidance and control team wanted to make a full "closed-loop" descent. Closed-loop means that it would have used the onboard laser altimeter to sense the altitude above the surface and use that information to continuously control the thrusters and ensure a relatively soft touchdown. However, the software development and testing for this closed-loop approach was ruled to be too costly and time consuming. Therefore, an open-loop descent was chosen in which the spacecraft would free-fall toward the surface and the thrusters would be fired at three predetermined times based on the predicted rate of descent. The open-loop descent had a much lower probability of arriving at the surface at a low rate of speed.

The landing occurred on February 12, 2001. The plan was to take a number of images on the way down, but the no-moving-parts design meant that the spacecraft had to reorient itself to see the surface and to point its antenna toward Earth. Therefore, on the way down, the spacecraft would take an image, and then the whole spacecraft would have to be turned to point its antenna toward Earth; this took minutes. Then it took another a few minutes to transmit the image, and the spacecraft would take more time to turn back to face Eros and take another image. This process

FIGURE 20.2 The final four images sent by the NEAR spacecraft during its descent to the surface of 433 Eros. The last image, on the left, was only partially transmitted before landing. It was taken at an altitude of about 150 m and has a resolution of 1 cm per pixel. (Courtesy of Johns Hopkins University Applied Physics Laboratory.)

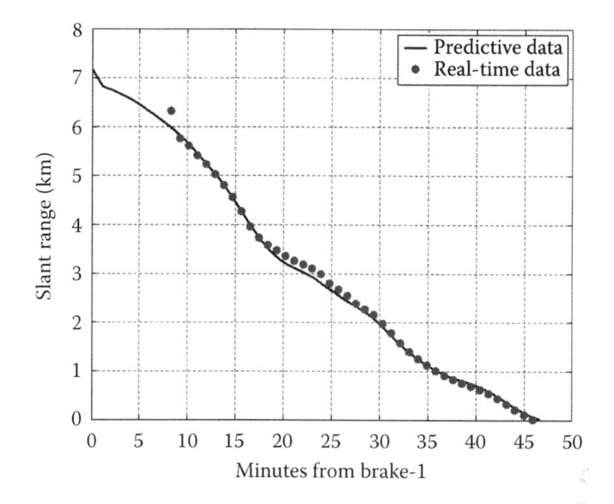

FIGURE 20.3 The predicted and actual altitude profile of the NEAR spacecraft while landing on 433 Eros. The solid line predictions dip at the times of the three planned thruster firings to slow the descent. The dots of actual data from the laser altimeter show excellent agreement. (Courtesy of Johns Hopkins University Applied Physics Laboratory.)

was repeated all the way down, and the final image of the surface, taken from an altitude of about 150 m, has a resolution of about 1 cm per pixel (see the left-hand segment of Figure 20.2). The altitude profile of the open-loop descent matched the predictions almost exactly (Figure 20.3), and NEAR touched down with a velocity of about 1.5 m/s (~3.4 mi/h). Because of the low speed at touchdown, the spacecraft was not damaged, and it remained operational for about 2 weeks until it was turned off and placed in hibernation by ground commands. The hibernation status meant that all of the subsystems with power switches were turned off. Some essential systems, such as the solar panels and the radio receiver, were not switchable and could be used to revive the spacecraft if the conditions were right.

The 2 weeks of operations on the surface of Eros were vital to understanding the composition of the asteroid from data gathered by the gamma-ray instrument and the magnetometer. Figure 20.4 is an artist's conception of NEAR on Eros. The approximate spacecraft orientation in the drawing was confirmed by the onboard sensors and the solar panel power output following the landing.

NEAR landed near the pole of Eros when it was in continuous sunlight, and the approximate landing site is shown in Figure 20.5. By the end of March, 2001 the landing site was experiencing sunsets every 5.3 hours. From August through mid-November, the spacecraft was in continuous darkness and very cold. Then came another series of sunrises and sunsets, until it was once again in full sunlight in late August of 2002. Planetary spacecraft are typically designed to withstand temperatures between –35°C and +65°C, so the extreme thermal cycles to very cold temperatures that NEAR experienced during the light and dark seasons on Eros probably damaged a number of its subsystems. There was an attempt to contact the NEAR spacecraft when it returned to full sunlight, but there was no response. Therefore,

FIGURE 20.4 Artist's conception of the NEAR spacecraft on the surface of Eros. (Courtesy of Johns Hopkins University Applied Physics Laboratory.)

NEAR remains on Eros to be rediscovered by a future mission. The empty spacecraft on Eros has a mass of 468.1 kg.

Hyabusa Landing Target on 25143 Itokawa

On May 9, 2003, the Japan Aerospace Exploration Agency launched a very ambitious mission to land a spacecraft on an asteroid, take a sample of the surface material, and bring the sample back to Earth. The mission, originally known as MUSES-C and later renamed Hyabusa, was only the second interplanetary spacecraft to use solar-electric propulsion. It was designed to fly to asteroid 25143 Itokawa and use a ballistic sampling technique to obtain a small sample of the surface rocks.

Itokawa is a small, irregularly shaped body, about 500 m long. Objects this small do not have a gravitational sphere of influence large enough to provide a stable orbit for a spacecraft. Therefore, it was planned that the spacecraft would "station keep" with the asteroid: it would put itself in a heliocentric orbit almost identical to that of the asteroid and stand off from the asteroid a few kilometers. The spacecraft would make small corrections to its orbit to maintain its relative position to Itokawa. The Hyabusa mission planned to leave two objects on Itokawa. One was a microrover named *Minerva* that would actually hop around the asteroid. The other was a small target marker that would be used to assist in navigating the spacecraft to the landing site.

The mission experienced a number of problems en route to Itokawa. In 2003, a large solar flare damaged the spacecraft solar cells. This reduced the electrical power available for the ion engines and delayed the arrival of the *Hyabusa* spacecraft

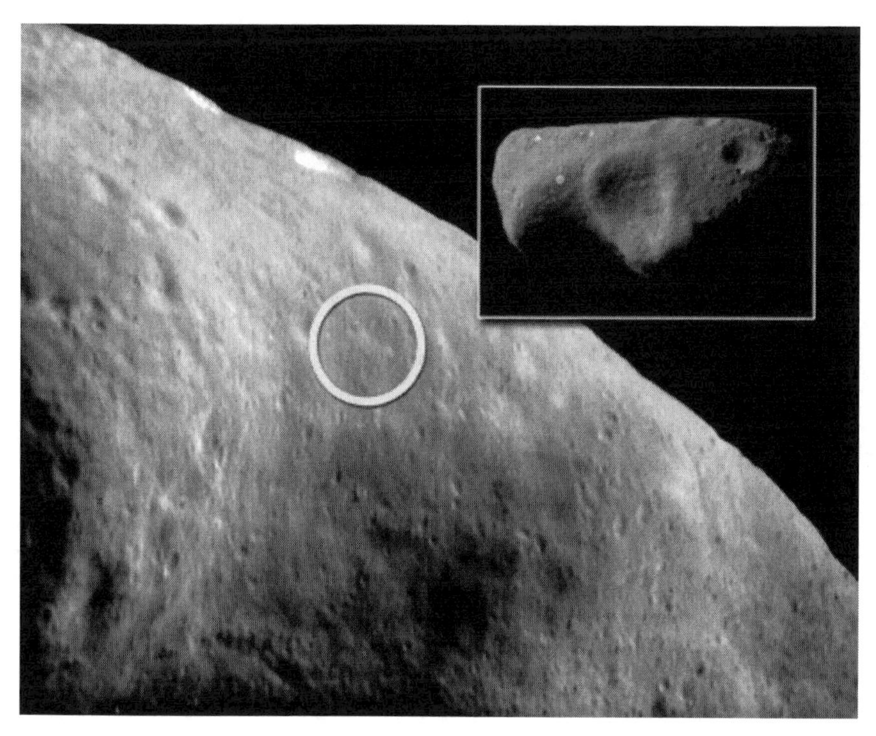

FIGURE 20.5 The location of the NEAR spacecraft on 433 Eros. The dot on the inset image shows the location of the landing zone. The circle in the large image encloses the estimates of the final position of NEAR. (Courtesy of Johns Hopkins University Applied Physics Laboratory.)

at Itokawa by 3 months to September 2005. However, orbital mechanics required that the departure date from Itokawa remain unchanged. This significantly reduced the planned time in the vicinity of the asteroid.

Hyabusa reached Itokawa in September 2005 (Figure 20.6). Additional problems plagued the mission. In 2005, two of the momentum wheels that controlled the pointing of the spacecraft failed. After that, attitude changes had to be accomplished with thrusters. On November 12, 2005, the *Minerva* microrover was released, but the spacecraft was actually drifting away from the asteroid at the time, rather than toward it as the mission controllers believed. Therefore, *Minerva* (Figure 20.7) never reached Itokawa, and it is still drifting in its own heliocentric orbit.

After an earlier landing attempt was aborted, the *Hyabusa* spacecraft descended to the surface of Itokawa in the relatively smooth region known as the Muses Sea to take its sample. It released the target marker (Figure 20.8), which is a spherical object with a surface pattern that would be recognized in images of the surface by the guidance software and used to keep the spacecraft from drifting relative to the landing site. Figure 20.9 is an image taken by the *Hyabusa* spacecraft during the descent. The right-hand panel shows the shadow of the spacecraft on the asteroid, and the left-hand panel is a close-up that clearly defines the outline of the spacecraft

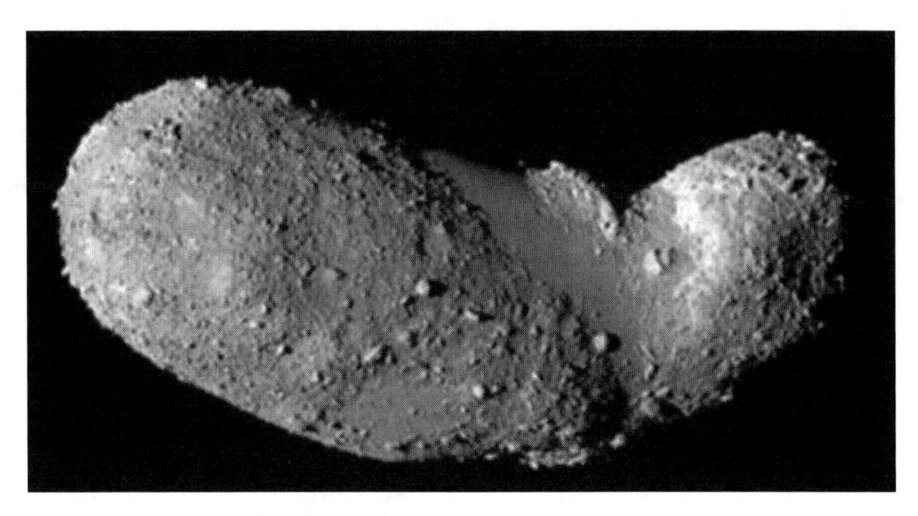

FIGURE 20.6 Image of asteroid 25143 Itokawa taken by the *Hyabusa* spacecraft. (Courtesy of Japan Aerospace Exploration Agency.)

and the bright target marker with a black circle drawn around it. The target marker remains on *Hyabusa*.

Unfortunately, later analysis showed that this sampling attempt had also failed, but the target marker remains on the surface of Itokawa.

Deep Impact Mission to a Comet

The NASA Deep Impact mission consisted of a primary spacecraft and a probe (impactor) that was carried as a daughter spacecraft. The impactor was released to collide with comet Tempel 1 at very high velocity. The goal of the mission was to learn more about the composition of comets by observing the impact ejecta and the impact crater. The impact occurred on July 4, 2005, when the 370-kg impactor hit Tempel 1 at a speed of 10.3 km/s (23,040 mi/h). The large ejecta plume that resulted was imaged by the primary spacecraft and by other telescopes in space and on the ground. At these very high impact speeds, the energy released during the impact (approximately 20B W/s) almost certainly vaporized all of the impactor and a great deal of cometary mass as well. Therefore, a large crater is probably the only feature of the Deep Impact mission left on Tempel 1.

Cassini-Huygens Missions to Saturn's Moon Titan

There have been several missions to study planetary moons of Mars, Jupiter, and Saturn. However, only the *Huygens* probe on the Cassini-Huygens mission has actually landed on one of these bodies. The Cassini mission was engineered to remotely study Saturn and its rings and moons. The *Huygens* probe, which was carried to Saturn on the *Cassini* spacecraft, was designed to penetrate through the very thick atmosphere of Saturn's moon Titan and land on its surface.

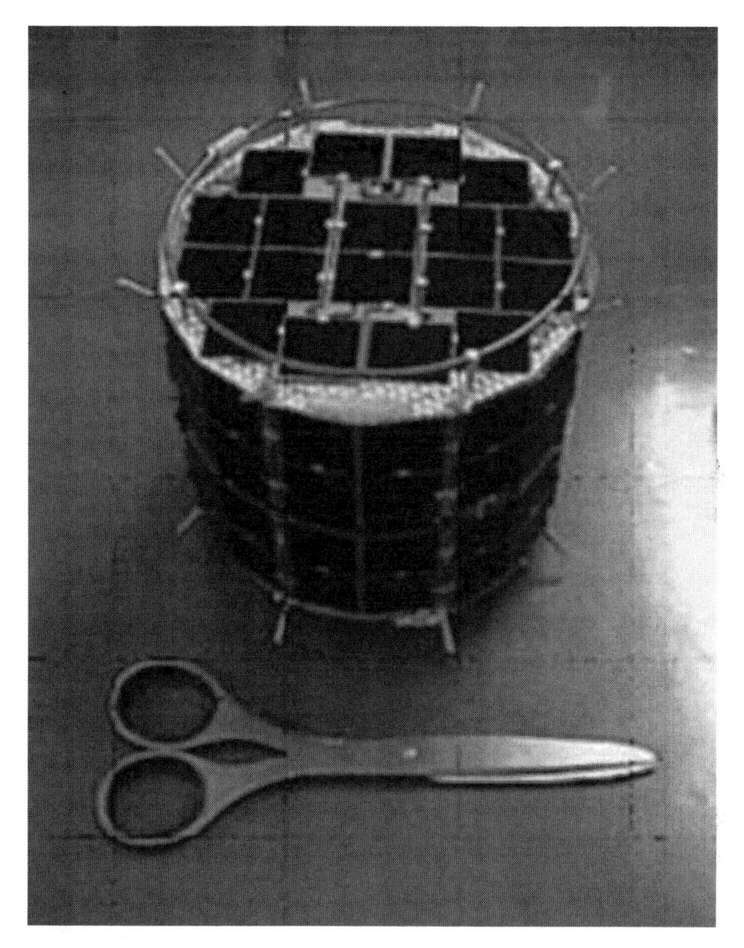

FIGURE 20.7 The *Minerva* microrover was intended to hop about on asteroid 25143 Itokawa. It did not reach the surface and remains in heliocentric orbit. (Courtesy of Japan Aerospace Exploration Agency.)

The *Cassini* spacecraft, built by the Jet Propulsion Laboratory, is one of the largest interplanetary spacecraft ever built. It is 6.7 m high and more than 4 m wide, with an 11 m boom extending out to carry some of its scientific instruments. The 2,125 kg spacecraft carried the 349-kg *Huygens* probe to Saturn and targeted it at Titan to study both the atmosphere and surface of this moon that has prebiotic compounds in its atmosphere. *Cassini-Huygens* was launched on October 15, 1997, and entered Saturn orbit on July 1, 2004. Prior to this time, observations of Titan had revealed only a body enshrouded by a thick haze. However, during the initial flyby of Titan, the *Cassini* onboard imaging radar showed pieces of the surface with varied topography and small altitude variations of less than 50 m.

The *Huygens* probe was built for the European Space Agency by Aerospatiale, now absorbed as part of Thales Alenia Space. On December 25, 2004, *Cassini*

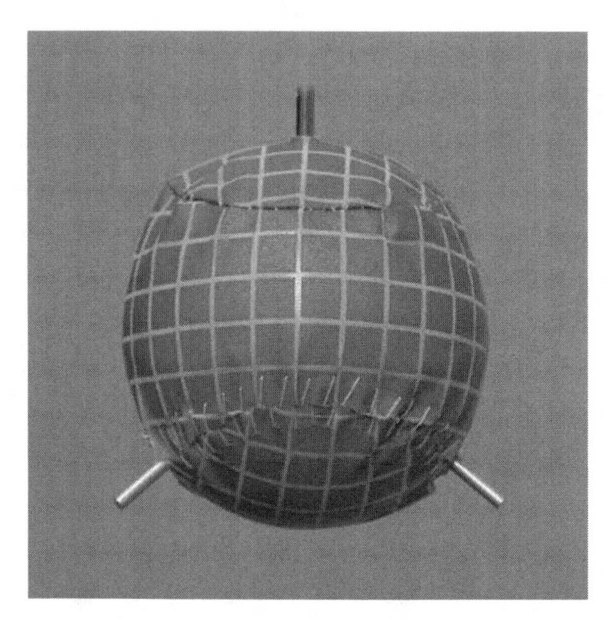

FIGURE 20.8 This 10 cm landing marker containing 888,000 names from 149 countries was placed on the surface of Itokawa to help guide the landing of the *Hyabusa* spacecraft. (Courtesy of Japan Aerospace Exploration Agency.)

released the *Huygens* probe, which landed on Titan on January 14, 2005, at latitude 10.2936° S and longitude 163.1775° E. The full probe assembly released by *Cassini* consisted of a 1.3 m diameter probe craft sandwiched between a 2.7 m diameter front heat shield and a backshell (Figure 20.10). The heat shield protected the probe

FIGURE 20.9 The *Hyabusa* spacecraft cast a shadow of itself during the landing on asteroid 25143 Itokawa. The panel on the left is a close-up of the shadow, with the bright reflection of the landing target marker seen in the black circle. (Courtesy of Japan Aerospace Exploration Agency.)

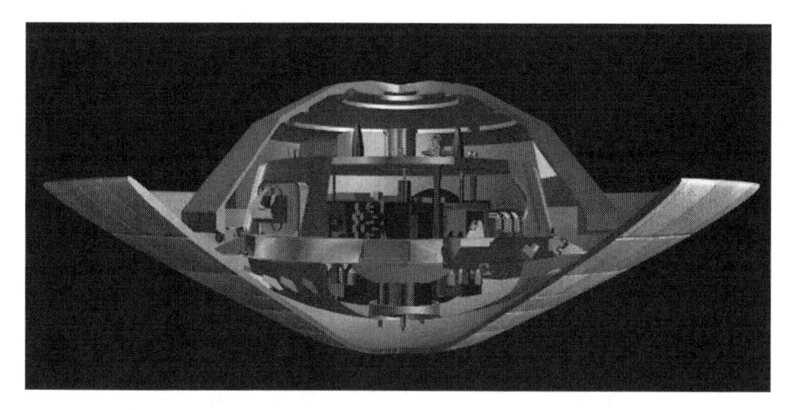

FIGURE 20.10 Artist's cross-section of the *Huygens* probe package as released from the *Cassini* spacecraft, with the heat shield and backshell enclosing the descent module. (Courtesy of European Space Agency.)

through the high-dynamic-heating portion of its entry into the Titan atmosphere. As the probe slowed in the atmosphere, the heat shield and backshell were ejected, and a parachute was deployed from the probe to further slow its descent to the surface. Figure 20.11 is a picture of a full-scale replica of the probe.

The probe took images and used its other instruments on the way down, and when it reached the surface, it probably looked like the artist's rendition in Figure 20.12. The figure does show the parachute lying near the probe, but it does not include the other objects left on the surface of Titan, the heat shield and the backshell. Since the heat shield and backshell were ejected well above the surface, they may reside quite far from the probe and parachute.

FIGURE 20.11 This replica of the 1.3 m diameter descent module shows the many sensors on the *Huygens* probe. (Courtesy of European Space Agency.)

FIGURE 20.12 Artist's rendition of the *Huygens* probe on the surface of Titan. (Courtesy of European Space Agency.)

Venus Missions

In the early years of planetary exploration, the United States concentrated most of its efforts on Mars, while the Soviet Union sent many missions to Venus. Venus has a very hostile environment and in many ways it is far more challenging than Mars. The atmosphere of Venus is very dense and is composed of 96.5% CO_2 and 3.5% N_2. The pressure at the surface is about 90 atmospheres (9.12M Pa; 1,322 psi) and the temperature is about 480°C (896°F). These high pressures and temperatures have severely limited the operational lifetime of all landed missions to Venus. Table 20.3 lists all of the missions and their primary objects sent to the surface of Venus by the USSR and the United States. Together, there are 21 primary objects on the surface. There are also numerous secondary objects from heat shields to parachutes. Venus, with these twenty-one primary objects, is second only to the Moon in the number of items that now reside on its surface. However, the high temperature and pressure may have significantly modified many of the objects so that they probably no longer appear as they did when launched.

Venera 3 through *Venera 8*

The *Venera 3* through *Venera 8* spacecraft launched by the Soviet Union to Venus were all of similar design. *Venera 3* was a cylindrical spacecraft with a launch mass of 958 kg, and it carried a 0.9 m diameter pressurized spherical descent capsule weighing 337 kg (Figure 20.13). The capsule descended to the surface on a parachute, and it was designed to measure the temperature, pressure, and composition of the atmosphere on the way down and send the data to the *Venera* spacecraft for relay back to Earth. Unfortunately, communications with the capsule failed, and no

TABLE 20.3
All Missions Sent to the Surface of Venus

Body	Nation	Mission or Object	Date Placed/Last Contact	Latitude	Longitude
Venus	USSR	*Venera* 3	March 1, 1966	Unknown	Unknown
		Venera 4 (capsule)	October 18, 1967	19° N	38° E
		Venera 5 (capsule)	May 16, 1969	3° S	18° E
		Venera 6 (capsule)	May 17, 1969	5° S	23° E
		Venera 7 (capsule)	December 15, 1970	5° S	9° W
		Venera 8 (capsule)	July 22, 1972	10° S	25° W
		Venera 9 (capsule)	October 20, 1975	32° N	69° W
		Venera 10 (capsule)	October 23, 1975	16° N	69° W
		Venera 11 lander	December 25, 1978	14° S	61° W
		Venera 12 lander	December 21, 1978	7° S	66° W
		Venera 13 lander	March 1, 1982	7.5° S	57° W
		Venera 14 lander	March 5, 1982	13.25° S	50° W
		Vega 1 descent unit	June 11, 1985	7.5° N	177.7° E
		Vega 1 balloon gondola	June 11, 1985	Unknown	Unknown
		Vega 2 descent unit	June 15, 1985	8.5° S	164.5° E
		Vega 2 balloon gondola	June 15, 1985	Unknown	Unknown
	United States	*Pioneer Venus Bus*	December 9, 1978	37.9° S	69.1° W
		Pioneer Venus Large Probe	December 9, 1978	4.4° S	56° W
		Pioneer Venus Small Probe—North	December 9, 1978	59.3° N	4.8° E
		Pioneer Venus Small Probe—Day	December 9, 1978	31.3° S	43° W
		Pioneer Venus Small Probe—Night	December 9, 1978	28.7° S	56.7° E

atmospheric data were returned. *Venera 3* landed on the surface of Venus on March 1, 1966. Although no data were returned, *Venera 3* is a very important historical object since it was the first human-made object to land on another planet.

Venera 4 was a much more successful mission than *Venera 3*. It arrived at Venus on October 18, 1967, and made the first ever in situ analysis of the atmosphere of another planet. It carried thermometers, pressure sensors, gas analyzers, and a radio altimeter. *Venera 4* provided the first direct measurement of the heat within the atmosphere and at the surface of Venus. The descent capsule (Figure 20.14) floated down on its parachute and returned data within the atmosphere for 94 minutes, to

FIGURE 20.13 The *Venera 3* spacecraft, with the spherical descent module at the bottom of the stack. (Courtesy of NASA National Space Science Data Center.)

FIGURE 20.14 The *Venera 4* descent module. (Courtesy of NASA National Space Science Data Center.)

an altitude of 24.96 km, when communications failed. Although there were no data returned during the final portion of the descent, the estimated location of the landing site is near 19° N and 38° E.

Venera 5 and *Venera 6* were a pair of nearly identical spacecraft launched separately in 1969. With the knowledge gained from the *Venera 4* results, *Venera 5* and *Venera 6* were fitted with strengthened descent capsules weighing 405 kg each. They also had smaller parachutes that would let them descend more quickly through the very dense atmosphere, allowing them to reach their "crush" depth before they ran out of power and lost communications. During the descent, the instruments confirmed that the atmosphere below the cloud layer was 97% CO_2, and they measured the N_2 and H_2O content. They landed near the Venus equator, as listed in Table 20.3.

Venera 7, launched in 1970, became the first spacecraft to transmit data back from the surface of another planet. The descent capsule (Figure 20.15) reached the surface after 35 minutes and continued to transmit for another 23 minutes. This highly strengthened descent capsule weighed 495 kg and had very limited room for scientific instruments. It reported surface temperatures between 457°C and 474°C. Doppler measurements of the signal shortly before touchdown showed that the surface winds were less than 2.5 m/s (5.6 mi/h).

Since *Venera 7* had reached the surface and shown that the surface pressure was about 90 atmospheres, it was realized that the *Venera 7* pressure vessel was overdesigned. Therefore, the walls of *Venera 8* could be thinned somewhat, to allow more mass and volume to accommodate scientific instruments. *Venera 8* reached Venus on

FIGURE 20.15 The *Venera 7* descent capsule was the first to transmit data from the surface of another planet. (Courtesy of NASA National Space Science Data Center.)

July 22, 1972, with a 495-kg descent capsule that contained altimeter, temperature, pressure, and light sensors, a gamma-ray spectrometer, and gas analyzers. *Venera 8* was the first of the Venera craft to land on the daylight side of Venus. During the descent through the atmosphere, its light sensors showed that there was a bottom to the cloud deck and that the atmosphere below about 35 km altitude was clear. When the capsule reached the surface, it continued to operate for another 50 minutes and reported a temperature of 470° C and a pressure of 90 atmospheres. The gamma-ray spectrometer measured the ratios of radioactive potassium, thorium, and uranium and reported that they were similar to those of granite rocks on Earth.

Venera 9 and *Venera 10*

The *Venera 9* and *Venera 10* missions used a new class of spacecraft that was much larger than the previous Venera craft. They had both an orbiter and a lander. The lander entry sphere of this new spacecraft was 1,560 kg, and the surface payload was 660 kg. Venera 9 landed on October 22, 1975, and was the first spacecraft to return images of the surface of Venus. Figure 20.16 shows the enlarged lander that includes a spherical pressure vessel evolved from the descent capsules of the earlier Venera craft, with a large drag disk to slow its descent and a cushioned circular landing ring to absorb the shock of the touchdown on the surface.

FIGURE 20.16 The *Venera 9* lander was much larger than the previous descent modules. The large disc provided drag during the descent through the dense atmosphere of Venus, and the ring at the bottom cushioned its landing. (Courtesy of NASA National Space Science Data Center.)

Venera 9 and *Venera 10* landed in daylight and used a scanning photometer system to take panoramic images of the surface at the base of the lander. Each spacecraft had two cameras, in order to take a full 360° panorama. However, one of the camera covers on each spacecraft failed to open, producing a 180° panoramic image from each lander. Both landers touched down in regions with relatively unweathered rocks that were about 30 to 40 cm in diameter. The *Venera 10* image shows somewhat smoother rock surfaces than that taken by *Venera 9*. Figure 20.17 is the officially released picture from *Venera 9*. A recent reconstruction of this image from the original data tapes using modern image processing techniques[1] has improved the gray scale and the apparent resolution of the image in the region near the horizon (see Chapter 15 of this volume), but the essential picture of a rock-strewn plane remains the same.

Venera 11 and *Venera 12*

Venera 11 and *Venera 12* were identical spacecraft with landers of a design similar to *Venera 9* and *Venera 10*. The carrier spacecraft were not orbiters, and they flew past Venus as the landers descended through the atmosphere. Therefore, they could receive data from the landers for only a limited time. The landers touched down on Venus on December 25, 1978, carrying many instruments, including color cameras. However, all the camera covers on the two landers failed to open, and no images were returned. Their landing sites are listed in Table 20.3. The landers did report evidence of lightning and thunder in the Venusian atmosphere and the existence of carbon monoxide at low altitudes.

Venera 13 and *Venera 14*

Venera 13 and *Venera 14* were identical spacecraft, similar in design to *Venera 9* through *Venera 12*, that were launched in October and early November 1981 (Figure 20.18). They had a large variety of instrumentation that was updated from the earlier Venera missions. They arrived at Venus on March 1 and March 5 of 1982 carrying 15 types of instrumentation in the descent modules for studying both the atmosphere during the descent and the surface structure and composition after landing. These large descent modules, weighing 760 kg each, used parachutes to slow

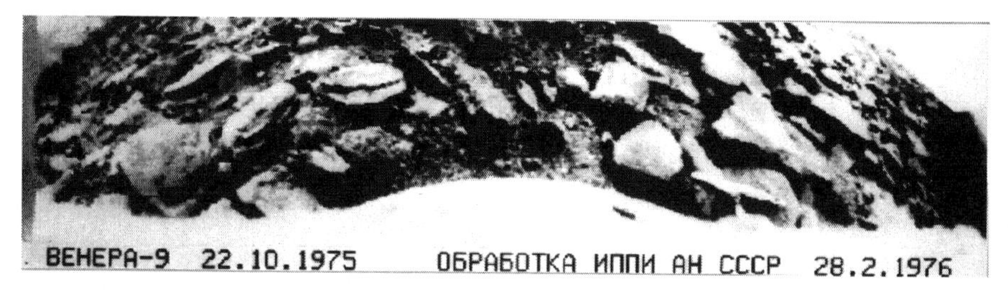

FIGURE 20.17 The first image of the surface of another planet, taken by *Venera 9*. (Courtesy of NASA National Space Science Data Center.)

FIGURE 20.18 The 760-kg *Venera 13* descent module was an evolution of the design first used for *Venera 9*. (Courtesy of NASA National Space Science Data Center.)

their initial descent and large atmospheric drag disks to control their descent in the lower atmosphere. The parachutes were released at an altitude of 50 km, and it is not known where they eventually landed. The descent module landing sites are listed in Table 20.3.

The thermal management system of the landers included a provision for precooling the lander prior to atmospheric entry; there was thick insulation in the pressure vessel, and a liquid circulation system distributed the heat uniformly. The thermal control system and the batteries were expected to enable the landers to survive on the Venusian surface for at least 32 minutes after touchdown. *Venera 13* actually operated on the surface for 127 minutes. The surface conditions reported by the *Venera 13* and *Venera 14* landers (457°C and 465°C; 89 and 94 atmospheres pressure) were similar to those reported by previous missions. Both landers had a pair of cameras that returned the first color photographs from the Venusian surface. Figure 20.19 is the 170° panoramic image that was made from separate exposures through blue, green, and red filters by the two cameras on *Venera 13*. The apparent orange tint to the rocks and the atmosphere off in the distance is the result of atmospheric absorption of blue wavelengths and Rayleigh scattering of sunlight in the thick atmosphere. After landing, *Venera 13* extended an arm with a drill that took a subsurface sample and delivered it for analysis by an x-ray fluorescence spectrometer. The atomic composition was similar to that of basaltic rocks on Earth.

Venera 14 was the last of the Venera series to land on Venus. There were two more missions in the series (Venera 15 and Venera 16), but they were orbital missions designed to map the planet with synthetic aperture radars.

FIGURE 20.19 *Venera 13* took these first color pictures (now in black and white) of the surface of Venus. The orange tint results from Rayleigh scattering and atmospheric absorption of the blue light in the visible spectrum. (Courtesy of NASA National Space Science Data Center.)

VEGA 1 AND *VEGA 2*

Following the cancellation of the American mission to Halley's comet, the Soviet Union modified its next planned Venera mission to fly by Venus and continue on to observe comet 1P/Halley. The mission name was changed to Vega, which combined the Russian names for Venus and Halley's comet. Once again, the Soviet Union launched two identical spacecraft to help ensure mission success. The basic spacecraft and lander designs were a development of the *Venera 9* through *Venera 16* spacecraft. The Vega missions were more international than the previous Venera series, with significant contributions from Austria, Bulgaria, Hungary, the German Democratic Republic (East Germany), Poland, Czechoslovakia, France, and the Federal Republic of Germany (West Germany).

The Venus portion of the Vega missions had a lander of a similar design to *Venera 13*, plus a balloon that was released to drift through the atmosphere of Venus. The two spacecraft were launched a few days apart in December 1984, and they arrived near Venus in June 1985. As each spacecraft neared Venus, its lander separated from the spacecraft 2 days before entry into the Venus atmosphere. A parachute slowed it down during the initial descent. At about 63 km altitude, the initial parachute, which was attached to the cap of the lander, was released with the cap. A second parachute then deployed, and it pulled the balloon out of the lander. About 100 s later, the balloon was inflated, and the parachute and inflation system were then jettisoned. At about 50 km altitude, ballast was jettisoned, and the balloon was able to float to its equilibrium altitude of 53.6 km. The lander continued to descend to the surface, while the balloon then drifted on the zonal winds at its equilibrium altitude, which is in the middle of most active of the three cloud layers at Venus. Although the surface of Venus is about 480°C and 90 atmospheres, at the balloon's altitude, conditions were about 0.7 atmospheres and room temperature . The balloon was 3.4 m in diameter and supported 25 kg of attached mass, 5 kg of which was the science payload. The scientific instruments measured the pressure, temperature, vertical wind velocity, and the density of aerosols in the clouds.

The balloon entered the atmosphere on the night side of Venus and drifted westward 8,500 km at an average speed of 69 m/s (154 mi/h) for nearly 2 days. At that point it crossed the terminator into daylight, and the added solar heating greatly expanded the balloon until it burst. Table 20.3 lists the position of the lander on the surface, but the final locations of the multiple parachutes, descent module cap, balloon, and balloon gondola are not known.

Pioneer Venus Multiprobes

Pioneer Venus was a NASA mission to study the solar wind in the vicinity of Venus, map the surface with radar, and study the upper atmosphere and ionosphere of Venus. The mission was launched in two parts, a Venus orbiter in May 1978 and the Multiprobes in August 1978. The orbiter carried the radar and a variety of instruments to study the atmosphere remotely. It continued to operate for 14 years, until its orbit decayed and it entered the atmosphere of Venus on October 8, 1992.

The Pioneer Venus Multiprobes consisted of the carrier spacecraft, referred to as the Bus, and four separates probes to make measurements as they descended through the atmosphere. Although none of them was designed for soft landing, all five objects entered the atmosphere on December 9, 1978, and descended to the surface of Venus. The probes were of two types: one Large Probe, and three small probes that were targeted to different areas of Venus.

The Bus, a cylindrical spacecraft 2.5 m in diameter and weighing 290 kg, carried the probes to Venus and released them. It was designed to enter the Venusian atmosphere at a low angle and measure the upper atmosphere. It had no heat shield, and it transmitted data until it was destroyed by the high temperatures generated by atmospheric friction. It had both an ion mass spectrometer and a neutral mass spectrometer that provided the only direct measurements of the upper atmosphere. The probes did not take any data until they were at much lower altitudes. The Bus continued to transmit down to an altitude of 110 km before signal was lost. Because there was no signal down to low altitudes, the latitude and longitude for the Bus position in Table 20.3 are approximate.

The 315-kg Large Probe carried seven instruments to study the composition and structure of the atmosphere and clouds. It had an approximately 1.5 m diameter conical heat shield to protect it during initial entry into the atmosphere and a central 0.73 m spherical titanium pressure vessel that contained all of the electronics. The heat shield and rear protection cover were jettisoned at 47 km altitude, and a parachute was deployed to slow its descent.

The three identical small probes weighed 90 kg each and were 0.8 m in diameter. Each had a heat shield and a spherical pressure vessel, but unlike the Large Probe, the heat shield did not separate and there was no parachute. Each small probe measured temperature, pressure, acceleration, cloud particulates, and solar illumination. The North Probe was targeted to about 60° N latitude on the day side of the planet. The other two small probes, the Night Probe and the Day Probe, measured atmospheric differences with and without solar illumination. The Day Probe was the only one to continue sending data all the way to the surface. It continued to operate for about an hour after touchdown.

CONCLUSION: OBJECTS FOR REDISCOVERY

The human-made objects left on other planets remain opportunities for rediscovery on future missions. They are both historically significant and of strong scientific interest. Space weathering is the general phrase used to describe the long-term effects of the space environment on physical objects. For the NEAR spacecraft, this includes exposure to vacuum; very large thermal swings every 5.3 hours during most of the Eros year; exposure to high-energy cosmic rays, the solar wind, and strong fluxes of solar energetic particles from solar flares; and interplanetary shock waves. It also includes any redistribution of regolith particles on the surface of the asteroid from seismic disturbances caused by impacts around the asteroid. This is a very different environment from the extreme but relatively constant temperature and pressure for objects left on the surface of Venus. The Apollo astronauts visited an old *Surveyor* spacecraft on the Moon to examine its deterioration in the lunar environment, and the Long Duration Exposure Facility was returned to Earth after years in low Earth orbit for study of its deterioration; likewise, humans will learn a great deal of vital information about building for long-term planetary outposts by examining some of these great historic objects on other planetary surfaces. I expect that there will eventually be a debate as to whether some of the objects should be returned to Earth for detailed study or should be left undisturbed because of their historical significance.

REFERENCES

1. http://www.mentallandscape.com/V_DigitalImages.htm.

21 Culture of Apollo: A Catalog of Manned Exploration of the Moon

P.J. Capelotti

CONTENTS

INTRODUCTION

Of the thirty-eight expeditions to the Moon that achieved impact on the surface, twenty-three were launched by the United States, and of these, eight were manned Apollo missions (Figure 21.1). The archaeological evidence of manned missions to the Moon is exclusively American.

From 1969 to 1972, the Apollo missions left behind on the Moon twenty-three large-scale artifacts. These large-scale or main-body artifacts fall into five categories: lunar module ascent stages (Figure 21.2), lunar module descent stages (Figure 21.3), Saturn V third-stage rockets (S-IVB) (Figure 21.4), subsatellite science probes (Figure 21.5), and lunar rovers (Figure 21.6).

Around these main-body archaeological assemblages, or collections of similar technologies, smaller artifacts include scientific instrument packages and their power generators, personal artifacts, and the only piece of artwork brought to the Moon and left there. Other evidence of human exploration includes footprints and rover tread pathways created by astronauts and the vehicles they brought to the Moon to extend their exploring range beyond walking distance from the lunar module base camps.

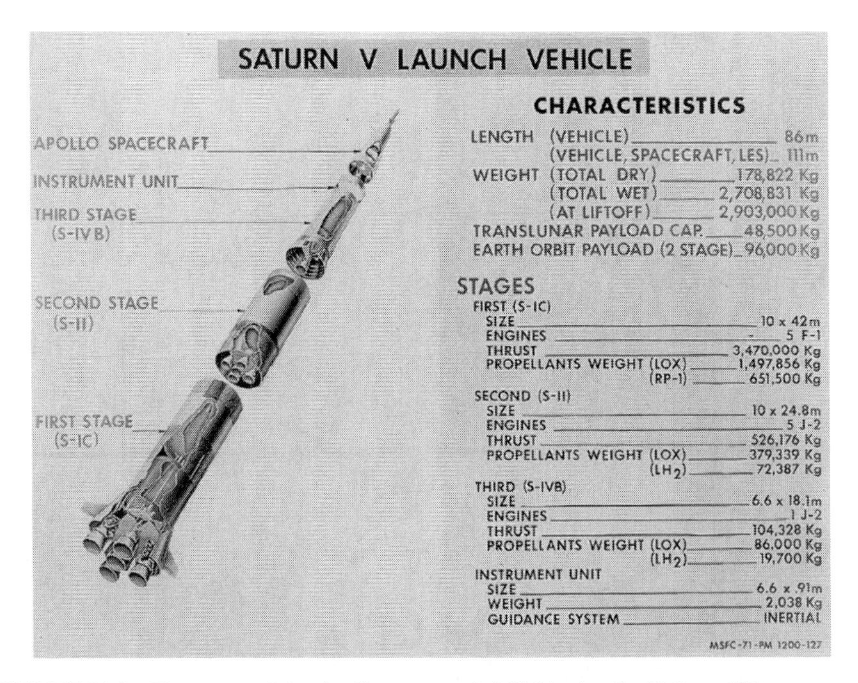

FIGURE 21.1 The stages of the Apollo spacecraft (NASA: *Apollo 10 Press Kit*).

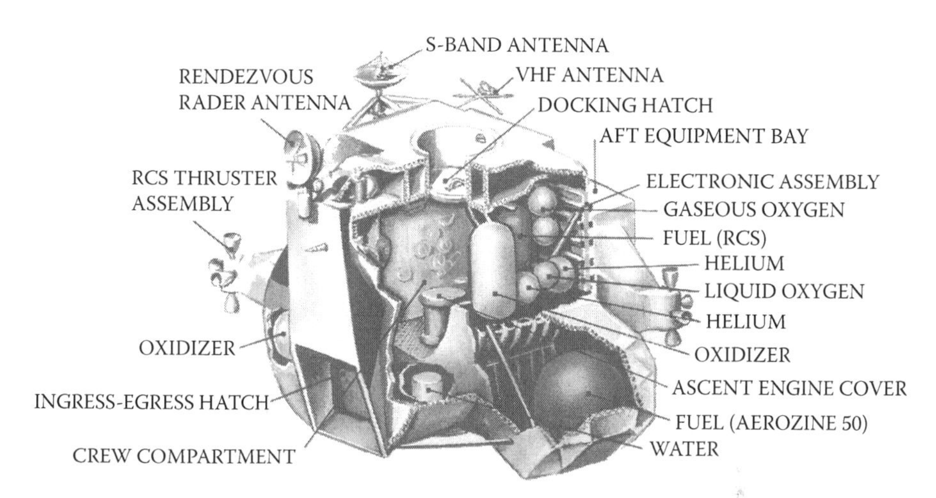

FIGURE 21.2 The lunar module, showing the ascent or upper stage attached to the descent stage (NASA: *Apollo 10 Press Kit*).

These assemblages transitioned at different moments from what the archaeologist Michael Schiffer calls their *systemic context* (operating as part of an active human enterprise), to enter the realm of archaeology, where they became relics embedded in the natural landscape. There they found in Schiffer's terms, their *archaeological context*.[1] Most of the areas of the sites made this transition as soon as the humans vacated them. Others, like the Apollo Lunar Scientific Experiment Package (ALSEP)

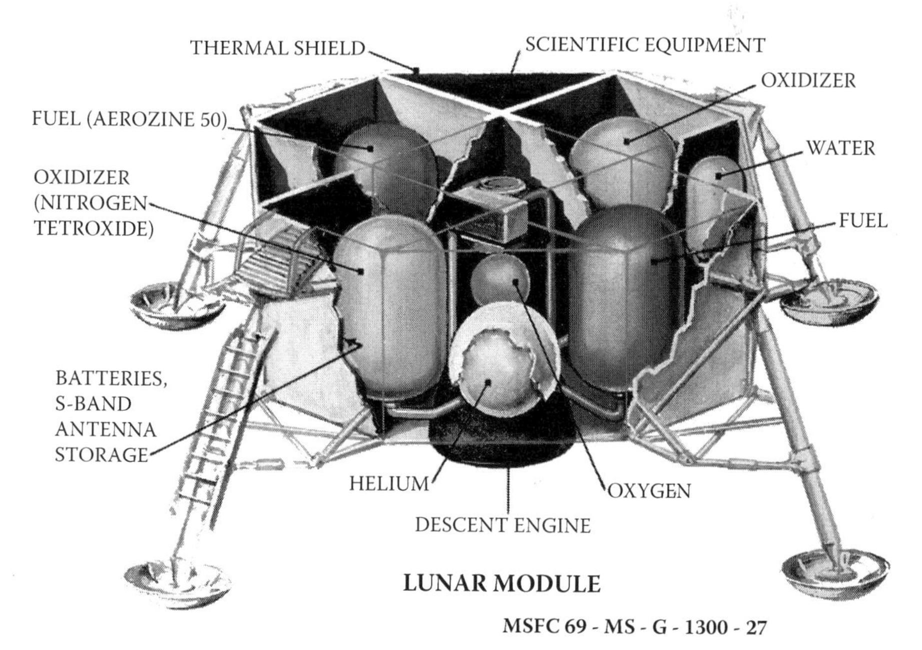

FIGURE 21.3 The lunar module descent stage (NASA: *Apollo 10 Press Kit*).

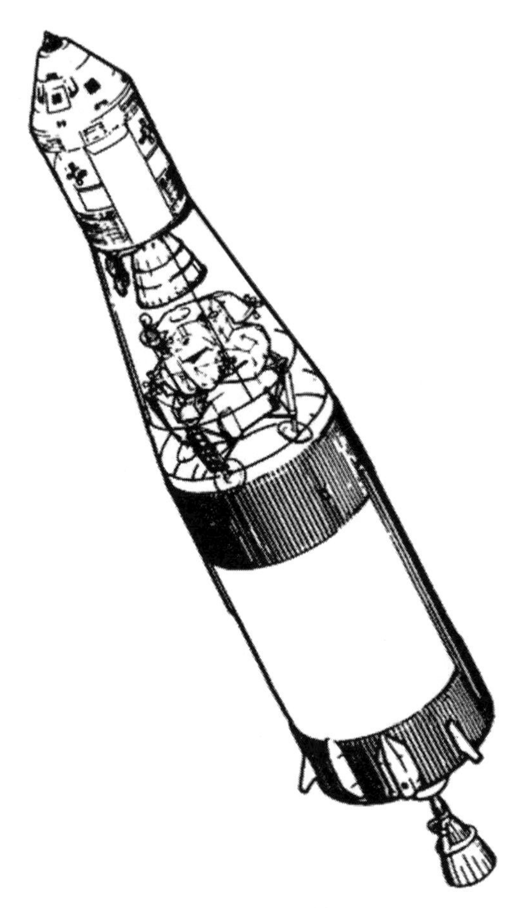

FIGURE 21.4 The Saturn V third-stage S-IVB rocket consisted of the entire area below the enclosed lunar module (NASA: *Apollo 10 Press Kit*).

and its power generators, remained in systemic context until the generators were shut down by commands sent from Earth several years later.

Taken together, the several sites of manned exploration of the Moon can be thought of in archaeological terms as a culture. In this case, they are the *culture of Apollo*. This chapter explores the historical archaeology of these assemblages in order to create a descriptive catalog of the culture of Apollo, offers some thoughts on how cultural resource protections might be applied to them, and places them within a theoretical framework for aerospace archaeology.

ARCHAEOLOGICAL CATALOG

Apollo 10[2]

History

The Apollo 10 expedition lifted off from Cape Canaveral on the afternoon of May 18, 1969, closely orbited the Moon in a test of the full Apollo spacecraft, and returned

FIGURE 21.5 Artist's drawing of the *Apollo 16* subsatellite science probe. (Courtesy of NASA.)

to Earth almost exactly 8 days later, on May 26. The objective of the mission was to evaluate the complete complement of manned Apollo technologies and their associated Earth-based support systems prior to an actual landing on the Moon planned for the Apollo 11 mission later that summer. All operations of the later Apollo 11 mission were simulated with the exception of the actual landing on the Moon. As in all Apollo missions, the expedition consisted of a three-man crew, led by mission commander Thomas P. Stafford, command module pilot John W. Young, and lunar module pilot Eugene A. Cernan. The technology that reached the Moon included the command and service module (CSM) and the lunar module (LM).

After a 3-day cruise to the Moon, the lunar module was separated from the command module and flown to within 14 km of the lunar surface. Returning toward the command module, the descent stage of the lunar module was disconnected and jettisoned into lunar orbit. The two-person LM crew returned to the command module, and then the ascent stage of the lunar module was jettisoned into solar orbit. The crew returned to Earth on board the command module. The crew capsule reentered the Earth's atmosphere and splashed down 643 km east of American Samoa on May 26, 1969 where both crew and capsule were recovered by the USS *Princeton*.

Archaeology

The *Apollo 10* lunar module descent stage is the first object of manned exploration humans left behind on the Moon. It was jettisoned into lunar orbit after the LM made its closest inspection of the lunar surface (Figure 21.7). Immediately after its separation, the crew in the ascent stage of the LM made a quick maneuver to avoid hitting

FIGURE 21.6 The lunar roving vehicle during the *Apollo 17* expedition. (Courtesy of NASA.)

the descent stage. NASA mission records do not record what happened to the descent stage after separation from the ascent stage.[3] When official records speak of jettisoning operations, it is with regards to the jettison of the ascent stage after rendezvous with the command service module[4] since, ordinarily, the descent stage would be left on the lunar surface and only the ascent stage would lift off the Moon to return the two crew members to the CSM.

The discarded *Apollo 10* LM descent stage would have continued in orbit around the equator of the Moon until the orbit decayed and the artifact impacted the surface, transitioning as it did so from systemic to archaeological context. The crash site, as well as the degree of damage done to the descent stage during impact, is unknown. All known archaeological sites created by the Apollo program exist on the near side of the Moon. The possibility exists that the wreck of the *Apollo 10* LM descent stage—along with the wrecks of the *Apollo 11* and *Apollo 16* ascent stages and the *Apollo 16* subsatellite—comprise the only evidence of manned exploration to have been discarded on the far side of the Moon.

FIGURE 21.7 The *Apollo 10* lunar module descent and rendezvous operation, showing the discard of the descent stage (upper middle panel) onto the surface of the Moon (NASA: *Apollo 10 Press Kit*).

Apollo 11[5]

History

The Apollo 11 expedition left Earth on the morning of July 16, 1969, with the purpose of landing the first humans on the Moon and returning them safely. The crew consisted of commander Neil A. Armstrong, command module pilot Michael Collins, and lunar module pilot Edwin E. (Buzz) Aldrin, Jr.

At the launch point in the coastal estuaries of eastern Florida, roads and parking areas overflowed with an estimated one million eyewitnesses, including the author. One did not see so much as feel the launch of the Saturn V rocket, as the 7.6-million-pound thrust of its main engines shook the ground for several miles around.

Three days later, during the afternoon of July 19, the spacecraft—now consisting of the command service module attached to the lunar module (the three stages of the Saturn rocket itself having been each jettisoned in turn as fuel was exhausted)—entered lunar orbit and passed to the far side of the Moon. Each lunar orbit required about 2 hours.

On July 20, Aldrin and Armstrong entered the lunar module, separated the lander from the CSM, and descended toward the surface of the Moon. In the late afternoon of July 20, the lunar module was set down on the southwestern edge of the Sea of Tranquility, about 4 miles from its target point. Armstrong's radio message to Mission Control that "the *Eagle* has landed" marked the first time humans established a manned base camp on another planetary body. A few hours later, Armstrong crawled backward through the lunar module hatch, descended the 10-ft.-tall ladder, and stepped off onto the surface of the Moon. Armstrong, soon joined by Aldrin, began a rapid series of both scientific observation, recording, and sampling and the emplacement and photography of symbols such as a flag and memorial plaque. Less than 3 hours later, with their exploratory and symbolic work accomplished, the two humans climbed back on board the lunar module.

Sleeping overnight in the module, the explorers left the Moon the following afternoon after less than a day on the surface. The ascent engine of the ascent stage of the lunar module was fired. Using the platform of the descent stage as a launch pad, the spacecraft lifted off, and the first human base camp on the Moon was abandoned. The expedition returned to Earth on July 24, 1969.

Archaeology

More than 5,000 pounds of spacecraft and other exploratory gear were left behind at the landing site/base camp, along with hundreds of footprints in the soft, dusty soil. These artifacts include the descent stage/launch pad, video and still cameras, scientific sampling tools, discarded life support systems, an American flag, and several remotely operated scientific instruments, including a laser beam reflector, seismic detector, and a gnomon, a device to verify colors of objects photographed.

The base camp is located just north of the lunar equator, at 0° 40′ 26.69″ N, 23° 28′ 22.69″ E. What is unknown is the location of the *Apollo 11* lunar module ascent stage. Two hours after the lunar module ascent stage redocked with the command service module, the ascent stage was jettisoned. As with the *Apollo 10* descent stage released 2 months earlier, the location of the *Apollo 11* ascent stage was not recorded. The surviving command module is on display at the National Air and Space Museum in Washington, D.C.

Apollo 12[6]

History

Four months after Apollo 11, on November 14, 1969, the second mission to land humans on the lunar surface lifted off from Earth. The Apollo 12 expedition consisted of mission commander Charles P. "Pete" Conrad, lunar module pilot Alan L. Bean, and command and service module pilot Richard F. Gordon. On the morning of November 19, 1969, Bean and Conrad descended to Oceanus Procellarum (the Ocean of Storms) on the Moon. The lunar module touched down on the rim of a crater that was named after an unmanned *Surveyor 3* probe that had landed on the Moon 2 1/2 years earlier.

Near the lunar module base camp area, Bean and Conrad emplaced an ALSEP and a solar wind collector on the surface of the Moon, which were connected by cables to a central power unit. The power unit used a small amount of plutonium-238 to create heat and convert this heat into electricity.

During a long (1.3 km) walk away from the lunar module on November 20, the astronauts reached the *Surveyor 3* probe inside the crater and removed several components of it. These components, weighing some 10 kg, were returned to Earth for study. This research—an analysis of the transformation processes at work on an artifact during its years in the lunar environment—can be appropriately considered the first example of extraterrestrial archaeology.

After more than 31 hours on the lunar surface, the two astronauts lifted off from the Moon in the lunar module ascent stage and rendezvoused with the command service module. The ascent stage was jettisoned and deliberately crashed back into the lunar surface to create an artificial quake recorded by the ALSEP's seismometer. The expedition returned to Earth on November 24, 1969.

Archaeology

In terms of material culture, Apollo 12 both conducted archaeological research in the form of retrieving components of the *Surveyor 3* probe and created distinct archaeological assemblages in the form of the lunar module descent stage base camp and its associated scientific experiments and footprints, and the wreck of the lunar module ascent stage. The descent stage launch pad/base camp with the associated ALSEP experiment package is located south of the lunar equator, at 2.99° S, 23.34° W. The wreck of the ascent stage is located slightly south and east of the base camp, at 3.94° S, 21.2° W. The ALSEP instruments, powered up on November 20, 1969, were powered down along with other ALSEP instrument assemblages on September 30, 1977, making them the last of the *Apollo 12* artifacts to make the transition from systemic to archaeological context.

The surviving command module was placed on display at the Virginia Air and Space Center in Hampton, Virginia, while the television camera recovered from the *Surveyor 3* probe was placed on display at the National Air and Space Museum.

Apollo 13[7]

History

The Apollo 13 mission, April 11–17, 1970, was intended to explore the Fra Mauro region of the Moon. The expedition was aborted when one of the oxygen tanks on board the CSM exploded en route to the Moon. The cause of the explosion was later determined to be a short caused by damaged wires. With nearly half the service module's oxygen supply gone, the astronauts, led by Captain James A. Lovell, Jr., and including command module pilot John L. Swigert, Jr., and lunar module pilot Fred W. Haise, Jr., transferred to the lunar module. They used its supplies for survival until and after the whole command service module–lunar module assembly could orbit the Moon and make a return voyage to Earth on April 17, 1970.

Archaeology

The only evidence of the Apollo 13 mission that survives on the Moon is the wreck of the Saturn V third-stage S-IVB (S-IVB-508), which lies at 2.75° S, 27.86° W, just west of the base camp of the Apollo 12 mission. After separation from the command service module, the S-IVB was placed on a lunar impact trajectory and impacted the Moon on April 14, 1970.

Possible remains of the lunar module exist on Earth, in the Pacific Ocean off the coast of New Zealand, where they would have fallen if they survived reentry in Earth's atmosphere. The command module was eventually placed on display at the Kansas Cosmosphere and Space Center in Hutchinson, Kansas.

Apollo 14[8]

History

The Apollo 14 expedition left Earth on January 31, 1971, on a mission to land on the Moon near the Fra Mauro crater. The expedition was led by commander Alan B. Shepard, Jr., and included lunar module pilot Edgar D. Mitchell and command module pilot Stuart A. Roosa. The Saturn V third-stage S-IVB was powered into a lunar impact trajectory during the flight to the Moon and crashed into the lunar surface on February 4.

Shepard and Mitchell landed on the Moon in the lunar module, 24 km from Fra Mauro, on February 5, 1971, and emplaced a series of scientific experiments, including an ALSEP, took photographs, and collected more than 42 kg of geological samples from the lunar environment. Before concluding the second of their two walks on the Moon, Shepard teed up two golf balls and hit them away into space using a 6-iron attached to the shaft of a scientific sampling device.

The lunar module ascent stage left the Moon on February 6, and the astronauts returned to Earth on board the command module on February 9.

Archaeology

The *Apollo 14* S-IVB Saturn V third stage was deliberately crashed into the lunar surface to trigger a Moon quake that could be recorded by the Apollo 12 ALSEP. The impact point, 8.09° S, 26.02° W, places the wreckage in a cluster of S-IVB discards surrounding the Apollo 12 and Apollo 14 base camps. After being jettisoned by the crew, the *Apollo 14* lunar module ascent stage crashed into the Moon on February 7, 1971, at 3.42° S, 19.67° W, a spot roughly midway between the Apollo 12 and Apollo 14 base camps.

The base camp itself, with the lunar module descent stage and the associated walking area of the astronauts, is located at 3° 38′ 43.08″ S, 17° 28′ 16.90″ W. The ALSEP station, which recorded the impact of the lunar module ascent stage, was activated on February 5, 1971. It was powered down by commands from Earth and entered its archaeological context along with the other ALSEP stations on September 30, 1977. The command module was placed on display at the Astronaut Hall of Fame in Titusville, Florida.

Apollo 15[9]

History

The Apollo 15 expedition left Earth on July 26, 1971, on a mission to explore the Hadley Rille/Apennines region of the Moon, the northernmost landing site in the Apollo series of lunar explorations. The mission was led by commander David R. Scott and included lunar module pilot James B. Irwin and command module pilot Alfred M. Worden.

The Saturn V third-stage S-IVB separated from the command service module when the spacecraft reached orbit around the Earth and was then sent on a collision course with the Moon.

Scott and Irwin landed on the Moon on July 30, 1971, at the foot of the Apennine mountain range. During the course of their time on the surface, they deployed a lunar roving vehicle from its storage "garage" on board the descent stage of the lunar module. This was the first time such human had driven a wheeled vehicle on the Moon. With the rover, the astronauts were able to journey nearly 28 km over the lunar surface while collecting more than 77 kg of geological samples, taking photographs, and setting up an ALSEP instrument package.

The lunar module descent stage, the rover, and the array of scientific instruments were left behind when the lunar module ascent stage left the Moon on August 2. The mission concluded when the astronauts returned to Earth on board the command module on August 7.

Archaeology

The *Apollo 15* S-IVB Saturn V third stage was crashed into the lunar surface at 1.51° S, 17.48° W, a position north-northwest of the Apollo 12 and Apollo 14 base camps and where the ALSEPs emplaced nearby could record its impact. Apollo 15's ALSEP was activated on July 31, 1971, and entered its archaeological context when it and the other ALSEP stations were powered down on September 30, 1977.

The Apollo 15 expedition was the first manned lunar mission to employ a lunar roving vehicle. This four-wheeled machine allowed the astronauts to extend their range up to 5 km away from the lunar module base camp site. The rover and the *Apollo 15* lunar module descent stage base camp area are located at 26° 7′ 55.99″ N, 3° 38′ 1.90″ E. It is also here on the lunar surface that a small work of art—a stylized astronaut—was left behind with a plaque dedicated to the Americans and Russians who had died in the development of their nations' respective space exploration programs.

When the *Apollo 15* lunar module ascent stage was jettisoned, it crashed into the Moon at 26.36° N, 0.25° E, within 100 km of the expedition's base camp. Another archaeological element of the Apollo 15 mission was a so-called subsatellite that was released into orbit around the Moon from the expedition service module. The subsatellite measured plasma, particle, and magnetic fields of the Moon and mapped the lunar gravity field. When its orbit decayed, the cylindrical satellite with its three boom antennae crashed at an unrecorded location on the lunar surface.

Apollo 16[10]

History

The Apollo 16 expedition left Earth on April 16, 1972, on a mission to explore the Descartes region of the Moon. The expedition was led by commander John W. Young and included Charles M. Duke, Jr., as lunar module pilot and Thomas K. Mattingly II as command module pilot.

The Saturn V third-stage S-IVB separated from the command service module after the spacecraft reached Earth orbit. The S-IVB was then sent on to an impact on the Moon, where it crashed on April 19 at 1.3° N, 23.8° W, continuing the cluster of S-IVB wreckage around the landing sites/base camps of the Apollo 12 and Apollo 14 missions.

Young and Duke landed on the Moon in the lunar module on April 21. During the course of three excursions away from the lunar module base camp, they drove a total of 27 km with a new lunar roving vehicle. Another ALSEP package was emplaced on the Moon, and as with all other missions save for Apollo 13, another sampling of lunar rocks and soil was collected for return to Earth.

The LM took off from the Moon on April 24, and the astronauts returned to Earth on April 27.

Archaeology

The *Apollo 16* S-IVB Saturn V third stage was crashed into the lunar surface at 1.3° N, 23.9° W, just north of the lunar equator and within the general cluster of S-IVB wrecks from *Apollo 13–17*. The Apollo 16 base camp, with its lunar roving vehicle and ALSEP station, is located at 8° 58′ 22.84″ S, 15° 30′ 0.68″ E, southwest of the original *Apollo 11* landing site.

A fault in *Apollo 16*'s lunar module ascent stage meant that the planned crash into the lunar surface had to be abandoned. Like the expedition's subsatellite, both machines were placed in orbit around the Moon. It is estimated that the ascent stage remained in orbit for a year before it fell from orbit to an unknown location on the lunar surface. The expedition's subsatellite experienced mechanical problems as well, and its planned lunar orbit of 1 year was shortened to 1 month, before it too crashed into an unrecorded location on the lunar surface. The expedition's ALSEP station was the final piece of the expedition to transition from systemic to archaeological context. Activated on April 21, 1972, it was powered down with the other ALSEP stations on September 30, 1977.

Apollo 17[11]

History

The Apollo 17 expedition left Earth on December 7, 1972, on a mission to explore the Taurus-Littrow region of the Moon. The expedition was led by commander Eugene A. Cernan, with lunar module pilot Harrison H. Schmitt and command module pilot Ronald E. Evans. The mission was both the final Apollo mission and the last time humans reached and walked on the surface of the Moon.

The Saturn V third-stage S-IVB separated from the command service module soon after the spacecraft entered orbit around the Earth and was sent on course to the Moon.

Cernan and Schmitt landed on the Moon in the lunar module on December 11, 1972, on the southeastern rim of Mare Serenitatis (the Sea of Serenity). There they deployed another lunar roving vehicle from its storage bay on board the lunar module descent stage. Over the next 75 hours, they explored the area around the base camp, extending their range by driving the rover a total of 30 km. The astronauts collected more than 110 kg of lunar samples and emplaced an ALSEP instrument package.

The lunar module ascent stage lifted off from the Moon on December 14, and the astronauts returned to Earth on board the command module on December 19.

Archaeology

The *Apollo 17* Saturn V third-stage S-IVB rocket crashed into the lunar surface on December 10 at 4.21° S, 12.31° W, completing a cluster of S-IVB wreckage from *Apollo 13, 14, 15, 16,* and *17* arrayed around the descent stage/base camps of Apollo 12 and 14.

The *Apollo 17* lunar module descent stage/base camp, with its associated lunar roving vehicle and ALSEP instrument array, is located at 20° 11′ 26.88″ N, 30° 46′ 18.05″ E. After being jettisoned from the command module, the ascent stage of the expedition's lunar module crashed into the lunar surface at 19.96° N, 30.50° E.

The final element of the expedition to transition from systemic to archaeological context was the ALSEP array. Activated on December 12, 1972, it was closed down with the other ALSEP stations on September 30, 1977. The primary artifact to return to Earth, the *Apollo 17* command module, was placed on display at the Johnson Space Center in Houston, Texas.

LUNAR ARCHAEOLOGY OF THE APOLLO MISSIONS AS CULTURAL RESOURCE

Methodological Catalog

Eight Apollo missions left archaeological evidence behind on the lunar surface (Figure 21.8).[12] Six of these included a human presence on the Moon. As such, the Apollo expeditions comprise a crucial cultural resource database within the lunar archaeology of *Homo sapiens*. The progress of aerospace technology and the projection of privately funded space vehicles carrying paying passengers into orbit have combined to motivate both theoretical discussions[13] and actual (if remote) studies of Apollo sites,[14] and discussions of how this database could be both preserved and utilized for studies within the field of aerospace archaeology.

Cultural resource management of such aerospace sites in extreme environments as those of the Apollo program on the Moon will likely be founded on many of the same principles that currently guide the attempts at stabilization, preservation, and study of historic sites in places such as Antarctica, where scientific exploration during the International Geophysical Year of 1957–1958 led directly to the Antarctic Treaty

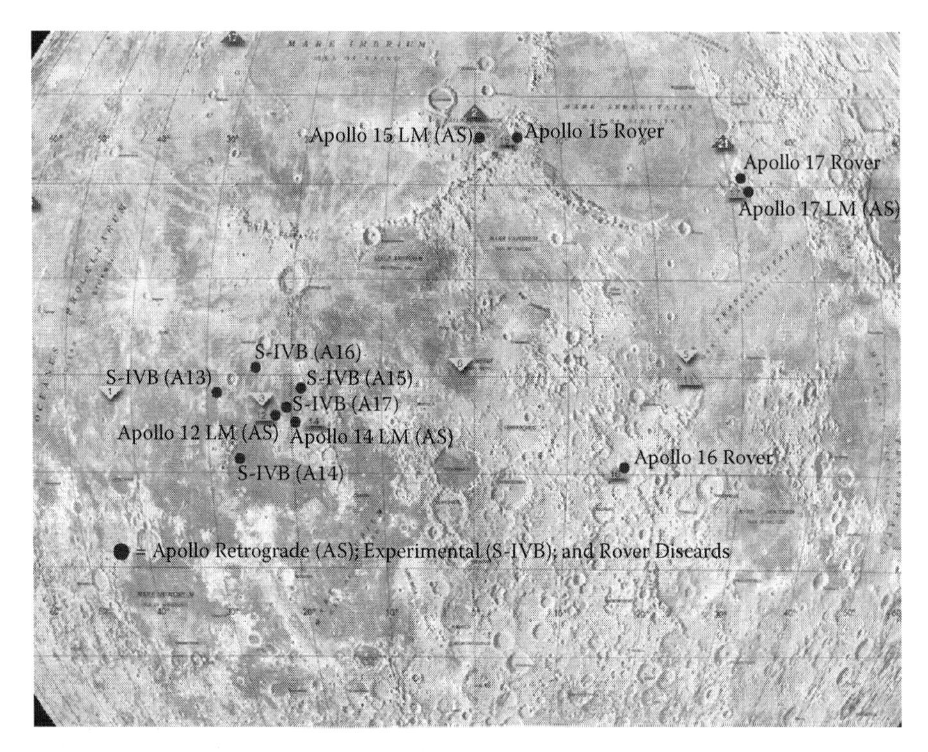

FIGURE 21.8 Locations of the Apollo landing sites, in relation to the locations of Apollo retrograde (lunar module ascent stage), experimental (S-IVB), and lunar rover discards. (Modified by the author from National Space Science Data Moon Landing Map, http://nssdc.gsfc.nasa.gov/planetary/lunar/moon_landing_map.jpg.)

of 1961, which in turn became "a model for subsequent international treaties about space."[15]

Under the treaty, nations such as New Zealand have established protection and study regimes for historic sites in Antarctica,[16] and archaeological techniques and technologies have advanced for research in this extreme environment.[17] These techniques are episodic and directed toward particular sites during brief summer field seasons, just as one might anticipate in any planning of a mission to a particular site of extraterrestrial archaeology. (The more theoretical considerations that might govern such archaeological research at lunar and other sites in space are discussed in Chapter 45.)

The extreme environment in which the Apollo database is embedded is a temporary advantage in its preservation. It should preclude most cultural transformational processes for several more decades, until the expected advent of regular tourist or development travel to the Moon. Once humans revisit the Moon, one of the first priorities should be the active preservation of the culture of Apollo. Natural transformation processes of extreme temperatures and radiation can be partially ameliorated through a program of shielding the major artifact concentrations. (Other natural processes, such as potential damage or destruction by meteorites, cannot be accounted for or fully prevented.) Cultural transformation processes such as tourism and souvenir collection should be accounted for prior to any opening of the lunar surface to the general public.

While the static base camps of the Apollo missions would comprise the main foci of cultural resource protection regimes, equally important will be the preservation of the Apollo wreck sites. These are artifacts that demonstrate, like the impact zones of the Saturn V S-IVB third-stage rockets, reuse of one form of technology for a distinct scientific experimental series, or, like the undiscovered wreckage of the *Apollo 10* descent stage and the undiscovered ascent stages of *Apollo 11* and *Apollo 16*, techno-cultural failure in the course of exploratory voyaging.

For example, the site where the Soviet *Luna 5* probe crash-landed onto the surface of the Moon with such force "that a German observatory claimed to have seen a 135 by 49 mile dust cloud where it impacted"[18] may one day provide excellent archaeological opportunities for the study of the secretive Luna series of unmanned probes launched in an era of intense superpower competition for priority on the Moon.

While it seems rather extraordinary to consider the possibilities of archaeological field research on the Moon, we can take note of the fact that it has already taken place. In fact, NASA has been conducting formational archaeology research on lunar sites at least since 1969.[19] As noted above, on November 20, 1969, astronaut Charles Conrad, Jr., recovered pieces (the television camera, remote sampling arm, and sections of tubing) from the unmanned *Surveyor 3* probe that had soft-landed in the Moon's Ocean of Storms on April 19, 1967 (Figure 21.9). These artifacts were brought back to Earth so that the Johnson Space Center and the Hughes Air and Space Corporation in El Segundo, California, could analyze the natural transformational processes operating on aerospace artifacts left on the Moon (e.g., National Aeronautics and Space Administration 1972[20]).

A study of the archaeological catalog left by the Apollo program on the lunar surface suggests that its deposits can be arranged into five cohesive geographic areas on the near side of the Moon. The areas (Figure 21.10)[21] can be thought of as notional archaeological preserves, areas of the Moon specially demarcated in order to place the entire database (minus the artifacts yet to be located) under a protective regime that will shield them both from environmental deterioration and from the effects of future exploration, visitation, and potential exploitation.

The boundaries of the first of these notional archaeological parks would surround the landing site of *Apollo 11*, from 20° to 30° E and from 5° S to 5° N. This area would allow for the inclusion of the earlier unmanned *Surveyor 5* probe and perhaps enclose the undiscovered wreckage of the *Apollo 11* ascent stage as well.

If the first such lunar archaeological preserve is so designated for its obvious historical value and associations, the second would encompass by far the largest concentration of associated remains of the Apollo program. This second area, extending from 15° to 30° W and from 10° S to 3° N, would enclose the base camps of both the Apollo 12 and Apollo 14 missions, the wrecks of the discarded ascent stages of both missions, the deliberately crashed wreckage of the S-IVB third-stage rockets of the Apollo 13, 14, 15, 16, and 17 expeditions, and also the area around Surveyor crater and the *Surveyor 3* spacecraft and its associations with the Apollo 12 mission. Such a preserve would offer an extraordinary database of for aerospace archaeology and the history of science, exploration, and technology.

FIGURE 21.9 The *Apollo 12* lunar module (in the distance on the edge of *Surveyor* crater), with the *Surveyor 3* probe in the foreground. (Courtesy of NASA.)

Next in importance would be the areas surrounding the missions of Apollo 15 and Apollo 17. Both of these demarcated areas include not only the respective expedition base camps but, for the first time in the series (with Apollo 15), the additional technology of the lunar rover. Also associated within the Apollo 15 preserve would be the remains of the unmanned Russian spacecraft *Luna 2*. The Apollo 17 preserve likewise would include the remains of a Russian probe, in this case *Luna 21*. The Apollo 15 site is also the highest-latitude base camp of all of the Apollo missions. The final site is that of Apollo 16, which also has an associated lunar roving vehicle.

Ranking these areas according to cultural, archaeological, and historical value, and assuming the potential that each area might not receive complete protective quarantine, these preserves should be established in the following order:

1. Apollo 11/Surveyor 5
2. Apollo 12/Surveyor 3/Apollo 14/S-IVB cluster
3. Apollo 15
4. Apollo 17
5. Apollo 16

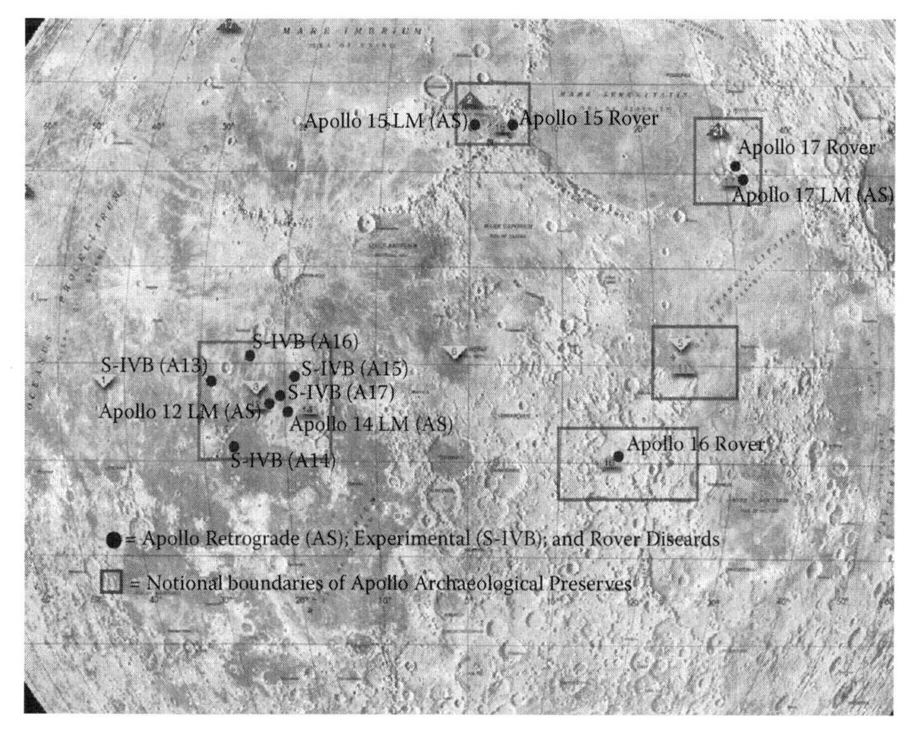

FIGURE 21.10 Notional boundaries of Apollo Archaeological Preserves. (Modified by the author from National Space Science Data Moon Landing Map, http://nssdc.gsfc.nasa.gov/planetary/lunar/moon_landing_map.jpg.)

Shielding the Apollo base camps will require the development of a dome-like structure specifically engineered to address a combination of natural and cultural transformational processes. Such an "archaeodome" would first and foremost offer protection against the gradual natural decay of extreme temperatures and solar radiation. At the same time, connected to a nearby landing platform, the dome would allow for tourist visits while shaping the access pathways to the site. Tourist pathways across the dome—complete with handholds, interpretive panels, and emergency life support stations—would allow for both oblique and plan views of the archaeological sites while leaving sensitive areas such as the original Apollo footprints and rover treads untouched. Such a concept, developed for the protection of the Apollo base camps and wreck sites on the Moon, could be adapted for similar cultural resource protection regimes for such sites as the Viking landers on Mars.

Theoretical Catalog

Prosaic questions of the survey, stabilization, and preservation of archaeological sites on the Moon need not necessarily be examined in a theoretical, shall we say, vacuum. The archaeological study of both earthbound and planetary sites from the history of aerospace exploration leads neatly into cultural and biological considerations inherent in humanity's exploration of and expansion into space.

Hundreds of sites exist where operational spacecraft have been discarded, whether by mission requirements, accident, or obsolescence. Abandoned space *ports* testify to the base camps necessary to operate in the aerospace environment. The Apollo lunar module descent stages and the Saturn V third-stage S-IVBs on the Moon likewise testify to the extraordinary dual use of much of the Apollo technological catalog. When wrecks such as those of the S-IVBs occur, or the Apollo base camp sites are abandoned, the artifact leaves history, where it was a part of people's thoughts and actions (Schiffer's systemic context), and enters Schiffer's archaeological context.[22]

In order to make credible use of the material record in the construction of screens through which to filter the words of the historical record, primary attention must be given to accounting for how what exists as the present-day (archaeological) material record arrives in our time from the (systemic) past. This idea has no doubt long been an intuitive one on the part of archaeologists, but its explicit application as a primary mechanism in the analysis of material remains has come only recently. As Schiffer writes, "the past—manifest in artifacts—does not come to us unchanged."[23]

It was Ben Finney, who for much of his career had explored the technology and techniques used by Polynesians to settle the Pacific Ocean, who first suggested that it would not be premature to begin thinking about the archaeology of aerospace sites on the Moon and Mars, "the remains or imprint of Russian and American space ventures there."[24] Finney was certainly clear-eyed about the improbability of anyone being able to conduct fieldwork any time soon, yet he was equally convinced that one day such work would be done. This path of this emerging field could be seen, by way of illustration, as early as the early to mid-1990s in the examinations made of aerospace base camps and other sites as recent as those remaining from the years of the Cold War.[25]

While the archaeologist O.G.S. Crawford "referred to obsolete aircraft as strictly archaeological"[26] and archaeologist James Deetz[27] referred to "interplanetary space vehicles" as a complex example of material culture, the first suggestion that aircraft wrecks might yield important anthropological data was made as early as 1983 by Richard A. Gould. Gould suggested that debates originating in the historical record could be evaluated through "the explanatory potential of archaeology ... [where] differing historical interpretations can be regarded as a source of alternate hypotheses, with archaeological evidence being used to test each alternative."[28]

Studies of the culture of Apollo must combine appropriate material recording and intensive historically particular documentation with anthropological approaches on the model of Gould's "trend analysis."[29] Gould suggests that there is "an urgent need to examine the relationship of specific kinds of behavior to the materials that are deposited at wreck sites," and he goes on to suggest that such sites possess "signatures" and "distinctive patterns," "akin to what might appear as a particular configuration of sounds on sonar or other underwater listening apparatus to identify a particular submarine or class of submarine and what it might be doing."[30]

Consideration of the nature of the mating of *Homo sapiens* to our machines throughout aerospace history inevitably leads to cultural and biological considerations inherent in humanity's exploration of and expansion into space, a move that "represents a continuation of our terrestrial behavior, not a radical departure from it."[31] "Our terrestrial behavior," of course, includes the triumphal (*Apollo 11*,

Apollo 12), the precarious (*Apollo 13*), and the catastrophic (*Apollo 1*, *Challenger*, *Columbia*).

Eventually, this model can serve as an archaeological basis by which to examine Finney's thesis that humankind "evolved as an exploratory, migratory animal."[32] Not only might the human exploration of space supersede the movement of *Homo erectus* out of Africa or *Homo sapiens* across the Wallace Line, Finney argued that it could "be studied directly without having to reconstruct and test the vehicles involved or interpret ambiguous texts."[33]

Although at first it may appear that Finney's notion of *Homo sapiens* as an explorer is ultimately too large and vague to test archaeologically, it contains several archaeological implications that can be applied to the Apollo sites on the Moon. Exploration is a uniquely human behavior and produces, as in the Apollo lunar archaeological preserves, a novel material component. As such it falls within the hard boundaries of credible archaeological analysis.

The testing of Finney's hypothesis should be directed at establishing links between the cultural responses to the need to explore read in the archaeological record and the suggestion that *Homo sapiens* has evolved as an inherently exploring species, one "that spread from [a] tropical homeland through developing technology to travel to and survive in a multitude of environments for which [it was] not biologically adapted."[34]

CONCLUSION

The Apollo program left six lunar module descent stages fixed at base camps on the Moon, another descent stage at an undiscovered site, and another six ascent stages deliberately discarded and impacted on the lunar surface after they had delivered their crews back to the mission's command module. These spacecraft, dead and discarded, have all found Schiffer's archaeological context.

The scientific instruments packages (ALSEPs), left behind on the Moon and powered by small amounts of plutonium-238, were the last Apollo artifacts to leave their systemic context. When, nearly 5 years after the last human walked on the Moon, all of the ALSEP stations were powered down on September 30, 1977, they too entered their archaeological context along with the rest of the Apollo archaeological database.

The goals of aerospace archaeologists with regard to the culture of Apollo from this moment forward should be twofold. First, a protective regime should be adopted to preserve this database as the vital cultural and behavioral catalog of scientific exploration that it is. Second, the database should be seen as a primary tool in the examination of *Homo sapiens* as an exploratory, migratory species.

Archaeology, unlike history, cannot exist without the material residue of human triumph and failure, be it ruins of abandoned settlements, shipwrecks at the bottom of the sea, or crash sites of Soviet robot probes on the surface of the Moon. If we can study and isolate the processes by which humans adapt such technology to explore such extreme environments as space, we can gain insight into human behavior that will aid or even propel our movement as a species into the solar system and beyond.

REFERENCES

1. Schiffer, M. 1987. *Formation Processes of the Archaeological Record.* Salt Lake City: University of Utah Press.
2. NASA, National Space Science Data Center. *Apollo 10*. http://nssdc.gsfc.nasa.gov/nmc/masterCatalog.do?sc=1969-043A.
3. NASA. 1969. *Apollo 10 Press Kit.* Release No. 69-68. Washington, DC: NASA. See also: NASA. 1969. *Apollo 10 Mission Report*; MSC-00126. Houston, TX: Manned Spacecraft Center.
4. NASA. 1969. *Apollo 10 Press Kit.* Release No. 69-68. Washington, DC: NASA. See also: NASA. 1969. *Apollo 10 Mission Report; MSC-00126.* Houston, TX: Manned Spacecraft Center (specifically pp. 9–18 and 9–19).
5. NASA. National Space Science Data Center, *Apollo 11*. http://nssdc.gsfc.nasa.gov/nmc/spacecraftDisplay.do?id=1969–059A.
6. NASA. National Space Science Data Center, *Apollo 12*. http://nssdc.gsfc.nasa.gov/nmc/spacecraftDisplay.do?id=1969–099A.
7. NASA. National Space Science Data Center, *Apollo 13*. http://nssdc.gsfc.nasa.gov/nmc/spacecraftDisplay.do?id=1970–029A.
8. NASA. National Space Science Data Center, *Apollo 14*. http://nssdc.gsfc.nasa.gov/nmc/spacecraftDisplay.do?id=1971–008A.
9. NASA. National Space Science Data Center, *Apollo 15*. http://nssdc.gsfc.nasa.gov/nmc/spacecraftDisplay.do?id=1971–063A.
10. NASA. National Space Science Data Center, *Apollo 16*. http://nssdc.gsfc.nasa.gov/nmc/spacecraftDisplay.do?id=1972–031A.
11. NASA. National Space Science Data Center, *Apollo 17*. http://nssdc.gsfc.nasa.gov/nmc/spacecraftDisplay.do?id=1972–096A.
12. National Space Science Data Moon Landing Map. http://nssdc.gsfc.nasa.gov/planetary/lunar/moon_landing_map.jpg.
13. O'Leary, B.L. 2006. "The cultural heritage of space, the Moon and other celestial bodies." *Antiquity* 80 (307), March 2006.
14. O'Leary, B.L., J. Hunner, J. Versluis, R. Gibson, and J. Culp. 2003. Lunar Legacy Project. http://spacegrant.nmsu.edu/lunarlegacies/.
15. Gillmor, C.S. 1994. Science and Travel in Extreme Latitudes. *Isis* 1994: 482–485.
16. Harrowfield, D.L. 1988. Historic Sites in the Ross Dependency, Antarctica. *Polar Record* 24: 277–284.
17. Ritchie, N.A. 1990. Archaeological Techniques and Technology on Ross Island, Antarctica. *Polar Record* 26: 257–264.
18. Burrows, W.E. 1990. *Exploring Space*. New York: Random House (see specifically p. 162).
19. Capelotti, P.J. 2004. Space: The Final (Archaeological) Frontier. *Archaeology* 57: 46–51. Capelotti, P.J. 1996. A Conceptual Model for Aerospace Archaeology: A Case Study from the Wellman Site, Virgohamna, Danskøya, Svalbard. PhD diss., Rutgers University. University Microfilms #9633681.
20. NASA. 1972. *Analysis of Surveyor 3 Material and Photographs Returned by Apollo 12.* NASA SP-284.
21. National Space Science Data Moon Landing Map. http://nssdc.gsfc.nasa.gov/planetary/lunar/moon_landing_map.jpg.
22. Schiffer, Michael. 1987. *Formation Processes of the Archaeological Record.* Salt Lake City, UT: University of Utah Press.
23. Ibid. (see specifically p. 5).
24. Finney, B., personal communication, December 3, 1993.

25. Delgado, J.P., D.J. Lenihan, and L.E. Murphy, 1991. *The Archaeology of the Atomic Bomb: A Submerged Cultural Resources Assessment of the Sunken Fleet of Operation Crossroads at Bikini and Kwajalein Atoll Lagoons*. Southwest Cultural Resources Center Professional Papers, no. 37. Santa Fe, NM: National Park Service. Center for Air Force History. 1994. *Coming in from the Cold: Military Heritage in the Cold War*. Washington, DC: Center for Air Force History.
26. Dymond, D.P. 1974. *Archaeology and History: A Plea for Reconciliation*. London: Thames and Hudson.
27. Deetz, J. 1977. *In Small Things Forgotten*. Garden City, NY: Anchor Books (see specifically p. 24).
28. Gould, R.A. 1983. The Archaeology of War. In *Shipwreck Anthropology*, ed. R.A. Gould, 105–142. Albuquerque, NM: University of New Mexico Press.
29. Gould, R.A. 1990. *Recovering the Past*. Albuquerque, NM: University of New Mexico Press (see specifically pp. 178–179).
30. Gould 1983 (note 27).
31. Finney, B.R. 1992. *From Sea to Space*. Palmerston North, New Zealand: Massey University Press (see specifically p. 105).
32. Ibid.
33. Ibid.
34. Ibid.

Section V

Spacecraft Forensics and Mystery Solving

Searching for Life on Mars.

22 Space Hardware: Mystery Solving

Ann Garrison Darrin and Patricia M. Prettyman

CONTENTS

INTRODUCTION

This chapter discusses failures in space: What are the causes and effects of space failures? How do anomalies in space cause such failures? Do they provide any insight? Also discussed in this chapter are space phenomena that create space failures, socioeconomic-related space failures, and other unusual anomalies found in space. Determining the cause of failures and anomalies is of utmost importance to prevent recurrence. Many failures are eventually attributable to human errors, including design, software, processing, and material concerns. In discussing failures, it is important to address launch vehicle anomalies, which are catastrophic. From the perspective of the spaceflight community, catastrophic launch failures and ghosts of missions that never materialized have a significant impact. Ghosts of missions lost may explain the dedication to failure avoidance and risk mitigation seen in mission planners, managers, and designers. This conservative stance by the fiscal stakeholders, mission and program managers, designers, and engineers is often in direct opposition to mission needs, such as the need to address exploration, scientific requirements, military, or commercial goals.

From the specter of mission launch, this chapter will address mission failures once preliminary orbital requirements were achieved. Unsolved mysteries and unanswered questions related to artifacts of missions past will also be discussed. These unanswered questions leave a void for future researchers from all fields to pursue. A failure is that which has not fully been explained or cannot be fully understood by reason or, less strictly, whatever resists or defies explanation. Space-related mysteries relative to human explorations are found in orbit, on planetary and lunar soils, and terrestrially, often in museums. An introduction to a few of these unsolved mysteries is provided here. Often the term "unsolved mystery" is used relative to long discussions of UFOs and alien visits or abductions, providing a neverending need for debunking these theories. Unsolved mysteries relative to failures relate to not fully understanding the cause and effect of errors and mishaps. Other unsolved mysteries pertain to phenomena on missions yet to be explained. Solution sets are often difficult when sites are remote and can rarely, if ever, be visited. In addition, the ability to gather data and information is constrained by a myriad of reasons. Some of these solutions will need to wait until opportunities for further investigation are available. Many may remain speculations because of lack of data and lack of the ability to repeat exact conditions. This section includes a discussion by Dr. Dirk Spenneman and Guy Murphy on Mars landers, the Beagle 2 case study, and the importance of preserving space archaeological sites for space heritage (see Chapters 23 and 41 of this volume). Later in this section, there is a continued discussion by Dr. Ralph Lorenz on space hardware artifacts (see Chapter 25 of this volume), followed by Dr. Philip Stooke discussing lost spacecraft and the knowledge behind finding Moon landers and their locations, if known (see Chapter 24 of this volume).

WHAT IS FAILURE?

At full performance, if a space system is performing nominally, and at the onset of an event that degrades that performance in any manner, the term *anomaly* is used. *Nominal* is defined in the aerospace community as performing or achieved within expected, acceptable limits, or normal and satisfactory (e.g., "the mission was nominal throughout"). A *failure* in the aerospace world is the condition or fact of not achieving the desired end or ends, the failure of an experiment.[1] The term *anomalies* relates in general to deviations from the norm. In engineering jargon, a very small anomaly, often temporary, is often referred to as a glitch. When one speaks of launch failure, it implies that the desired final orbit was not achieved. It is notable that the Hubble Space Telescope never had a repair mission; rather, it has had servicing missions.

Fifty years of experience in building spacecraft and measuring the space environment has yielded a refined understanding of how to avoid failure. This experience is reflected in the following facts:

- The number of spacecraft failures has been steadily decreasing.
- Failures, when they do occur, are less severe.

Within these trends, however, are some significant areas of concern that could affect continuous improvement in mission performance.

- Design-related failures are playing a more significant role as the total number of failures diminishes.
- Mechanical failures contribute significantly to reduced performance or loss of spacecraft.[2]

There are many glossaries of failure terminology and thousands of cause-and-effect reviews. Unfortunately, as we move forward, we often repeat the mistakes of the past.

OVERVIEW OF CAUSES OF FAILURES

Failures range from small, recoverable upsets to catastrophic. The following terms are often used in referring to analyzing an event with a spacecraft.

- *Contributing factor*: An event or condition that may have contributed to the occurrence of an undesired outcome, but whose elimination or modification would not by itself have prevented the occurrence.
- *Proximate cause*: An event that occurred, including any conditions that existed immediately before the undesired outcome and that directly resulted in its occurrence, whose elimination or modification would have prevented the undesired outcome. Also known as the *direct cause*.
- *Root cause*: One or more factors (events, conditions, or organizational factors) that contributed to or created the proximate cause and subsequent undesired outcome and whose elimination or modification would have prevented the undesired outcome. Typically, multiple root causes contribute to an undesired outcome.
- *Root cause analysis*: A structured evaluation method that identifies the root cause(s) of an undesired outcome. Root cause analysis should continue until organizational factors have been identified or until data are exhausted.
- *Observation*: A factor, event, or circumstance identified during the investigation that did not contribute to the mishap or close call but that, if left uncorrected, has the potential to cause a mishap or increase the severity of a mishap; or a factor, event, or circumstance that is positive and should be noted.[3]

Table 22.1 lists spacecraft events noted in 2007. Some of these events shortened the crafts' functional life expectancy, such as RASCOM-QAF1's helium leak, and/or reduced functionality, such as the GOES-12 and MetOp-A events. FUSE, which was launched with a 3-year life expectancy in 1999, continued to fulfill some aspect of mission demands till true death, although it was plagued through its lifetime with reaction wheel issues. For Meteosat-8, various problems were found, but work-arounds allowed for recovery to nominal operations. *Universitetsky*, a student-built satellite, has an unknown but nonrecoverable failure, perhaps from a bad uplink command. GeoEye experienced technical difficulties and is not sending imagery. The infamous Chinese ASAT test of January 11, 2007, virtually disintegrated the

TABLE 22.1
Short List of Anomaly-Related Satellite Events for 2007

Date	Satellite	Event
December 30, 2007	Arirang 1	Contact lost; mission terminated
December 29, 2007	RASCOM-QAF1	In Orbit Test (IOT) Leak in helium subsystem detected
December 4, 2007	GOES-12	Three-day attitude loss after routine stationkeeping maneuver
December 2007	Ekspress-AM22	Report: all gyros lost, possibly total loss (unconfirmed)
October 2007	Landsat 5	Diminished battery capacity after cell failure
September 26, 2007	Arabsat 2A	Two-hour outage
September 17, 2007	MetOp-A	One and a half day outage
August 2007	FUSE	Last reaction wheel lost; mission terminated
August 2007	Anik F3	Ka-band payload could not be activated
July 4, 2007	MetOp-A	Advanced High Resolution Picture Transmission failed
July 3, 2007	Solidaridad II	Report: 10-hour outage
June 2007	EchoStar V	Eighth solar array strip lost
May 22, 2008	Meteosat-8	Orbit change, likely hit by unknown object
May 21–22, 2007	XM 3	Software glitch, 1-day outage
April 5, 2007	Yamal 201	Attitude loss, 6-hour outage
March 25, 2007	IGS Radar 1	Shutdown after power system problem
March 19, 2007	New Horizons	Temporary safe mode after memory error, recovered
March 14, 2007	Hotbird 2	Anomaly in power subsystem
March 6, 2007	Universitetsky	Stopped transmitting for unknown reason
March 4, 2007	Orbview 3	Stopped sending usable imagery
March 2007	Orbital Express/ASTRO	IOT: Guidance problem, resolved by software update
February 2007	EchoStar II	Rotation of north solar panel failed, switched to backup
February 2007	Beidou 1D (2A)	Failed to deploy solar arrays, problem resolved in April 2007
February 2007	Kiku No. 8	IOT: part of payload could not be activated
January 27, 2007	Hubble Space Telescope	Safe mode after permanent failure of Altitude Control System instrument, recovered
January 11, 2007	FY-1C	Satellite disintegrated as result of Chinese anti-satellite weapon test, total loss
January 2, 2007	INSAT-4A	Thirty-minute outage owing to solar disturbances

Chinese defunct weather satellite FY-1C.[4] In some instances, such as the *Landsat 5* event, full mission recoverability is possible. This list is cursory but typical of a year in flight, which of course includes launch failures, such as the Zenit-3SL and lost rovers. In 2007, the Russian Zenit-3SL (no. SL24) rocket exploded at liftoff from the Sea Launch platform stationed in the Pacific Ocean. On February 3, 2007, the first images of the top deck of the Odyssey platform and its underside, as they looked after the accident, surfaced on the Internet. They revealed a missing flame deflector below the launch platform, sliding doors of the hangar blown off their tracks, and a moderate amount of burn marks around the pad. However, the rest of the structures looked intact, including fragile lights and radar antenna on the top deck. It was also obvious from the photos that the interface plate connected to the side of the Zenit at launch had not jettisoned at liftoff, which was evidence that the rocket's upward movement during the accident did not exceed 85 mm. According to the latest unconfirmed reports from Russia, a preliminary analysis of the failure showed that the rocket lifted around 10–15 cm above the launch pad, after which pressurization of the oxidizer tank was lost, apparently leading to the main engine shutdown.[5]

EMOTIONAL TOLL OF FAILURE

It is important to discuss the emotional toll of working in and around the space industry. Few vocations are so rife with failures, yet there is a sense of determination that after each failure one will voluntarily sign up to try again. The Mars 96 project took 10 years and $300M to complete; the spacecraft carried 23 instruments supplied by 22 countries. *Mars 96* plunged into the Pacific Ocean 1 day after its launch from Kazakhstan. "I'm looking at a desk full of my Mars stuff, and I feel absolutely gutted to think all my work has ended in failure," says Adrian James, who worked on one of the *Mars 96* instruments, the Fast Omnidirectional Non-scanning Energy Mass Analyser, since 1988 and had been project manager for 5 years. "It's devastating. You work all through your weekends, you tear your hair out over all the problems, you have hassles with the Russians, and then when you finally get your instrument on its way to Mars, it all ends in tears," he says.

It is important to note that entire careers are dedicated to the success of space missions. It is little wonder that crews from failed missions will never forget and have an ingrained dedication to failure avoidance. Failure is a tough training ground. The *Mars Observer* is a case in point. Mars Observer was a NASA mission designed to study the surface, atmosphere, interior, and magnetic field of Mars from orbit. The mission was to last 1 Martian year (687 Earth days), to allow observations during all four seasons. *Mars Observer* was launched on September 25, 1992, aboard a Titan II rocket. The journey to Mars took 11 months. Communications were lost with the spacecraft on August 22, 1993, as the *Mars Observer* was preparing to go into orbit around the planet. The mission's total price tag exceeded $800M. The hardware was a small portion of the costs; however, in spite of the mission lost, many people earned advanced degrees, and the knowledge base was significantly expanded. Many of the personnel involved in Mars Observer went on to become part of the teams that designed the successful Mars Pathfinder and Global Surveyor missions to Mars. These emotional tolls pale in comparison to catastrophic failures involving loss of life. Tragic failures resulting in the loss of life are most poignant and leave us scarred from the sense of corporate loss. Both the Soviet and the American communities have suffered fatal losses, such as those caused by the oxygen-rich flash fires inside a capsule prior to launch suffered in a 1961 Soyuz and the 1967 *Apollo 1*. In 1967, *Soyuz 1* had a catastrophic vehicle crash from parachute failure on reentry. In 1971, a fatal depressurization of a Vostok capsule was caused by a valve prematurely opening during reentry. In 1986, the tragic explosion of the space shuttle *Challenger*

STS-51L was traced to an O-ring failure on the booster rocket that initiated catastrophic destruction of the entire vehicle seconds after launch. In 2003, a heat tile failure on space shuttle *Columbia* STS-107 initiated a catastrophic destruction of the vehicle on reentry. These failures permanently mar our space programs and remind us of the risk of exploration.

GETTING PAST LAUNCH FAILURES

Overly simplified, if a successful launch is not achieved, then the mission story normally ends. Therefore, an overview of failures will begin with launch systems. There have been over 5,000 launches into the exoatmosphere conducted by Commonwealth of Independent States/USSR, the United States, Europe, China, Japan, India, Israel, Brazil, North Korea, France, the United Kingdom, and Australia. Over the years, the failure rate has been close to 10%, with earlier launches running with higher failure rates than the most recent events.[6] Launch failure statistics do not truly reflect the impact to space exploration and exploitation, as numerous launch vehicles carry several payloads, and it can be assumed that a launch failure is catastrophic to all payloads. Teams of hundreds of researchers, engineers, program managers, sponsors, and customers with vested interest in the success of a mission are truly affected after investing years of work in a mission to watch the effort disintegrate because of a failure. Statistics show that among the causes of failure for worldwide space launches in the 25-year period from 1980 to 2004, propulsion subsystem problems predominated. The propulsion subsystem appears to be the Achilles' heel of launch vehicles, and because of the significance of these system failures, it will be reviewed further. The propulsion subsystem, the heaviest and largest subsystem of a launch vehicle, consists of components that produce, direct, or control changes in launch vehicle position or attitude. Its many elements include different stages of main components of rocket motors, liquid engines, and thrusters; combustion chamber; nozzle; propellant (both solid and liquid); propellant storage; thrust vector actuators and gimbal mechanisms; fuel and propulsion control components; feed lines; control valves; turbo pumps; igniters; and motor and engine insulation. Similar components are also used as separation mechanisms in the separation/staging subsystem. Propulsion subsystem failures can be divided into failures in solid rockets (SRs) and liquid rockets (LRs). Of the 76 propulsion failures in 1980–2004, 16 were SR failures and 60 were LR failures. There were 574 launches with SRs and 2,257 launches with LRs. Therefore, the success rate is 97.2% for SRs and 97.3% for LRs. The success rate for SRs and LRs in the 25-year period 1980–2004 was the same. The statistical results essentially settle the ongoing argument about whether LRs are more reliable than SRs or vice versa.[7]

UNSOLVED PHENOMENA: WHAT IS AFFECTING PIONEER?

Failure is not the only producer of unsolved mysteries. The *Pioneer anomaly* or *Pioneer effect* is the observed deviation from expectations of the trajectories of various unmanned spacecraft visiting the outer solar system, notably *Pioneer 10* and *Pioneer 11*.

The *Pioneer effect* causes satellites to speed up or slow down slightly as they conduct Earth flybys, and as yet the physical explanation for the anomaly remains a mystery.[8] More recently, teams at the Jet Propulsion Laboratory have built detailed models of *Pioneer 11*, including a thermal model which predicts how heat is distributed. These models lend credence that uneven heat emission could potentially account for 28%–36% of the anomaly detected.

At present, there is no universally accepted explanation for this phenomenon; while it is possible that the explanation will be mundane—such as thrust from gas leakage—the possibility of entirely new physics is also being considered. Explanations for the discrepancy that have been considered include the following.

- Observational errors, including measurement and computational errors, in deriving the acceleration.
- A real deceleration not accounted for in the current model, such as
  - Gravitational forces from unidentified sources, such as the Kuiper belt or dark matter; drag from the interplanetary medium, including dust, solar wind and cosmic rays;
  - Radiation pressure of sunlight, the spacecraft's radio transmissions, or thermal radiation pressure; or
  - Electromagnetic forces due to an electric charge on the spacecraft.
- New physics
  - Relative to clock acceleration?
  - Relative to the law of gravity, which could also be interpreted as a modification of inertia?
  - Relative to down-scaling of photon frequency?
  - Relative to extending the Hubble law [which relates the increase (redshift) of the wavelength of a photon from another galaxy to the expansion of the universe]?

There are several remaining options for further research:

- Further analysis of archived Pioneer data could be conducted.
- The *New Horizons* spacecraft to Pluto is spin-stabilized for much of its cruise, and there is a possibility that it can be used to investigate the anomaly.
- A dedicated mission has also been proposed. Any such mission would probably need to surpass 200° AU from the sun in a hyperbolic escape orbit.
- Observations of asteroids around 20° AU may provide insights if the anomaly's cause is gravitational.[9]

UNSOLVED MYSTERIES

Lost and Found: Camera Tracking and Sighting

A puzzling object was discovered in orbit around Earth on a voyage through the solar system on September 20, 2002. It is small, only perhaps 30 m long, and rotates once every minute or so. Amateur astronomer Bill Yeung first spotted the 16th-magnitude speck of light in the constellation Pisces on September 3. He named it J002E3. Analysis of the observations by the Minor Planet Center established that the object was in a geocentric rather than heliocentric orbit. This extremely unusual orbit prompted a flurry of observational and analytic interest. Detailed analysis of incoming position observations at the Jet Propulsion Laboratory Near-Earth Object Program Office determined that the object was in an unstable, 42-day orbit about the Earth. Researchers at the laboratory concluded that this orbital behavior was inconsistent with a natural solar system body such as an asteroid but was very consistent with a human-made body launched from the Earth. Their analysis of the orbital characteristics and the timeline pointed to the upper S-IVB stage of the *Apollo 12* launch in November 1969. The measured spectra of J002E3 matched white paint better than asteroid surface materials, and the measured photometric characteristics indicated that J002E3 did not appear to be an asteroid. Photometric and high-resolution visible spectral observations of J002E3 were made using sensors on the 3.67-m Advanced Electro-Optical System telescope and were in quantitative agreement with the astronomical observations. Detailed analyses of these observations were conducted with the application of algorithms designed for use with artificial satellites. The configuration, size, and dynamics of J002E3 were determined using photometric lightcurves. The high-resolution visible spectra were compared with the spectra of aged white paint observed on other rocket bodies.[10] Speculations include that it might be one of the Spacecraft-Lunar Module Adapter panels, or an S-IVB from *Apollo 12*. Unlike that of *Apollo 14*, *Apollo 12*'s S-IVB did not crash into the Moon. The crew jettisoned it on November 15, 1969, when it was nearly out of fuel. Once the astronauts were safely away, ground controllers ignited the S-IVB's engine. They meant to send the 60-ft.-long tank into a sun-centered orbit, but something went wrong; the burn lasted too long. Instead of circling the sun, the S-IVB entered a barely stable orbit around the Earth and Moon. The object is as bright as a 30-m-wide space rock, and it is moving about as fast as an asteroid should move. It loops around Earth once every 48 days or so, coming as close to our planet as the Moon and ranging as far away as two lunar distances. There is no evidence that the speck is moving under its own power. The orbit is constantly changing because of gravitational perturbations by the Sun and Moon. J002E3 left Earth orbit in June 2003 to resume its orbit around the sun. It is predicted to return in about 30 years, when this mystery may well be solved.[11] Another mysterious orbiter is 6Q0B44E, sometimes abbreviated to B44E; it is a small object, probably an item of space debris, currently orbiting the Earth outside the orbit of the Moon. 6Q0B44E was first observed by Catalina Sky Survey researchers at the Lunar and Planetary Laboratory of the University of Arizona on August 28, 2006. The object is just a few meters across and has been provisionally classified as artificial. B44E orbits the Earth between 585,000 and 983,000 km, which is 2 to 3 times the distance of the Moon's orbit, over a period of 80 days. Calculations from the observations suggest that B44E probably entered the Earth-Moon system between 2001 and 2003, although it may have arrived up to a decade earlier. Similarities between the discoveries of B44E and J002E3, now believed to be part of the *Apollo 12* rocket, have led some astronomers to speculate that B44E may be another relic of human space exploration that has returned to Earth orbit.[12]

The *Mars Polar Lander* was part of the NASA Mars Surveyor '98 program, which consisted of two spacecraft launched separately, the *Mars Climate Orbiter* (formerly the *Mars Surveyor '98 Orbiter*) and the *Mars Polar Lander* (formerly the *Mars Surveyor '98 Lander*). The two missions were designed to study the Martian weather, climate, water, and carbon dioxide levels in order to understand the reservoirs, behavior, and atmospheric role of volatiles and to search for evidence of long-term and episodic climate changes. Communication with the lander was lost prior to atmospheric entry. The *Mars Polar Lander* was to touch down on the southern polar layered terrain, between 73° S and 76° S in a region called Planum Australe, near the south pole, near the edge of the carbon dioxide ice cap in Mars' late southern spring. The last telemetry from *Mars Polar Lander* was sent just prior to atmospheric entry on December 3, 1999. No further signals have been received from the lander. According to the investigation that followed, the most likely cause of the failure of the mission was a software error that mistakenly identified the vibration caused by the deployment of the lander's legs as being caused by the vehicle touching down on the Martian surface, resulting in the vehicle's descent engines being cut off while it was still 40 meters above the surface, rather than on touchdown as planned. Another possible reason for failure was inadequate preheating of catalysis beds for the pulsing rocket thrusters: hydrazine fuel decomposes on the beds to make hot gases that throttle out the rocket nozzles; cold catalysis beds caused misfiring and instability in crash review tests.

Attempts were made in late 1999 and early 2000 to search for the remains of the *Mars Polar Lander* using images from the *Mars Global Surveyor*. These attempts were unsuccessful, but re-examination of the images in 2005 led to a tentative identification, described in the July 2005 issue of *Sky and Telescope*.[13] However, higher-resolution photos taken later in 2005 revealed that this identification was incorrect and that *Mars Polar Lander* remains lost.[14] NASA is hoping that the higher-resolution cameras of the *Mars Reconnaissance Orbiter*, currently in Martian orbit, will finally locate the lander's remains.[15] Recently, the HiRISE camera on the *Mars Reconnaissance Orbiter* has taken high-definition images of the probable crash area, and the public is invited to help solve this mystery by scanning these images for signs of the *Mars Polar Lander*. A typical screen would require 1,600 screenshots to view one image, and there are 18 images total to scan.[16] This becomes a scavenger hunt on a global (cosmic?) scale.

Sociopolitical Motivators: Mystery Solving and the Cold War

James Oberg, former NASA mission manager and scholar of the Soviets in space during the Cold War, has compiled an enticing list of questions and unsolved mysteries relative to the USSR and the Cold War, where very little information—or at least extremely "scrubbed" data—was released. I list only a few here relative to this handbook to be solved through further research into archives and oral histories.

- The early payloads of the Zenit booster (the so-called "hulks") in the late 1980s: What were they really? Why was one sent into a sun-synchronous retrograde orbit with no follow-on? And what were those debris that wound up flung into a higher orbit at insertion and that, on the first launch, were the only objects to reach orbit? (The Soviet failure to register them with the United Nations was their most egregious violation of the Convention on Registration of Outer Space Objects.)

- Relative to a jettisoned old spacesuit on an early *Mir* extravehicular activity: there is a wonderful story of cosmonauts saluting the "fallen comrade" as "he" drifts away. When did this happen, and are there pictures?
- Did the Soviet military recover any lost *Discoverer* capsules? What other U.S.-origin space debris have they found, studied, and preserved? Did they ever have to go outside their own country—say, into Iran or China—to retrieve errant space objects? And what about that reported late-1960s Soyuz-class vehicle on display in the People's Army Museum in Beijing?[17]

CONCLUSION

Failures in space are caused by a wide variety of events and may range from a small, recoverable upset to catastrophic incidents and traumatic impact on all the parts of a spacecraft mission. As future research is conducted, we will be able to determine the greater significance of causes of failures. Seeking out reasons behind unresolved anomalies is one reason to revisit space hardware—that is, the material remains. Archaeology is the science that studies human cultures through the recovery, documentation, analysis, and interpretation of material remains and environmental data, including architecture, artifacts, features, biofacts, and landscapes.[18] Within the realm of space archaeology, some of these mysteries will remain until the opportunity for crewed or robotic field work is available. As is noted above in some of James Oberg's work and later in this volume, in Ralph Lorenz's Chapter 25, much of this field work may be done terrestrially.

NOTES AND REFERENCES

1. Dictionary.com Unabridged (v 1.1). Random House, Inc. http://dictionary.reference.com/browse/nominal.
2. Sarsfield, L. 1998. Failure in Spacecraft Systems. Cosmos on a Shoestring Appendix B. Santa Monica, CA: Rand Corporation.
3. Mishap Reporting, Investigating, and Recordkeeping, NASA Procedural Requirements for, February 11, 2004 NPR 8621.1A. Appendix A.
4. http://www.sat-nd.com/failures/. From the Web site's author: "On these pages, I've tried to collect facts and figures relating to on-orbit satellite failures or outages."
5. http://www.russianspaceweb.com/zenit_nss8.html.
6. http://www.sat-nd.com/failures.
7. Chang, I.-S. Space Launch Vehicle Reliability. Paper presented at The Aerospace Corporation 41st AIAA-2005-3793 AIAA/ASME/SAE/ASEE Joint Propulsion Conference & Exhibit, July 10–13, 2005, Tucson, AZ.
8. Chang, I.-S., and E.J. Tomei. Solid Rocket Failures in World Space Launches. Paper presented at The Aerospace Corporation 41st AIAA/ASME/SAE/ASEE Joint Propulsion Conference & Exhibit, July 10–13, 2005, Tucson, AZ. AIAA 2005-3793.
9. Daily Launch Today's News for AIAA from Newspapers, TV, Radio, & Journals. March 5, 2008. See http://aiaa.custombriefings.com or http://www.aiaa.org/content.cfm?pageid=4 for archived material.

10. http://science.nasa.gov/headlines/y2002/20sep_mysteryobject.htm.
11. Lambert, J.V., K. Hamada, D.T. Hall, J.L. Africano, K. Luu, P. Kervin, M. Giffin, and K. Jorgensen. 2004. Photometric and spectral analysis of MPC object J002E3. In *IEEE Aerospace Conference Proceedings*, vol. 5, 2866–2873. Big Sky, MT: IEEE Aerospace Conference Proceedings.
12. Mystery Object Orbits Earth: A Puzzling Object Just Discovered in Orbit around Earth Might Be an Apollo Rocket on a Fantastic Journey through the Solar System.
13. http://ssd.jpl.nasa.gov/?horizons_news.
14. Editorial. Mars Polar Lander Found at Last? 2005. *Sky & Telescope*. http://www.skyandtelescope.com/news/3310281.html?page=1&c=y.
15. http://en.wikipedia.org/wiki/Mars_Polar_Lander#cite_note-1#cite.
16. Muir, H. 2008. http://www.newscientist.com/article/dn13884.
17. Oberg, J. 1995. Soviet Space Secrets. *Spaceflight (British Interplanetary Society Monthly)* 37: 254–255.
18. Renfrew, C., and P.G. Bahn. 1991. *Archaeology: Theories, Methods, and Practice*. London: Thames and Hudson Ltd.

23 Failed Mars Mission Landing Sites: Heritage Places or Forensic Investigation Scenes?

Dirk H. R. Spennemann and Guy Murphy

CONTENTS

LOST IN SPACE: FAILED MARS MISSION LANDING SITES AS FORENSIC INVESTIGATION SCENES

Humanity is busily leaving traces on various celestial bodies, such as the Moon and, more recently, Mars. While several of the missions have been successful, some of them have failed, on occasion quite spectacularly. There is an interest in understanding what went wrong in the mission profile that lead to the failure. It can be posited that some of these sites, such as the as yet undiscovered crash site of the *Beagle 2* mission, will become prime objectives of future recovery operations. This chapter explores the management and ethics of crash analysis methodology and heritage conservation, as it applies to Mars mission crash sites. What future heritage value will the crash site of a failed mission site have compared with a successful one, and how should we manage this?

INTRODUCTION

The enterprise of sending spacecraft to the surface of Mars is a highly risky one, with more than half of all launched missions ending in failure. Each failed mission represents a costly investment in human energy, money, and national prestige, which has typically ended with a sudden radio silence that then became a mystery. Why did the mission fail, and where is the spacecraft now? Confronted with a failure to signal home after atmospheric entry or a sudden halt in communications, mission controllers have been left to only speculate as to what might have gone wrong.

With the arrival of new successive orbiting probes with ever-higher surface imaging capacity, it should soon be possible to identify the landing (or crash) sites of failed Mars missions on the planetary surface. Human curiosity being what it is, at some point in the distant future, people will eventually visit these sites, and that they have value as heritage places is generally recognized.[1] As future missions to Mars are conceptualized, and as the previously landed spacecraft have heritage value of varied degrees of significance, there is a need to develop procedures that govern the disturbance and removal of parts from landed spacecraft. This chapter examines how failed mission surface sites may be characterized as potential forensic investigation scenes and considers what policy implications may flow from this when humans eventually arrive on the scene. It uses the *Beagle 2* lander as a case study from which to draw a series of more general policy conclusions.

There are two main reasons to pursue this line of examination. A systematic forensic investigation of why missions failed will inform future mission planning. It will also fill in gaps in the history of space exploration.

It may seem a premature to be considering the possibility of the human disturbance of heritage artifacts on other planetary bodies, but there already exists a precedent for this on the lunar surface, when *Apollo 12* landed just 180 m away from *Surveyor 3*, which had touched down 31 months earlier.[2] Astronauts Pete Conrad and Alan Bean visited the spacecraft on their second moonwalk on November 20, 1969, examining *Surveyor 3* and its surroundings, taking photographs of the spacecraft and its surroundings, and then removing about 10 kg of parts from the spacecraft for later examination back on Earth. Items removed from the craft included the television camera, electrical cables, the sample scoop, and pieces of aluminum tubing.[3]

It has been speculated that future human explorers on Mars may wish to visit historic lander sites in order to collect artifacts as souvenirs or trophies,[4] and this could result in the loss of critical information as to why the mission failed.

POTENTIAL FOR FORENSIC SITES ON MARS

Commencing with the crash landing of *Mars 2* some 35 years ago, humanity has left a range of traces on the Martian surface. Ignoring the flybys and launch failures, to date, twelve surface missions have been launched to Mars, eleven of which entered the Martian atmosphere and five of which can be regarded as successful. A summary of all failed missions known or suspected to have left remnants on the Martian surface is shown in Table 23.1.[5] In addition to these, two other missions warrant mentioning, Phobos 2 and Mars Climate Orbiter.

Phobos 2

Phobos 1 and Phobos 2 were the next generation in the Soviet Venera-type planetary missions. The objectives of the Phobos missions were to (1) conduct studies of the interplanetary environment, (2) perform observations of the sun, (3) characterize the plasma environment in the Martian vicinity, (4) conduct surface and atmospheric studies of Mars, and (5) study the surface composition of the Martian satellite Phobos. To achieve the latter objective, the Phobos craft were to approach within 50 m of Phobos' surface and release two landers, one a mobile "hopper," the other a stationary platform.

Although *Phobos 1* was lost en route to Mars because of a communications failure, *Phobos 2*, launched on July 12, 1988, was successfully inserted into the Martian orbit on January 29, 1989. *Phobos 2* operated as scheduled throughout its cruise and Mars orbital insertion phases. Shortly before the final phase of the mission, when it was approaching Phobos' surface, contact with the spacecraft was lost. The mission ended

TABLE 23.1
Possible Crash Sites of Spacecraft on the Martian Surface

	Mission							
	Mars 2	***Mars 3***	***Mars 6***	***Phobos 2***	***Mars Climate Orbiter***	***Mars Polar Lander***	***Deep Space 1 and 2***	***Beagle 2***
Country	USSR	USSR	USSR	USSR	United States	United States	United States	ESA
Launch date	May 19, 1971	May 28, 1971	August 5, 1973	July 12, 1988	December 11, 1998	January 3, 1999	January 3, 1999	June 2, 2003
Landing date	November 27, 1971	December 2, 1971	March 12, 1974	March 27, 1989	September 23, 1999	December 3, 1999	December 3, 1999	December 25, 2003
Location	45° S, 313° W	45° S, 158° W	23° S, 19° W	?	?			
Item	Descent module, tethered robot	Descent module, tethered robot	Descent module, tethered robot	Lander 1, Lander 2	Orbiter	Descent module	Impact probe A, Impact probe B	Descent module, rover
Mass (kg)	1,210	1,210	?	2,600	338*	?	3.5 each	?
Mass of lander (kg)	350	350	635	?	?	290	?	?
Dimensions	?	?	?	?	2.1 × 1.6 × 2 m	?	?	?
Nature of failure	Crash, entry angle too steep	Failure 20 s after landing	Failure just before landing	?	Crash due to altitude error	Crash, premature engine shut down?	Failure to separate, failure to release	?

* Not counting propellant.

when the spacecraft signal failed to be successfully reacquired on March 27, 1989. The cause of the failure was determined to be a malfunction of the onboard computer.[6] It is not known whether or not the lander touched down or crashed onto Mars' surface.

Mars Climate Orbiter

Mars Climate Orbiter was launched on December 11, 1998, with the mission of studying the Martian weather, climate, and water and carbon dioxide budget. The spacecraft reached Mars and commenced an orbit insertion main engine burn on September 23, 1999. Soon afterward, the spacecraft passed behind Mars. Radio contact was never reestablished, and no signals from the spacecraft were obtained.

Subsequent investigations revealed that some spacecraft commands had been sent in Imperial units instead of metric. As a result, the spacecraft entered the Martian atmosphere at about 57 km instead of the intended 140–150 km altitude above Mars. The lower insertion altitude would have resulted in the spacecraft either reentering heliocentric space or being destroyed by atmospheric stresses and friction.[7] It is unclear whether any debris from the craft would have reached the Martian surface. Given that there is a high degree of confidence that the cause of the mission failure has been identified and that surface remnants (if any) are likely to consist only of burnt debris buried in impact craters, the debris are presumably of little or no forensic value.

Physical Characteristics of The Sites

Depending on the point in the mission cycle at which these missions failed, the tangible evidence of these missions on the Martian surface may range from landed and thus essentially intact (albeit not operationally successful) spacecraft with associated secondary artifacts to mere impact craters containing debris.

Artifacts associated with successful surface missions have been described by the authors in a previous paper[8] as falling into three categories: (1) *Technological artifacts:* those human-made objects that were distributed onto the Martian surface during the landing sequence. These include the landers, heat shields, back plates, parachutes, and so forth. (2) *Environmental artifacts:* those marks left on the Martian surface by the technological artifacts, including impact marks, bounce marks, rover tracks, trenches and abrasion marks on rocks, and so forth. (3) *Cultural artifacts:* these arbitrarily exist primarily in the human mind. They do not necessarily have a physical manifestation on the surface but nevertheless contribute to the significance of the site. Missions that landed successfully will have all three categories of artifacts present.

If the mission crashed, however, the presence and intactness of technological artifacts will be compromised by the force of impact and possibly also heat from atmospheric friction. In general, the higher the altitude of the craft when mission sequence failed, the less intact any surviving surface artifacts will be, and the larger (and potentially more scattered) will be any impact craters. With impactor missions such as Deep Space 2, surface probes were designed to crash into the planetary surface at high velocity and continue operating after impact.

Depending on when the last status signal was received from the lost craft, mission planners will typically be left with a range of possible failure scenarios, which will each in turn have distinct outcome in terms of surface artifacts.

In at least one case (*Mars 6*), the lander was designed to communicate with Earth via another spacecraft in Martian orbit (*Mars 5*). Communication with this lander ceased several seconds before it reached the surface. It is not clear whether this could have occurred because of a failure in the orbiter. If this was the case, then critical forensic information may still be present within that craft if it is still in Martian orbit. While the forensic examination of spacecraft still in space warrants further discussion, the discussion in this chapter will be restricted to artifacts on the Martian surface.

WHAT MAKES THESE SITES FORENSIC?

In a formal sense, *forensic* is a legal term referring to things that are used or applied in the investigation and establishment of facts or evidence in a court of law. In the context of this chapter, it is assumed to refer more broadly to the processes, procedures, and physical evidence that may be used to establish presently unknown or inconclusive facts about the causes and nature of the failure of a Mars surface mission.

Integral to the act of forensic investigation is the piecing together of undisturbed, primary evidence of an unobserved event to determine as accurately as possible what has occurred. In terrestrial contexts, this is inherent to archaeological investigation. Archaeological methodology has been employed in a forensic science context to document exhumations,[9] crime scenes,[10] and graves of victims of war crimes.[11] Another investigative paradigm is the examination of air crashes, where in order to understand the cause of the air crash, investigation teams spatially record and then collect and analyze all pieces of the aircraft wreckage. A related field is the area of shipwreck investigation. Paleontology shares a concern with recording the position and condition of material. In all such situations, a diligent and detailed documentation is made of the extant remains in order to fully and accurately collect and interpret the information contained at the site.

The failed Mars mission crash sites identified in the previous section may be defined as forensic sites because their currently largely undisturbed physical condition contains evidence from which unique information about why and how the mission failed may be obtained. These differ from successful missions, where the descent, landing, and operational sequences of the missions proceeded as planned and where there will typically be data about the landing site and surface artifacts that have been collected by the probes themselves and returned to Earth.

They are also forensic sites in the broader sense that the precise location on the surface where the craft's descent occurred and where remnant artifacts now remain is, at the time of writing, unknown for all missing lander craft. Like a lost archaeological site or an unknown location associated with a crime, a process needs to be undertaken first to identify the site where part of the story or event occurred and where physical evidence may now remain.

The lack of knowledge of the location of missing spacecraft is a reflection of a number of factors. The earliest probes were sent to a planet for which there were no detailed surface maps or images. There was no global positioning system in operation on Mars to allow specific sites on the surface to be pinpointed using satellites as may be done on Earth. The distance from Earth meant the descent of spacecraft could not be controlled in real time, and not all landers were designed to transmit data back to Earth during descent. Mission planners sending surface probes to Mars on a trajectory involving direct entry into the Martian atmosphere had to content themselves with selecting a broad geographic region where the landing would occur, the actual landing site being a random location within large ellipse measuring tens of kilometers across, with an area of hundreds of square kilometers. Until recently, orbiting spacecraft have not had the sufficiently high-resolution imaging capacities to identify features as small as spacecraft artifacts, and small objects are still beyond the range of current apparatus.

The central initial issue with the management of landing or crash sites as forensic scenes is trying to determine from the surface artifacts a definitive explanation as to why the mission failed. Understanding this surface information presupposes the availability of a complete and comprehensive data set about the mission hardware, mission sequence, and returned postlaunch data. With both successful and unsuccessful missions, the increasingly high resolution imaging capacity of orbiting spacecraft will make it possible to obtain aerial imagery of landing (or crash) sites. Although this is valuable, it is only part of a much more comprehensive data set potentially available on the Martian surface, which can be fully obtained only by direct onsite investigation.

Relevant forensic information includes different types of data, which are described below. The potential for environmental conditions on the Martian surface to cause information loss is then outlined.

Types of Forensic Data

Spatial Distribution of Artifacts

This category includes the number of artifacts to have reached the surface, their position on the surface relative to each other and the surrounding environment, and also any surface markings. These give an indication of the point at which mission failure occurred when assessed against the predicted distribution resultant from a successful landing.

Configuration of Artifacts

The configuration of individual hardware elements at a successful landing would give further indication of the mode of mission failure. This might include whether elements such as parachutes or solar panels were deployed.

Physical Condition of Artifacts

The physical condition of surviving landed artifacts can be a source of information about the descent phase and landing process. This could include evidence of heat damage from atmospheric entry, physical damage resultant from hard impacts or be the absence of damage, which would be indicative of a successful landing sequence.

Stored Data

There may be electronically stored data within surface artifacts that can be used to piece together the story of the mission. Some landers may have been designed to collect data during the descent phase. If the mission failure was caused by a breakdown of the craft's earthbound communications system, the lander may have commenced collecting surface data, which could still be held within its electronic systems.

Biological

This category is described further below. It consists of surviving microbial life that may have been accidentally transported to Mars on the landers, and also evidence of biological activity that may exist in samples taken by landers.

Environmental Conditions as a Source of Information Loss

An important consideration that needs to be taken into account when interpreting forensic information from lander crash sites is the potential for environmental conditions on the Martian surface to cause disturbance of the original distribution and condition of surface artifacts after the crash or landing has occurred. This may arise because of things such as wind action, abrasion from surface dust, extreme daily temperature differentials, effects of moisture, ultraviolet light exposure, reactive chemicals in the surface regolith, and electrostatic action.

The longer the time period between the initial landing or crash and the forensic investigation, the greater effect these forces will have and the greater the potential loss of forensic information will be. Artifacts that have become buried or semiburied could also be made more vulnerable through corrosion through contact with the regolith. Damaged spacecraft components may have interior elements exposed, which could make them more vulnerable to deterioration.

Planetary Contamination as an Additional Consideration

Although independent of what caused the mission failure, the issue of planetary contamination at crash sites may also be classed as forensic.

Planetary contamination is the introduction (intentional or unintentional) to a planetary body of non-native biological organisms from another body (Earth), which may then either damage or displace native ecosystems or make it difficult to subsequently determine whether the organisms had arrived by natural processes. Over recent decades there has been a realization that life is capable of persisting in much more extreme environments on Earth than had previously been thought possible, meaning it may also be able to persist in a greater range of places in the solar system. Earth organisms, known as extremophiles, are theoretically capable of surviving the extremes of Martian surface conditions such as cold, acidity, and ultraviolet light, particularly if they are subject to some localized protection, such as shielding or becoming buried. Over long periods of time, such organisms have the potential to cause significant changes to the environment on a local or wider scale through biological action.

Although it has been theorized that there may be (or have been in the past) a small degree of natural exchange of biological material between Mars and Earth through natural processes involving meteor impact ejecta,[12] it is generally agreed that the deliberate contamination of the Martian surface with Earth organisms should be avoided. If biological organisms exist on Mars, it will be of great scientific interest to determine whether they evolved independently from life of Earth or were transported to Mars from Earth in the recent or distant past. Deliberately or accidentally introduced Earth organisms may potentially confuse an understanding of any life that exists on Mars on account of natural processes and could potentially damage native ecosystems.

There has been agreement among various national space agencies since at least the 1980s to prevent unintentional contamination of the Martian surface by applying sterilization procedures to Mars-bound landers. More investigation is required into the planetary protection procedures applied to all surface probes launched to date. With the early Soviet landers, it is unclear whether sterilization procedures were applied. Also, if sterilization procedures were not applied to the interiors of all components, the compartments may have been exposed if the craft was exposed to violent impact forces on impact, thus becoming a source of contamination. Microbes may survive at forensic crash sites on Mars, and they may be destroyed by subsequent human-induced exposure to ultraviolet light or chemicals in the regolith, so their possible presence ought to be addressed in forensic management policy.

Regardless of whether lander craft have been subjected to sterilization procedures, there would be scientific interest in trying to determine whether Earth organisms have arrived on Mars via landed/crashed Mars probes and survived. This would apply to successful missions as well as unsuccessful ones.

A slightly different future scenario might involve examining samples taken by successful lander missions. The Viking missions that landed in 1976 undertook a series of experiments on surface regolith samples that were designed to detect evidence of life. The results were ambiguous to an extent that scientific debate continues about their correct interpretation, particularly as a deeper understanding of the biochemistry and growth habits of extremophiles has emerged. It may be possible at a future date to retrieve samples taken from the Viking and other landers for further analysis, and careful treatment will be required at the time of disturbance to ensure that delicate chemical and possibly biological evidence is not destroyed.

When Does the Science Stop?

The question of whether there may be Earth microorganisms present at forensic crash sites touches on a larger issue that applies to both successful and unsuccessful Mars surface missions. When does the landing/crash site cease to operate in subjugation to the needs of scientific investigation and become a place governed by primarily the principles of heritage management? This point could be defined as when the craft is no longer able to directly or indirectly send signals back to Earth, or alternatively when it stops collecting data about the surrounding environment. A more formal definition could simply be when the craft ceases to fulfill any of its mission objectives. Assuming one or more of these definitions are followed, there is still likely to be scope for opportunistic scientific investigations when humans visit the site that were not part of the original mission plan. Microbial investigation of the artifacts and landing site is one such field of investigation. Given the rarity of opportunities to undertake scientific research of this nature on the Martian surface, the heritage management of lander sites should try to allow for the undertaking of opportunistic science, and their forensic management should seek to ensure that opportunity is not lost when the site is first disturbed.

BEAGLE 2: A CASE STUDY

It is the argument of this chapter that the management of crashed lander mission sites differs from that of successful missions, in that information confirming the mode of mission failure is likely to be present in physical evidence on the planetary surface and that preserving and obtaining this information will require additional specific management procedures in addition to those heritage management policies advisable for successful landing sites. The following analysis intends to demonstrate this using the lost Beagle 2 mission as a case study. It identifies the probable causes of mission failure and the likely physical evidence produced by each of them and then, in conclusion, suggests management procedures to optimize the survival and collection of this evidence (Figure 23.1).

History of the Mission

Beagle 2 was launched with the *Mars Express* orbiter at the Baikonur Cosmodrome on June 2, 2003. It was a British-designed and British-built surface lander, the first surface mission to be sent to Mars by the European Space Agency. Its primary science objective was to detect "extinct and/or extant life on Mars, or at least to establish if the conditions on the planet were ever suitable for life to have evolved."[13] It was to investigate the Martian atmosphere and surface geochemistry and mineralogy, undertaking analysis of surface and subsurface samples.[14]

A distinguishing feature of the probe was its small size, its dimensions in a closed configuration being 523 by 924 mm, with a weight of 33.2 kg. Compared with previous Mars surface missions, *Beagle 2* was physically the smallest lander, its maximum dimension reaching 1.9 m with its lander lid open and solar panels folded out. By contrast, the *Viking 1* and *Viking 2* craft measured almost 3 m across, and the *Pathfinder* lander was approximately 2.75 m across in its open configuration. The Mars exploration rovers (MER) *Spirit and Opportunity* measure approximately 2.3 by 1.6 m.[15] The failed *Mars Polar Lander* measured 3.6 m across when fully deployed.[16] *Beagle 2* was also distinguished by the ratio of its science payload (9 kg)

FIGURE 23.1 The *Beagle 2* lander in its surface operation configuration. (Model used in compliance with the *Beagle 2* team requirements; all rights reserved.)

to support systems, understood to be the highest of any planetary lander. For its size, it had an extraordinary ability to investigate its surroundings. The lander contents consisted of three parts, the electronics module, gas analysis package, and a platform for tools and instrumentation called the Position Adjustable Workbench (PAW), which was located at the end of the Anthropomorphic Robotic Manipulator (ARM), a robotic sampling arm. There was also PLUTO (Planetary Underground Tool, also know as the Mole), a small burrowing device attached by cabling, with the capability to bring soil samples back to the lander (Figure 23.2).[17]

Mars Express and *Beagle 2* arrived at Mars early on December 25, 2003, the lander being scheduled to touch down on the Martian surface at 2:54 Universal Time, on Isidis Planitia at approximately 10.6° N, 270° W. The planned landing sequence involved initial deceleration due to friction on the heat shield as the landing pod entered the atmosphere, with further slowing provided by parachutes, which were to start opening 4.5 minutes before landing. At about 1 km above the surface, gas bags were to inflate around the landing unit, the parachutes then being released. The lander would come to rest on the surface, having bounced up to fifteen times while protected by the gas bag system. Once stationary, these bags were to deflate and detach.

The lander would then undergo deployment, which involved its lid of the lander flipping open and then four petal-like solar panels unfolding from the lid in a radiating pattern. Twenty minutes after deployment was complete, the camera system was to use its right camera to take a compressed, black-and-white 360° image, subsequently taking a second image, and a third the next day. During the first night it was to attempt photograph Phobos passing within view of its right camera. On the second

FIGURE 23.2 Computer-generated picture of the *Beagle 2* lander leaving *Mars Express*. (Model used in compliance with the *Beagle 2* team requirements; all rights reserved.)

day, the bolts holding the ARM and PAW were to be released, and also the Mole's lock pin, the craft then awaiting further instruction from Earth.[18]

Beagle 2 was not designed to broadcast signals during its descent phase. It was capable of communicating indirectly with Earth via signals by relayed through the European Space Agency's *Mars Express* and NASA's *Mars Odyssey* spacecraft from Martian orbit, and also directly via radio telescopes on Earth such as Jodrell Bank. The first opportunity to receive a signal indicating a successful landing was to occur at 5:30 a.m. GMT on December 25 via a relay from the *Mars Odyssey* orbiter. No signals were ever received from the lander from the surface over the following months by either the *Mars Express* or *Mars Odyssey* orbiters or by Earth-based receivers.

Had *Beagle 2* landed and deployed successfully, the expected surface distribution of technological artifacts would have consisted of the lander unit, with the airbags and parachutes nearby, and the heat shield a further distance away. Environmental artifacts may consist of a surface impact mark associated with the heat shield, and a series of bounce marks on the surface from the lander and airbags, these being closer together and lighter nearer to the lander. The airbags would have deflated. Both the airbags and parachutes would have initially been deposited on the surface intact, although they would deteriorate over time as they were subjected to environmental forces. The lander would be undamaged externally and internally and would have deployed itself by flipping open the lid and unfurling the solar panels. Had the first 2 days on the surface proceeded as planned, the ARM, PAW, and Mole would have all been unlocked, and several surface images would have been taken.

Possible Failure Causes

The last information received on the condition and status of the Beagle consisted of photographs taken by the *Mars Express* orbiter as the probe detached and moved away from the orbiter on its planned trajectory for atmospheric entry (Figure 23.3). The lander was not designed to broadcast signals during its descent phase. No more information is known about the fate of the craft, leaving mission planners only able to speculate about when during the descent sequence that failure occurred and what was its cause. Unless the craft completely burned up during atmospheric entry, there should be physical evidence of the lander on the Martian surface, and this will constitute forensic evidence that is suggestive (if not conclusively indicative) of the mode of failure.

An outcome of the formal investigation into the mission failure was the publication of the *Beagle 2* Mission Report in 2004. A series of likely causes was identified in the executive summary of the report, and these are described in detail in the following section, with comment on the likely forensic evidence resultant from each as deduced from an understanding of the overall mission plan and mode of failure.[19] These have also been summarized in Table 23.2.[20] It should be noted that this list does not comprehensively cover every possible failure mode and that other, low-probability events may have been the cause of failure. If so, these will also have characteristic outcomes of surface artifact distribution and contribution. It is beyond scope of this chapter to identify and describe every possible cause of mission failure.

Low Atmospheric Density

An unusually low atmospheric density was detected at 20–40 km over the Martian surface in early January. If this existed when *Beagle 2* descended, the craft may have descended too fast and parachute deployment failed. The parachutes may not have

FIGURE 23.3 The last communications received concerning the status of *Beagle 2* consist of photographs taken by *Mars Express* as the lander detached and commenced its descent trajectory on December 19, 2003. (ESA; *Beagle 2*; all rights reserved.)

TABLE 23.2
Possible Causes of Mission Failure for *Beagle 2*, with Associated Forensic Evidence

Failure Cause	Artifact Status					
	Lander	Airbags	Parachutes	Heat Shield	Bounce Marks	Impact Crater
Low atmospheric density	Possibly debris	Possibly debris	Possibly debris	Possibly debris	Absent	Possibly present
Heat shield malfunction	Possibly debris	Possibly debris	Possibly debris	Possibly debris	Absent	Possibly present
Electronics malfunction (entry induced)	Present	Present	Present	Present	Present	Absent
Lander electronics malfunction	Present	Present	Present	Present	Present	Absent
Collision with parachutes/ airbags	Present	Present	Present	Present	Modified	Absent
Gas bag malfunction	Damaged	Damaged	Present	Present	Modified	Absent
Landing site hazards	Possibly damaged	Present	Present	Present	Present	Absent
Landing damage preventing deployment	Present	Present	Present	Present	Present	Absent
Damage to antenna during landing	Present	Present	Present	Present	Present	Absent

separated. The descent payload may have superheated and burned up or disintegrated on atmospheric entry.

Surface forensic evidence would depend on the extent to which hardware elements survived the stresses of atmospheric entry. At maximum, it may consist of technological artifacts including lander debris, heat shield, and parachute remnants and environmental artifacts consisting of one or more impact craters. The crater from the craft impacting on the Martian surface would be an estimated 5–6 m in diameter, with a maximum feature size dimension of 9 m.[21] If the heat shield separated successfully, it would have landed a short distance from the main impact site in the same manner as in a successful mission landing. At the other extreme, the spacecraft may have completely burned up or may have showered onto the surface as very fine scattered debris, which cannot be remotely detected.

Heat Shield Malfunction

The heat shield may have malfunctioned on account of the failure of thermal protection tiles, the structural failure of the heat shield, or incorrectly calculated aerodynamics. The descent payload may have superheated and burned up or disintegrated on atmospheric entry.

The range of possible surface forensic evidence would be the same as for the previous possible failure cause. No surface artifacts would be consistent with a successful landing.

Entry-Induced Electronics Malfunction

Beagle 2's electronics may have malfunctioned while undergoing shocks and other stresses of during the decent and landing sequences. It is assumed that the heat shield, parachutes and airbags would have functioned correctly in this scenario.

The distribution of the lander, airbags, parachutes, heat shield, and surface markings would be consistent with a successful landing, although the lander's configuration may not. Either the communications system or the craft itself failed, and identifying the malfunctioning electronic component, if this is still possible, could be done only through physical examination and possibly testing the landing craft. If it is possible to determine which electronic component failed through physical investigation, there may be no way of establishing whether this occurred because of descent/landing stresses, or random component failure as described below. Failure caused by descent/landing stresses may be more likely to have a physical manifestation, such as detached components or broken wire connections. Surface conditions may have caused further deterioration of the electronics since landing. There may be stored electronic data that are of interest.

Lander Electronics Malfunction

Electronic components and/or systems in *Beagle 2* may have failed for reasons unrelated to the physical stresses of entry and landing, such as random component failure. Again, it is assumed that the heat shield, parachutes, and airbags would have functioned correctly in this scenario.

The relative distribution and condition of the heat shield, parachutes, air bags, and surface markings would be consistent with that expected in a successful mission. Forensic evidence supporting this as a mode of failure would be identical to that resultant from entry-induced electronics malfunction. If physical investigation of the lander's electronic systems can identify specific components as the cause of failure, it may be the absence of physical damage suggestive of descent/entry stresses that is indicative of electronics malfunction as the cause of failure.

Collision with Parachutes

Collision of *Beagle 2* within its airbags with the braking parachutes could not be ruled out as a cause of mission failure.[22] The lander may have become entangled in the parachute after bouncing on the surface and been unable to separate from the gas bags and deploy once stationary.

The main parachute, the airbags, and the lander may still be entangled on the surface, the heat shield being located a short distance away, as it would be in a successful mission. There may be fewer surface impact bounce marks, and possibly impact and drag marks from the movement of the entangled parachutes and lander on the surface. The fabrics of the parachutes and bags may have subsequently deteriorated with environmental exposure and become detached from the lander.

Gas Bag Malfunction

Simulations suggest the lander would bounce up to 15 times before coming to rest on the surface.[23] The gas bag system may have failed upon initial impact or subsequent bounces because of the high velocities involved, or the bags possibly being punctured by rocks, the lander and/or bag separation system then becoming damaged.

The parachutes and heat shield would be located a short distance from the lander in a manner consistent with a successful landing. The bounce marks on the planetary surface may be closer together, fewer in number, or more pronounced. Both gas bags and the lander may show evidence of physical damage, although the gas bags may have also been subject to environmental degradation. The gas bags may not have separated. The lander may have failed to deploy or only partially deployed.

Landing Site Hazards

Beagle 2 may have encountered hazards on the surface such as craters, large rocks, and so forth, resulting in damage to the unit or it resting in a configuration preventing communication. Failure would have been triggered as or when lander came to rest rather than during the bouncing phase.

The relative distribution and condition of the heat shield, parachutes, air bags, and most surface markings would be consistent with that expected in a successful mission. The condition and configuration of the lander plus the presence of hazardous natural element in the immediate vicinity would be indicative of this mode of failure.

Landing Damage Preventing Deployment

The parachute and airbag systems may have functioned as planned, but damage could have nevertheless occurred during landing that prevented the lander from folding out its lid and solar panels correctly.

The relative distribution and condition of the heat shield, parachutes, and air bags would be consistent with that expected in a successful mission. The lander may have failed to deploy or only partially deployed. The absence of obvious landing site hazards would distinguish this from the previous cause of failure.

Damage to Antenna during Landing

While correct deployment of *Beagle 2* may have occurred, damage to the antenna during landing would prevent it from communicating with Earth.

The surface configuration of artifacts would be entirely consistent with that expected from a successful landing and deployment. The scene would be of a successfully deployed lander, possibly with several surface images stored in its memory, and with its ARM and Mole unlocked and ready to explore the surroundings. Otherwise fully functional, it would have ceased operation after a period, without having sent images back to Earth or used the ARM or the Mole. This is a scenario where the potential presence of stored data may be of most interest. *Beagle 2* may have collected images and other data relating to the Martian surface, and it may be possible at some future time to recover this information from the lander itself (Figure 23.4).

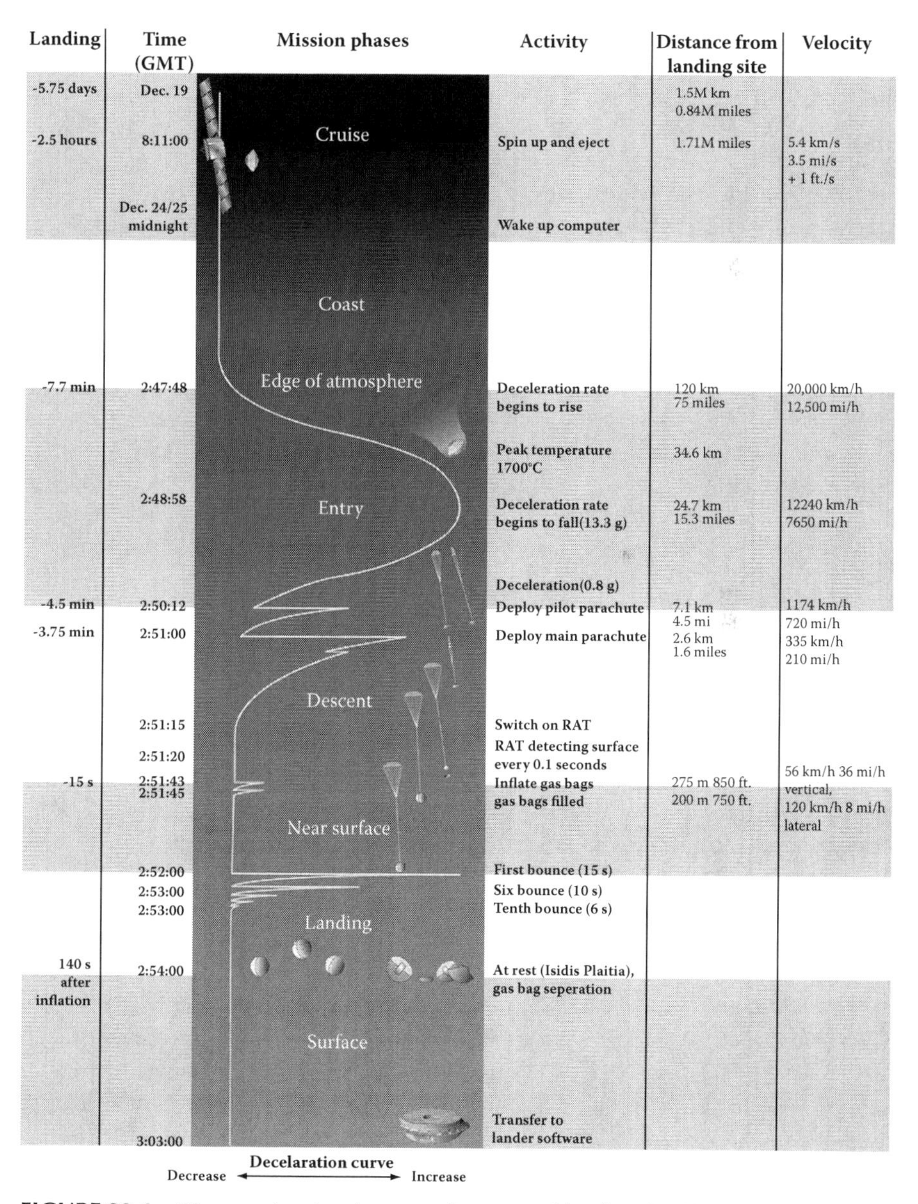

FIGURE 23.4 Diagram showing the entry, descent, and landing timeline showing stages of deceleration, distances above the surface, and velocities. Letters indicate stage in the descent at which each possible cause of failure would have occurred. (*Beagle 2*; all rights reserved.)

STATE OF FORENSIC INVESTIGATIONS

Successful lander missions to date have had the unusual characteristic of being self-documenting artifacts, in that they have returned to Earth photographic and other information about surface artifacts and the landing site. With unsuccessful missions, and in the absence of crewed or automatic missions to Mars specifically tasked with the search for and investigation of the crashed landers, the current state of investigation is confined to photographic observations by orbiting camera platforms.

Efforts to locate *Beagle 2* using orbital imagery has so far proved unsuccessful. In December 2005, a possible landing site was reported as identified in photographs by NASA's *Mars Global Surveyor*. The small size of the lander was slightly beyond the resolution of the best available images, making this identification at best conjectural. Higher-resolution images obtained of the site in 2007 by NASA's newly arrived *Mars Reconnaissance Orbiter* (MRO), however, ruled out this location as showing any firm evidence of *Beagle 2*. While MRO has successfully identified the Viking, Pathfinder, and MER landing sites with their associated artifacts in a series of extraordinary images, the search for *Beagle 2* is probably still being hampered by its smaller physical size, identification of the lander in its unopened configuration being still beyond its resolution capacities. This visual search is likely to continue over coming years as successive new orbiters are launched with better imaging capabilities. Carl Sagan once commented that in science the absence of evidence is not the evidence of absence. In this case, as noted, it is possible that physical evidence of *Beagle 2* may never be found if it burned up during atmospheric entry, but this cannot be yet be concluded.

DISCUSSION AND ANALYSIS

Having identified the likely artifact distribution resultant from different failure modes, a comparison and analysis of these can now be made.

If the heat shield malfunctioned, *Beagle 2* may have burned up or left a crater containing debris. A crater would have also been left if low atmospheric density was the cause. If there was a parachute collision, the parachute and lander may still be entangled, with a modified bounce mark pattern and the heat shield nearby. Gas bag malfunctioning would be evident from examining the gas bags, the lander may be damaged, and there may be a modified bounce mark pattern. Landing site hazards would result in the expected distribution of heat shield and parachute, a possibly modified bounce mark pattern, and the partially or fully deployed lander and airbags nearby, with identifiable physical hazards such as rocks, craters, or crevices.

Forensic information from the remaining modes of failure will be more difficult to distinguish, and it would be necessary to inspect the condition and/or configuration of artifacts, not just their distribution. If electronics malfunctioned because of landing damage or other reasons, there would be the expected surface distribution of lander, gas bags, parachutes, heat shield, and bounce marks, and the lander may be partially or fully deployed. Examination and testing of electronics would allow the failure cause to be further narrowed. If there was landing damage to *Beagle 2*, artifact distribution would be the same, but the undeployed or partially deployed lander would show evidence of this, and the airbags would probably be undamaged. If the antennae had landing damage, the artifact configuration would be the same as for a successful mission.

A gradient thus becomes apparent between scenarios leaving little or no forensic evidence at one extreme, such as with low atmospheric density or heat shield malfunction, and a very rich assortment of artifacts and forensic information at the other, such as in the case of antenna malfunction. The situation of various failure scenarios along this gradient is shown in Table 23.2. Heat shield failure and low atmospheric density and parachute would appear to be most readily identifiable with the aid of aerial imagery, and to a lesser degree possibly also evidence of a parachute collision or landing site hazards. Other failure modes would really require examination and/or testing of lander hardware, including the deployment mechanisms, electronics, antenna, and also the bags. The scenario resulting in the richest volume of forensic data would be if *Beagle 2* successfully deployed and collected surface data but failed to communicate because of a damaged antenna. This is the scenario that would warrant the most sophisticated investigation, particularly as it may involve the recovery of electronic data.

The scope for undertaking forensic investigation directed toward identifying the causes of mission failure will this vary depending on the mode of failure, as also will the degree of policy guidance required to maximize the recovery of forensic information from the site.

FORENSIC MANAGEMENT POLICY

The preceding analysis indicates that there is likely to be forensic information indicative of the mode of failure of *Beagle 2* on the Martian surface. This information will include the spatial distribution of technological and environmental artifacts, the physical condition of individual artifacts, and possibly also data stored electronically within lander components. Collection of this information is critical to understanding the cause of mission failure. Some of the information can be gathered remotely, particularly from high-resolution images taken by orbiters. Some can be gathered only by a close, in situ examination of artifacts and possibly also their offsite analysis. The types of forensic information and methods of collection are also applicable to the investigation of other failed Mars mission sites.

It has been an assumption of this discussion so far that comprehensive information about the mission's hardware design, mission plan, and mission outcome would be available to any future investigators of the crash site. This is a critical precondition for undertaking an optimal forensic investigation and analysis of the crash site. Crash sites on Mars will not necessarily be investigated by the nation that sent them. In the case of *Beagle 2*, such detailed information is in the public realm or is otherwise accessible, and it is likely to remain so in future. This is not necessarily the case for all other lost Mars surface missions, however, particularly the early Soviet probes. There was a high degree of secrecy surrounding the Soviet space program, and not all Soviet space archives have been opened to the public. Not all Soviet material has been translated into English. It is not clear that comprehensive archives have been kept relating to all failed Mars missions to an extent that would be desirable to forensic investigators. Information may also be lost as individuals involved the missions retire or pass away.

SIGNIFICANCE OF FORENSIC DATA

It is worth considering the relative significance of forensic information in a broader context. It may be reasonably asked how important obtaining precise information on the causes of past mission failures is compared with the fundamental priority of undertaking basic planetary science, given the extraordinary financial cost of undertaking activities of any kind of the Martian surface. Obtaining scientific data is clearly the overwhelming priority. Basic questions still remain unanswered about a planet with a surface greater than that of all the Earth's continents combined. Definitively answering the basic question of whether there is life on Mars is likely to take decades of robotic and human exploration. Writing in 2007, the authors would not suggest there is justifiable value in undertaking a Mars mission for the sole purpose of determining the fate of lost (albeit historically significant) surface probes. Consideration involving policy largely focuses on projecting forward to the far future, when there is a human presence on the planet.

Solving the mysteries of what happened to these missing craft is still of interest, however. Realistically, circumstances that arise to further the forensic investigation are most likely to be through the incidental collection of relevant data, such as the continued imaging of the surface with more powerful orbiting instrumentation. At the same time, it is still worth giving detailed consideration to issues of forensic data collection given that all the crash sites are presently completely undisturbed by human activity. Collecting and preserving forensic data by future explorers is not necessarily a complex activity. It would seem very unfortunate to unnecessarily lose historically important information through a lack of forethought.

FUTURE DIRECTIONS

In considering the possibility that humans may one day visit the *Beagle 2* and other crash sites on Mars, we would propose the following general guiding principles to ensure the protection and maximum collection of all available forensic data. It should be noted that principles guiding the optimal collection of forensic data are not the same as policy guidelines for the overall management of these places as heritage sites, which will be covered elsewhere. The collection of forensic data should be nevertheless be addressed in any broader management program devised for individual sites. We consider the following aspects essential for a comprehensive and ethically responsible policy.

General Guidelines

Identifying the location of all landing or crash sites on the Martian surface of missing spacecraft is the first priority. This includes the location of secondary artifacts such as heat shield and parachutes, which may be some distance from the lander itself.

Addressing forensic issues should be seen as the first step of a broader program to manage the landing site as a heritage place.

International agreement should be reached between current (and future) spacefaring nations laying down principles and protocols to be followed in the treatment of forensic crash sites and also the ownership of artifacts.

Artifacts should not be disturbed unless in the context of a forensic investigation and heritage management program.

Spatial Distribution and Configuration of Artifacts

The original distribution and configuration of technological and environmental artifacts at the crash site should be precisely recorded prior to any human disturbance. This should be done with the aid of aerial photography and may also involve surface excavation of craters or embedded artifacts.

Physical Condition of Artifacts

The condition of surface artifacts could be assessed through inspection after the initial documentation of artifact distribution at the landing site. Artifacts may need to be taken off-site for further analysis.

Assessment of the condition of artifacts needs to take into account damage caused by environmental deterioration.

Stored Data

Electronically stored data may be highly vulnerable to deterioration, so its collection should be given high priority.

The possible hazards to artifacts of surface electrostatic forces need to be considered in any interventions involving surface disturbance.

Planetary Contamination

Any investigation for surviving Earth organisms is best made at the time of first intervention to minimize the risk of any organisms being unnecessarily disturbed or killed prior to study and of additional Earth organisms being introduced at this time.

It is incumbent on the present generation of heritage managers and space scientists to develop management strategies for the investigation and documentation of both successful and unsuccessful Mars landing sites. At this juncture, before any actual recovery missions are planned in any detail, we have the unique opportunity of ensuring that the heritage sites on Mars can be safeguarded from inadvertent diminishment of forensic and/or heritage values.

Clearly, in the meantime, attempts to locate the missing lander crash sites using orbital imagery should continue. Ideally, techniques would be developed that allow the acquisition of a maximum of information without actually setting foot, human or robotic, onto the crash sites.

In preparation for future forensics and/or heritage investigations, efforts need to be made to establish a comprehensive, and open, international archive of all Mars missions, successful as well as failed. As a corollary, it is incumbent upon currently planned missions to develop procedures that ensure that mission information is archived in a fashion that will make it available (and readable/usable) to future generations. The history of managing cultural heritage sites is littered with examples of inadequate or insufficient documentation contemporary with the creation of the sites. Today, as we are shaping our interplanetary futures, we have the opportunity to correct such mistakes and to provide to future generations of heritage managers the tools and data they will require.

NOTES AND REFERENCES

1. Spennemann, D.H.R., and Murphy, G. 2007. Technological Heritage on Mars: Towards a Future of Terrestrial Artifacts on the Martian Surface. *Journal of the British Interplanetary Society* 60(2): 42–53.
2. National Space Science Data Centre. NSSDC Master Catalog: Spacecraft. Surveyor 3. NSSDC ID: 1967-035A. http://nssdc.gsfc.nasa.gov/nmc/spacecraftDisplay.do?id=1967–035A.
3. NASA. 2002. Experiments: Operations During Apollo EVA's. http://www.ares.jsc.nasa.gov/HumanExplore/Exploration/EXLibrary/docs. NASA. 2003. Apollo XII Surveyor III Analysis. http://www./pi.usra.edu/expmoon/Apollo12/A12_Experiments_III.html.
4. Baxter, S. 2004. Trophy Fishing: Early Expeditions to Spacecraft Relics on Mars. AAS 03-307. *Journal of the British Interplanetary Society* 57: 99–102.
5. For a complete listing including detailed descriptions of mission sequences and outcomes, see Spennemann and Murphy 2007 (note 1).
6. National Space Science Data Centre. NSSDC Master Catalog: Spacecraft. Phobos 2. NSSDC ID: 1988-059A. http://nssdc.gsfc.nasa.gov/nmc/spacecraftDisplay.do?id=1988–059A.
7. NASA. 1999. Mars Climate Orbiter Mishap Investigation Board, Phase 1 Report, November 10, 1999. Washington, DC: NASA (see specifically p. 7).
8. Spennemann and Murphy 2007 (note 1).
9. See, for example, Spennemann, D.H.R., and B. Franke. 1995. Archaeological Techniques for Exhumations: A Unique Data Source for Crime Scene Investigations. *Forensic Science International* 74(1): 5–15.
10. See, for example, Boddington, A., A.N. Garland, and R.C. Janaway. 1988. *Death, Decay, and Reconstruction: Approaches to Archaeology and Forensic Science*. Manchester, U.K.: Manchester University Press.
11. See, for example, Haglund, W.D. 2002. Recent Mass Graves: An Introduction. In: *Advances in Forensic Taphonomy: Method, Theory and Archaeological Perspectives*, ed. W.D. Haglund and M.H. Sorg, 243–261. New York: CRC Press. Skinner, M., D. Alempijevic, and M. Djuric-Srejic. 2003. Guidelines for International Forensic Bio-archaeology Monitors of Mass Grave Exhumations. *Forensic Science International* 134: 81–92.
12. Davies, P. 2000. *The Fifth Miracle*. New York: Simon & Schuster (see specifically Chapter 9, pp. 221–244).
13. Pillinger, C.T. 2003. *The Guide to Beagle 2*. Milton Keynes, UK: The Open University (see specifically p. 116).
14. Pillinger 2003 (note 13) (see specifically pp. 2–14, 80, 116).
15. NASA. 2004. *Mars Exploration Rover Landings Press Kit*. Washington, DC: NASA (see specifically p. 6).

16. NASA. 1999. *Mars Polar Lander/Deep Space 2 Press Kit.* Washington, DC: NASA (see specifically p. 8).
17. Pillinger 2003 (note 13) (see specifically pp. 2–14, 80, 116).
18. Pillinger 2003 (note 13) (see specifically pp. 196–200).
19 Sims, M., ed. 2004. *Beagle 2 Mission Report.* Leicester, U.K.: Lander Operations Control Centre, University of Leicester (see specifically pp. 1–2).
20. They have been listed in approximate order of occurrence in the mission cycle, rather than order of probability.
21. Simms 2004 (note 19) (see specifically pp. 247–256).
22. Simms 2004 (note 19) (see specifically pp. 223–225).
23. Pillinger 2003 (note 13) (see specifically pp. 49–50).

24 Lost Spacecraft

Philip J. Stooke

CONTENTS

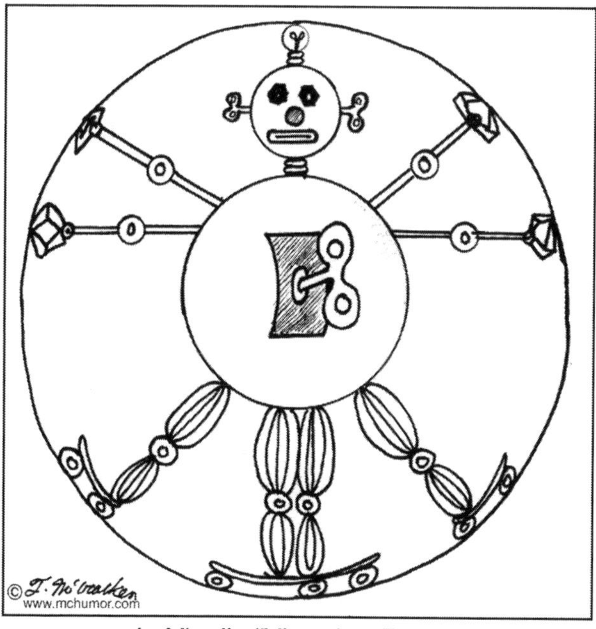

da Vinci's "Vitruvian Rover."

INTRODUCTION

A spacecraft can be considered lost if we do not know its position well enough to image it from orbit or drive a rover to it. The position of an object can be described either in latitude-longitude coordinates or relative to surface features, and it must be emphasized that these are not the same. On Earth we are accustomed to regard them as interchangeable, so that if we know one we can convert it to the other by consulting a suitable map. This is possible because two centuries of geodetic surveying have enabled us to tie surface features to the geographic grid with high precision.

On the Moon or Mars, by contrast, our maps are less accurate. The locations of craters and hills are not known in latitude-longitude coordinates to the terrestrial level of precision. A spacecraft may be located precisely among nearby craters, even imaged from orbit (e.g., *Surveyor 1* on the Moon), while its coordinates remain uncertain. Conversely, coordinates may be measured to high precision by radio signal tracking (e.g., *Viking 2* on Mars), but the location of those coordinates in an image, or rather the location of the image relative to the coordinate grid, may be poorly constrained. This has another effect on coordinates. As our ability to locate features on the coordinate grid improves, map accuracy improves. The location of *Surveyor 1* might be known to within a few meters in orbital images, but when that location is plotted on maps of different dates, from the early 1960s to today, different coordinates may be read off the maps. Thus, sources of coordinates from different dates may vary, but more problematic is the fact that no effort has been made to bring old values up to date, except for Apollo sites, which are the main control points for lunar mapping. The situation will improve when very precise global positional data are obtained from the laser altimeter on the *Lunar Reconnaissance Orbiter*, due to launch in May 2009, and on other contemporary missions. Then it will be possible to bring all site coordinates into one consistent system for the first time.

Adding to this problem is that many locations quoted in the literature are given without any measure of their uncertainty. It is common to find the location of a spacecraft such as *Luna 9* given to within 0.01° (300 m on the Moon), when its location is in fact uncertain by at least 1° (30 km). Thus, the literature on lunar exploration is full of inconsistencies in spacecraft locations. Here, I summarize the state of knowledge, highlighting some specific cases, and provide a table of spacecraft locations on the Moon and Mars with an assessment of the accuracy with which we know them (Tables 24.1 and 24.2). The lunar locations are taken from research conducted by the author[1] and are based mainly on data provided by the National Space Science Data Center (NSSDC). Some of the modifications to NSSDC values are described below. Mars locations are also from NSSDC unless updated for this book.

LOCATING LUNAR AND PLANETARY SPACECRAFT

The first guide to locating a spacecraft is knowledge of its trajectory, leading to a prediction of the landing or impact location. Uncertainties were on the order of tens of kilometers (about 1°) in the 1960s and perhaps a tenth of that today, for the Moon, and ten times that at Mars. There were reports of visual detection of impacts in the

TABLE 24.1
Spacecraft Locations on the Moon

Spacecraft	Date of Contact	Location	Uncertainty
Luna 2	September 13, 1959	29° N, 0° E, or 26.42° N, 2.08° E	30 km (from tracking) 15 km (reported observation of impact)
Upper stage	September 13, 1959	25° N, 70° E	300 km (very rough estimate)
Ranger 4	April 26, 1962	15.5° S, 130.7° W	30 km (from tracking)
Ranger 6	February 2, 1964	9.4° N, 21.5° E, or 9.5° N, 21.3° E	30 km (from tracking) 100 m (candidate site in Clementine images)
Ranger 7	July 31, 1964	10.63° S, 20.55° W	100 m (crater seen in Apollo 16 image)
Ranger 8	February 20, 1965	2.72° N, 20.55° E	100 m (crater seen in Lunar Orbiter 2 image)
Ranger 9	March 24, 1965	12.82° S, 2.42° W	100 m (crater seen in Apollo 16 image)
Luna 5	May 12, 1965	1.6° S, 25° W	30 km (from tracking)
Upper stage	May 12, 1965	32° S, 8° W	30 km (reported observation of impact)
Luna 7	October 7, 1965	9.8° N, 47.8° W 9° N, 51° W	30 km (from tracking) 15 km (reported observation of impact)
Luna 8	December 6, 1965	9.6° N, 62° W	30 km (from tracking)
Luna 9	February 3, 1966	8° N, 64° W	30 km (based on surface images: reported site at 7.08° N, 64.37° W does not match images)
Surveyor 1	June 2, 1966	2.55° S, 43.4° W	5 m (seen in Lunar Orbiter 3 image)
Surveyor 2	September 22, 1966	4° S, 11° W	50 km (tracking before thruster failure)
Lunar Orbiter 1	October 29, 1966	7° N, 161° E	30 km (from tracking)
Luna 12	Unknown	Unknown	Within 20° of equator, uncontrolled, untracked
Luna 13	December 24, 1966	18.9° N, 62.1° W	30 km (from tracking)
Surveyor 3	April 20, 1967	2.94° S, 23.34° W	10 m (Lunar Orbiter and Apollo 12 images)
Surveyor 4	July 17, 1967	0.43° N, 1.62° W	5 km (from tracking)
Surveyor 5	September 11, 1967	1.41° N, 23.18° W	300 m (estimate from surface and orbital images)
Lunar Orbiter 3	October 9, 1967	14.3° N, 92.7° W	30 km (from tracking)
Lunar Orbiter 2	October 11, 1967	3° N, 119.1° E	30 km (from tracking)
Lunar Orbiter 4	No later than October 31, 1967	0° N, 25° W	150 km (E-W), 300 km (N-S) (predicted)
Surveyor 6	November 10, 1967	0.47° N, 1.48° W	10 m (from surface and orbital images)
Surveyor 7	January 10, 1968	40.92° S, 11.45° W	20 m (from surface and orbital images)
Lunar Orbiter 5	January 31, 1968	2.79° S, 83.04° W	30 km (from tracking)
Apollo 10 LM ascent stage	Unknown	Unknown	Within 2° of equator, uncontrolled, untracked
Apollo 11	July 20, 1969	0.67° N, 23.47° E	10 m (from surface and orbital images)
LM ascent stage	Unknown	Unknown	Within 2° of equator, uncontrolled, untracked
Luna 15	July 21, 1969	17° N, 60° E	100 km, inconsistent reports

(*Continued*)

TABLE 24.1 *(Continued)*

Spacecraft	Date of Contact	Location	Uncertainty
Apollo 12	November 19, 1969	2.94° S, 23.35° W	10 m (from surface and orbital images)
LM ascent stage	November 20, 1969	3.94° S, 21.20° W	5 km (from tracking and seismic signal timing)
Apollo 13 SIVB upper stage	April 14, 1970	2.54° S, 27.79° W	20 m (crater seen in Apollo 14 images)
Luna 16	September 20, 1970	0.68° S, 56.3° E	10 km (from tracking)
Luna 17	November 17, 1970	38.22° N, 35.20° W	300 m (possible candidate sites based on surface and orbital images, unconfirmed, 10 km uncertainty from tracking if incorrect)
Lunokhod 1 End of drive	September 14, 1971	38.29° N, 35.19° W	
Apollo 14	February 5, 1971	3.67° S, 17.46° W	10 m (from surface and orbital images)
SIVB upper stage	February 4, 1971	8.17° S, 25.95° W	20 m (crater seen in Apollo 16 images)
LM ascent stage	February 8, 1971	3.37° S, 19.4° W	300 m (ejecta seen in Apollo 16 images)
Apollo 15	July 30, 1971	26.1° N, 3.6° E	10 m (from surface and orbital images)
SIVB upper stage	July 29, 1971	1.51° S, 11.81° W	10 km (from tracking and seismic signal timing)
LM ascent stage	August 4, 1971	26.36° N, 0.25° E	10 km (from tracking and seismic signal timing)
Subsatellite	After January 1973	Unknown	Within 30° of equator, uncontrolled, untracked
Luna 18	September 11, 1971	3.6° N, 56.5° E	5 km (from tracking)
Luna 19	Unknown	Unknown	Within 40° of equator, uncontrolled, untracked
Luna 20	February 21, 1972	3.53° N, 56.55° E	1 km (from surface and orbital images)
Luna 21	January 15, 1973	25.85° N, 30.45° E	300 m (from surface and orbital images)
Lunokhod 2 End of drive	May 10, 1973	25.72° N, 30.93° E	500 m (from surface and orbital images)
Luna 24	August 18, 1976	12.8° N, 62.2° E	5 km (from tracking)
Hiten	April 10, 1993	34° S, 55.3° E	5 km (tracking and telescopic observation)
Lunar Prospector	July 31, 1999	87.7° S, 42.35° E	10 km (from tracking, attempts to observe impact failed)
SMART-1	September 3, 2006	34.4° S, 46.2° W	2 km (from tracking and telescopic observation)
Chandrayaan-1 Moon Impact Probe	November 14, 2008		Location not yet announced
Okina	February 12, 2009	28.2°N, 159.0°W	5 km (from tracking)

early years of lunar exploration (*Luna 2, Luna 5, Luna 7*), but these have always been controversial. More recently, there have been uncontested observations in the infrared (*Hiten, SMART-1*). These observations allow impacts to be located to within

TABLE 24.2
Spacecraft Locations on Mars

Spacecraft	Date of Contact	Location	Uncertainty
Mars 2	November 27, 1971	45° S, 302° W	300 km (from tracking)
Mars 3	December 2, 1971	45° S, 158° W	300 km (from tracking)
Mars 6	March 12, 1974	24° S, 19.5° W	300 km (from tracking)
Viking 1	July 20, 1976	22.48° N, 49.97° W	5 m (from MRO imaging)
Viking 2	September 3, 1976	47.97° N, 225.74° W	5 m (from MRO imaging)
Mars Pathfinder	July 4, 1997	19.33° N, 33.55° W	5 m (from MRO imaging)
Mars Climate Orbiter	September 23, 1999	34° N, 170° W	1,000 km (tracking, with many uncertainties; may have escaped Mars without impacting)
Mars Polar Lander	December 3, 1999	76.3° S, 194.5° W	15 km (from tracking)
Deep Space 2	December 3, 1999	75.0° S, 195.6° W	15 km (from tracking)
Beagle 2	December 25, 2003	11.6° N, 269.6° W	30 km (from tracking)
MER-A lander	January 4, 2004	14.57° S, 175.48° E	5 m (from MRO imaging)
Spirit rover (end of drive)		14.52° S, 175.55° E	
MER-B lander		1.95° S, 354.47° E	5 m (from MRO imaging)
Opportunity rover (end of drive)		2.06° S, 354.48° E	
Phoenix	April 30, 2008	68.22° N, 234.25° E	5 m (from MRO imaging)

Longitudes on Mars were usually measured west from the prime meridian before 2004, but they are now usually measured to the east. This table reflects the change in usage beginning with MER-A (Spirit) in 2004.

1 or 2 km on the lunar surface. In some cases, the impact crater made by an artificial object has been imaged (*Ranger 7, 8,* and *9*, some Apollo hardware), assuming that the identification is correct. Ewen Whitaker[2] illustrated several examples and showed that spacecraft ejecta often appears dark, unlike fresh meteorite impact ejecta. The reason for this is not clear, and it is not consistent among sites, but escape of residual fuel may be implicated.

Some landed spacecraft have been imaged on the surface of the Moon (*Surveyor 1*, lunar modules of *Apollo 15, 16,* and *17*) and Mars (all successful landers). In these cases, a comparison of tracking and imaging reveals typical discrepancies of a few kilometers, representing a combination of tracking and map errors. Where no imaging is yet available, another approach is necessary. If surface images are available, features in the vicinity of a lander can be matched with those seen in orbital images. In areas of suitable relief and adequate orbital imaging, this works well, and the method was used to find or constrain the locations of most of the lunar Surveyors, some Luna spacecraft, and all successful Mars landers. For Mars, the locations suggested by feature matching were used to target orbital high-resolution images, in which the landers were found. The process fails where the local horizon lacks identifiable features, where orbital images are of low quality, or both (*Luna 13, Surveyor 5, Viking 2*). If a spacecraft is found by feature matching, its location can be converted to coordinates only as accurately as the local map control allows. This may still result in uncertainties of several kilometers in that coordinate space.

SOVIET LUNA SPACECRAFT

No Soviet-era spacecraft of the Luna series can be located precisely. The locations known most reliably are those of *Luna 20* and *Luna 21/Lunokhod 2*, where surface images of local relief can be matched with orbital images. Most Luna locations are based on tracking from Earth, with uncertainties of several tens of kilometers. A good example of the unreliability of tracking is *Luna 9*, the first successful lunar lander. It is usually said to have landed at 7.08° N, 64.37° W on the western edge of Oceanus Procellarum based on Soviet-era reports. U.S. Air Force maps place this location immediately adjacent to hills at least 800 m high, based on measurements of their shadows. The horizon to the south and west should resemble the *Apollo 17* landing site. But *Luna 9* returned panoramic images showing a flat horizon, indicating that it is far enough from those hills and others to the west that they are hidden below the local horizon. The most likely location for *Luna 9*[3] is near 8° N, 64° W, as seen in Figure 24.1.[4] The best hope of finding its exact location would be very high resolution imaging, in which the pattern of small craters around it might be identified.

Every mission has unique circumstances. *Luna 2*, the first human artifact to reach the Moon, was tracked to about 29° N, 0° E, but observers from Earth reported a dust cloud at 26.42° N, 2.08° E.[5] This observation is not universally accepted, but clearly two alternate sites have to be considered. In addition, the upper stage of the *Luna 2* launch vehicle struck the Moon 30 minutes after *Luna 2* itself. This is often reported, but no impact site has been described. If the two vehicles traveled on the same trajectory, the second impact site would be displaced by the Moon's orbital motion, leading to the very rough estimate of its position given in Table 24.1. *Luna 5* struck the surface at 1.6° S, 25.0° W, but its upper stage apparently crashed near 32° S, 8° W, raising a large dust cloud that was imaged from Earth.[6] If the imaging report is correct, it is presumed that residual fuel in the upper stage rocket vaporized and lifted the dust. *Luna 17* carried a rover, *Lunokhod 1*, to a site near 38.3° N, 35.0° W, which lacks high-resolution orbital images. Although *Lunokhod 1* obtained many images over a 1 × 4 km area, enabling a map of the local craters to be compiled, the pattern of craters cannot yet be unambiguously matched with orbital images. One candidate site for *Luna 17*, based on an uncertain topographic match, is 38.22° N, 35.20° W, shown in Figure 24.2. This could be confirmed from orbit with high-resolution imaging and possibly also by obtaining reflections from its laser reflector, which has not been detected since 1970.

Several Lunas (*Luna 11, 12, 14, 19,* and *22*) were orbiters that would have crashed at some unknown time and place, limited only by their orbital inclinations. All low

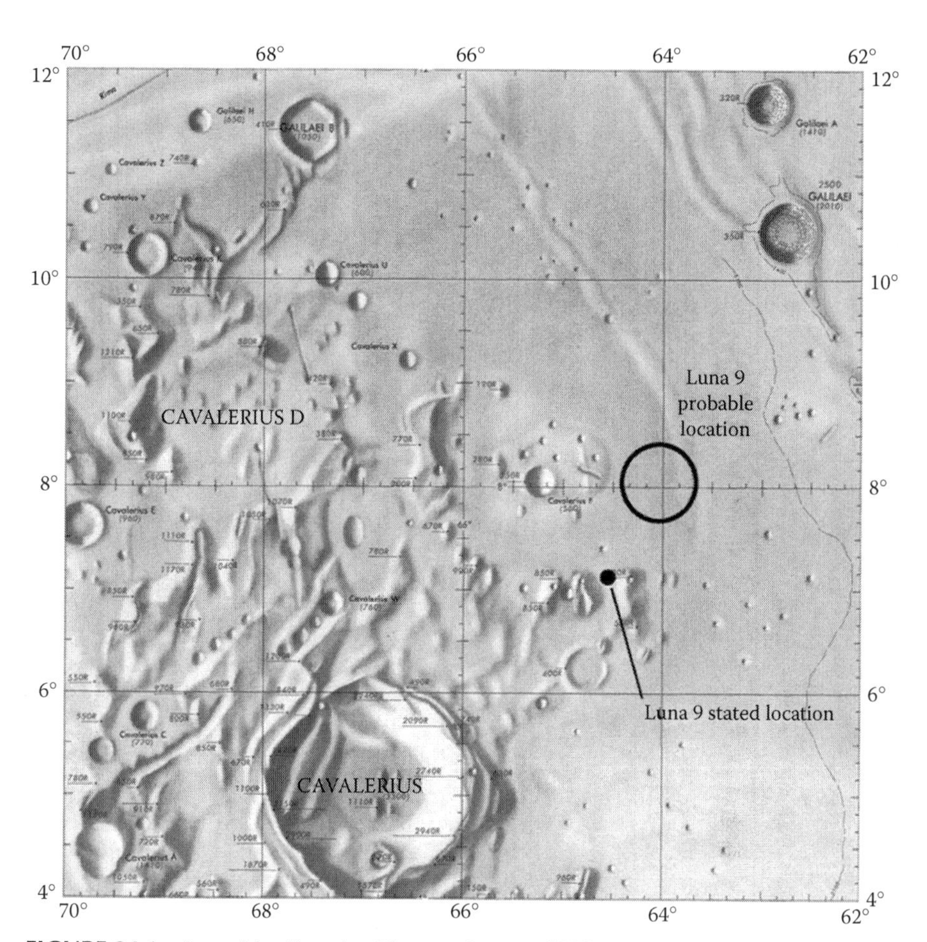

FIGURE 24.1 *Luna 9* landing site. The map is part of U.S. Air Force Lunar Astronautical Chart LAC 56, 1963.

lunar orbits are unstable and eventually evolve until they strike the surface. An earlier orbiter, *Luna 10*, was in a higher orbit and has probably not crashed. *Luna 21* landed in a region with hills around much of the horizon, and its rover, *Lunokhod 2*, drove for more than 30 km through a landscape with many topographic features visible in *Apollo 15* images. The landing site and rover route are fairly well constrained by these images, but orbital resolution was reduced by oblique viewing, and the number of surface panoramas was relatively small. Images of the end of the route have not been published, so the final parking place is uncertain by about 200 m. Other Luna sites are listed in Table 24.1.

RANGER IMPACT SITES

The successful Ranger impact sites can be predicted from trajectory analysis using their own images. Those sites can be inspected in images from subsequent missions, and the expected craters are indeed found.[7] It would be difficult to demonstrate

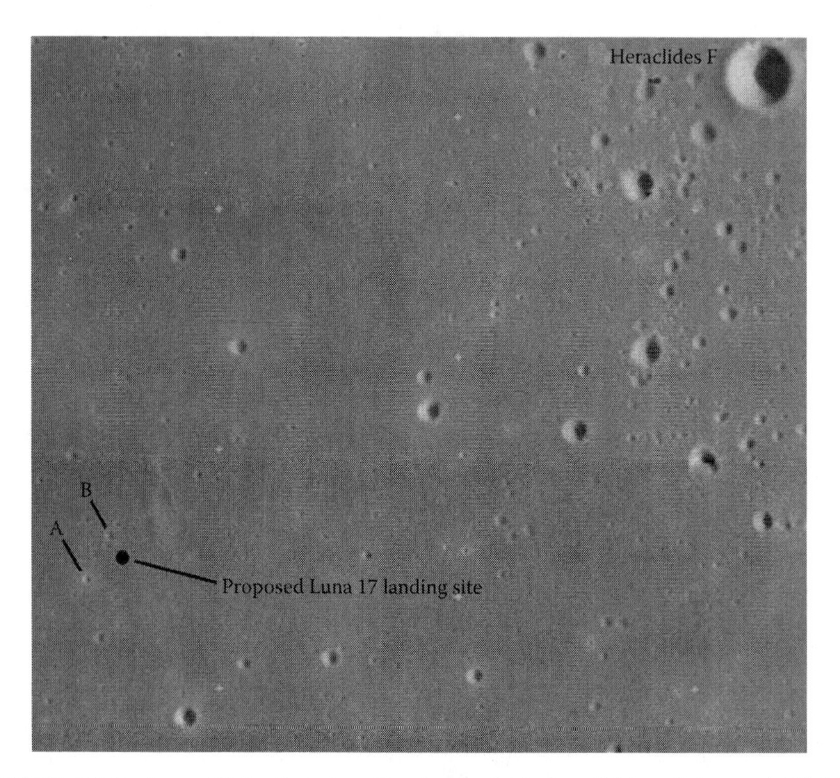

FIGURE 24.2 *Luna 17* landing site. The image is *Lunar Orbiter 4* frame 4-145-H2. Heraclides F (upper right) is located at 38.5° N, 33.75° W. The proposed location is based on the identification of the craters marked A and B, seen in surface images. Despite the obvious uncertainties, there are few if any other possible candidates. The image covers an area 40 km wide, and north is a little left of the top of the image.

that the craters were artificial without the prior predictions. Some of these craters and three Apollo impact sites are surrounded by dark ejecta deposits, which are otherwise rare on the Moon. That serves as another clue to the artificial nature of the craters. Using this idea, a possible candidate for the *Ranger 6* impact (in which the cameras failed) has been found in *Clementine* data.[8] High-resolution images from future missions might confirm its identity. *Ranger 4* struck the lunar far side, returning no imaging, and its site is known only from tracking. At the *Ranger 8* site, Whitaker showed that two previously reported candidate craters with bright ejecta in *Lunar Orbiter 2* images were not produced by the spacecraft. A dark ejecta crater lying between the candidates is now regarded as the *Ranger 8* impact crater shown in Figure 24.3. *Ranger 8* flew an oblique descent trajectory and did not image its impact point.

SURVEYOR LANDING SITES

All successful Surveyor sites except that of *Surveyor 5* are known to within a few meters in orbital images. *Surveyor 1* was imaged on the surface by *Lunar Orbiter*

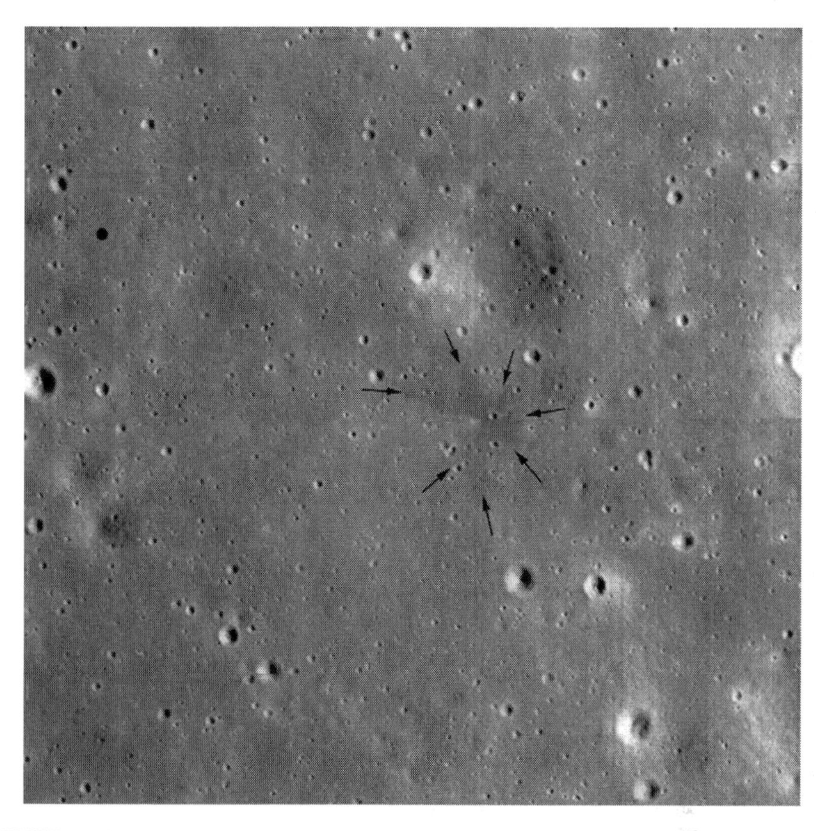

FIGURE 24.3 *Ranger 8* impact site. The dark ray pattern (arrowed) is ejecta from the *Ranger 8* impact, which approached the surface from the lower left. The 10-m-wide crater at the center of the dark ray pattern is the *Ranger 8* impact crater. The image is part of *Lunar Orbiter 2* frame 2-070-H2, covering an area 1 km wide, with north at the top.

3, after landform matching using distant hills revealed its approximate location. *Surveyor 3, 6,* and *7* were found by Whitaker[9] by matching features in their surface panoramas with orbital views. *Surveyor 5* could not be located in this way because it landed in a very flat area outside high-resolution image coverage. Nevertheless, the surface images place some constraints on the location, since *Surveyor 5* cannot be inside or within view of any significant craters. One small crater rim glimpsed on the horizon may provide a better indication, from its azimuth and angular size, and the location in Table 24.1 is based on that observation. High-resolution imaging should suffice to locate *Surveyor 5* or the pattern of small craters around it, a process made easier by the relative accuracy of the existing prediction. Cases like that of *Luna 13* are more difficult because a large area would have to be imaged to find the crater pattern.

LUNAR ORBITER IMPACT SITES

The five Lunar Orbiters helped to locate the Apollo landing sites and survey the entire lunar surface in more detail than was possible before. After their photographic missions ended, they were crashed on the lunar surface. *Lunar Orbiter 4* lost contact with Earth, and its location is only approximately known, perhaps to within a few hundred kilometers. The other four orbiters were deliberately crashed and could be located within 30 km of the coordinates listed in Table 24.1. These uncertainties would make finding the impact sites very difficult. One major source of uncertainty in these figures is knowledge of surface elevations in the impact region. The coordinates are based on orbit predictions and assume a smooth surface of mean lunar elevation, as topography was unknown at the time. It might be possible to recalculate the impact locations based on better topographic data, especially the *Lunar Reconnaissance Orbiter* altimetry expected in 2009.

APOLLO LANDING AND IMPACT SITES

All Apollo landing sites are known precisely because high-resolution orbital image and surface panoramas are available for each location. At the *Apollo 11, 14,* and *15* sites, these are supplemented by very accurate positions from laser ranging to reflectors deployed by the astronauts. In fact, the link between coordinate positions defined by laser data and the known locations relative to surface features are important tie points for the lunar map control system. For *Apollo 15, 16,* and *17*, the lunar module (LM) itself and soil disturbed by rover tracks and footprints were imaged from orbit. Surface photography also helps locate equipment and experiments set out around the Apollo sites. But Apollo missions also left other artifacts on the Moon, spacecraft components that struck the lunar surface when not needed any more. The Saturn rocket upper stages for *Apollo 13–17* were deliberately crashed to provide signals for the seismometers. The impact craters made by the *Apollo 13* and *14* upper stages were found by Whitaker[10] in Apollo orbital images. The other upper stages fell in areas without high-resolution image coverage from orbit, but they might be found in future data. Most of the Apollo lunar module ascent stages also struck the lunar surface, providing seismic signals. Only *Apollo 14*'s lunar module ascent stage has been located from its distinctive ejecta deposit in an orbital Apollo image.[11] The *Apollo 11* and *16* ascent stages were abandoned in orbit and made uncontrolled impacts at unknown times and locations, the possible sites limited only by their orbital inclinations. *Apollo 17*'s LM ascent stage is thought to have struck a mountainside south of the landing site, in view of a television camera mounted on the lunar rover left at the site. Unfortunately, the television failed before the impact. Efforts to locate the impact site in images from *Clementine*, the Hubble Space Telescope, and terrestrial radio telescopes have all failed.

OTHER SPACECRAFT

After Apollo and the end of the Soviet Union's Luna program in 1976, there was a long hiatus in lunar exploration. The next human artifact to reach the lunar surface was *Hiten*, a Japanese spacecraft launched in 1990. It crashed on the Moon on April 10, 1993, and for the first time there was an uncontested observation of the impact from Earth. *Hiten* crashed just on the dark side of the terminator, the line separating sunlit and shaded halves of the Moon. A flash was observed in infrared wavelengths at an Australian observatory, allowing the impact site to be located within a few kilometers. Figure 24.4 shows that site and the direction of the spacecraft's low-angle approach to the surface. A small bright spot just short of the suggested location might possibly be the *Hiten* impact point, but this is impossible to confirm with existing images. Future spacecraft should be able to locate the crater unambiguously. A small separate orbiter, *Hagoromo*, released by *Hiten*, is thought to remain in lunar orbit.

The next lunar visitor from Earth, the U.S. *Clementine* spacecraft, orbited the Moon but left it to pursue an asteroid before failing. Then, in 1998 and 1999, the *Lunar Prospector* spacecraft orbited for 18 months, ending its mission with a deliberate impact. *Lunar Prospector* was intended to resolve a controversy concerning the presence of ice in permanently shaded lunar craters. It crashed in a shaded crater near the Moon's south pole, and observers on Earth searched for a plume of debris that might have ejected ice or water vapor above the crater rim. No ejecta plume was observed, but the impact presumably took place as expected. The crater is hidden in permanent shadow, but future observations might reveal it. *Lunar Prospector* carried a small capsule containing a portion of the cremated remains of Eugene Shoemaker, a lunar geology pioneer in the earliest days of lunar exploration. Thus its crater is the first lunar "burial" or memorial site, and the shaded crater is now named Shoemaker as another memorial.

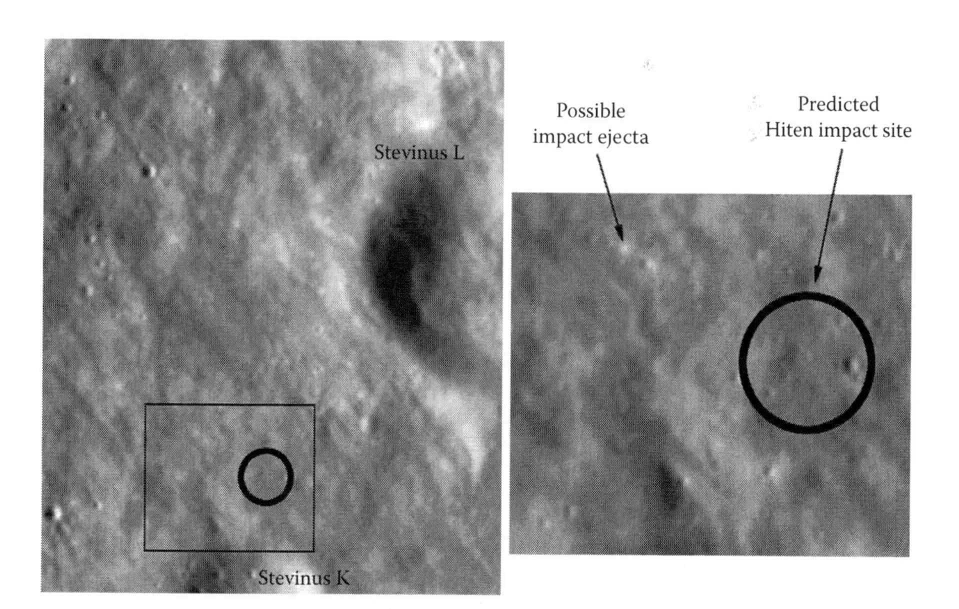

FIGURE 24.4 The *Hiten* impact site. *Hiten*, a Japanese spacecraft, approached the surface from the upper left and crashed between the craters Stevinus K and Stevinus L. The inset shows the predicted impact site (the circle is 5 km across, for scale), and a small, bright ejecta deposit. Although it is tempting to suggest that this is the actual impact site, only higher-resolution images could confirm the identification. (From *SMART-1*. Courtesy of H. Svedhem, ESA, and SPACE-X, Space Exploration Institute.)

The next lunar impact was a European spacecraft called *SMART-1*. (The name is an abbreviation of Small Missions for Advanced Research in Technology.) At the end of its remote sensing mission, *SMART-1* was deliberately crashed into a hillside at 34.4° S, 46.2° W, and again the impact and its spray of ejecta were observed from Earth in infrared images. The impact was on the dark side of the terminator, which benefits infrared observations. The crater location is known within a few kilometers and should be easy to find in future high-resolution images. The same should apply to India's first impact probe (the Moon Impact Probe of the Chandrayaan-1 mission), which crashed near the south pole on November 14, 2008. Its location has not yet been announced.

MARS SPACECRAFT

The situation for Mars is very different from that for the Moon. Tracking uncertainties are necessarily larger at this more distant target, and the effects of atmospheric friction on a descending spacecraft introduce a variable not encountered at the Moon. Set against this is the easily observed rotation of Mars, which introduces a Doppler shift in radio transmissions that can be used to measure a landed spacecraft's latitude very precisely. This cannot be done effectively on the Moon because its rotation is synchronous with its orbit. Thus, for *Viking 2*, a very precise location, especially in latitude, could be stated, while the position of the landing site among the surface features of Mars remained uncertain. Only with the advent of extremely high resolution imaging from orbit with the *Mars Reconnaissance Orbiter* (MRO) has it been possible to locate all successful landers on Mars, finally establishing tie points between coordinates and surface features. Attempts to identify sites before MRO were compromised by an incorrect identification of the *Viking 1* site and the initial failure to find *Viking 2*. Mars locations are listed in Table 24.2.

The Soviet Union placed three landers on Mars. Two crashed (*Mars 2, Mars 6*), and one (*Mars 3*) landed successfully but failed before sending useful data to Earth. Because no surface images or signal tracking from the surface could be examined, only prelanding tracking is available to locate these sites. They are thought to be known within about 5°, or 300 km on Mars. The likelihood of locating these spacecraft in orbital images is very small.

The United States has landed six spacecraft successfully on Mars: *Viking 1, Viking 2, Mars Pathfinder,* the Mars Exploration rovers A (*Spirit*) and B (*Opportunity*), and *Phoenix*.

In every case, an approximate location could be determined from a comparison of surface panoramic images and orbital images. The method is not without problems, as the Viking experience shows. *Viking 1*'s site was misidentified by Morris and Jones,[12] and the problem was corrected only after its coordinates could not be made to match those of the nearby *Mars Pathfinder*, thus prompting a reassessment. The process of matching surface and orbital images can be compromised by ambiguous landforms and by our uncanny ability to explain discrepancies. If an expected feature is not seen, some explanation can often be devised from an unusual combination of landforms. This author's attempt to find *Viking 1*, using a location suggested by

Merton Davies of the RAND Corporation during an early stage of the reassessment of its position, was hopelessly wrong for this reason.[13] *Viking 2* landed in a very bland landscape covered by very poor orbital images, and it was not found unambiguously on the surface until MRO images became available. All later spacecraft landed in areas of better image coverage and higher relief and were quickly located by Tim Parker at the Jet Propulsion Laboratory, who also found the correct *Viking 1* location.[14]

Two other U.S. sites deserve mention here. *Mars Climate Orbiter* (MCO) failed as it entered the atmosphere of Mars on September 23, 1999. Some of its fragments might have reached the ground, probably in the vicinity of 34° N, 170° W, but it is also possible that the disabled spacecraft was not captured by Mars and entered solar orbit. At any rate these coordinates are very uncertain, perhaps by as much as 15° or 1,000 km. It should be noted here that all other Mars orbiters are still in orbit, having been designed not to impact Mars for decades after their missions ended to prevent accidental contamination of the planet. Only a few months after the demise of MCO, the *Mars Polar Lander* and two small probes carried with it crashed near 76° S, 195° W, as shown in Figure 24.5. Searches by orbiting spacecraft have not yet located the crash site, but the very high resolution provided by MRO may eventually reveal it. In this case the location is known from tracking to within about 15 km.

The European Space Agency attempted to land a small spacecraft, *Beagle 2*, on Mars on December 25, 2003. No signal was received from the surface, and it is presumed to have crashed. Tracking suggests a location to within 30 km (Table 24.2). One suggested impact site, a small dark area in a crater, was shown to be an entirely natural feature in MRO images. In this case, imaging may eventually locate the impact site, but searching for sites by taking high-resolution images is very resource-intensive and can be hard to justify.

The *Phoenix* lander reached Mars on May 25, 2008. Its landing site is in an extremely bland region of the northern high-latitude plains, but it was located immediately in images taken by *Mars Reconnaissance Orbiter* within hours of landing. In fact, in a remarkable demonstration of our new remote sensing abilities, *Phoenix* was also imaged by MRO during its descent to the surface on a parachute.

CONCLUSION

This survey of spacecraft locations on the Moon and Mars illustrates our knowledge up to the second half of 2008. New, very high resolution imaging and greatly improved cartographic coordinates will improve our knowledge of these locations in the next few years, though there is no plan at present to update spacecraft locations in a systematic way. Where a spacecraft location is known (e.g., *Surveyor 1*) or may be found quite easily (*Luna 17*) the current estimated coordinates should be improved. Conversely, if a site is poorly known (*Luna 9*, *Mars 3*), there is little chance that we will be able to find it in the limited areas of new high-resolution imaging, and its coordinates will probably remain uncertain. The experience with *Phoenix* suggests that some future spacecraft sites on the Moon and Mars can be located precisely and rapidly, but this will not always be the case. As a postscript, on February 12, 2009, a subsatellite of the Japanese Kaguya mission crashed on the far side, unseen from

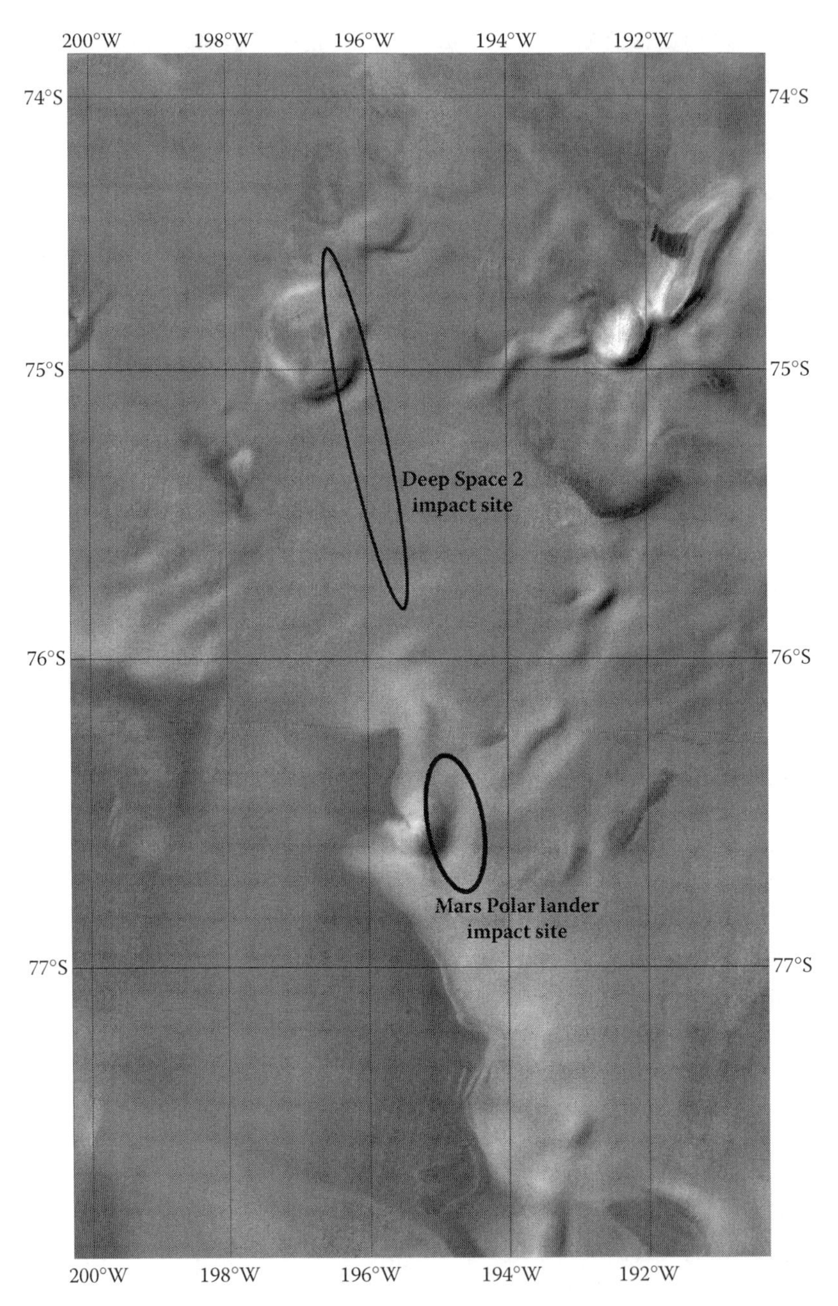

FIGURE 24.5 *Mars Polar Lander* and *Deep Space 2* probe impact sites. The image background is a combination of Mars Odyssey and Viking data.

Earth. Its coordinates are uncertain by at least several kilometers, and it may be hard to find in future images. There will still be puzzles to solve for a long time to come.

REFERENCES

1. Stooke, P.J. 2007. *The International Atlas of Lunar Exploration.* Cambridge, U.K.: Cambridge University Press.
2. Whitaker, E.A. 1972. Artificial Lunar Impact Craters: Four New Identifications. *Apollo 16 Preliminary Science Report.* NASA SP-315. Washington, DC: NASA (see specifically pp. 29–39 to 29–45).
3. Stooke, P.J. 2006. Locating Landed Spacecraft and Artificial Impact Craters in LRO Images. Abstract no. 1341. 37th Lunar and Planetary Science Conference, Houston, Texas, March 2006.
4. U.S. Air Force. 1963. Lunar Astronautical Chart LAC 56.
5. Anonymous. 1960. Lunik 2's Landing on the Moon. *Sky and Telescope* 20(5): 265.
6. Anonymous. 1965. Photos of Dust Raised by Moon Probe. *New Scientist* 26(449): 842.
7. Whitaker 1972 (note 2).
8. Stooke 2006 (note 3).
9. Whitaker 1972 (note 2).
10. Whitaker 1972 (note 2).
11. Whitaker 1972 (note 2).
12. Morris, E.C., and K.L. Jones. 1980. Viking 1 Lander on the Surface of Mars: Revised Location. *Icarus* 44: 217–222.
13. Stooke, P.J. 1999. Revised Viking 1 Lander Site. Abstract no. 1020. 30th Lunar and Planetary Science Conference, Houston, Texas, March 1999.
14. Parker, T.J., R.L. Kirk, and M.E. Davies. 1999. Location and Geologic Setting for the Viking 1 Lander. Abstract no. 2040. 30th Lunar and Planetary Science Conference, Houston, Texas, March 1999.

25 Space Hardware: Models, Spares, and Debris

Ralph D. Lorenz

CONTENTS

INTRODUCTION

The nature of space exploration is such that it is experienced directly by only a few people. The images or other data sent back to Earth as a remote extension of our own senses give an element of "being there," but otherwise our only connection to space is via the tangible hardware associated with space projects. By definition, most spacecraft are inaccessible, but there are a number of objects that we can see and touch.

There are two main categories of artifacts that can be inspected on Earth. First are the rare instances of hardware that returns from space. Second are items constructed during the development of a space project, alongside the vehicle that actually flies. In some cases these are identical copies, although many development and engineering models also exist that reproduce only some aspects of the flight vehicle: these are used in the design and debugging of a mission. We will consider these two broad classes in turn.

RETURNED HARDWARE

Parts that can return from orbit must either have a very low mass-to-area ratio (such that their heating rate is modest enough to keep the item below its melting point) or be protected from heating. Earth-return capsules for astronauts (such as those for *Apollo*, *Soyuz*, and *Shenzou*) and unmanned capsules for microgravity experiments and photographic film are examples here.

Because the technology for returning equipment from space is essentially identical to that required to deliver equipment (such as nuclear warheads) ballistically over intercontinental distances, the hardware associated with the aerodynamic and aerothermodynamics of hypersonic flight is often considered strategic. Even when the materials themselves are low technology (such as powdered cork in an epoxy matrix), the formulation and testing data that make the material useful may be treated at least as commercial, if not national, secrets, and so the dull, dirty-looking heat shield materials are typically less available for intimate inspection than are shiny, machined metal structures (i.e., the material itself, rather than the fabrication of the article from the material, is considered proprietary). Entry from orbit entails the dissipation as heat of the same amount of energy that was required to deliver the vehicle into orbit in the first place: this violent process, and subsequent exposure to the elements (capsules perhaps sitting on the Khazak steppes or bobbing in the ocean for hours before recovery) means that reentered spacecraft are often irregularly stained or scorched (Figure 25.1).

Some items that reenter survive only inadvertently. While dense objects will burn up on entry, empty fuel and pressurant tanks are thin-walled shells with a low mass-to-area ratio. They therefore decelerate at high altitude. This permits the kinetic energy to be dissipated more gradually in the thin air, and so heating is

FIGURE 25.1 A slightly space-worn *Shenzou 5* capsule on display in Beijing. This vehicle launched and returned China's first taikonaut in 2003. (Photo by author.)

modest, permitting their survival (see, e.g., Figure 25.3). A particular set of objects in this class consists of the titanium gas tanks used on Russian Molniya and other satellites. These are spherical, thin-walled tanks that hold helium pressurant. During entry, as the satellite breaks up, these tanks are ripped off, and being made of titanium (which has a rather higher melting point than aluminum alloys) they are able to survive entry heating. One such tank, which was found on an Australian beach, was obtained by a German meteorite collector, who noted the metric thread on the tank's nipple, as well as some evidence of melting and several tiny hypervelocity impact craters. The present author determined[1] the composition of a small fragment to be titanium alloy using x-ray fluorescence in an electron microscope and established the density of craters to be consistent with an exposure age in orbit of several years, consistent with the Molniya origin (Figure 25.2).

A few other, unfortunate artifacts exist that have fallen to Earth during the failure of a launch vehicle.[2] Since this may occur at lower speed and lower altitude than are typical for reentry, these may not endure significant aerodynamic heating, although they may be damaged by impact with the ground or effects during the launch vehicle explosion (if control will be lost, damage on the ground is minimized by detonating charges on the launch vehicle to ensure dispersal of the fuel and oxidant.) A celebrated example is the failure of the first Ariane 5 launcher, carrying the four identical Cluster magnetospheric research satellites. The vehicle failed very soon after launch, and the swampy terrain around the Kourou launch site in French Guyana softened the impact of some debris, which was then recovered. Figure 25.3 shows

FIGURE 25.2 The 250-kg main propellant tank for the second stage of a Delta 2 launch vehicle that landed in Texas in 1997. Because the empty tank has a low ratio of mass to area and because it is made of a high-melting-point material (stainless steel), it survived entry relatively intact. (Courtesy of NASA.)

FIGURE 25.3 Electronics box from Cluster on exhibit at the European Space Research and Technology Centre, recovered from a South American swamp near the launch site after the failure of Ariane 501. While the metal box itself is relatively intact, the torn honeycomb panel to which it was bolted and the shredded cabling at left attest to the violent disassembly of the spacecraft. (Photo by author.)

several parts exhibited at the European Space Technology and Research Centre in the Netherlands. A few small mechanical parts such as screws were in fact reflown on a subsequent launch (Cluster-Phoenix), more for symbolism than for practical economy.

A final class of objects returned from space are those that are actually brought back intentionally, almost invariably by astronauts within the protective confines of another spacecraft. Typically these are items that have been exposed to the space environment for some years, either in order to study those effects specifically, or to replace them with improved parts. Three examples are worth calling out. First, the *Apollo 12* landing site was targeted very close to the *Surveyor 3* lander, which had arrived 2 years earlier. Astronauts walked to *Surveyor 3* and removed components for analysis of how the lunar environment had affected materials. The camera of *Surveyor 3*, now on display in the Smithsonian's National Air and Space Museum, was reported to contain bacteria, suggesting that biota may survive long periods. However, it seems more likely that contamination after recovery was the source of these bacteria. A second example is the Long Duration Exposure Facility. This was a large (but essentially passive) satellite that was released from the space shuttle *Challenger* in 1984 in order to measure space environment effects on various materials. It was recovered by the space shuttle *Columbia* in 1990 and revealed much data on atomic oxygen degradation, radiation effects, and micrometeoroid and space debris impacts (including tiny craters with trace residues of nitrogen, suggesting that they had been formed by impacts with ice crystals from the space shuttle's toilet!). A final example is the solar array from the Hubble Space Telescope (Figure 25.4):

FIGURE 25.4 The Hubble Space Telescope, anchored by the remote manipulator system robot arm above the payload bay of the space shuttle. The solar arrays at either side of the telescope tube were recovered and brought back to Earth for analysis. (Courtesy of NASA.)

launched in 1990, the telescope experienced some attitude perturbations on crossing from day into night each orbit due to the thermal shock on the flexible arrays and the masts that held them outstretched. These arrays were designed to be replaced on-orbit, and one of the two arrays was recovered in December 1993 (the other failed to retract properly and had to be jettisoned). This recovered array was examined in detail for micrometeoroid impacts (Figure 25.5), the large area of the array (60 m^2) giving good statistics on some 3,600 impacts. The replacement array was itself recovered and replaced in 2002 with smaller and more efficient arrays.

SPACECRAFT ASSEMBLY AND TEST: ENGINEERING AND OTHER MODELS

Although many spacecraft are hand-built "one offs," the manufacture and test of a space system usually involves the creation of several versions or models of the system. These various models allow builders to rehearse construction operations and more particularly reproduce certain aspects of the final system's function in order that problems may be identified and corrected before the final flight version.

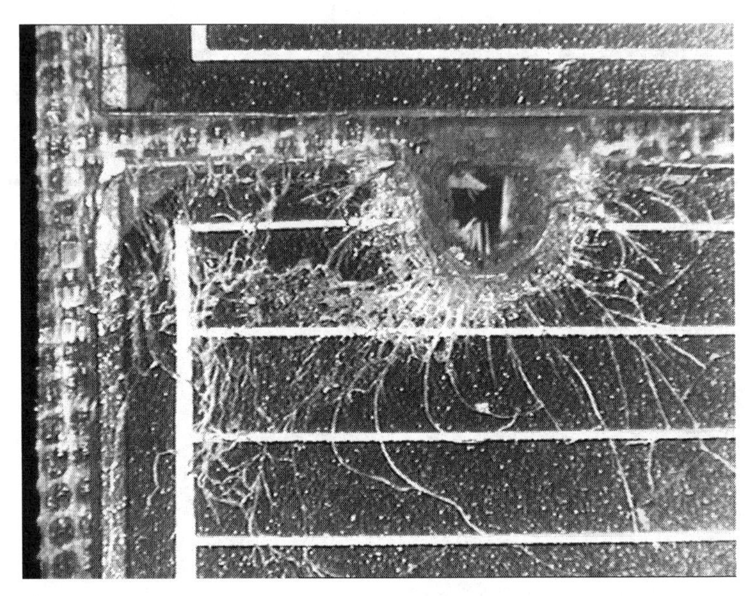

FIGURE 25.5 A closeup image of a small impact crater in the second Hubble Space Telescope array, recovered in 2002 after 8.25 years in space. Analysis of the density of craters on such returned materials allows the deleterious effects of the space environment to be better understood. (Photo: European Space Agency.)

First, for mechanical tests (such as interface checks—to check that bolts line up with holes on other systems), vibration tests, and thermal tests, a structural and thermal model is often constructed. This will have major structural elements, including deployable mechanisms, but will typically lack any electronics. To test mechanism deployments, often by pyrotechnic actuators, this model may sometimes be called a structural thermal and pyrotechnic model.

Because such models are visually representative of the overall layout of a spacecraft, these models often find their way into museums after a mission, in some cases soon after launch (when the major structural loads are typically encountered). The same is usually not the case of so-called engineering models (EMs) of spacecraft. In some cases these are structurally representative, but usually they are laid out on some sort of bench or fixture that can be readily worked on by engineers and technicians. The main function of an EM is to verify the electrical and software interfaces between subsystems, so a major activity is the mate and demate of connectors between subsystems. Initially, this may involve engineering models of these subsystems themselves, during which time one hopes major problems are resolved, but it can also serve as a way of comprehensively testing some flight model (FM) units before others. (For example, if an FM power unit were to have been manufactured with an undetected wiring error, plugging it straight into the FM spacecraft might lead to damage to the other FM units; by plugging it into an EM spacecraft first, the unit can be comprehensively tested, and the worst that happens is it damages other EM systems.) EM units should have parts that are electrically representative but may, for example, use less expensive commercially rated components rather than military- or space-rated radiation-tolerant or large-temperature-range parts. The latter may be expensive and/or have long lead times for procurement, so repairing an EM unit is likely to be much less damaging to a project than having to repair the FM. EM cabling, because it may have encountered more adjustments during assembly and testing, may often be more prominently labeled than the tight bundles of secured cabling that are more typical of a flight cable harness.

Because of the utility in diagnosing problems and in, for example, verifying the correct operation of new software, engineering models are usually retained under project control for the operational duration of a project. Depending on their visual appeal (functionally they are often a set of featureless electrical boxes plugged together; Figure 25.6), they may be exhibited in museums, but sometimes they are simply disposed of.

FIGURE 25.6 *Huygens* probe engineering model at left. Most of the equipment, including the large wire harness, is mounted in a representative layout, but it is mounted on a fixture that enables the equipment platform to be flipped over. This, and the lack of a protective structure, permits the easier addition, removal, adjustment of, and connection of test equipment to the units. The probe's batteries are replaced by a laboratory power supply to permit continuous operation, and in the "shirtsleeve" laboratory environment, the probe's thermal insulation is not necessary. The model on the floor at right is an electrically representative mockup of the probe structure, so that the radiation pattern of energy from the radio antennas (black cones) can be measured. This will differ from their theoretical performance because of reflections from parachute box on the roof of the probe. (Photo at the European Space Operations Centre by author.)

In some applications there may be special models constructed for specific tests. An example is the Special Model-2 (SM2) of the *Huygens* probe, constructed as an aerodynamically representative article for a drop test from a stratospheric balloon over Arctic Sweden, in order to verify the parachute deployment sequence. Although constructed exclusively for a single flight test (the descent lasting only an hour) in 1994, this article served as the centerpiece prop for media interviews and so forth during the actual *Huygens* descent on Titan in 2005 (Figure 25.7). Spacecraft with large optical systems, such as telescopes, may often have dedicated optical models.

Another class of spacecraft is the flight spare (FS). This is typically identical or nearly identical to the FM. In some projects, problems during testing or delivery may lead to the late substitution of the FS for the FM. Flight spare equipment may eventually appear in a museum, and there have even been instances of such exhibits being pressed back into flight service or cannibalized for parts.

Occasionally, some projects produce many identical spacecraft. This is rare for the specific and expensive requirements of interplanetary exploration, but for applications in Earth orbit that require the provision of a continuous and global service, a series of identical spacecraft may be produced in a production-line environment. Examples here include the global positioning system navigation satellites, and especially the

FIGURE 25.7 *Huygens* probe Special Model-2 (SM2), with the author. This is a physically much more appealing item than the engineering model in Figure 25.6, and it is very similar in external appearance and aerodynamics to the article that actually flew on Titan a billion miles away. There are some differences between this and the flight unit—for example, this SM2 lacks the science payload, but it does have a small square window on the upper surface for a camera to film the parachute deployment during the balloon drop test over Sweden. (Photo by author.)

commercially unsuccessful low Earth orbit communications constellations such as Iridium (so named because it was to have 78 orbiting satellites, this being the atomic number of the element iridium). In addition to the initial operating complement, a series of replenishment satellites (to make up attrition losses in orbit) may also be constructed. At the conclusion of the project, perhaps because of a series being superseded by more capable or economical systems, one or more of the series may remain on the ground and be available for exhibition. Programmatic changes, such as loss of political support for a project, can leave hardware unlaunched—Apollo hardware, such as a full-scale, flight-ready *Skylab* space station on display at the National Air and Space Museum, is a prime example (this was one of two built—the other was launched in 1973, visited by astronauts in 1973 and 1974, and reentered over Australia in 1979). In extreme cases, a program may be terminated altogether before anything has flown, even if the spacecraft was ready to fly. A notable example here is the Deep Space Climate Observatory, formerly known as *Triana*. This spacecraft was the 1998 idea of then-Vice President Al Gore, intended to continuously observe the sunlit Earth for an extended period from the L_1 Lagrange point. As well as providing iconic images, the vehicle would be well placed to monitor climate change. The project became deeply politicized, not least by a change in administration, and its intended 2003 launch was blocked. However, it enjoyed scientific support that made it difficult to cancel the project altogether, and thus the full-fledged satellite remains in storage (at a reported cost of $1 million per year) on the ground.

CONCLUSION

An important legacy of the space age is the hardware left behind on, or occasionally returned to, Earth. This equipment can vary in appearance from the ugly charred and melted metalwork that may occasionally survive the rigors of reentry, to delicate hardware recovered from space at great expense by astronauts, to the collateral mockups and testbeds developed during the construction and testing of a space system. Each of these classes of object has its own aesthetic appeal, and each has a story to tell.

REFERENCES

1. Harland, D.M., and R.D. Lorenz. 2005. *Space System Failures*. Berlin: Praxis-Springer.
2. Lorenz, R.D., and W. Zeitschel. 1999. An Intact Item of Soviet Space Debris with an Exposure Age of 10 Years. *Journal of the British Interplanetary Society* 52: 439–442.

Section VI

Environmental Effects and the Material Record

26 Natural Formation Processes and Their Effects on Exoatmospheric Objects, Structures, and Sites

Edward Staski and Roger Gerke

CONTENTS

INTRODUCTION: CULTURAL AND NATURAL FORMATION PROCESSES

Archaeological formation processes are all the factors responsible for creating the archaeological record, whether on the Earth or not.[1] Cultural formation processes are those that result from human behavior; while not the subject of this chapter, the potential of certain cultural formation processes to have adverse impacts on archaeological resources in space is what initiated much of what we currently identify as space archaeology (see Chapter 2). Natural formation processes are those that result from the conditions and forces of nature; the natural formation processes that operate in various exoatmospheric environments and that have impacts on various exoatmospheric archaeological resources, are the subject of this chapter.

There are four general types of cultural formation processes: deposition, reclamation, disturbance, and reuse.

Deposition includes all cultural formation processes that result in the movement of material culture from systemic context (participating within a living culture and society) to archaeological context (no longer participating within a living culture and society, i.e., part of the archaeological record). Examples of deposition include *discard* (the intentional throwing away of items no longer considered useful), *loss* (the unintentional movement of materials into archaeological context), *caching* (the purposeful burying of materials with the intent to recover them at a later time), *burial* (placing the remains of the dead into the archaeological record), and *abandonment* (when a place rather than a particular object enters the archaeological record). Space debris, exoatmospheric material culture that has been deposited for various reasons and in various ways, is the archaeological record of interest in this chapter, and in this book. Some of this debris can be found on the Moon, Mars, Venus, and a few other celestial bodies. It reached these bodies either by intentionally landing on them, fulfilling a mission, and then remaining in place (*landing sites*, an example of discard), or by unintentionally crashing into them (*crash sites*, an example of loss).

Most of this debris, however, is flotsam: free-floating debris deposited in space, just as flotsam on Earth is deposited and free-floating in water. These objects are also an example of discard.

Reclamation is the opposite of deposition, at least so far as the direction objects move. It includes all cultural formation processes that result in the movement of material culture from archaeological context back to systemic context. Examples of reclamation include *scavenging* (the retrieval of a community's archaeological materials by residents of the same community), *collecting* (the retrieval of a community's archaeological materials by residents of a different community—scavenging and collecting can sometimes be difficult to differentiate), *treasure hunting* (the retrieval of a community's archaeological materials *from the subsurface* by residents of a different community), *salvage* (the reclamation of materials from a place previously occupied and then abandoned by either the people who once lived there or other people) and, of course, *archaeological recovery* (see Chapters 2 and 40 of this volume). Space scientists talk of *cannibalizing* exoatmospheric material culture that has become space debris and *recovering* salvageable parts of this debris for reuse elsewhere (see below). These processes are special, exoatmospheric examples of reclamation.

Disturbance is that category of cultural formation processes that change either the condition or location of material culture while not removing it from archaeological context. On the Earth, numerous, varied events and activities result in disturbance, including the plowing of agricultural fields and the construction of buildings, highways, and other aspects of the modern infrastructure, and the result is that disturbance is widespread and significant. In contrast, there is presently only one exoatmospheric site where disturbance has occurred, the *Surveyor 3* site on the Moon, visited by *Apollo 12* astronauts (see discussion below). This situation is most likely going to change in the next several decades, with return manned missions to the Moon, manned missions to Mars, and most significantly with the emergence of space tourism. Programs such as the X Prize competition and plans to create spaceports through public-private cooperative ventures have made space tourism a very real, not-so-distant possibility. Indeed, the current X Prize competition calls for placing a privately developed probe on the Moon and offers a bonus if objects previously deposited there can be videotaped or photographed (see http://www.xprize.com). Such an incentive makes disturbance a likely outcome, and similar future activities will make disturbance of the exoatmospheric archaeological record common.

Finally, *reuse* includes all cultural formation processes that prolong the time that material culture remains in systemic context by changing the user, use, or form of the material, or some combination of these features. Examples of reuse include *lateral cycling* (a change in the user but not the use or form of an object), *recycling* (a significant change in the use and form of an object, and usually the user as well), *secondary use* (a change in use but not a profound change in form, and usually not a change in user), and *conservation* (the conscious effort to preserve an object because of its perceived importance or value).

Natural formation processes cannot be categorized as easily as cultural formation processes. There are some natural formation processes that move material culture from systemic context to archaeological context, and so are analogous to deposition. Likewise, and much more significantly, there are many more natural formation processes that alter the condition or location of materials while not removing them from archaeological context, and thus are analogous to disturbance. No natural formation process moves material culture from archaeological context to systemic context, so there is no analogy to reclamation (even though certain natural processes increase or decrease the chance that reclamation will occur), and no natural formation process moves material within systemic process, so there is no analogy to reuse (even though certain natural processes have impacts on materials that nonetheless remain in systemic context).

Among the innumerable natural formation processes analogous to disturbance are the better-known and better-studied examples of *faunalturbation* (the impacts that nonhuman animals can have on the archaeological record) and *floralturbation* (the impacts that plant life can have on the archaeological record). There is also *perturbation* and *pedoturbation* (changes in soils and sediments, which can have impacts on the archaeological record). The forces of wind and water cause significant changes in soils and sediments, and thus are ultimately responsible for the occurrence of these natural formation processes.

NATURAL FORMATION PROCESSES UNDER THE WATER AND OFF THE EARTH

It should be obvious to everyone that many of the natural formation processes acting on Earth and studied by archaeologists on Earth have no or little impact on material culture off the Earth. Those exoatmospheric natural formation processes that do exist appear to have little impact on material culture, except in rare instances when their impacts are stupendous (e.g., asteroid and comet impacts and volcanism, which of course can occur on the Earth as well). Indeed, exoatmospheric environments appear to be generally benign when it comes to impacts on the archaeological record (see our discussion below), yet surprisingly, the general, uninformed consensus seems to be that these environments are extremely damaging.

Our lack of interest in and resulting ignorance of exoatmospheric natural formation processes, combined with our ill-conceived perceptions of the exoatmospheric environment, lead us to exaggerate the adversity of outer space. Not so long ago the same could have been said about the underwater environment, and because of these and other similarities, and because our knowledge about and perception of the underwater world have been changing and improving recently (as our knowledge and perception of outer space must change and improve in the future), a very useful comparison can be made between the two settings.[2]

The assumption that both the underwater and space environments have many significant adverse impacts on archaeological resources results from the fact that in many ways these are foreign and hostile surroundings for humans. These are exotic and dangerous places where most people spend little or no time. Furthermore, those few people who do go under the water or out in space are mere visitors to these environments. They do not inhabit them in any permanent sense. Special precautions need to be taken for survival while they are there.

Yet the assumption is wrong in both environments, as underwater archaeologists have been learning more and more in recent years. Formation processes are different under the water and out in space, but their impacts on archaeological resources are generally inconsequential. Indeed, both are relatively benign environments when it comes to the preservation and integrity of the archaeological record. They are not benign everywhere, of course—in some underwater and exoatmospheric locations, archaeological resources are obliterated, either in an instant or over some length of time. But in many other places both underwater and exoatmospheric natural site formation processes serve to maintain or even enhance preservation.

Take the case of the *Vasa*, a seventeenth-century Swedish warship that sank in Stockholm Harbor in 1628 and was raised from the water in the early 1960s. It was found to be "spectacularly well-preserved and intact," largely because the salinity of the Baltic Sea is too low for shipworms (e.g., *Teredo navalis*) to thrive and because there was little turbulence in the surrounding water. Otherwise, perishable materials such as wood and cloth were essentially undamaged when the warship was recovered, allowing archaeologists and historians ample opportunities to learn about seventeenth-century shipbuilding technology, maritime military capabilities, and cultural expression.[3]

Or take the case of the nineteenth-century American steamboat *Bertrand*, which sank after striking a snag in the Desoto bend of the Missouri River on April 1, 1865, on its way to supply Fort Benton, a frontier military outpost at the time in present-day Montana. This wreck was located beneath the water table in 20–28 ft. of sediment, and after much effort it was raised in the late 1960s. To everyone's delight (and astonishment) the integrity of the vessel was preserved, and more importantly, the cargo on the vessel was intact. This vast and diverse cargo included foodstuffs, textiles, household goods, mining supplies, hardware, tobacco supplies, military ordinance, and the personal belongings of passengers. The underwater environment had done little to compromise the integrity of either the ship or its contents, and much has been learned from various analyses.[4]

Numerous additional examples of well-preserved underwater archaeological sites and resources could be cited. In all cases, it is the underwater environment that has enhanced preservation. Indeed, both the *Vasa* and the *Bertrand* did not experience significant deterioration, or the threat of significant deterioration, until after they were raised out of the water! In the case of the *Vasa*, which remains today on display at the Vasa Museum in Stockholm, extraordinary measures are needed to even slow the rate of destruction. It enjoyed much better protection when it was at the bottom of Stockholm Harbor.

The real danger to both underwater and exoatmospheric archaeological resources comes not from nature but from humans. The adverse impacts of cultural formation processes—various forms of reclamation, including scavenging, collecting, treasure hunting, and salvage—are what require our attention and mitigation, not the supposed adverse impacts of natural formation processes. The latter are working in our favor.

Despite the threat, both underwater and exoatmospheric archaeological resources have been given little legal protection. Ultimately, this is due to the ill-conceived perceptions of these environments discussed previously, but more directly it is due to the absence of consistent national policies for the preservation of both underwater and outer space archaeological resources. The absence of such policies has made it impossible to reconcile maritime and space law with preservation law, which remain principally at odds. The legal contrast is particularly striking when it comes to questions regarding jurisdiction over and ownership of both underwater and exoatmospheric resources. Determinations have been vague, and decisions have been and will continue to be unclear.[5]

This situation cannot continue. As the technology of reclamation advances, both in the underwater and outer space environments, and becomes increasingly available to those who can afford it and who would do harm to these resources, the threat only grows. Costly, complex reclamation technology is a double-edged sword in this context; while it enhances the opportunities for harm, by those who can afford it, its exorbitant cost simultaneously stymies the attempts of archaeologists to locate, identify, inventory, evaluate, and protect both underwater and outer space archaeological data.[6]

Underwater and outer space archaeological resources are not just similar in legal and managerial ways. They are also similar in some substantive ways. Because of the means by which many of these resources find themselves in either an underwater or outer space environment, and because of the generally benign effects of natural formation processes in these environments, a significant number of them can be found in what archaeologists would recognize as *time capsule* sites. Archaeological time capsules result from single episodes or very short-term periods of deposition. Many of these time capsules are subsequently well preserved, with the materials so deposited being well protected from the impacts of many formation processes. As a result, when archaeologists identify and excavate them, a great amount and range of significant information about one moment in time—the time of site formation—can be recovered. And since these are archaeological time capsules and not ceremonial ones, their creation can be described as unintentional, and thus the nature of their contents is generally less selective and more representative of life at the time of deposition.[7] Shipwrecks are good examples of archaeological time capsules, and so are the Apollo landing sites on the Moon.

CASE STUDY OF EXOATMOSPHERIC NATURAL FORMATION PROCESSES: THE MOON

All of the natural formation processes found on Earth operate within (and largely because of) an atmosphere composed of nitrogen, oxygen, and small amounts of other gases. On the Moon (and in many other exoatmospheric locations) there is no atmosphere, and only traces of these gases can be found. This absence of an atmosphere enhances certain environmental variability (e.g., our atmosphere regulates temperature, which on the Moon can vary from 260°F in the daytime sun to −280°F at night[8]), and this enhanced variability might have impacts on archaeological resources. However, the most profound outcome of this absence of an atmosphere is the absence of all the most significant natural formation processes impacting archaeological resources on Earth. No atmosphere means no animals or plants on the Moon or elsewhere in outer space, for example, and so faunalturbation and floralturbation are nonexistent. Likewise, no atmosphere means that water cannot maintain a liquid state, and it also means there is no wind. Thus, the primary causes of pedoturbation are also nonexistent on the Moon.

Volcanism has occurred on the Moon in the past. This has been confirmed by comparing lunar features to terrestrial volcanic features.[9] Marias, rilles, and volcanic craters can all be seen on the Moon. Evidence for recent eruptions and lava flows does not exist, however. There has been no volcanic activity on the Moon's surface since about 1 billion years ago.[10]

Nevertheless, the lack of an atmosphere means that the Moon is at the mercy of the outer space environment, meaning that outer space itself is the source of natural formation processes operating on archaeological resources on the Moon. Various possible causes of these formation processes, including comets, meteors, micrometeoroids, static electricity, radiation, and the absence or diminished (or, in some rare instances, increased) force of gravity, must be considered. Still, even after such consideration, it is concluded that the outer space and Moon environments are generally benign when it comes to impacts on the archaeological record.

In order to adequately evaluate natural formation processes on the Moon, we need to physically examine evidence from the Moon, and at present there is only so much that can be determined or even speculated upon. We do not have easy access to the Moon, and evaluating lunar archaeological sites directly is currently impossible. Any such study will need to wait until crewed lunar missions resume. Until then, archaeologists are limited to learning and evaluating what NASA scientists have concluded from their assessments of various artifacts either on or returned from the lunar surface.

One very important assemblage of such artifacts comes from *Surveyor 3*, an automated probe that accomplished one of the first soft landings on the Moon, as shown in Figure 26.1. It was launched on April 17, 1967, and landed on the Ocean of Storms on April 19.[11] Thirty-one months later, the *Apollo 12* lunar module landed 600 ft. from this probe,[12] and during their second extravehicular activity, astronauts Alan Bean and Pete Conrad began their evaluation and collection of parts from the *Surveyor 3* site. As the NASA *Surveyor 3* report states, "... the astronauts examined and photographed the spacecraft and removed selected parts and enclosed soil for return to Earth."[13] This exercise was certainly done to test hardware and materials after being off the Earth for some time, but it can nevertheless also be considered data recovery from an archaeological site on the Moon. As such, it can be seen as the first archaeological field work conducted at

FIGURE 26.1 *Surveyor 3* on the Ocean of Storms, 1969. This is the site of the first and only archaeology on the Moon. (Courtesy of NASA.)

an exoatmospheric location, a fact clearly recognized by Penn State University anthropologist P.J. Capelotti:

> As such, the mission of Apollo 12 provided the first example of aerospace archaeology, extraterrestrial archaeology and—perhaps more significant for the history of the discipline—formational archaeology, the study of environmental and cultural forces upon the life history of human artifacts in space.[14]

However, the astronauts did not follow proper archaeological procedure. If they had, different techniques might have been applied. Detailed maps might have been drawn, a survey surrounding the site might have been conducted, and an accurate, detailed description of how the artifacts were collected might have been included with the report. None of these tasks were undertaken, and that resulted in valuable archaeological data being lost.

As a result, little evidence regarding natural formation processes affecting either the material observed or returned can be found in the *Surveyor 3* report.[15] The study indicates that the material looked essentially the same as when it was launched, with only one or two minor changes, such as the appearance of tiny dust impacts and some minor element modifications to the metal casing around the camera, but that is all. One might reasonably ask, What is there to evaluate?

Indirectly, quite a lot can be determined and inferred. Indeed, what the *Surveyor 3* report offers regarding natural formation processes and the lunar archaeological record is unique. Micrometeorite impacts, radioactivity damage from sunlight and gamma rays, the effects of lunar dust attached to the equipment, and even the possible impact of organic matter (see below) are all covered in the report as significant problems.[16] Of course, the sole purpose of this NASA investigation was to evaluate the probe with hopes of improving and generating longer-lasting space equipment, but by using their data and results it might be possible to better deduce what affects the archaeological record on the Moon, and how.

Apollo 12's astronauts removed and returned a television camera, an unpainted aluminum tube, a painted aluminum tube, a surface sampler scoop, and a television cable for analysis.[17] Because of the sensational impacts its existence might have had, NASA's evaluation of the organic material on *Surveyor 3* was a priority in the report. The presence of organic matter was suspected because of discoloration on the outside of the probe and because of the existence of a mass on the exterior of the camera lens that some thought might be organic material from the Moon.[18] After a series of tests conducted on the returned material, the investigators determined that any organic materials on *Surveyor 3* were from Earth. The discoloration was caused by lunar dust settling on the probe during both the *Surveyor 3* and *Apollo 12* landings and organic material from Earth.[19] The mass on the lens could have been from a sneeze on Earth months before the probe was launched or landed on the Moon.

Another important environmental cause of natural formation processes, analyzed by the *Surveyor 3* team, was micrometeorites. Although NASA researchers thought that tiny impacts on the returned camera's metal casing were from micrometeorites, occurring naturally and continuously, they later determined that they were caused by particles disturbed by the landing events of both *Surveyor 3* and *Apollo 12*. There are no definite impacts of micrometeorite origin found on the returned *Surveyor 3* equipment.[20] Researchers found that the actual landings were more damaging to the *Surveyor 3* camera than was exposure to free-floating micrometeorites.

The evaluation of radiation, solar wind, or cosmic rays effects on *Surveyor 3* showed no evidence of serious impact. While there was an increase in the element sodium, the actual changes to the structures were negligible.[21] The analysis determined that there was no evidence of microstructure effects caused by solar wind, solar flares, or cosmic rays.[22]

Surveyor 3 was not evaluated for archaeological purposes. NASA's goal was to learn how to design better, longer-lasting equipment. By using the *Surveyor 3* analysis as a starting point, however, we get a glimpse of what might be awaiting archaeologists in space. And there are other sources of this type of presaging information. Some derive from the first Moon landing in history, *Apollo 11*.

While the primary goal was to get people on the surface, one of the additional objectives of the Apollo 11 mission was to place a range of scientific equipment on the Moon. Doing so would increase understanding of a myriad of lunar functions. One experiment involved placing a lunar ranging retroreflector on the lunar surface to study gravitation, relativity, and lunar physics, as shown in Figure 26.2. This instrument was designed for use by the Lick Observatory and MacDonald Observatory and eventually used by Apache Point Observatory Lunar Laser-Ranging Operation project for determining lunar distance and precisely evaluating the Moon-Earth-Sun system. It was not expected that the retroreflector would continue functioning for 38+ years, but it has. It still reflects back laser beams sent from Earth.

The surprising longevity of the retroreflector suggests that lunar natural formation processes may not have had many (or any) adverse impacts on this particular object. Is that the result of the benign nature of the lunar environment, or did the design

FIGURE 26.2 Lunar laser retroreflector left by *Apollo 11, 14,* and *15*. These are still in use today. (Courtesy of NASA.)

of the object protect it effectively from damage? The retroreflector was designed to contain one hundred corner-cube prisms made of fused quartz.[23] This array was designed to minimize thermal gradients as the Moon rapidly changes temperatures from day to night and from night to day,[24] in order to ensure that a clear signal was relayed back to Earth. Thus, there is a possibility that this design protected the retroreflector from extreme temperature changes.

The instrument was also set about 14 m away from the lunar module, which might have protected it from the dust and static discharge caused by takeoff.[25] As for radiation and micrometeoroids affecting the retroreflector, since observatories have been receiving data from it for more than 38 years, it unlikely that either of these is having a significant negative impact. Dust on a retroreflector can block signals being transmitted to it. In contrast to the situation of Mars, however, the only dust that could settle on the Moon's retroreflector would come from the results of lunar termination or from impacts of matter floating in space. As long as there are no major meteoroid impacts or volcanic/seismic activities in the vicinity of its environment, space and the lunar environment should preserve the retroreflector for years to come.

One of the more significant moments of the Apollo 11 mission occurred when the astronauts planted the American flag on the lunar surface, as shown in Figure 26.3. At that moment, the flag was a powerful symbol of the American victory over the Soviet Union in the Space Race.[26] Yet when the lunar module launched from the Moon, the thrust from its propulsion probably knocked the flag over since it was placed in such proximity to the craft.[27] There are several questions to be asked about the flag's preservation, then: Did the launch wreck or change any qualities of the flag? Has it since survived intact on the surface? To answer these and similar questions, we can look at NASA's experiments and evaluations of materials.

FIGURE 26.3 Edwin "Buzz" Aldrin on the Sea of Tranquility with the first flag planted on the Moon. (Courtesy of NASA.)

The flag was made of nylon. It is safe to assume that a nylon flag can handle short-term exposure,[28] and as the flag presumably is lying on the ground near the lunar module decent stage, it is reasonable to consider it somewhat protected from bombardment or minor impact from debris. The only force that would significantly affect it has been intense illumination from the sun. Without an atmosphere's protective ozone layer to protect the flag, it may be faded to the point of being white or colorless.

Finally, consider the famous and historic astronaut footprints on the Moon, known to nearly everyone. The possible impacts of static electricity were not considered significant until Moon missions started, yet it has since been suggested that a large buildup of static conductivity between lunar terminations could raise soil particles in a dust storm that could result in the movement of significant amounts of lunar soil.[29] If this has been the case, there is then the possibility that these footprints have been disturbed or even obliterated. Static electricity in the lunar soil might also have made the nylon and the aluminum pole of the flag statically charged, and this would have caused lunar dust to attach to the flag from the regolith.

Has the cultural and historical significance of the flag or the footprints been changed because of serious alterations to these artifacts? No. Even if every star and stripe on the flag has been ruined, it would still be the first flag on the Moon, and that will never change. Even if every footprint now is gone, the image of them will last forever as a cultural icon. Regardless, until an investigation is done on the site, an evaluation of significance cannot even be broached.

CONCLUSION

Archaeological formation processes—all the factors responsible for creating the archaeological record—are either cultural or natural in origin. Cultural formation processes are those caused by humans, and they operate wherever people are found. Natural formation processes are those caused by nature, and they operate everywhere material culture exists. Exoatmospheric natural formation processes have been operating in recent years, because an exoatmospheric assemblage of material culture has been accumulating.

The natural formation processes operating on our planet are fairly well understood. Those operating in outer space are remarkably different, however, and only now are we beginning to consider what their impacts on material culture might be. Our ignorance up until now has resulted in a false assumption that all exoatmospheric environments must have highly adverse impacts on all exoatmospheric archaeological materials. This assumption has no empirical support and is likely at least an exaggeration if not an outright falsehood. Indeed, initial consideration of the issue leads to the reasonable conclusion that outer space environments are generally benign when it comes to the archaeological record. It is illuminating to consider the underwater environment, a similarly hostile, exotic, and largely unexplored setting until recently, when attempting to understand better the impacts of natural formation processes operating in outer space. Contrary to widespread suppositions, archaeological study has revealed that underwater environments oftentimes enhance the integrity and preservation of archaeological resources. Chances are good that the same is true for resources located off the Earth. Space archaeology has great investigative potential.

NOTES AND REFERENCES

1. Staski, E. 2000. Archaeological Sites, Formation Processes. In *Archaeological Method and Theory: An Encyclopedia*, ed. L. Ellis, 39–44. New York: Garland Publishing. For complete consideration of the topic, see Schiffer, M.B. 1987. *Formation Processes of the Archaeological Record*. Albuquerque, NM: University of New Mexico Press.
2. For general background information on underwater and maritime archaeology, see Babits, L.E., and H. Van Tilburg, eds. 1998. *Maritime Archaeology: A Reader of Substantive and Theoretical Contributions*. New York: Springer. Delgado, J.P. 1997. *Encyclopedia of Underwater and Maritime Archaeology*. New Haven, CT: Yale University Press. Gould, R.A. 2000. *Archaeology and the Social History of Ships*. Cambridge, U.K.: Cambridge University Press. Muckelroy, K. 1979. *Maritime Archaeology*. Cambridge, U.K.: Cambridge University Press. Ruppe, C.V., and J.F. Barstad, eds. 2002. *International Handbook of Underwater Archaeology*. New York: Springer. Spirek, J.D., and D.A. Scott-Ireton, eds. 2003. *Submerged Cultural Resources Management: Preserving and Interpreting Our Sunken Maritime Heritage*. New York: Springer. U.S. Congress, Office of Technology Assessment. 1987. *Technologies for Underwater Archaeology and Maritime Preservation—Background Paper OTA-BP-E-37*. Washington, DC: U.S. Government Printing Office.
3. Cederlund, C.O., and F.M. Hocker, eds. 2007. *Vasa I: The Archaeology of a Swedish Royal Ship of 1628*. Stockholm, Sweden: National Maritime Museum of Sweden. Also, Saunders (1962) presents the story for a general audience: Saunders, R. 1962. *The Raising of the Vasa: The Rebirth of a Swedish Galleon*. London: Oldbourne.
4. See Peterson, L.P. 1997. *The Bertrand Stores: An Introduction to the Artifacts from the 1865 Wreck of the Steamboat Bertrand*. Missouri Valley, IA: Midwest Interpretive Center. Petsche, J.E. 1974. *The Steamboat Bertrand: History, Excavation, and Architecture*. Washington, DC: National Park Service, U.S. Department of the Interior. Switzer, R.R.

1974. *The Bertrand Bottles: A Study of 19th Century Glass and Ceramic Containers.* Washington, DC: National Park Service, U.S. Department of the Interior. Tuthill, T. 2008. *Artillery Technology from Defacto Sites as an Indicator of Technology in Frontier Conflict Zones.* Manuscript on file, Department of Sociology and Anthropology, New Mexico State University, Las Cruces, NM.

5. U.S. Congress, Office of Technology Assessment. 1987. *Technologies for Underwater Archaeology and Maritime Preservation—Background Paper OTA-BP-E-37.* Washington, DC: U.S. Government Printing Office (see specifically pp. 3–37).
6. U.S. Congress, Office of Technology Assessment 1987 (note 6) (see specifically p. 8).
7. Jarvis, W.E. 2002. *Time Capsules: A Cultural History.* Jefferson, NC: McFarland.
8. World Book at NASA: The Moon. http://www.nasa.gov/worldbook/moon_worldbook.html.
9. Mutch, T. 1970. *Geology of the Moon: A Stratigraphic View.* Princeton, NJ: Princeton University Press (see specifically p. 176).
10. NASA. 2007. *World Book at NASA—The Moon.* http://www.nasa.gov/worldbook/moon_worldbook.html.
11. McNamara, B. 2001. *Into the Final Frontier: The Human Exploration of Space.* Philadelphia, PA: Harcourt College Publishers.
12. McNamara 2001 (note 12) (see specifically pp. 156–157). World Book at NASA (note 9).
13. Carroll, W.F., R. Davis, M. Goldfine, S. Jacobs, L.D. Jaffe, L. Leger, B. Milwitzky, and N.L. Nickle. 1972. Introduction. In *NASA SP-284: Analysis of Surveyor 3 Material and Photographs Returned by Apollo 12*, 1–8. Washington, DC: National Aeronautics and Space Administration.
14. Capelotti, P.J. 2004. Space: The Final (Archaeological) Frontier. *Archaeology Magazine*, 47–51. http://www.archaeology.org/0411/etc/space.html.
15. Nickle, N.L., and W.F. Carroll. 1972. Summary and Conclusions. In *NASA SP-284* (note 14), 9–13.
16. Ibid.
17. Ibid.
18. Ibid. See also Simoneit, B.R. and A.L. Burlingame. 1972. High Resolution Mass Spectrometric Analysis of Surface Organics on Selected Areas of Surveyor 3. In *NASA SP-284* (note 14), 127–142.
19. Simoneit and Burlingame 1972. In *NASA SP-284* (note 14) (see specifically p. 142).
20. *NASA SP-284* (note 14) (see specifically pp. 11, 150).
21. Nickle and Carroll 1972 (note 14).
22. Ibid.
23. Department of Physics, UCSD. 2007. Lunar Retroreflectors. http://physics.ucsd.edu/~tmurphy/apollo/lrrr.html.
24. Ibid.
25. NASA. 1971. *NASA SP-238: Apollo 11 Mission Report.* Washington, DC: National Aeronautics and Space Administration (see specifically p. 154).
26. Gorman, A.C. and B.L. O'Leary. 2007 An Ideological Vacuum: The Cold War in Outer Space. In *A Fearsome Heritage: Diverse Legacies of the Cold War*, eds. J. Schofield and W. Cocroft, 73–92. Walnut Creek, CA: Left Coast Press.
27. McNamara 2001 (note 12).
28. NASA. 2002. *TIP No. 015 Antistatic Properties of Transparent Films.* http://code541.gsfc.nasa.gov/documents/materials_tips_PDFs/TIP%20015R.pdf.
29. Heiken, G., D. Vanimen, and B. French, eds. 1991. *Lunar Sourcebook: A User's Guide to the Moon.* Cambridge, U.K.: Cambridge University Press.

27 Space Environmental Effects

Jennifer L. Sample

CONTENTS

INTRODUCTION

Engineered structures existing in space for any period of time will be subject to environmental conditions very different from those on Earth. The effects of vacuum and radiation, for example, on various materials used for spacecraft design depend on the material properties, such as density, atomic number, conductivity, and so forth, and length of exposure, and they may affect performance. Earth's atmosphere shields us from radiation. The radiation exposure encountered by structures in space causes damage by a variety of mechanisms, including charging, electronics degradation, and radiation-induced sublimation. Plasma, ionizing radiation, micrometeor/orbital debris, neutral gases, and the solar and thermal environments in space have their own effects on materials and structures. These environmental conditions and their effects on engineered structures in space will be explored in this and the following chapters.

The space environment is quite different from that on Earth. It is in many ways harsher, and thus spacecraft are designed with the specific environment(s) they will encounter in mind. Proper design, including shielding critical instruments or parts of the spacecraft, can greatly minimize radiation effects. Environmental effects can cause failures or contribute to degradation after unrelated spacecraft failures. In general, the spacecraft can interact with the space environment in ways that can cause failure or damage or can affect the integrity of measurements being taken by the spacecraft's instruments. Sometimes the data disruption can be a way of learning indirectly about spacecraft-environment interactions.

Much attention has been paid to defining and understanding the space environment, specifically for the purposes of spacecraft design.[1] Here we seek to also understand the effects of space environment on spacecraft decay as it pertains to archaeology. These effects may be unintended but are instructive, as will be outlined in subsequent chapters in this section.

SPACE ENVIRONMENT AND INTERACTIONS WITH SPACECRAFT

For the purposes of this section, the space environment will be discussed as it pertains to spacecraft degradation. The relevant environmental factors discussed will be plasma, ionizing radiation, micrometeor/orbital debris, neutral particles, solar environment, and thermal environment.

The space radiation environments have been modeled and measured for decades.[2] Chapter 28 provides an excellent summary of the available models and data on solar protons, heavy ions, and trapped radiation belts, including trapped particle model development over several decades and the Combined Release and Radiation Effects Satellite mission magnetic storm data. Additional chapters cover the thermal environment, atmospheric effects, electrical system degradation due to the space environment, contamination, and studies of aging in greater detail. This chapter will serve to introduce the various environmental factors and their potential effects on spacecraft.

PLASMA

The plasma environment that a spacecraft encounters depends on its orbit and primarily affects the spacecraft through surface charging processes. Spacecraft charging is a process by which a current balance is achieved between currents into and out of the spacecraft. Current in is due to plasma striking the spacecraft surface, and current out can be in the form of various currents, the magnitudes of which depend on the specific materials emitting them. Currents out of the spacecraft include secondary electrons emitted because of electron and ion impacts, backscattered electrons, and photoemission. Effects on spacecraft include the possibility of arc discharging when two regions of the spacecraft (e.g., sunlit and shaded) charge to different voltage potentials, along with enhanced contamination, shifted spacecraft electrical ground (a problem for measurement integrity) and effects on drag and electromagnetic torque experienced by the spacecraft. Surface charging due to plasma interactions with the spacecraft is typically minimized by making spacecraft surfaces all electrically conductive. For example, Kapton is metalized to ensure that there is a Faraday cage effect on the surface of a spacecraft. However, in some cases, nonconductive surfaces may be used, and charging may be estimated using modeling software, such as NASCAP-2K.[3]

Ionizing Radiation

Charged particles including protons, electrons, and heavy ions have the potential to cause spacecraft charging, degrade electronics, and cause single-event upsets (SEUs). An SEU can result from a high-energy particle impacting and ionizing a sensor element or electronic circuit. Shielding is used to mitigate the effects of ionizing radiation, and its effectiveness can be modeled and measured postmission. This radiation can also cause dielectric charging and breakdown via arc discharging, can degrade optical materials by various processes including color center formation, and can have a negative impact on humans in space. Solar arrays are particularly susceptible to radiation effects, which in turn can affect the mission via loss of power generated by damaged solar panels.

Micrometeor/Orbital Debris

Macroscopic particles exist near planets because of interplanetary meteoroids and debris. Such debris exists near Earth because of decades of human presence in the space surrounding Earth. Material from previous satellites, launches, spacecraft, and so forth surrounds Earth and can cause physical surface damage when impacting a surface. These effects can be understood through modeling, and various accounts of damage due to particulate impacts have been recorded in past missions.[4] Human-made debris or discarded materials or structures are perhaps what first come to mind when thinking of space archaeology, as these are the material remains left by humans in outer space.

Neutral Gases

Neutral gases encountered by a spacecraft predominantly in low Earth orbit (LEO) and polar Earth orbits can cause atmospheric drag, surface erosion, and spacecraft glow, as well as contamination due to neutral gases emitted by spacecraft. The impacting of neutral gas molecules on spacecraft in LEO transfers momentum and energy to the spacecraft, resulting in drag, which can impact orbital precision.

Contamination due to neutral gases can originate from outgassed material from the spacecraft itself or in the form of propellant molecules from solid rocket motor propulsion systems condensing on the spacecraft. This material can degrade performance of the spacecraft or its instruments. Great care is taken in spacecraft design to use materials that are *flight qualified:* that is, whose outgassing properties under stringent limitations are known. Materials can also be thermally pretreated on Earth to minimize outgassing in space.

Neutral gases also include atomic oxygen, which degrades spacecraft surfaces via erosion and recession. This effect was seen on space shuttle flights in the 1980s.[5] Oxygen is a strong oxidizing agent and is a significant component of LEO atmosphere. Erosion is enhanced by ultraviolet (UV) exposure and is most significant with many organic materials. Spacecraft glow is also a neutral gas effect related to contamination and spectrally can affect optical sensor system measurements.

Solar Environment

The sun's electromagnetic flux and emitted charged particles are its main contributions to space environmental effects. The sun emits high-energy protons and lower-energy plasma, known as the solar wind. Solar activity variations result in solar flares and geomagnetic storms. Missions are often planned so as to avoid major solar events that could compromise the spacecraft, spacewalker health, ability to acquire data, or data integrity. The sun's photons also cause environmental effects in the form of photoemission and electron generation via processes such as Compton scattering and pair production.

Thermal Environment

The thermal environment and its effects on aging are discussed in detail in Mehoke's Chapter 29 of this volume. Briefly, space instruments and electronics have operational temperature ranges close to ambient Earth temperatures. Care must be taken in space to keep these electronics within their operational temperature ranges, which is accomplished both by spacecraft design and via materials such as thermal blankets. These materials are susceptible to aging and degradation in space.

Natural versus Induced Environments

The environments covered thus far are natural, defined as the space environment as it occurs independently of a spacecraft.[6] Orbital debris, while manufactured by humans at some point, may be present as part of the natural environment to a new mission. Spacecraft operation may also *induce* a local environment not representative of that which the craft passes through. Many examples of this exist throughout the history of our space exploration, ranging from the somewhat mundane wastewater dumps from the space shuttle to plasma cloud seeding experiments designed to generate plasma for study. Retrorocket fire from the space shuttle approaching a space station is another source of induced environment, as well as any local thermal perturbations due to the presence of the spacecraft. The classic example of an induced environment is contamination. Contamination can be molecular, such as outgassed molecules recondensing onto windows or solar cells, or particulate, such as particles released into the local environment by vibrations of the spacecraft or its parts. This topic is addressed in detail in Donegan's Chapter 33 of this volume.

Combined Effects

All of the space environmental effects described here may coexist together during a given mission. Many combine leaving the spacecraft more vulnerable to problems; for example, outgassing due to vacuum (microgravity) environment is typically exacerbated by elevated temperature. However, other combinations of effects may mitigate overall environmental effects. For example, increased temperature typically renders insulating materials more conductive, thereby possibly providing a mitigating effect on differential spacecraft charging. Combinations of these effects are obviously as specific to the mission as the particular environments encountered and materials used. However, it is necessary to keep in mind that effects may combine, producing unanticipated effects on the spacecraft or mission data integrity.

The combined space environmental effects on potential Apollo window materials were extensively studied in 1965.[7] The study identified that the primary environmental factors most likely to cause degradation of the optical windows were UV and extra-UV photons, the solar wind protons, any solar flare protons and electrons, and protons and electrons trapped in the Van Allen belts. Depending upon the duration of exposure, optical transmissivity was decreased through many of the potential window materials. Often, terrestrial studies such as this one can be useful to help spacecraft engineers determine which materials to use for a given mission.

Combined space environmental effects are particularly relevant to space travelers such as astronauts or tourists. Just as spacecraft materials are exposed to the environment, so is the human body in outer space, particularly on a spacewalk. Periods of weightlessness (due to the microgravity environment) are known to reduce the body's bone mass, and they also may depress the immune system and lead to changes usually associated with aging.[8] In addition to the microgravity environment, the radiation environment presents many potential hazards. Ionizing radiation can damage cell DNA, potentially preventing cells from being able to self-repair, or may cause mutations. Radiation can cause acute sickness, depending on the dose, and solar flare incidents would be particularly hazardous to the unprotected astronaut. Humans are generally protected from many types of radiation exposure by the Earth's environment and from the shielding capability of structures such as a spacecraft or space station.

FUTURE

Various emerging materials and materials systems discovered or fabricated in recent decades, such as carbon nanotubes, nanocomposites, and microelectromechanical systems (MEMS), show great promise for space applications. For example, carbon nanotubes are a relatively new form of carbon in the form of nanoscale sheets of graphite rolled into tubes. These are remarkably strong and lightweight, and they are similar to nanoscale carbon fibers in some respects. Because of their very light weight and high strength, they have a high strength-to-mass ratio, making them ideally suited for space applications (providing strength while minimizing launch costs, for example). These tubes are being fabricated into composites that are currently under investigation for space applications.[9] Carbon nanotubes also emit electrons with very low applied voltage and without heating, so they are also under investigation for space instruments requiring an ionization source.

In addition to new nanomaterials, MEMS may positively impact space missions, helping to achieve scientific objectives in new and potentially better ways. MEMS are actual systems capable of performing entire functions, such as analyzing energy of incident plasma, but because of their miniature size, they are lightweight, have low power consumption, and less expensive to launch.[10]

Materials used in space undergo extensive testing, rendering them space qualified. Space qualification involves studies such as thermal vacuum, irradiation, and temperature cycling. Thus, in addition to the challenges of fabricating and understanding the properties of new materials, such as nanomaterials and MEMS, space qualification is also necessary. As these materials are relatively new and may possess variations from batch to batch, passing the stringent requirements to be deemed space qualified can be a significant challenge. Space qualification of these new materials and material systems is currently undertaken on an as-needed basis.

Spacecraft designers and materials scientists focus on the effects of space on materials from the perspective of mechanical failure: failure to function, failure to provide the data needed, or other failure to achieve designed purpose. Archaeologists focus on the significance of spacecraft and objects in a historic context and how to preserve those that are important to future generations. The two synergistically combine, however, as archaeological investigations are necessarily concerned with the physical and locational integrity of material culture. Additionally, the space environmental effects from both natural and induced environments become increasingly important to the field of space heritage and preservation, as space becomes the place for more and different missions, such as space tourism in the future.

CONCLUSION

Spacecraft materials interact with the space environment in various ways, depending on the specific environment (orbit-dependent) and material properties (e.g., density and conductivity), which govern the susceptibility of those materials to the environment. Overall, interactions can be categorized into physical interactions, such as erosion due to micrometeor impacts with external spacecraft surfaces, and chemical interactions, such as contamination due to outgassing in the microgravity environment or color center formation due to interactions with radiation.

Space environmental effects can also be short term or cumulative over relatively long periods. Short-term spacecraft effects include spacecraft charging and outgassing, and longer-term spacecraft effects include general collective interactions known as aging. Aging can include long-term effects from any of the above processes. Typically, the longer the time spent in space the greater the damage to materials due to aging processes. Chapter 34 provides a summary of predictive environmental effects studies, including Long Duration Exposure Facility testing, and postmission case studies, including the Materials International Space Station Experiment material exposure tests.

REFERENCES

1. Bedingfield, K.L., R.D. Leach, and M.B. Alexander. 1996. *Spacecraft System Failures and Anomalies Attributed to the Natural Space Environment.* NASA Reference Publication 1390. Huntsville, AL: NASA.
2. Hastings, D.E., and H.B. Garrett. 1996. *Spacecraft Environment Interactions.* New York: Cambridge University Press.
3. Sample, J.L., M. Donegan, T. Wolf, D. Drewry, and D. Mehoke. 2007. Charging of Ceramic Coatings in Space. Paper presented at the 48th AIAA/ASME/ASCE/AHS/ASC Structures, Structural Dynamics, and Materials Conference Proceedings, Honolulu, HI, April 23–26, 2007.

4. Sample et al. 2007 (note 3).
5. Tribble, A.C. 2003. *The Space Environment.* Princeton, NJ: Princeton University Press.
6. James, B.F., O.W. Norton, and M.B. Alexander. 1994. *The Natural Space Environment: Effects on Spacecraft.* Alabama: NASA Reference Publication 1350. Huntsville, AL: NASA.
7. Jones, R.H. 1965. *Combined Space Environment Effects on Typical Spacecraft Window Materials.* NASA-CR-65142. Tulsa, OK: Avco Corporation.
8. Newman, D.J. 2000. Life in Extreme Environments: How Will Humans Perform on Mars? *Gravitational and Space Biology Bulletin* 13(2): 35.
9. Hofstra, A.A., M.L. Morris, J.L. Sample, and W.D. Powell. 2007. Non-covalent Functionalized Nanotubes in Nylon 12. *Proceedings of SPIE*, 6643, 664305.
10. Wesolek, D.M., R. Osiander, A. Darrin, J. Lehtonen, and F.A. Hererro. 2005. Micro Processing a Path to Aggressive Instrument Miniaturization for Micro and Pico Sats. Paper presented at the 2005 IEEE Aerospace Conference, Big Sky, MT, March 2005.

28 Space, Atmospheric, and Terrestrial Radiation Environments

Janet L. Barth

CONTENTS

INTRODUCTION

The complex environment of Sun-Earth space consists of time-varying ultraviolet, x-ray, plasma, and high-energy-particle environments. Variations depend on location in space and on the year in the solar cycle, both somewhat predictable. However, large variations that depend on events on the sun are not predictable with reasonable certainty. Our knowledge of large variations is statistically based and is useful in the design phase of space systems to determine upper limits on the severity of the space environments. However, knowledge based on statistics has limited application for forecasting the real-time space environment for operational systems.

The natural space environment and its solar-induced variability pose a difficult challenge for designers of technological systems. The effects of environment interactions include degradation of materials, thermal changes, contamination, excitation, spacecraft glow, charging, communication and navigation errors and dropouts, radiation damage, and induced background interference. This chapter will focus on the radiation environments that produce radiation effects.

OVERVIEW OF THE RADIATION ENVIRONMENT

First, a brief overview of space and atmospheric radiation environments will be presented. More complete descriptions of the radiation environments can be found in other sources.[1,2,3,4]

SPACE RADIATION ENVIRONMENTS

The natural space radiation environment can be classified into two populations:

1. Particles trapped by planetary magnetospheres in "belts," including protons, electrons, and heavier ions
2. Transient particles, which include protons and heavy ions of all of the elements of the periodic table. The transient radiation consists of galactic cosmic ray (GCR) particles and particles from solar events, such as coronal mass ejections and flares. These two types of solar eruptions periodically produce energetic protons, alpha particles, heavy ions, and electrons that are orders of magnitude higher in abundance than the background GCRs

Table 28.1 lists the maximum energy of space radiation particles. The table shows that much of the environment is high energy; therefore, shielding is not always effective in protecting space systems from many types of radiation environments. When modeling, particles are treated as isotropic and omnidirectional, with the exception of plasma, low-altitude trapped protons (<500 km), and cosmic radiation on the ground.

TABLE 28.1
Maximum Energies of Particles

Particle Type	Maximum Energy
Trapped electrons	Tens of MeV
Trapped protons and heavy ions	Hundreds of MeV
Solar protons	GeV
Solar heavy ions	GeV
Galactic cosmic rays	TeV

Trapped Populations

James Van Allen is credited with discovering the trapped proton and electron regions around the Earth. Figure 28.1 is an artist's drawing that shows the belt-like structure of these particle regions. The tilt of the Earth's magnetic pole from the geographic pole and the displacement of the magnetic field from the center cause a dip in the magnetic field over the South Atlantic Ocean, resulting in a bulge in the underside of the inner belt. This region (~300 to ~1,200 km), not shown in Figure 28.1, is called the South Atlantic Anomaly (SAA). In spite of the SAA's reputation for plaguing spacecraft, the flux levels there are actually much lower than those at higher altitudes. The E > 30 MeV proton fluxes peak at approximately 2,500 km altitude at the equator. The electrons are trapped into two regions, the inner and outer zones. The E > 2 MeV peak electron fluxes at the equator are at approximately 2,500 km altitude in the inner zone and at 20,000 km altitude in the outer zone. Heavy ions are also trapped in planetary magnetic fields. For most shielded spacecraft systems, the

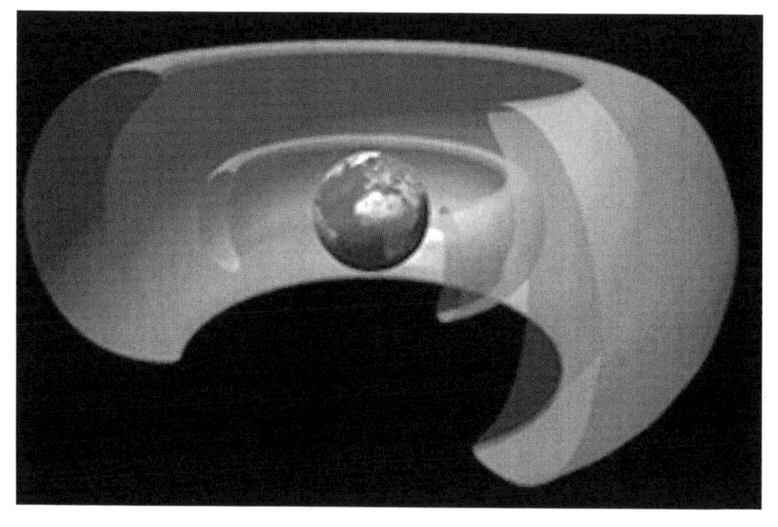

FIGURE 28.1 Artist's drawing of Earth's Van Allen radiation belts. The drawing does not show the South Atlantic Anomaly. (Courtesy of NASA.)

abundances of these ions at energies high enough to penetrate spacecraft materials are too low to be a dominant factor in the rates of single-event effects in electronics systems.

The trapped particle levels and locations are highly dependent on particle energy, altitude, inclination, and the activity level of the sun, and the levels are highly dynamic. The slot region between the inner and outer zones ($2 < L^5 < 2.8$) and the outer zone ($L > 2.8$) populations can increase above averages by several orders of magnitude because of changes in the magnetosphere induced by solar and magnetic storms. Figure 28.2, a plot of measurements of trapped electrons over a 1-year period, shows changes in the extremely dynamic outer zone and the slot region filling periodically with storm electrons. Because of their complex distribution and dependence on long- and short-term solar variability, the trapped particle populations are difficult to model and forecast.

The minimum requirement for the existence of a planetary radiation belt is that the planet's magnetic dipole moment must be sufficiently great to arrest the flow of the solar wind before the particles reach the top of the atmosphere, where the particles will lose their energy because of collisions. Venus and Mars do not have magnetospheres and therefore cannot support particle trapping. The magnetic fields of some of the other planets are similar to the Earth's; however, they vary in strength. Mercury has a weak magnetic field, so it is expected that it has a trapped particle population proportionally lower than that of the Earth. Saturn, Uranus, and Neptune have magnetic fields with strength similar to that of the Earth, but measurements indicate that the intensities of the trapped radiation environments of Saturn and Uranus are much lower than the Earth's and do not pose serious problems to the design of spacecraft systems. On the other hand, Jupiter's enormous magnetic

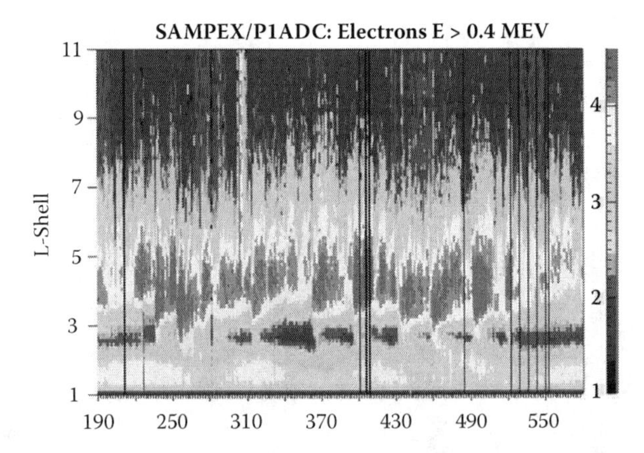

FIGURE 28.2 Radiation belt electrons (E > 0.4 MeV) measured by the Sampex spacecraft. The measurements are plotted as a function of L and the day of year. The outer zone (L > 3), slot region (L = 2–3), and the inner zone (L < 3) can be clearly seen. The plot shows the dynamic electron population in the outer zone and the numerous episodes of the slot region being filled with storm electrons (S. Kanekal, NASA/Goddard Space Flight Center).

dipole (428,000 nanotesla, compared with 30,760 nanotesla for Earth) can support an intense particle environment. Its magnetosphere is the largest object in the solar system. Measurements have shown that Jupiter's radiation environment is considerably more intense than the Earth's and is more extensive; therefore, mission planning for spacecraft that will spend even short times in the trapping regions of Jupiter must include careful definitions of the radiation environment. The *Phobos* probe showed that Mars has a radiation environment; however, it is due to the thin atmosphere of Mars, which allows interplanetary GCRs and solar particles to penetrate to the surface. Interaction of these particles with the atmosphere produces neutrons, which penetrate to the planetary surface and then reflect back.

Transient Populations

The GCR population is a continuously present, slowly varying population of ions from all elements of the periodic table. The levels of GCRs are modulated by the 11-year solar cycle, with the peak GCR populations occurring near solar minimum. The modulation occurs because the solar magnetic field modifies the interstellar GCR environment, suppressing GCR levels, during periods of increased solar activity. Superimposed on the GCR levels are unpredictable, sudden rises in the flux levels due solar energetic particles from solar storms. Galactic and solar particles have free access to spacecraft outside the magnetosphere. Because the transient particles penetrate the Earth's magnetosphere, they can reach near-Earth orbiting spacecraft and are particularly hazardous to satellites in polar, highly elliptical, and geostationary orbits. Figure 28.3 shows measurements from the IMP-8 spacecraft, which illustrates variations in the two populations over a 20-year period. The slowly varying, low-level GCR background population is seen to be approximately anticorrelated with the sunspot number. This is because the GCRs originate outside our solar system and must "fight" against the solar wind to reach our planetary systems. Figure 28.3 also shows the solar particle events, seen as spikes superimposed on top of the GCR measurements. These are the sudden increases in particle populations due to coronal mass ejections and/or solar flares.

Atmospheric Environment

As cosmic ray and solar particles enter the top of the Earth's atmosphere, they are attenuated by interaction with nitrogen and oxygen atoms. The result is a shower of secondary particles and interactions created through the attenuation process. Products of the cosmic ray showers are protons, electrons, neutrons, heavy ions, muons, and pions. Our knowledge of neutron levels comes from balloon, aircraft, and ground based measurements. Taber and Normand[6] give an overview of the neutron in the atmosphere as a function of altitude. The energies of neutrons in the atmosphere is in the hundreds of megaelectronvolts (MeV).

TRAPPED RADIATION BELTS

Kristian Birkeland theorized the existence of radiation trapping in planetary magnetospheres in 1895 when he performed vacuum chamber experiments to study aurora.

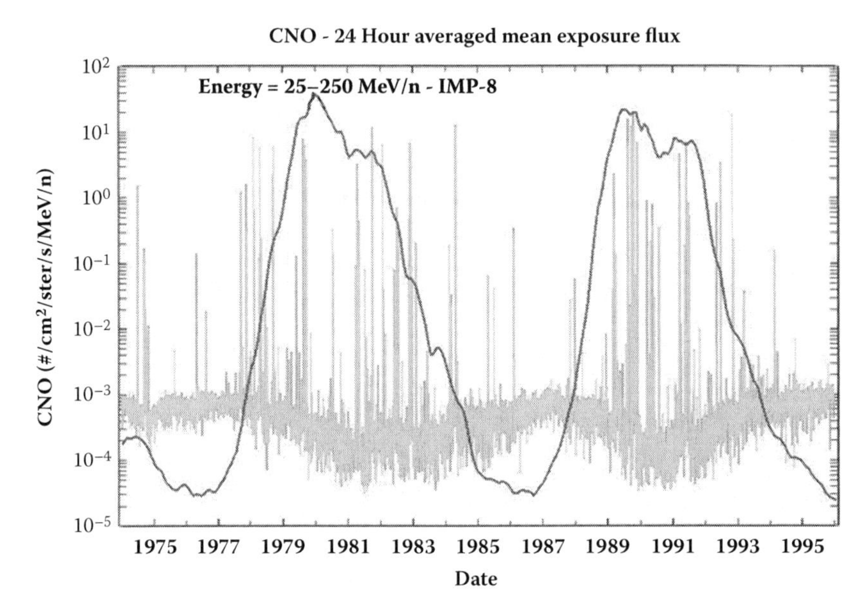

FIGURE 28.3 The IMP-8 measurements of heavy ions in interplanetary space. Note the sudden rises in the counts due to solar particle events and the anticorrelation of the galactic cosmic ray levels with solar activity. The solid line is the sunspot number.

With Henri Poincaré, he showed that charged particles spiraled around field lines and are repelled by strong fields. Later, Carl Stöermer continued work of Birkeland on aurora and made calculations that led to the theory that there was a belt-like area around the Earth in which particles were reflected back and forth between the poles. However, it was felt that the magnetic field was not strong enough to hold the particles. In 1957, S. Fred Singer proposed that ring current could be carried by lower energy particles injected by into trapped orbits by magnetic storms. A complete history of radiation belt science can be seen in *Space Storms and Space Weather Hazards*.[7]

EARTH'S VAN ALLEN BELTS

The launch of the Sputnik Earth orbiter by the Russians on October 4, 1957, sparked intense interest in developing a U.S. space program. In 1955, James Van Allen and several other American scientists had proposed the launch of a scientific satellite as part of research programs to be conducted during the International Geophysical Year of 1957–1958. The success of *Sputnik* led to the approval of Van Allen's proposal for the *Explorer I* spacecraft. Van Allen's interest in getting instruments into space was to study cosmic rays and their origin. He designed a cosmic ray detector to measure the low background cosmic rays, which was launched on *Explorer I* on January 31, 1958, from Cape Canaveral, Florida. In analyzing the data from his instrument, Van Allen was puzzled by zero readings. At first he thought that the instrument had malfunctioned, but later he realized that the instrument was being "flooded" with radiation measurements. Van Allen determined that his instrument was measuring intense radiation surrounding the Earth, and he announced his discovery on May 1, 1958.[8,9]

Most of the scientific instruments flown in space in the late 1950s and early 1960s were designed to detect energetic protons and electrons. With the results, scientists gained a general understanding of the near-Earth radiation environment but found differences up to a factor of 10 when making quantitative comparisons between measurements.

It was also during the early 1960s that spacecraft electronics were found to be unreliable. Problems from differential charging from the solar wind and from noisy data transmission to the Earth from soft fails were noted. These problems were largely dealt with by building redundancy into systems. As the first scientific satellites were being launched in the late 1950s and early 1960s, the USSR and United States also detonated nuclear devices at altitudes above 200 km. The most dramatic of these tests was the U.S. Starfish detonation on July 9, 1962. Ten known satellites were lost because of radiation damage, some immediately after the explosion (see Vette, reference 9). The Starfish explosion injected enough fission spectrum electrons with energies up to 7 MeV to increase the fluxes in the inner Van Allen belt by at least a factor of 100. Effects were observed out to 5 Earth radii. The Starfish electrons that became trapped (modeled by Teague and Stassinopoulos[10]) dominated the inner zone environment (~2.8 Earth radii at the equator) for 5 years and were detectable for up to 8 years in some regions.

BEGINNINGS OF THE RADIATION BELT MODELING PROGRAM

The production of enhanced radiation levels by the Starfish explosion and others and the ensuing problem of shortened spacecraft lifetimes emphasized the need for a uniform, quantitative description of the trapped particle environment. Wilmont Hess of the National Aeronautics and Space Administration's (NASA's) Goddard Space Flight Center (GSFC) developed the first empirical models of the trapped radiation belts. Using data from several satellites, he began constructing quantitative radiation models for inner zone protons and electrons. These models were designated P1, P2, and so forth, and E1, E2, and so forth, where "P" indicates protons and "E" indicates electrons. Starting in 1962 and continuing through the late 1960s, several series of satellites were launched with instruments designed to measure the effects of Starfish, providing a large volume of particle data. In late 1963, James Vette of The Aerospace Corporation and later of NASA/GSFC was appointed to lead a trapped radiation environment modeling program jointly funded by NASA and the United States Air Force. At that time, there were several groups actively involved in trapped particle measurements, including The Aerospace Corporation, Air Force Cambridge Research Laboratory (now Phillips Laboratory), Johns Hopkins University Applied Physics Laboratory, Bell Telephone Laboratories, GSFC, Lawrence Livermore Laboratory, Lockheed Missile and Space Corporation, the University of California at San Diego, and the University of Iowa. Each organization agreed to make its measurements available to the modeling program.

In 1965, James Vette of The Aerospace Corporation presented an overview paper of the efforts to model the trapped radiation environment.[11] He referenced 31 papers from various scientific meetings and journals (American Geophysical Union, *Journal of Geophysical Research*, etc.[11]) where measurements and modeling results had been presented. Vette reviewed the efforts to model the outer zone particles, stressing the difficulty of developing static maps of the dynamic electrons in that region. He compared the lifetimes of the outer particles (minutes) to those of the inner zone (years). As his paper demonstrates, the modulation of the electrons driven by the 27-day solar rotation period and the fluctuations connected with magnetic activity were known at that time. Rather than presenting maps of the dynamic outer zone, he gave "typical" integral spectra. It was known at the time that the protons in the outer zone are more stable, are more closely confined to the magnetic equator, and have a "soft spectrum." With respect to damage to spacecraft, he noted that outer zone protons will affect only unshielded devices, but that electron exposure during long missions would result in measurable effects.

Vette presented the AE-1[12] map of the inner zone electron model and the newly developed AP maps for protons with energies greater than 4 MeV. He noted that the natural inner zone electron population was not well known before Starfish and that Starfish electrons dominated the population levels in regions below an L of 1.8. He also stated that, before Starfish, protons up to several hundred MeV dominated the inner zone. Interestingly, as an aside, Vette made reference to a then-recent observation by McIlwain that a redistribution of protons at E>34 MeV followed a large magnetic storm in the L = 2–3 region.[13]

TRAPPED PARTICLE MODEL DEVELOPMENT: 1960s, 1970s, AND 1980s

Eight trapped proton models, eight trapped electron models, and one Starfish decay model were released during the 27 years that the trapped radiation modeling program was operative. The trapped particle models that are most often used at this time are the AP-8[14] for protons and the AE-8[15] for electrons. The AP-8 model, released in 1976, was the culmination of a long-term effort to include all of the previous models under one common approach and to include all of the data after 1970. After 1977, the modeling budget was significantly reduced, so a similar effort to consolidate the electron models into the AE-8 model was not completed until 1983. The formal documentation of that model was released in 1991.

The AP-8 and AE-8 models include data from 43 satellites, 55 sets of data from principal investigator instruments, and 1,630 channel-months of data. By the 1970s, scientific interest had shifted from trapped particles to the plasma regime to determine the physical mechanisms of particle energization and transport. As a result, the number of new data sets available for trapped radiation environment modeling was drastically reduced. It was not until the measurement of storm belts by the Combined Release and Radiation Effects Satellite (CRRES) mission in 1991 that concerns were renewed about the ability to model the trapped radiation belts to sufficient accuracy for using modern microelectronics in space.

STARFISH EXOATMOSPHERIC, HIGH-ALTITUDE NUCLEAR WEAPON TEST[16]

In 1958, the United States conducted the Hardtack series of nuclear weapons tests over the Pacific Ocean and the Argus series in the South Atlantic Ocean. In 1962, additional tests were conducted in the Fishbowl series. One test, the Starfish Prime, with a yield of 1.4 megatons TNT equivalent, was exploded on July 9, 1962, at a very high altitude (approximately 400 km) over Johnson Island in the Pacific Ocean (about 700 miles from Hawaii). This exoatmospheric nuclear explosion released about 10^{29} fission electrons into the magnetosphere, creating an artificial radiation belt and raising the intensity of the inner zone electron population by several orders of magnitude. This additional radiation increased the radiation damage on spacecraft flying in that region to critical levels. The first failure (due to total ionizing dose) was the *Telstar* satellite, which was launched 1 day after the Starfish. It was estimated that the spacecraft experienced a total dose from the explosion that was 100 times larger than was planned for the spacecraft lifetime. Within months after the tests, seven satellites failed, primarily because of solar cell damage.

Initial predictions of the longevity of the Starfish debris ranged from the overly optimistic predictions of some months to the more realistic ones of a few years. Studies conducted in the late 1960s[17,18,19,20] attempted to define the rate of decay with varying results. An in-depth evaluation was performed in 1970–1971[21] using data from the 1963-38C satellite that covered the time span from September 1963 to December 1968. The researchers identified three distinct regions within the inner zone domain that were populated by the artificial electrons and established that their decay lifetime τ in days could be best expressed by a function of three variables:

$$\tau_1 = \tau\,(B,L,E) \text{ in Region 1} \tag{28.1}$$

$$\tau_2 = \tau\,(L,E) \text{ in Region 2} \tag{28.2}$$

$$\tau_3 = \tau\,(E) \text{ in Region 3} \tag{28.3}$$

where B is the field strength in gauss, L is the magnetic shell parameter, and E is the energy in MeV. In 1972, a more thorough approach produced a model of the Starfish flux for September 1964 based on data from several spacecraft (OGO-1, OGO-3, OGO-5, OV3-3, and 1963-38C) (see Teague and Stassinopoulos, reference 10). That model distinguished between artificial and natural electrons and provided the artificial flux as a function of equatorial pitch angle, energy, and L value. The decay times for this flux were determined by two separate methods, which were combined to yield average values that are appropriate for the evaluation of the long-term loss process of the artificials. Numerical values relating to nuclear explosions must include a substantial margin of error. In addition to the difficulties in making measurements of these events at the time of their occurrence, the results are dependent on circumstances that cannot be predicted. Two nuclear weapons of different design may have the same explosive energy yield, but the effects could be markedly different. In the case of the Starfish, estimates of errors associated with the cutoff-time model are given in Table 28.2.

It is interesting to compare the Starfish effects with those of a Soviet high-altitude test of a low-yield weapon that was performed on October 28, 1962, over Semipalatinsk in Kazakhstan. Figure 28.4 shows the integral Van Allen belt electron fluxes before and after in the regions from L of ~1.8–4.0 for particles with energies of E>0.5 and E>1.9 MeV. This L region covers the outer edge of the inner zone, the slot region, and

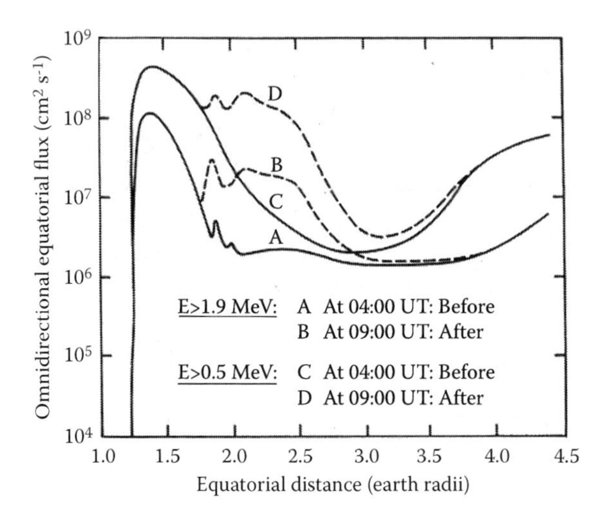

FIGURE 28.4 Integral Van Allen belt electrons before and after the Soviet event.

the beginning of the outer zone. The bulk of the fission electrons are concentrated between L of ~1.8 and 2.7, which is to be expected because their injection occurred at a high latitude location. In contrast, the Starfish debris was concentrated in the inner zone because of the low latitude of the Johnson test site. A schematic of the distribution of the fission electrons from these two tests is shown in Figure 28.5 in terms of magnetic shell parameter L and magnetic latitude. Figure 28.6 is an attempt to show the average lifetimes of the E > 2 MeV electrons from the Starfish and Soviet experiments. Although it is difficult to draw conclusions from only two tests, the data suggest that longevity is maximum at low L values, decreases rapidly toward the slot region, and is in the range of months and perhaps weeks at larger L values.

STORM BELTS: THE CRRES MISSION

A discussion of developments in our knowledge of the trapped radiation environment cannot be complete without including the contributions that were made by the

TABLE 28.2
Accuracy of the Starfish Decay Model in Months

	Energy (MeV)						
L	**0.03**	**0.05**	**0.1**	**0.3**	**0.5**	**1.0**	**3.0**
1.3	±4	±4	±4	±4	±5	±8	±10
1.4	±3	±3	±3	±3	±4	±8	±10
1.5	±3	±3	±3	±3	±4	±8	±10
1.6	±3	±3	±3	±3	±4	±6	±7
1.7	±3	±3	±3	±3	±4	±8	±8
1.8	±3	±3	±3	±3	±6	±8	
1.9	±8	±7	±5	±5	±7	±8	
2.0	±8	±7	±5	±6	±8	±9	

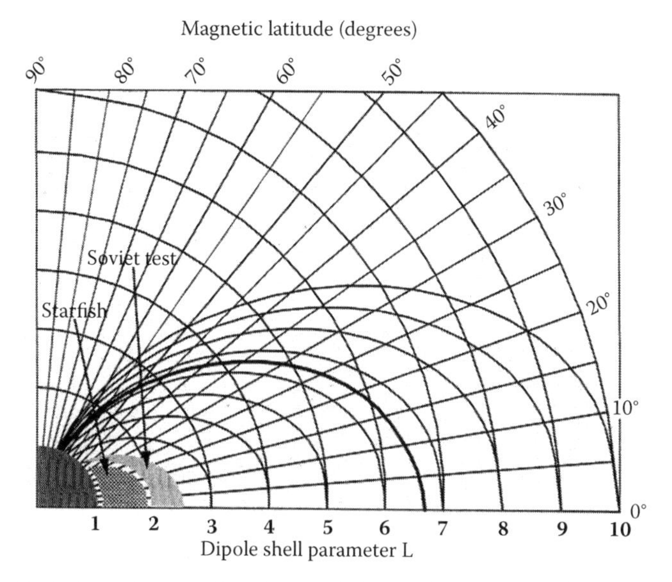

FIGURE 28.5 Schematic of the distribution of fission electrons from the Starfish and Soviet tests in magnetic coordinates.

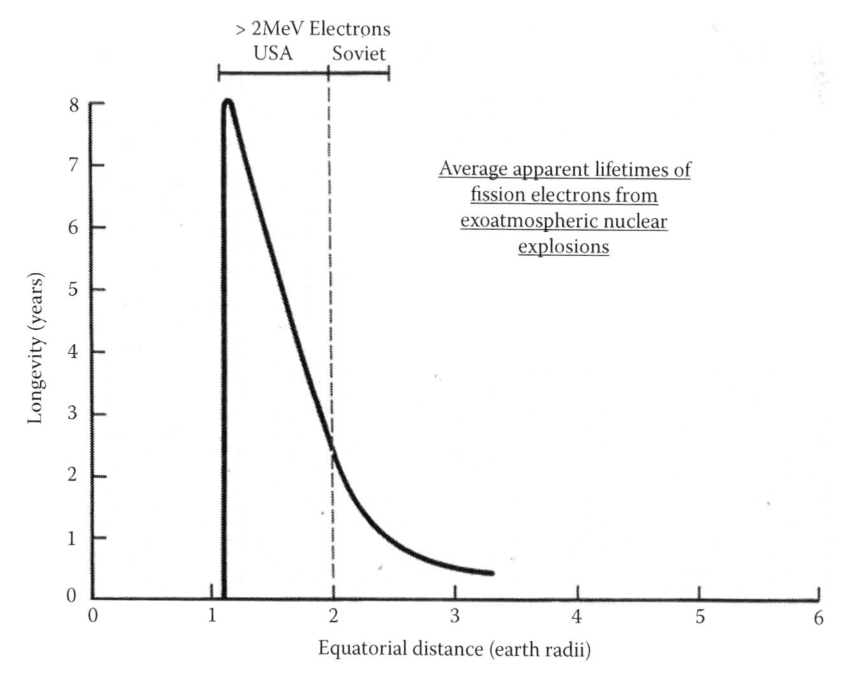

FIGURE 28.6 Comparison of the average lifetimes of >2 MeV electrons for the Starfish and Soviet tests.

CRRES mission. Not only did the mission provide data for radiation belt modeling, but it also rekindled interest in radiation belt science. The CRRES satellite carried a full complement of particle detectors, and its geosynchronous transfer orbit of 350 km perigee and over 30,000 km apogee at 18° inclination was ideal for measuring radiation belt particles. Unfortunately, the CRRES launch was too late to measure the large September and October 1989 solar events. However, CRRES was in operation for the second peak of solar cycle 22 and observed the largest magnetic event ever recorded in March of 1991. Analysis of data from the instruments on CRRES before and after the March 1991 magnetic storm showed extremely large redistributions of the trapped particle populations. Mullen et al.[22] reported that the CRRES proton instrument had measured a new proton storm belt in the slot region (L = 2–3). In the 1996, Gussenhoven et al.[23] summarized all of the CRRES results, including a review of this proton belt formation and the production of an electron storm belt during the same storm. Recall from the previous section "The Beginnings of the Radiation Belt Modeling Program" that in 1965 McIlwain had previously reported that he had observed a redistribution of protons at E > 34 MeV following a large magnetic storm in the L = 2–3 region.

One of the most important conclusions that resulted from the CRRES program was that existing theory cannot explain the particle penetrations deep into the magnetosphere observed by McIlwain and Mullen et al. The frequency of occurrence of these atypical events is also unknown; therefore, applying current models to setting design and operational radiation hardening requirements creates uncertainties that are impossible to quantify. Dynamic environment simulations are a method that could be used to address this issue. Boucher et al.[24] reviewed modeling techniques that provide computer generated models of trapped particle transport during storms. Bourdaire, Boucher, and others[25] are developing a 4-D diffusion code to calculate the transport of particles throughout the inner magnetosphere. They have applied the code to several problems, including calculation of the transport of existing and injected particles during storms, determination of the ring current growth using calculations for both protons and electrons, and discovery of the process by which high-energy particles are totally removed from the internal magnetosphere. Case studies have been validated using CRRES and Space Technology Research Vehicle-1b measurements.[26] In the future they plan to drive the model with interplanetary magnetic field strength, thereby greatly increasing the database for running the code and providing statistics for a range of conditions for a given time period.

EUROPEAN SPACE AGENCY MODEL IMPROVEMENTS

Daly et al.[27] identified errors in the NASA models and documentation, including a source code error in AP-8-MIN and the fact that the AZUR data set on which the AP-8-MAX is based covered a time span of 3 months, not 6 months. They also noted that the space shuttle and the Long Duration Exposure Facility satellite measured environments in low altitudes (300–500 km) were from 60% to 100% higher than those predicted by the AP-8 models. The authors determined that a large source of this error is due to the method used to interpolate between the B/B0 values in the regions near the atmospheric cutoff. They developed an alternate interpolation method that increased the estimated fluxes by 10% to 40%, bringing them closer to the measured levels. They also recommended that an additional L increment at the low L values be included in the models to give better resolution at steep gradients. This increased the estimated fluxes by about 40%. When these two refinements to the interpolation scheme were combined, the revised flux levels were much closer to the measured values. The work of Daly et al. renewed interest in improving the radiation belt models for applications to enabling and commercial off-the-shelf technologies, which can have very low radiation tolerance.

RADIATION BELT MODELS WITH IMPROVED TIME RESOLUTION

Analysis of the CRRES instrument and experiment data showed that not only is the environment extremely dynamic but also that electronic parts respond to the short-term changes. The AP-8 and AE-8 radiation belt models, with their 4–6-year averages, were adequate for long mission durations and for long-term degradation effects. However, modern spacecraft and instrument systems have serious problems with short-term effects, such as interference and data corruption. With a time resolution of 4–6 years, the radiation belt models place a serious restriction on the ability to address design and operation issues of systems.

Several researchers have recognized the need for trapped particle models with finer time resolution. Four empirical models were developed using CRRES data to estimate short-term dynamic changes in the particle population, the CRRESPRO,[28] CRRESELE,[29] CRRESRAD,[30] and APEXRAD[31] models. While the models are based on data collected over a short 14-month period and during solar maximum conditions only, they give the most comprehensive picture available of the environment resulting from a geomagnetic storm.

The time resolution of the low-altitude portion of the AP-8 model has improved in recent years. Pfitzer[32] plotted predicted flux values for low inclination orbits as a function of the average atmospheric density. From this index, he interpolated and extrapolated fluxes from the AP-8 and AE-8 models for solar activity conditions. Later, Huston and Pfitzer[33] analyzed proton instrument data from the CRRES and Television Infrared Observation Satellite/National Oceanic and Atmospheric Administration (NOAA) satellites with the goal of developing a low-altitude (<850 km) trapped proton model with variation over an entire solar cycle as a function of solar activity indices. In 1998, Huston and Pfitzer presented the first trapped proton model with true solar cycle variation.[34] The proton flux levels were determined by using the solar radio flux proxy for atmospheric heating and included the phase lag between rise and fall of flux levels and solar activity.

Recently, the Huston team joined efforts with Xapsos and others at GSFC to add statistical variations to the solar cycle-driven model, thereby adding confidence level information to proton levels.[35] This increases the ability to address trapped proton variations for spacecraft design and mission planning.

Heynderickx and Lamaire[36] also plotted model fluxes as a function of the average weighted density of the atmosphere (ns) that is encountered by a particle on its drift shell. They found that the relationship between ns and the AP-8 and AE-8 models fluxes is well ordered, especially at low L values. If a practical form for the ns calculation is developed, this method could lead to replacing B0 with B0(ns) when accessing the AP-8 and AE-8 models, thereby reflecting solar activity effects. Using data from the proton instrument on the Sampex spacecraft, they also developed a trapped proton model with improved time resolution.[37]

In 2003, Boscher et al. published a model for electrons in geostationary regions of the Earth's radiation belts.[38] This model is the result of an international collaboration between Onera, the French Research Laboratory, and the U.S. Los Alamos National Laboratory. The model is based two and a half solar cycles of electron outer radiation belt measurements from Los Alamos National Laboratory geostationary satellites, covering the period 1976–2001. The measurements were cross-calibrated between the satellite instruments and referenced to CRRES observations. The model takes into account the solar cycle variation.

In October 2004, the National Aeronautics and Space Administration's Living with a Star Program sponsored the Working Group Meeting on New Standard Radiation Belt and Space Plasma Models for Spacecraft Engineering (October 2004, College Park, Maryland).[39] The workshop was organized and led by Barth of NASA, with coleaders from The Aerospace Corporation, the Air Force Research Laboratory, and the European Space Agency. Members of the international radiation belt and space plasma modeling community (industry, National Aeronautics and Space Administration, Air Force Research Laboratory, the Department of Defense, and the European Space Agency) attended, focusing on the development of the next generation of radiation belt models to replace the AP-8 and AE-8 models; the primary purpose for the development of new models is the reduction of risk and costs associated with the exposure of spacecraft components and instruments to the space radiation environment. The goals of the workshop were to report and document recent progress on radiation belt model development and to complete a road map for the development of new standard radiation belt and space plasma models for spacecraft engineering.

The most important result of the workshop was user community–developed model requirements that were derived from the United States Space Technology Alliance's Space Environment and Effects Working Group and from the European Space Agency's Space Environment and Effects Program. Plenary sessions reviewed new developments in radiation belt models for protons and electrons, space plasma models, data set availability and calibration, and current and planned missions and instrumentation for space radiation environment modeling purposes. In all, thirty-five presentations were given over 3 days. The workshop provided inputs to agencies for future investments, agreements on interagency cooperation, long-term modeling goals of the participants and their institutions, and community agreement on data set management and the model standardization process. Also, two new models were proposed for data set review and standardization. The models cover critical regions in space for space agency, military, and commercial spacecraft missions, namely, low-altitude protons [a combination of the Huston and Pfitzer (see reference 34) and Heynderickx et al. (see reference 37) models] and geostationary electrons [the Onera/Los Alamos POLE model (see reference 38)]. These two new models have progressed through committee review and have been opened up for comments from the modeling community as a final step for standardization.

Recently, the National Reconnaissance Office, National Aeronautics and Space Administration, the Air Force Research Laboratory, The Aerospace Corporation, Los Alamos National Laboratory, and the Naval Research Laboratory have embarked on a project to produce the next generation radiation belt model, AP(E)-9. This model upgrade will offer significant improvements in terms of the radiation hazards specified, accuracy and uncertainty quantification, spectral and spatial coverage, and time-correlated probability of occurrence statistics. Preliminary requirements have been gleaned from participation in the Department of Defense–sponsored Space Environment Effects Working Group, the NASA Living with a Star Workshop (see Lauenstein et al., reference 39), and the NOAA Space Weather Workshop. A goal of the NASA's Living with a Star Program is to increase our knowledge of the radiation belts surrounding the Earth. NASA will launch two satellites (Radiation Belt Storm Probes) into the radiation belts to discover the fundamental physics underlying the source, loss, and transport processes that govern the levels of particles in the radiation belts. The data will be used to improve the AP-8 and AP-9 models used by engineers to design radiation-hardened spacecraft.

TRAPPED RADIATION AT OTHER PLANETS

After numerous missions to explore the Earth's radiation belts and interplanetary space, scientists became interested in exploring the outer planets. In 1971, J.W. Haffner contributed a paper, "Natural Radiation Environments for the Grand Tour Missions"[40] that presented an overview of the solar "flare," galactic, and magnetically trapped particle radiation expected during missions to the four outer planets, Jupiter, Saturn, Uranus, and Pluto. He estimated the environment levels from data and models and predicted mission doses. The ability to predict the environments at the outer planets was greatly hampered by the lack of measurements of the interplanetary environment beyond 1 astronomical unit (AU) and by the lack of any measurements of the trapped radiation environments of Jupiter and Saturn.

Haffner presented estimates of the trapped radiation belts of Jupiter and Saturn. It was theorized that Jupiter and Saturn had belts similar to those of the Earth based on decametric (bursts) and decimetric (quasi–steady state) radio frequency radiation emitted by Jupiter and Saturn. Similar radio frequency radiation emission was measured from the Earth's belts. The decimetric radiation is due to synchrotron emission of the electrons trapped in the magnetic field, and the decametric radiation is associated with one of Jupiter's moons, Io. On the basis of assumptions about the limiting particle fluxes, similar relationships in Jupiter's and Earth's magnetic fields, and particle and plasma densities and by ignoring the effect of the planet itself, Haffner derived relationships between the magnetic field at Jupiter's equator, the particle density relative to the plasma stability limit, and the effective inner radius of the Jovian belts. From that set of parameters, the electron dose rates for mission flybys were estimated.

The same methodology could not be applied to trapped protons because they do not radiate as the electrons do. Haffner pointed out that theory explaining the source and loss mechanisms for protons or electrons in the Earth's belts that could be applied to estimating the proton belts of Jupiter did not exist. Therefore, he had to base the proton estimates for Jupiter and Saturn on the ratios of the protons/electrons in the Earth's Van Allen belts. He used these estimates to calculate doses for the missions.

In 1972, Kase followed up Haffner's work with a presentation that focused on concerns about displacement damage on spacecraft electronics due to proton and neutron environments.[41] In addition to natural sources, Kase presented the problem posed by having neutron-emitting radioisotope thermoelectric generators on board spacecraft. Kase also revisited the problem of modeling the proton belts of Jupiter. By that time, Divine had developed nominal and upper limit models of the proton belts.[42] The three orders of magnitude difference between the two models was an indication of the inaccuracy inherent in the unvalidated theoretical approach. It was expected that the proton spectra were "very hard" near the surface of Jupiter and softer at great distances.

MEASUREMENTS OF THE GRAND TOUR MISSION

The Pioneer missions to the outer planets carried instruments to measure the radiation environment. The measurements showed that the radiation was orders of magnitude higher than expected. Peak intensities of electrons in the belts, as measured by *Pioneer 10*, were 10,000 times greater than Earth's maximum. Also, the electron energies were found to be greater than 20 MeV. Protons were several thousand times as intense as Earth's belts. The inner radiation belts of Jupiter, as measured by *Pioneer 10*, had the highest radiation intensity so far measured, comparable to radiation intensities following an explosion of a nuclear device in the upper atmosphere. *Pioneer 11* confirmed these high intensities. In the inner region of the magnetosphere, high-energy protons exceeding 35 MeV appear to peak in two shells; the outer shell was detected at 3.5 Jovian radii by *Pioneer 10* (confirmed by *Pioneer 11*), and an inner shell, discovered by *Pioneer 11*, has a peak at 1.78 radii of Jupiter. *Pioneer 11* also found that there is a greater flux of energetic particles at high Jovian latitudes than would have been expected from the measurements made by *Pioneer 10*. It also discovered that the flux of energetic particles peaks on either side of the dipole magnetic equator.[43] This discovery led to the need to retrofit the Galileo spacecraft with radiation-hardened bipolar processors, because the spacecraft design with unhardened processors had been fixed before the arrival of the Pioneer spacecraft at Jupiter.[44] The models of the Jupiter radiation environment were updated using data from the missions to the outer planets.[45]

Saturn also has radiation belts; however, they are not nearly as intense as those of Jupiter because the rings around Saturn deplete the particle levels near where their peak would occur. Measurements from instruments on the Cassini spacecraft were used to improve the SatRad model, which was originally based on data from the Pioneer and Voyager missions.

SOLAR PROTON EVENT MODEL DEVELOPMENT

EARLY ESTIMATES OF SOLAR PROTON LEVELS

In 1965, Vette compared the levels of solar "flare" protons and galactic cosmic rays based on work by Malitson and Webber[46] and noted that the solar protons are low during periods of low solar activity. He also pointed out that they were considered a serious hazard for astronauts on Apollo missions. To predict the solar proton environment, Haffner compiled all of the existing solar proton data at 1 AU for 1956–1968, including data from Webber, McDonald, Lewis, Modisette, and Mosley (see reference 46). He developed fitting functions to fill in missing data that described the onset, rise time, and decay of solar events as a function of time. He then reconstructed probable values based on sunspot numbers by using predictions of future sunspot numbers developed by Weddell and Haffner.[47] To estimate the diffusion of the solar protons throughout interplanetary space, Haffner assumed a spatial dependence of $1/r^2$, where r is measured in AU, which is an approximation still used today. The method was used to predict the expected solar particle environment for each year from 1970 to 1989. Haffner points out the large uncertainties in predicting solar particle levels. In fact, as seen in Table 28.3, the differences between the solar proton levels predicted for 1972 and 1989 at 1 AU and those measured by spacecraft are large.

First Statistical Models

As concern grew over exposure of electronics and human to solar protons and as the amount of available on-orbit particle measurements increased, the use of compilations of data for estimating solar proton levels was replaced by modeling efforts. In 1974, King[48] published the first statistical model for solar proton events using Poisson distributions. He concluded from his analysis of proton data from the twentieth solar cycle that solar proton events could be classified into "ordinary" and "anomalously large." This was based on the fact that only one anomalously large event occurred in the twentieth solar cycle—the August 1972 event. That event alone accounted for 84% of the total proton fluence in the solar cycle at energies $E > 30$ MeV. However, when Feynman et al.[49] added data from cycles 19 and 21 to the solar proton event database, they were able to conclude that individual solar proton events actually form a continuum of event severity from the smallest to the largest, blurring the distinction between ordinary and anomalously large events.

Engineering-Oriented Statistical Models

Many large events similar to the August 1972 event occurred in cycle 22, increasing concern about the validity of the solar proton models. With the goal of improving the ability to address practical aspects of spacecraft reliability, a team led by Xapsos began compiling solar proton data for solar cycles 20, 21, and 22 and using statistical

TABLE 28.3
Solar Protons >10 MeV

Year	Predicted Protons/cm^2 per Year	Actual Protons /cm^2 per Year
1972	7.0×10^9	2.4×10^{10}
1989	8.0×10^8	4.5×10^{10}

techniques to derive probability distributions of cumulative solar proton fluences. A review of the data sets can be found in Xapsos et al.[50] In 1996, Xapsos et al.[51] presented a paper that described the application of extreme value theory to determine probability of encountering a single large event over the course of a mission. They also used compound Poisson process theory to describe the probability of encountering various fluence levels during a mission. The work of the Xapsos team confirmed the Feynman conclusion that a "typical event" cannot be defined.

The Xapsos team then turned its focus to understanding how to define the peaks of solar proton events. To accomplish this, the team's members applied the maximum entropy principle (MEP) to select the least biased event probability distribution. The MEP, used for earthquake predictions, is valuable for analyzing incomplete data sets. They validated the results with lunar rock records dating back to ancient times.[52] The Xapsos team members continued their work by establishing worst-case solar proton spectra for solar events.[53] When comparing their model with the CREME96[54] "worst-week" solar proton model, which was based on the October 1989 solar particle event, they found that, statistically, the CREME96 model is closer to a 90% worst-case event model.[55] Xapsos et al. have combined the model elements into the emission of solar protons (ESP) model, which is available as a computer code (see Xapsos et al., references 50 and 55).

Solar particle intensity also varies with distance from the sun. As particles move away from the sun, the particles diffuse into the larger volume of space. Spacecraft near the Earth have provided us with a large data set of particle measurements, so particle intensities are scaled as a function of distance from the Earth. (The Earth is at 1 AU, or the distance between the sun and the Earth.) We know from Pioneer and Voyager measurements that the particle intensity varies by $1/r^2$ at distances greater than 1 AU from the Earth, where r is in units of AU. Because we have so few measurements as we move closer to the sun (less than 1 AU), we are not certain about the variation of particle intensity. Estimates range from $1/r^2$ to $1/r^3$.

HEAVY IONS

Heavy Ions Prior to 1975

In the early 1900s, scientists were puzzled by ground "charge" on electroscope leaves. During laboratory experiments, the leaves of an electroscope repelled each other without the presence of charge. Scientists assumed that this was the result of the ionization of air by the natural radiation present on Earth (see previous section, "Atmospheric Environment"). Efforts were made to eliminate the radiation by using radiation pure materials; however, the problem persisted. In 1913, an Austrian scientist, Victor Hess, devised an experiment to put an electroscope in a balloon to get it away from the Earth's radiation. As Hess and his experiment ascended in the balloon, he observed that the radiation source did not go away; rather, as the altitude of the balloon increased, the radiation increased. Hess concluded that the source of this radiation was outer space. In the summer of 1925, Millikan showed with his lake experiments that the radiation source was outer space. When he presented his lake experiment findings, he called the radiation "cosmic rays."[56] Hess and C.D. Anderson received the Nobel Prize for their discovery of cosmic rays in 1936.

Intense interest in understanding cosmic rays continued into the 1950s. Recall from the previous section, "The Earth's Radiation Belts," that Van Allen's interest in getting instruments into space was to study cosmic rays and their origin. In 1965, Vette (see reference 11) compared the levels of solar "flare" protons and galactic cosmic rays based on work by Malitson and Webber (see reference 46). The primary concern was dose levels on spacecraft components, which was considered a nonissue in the presence of protons and electrons.

In the late 1960s, the interest in cosmic rays of solar and galactic origin went beyond basic scientific research and became a safety issue when astronauts on Apollo missions reported visual light flashes. McNulty[57] proposed that Cerenkov radiation generated by individual cosmic ray ions traversing the vitreous of the eye were responsible for the flashes. He and his colleagues proceeded with a series of experiments that exposed human subjects, P. McNulty, V.P. Pease, V.P. Bond, and L. Pinsky, to energetic heavy ions at accelerators. Understanding of the source and mechanism of the light flashes raised the concern for astronaut safety, which in turn generated interest in measuring and modeling the galactic and heavy ion space environment. In fact, the IMP-8 spacecraft, which has provided the best long-term data set of heavy ions, was planned as a result of these concerns.

Haffner discussed the concern for the GCR contribution to total dose on spacecraft in his 1971 paper (see reference 40). He presented values for the levels of galactic cosmic rays[58] and estimated that the expected GCR dose for the Grand Tour missions was in the range of a few hundred rads. He also assumed $1/r^2$ spatial dependence for the galactic cosmic rays; however, in 1973, we learned from the Pioneer 10 and 11 missions to the outer planets that the GCR flux levels are relatively invariant over large distances from the sun.

Newer Technologies Increase Concerns about Heavy Ions

Prior to 1975, heavy ion populations, whether of galactic or solar origin, were not considered a major concern for the reliability of spacecraft electronics. Regardless of the region of space that missions visited, the contribution of heavy ions to spacecraft charging, ionizing dose, or displacement damage effects were insignificant compared with other sources of radiation, such as the trapped radiation belts or protons from solar events. That changed in 1975 when Binder et al.[59] reported, "Anomalies in communication satellite operation have been caused by the unexpected triggering of digital circuits. Although the majority of these events have been attributed to charge buildup from high temperature plasmas, some of the events appear to be caused by another mechanism." In addition to an analysis of the circuit effects and the basic mechanism of these events, the authors presented cosmic ray spectra of Meyer,[60] and they calculated the intensities using abundances from various authors, including Burrell and Wright,[61] who investigated dose rates of GCRs for astronaut exposure. Previous research on galactic cosmic rays by magnetospheric physicists interested in basic scientific research and by nuclear physicists concerned with astronaut dose became significant for the space community. As explained by McNulty (see reference 57) the interaction models used to explain the light flashes observed by astronauts eyes were modified slightly to explain the upsets observed in microelectronic circuits. The Binder paper sparked intense interest in modeling heavy ion environments and interactions.

COSMIC RAY EFFECTS ON MICROELECTRONICS (CREME) CODE

A team at Naval Research Laboratory, led by Jim Adams, recognized the need for a comprehensive software package to calculate single-event upset rates in space that integrated environment predictions with particle interaction models. They embarked on the task of developing a comprehensive computer tool that could be used by researchers and engineers. Because of the extent of the upset problem, the effort had to include compilation of data sets for GCR and solar heavy ion populations, development of GCR and solar heavy ion models, evaluation of solar proton data and models, development of magnetospheric cutoff calculations, analysis of spacecraft shielding effects, and development of interaction models for upsets due to heavy ions and protons. Adams et al. produced two Naval Research Laboratory (NRL) Memorandum Reports, which reported on topics related to this task.[62,63] The first report, published in 1981, contained a comprehensive review of the near-Earth particle environment.

CREME HEAVY ION ENVIRONMENT MODELS

The GCR environment model was based on data from several researchers collected through 1980 (see Adams, reference 62). Because of the dissimilar shape of their energy spectra, the hydrogen, helium, and iron ion distributions were treated as separate cases, and the other elements were scaled to one of the three spectra, as appropriate, using the relative abundances of the elements. Four different models of the GCR environment were developed, one of them being the well-known "90% worst case environment."

The solar heavy ion environment was more difficult to model because of the unavailability of a good data set from spacecraft instrumentation. Adams et al. assumed that the solar particle events with the highest proton fluxes are always heavy ion rich and estimated fluence levels for the higher energy solar heavy ions (>1 MeV) by scaling the abundances to protons. Eight different models of the solar heavy ion environment were developed because, without the benefit of comprehensive space measurements, the authors had to account for all possible solar activity conditions.

EXTENDING THE MODELS TO A RATE PREDICTION TOOL: CREME86

In 1982, Adams et al. presented a paper[64] that, with a paper by Petersen et al.,[65] essentially laid the groundwork for the CREME86 code. This code was the first end-to-end desktop capability to calculate radiation environments throughout near-Earth regions and to use laboratory test parameters from devices to calculate the rate of a radiation effect. The simple title, "The Natural Radiation Environment Inside Spacecraft," does not reflect the complexity and groundbreaking nature of the work. The paper reviewed the work on the development of the environment models (see Adams, reference 62) and presented methods to calculate the transport of particles through the magnetosphere and through spacecraft shielding. The authors also showed the utility of using linear energy transfer (LET) spectra to represent the heavy ion environment in a form that condenses the energy spectra of all ions into a compact expression and that can be applied to calculating energy transfer in microvolumes. Heinrich constructed the first LET spectrum describing the ion environment in space in 1977 for biomedical purposes[66] (see also Peterson[67]). In 1978, Pickel and Blanford showed the applicability of the LET spectrum to the single-event upset problem in microelectronics.[68] The resulting CREME[69] code was first released in 1986.

UPDATES TO THE CREME CODE

For 10 years, the CREME86 code was a standard for calculating heavy ion environments. As knowledge of the radiation environment increased, it became apparent that the CREME86 models could be improved. Using International Sun-Earth Explorer 3 spacecraft data, Reames et al.[70] found an inverse correlation between proton intensity and the iron/carbon heavy ion abundance ratio and that the composition of a solar particle event was a result of the location of the event on the sun. This contradicted Adams' assumption that all solar events are helium-rich, meaning that the solar heavy ion models were probably overpredicting. Dyer et al.[71] measured LET spectra during the March 1991 event with the Cosmic Radiation Environment and Dosimetry instrument on University of Surrey Satellite-3. When they compared the measurements with LET calculated using the CREME86 solar particle models, they found that in the LET range important for single-event effects analyses, all of the models severely overpredicted the LET levels.

In the 1990s, NRL recognized the need to update the environment models in the CREME86 code. The most important update to the code was the solar heavy ion model. Dietrich, from the University of Chicago, analyzed the solar heavy ion data from the IMP-8 satellite, providing the most comprehensive set of solar heavy ion space data to date.[72] The data set is especially important for modeling the fluences at higher energies. A team led by Tylka used the results to model the solar heavy ions based on the October 1989 solar particle event. An analysis of one hundred solar heavy ion events in the Dietrich database showed that this event could be used as a representative of a "worst-case" environment (see Sims et al., reference 71). The CREME96 solar heavy ion estimates are significantly lower than the heavy ion models in CREME86. Figure 28.7 compares the LET energy spectra for the CREME86 and CREME96 solar heavy ion models.

The GCR environment model was also updated to include the analysis of the Sampex measurements of anomalous cosmic rays (ACRs). With the finding that the ACRs are not singly charged over energies of 20 MeV/n, four models in CREME86 were replaced with one GCR model in CREME96. The model change significantly reduced the calculated single-event effects rates for high-threshold devices in low-altitude orbits (see Sims et al., reference 71).

A team led by Xapsos identified the concern that the solar heavy ion model currently in CREME96 does not have a true statistical basis and may misrepresent a worst-case environment (see previous section, "Engineering Oriented Statistical Model"). This team is applying the methodology used in the ESP model to derive a statistically based solar heavy ion model. Initial estimates are that the CREME96 model is in the range of a 90% worst-case model.

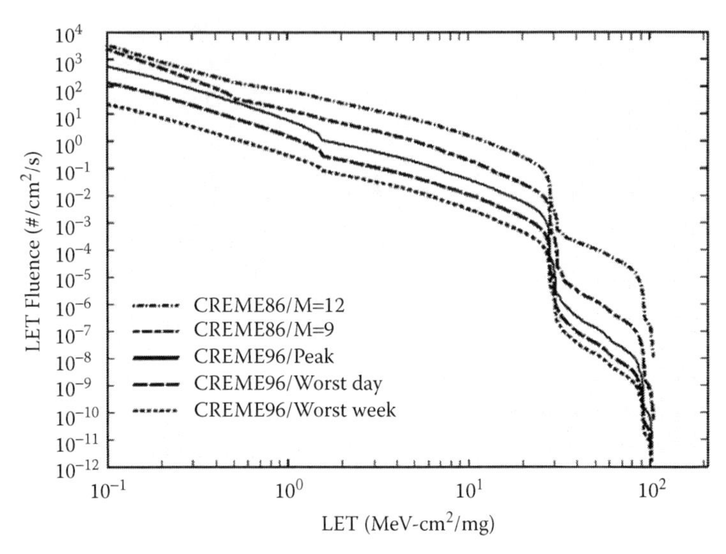

FIGURE 28.7 Comparison of CREME86 and CREME96 solar heavy ions models for a geostationary orbit.

ATMOSPHERIC ENVIRONMENTS

In the last 20 years it has been discovered that electronics in aircraft systems, which are subjected to increasing levels of cosmic radiation and their secondaries as altitude increases, are also sensitive to single-event effects. In 1984, Tsao et al.[73] showed that, below altitudes of about 60,000 ft., secondary neutrons from cosmic ray heavy ion fragmentation are the most important contributor to single-event upsets (SEUs). Silberberg et al.[74] gave a method for calculating the neutron SEU rate. They also predicted that SEU rates would increase with enhanced solar particle backgrounds.

An increasing body of data on upsets in avionics systems was accumulated that pointed to neutrons as the being the primary cause of SEUs on aircraft. In an unintentional experiment in 1993, reported by Olsen et al.,[75] a commercial computer was temporarily withdrawn from service when bit-errors were found to accumulate in 256 Kbit Complementary Metal-Oxide Semiconductor (CMOS) static random-access memory (SRAMs) (D43256 A6U-15LL). Following ground irradiations by neutrons, the observed upset rate of 4.8×10^{-8} upsets per bit-day at conventional altitudes (35,000 ft.) was found to be explicable in terms of SEUs induced by atmospheric neutrons. In an intentional investigation of single-event upsets in avionics, Taber and Normand[76] flew a large quantity of CMOS SRAM devices at conventional altitudes on a Boeing E-3/AWACS aircraft and at high altitudes (65,000 ft.) on a NASA ER-2 aircraft. Upset rates in the IMS1601 64Kx1 SRAMs varied between 1.2×10^{-7} per bit-day at 30,000 ft. and 40° latitude to 5.4×10^{-7} at high altitudes and latitudes. Reasonable agreement was obtained with predictions based on neutron fluxes.

Our knowledge of neutron levels comes from balloon-, aircraft-, and ground-based measurements. These studies show that the energies of the neutron flux range from keV to hundreds of MeV. For SEU applications in aircraft where shielding stops low-energy particles, only the energies greater than 1 MeV are significant for modeling the environment external to the spacecraft. It is also known that the flux peaks at an altitude of about 60,000 ft., which is the same altitude of the peak of observed SEU rates. Because the shape of the neutron spectrum varies little over altitude, models can be greatly simplified.

Two coordinate systems are commonly used to define the neutron distributions, energy-altitude-latitude and energy–atmospheric depth–magnetic rigidity. Taber and Normand (see reference 6) have developed an empirical model in the energy-altitude-latitude system based on studies by Mendall and Korff,[77] Armstrong et al.,[78] and Merker et al.[79] A model by Wilson-Nealy (see reference 6), based on the other system, is more recent and more comprehensive, but it is not as easy to use as the older model. Taber and Normand believe that the older energy-altitude-latitude model is sufficiently accurate for microelectronics applications.

Since the discovery of SEUs on at aircraft altitudes, researchers have made significant efforts to monitor the environment (see Barth et al., reference 16). Dyer et al. flew a version of their Cosmic Radiation Environment and Activation Monitor (CREAM) on regular flights on board Concorde G-BOAB between November 1988 and December 1992. In 1989, Dyer et al. first reported on the results of measurements aboard the Concorde aircraft.[80] Results from 512 flights have been analyzed of which 412 followed high latitude transatlantic routes between London and either New York or Washington, D.C. (see Sims et al., reference 71). Thus, some 1,000 hours of observations have been made at altitudes in excess of 50,000 ft. and at low cut-off rigidity (<2 GV), and these span a significant portion of solar cycle 22. Figure 28.8 shows the count rate in CREAM

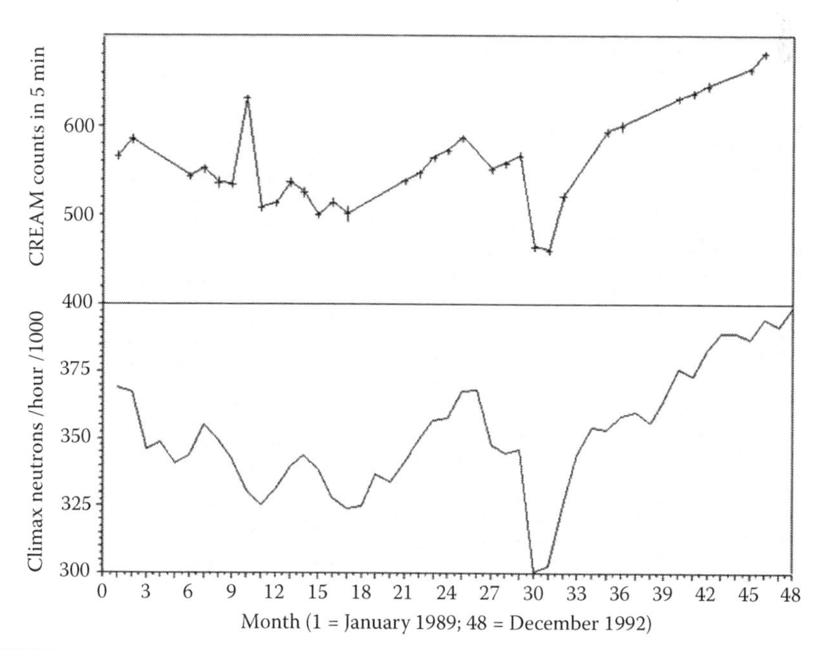

FIGURE 28.8 Monthly mean count rates from the CREAM instrument on Concorde from January 1989 to December 1992 compared with ground-level neutron monitor at Climax.

channel 1 (19fC to 46fC, LET 6.1 MeV cm^2 g^{-2}) plotted as monthly averages for the ranges 54,000–55,000 ft. and 1–2 GV. The rates show a clear anticorrelation with the solar cycle and track well with the neutron monitor at Climax Colorado (altitude 3.4 km, cut-off rigidity 2.96 GV). The enhanced period during September and October 1989 comprised a number of energetic solar particle events observed by ground-level, high-latitude neutron monitors, and the Concorde observations are summarized in Table 28.4 (see Dyer et al., reference 80),[81] which gives the enhancement factors compared with adjacent flights when only quiet-time cosmic rays were present.

More recently, the CREAM detector has been operated on a Scandinavian Airlines Boeing 767 operating between Copenhagen and Seattle via Greenland, a route for which the cut-off rigidity is predominately less than 2 GV. Approximately 540 hours of data accumulated between May and August 1993 have been analyzed, and these are combined with Concorde data from late 1992 to give updated altitude profiles.[3] Using the AIRPROP code,[82] they have also shown that cosmic rays and their secondary fragments are not the major contribution to SEUs at aircraft altitudes. Recent work[83] has concentrated on explaining both the altitude dependence and the energy deposition spectra using a microdosimetry code extension to the Integrated Radiation Transport Suite. Figure 28.9 shows that atmospheric secondary neutrons are the major contribution but that ions start to become important at the highest altitudes. Figure 28.10 shows that at 30,000 ft. the charge deposition spectrum is dominated by neutron interactions at the high end, while energetic secondary electrons and muons contribute to the low channels. The work of Normand et al.[84] arrived at similar conclusions on the neutron contribution by scaling results of irradiation of silicon detectors obtained at a spallation neutron source.

CONCLUSION

The state of the knowledge of the space, atmospheric, and ground-level radiation environments was reviewed. We saw that basic science research often is used to derive the definitions required for understanding radiation effects and for developing models that are useful for designing radiation-hardened systems. Table 28.5 shows a

TABLE 28.4
Enhancement Factors for CREAM on Concorde during Solar Particle Events

Channel Number and Charge Deposition Threshold	September 29 1406–1726	October 19 1420–1735	October 20 0859–1204	October 22 1814–2149	October 24 1805–2135
1, 19 fC	3.7 ± 0.02	1.6 ± 0.01	1.4 ± 0.01	1.5 ± 0.01	3.4 ± 0.01
2, 46 fC	4.9 ± 0.1	1.9 ± 0.04	1.6 ± 0.04	1.8 ± 0.04	4.5 ± 0.06
3, 110 fC	5.7 ± 0.1	2.1 ± 0.07	1.8 ± 0.07	1.9 ± 0.07	5.2 ± 0.1
4, 260 fC	5.9 ± 0.2	2.0 ± 0.1	1.8 ± 0.1	2.0 ± 0.1	5.7 ± 0.2
5, 610 fC	5.6 ± 0.6	2.0 ± 0.3	2.0 ± 0.4	2.1 ± 0.3	4.9 ± 0.4
6, 1.50 pC	6.1 ± 1.5	3.0 ± 0.7	1.1 ± 0.8	1.0 ± 0.6	4.3 ± 1.1
7, 3.40 pC	(17.4 ± 17.4)		(30.4 ± 30.4)		
8, 8.10 pC					
9, 19.3 pC					

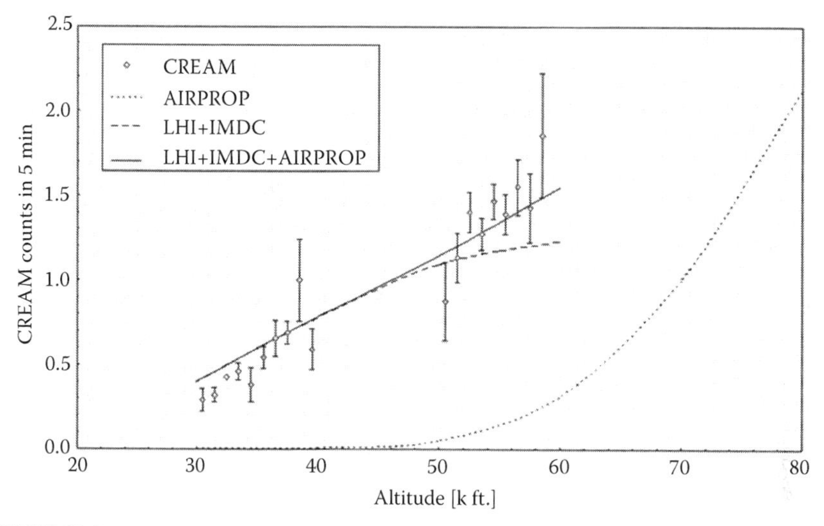

FIGURE 28.9 Average CREAM channel 5 count rates as a function of altitude at 1–2 GV from SAS and Concorde flights. Also shown are the predictions from AIRPROP and from neutron interactions as calculated using radiation transport and microdosimetry codes [Light Heavy Ion (LHI) + Ion Microdosimetry Code Extension (IMDC)]. Neutrons dominate at 3,000–40,000 ft., but cosmic ray ions contribute at supersonic altitudes.

timeline of important discoveries that led to the development of engineering models of the radiation environment. Often, our ability to increase system capability is limited by appropriate models of the radiation environment. Table 28.6 shows a timeline of engineering model development. Large uncertainty factors in environment

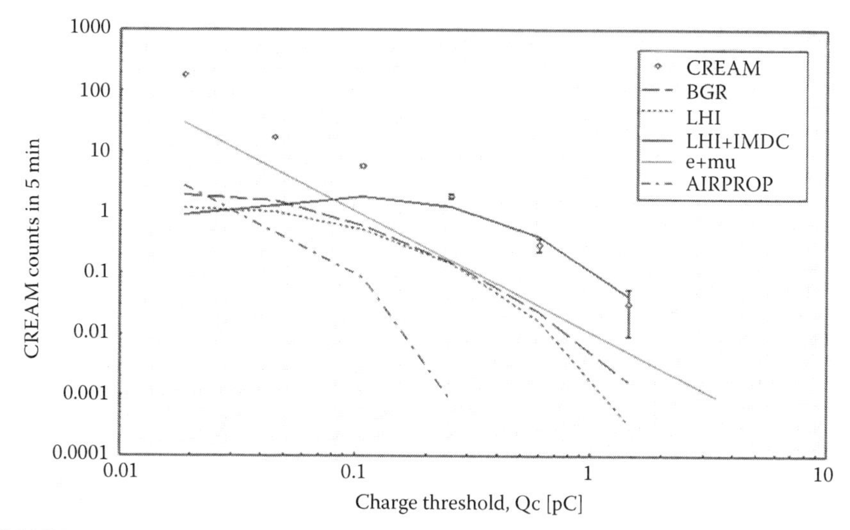

FIGURE 28.10 The spectrum of charge deposition from CREAM at 30,000–31,000 ft. compared with predictions. Neutrons dominate at high values, while electrons and muons contribute at the low end.

TABLE 28.5
Timeline of Events Related to Discoveries and Data Acquisition

Date	Event
1895	Birkeland theorizes particle trapping in the Earth's magnetic field.
Early 1900s	Electroscope experiments suggest that there is a source of radiation on Earth.
1904	Stöermer makes calculations that lead to the theory that there is a belt-like area of radiation around the earth in which particles are reflected back and forth between the magnetic poles.
1913	Hess' electroscope balloon experiment shows that radiation from outer space ("cosmic rays") is reaching Earth.
1925	Millikan Lake experiments support Hess' theory of cosmic ray radiation from outer space.
1936	Hess and Anderson receive Nobel Prize for discovery of cosmic rays.
1955	James Van Allen proposes to launch a satellite to study cosmic rays to mark the International Geophysical Year.
1957	Singer proposes that high-energy particles can be injected by magnetic storms be trapped on magnetic field lines around the Earth.
1958	Van Allen's experiment on *Explorer 1* measures high-energy particles in the Earth's radiation belts.
1958	HARDTACK series of nuclear tests started.
Early 1960s	"Upsets" in electronics on spacecraft reported.
1962	Starfish atmospheric nuclear detonation leaves nuclear debris in the Earth's radiation belts.
Early to late 1960s	Spacecraft are launched to measure the effects of Starfish.
Early 1960s	Astronauts report visual light flashes.
1965	McIlwain notes enhancements in the Earth's radiation belts from a magnetic storm.
1973	*Pioneer 10* and *Pioneer 11* measure Jupiter's radiation belts.
1975	McNulty et al. conduct particle accelerator experiments to validate the theory that light flashes are induced by charge deposited cosmic rays.
1975	Binder et al. report on increased anomalies on communication satellites due to radiation.
1977	Heinrich constructs cosmic ray Linear Energy Transfer spectra for biomedical applications.
1978	Pickel and Blanford use Linear Energy Transfer spectra to calculate upset rates on electronics.
1984	Tsao et al. report upsets on aircraft avionics.
1988–1992	Dyer measures upset rates in electronics on Concorde flights, reports anticorrelation with solar cycle.
1989	*Phobos* measures Mars radiation.
1990	Reames et al. contradicts Adams' assumption that all solar particle events are helium rich.
1990	CRRES spacecraft launched.
March 1991	"Storm" radiation belts measured by CRRES after largest magnetic event ever recorded.
1992	Radiation belt enhancements from magnetic storm reported.

TABLE 28.6
Timeline of the Development of Models of the Radiation Environment

Date	Engineering Model Development
Early 1960s	Hess releases first empirical models of the Earth's radiation belts.
1963	Vette appointed to lead coordinated Earth's radiation belt modeling effort.
1965	Vette presents overview paper of the first electron map (AE-1) and proton maps (AP).
1966	Weddel and Haffner perform statistical analysis of solar proton measurements.
1971	Stassinopoulos and Verzariu publish a model of Starfish electrons.
1971	Haffner presents measurements of galactic cosmic rays.
1971	Haffner presents review of estimated radiation belt intensities for outer planets.
1974	King releases the first statistical model of solar protons based on solar cycle 20 measurements.
Mid-1970s	AE-8 and AE-8 model maps of the Earth's radiation belts are released.
1977	Budgets to support modeling of the Earth's radiation belts are greatly reduced.
1981	Adams et al. release comprehensive galactic cosmic ray and solar proton and heavy ion models.
1983	Divine and Garret update models of Jupiter's radiation belts based on *Pioneer* measurements.
1986	CREME86 (cosmic ray effects on microelectronics) package released.
1991	CRRES spacecraft measurements of "storm" belts increase concerns over the AE-8 and AP-8 model accuracy.
1992–1993	Earth radiation belt models based on CRRES measurements are released, CRRES-PRO and CRRES-ELE.
1993	Feynman et al. release a statistical model of solar protons based on solar cycle 19–21 measurements.
1995	Taber and Normand release an atmospheric neutron model based on balloon measurements.
1996	CREME96 (cosmic ray effects on microelectronics) package released, updating galactic cosmic ray and solar proton and heavy ion models to account for Reames' work.
1997	Earth radiation belt dose maps are released based on the APEX satellite measurements.
1998	Huston and Pfitzer release the first proton radiation belt model with solar cycle variation based on NOAA measurements.
1999	Heynderickx et al. release a proton belt model with solar cycle variation based on SAM PEX measurements.
1999	Xapsos et al. present the Emission of Solar Protons model based on solar cycle 19–22 measurements.
2003	Boscher et al. publish a model for electrons in geostationary regions of the Earth's radiation belts.
2004	NASA sponsors a meeting to address a path forward to construct new electron and the proton models of the Earth's radiation environment.
2007	U.S. agencies announce new modeling effort to develop AE-9 and AP-9.

definition translate to large design margins. The direct result is reduced system resources due to increased shielding, higher mitigation overhead, and/or the use of less capable components.

The availability of useful models of the radiation environment is dependent on our knowledge of the environment, the availability of appropriate data for modeling, and funds for modeling and validation. The use of radiation sensitive modern

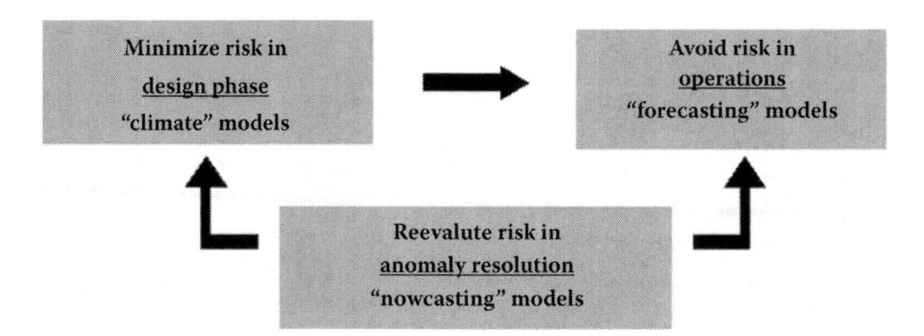

FIGURE 28.11 Environment model types that are required in each phase of risk management for missions.

microelectronics in systems requires models that can meet the needs of system designers and operators through all phases of the life cycle. Figure 28.11 shows three areas that require environment models to address risk issues in systems. Statistically based specification or "climate" models are used for the design phase, which includes system design and mission planning. The minimum requirements of the models are that they should represent long-term variation over the solar cycle with at least 1-month resolution, provide worst-case estimates, provide confidence levels, and represent the environment in a broad energy range so they are applicable for surface materials to deeply embedded sensors. The operation phase requires "forecasting" models for protecting the systems investment, for mission planning, and for personnel scheduling. The minimum requirements of these models are spatial coverage from interplanetary to low Earth orbits, information about the level of severity of storms, forecasts of quiet times for maneuvers and onboard operations, and specificity to effects on technologies. The third type of model for "nowcasting" the environment is used to resolve anomalies so risk can be reassessed both for the operating systems and for other systems that are in development. The minimum requirements of these models are spatial coverage from interplanetary to low Earth orbit, time profile of the radiation event, adequate energy range and ion composition/LET range, and specificity to the effect on technology. If anomaly resolution is critical to the mission and must be performed in near real-time, monitors that are in close proximity to the system may be required.

NOTES AND REFERENCES

1. Stassinopoulos, E.G. IEEE NSREC Short Course: Microelectronics for the Natural Radiation Environments of Space. Chapter I: Radiation Environments of Space, Reno, NV, July 16, 1990.
2. Barth, J.L. IEEE NSREC Short Course: Applying Computer Simulation Tools to Radiation Effects Problems. Section II: Modeling Space Radiation Environments I 1–83, Snowmass Village, CO, July 21, 1997.
3. Dyer, C.S. IEEE NSREC Short Course: Radiation Effects in the New Millennium—Old Realities and New Issues. Section II: Space Radiation Environment Dosimetry II 1–76, Newport Beach, CA, July 20, 1998.

4. Mazur, J. IEEE NSREC Short Course: Radiation Effects—From Particles to Payloads. Section II: The Radiation Environment Outside and Inside a Spacecraft II 1–69, Phoenix, AZ, July 15, 2002.
5. L is most simply described as the value that marks the particle drift shells by their magnetic equatorial distance from the center of the Earth.
6. Taber, A.H., and E. Normand. *Investigation and Characterization of SEU Effects and Hardening Strategies in Avionics.* DNA-TR-94-123. Alexandria, VA: Defense Nuclear Agency. 1995.
7. Daglis, I.A. *Space Storms and Space Weather Hazards*, vol. 38, chapter 3. Dordrecht, The Netherlands: Kluwer Academic Publishers, 2001.
8. Van Allen, J.A., C.E. McIlwain, and G.H. Ludwig. Radiation Observations with Satellite 1958. *Journal of Geophysical Research* 64: 271–286, 1959.
9. Vette, J.I. *The NASA/National Space Science Data Center Trapped Radiation Environment Model Program (1964–1991).* NSSDC 91-29. Greenbelt, MD: NASA/Goddard Space Flight Center, National Space Science Data Center, 1991.
10. Teague, M.J., and E.G. Stassinopoulos. *A Model of the Starfish Flux in the Inner Radiation Zone.* X-601-72-487. Greenbelt, MD: NASA/Goddard Space Flight Center, 1972.
11. Vette, J.I. The Space Radiation Environment. *IEEE Transactions on Nuclear Science* 12(5): 1–17, 1965.
12. The “A” is for Aerospace Corporation, where Vette first worked.
13. McIlwain, C.E. Forthcoming. *The Redistribution of Trapped Protons during a Magnetic Storm.* University of California at San Diego.
14. Sawyer, D.M., and J.I. Vette. *AP-8 Trapped Proton Environment for Solar Maximum and Solar Minimum.* NSSDC/WDC-A-R&S, 76-06. Greenbelt, MD: NASA/Goddard Space Flight Center, 1976.
15. Vette, J.I. *The AE-8 Trapped Electron Model Environment.* NSSDC/WDC-A-R&S 91-24. Greenbelt, MD: NASA/Goddard Space Flight Center, 1991.
16. Barth, J.L., C.S. Dyer, and E.G. Stassinopoulos. Space, Atmospheric, and Terrestrial Radiation Environments. *IEEE Transactions on Nuclear Science* 50(3, part 3): 466–482, 2003.
17. Beall, D.S., C.O. Bostrom, and D.J. Williams. Structure and Decay of the STARFISH Radiation Belt, October 1963 to December 1965. *Journal of Geophysical Research* 72: 3403–3424, 1967.
18. Bostrom, C.O., D.S. Beall, and J.C. Armstrong. Time History of the Inner Radiation Zone, October 1963 to December 1968. *Journal of Geophysical Research* 75: 1246–1256, 1970.
19. Brown, S.L. Observations of the Artificial Radiation Belts. In *Radiation Trapped in the Earth's Magnetic Field.* Astrophysics and Space Science Library. Vol. 5. Dortrecht, Holland: D. Reidel, 1966.
20. Van Allen, J.A. Spatial Distribution and Time Decay of the Intensities of Geomagnetically Trapped Electrons from the High Altitude Nuclear Burst of July 1962. In *Radiation Trapped in the Earth's Magnetic Field.* Astrophysics and Space Science Library. Vol. 5. Dortrecht, Holland: D. Reidel, 1966.
21. Stassinopoulos, E.G., and P. Verzariu. General Formula for Decay of Starfish Electrons. *Journal of Geophysical Research* 76: 1841–1844, 1971.
22. Mullen, E.G., M.S. Gussenhoven, K. Ray, and M. Violet. A Double-Peaked Inner Radiation Belt: Cause and Effect as Seen on CRRES. *IEEE Transactions on Nuclear Science* 38(6): 1713–1717, 1991.
23. Gussenhoven, M.S., E.G. Mullen, and D.H. Brautigam. Improved Understanding of the Earth's Radiation Belts from the CRRES Satellite. *IEEE Transactions on Nuclear Science* 43(2): 353–368, 1996.

24. Boucher, D., S. Bourdaire, and T. Beutier. Dynamic Modeling of Trapped Particles. *IEEE Transactions on Nuclear Science* 43(2): 416–425, 1996.
25. Bourdaire, S., D. Boucher, and T. Beutier. Modeling the Charged Particle Transport in the Earth's Internal Magnetosphere. In Workshop of the GdR Plasmae, Tournon, France, April 10–12, 1995.
26. Bourdarie, S., personal communication, Onera, Toulouse, France, September 2002.
27. Daly, E.J., J. Lemaire, D. Heynderickx, and D.J. Rodgers. Problems with Models of the Radiation Belts. *IEEE Transactions on Nuclear Science* 43(2): 403–415, 1996.
28. Gussenhoven, M.S., E.G. Mullen, M.D. Violet, C. Hein, J. Bass, and D. Madden. CRRES High Energy Proton Flux Maps. *IEEE Transactions on Nuclear Science* 40(6): 1450–1457, 1993.
29. Brautigam, D.H., M.S. Gussenhoven, and E.G. Mullen. Quasi-Static Model of Outer Zone Electrons. *IEEE Transactions on Nuclear Science* 39(6): 1797–1803, 1992.
30. Gussenhoven, M.S., E.G. Mullen, M. Sperry, and K.J. Kerns. The Effect of the March 1991 Storm on Accumulated Dose for Selected Orbits: CRRES Dose Models. *IEEE Transactions on Nuclear Science* 39(6): 1765–1772, 1992.
31. Gussenhoven, M.S., E.G. Mullen, J.T. Bell, D. Madden, and E. Holeman. APEXRAD: Low Altitude Orbit Dose as a Function of Inclination, Magnetic Activity and Solar Cycle. *IEEE Transactions on Nuclear Science* 44(6): 2161–2168, 1997.
32. Pfitzer, K.A. *Radiation Dose to Man and Hardware as a Function of Atmospheric Density in the 28.5 Degree Space Station Orbit.* MDSSC Rep. H5387. Huntington Beach, CA: McDonnell Douglas Space Systems Co., 1990.
33. Huston, S.L, G.A. Kuck, and K.A. Pfitzer. Low Altitude Trapped Radiation Model Using TIROS/NOAA Data. In *Proceedings from Model Workshop*, D. Heynderickx, ed. Belgisch Instituut voor Ruimte-Aëronmie/Institut d'Aéronomie Spatiale de Belgique (BIRA/IASB), 1997. http://magnet.oma.be/trend/trend2.html.
34. Huston, S.L., and K.A. Pfitzer. A New Model for the Low Altitude Trapped Proton Environment. *IEEE Transactions on Nuclear Science* 45(6): 2972–2978, 1998.
35. Xapsos, M.A., S.L. Huston, J.L. Barth, and E.G. Stassinipoulos. Probabilistic Model for Low-Altitude Trapped Proton Fluxes. *IEEE Transactions on Nuclear Science* 49(6): 2776–2781, 2002.
36. Heynderickx, D., and J. Lamaire. Coordinate Systems for Mapping Low-Altitude Trapped Particle Fluxes. In *Proceedings of the Taos Workshop on Earth's Trapped Particle Environments,* August 14–19, Taos, NM, 1994.
37. Heynderickx, D., M. Kruglanski, V. Pierrard, J. Lemaire, M.D. Looper, and J.B. Blake. A Low Altitude Trapped Proton Model for Solar Minimum Conditions Based on SAMPEX/PET Data. *IEEE Transactions on Nuclear Science* 46(6): 1475–1480, 1999.
38. Boscher, D.M., S.A. Bourdarie, R.H.W. Friedel, and R.D. Belian. Model for the Geostationary Electron Environment: POLE. *IEEE Transactions on Nuclear Science* 50(6): 2278–2283, 2003.
39. Lauenstein, J.-M., J.L. Barth, and D.G. Sibeck. Introduction to Special Section on International Working Group Meeting on New Standard Radiation Belt and Space Plasma Models for Spacecraft Engineering. *Space Weather* 3(S07B01): doi:10.1029/2005SW000159, 2005.
40. Haffner, J.W. Natural Nuclear Radiation Environments for the Grand Tour Missions. *IEEE Transactions on Nuclear Science* 18(6): 443–453, 1971.
41. Kase, P.G. The Radiation Environments of Outer-Planet Missions. *IEEE Transactions on Nuclear Science* 19(6): 141–146, 1972.
42. Divine, N. *The Planet Jupiter (1970).* NASA SP-8069. Pasadena, CA: Jet Propulsion Laboratory, 1971.
43. *History of NASA.* SP-349/396 Pioneer Odyssey. http://history.nasa.gov/SP-349/ch5.htm.

44. T. Oldham, personal communication regarding the talk by G. Lee at the 1983 Single Event Effects Symposium, QSS Group at Goddard Space Flight Center, Code 561, Greenbelt, MD, January 2003.
45. Divine, N., and H.B. Garrett. Charged Particle Distributions in Jupiter's Magnetosphere *Journal of Geophysical Research* 88(A9): 6889–6903, 1983.
46. Malitson, H.H., and W.R. Webber. A Summary of Solar Cosmic Ray Events. *Solar Proton Manual*. Greenbelt, MD: NASA TR R-169, 1963.
47. Weddel, J.B., and J.W. Haffner. *Statistical Evaluation of Proton Radiation from Solar Flares*. Report SID 66-421. North American Aviation, July 1966, Inglewood, CA.
48. King, J.H. Solar Proton Fluences for 1977–1983 Space Missions. *Journal of Spacecraft and Rockets* 11(6): 401–408, 1974.
49. Feynman, J., G. Spitale, J. Wang, and S. Gabriel. Interplanetary Fluence Model: JPL 1991. *Journal of Geophysical Research* 98: 13281–13294, 1993.
50. Xapsos, M.A., J.L. Barth, E.G. Stassinopoulos, G.P. Summers, E.A. Burke, and G.B. Gee. Model for Prediction of Solar Proton Events. Proceedings from the Space Radiation Environment Workshop, Farnborough, United Kingdom, November 1–3, 1999.
51. Xapsos, M.A., G.P. Summers, P. Shapiro, and E.A. Burke. New Techniques for Prediction Solar Proton Fluences for Radiation Effects Applications. *IEEE Transactions on Nuclear Science* 43(6): 2772–2777, 1996.
52. Xapsos, M.A., G.P. Summers, and E.A. Burke. Probability Model for Peak Fluxes of Solar Proton Events. *IEEE Transactions on Nuclear Science* 45(6): 2948–2953, 1998.
53. Xapsos, M.A., G.P. Summers, J.L. Barth, E.G. Stassinopoulos, and E.A. Burke. Probability Model for Worst Case Solar Proton Event Fluences. *IEEE Transactions on Nuclear Science* 45(6): 1481–1485, 1999.
54. Tylka, A.J., F. Dietrich, and P.R. Boberg. CREME96: A Revision of the Cosmic Ray Effects on Microelectronics Code. *IEEE Transactions on Nuclear Science* 44(6): 2150–2160, 1997.
55. Xapsos, M.A., J.L. Barth, E.G. Stassinopoulos, E.A. Burke, and G.B. Gee. Model for Emission of Solar Protons (ESP)—Cumulative and Worst Case Event Fluences. NASA-Marshall Space Flight Center SEE Progra. http://see.msfc.nasa.gov.
56. Millikan, R.A. Presentation before the National Academy of Sciences, November 9, 1925, Madison, WI.
57. McNulty, P.J. Single Event Effects Experienced by Astronauts and Microelectronic Circuits Flown in Space. *IEEE Transactions on Nuclear Science* 43(2): 475–482, 1996.
58. The source of the data is not referenced.
59. Binder, D., E.C. Smith, and A.B. Holman. Satellite Anomalies from Galactic Cosmic Rays. *IEEE Transactions on Nuclear Science* 22(6): 2675–2680, 1975.
60. Meyer, P., R. Ramaty, and W.R. Weber. Cosmic Rays—Astronomy with Energetic Particles. *Physics Today* 27(10): 23, 1974.
61. Burrell, M.O., and J.J. Wright. *The Estimation of Galactic Cosmic Ray Penetration and Dose Rates*. NASA TN D-6600, Greenbelt, MD, 1972.
62. Adams, J.H., Jr., R. Silberberg, and C.H. Tsao. *Cosmic Ray Effects of Microelectronics, Part I: The Near-Earth Particle Environment*. NRL Memorandum Report 4506, Washington, DC, August 25, 1981.
63. Adams, J.H., Jr., J.R. Letaw, and D.F. Smart. *Cosmic Ray Effects of Microelectronics, Part II: Geomagnetic Cutoff Effects*. NRL Memorandum Report 5099, May 26, Washington, DC, 1983.
64. Adams, J.H., Jr. The Natural Radiation Environment Inside Spacecraft. *IEEE Transactions on Nuclear Science* 29(6): 2095–2100, 1982.

65. Peterson, E.L., P. Shapiro, J.H. Adams, Jr., and E.A. Burke. Calculation of Cosmic-Ray Induce Soft Upsets and Scaling in VLSI Devices. *IEEE Transactions on Nuclear Science* 29(6): 2055–2063, 1982.
66. W. Heinrich. Calculation of LET Spectra of Heavy Cosmic Ray Nuclei at Various Absorber Depths. *Radiation Effects* 34(143), 1977.
67. Peterson, E.L. Approaches to Proton Single-Event Rate Calculations. *IEEE Trans on Nucl. Sci.* 43(2): 496–504, 1996.
68. Pickel, J.C., and J.T. Blanford, Jr. Cosmic Ray Induced Errors in MOS Memory Cells. *IEEE Transactions on Nuclear Science* 25(6): 1166, 1978.
69. Adams, J.H., Jr. *Cosmic Ray Effects on Microelectronics, Part IV.* NRL Memorandum Report 5901. Washington, DC: Naval Research Laboratory, December 31, 1986.
70. Reames, D.V., H.V. Cane, and T.T. von Rosenvinge. Energetic Particle Abundances on Solar Electron Events. *The Astrophysical Journal* 357: 259–270, 1990.
71. Sims, A., C. Dyer, C. Peerless, K. Johansson, H. Pettersson, and J. Farren. The Single Event Upset Environment for Avionics at High Latitude. *IEEE Transactions on Nuclear Science* 41(6): 2361–2367, 1994.
72. Tylka, A.J., F. Dietrich, and P.R. Boberg. Probability Distributions of High-Energy Solar-Heavy-Ion Fluxes from IMP-8: 1973–1996. *IEEE Transactions on Nuclear Science* 44(6): 2140–2149, 1997.
73. Tsao, C.H., R. Silberberg, and J.R. Letaw. Cosmic Ray Heavy Ions at and above 40,000 Feet. *IEEE Transactions on Nuclear Science* 31(6): 1066–1068, 1984.
74. Silberberg, R., C.H. Tsao, and J.R. Letaw. Neutron Generated Single Event Upsets. *IEEE Transactions on Nuclear Science* 31(6): 1183–1185, 1984.
75. J. Olsen, P.E. Becher, P.B. Fynbo, P. Raaby, and J. Schultz. Neutron-Induced Single Event Upsets in Static RAMS Observed a 10 km Flight Attitude. *IEEE Transactions on Nuclear Science* 40(6): 74–77, 1993.
76. Taber, A., and E. Normand. Single Event Upset in Avionics. *IEEE Transactions on Nuclear Science* 40: 120, 1993.
77. Mendall, R.B., and S.A. Korff. Fast Neutron Flux in the Atmosphere. *Journal of Geophysics* 68: 5487, 1963.
78. Armstrong, T.W., K.C. Chandler, and J. Barish. Calculation of Neutron Flux Spectra Induced in the Earth's Atmosphere by Galactic Cosmic Rays. *Journal of Geophysical Research* 78: 2715, 1973.
79. Merker, M., E.S. Light, H.J. Verschell, R.B. Mendell, and S.A. Korff. Time Dependent World-Wide Distribution of Atmospheric Neutrons and Their Products. *Journal of Geophysical Research* 78: 2727, 1973.
80. Dyer, C.S., A.J. Sims, J. Farren, and J. Stephen. Measurements of the SEU Environment in the Upper Atmosphere. *IEEE Transactions on Nuclear Science* 36(6): 2275–2280, 1989.
81 Dyer, C.S., A.J. Sims, J. Farren, J. Stephen, and C. Underwood. Comparative Measurements of the Single Event Upset and Total Dose Environments Using the CREAM Instruments. *IEEE Transactions on Nuclear Science* 39(3): 413–417, 1992.
82. Tsao, C., R. Silberberg, J. Adams, Jr., and J. Letaw. *Cosmic Ray Effects on Microelectronics: Part III: Propagation of Cosmic Rays in the Atmosphere.* NRL Memorandum Report 5402, Washington, DC, August 1984.
83. Dyer, C.S., and P.R. Truscott. Cosmic Radiation Effects on Avionics. In *ERA Technology Conference Volume for 1997 Avionics Conference*, 6.3.1–6.3.10. Heathrow, United Kingdom, November 1997.
84. Normand, E., D. Olberg, J. Wert, J. Ness, P. Majewski, S. Wendron, and A. Gavron. Single Event Upset and Charge Collection Measurements Using High Energy Protons and Neutrons. *IEEE Transactions on Nuclear Science* 41(6): 2203–2209, 1994.

29 Thermal Environment and the Effects on Aging

Douglas Mehoke

CONTENTS

INTRODUCTION

The costs and constraints associated with putting hardware into space are so restrictive that vehicles are specifically designed for their particular orbit, attitude, payload, and mission. One of the restricting elements that drive spacecraft configurations is the need to keep the electronics and sensor systems within their operational temperature ranges. To understand the interaction between space vehicles and their temperatures requires a discussion of temperature itself, the heat flows that drive temperatures in space, and how these heat flows are controlled using the space vehicle thermal design.

Major heat flows in space come from the sun and planetary bodies. This heat flows through the spacecraft to cold space. Solar and planetary heat sources differ from each other in magnitude, direction, and wavelength range. How these heat sources vary with location and attitude is an important feature in the thermal designs. The designs of most of the large external features of any space vehicle are influenced by thermal issues. Such features include spacecraft radiators, insulation systems, solar panels, and antennas.

TEMPERATURE

The ambient environment in space is very different from what we normally think of in terrestrial situations. Typically, the ambient environment refers to the local air temperature. Air and air movement is an effective way of keeping the surface temperatures in one area within a relatively small range. In space, there is no such unifying mechanism, and as a result space hardware will experience very wide temperature extremes of 200°C (360°F) over short distances. Rather than an ambient environment, in space we refer to the thermal environment as the driving mechanism that determines temperatures.

Temperature changes occur as heat flows through a system from the hot areas to the cold ones. In vacuum situations, there is no air to serve as a mixing agent, so temperatures are set by the magnitude of the local hot and cold sources. The space environment can harbor huge temperature extremes because the hot and cold sources are very hot and very cold. The principal source of heat input to most space objects is the sun. Other inputs come from nearby planetary bodies. The cold source in the space environment is space itself.

While discussing hot and cold, it is worth taking a moment to talk about temperature itself. Historically, temperature is a relative thing. It is measured relative to some known condition (e.g., 20°F above freezing). The common Fahrenheit scale dates to 1724, when Daniel Gabriel Fahrenheit developed a system with three reference point temperatures over a 96-point scale. By mixing salt into a water-ice mixture, he achieved the cold reference point, about 0°F. The ice water bath without the salt produced a midpoint, 32°F. The hot reference point used the human body temperature, about 96°F. He then divided his scale into twelve divisions and each of those into eight subdivisions, for a 96-point scale. The reason for choosing the coldest possible reference point was to eliminate the need for the negative numbers used in other contemporary temperature scales. The important issue for the reference points is that they can be generated in a repeatable manner in different locations, using a simple approach. Temperatures are measured using a material whose volume changes significantly with temperature, such a mercury in a thin glass tube. By noting the hot and cold positions of the mercury in the tube, a scale can be generated that measures how many degrees the thermometer is above the reference condition.

Another common temperature scale is Celsius (also known as centigrade). While it is much like the Fahrenheit scale, the cold and hot reference points were adjusted to more easily repeated end points. The ice/water bath is used for the coldest reference point (0°C), and boiling water is used to determine the hot reference point (100°C).

Both of these scales are relative, in that they define how much warmer something is relative to a standard reference point. Temperature also has an absolute measure. Quantum physics tells us that matter has a minimum energy state. If this condition is used as our cold reference point, temperature takes on the normal attributes of a measured quantity: zero means none. The absolute form of the Celsius scale is the Kelvin scale: 1°C change = 1 K change, but 0 K is –273.15°C. On the Fahrenheit side, the absolute scale in the Rankine scale: 1°F change = 1°R change, but 0°R = –459.67°F. This absolute feature will become important when we talk about how heat flows in space.

HEAT TRANSFER

Conservation of energy is the basis of the discussion for heat flow in a system. There is a difference between energy and power that is important to remember. Power is defined as energy per unit time, or an energy flow. Heat and work are defined as terms of energy. Heat and work are important in thermodynamics. When discussing temperatures, heat flow, or power, is used.

Temperatures results from the movement of heat thorough a system. Using an electrical analogy, hot and cold temperatures represent driving potential or voltage. Heat flow is the moving energy or current that flows between the two potentials. Thermal resistance is the measure of how hard it is for the energy to flow in a material. The electrical analog is electrical resistance.

To be able to talk about temperatures and heat flow in a system, we have to define a system where heat flow can be related to something that can be measured, temperature, and some inherent properties of the system, the materials and their configuration. The study of heat flow is known as heat transfer, and it has three major subdivisions: conduction, convection, and radiation.

Conduction is the study of how heat moves in a solid. The basic equation of conduction is known as Fourier's law.

$$Q = -k\,A\,dT/dx \qquad (29.1)$$

The law states that the heat flow through a solid is proportional to the temperature gradient in the material (dT/dx), the cross-sectional area through which the heat flows (A), and a factor known as the thermal conductivity (k), which is an inherent property of the material. The thermal conductivity of materials is measured, and it does vary over wide temperature ranges but can be treated as a constant near room temperature. In a particular instance, the cross-sectional area through which the heat is flowing is defined by the geometry of the item in question. Therefore, for solid materials, we can define the heat flow in terms of the temperatures, materials, and geometry of the system.

Convection takes the same heat flow mechanism used in conduction and extends it to non-solids. The simplifying assumption in conduction is that the various parts of the system remain in the same shape throughout the process. If we look at conduction between a solid and a nonsolid, such as air or water, there is an associated complicating mechanism that occurs, buoyancy. Buoyancy refers to the mechanism that causes parts of a fluid to move relative to each other as their local density changes. Heat is transferred at the solid-fluid interface by conduction, but then the fluid density changes as its temperatures increases. Since the local fluid segment is less dense than the fluid around it, the segment rises due to buoyancy. As the fluid segment moves away from the heated surface, it is replaced by a lower-temperature piece of fluid that is heated and rises, so that the fluid is in continuous motion. Convection is the study of the heat transfer when the participating fluid is moving away from the hot surface. The basic convection equation is as follows:

$$Q = h\,A\,(T_w - T_{fs}). \tag{29.2}$$

The heat flow from a surface is proportional to the temperature difference between the surface and the fluid temperature well away from the wall, the surface area (A), and a factor known as the convective heat transfer coefficient (h). Like the thermal conductivity, the convection coefficient is a measured quantity, but unlike the conductivity, it is a function of many parameters: the flow speed, surface geometry, and fluid properties. However, for a given geometry and given materials, the heat transfer can be expressed in terms of temperatures, materials, and the geometry of the system.

The final heat transfer mechanism is radiation. Radiation heat exchange is based on the electromagnetic energy that is radiated from all materials. In 1900, Max Planck determined that the energy is radiated from a perfect, or blackbody, source according to the equation:

$$E_{b\lambda} = C_1\,\lambda^{-5} / \{\exp(c_2/\lambda \mathrm{T}) - 1\} \tag{29.3}$$

where T = temperature (K), $C_1 = 3.743 \times 10^8$ W μm^4/m^2, and $C_2 = 1.4387 \times 10^4$ μm K. From this equation it can be seen that there is a spectral distribution of the radiated energy that is a strong function of the temperature. Figure 29.1 shows the spectral distribution of energies at different temperatures. Not surprisingly,

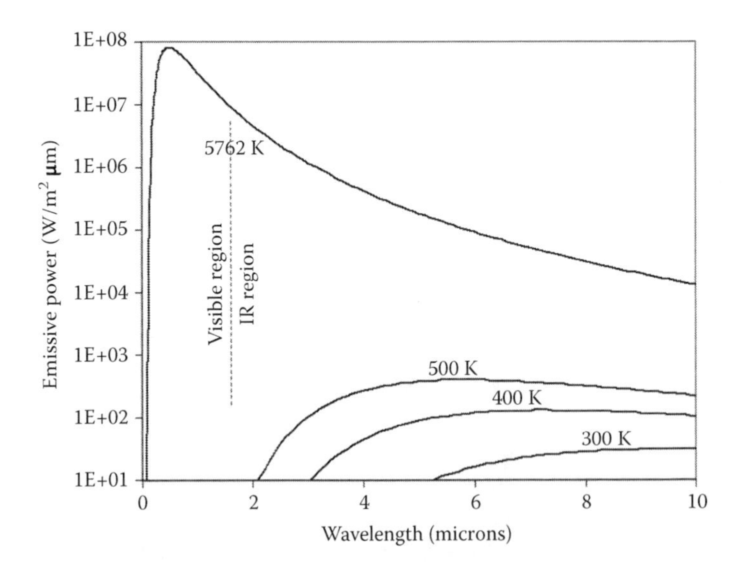

FIGURE 29.1 Spectral distribution of energies at room temperature.

virtually all of the energy is radiated in the infrared (IR) region of the spectrum at room temperature. Those things that emit visible radiation, like the sun, operate at much higher temperatures.

By integrating the energy flux over all wavelengths, and including the Stefan-Boltzmann constant, the total energy radiated from a blackbody can be calculated using the following equation:

$$Q(\mathrm{W/m^2}) = \sigma T^4 \tag{29.4}$$

where $\sigma = 5.6704 \times 10^{-8}$ W/m^2 K^4.

The discussion of radiation, so far, has used idealized or blackbody surfaces. In real life, surfaces that are not perfect are known as gray bodies. These surfaces may absorb, reflect, or transmit the energy incident on them. To account for these properties on actual surfaces, new factors are defined that combine the wavelength dependent energy profile with the wavelength dependent surface optical property. These factors are the absorptivity, α, reflectivity, ρ, and the transmissivity, τ. In use, they represent the fraction of the incident energy striking a surface that is absorbed, reflected, or passes through the surface.

These terms define the heat flow actually interacting with the surface in terms of a material property and its incident energy. Since these optical properties are integrated over all wavelengths, they are dependent on the spectral distribution of the incoming energy. This complication seems to require that the radiative energy interaction for a surface requires knowledge of the temperatures of all the contributing sources and their wavelength distribution. Clearly, this requirement would make the calculation of radiative heat exchange very complicated.

A solution to this problem is found by noting that radiative interchanges in space tend to occur only between two types of sources. The first source is the sun. It is very hot, 5,762 K, and emits most of its energy in the visible wavelength range. The second source is everything else. Virtually all nonsolar surfaces emit most of their energy in the IR range. Therefore, by defining two optical properties—(1) α_s = fraction of the incident visible energy that is absorbed or emitted by the surface, and (2) ε_{IR} = fraction of the incident IR energy that is absorbed or emitted by the surface—the calculation of the radiation interchange between surfaces can be done without worrying about the spectral nature of radiation.

SOLAR ENVIRONMENT

The sun is typically the largest incident energy source for space hardware. The sun acts as a blackbody, emitting energy primarily in the visible range. By the time this energy reaches the Earth, it is functionally collimated and has a uniform intensity of 1,367 W/m^2 [solar constant at 1 astronomical unit (AU)]. The solar flux at different sun distances can be calculated by assuming that the sun's energy is expanding spherically from a central point, or

$$Q(R_2) = Q(R_1) \times R_1^2/R_2^2 \tag{29.5}$$

This relationship can be used to within a few solar radii of the surface of the sun with acceptable accuracy.

Solar-powered spacecraft operate between Mercury and Jupiter. The closest a spacecraft has come to the sun is 0.3 AU. At 0.3 AU, the solar flux is 11 times higher than at the Earth. Jupiter is at 5 AU, and the solar flux is 25 times less than at the Earth. At the Earth, solar-facing temperatures can vary between 0°C and 125°C. At sun distances closer than about 0.8 AU, a sun shield is required to keep the direct sun off the spacecraft.

At distances farther than about 2 AU, the sun's energy alone is not enough to keep an exposed spacecraft surface above room temperature. Beyond about 5 AU, nuclear power becomes the usual power source for space hardware. At these distances, the spacecraft relies on its own internally generated power to keep itself warm. A Cassini heritage Radioisotope Thermoelectric Generator generates about 300 W of electricity (We) but about 4,000 W of heat. The newer versions, the Multi-Mission Radioisotope Thermolelectric Generator or the Advanced Stirling Radioisotope Generator, generate about 100 We each. When you note that a 1-m^2 surface, at room temperature, and a high IR emissivity radiates 450 W, and that the surface area of a spacecraft will include several square meters, controlling the heat loss on deep space missions becomes a critically important issue.

PLANETARY HEAT INPUTS

While the sun represents the largest heat source for most space hardware, planets also can be significant heat sources for space hardware. Energy from planets comes in two forms, reflected solar and emitted IR. The Earth has an average reflectance of about 30%. It may seem surprising, but the Moon's reflectivity is about 5%. Reflected solar energy is known as albedo. This energy is still in the visible wavelengths. It peaks at the point on the planet closet to the sun, the subsolar point, and falls off as the cosine of the angle from the subsolar point.

The planet's surface also radiates energy to space, some of which hits the spacecraft. This energy comes from the warm surface, so it is in the IR portion of the wavelength. The Earth, because of its atmosphere, has a relatively constant surface temperature. Other planets and the Moon have very large differences in the surface temperatures. In the design of the Messenger mission to Mercury, the IR energy emitted by the planet was the driving heat input to parts of the spacecraft, because they could not be shielded by the sunshade.

Unlike the energy from the sun, emissions from the planet surfaces are not collimated. They radiate diffusely from the surface up into space. A spacecraft looking down at the planet would receive heat not only on its planet-facing side, but on surface normal to the planet direction. For example, a cube flying over the Earth would have the largest input on the surface facing the planet. However, the surfaces perpendicular to the planet's surface would see a flux of about a third of the planet-facing surface.

SPACECRAFT THERMAL DESIGN

Spacecraft thermal designs combine the hardware, environmental heat inputs, and internal power dissipation into a configuration that produces component temperatures that are typically near room temperature. Because of the importance of radiation heat transfer in space, as both a heat input and loss, the thermal designs of space vehicles focus on the external surfaces of the system.

The specific design represents the thermal requirement particular to that mission, but there are generic thermal limits that are common to most spacecraft. The general support electronics usually operate well between –35°C and 60°C. Propulsion systems, including the tanks, lines, and thrusters, must be kept above about 5°C to keep the propellant from freezing. They usually have a maximum temperature near 50°C. Scientific instruments tend to like to operate colder than room temperature, –20°C to 0°C. Cryogenic sensors need very cold temperatures to separate their signal from the surrounding background. The sensors operate between 70 and 120 K, and the supporting hardware needs to be kept as cold as possible to minimize leakage of heat into the cold focal planes.

In hardware designed to operate in space, most of the exposed surfaces are involved in the thermal control of the item. All spacecraft designs are based on balancing the heat dissipated internally and absorbed from external sources with the heat lost to cold space. It is not usually a good idea to have the spacecraft move large amounts of heat from point to point, because large temperature gradients come with this tactic. Therefore, areas with large heat sources tend to be kept near the place where that energy is rejected to space by radiators. The rest of the exposed surfaces are covered with thermal insulation.

SPACECRAFT EXTERNAL SURFACES

All of the external surfaces of a spacecraft are part of the overall thermal design. Two of the largest spacecraft surfaces that are externally visible are the insulation blankets and the radiators. They are used to control the spacecraft's temperatures. Other large surfaces include the antennas, solar arrays, and instrument apertures. Each of these surfaces has particular characteristics that are related to both the component's operational function and the thermal control of the item.

Most of the exposed area on spacecraft not covered with the mirrored radiator, or instrument apertures, is covered by insulation. Spacecraft insulation is different from atmospheric insulation. On Earth, insulation is made up of low-conductivity materials spun into fibrous batts. Their purpose is to restrict the movement of air while providing a low-conductivity path from one side of the batt to the other. Having more insulating capability means adding more thickness. In space, the dominant heat transfer mode is radiation. In the discussion on radiation, the only property we can change is the emissivity. Lowering the emissivity lowers the heat loss. Spacecraft insulation lowers the effective emissivity by layering low-emissive sheets of material into insulation blankets. These Multi-Layer Insulation (MLI) blankets are very thin, less than a quarter of an inch thick, but reduce the heat loss by the spacecraft by 97% to 99%.

Radiators can be specially designed structures or just exposed areas of the spacecraft body. Radiators have a coating that is usually designed to reflect the most solar energy but emit the most IR energy. These surfaces are said to have small $\alpha_s/\varepsilon_{IR}$ ratios. These surfaces appear as shiny mirrors or white surfaces distributed around the body of the spacecraft. By looking at where the radiator surfaces are located, some interesting inferences can be drawn about the spacecraft. Radiators mark the location of the general spacecraft support electronics, which typically are located within main body. Except for deep space vehicles, radiators need to be located where they do not have extended views of the sun. Therefore, the radiator side of the spacecraft is the side normally facing away from the sun.

Spacecraft radiator designs fall within large categories based on where they fly. Vehicles in low Earth orbit typically see a large range of solar angles. Vehicles with radiators spread around the body have been designed to fly in an Earth-oriented mode. If the radiators are all on one side, the spacecraft will probably fly in a sun-oriented mode. Spacecraft in geosynchronous orbits tend to have their main radiator panels on the upward and downward sides (viewed from the orbital plane). If the spacecraft is a spinner, the radiators tend to be spread around the spin axis. A spacecraft with very minimal radiators is probably heading to Mars or beyond. A larger area of radiators implies higher power in the vehicle.

EVOLUTION OF SPACECRAFT THERMAL DESIGN

The evolution of spacecraft thermal designs has paralleled the progression of spacecraft in general. They have become much bigger and more complex. This progression is seen in the materials and the analysis tools. As the experience base has increased, the generation of standard buses has made some spacecraft assemblies into a nearly production-line process. At the same time, missions are being sent across the solar system into environments where their designs are pushing out our knowledge envelope.

While many older spacecraft materials are still used, they are used in ways that would never have been attempted a decade ago. Aluminum honeycomb structures are still widely used, but composite structures are becoming more common as a way to lower the overall mass of the system. Louvers and phase change materials are thermal control components widely used to allow higher power electronics. As spacecraft get larger, the paths between the high-power units and the external radiators require more efficient heat paths, such as heat pipes. Insulation materials have remained much the same, but the testing and analysis of these materials have allowed more precise knowledge of their performance, allowing for designs such as the tennis-court-sized James Webb Space Telescope sunshield, which blocks essentially all of the incident solar energy from reaching the cold optics.

These new design ideas have been supported by a huge increase in the capabilities of the supporting analytical programs. Early spacecraft thermal designs were supported by hand calculations. This approach severely limited the size and shape of spacecraft that could be attempted. In the 1970s, the push for larger spacecraft drove the development of specialized thermal analysis codes that predicted the external environment and the radiation interchange between spacecraft surfaces. These codes were run on large main frame computers. With limited input/output, the level of complexity that could be achieved was restricted by the physical limitations of creating and checking the spacecraft thermal model. Each surface needed to be identified and input into a computer model. The time required for model input, checking, and run-time was prohibitive for models larger than a thousand nodes. In the 1990s, the move to the personal computers and the development of graphical user interface front ends to the analysis codes allowed the model size to increase from a few hundred to a few thousand nodes. Recently, the connection of the thermal analysis codes with existing mechanical computer-aided design systems has allowed much larger spacecraft to be built. The present James Webb Space Telescope thermal model has more than 300,000 elements. While the increasing complexity of the thermal analysis codes does allow much larger thermal models to be built and run, the capability to check these results is still limited by our ability to test the system in a space-like environment. Testing is now the limiting item for the expansion of spacecraft thermal designs.

SPACE ENVIRONMENT

RADIATION

Space is both a harsh environment in some areas and a gentle one in others. Radiation, particle debris, and contamination can have a greater effect on materials in space than on Earth because of the lack of the protective atmosphere. However, materials in space are not subjected to wind damage or sagging due to gravity. Thus, the space environment impacts materials and structures both more and less than an Earth environment.

Two of the most harmful radiation sources in space include ultraviolet (UV) and ionizing radiation. Harmful effects from radiation on materials, especially organic ones, are a decrease in material cohesion, the development of color centers in the material, and increase in brittleness.

Frequently studied materials are Kapton and Teflon, which are used as the outer surface for MLI and radiators and are typically used with a metalized backing surface. Silver-backed Teflon is the external material on the MLI blankets on the Hubble Space Telescope (HST). With the HST servicing missions, NASA had a chance to inspect these materials after exposure for years in a low Earth orbit. When exposed to the sun for long periods, Teflon has the tendency to develop an extensive serious of very fine cracks, known as crazing. Visually, the surface changes from a mirror-like finish to a milky finish. Crazing is caused by UV radiation breaking down the bonds within the film material. While these surfaces become more fragile, the lack of stress on the material means the structural impact can be minimal. Another commonly used material is Kapton. Widely used as the outer layer of MLI blankets, Kapton is generally very stable in the space environment. Measured data from space experiments have shown no detrimental mechanical effects on the creep behavior of Kapton. It is sensitive to erosion at low altitudes because of atomic oxygen bombardment.

Another effect of UV radiation is the darkening or degradation of thermal control surfaces. As noted earlier, popular surfaces used in space applications have a small $\alpha_S/\varepsilon_{IR}$ ratio. Some of these surfaces have a clear outer layer, and some are a white paint or material.

CONTAMINATION

Contamination is a general term referring to the deposition on spacecraft surfaces of very thin films and the deposition of particulates. The films are made up volatiles escaping from the paints, adhesives, and electronic assemblies and from the vaporization of materials due to the vacuum and temperature environment. Particulate contaminants come from preflight exposure sources, launch vehicle sources, and on-orbit material degradation.

Film contamination affects the performance of optical and thermal systems. UV optics are especially sensitive to film contamination. Particle contamination is a source of scattered stray light in optical systems. Contamination sensitivity is quantified in terms of film thickness or particles per surface area. Another important effect of contamination is the degradation of thermal control surfaces. Thermal control surfaces rely on a low solar absorptivity, α_s. Contamination causes these surfaces to darken or become more opaque, raising the α_s and increasing the temperature of the structure behind them.

One of the driving forces in the world of contamination control has been the telecommunications industry. Relay satellites use large numbers of powerful transmitters. The life of these spacecraft is directly related to their ability to keep the transmitters cool. Increasing the temperature of electronics systems decreases their operational life. Satellite systems are a high-cost, high-return business. The cost is fixed, but the return is directly proportional to how long the system lasts. Transmitter panels are kept cool using large radiators. Great care is taken to keep contamination from degrading the radiator panels by directing the outgassing paths away from the sensitive surfaces.

Atomic Oxygen

One of the surprises of the early shuttle flights was the striking change occurring in organic films exposed in the forward velocity direction. Resembling weathering, the effect is attributed to the impact of high-velocity atomic oxygen. The surface bombardment results in physical erosion over time and a measurable reduction in the thickness of the material.

Atomic oxygen flux varies with altitude and the solar cycle. For example, the Long Duration Exposure Facility (LDEF) experiment (described below) flew in low Earth orbit for 69 months. The mission began near solar minimum conditions and ended close to solar maximum. This increase in solar activity, coupled with the decrease in altitude over the mission, resulted in the majority of the oxygen exposure occurring at the end of the mission. About 57% of the atomic oxygen exposure accumulated during the last 6 months of the LDEF mission. The last year of the flight accounted for roughly 75% of the total exposure.

Some materials, such as carbon and organics, are susceptible to atomic oxygen (AO) erosion, while others, such as fiberglass are not. Data from on-orbit studies have shown AO "scrubbing" of forward-facing surfaces, where the effects from contamination are removed by the low-level erosion of the surface.

CONCLUSION

Unlike terrestrial systems, the thermal environment in space drives the configuration, attitude, and external surface materials used in space hardware designs. The lack of an atmosphere allows the generation of very high and very low temperatures that are detrimental to the intended operation of equipment and systems. These temperatures are the result of the heat flowing through the spacecraft from the hot solar or planetary sources to the cold of deep space. It is the aim of spacecraft thermal designers to control and direct these heat flows to provide the necessary temperatures that a particular system needs to operate successfully. Over time, the evolution of spacecraft thermal designs has been driven by the increased performance requirements of space systems and the advances in the analytical programs used to analyze the hardware.

A significant part of the spacecraft design process is understanding the impacts that the space environment will have on the hardware during the particular mission. Subsystem damage resulting from ionizing radiation, UV exposure, micrometeoroids, atomic oxygen, and contamination can jeopardize the proper operation of systems in ways that are not a concern for Earth-based designs. As little space hardware is brought back for study, our understanding of the importance of these space environments has changed and developed over time based on data received and inferred from on-orbit performance of the hardware.

FURTHER READING

Burmeister, Louis C. 1993. *Convective Heat Transfer, 2nd Edition.* New York: John Wiley & Sons.

Campbell, W.A., and J.J. Scialdone. 1984. *Outgassing Data for Selected Spacecraft Materials.* Greenbelt, MD: NASA Reference Publication 1124.

Carslaw, H.S., and J.C. Jaeger. 1980. *Conduction of Heat in Solids.* New York: Oxford University Press.

Gaski, J.H.D. 1987. *Sinda/1987/ANSI.* Fountain Valley, CA: Network Analysis Associates, Inc.

Heaney, J.B. 1971. Suitability of Metalized FEP Teflon as a Spacecraft Thermal Control Surface. Paper no. 71-Av-35, presented at the SAE/ASME/ASAA Life Support and Environmental Control Conference, San Francisco, July 12–14, 1971.

Holman, J.P. 1963. *Heat Transfer.* New York: McGraw-Hill.

IIT Research Institute. 1971. *Ultraviolet Irradiation of Thermal Control Coatings.* IITRI Report No. IITRI-C6232-5, APL/JHU Contract No. 341785, June 1.

Kays, W.M., and M.E. Crawford. 1980. *Convective Heat and Mass Transfer.* New York: McGraw-Hill.

Levine, A., ed. *LDEF—69 Months in Space.* In the Proceedings of the First Post-Retrieval Symposium, Kissimmee, Florida, June 1991. Hampton, VA: NASA Conference Publication 3134.

Minkowycz, W.J., E.M. Sparrow, G.E. Schneider, and R.H. Pletcher. 1988. *Handbook of Numerical Heat Transfer.* New York: John Wiley & Sons.

Piscane, V.L., and R.C. Moore. 1994. *Fundamentals of Space Systems.* New York: Oxford University Press.

Sparrow, E.M., and R.D. Cess. 1978. *Radiation Heat Transfer.* Washington, DC: Hemisphere Publishing Company.

Van Vliet, R.M. 1965. *Passive Temperature Control in the Space Environment.* New York: The Macmillan Company.

30 Potential Effects: Atmospheres of Space Bodies and Materials

William S. Heaps

CONTENTS

INTRODUCTION

For most archaeological artifacts, features, and sites the principal form of deterioration arises from interactions with the atmosphere. Examples of this abound for structures on earth. The pyramids, for example, show the effects of millennia of erosion primarily by wind-borne particles. In addition, many limestone structures have been harmed by the acidic properties of the air, particularly as this affects the pH of rain. Finally, the degradation of many construction materials arises from oxidation. High levels of tropospheric ozone arising from urban air pollution greatly accelerate this process. In attempts to reduce the impact that this may present, many industrialized nations have undertaken various mitigation steps to combat this issue.

In the extraterrestrial realm archaeology is typically confined to orbiting objects plus a limited amount of hardware that has successfully landed on the surface of the

Moon, the inner planets, the moons of planets, and possibly asteroids. The larger planets—Jupiter, Saturn, and Neptune—do not possess a clearly defined surface. As such, the likelihood of ever recovering an object that enters their atmosphere is minimal.

ATMOSPHERE

An atmosphere is an assemblage of molecules bound to the surface of a heavenly body by gravity. The properties of the atmosphere depend on a number of other properties of the object, including but not limited to composition, albedo, and even, possibly, the magnetic field. On Earth the atmosphere is about four-fifths nitrogen and one-fifth oxygen. Water vapor, argon, carbon dioxide, methane, and a host of other trace gases are also present. The density of the atmosphere at any point is given by an equilibrium between the pressure of the gas and the weight of the gas above that point. The pressure depends on the temperature of the gas, and the temperature depends on the balance of a number of processes, including heating at the planet's surface, where sunlight is absorbed, as well as absorption of sunlight and surface radiance by gases in the atmosphere itself. Most atmospheres get cooler with altitude as the distance from the warm surface increases; however, the Earth's atmosphere in fact exhibits a second maximum temperature arising from the absorption of sunlight by ozone in the upper atmosphere. The upshot is that the density of a planetary atmosphere decreases with height as an exponential function.

$$N(z) = N_0 \exp - [z/(kT/mg)] \quad (30.1)$$

Here N is the atmospheric density, N_0 is the density at the surface, z is the height above the surface, k is Boltzmann's constant, T is the temperature, m is the mass of the atmospheric molecule, and g is the acceleration of gravity at the surface. The quantity (kT/mg) represents the altitude change over which the density is reduced by $1/e$. It is called the scale height of the atmosphere. For Earth it is about 7 km. It is a useful concept despite the fact that the atmospheric temperature is different at different times and locations. The formula also implies that various gases with different masses should fall off in density at different rates. In fact, the effects of collisions among different molecular species within the atmosphere dominates over gravity to such an extent that the mixing proportions of the Earth's atmosphere are constant up to considerable heights. This is to say that the weight of the column of both nitrogen and oxygen molecules is balanced by the pressure (change of momentum in molecular collisions) arising from both types of molecule.

The composition of an atmosphere is also influenced by photochemical effects. In the upper reaches of an atmosphere (more than approximately 80 km for Earth), the ultraviolet radiation from the sun consists of photons with enough energy to split apart a molecule of nitrogen (N_2) or oxygen (O_2), yielding the atomic forms of these elements. These ultraviolet (UV) photons are also able to knock an electron off the molecule to form charged particles called ions (N^{2+} and O^{2+}). The ions and the elemental forms of nitrogen and oxygen are much more chemically active than the molecular forms, giving rise to a host of additional trace species in the upper atmosphere. It is the reaction of atomic oxygen with molecular oxygen that gives rise to the formation of an ozone layer in the upper atmosphere:

$$O + O_2 \rightarrow O_3 \tag{30.2}$$

which has important consequences for Earth. These ionization and photodissociation processes absorb all of the very short wavelength radiation from the sun. Longer wavelengths (230–400 nm) are transmitted by nitrogen and oxygen, but they are absorbed by ozone. On planets without a significant ozone layer, objects at the planet's surface are subject to considerable degradation arising from solar UV, but that is not the subject of this chapter.

At greater altitudes it is no longer valid to assign a constant value, *g*, for the acceleration of gravity. The gravitational force is proportional to the inverse square of the distance between masses. For a large, spherical object, such as a planet, this means that the distance in the formula is effectively equal to the distance from the center of the sphere. The solution to the equation for density with altitude is no longer an exponential function. At these altitudes, temperatures become more extreme and the molecular collision rate is such that fractionization of constituents by mass takes place (more than approximately 100 km for Earth). For the Earth, even at altitudes as great as 500 km, the number density for atoms is assumed to be on the order of 10^{10}–10^{11} particles per cubic centimeter. At these altitudes, the density is highly variable, depending largely on solar activity. Numbers of this magnitude still represent a substantial drag force on a satellite and will bring it down in the end. Eventually the density of the planetary atmosphere becomes comparable to the density of the solar wind (~7 protons per cubic centimeter at the Earth's orbit), and the atmosphere of the planet is no longer a valid concept.

AEROSOLS

Finally it should be noted that atmospheres also contain small solid or liquid particles with diameters up to tens of microns. Aerosols can arise from a number of mechanisms. In the lower atmosphere, soil particles can be lofted by winds. Fires provide a mechanism to carry particles of carbon or other combustion products into the atmosphere. Chemical processes can condense or precipitate a particle as well. In fact, there exists a layer of particles in the lower stratosphere on Earth caused by the formation of sulfate compounds arising from injection of sulfur dioxide into the stratosphere by volcanoes. The most significant consequence of aerosols from the archaeological perspective is probably the effects of erosion by soil particles carried by a high wind.

DAMAGE MECHANISMS

There are two principal mechanisms by which an atmosphere can give rise to the deterioration or destruction of an archaeological artifact or feature. These are physical processes, as exemplified by erosion, and chemical processes, as exemplified by oxidation. Erosion, or "weathering," is a process of physical damage to an object arising from mechanical forces. Clearly the most straightforward of these is just abrasion by hard particles carried by wind. The magnitude of this effect depends on several factors, including the physical properties of the blowing particles, the size and abundance of the particles, and the wind velocity. Generally, this is a slow process on Earth capable of profound effects over long periods of time but probably not significant over the time scales that embrace the Space Age. Another weathering mechanism is the freeze-thaw process. This is not strictly an atmospheric mechanism, taking place only in the presence of an atmosphere, which is necessary to carry the operational fluid and deposit it on and in the weathering object. This mechanism is most powerful when operating with water. This is because of water's rather unusual property of expansion on freezing. This permits liquid water to flow into a crack or fissure on the weathering object and conform perfectly to the space in size and shape. When the water freezes it expands and is capable of exerting tremendous forces on the confining walls, giving rise to deformation or even breaking. Most other liquids do not expand during freezing and so do not exert these great forces.

The atmosphere can also cause significant deterioration of archaeological artifacts through chemical action. This process is enhanced by the presence in the atmosphere of highly reactive chemical species. On Earth oxygen itself is capable of causing significant deterioration of objects, particularly metals. Ozone is even more destructive because of its ability to yield the highly reactive oxygen atom. Halogen atoms and weak acids are also present in the Earth's atmosphere, arising from both natural processes and human-made pollution. Planets with thin atmospheres that transmit significant amounts of UV tend to have chemically reactive atmospheres because of the ability of UV to initiate photochemical processes that produce chemically active compounds.

For organic species, such as plastics, the initial step in the degradation process is often attack by the hydroxyl radical OH. Radicals are compounds that have unpaired electrons. Having an unpaired electron makes a species particularly reactive. The hydroxyl radical can be formed by the direct action of UV light on water vapor. In the lower atmosphere of Earth, it is more commonly produced by a chain of reactions starting with the photolysis of ozone yielding an excited-state oxygen atom that subsequently reacts with water vapor to form two OH radicals. OH radicals are strongly reactive with hydrogen atoms, re-forming water vapor. Once an organic compound has lost a hydrogen atom it is susceptible to many additional forms of chemical attack that lead to its destruction.

The mechanisms described above are operational to a greater or lesser extent in the atmospheres of all the bodies in the solar system. The next section considers the details for some of the individual components of the solar system.

EARTH

The atmospheric effects for the Earth have already been detailed to a great extent, as Earth served to provide examples that are familiar. Most of the objects sent into space, thus far, have remained in close proximity to the Earth. This means that the majority of artifacts will be found on Earth or still in orbit around Earth. For many objects the most profound atmospheric effect of all has been and will be destruction by heating from friction upon reentry. Once an object has successfully completed this most difficult stage, it becomes subject to additional damage from the atmosphere. It should be mentioned that the atmosphere does have limited effects on objects still in orbit. Even at altitudes as great as 500 km, the atmospheric pressure is still on the order of 10^{-6} mB, and the atmosphere is composed of fairly reactive particles moving at the rather significant velocity of 5 km/s relative to a satellite. Still, this pressure is so low that most objects designed for operation in space are little damaged by the atmosphere during their time in orbit.

MERCURY

The atmosphere of Mercury is pretty much negligible. Whatever atmosphere Mercury ever had has been blasted away by the solar wind. The surface pressure on Mercury is believed to be on the order of 10^{-12} mB, which is a million times lower than the pressure of the Earth's atmosphere 500 km up in space. Mercury's atmosphere is thought to consist mostly of oxygen, sodium, hydrogen, and helium, but it is so rarefied that even these chemically active species could have little effect on a human-made object on the surface.

VENUS

Venus has the potential to be very rich for space archaeology. There have been more than forty probes of Venus, particularly by the Soviet space agency. Many of these missions were landers, so the possibility of human-made objects on the surface of Venus is high. The atmosphere of Venus, however, is extremely harsh and may very well have caused significant damage over time to any hardware that survived to reach the surface. The atmosphere of Venus is in a condition known as a "run-away greenhouse." The atmosphere is thick and composed largely of carbon dioxide. This means that solar heating of Venus is not effectively reradiated into space, with the result that the surface temperature exceeds 900°F. The surface pressure is on the order of 90–100 B. The scale height is about 16 km. Although there is little free oxygen on Venus, there are substantial amounts of carbon monoxide, which is somewhat reactive, as well as sulfuric acid in cloud droplets, which would attack most metal objects during an descent to the surface.

MARS

Mars is an excellent source of space archaeological artifacts. There have been any number of orbiters around Mars, although the status of most of them is unknown because the missions end when the instruments cease to communicate. There are a number of landers on the surface of Mars, and their locations are well known so their eventual recovery is more likely. These include the Viking landers from 1976, the *Sojourner* rover from 1996, and the *Spirit* and *Opportunity* rovers that continue to operate on the surface of Mars today. The atmosphere of Mars does represent a threat to these landers but is not likely to cause total destruction in a short time frame. The Martian atmosphere is about 97% carbon dioxide with trace amounts of a number of other species. Its scale height is about 11 km. The oxygen and ozone abundance are low, as is the water vapor. This means that attack by chemicals in the atmosphere is a small worry. Mars does experience very significant dust storms of long duration and with high wind velocities. Particle sizes are estimated to be small—on the order of 1.5 microns in diameter. The wind velocities, driven by the summer-winter freezing and thawing of significant portions of the atmosphere, can reach 400 km/h (250 mph). However, the surface pressure on Mars is only about 7 mB, and so the amount of abrasive dust that could be lofted by such a tenuous atmosphere is probably small. Physical erosion should be slow. On the other hand the thin atmosphere does mean that the surface of Mars is exposed to intense solar UV radiation, and the temperature extremes of the atmosphere are considerable. These latter two problems are more likely to cause stress to a human-made object on Mars.

PLUTO

Pluto is the last planet (or almost planet) that has both an atmosphere and a surface. The surface pressure is about .003 mb, with a scale height of 60 km. No human-made object has even approached Pluto to date.

THE MOON

There are a number of human-made artifacts on the surface of the Moon—the largest being the remains of landers from the Apollo missions. The atmosphere of the Moon is also pretty much negligible with surface pressures on the order of 10^{-12} mB. This thin atmosphere does mean that surface artifacts are exposed to extreme UV radiation, extreme temperatures from the monthly cycles of exposure to solar heating and then deep space, and finally to the solar wind. There are reactive chemical species in the atmosphere because exposure to extreme UV initiates photochemistry that forms radicals, but the number density of these reactants is too low for them to be factors in the deterioration of archaeological objects.

THE REST

The only remaining objects in the universe that possess a surface and an atmosphere are the asteroid Ceres, Jupiter's moon Io, Saturn's moon Titan, and Neptune's moon Triton. To date, no human-made object has landed on any of these smaller inhabitants of the solar system, but the fact that they exhibit an atmosphere makes them of greater interest to planetary scientists. Of course there are other motivations for the exploration of a moon besides an atmosphere. The Galileo mission to Jupiter discovered the existence of a subsurface ocean on one of the moons, Europa. Such an ocean may provide a habitat for life, making Europa a become a strong candidate for a planetary mission. At the end of its service life, the *Galileo* probe was deliberately destroyed by flying it into the atmosphere of Jupiter. This was done largely to prevent the possibility of its crashing into Europa or one of the other moons and contaminating the surface. A mission to one of these destinations may be in the offing in the future.

CONCLUSION

The field of space archaeology is in its infancy, and the good news is that the Space Age is a short era and that many of the objects that humanity has deposited in outer space may yet survive. Atmospheric effects are only one of several mechanisms by which space probes may deteriorate. Moreover, missions to bodies with harsh atmospheres have been engineered to survive these conditions. If efforts are begun soon, it is likely that the rich heritage of the Space Age can be documented successfully through its artifacts.

FURTHER READING

Arnett, W.A. 2008. *The Nine 8 Planets*. http://www.nineplanets.org.

Atmosphere of Mars. http://en.wikipedia.org/wiki/Atmosphere_of_Mars.

Bauer, S.J., and H. Lammer. 2004. *Planetary Atmospheres*. Berlin: Springer.

Galileo: Journey to Jupiter. 2003. Jet Propulsion Laboratory. http://www2.jpl.nasa.gov/galileo.

Seinfield, J.H., and S.N. Pandis. 2006. *Atmospheric Chemistry and Physics*. Hoboken, NJ: Wiley.

31 Wear and Tear: Mechanical

Theodore S. Swanson

CONTENTS

INTRODUCTION

The focus of this chapter is on the long-term wear and tear, or aging, of the mechanical subsystem of a spacecraft. The mechanical subsystem is herein considered to be the primary support structure (as in a skeleton or exoskeleton) upon which all other spacecraft systems rest, and the associated mechanisms. Mechanisms are devices that have some component that moves at least once in response to some type of passive or active control system. For the structure, aging may proceed as a gradual degradation of mechanical properties and/or function, possibly leading to complete structural failure over an extended period of time. However, over the 50 years of the Space Age, such failures appear to be unusual. In contrast, failures for mechanisms are much more frequent and may have a very serious effect on mission performance.[1]

Just as on Earth, all moving devices are subject to normal (and possibly accelerated) degradation from mechanical wear due to loss or breakdown of lubricant, misalignment, temperature cycling effects, improper design/selection of materials, fatigue, and a variety of other effects. In space, such environmental factors as severe temperature swings (possibly hundreds of degrees Celsius while going in and out of direct solar exposure), hard vacuum, micrometeoroids, wear from operation in a dusty or contaminated environment, and materials degradation from radiation can be much worse. In addition, there are some ground handling issues, such as humidity, long-term storage, and ground transport, which may be of concern.

This chapter addresses the elements of the mechanical subsystem subject to wear and identifies possible causes. The potential impact of such degradation is addressed, albeit with the recognition that the impact of such wear often depends on when it occurs and on what specific components. Most structural elements of the mechanical system typically are conservatively designed (often to a safety factor of greater than ~1.25 on yield for unmanned spacecraft) but do not have backup structure because of the added mass this would impose, and also because structural elements can be accurately modeled mathematically and in testing.[2] Critical mechanisms or devices may have backups, or alternate work-arounds, since characterization of these systems in a 1*g* environment (e.g., 100% of normal Earth gravity) is less accurate than that of structure in a 1 *g* environment, and repair in-space is often impossible.

CAUSES OF MECHANICAL AGING

Mechanical aging is not necessarily the same phenomenon as events that are typically classified as "failures." Both aging and failures will be addressed in the following discussion, as failures may be viewed as an "infant mortality" type of aging.

Some events that are traditionally classified as failures are due to poor design, poor quality, or improper selection of materials. One proximate cause of such mechanical failure is the stresses induced by acceleration forces, such as those experienced during launch, orbital maneuvers, deployment of antennas and other such structures, and planetary entry and landing. Other failures are caused by poor ground handling (e.g., contamination, excessive testing for vibration or temperature extremes, etc.) or workmanship issues. Many mechanical or mechanism failures are caused by exposure to and continuous operation within the space environment. Many of these failures may be more commonly termed aging effects. Space environmental effects include the thermal environment, plasma, micrometeoroids and space debris, solar thermal effects, magnetic fields, and changes in the gravitational environment.[3] Some of the fundamental causes for failure by space environmental effects are discussed in later in this chapter and elsewhere in this book and will thus be addressed in this chapter only as they apply to mechanical subsystems.

It should be recognized that there are several classifications of what constitutes a failure. Different groups define and classify failures in different ways. For the purposes of the following discussion, only significant failures, which are generally defined as being an event that results in a loss of 33% or more of a mission's or instrument's objective, will be considered.[4]

FAILURE AND AGING DUE TO POOR DESIGN, POOR QUALITY, OR IMPROPER MATERIALS

Spacecraft mechanical systems, and especially mechanisms, are often a one-of-a-kind design.[5] This is especially true for spacecraft/instruments intended for missions of exploration or for science missions. Some components that are common to many spacecraft, such as solar drive mechanisms, gyroscopic reaction wheel assemblies, motors, mechanical louvers, and so forth, may share a common design and be manufactured in lots. But there are often variations in how they are used and the environment to which they are exposed that create unique design challenges. Even communication satellites are typically made in lots of no more than a few dozen. Hence, mechanical subsystems may lack the design heritage and wider use that other subsystems might enjoy.[6,7] This lack of a broader data base means less experience with a given design, and this may contribute to design flaws. In addition, development of a reliability model to accurately predict the probability of failure is nearly impossible because of the low sample size that can be used in the analysis.

The effects of poor design, poor quality, or improper material choice may manifest themselves very quickly, such as at launch, or result in a degradation (possibly leading to failure) over time. A not uncommon example of long-term degradation would be failure of a gyroscopic stabilizer wheel due to problems with the bearings. Moving parts, such as bearings, motors, and gears often have lubrication that may eventually outgas to space, become dislocated from the moving parts (because of migration, lubricant is driven away from the hot spots). Moving parts may also become stuck because of friction from wear or thermally induced expansion/contraction of the materials. These have been identified as common causes of failure and for this reason moving parts are often avoided, if possible. Deployment mechanisms have sometimes failed to engage fully or at all, leading to problems with the power, thermal, or communications subsystem. A classic example would be the failure of the solar array/micrometeoroid shield to properly deploy on *Skylab* in 1973, which caused power problems and excessive temperatures until makeshift repairs could be performed. A more recent example would be the failure of the solar array on the *Mars Global Surveyor* to latch properly (due to a damper arm failure), which resulted in a flight plan change.[8]

Data collected over the last few decades shows that design flaws are becoming fewer, at least for missions and applications for which we are gaining experience.[9] Ground testing and a sound quality control program have generally mitigated mechanical design and quality issues. However, as humans move further out into the solar system and beyond, we will be exposing our spacecraft to ever more challenging environments, some of which are not as well understood as the near-Earth environment. The near-Earth environment, as harsh as it may seem when compared with terrestrial applications, is actually relatively benign compared with the Moon, outer planets, or nearer to the sun. This will complicate the design of all subsystems, including mechanical, and often demand more sophisticated and complex designs that have fewer margins for error.

Ground testing and verification is another issue that may be complicated because of the difficulty in precisely replicating long-term space exposure via traditional ground testing techniques. Space environmental simulation testing, such as thermal cycling in a vacuum, launch vibration loads, acoustic loads, radiation effects, and so forth, is normally performed in a piecemeal fashion because of practical and cost limitations. This approach has proven to be very effective when done properly, but it does not really address long-life issues that may only manifest themselves after decades, centuries, or millennia of time.

It should be noted that it important to test the spacecraft to the limits of what it will be exposed to in space, with some margin, and to avoid either under- or over-testing. This is often referred to as the "test as you fly, fly as you test, with some margin" philosophy.[10,11] Testing to a less stringent environment can clearly lead to premature failure due to the actual use exceeding the mechanical and/or thermal environment the spacecraft was designed for, and qualification tested to. However, overtesting by either going to excessive levels or excessive cycles, beyond a reasonable margin above the expected environmental limits, can lead to stressing the mechanical subsystem beyond its design limits. This can lead to performance deterioration or outright failure. Accordingly, it is vital to be able to accurately predict the mechanical stress that the spacecraft will be exposed to in its intended service. A great deal of sophisticated analysis is typically employed to ensure accuracy, and ground testing is intended to "qualify" a given piece of hardware to the calculated exposure limits.

Improper material selection is really a design issue that should be mitigated through experience with similar applications, knowledge of material performance in the anticipated environment, and a thorough ground test program. However, examples of failure through improper material selection continue to occur, albeit at a reduced rate due to the maturing of the industry.

FAILURE AND AGING DUE TO ACCELERATION FORCES

Acceleration forces can have a significant impact on both the spacecraft's structure and mechanisms, and careful design is needed to mitigate these effects.[12] The result of improper design or poor quality may be manifested as a catastrophic failure or through a delayed failure caused by material fatigue. Fatigue is a particularly difficult issue, as it is more difficult to predict and can be significantly impacted by microcracks and other small flaws.

Mechanical loads may be either static or dynamic. Static loads may be imposed externally, such as by gravity during spacecraft assembly/integration/testing, or they may be self-contained, such as from stored propellants that are under pressure (typically temperature dependent), preloads from tightening a bolt, or thermoelastic stresses from temperature changes that cause materials to expand/contract. Dynamic loads include launch thrust vibration, air pressure waves during launch while in the atmosphere, shock impulse loads from the pyrotechnic devices used to perform stage separation, stresses from orbital maneuvering, and reentry/landing loads. It should be noted that these dynamic loads are expressed over a fairly wide frequency range and may vary in intensity during the acceleration/deceleration event.[13] These transient dynamic loads may initiate a fatigue failure, which might be completed by cyclical stresses occurring during normal operations while in space. Cyclical stresses while in orbit or in transit may be due to thermally induced cycling (e.g., the spacecraft going in and out of exposure to the sun), operation of equipment that has some vibration, or some other cause. The combined effect of thermally induced stresses with additional mechanical stress can be particularly damaging for structures made of composite materials, as it may cause microcracking.[14] For many modern rockets, peak axial acceleration is on the order of 3 to 5 g, meaning that the mechanical stress applied to the structure is 3–5 times that normally applied by gravity. However, depending on how and where a component is mounted, it may experience significantly higher transient acceleration forces, on the order of tens of *g*. For example, many components are, in effect, mounted in a cantilever fashion, which can significantly increase their *g* loading.

In addition to the basic structure, which clearly must survive launch without functional damage, there are typically mechanisms, such as gyroscopic stabilizer wheels, deployment devices, turntables for scanning instruments, filter wheels, motors, shutters, and so forth, that need to operate once launched, either as a single event or routinely. Depending on their design, some mechanisms may require "launch locks" to constrain moving parts or bypass the load path during the violent launch event. Failure of the launch locks to properly engage, failure to release when so commanded, and early release are additional causes of an infant mortality type of aging. Additionally, mechanical components that contain pressurized fluids, such as propulsion lines/tanks and heat pipes, are at risk for developing leaks (due to weld or seal failure, collisions with micrometeoroids or space junk, etc.), which can also lead to early failure.

FAILURE AND AGING DUE TO POOR GROUND HANDLING OR WORKMANSHIP

There have been a variety of premature failures caused by poor handling during integration and testing, or while the spacecraft is transit between assembly, testing, and/or launch facilities. Other failures have been caused by simple poor-quality workmanship. Stringent quality control procedures are typically employed to prevent such unnecessary problems, but the problems continue to occur. Part of the issue is the complexity and interrelationship between the spacecraft's various subsystems, which can result in unintentional and unrecognized damage to a component not obviously associated with one that is being worked on. Sometimes cost and/or schedule will drive the management to take risks that later prove to be poor decisions.

Typical problems in ground handling tend to focus around excessive vibrations during transit, exposure to excessive temperatures or humidity during transit, and contamination from particulate or molecular sources. From a mechanical subsystem perspective, excessive vibrations while still on the ground risk the loss of lubricant in gears or sliding components, or misalignment of critical components such as optics and lasers. For example, the high-gain antenna on *Galileo* got stuck and could not be released while in transit to Jupiter. It is conjectured that this might have been caused by misalignment/loss of lubricant during its numerous transits on highways to and from the launch facility.

The presence of excessive contamination on a spacecraft, particularly molecular buildup on sensitive optics and precise mechanisms, is another risk to long-term life. This is typically caused either by assembly and/or testing in a dirty (i.e., noncontrolled) environment or molecular out-gassing of volatiles from within the materials used to assemble the spacecraft. This will occur under the hard vacuum of space (or a thermal vacuum space simulation test) and is accelerated by higher temperatures. Proper material selection and an appropriate approach to contamination control during assembly/testing/shipping are needed to avoid this potentially serious problem.[15]

FAILURE AND AGING DUE TO THE NATURAL SPACE ENVIRONMENT

The natural space environment can have a very significant impact on a variety of spacecraft subsystems, including mechanical. *Natural space environment* means the environment that is present in space independent of the presence of the particular spacecraft in question. It includes both naturally occurring phenomena, such as radiation and solar illumination, and human-made objects such as space debris. More specifically, it includes the following nine environments: the neutral thermosphere, the thermal environment, plasmas, meteoroids and human-made space debris, the solar environment, ionizing radiation, the geomagnetic field, the gravitational field, and the mesosphere. These environments may cause either sudden or, more likely, gradual deterioration of a spacecraft subsystem, leading to eventual failure. Since such degradation is the result of the natural space environment, such a failure may be better termed true aging.

The space environments that have the most impact on the mechanical subsystem include the thermal environment, plasma, micrometeoroid/space debris, and the solar environment. Magnetic fields, gravitational effects, and the mesosphere may have secondary effects. The thermal environment is coupled to the solar environment and is driven by the presence or lack of solar illumination and by any planetary infrared radiation or reflected solar radiation that the spacecraft may be exposed to. The thermal properties of the exposed surfaces of the spacecraft will determine how hot or cold each surface gets in response to this incident radiation. This has a significant impact on the placement of various spacecraft components, such as radiators, so that they may be maintained within acceptable temperature limits. Additionally, if the spacecraft is exposed to a variable thermal environment, such as will occur for spacecraft in low Earth orbit, on the lunar surface, or on a planetary surface, then it will be alternately heated and cooled. For example, a spacecraft on the surface of the Moon may be exposed to an effective thermal environment of more than 100°C during the lunar day, which falls to perhaps –200°C during the lunar night. This temperature differential of 300°C or more can cause serious problems because of the natural expansion and contraction of materials in response to temperature changes. This may lead to fatigue cracks, which can ultimately cause a structure to fail, especially when under an additional load.[16] Composite materials may be particularly susceptible to this phenomenon.[17]

Plasma occurs naturally in the upper atmosphere (above 90 km) as a result of solar radiation.[18] Plasma can cause mass loss from erosion, arcing, and sputtering, and thus change the structural dimensions of an object. A common phenomenon is the splitting of the O_2 molecule into atomic oxygen, which is highly reactive with certain polymers used in spacecraft construction. An example is the aluminized Kapton commonly used for insulation/thermal control. The resulting deterioration of such materials due to atomic oxygen exposure can have significant impacts on the thermal control subsystem, leading to problems with the mechanical subsystem, especially mechanisms. Another common effect of solar radiation in the upper atmosphere is the stripping of electrons off an atom's outer shell. This creates plasma of positively charged atoms and negatively charged electrons, which can cause a charge differential across a spacecraft. Discharges move material from one location to another, creating dimensional changes, as well as having significant impacts on the spacecraft's electrical subsystem.

Micrometeoroid/space debris is an increasing problem due to the long life of some "space junk." Space junk is material left over from launches or generated by the intentional or unintentional breakup of spacecraft. While these particles may be small, they are traveling at a very high velocity differential relative to a given spacecraft. For example, a small, 90-gram particle will impart over 1 MJ of energy from an impact.[19] There are numerous examples of such micrometeoroid/space debris hits.[20] Just through September of 1993, the shuttle program had to replace more than forty-six orbiter windshields because of impact damage. After the December 1993 Hubble Servicing Mission, the retrieved solar array showed more than 5,000 micrometeoroid impacts over its 4-year life in space. And late in 1989, when the shuttle was retrieving the Long Duration Exposure Facility (LDEF) spacecraft, a picture was taken showing a large piece of debris, with a relative velocity of about 170 km/h, passing between LDEF and the shuttle when they were only 660 ft. apart. A collision might have been catastrophic.[21]

FAILURE AND AGING DUE TO PLANETARY ENVIRONMENTS

Entry into, landing, and long-term survival on a planetary body may impose severe stresses on the mechanical subsystem. This is particularly true if the planet or moon has an atmosphere. Depending on how it is accomplished, reentry into a planetary atmosphere may be at hypersonic speeds and incur high *g* forces from deceleration. The energy associated with the change in speed will be dissipated as heat. Both the resulting mechanical stress and the weakening of material properties from the elevated temperatures will place significant demands on the mechanical structure. To help mitigate these effects, a variety of techniques are used, such as an ablative heat shield that may be jettisoned during landing, parachutes, and some sort of landing technology such as reverse thruster rockets, inflatable balloons, and/or collapsible structure to absorb the landing shock.

Once on the surface, the spacecraft will be fully exposed to the ambient environment, which may be extremely harsh. These conditions vary tremendously depending on the planet or moon involved, and the spacecraft will have to be designed to survive (for whatever time is intended) in this environment. For example, the atmospheric pressure on the surface of Venus is 90 times that of Earth, and the temperature is approximately 430°C. The atmosphere is nearly pure carbon dioxide, but higher in the atmosphere there are significant quantities of sulfuric acid. The moons of the outer planets, which are of significant interest as they may have subsurface water oceans or prebiotic compounds, are extremely cold (about –200°C). For example, Titan, Saturn's largest moon, has a thick atmosphere at –180°C composed primarily of nitrogen, with clouds of methane and ethane. It has atmospheric circulation with rain and lakes of liquid methane. Some moons, such as those of Jupiter, are exposed to an intense radiation environment. The longevity of mechanical systems in such environments is clearly a function of the design and material choices and will no doubt be affected by unknown corrosive and/or other environmental forces particular to a given planet or moon.

HUBBLE SPACE TELESCOPE TO JAMES WEBB SPACE TELESCOPE: MECHANICAL EVOLUTION

The comparison of the Hubble Space Telescope (HST) with its scientific replacement the James Webb Space Telescope (JWST) is representative of the ever-evolving mechanical structures in space. The model of the JWST with the program team, as shown in Figure 31.1, demonstrates the growth of mechanical complexity and team size to support these large spacecraft developments.

Mechanical engineering in space has evolved in complexity and size and allows greater support for increase mission functionality. Logarithmic increases in launch lift, combined with years of experience in materials in space applications, enable greater functionality to accomplish significantly more tasking science drivers.

The Hubble Space Telescope orbits around the Earth at an altitude of ~570 km above it.

JWST will not actually orbit the Earth—instead it will sit at the L_2 Lagrange point, 1.5 million kilometers away. Because HST is in Earth orbit, it was able to be launched into space by the space shuttle. JWST will be launched on an Ariane 5 rocket. Larger lift capability enables larger structures, as shown in Figure 31.2.

SIZE

HST is 13.2 m (43.5 ft.) long, and its maximum diameter is 4.2 m (14 ft.), where as JWST's sunshield is about 22 by 12 m (72 by 39 ft). That is about the size of a tennis

FIGURE 31.1 Model of the James Webb Space Telescope and team. (Courtesy of NASA.)

court. HST's mirror is a much smaller 2.4 m in diameter, and its corresponding collecting area is 4.5 m^2. Figure 31.3 demonstrates the larger primary mirror area.

Mission Science and Wavelength

JWST will have a 6.6-m-diameter primary mirror, which will give it a significant larger collecting area than the mirrors available on the current generation of space telescopes; about 7 times more collecting area than HST.

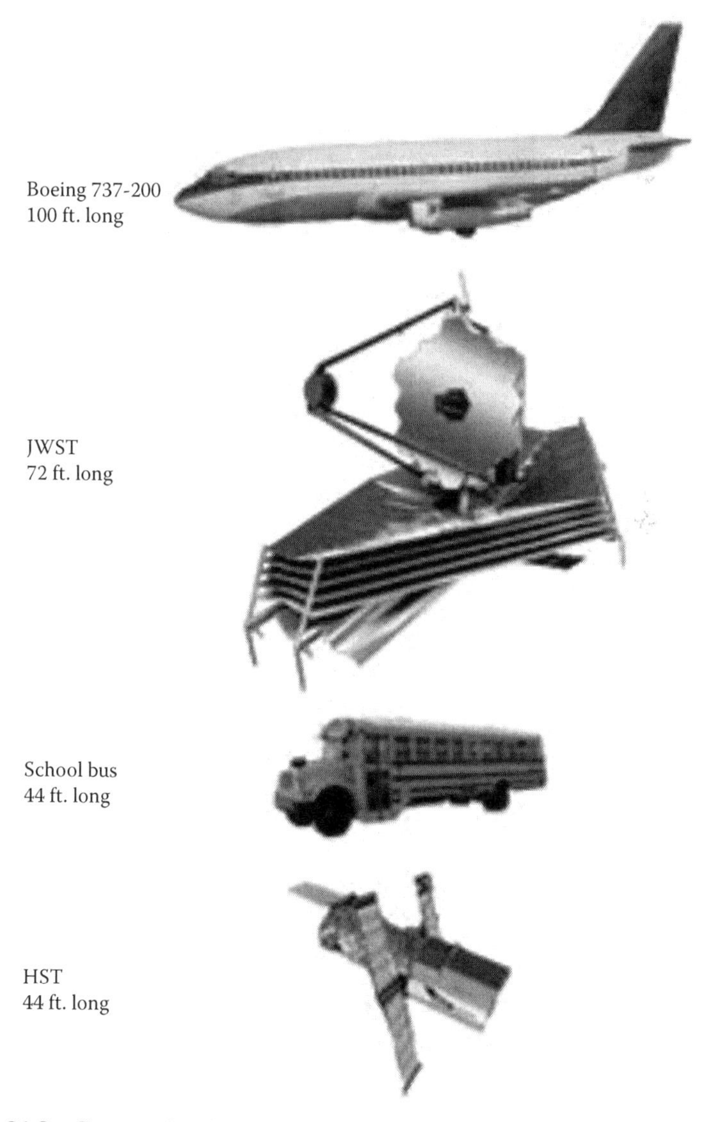

FIGURE 31.2 Comparative sizes of the Hubble Space Telescope and the James Webb Space Telescope. (Courtesy of NASA.)

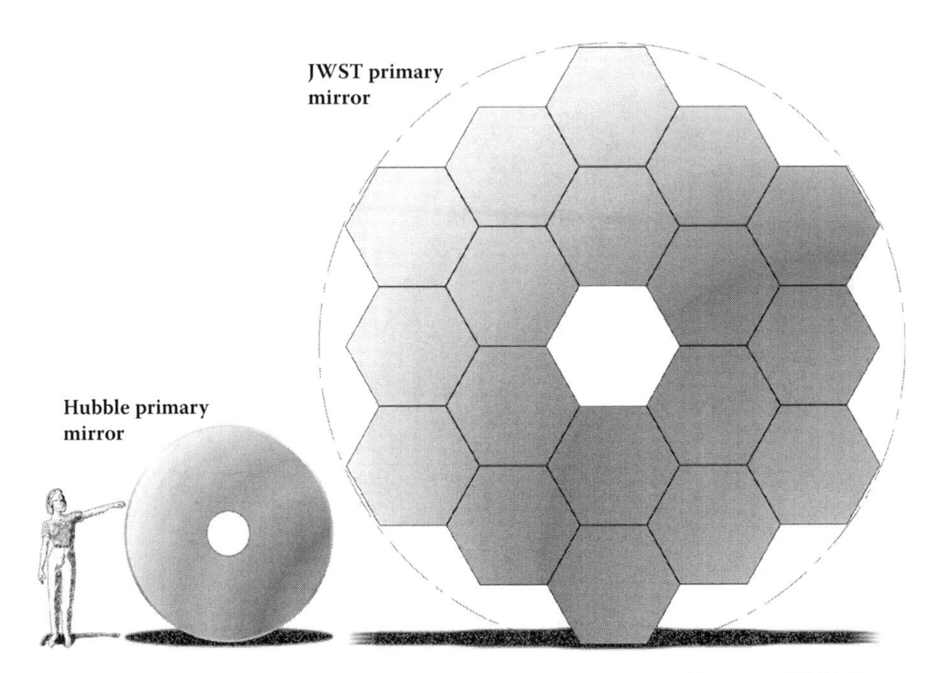

FIGURE 31.3 Comparative sizes of the Hubble Space Telescope and the James Webb Space Telescope primary mirror. (Courtesy of NASA.)

JWST will also have significantly larger field of view than the Near Infrared Camera and Multi-Object Spectrometer camera on HST (covering more than ~15 times the area). Significantly, JWST will observe primarily in the infrared in order to study the origin and evolution of galaxies, stars, and planetary systems by providing infrared imagery and spectroscopy. JWST's four instruments will provide wavelength coverage from 0.6 to 28 micrometers (or microns). The instruments on HST can observe a small portion of the infrared spectrum from 0.8 to 2.5 microns, but its primary capabilities are in the ultraviolet and visible parts of the spectrum, from 0.1 to 0.8 microns.[22]

TRENDS IN SPACECRAFT DESIGN AFFECTING AGING

Over the 50 years of the Space Age, we have learned a great deal about how to design and operate spacecraft to enhance their odds for mission success and long life. The major driver for mechanical aging, the space environment, has not changed, but our knowledge of this demanding environment has improved greatly; hence, we have changed the materials and techniques used to build spacecraft. Our technologies have also advanced in many areas. This improved understanding of the environment and advanced materials/technology have led to reduced mechanical failures for the typical spacecraft in low Earth orbit.

As composites offer the opportunity for reduced structural mass and the ability to tailor thermal, electrical, and radiation transparency properties, the trend is for their increased use on future spacecraft. While generally robust in the space environment, they can fail in manners that solid metals do not. Composite properties may vary greatly depending on material selection and fabrication, and they are often nonisotropic in mechanical, thermal, and electrical properties. For example, they are more subject to microcracks, which may lead to delamination or fatigue failure. Failures or design/fabrication flaws may also be more difficult to detect using traditional inspection techniques, and also more difficult to repair. Another emerging trend, again to save mass and also packing volume, is toward inflatable structures to be used as habitats, reflectors, storage containers, shields, balloons, or structural members such as beams. Such subsystems may fail because of improper inflation, improper or incomplete hardening (if employed), radiation damage, or leaks of the pressurant (if employed), as well as because of the harsh environment of space.

CONCLUSION

As we send spacecraft to ever more demanding environments, such as very near the sun, to our moon, or to the moons of the outer planets, the stresses induced by the environment will increase. To accommodate this increasingly severe environment and the typically more challenging mission objectives, these spacecraft will become more sophisticated and advanced in their design and materials choices. Additionally, they will often need to last decades, instead of years. This might even lead to material deterioration (and hence mechanical deterioration) from the expansion/contraction caused by thermal cycling and extreme temperatures. The possibility of long-term radiation damage, especially for some composites, may also increase with long exposure. Hence, a future in which spacecraft go to ever more environmentally demanding locations, carrying more sophisticated equipment and operating for a much longer timeframe, will increase the risk of failure from environmental factors, material selection, new technologies, and a combination of such considerations. Clearly, our designs and materials must improve, and with such improvement will come improved performance and durability. But failures, whether from inadequate design, poor material choices, misunderstanding of the environment, inadequate testing, poor workmanship, or human error, will continue to occur, as space is a harsh and very unforgiving environment. Our goal is to anticipate and meet the challenges with an ever-increasing capability.

REFERENCES

1. Newell, J., and D. Oberhettinger. 2000. NASA Engineering Network—Lessons Learned; Entry 0913; Space Mechanisms Reliability. http://www.nasa.gov/offices/oce/llis/0913.html.
2. Harper, A. 2005. *General Environmental Verification Standard (GEVS) for Goddard Flight Programs and Projects*. GEVS-STD-7000. Greenbelt, MD: NASA Goddard Space Flight Center.
3. Bedingfield, K.L., and R.D. Leach. 1996. *Spacecraft System Failures and Anomalies Attributed to the Natural Space Environment,* ed. M.B. Alexander. Huntsville, AL: National Aeronautics and Space Administration.
4. Sarfield, L.P. 1998. *The Cosmos on a Shoestring: Small Spacecraft for Space and Earth Science; Appendix B: Failure In Spacecraft Systems*. Santa Monica, CA: Rand Corporation.

5. Newell and Oberhettinger 2000 (note 1).
6. Sarfield 1998 (note 4).
7. Sarfield 1998 (note 4).
8. Sarfield 1998 (note 4).
9. Sarfield 1998 (note 4).
10. NASA. 2006. Rules for the Design, Development, Verification and Operation of Flight Systems. GSFC-STD-1000. http://gsfcrules.gsfc.nasa.gov/.
11. Oberhettinger, D. 2002. NASA Engineering Network—Lessons Learned; Entry 1196; Test as You Fly, Fly as You Test, and Demonstrate Margin. http://www.nasa.gov/offices/oce/llis/1196.html.
12. Sarafin, T.P., and W.J. Larson. 1995. *Spacecraft Structures and Mechanisms.* Torrance, CA: Microcosm Inc. (see specifically pp. 38–56).
13. Harper 2005 (note 2).
14. Issoupov, V., O.V. Startsev, et al. 2004. Combined Effect of Thermal and Mechanical Stresses on the Viscoelastic Properties of a Composite Material for Space Structures. In *Space Technology Proceedings*, Vol. 5, pp. 271–280. The Netherlands: Springer.
15. NASA 2006 (note 10).
16. National Academy of Sciences. 1996. *Accelerated Aging of Materials and Structures: The Effects of Long-Term Elevated Temperature Exposure; Degradation Mechanisms.* Washington, DC: The National Academies Press.
17. Issoupov et al. 2004 (note 14).
18. Sarafin and Larson 1995 (note 12).
19. Sarafin and Larson 1995 (note 12).
20. Sarafin and Larson 1995 (note 12).
21. Sarafin and Larson 1995 (note 12).
22. NASA. The James Webb Space Telescope Home Page. http://www.jwst.nasa.gov.

32 Space Environment and Spacecraft Electrical Systems

Ann Garrison Darrin

CONTENTS

INTRODUCTION

Three main concerns of operating in the environment of space on electrical systems are vacuum, thermal, and radiation effects. In addition, other impacts from operating in the low- to zero-gravity environment and particular constraints to designs will be discussed in this chapter. The space environment basically does not change; rather, our approach to performance in the space environment has changed over the years. Therefore, beyond understanding the space environmental impacts there are discussions of the evolution of electrical systems elements and how to track and understand pedigrees and dates. It is important to understand the dramatic difference between electric systems deployed on Earth or near Earth where servicing is a built-in factor compared with space applications. There is, with the exception of the euphemistic term "servicing" of Hubble Space Telescope (HST), limited to no servicing of systems, which couples with the incredible cost of placing useable satellites, manned missions, space station, and launch vehicles into space. This pairing drives distinct differences in the design of systems that are unrecoverable and irreparable. For many critical systems the equipment is redundant; that is, there are two (or more) of each system, providing backup. Where full redundancy is impractical, other techniques are used, such as paralleled converters, fused bus capacitors, majority voting, and so forth. Beyond redundancy, electrical systems for space have relied on heritage systems and derating. *Heritage* refers to systems and components that have been flown before, and *derating* refers to driving systems well within their performance range. In order to understand the electrical systems, a description of the various components of these systems is also included and discussed in relationship to variety of impacts of the space environment. Additionally, an understanding of the various power systems is included in this chapter to emphasize its necessity and vulnerability—the Achilles heel. The electrical system is the lifeline of all electromechanical space systems. Although failures in space are covered in separate chapters, a discussion of the electrical failures and anomalies as they relate electronics in the space environment are discussed as a representative case study in a discussion of the Hubble Space Telescope.

BACKGROUND

The importance of the electrical system for power and communication to space systems must not be underestimated. The single most common reason that end of life comes to any satellite, probe, shuttle, station, or rover is the end of its power.

The space environment, in combination with confined volume and (in the case of space stations and vehicles) human factors, drives the requirements for careful material and electronic component selection. In a vacuum, outgassing and offgassing can contaminate critical components, including the clouding of optical lenses. In these environments components can implode or explode without proper venting. In addition, component functions often change in a vacuum. This is particularly true of electromechanical devices, such as relays. Thermal issues are exacerbated, as traditional heating and cooling methods such as convection are not the same in zero gravity. The zero gravity imposes novel mechanical challenges, for example, a CD-ROM drive will not function where the CD goes in to free float and is not in an air volume. Ionizing radiation can latch up a component or cause disabling upsets. In the area of stations and human space travel, additional concerns of toxicity and flammability are noted.

HUBBLE'S TROUBLES BEYOND OPTICS: ELECTRONICS

The HST was beset by troubles in construction and budget overruns. Shortly after its 1990 launch, it was found that the main mirror suffered from spherical aberration due to faulty quality control during its manufacturing, severely compromising the telescope's capabilities. After the Mr. Magoo cartoon jokes had subsided and the successful servicing mission in 1993, the telescope was restored to its intended quality and became a vital research tool, as well as a public relations boon for astronomy. However, design, manufacturing, and effects from the environment continued to plague the satellite. The myopic start of the telescope was continued with a battery of environmental effects from radiation, overheating, micrometeoroids, and other abusive treatments.

Less than 1 month after launch, radiation-induced upsets began to occur. On May 7, 1990, bit flips occurred in the random access memory of the fine guidance electronics and affected the guidance system while HST was passing through the South Atlantic Anomaly (SAA). Onboard software was modified to compensate for the flips. On June 20, 1990, the SAA also caused high photomultiplier tube counts in the fine guidance system. This resulted in guide star acquisition failures. Both incidents are suspected to have been due to increased radiation effects. These radiation effects are discussed later in this chapter. The SAA is discussed in greater detail with the environmental descriptors of the radiation environment.

Temperature cycling–induced anomalies were noted before December 1993 on the HST service mission, where the solar arrays vibrated severely each time the observatory emerged from shade into sunlight. Active vibration cancellation using the onboard gyros was implemented before the service mission to minimize the problem. Thermal expansion of the support poles (also called bistems) was blamed for the vibrations, which interfered with deep-space observations. The new arrays installed during the service mission had sleeves over the bistems to provide jitter-free imagery.

During the December 1993 Hubble service mission, the HST original arrays were swapped out for replacements and returned to Earth. British Aerospace inspection of the retrieved HST array revealed that the whole wing suffered between 5,000 and 6,000 micrometeoroid impacts in its 4-year life. The effects of these impacts range from slight grazing to puncture of cells and blankets. The physical appearance and durability of the materials on the solar array and the physical position of hardware indicators were investigated. Evidence of micrometeoroid damage and the darkened silicone coatings were apparent.[1] The Room Temperature Vulcanized Silicones (RTV) adhesive covering the bus bars and diodes had bubbled and cracked exposing bus bars as the RTV contracted. On the exposed cell cover slides, marks of micrometeoroid collisions were present. A few slides have chipped holes off the cover slide or continued into the blanket substrates. Some holes went through the blanket. The stainless steel on the bistems was a gray ash color, and the red knobs on the end of the bistem were a dull brown color. The ultraviolet-exposed Kapton was considerably darkened.

The material used to cover the secondary deployment mechanism motor housings was metalized Teflon and Mylar. Some of the silver-colored tape strips had changed to dull gray, while other strips are were as shiny as the solar array coverings. The material used to hold the wire harness on the exterior of the primary deployment mechanism mast had discolored and cracked around the screw holes.[2]

Over the first 148 months that HST was in orbit, vehicle configuration, charge system degradation, and failures, together with thermal design limitations, had a significant effect on the ability to optimally charge the NiH batteries. As a result, battery capacity as measured during deep discharge capacity tests was degraded as noted in Table 32.1.[3]

SPACE ENVIRONMENT

ATOMIC OXYGEN

Atomic oxygen (AO) exists in significant amounts around low Earth orbits and around Mars. Atomic oxygen is highly reactive and will react differently depending on the nature of the materials involved. Atomic oxygen effects were first detected during shuttle missions. Exposure to AO tends to cause metals to develop an oxide on their surface, whereas polymers tend to lose mass and undergo a change in surface morphology. Because of their high reactivities with AO, polymers and other composites need to be protected. On an order of magnitude of scale, surfaces such as the solar arrays will be exposed to a stronger AO flux field than inboard components. Exposure to AO is a known detriment to Kapton wire, as AO reduces the thickness

TABLE 32.1
Hubble Battery Issues

History of Battery Events in Hubble

- December 1993: A solar panel trim relay failed in the power control unit. The charge cutoff levels then became too high for the heat dissipation capacity of the battery. The battery capacity began to decrease at a rate of 4.8 ampere-hours per battery per year.
- February 1997: Another solar array trim relay failed. The trickle-discharge frequency grew from an occasional event to an average of one orbit in five, with discharge rates of 5 amperes per battery. The battery-capacity test indicated a declining capacity.
- January 1999: A bus fault in the power-control unit made battery-capacity testing too risky. Batteries 5 and 6 accepted slightly increased charge while serving less of the spacecraft's load.
- December 1999: A relay failed open, resulting in a loss of two solar panels.
- January 2000: During the third service mission by astronauts, the bus fault went into remission and did not resume until June 2000.
- April 2000: Closing a relay that bypassed the failed charge control relay mitigated the loss of the solar array current caused by the previous failure. Battery capacity testing was to resume, but it was halted by the return of the bus-fault impedance in July 2000.
- March 2002: During a service mission, the astronauts replaced the faulty power control unit. They also replaced the radiation-degraded solar array with a 30% more powerful solar array for carrying Hubble's electrical loads and charging batteries. The power system was optimized for the new flight configuration and allowed to stabilize.[4]

of insulation materials and degrades their insulating properties. A thin protective coating of silicon oxide is often used on Kapton solar array substrates for protection against AO threats.

Vacuum Operation

The vacuum 100 miles above Earth (considered to be a low orbit) is quite hard, a pressure of less than 10^{-10} torr. In comparison, the pressure at sea level is about 760 torr. Note that the best vacuum chambers on the ground can only get to about 10^{-7} torr, although anything beyond about 10^{-4} torr has little additional effect. The effects of operating in a vacuum are related to both pressure differential and the absence of air itself.

Surprisingly, the effects on most electronics of going from sea level into a space vacuum are not as profound as one might expect. Sealed parts, such as tantalum capacitors and nickel-cadmium batteries, are typically constructed to withstand internal pressures of several atmospheres. Going into a space vacuum from sea level represents a pressure delta of only 1 atmosphere. Nonhermetic components, such as aluminum electrolytic capacitors and lead-acid batteries, are not normally used. Film resistors and ceramic capacitors work fine. A more likely area of concern is material that can trap air bubbles, such as potting materials and closed-cell foam, both common in electronic packaging. Potting materials should go through a vacuum treatment before curing to remove air bubbles.[5]

In a vacuum there is no forced air cooling, there is no convective cooling, there is only conduction and radiation. The only cooling is through thermal radiation to deep space; however, the space vehicle will usually have a cooling system, providing a cooled surface for power circuits to mount to. A note of caution in conductive cooling: the conductive interface does not have the same thermal resistance in a vacuum as in air.[6]

SPACECRAFT EFFECTS

Thermal

High Temperature

High temperature causes adverse effects on spacecraft system parts and components such as cracking, separation, wear-out, corrosion, and performance degradation. These temperature-related defects may affect the electronic parts, the mechanical parts, and the materials in a spacecraft.

Low Temperature

Spacecraft environments rarely expose electronic parts to temperatures below −55°C. A few spacecraft applications can involve extremely low temperatures; these cryogenic applications may be subjected to temperatures as low as −190°C. Cryogenic environments may be experienced by the electronics associated with solar panels or with liquid nitrogen baths used with ultrasensitive infrared detectors. The reliability of many electronic parts improves at low temperatures, but their parametric characteristics can be adversely affected. At these low temperatures many materials strengthen but become brittle.

Temperature Cycle

In many cases, space environment presents an extreme thermal stress on the spacecraft. High temperature extremes result from the exposure to direct sunlight, and low temperature extremes result because there is no atmosphere to contain the heat when not exposed to the sun. This cycling between temperature extremes can aggravate thermal expansion mismatches between materials and assemblies. Large cyclic temperature changes in temperature can cause cracking, separation, and other reliability problems for temperature-sensitive parts. Temperature cycling is also a major cause of fatigue-related solder joint failures.

Charging

One of the most common anomalies caused by radiation hazards is spacecraft or satellite electrical charging. An electrostatic discharge can produce spurious circuit switching; degradation or failure of electronic components, thermal coatings, and solar cells; or false sensor readings. In extreme cases, a satellite's life span can be significantly reduced, necessitating an unplanned launch of a replacement satellite.[7] An electrical charge can be deposited either on the surface or deep within an object. Normally, electrical charging will not (in itself) cause an electrical upset or damage. It will deposit an electrostatic charge that will stay on the vehicle (for perhaps many hours) until some triggering mechanism causes a discharge or arcing. Such mechanisms include a change in particle environment, a change in solar illumination (such as moving from eclipse to sunlit), or onboard vehicle activity.[8] Charged particles accumulate on spacecraft surfaces, creating differential charging and strong local electric fields. If a surface builds up sufficient electric potential, a high-energy discharge (arc) can blow away material and deposit it on optical or other sensitive equipment. The hot, thin plasma of the magnetosphere creates more devastating problems at the geosynchronous altitude. In the region above 1,000 km, the electromagnetic influence of plasma particles extends over a kilometer or more. High-energy (>100 keV) electrons from plasma penetrate external spacecraft surfaces, accumulating inside on grounded conductors, insulators, and cables, causing strong electric fields and ultimately breakdown. Because of their high resistivities, dielectric surfaces can be charged to different potentials than metallic surfaces (which should be at spacecraft ground potential). Considering the effects of internal discharges is important when a system is expected to operate in an environment where penetrating radiation causes charging inside the system. Internal discharges occur when ungrounded metal or dielectric surfaces collect enough charge from the plasma field that the electric field generated exceeds the breakdown strength from the point of the deposited charge to a nearby point. Internal discharges have been suspected as the cause of a number of spacecraft performance anomalies. The conditions for discharging are dependent on the environment, the shielding provided by the spacecraft, the material that is charging, and the geometry of the charged materials. System response to internal charging depends on the location of the discharge and the sensitivity of the circuits. Charges that would go unnoticed on the exterior of a space system can be significant when they occur internally.

Meteoroids/Orbital Debris

The impact of debris and micrometeoroids is potentially catastrophic to spacecraft. The effects may degrade such elements as the solar arrays. Nonorbital debris such as dust on the surface of Mars will also have a degradation effect that must be considered in developing end of life predictions. On each *Mars Exploration Rover*, the solar panels fold up to fit inside the lander for the trip to Mars and deploy to form a total area of 1.3 m^2 (14 square feet) of three-layer photovoltaic cells forming the solar arrays. Each layer is of different materials: gallium indium phosphorus, gallium arsenide, and germanium. The array can produce nearly 900 watt-hours of energy per Martian day, or sol. However, by the end of the 90-sol mission, the energy generating capability is reduced to about 600 watt-hours per sol because of accumulating dust and the change in season. The solar array repeatedly recharges two lithium-ion batteries inside the warm electronics box.[9]

Solar Environment

Modern electronics are becoming increasingly sensitive to ionizing radiation. The parts today are more vulnerable, as much as there are either fewer parts designed for the space environment and/or we are now choosing to use more commercial state-of-the-art parts.

In semiconductor devices, total dose effects can be time-dependent threshold voltage shifts, adversely affect current gain, increase leakage current, and even cause a loss of part functionality. A single-event phenomenon, which is caused by a single high-energy ion passing through the part, can result in either soft or hard errors. Soft errors (also referred to as single-event upsets) occur when a single high-energy ion or high-energy proton causes a change in logic state in a flip-flop, register, or memory cell of a microcircuit. Also, in low-power, high-density parts with small feature sizes, a single heavy ion may cause multiple soft errors in adjacent nodes. Soft errors may not cause permanent damage. A hard error is more permanent.

While the total dose of radiation varies considerably with the amount of shielding between the part and the outside environment, the susceptibility to single-event upset does not change significantly with additional shielding, and latch-up problems are most critical for digital parts, such as memories and microprocessors, which have a large number of memory cells and/or application registers.

Geomagnetic Field

Surface charging occurs predominantly during geomagnetic storms. It is usually more severe in the spacecraft local times of midnight to dawn but can occur at any time. Night-to-day and day-to-night transitions are especially problematic during storms since the photoelectric effect is abruptly present or absent, which can trip discharges. Additionally, thruster firings can change the local plasma environment and trigger discharges.[10]

GENERIC ELECTRICAL SYSTEM DESCRIPTION

The generic electrical system has a framework of metal or composites, in order to get through the launch stresses. The bus (power and communication) is the lifeblood the spacecraft system. At the end of power or at a level where systems will no longer perform their respective functions, subsystems and systems fail. A typical wiring harness to a system box is shown in Figure 32.1. Generically, all spacecraft have a source of power that typically comprises solar cells and batteries for storage. Additional discussions on power systems will detail alternate systems, including fuel cells and, historically, nuclear systems for space probes to outer planets. Power systems are constantly monitored, and data on power and all other onboard systems are sent to relay or ground stations in the form of telemetry signals. This communication is typically wireless: most commonly radio systems, but there may be optical communication that implements a data link for large, complex systems. The telemetry system allows automatic monitoring, alerting, and recording of vital space systems, including the status of the power systems during data downlinks and uplinks via the telemetry data systems. Standards for this activity are set by the Consultative Committee for Space Data Systems.

FIGURE 32.1 Wiring harness to system box. (Courtesy of Johns Hopkins University Applied Physics Laboratory.)

Project Mercury was the first U.S. manned orbital space flight; this flight occurred on February 20, 1962, and was groundbreaking for the evolution of real-time telemetry. The project lasted 55 months, involved more than 2 million people, and cost more than $400 million. In retrospect, Project Mercury's real-time computer, programming, and data-processing aspects seem a minor element of the total project. However, it was the computer and real-time telemetry that paved the way for future manned space projects and gave the computer profession its first glimpse of a real-time computing system that was predicated on the safety requirements of a man in-the-loop.[11] The two computers tracking at NASA Goddard Space Flight Center were duplexed 7,090 transistorized computers supplemented by one 709-vacuum-tube computer at the Bermuda tracking station. In contrast, NASA's Mercury capsules of the early 1960s had no computers at all. For reentry, retrorocket timing and attitude information was radioed to the spacecraft from a tapes-and-teletypes computer center on the ground. The later two-man Gemini capsules had their own rudimentary computer (capable of seven thousand instructions per second), which helped with tricky tasks such as rendezvous operations. Without that, the Apollo Moon missions would not have been possible. The Apollo missions themselves were a computing landmark, too. The Apollo system computer prototypes consumed, at the time, roughly two-thirds of the world's total supply of integrated circuits.[12]

Just as the communication and power buses are a spacecraft's lifeblood, the onboard computer is the brain; it is a component of the command and data handling system. To demonstrate the generic elements of the bus, the James Webb Space Telescope Observatory bus is home to six major subsystems:

- Electrical power subsystem
- Attitude control subsystem
- Communication subsystem
- Command and data handling subsystem
- Propulsion subsystem
- Thermal control subsystem

POWER

The power systems on space vehicles are almost invariably direct current (DC). This is because the power sources—solar cell arrays, fuel cells, and batteries—are DC. The most standard system bus voltages are 28 and 120 V DC (space station).[13,14] Because of the various power busing architectures, the main bus voltage may not be regulated and could vary widely, such as from 18 to 36 V for a nominal 28 V DC bus. Power levels go from less than 100 W for a small satellite to more than 10 kW in a very large satellite. The nominal 28 V was the original specification for most avionics in planes, and this specification migrated to the space industry. Recently, lower-voltage buses have been used, such as the 7 V bus on ST-5 of the New Millennium Program. Typical loads are what one would expect in most commercial electronic system: 5 and 3.3 V logic power, ± 15 V analog power, and so forth. Note that most space platforms, whether they are scientific experiment packages or communications satellites, are going to spend a great deal of time transmitting information, so power to radio transmitters is a given. Other power loads in a satellite system are heaters, battery chargers, reaction wheel motors, solar array drive motors, and electromagnets (for alignment control).

ELECTRONICS IN SPACE HISTORY

In the 1950s and early 1960s there were no space-quality parts, only standard military specification parts (of pretty low reliability) and commercial-quality parts. The famous *Sputnik 1* was useful both scientifically and politically. The satellite's power included a chemical battery—a silver-zinc accumulator—which accounted for 38% of its total mass. It allowed transmissions to continue for 22 days.[15] The first practical photovoltaic cell was developed at Bell Telephone Laboratories (United States) in 1954, and *Vanguard 1* presented the first opportunity to fly it in space (Figure 32.2). The spacecraft carried two radio transmitters, one powered by mercury batteries, the other by the 34 solar cells. Protected by six tiny windows spaced around the sphere, the cells produced just 10 mW of power.[16] The parts list for the Vanguard rocket would have consisted mostly of alloy and grown junction transistors (2N45 and 2N335), alloy diodes, aluminum electrolytic, and paper capacitors, with a few ceramics, carbon composition resistors, and lots of relays.

The U.S. Army had built the Redstone Missile as a derivative from the German V-2 rocket. The Jupiter and Juno rockets were derived from the Redstone with

FIGURE 32.2 Breadboard of miniaturized components of *Vanguard*. (Courtesy of NASA.)

upper stages capable of putting small payloads into Earth orbit. However, President Eisenhower had given the job of launching the first U.S. satellite to the Navy on the Vanguard. The Army (Dr. von Braun and his Redstone staff, later to become NASA–Marshall Space Flight Center) had very explicit orders that anything they launched had better be only ballistic and return to Earth without going into orbit. After three failures of the Vanguard, the President directed the Army to launch the satellite that they had stored in a warehouse. This became *Explorer 1*, which was built by Jet Propulsion Laboratory. It was a very simple satellite with a gyro, Geiger counter, and two transmitters; however, the launch vehicle and the satellite worked perfectly, and the United States finally had a satellite in space. The electrical systems used in these systems and the other early Explorers and launch vehicles were mostly composed of commercial and a few military-specification parts relying on incoming testing and inspection to weed out infant mortality–related concerns. They were more electromechanical and electrical than solid-state parts. The designs were simple with lots of relay logic and diodes; mostly for isolation. Most circuits operated with predetermined sequencing and timing.[17] *Explorer 1*, the first U.S. Earth satellite, was successfully launched on February 1, 1958. The *Explorer 1* instrumentation payload used transistor electronics, consisting of both germanium and silicon devices. This was a very early time frame in the development of transistor technology and represents the first documented use of transistors in the U.S. Earth satellite program. In addition to the historic use of transistors, the Explorer satellite instrumentation package achieved other major scientific breakthroughs: the discovery of the Van Allen radiation belts is one of the most famous.[18]

Russian spacecraft traditionally used pressurized vessels with a gas environment inside. These gas-filled vessels provide convection heat transfer for cooling the electronics and transferring the heat to the outside environment via radiation. The use of pressure vessels greatly simplifies the thermal management but requires much more mass and volume to implement than Western conduction/radiation designs. The reactor control system that was originally designed using the Russian philosophy is based on small-scale integration discrete logic, which made the unit large and heavy. Also, it relied on a pressure vessel with convective heat transfer for cooling its electronics. The unmanned U.S. spacecraft typically do not use pressure vessels and instead use conduction and radiation techniques for thermal control of their electronics. Because of mass and volume limitations and the desire to avoid building a pressure vessel, most electronics have since been replaced with a much smaller, microprocessor-based unit that retains all the functionality of the original design but uses Western thermal techniques.[19]

In the early 1960s, a common approach in military avionic systems was to use a central minicomputer independently connected to a variety of different subsystems. It was realized that the complexity of the mechanism used to transfer data could be greatly reduced if a common approach could be established. A widely selected method was to use multiplexing techniques, which allowed the mass of cabling used previously to be replaced with a single pair of wires to form a bus over which data could be transmitted serially, with each transaction being allotted a given time frame. Large systems were interconnected with bulky harnesses and perhaps hundreds of point-to-point connections. Airborne digital computers began to be introduced to replace analog systems in the late 1960s; they required large interface units to collect, sample, and analog/digital-convert individual input/output signals from around the aircraft. Erv Gangl, employed as a civilian by the U.S. Air Force (USAF), who has long been considered the father of the Mil-Std-1553 data bus, has described his difficulties convincing management of the benefits of digital data transmission.[20] As a student, he became aware of all-digital interfaces used by the mainframe computers of the day. As a new U.S. Air Force employee, he proposed simplifying avionics integration using the data bus concept. He was informed that analog interconnection was a proven technology and "digital was just too risky." When the initial F-15 avionics design came in a couple of hundred pounds overweight, however, digital suddenly became attractive to reduce the wiring. It is ironic that the digital bus was accepted to reduce weight rather than for its flexibility. Because many failures had been traced to wiring, drastically reducing the number of interconnections also improved reliability.[21] However, no standard approach existed until the release of MIL-STD-1553 in 1973, followed by Revision A in 1975 and finally Revision B in 1978. In 1980, USAF issued Notice 1, which limited the options that could be implemented in systems designed for USAF, and in 1986 Notice 2, which is intended to enhance interoperability and interchangeability among the U.S. tri-services, was issued. Since its release, 1553B has come to be regarded as the preferred standard for medium-speed transmission of data within aircraft. It has been adopted by the United Kingdom as DEF-STAN 00-18 (Part 2) and by the North Atlantic Treaty Organization as STANAG.[22]

Considering the mass and volume limitations of spacecraft, space exploration as we know it would have been impossible without the transistor and other miniature components. This is not to say, however, that space exploration per se would have been impossible. Given the political impetus behind the first launches, it seems likely that development would have continued to the point where large space stations, stacked with valves and vacuum tubes, were commonplace. Indeed, Arthur C. Clarke and other space advocates, writing before the Space Age began, assumed that communications satellites would be large manned space stations, mainly because the necessary levels of component reliability and control system autonomy were unimaginable before transistorization. "Men," Clarke said, "would have to be on board to operate and maintain the spacecraft." The field of satellite communications would have been very different if component miniaturization had not occurred.[23]

TYPES OF SPACECRAFT ELECTRIC POWER SOURCES

The classic spacecraft electrical power system consists of a power source (usually photovoltaic but sometimes nuclear), an energy storage device (usually rechargeable batteries but sometimes fuel cells, with a future possibility of using capacitors and flywheels), and the power management and distribution subsystem (sometimes referred to as power conditioning and control subsystem). The choice of the subsystems comprising the electrical power system is largely dictated by the mission requirements.

- Photovoltaic power sources provide a long-term source of power at a known degradation rate. Photovoltaic power sources coupled with rechargeable batteries are the usual choice for spacecraft operating in the inner solar system.

- For short missions, energy storage systems may be sufficient to power the space system. This was the practice in the early days of the space program, and this is how the crewed U.S. missions (e.g., Mercury, Gemini, Apollo, space shuttle) are powered.
- Nuclear power sources provide a long-term source of power and are very attractive for missions operating where there is very little sunlight (e.g., outer solar system, polar regions of Mars, lunar nights, surface of Venus) or in hostile environments (e.g., radiation belts, very close to the sun).

In order to complete the bus system, electrical components in space include but are not limited to capacitors, connectors and contacts, crystals, crystal oscillators, fiber optics, filters, fuses, heaters, magnetics, hybrid microcircuits, monolithic microcircuits, electromagnetic relays, resistors, and semiconductor devises (transistors and diodes, switches, thermistors, and wire and cable). All of these components are affected by the space environment.

COMMERCIAL VERSUS SPACE-GRADE COMPONENTS

In general, space hardware does not use commercial off-the-shelf components and subsystems. The major exceptions are the amateur satellites and many university-grade experimental spacecraft. In fact, an off-the-shelf commercial power supply, if placed in a shielded, air-filled container, would most likely function correctly if placed in orbit, at least for a short time. The key for designing space-quality power circuits is to make the total product tolerant of radiation and vacuum operation, which commercial and industrial applications do not demand, individually or in combination.

WIRING FAILURES

In the aerospace arena, wiring system failures have proven to be very costly in terms of loss of very expensive equipment, imperilment of missions, and loss of lives. Often a wiring system failure is not simply the result of inadequate insulation but is due to a combination of wiring system factors. These include mishandling of wiring insulation, system designs that expose wires to abnormal stresses, and exposure to fluids that degrade the insulation. Some of the NASA missions with wiring system failures are shown in Table 32.2.[24]

CONCLUSION: DEVELOPMENT OF THE TRANSISTOR—THE TECHNOLOGICAL PERFECT STORM

Without transistors, there would not have been a space program. *Sputnik* appeared as a fourth element creating a perfect technology storm, starting with the transistor, solar cell, and maser in the 1950s. Even without *Sputnik* as an incentive, AT&T had the means and the economic motivation to develop communications satellites.

TABLE 32.2
Sample Space Missions with Wiring System Failure

Mission	Cause	Result
• Gemini 8	• Electrical wiring short	• Shortened mission—near loss of crew
• Apollo 204	• Damaged insulation, electrical spark	• 100% O_2 fire, three astronauts lost
• Apollo 13	• Damaged insulation/short circuit/flawed design	• Oxygen tank explosion, mission incomplete
• STS-6	• Abrasion of insulation/arc tracking	• Wire insulation pyrolysis, six conductors melted
• STS-28	• Arc tracking	• Teleprinter cable insulation pyrolysis
• Magellan	• Wrong wiring connection, wiring short	• Wire insulation pyrolysis—ground processing
• Spacelab	• Damaged insulation arc tracking	• Wiring insulation pyrolysis during maintenance
• Delta 178_OES-G	• Mechanical or electrochemical insulation damage	• Loss of vehicle
• STS-48	• Insulation breakdown—fluid exposure	• Solid rocket booster fuel isolation valve failure
• European Space Agency: Olympus	• Insulation breakdown—electrical wiring short	• Loss of solar array

Engineers from Bell Laboratories first developed critical electronic components used in active satellites and ground stations, including the maser, the transistor, solar cells, and the traveling-wave tube. A traditional high-gain receiver in a ground station will add substantial noise to the transmissions relayed from a satellite. The solid-state maser could amplify the weak reflected signal without adding background noise. The traveling-wave tube allowed operators to make microwave communications over a wide range of frequencies.[25] *Sputnik*'s societal effect was immediate and electric—or perhaps more correctly, electronic. Congress quickly funded a billion-dollar educational aid bill, targeted to boosting science and technology education in the name of better ensuring national defense. School programs newly emphasized science and engineering. Science fairs for teens started joining football and basketball as after-school activities. Science clubs made amateur rockets and radio telemetry systems. Science clubs sprang up by the thousands nationwide! Hobby stores started carrying more electronic remote control systems for cars, boats, and model aircraft. Prominent toy companies provided electronics and communications kits to their young science series. In coming years, Arrow Electronics, Schweber Electronics, LaFayette Radio, and Harrison Electronics would supply the electronics revolution. A decade later, those educational programs and scholarship assists paid off well. America's electronics edge led directly to the large scale integration circuits and integrated circuit chips that enabled the Apollo command module and Grumman lunar module to calculate space rendezvous and landing.[26] Advances in transistors from 1 to 50 million are shown in Figure 32.3.[27,28]

FIGURE 32.3 Texas Instruments transistor type flown on *Explorer 1* (on the left) versus early 2000 Pentium technology containing 50 million transistors. (From Historic Transistors. Courtesy of http://semiconductormuseum.com.)

One of those inspired in the late 1950s to pursue science was the nanotech pioneer Dr. Richard Smalley. Smalley, a Nobel Prize-winning chemist, believed the emerging field of nanotechnology might reverse the long slide in the number of new students choosing careers in science, just as the Space Race inspired an earlier generation. "It was Sputnik that got me into science," Smalley says. "Of all the impacts of [nanotechnology's rise], the most important impact—and one that I dearly hope will happen—will to be to get more American girls and boys interested in science."[29]

REFERENCES

1. Winslow, C. 1995. Hubble Space Telescope Solar Array Change-Out, Mission Anomalies and Returned Flight hardware. *Aerospace and Electronic Systems Magazine, IEEE* 10(4): 3–13.
2. Krol, S.J., Jr. and G.M. Rao. 2003. Hubble Performance On-Orbit. *Aerospace and Electronic Systems Magazine, IEEE* 18(2): 13–17.
3. Rao, G., H. Wajsgras, and S.J. Krol, Jr. Hubble Space Telescope On-Orbit NI& Battery Performance (1990 to 2002). *Proceedings of the 37th ECEC*, paper no. 20068.
4. Oman, H. 2003. Space Telescope at Twelve Years of Age. *IEEE AES Systems Magazine*. June 2003: 31–37.
5. Mulkern, J.H., and G.T. Lommasson. 1999. Out of This World Products-Designing for Space; *Applied Power Electronics Conference and Exposition, 1999. APEC '99. Fourteenth Annual* Vol. 1: 129–134.
6. Steinberg, D.S. 1980. *Cooling Techniques for Electronic Equipment*. New York: John Wiley and Sons.
7. Mulkern and Lommasson 1999 (note 5).
8. Space Environment Center. http://sec.noaa.gov/SatOps/.
9. The Mars Exploration Rover Spacecraft. *Spaceflight Now*, June 4, 2003. http://www.spaceflightnow.com/mars/mera/030604mer.html.
10. Space Environment Center (note 8).
11. Gass, S.I. 1999. Project Mercury's Man-in-Space Real-Time Computer System: "You Have a Go, at Least Seven Orbits" (University of Maryland). *IEEE Annals of the History of Computing* 21: 37–48.

12. Computers in Space. 2004. *Atomic: Maximum Power Computing*. http://www.dansdata.com/spacecomp.htm.
13. NASA. 1997. *Electric Power Specifications and Standards, Volume 1: EPS Electrical Performance Specifications*. SSP 30482 Volume 1, Revision C. NASA Space Station Program Office, July 1997.
14. NASA. 1994. *Electric Power Specifications and Standards, Volume 2: Consumer Restraints*. SSP 30482 Volume 2, Revision A. NASA Space Station Program Office, January 1994.
15. Allward, M., ed. 1968. *The Encyclopedia of Space*. Feltham, UK: Paul Hamlyn.
16. Shelton, W. 1968. *Soviet Space Exploration: The First Decade*. New York: Washington Square Press.
17. Hamiter, L. 1991. *The History of Space Qualified Parts in the United States: Leon Hamiter Components Technology Institute Inc*. Presented at the ESA Electronic Components Conference, ESTEC, Noordwijk, The Netherlands, November 12–16, 1990; ESA SP-313, March 1991.
18. http://semiconductormuseum.com/Museum_Index.htm.
19. Reynolds, E., E. Schaefer, G. Polansky, J. Lacy, and A. Bocharov. 1994. Utilizing a Russian Space Nuclear Reactor for a United States Space Mission: Systems Integration Issues. Presented at the 11th Symposium on Space Nuclear Power and Propulsion, January 9–13, Albuquerque, NM.
20. Adams, C. 2002. Interview with Erv Gangl. *Avionics Magazine (now Aviation Today)*. www.aviationtoday.com.
21. Aerospace and Electronic Systems Magazine. 2003. Plug and Play Digital Avionics Traced to Erv Gangl. *IEEE* 18(2): 41–42.
22. Stone, F.W. 1989. 1553 Overview; Buses for Instruments: VXI and Beyond. IEEE Colloquium, April 19, 1989: 8/1–8/7, London.
23. Williamson, M. 2001. The Early Development of Spacecraft Electronics. *Engineering Science and Education Journal* 10: 68–74.
24. Stavnes, M.W., A.N. Hammoud, and R.W. Bercaw. 1994. *NASA Technical Memorandum 106655 Operational Environments for Electrical Power Wiring on NASA Space Systems* (see specifically Table 2-1). Cleveland, OH.
25. Slotten, H.R. 2002. *Satellite Communications, Globalization, and the Cold War Technology*. Baltimore, MD: Johns Hopkins University Press.
26. Doherty, R. 2007. After Sputnik, "Tech Island's" Revolution Spurred Today's Microelectronic World 10/04/2007. http://www.technologypundits.com/index.php?article_id=442.
27. http://upload.wikimedia.org/wikipedia/en/34/pentium/ds-jpg.
28. The Transistor Museum. http://www.semiconductormuseum.com.
29. Service, R.F. 2000. Atom Scale Research Gets Real. *Science* 290(5496): 1524–1531.

33 Contamination Control and Planetary Protection

Michelle M. Donegan

CONTENTS

INTRODUCTION

Contamination is an ever-present influence on a spacecraft throughout its life, from sensor fabrication, system integration, prelaunch preparation, the launch itself, and mission performance, through the end of the mission. Concerns about contamination even continue after the conclusion of the mission. This chapter will examine the contamination problem from all angles, including its sources; methods of prevention; ways to detect, measure, and control contamination; and its effects on the spacecraft's sensors. Beyond the sensors, the issues involved in contamination control include the artificial objects placed into space as artifacts and debris and also the effects of space exploration on the larger environment.

Control of contamination is an essential element of spacecraft design. It can occur during the building or integration of spacecraft, during launch, and in the space environment. Despite efforts at control, particulate and/or molecular contamination may occur before the spacecraft launch. The spacecraft may also experience contamination on-orbit from launch acceleration, outgassing, venting, leaks, or thruster firings; spacecraft on planetary surfaces will also experience dust deposition. One serious effect is surface degradation, including obscuration of optical elements such as lenses and mirrors, or even of solar cells, which can lead to power loss. Other consequences of contamination include disruption of the payload, electrostatic discharge, and disruption of data collection. In some cases, contamination may lead to spacecraft failure. One example of a mission impacted by contamination is the 1970s satellite *Landsat 3*,[1] which experienced contamination of its infrared sensors by gas molecules, leading to a loss of data. The effects of solar radiation on spacecraft surfaces have been observed, such as water vapor bursts from multilayer insulation (MLI) during emergence from dark to sunlight.[2] Particulate contamination can be caused by moving parts, as when the instrument doors on the *Midcourse Space Experiment* were opened, releasing a virtual "snowstorm" of particles that blinded the optical sensor for a time.[3] Finally, the possibility of biologically contaminating the very environment that the spacecraft is visiting must be taken into consideration whenever possible. Anaerobic bacteria have been reported to grow inside hydrazine tanks slated for an exoplanetary mission;[4] this is an issue that must be considered in order to avoid contamination of lunar and planetary environments.

This chapter covers the sources of spacecraft contamination, its effects on missions, and its role in the study of space archaeology and heritage. It will discuss the diagnosis of a contamination problem after the fact, the challenges that would be faced in investigating a spacecraft because of contamination it has suffered, and what we can learn about a spacecraft's history from its current state. It will also cover efforts to prevent environmental contamination, including planetary protection programs.

BACKGROUND

Contamination is a leading cause of spacecraft anomalies. An analysis by Landis [5] found that 14% of on-orbit failures were due to darkening or optical contamination, for a total cost of over $1 billion, over the period 1990–2005.

The *Chandra* x-ray observatory, for example, experienced a gradual degradation in the performance of one of its instruments from its 1999 launch through 2003. The optical filter in front of *Chandra's* Advanced CCD Imaging Spectrometer (ACIS) accumulated a grease-like coating, causing the instrument to lose sensitivity to low-energy x-rays.[6] (Note that CCD here stands for "charge-coupled device" but is used officially as an acronym-in-an-acronym in ACIS.) The nature of the contaminant was determined by taking high-resolution spectra of known bright sources and measuring absorption edges. The coating was found to be composed of carbon, oxygen, and fluorine, implicating a lubricant on the spacecraft.[7] The *Chandra* team concluded that the fluorocarbon lubricant had been broken down by radiation exposure and that the constituents had then accumulated on the optical filter. The team considered a bake-out in order to boil away the film, but this carried significant risk that the contaminant would migrate to other parts of the spacecraft and settle there, as well as the possibility of damage to the instrument.[8,9] As of this writing, the bake-out has been postponed indefinitely, and for now, scientists are just living with the lower sensitivity.

Similarly, the *Cassini* narrow-angle camera acquired a haze, probably from engine exhaust, when it was heated to 30°C for maintenance and then cooled back to its normal operating temperature of –90°C. Images taken while the haze was present were blurry, about 70% of light from a given star being diffused, rather than concentrated at the center (Figure 33.1). To address the problem, the spacecraft went through a series of controlled warmings to temperatures such as –7°C and 4°C,

FIGURE 33.1 Images of single stars, taken with the narrow-angle camera on NASA's *Cassini* spacecraft, showing the effects of haze on the camera's optics and the improvement after the successful removal of the haze. The image on the left predates the haze problem; the second image from the left is after the accumulation of the haze; and the subsequent three images are from successive warming treatments to decontaminate the optics (NASA/JPL).

followed by cooling back to the operating temperature. One of the major types of contamination is the formation of molecular films on spacecraft surfaces. Such surfaces may reflect, transmit, and/or absorb incident electromagnetic radiation; conservation of energy requires that their respective coefficients obey the relation $p + \tau + \alpha = 1$. These coefficients are functions of the material composition, the angle of incidence and polarization of the incident radiation, and the incident wavelength.[10] Surfaces are often designed to be either mostly reflecting (e.g., for thermal control in hot environments) or mostly transmitting (e.g., solar array coverslides.)

Thermal control systems that are contaminated by molecular films will experience a change in surface properties that can adversely effect the system performance. For a given wavelength, the emittance, ε, of a surface is equal to its absorptance, α. However, the ability of a radiating surface to lose heat is generally expressed by the ratio α/ε, where α is taken at the incident (optical) wavelength and ε is taken at the radiating (infrared) energy. Contamination of a thermal control surface may lower α, increase ε, or both, thereby causing the spacecraft to maintain a higher temperature than that for which it was perhaps designed.

For the case of transmitting surfaces, such as solar array coverslides, any molecular film that forms on a surface will have the effect of decreasing the transmittance, τ, reducing the energy incident on the solar cells and ultimately lowering the amount of power available to the spacecraft and its instruments.

Molecular contamination may be the result of the space environment, but contamination due to the outgassing of materials from the spacecraft itself is a major concern. Such material can be deposited on spacecraft surfaces; for this reason, it is critical that care be taken to select materials with low outgassing rates. Additionally, material from thruster plumes may be reattracted to the spacecraft and deposited. It should be noted that attraction and reattraction of molecular contaminants will be worsened if the spacecraft is significantly charged relative to the plasma ground.

Contamination by particulate matter (also known as dust) includes the possibility of obscuration of reflecting or transmitting surfaces. Additionally, the particles may cause scattering of incident radiation, which could wreak havoc with optical systems and solar arrays. Particulate contamination has had a major impact on the operations of the Mars exploration rovers *Spirit* and *Opportunity*. On multiple occasions, the rovers have been forced to pause and conserve power while waiting for a dust storm to pass; this was necessary because contamination of the surfaces of their solar arrays by dust greatly limited the power available to the rovers. In August 2007, dust storms nearly brought about a permanent end to the rovers' missions, since the dust accumulation drastically limited the power available to the solar arrays. Figure 33.2 shows a photo of the deck of *Spirit* as of December 2007, at which point the rover had been on Mars for nearly 4 years. The buildup of dust on the solar panels has rendered the contrast between the panels and the surroundings to be minimal. As of April 2008, the fraction of sunlight that manages to get through the dust coating and reach the solar panels is only about one-third, which will continue to severely restrict the power available to the rovers. Unfortunately, the only possible way for this fraction to improve is from a strong and sustained wind, which is not something that will occur on demand.

Finally, the study of any recovered space artifact should recognize the destructive effect of solar radiation on polymeric materials such as MLI, which can become brittle because of bond cleavages caused by ultraviolet irradiation. Handling must

FIGURE 33.2 Photo of the deck of the *Spirit* Mars rover showing the dust accumulation on the rover's solar panels: the accumulation reached a point where the panels blended with the dusty background (NASA/JPL-Caltech/Cornell).

be performed delicately so as not inadvertently destroy the embrittled surfaces surrounding the spacecraft.

CONTAMINATION CONTROL

A number of measures are possible to limit the effects of contamination on spacecraft systems. Prevention of contamination begins with the spacecraft design, normally documented as the Contamination Control Plan. For example:

- A contamination budget starting with the cleanliness level required for the mission, which is then derated for the various operational aspects of the spacecraft, such as launch
- Selection of materials with low outgassing rates (adhesives, tapes, and coatings are of special concern)
- Designing to prevent spacecraft charging
- Placement of thruster plumes such that they are unlikely to impinge on sensitive surfaces
- Placement of volatile materials such as oils and greases with no line-of-sight or away from sensitive optical elements. Cryogenically cooled long-wave infrared detectors may need to be isolated from sources of water vapor
- Designing to allow proper functioning through the lifetime of the spacecraft, taking predicted end-of-life conditions into account

Of course, care must be taken during manufacturing, integration, and testing to prevent contamination from terrestrial sources. Once contamination control requirements have been established for the project at hand, the level of clean room necessary can be identified and used. Sterilization of some spacecraft components may also be a part of the assembly process in order to comply with planetary protection policies. Specific procedures are also necessary for launch, lest the spacecraft become contaminated at the launch facility or during the launch itself.

A full description of contamination control procedures would be a book unto itself, but a brief summary is attempted here.[11]

- Spacecraft assembly should take place in properly maintained cleanrooms, and optical components should be handled in class 100 or better cleanrooms only. (Class 100 cleanrooms have no more than 100 particles of at least 0.5 μm in diameter per cubic foot.) Optical payloads are generally the most contamination-sensitive components of the spacecraft.
- Cleaning methods include vacuuming, dry wiping, solvent wiping, and ultrasonic cleaning.
- Hardware and materials that can tolerate baking should be baked at high temperature in order to remove molecular contamination. In particular, they should be baked at a temperature higher than the highest temperature expected during the mission in order to force outgassing.
- Optical payloads obtained from vendors should be assembled and tested, and then disassembled and cleaned. Before launch, their exterior surfaces

should also be thoroughly cleaned. After cleaning, handling should include double bagging with electrostatic protective containers and even purging with dry nitrogen during storage.
- Solar arrays should be shielded and mounted vertically in order to minimize contamination during launch.

CONTAMINATION CONSEQUENCES

Failures from contamination can occur at any time, from launch failures to degradation of an instrument many years into the mission. For example, in March 2006, a Russian-built Proton rocket failed upon launch, carrying *Arabsat4A*. The ensuing investigation concluded that the oxidizer supply system unexpectedly shut down the upper stage main engine early. The mostly likely cause was determined to be a foreign particle that obstructed a nozzle of the booster hydraulic pump.[12]

On the other hand, sometimes action can be taken during a mission to prevent additional contamination. For example, when the space shuttle *Atlantis* visited *Mir* in 1995, *Mir* turned its solar arrays edge-on to the shuttle's approach in order to minimize the degree of contamination from *Atlantis*' thruster plumes.[13] This was an important precaution to take, especially since *Mir* suffered from chronic power problems throughout its lifetime, largely because of contamination-aggravated atomic oxygen erosion of its solar arrays.[14]

Among the earliest examples of mission degradation due to contamination are the Orbiting Solar Observatory (OSO) satellites, which were launched in the 1960s and 1970s. One of the instruments aboard *OSO-8*, launched in 1975, had its sensitivity reduced to 0.1% of the original value at the astronomically important Lyman-α wavelength because of outgassing from electronics.[15] With careful calibration, the instrument was still usable for scientific purposes, but a loss of 3 orders of magnitude in sensitivity must certainly have been a disappointment.

PLANETARY PROTECTION

So far, this chapter has dealt with contamination of human artifacts by the space environment. We will now consider the reverse problem—for example, the contamination of a planetary surface by substances aboard a spacecraft. This field, formerly known as planetary quarantine, now goes by the name of planetary protection. Concern about planetary protection is twofold: on the one hand, the integrity of scientific measurements depends on the noncontamination of the environment, and biological contamination could interfere with as-yet-undetected life-forms; on the other hand, any sample return missions need to avoid contaminating the Earth with any extraterrestrial life.

The basis for the planetary protection policies of NASA[16] and the Committee on Space Research (COSPAR)[17] derives from Article IX of the 1967 United Nations "Moon Treaty."[18] These policies have been further developed over the years, taking a risk-reduction approach to the requirements for different mission types. The modern COSPAR policy subdivides missions into five categories:

- Category I missions are those that interact only with bodies that do not contain life and are not expected to provide information about the origin and evolution of life (e.g., asteroids). Therefore, these missions have no planetary protection requirements.
- Category II missions involve target bodies that are interesting from a biological perspective but are highly unlikely to support the spread of any Earth-originating biological contamination (e.g., comets). The requirements here are for documentation only.
- Category III missions encounter bodies of interest from a biological perspective, but only flyby or orbit rather than landing (e.g., Mars orbiters). The requirements here are more stringent than for Category II.
- Category IV missions are similar to Category III but contain a landing component (e.g., the Mars rovers.) In addition to detailed documentation, there may be special processes required prior to launch, such as sterilization of certain components.
- Category V missions are any missions that return to the Earth-Moon system (e.g., proposed Mars sample returns). Requirements include full containment of any materials from the body that was visited and prompt analysis of any samples returned from the mission.

Note that none of these categories requires sterilization of the entire spacecraft. This method, which was attempted for the Lunar Ranger spacecraft, turned out to be very difficult. For example, a gaseous sterilization would reach only exposed surfaces, failing to decontaminate inner surfaces and joints, while radiation sterilization would have required extremely high doses to ensure lethality to microorganisms. Heat sterilization carried the risk of damaging essential spacecraft components at high temperatures and of failing to kill all contaminating microorganisms at low temperatures. In the end, NASA settled on a modified version of dry heat sterilization, exempting certain heat-sensitive components so that they would not be damaged.[19,20] Since then, planetary protection policies have become and remained probabilistic in nature.

While early missions such as Lunar Ranger were concerned with contamination of surfaces by Earth-originating material, the Apollo missions were the first to have to worry about contamination in the other direction. Upon return from the Moon, the lunar samples were kept in isolation, and the astronauts were quarantined for 31 days. However, other contamination issues, such as the probe landing in the ocean, the spacesuits that had been used for moonwalks, and so on, were not addressed, so it is just as well that lunar materials have apparently not posed a threat to the Earth.

The likelihood of discovering materials of biological interest on the Moon was relatively low, but Mars is a whole other story, so the Viking missions represented a large step in planetary protection efforts. The Viking project set the requirements that the total probability of biological contamination of Mars over the entire system and mission was not to exceed 10^{-4}, and the probability of the lander causing such contamination was not to exceed 2×10^{-5}. The lander was therefore dry-heat-sterilized before launch.[21]

Planetary protection requirements have continued to evolve. Current requirements for a Mars lander include, for example, the maximum allowable number of spores that may be present on the spacecraft. The probability of accidental impacts also considered (such as impacts due to a major spacecraft failure.[22] A mission to Europa will present some unique challenges. The spacecraft may have to be sterilizable at the system level, a feat not attempted since the Lunar Ranger program. The spacecraft will need to be extremely radiation-tolerant in order to survive in the Jovian radiation environment, as well as capable of surviving very cold temperatures. On the flip side, that same radiation environment may provide some degree of sterilization. The possibility of ice and liquid water on Europa's surface adds an additional degree of complication.[23]

Looking ahead to humanity's continued exploration of the solar system (and beyond), mission concepts to worlds such as Titan are under consideration as of this writing. Planetary protection requirements for Titan have not yet been established; one might expect them to be similar to those for Europa, although Titan's thick atmosphere complicates matters somewhat. In fact, we have already begun to protect Europa from contamination by human artifacts. The *Galileo* spacecraft, which had spent nearly 8 years orbiting Jupiter, was deliberately steered into a controlled impact with that planet in 2003, largely in order to eliminate the possibility that it would one day contaminate Europa.

CONTAMINATION FORENSICS

What happens if we come across a spacecraft or other artifact and want to find out what happened to it with respect to contamination? Of course, the approach depends on whether the spacecraft is still functioning and communicating, and which components are still functioning. For the example of *Chandra's* ACIS instrument given earlier in this chapter, it was possible to diagnose the issue from afar by looking at the instrument response to known standard sources.

In other cases, the spacecraft may be inspected at closer range, whether by a human being or by a robotic tool. Some information can then be gleaned from some distance. For example, the spectral characteristics of the solar arrays could be measured, providing data about what type of contamination they had suffered. If the spacecraft's environment is consistent over time, the impact on the power system might then be assessed.

In some cases, investigating the contamination of an artifact many years after it stopped functioning could be quite challenging. In particular, planetary landers may have accumulated large amounts of dust/particulate contamination over the course of many years, which would not necessarily have been the cause of any failures.

Finally, if the artifact is being examined at close range, many of the same issues applicable to planetary protection will need to be considered, so that the investigation does not itself contaminate the spacecraft and corrupt any information that might be recovered.

CONCLUSION

Contamination is a factor throughout the entire course of a spacecraft's mission from the earliest design phases. Its sources include the space environment, as well as elements of the spacecraft itself via outgassing, thruster plumes, etc. The consequences of contamination are myriad and especially include loss of power due to solar cell obscuration, degradation of instrument sensitivity, and, in some cases, complete spacecraft failure. Therefore, it is necessary to develop and utilize contamination control procedures for use during the design, manufacture, integration, and test phases of the spacecraft development. When contamination does occur, its effects can often be observed even if the spacecraft is no longer directly accessible. In some cases, techniques such as controlled bake-outs can be used to address the situation.

The reverse problem of planetary protection is also considered. In order to preserve the integrity of the places we explore, particularly with regards to the possibility of non-Earth-based life-forms, it is necessary to prevent contamination of planetary surfaces by human artifacts and life-forms. The degree of precautions necessary for planetary protection ranges from none at all (for missions to nonhospitable locations such as asteroids) to very serious (such as for a Mars sample return mission).

REFERENCES

1. Bedingfield, K.L., and R.D. Leach. 1996. *Spacecraft System Failures and Anomalies Attributed to the Natural Space Environment.* NASA Reference Publication 1390. Huntsville, AL: NASA/George C. Marshall Space Flight Center.
2. Boies, M.T., B.D. Green, G.E. Galica, O.M. Uy, R.C. Benson, D.M. Silver, B.E. Wood, J.C. Lesho, D.F. Hall, and J.S. Dyer. 2004. The 8-Year Report on MSX Thermal Blanket Outgassing: An Inexhaustible Reservoir. In *Optical Systems Degradation, Contamination, and Stray Light: Effects, Measurements, and Control*, ed. P.T.C. Chen, J.C. Fleming, and M.G. Dittman, 20–31. Proceedings of the SPIE 5526. Bellingham, WA: SPIE—The International Society for Optical Engineering.
3. Uy, O.M., R.C. Benson, R.E. Erlandson, M.T. Boies, R.P. Cain, J.F. Lesho, et al. 1997. Molecular Densities Measured During Early Operations Around Midcourse Space Experiment Satellite and a Comparison with Pre-Launch Models. In the Proceedings of the 7th International Symposium on Materials in Space Environment, Toulouse, France, June 16–20, 1997: 321–329. Noordwijk, Netherlands: ESA.
4. Hogue, P. 2006. Potential Biofouling of Spacecraft Propellant Systems due to Contaminated Deionized Water. In *Optical Systems Degradation, Contamination, and Stray Light: Effects, Measurements, and Control II*, ed. O.M. Uy, S.A. Straka, J.C. Fleming, and M.G. Dittman. Proceedings of the SPIE 6291: 62910N. Bellingham, WA: SPIE—The International Society for Optical Engineering.
5. Landis, G.A. 2006. Causes of Power-Related Satellite Failures. *Conference Record of the 2006 IEEE 4th World Conference on Photovoltaic Energy Conversion*, ed. R. Tischler and S.G. Bailey. Vol. 2: 1943–1945. Piscataway, NJ: IEEE.
6. Plucinsky, P.P., N.S. Schulz, H.L. Marshall, C.E. Grant, G. Chartas, D. Sanwal, et al. 2003. Flight Spectral Response of the ACIS Instrument. In *X-Ray and Gamma-Ray Telescopes and Instruments for Astronomy*, ed. J.E. Truemper and H.D. Tananbaum, 89–100. Proceedings of the SPIE 4851. Bellingham, WA: SPIE—The International Society for Optical Engineering.

7. Marshall, H.L., A. Tennant, C.E. Grant, A.P. Hitchcock, S.L. O'Dell, and P.P. Plucinsky. 2004. Composition of the Chandra ACIS contaminant. *X-Ray and Gamma-Ray Instrumentation for Astronomy XIII*, ed. K.A. Flanagan and O.H.W. Siegmund, 497–508. Proceedings of the SPIE 5165. Bellingham, WA: SPIE—The International Society for Optical Engineering.
8. O'Dell, S.L., D.A. Swartz, P.P. Plucinsky, M.A. Freeman, M.L. Markevitch, A.A. Vikhlinin, et al. 2005. Modeling Contamination Migration on the Chandra X-Ray Observatory. *UV, X-Ray, and Gamma-Ray Space Instrumentation for Astronomy XIV*, ed. O.H.W. Siegmund, 313–324. Proceedings of the SPIE 5898. Bellingham, WA: SPIE—The International Society for Optical Engineering.
9. Plucinsky, P.P., S.L. O'Dell, N.W. Tice, D.A. Swartz, M.W. Bautz, J.M. DePasquale, et al. 2004. An Evaluation of a Bake-Out of the ACIS Instrument on the Chandra X-Ray Observatory. *UV and Gamma-Ray Space Telescope Systems*, ed. G. Hasinger and M.J.L. Turner, 251–263. Proceedings of the SPIE 5488. Bellingham, WA: SPIE—The International Society for Optical Engineering.
10. Tribble, Alan C. 2000. *Fundamentals of Contamination Control*. Bellingham, WA: SPIE—The International Society for Optical Engineering.
11. See Tribble, note 10.
12. Ray, J. Proton Launch Report: Mission Status Center. *Spaceflight Now*. http://www.spaceflightnow.com/proton/hotbird8/status.html.
13. McDonald, S. 1998. *Mir Mission Chronicle*. NASA TP-1998-208920. Houston, TX: National Aeronautics and Space Administration, Lyndon B. Johnson Space Center.
14. See Bedingfield, note 1.
15. Bonnet, R.M., P. Lemaire, J.C. Vial, G. Artzner, P. Gouttebroze, A. Jouchoux, A. Vidal-Madjar, J.W. Leibacher, and A. Skumanich. 1978. The LPSP Instrument on OSO 8. II—In-Flight Performance and Preliminary Results. *Astrophysical Journal* 221: 1032–1053.
16. Barengoltz, J. 2005. A Review of the Approach of NASA Projects to Planetary Protection Compliance. Paper presented at the 2005 IEEE Aerospace Conference, Big Sky, MT.
17. Committee on Space Research (COSPAR). 2005. COSPAR Planetary Protection Policy. http://cosparhq.cnes.fr/Scistr/Pppolicy.htm.
18. United Nations. 1967. Treaty on Principles Governing the Activities of States in the Exploration and Use of Outer Space, Including the Moon and Other Celestial Bodies. In *United Nations Treaties and Principles on Outer Space*. New York: United Nations.
19. Hall, L.B. 1973. *The Planetary Quarantine Program*. NASA, SP-4902. Washington, DC: NASA.
20. Hall, R.C. 1977. *Lunar Impact: A History of Project Ranger*. NASA, SP-4210. Washington, DC: NASA.
21. See Barengoltz, note 16.
22. Gershman, R., M. Adams, R. Mattingly, N. Rohatgi, J. Corliss, R. Dillman, J. Fragola, and J. Minarick. 2004. Planetary Protection for Mars Sample Return. *Advances in Space Research* 34(11): 2328–2337.
23. Belz, A.P., and J.A. Cutts. 2006. Planning for Planetary Protection and Contamination Control: Challenges beyond Mars. Paper presented at the 2006 IEEE Aerospace Conference, Big Sky, MT.

34 Studies in Aging

Paul J. Biermann

CONTENTS

Slowing spacecraft aging.

INTRODUCTION

The space environment provides numerous challenges for materials. Some of the primary concerns involve ultraviolet (UV), radiation, micrometeorites, atomic oxygen, outgassing contamination, temperature cycling, and effectiveness of protective coatings. Each of these areas could, and has, generated chapters of data and analysis. As our time spent in space increased, the need for long-term data grew. The result of this was a series of tests designed for exposure to the conditions in space and then retrieved for study here on Earth. Each of them was planned for a specific duration of exposure, but events that altered the retrieval mission's schedule resulted in extended missions. This chapter will focus on long-term tests conducted over various lengths of exposure, some of them for much longer than originally intended. We will list the various analytical instruments used to examine the samples once they are recovered. Finally, after looking at the individual programs, we will provide references to test results from these programs categorized by the area of study, such as UV, radiation, micrometeorites, and so forth.

The major difference between spacecraft manufacturers and space archaeologists is time scale. Both have a common interest in determining the long-term effects of the space environment on an article placed in that environment, but the duration of the longest-working spacecraft is currently less than 35 years, while an archaeologist may typically look at items hundreds to thousands of years old. The following studies point toward the types of degradation that can be expected, but the degree of degradation over time may be difficult to accurately extrapolate.

STUDIES

Probably the most well-known study is the Long Duration Exposure Facility (LDEF),[1] as shown Figure 34.1. NASA designed the LDEF for an 18-month mission to expose thousands of samples of candidate materials that might be used on a space station or other orbital spacecraft. The LDEF was launched in April 1984 and was to have been returned to Earth in 1985. Changes in mission schedules postponed retrieval until January 1990, after 69 months in orbit. Analyses of the samples recovered from LDEF have provided spacecraft designers and managers with the most extensive database on space materials phenomena. Many LDEF samples were greatly changed by extended space exposure. Among even the most radically altered samples, NASA and its science teams are finding a wealth of surprising conclusions and tantalizing clues about the effects of space on materials.[2] The overall goals of the LDEF investigations, established by the Materials Special Investigation Group prior to LDEF retrieval, are to provide useful engineering data to people designing and building spacecraft, and secondarily, to obtain data of potential interest to materials researchers. The specific objectives are to support predictions of materials lifetimes under the various low Earth orbit (LEO) environments to determine how long the material will physically survive; to estimate the engineering performance lifetimes of these same materials under specific LEO exposures; to identify materials and processes by which given materials degrade; and to provide insights into development of new, more inherently LEO environmentally resistant materials. To achieve

FIGURE 34.1 Long Duration Exposure Facility. (Courtesy of NASA.)

the established objectives, two criteria were established to select which materials had the highest priorities for analyses. The first criterion was to examine materials that are still being used on new spacecraft. The second priority was to examine materials with multiple exposure locations on LDEF, because this provided the opportunity to develop predictions on how a material will behave as the exposure environment varies. The goals as defined led to the identification of silverized Teflon (Ag/FEP), chromic acid anodized aluminum, and certain thermal control paints used on LDEF as the material types of highest priority for examination. Ag/FEP was chosen because of an excellent absorptivity/emissivity ratio or heat flux (extremely low), and extensive flight history. Also influencing the decision was the fact that the Hubble Space Telescope (HST) uses Ag/FEP as its primary passive thermal control system (and was launched shortly after LDEF retrieval) and that a number of other spacecraft are using or considering this material for use. The various forms of Ag/FEP were used on virtually every side of the LDEF except the Earth end. The potential uses of this material, and the location distribution and therefore the exposure conditions, have provided the rationale and opportunity for a comprehensive study of this material.[3]

Because of the vast amounts of information generated by this program, NASA developed an archive system. The LDEF Archive System is designed to provide spacecraft designers and space environment researchers single-point access to all available resources from LDEF. These include data, micrographs, photographs, technical reports, papers, hardware, and test specimens, as well as technical expertise.[4]

The Hubble Space Telescope repair mission of December 1993 was first and foremost a mission to improve the performance of the observatory,[5] as shown in Figures 34.2 and 34.3. But for a specialized segment of the aerospace industry, the primary interest is in the return to Earth of numerous pieces of the HST hardware, pieces that have been replaced, repaired, improved, or superseded. The returned hardware is of interest because of the information it potentially carries about the effects of exposure to the space environment for 3 ½ years. Like the LDEF retrieval mission 4 years ago, the HST repair mission is of interest to many engineering disciplines, including all of the disciplines represented by the LDEF Special Investigation Groups. There is particular interest in the evaluation of specific materials and systems in the returned components. Some coated surfaces have been processed with materials that are newer and still in use by, or under consideration for, other spacecraft in a variety of stages of development. Several of the systems are being returned because a specific failure or

FIGURE 34.2 Hubble Space Telescope servicing mission. (Courtesy of NASA.)

FIGURE 34.3 Hubble Space Telescope servicing mission. (Courtesy of NASA.)

anomaly has been observed and thus there is, at the outset, a specific investigative trail that needs to be followed.[6]

The European Space Agency also placed an experiment called the European Retrievable Carrier (EURECA) in LEO for 11 months, commencing in August 1992.[7,8,9] EURECA was the first spacecraft to be retrieved and returned to Earth after the recovery of LDEF. The primary mission objective of EURECA was the investigation of materials and fluids in a very low microgravity environment. In addition, other experiments were conducted in space science, technology, and space environment disciplines.

Hundreds of material samples were passively exposed to the space environment for nearly 4 years as part of the Materials on International Space Station Experiment (MISSE) units 1–4.[10] The experiment was planned for 1 year of exposure, but its return was delayed by the *Columbia* accident and subsequent grounding of the space shuttle fleet. The experiment was attached externally to the Quest Airlock. Atomic oxygen fluence and ultraviolet radiation dose varied across the experiment because of shadowing and space station orientation. More than a hundred meteoroid/space debris impacts were found. Many polymer film samples were completely eroded by atomic oxygen.[11]

MATERIALS ANALYSIS TECHNIQUES FOR SPACE ARCHAEOLOGY

There are many modern instruments that can be used to analyze artifacts from space exploration. There are several spectroscopic techniques that can be employed in a nondestructive fashion. Fourier transform infrared (FTIR) microscopy and Raman microscopy are widely used to examine materials on the basis of their molecular or crystallographic structure. Either technique can be used to identify the class of material under investigation and in some cases can be used to positively identify a material if appropriate standards are available. One of the benefits of FTIR or Raman microscopy is the ability to focus on specific sites on a sample and determine whether there is a variation of the material across the surface. These microscopic techniques are limited to surface analysis only.

Scanning electron microscopy with energy dispersive spectrometry is another widely used tool. In this case x-rays are used instead of photons to image the sample. Using x-rays instead of photons enables the analyst to achieve much higher levels of magnification than a light microscope. When this technique is coupled with energy dispersive spectrometry, the analyst is able to determine elemental composition of small regions of a sample. One of the limitations of these techniques is that the surface of the sample is bombarded with x-rays. Nonconducting samples can generate a static charge, which can affect the quality of the images obtained. To counteract this, samples are typically overcoated with a thin layer of conductive material. After being coated, the sample has been altered from its original form.

Also of interest are thermoanalytical techniques. Differential scanning calorimetry can provide information about a material by examining the amount of heat required to alter the temperature of a sample compared with a standard material. This information can provide insight into the phase transitions a material undergoes. Another thermoanalytical technique is thermal gravimetric analysis, in which the mass of a sample is monitored during experiments where the temperature can be held constant (evaporation rate) or changed at a known rate. The resulting data can provide insight into the performance of the material at different temperatures. One drawback of these techniques is that the sample is typically altered from its initial state or consumed. However, a relatively small portion of material is necessary to run these tests.

There are other spectroscopic techniques available, which are more destructive. The simplest is a flame test, where a small portion of the material of interest is burned in a flame. The color produced in the flame will correspond to the metal content. For example, when salt is burned in a flame, the orange color of the flame shows the presence of sodium. More elaborate versions of the flame test, atomic absorption spectroscopy (AAS) and inductively coupled plasma-optical emission spectroscopy (ICP-OES), use these principles on a grander scale. AAS and ICP-OES are two distinct tests that fall in the same broad family of tests that originated with the flame test. Basically, in AAS a sample is introduced into a flame and a lamp made of the element of interest is used to determine the concentration of the element in the material. The intensity of the absorbance will be proportional to the element's concentration. This technique is widely used because of the specificity of the analysis. Two drawbacks are that the sample is consumed and that only a single element may be screened at a time. In ICP-OES, the material is introduced into plasma, instead of a flame, which because of a much greater amount of energy is able to excite the elements to emit several discrete wavelengths of light. The amount of light emitted can be related to the element's concentration in the material. However, very few materials are composed of a single element, which can lead to the overlap of the wavelengths of light emitted. Choosing appropriate wavelengths for analysis and employing proper analytical techniques can offset this problem. While the overlap of spectral lines can be a problem, the presence of multiple wavelengths of light allows for the analysis of several elements simultaneously, decreasing the amount of time and material necessary for an analysis. These techniques provide information about the elements present in a material in increasing sensitivity. A flame test is as sensitive as the observer's eye and can only confirm the presence of an element at significant concentrations. AAS has been demonstrated to be effectively sensitive to detect elements at concentrations in the sub-milligram per liter or sub-part per million levels. ICP-OES has been demonstrated to be sensitive to concentration levels of several magnitudes lower, reaching the nanogram per liter or part per trillion range.

Mass spectrometry can also be employed to analyze a material by examining the mass spectrum of the analyte. There are several mass spectrometric techniques that can be employed. In secondary ion mass spectrometry (SIMS), a sample's surface is bombarded with an ion, typically argon or oxygen, which then dislodges other secondary ions from the surface. These secondary ions are then analyzed by the mass spectrometer. SIMS is a surface technique capable of looking directly at material on the surface of a sample. By dwelling on a portion of the sample for a period of time, the sample can be depth profiled as the primary ions sputter the ions on the surface, in effect drilling into the surface. One of the drawbacks of SIMS or any other mass spectrometric technique is that if the material is not suitably pure, the resulting spectra can be very difficult to interpret. One way to overcome this limitation is to use a separative technique in combination with mass spectrometry.

Gas chromatography mass spectrometry (GC/MS) is such a technique where a sample is introduced into a gas chromatograph, and its volatile chemical components are separated and then analyzed by the mass spectrometer, providing mass spectral data for each compound separated by the gas chromatograph. The amount of time the chemical is retained on the column is also characteristic of that species. The combination of these techniques can provide a significant amount of information about the material of interest. Pyrolysis gas chromatography mass spectrometry allows for the analysis materials that are typically not volatile enough to be analyzed by GC/MS, by pyrolizing or burning the material in controlled conditions and analyzing the volatile components (pyrolyzates) by GC/MS. The resulting pyrolysis chromatograms (programs) will be unique for a material based on the pyrolysis parameters and the material's composition.

Liquid chromatography mass spectrometry (LC/MS) is another separative technique that can be employed when it is detrimental for the analysis to use gas chromatography. LC/MS will provide data similar to that provided by GC/MS; however, the analysis is performed in the liquid phase instead of the gaseous phase.

Inductively coupled plasma mass spectrometry (ICP-MS) uses plasma to separate a sample into its elemental components. In ICP-MS, the plasma is passed into an expansion chamber and then into the vacuum of the mass spectrometer, where the elements are analyzed on the basis of their atomic mass. One of the benefits of ICP-MS is the ability to do isotopic analysis. Typically, an argon plasma is used that can make it difficult to analyze elements that posses a similar atomic mass, such as iron and calcium. Also, because it is nearly impossible to remove the constituents in air from the plasma, components such as silica are equally difficult to analyze.

These are but a few of the instruments and techniques currently available to modern researchers, and they should by no means be considered the only methods to be used to analyze space artifacts.

SPECIFIC STUDY RESULTS

Time frame and location for each test group of materials are as follows:

LDEF	5.75 years	LEO (275 miles)
EURECA	11 months	LEO (316 miles)
Hubble Space Telescope	3.5 years	LEO (360 miles)
MISSE 1 and 2	3 years	LEO (212 miles)
MISSE 3 and 4	1 year	LEO (212 miles)

- UV effects:
  - The effect on solar cells due to UV exposure[12]
  - Smart materials—polymer piezoelectric behavior under UV[13]
  - Effects on spacecraft paint of combined atomic oxygen (AO)/UV[14]
  - Solar cell data from LDEF[15]
- Radiation effects:
  - Ionizing radiation effects on star camera charge-coupled devices[16]
  - Teflon degradation on orbit[17,18]
  - X-ray solar flare effects on Teflon[19]
  - Radiation effects on polyimide film[20]
- Micrometeorite effects:
  - Chemical and isotopic analysis of micrometeoroids on LDEF[21]
  - Solar cell data from LDEF[15]
- Atomic oxygen effects:
  - The effect on solar cells due to AO exposure[12]
  - Smart materials—polymer piezoelectric behavior under AO[13]
  - Survivability of polyhedral oligomeric silsesquioxane (POSS) polyimides[22]
  - Effects on super-smooth (0.1 nm roughness) optical surfaces[23]
  - Teflon degradation on orbit[24,25]
  - AO and silicone contamination interaction[26]
  - AO-induced outgassing and surface changes[27]
  - Oxygen interaction with materials in LEO[28]
- Outgassing contamination:
  - Contamination effects on optical transparencies and first surface mirrors[29]
  - Contamination model versus actual LDEF data[30]

- Contamination induced effects on optics in the HST interior space[31,32]
- Solar cell data from LDEF[15]

- Thermal cycling in vacuum effects:
  - Smart materials—polymer piezoelectric behavior at elevated temperature[13]
  - LDEF experience applied to space station thermal coatings[33]
  - Solar cell data from LDEF[15]
- Protective coating performance:
  - Protective silicone/siloxane coatings[34]
  - Effects on coated tether materials[35]
  - Properties of POSS polyimides[22]
  - Coating degradation mechanisms on the HST[36]

CONCLUSION

At over 30 years of flight, *Voyager 1* and *Voyager 2* currently hold the record for active operational performance in space, and they can be expected to function for 5 or 6 more years. Their trajectories make it unlikely that we will recover them, but many other space objects will return to our neighborhood on a cyclical timeline. Eros comes close enough to retrieve the Near Earth Asteroid Rendezvous spacecraft every 7.5 years. The satellite orbited the asteroid Eros for a year before it was allowed to "land" (controlled low velocity impact) on the surface of Eros. Any number of Earth-orbiting and geostationary platforms could be recovered in the future, each providing a different snapshot of aging effects over time. Since the launch of *Sputnik* in 1957, more than 31,102 human-made objects have been catalogued, many of which have since reentered the atmosphere. Currently, the Joint Space Operations Center tracks more than 17,000 human-made objects orbiting Earth that are 10 centimeters or larger. About 10% of those being tracked are functioning payloads or satellites, 15% are rocket bodies, and about 75% are fragmentation and inactive satellites. Between 2,000 and 2,500 objects, including geostationary communications satellites, are in deep space orbits more than 22,500 miles from Earth.[37] As we return to the Moon we will have the entire catalog of lunar landers, both unmanned and manned to visit and explore. And if you have really long vision, *Deep Space 1* is coming back in 3012—imagine what we could learn.

REFERENCES

1. http://setas-www.larc.nasa.gov/LDEF/index.html
2. Whitaker, A.F., and D. Dooling. *LDEF Materials Results for Spacecraft Applications: Executive Summary.* Conference sponsored by the University of Alabama, the Alabama Space Grant Consortium, and the Society of Advanced Materials and Process Engineers (SAMPE), October 27–28, Huntsville, AL. NASA-CP-3261; M-745; NAS 1.55: 3261, 1992.
3. Pippin, H.G., and H.W. Dursch. Recent Results from Long Duration Exposure Facility Materials Testing. NASA. *Langley Research Center, LDEF: 69 Months in Space.* Third Post-Retrieval Symposium, Part 2, November 8–12, 1993, Williamsburg, VA, pp. 555–565 (see N95-23896 07-99), 1995.

4. Wilson, B. NASA. *Langley Research Center, LDEF: 69 Months in Space.* Third Post-Retrieval Symposium, Long Duration Exposure Facility (LDEF) Archive System, Part 3, November 8–12, 1993, Williamsburg, VA, pp. 1249–1262 (see N95-27629 09-99), 1993.
5. http://www.stsci.edu/hst/.
6. Edelman, J., and J.B. Mason. Space Environmental Effects Observed on the Hubble Space Telescope. NASA. *Langley Research Center, LDEF: 69 Months in Space.* Third Post-Retrieval Symposium, Part 3, November 8–12, 1993, Williamsburg, VA, pp. 1231–1236 (see N95-27629 09-99), 1995.
7. http://heasarc.nasa.gov/docs/heasarc/missions/eureca.html.
8. Vaneesbeek, M., M. Froggatt, and G. Gourmelon. Material Inspection of EURECA First Findings and Recommendations. NASA. *Langley Research Center, LDEF: 69 Months in Space.* Third Post-Retrieval Symposium, Part 1, November 8–12, 1993, Williamsburg, VA, pp. 71–86 (see N95-23796 07-99), 1993.
9. Dover, A., Aceti, R., and G. Drolshagen. EURECA 11 Months in Orbit: Initial Post Flight Investigation Results. NASA. *Langley Research Center, LDEF: 69 Months in Space.* Third Post-Retrieval Symposium, Part 1, November 8–12, 1993, Williamsburg, VA, pp. 23–35 (see N95-23796 07-99), 1995.
10. http://missel.larc.nasa.gov/.
11. Finckenor, M. The Materials on International Space Station Experiment (MISSE): First Results from MSFC Investigations. 44th AIAA Aerospace Sciences Meeting and Exhibit, January 9–12, 2006, Reno, NV; pp. 1–9, 2006. AIAA-2006-472. AIAA.
12. Okumura, T., S. Hosoda, J. Kim, M. Iwata, and M. Cho. Optical Characteristic Change of Transparent Film due to Combined Exposure to Atomic Oxygen and Ultraviolet Ray. *Journal of the Japan Society for Aeronautical and Space Sciences* (Japan) 54(628): 232–242, 2006.
13. Dargaville, T., M.C. Celina, P.M. Chaplya, and R.A. Assink. Smart Materials for Gossamer Spacecraft—Performance Limitations. 46th AIAA/ASME/ASCE/AHS/ASC Structures, Structural Dynamics, and Materials Conference, April 18–21, 2005, Austin, TX; pp. 1–13, 2005. AIAA 2005-2364. AIAA.
14. Tong, J., W. Jihui, L. Xiangpeng, and L. Jinhong. Synergistic Effects of Simulated AO/UV on Some Thermal Control Paints of Spacecrafts. *Advances in the Astronautical Sciences* 117: 945–950, 2004.
15. Hill, D., and F. Rose. *Summary of Solar Cell Data from the Long Duration Exposure Facility (LDEF) (Final Report, 21 Jul. 1993–19 Aug. 1994).* NASA-CR-196539; NAS 1.26:196539, 1994. Huntsville, AL: NASA Space Power Institute.
16. Marshall, P., C. Marshall, E. Polidan, A.Wacyznski, and S. Johnson. Single Particle Damage Events In Candidate Star Camera Sensors. *Advances in the Astronautical Sciences* 121: 371–380, 2005.
17. Townsend, J., P.A. Hansen, and J.A. Dever. On-Orbit Teflon FEP Degradation. Space Simulation Conference, 20th—The Changing Testing Paradigm, October 27–29, 1998, Annapolis, MD; pp. 219–232, 1999.
18. Townsend, J., P.A. Hansen, M.W. McClendon, J.A. Dever, and J.J. Triolo. Evaluation and Selection of Replacement Thermal Control Materials for the Hubble Space Telescope. Materials and Process Affordability—Keys to the Future; Proceedings of the 43rd International SAMPE Symposium and Exhibition, May 31 to June 4, 1998, Anaheim, CA; pp. 968–982, 1998.
19. Milintchouck, A., M. Van Eesbeek, T. Harper, and F. Levadou. Influence of X-Ray Solar Flare Radiation on Degradation of Teflon in Space. *Journal of Spacecraft and Rockets* 34(4): 542–548, 1997.

20. Forsythe, J.S., D.J.T. Hill, F.A. Rasoul, J.S. Forsythe, J.H. O'Donnell, P.J. Pomery, G.A. George, P.R. Young, and J.W. Connell. The Effect of Simulated Low Earth Orbit Radiation on Polyimides (UV Degradation Study). NASA. *Langley Research Center, LDEF: 69 Months in Space.* Third Post-Retrieval Symposium, Part 2, pp. 645–656 (see N95-23896 07-99), 1995.
21. Zinner, E. *Analysis of LDEF Experiment AO187-2 Chemical and Isotopic Measurements of Micrometeoroids by Secondary Ion Mass Spectrometry (Final Report).* St.Louis, MO: Washington University Department of Physics, 1995.
22. Tomczak, S., S.J. Tomczak, D. Marchant, S. Svejda, T.K. Minton, A.L. Brunsvold, I. Gouzman, E. Grossman, G.C. Schatz, D. Troya, L. Sun, and R.I. Gonzalez. Properties and Improved Space Survivability of POSS (Polyhedral Oligomeric Silsesquioxane) Polyimides. *Materials for Space Applications.* Materials Research Society Symposium Proceedings, vol. 851, pp. 395–406, 2005.
23. Schmitt, D.-R., H. Toebben, G.A. Ringel, P. Weissbrodt, M. Schrenk, L. Raupach, and E. Hacker. Degradation of Supersmooth Surfaces for UV/EUV/X-Ray Applications in Space. *Proceedings of SPIE, the International Society for Optical Engineering, UV/EUV and Visible Space Instrumentation for Astronomy and Solar Physics,* O.H. Siegmund, S. Fineschi, M.A. Gummin, eds, vol. 4498, pp. 111–120, 2001.
24. Townsend, J., P.A. Hansen, and J.A. Dever. On-Orbit Teflon FEP Degradation. 20th Space Simulation Conference —The Changing Testing Paradigm, October 27–29, 1998, Annapolis, MD; pp. 219–232, 1999.
25. Townsend 1998 (note 18).
26. Banks, B., K.K. de Groh, S.K. Rutledge, and C.A. Haytas. Consequences of Atomic Oxygen Interaction with Silicone and Silicone Contamination on Surfaces in Low Earth Orbit. Proceedings of the Conference on Rough Surface Scattering and Contamination, July 21–23, 1999, Denver, CO; pp. 62–71, 1999.
27. Linton, R., M.M. Finckenor, and R.R. Kamenetzky. Further Investigations of Experiment A0034 Atomic Oxygen Stimulated Outgassing. NASA. *Langley Research Center, LDEF: 69 Months in Space.* Third Post-Retrieval Symposium, Part 2, pp. 727–736 (see N95-23896 07-99), 1995.
28. Koontz, S.L., L.J. Leger, S.L. Rickman, J.B. Cross, C.L. Hakes, and D.T. Bui. Evaluation of Oxygen Interactions with Materials 3: Mission and Induced Environments. NASA. *Langley Research Center, LDEF: 69 Months in Space.* Third Post-Retrieval Symposium, Part 3, pp. 903–916 (see N95-27629 09-99), 1995.
29. Pippin, H.G., and M. Finckenor. Measurements of Optically Transparent and Mirrored Specimens from the POSA, LDEF A0034, and EOIM-III Space Flight Experiments. *Proceedings of SPIE, the International Society for Optical Engineering,* SPIE-4774, 2002.
30. Rantanen, R., T. Gordon, M.M. Finckenor, and G. Pippin. Comparison of Contamination Model Predictions to LDEF Surface Measurements. Proceedings of the Meeting on Optical Systems Contamination and Degradation, July 20–23, 1998, San Diego, CA; pp. 260–271, 1998.
31. Tveekrem, J., L. June, D.B. Leviton, C.M. Fleetwood, and L.D. Feinberg. Contamination-Induced Degradation of Optics Exposed to the Hubble Space Telescope Interior. Proceedings of the Conference on Optical System Contamination V and Stray Light and System Optimization, August 5–7, 1996, Denver, CO; pp. 246–257, 1996.
32. MacKenty, J., S.M. Baggett, J. Biretta, M. Hinds, C.E. Ritchie, L.D. Feinberg, and J.T. Trauger. Effects of the Space Environment on the HST Wide Field Planetary Camera-I. Proceedings of the Meeting on Space Telescopes and Instruments, April 18–19, 1995, Orlando, FL; pp. 160–166, 1995.

33. Babel, H. LDEF's Contribution to the Selection of Thermal Control Coatings for the Space Station. NASA. *Langley Research Center, LDEF: 69 Months in Space.* Third Post-Retrieval Symposium, Part 3, pp. 1273–1283 (see N95-27629 09-99), 1995.
34. Dworak, D., B. Banks, C. Karniotis, and M. Soucek. Evaluation of Protective Silicone/Siloxane Coatings in Simulated Low-Earth-Orbit Environment. *Journal of Spacecraft and Rockets* 43(2): 393–401, 2006.
35. Gittemeier, K., C. Hawk, M. Finckenor, and E. Watts. Space Environmental Effects on Coated Tether Materials. 41st AIAA/ASME/SAE/ASEE Joint Propulsion Conference & Exhibit, July 10–13, 2005, Tucson, AZ; pp. 1–10, 2005.
36. Triolo, J., and Y. Yoshikawa. Investigation of Coating Degradation Mechanisms on the HST Spacecraft. 7th International Symposium on Materials in Space Environment, June 16–20, 1997, Toulouse, France; pp. 357–365, 1997.
37. http://www.stratcom.mil/fact_sheets/STRATCOM%20Space%20and%20COntrol%20Fact%20Sheet%20--%2025%20Feb%2008.doc.

Section VII

Preservation of Space Objects and Case Studies

35 Space Technology: *Vanguard 1*, *Explorer 7*, and *GRAB*—Materials and Museum Concerns

Hanna Szczepanowska

CONTENTS

"Can't keep it all."

INTRODUCTION

Today's life is unimaginable without the involvement of satellites—from space exploration, weather tracking, and navigation to cellular phone communication, to name just a few applications. The early satellites in the collection of the Smithsonian's National Air and Space Museum in Washington D.C., the subject of this study, exemplify the technological developments of space systems from the very inception of space exploration.

Vanguard 1 (1958), *Explorer 7* (1959), and *GRAB* (1960), the iconic artifacts of the Space Race era, provide insight through analysis of their technology to the culture, work practice, and sociopolitical base of a specific period, enabling their historiographic study. Thus, preservation of a tangible product, an artifact itself, is essential and is a primary objective of a museum conservator's work.

Particularly at the early stage of spacecraft development, there was reliance on empirical experience, planning for the unknown—space and its unexplored environment. Numerous failures at the initial stages of the Vanguard program indicated that struggle of the early days. Experimentation with a variety of materials is evident even within one class of the satellites, such as Vanguard.[1] As John Naugle, Deputy Associate Director of NASA, remarked in a foreword to *Scientific Satellites* in 1967:

> In the hustle of progress, few men have had opportunity to describe what they have done or used. There is grave danger that the line of development of space equipment and instrumentation may be lost if care is not given to its preservation. Much information is contained in in-house reports, but, as in all active fields, the records are scattered, often incomplete, and sometime[s] silent on important points. As future investigators try to assess past results and to combine them with their own, they will need to know accurately how the results were obtained.[2,3]

Three main objectives are outlined in John Naugle's summary: the need to capture the evidence of experimentation, the documentation of the final product, and its preservation. These objectives served as a premise for developing a protocol of the museum care of space objects.

Satellite technology is a huge and multifaceted subject; therefore, the focus of this chapter is limited to power systems, in particular to solar cells and potting cements. Two factors decided this selection—first, the critical role of solar cells for the operation of satellites, and second, clear evidence of their deterioration, manifested as efflorescence and cracking of potting material and oxide-like deposits on solar cells.

Materials used in the composition of the solar panels had to be identified to understand the processes of their deterioration before systematic care could be proposed. To the best of the author's knowledge, until now no study has been undertaken of the deterioration of materials in the early satellites in a museum context. Design of a testing protocol was drawn from the practice applied to artworks, ceramic, pottery, metal sculptures and contemporary composite artifacts. The process of designing the analysis and diagnostic methodology for the spacecraft is the subject of this chapter.[4]

PROJECT BACKGROUND

The space artifacts at the National Air and Space Museum (NASM) are testimony to relatively recent events in space exploration. The documentation associated with these artifacts is incomplete and in some cases nearly nonexistent, illustrating the comment, cited above, of John Naugle in 1967. To supplement the historical records and compare their accuracy with the actual objects, the author, in collaboration with the curators of the Space History Department of the NASM, initiated diagnostic studies of the selected artifacts. The results not only served as basis for assessment of materials used in the construction of the satellites but also contributed to an intellectual control of the collection. Discoveries revealed in the process of detailed examination of satellites' instrumentation and analysis of the materials often confirmed the authenticity of the artifacts, placing them in a different rank in the museum collection.

The project began in June 2005 in collaboration with several analytical laboratories. Tests were carried out at the Johns Hopkins University Applied Physics Laboratory, the Department of Scientific Research at the Freer and Sackler Galleries, and the Smithsonian Museum Conservation Institute.[5] The Scanning Electron Microscopy Laboratory at the Smithsonian National Museum of Natural History assisted with topographic high-resolution digital imaging.[6]

PROJECT METHODOLOGY

Vanguard 1, *Explorer 7*, and *GRAB*, the artifacts themselves, served as the primary sources of information. Examination of their structure, analysis of materials, and comparative research of designs supplied by scarce engineering drawings were considered in this project. The final shape of the satellites, the design choices, and selection of materials were dictated by the objectives of the program missions. Therefore, a brief overview of each one is provided for the artifacts studied, with a focus on the power systems, in particular construction and design of solar arrays. Technological developments of solar cells and raw materials processing were traced from the first patents dating to 1941.

A survey of the NASM collection database provided examples of artifacts suitable for this study. Faced with unknown materials, we focused the examination and laboratory analysis on establishing the category of materials in order to select the appropriate testing methodology. The analytical tests are described in test methodology.

ARTIFACTS: SELECTION CRITERIA

In a group of early satellites, the selection for case studies was based on the following criteria:

- Time frame of launching
- Mission objective, with an emphasis on solar experiments
- Use of solar cells as the source of energy
- Similarities in deterioration of solar panels

The Museum System (TMS) artifact database was surveyed for satellites that were launched in the 1958–1960 time frame. The selection was narrowed by searching for similarities in structure, materials used in solar panels, and their state of preservation. *Vanguard 1* (1958), *Explorer 7* (1959), and *GRAB* (1960) met the above criteria (Figures 35.1–35.3). These satellites are flight-qualified and engineering test models currently on exhibit at the main NASM location on the National Mall, Washington D.C., and at the Udvar-Hazy Center, Chantilly, Virginia, a recent expansion of the NASM. It is important to note that the artifacts on display were never launched into space, although some were equipped as backups and were ready to be launched if needed. Some of the first-generation satellites launched into space are still orbiting (*Vanguard 1*), while others have burnt out and disintegrated upon re-entry.

VANGUARD 1, EXPLORER 7, AND *GRAB*: DESIGN AND MATERIALS

Design criteria used in building the satellites focused on their operation in space; the materials had to endure the effects of space environment to ensure their reliable performance. The satellites orbiting in full-sun periods and eclipse seasons were exposed to extreme fluctuations of temperature and immense light intensity. Solar winds and meteorites coming in contact with solar cells would cause physical damage on impact. Solar cells are essential to a satellite, as they generate energy

FIGURE 35.1 *Vanguard 1*. National Air and Space Museum collection (TMS 1976–1019), displayed at the Udvar-Hazy Center, Chantilly, Virginia (the new extension of the National Air and Space Museum, Smithsonian Institution).

permitting transmission of data to the receivers on the ground. Thus, damage of solar cells means termination of their functionality.[7,8]

The design of satellites was determined by the experimental objectives of their missions. Typically, in a simplified definition, each scientific, military, or communication satellite contains instrumentation and operation systems packaged in a shell. There are only four basic structures utilized in scientific satellites: spheres, cylinders, polygons, and cones.[9] A shell of spherical shape, such as that of *Vanguard 1* and *GRAB*, was considered optimal as it can assume any orientation with respect to the sun, receiving equal light exposure. *Explorer 7*, shaped as two cones connected by a cylinder, had an additional surface of solar arrays on its equator to gather data on solar cell erosion caused by solar radiation.

Examination of the solar panels on the selected satellites indicated deterioration—efflorescence of cement and deposits that resemble metal-like corrosion on solar cells (Figures 35.4–35.6).

FIGURE 35.2 *Explorer 7*. National Air and Space Museum collection (TMS 1976-1109), in preparation for display at the Udvar-Hazy Center.

The design of the satellites is briefly described in relation to the missions' objectives. However, it is limited to experiments that directly utilized solar cells as a main source of power.

Vanguard 1, Launched on March 17, 1958

The NASM collection has seven objects identified as *Vanguard 1* backup models, test models, replicas, and display models.[10] Three of this series were analyzed. Vanguard evolved in the early 1950s as a scientific mission, and in 1955, it was presented by James van Allen as performing cosmic ray observation.[11]

FIGURE 35.3 *GRAB* (TMS 2002-0087) satellite on display at the National Air and Space Museum on the National Mall in Washington, D.C.

FIGURE 35.4 *Vanguard 1* (TMS 1976-1019): closeup of a solar panel illustrating efflorescence, powdery deposits, and crackelure of cement.

FIGURE 35.5 *Explorer 7*: solar panel with deposits of powdery deterioration, crackelure, and losses of cement.

Structure

Each satellite in the *Vanguard 1* series is constructed of a rigid, pressurized aluminum sphere 16.5 cm in diameter, with six antennas and six solar panels (Figure 35.7). Although the main source of energy was generated by solar cells, some energy was stored in mercury batteries and supplied during eclipse periods.[12] All three examined models of *Vanguard 1* in the NASM collection offer a variety of technical solutions to the problem of mounting solar cells and joining two units of the sphere. Solar cells were mounted with adhesive, resin, or an unidentified substance. Hemispheres were joined with rivets or snap-like pins.

Explorer 7, Launched on October 13, 1959 (Also Known as 1959 Iota)

The scientific payload for Explorer was designed to furnish data on the radiation balance of the Earth, total cosmic ray intensity, and the performance of solar cells unprotected from micrometeorite impact, just to name a few. The purpose of the solar experiment was to measure the deterioration of solar cells caused by direct exposure to the radiation in the Van Allen belts and to determine the effectiveness of glass filters in prevention of that deterioration.[13] The solar cell erosion experiment had a direct impact on the design of solar panels, which used various configurations of solar arrays, protecting some by glass and leaving others unprotected by any shielding material. To obtain a fairly smooth, continuous power output, the cone was divided into eighteen panels, of which six, when oriented perpendicular to the

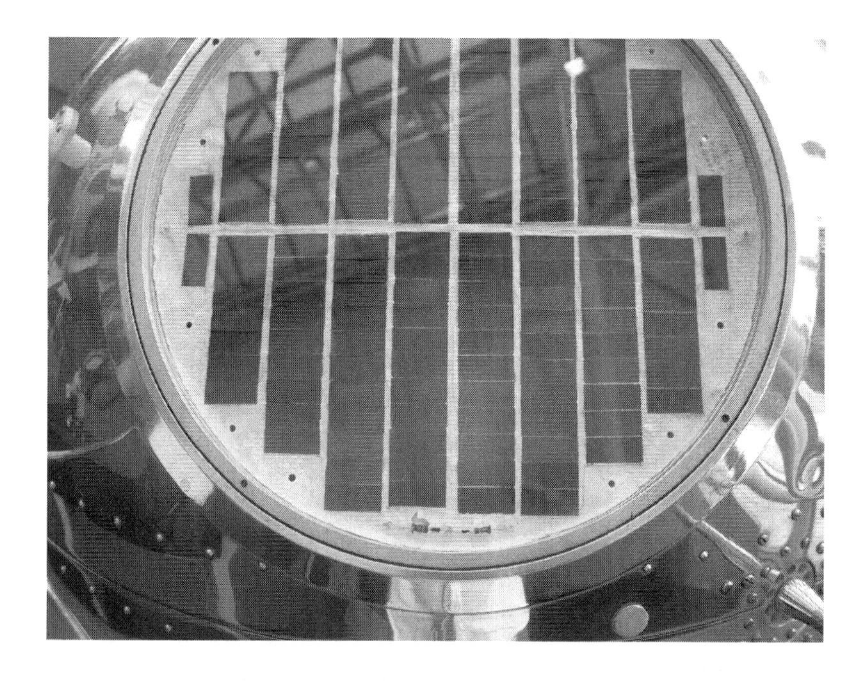

FIGURE 35.6 *GRAB*: solar panel, detail illustrating deposits of deterioration on potting material.

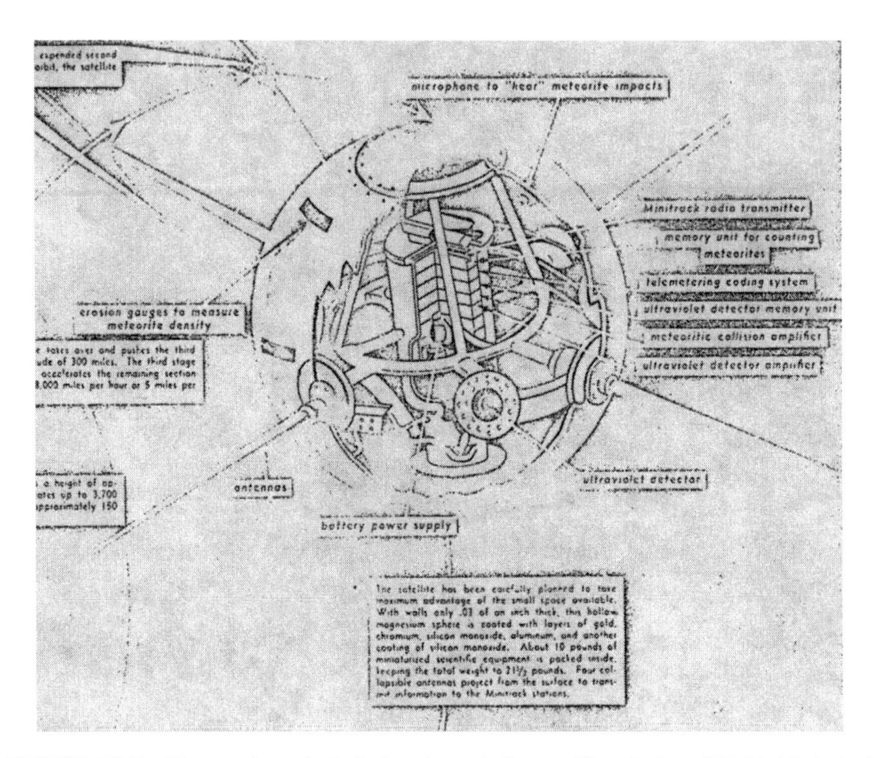

FIGURE 35.7 Vanguard: technical drawing of the satellite design (NASA Technical Notes).

incident solar energy, were capable of furnishing power to operate the complete electronics system (Figures 35.8, 35.9).

Structure

The satellite's shape is a combination of two cone frusta connected by a cylinder (Figure 35.10). Fiberglass covered with sandblasted aluminum foil was used as the skin material.[14] The satellite's antenna is a crossed dipole. The nickel-cadmium batteries stored the energy generated by solar cells and provided continuous power during eclipse period.[15]

The *Explorer 7* that was restored at the Paul Garber Restoration Facility of the National Air and Space Museum revealed, in the process of examination, the original components of the flight backup unit. The glass covers on the solar panels were fused silica. The presence of Sauereisen cements, the only referenced potting material in technical notes, was confirmed as such by analytical examination, and gray paint applied on the panels was evidence of passive thermal stabilization. Micrometeoroid detectors marking the first attempt to record a meteoroid impact were located on the flat surface of the satellite's equator.[16]

GRAB, Launched on June 22, 1960

The Galactic Radiation and Background satellite (GRAB) was also known under other names as SOLRAD 1, Greb 1, SR 1, Solar Radiation Satellite 1, Solar

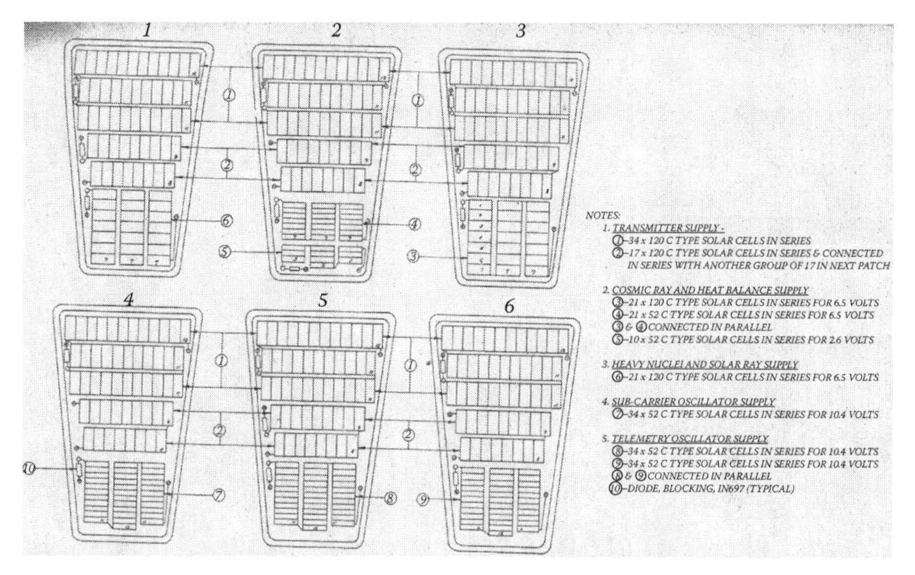

FIGURE 35.8 Explorer: pattern of solar panel configurations (NASA, D-608, 1961).

Monitoring Satellite, and Sunray.[17,18] The name GRAB does not appear in technical notes until its declassification in 1995. It was the first satellite system for signal surveillance and operated until August 1962, gathering information on Soviet air defense radars located in the area inaccessible to the Air Force and Navy aircraft.[19] Sunray 1 was the name of the scientific (nonclassified) part of the satellite's program, and its objective was to study solar activity and x-radiation.[20]

Structure

GRAB is constructed of aluminum as a 51-cm-diameter spherical shell, powered by energy from solar cells arranged in six patches and supplemented by energy stored in nickel-cadmium batteries. Four antennas and sensors protrude from the shell. Solar patches are covered with screw-on shielding glass (Figure 35.11).

SOLAR PANEL TECHNOLOGY IN THE EARLY SATELLITES

The success of a space mission depends on a properly working power supply. Nearly all spacecraft, including present-day satellites, use solar radiation as the primary source of power, which is stored in battery cells to keep the spacecraft alive during eclipse.[21]

SOLAR CELL STRUCTURE

A solar cell works on the principle of photovoltaic effect and converts incident solar radiation into electrical energy. Since the invention of "solar converters" in the early 1940s, silicon remains the most popular material in the production of solar cells.[22,23] It was used for the first time in a space application on *Vanguard 1*. The output of

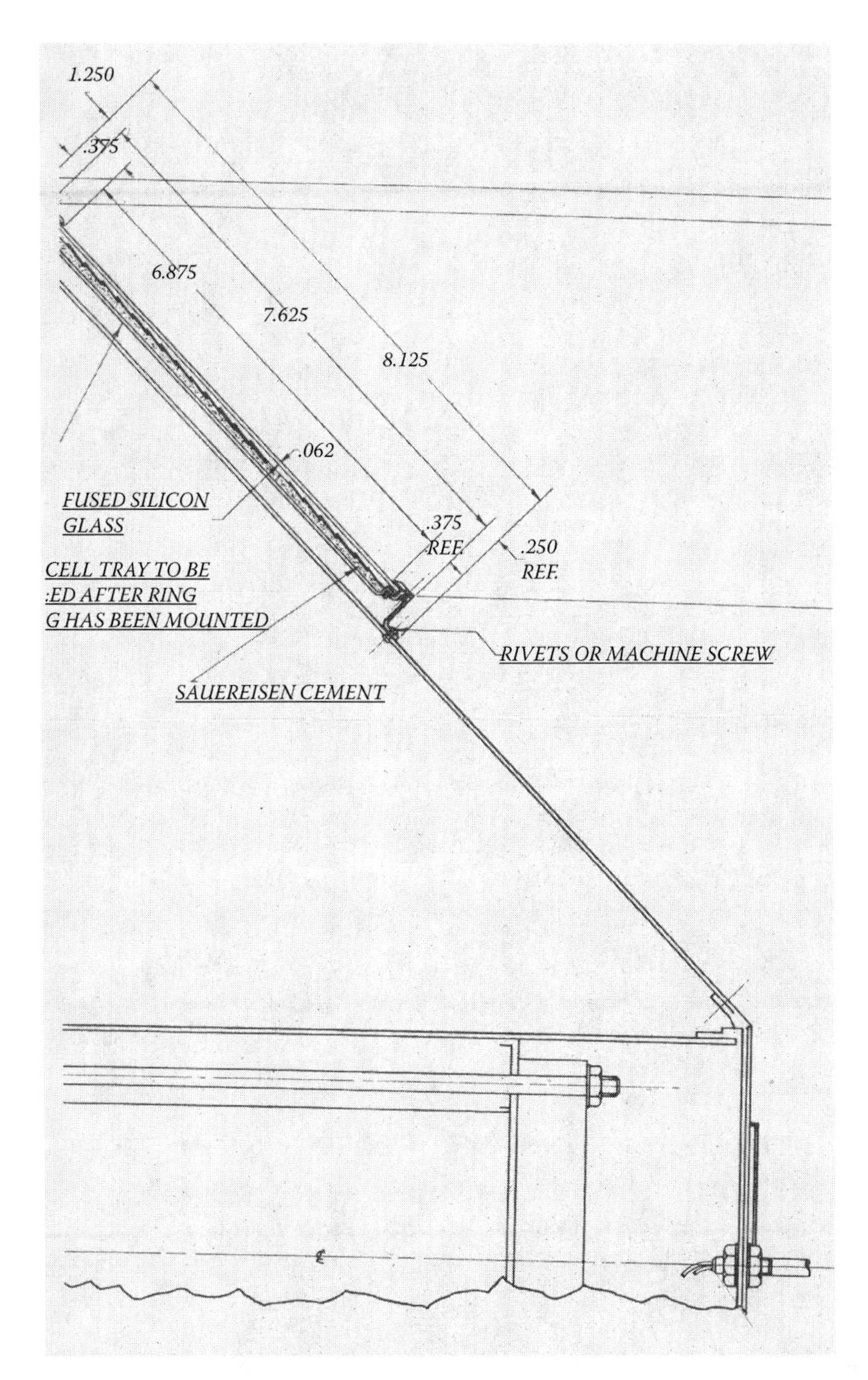

FIGURE 35.9 *Explorer 7*: technical drawing of solar panel; cross-section of the satellite design.

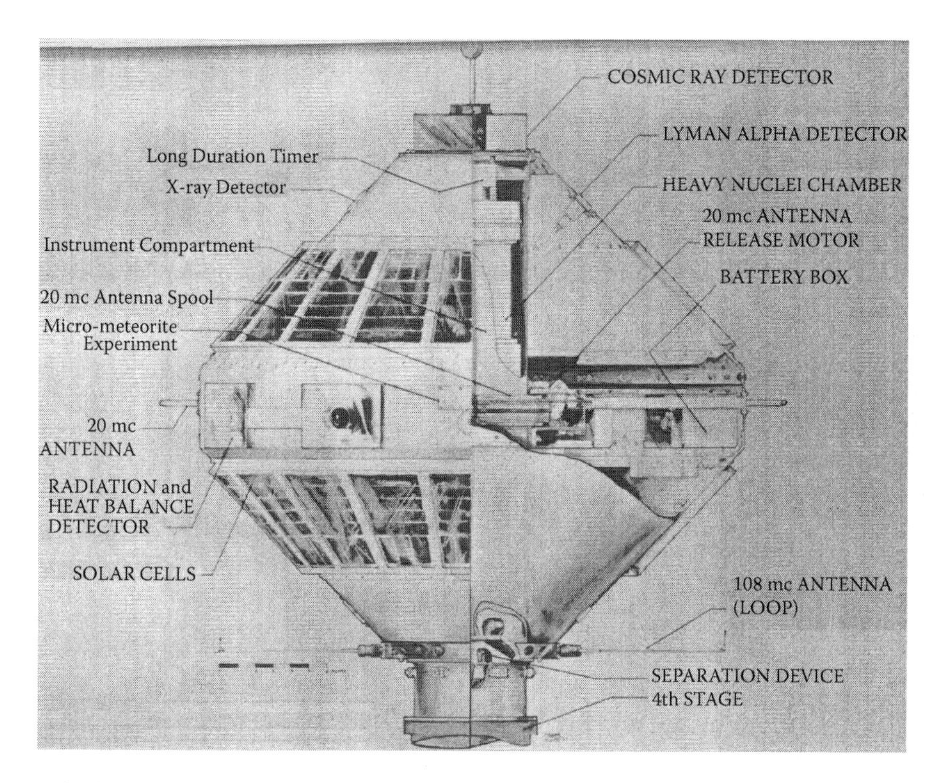

FIGURE 35.10 *Explorer 7*: technical drawing of the Explorer satellite; cross-section of the structure (NASA Technical Notes, D-608).

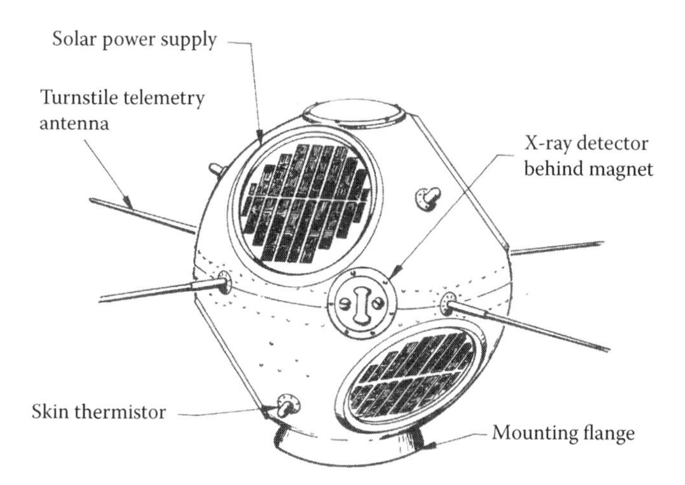

FIGURE 35.11 *GRAB*: drawing of SOLRAD structure (the same design was used in both satellites). (Adapted from Corliss, W.R., *Scientific Satellites*, NASA SP-133, NASA Scientific and Technical Information Division, Office of Technology Utilization, Washington DC, 1967.)

energy depends on the purity of the silica. Metallic inclusions present in raw silica play an important role not only in the energy output but also in maintaining the configuration of n-type and p-type zones within a single cell. The chemical composition of a material used in the production of solar cell is as follows: silicon, 99.85%; iron, 0.031%; aluminum, 0.020%; phosphorus, 0.011%; and trace amounts of carbon, calcium, nitrogen, magnesium, and manganese.[24,25,26] These inclusions may be also responsible for deposits on the solar cell surface that resemble metal oxides.

Fused silica glass, 1/16 inch thick, protected solar cells on *Vanguard 1* and *Explorer 7*. A bare solar cell patch on *Explorer 7* was used to measure solar aspect, the effects of micrometeorites, and radiation.

A solar array comprises solar cells connected to yield the voltage and power required by a particular mission. Cells can be arranged in a shingle-type mounting, overlapping each other and connected to adjacent cell, or in a flat-type mount, side by side (Figure 35.12).

Solar arrays on the studied satellites are shingle-type, mounted to the satellite body as patches (*Vanguard 1* and *GRAB*) or arrays (*Explorer 7*). In all three cases, the solar cells are mounted to a lightweight metallic sheet fastened to the satellite's shell. The instantaneous power output of this array is proportional to the area exposed to light.

COATINGS AND SHIELDS ON SOLAR CELLS

Light reflection, among other factors, was one of the major obstacles in producing a desired energy output. Historically, several methods were tried to overcome that problem—using a glass shield directly on the surface of cells, applying coating to the cells' surface or abrading it to change the reflection angle, and electroplating (Figure 35.13).

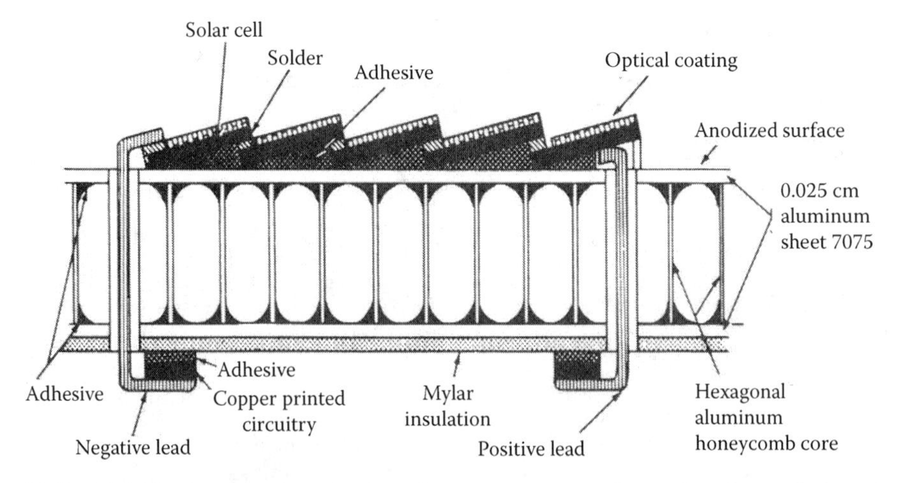

FIGURE 35.12 Cross-section of solar cells arrangement, as observed on the studied satellites. (Adapted from Corliss, W.R., *Scientific Satellites*, NASA SP-133, NASA Scientific and Technical Information Division, Office of Technology Utilization, Washington DC, 1967.)

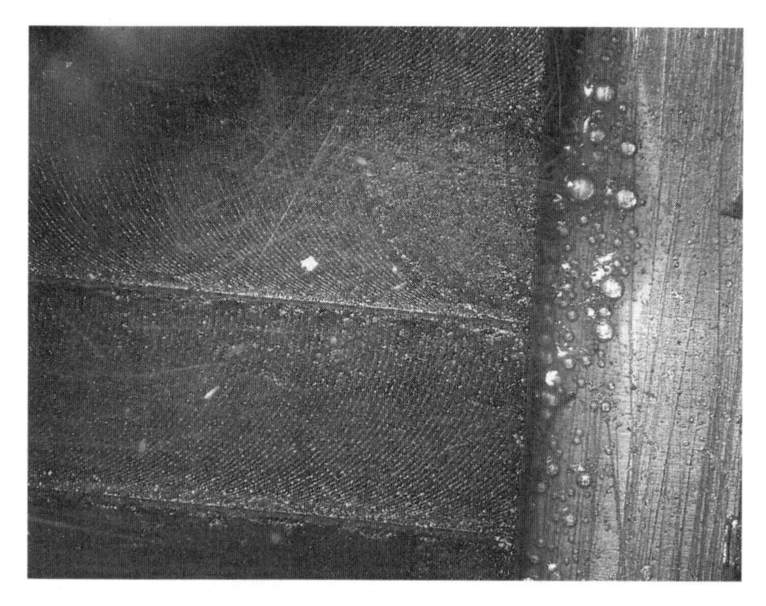

FIGURE 35.13 *Vanguard 1* series: closeup of a grooved solar cell which was one of the methods to diffract light reflection. Brown resin coating is used as adhesive and coating on solar cells.

The cover glass reduced the degradation of the cells. Antireflective coating on the glass minimized the reflection. One product only, Victron lacquer, was mentioned as a coating in the first patent disclosed in 1941.[27] However, there is no information regarding its chemical composition. The natural tendency of silica is to acquire a transparent oxide coating when exposed to the atmosphere; the oxide coating minimizes reflection losses. Alternatively, to accelerate that natural process, the silicon surface can be oxidized by heating in water vapor.[28]

CEMENT USED AS ADHESIVE IN SOLAR PANELS

When the spacecraft passes from full sunlight to total eclipse, the cells' temperature may drop 150°C. Thus, both cells and adhesive systems have to withstand that rapid change in temperature.

Sauereisen cement was mentioned in the *Explorer 7* technical note as an adhesive used for securing solar cells in panels. However, even at that time, during the temperature-vacuum test the cement buckled, loosing its adhesive quality. The technical note states that "the cement developed a chalky texture and, subsequently, became more moisture absorbent."[29] To insure proper bonding, epoxy resin cement was used to supplement the Sauereisen cement when adhesion was doubtful. However, no reference was found as to the type of resin or adhesive used.

Examples of three *Vanguard 1* satellites illustrate three variants of solar panel assembly: one with cells mounted with Sauereisen cement, another with the textured cells covered with brown, transparent coating serving as adhesive, and the third one, in the best condition, which does not indicate the presence of either (Figure 35.14, A and B). The conversion of a vision from a drawing board to a three-dimensional object is a creative process, and particularly in the time of fabricating new instruments, such as early satellites, not all ideas are captured in the form of technical notes, leaving room for experimentation with materials and designs. This explains the disparity between seemingly identical engineering models of *Vanguard 1*, and it also explains their different states of preservation.

FIGURE 35.14 Vanguard 1 series: example of different mounting of solar cells. Top, Sauererisen (TMS 1976-1019); Bottom, resin-type adhesive (TMS 1959-0212).

TEST METHODOLOGY

Examination and analysis of materials used in solar panels aimed at determining what materials were used in solar panels, particularly the adhesive system, in order to understand the reason for efflorescence formation. Samples of deposits on solar panels and cells were examined under a high-resolution stereomicroscope to establish a general category of materials and to select the most appropriate laboratory techniques. The microscope used in this examination was Olympus SZX12 with camera attachment QImaging U-TVO.5xc-2 (Figure 35.15, A and B).

SAMPLE PROCUREMENT

The chalky powder or efflorescence present on the cements in all three case-study satellites was collected for analysis. Relying on technical notes associated with *Explorer 7*, which referred to Sauereisen cement as the mounting or potting cement, the sources of the same material were traced for comparative analysis.[30] The Sauereisen company, established in 1899, is still in existence and supplied samples of a material whose chemical composition was determined to be as close as possible to that used on the satellites studied. Results of analysis of samples prepared from the new Sauereisen material were compared with tests of the old ones, proving that in all cases the material was the same. Sauereisen cement is a two-part, chemically setting compound. Samples were prepared in 20%, 25%, and 30% ratios of powder to solvent.

Examination of solar cells was limited at this time to evaluation of the deposits with a binocular surface analysis microscope in polarized and raking light. Results of the surface examination revealed oxides and metal-like corrosion, indicating the most likely impurities in the raw materials used for the production of solar cells. Analytical tests of solar cells were not carried out at this time because of limited access to solar panels and scarcity of individual cell material. In most cases, the satellites need to be open to gain access from the inside to release the mounting mechanism of the solar panels.

LABORATORY TECHNIQUES

Laboratory examination of an artifact is carried out on a macro or micro level. *Macro* refers to a visual observation, and *micro* to microscopic analysis. The goal is to understand the fabric of an artifact and alterations that occurred to it over a period of time. Scarcity of sampling material, particularly in the case of space objects, demands a rigorous analytical program to obtain as much information from a sample as possible without the necessity of returning for more. There is usually a progression of analytical protocol, starting with a broad evaluation of the material and proceeding to selection of more precise analytical techniques and tools that will answer specific questions by targeting elements of the materials from which the artifact is made. Surface examination with a stereomicroscope provided information about the topography of the cement samples and characteristics of the deposits on solar cells.

FIGURE 35.15 Closeup of a Sauererisen cement chalky texture, as observed on sample collected from *Explorer 7* (top) and *GRAB* (bottom). Surface analysis microscope, SEM Lab National Museum of Natural History.

The following analytical techniques were applied to the examination of each sample:

- Second ion mass spectroscopy (SIMS)
- Scanning electron microscopy coupled with energy-dispersive x-ray spectrometry (SEM-EDS)

- X-ray diffraction (XRD)
- Fourier transform infrared spectroscopy (FTIR)
- FTIR coupled with attenuated total reflectance (FTIR-ATR)

SI MS is an analytical spectroscopic tool primarily concerned with the separation of molecular (and atomic) species according to their mass and charge.

Scanning electron microscopy (SEM) is often combined with energy-dispersive x-ray spectroscopy. It is a tool that provides elemental analysis of the surface. The SEM microscope can magnify the sample up to 100,000×, revealing features of the surface. In addition, analysis of the elements is carried out by energy-dispersive x-ray spectroscopy. The results are captured as images and spectra produced by the elements present. The technique is carried out in a vacuum, and the sample requires preparation and mounting on small studs.

X-ray fluorescence spectrometry (XRF) has been one of the most widely used techniques for the past 50 years in the diagnosis of museum art collections. This technique has the ability to survey virtually all the elements in the periodic table. To simplify the analytical process, it can be described as measuring the energy of characteristic x-rays of atoms of the examined sample. Peaks in a spectrum are graphic illustrations of energy changes on the atomic level that are characteristic of each element.

XRD is an analytical tool used primarily for an identification of inorganic crystalline materials, such as pigments. A material's crystalline structure reflects or changes the angle of reflection of x-rays. The recorded angle and intensity of the reflection and refraction are recorded as a spectrum and compared with references of known materials. This technique requires a very small sample size and produces accurate identification results.

FTIR is an analytical tool that produces a characteristic spectrum of different chemical compounds found in the analyzed material. It is particularity useful in the characterization of organic materials that comprise paint bindings, coatings, finishes, or adhesives. Radiation/absorption in the infrared region of electromagnetic spectrum is characteristic of each chemical compound, thus allowing accurate assessment of chemical composition of the examined materials. Comparison of the spectrum produced by an unknown compound with that of a known compound provides a positive identification. Spectrum reference libraries are available online.

Energy-dispersive x-ray fluorescence spectrometry (XRF) is another technique used in identification of inorganic compounds, such as glass, ceramics, pigments, and metal alloys. An x-ray beam directed at the sample causes excitation of materials, resulting in the emission of a characteristic fluorescence radiation. That radiation is recorded as a spectrum, providing identification of the elements present.

SI MS analysis was carried out on the green deposits collected from the solar panels of *Vanguard 1* (TMS 1976-1019), and it indicated the presence of copper. That points, most likely, to an interaction between the metal components connecting individual cell shingles and interaction with the product of Sauereisen deterioration (Figure 35.16).

SI MS and XRD were performed on cement samples from *Explorer 7* and *GRAB*; the data were compared with spectral analysis of the new cement. XRD

FIGURE 35.16 *Vanguard 1* (TMS 1976-1019) solar panel; blue deposit of deteriorated Sauereisen and blue-pigmented products of metal corrosion. Inset, scanning electron microscopy micrograph of the corrosion material and plotting resulted from second-ion mass spectroscopy analysis of the blue-white deposit. Characteristic blue-green color (not shown here) indicates the presence of copper. The residue is a product of chemical reaction that occurred inside the solar panel, most likely between the copper wire connecting solar cells and the potting cement, in the presence of humidity in the environment.

was carried out on samples from *Vanguard 1*, and it narrowed down the findings of other techniques, confirming the presence of sodium, aluminum, and silica, which are characteristic of the cement Sauereisen and confirmed its presence in both satellites.

Furthermore, SEM-EDS analysis of powdery deposits from *Vanguard 1* confirmed content rich in sodium and silica. That was in agreement with the FTIR analysis of samples from *Explorer 7*. The cement on both satellites was identified as Sauereisen.

FTIR spectroscopy was used to determine the chemical components. Results of the analysis of *Explorer 7* material were compared with those of three samples of the new cement. That method, by transmitting the infrared energy through the material, provides information on chemical compounds in the mass of the material. Changes that occurred throughout may indicate that the mechanism of degradation is related to aging of the material itself. FTIR-ATR provided information on the chemical characteristics of the surface (Figure 35.17).

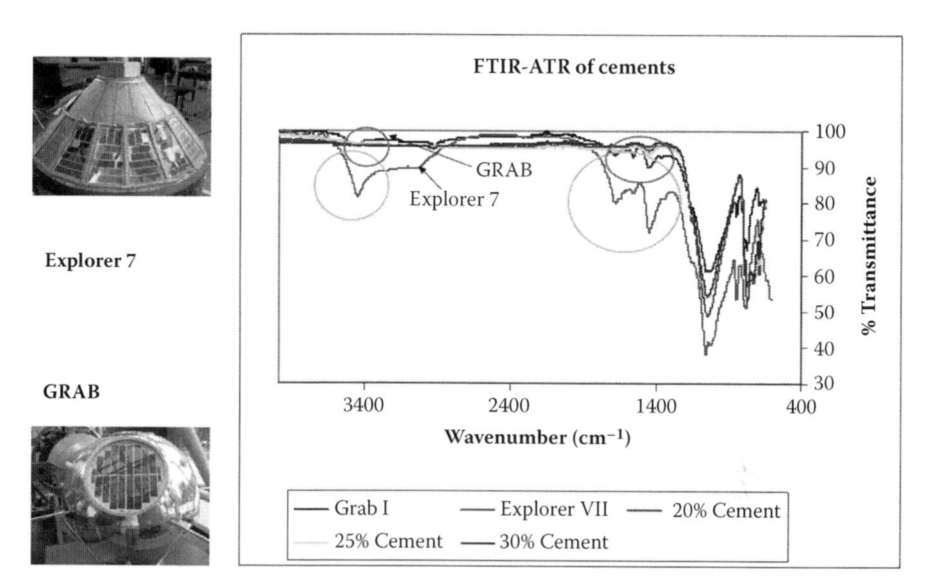

FIGURE 35.17 Comparative analysis of cement samples collected from two satellites (*Explorer 7* and *GRAB*) and the samples prepared from new Sauereisen. Plotting resulting from Fourier transform infrared spectroscopy–attenuated total reflectance analysis illustrates similarities of elements in the analyzed samples. The old materials showed similarities that set them apart from the new sample, such as the presence of peaks in similar spectral ranges. That feature can be attributed to the aging of material and will be evaluated.

TEST RESULTS

SI MS positive ion results are qualitative in nature and indicated that the principle species were silicon at 28 and 29 atomic mass units (AMU); oxygen (O_2) at 32 AMU; carbon at 24, 25, and 45 AMU; and chlorine at 35 and 37 AMU. Chlorine was present in the sample of new cement and not in the samples from the satellites.

The FTIR spectra of the three new cement samples (mixed at ratios of solid to liquid of 20%, 25%, and 30%) were indistinguishable. However, there is a slight difference between them all and the one from *Explorer 7*. As described above, presence of chlorine was confirmed in all new samples (indicated by the shoulder near 734.86) and not in the old one, which would eliminate a theory that acidity from the environment may be responsible for the degradation, as illustrated in Figure 35.18.

By FTIR-ATR, the chemistry of the surface of our samples was determined. The strongest peaks are evident in the older sample from *Explorer 7* and, in the same regions, in samples from *GRAB*, but they are absent in samples from the new cement. This indicates similarities among the samples of older material and sets them apart from the new material.

It is important to mention that none of the new samples were aged to match the sampled materials from the satellites that are 50 years old. Therefore, there is a chance that the chemical changes could be related to the material aging.

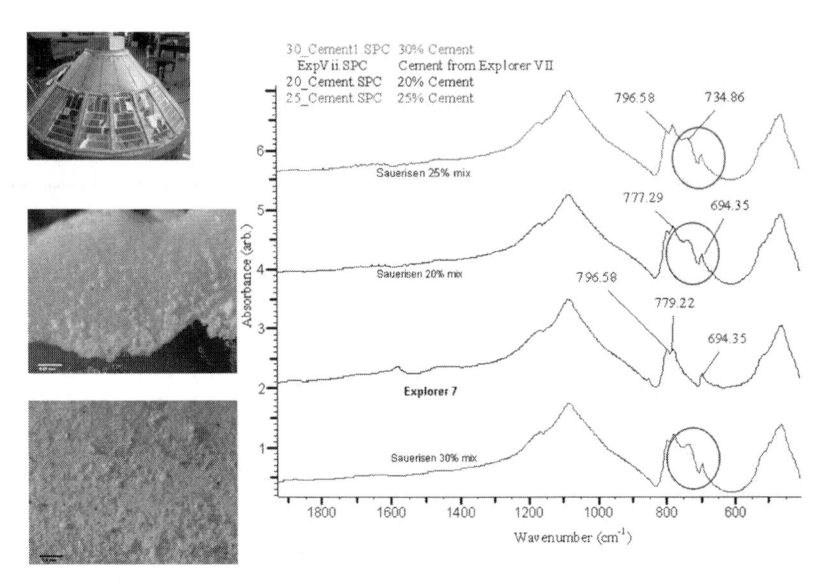

FIGURE 35.18 Comparison of Fourier transform infrared spectroscopy plotting produced from analysis of Sauereisen samples of new cements with a sample of that from *Explorer 7*. The new material was prepared in the following ratios of dry (powder) to wet: 20%, 25%, and 35%. Absence of chlorine in the old sample and its presence in all new samples was observed in the future aging studies.

IMPLICATIONS OF FINDINGS

1. Sauereisen cement was used on all three satellites as cementing material in solar panels, indicated by presence of sodium-aluminum-silicate, predominant in all samples. Laboratory tests were supported by surface analysis microscopy and historical reference. That was a particularly valuable conclusion in regard to the GRAB reconnaissance satellite, since no technical information was available for this artifact. The presence of that cement confirms that the artifacts in our collection are identical to the flown spacecraft.
2. Changes in the chemistry in the bulk of the material correlate with the changes on the surface, indicating that the terrestrial environment and immediate environment of an enclosure are most likely contributing to chemical alterations within the material itself. This information has a direct impact on designing display cases and selection of materials remaining in close contact with space objects.
3. Oxide formations on the solar cells may be attributed to the early technologies of solar cells production, metallic impurities in the cells' material, and coatings aiming to reduce the light reflection. Further analysis of the solar cells surface will be undertaken as soon as samples of the material become available.

The proposed testing protocol is a work in progress. The principles of the approach, which include historical research of available data, comparative analysis of technical notes, and laboratory testing of artifacts and material, are applicable to a wide range of complex artifacts and composite materials found in space objects. However, each case needs an individual evaluation and tailoring of tests to obtain meaningful results.

CONCLUSION

A collection of early satellites exemplifies historically important artifacts that are the icons or symbols of space exploration. Their unique technical detail, revealed through the process of examination and testing, supports their scholarly assessment and provides insights into the social, cultural, and institutional contexts of these artifacts. The proposed systematic assessment of the materials and satellite construction supports development of relevant long-term preservation guides to their exhibit and affects the choice of storage systems. From a sociopolitical standpoint, technical solutions and choices illustrated by the technical satellites' designs are representative of events and trends in society that produced them. Preservation efforts, substantiated by a systematic testing and examination protocol, not only support one the main museum mandates but also are critical in ensuring continuity in the history of the beginning of space exploration, represented by these tangible artifacts.

NOTES AND REFERENCES

1. Class or type of satellites is referred to in the Evolution of Scientific Satellites Type. Five types are or classes are identified as follows: I: Explorer class, to which Explorer and Vanguard belong; II: Observatory class; III: Recoverable spacecraft; IV: Piggyback type, to which GRAB belongs; and V: All other satellites. Adapted from Corliss, W.R. 1967. *Scientific Satellites.* NASA SP-133. Washington DC: NASA Scientific and Technical Information Division, Office of Technology Utilization.
2. Corliss 1967 (note 1).
3. J.E. Naugle. Forward to Corliss 1967 (note 1).
4. Acknowledgments: David DeVorkin, James David, and Roger Launius, curators of the NASM Division of Space History for their generous contribution regarding the satellites history. Paul Biermann, principal professional staff, composite M&P engineer, Applied Physics Laboratory, JHU, for making tests at APL possible. At the Smithsonian Institution, Dr. Paula DePriest and Dr. Robert Koestler, directors of the Museum Conservation Institute, and Paul Jett, head of the Department of Conservation and Scientific Research at the Freer and Sackler Galleries, for facilitating analytical tests in their laboratories. Elisabeth West FitzHugh, research associate, Freer and Sackler Galleries, for her invaluable editorial and contextual input.
5. Analytical testing as carried out by Edward Ott, associate chemist, Applied Physics Laboratory, Johns Hopkins University; Janet Douglass, senior scientist, Department of Conservation and Scientific Research, Freer and Sackler Galleries; Jeff Speakman, Museum Conservation Institute, Smithsonian Institution.
6. Scott Whittaker, head of the SEM Laboratory, National Museum of Natural History, Smithsonian Institution.
7. Kennedy, G.C., and M.J. Crawford. 1998. Innovations Derived from the Transit Program. *Johns Hopkins APL Technical Digest* 19(1): 27–36.

8. Bostrom C.O., and D.J. Williams. 1998. The Space Environment. *Johns Hopkins APL Technical Digest* 19(1): 43–52.
9. Corliss 1967 (note 1).
10. DeVorkin, D. 2007. Preserving the Origins of the Space Age: The Material Legacy of the International Geophysical Year (1957–1958) at the National Air and Space Museum. In *Smithsonian at the Poles: Contributions to International Polar Year Science*, eds. Igor Krupnik, Michael A. Lang, and Scott Miller, 35–48. Washington DC: Smithsonian Institution Scholarly Press.
11. McLaughlin Green, C., and M. Lomask. 1971. *Vanguard: A History.* Washington, DC: Smithsonian Institution Press.
12. Corliss 1967 (note 1).
13. Richter, L.H., ed. 1966. *Space Instrument Survey, Instruments and Spacecraft, October 1957–March 1965.* Prepared for NASA by Electro-Optical Systems, Inc., Scientific and Technical Information Division. Washington, DC: NASA.
14. Corliss 1967 (note 1).
15. NASA Technical Note D-608. 1961. *Juno Summary Project Report.* Vol. 1. Explorer VII Satellite. George C. Marshall Space Flight Center Huntsville, Alabama. Washington, DC: NASA.
16. Cherrick, I.L., ed. 1960. NASA Technical Note D-484. *Telemetry Code and Calibrations for Satellite 1959 IOTA (Explorer VII).*
17. DeVorkin, David (Senior Curator, Space History Department, NASM), personal communication, March 2005.
18. NRL. 2000. GRAB: *Galactic Radiance and Background, First Reconnaissance Satellite.* Washington, DC: Naval Research Laboratory (NRL).
19. Corliss 1967 (note 1).
20. Richter 1966 (note 13)
21. Chetty, P.R.K. 1988. *Satellite Technology and its Applications.* Fairchild Space Company, Germantown, Maryland. Blue Ridge Summit, PA: TAB Books Inc.
22. Chetty 1988 (note 21).
23. Cruise, A.M., J.A. Bowles, T.J. Patric, and C.V. Goodall. 1998. *Principles of Space Instrument Design.* Cambridge Aerospace Series 9. Cambridge, United Kingdom: Cambridge University Press.
24. Ohl, R., and L.N. Silver. 1941. Light-Sensitive Electric Device. US Patent 2,402,662, filed May 27, 1941, patented June 25, 1946.
25. Bogachev, E., I. Abdjukhanov, and A. Timofeev. 2000. Method for Producing Metallic Silicon. International Patent Bureau WO/2000/047784.
26. www.semiconductor.net.
27. Ohl, R., and L.N. Silver. 1941. Light-Sensitive Electric Device. US Patent 2,402,662, filed May 27, 1941, patented June, 25 1946.
28. Chapin, D., and B. Ridge. 1954. Solar Energy Converting Apparatus. US Patent 2,780,765, filed March 5, 1954, patented February 5, 1957.
29. NASA 1961 (note 15).
30. Determination of the type of contemporary Sauereisen cement was made by the APL Principal Investigator, Composite Materials, Paul Biermann.

36 *CORONA* KH-4B Museum Preservation of Reconnaissance Space Artifacts: A Case Study

Hanna Szczepanowska

CONTENTS

INTRODUCTION

The highly classified Central Intelligence Agency (CIA)–Air Force *CORONA* photoreconnaissance satellite program operated from 1958 to 1972 and was America's and the world's first successful space-based imagery collection platform. During its existence, 145 *CORONA* satellites were placed in orbit and brought back more than 800,000 usable photographs of the USSR and other nations' territories that provided invaluable intelligence, which could not be obtained by any other means. The cameras and film return capsules were carried on Agena upper stage rocket, which was launched by a Thor booster into a near-polar orbit. Once all the film had been exposed, the film return capsule and its protective heat shield separated from the satellite, reentered the atmosphere, and at about 12 miles altitude deployed a parachute that was snagged by an Air Force plane. The Agena with the cameras eventually reentered the atmosphere and burned up (Figure 36.1).

In 1995, the *CORONA* program was declassified, and the National Reconnaissance Office and CIA transferred a KH-4B camera and its film return capsule made from

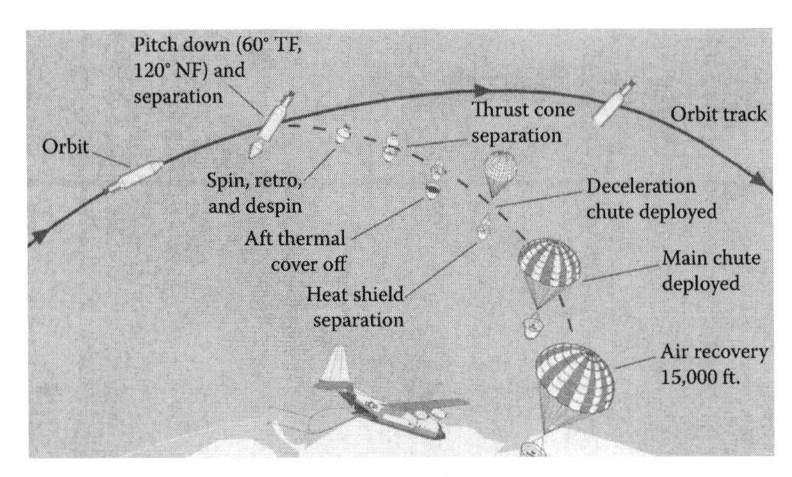

FIGURE 36.1 Schematic drawing of the capsule recovery path, published online by several sites. (*CORONA* Summary. http://www.globalsecurity.org.)

spare parts to the Smithsonian Institution's National Air and Space Museum, Washington D.C. (Figure 36.2). This was the last and most advanced camera used in the program and was capable of achieving a ground resolution as high as 6 ft. The museum's KH-4B is the only camera to survive from the program. However, little information about the artifact—its structure, materials, and film assembly—accompanied the artifact or is set forth in available program records. Although exhibiting the camera fulfills the educational aspect of museum's mission, its proper preservation,

FIGURE 36.2 *CORONA* camera on display at the National Air and Space Museum, Smithsonian Institution, Washington D.C.

which is a primary responsibility of the museum, depends on understanding the materials used in its construction. The need for conservation of an artifact remaining on exhibit for 13 years presented an opportunity to access the return capsule interior and removable film rewinding mechanisms, allowing for the first and most complete photodocumentation since the artifact was bestowed into public care. As a backdrop, the geopolitical climate of that period, even if surveyed only briefly, contributes to the understanding of the tangible product that is the *CORONA* camera itself. The process of developing a preservation plan for it is the subject of this chapter.[1]

CORONA: HISTORY AND MEANING OF THE PROGRAM

The *CORONA* program was conceived in deep secrecy some time at the end of 1957 and at the beginning of 1958, under a joint CIA–Air Force management team, with the purpose of furnishing intelligence without a high risk of human-operated overhead reconnaissance.[2] The *CORONA* program played a crucial role in the Cold War era, providing intelligence that, as many believe, prevented a world nuclear disaster.[3] The *CORONA* return vehicle, or capsule, had pioneered the way for satellite reconnaissance contributing to space technology and consequently the nation's security, particularly during the unsettling times of the Cold War. Reentry experiments laid the foundation for the development of materials used in ablative shields, which in subsequent years protected men on their return from space.

The geopolitical climate of the period fueled intellectual efforts of experts in myriad branches of science, bringing together academics and clandestine constituencies of the Department of Defense and the intelligence community. Thus, the success of the *CORONA* program was a result of collaboration between experts from the private sector, scientists, and engineers on one side and military strategists with a vision on the other. The multilevel collaboration resulted in technological advances in Earth science and imaging, providing the basis for today's global mapping, which have branched out into novel applications that are enhancing our daily life.[4]

CORONA was an ambitious and complex program that pioneered in technical fields about which little was known. It was the first, longest, and most successful of the nation's space recovery programs. Several years preceded the *CORONA*'s success. The code word "*CORONA*" referred to a classified part of the program known to the public as DISCOVERER, particularly during 1960–1962. The DISCOVERER XIII satellite recovery vehicle was the first object orbited in space and recovered according to plan.[5] The capsule from the diagnostic flight, on August 10, 1960, without a camera or film, was recovered from the ocean on August 11, and the President and the Smithsonian received the actual hardware from that first historic event.[6]

The first launch of *CORONA*, with a camera, film, and telemetry payload as DISCOVERER XIV was on August 18, 1960.[7] DISCOVERER XIII is currently on display at the National Air and Space Museum (NASM) (Figure 36.3). The capsule was opened in June of 2006 for the first time since its arrival to the museum in 1961,[8] revealing most of the instrumentation intact and in its original position. Markings that are usually placed on matching components, in this case red paint dabs, were perfectly aligned (Figure 36.4, A and B). The capsule's structure, particularly the shape of its lid-cover, provided an explanation for how the parachute

FIGURE 36.3 DISCOVERER capsule during conservation in the preservation workshop of the Paul Garber Restoration Facility, National Air and Space Museum, Suitland, Maryland.

that enabled a midair retrieval was stored away on *CORONA* (Figure 36.5). No accompanying documentation was transferred to the museum along with the artifact, and most of the electronic components remain unidentified as to their purpose and function.

DISCOVERER XIV/*CORONA* was launched on August 18, 1960. The capsule contained the 20-lb load of film.[9] It delivered substantial photographic intelligence of the Soviet Union territories, more than all previously gathered by twenty-four manned U-2 overflights. Not all was perfect. The developed film was marked with density bars running diagonally, as a result of electrostatic discharge (known, ironically, as the *CORONA* phenomenon). An official announcement to the public regarding the capsule retrieval noted

> The satellite has been placed into orbit with a 77.7 degree of inclination, an apogee of 502 miles, a perigee of 116 miles, and orbital period of 94.5 minutes. A retro-rocket had slowed the capsule to reentry velocity, and a parachute had been released at 60,000 ft. The capsule, which weighed 84 pounds at recovery, was caught at 8,500 ft. by a C-119 airplane on its third pass over the falling parachute.[10]

(A)

(B)

FIGURE 36.4 (A) DXIII: An overall view of the instrumentation. (B) A detail of alignment markings.

Of the following ten launches, extending through August 3, 1961, only four were *CORONA* missions.[11] The DISCOVERER program officially ended on January 13, 1962, with the failure of an Agena launch.

Ten years later, the *CORONA* program terminated with its last launch on May 25, 1972, and recovery on May 31; available hardware from developmental models was assembled to recreate the only existing model of the *CORONA* artifact, which is currently on exhibit at the National Air and Space Museum.

Technological developments of the *CORONA* space vehicle and camera, although evolved for military applications, benefited civilians in numerous ways. One that has far-reaching implications is the contribution to the art and science of photogrammetry and remote sensing of geographic information systems. Would the global

SATELLITE RECOVERY VEHICLE CONFIGURATIONS

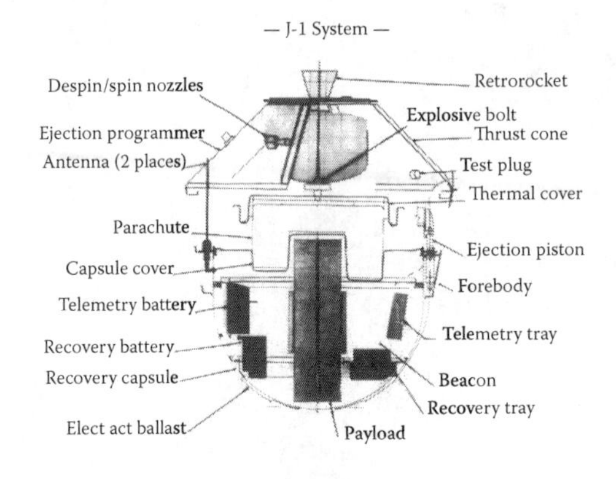

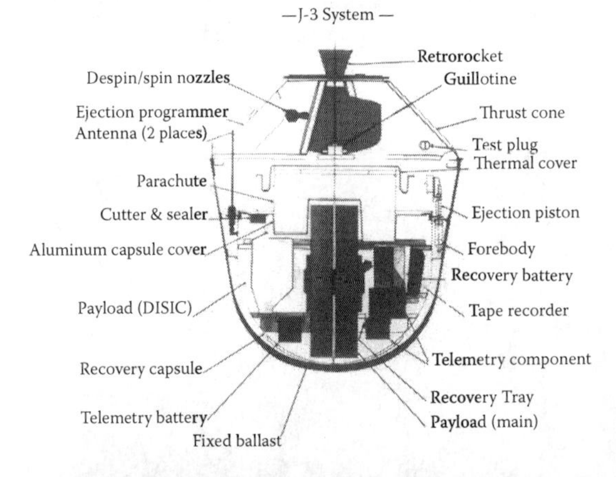

Bye 15274-74
Handle via Byeman
Controls Only

FIGURE 36.5 Top and middle: Engineering drawing of the capsule. Bottom: DXIII lid cover during conservation at the Paul Garber Restoration Facility, National Air and Space Museum, Suitland, MD. (Adapted from *Corona History*, Vol. IV, declassified documents BYE 15274-74. Available online.)

positioning system have been possible without *CORONA*? Most likely it would have been, but one could argue when and how. The *CORONA* program left a technological and information heritage that was a catalyst for today's achievements in space-based information technology, providing new tools and new insights to study a myriad of environmental and historical occurrences.[12] The *CORONA* camera at the National Air and Space Museum embodies the message that preservation of the artifact, as the only tangible remnant of the program, ensures preservation of the message that it conveys.

Development of the preservation program of the *CORONA* camera is illustrated by tracing the *CORONA* program's objectives, reconstructing technology chosen to meet those objectives, with particular emphasis on the camera's optics. Scarce technical drawings available in partially declassified documents of the *CORONA* and DISCOVERER programs have been used as references in identifying the components of our artifact tangible representations of its time (Figures 36.6 and 36.7).

CORONA CAMERA EXHIBIT AT CIA HEADQUARTERS

The first exhibit of *CORONA* prior to the declassification of the program was set up at as a small exhibit in Washington D.C. Richard Helms,[13] the Director of Central Intelligence, said in his 1970 dedication speech,

> *CORONA*, the program which pioneered the way in satellite reconnaissance, deserves the place in history which we are preserving through this small Museum display. 'A Decade of Glory' as the display is titled, must for the present remain classified. We hope, however, that as the world grows to accept satellite reconnaissance, it can be transferred to the Smithsonian Institution. Then the American public can view this work, and then the men of *CORONA*, like the Wright Brothers, can be recognized for the role they played in the shaping of history. (Figures 36.8 and 36.9)

Not until the 1995, when the program was declassified, was the *CORONA* camera transferred to the Smithsonian as the representation of the last and most advanced stage of the program KH-4B. What this means, how the technologies evolved to reach that stage, and which of those elements are embodied by the artifact will be revealed in the course of this chapter.

Tracing the path of manufacturers and companies originally involved in producing components of the artifact has contributed to the understanding of the ever-changing technology and evolution of *CORONA*'s two most significant components, the camera system and the return capsule, both of which are examined in subsequent sections.

CORONA: CAMERA

The *CORONA* program's most important objective was high-resolution photography of the territories not accessible by land using panoramic cameras. Thus, *CORONA*'s camera was the essential component of the program. It provided images of the territories behind the Iron and Bamboo Curtains, which kept the peace negotiations in

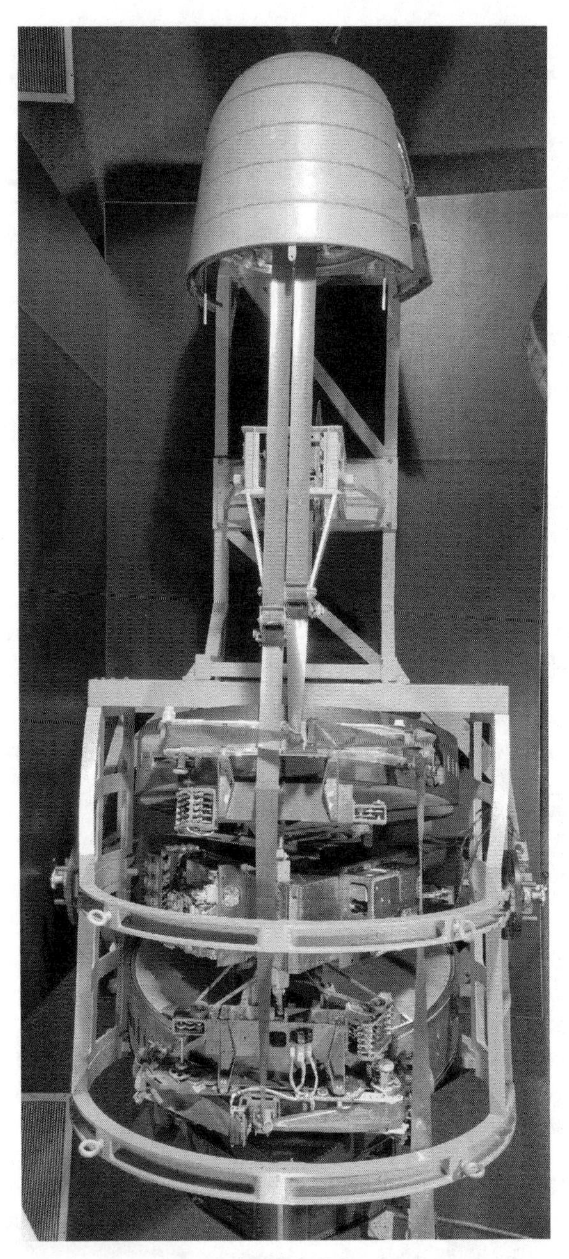

FIGURE 36.6 Bird-eye view of *CORONA*, on display at the National Air and Space Museum (NASM) next to the line drawing. (Photo credit: Eric Long, NASM staff photographer.)

balance. More than 800,000 images gathered by *CORONA* camera were sealed in capsules, jettisoned from space to Earth orbit, and captured in midair by military aircraft. The concept, operation, and execution seem, even by today's standards, like a realization of an unthinkable dream at a time when technology was just embracing computers and transistors as part of everyday life.

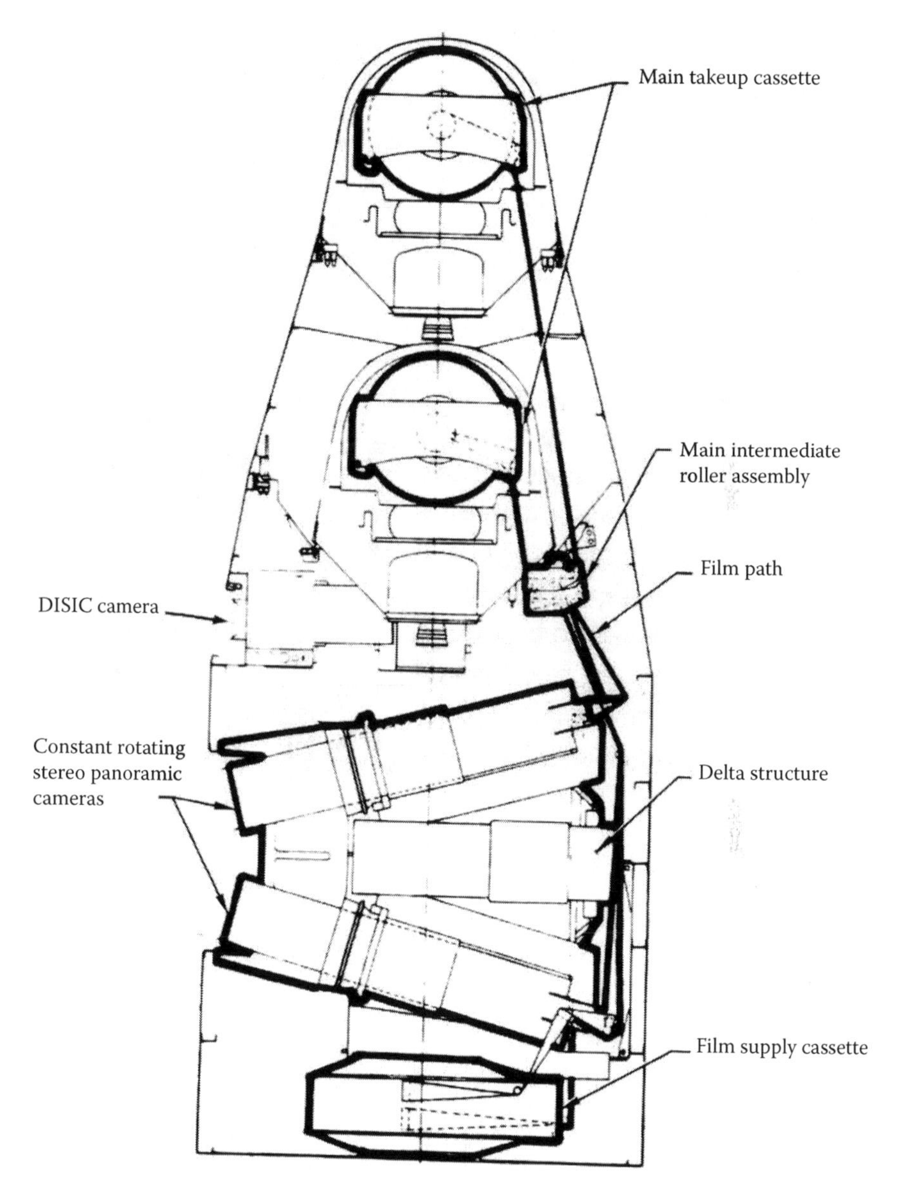

FIGURE 36.7 Line drawing of major components of KH-4B camera. (Published on several Web sites; this one was adapted from http://declassified-top-secret-cold-war-documents.com/images/Line%Drawing%20KH.)

The imagery acquired from the satellites and cameras that composed the *CORONA* program was known by the codeword Talent-Keyhole—where the first part (Talent) was initially used for images collected by aircraft, and the second part (Keyhole) referred to satellite collection. Thus, the abbreviation KH was coined. Consecutive numbers next to the KH denotation, such as KH-1 and KH-2 through KH-6,

FIGURE 36.8 Former vice president Al Gore at the CIA headquarters' opening ceremony of the *CORONA* exhibit. (Adapted from McDonald, E., ed. *CORONA between the Sun and the Earth*. p. 175.)

referred to improvements of the original camera. The KH-4 mission launched in 1962 brought a major breakthrough in photographic technology with the installation of the Mural camera, to provide stereoscopic imagery; two images of one area were obtained from the side-oriented cameras and a third one from a horizontal camera.

Initially, two companies provided camera systems designs: Fairchild Corporation and Itek Corporation, formed by scientists who originally worked at Boston University. Itek's system configuration, relying on an Earth-center-stabilized approach, won the contract. Acceptance of the civilian base technology was another unprecedented element of that time—it changed the boundaries of clandestine and public involvement.[14]

Over the entire period of its operation, the camera, as the most essential component of the mission, was undergoing continuous improvements aimed at enhancing image resolution. To identify which elements of the museum's artifact are original, a survey of the reconnaissance camera systems of that period was undertaken (Figure 36.10).

FIGURE 36.9 *CORONA*, as displayed at National Air and Space Museum. View of the capsule partially covered with the heat shield. Gold-plated surface of the capsule is visible through the cut-out in the heat shield.

The first stage, which evolved over the years, was camera C, a scanning, panoramic camera in which the cycling rate and the velocity-over-height ratio were constant and established prior to launching. Image motion was fixed mechanically. Those measures were aimed at matching the number of film exposures in a given period of time (camera cycling rate) with the varying ratio between vehicle altitude and velocity on orbit (velocity-over-height) to prevent gaps or overlapping of photographs.[15]

The upgraded version, known as C‴ Triple Prime, built by Itek Corporation, was a reciprocating camera, with lager aperture lens, improved film transport mechanism, and a greater flexibility in commanding the vehicle operations, particularly regarding the control of velocity-over-height factor. An improvement was even greater when two C‴ cameras were combined in a system mentioned earlier and known as Mural camera system. It was the first system to provide stereoscopic coverage of the photographed territory. In addition, counter-rotating lenses further improved the camera stability.[16]

One camera pointed forward, the other one backward (Figures 36.11–36.13). The load of film was increased to two 40-pound rolls, carried in a double-spool film supply cassette (Figure 36.14, A and B). The two film webs were fed separately to the two cameras, where they were exposed during lens-cell rotations, and next fed to a double take-up cassette in the satellite recovery vehicle (Figure 36.15). The system was designed for up to 4 days of operation in space.

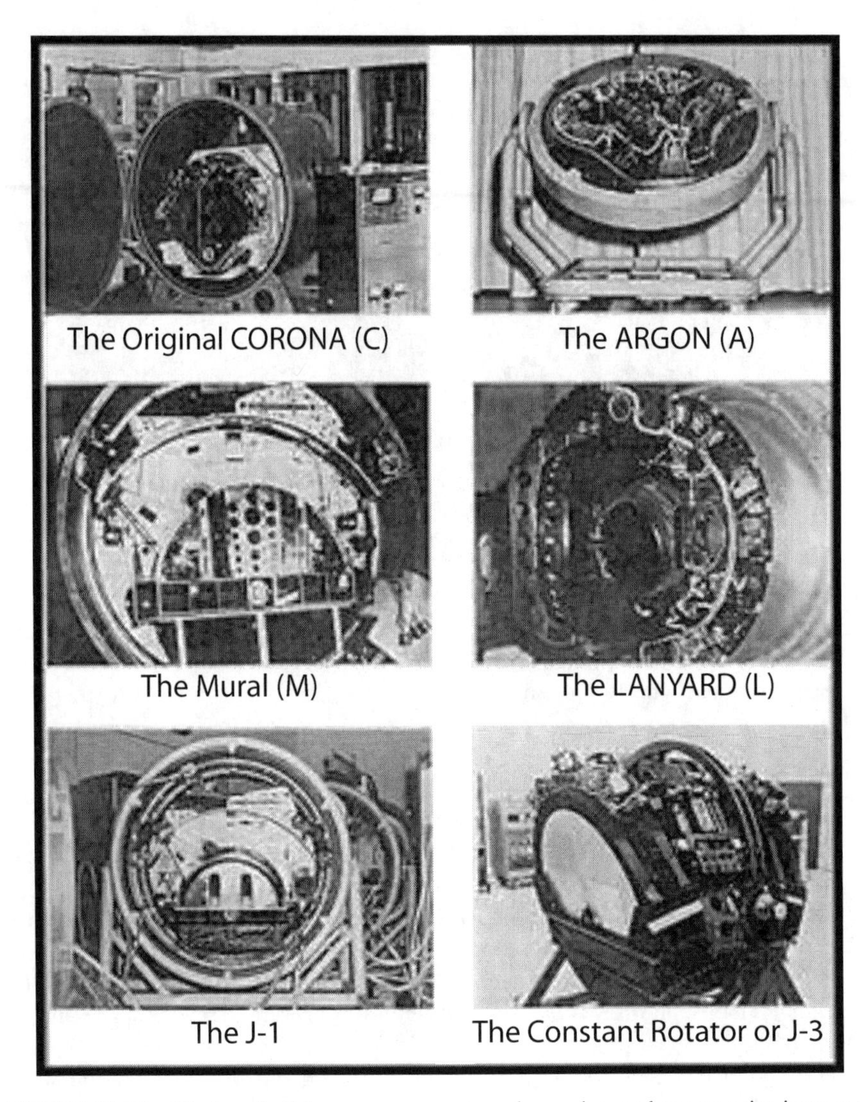

FIGURE 36.10 Black-and-white camera systems and overview and comparative images of the camera configuration. (Source: GlobalSecurity.com).

The first Mural camera was launched as program flight number 38 in February 1962.[17] An anomaly that occurred on the return vehicle, failure of the heat shield separation, unexpectedly provided a valuable diagnostic data on reentry effects. The diagnostic data furnished by the heat shield exposed to the reentry impact, friction, and heat deserves its own publication.

Enhancements of imaging performance on the Mural system were fulfilled by the index camera and a stellar camera. The latter, taking photographs of stars in

FIGURE 36.11 Frontal view of the camera lens on National Air and Space Museum's object

combination with those taken of the horizons, provided a point of reference for a more precise means of determining vehicle attitude on orbit.

CORONA: THE CAPSULE

The capsules are the only objects that returned intact from space. Two capsules (or buckets), one from the *CORONA* program and one from DISCOVERER XIII, provide complementary insights into their structure and retrieval mechanism. The buckets, which are gold-plated containers, housed diagnostic instrumentation on the DISCOVERER XIII mission, which was replaced by intake blades for collecting exposed film on the *CORONA* flight (Figure 36.16). Items 2, 3, 4, 6, 8, and 10 were identified on the original capsule (see Figures 36.17 and 36.18). The intake blades inside the return capsule, once filled, triggered a sheering mechanism that cut the film and subsequently sealed off the capsule. In a simplified understatement of the events' sequence, that in turn activated the mechanism of the capsule separation.[18] The return *CORONA* capsules were initially designed to float for up to 3 days and to self-destruct by sinking if not retrieved by then. Later, the flotation time was extended to 4 days. A capsule sink valve containing compressed salt, which would dissolve in a predicted period of time, controlled the duration of the flotation period (Figure 36.19). Once water entered the interior, exposed film was destroyed.[19] Some of the capsules sank, but none were captured by the enemy.

FIGURE 36.12 Lens closeup.

Program flight number 69, launched on August 23, 1963, introduced a two-bucket configuration. The second container increased the film capacity to 160 pounds. By the end of 1966, thirty-seven KH-4B systems had been launched, and thirty-five of them successfully reached orbit, furnishing sixty-seven buckets of usable, recoverable film.[20] By 1967, the mission life had increased from the original 1 day to 15 days in space.

CORONA: THE ARTIFACT AND ITS PRESERVATION

On February 22, 1995, President William J. Clinton issued an executive order to release space-based reconnaissance imagery.[21,22] On February 24, just 2 days later, Vice President Al Gore announced declassification of the *CORONA* program at the CIA headquarters. Soon afterward, a Memorandum of Understanding was issued between the Director of the National Reconnaissance Office, on behalf of its program elements, and the Smithsonian Institution, on behalf of the National Air and Space Museum. Among other stipulations, it was agreed that "NASM is dedicated to the collection, preservation, study and display of objects of historical significance documenting the accomplishments of the nation's space program including its art in the

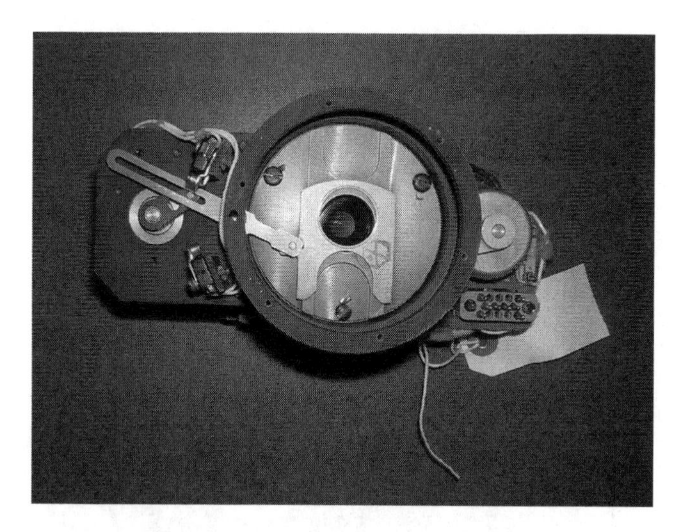

FIGURE 36.13 Back camera. The lenses were removed for the restoration purposes; metal frame showed corrosion deposits that needed to be stabilized.

preservation of peace and is the most appropriate repository for the DISCOVERER/ *CORONA* camera and related artifacts."[23,24] Pursuant to section 77d of title 20 of the United States Code, the transfer of *CORONA* became effective as of February 24, 1995. The lengthy declassification process revealed technical notes about some aspects of the mission and choices of design to meet those objectives. As illustrated above, these documents enabled understanding of the artifact, its components, its functions, and their interrelations, which in turn served as the basis for developing a careful preservation plan.

The preservation of any artifact aims toward stabilization of the original structure and materials with minimum interference with the original fabric and original concept; any alteration may inadvertently change the message that the artifact was intended to convey. Thus, familiarity with the history and purpose of the artifact's creation is the starting point for development of the preservation plan. The artifact's purpose and function, and the conditions in which it was intended to perform, justify the choice of materials, structure, and design of components. Intricate space artifacts, and in particular those specifically designed for reconnaissance missions, present a particular challenge for a museum specialist. By virtue of their nature, the reconnaissance space vehicles are transferred to a museum without documentation. Loss or destruction of some records and slow declassification of the remainder present a challenge to a curator, who is responsible for historical interpretation, and hinders the transfer of the information that usually serves as base for designing proper museum care.[25] Evolving materials technologies are obscured by changes of polices, so the object itself becomes an icon of past achievements and an archaeological specimen.

The *CORONA* camera, which was accepted by NASM in 1995, has remained on exhibit ever since. After 14+ years of continuous open display (without a

(A)

(B)

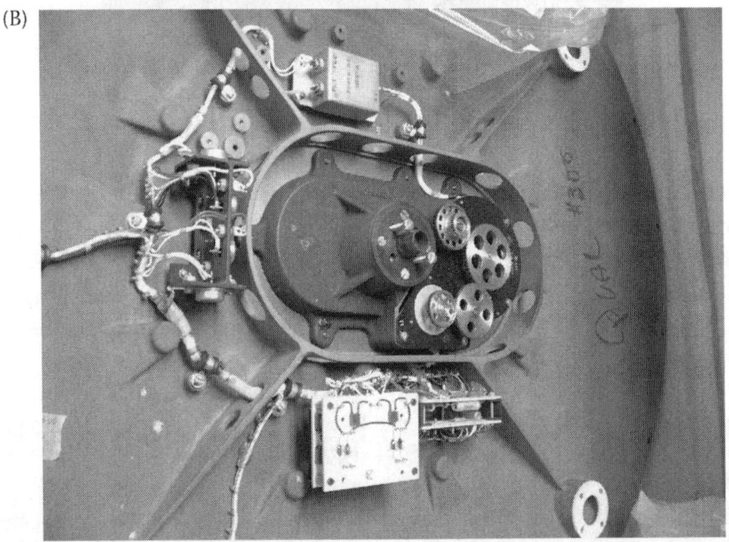

FIGURE 36.14 Film cassette, closed (A) and open (B), illustrating film winding mechanism on the interior of the cassette cover. (National Air and Space Museum object.)

FIGURE 36.15 Illustration of a double-feed film system, at various stages of film movement, including closeup of the intermediate roller assembly.

protective case) in the one of the most popular Smithsonian museums (with a visitorship of more than 90,000 on a holiday weekend), the artifact was in great need of preservation. For curator and conservator alike, it presented an unparalleled opportunity to learn about the structure, materials, and intricate components of the artifact.

The historical background of the *CORONA* mission and the function of the camera provides an understanding of how the materials were expected to perform in space. Comparing the artifact's structure with the technical notes carried out in parallel with the disassembling of parts guided separation of removable elements from the camera body. The existing technical notes were retrieved from the National Archives of National Reconnaissance Office after a lengthy process of the curatorial research.

The process of developing the preservation plan relied on the following:

1. Historical research into the mission and program
2. An understanding the artifact's components and choices made to fulfill the mission objectives

FIGURE 36.16 *CORONA* capsule with intake blades.

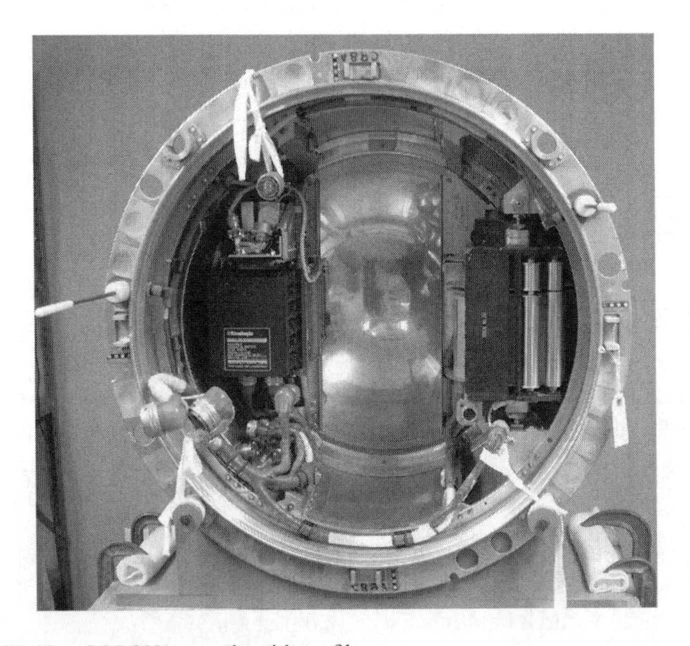

FIGURE 36.17 *CORONA* capsule without film.

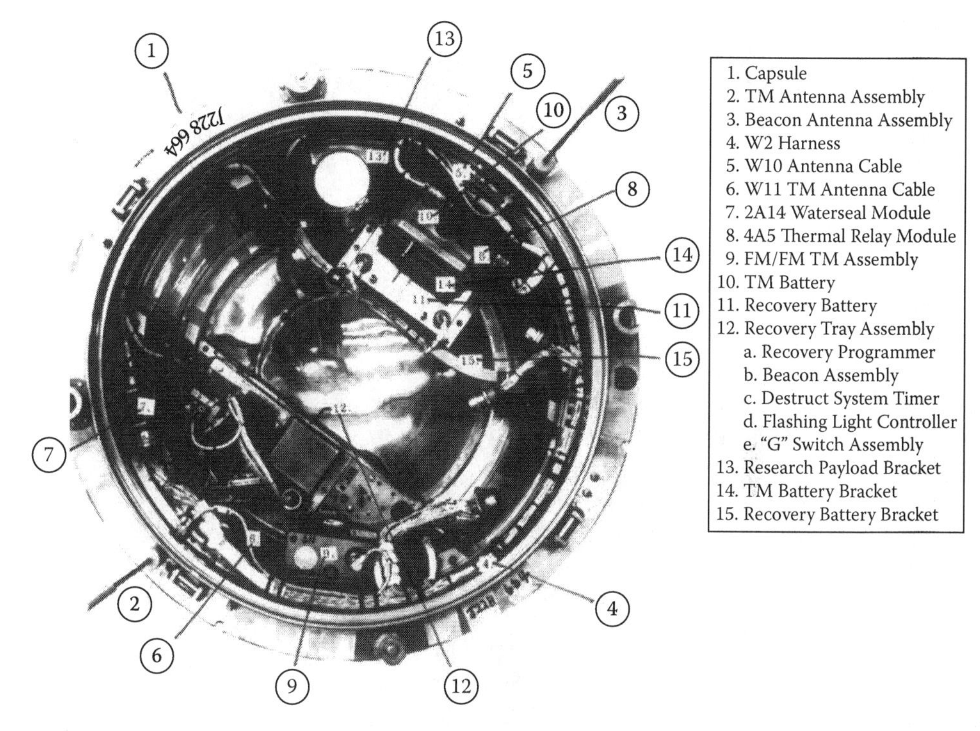

FIGURE 36.18 Engineering diagram of the *CORONA* capsule and its components.

3. Designing the least invasive conservation process that chemically stabilized the artifact in situ, particularly arrested metal corrosion
4. Photographic documentation of every component that underwent conservation/stabilization
5. Carrying out the actual stabilization process
6. Selection of materials accepted by the conservation community for the treatment of museum artifacts, such as metal sculptures or composite materials
7. Written and photographic documentation of the procedures applied to the artifact and discussion of materials selected for preservation treatment

CONCLUSION

The final product—a preserved artifact—represents a collaborative effort of a curator's historical research and a conservator's interpretive analysis of structure and materials. *CORONA* as a museum preservation case study is particularly interesting because, in spite of the extensive work carried out over the past several years, many areas are left to be illuminated in the future with the progression of the declassification process, which will reveal details about the products of the program and its instruments.

FIGURE 36.19 Top: Exterior of the capsule and elements identified as indicated on the engineering drawing. Bottom: Valves on capsule and drawing-diagram.

NOTES AND REFERENCES

1. Acknowledgments for contributions and support: James David, curator of the NASM, Division of Space History, for his contextual editorial contribution and generously sharing information regarding the history of *Corona*; Roger Launius, senior curator at NASM, Space History Division, for his encouragement and support; and Elisabeth West FitzHugh, research associate, Freer and Sackler Galleries for her invaluable editorial and contextual input.
2. Greer, K. 1973. *Corona*, in Ruffner, K.C., ed. *CIA Cold War Records. CORONA: America's First Satellite Program*, pp. 3–39. Washington DC: CIA History Staff, Center for the Study of Intelligence.
3. Studeman, W.O. 1995. Remarks at the Signing of the Executive Order Declassifying Early Satellite Imagery. ADCI Speech 2/24/1995. www.GlobalSecurity.org.
4. Cloud, J. 2001. Imaging the World in a Barrel: *CORONA* and the Clandestine Convergence of the Earth Sciences. *Social Studies of Science* 31(April): 231–251.
5. *CORONA* Program History. 1976. *Recovery from Orbit; Re-entry Vehicle Personnel and Development Testing.* Vol. VI, Part II. Directorate of Science and Technology, Central Intelligence Agency; HQ Air Force Special Projects Production Facility. Declassified and released by NRO in Accordance with E.O. 12985 on November 26, 1997.
6. Greer 1973 (note 2). Ruffner (note 2).
7. Oder, F., and M. Belles. 1997. *CORONA*: A Programmatic Perspective. In *CORONA, between the Sun and the Earth: The First NRO Reconnaissance Eye in Space,* ed. R. McDonald, 75–84. Bethesda, MD: American Society for Photogrammetry and Remote Sensing.
8. David, J. personal communication, 2008.
9. Greer 1973 (note 2). Ruffner (note 2).
10. Ruffner, K., ed. 1975. Part I: History of the *CORONA* Program. In Ruffner (note 2), 1–3.
11. Ibid.
12. Harris, J.K. 1997. A Look Back to *CORONA* and A Look Forward to the Information Era. In McDonald 1997 (note 7), 277–287.
13. Helms, R. A DCI Perspective. (Foreword.) In McDonald 1997 (note 7), v–vii.
14. Hall, C. 1997. Post War Strategic Reconnaissance and the Genesis of Project *CORONA*. In McDonald 1997 (note 7), 25–60.
15. Greer 1973 (note 2). Ruffner (note 2).
16. Smith, D. 1997. The Design and Engineering of *CORONA*'s optics. In McDonald 1997 (note 7), 111–120.
17. Greer 1973 (note 2). Ruffner (note 2).
18. *CORONA* Program History 1976 (note 5).
19. Greer 1973 (note 2). Ruffner (note 2).
20. Greer 1973 (note 2). Ruffner (note 2).
21. After McDonald 1977 (note 7).
22. Clinton, W.J. 1995. Executive Order 12951: Imagery Acquisition by Space-Based National Intelligence Reconnaissance Systems, February 22, 1995. In McDonald 1997 (note 7), 359–360.
23. Memorandum of Understanding between Director of NRO and Secretary of the Smithsonian Institution, February 24, 1995. In McDonald 1997 (note 7), 326–327.
24. McDonald, R. 1997. Introduction: The Impossible Dream. In McDonald 1997 (note 7), 1–13. McDonald, R. 1997. *Corona*, Argon, and Lanyard: A Revolution for US Overhead Reconnaissance. In McDonald 1997 (note 7), 61–75. McDonald, R. 1997. Lessons and Benefits from *CORONA*'s Development. In McDonald 1997 (note 7), 255–269.
25. David, J. 2008. Two Steps Forward, One Step Back; Mixed Progress under the Authoritative System Declassification Review Program. *The American Archivist* 70: 219–251.

37 In Situ Preservation of Historic Spacecraft

Robert Barclay and Randall C. Brooks

CONTENTS

INTRODUCTION

The deliberate deorbiting of the *Mir* space station provides a focus for discussion of a new consciousness of the value of space artifacts. This chapter examines the accretive technical and social processes whereby such artifacts as *Mir* become historic objects worthy of preservation. Protocols for decision-making are discussed in view of comparative technical and historical value and feasibility of conservation. Preservation of space vehicles in situ is discussed, with particular reference to safety, monitoring, potential future public access, and long-term costs. An argument is made for a wider definition for World Heritage designations to include material beyond the surface of the Earth, and for international bodies to assess, designate, monitor, and oversee such projects.

When space station *Mir* was deorbited in a shower of burning wreckage over the Pacific Ocean on March 23, 2001, as shown in Figure 37.1, it became, perhaps unwittingly, a quintessential symbol of the ugly side of modern industrial society. It was obsolete, beyond economic repair, and technically outdated. And in the modern world this means only one thing: disposal. But what came down that day was, in fact, an irreplaceable example of world heritage. *Mir* symbolized humankind's exploration of space in a way that other, more evanescent objects did not. It became the first permanent manned presence in Earth orbit, its personnel set records for long-duration spaceflight, and it survived 15 years in orbit—5 years past its planned working life. It was the site of both danger and triumph. Cosmonauts and astronauts who have lived on *Mir* describe the experience as one in which they were left feeling a close affinity for the station; it embodied the emotional difference between a habitation and a home. This was in spite of the near catastrophes of 1997, when *Mir* suffered a fire when an oxygen canister caught fire, and later when a Progress module full of garbage collided with the Spectr module, causing loss of pressure and power for several hours (the damaged *Mir* is shown in Figure 37.2). In fact, it is crises such

FIGURE 37.1 Space station *Mir*'s descent into the Pacific over Fiji on March 23, 2001 (From Reuters).

as these that actually bring people closer to one another. A number of other minor technical problems caused power failures, loss of computer control, and loss of radio communication with ground stations. Though *Mir* was battered, the station's crew always averted worse disasters and kept it flying. However, such accidents were offset by triumphs when *Mir* set records for duration and when astronauts from states that had been Cold War enemies worked together. Even the United States recognized *Mir*'s utility when planning the next phase of space exploration, the International Space Station. Physicists, astronomers, biologists, and other specialists from many countries around the world were invited to participate in its multitude of scientific and exploratory activities.[1] *Mir* was the meeting place of countries and cultures. It is through such human activities that objects acquire meanings that transcend their materials and techniques of fabrication. *Mir* was much greater than the sum of its parts. In this chapter we explore why objects such as *Mir* should be preserved and how this might be accomplished, and we conclude with some speculations upon the future of museums beyond the atmosphere of Earth.

FIGURE 37.2 A view of space station *Mir* showing damage to solar panels. This picture evokes the beauty of many pragmatically designed space structures. (Courtesy of NASA.)

MUSEUM PERSPECTIVE

Museums throughout the world boast examples of early steam engines, pioneering cars, and antique aircraft, all conscientiously preserved and celebrated for what they are and represent. To the general public they are touchstones to historic technical developments or design values, while to students, historians, and technical specialists they are sources of valuable information. But unlike all land-based vehicles, historic spacecraft present new and challenging problems. Hitherto, space exploration has been essentially disposable; all that museums can exhibit are test pieces, mock-ups, and backup units, because the real things are almost always lost. The only pieces of the Mercury, Gemini, and Apollo craft that returned to Earth were the astronauts' capsules, or command modules. The Canada Science and Technology Museum in Ottawa had the command module of *Apollo 7* on display for 30 years, as shown in Figure 37.3, with all its scorch marks from reentry, and other evidence of subsequent dismantling and testing to assess its performance. But until the advent of the space shuttle, the remaining parts were always disposed of in a process that was the epitome of planned obsolescence. This often resulted in the loss of culturally and historically valuable items. For example, the lunar excursion module (LEM) that was used as a lifeboat for the *Apollo 13* crew is a prime example of the consequences of such a loss. Thirteen hours into the voyage, an explosion rendered the mission impossible to continue and necessitated a loop around the Moon, using the LEM instead of the service module to keep the astronauts alive. After a highly eventful and near fatal voyage, that heroic little spacecraft, like the other LEMs that had set foot upon the Moon, ended its life in flames over the Pacific Ocean. It is noteworthy here that we apply the word *heroic* to an inanimate object, witness to the power of subjective personification.

As museum professionals, we feel very strongly that representative examples of all technologies must be rescued and preserved. Space travel has now reached a maturity that encourages and demands retrospection. The loss of *Mir*—but more importantly, the attitude behind its disposal—strengthened and focused our feelings and made us realize that the definition of heritage, and the range of material that humankind should preserve, must be widened. We also understand the enormous technical and financial challenges that such a broadening of heritage definitions would incur. There are strong arguments to suggest that this is not a viable objective, but who, until the 1960s, would have considered the raising and preservation of a lost seventeenth-century Swedish vessel technically possible?[2] But now the *Vasa*, preserved in Stockholm, impresses all who see her, and the ship and its supporting exhibits provide information on the culture, trades and politics of the time (1628) that is rarely found by other means (as shown in Figure 37.4). More recently, and building upon the techniques learned in Stockholm, the *Mary Rose*, flagship of the English King Henry VIII, has been raised from the sea and has provided a wealth of minute detail on everyday life aboard a warship of the sixteenth century. *Mir* would have been a similar unique window on the workings of society of our times for those in the twenty-fifth century and beyond. As the first permanently occupied space vehicle, its position in space history is unique. This is not, of course, to say that *Mir* could have been saved indefinitely—this was beyond technical capability at

FIGURE 37.3 The command module from *Apollo 7* being mated to the service module, with the ablation of the heat shield shown in the inset (From NASA and Canada Science & Technology Museum.).

the time, and will be perhaps for decades to come—but there is no reason to suppose that a temporary solution would not have been available given the necessary political will. This case serves as a strong example of the parallel directions in which cultural thought and technical feasibility travel. If there is sufficient cultural drive, technology will be harnessed.[3]

We deliberately make the above association with underwater archaeological sites as there are parallels between ships and vehicles in a space environment. A draft proposal was placed before the General Conference of the United Nations Educational, Scientific and Cultural Organization (UNESCO) in late 2001 concerning the preservation of marine sites.[4] The introduction states that

> At present, there is no international legal instrument which adequately protects the underwater cultural heritage, which is increasingly threatened by pillage and natural

FIGURE 37.4 The *Vasa* was so well preserved that, after being salvaged, she was able to float unaided. (Photo by Sture Silvander. Courtesy of Vasamuseet.)

> damage. This has led to the irretrievable loss of a vast part of our collective cultural heritage.

The authors go on to state

> Shipwrecks are invaluable in restructuring life-styles no longer existing and represent a buried treasure in terms of knowledge about life on board, boat construction and trade routes. A shipwreck is a time capsule waiting to be unlocked since time stops when a vessel founders.

Preserved pioneering spacecraft will provide future generations with the same glimpse into our times, and although surviving spacecraft are not in imminent danger of plunder, the fact that groups are actively participating in competitions to put privately funded manned craft into space suggests that this danger may not be too far off. More importantly, though, gravity is a slow but unrelenting plunderer of craft in low Earth orbit.

HISTORIC OBJECT

WHAT MADE *MIR* HISTORIC?

Some people considered *Mir* to be a historic object; others saw it as a mere disposable item. What factors in society drive such diverse viewpoints regarding the same object? To answer this question, we need to digress into an area perhaps not normally encountered by readers of technical and scientific publications. We move from the technical history of objects to their social history. Like all other artifacts, space vehicles are commodities; the process of commodification is conferred upon them as a result of their manufacture to suit a specific need. However, with some particular commodities, a process called singularization takes place. It is a social/historical view that society needs to set apart some portion of objects, or commodities, as "sacred" and that "culture ensures that some things remain unambiguously singular."[5] In a state society, such objects may belong to the "symbolic inventory of a society: public lands, monuments, state art collections" and so on.[6] The process of singularization, of turning a disposable commodity into an object of unique value, is a social transaction.

Michael Thompson's work *Rubbish Theory: The Creation and Destruction of Value* has provided new insights into social transactions from the point of view of the collector and connoisseur. He argues that artifactual commodities are generally assigned to the categories of either transient or durable, both of which exist in a region of fixed assumptions where worldview precedes societal action. In other words, the object's value—monetary, aesthetic, and cultural—in both these categories is clearly circumscribed, and society's action toward it is therefore predetermined.[7] A transient object is one whose value is falling. It is a utensil, like *Mir*, that is in the process of being used up, and at some stage its value, both monetarily and culturally, will approach zero. Typical examples of such objects are disposable ballpoint pens, cigarette lighters, and, sadly, such sophisticated and intricate items as cell phones. A durable object, on the other hand, is one that has been assigned aesthetic, monetary, or other values by society, and these values are either stable or increasing. Should a head of state use the ballpoint pen to sign a peace treaty, should the cigarette lighter be applied to that head of state's celebratory cigar, and should the cell phone have been used to cement the deal, these objects would then (with sufficient documentation) make the leap into singularization. If such trivial items can become valued by society, what then of whole space stations? It is clear that any one of many incidents aboard *Mir* had the potential to provide the necessary "societal push"; together they made it inevitable.

Interestingly, both Thompson and Igor Kopytoff, another commentator on the meanings of objects in society, use automobiles to illustrate the social transactions that may take place around a complicated functioning artifact subjected to intricate and varied interventions.[8] Thompson uses the example of the car to illustrate variations in market value, while Kopytoff demonstrates the diversity of potential social biographies such an artifact may encourage.[9] In both cases, the significance of the choice of the car lies in the combination of its intricacy of operation and its social symbolism. Because it is both a functioning machine kept in working condition by technical intervention and a focus for social transactions, it carries information of both technical and social value. A space station provides an equally valid example. Like a motor car, it is supplied from the manufacturer in new condition, must be brought into a working state by its users, and is then serviced, adapted, repaired, and altered to suit the exigencies of continuing use. In the process it accumulates both social and technical biographies.

Objects that have been singularized by societal transactions in this way—those that have passed from transient to durable—are often referred to as historic. Until the twentieth century, history had been confined to the study of events far enough in the past that academic perspective could be ensured. In the twentieth century, the definition of history itself underwent a metamorphosis, and it is now quite legitimate to consider events of only a few years ago as genuinely historic. The same is true for museum objects, and collections now comprise many categories of recently produced artifacts. An object may be designated by society as historic by the application of any one of an open-ended list of cultural markers relating to such features as ownership, antiquity, beauty, historical and monetary value, and, of course, events that the object may have "witnessed" or, as in the case of *Mir*, influenced.

To Destroy or to Preserve?

If *Mir* could indeed have been defined by societal pressure as a historic object—one worthy of addition to the world's "symbolic inventory"—why was its destruction seen by some as necessary and expedient and by others as a tragedy? It all depends upon how one looks at it. Any object signifies different things to different people. What exists in the mind of the observer is the semiological idea of the object, and this idea is a social construct compounded of knowledge about the object and attitudes and assumptions toward it. People's reactions to such objects are necessarily integrated; their perception is selective and conditioned by both immediate surroundings and past and ongoing experiences. Thus, such an object is polysemic in nature, in that different people react in different ways to it. At a primary level, *Mir* was an obsolete piece of space hardware. All agree on this. Because such primary level significations are shared by all viewers, they carry no emotional implications. However, at the secondary level of signification, which Roland Barthes names *connotation*, contemplation of the future disposition of the object leads viewers into a cognitive level of subjectivity. It is at this level that "myth is created and consumed."[10] To one viewer *Mir* might be an orbiting hazard, while to another it might provide the touchstone to fantasies upon the nature of space exploration. In both "readings" the subjective extrapolations go far beyond the bounds of the physical object. The widely divergent readings of the object, and the equally divergent potential actions, are thus channeled by the cultural predisposition of the viewer and programmed by the social milieu in which the observation takes place.

Such conflicting opinions on historic objects are generally confined to modern material. The continuing preservation by society of those artifacts that have existed for considerable periods of time is rarely held in question. They are solidly durable. Russian officialdom, for example, would never have dreamt of destroying an out-of-date artwork or an ancient building that had lost its function. The Hermitage and its contents are safely ensconced in familiar cultural and social territory. Similarly, humankind would hardly condone the destruction of the historic spacecraft that dot the surfaces of the Moon, Mars, and Venus. This demonstrates that the "test of time" is still valid, but it also shows that those who champion the preservation of modern material, especially technical artifacts, do so under significant resistance from a large sector of society.[11]

SPACE TOURISM

The future of spaceflight for the general public is tied closely to commercial interest; if people are willing to spend the money, a way will be found to help them spend it. The pioneering 7-day, US$20M sojourn of millionaire Dennis Tito on *Mir* showed that private individuals could aspire to become space tourists. South African Mark Shuttleworth and others have since shown how far it is possible for an individual to go.[12] A recent study by Azabu University in Tokyo has shown keen interest for space travel from the general public, with 80% of respondents under the age of 30 stating that they would willingly spend large sums for the opportunity. In 2001, the Japanese tourist company Kinki Nippon formed a branch dedicated to developing and marketing space tourism. The company estimated (perhaps over-optimistically) that by 2030 as many as five million people per year could take trips into space. Kinki Nippon has since merged with JTB Corp., which is partnering with Space Adventures Ltd. to bring space tourism to consumers. The winning of the X Prize in 2004 by *SpaceShipOne* (Figure 37.5), designed and built by Burt Rutan's company Scaled Composites, signaled a new era in private spaceflight. An agreement has been signed between Scaled Composites and Sir Richard Branson (founder of the Virgin Group) to form a new aerospace production company to build a fleet of commercial suborbital spaceships and launch aircraft. *SpaceShipTwo* and the *White Knight Two* launch system will be manufactured and marketed to space line operators, including Branson's Virgin Galactic.[13] Spaceport America near Las Cruces, New Mexico, is slated to open in late 2009 or early 2010, and it is here that Virgin Galactic's *SpaceShipOne* will make its home. Although these adventures are

FIGURE 37.5 Burt Rutan's *SpaceShipOne* is slung below its mother ship *White Knight* during a test flight in 2003. Virgin Galactic's *SpaceShipTwo* will be based on a similar design (Courtesy of Scaled Composites).

still suborbital, it is clear that space tourism is rapidly becoming more than mere futuristic speculation.

Visits to preserved space vehicles may well become a highlight of space tours in the later twenty-first century. In the eighteenth century it was the fashion to take the grand tour and to visit the sites of ancient Egypt, Greece, and Rome. In the nineteenth century the museums, galleries, and historic sites of the world burgeoned under new waves of historic cultural awareness. The twentieth century saw the tourist industry become the main economic driver of the globe. In the twenty-first century we can be sure that people will be visiting heritage sites beyond the confines of our atmosphere. It is to be hoped that they will find something worth visiting when that time comes. In the following section we lay the speculative groundwork for such cultural tourism.

STRATEGIES FOR PRESERVATION

There are two classes of space objects, those that are small enough to retrieve and those that, if they are to be preserved, have to be maintained in space because of their size or fragility. On *Mir*, when experiments were completed, old equipment was stuffed into nooks and crannies of one of the modules or was junked on Progress modules that burned up on reentry. For a curator struggling with the problems of preserving space technology, it is not an easy choice to resolve—that is, to preserve all or just parts, such as the scientific equipment, of a significant artifact. In the early days of space science, engineering models provided one way to preserve a three-dimensional representation of the heritage object that later flew but was rarely recovered. With years of design experience and the advent of computer-aided design/computer-aided mapping techniques, these models have generally fallen by the way-side and we are being left with a tiny fraction of space hardware being preserved, hence the desire and need to go beyond preservation of representative objects.

We consider it beyond argument that larger spacecraft and their components should be preserved in space. Accretive structures such as space stations are far too extensive and were never intended to be self-supporting, requiring periodic servicing even during their useful lives. Nor could they be engineered to survive reentry. Dismantling for recovery by a space shuttle would, of course, be possible but very expensive and demanding of high-end technologies. In situ preservation of spacecraft is, on the other hand, respectful of the historic object's context, regardless of whether or not it would be feasible to return it to Earth.

One could contemplate returning to Earth just some of the many pieces of scientific experimental apparatus that normally burn up in a fiery descent, but such a strategy is a second-best solution. It is akin to removing just the Elgin Marbles to the British Museum, and leaving the Parthenon without them (a situation many are trying to reverse after more than 150 years). Fortunately, the Parthenon survives. What we are looking for is to maintain objects in their entirety in Earth orbit, but in a place that is both safe and accessible.

LOW EARTH ORBITS

Satellites in low Earth orbit (less than 500 km) pose the greatest challenge for preservation as there is constant atmospheric drag on these satellites, slowing their motion and ultimately pulling them toward the Earth. Given the cost of maintaining such satellites, an orbit in excess of 1,000 km is most desirable for a parking orbit, as the added altitude then decreases the ongoing costs and frequency of visits for reboosting the satellite. However, the technical problem of boosting a massive object like *Mir* to such a high orbit is anything but insignificant. Add-on boosters to raise and continuously maintain the orbit is clearly a more practical approach than using the small on-board rockets used to maintain the craft's orbital station. However, such major additions require consideration of this potential need before the craft is even launched.

Under current practices, satellites in low Earth orbit are, in the vernacular of the industry, deorbited into the Pacific, but in the end, if a station like *Mir* is to be preserved for posterity, strategies from the early design stages will need to be instigated. For example, mountings and connections for externally mounted fuel cells and computer modules, with backups, to monitor and control the satellite independently of the original hardware would be required. Of course, these strap-on modules would need to be replaced as they failed or as the fuel was exhausted.

PARKING AT A LAGRANGE POINT

Geosynchronous communications or meteorological satellites (in orbits 35,786 km from Earth), on the other hand, are pushed out from the Earth on disposal, and they gradually spiral outward, to be lost from Earth orbit. To preserve these, and perhaps for some of the most significant low Earth orbit satellites, the logical spot is at one of the Lagrange points. Here, only small amounts of fuel would be required to maintain a satellite's position for long periods.

The eighteenth-century mathematician Joseph Lagrange discovered, when solving problems in orbital mechanics, that there are five points at which conditions of gravity between two bodies are balanced (except for small disturbing forces from other more distant objects) (Figure 37.6). For the Earth-Moon system, there are three points on the line between the two bodies. The L_{M1} and L_{M2} points are stable, while L_{M3} is less so, and being on the far side of the Moon, L_{M3} would make a less desirable parking location. A satellite placed at L_{M1} or L_{M2} would stay there with relatively little energy required to keep it on station. A craft placed at the L_{M4} or L_{M5} points, which lead and trail the Earth-Moon in their orbit around the Sun, are stable as well, though they also "orbit" as the Moon goes though its monthly orbit around Earth. For this reason, they may not be as desirable for space tourists. However, these points have been proposed as locations of space colonies, in which case parked historic satellites might be one of the first cultural attractions for an otherwise desolate region of space.

We should also mention that there is a similar set of five Lagrange points for the Earth-Sun system. The solar observation satellite *SOHO* is in a "halo" orbit slowly moving around L_{S1} in approximately 178 days. The *MAP* satellite (Microwave Anisotropy Probe), launched in June 2001, is currently near the L_{S3} point, and NASA is considering placing the replacement for the Hubble Space Telescope at L_{S3} as well. After the two-and-a-half month voyage to L_{S1}, *SOHO* is maintained in its halo orbit

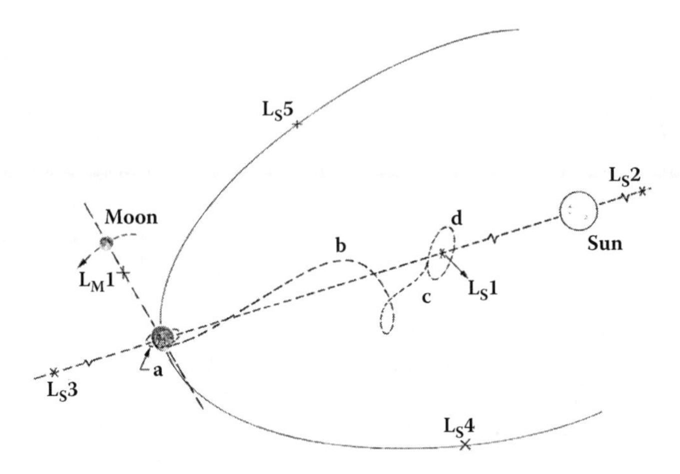

FIGURE 37.6 The Lagrange points for the Earth-Sun system (not to scale) showing the orbital path of the *SOHO*, which uses point L_{S1}, which is approximately 1.5M km from Earth. L_{S4} and L_{S5} are 150M km. (i.e., the same distance as the Sun) and thus impractical for our purposes. L_{M1} is approximately 300,000 km from Earth. The velocity required to get an object from Earth orbit to L_{S1} is 3.2 km/s.

with hydrazine thrusters, and it has been there since February 1996, with several years left in its useable life. Because of their relatively large distances from Earth, (1.5M km, or approximately four times the distance between the Earth and the Moon for L_{S1}) and nearly 150M km for the L_{S4} and L_{S5} points, these regions probably do not make as much sense for "visible storage" of historic spacecraft for quite a long time to come.

As an example of the challenges, the *SOHO* craft, launched with an Atlas/Centaur rocket, reached its transfer orbit at 180 km with a speed of 8 km/s. To escape Earth orbit, a craft requires a velocity of 11.2 km/s, which means that to park a satellite at a Lagrange point one needs to add approximately 3.2 km/s to an orbiting satellite's velocity—clearly a manageable feat even today. These numbers do not preclude speculation on long-term possibilities. If one assumes the continued pace of humankind's technical development, the Lagrange points could become, within decades, an extremely useful holding place for disused but potentially "collectible" space vehicles. A kind of space scrap yard would be established, which could be visited by tourists from Earth. Thus, potentially dangerous objects that exist in increasingly crowded low orbits would be towed out by space tugs to a final resting place that was safely out of harm's way.

Until that time comes, lower orbital slots must continue to be used. Presently, such slots are assigned to the most appropriate use by international committees, and perhaps it is time that international space law experts tackled this issue on behalf of World Heritage.

"WEATHERING"

Damage due to weathering is a common concern in Earth-based outdoor artifacts and structures. The vacuum of space might be considered the most benign of preservation media, but this is far from the case. Space objects suffer continual bombardment from micrometeorites, subatomic particles, and hard radiation. In 1984 a satellite termed the Long Duration Exposure Facility was deployed from the space shuttle. Its fifty-seven experiments involved more than 200 investigators from ten countries. The experiments covered long-term effects on materials, coatings, thermal systems, power and propulsion systems, science experiments, and electronics and optics. The satellite was retrieved in 1990 and the results analyzed. Among other experimenters, the University of Toronto Institute for Aerospace Studies provided a series of test samples of epoxy matrix composite materials using graphite, boron, and Kevlar fiber reinforcements. One of these samples (now in the collection of the Canada Science and Technology Museum) is illustrated in Figure 37.7. It is clear that degradation and damage take place upon various materials in relatively short periods of time. While action is being taken to protect materials used in working spacecraft, or to use materials of known durability, there is little that can be done for potentially historic material. It is sufficient at this stage to be aware of this issue and to plan for remedial action when it becomes feasible.

In addition to the effects of a harsh environment, there is the issue of materials that were never intended for long-term stability and that may have been chosen for other characteristics of a more immediately useful nature. Many modern synthetics and composites possess properties designed for specific functions, and in a world of planned obsolescence there is no need for long-term stability, especially if it will compromise initial function. As with the weathering referred to above, it is sufficient to record this information.

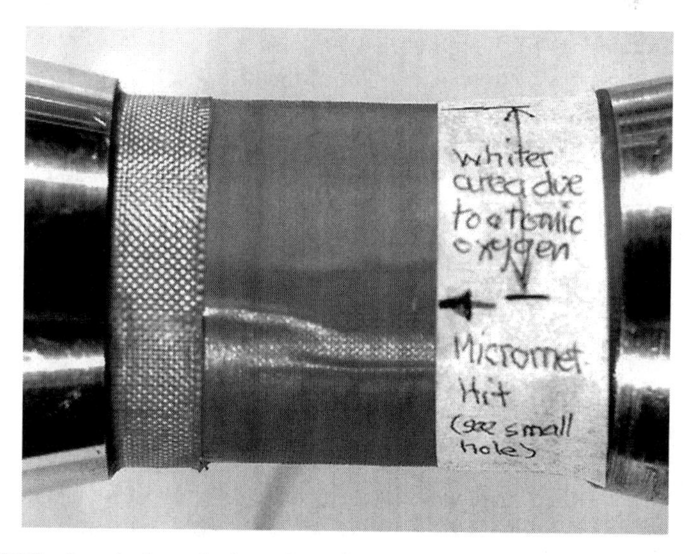

FIGURE 37.7 Sample from the Long Duration Exposure Facility. This sample shows the effects of both micrometeoroids and atomic oxygen. The latter caused the lighter area, while the darker region resulted from ultraviolet radiation affecting the plastic material. (From Canada Science & Technology Museum.)

DECISION MAKING

Assessing Cultural Value

Financial constraints were a primary reason for the decision to deorbit *Mir*, a decision that might have been different in better economic or political times, or had *Mir* been a NASA or combined NASA/European Space Agency craft. This decision was made in the absence of an international structure for assessing and assigning historical and cultural value. For many years museums and other cultural organizations have had criteria by which they judge the worthiness of an object for preservation. However, within those agencies, scientific and technological objects often take a back seat to paintings or household furniture, even though in the overall scheme of the evolution of society, scientific and technological artifacts have had a much more fundamental impact on our modern world.

In establishing the worthiness of spacecraft for preservation, as for any scientific or technological object, contributions to "firsts" are always primary considerations. This was clearly the case for *Mir*. Among many firsts, it was the first manned program that involved many nations cooperatively; it established the viability of a permanent orbiting space laboratory; and it influenced the design of the next-generation craft, the International Space Station. Not only did the craft itself contribute to the advancement of technology, but many of the experiments carried on board incorporated leading-edge technologies. Many of these were lost over the Pacific, having been stored on completion of their experiments. On the social and political side, *Mir* was in its early years a symbol of USSR pride, and with the changing political climate of the 1990s, *Mir* helped bridge and influence political and cultural differences.

A similar series of historical and technical firsts is associated with the Hubble Space Telescope (HST)[14] (Figure 37.8). Launched in April of 1990, the HST immediately expanded astronomers' view of the heavens, and from a cultural perspective, it gained an enormous popular following with the advent of the World Wide Web and digital imaging technology. Space shuttle service calls to the telescope (the first of which was to correct an embarrassing optical error) became television dramas and captured the imagination of viewers around the globe. The near-disastrous optical problem was turned into a public relations triumph. To date, the HST has had more than 129 hours of extravehicle servicing from four space shuttle visits (STS-61 in 1993, STS-82 in 1997, STS-103 in 1999, and STS-109 in 2002),[14] and the final servicing mission was in 2008. The telescope is in a 600-km-high orbit, so orbital decay due to atmospheric drag is slow, but it has required one boost to its original orbit. Nevertheless, there will come a time when the technology is obsolete and servicing is too costly to continue. The current objective is a planned retirement in 2013. The Hubble Space Telescope will then sit on the knife edge between transient and durable; it will be either rubbish to be discarded, or a historic object to be preserved. It is to be noted that while the HST was small enough to lift into orbit in one space shuttle mission, it could not be returned to Earth by the same means. Thus, its appearance in a conventional terrestrial museum is not at present an option.

Given all of these features of *Mir* and Hubble as historic objects, their preservation would be guaranteed had the decision been made by a typical museum committee

FIGURE 37.8 The Hubble Space Telescope before deployment of its solar cells. A historic object in the making. (Courtesy of NASA.)

at a local level. Following the procedures used at the Canada Science & Technology Museum, for instance, we have an established collection policy and thematic framework that helps curators assess an artifact's significance. The curators as a group assess the arguments and make carefully weighted decisions. The fate of *Mir* and the future of Hubble should focus our attention, not just on defining what technology deserves to be preserved but how we can go about it on an international scale. In particular, the cost of preservation of large industrial heritage sites or spacecraft demands a coordinated effort. The implications, costs, and preservation strategies are an order of magnitude greater. The scientific, technical, and cultural significance of such artifacts must be weighed on a global scale by a wide range of experts from diverse backgrounds.

Need for a Coherent Collection Development Strategy

Because of the costs and collaborative efforts that will be required to preserve space artifacts that have been launched, or historically significant sites from our early years of exploration, a well-developed decision making process is essential. The model used by the Canada Science and Technology Museum (based on that developed by, though subsequently abandoned by, the Ford Museum) is an excellent guide to a rational decision-making process. It requires a dedicated and prolonged commitment in both financial and human resources. However, the upfront costs for decision making are paid for many times over in the long-term improvements in the quality of the collection that is developed as a result of the process.

The Canada Science and Technology Museum has had and follows one of the most rigorous collection development strategies (CDSs) of any comprehensive science and technology museum. Its CDS has been in use for almost 20 years and has resulted in a massive amount of research—the basis of its collecting activities and an excellent model for the systematic development of a space technology collection, whether on the ground or preserved in space.[15]

The principle objectives of the CSTM strategy are as follows:

- To address the future scope and direction of collection and research
- To identify and prioritize issues, ideas, subject areas, or themes requiring research and development attention
- To provide access to the collection

The strategy outlines an intellectual framework upon which collection development decisions are based on rigorous research, peer review, curatorial assessment of artifacts for addition to the collection based on the research, and periodic review of each subject area collection. The following proposes a similar framework with appropriate themes for an international effort to define and preserve space hardware, significant sites, and so forth.

GUIDING PRINCIPLES FOR A COLLECTION DEVELOPMENT STRATEGY

The primary purpose of the collection will be to help people understand the transformation of human activities carried out in space resulting from the application of science and technology. It is accepted that ground-based collection development will continue to be the purview of national bodies and traditional museums. Collection activities will focus on the relationships between people and science and technology. The principal criterion for evaluating an object or site is the set of stories that the object or site tells, and therefore, its ability to foster understanding and engagement with all humans.

The following guiding principles will also be applied to the CDS process and decision making.

- A focused collection is achieved by identifying and acquiring those objects and protecting those sites and supporting documentation that best reflect the framework. Deaccessions occur with objects or delisting of sites that are not consistent with the Collection Development Strategy.
- Collection development decisions are consistent with established International Council of Museums (ICOM) museum professional standards and ethics.
- Curatorial research is fundamental to the development and management of the collection and public programming and education, as well as physical and intellectual access.

- Once an artifact or site is officially confirmed as international space heritage, its conservation is automatically committed to providing long-term preservation for all objects associated with the artifact or site.
- Conservation and restoration resource requirements will be assessed in the decision to acquire an object.

CDS THEME STATEMENT: THE HUMAN AND ROBOTIC EXPLORATION OF SPACE

"The Human and Robotic Exploration of Space" theme and its subthemes provide our intellectual framework to guide all aspects of collection development, including research. The following theme statement will guide future priorities, decisions, and activities of collection and research:

> *Scientific and technological endeavor has transformed humans and their understanding of their place and limitations in space.*

The main theme of "The Human and Robotic Exploration of Space" embodies the following subthemes:

- *Human context*: The human scientific and technological endeavor is reflected in the challenges encountered and the choices made in the development of space.
- *Finding new ways*: Science and technology play key roles in efforts to find new ways of living, learning, and working in space and, through robotic devices, the exploration of our planetary system and beyond.
- *How things work*: Developing a knowledge of how "things" work can help people better understand the factors that have contributed to the human and robotic exploration of space.
- *People, science, and technology*: Our culture and lives are shaped and influenced by scientific and technological change. At the same time, individually and collectively, people shape the evolution of science and technology through their decisions and actions.

WORLD HERITAGE DESIGNATION

We propose that submissions be made to a new purpose-specific Commission of the International Union for the History and Philosophy of Science (IUHPS), following the general principles used to assess UNESCO's World Heritage sites. We further propose a new body to assess space hardware because of the unique problems and the special expertise required in weighing submissions for space-based heritage sites (see also O'Leary, Chapter 43 of this volume). This body would comprise a wide range of international expertise, including scientists, engineers, curators, historians, conservators, and any others whose particular areas of specialization could be called upon as needed.

The following provides a very quick summary of how the UNESCO's World Heritage Committee and Fund oversees the assessment of sites and natural environments for designation with the International Council on Monuments and Sites (ICOMOS). From the UNESCO Operational Guidelines (1997), we learn:

> The cultural heritage and the natural heritage are among the priceless and irreplaceable possessions, not only of each nation, but of humankind as a whole. The loss, through deterioration or disappearance, of any of these most prized possessions constitutes an impoverishment of the heritage of all the peoples in the world. Parts of that heritage, because of their exceptional qualities, can be considered to be of outstanding universal value and as such worthy of special protection against the dangers which increasingly threaten them.[15]

Historic space vehicles would clearly fit into such a definition. The Operational Guidelines state further:

> The World Heritage Committee has four essential functions:
>
> - To identify, on the basis of nominations submitted by States Parties, cultural and natural properties of outstanding universal value which are to be protected under the Convention and to list those properties on the "World Heritage List"
> - To monitor the state of conservation of properties inscribed on the World Heritage List, in liaison with the States Parties
> - To decide in case of urgent need which properties included in the World Heritage List are to be inscribed on the "List of World Heritage in Danger" (only properties which require for their conservation major operations and for which assistance has been requested under the Convention can be considered)
> - To determine in what way and under what conditions the resources in the World Heritage Fund can most advantageously be used to assist States Parties, as far as possible, in the Protection of their properties of outstanding universal value.

Other than describing the exact position on the Earth, these criteria are suitable for assessing proposals on space vehicles, although as noted above, the skills required for assessment are beyond those of ICOMOS. Incidentally, review of proposals takes 1.5 years from receipt of nomination—hence, planning well in advance would be required for space vehicle preservation. This is an apposite observation in view of the rapidity with which decisions about *Mir* had to be made.

FUNDING

One of the chief reasons for not performing salvage or preservation strategies on *Mir* was the potential cost. In view of the financial situation facing Russia's space program at that time, this was a genuine concern. Lack of money is an argument that will doubtless be used in the future. However, it is clear from past large-scale preservation projects, such as the dismantlement and raising of the Temples of Abu Simbel in southern Egypt or the recovery of the *Vasa* in Stockholm, that if the project is deemed worthy by society, financing will be found.

The basic question that must be posed is whether it is the responsibility of nations to preserve such material, or whether there should not be an international organization with member states that provides disbursements. In current practice, once a proposal for World Heritage status has been granted, funding is provided via UNESCO's World Heritage Fund (WHF), which accepts contributions from a broad range of sources. Contributions may include those made by the member countries to the Convention; contributions, gifts, or bequests that may be made by other countries; UNESCO itself; and other organizations of the United Nations system, particularly the United Nations Development Programme. UNESCO's policies also allow other interdepartmental organizations, public or private bodies, or individuals to make contributions. Funds for the WHF may also be raised by collections and receipts from events organized for the benefit of the Fund or from other resources acceptable to the World Heritage Committee. Hence, a new Commission of the IUHPS could also be empowered to raise funds for preservation of space technology.

CONCLUSION

With *Mir*, humankind had a unique opportunity to preserve for posterity a wonderful example of its technology, ingenuity, and spirit, one that combined scientific and technical achievement with social and political impact and had great influence on future developments. We have come to regret disposing of it as if it were mere space junk. *Mir* was more than the sum of its age and battered parts; like all museum pieces, it was a symbol and a touchstone to a historic past. It provided a door through which we could step into a past cultural landscape. Nothing can replace the lost technical intricacy. Nothing can replace the evidence of human ingenuity in surviving in the most hostile of environments. And nothing can equal the sensation that is provided by "the real thing," which is why we develop and maintain museums and world heritage sites. But *Mir*'s passing can have the positive side of providing a wake-up call. Now, because of the example that *Mir* gave us, we must begin to think of world heritage on a much wider canvas. Our space vehicles are the twentieth and twenty-first centuries' legacy of antiquity; they are our counterpoint to Stonehenge, the Pyramids, Harrison's chronometers, and the telescopes of Galileo and Newton.

APPENDIX 1: COLLECTION DEVELOPMENT STRATEGY PROCESS: KEY ELEMENTS

A successful CDS begins with the adoption of a focused and relevant conceptual theme. "The Human and Robotic Exploration of Space" will provide a framework within which focused historical research may be conducted and decisions made. The CDS comprises several steps.

HISTORICAL ASSESSMENTS

The products of historical research are *historical assessments*, which identify and analyze important concepts, ideas, objects, and issues key to the historic development of our collective scientific and technological endeavor and our understanding of them. The historical assessment places the concepts within the Human and Robotic Exploration of Space theme and subthemes, and latitude is exercised in identifying and developing those historical concepts reflecting the theme.

Following the completion of a series of historical assessments, a *collection assessment* is prepared. Collection assessments comprise three sections: 1) the ideal collection; 2) a profile of the existing collection; and 3) collection needs, which are obtained by comparing the ideal collection with the existing collection profile. This process identifies artifacts or classes of artifacts or historic sites to be acquired and artifacts to be deaccessioned.

Collection Assessments

The collection development strategy comprises three elements and therefore includes vital functions, which permit making informed decisions on collection acquisitions, content, and, where justified, the selection of historic sites for preservation.

1. An *ideal collection* identifies, from the historical assessment and associated documentation, categories of objects and sites and supporting media that effectively represent the historical concepts, ideas, and issues in a subject area. The ideal collection is defined according to subject area, subarea, and/or time period. A preferred strategy for developing an ideal collection for a curatorial area is identified and substantiated a priori by the curator and may be one or more of the following types:
   - A *representative collection,* that is, one that includes examples of significant developments in science and technology, is in most cases the preferred form of an ideal collection for a curatorial area. For specific topics and/or time periods, the curator recommends the development of an iconographic or study collection. It illustrates innovations, developments, and evolution within a broadly defined area.
   - An *iconographic collection* includes one or more artifacts and supporting media to represent a broad thematic area where, for some reason, it is inappropriate or difficult to build a representative collection.
   - A *study collection* is an in-depth collection developed to represent an important aspect of science and technology in Canada. It illustrates innovations, developments, and evolution within a narrowly defined area.
2. *Collection profiles* document the context and scope of the existing artifact collection and supporting documentation. Artifacts or physical sites in an area of the collection are described sufficiently to determine whether they are consistent with artifact or sites needs as identified in the ideal collection.
3. As part of the collection profile, *collection needs* are identified by comparing the ideal collection with the collection profile for each subject area. This identifies artifacts, classes of artifacts, or physical sites to be acquired and

preserved and, if not consistent with historical concepts or collection objectives, artifacts to be deaccessioned or physical sites to be delisted. At the Canada Science and Technology Museum, we review the holdings of other Canadian museums to avoid unnecessary duplication—a prudent approach given the long-term costs of documenting, conserving, and storing large, technically sophisticated materials. For a global CDS, coordination and recognition of objects in national collections will also be important and will ensure that the broader contributions to space exploration are not focused solely on the achievements of Soviet and American programs.

Committee Process

In the CSTM collection development strategy, two committees play an important role in the implementation of the strategy, the Acquisition Committee and a Collection Development Committee. For an international process focused on space exploration, only the constituting body, reporting structure, and committee membership will need to be negotiated. In brief:

- The Collection Development Committee's purpose is to plan, review, and approve historical assessments and collection assessment documents. This committee will be a smaller committee composed of a subset of the Acquisition Committee plus representative of the funding bodies, namely UNESCO and/or ICOM.
- The Acquisition Committee's purpose is to bring appropriate staff together in a regular forum and, using the results of the historical and collection assessments, to review acquisition proposals and to recommend acquisitions and deaccessioning of artifacts. It will be composed of an international panel of curators and historians of space science (including astronomy) and technology.

Other aspects of the CSTM process and supporting documents (e.g., committee terms of reference) may be found at: http://www.sciencetech.technomuses.ca/english/collection/pdf/collection_development_strategy_2006.pdf

NOTES AND REFERENCES

1. Comprehensive coverage of *Mir* can be found in Hall, R., ed. *The History of Mir: 1986–2000.* London: The British Interplanetary Society, 2000.
2. Recovery and preservation of the *Vasa* was a massive national project that began with the ship's discovery in 1956. In 1961 the entire 70-m-long hull, cradled between floatation barges, broke the surface and was towed into dry dock. The formidable problems of preserving this enormous wooden structure are detailed in Barkman, L. The Preservation of the Warship Wasa. In *Problems of the Conservation of Waterlogged Wood*, ed. A. Oddy, 65–105. Maritime Monograph and Reports 16. Greenwich, United Kingdom: National Maritime Museum, 1975
3. A brief scan of Web sites on the subject of *Mir* and MirCorp will give an idea of the political factors that may or may not have played a part. For the purposes of this chapter, we stick to the technical and cultural aspects of preservation.

4. See UNESCO's Web page, Culture and UNESCO/Legal Protection and Heritage/Underwater Cultural Heritage. http://www.unesco.org/culture/legalprotection/water/html_eng/index_en.shtml.
5. Kopytoff, I. The Cultural Biography of Things. In *The Social Life of Things: Commodities in Cultural Perspective*, ed. A. Appadurai, 73. Cambridge, United Kingdom: Cambridge University Press, 1986.
6. Ibid.
7. Thompson, M. *Rubbish Theory: The Creation and Destruction of Value*. Oxford, United Kingdom: Oxford University Press, 1979 (see specifically p. 8).
8. Kopytoff 1986 (note 5), p. 64.
9. Kopytoff 1986 (note 5), p. 68. Thompson 1979 (note 7), pp. 19–20.
10. Storey, J. *An Introductory Guide to Cultural Theory and Popular Culture*. Athens, GA: University of Georgia Press, 1993 (see specifically p. 78).
11. Indeed, the very durability of this publication is tied to societal pressure. Some will deem it rubbish, and others will see merit in it. Time will tell whether its message becomes durable, or remains transient and is discarded. The editorial decision to publish it represents its first challenge.
12. Tito's space trip, and the political negotiations surrounding it, also indicate the state of financing of Russia's space program.
13. http://www.scaled.com/news/2005-07-27_branson_rutan_spaceship_company.htm.
14. A summary of the astronauts performing the servicing and their activities may be found at http://www.cbsnews.com/network/news/space/hstevas.gif.
15. The museum's full CDS may be found at http://www.sciencetech.technomuses.ca/english/collection/pdf/collection_development_strategy_2006.pdf.

38 Archaeology of the Putative Roswell UFO Crash Site: A Case Study

William Doleman

CONTENTS

INTRODUCTION

As the story goes, one morning in early July 1947, William "Mack" Brazel, foreman of the J.B. Foster sheep ranch, located some 30 miles southeast of Corona, New Mexico, discovered "strange metallic debris" spread across a part of the Hine Pasture (Figure 38.1). On or before July 7, Brazel took samples of the material into Roswell to the sheriff's office. On July 8, 1947, an article under the front-page headline "RAAF Captures Flying Saucer," the *Roswell Daily Record* newspaper led off with, "The intelligence office of the 509th Bombardment group at Roswell Army Air Field announced at noon today, that the field has come into possession of a flying saucer." On July 9, the paper published a report with the lead "Ramey [of the RAAF] Says Excitement Is Not Justified. General Ramey Says Disk is Weather Balloon." And thus was launched one of the most interesting, if controversial, chapters in New Mexico's colorful history.

On September 16–24, October 10–11, and October 15–17, 2002, archaeologists from the University of New Mexico (UNM) Office of Contract Archeology (OCA) conducted archaeological testing and related research at the reported location where Mack Brazel found the debris. The research was privately funded by a subsidiary of USA Cable (now NBC-Universal), the Sci Fi Channel, with Larry Landsman, special projects director, acting as Sci Fi's contact. Technical advisors to the project were Donald Schmitt and Thomas Carey, well-known independent Roswell UFO researchers. The author served as principal investigator. The project, as well as the background of the Roswell incident and new evidence derived from a 1947 photograph, was documented in the Sci Fi Channel's *The Roswell Crash: Startling New Evidence*. First broadcast on November 22, 2002, the program is occasionally rebroadcast and remains Sci Fi's highest-rated special. In addition, the final report[1] was published as a *Sci Fi Channel Book*, together with diaries kept by project participants, a Roper poll on American attitudes about UFOs, and a foreword by New Mexico Governor Bill Richardson.[2]

Archaeological excavations were accomplished by volunteer excavators under the supervision of Doleman and OCA staff members J. Robert Estes and Louis Romero. In addition, geophysical prospection surveys (electromagnetic conductivity and metal detection) were conducted by David Hyndman of Sunbelt Geophysics in Socorro, New Mexico, for the purposes of locating subsurface anomalies warranting archaeological investigation. Finally, backhoe trenches were excavated across the reported location of a "furrow" reportedly created by the vessel impact and in two geophysical anomalies identified by the geophysical prospection.

Prior to the investigation reported herein, evidence of the Roswell Incident had been limited largely to early newspaper reports, eyewitness accounts, signed affidavits, ancillary documents, and interviews—many conducted long after the event itself. Much of this evidence has been summarized and critiqued in numerous articles and books, not to mention being an ongoing subject of discussion and dispute at

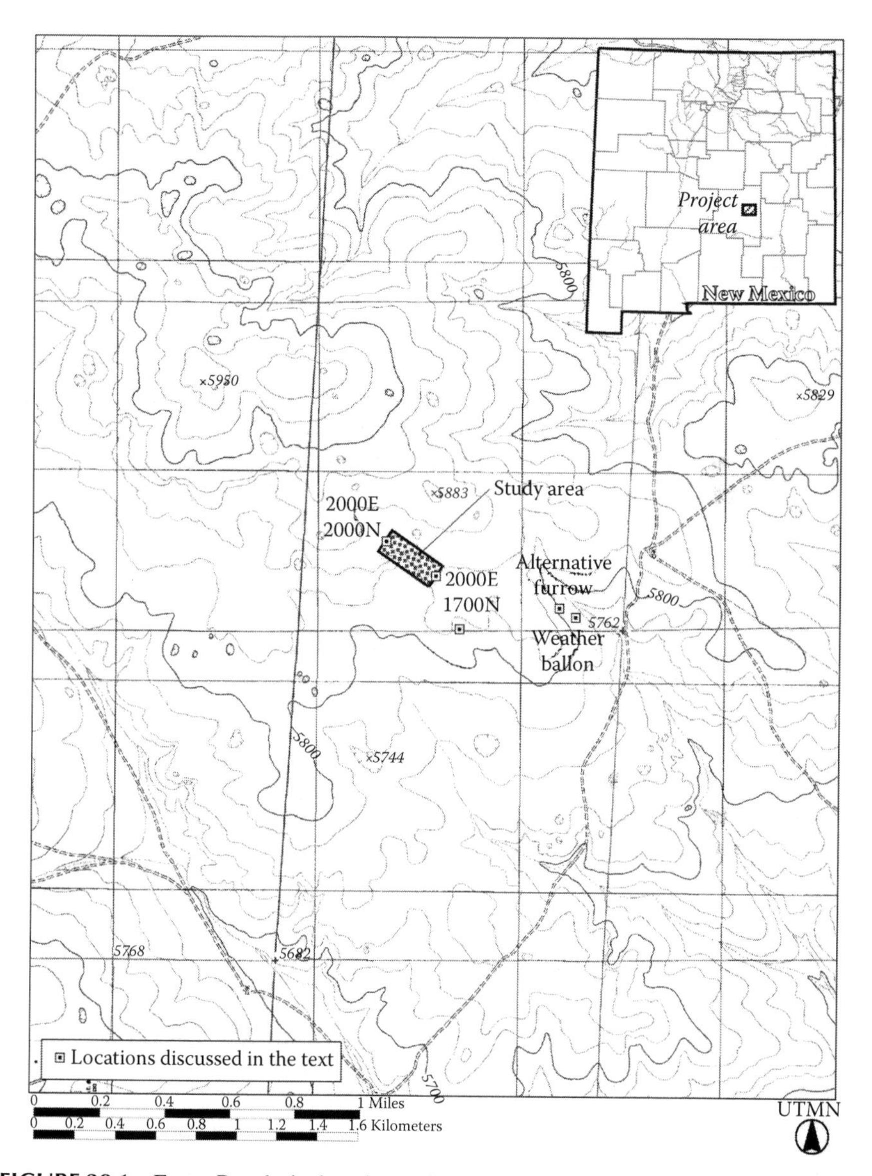

FIGURE 38.1 Foster Ranch site location and topographic setting. Note southeastward trend of study area drainages.

conventions and in Internet newsgroups, chat rooms, and mailing lists. Many of those who were first involved as eyewitnesses or investigators are now dead. Throughout the 55-plus years that have ensued since the summer of 1947, no evidence other than written declarations, recorded verbal testimony, and theorizing (fact-based and otherwise) has been offered to support claims that something—many believe an extraterrestrial vessel—crashed on the high-desert grasslands of southeastern New Mexico in 1947. In other words, despite reported accounts of one or more crashed vessels, nonhuman bodies, impact marks on the ground, and unusual crash debris, none of this reported physical evidence has survived to the present. Or if it has, it has not been made available for public scrutiny. Thus, the importance of the project reported herein lies in the fact that it represents the first-ever comprehensive attempt to discover physical evidence of the event and to apply modern scientific methods to its discovery and evaluation.

The specific area chosen for the investigation was determined by the technical advisors based on eyewitness accounts provided to them in the course of past research[3] and is purportedly the location at which Mr. Brazel discovered the "debris." It is important to note that the UNM archaeologists relied entirely on the technical advisors' knowledge of the site's location because there is no—easily discernible at least—evidence of the reported impact or debris on the surface at the present time.

This site—referred to herein as "Foster Ranch site"—is commonly known as the "debris field" and/or "skip site," and even the U.S. Air Force has acknowledged the site as the location where something came down in 1947 (albeit a Project Mogul balloon train and not a UFO[4]). According to various testimony-based reconstructions of events, this location was the initial impact point of an object that hit the ground with a glancing impact and then rose into the air and crashed again some 25–40 km (15–20 miles) away. The specific area subjected to intensive study during the present project is centered on a reportedly once-visible impact mark,[5] measures 300 × 120 m (984 × 394 ft.), and comprises an area of 3.6 hectares (8.9 acres). The land where the investigation took place is owned and administered by the United States Department of Interior, Bureau of Land Management (BLM), Roswell Field Office. Pursuant to BLM regulations, the work was conducted under Cultural Resource Use Permit No. 05-8152-02-01, which was issued by BLM's Roswell Field Office upon approval of a testing plan submitted to BLM by OCA in early September 2002.

The possibility of using archaeological and geophysical methods to investigate the Foster Ranch site was first broached with OCA personnel by Schmitt and Carey in 1999. Having secured a commitment for funding from the Sci Fi Channel in the spring of 2002, Schmitt and Carey, together with Mr. Landsman of Sci Fi, met with OCA Director Richard Chapman and OCA principal investigator Doleman in June to discuss appropriate research methods. The basic components of the research reported herein, including the use of subsurface geophysical prospection, were hammered out in that meeting. The specific geophysical prospection technologies used were determined by Hyndman. The general methodology bears striking similarities to that proposed in a 1998 paper by Greg Fewer,[6] but the similarities are entirely coincidental, as the paper was not known to the author or other project participants at the time.

The use of archaeological methods is entirely appropriate to the search for physical evidence of whatever happened on the Foster Ranch in 1947 because archaeology is essentially the forensic science of the past. This is true because the primary goal of archaeologists is to find and interpret physical evidence of past events, and archaeological methods are designed to do just that. These methods include modeling the dynamic processes that create the static archaeological "sites" we see today, as well as those processes that subsequently modify them. Together, these processes are referred to as *archaeological formation processes*. In addition, geophysical prospection methods and technology—originally developed for detecting the subsurface presence of geological phenomena—were incorporated into the archaeological fieldwork as an aid in choosing the locations of archaeological explorations and were warranted by the nature of the physical evidence being sought. OCA relied entirely on the knowledge of the project's technical advisors and their familiarity with eyewitness reports both to choose the specific location where the investigation would be pursued and to determine what physical evidence might be present.

The research strategy and methods used in the Foster Ranch site investigation are described in the following section. The core of this research approach is based on the fact that, at present, no obvious evidence of either debris or a furrow is visible on the ground surface at the Foster Ranch site. Hence, the approach involved combining reports of the physical evidence once present and the alleged removal of most of it, together with knowledge of the natural processes that may have further affected it, for the purposes of designing an appropriate set of investigative methods for discovering any remnants of that physical evidence that might remain—albeit buried—at the site.

In terms of the introduction to archaeology presented by Edward Staski in Chapter 2, the present study is methodologically closest to his fourth definition of space archaeology: "The perceived future role archaeologists might play in studying the material culture of extraterrestrials, living or not, once such material is discovered and verified as genuine." Furthermore, the present study conforms to Staski's "traditional archaeology" (studying the past to understand the past) in that both the events (behavior) of interest and the data being sought are of the past. But the normal methodological trajectory is reversed here: instead of using what is found by means of excavation to infer past events from recovered archaeological data, we are attempting to infer what evidence we might find from postulated events in the past. As a result, the methods used are focused on discovering evidence in the archaeological record rather than on uncovering what is known to be embedded in the depositional record.

For further information on the Roswell Incident, the interested reader is referred to books coauthored by one of the project's technical advisors.[7] Information concerning this project, as well as a number of Roswell Incident-related links, can be found at http://www.scifi.com/ufo/ and http://www.scifi.com/roswellcrash.

RESEARCH RATIONALE AND METHODS

The fundamental objective of the Foster Ranch site project was to determine whether any remnants of the originally reported physical evidence at the site remain today. In order to accomplish this goal, a staged research strategy was implemented. The first stage entailed reviewing the physical evidence that was reportedly produced by the 1947 event. In the second, the various human and natural postevent processes and actions that could have affected the nature, distribution, and detectability of that evidence were detailed. In the third, three techniques commonly used by geologists and archaeologists to detect buried phenomena were applied to the site in an attempt to locate possible buried evidence. Finally, areas pinpointed using these geophysical prospection methods were targeted with standard archaeological test excavations. Evidence recovered from the test excavations dictated the nature of postfield analyses.

Reported Physical Evidence and Postevent Activities

On the advice of the project's technical advisors, research at the Foster Ranch site focused on discovering relict evidence of two things reported by eyewitnesses. Assorted accounts indicate that witnesses saw (a) debris spread out across an area measuring about 300 ft. wide by 0.75 miles long, and (b) a shallow, linear depression in the ground surface variously described as a "furrow" or "gouge" and measuring about 10 by 500 ft. long, the latter presumably left by an impact of some sort.[8] One eyewitness described the furrow as being up to 2 ft. deep at the impact end, but shallower (a few inches) for most of its length.[9] Finally, some accounts suggest that the concentration of debris was greatest at the presumed impact point and hence the point where the furrow was deepest.[10] In addition, reports indicate that within days of the original reporting of the debris field, a large team of military personnel proceeded to the site for the purposes of removing all the debris,[11] walking—it is said—shoulder-to-shoulder across the entire debris field until all visible remnants had been acquired. Finally, all recovered debris, as well as the remains of the craft and its occupants, was allegedly removed to secret facilities for further study by the Army Air Force and/or other government agencies, which explains the lack of any material being available today for inspection or analysis.

Implications of Foster Ranch Site Characteristics for Testing Strategies

Several aspects of this story are relevant to the research methods implemented at the Foster Ranch site investigation. Paramount among these is the fact that the two forms of physical evidence being sought by testing activities at the Foster Ranch site are (a) any residual debris not removed by the military, and (b) vestiges of the observed impact mark, gouge, or furrow (the term *furrow* will be used throughout the remainder of this report). The descriptions of the debris field and furrow played a crucial role in the project's overall strategy, as well as in determining more specifically where project activities would be focused.

Debris Field

The size of the originally reported debris field, together with the reported nature of the debris itself (light and easily moved by the wind[12]), and the alleged postevent cleanup suggest the characteristics to be expected of any debris that might yet be present at or near the site.

First, the amount of debris was reportedly considerable, and—under the presumption that the debris was produced by partial destruction and breakup of the outer portions of whatever struck the ground—it was hypothesized that more small pieces were produced than large ones, with the most abundant pieces being the smallest. This is because things that break usually break into more small pieces than large, a fact familiar to anyone who has found a small but painful glass sliver in a bare toe long after having thoroughly cleaned up the glass debris that produced it. Call this the "broken glass" theory of material destruction. Of course, different materials exhibit different breakage patterns depending on their strength in compression, tension, and shear, as well as the nature of the destructive force. Furthermore, nothing is known about the nature of the material(s) that comprised the hull of the vessel that reportedly struck hard enough to produce debris. Nonetheless, based on reports of abundant debris, it was assumed that the numerical bulk of the debris would have consisted of smaller pieces that would be the hardest to find during any cleanup effort and hence the most likely to remain at the site 56 years later. No doubt, many of the contributors to this volume could shed valuable light on this question.

Second, the reported light weight of the material suggests that it may have blown considerable distances, both between the time it was originally found and the cleanup, and subsequently. Together, these conclusions reveal an important fact about the search for remnant debris: namely that it is equivalent to the old aphorism "looking for a needle in a haystack." This potentially broad distribution, together with the likelihood of postevent burial by a variety of possible processes (see "Post-Event Effects of Natural Processes" below), implies a "haystack" that is not only broad but deep. Given reports suggesting that the debris was densest near the impact point, the Foster Ranch site testing project focused on the reported location of the furrow—in essence, looking in the densest part of the haystack. Despite the fact that this area was presumably the focus of the most intensive cleanup efforts, this strategy was deemed preferable to searching a much broader area in which the density of remnant debris would be exponentially lower.

Post-event Effects of Natural Processes

The reported lightness of the debris also suggests that any material still present would have to have been buried or otherwise secured to the ground surface not long after the impact; otherwise it would have eventually blown away. At least two natural processes could have contributed to burial or securing and obscuration of small, undetected debris. The first is burial by long-term natural erosion and deposition—that is, slopewash. In this process, surface sediments are moved from upslope downward by a combination of water (which runs across the surface when the rainfall rate exceeds the rate at which the soil absorbs it) and gravity. Slopewash tends to move light things across the surface and eventually bury them in low areas where the eroded material accumulates.

Another burial mechanism is burrowing animals. Archaeologists have long studied the effects of animal burrowing—or faunalturbation—on archaeological deposits. Burrowing animals ranging in size from ants to coyotes can churn sediments and translocate artifacts horizontally and vertically, thus altering their distribution and affecting interpretations based thereon.[13] In the desert, diurnal temperature variations, together with a general lack of vegetation large enough to hide in, mean that burrowing is very common among small to medium animals. Burrowing by coyotes or foxes and smaller rodents is evident at the site, suggesting that even in the 56 years that passed between 1947 and 2002, a considerable amount of subsurface deposits may have been excavated and moved to the surface, while surface materials became slowly buried.

Furrow

As noted above, available descriptions of the furrow suggest a linear feature measuring 10 by 500 ft., only a few inches deep for much of its length, but as deep as 2 ft. (about 60 cm) along the main impact point. Descriptions of the furrow's profile are less specific, but most suggest a concave cross-section. As there is no obvious surface evidence of the furrow at present, it has presumably been buried and/or obliterated by erosional and bioturbational processes. This fact led to the incorporation of geophysical prospection technology in the testing project. The absence of a surface-visible furrow also required that the originally observed furrow's location be determined by the project's technical advisors, one of whom had been shown the location by two separate eyewitnesses.

On the first day of the project (September 16, 2002), the OCA archaeological staff met with technical advisors Schmitt and Carey for the express purpose of selecting the specific location where the testing activities would be focused. As noted above, the location was to be centered on the furrow, as the area most likely to contain remnant debris, as well as being the only place to search for evidence of the furrow itself. Mr. Schmitt designated the reported initial impact point as well as the furrow's alignment (reportedly more or less straight). Based on these consultations and Mr. Carey's statement that uncertainty concerning the actual locations of the furrow was no greater than 200 ft., a central baseline and study area of 120 by 300 m was designated. This area represents 36,000 m^2 (36 hectares, 15 acres), or about one-third the total area of the reported debris field. The baseline and grid served as the basis for all subsequent provenience determinations and topographic mapping of the site (Figures 38.2 and 38.3). In Figure 38.3, the grid's coordinate system [expressed in meters north (N) and east (E)] is arbitrary, and grid south is oriented at an angle of 124° 43′ from true north, meaning that grid north is oriented approximately 56° west of true north. Throughout this report, directional information will be expressed in the grid system, with cardinal directions noted as such (e.g., "true north").

Staged Research Methods

Three geophysical prospection methodologies were used in an effort to locate subsurface evidence of the furrow and debris. The research was conducted by David Hyndman of Sunbelt Geophysics in Socorro, NM. Detailed descriptions of the methodologies and results of all three studies are presented in a report prepared by Hyndman and submitted separately to the Sci Fi Channel.[14]

Aerial Photograph Study

Because aerial photographs contain a wealth of information about Earth's surface that is often not readily apparent from a surface perspective, they are a commonly used tool among earth scientists, including geologists, biologists, and archaeologists. Hyndman acquired aerial photographs of the Foster Ranch project area that bracketed the "event date" of July 1947 as closely in time as possible. The acquired aerial photographs were made by the U.S. Department of Agriculture's Soil Conservation

FIGURE 38.2 View south along grid system centerline from the north end of grid system. The deepest part of the reported furrow is located just beyond two foreground backdirt piles (backhoe study units 101 and 102).

Service (now the Natural Resources Conservation Service) and are dated November 19, 1946 (1:31,680 scale, or 1 inch equal to 0.5 mile), and February 3 and 4, 1954 (1:54,000 scale, enlarged in printing to 1:27,000, or 1 inch equal to 0.43 mile). In addition, a single aerial photograph taken October 12, 1996 (1:40,000 scale, or 1 inch equal to 0.63 mile), by the United States Geological Survey's National Aerial Photography Project was also acquired for comparison with the earlier imagery.

The study of the aerial photographs served two purposes. The first was to see whether any linear features that might represent a once-visible furrow were present in 1954 but not in 1946. The presence of such a feature in the 1954 photographs but not in a 1946 image would be consistent with a furrow—whatever its origin—having been created in 1947. The second purpose was to determine—should such a suspicious linear feature be visible in the 1954 photography as well as in the 1996 photograph—to use the photographs to precisely establish the feature's location on the ground so that it might be further investigated.

The aerial photography from all three dates also proved valuable in addressing a number of other, unanticipated questions that arose in the course of the project.

Geophysical Prospection

Two related geophysical prospection technologies were used at the Foster Ranch site to obtain evidence of possible buried features (the furrow) and buried debris. Both acquire information on the subsurface characteristics of the soil by projecting a signal into the ground and measuring the return signal. The methods

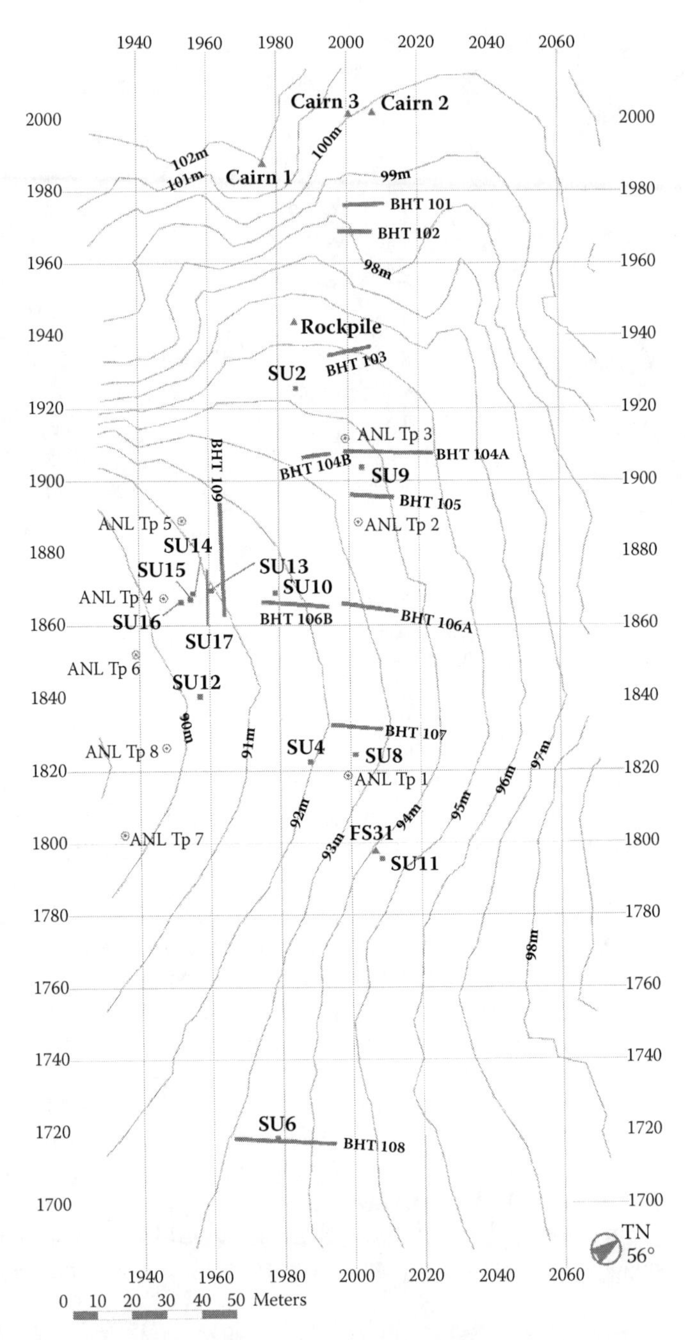

FIGURE 38.3 Transit-based topographic map of site grid system, showing excavated study units (SU), backhoe trenches (BHT), documented rock features (cairns, rockpile), Center for UFO Studies test pits (ANL Tp), and one surface find (FS 31; see text).

utilized—electromagnetic conductivity (EMC) and high-resolution metal detection (HRMD)—were chosen on the basis of the anticipated properties of the evidence being sought.

Electromagnetic Conductivity Survey

The EMC survey was used for the purposes of finding any buried features that might represent a buried furrow. At the heart of the method lies the fact that any furrow that was present in 1947, but that was subsequently buried by human or natural processes, would contain deposits that are less compact and hence capable of holding more moisture than the surrounding undisturbed and naturally consolidated sediments. These differences in consolidation or compaction would produce a higher moisture content in the feature's fill, and the higher moisture would in turn yield a higher electrical conductivity. In contrast, more compact soils can be expected to yield lower conductivity signals, and the limestone bedrock the lowest strength signal of all, owing to its low porosity. The EMC survey was performed with a Geonics EM-31 ground conductivity meter, which is capable of detecting patterned variations in soil conductivity.[15] The spatial resolution of the EMC survey was limited by the sampling interval (0.5 m) and the distance between transects (10 m). The use of 10-m transect intervals allowed the broad coverage required to survey the entire site area.

Because the EMC survey transects were tied to the grid system laid out by the archaeological team and marked at regular grid intervals by rebar stakes and pinflags, the resulting data could be used to prepare contour maps of both the conductivity signal and the inferred electromagnetic conductivity across the site. The EMC study was conducted first and implemented across the entire 120 × 300 m study area. Once the data processing was complete (2 days later), Hyndman returned to the field with the conductivity map, which revealed three or four anomalously high-conductivity locations. Two of these anomalies were elongated, located within 10 m of the projected furrow, and were parallel to the furrow alignment as determined by the project's technical advisors. This fact was a considerable surprise, as elongated high-conductivity anomalies are exactly what would be expected of a buried furrow. These two anomalies, one each located at the north and south ends of the study area, were targeted with archaeological test excavations. One was also investigated with a backhoe trench.

High-Resolution Metal Detection Survey

The high-resolution metal detection survey was conducted for the purposes of locating any buried or obscured metallic material that might represent remnants of the "metallic debris" originally reported in 1947. Almost all metallic elements are at least moderately conductive. Thus, if any metallic debris were present on or below the surface, the HRMD survey should reveal it—within the limits of the device's sensitivity. The metal detection survey was conducted using a Geonics EM-61 (HH) high-precision metal locator.[16] The EM-61 detects the presence of high-conductivity materials through a process that is similar to that of the EM-31 EMC meter but that utilizes a magnetic pulse in place of a radio frequency signal. The spatial resolution of the HRMD survey is determined by the sampling interval (0.2 m) and the distance between transects (1 m). These close intervals provided a much higher spatial resolution than that of the EMC survey, but they also limited the survey to an area of about 2,025 m^2, or roughly 6% of the overall site area.

The HRMD survey was focused on parts of the site deemed most promising on the basis of results from the EMC survey and a knowledge of the site's erosion and deposits. The HRMD survey was conducted after the first 2 days of archaeological testing, when the EMC survey results were available. Two areas were targeted, one in the area of the most prominent electromagnetic conductivity anomaly and the other covering the topographically low areas, where deposition would be most likely to have buried any remnant debris. When available after computer processing and map preparation, the results of the HRMD survey were used as the basis for placing a number of archaeological testing units, as well as a backhoe trench in a distinct anomaly.

Archaeological Activities

Archaeological test excavations were conducted for the purposes of exploring subsurface anomalies revealed by the electromagnetic conductivity and metal detection surveys, as well as to search for any buried materials of possible extraterrestrial origin that might remain at the site. In addition to geophysical prospection anomalies, the locations of archaeological test excavations were chosen on the basis of a topographically based erosion-deposition model for the site, as well as on evidence of past, but not recent, bioturbation. The testing program incorporated standard archaeological excavation methods and included photography, mapping, and excavation of exploratory test pits to determine subsurface stratigraphy, as well as to investigate anomalous localities. All excavations were documented in the field using standard excavation forms.[17]

All recovered artifacts were collected. Volunteers were trained as much as possible in the recognition and distinction between natural materials and Native American material culture (artifacts), and instructed to record, bag, and catalog any item not identifiable as either natural or Native American as "historic material of uncertain origin" or HMUO (pronounced "Huh-muo"). Plans for the disposition and analysis of any suspected extraterrestrial artifacts were determined by prefield consultation with BLM personnel and the project's sponsors. It was presumed that any such materials would be transported to an appropriate laboratory for scientific evaluation. At the end of the project, all silt samples, artifacts, and HMUOs were transported to a secure storage facility in Roswell (see "Identification of HMUOs").

Backhoe Trenching

A backhoe equipped with a specialized "archaeology blade" was used to excavate trenches at the site for three purposes. The backhoe was provided and operated by Mr. Eligio Aragon of Alleycat Excavating in Los Lunas, New Mexico. Mr. Aragon—known to New Mexico archaeologists as "Alley"—has more than 20 years of experience using a backhoe in archaeological applications. The first use of the backhoe trenches was to search for evidence of the buried furrow by excavating a series of trenches across the furrow's presumed alignment (Figure 38.3). Just as subtle subsurface archaeological features, such as hearths and pithouses, are revealed and easily identified in cross-sectional profiles created by excavations, it was expected that any buried furrow would similarly be evident in the sides of backhoe trenches, provided they intersected the now-buried feature.

Secondly, backhoe trenches were used to investigate the anomalies revealed by the electromagnetic conductivity and metal detection surveys—one of each—and allowed investigation of depths greater than those possible in 50 × 50 cm test pits. Finally, the backhoe trench provided a more comprehensive record of the site's stratigraphy, with the deeper and more elongated profiles providing a better picture of important lateral variations.

Soil-Stratigraphy Study

A soil-stratigraphy study was conducted at the site by evaluating soil profiles exposed in the walls of backhoe trenches and archaeological test pits. Soil-stratigraphy studies are an important geomorphological methodology that focuses on both the depositional stratigraphy of sediments and the ways in which the deposits have been altered through the various biological and geochemical processes associated with pedogenesis—the formation of soil horizons. Because pedogenesis occurs at regular rates, the general age of deposits can often be inferred from the soil horizons present, thus making soil stratigraphy a valuable tool for archaeologists.[18]

The value of the soil-stratigraphy study to the Foster Ranch project was thus threefold. First, it provided information on the natural soil stratigraphy and estimated ages of intact strata and soils. Once the site's natural stratigraphy was documented through excavation of a "strat pit" and the earlier test pits, it was possible to determine the depth at which test excavations could be curtailed because soils—and hence deposits—older than 55 years had been reached. Second, it supplied a baseline against which to evaluate "anomalous" stratigraphy, such as might have been produced by a ground-disturbing event in the past. Finally, the study was intended to yield a better understanding of intrasite variations in stratigraphy and their implications for long-term geomorphic and soil-forming processes at the site.

Geomorphic processes in the Foster Ranch site area reflect the region's geological context. The Foster Ranch site lies in the middle of a vast, undulating plateau underlain by limestone of the Permian-age San Andres Formation. Because the plateau is more or less level and much of the precipitation that falls enters the ground instead of flowing across the surface, the soils in the area are classified as residual soils. Other processes that contribute to the project area's deposits are (in order of importance): (a) alluvial (water-caused) processes consisting primarily of "sheetwash", and (b) blowing wind, which imports both sand-sized particles and airborne dust particles from outside the area. Soils in the study area are correspondingly fine-grained (loamy texture) and highly calcareous, consisting largely of in situ weathering products, including limestone clasts, and rounded fine sand particles of clearly eolian origin.

RESULTS OF THE INVESTIGATION

INVESTIGATIONS BY SUNBELT GEOPHYSICS

As noted, three studies that provided crucial information to guide the archaeological testing program were conducted by David Hyndman and Sidney Brandwein of Sunbelt Geophysics of Socorro, New Mexico. These were as follows:

1. A study of aerial photographs from before and after July of 1947
2. An electromagnetic conductivity survey of the entire gridded site area
3. A high-resolution metal detection survey of selected portions of the site

All three studies were performed under contract to the Sci Fi Channel and are detailed in a report delivered separately to Sci Fi.[19] The results are summarized here, together with their use in archaeological testing.

Aerial Photograph Study

Both the November 1946 and February 1954 aerial photography was acquired in stereo pairs, thus allowing inspection through a magnifying stereoscope designed for just such viewing, which exaggerates topography and makes non-natural features more easily detectable. The 1996 aerial photograph was acquired in a single image. Figure 38.4 is an enlarged portion of the 1946 aerial photograph from Hyndman and Brandwein's report showing the grid system outline.

No linear trace that is visible in the 1954 aerial photographs but not visible in the 1946 photographs was found during inspection of the aerial photography. Thus, no evidence of a furrow-like feature that was present in 1954 but not in 1946 was detected in the aerial photograph study.

A segmented linear trace was observed within the study area that represents a southeast-trending drainage or ephemeral channel that enters the study area at the east end of the north end on a bearing of about 125° from true north. In Figure 38.4, the drainage is a faintly darker line and area, whose appearance is caused by vegetative differences related to associated moisture variations. About a quarter of the way into the study area, the drainage turns true south and spreads out somewhat into the swale in the middle (compare Figure 38.4 with the topography in Figure 38.3). Observations of local bedrock suggest that the drainage's linear course in the northwestern part of the study area—as opposed to a more "natural" winding and erratic one—may be the result of underlying bedrock fracturing and subsidence (common in karst terrains). This is particularly true in the case of the south-trending segment, which lies between two inward-tilted blocks of bedrock.

A linear feature that is evident in the 1996 aerial photograph, but not in the 1946 or 1954 photographs, offers some insight into the natural processes that may have affected the site since 1947 and that may have played a role in obscuring evidence that was once present and visible to eyewitnesses. The feature is an apparent vehicle track that extends from the western end of the study area almost due east and out of the area. The track is considerably less well-defined than other, well-used two-track roads in the 1996 photograph, suggesting that is was not regularly used. In addition, the track was not detectable on the ground in September of 2002, 6 years later.

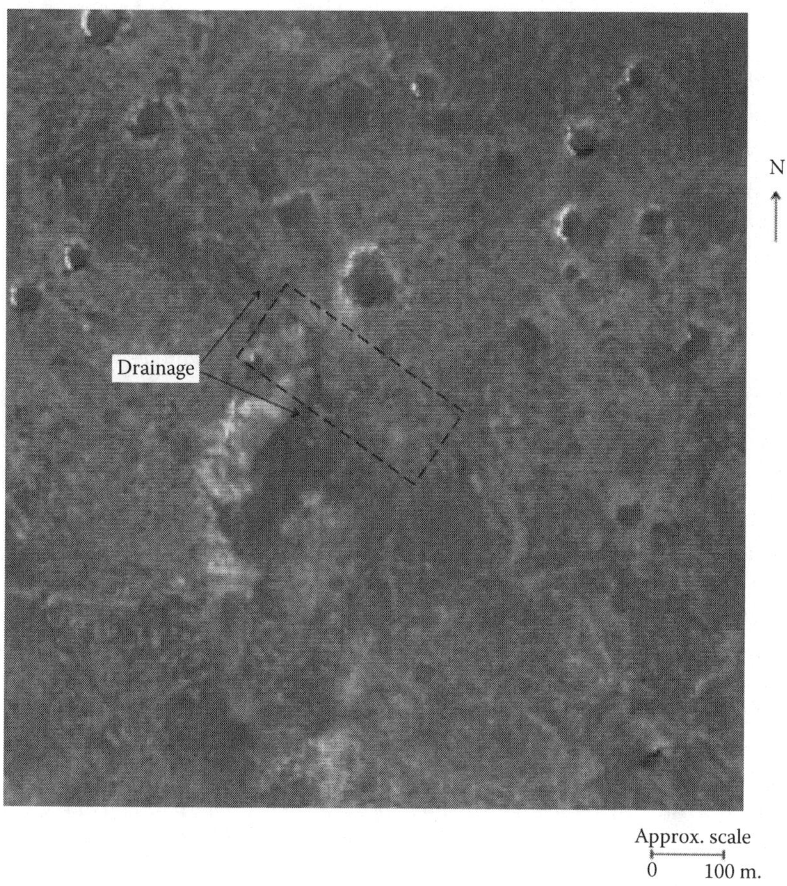

FIGURE 38.4 Enlarged portion of a 1946 aerial photograph showing the grid system outline and drainage traces. Circular features are natural sinkholes.

These facts shed some light on the question of how long a shallow linear feature—such as a furrow—might last in this landscape. The time required to obscure a linear depression through natural processes of erosion and revegetation would be a function of several factors. Not counting climate variations, the most important of these factors are (a) the orientation of the feature with respect to local slopes (approximately perpendicular), and (b) the depth and width of the original feature (assumed to be the width of a vehicle and no more than a few inches deep, although we have no way of knowing how pronounced the feature was in 1996). Given this, it seems quite possible that the shallow furrow described by eyewitnesses in July 1947 could have been erased in the 6 ½ years that passed between then and February 1954, when the postevent photograph was taken. Furthermore, the 55 years that passed between 1947 and 2002 may have been more than sufficient time to obscure or even completely erase much or all of the furrow described by eyewitnesses. The crucial question thus becomes: how deep was the deepest portion of the furrow? As noted earlier, it may have been as much as a couple of feet deep at the initial impact point, while most of it was reportedly but a few inches deep.[20] Thus, if a furrow was once present and visible, natural processes could have obscured all but a small, deep portion of it. If so, then the likelihood of detecting the feature on aerial photographs—although greater than on-the-ground inspection—would decline.

Electromagnetic Conductivity Survey

As noted, the goal of the EMC survey was to use lateral variations in the electrical conductivity of the ground to detect any buried feature that might represent a buried or obscured furrow. If present, the furrow should present itself as an abnormally high-conductivity signal. In the following discussion, the locations of specific areas on the site are expressed in terms of the grid coordinate system described earlier under "The Furrow," with coordinates given as measured in meters north (N) and east (E) from the arbitrary 2000 N/2000 E datum.

Figure 38.5 shows the computer-processed results of the EMC survey as a shaded map with the site grid system and investigative units from Figure 38.3 superimposed. The black dashed-line arrows show the approximate location of the drainage documented in the aerial photographs. In the shaded map, the darkest shades (red and blue in the color-shaded original) represent the highest and lowest conductivity measurements, respectively, while the lighter ones represent intermediate values (yellows and greens in the original). The broad, dark areas in the northwest, the northeast between ca. 1940 and 1960 N, the southeast between ca. 1740 and 1860 N, as well as a small area 1700–1720 N and 2000–2040 E are all lows that represent slopes or topographic highs where limestone bedrock lies at or just below the surface. The dark area defined by 1900–1940 N and 1960–1990 E (identified as Anomaly 1) as well as that defined by 1700–1740 N and 1960–1990 E (Anomaly 2) are the strongest conductivity highs identified by the conductivity survey. At least some of the conductivity highs appear to be correlated with bedrock variations as well.[21]

A third anomaly lies in the southwest part of the study area and occupies the area defined by 1760–1780 N, 1940–1960 E and some of the areas to the north and south. The third (anomaly 3) lies at the south end of the study area and is centered just southwest of 1720 N, 1980 E. Three aspects of anomalies 1 and 3 make them particularly interesting. First, they are the strongest and largest of the three main anomalies. Second, the third is oriented almost exactly parallel to the axis of the reported furrow, while the first is oriented on a line between that of the drainage and the furrow. Third, each lies from 15 to 25 m west of the centerline of the furrow, so that together the line connecting them is also nearly parallel to the reported furrow's centerline.

In contrast, anomaly 2 lies 40–60 m west of the supposed furrow axis, and its alignment can be projected to the northeast to connect with that of the bedrock anomaly discussed above. Given its alignment and its distance from the centerline, this anomaly was assumed to have an origin similar to that of the bedrock anomaly, although this assumption was not tested.

Based on these observations, anomalies 1 and 3 were targeted for archaeological test excavations (study units 2 and 6, respectively). The south anomaly was also investigated with a backhoe trench (BHT 108).

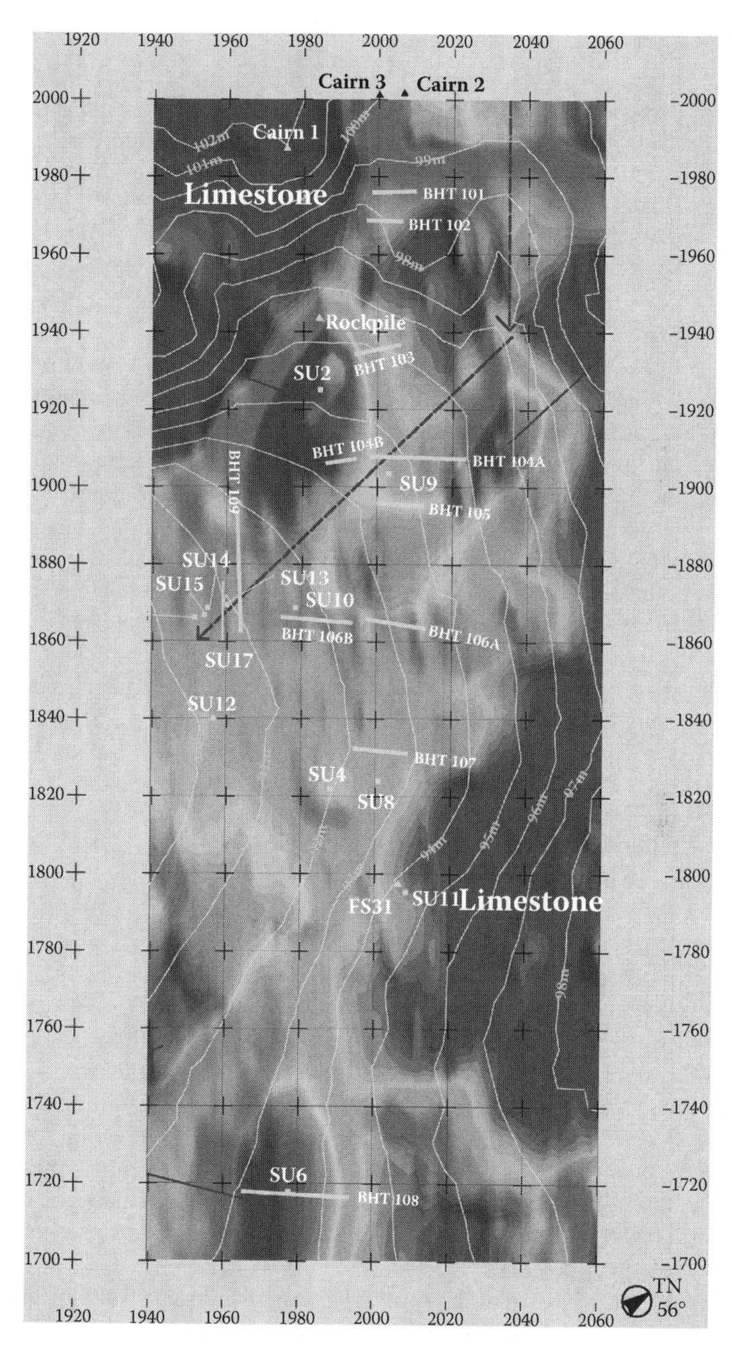

FIGURE 38.5 Computer-generated shaded contour map of electromagnetic conductivity survey readings.

High-Resolution Metal Detection Survey

As noted, the HRMD survey was conducted largely for the purposes of detecting buried debris resulting from the purported extraterrestrial vessel impact in 1947. Owing to the time constraints of a 4-day project, the areas targeted for the HRMD survey were based on information gathered during the first 2 days of test excavations, as well as the likelihood that any buried debris would be most likely to occur in deposits laid down by the prevailing drainage.

The "late-time" results of the study (strictly metal detection) detected only surface metal objects of known origin, including the study area grid system's baseline nails and pinflags, as well a few pieces of rusted iron-bearing metal on the surface, all undoubtedly the product of past ranching activities (primarily cans and wire).

The "early-time" results, which are more sensitive and are intended to detect both metal and highly conductive soil constituents, such as iron- and manganese-bearing (ferromagnesian) minerals, are presented as a shaded map in Figure 38.6. In the map, the dark gray shading (blue in the color-shaded original) represents areas where no signal was detected, while intermediate and strong signals (green and red, respectively, in the original) appear as light gray, including centerline nails, corner nails for test pits that were present at the time of the HRMD survey, and metal pinflags marking the study area boundary. Also present is an unmistakable linear trace of intermediate signals about 100 m long that is exactly aligned with the southwest-trending segment of the area's main drainage. This trace terminates in a roughly elliptical area measuring about 30 by 40 m north-south. In addition, several isolated higher-strength signals (yellow) occur along the main drainage-related trace, as well as near the center of the elliptical anomaly. Another faint linear trace also enters the HRMD-surveyed area from the east and also ends at the elliptical feature. No material or other phenomena were visible on the surface in the area of these HRMD anomalies that could account for them, indicating that whatever caused the signal is either not visually detectable or—as seemed more likely—is buried.

Both linear traces appear to correlate with possible drainages as indicated by slight U-shaped bends in the contour lines. The main trace even appears to bend to the north, where the two drainage segments meet. Two conclusions seem warranted by the HRMD survey results. First, whatever material is responsible for the distinct but moderate signals appears to be correlated with local drainage patterns and thus has quite likely been transported, concentrated, and buried by surface or subsurface water flow associated with the drainages. Second, the moderate signal strength is due either to the low relative conductivity of the signal-generating material or to the material's being buried at some depth (assuming and inverse correlation between signal strength and depth).

At the time the HRMD survey results became available, results of other test excavations had led to a growing feeling that the lower parts of the local drainages might offer the best target for finding buried debris if any were present. The HRMD survey results buttressed this conclusion considerably, and several test pits and a stripping trench [study units (SUs) 12–17] were excavated in the elliptical anomaly, as was a backhoe trench (BHT 109).

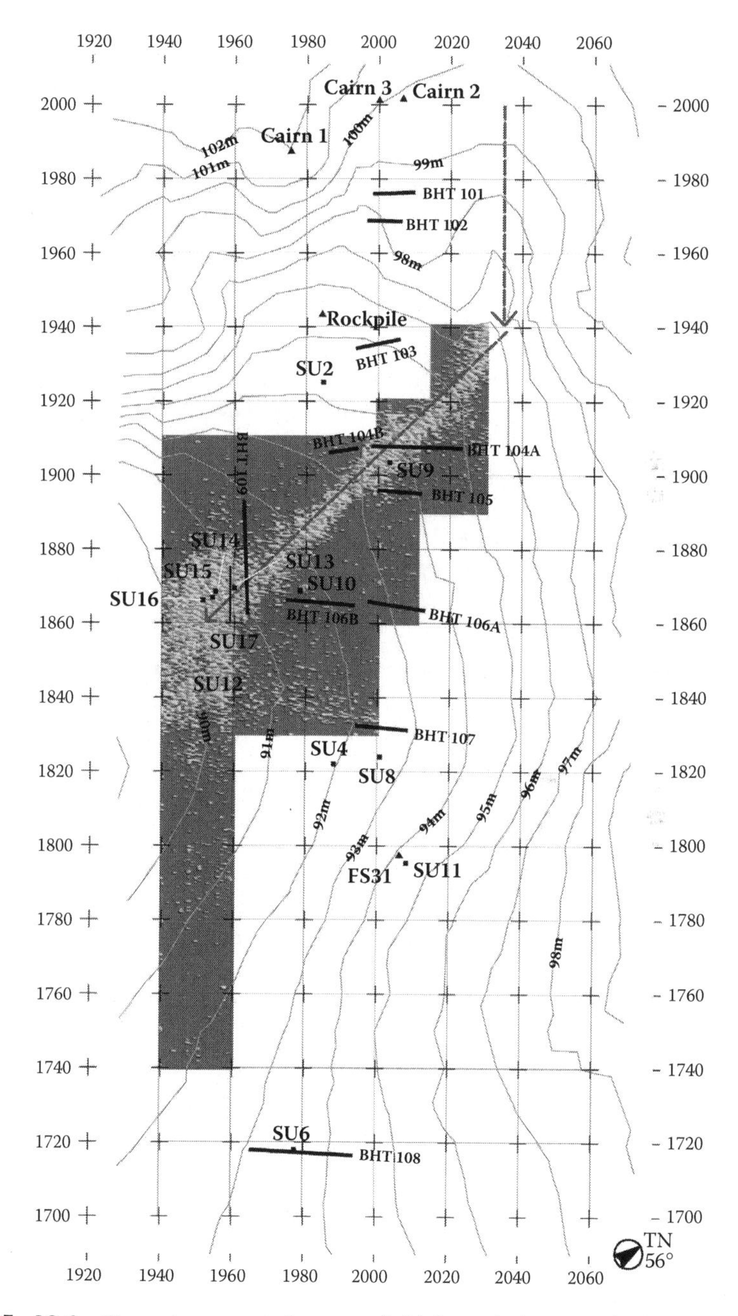

FIGURE 38.6 Computer-generated map of high-resolution metal detection survey readings.

Archaeological Testing Results

A total of 12 test pits measuring 50 × 50 cm and ranging in depth from 10 to 90 cm were excavated. These were designated SUs 2, 4, 6, 8, and 10–16 (in addition, SUs 1, 3, 5, 7, and 9 were designated and laid out, but time did not allow their excavation). In addition, two 10-m × 50-cm × 10-cm-deep stripping trenches were excavated as study unit 17. SUs 12–17 were excavated on the last day of the project as time was running out, and no detailed excavation notes were kept. Table 38.1 lists the excavated test pits under headings that reflect the rationale behind their choice, together with basic information concerning locations, dimensions, excavated levels (times 10 cm = depth) and numbers of soil samples and HMUOs recovered, plus comments. Together, these excavations resulted in removal and screening through one-eighth-inch mesh (to recover small artifacts) of a total of 2.38 m^3 of deposits from 13 m^2, or 0.036% of the somewhat arbitrarily defined 120 × 300 m site area haystack. Detailed results from the individual test pits are presented in the final report.[22]

From Table 38.1, it is clear that no obvious alien vessel parts or evidence of an impact was found in any of the test pits. Artifacts of any kind were rare, and most collected items were bagged as HMUOs (see "Identification of HMUOs" below). Nor did the test pits reveal obvious explanations for the conductivity or high-resolution metal detection anomalies. The soil-stratigraphy study and postfield analyses of soil samples did provide the basis for plausible explanations, however.

Identification of HMUOs

On April 15, 2003, the author conducted a preliminary inventory of the HMOUs, which were at the time being stored in safety deposit boxes at a bank in Roswell at the request of the Sci Fi Channel to ensure their security. The results of this inspection were published in the final report. Subsequently, the HMOUs and soil samples were transported to the offices of UNM-OCA for further analysis. (All excavated materials and sample remain at UNM-OCA as of August 2008, although is it anticipated that they will eventually be curated at the International UFO Museum and Research Center in Roswell.)

The HMUOs recovered from the Foster Ranch site fall into three categories: (a) definitely natural in origin (including five mineral specimens and one bone), (b) unidentified but probably of organic (i.e., biological) origin, and (c) apparently of manufactured origin. Appropriate forensic analyses of the HMUOs fall into two general and overlapping categories. One is *specialist-based analyses*, in which a scientist who specializes in a particular kind of material identifies specimens, often consulting a library of representative examples. The second category is *chemistry/physics-based*. This approach involves the use of a variety of analytical technologies to determine the chemical or elemental composition of unknown materials and use the resulting information to identify them. Both analytical approaches were used for the HMUOs, beginning with the author's visual inspection and use of an archaeologist's familiarity with the natural and human-made things archaeologists dig up regularly to make preliminary identifications. Table 38.2 lists the HMUOs by category, together with preliminary identifications and the results of subsequent

TABLE 38.1
Excavated Archaeological Study Units at the Foster Ranch Skip Site and Debris Field

Study Unit	Location and Purpose	SW Corner Coordinates	Dimensions (cm)	10-cm Levels	Soil Samples	HMUO Bags	Comments
Stratigraphy study unit							
4	Intermediate slope, explore stratigraphy	1820 N, 1991 E	50 × 50	7	7	0	Revealed typical arid-land soil profile, with upper A horizon, strong cambic B horizon more orange in color than other test pits; harder and more clay toward bottom
Electromagnetic conductivity survey anomaly study units							
2	EMC anomaly 1, W of north end of furrow	1924 N, 1985 E	50 × 50	3	4	2	Soil rocky with more clay in lower level 3, bedrock at bottom—possibly a reservoir for moisture; HMUOs from level 1
6	EMC anomaly 3, west of south end of furrow	1718 N, 1978 E	50 × 50	5	5	2	Revealed typical arid-land soil profile, with upper A horizon, cambic B horizon, calcium carbonate-rich Bk horizon; HMUOs from levels 1 and 2
Bioturbated area study unit							
8	Area of past but not recent bioturbaton	1824 N, 2001 E	50 × 50	7	7	3	Surface with small limestone pieces and lots of light-colored carbonate-rich material; recent coyote burrow nearby, one possible Native American artifact; HMUOs from levels 4 and 6
Possible ejecta impact area study unit							
11	Hillslope at distal end of furrow	1794.60 N, 2008.80 E	50 × 50	1	1	2	Very shallow, bedrock at 8 cm; 2 HMUOs from level 1

(Continued)

TABLE 38.1 *(Continued)*
Excavated Archaeological Study Units at the Foster Ranch Skip Site and Debris Field

Study Unit	Location and Purpose	SW Corner Coordinates	Dimensions (cm)	10-cm Levels	Soil Samples	HMUO Bags	Comments
Prevailing drainage study units							
9	Upper part, drainage, just east of furrow	1905 N, 2003.60 E	50 × 50	5	5	2	Limestone clasts throughout, larger with more clay in levels 3–5; near bedrock at bottom; 2 HMUOs from level 1
10	Center of drainage, west of furrow	1867.80 N, 1978.40 E	50 × 50	5	6	1	Loamy soils, few limestone clasts, more clay toward bottom; HMUO from level 3
12	Lower drainage, further west of furrow	1842.55 N, 1956.00 E	50 × 50	5	5	0	No notes
High-resolution metal detection survey anomaly units							
13	Center of elliptical anomaly	1870 N, 1960 E	50 × 50	5	4	0	No notes
14	Same as above	1869 N, 1955 E	50 × 50	3	3	0	No notes
15	Same as above	1866 N, 1955 E	50 × 50	5	5	4	No notes; HMUOs from levels 2 (1), 3 (2), and 4 (1)
16	Same as above	1866 N, 1952 E	50 × 50	4	3	0	No notes
17	North-south trench across elliptical anomaly	1860–1870 N, 1959 E 1875–1885 N, 1959 E	2 parts, each 50 cm × 10 m	1	0	8	Pair of surface-stripping units designed to screen broad area of surface deposits
Surface find							
n/a	Near SU 11	1797.70 N, 2006.61 E	n/a	n/a	n/a	1	FS no. 31, found on surface and mapped in with transit

EMC, electromagnetic conductivity; HMUO, historic materials of uncertain origin; n/a, not applicable

TABLE 38.2
Historic Materials of Uncertain Origin from the Foster Ranch Site: Current Analysis/Identification Status

FS Number	Item Number	Study Unit	Study Unit Type	Level	Original HMUO Description	Current Analytical Identification
Mineral materials						
11	1	8	Bioturbated area	4	Claw-shaped limestone pebble, about 10 mm—natural	Confirmed mineral
13	2	8	Bioturbated area	6	Two elongated pieces of calcareous mineral material, about 5 mm each—natural	Confirmed mineral, possible root casts
29	2	11	Possible ejecta impact area	1	"Flint" bagged as a HMUO is actually a natural red chert pebble, 9 mm	Confirmed mineral
49	1	15	High-resolution metal detection survey anomaly	2	Angular, non-limestone rock fragment, about 25 × 12 mm, with possible evidence of burning and some fine white fibers on one side—natural?	Confirmed mineral, fibers and burning not yet analyzed
50	1	15	High-resolution metal detection survey anomaly	3	Angular, non-limestone rock fragment, about 30 × 25 mm, with possible evidence of burning and some fine white fibers on one side—natural?	Confirmed mineral, fibers and burning not yet analyzed
50	2	15	High-resolution metal detection survey anomaly	3	Thin, curved calcareous mineral fragment, about 10 × 1 mm, probably a carbonate coating from a pebble	Confirmed mineral
51	1	15	High-resolution metal detection survey anomaly	4	Ten thin (1 mm), curved calcareous mineral fragment, about 10-25 mm, probably carbonate coatings from pebbles	Confirmed mineral
Apparent organic/biological origin						
3	1	2	Electromagnetic conductivity survey—north anomaly	1	Two pieces of organic (?) material dark green, 15–20 mm, possible algae (see also FS 56, Item 1, FS 57, Item 1)	Cryptogamic crust
3	2	2	Electromagnetic conductivity survey—north anomaly	1	One piece of light brown organic (?) material, about 20 mm in, possible insect pupa case	Confirmed as larval puparium

(Continued)

TABLE 38.2 *(Continued)*
Historic Materials of Uncertain Origin from the Foster Ranch Site: Current Analysis/Identification Status

FS Number	Item Number	Study Unit	Study Unit Type	Level	Original HMUO Description	Current Analytical Identification
3	3	2	Electromagnetic conductivity survey—north anomaly	1	Two pieces of unknown layered organic (?) material (about 15 mm), possibly lichen) on one side, mineral (?) (calcareous silt?) on other	Probable cryptogamic crust, requires final identification
13	1	8	Bioturbated area	6	One piece of light brown organic (?) material, about 20 mm, possible insect pupa case	Larval puparium or possible seed coat
29	1	11	Possible ejecta impact area	1	Three collapsed (?) ivory white, organic (?) items that were apparently originally hollow oblate spheres; material is thin and tough; possible reptile eggs	Confirmed as reptile eggs
23	1	9	Prevailing drainage—upper	1	Clump of unidentified fibers, slightly "curly," 15–20 mm, pale to white in color; possibly fur, probably not nylon	Not yet identified, requires forensic analysis
23	2	9	Prevailing drainage—upper	1	One piece of light brown organic (?) material, about 12 mm, possible insect pupa case	Confirmed as larval puparium
56	1	17	Stripping in high-resolution metal detection survey anomaly	1	Numerous small (15 mm) pieces of dark green, "papery" organic (?) material similar to FS 3, Item 3 (see also FS 56, Item 1)	Cryptogamic crust
56	2	17	Stripping in high-resolution metal detection survey anomaly	1	Small bone fragment with articular end (probably bird or small mammal)	Confirmed small animal bone
56	3	17	Stripping in high-resolution metal detection survey anomaly	1	A 15-mm piece of unknown organic (?) material, possibly part of a collapsed oblate "sac," tan-beige color, wrinkled, "leathery"	Not yet identified
57	1	17	Stripping in high-resolution metal detection survey anomaly	1	Three pieces (8–15 mm) of dark green, "papery" organic (?) material similar to FS 3, Item 3, and FS 56, Item 1	Cryptogamic crust

TABLE 38.2 *(Continued)*
Historic Materials of Uncertain Origin from the Foster Ranch Site: Current Analysis/Identification Status

FS Number	Item Number	Study Unit	Study Unit Type	Level	Original HMUO Description	Current Analytical Identification
Possible human-made materials						
5	1	6	Electromagnetic conductivity survey—south anomaly	1	One or two pieces of probable worn rubber shoe sole with molded tread (?), about 25 mm	Not yet identified, requires forensic analysis
5	2	6	Electromagnetic conductivity survey—south anomaly	1	One or two pieces of probable shoe leather, about 25 mm	Not yet identified, requires forensic analysis
6	1	6	Electromagnetic conductivity survey—south anomaly	2	One piece of probable worn rubber shoe sole with molded tread (?), about 30 mm	Not yet identified, requires forensic analysis
6	2	6	Electromagnetic conductivity survey—south anomaly	2	One piece of probable shoe leather, about 30 mm	Not yet identified, requires forensic analysis
32	1	10	Prevailing drainage—middle	3	Flattened clump 10–15 mm of probable clothing thread (cotton?), with an unknown substance adhering	Not yet identified, requires forensic analysis; unknown substance may be cryptogamic crust
57	2	17	Stripping in high-resolution metal detection survey anomaly	1	Fragment of apparent human-made white plastic tube [15 mm, wall thickness about 1.5 mm, estimated outside diameter 19 mm (0.75 in.)], flared at one end, dark stain on inside; probably not PVC	Assagai: "A white piece of plastic was observed that exhibited synthetic optical properties and appeared to be 'manmade'"
58	1	17	Stripping in high-resolution metal detection survey anomaly	1	Triangular piece (about 18 mm) of very thin (less than. 3 mm), translucent gray plastic (?), slightly shiny (between dull and glossy) on both sides; color of duct tape but lacks fibers; flexible and tough; reminiscent of gray refuse bag material	Assagai: "A piece of silver plastic was observed, is synthetic in nature and was felt to be possibly a piece of duct tape or a trash bag"

(Continued)

TABLE 38.2 *(Continued)*
Historic Materials of Uncertain Origin from the Foster Ranch Site: Current Analysis/Identification Status

FS Number	Item Number	Study Unit	Study Unit Type	Level	Original HMUO Description	Current Analytical Identification
59	1	17	Stripping in high-resolution metal detection survey anomaly	1	Two pieces (10, 12 mm) of very thin translucent white plastic (?), rolled, split and fragile; reminiscent of plastic drop cloth material	Assagai: "A piece of white fibrous material was observed that was determined to be cellulose"
31	1	n/a	Surface near study unit 11	n/a	Two pieces (originally 1?) of bright orange "plastic-like" material in the shape of flattened "blobs," about 12 and 23 mm by 5–8 mm thick; slightly flexible; surface has fine coral-like or lichen-like surface texture; fresh break is brighter and more vitreous; possibly modern plastic but otherwise unidentifiable	Assagai: "An orange agglomeration (blob) of material was observed and appeared to be an adhesive or glue material. It is partially translucent, weakly birefringent (anisotropic) and had a refractive index in the 1.53 to 1.54 range. This material is not fibrous and based upon the range of refractive indices mimics 'modacrylic,' a manmade synthetic material"

analyses. Additional forensic analysis was limited by the budget provided by the Sci Fi Channel.

Mineral Specimens

All of the mineral specimens were positively identified as such by the author and require no further analysis. To the extent that the possible burning and unidentified "white fibers" on two specimens might represent evidence of "something," however, they would benefit from further study.

Biological Specimens

The presumed-biological specimens were taken to the UNM Biology Department, where most of the author's identifications were unofficially confirmed by department faculty as several insect larva puparia, reptile eggs, a bone from a small (rat-sized) mammal, and a probable seed coat. Several examples of what came to be jokingly referred to as "dried green slime" were collected by the excavators, and all were identified as a typical desertic cryptogamic crust consisting of a desiccated mat of cyanobacteria, fungi, and algae that comes to life when it rains. Three items remain unidentified, of which two are most likely of terrestrial biological origin. The third, the "unidentified curly fibers" from SU 9, look like possible fur, but this identification has yet to be confirmed by forensic experts (Figure 38.7).

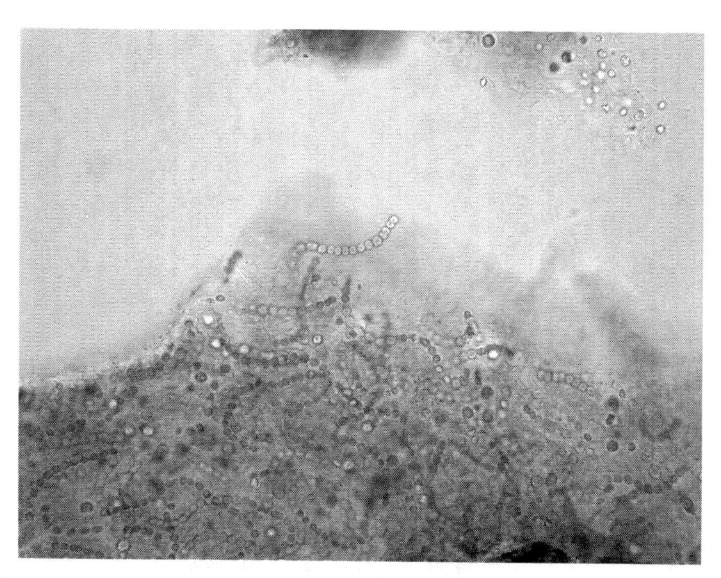

FIGURE 38.7 Microphoto of the "dried green slime."

Apparently Human-Made Specimens

Among the apparently human-made items, there is nothing that appears to be of obviously nonterrestrial origin (Figure 38.8, a–d). Only four, however were subjected to any analysis beyond the author's preliminary identifications. These are the "orange blob," the "white plastic tube" fragment, the "thin gray plastic" piece, and the several pieces of "very thin translucent white plastic." These items were submitted to Assaigai Laboratories in Albuquerque, New Mexico, for identification through

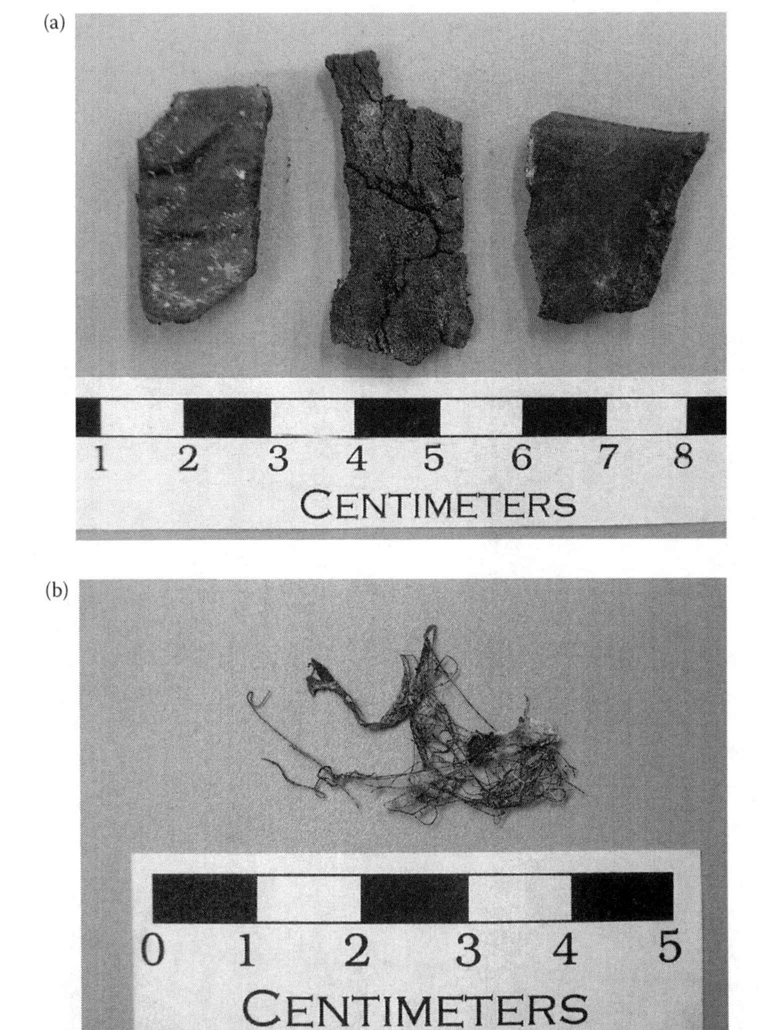

FIGURE 38.8 Apparently human-made HMUOs. a, Shoe sole fragments from study unit 6. b, Dark green (cotton?) thread from study unit 10. c, Plastic fragments from study unit 17. d, "Orange blobs" found on surface of site near study unit 11.

(c)

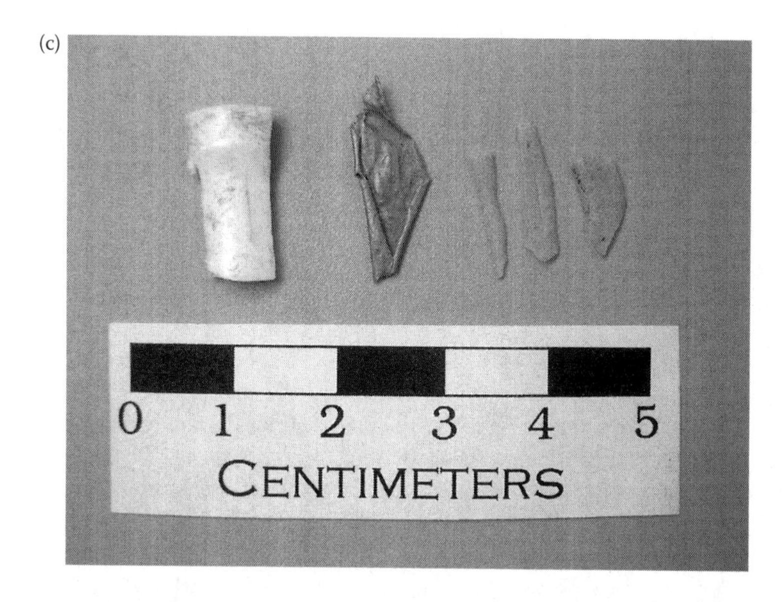

(d)

FIGURE 38.8 (Continued).

visual optical microscope identification.[23] Assaigai also subjected the orange blob to a refractive index (RI) analysis, in which the index of refraction is determined through the application of a series of oils with specific RIs until the correct one is found. The orange blob's RI suggests modacrylic to the Assaigai scientists, who said they believe it is a glob of "glue or adhesive" (*modacrylic* means modified acrylic, and both acrylic and modacrylic are plastics common in carpet fibers and some clothing). Because RI analysis is partially destructive, it was not performed on the other, smaller, apparently plastic specimens.

No analyses were performed on the four bags containing what appear to be shoe parts (rubber soles, leather) from SU 6, or on the "cotton thread" from SU 10. These items may have been left behind by postimpact visitors to the site. If so, analysis of these materials by the appropriate forensic specialists and specific identification of their age and origin might shed light on who visited the site and when.

Backhoe Trenching and Soil-Stratigraphy Studies

As discussed earlier, backhoe trenching was conducted at the Foster Ranch site for three purposes. The first goal entailed excavation of trenches across the reported furrow alignment (BHTs 101–107, Figure 38.3) in anticipation that—if present—deeper parts of the furrow would be evident as visually detectable features that crosscut or otherwise interrupted the site's natural stratigraphy. The second goal was to investigate anomalies revealed by the two geophysical prospection surveys (electromagnetic conductivity and high-resolution metal detection) at greater depths and in a more extensive fashion that was possible with the hand-excavated test pits (BHTs 108–109, Figure 38.3). Finally, together, all nine backhoe trenches provided extensive profiles for gathering soil-stratigraphic data from which to build a model of the site's natural stratigraphy and its variations. Of the nine backhoe trenches excavated, three yielded important results.

One backhoe trench, designated study unit 108, was excavated across the southern EMC anomaly just south of the test pit SU 6 (see Figures 38.3 and 38.5). The profile, which was recorded opposite the location of study unit 6, revealed unexpectedly deep soils with at least two depositional units and soil development representing ages on the order of 10,000 to more than 100,000 years. One aspect of the SU 108 profile is of particular interest and may be relevant to explaining the electromagnetic conductivity anomaly at this location. Four or five filled-in large-animal burrows were observed at a depth range of about 65–110 cm, all located within 2 m of one another at the point where the profile was recorded. The fill of each was noticeably softer and more porous. This cluster of buried, in-filled, and presumably old large animal burrows right at the peak of EMC anomaly 3 offers promising explanation for the anomaly. As sizable filled-in features they comprise a significant volume of more porous sediments that would naturally contain more moisture and thus be capable of producing a higher conductivity reading.

A single backhoe trench, designated study unit 109, was excavated across the main metal detection anomaly where study units 13–17 were excavated. This location is topographically the lowest area trenched and represents the bottom of the study area's prevailing drainage. The stratigraphic profile extended to a depth of 120 cm and exhibited a soil-stratigraphic sequence that is markedly different from that of SU 108 in two important ways: (a) much less pronounced calcareous horizons, and (b) clay is more common in all but the A horizon, and the soil colors are redder and darker overall.

Study unit 9's location in the bottom of the drainage—particularly in comparison with the topographic high location of SU 108—suggests that the associated increased surface and subsurface water flow might account for the differences in the soil profiles, as well as, possibly, the metal detection anomaly. Vadose zone water flow (i.e., transient flow that takes place between the surface and the saturated water table) in particular plays an important role in the characteristics of the soils and sediments. X-ray diffraction analyses of clays from this profile indicates the presence of two iron-bearing minerals—zeolite and hematite—which could account for the moderately strong metal detection anomalies in the area.[24] Although further testing would be required for confirmation, it is entirely possible that long-term vadose flow has both flushed pedogenic carbonates from the sediments and preferentially transported and deposited clay-sized particles and clay minerals into the deposits through a process similar to pedogenic illuviation. Finally, vadose flow may have imported iron- and manganese-bearing minerals dissolved from the local limestone bedrock and concentrated them in the area of the metal detection anomaly. Red stains on exposed limestone bedrock in the study area attest to the presence of trace iron minerals in the bedrock.

Finally, the characteristics of the site's natural stratigraphy, as revealed by the soil-stratigraphy study, indicated that—with the exception of animal burrowing and deep burial of artifacts—materials dating to the last 60 years should be largely confined to the upper 5–10 cm of the sites' deposits.

Trenching across the Reported Furrow Alignment

Seven backhoe trenches were excavated across the alignment of the impact furrow as reported to the project's technical advisors (Figure 38.3). The trenches were excavated on the afternoon of September 24, the last day of fieldwork. The locations were chosen according to the lead investigator's judgment, with an eye to sample different portions of the alignment while focusing on the presumably deeper section nearest the reported impact point, with three placed at the northern end of the alignment where initial impact occurred and the furrow was reportedly deepest, and an additional four placed along the remainder of the estimated 500-ft.-long feature. Study units 101 and 102 were placed at the estimated initial impact point, while study unit 103 was placed just on the other side of a limestone bedrock outcrop that eyewitnesses reportedly identified to the technical advisors as being a reference point. Study units 101 and 102 and the outcrop are visible in the foreground of Figure 38.2, while SU 103 lies just beyond the outcrop (the backhoe is visible excavating SU 104 in the distance).

Owing to their importance, study units 101–103 were inspected and photographed within an hour of excavation, while the other cross-furrow trenches were inspected during subsequent field visits to the site made by OCA personnel for the purposes of completing the transit-based topographic map and collecting soil samples. Only study unit 103 exhibited any evidence of a possible furrow, but the feature observed in the trench's south profile about 2.5 m east of the trench's west end—or approximately 3.5 m west of the estimated furrow alignment—provided the author with a considerable surprise. The feature was photographed immediately but was left otherwise untouched until subsequent visits, when it was analyzed and rephotographed (Figure 38.9a–c). The author's lengthy analysis of the feature appears in the final report, the main points of which are reviewed here.[25]

The first photograph taken of the "backhoe trench anomaly," as it came to be known, is shown in Figure 38.9a, which was made at 3:13 p.m. on September 24,

2003, with a digital camera and is unenhanced. Several aspects of the feature, which appears as an asymmetrical V shape with a rounded bottom, with the lowest point to the right of the trowel and just above the trench bottom, suggest that it marks a break or anomaly in the natural stratigraphy of the remainder of the profile. The feature is marked by a thin, intermittent line that starts below and to the right of the tip of the trowel and rises to the left at an angle of about 40°, continuing almost all the way to the surface. To the right of the bottom of the V, the feature is defined by a line marking a color difference that rises to the right at a steeper angle. To the right of this line, the soil appears slightly darker and redder in the color (not this version)

(A)

(B)

FIGURE 38.9 Photos of V-shaped stratigraphic anomaly in study unit 103 profile. (A) Photo taken at 3:13 p.m., September 24, 2003, about an hour after excavation. (B) Photo taken at 4:44 p.m., October 16, 2003, 22 days after discovery. (C) Photo taken after scraping and rewetting at 6:22 p.m., October 16, 2003, almost 2 hours after 9b.

(C)

FIGURE 38.9 (Continued).

photograph (confirmed by Munsell color analysis in the laboratory). This difference is more weakly evident to the left of the trowel, as well. Finally, the soil inside the V and above the trowel, where clods of grass and roots displaced by the backhoe can be seen hanging, appears to have slumped, as if the deposits inside the feature were softer and incapable of holding a vertical face such as that evident on either side of the anomaly.

These characteristics are essentially consistent with an in-filled and buried furrow or gouge resulting from an impact strong enough to significantly disturb the ground surface. The fine line defining the left side of the feature could represent either a layer of clay-dominated sediments washed in by the first precipitation event to occur after the feature was created, or a layer of sediments compressed by the impact. The other two characteristics that define the feature—color and soil compaction differences—are also consistent with the possibility that the feature represents a buried impact mark. Both represent apparent differences between the feature's fill and the surrounding undisturbed and natural deposits. That the fill appears to exhibit a lighter color than the slightly darker and redder color of the undisturbed sediments to the right and left would be consistent with light-colored surface deposits washing into a depression on either side of which the deposits retained the reddish shade characteristic of the natural soil's Bw horizon. Finally, recent fill would not be expected to exhibit the degree of compaction that is present in the undisturbed soils and that result from long-term soil-forming processes.

The fact that several characteristics of the feature are consistent with expectations of a furrow now buried (by natural or human actions), together with the feature's discovery within a few meters of the deepest part of the furrow's projected location, provided fertile ground for speculation that the anomaly might well be a preserved impact mark. The feature is not evident in the north profile of the backhoe trench, however, nor in any of the other cross-furrow trenches, although it should be noted that none of the others are closer than 30–35 m.

Most importantly, subsequent investigations cast doubts on the existence of the feature. When the author returned to the site on October 16—some 22 days after discovery—the feature was still visible (Figure 38.9b). But subsequent facing of the profile led to the near-complete disappearance of the feature (Figure 38.9c), leading to a puzzle: how was the anomaly created? It was determined that the anomaly was not an artifact of the trench excavation process, that is, either the outline of a single excavation "scoop" (not the operator's excavation method), or a "bucket wiggle" in which the side-to-side control was inadvertently moved during excavation by the backhoe operator, thus creating a bucket-shaped mark in the side of the trench. The "wiggle" hypothesis was negated by the fact that the bucket's bottom profile is a round 90° angle, while the angle of the anomaly's V was measured at 120°.

The apparent failure of the SU 103 anomaly to extend into the profile suggests that, whatever its origin, the feature was in fact superficial. Overall, the origin and "reality" of the study unit 103 anomaly remain uncertain. The feature visible in Figure 38.9a is unmistakable, yet it is essentially absent from the scraped profile in Figure 38.9c. Nonetheless, the soil color differences, although slight, have been independently established, and the backhoe operator, an archaeological specialist, claims to have observed the feature as well—in both sides of the trench. As noted above, the use of standard techniques of archaeological excavation, in which a grid system was laid out across the feature and horizontal excavations, such as those conducted in the test pits, would be required to fully evaluate the feature's existence and nature.

Other Discoveries

In addition to the results of the designed research, three unanticipated discoveries were made that bear on the project's research focus. The first of these are eight apparent test pit remnants that were probably excavated as part of a testing project sponsored by the Center for UFO Studies in 1989. Their presence attests to the chosen's sites having not changed since at least 1989. At the same time, the test pit's partially filled-in and vegetated condition offers valuable insight into the rates at which the natural processes of erosion, deposition, and revegetation can erase surface features on the high desert landscape and suggests that the reported furrow could possibly have been entirely erased by natural processes in the 55 years that passed between 1947 and 2002.

Two other discoveries are a linear, furrow-like feature found on a hillslope about ½ mile southeast of the Foster Ranch site, and a weather balloon (sic) found about 100 m to the southeast. For a variety of reasons, the feature—termed the "alternative furrow"—does not appear to represent any linear feature common in the area, including an abandoned two-track road, a cowpath, or even an entrenched erosional channel (which are, in fact, extremely rare). The feature appears on the aerial photographs used in the research, including those from 1946, indicating—unless the photographs' dates are erroneous—that the feature was present in November 1946. Mr. Charles Liles of the National Weather Service (NWS) Forecast Office in Albuquerque confirmed the balloon as a latex "high-altitude payload balloon" of the sort used by the NWS (he could not confirm it as being specifically a NWS balloon, however) and estimated its age at no more than 10 years. Given that the reported furrow was one of the two kinds of physical evidence being sought by the project, acquisition and study of additional aerial photography of the area seems warranted. Similarly, given the role of "weather balloons" in alternative explanations of the Roswell Incident, further analysis of the balloon is recommended for the purposes of shedding more light on the balloon's origins and its relationship, if any, to the alternative furrow.

CONCLUSIONS

The Foster Ranch archaeological testing project represents the first comprehensive attempt to locate extant physical evidence of whatever event occurred on the Foster Ranch site in July 1947. The project sought to uncover remnants of the two most commonly reported kinds of physical evidence: a furrow and unusual debris. No conclusive evidence of either was found, but two furrow candidates were discovered and should be further investigated. In addition, a number of unidentified items warranting forensic analysis were also recovered. Finally, soil samples collected during the project from both on- and off-site contexts were subjected to inductively coupled plasma mass spectrometry analysis by Assaigai Laboratories to determine whether microscopic evidence of debris is present in the form of traces of unusual metallic elements. Absent an extensive—and costly—background study of the occurrence and sources of rare metals in area soils, however, the results of this analysis cannot be effectively interpreted.[26]

In Chapter 2, Edward Staski rightly argues that archaeology is best defined as a science that attempts to link patterns (spatial or otherwise) in material culture to patterns in human behavior, rather than the study of particular ranges of space and time. The present study, however, can be said to operate at a level below the link between artifacts and behavior because it merely attempts to link evidence to putative events (the crash of an extraterrestrial vessel capable of interplanetary, even interstellar travel not being deemed a patterned behavior). In the modern, "processual" archaeological paradigm, considerable emphasis is placed on linking *facts* about the past with present-day observations using "middle-range theory." Inferences about behavioral patterns represented by the facts operate at a higher theoretical level.[27] Ethnoarchaeology—as defined by Staski—is in particular an important source of middle-range theory for archaeologists because it allows archaeologists to observe what kinds of physical evidence are produced by various behaviors.

A related methodology is *experimental archaeology*, in which archaeologists attempt to reproduce postulated past behaviors (such as various flint-working techniques) to determine what physical evidence (characteristic flaked-stone debris) might be produced by such behavior. Given that the present study reverses the inferential path of traditional archaeology and reasons from postulated events to hypothesized evidence, it might be desirable to use such an experimental approach to better determine what kinds of evidence should be sought at proposed UFO crash sites. Extraterrestrial vessels being in somewhat short supply, however, this approach is not at present a realistic option.

The humorous implications of such reasoning aside, as Staski points out, the burgeoning archaeological study of space exploration stands to inform us about ourselves through unique ways of studying present human behavior. With its focus on evidence of space exploration, as this new archaeological endeavor's methodologies evolve, it may even offer new ways to identify and investigate evidence of extraterrestrial use of space, be it on Earth or elsewhere. In this light, UNM-OCA's investigations at the Foster Ranch site can be viewed as a sort of precursor to Staski's fourth type of space archaeology: the future study of extraterrestrial material culture. Finally, the Foster Ranch project is a somewhat twisted version of Staski's sixth definition of space archaeology: "The archaeological study of material culture found in outer space, that is, exoatmospheric material that is clearly the result of human behavior." In contrast, the present study represents a focus on finding endoatmospheric material evidence that is clearly the result of extraterrestrial behavior. Such evidence was not found in the course of the investigations reported here. But, as evidence mounts that planets and the chemical precursors to life are the rule more than the exception in stellar systems throughout the universe, this *extraterrestrial archaeology* should become more than a curiosity (see Campbell, Chapter 46 of this volume).

Work at the Foster Ranch site is far from complete, especially given that only 0.036% of the study area was subjected to archaeological test excavations. The site represents a very large haystack, and the sought-after evidence may only be present as needles far and few between. Thus, continued strip excavations would be in order should funding become available. In addition, more in-depth forensic analysis of apparently human-made artifacts recovered might shed light on who visited the site and when. Finally, one important question remains: Is this the right place? Absent remnants of the surface evidence first reported—in particular the furrow—we may never know. Hence, in the spirit of all scientific archaeology, it can only be said that more data must be gathered and more research needs to be done.

NOTES AND REFERENCES

1. Doleman, W.H. 2003. Archeological Testing and Geophysical Prospection at the Reported Foster Ranch UFO Impact Site, Lincoln County, New Mexico. OCA/UNM Report No. 185-769. Albuquerque, NM: University of New Mexico Office of Contract Archeology.
2. McAvennie, M., ed. 2004. The Roswell Dig Diaries. Pocket Books, New York.
3. Carey, T.J. and D.R. Schmitt. 2007. *Witness to Roswell: Unmasking the 60-Year Cover-Up*. Career Press, Franklin Lakes, NJ. Randle, K.D., and D.R. Schmitt. 1991. UFO Crash at Roswell. Avon Books, New York. Randle, K.D., and D.R. Schmitt. 1994. The Truth about the UFO Crash at Roswell. Avon Books, New York.
4. Carey, Thomas, personal communication, July 28, 2003.
5. Schmitt, Donald, and Thomas Carey, personal communication, June 16, 2002.
6. Fewer, G. 1998. Searching for Extraterrestrial Intelligence: An Archaeological Approach to Verifying Evidence for Extraterrestrial Exploration on Earth. National Institute for Discovery Science. http://www.nidsci.org/pdf/fewer.pdf.
7. Berlitz, C., and W.L. Moore. 1980. The Roswell Incident. G.P. Putnam's Sons, New York. (Also published in 1980 by Berkeley Publishing Group, New York). Carey and Schmitt 2007 (note 3). Corso, P.J. 1997. The Day after Roswell. Pocket Books, New York. Eberhert, G.M., ed. 1991. The Roswell Report: A Historical Perspective. J. Allen Hynek Center for UFO Studies, Chicago, IL. Frazier, K., B. Karr, and J. Nickell, eds. 1997. The UFO Invasion: The Roswell Incident, Alien Abductions, and Government Coverups. Prometheus Books, Amherst, NY. Korff, K.K. 1997. The Roswell UFO

Crash: What They Don't Want You to Know. Prometheus Books, Amherst, NY. Leacock, C.P. 1998. Roswell Have You Wondered? Understanding the Evidence for UFOs at the International UFO Museum and Research Center. Novel Writing Publisher, Ann Arbor, MI. Pflock, K.T. 2001. Roswell: Inconvenient Facts and the Will to Believe. Prometheus Books, Amherst, NY. Randle, K.D. 1995. Roswell UFO Crash Update: Exposing the Military Cover-up of the Century. Global Communications, New Brunswick, NJ. Saler, B., C.A. Ziegler, and C.B. Moore. 1997. UFO Crash at Roswell: the Genesis of a Modern Myth. Smithsonian Institution Press, Washington, DC. United States Air Force. 1995. The Roswell Report: Fact versus Fiction in the New Mexico Desert. U.S. Government Printing Office, Superintendent of Documents, Washington, DC.
8. Schmitt, Donald, and Thomas Carey, personal communication, June 16, 2002.
9. Schmitt, Donald, personal communication, September 24, 2002.
10. Randle and Schmitt 1994 (note 3) (see specifically p. 42).
11. Ibid., p. 37.
12. Ibid., p. 31.
13. Doleman, W.H. 1995. Human and Natural Landscapes: Archeological Distributions in the Southern Tularosa Basin. Ph.D. diss., University of New Mexico, Albuquerque. Peacock, E., and D.W. Fant. 2002. Biomantle Formation and Artifact Translocation in Upland Sandy Soils: An Example from the Holly Springs National Forest, North-Central Mississippi, U.S.A. Geoarchaeology 17(1): 91–114.
14. Hyndman, D.A., and S.S. Brandwein. 2002. Geophysical Investigations of a Suspected UFO Crash Site near Corona, New Mexico. Sunbelt Geophysics, Albuquerque, NM.
15. For technical details and an exhaustive review of the results, see Hyndman and Brandwein 2002 (note 14).
16. For technical details and a discussion of the results, see Hyndman and Brandwein (2002) (note 14).
17. For details of the testing activities, see Doleman 2003 (note 1).
18. Birkeland, P.W. 1999. Soils and Geomorphology. Oxford University Press, New York.
19. Hyndman and Brandwein 2002 (note 14).
20. Schmitt, D., personal communication, September 24, 2002.
21. For details, see Hyndman and Brandwein 2002 (note 14) and Doleman 2003 (note 1).
22. Doleman 2003 (note 1).
23. Roark, A.A., and W.P. Biava. 2003. Unpublished letter report from Assaigai Laboratories on "Microscopic Analysis of Foster Ranch Samples," on file at the University of New Mexico Office of Contract Archeology, Albuquerque, NM.
24. Bloch, J. 2003. X-Ray Diffraction Analysis of Foster Ranch Samples. Unpublished report on file at the University of New Mexico Office of Contract Archeology.
25. Doleman 2003 (note 1).
26. Bloch, J., University of New Mexico Department of Earth and Planetary Sciences, personal communication, late 2005 or early 2006.
27. Binford, L.R. 1992. Seeing the Present and Interpreting the Past—And Keeping Things Straight. In Space, Time, and Archaeological Landscapes, ed. J. Rossignol and L. Wandsnider, 43–59. Plenum Press, New York.

Section VIII

Space Policy and Preservation

"Honey, this vacation let's
hit the Off World Heritage Sites."

39 International Space Community and Space Law

Stephen E. Doyle

CONTENTS

INTRODUCTION

This chapter briefly surveys the international space community and the law that has been establish for its management. It is intended as a source of references for any reader interested in learning more about the actors in outer space and the laws relevant to activities there. Provisions in the modern law of the sea offer potentially useful examples of archaeological protections that could provide a basis for considering archaeological provisions in space law. For this reason, short explications of relevant terms in maritime law are included. The chapter closes with a brief discussion of issues related to preservation of space artifacts and some suggestions for moving toward international agreement on archaeology in outer space.

SCOPE OF THE INTERNATIONAL SPACE COMMUNITY: THE PARTICIPANTS

The USSR's remarkable launch of the world's first orbiting satellite in October 1957 created widespread press notoriety and international discussion and consternation concerning what it would mean to world science, international security, and the prevailing Cold War's international balance of power.[1] Following a second launch by the USSR in November 1957 and additional launches by the USSR and the United States in 1958, substantial organizational and political reactions to these early spaceflights began to appear. A new, broadened appreciation of the significance of outer space entered global consciousness.

During the 1960s and 1970s, new organizations emerged,[2] some national governmental entities, some quasigovernmental/private entities, some international organizations, and some regional organizations, and all for the purpose of exploring and/or exploiting outer space.[3] By the end of the second half of the twentieth century, the establishment, operation, maintenance, and regulation of space systems, particularly those using space resources for provision of services on the earth, had become an annual multibillion-dollar industry. Systems emerged for national and international communication, remote sensing of the earth and its resources, meteorological services, navigation services, and national and regional military systems, including early warning systems and treaty verification systems.

The actors who made all this possible were initially nation-states. Governments of states undertook to organize and define space programs and to launch spacecraft for a wide variety of purposes. In the early years, launch services were provided by the United States and, later, the USSR on commercial bases. By the end of the century competitive commercial launch services were available from the United States, Russia, Europe (Ariane), Japan, China, and India. These launch providers and others could produce commercially competitive spacecraft for the use of countries who wished to conduct space activities but could not build their own satellites.[4]

Organizations created domestic satellite systems, regional systems, and global systems for communication services including telephony, telegraphy, data transmission, television relay, and eventually satellite broadcasting.[5] Major early participants included INTELSAT in Washington, D.C. (1964), the global communication satellite organization; INMARSAT in London (1976), the global maritime and mobile communication satellite system operator; ARABSAT in Riyadh, Saudi Arabia (1976), the Arabic language states' regional communication satellite system; INTERSPUTNIK in Moscow, Russia, the global organization of existing and former communist states, providing international communications of all kinds; and PALAPA (now SATELINDO) in Jakarta, Indonesia (1975), the Indonesian satellite system offering services among countries of the Asia Pacific region.[6] These few early systems are in addition to numerous additional national and regional systems.[7] As technology progressed during the third quarter of the twentieth century, more and more international cooperation arose in the planning and conduct of space flight programs on bilateral, regional, and broader international bases.[8]

In addition to the strong early emphasis on communication satellite systems,[9] systems emerged for meteorological services,[10] navigation services,[11] earth imagery from remote sensing satellites,[12] and space sciences,[13] and all these applications services became commercially and competitively available on the world market. Also, those governments with the capacity to do so established dedicated military satellite systems to strengthen their individual national defense situations.

In 1957, in response to the launch of *Sputnik 1*, the United Nations in New York took up the issue of use of outer space and urged that nations cooperate to ensure that outer space would be used exclusively for peaceful purposes.[14] Thereafter, the United Nations (UN) General Assembly created an Ad Hoc Committee on the Peaceful Uses of Outer Space.[15] Following its initial report, this ad hoc committee was succeeded by a permanent committee of the General Assembly.[16] In the early 1960s, only two states were regularly launching spacecraft and operating systems into space: the USSR and the United States.

Carl Christol notes correctly that for rules to be effective, these two space-resource states would "need to be in regular agreement on the formation of legal principles, rules, and policies for the space environment. In fact, many important developments in space law and policy down to the present [1982] have been the products of positions initiated by these two States."[17] To facilitate its work, the new UN committee established two suborgans: the Scientific and Technical Subcommittee, and the Legal Subcommittee of the Committee on the Peaceful Uses of Outer Space (COPUOS).[18] All three of these organs have met annually, continually since 1961.[19] By the early 1960s, it was apparent that there would have to be a body of international regulations and rules to manage the emerging mix of diverse space organizations and services. In 1963, COPUOS recommended a declaration of principles for the guidance of states' activities in outer space. The UN General Assembly adopted these recommended principles in a unanimously approved resolution.[20]

During the 1960s and 1970s, as the number of actors increased and the complexity of the relationships intensified, COPUOS addressed a wide range of issues, eventually producing five major treaties relating to the activities of states in outer space.[21] In parallel with these treaties, which were proposed to and promulgated by the UN General Assembly, the COPUOS also recommended guiding principles relating generally to activities of states in space (1963),[22] direct broadcasting by satellite (1982),[23] remote sensing of the earth from space (1986),[24] the use of nuclear power sources in outer space (1992),[25] and finally, a set of principles addressing international cooperation in the interest of developing countries (1996).[26]

In addition to these law-making actions of the United Nations, many states undertook the creation of national space program management organizations, as well as establishing enabling and regulatory national laws applicable to activities to gain access to and function in outer space. In the United States, for example, motivated by spaceflight during the 50 years from 1951 to 2000, seventeen significant new national laws were passed. More than sixty-five presidential policy directives and executive orders were issued, more than fifty new and reorganized national organizational structures were created inside the government, and at least twenty significant national organizations were created outside of government. At least thirty significant international treaties were signed and ratified, and many of these created new international organizations joined by the United States.[27] All of these new laws, regulations, policies, and organizations owed their genesis exclusively to spaceflight or space activity. None of them deal with preservation of materials of archaeological concern.

During the latter half of the twentieth century, a substantial amount of national and international regulation appeared addressing the need for management of cultural history, as for example the National Historic Preservation Act[28] in the United States and the UNESCO Convention Concerning the Protection of World Cultural and Natural Heritage, among others, which promotes protection of sites of "outstanding universal significance."[29] See the discussion of the implications of these laws by Beth Laura O'Leary in Chapter 40 of this volume.

By the end of the twentieth century at least twenty foreign national space organizations existed, with legal frameworks comparable to those established in the United States, to mange, facilitate, and regulate national space activities, and there were more than twelve international intergovernmental bodies engaged directly in some form of activity in outer space, with another ten or more private international cooperative organizations conducting or affecting space activities.[30]

AGREEMENTS AND LAWS IN PLACE

SITUATION ON LAND

There exist today substantial structures of law, policy, organizations, practices, international cooperation, and literature related to archaeology on and about the earth.[31] There is also in place an emerging community of institutions and people interested in space archaeology, and some nascent collections on earth include artifacts flown in outer space and in some cases artifacts recovered from the Moon.[32] With only U.S. and Russian round-trip visits to the Moon having been completed, there is only a small body of artifacts flown to the Moon and returned to Earth.[33] There are, in addition, the redoubtable Moon rocks, which include regolith materials collected on the Moon and brought to Earth by the United States and by the USSR.[34]

The United States' laws relating to historic preservation are numerous and long standing.[35] Many of these U.S. laws relate to and describe the roles, authority, and scope of activity of the Smithsonian Institution (SI), which is the primary national actor in the collection and preservation of historical materials and artifacts in the United States. Directly connected with and fortuitously physically close to the SI in Washington, D.C., is the headquarters of the National Aeronautics and Space Administration (NASA). Effective March 14, 1967, more than 2 years before NASA launched its first successful manned landing to the Moon, NASA entered into an agreement with the SI concerning the custody and management of NASA historical artifacts. Under this agreement, all NASA artifacts no longer needed by NASA or other governmental agencies for technical purposes are to be transferred to the Smithsonian Institution for custody, protection, preservation, and display. By terms of this agreement, displays may be at the Smithsonian or upon loan to NASA headquarters, NASA field centers, other federal agencies, museums, or other appropriate organizations. There are numerous well-organized and well-equipped museums of space and space artifacts about the United States, particularly in Alabama, Florida, New Mexico, and Texas.

Although national laws exist on abandonment of ships, protection of battlefields, preservation of native American artifacts and burial sites, selected and endangered species, coastal zones, continental shelves, wetlands, historical sites, some limited space-related Earth sites, federal records, marine sanctuaries, federal records, and other matters of historical interest, there has been as yet no congressional attention to the issues involved in preservation of evidence of human activities in outer space and sites and artifacts located or taken from there.

Pursuant to Article 2 of the *1967 Outer Space Treaty*, "outer space, including the moon and other celestial bodies, is not subject to national appropriation by claim of sovereignty, by means of use or occupation, or by any other means."[36] Thus, outer space stands in the law in a posture similar to that of the high seas, that is, beyond national territorial jurisdictions. In addition, Article 1 of the same treaty provides, in part, that "outer space, including the moon and other celestial bodies, shall be free for exploration and use by all states without discrimination of any kind, on a basis of equality and in accordance with international law, and there shall be free access to all areas of celestial bodies." See also the discussion of Article VIII of the treaty by O'Leary in Chapter 40 of this volume.

Among applicable provisions of the referenced international law are the terms of the 1970 UNESCO Convention on the Means of Prohibiting and Preventing the Illicit Import, Export and Transfer of Ownership of Cultural Property.[37] For this convention to have effect, there must be enabling national law providing applicable regulations and means for enforcement set up by signatory countries. Article 1 of the convention defines *cultural property* as property that

> … on religious or secular grounds, is specifically designated by each State as being of importance for archeology, prehistory, history, literature, art or science and which … [is]
>
> b. property relating to history, including the history of science and technology and military and social history, to the life of national leaders, thinkers, scientists and artists and to events of national importance.[38]

This definition is unarguably broad enough to embrace any and all things fabricated by humans and propelled into outer space, as well as any things people did, established, or left there, or anything brought back from space, provided only that all these things are designated cultural property by the originating or receiving state. The convention is not self-implementing; it requires governmental action by member states for its implementation.

Articles 9 and 10 of the 1970 UNESCO Convention on Cultural Property provide that

> Article 9
>
> Any State Party to this Convention whose cultural patrimony is in jeopardy from pillage of archaeological or ethnological materials may call upon other States Parties who are affected. The States Parties to this Convention undertake, in these circumstances, to participate in a concerted international effort to determine and to carry out the necessary concrete measures, including the control of exports and imports and international commerce in the specific materials concerned. Pending agreement each State concerned shall take provisional measures to the extent feasible to prevent irremediable injury to the cultural heritage of the requesting State.
>
> Article 10
>
> The States Parties to this Convention undertake:
>
> To restrict by education, information and vigilance, movement of cultural property illegally removed from any State Party to this Convention and, as appropriate for each country, oblige antique dealers, subject to penal or administrative sanctions, to

> maintain a register recording the origin of each item of cultural property, names and addresses of the supplier, description and price of each item sold and to inform the purchaser of the cultural property of the export prohibition to which such property may be subject;
>
> To endeavour by educational means to create and develop in the public mind a realization of the value of cultural property and the threat to the cultural heritage created by theft, clandestine excavations and illicit exports.[39]

The United States adopted the Convention on Cultural Property Implementation Act, approved January 12, 1983 and now found in Title 19, United States Code 2600. The law's definitions relate to materials within and subject to export control by States parties. The form and intent of the law may be argued as indicative of the kind of controls which would be useful and appropriate for dealing with space archaeological materials, but the law is silent with regard to outer space and materials located there or obtained there.

Limited attention has been given to date to protection of the history of human exploration and exploitation of space, but there is no comprehensive policy or body of regulations applicable to space artifacts, materials, and/or archaeology in or related to outer space. Although there are implementing guidelines established for various historical preservation laws, there are no published guidelines related to outer space, nor are there any relevant executive orders or presidential policy directives.[40] There are no provisions for tax incentives to preserve or support preservation of space related materials, artifacts, or sites.[41] Considered in contrast to the lands of the earth and the separately addressed areas of the high seas, archaeology in the increasingly active province of outer space has been substantially ignored by national legislatures, by the United Nations, and by UNESCO.

Situation at Sea

With regard to the situation of archaeological materials existing or found at sea, there are some relevant established agreements, but the United States is not a party to the two most relevant conventions dealing with archaeology in the oceans of the world. A brief historical recapitulation of the situation concerning archaeology beneath the world's oceans may be useful in considering what next steps, if any, should be taken concerning the preservation of history through the archaeology of outer space.

Nations negotiated many years to come to agreements modifying the freedom of the seas (*mare liberum*) which had prevailed on the world's oceans for centuries. In 1956, the first United Nations Conference on the Law of the Sea (UNCLOS I) met in Geneva, Switzerland, and produced four separate conventions, the negotiation of which was concluded in 1958.[42] These conventions entered into force at separate times, as provided by their respective terms. Some important questions of concern to states were left unresolved, such as the breadth of national territorial waters. A second UN Conference on the Law of the Sea (UNCLOS II) met for 6 weeks, also in Geneva in 1960, but reached no new agreement.

In 1973, UNCLOS III was convened in New York. This conference, seeking to avoid domination of the proceedings by selected groups of states, decided to employ a procedure that had been successfully practiced in the UN Committee on the Peaceful Uses of Outer Space since 1961. Participating states avoided voting by agreeing upon a consensus procedure. This provided basically that no proposed text would move forward to final agreement unless there was consensus among the participants (which meant basically an absence of explicit objection by any participating state, a tacitly agreed veto provision for every state). This conference formed a new, consolidated Convention on the Law of the Sea, which incorporated and superseded the earlier conventions agreed at UNCLOS I.[43] The new United Nations Convention on the Law of the Sea, of December 10, 1982, contained provisions related to protection of cultural heritage and archaeological interests of states in materials found in and taken from the high seas. Article 1 of the convention defines the term *area,* for purposes of the convention, as "the seabed and ocean floor and subsoil thereof, beyond the limits of national jurisdiction." Article 149 provides the following:

> Article 149—Archaeological and historical objects
>
> All objects of an archaeological and historical nature found in the Area shall be preserved or disposed of for the benefit of mankind as a whole, particular regard being paid to the preferential rights of the State or country of origin, or the State of cultural origin, or the State of historical and archaeological origin.[44]

It does not take a rocket scientist or a trained archaeologist to see that this provision is so general and vague as to have little value. It poses more questions than it answers. What constitutes disposal "for the benefit of mankind as a whole"? How shall the country of origin, cultural origin, or archaeological origin be discovered or determined? What are the preferential rights of states? There is another provision later in the same convention, containing more detail. Article 303 provides the following:

> Article 303—Archaeological and historical objects found at sea
>
> 1. States have the duty to protect objects of an archaeological and historical nature found at sea and shall cooperate for this purpose.
> 2. In order to control traffic in such objects, the coastal State may, in applying article 33, presume that their removal from the seabed in the zone referred to in that article without its approval would result in an infringement within its territory or territorial sea of the laws and regulations referred to in that article.
> 3. Nothing in this article affects the rights of identifiable owners, the law of salvage or other rules of admiralty, or laws and practices with respect to cultural exchanges.
> 4. This article is without prejudice to other international agreements and rules of international law regarding the protection of objects of an archaeological and historical nature.[45]

Herein we find more specific language, which refers in part to the locale of the discovery (Article 33 defines the "contiguous zone," nominally 24 miles outward from the nation maritime baseline) and declares a duty to protect objects found in that zone and to cooperate for the purpose of their protection. The provision in Article 303, paragraph 4, took on special relevance in 2001 because of the adoption of a new UNESCO Convention on the Protection of the Underwater Cultural Heritage, on November 2, 2001. Although UNCLOS III took tentative steps toward protection of oceanic archaeological interests, the 2001 convention was an instrument fully dedicated to that purpose. UNCLOS III had more than 150 states as parties, but not the United States. The 2001 convention has, at this writing, sixteen states that have ratified or accepted the convention. These are the states parties. When twenty states have ratified it, the 2001 convention will enter into force among ratified states parties. The United States is not a participant in this convention yet. Perhaps it will be in the future.

The 2001 convention requires states parties to preserve underwater cultural heritage for the benefit of humanity by appropriate action. For purposes of the convention, *underwater cultural heritage* means all traces of human existence having a cultural, historical, or archaeological character that have been partially or totally under water, periodically or continuously, for at least 100 years.

The primary principle of the convention requires that underwater cultural heritage be protected from commercial exploitation, trade, or speculation. This principle does not prevent professional archaeology, the deposition of objects recovered in research projects, nor does it prevent salvage activities or actions by finders, provided the requirements of the convention are fulfilled. A useful compromise is achieved in the convention between protection and operational needs. Any activity relating to underwater cultural heritage to which the convention applies is not subject to the law of salvage or law of finds. Authorized activity is approved by competent authorities as being in conformity with the convention, which ensures that a recovery of underwater cultural heritage will achieve maximum protection.

The convention establishes a preference for preservation in situ of underwater cultural heritage (i.e., the discovered location). That is the convention's first option. Archaeologists prefer in situ preservation because the objects preserved in this manner preserve their locational integrity. Alternative activities may be authorized for making a significant contribution to the protection or knowledge of underwater cultural heritage. The preference for in situ preservation underlines the importance of the historical context of the object and its scientific significance. In situ preservation also recognizes that such heritage is, under normal circumstances, well preserved by the low deterioration rate under water (due to the lack of oxygen) and therefore unlikely to be in danger.

The convention does not control ownership of wrecks or ruins between states concerned. The convention does set a high standard for preserving underwater cultural heritage, common to all states parties. The convention applies only among states parties, but each state so choosing may require even higher standards of protection than provided by the convention.

As a *lex specialis* (i.e., a special regulation), specific to underwater cultural heritage, the convention does not prejudice rights or duties of states under international law, including UNCLOS III. Any state may become a party to the 2001 convention, regardless of its status under the UNCLOS. States parties have the exclusive right to regulate activities in their internal and archipelagic waters and their territorial sea (nominally 12 miles from the maritime baseline), and within their contiguous zone (an additional 12 miles), they may regulate and authorize activities directed at underwater cultural heritage.

Within the exclusive economic zone (extending to 200 miles), or the continental shelf (of continental riparian states) and within the area (the high seas), a specific international cooperation regime encompasses notifications, consultations, and coordination in the implementation of protective measures pursuant to the 2001 convention.

The convention contains regulations against the illicit trafficking in cultural property and on training in underwater archaeology. It provides also that transfer of technologies and information sharing shall be promoted and that public awareness shall be raised concerning the value and significance of the underwater cultural heritage.

At the core of the 2001 convention, Article 33 provides the following:

> Article 33—The Rules
>
> The Rules annexed to this Convention form an integral part of it and, unless expressly provided otherwise, a reference to this Convention includes a reference to the Rules.[46]

Possibly the most important part of the convention, the annex sets forth thirty-six rules to guide the conduct of interventions underwater. This annex is internationally recognized as the primary reference document in the discipline of underwater archaeology. More significant than the rules thus formally placed before the international community is the fact that UNESCO, by virtue of adoption of this convention, has demonstrated the will of the international community to act for the preservation of archaeological materials even when found beyond national boundaries and in areas not subject to national appropriation. This is a significant precedent justifying eventual attention by the international community to the orphaned state of archaeology in outer space.

For its part, the United States Congress has already clearly indicated its willingness to formulate and adopt law applicable to national interest in the extraterritorial regions of outer space. On November 15, 1990, Congress adopted Public Law 101-580, which enacted the following provision:

> 35 USC Chapter 10—Sec. 105. Inventions in outer space
>
> (a) Any invention made, used or sold in outer space on a space object or component thereof under the jurisdiction or control of the United States shall be considered to be made, used or sold within the United States for the purposes of this title, except with respect to any space object or component thereof that is specifically identified and otherwise provided for by an international agreement to which the United States is a party, or with respect to any space object or component thereof that is carried on the registry of a foreign state in accordance with the Convention on Registration of Objects Launched into Outer Space.
>
> (b) Any invention made, used or sold in outer space on a space object or component thereof that is carried on the registry of a foreign state in accordance with the Convention on Registration of Objects Launched into Outer Space, shall be considered to be made, used or sold within the United States for the purposes of this title if specifically so agreed in an international agreement between the United States and the state of registry.[47]

The term *outer space* in this law includes the Moon and other celestial bodies and any other area beyond the tangible atmosphere of the Earth. Thus, there exist international precedents and at least this precedent in national law suggesting that it is appropriate and useful to consider the extension of law regarding archaeology to outer space, even to the establishment of appropriate safeguards and guidelines to regulate archaeological interventions in outer space, on the Moon, or on other celestial bodies.

We noted at the outset of this chapter that the primary actors in outer space are agencies of national governments or citizens (including legal entities) of states. Most if not all of the states involved in space activity today are parties to the *Treaty on Principles Governing the Activities of States in the Exploration and Use Outer Space, Including the Moon and Other Celestial Bodies.*[48] That treaty explicitly provides for the conduct of space activities in compliance with international law. Article 3 declares clearly:

> States Parties to the Treaty shall carry on activities in the exploration and use of outer space, including the moon and other celestial bodies, in accordance with international law, including the Charter of the United Nations, in the interest of maintaining peace and security and promoting international co-operation and understanding.[49]

Can it be assumed then that at least the spirit and general intent of existing provisions of international conventions relating to archaeology in areas beyond national jurisdictions should be considered applicable in outer space? The case is arguable and interesting but certainly not compelling. A far better result could be attained if the UN Committee on the Peaceful Uses of Outer Space would take up the question of appropriate safeguards for cultural materials in space or if UNESCO would take up the question, and one or both of these fora would produce either a proposed convention or a carefully constructed set of rules or guidelines, such as those annexed to UNESCO's 2001 Convention on the Protection of the Underwater Cultural Heritage. In addition, the World Heritage List formulated pursuant to the World Heritage Convention offers an example of what is possible through international cooperation in UNESCO.

REQUIRED NEXT STEPS

In the absence of action, particularly action of states wishing to assume leadership on this topic, nothing is likely to happen or be done about protection of archaeological materials in or from outer space. Museum operators, traders, collectors, and brokers will be content to have the existing situation prevail. It serves their economic interests. For any to whom the archaeology of outer space is a subject or interest and/or concern, little consolation can be offered, but some advice can be proffered.

Before the UN Committee on the Peaceful Uses of Outer Space will take up a subject for study, discussion, and possible action, it would be necessary that one or several member states propose that the topic of archaeology in outer space should be considered. The United Nations is an organization established to deal with the concerns of member states; it is not self-motivating. This is also true of UNESCO. Therefore, if one wished to see a formal international debate of the mater and desired to see decisions taken, one would be well advised to approach the United Nations

Division of their individual national Foreign Offices (in the United States, that is the State Department) and request that the matter be brought to the attention of the appropriate bodies and officials of the United Nations Organization.

To encourage and facilitate a meaningful, early, and substantive discussion, it would be useful to formulate a proposed approach, some recommended rules, and some form of final action recommended to be taken. The Foreign Service officers of the State Department are not the appropriate people to formulate proposals on such sophisticated issues to put before the United Nations. Relevant proposals should arise from the profession, be elaborated by the profession, and be presented to the State Department, or the appropriate Foreign Office for consideration, to be forwarded to the United Nations. A reader of this book is far more likely to appreciate what rules and guidelines are needed than a career Foreign Service officer of the State Department.

Thus, the future course of appropriate events appears to recommend itself.

1. Within one or more associations or groups concerned with archaeology in an interested nation, the archaeologists should establish a working group or a planning committee to examine archaeology in outer space. In the United States, this step has already been taken.[50]
2. One or more national workshops and/or possibly an international workshop should address (a) the proper national pathway(s) to approach the United Nations, (b) the issues involved, and (c) creation of a drafting group to prepare preliminary proposals.
3. National representative delegations should be encouraged to take the matter up with appropriate officials at the UN and/or at UNESCO, to begin an informal exploration of an appropriate path and mechanism to accomplish creation of an instrument dealing with archaeology in outer space.
4. Provide useful and credible written proposals to officials of national governments with requests that they be forwarded to appropriate international officials for consideration.
5. Identify and nominate qualified expert advisors who could meet with and counsel government employees during national preparatory meetings and potentially serve as advisors on delegations to international meetings dealing with the proposals.

CONCLUSION

In this chapter, I have attempted to survey the extant body of participants in outer space and the body of law that regulates the actions of those participants. I have discussed summarily the relevant international law dealing with activities in space, and I have considered in some detail the existing/emerging body of international law dealing with archaeology in and under the high seas, because therein appears to lie a body of potentially useful precedent. Finally, I have suggested some modest steps that could be taken, leading to possible international action to fill the need for rules or regulations in some form to encourage and facilitate the preservation of archaeological materials and to recognize the value of archaeology in outer space.

REFERENCES

1. See discussion and sources cited in Lay, S.H., and H.J. Taubenfeld. *The Law Relating to the Activities of Man in Space*. An American Bar Foundation Study. University of Chicago Press, Chicago, 1970 (see specifically pp. 11–12). See also Levy, L. (ed.). *Space: Its Impact on Man and Society*. W.W. Norton and Company, New York, 1965 (an anthology of commentary by political and technical leaders). Osgood, K.A. Before Sputnik: National Security and the Formation of US Outer Space Policy, 1953–1957. Paper presented at the Symposium Reconsidering Sputnik: Forty Years Since the Soviet Satellite, Smithsonian Institution, Washington, DC, September 30 to October 1, 1997 (a thorough, well-documented, and accurate assessment).
2. The United States established the world's first civil governmental organization devoted to the research in and development of space technology in National Aeronautics and Space Act, July 29, 1958, P.L. 85–568; 72 Stat 426, now contained, as amended, in U.S. Code, Title 42, Chapter 26, National Space Program. There are 20 nations that have established one or more national laws directly addressing activities in outer space. See the titles "Space Law" and "National Space Laws" at http://unoosa.org for a list of the twenty nations and texts of their laws.
3. An excellent general survey is Ley, W. *Rockets, Missiles and Men in Space*. The Viking Press, New York, 1968. See also McDougal, W.A. ... *The Heavens and the Earth: A Political History of the Space Age*. Basic Books, New York, 1985. For a more technical account, see Von Braun, W., F.I. Ordway III, and D. Dowling. *Space Travel: A History*. Harper & Row, New York, 1985. The early political context can be seen in Senate Committee on Aeronautical and Space Sciences. *Documents on International Aspects of the Exploration and Use of Outer Space, 1954–1962*, 88th Cong., 1st Sess., Doc. No. 18, May 9, 1963, Government Printing Office, Washington, DC.
4. For the national laws under which these various nations operate launch services, see the applicable laws at http://unoosa.org.
5. Martin, D.H. *Communication Satellites 1958–1995*. The Aerospace Press, El Segundo, CA, 1996.
6. Most entities listed maintain extensive Web sites, including histories. See also United Nations. *Space Activities of the United Nations and International Organizations*. United Nations, New York, 1992, and a revised edition, 1999. See also Levy 1965 (note 1).
7. Martin 1996 (note 5).
8. Frutkin, A.W. *International Cooperation in Space*. Prentice Hall, Englewood Cliffs, NJ, 1965.
9. NASA. *Conference on the Law of Space and of Satellite Communications*. NASA SP-44. Government Printing Office, Washington, DC, 1964.
10. For the formative years, see NASA. *Significant Achievements in Satellite Meteorology 1958–1964*. NASA SP-96. Government Printing Office, Washington, DC, 1966.
11. NASA. *Significant Achievements in Space Communications and Navigation 1958–1964*. NASA SP-93. Government Printing Office, Washington, DC, 1966.
12. National Academy of Sciences. *Resource Sensing from Space*. National Academy of Sciences, Washington, DC, 1977.
13. NASA. *Significant Achievements in Space Science 1965*. NASA SP-136. Government Printing Office, Washington, DC, 1967. NASA. *Satellite Geodesy 1958–1964*. NASA SP-94. Government Printing Office, Washington, DC, 1966.
14. UN General Assembly Resolution 1148 (XII), November 14, 1957; discussed in Christol, C.Q. *The Modern International Law of Outer Space*. Pergamon Press, New York, 1982 (see specifically p. 13). For an account of the U.S.-Soviet relations and negotiations at the threshold of the world's entry into outer space activities, see The Department of State. International Negotiations Regarding the Use of Outer Space 1957–1960.

Prepared by the Historical Office, Bureau of Political Affairs, Department of State, 1961. In *Legal Problems of Space Exploration: A Symposium,* Senate Committee on Aeronautical and Space Sciences, 87th Cong., 1st Sess., Doc. No. 26, March 22, 1961 (see specifically p. 985).

15. UN General Assembly Resolution 1348 (XIII), December 13, 1958; discussed in Christol 1982 (note 14) (see specifically p. 14).
16. UN General Assembly Resolution 1472 (XIV), December 12, 1959; discussed in Christol 1982 (note 14) (see specifically p. 15). For the foregoing three resolutions and more, see Senate Committee on Aeronautical and Space Sciences. *Documents on International Aspects of the exploration and Use of Outer Space 1954–1962*, 88th Cong., 1st Sess., Doc. No. 18, Government Printing Office, Washington, DC, May 9, 1963.
17. Christol 1982 (note 14) (see specifically p. 15).
18. Haley, A.G., *Space Law and Government.* Appleton Century Crofts, New York, 1963 (see specifically p. 320). See also Christol 1982 (note 14) (see specifically p. 16). A more esoteric, scholarly treatment of the development and nature of the COPUOS is in McDougal, M.S., H.D. Lasswell, and I.A. Vlasic. *Law and Public Order in Space.* Yale University Press, New Haven, 1963 (see specifically pp. 160–161, 209–212).
19. The United Nation Office of Outer Space Affairs, Vienna, Austria, offers a comprehensive Web site for UN documentation and historical reference materials at http://unoosa.org.
20. *Declaration of Legal Principles Governing the Activities of States in the Exploration and Use of Outer Space.* UN General Assembly Resolution 1962 (XVIII), December 13, 1963, adopted without a vote. In United Nations, *United Nations Treaties and Principles on Outer Space*, United Nations, New York, in multiple editions.
21. (1) *Treaty on Principles Governing the Activities of States in the Exploration and Use of Outer Space, Including the Moon and Other Celestial Bodies*; done at Washington, London, and Moscow on January 27, 1967; entered into force on October 10, 1967 (18 UST 2410; TIAS 6347; 610 UNTS 205). (2) *Agreement on the Rescue of Astronauts, the Return of Astronauts and the Return of Objects Launched into Outer Space*; done at Washington, London, and Moscow on April 22, 1968; entered into force on December 3, 1968 (19 UST 7570; TIAS 6599; 672 UNTS 119). (3) *Convention on International Liability for Damage Caused by Space Objects;* done at Washington, London, and Moscow on March 29, 1972; entered into force on September 1, 1972 (24 UST 2389; TIAS 7762; 981 UNTS 187). (4) *Convention on Registration of Objects Launched into Outer Space*; done at New York on January 14, 1975; entered into force on September 15, 1976 (28 UST 695; TIAS 8480; 1023 UNTS 15). (5) Agreement *Governing the Activities of States on the Moon and Other Celestial Bodies*; done at New York on December 18, 1979; entered into force on July 11, 1984 (18 ILM 1434; UN doc.A/RES/34/68 of December 14, 1979, Annex).
22. See note 16.
23. *Principles Governing the Use by States of Artificial Earth Satellites for International Direct Television Broadcasting* (General Assembly Resolution 37/92); adopted by a vote of 107-13-13 abstentions, on December 19, 1982; in United Nations, *United Nations Treaties and Principles on Outer Space*, United Nations, New York, in multiple editions.
24. *Principles Relating to Remote Sensing of the Earth from Outer Space* (General Assembly Resolution 41/65), adopted without a vote on December 3, 1986; in United Nations, *United Nations Treaties and Principles on Outer Space*, United Nations, New York, in multiple editions.
25. *Principles Relevant to the Use of Nuclear Power Sources in Outer Space* (General Assembly Resolution 47/68), adopted without a vote on December 14, 1992; in United Nations, *United Nations Treaties and Principles on Outer Space*, United Nations, New York, in multiple editions.

26. *Declaration on International Cooperation in the Exploration and Use of Outer Space for the Benefit and in the Interests of All States, Taking into Particular Account the Needs of Developing Countries* (General Assembly Resolution 51/122), adopted without a vote on December 13, 1996; in United Nations, *United Nations Treaties and Principles on Outer Space*, United Nations, New York, in multiple editions.
27. See Doyle, S.E., *The Impact of Spaceflight and Space Exploration on Laws and Governmental Structures of the United States*, a study done for and submitted to the NASA History Office, October 31, 2007, corrected April 1, 2008, available from the author.
28. 36 CFR 800; see also the Archaeological Resources Protection Act, 36 CFR 79, *et seq.*
29. World Heritage Convention, 1037 United Nations Technical Series U.N.T.S.151, 11 ILM 1358. As of June, 2003, 176 states are parties to the Convention. See also the *Convention on the Means of Prohibiting and Preventing the Illicit Import, Export and Transfer of Ownership of Cultural Property*, 823 U.N.T.S. 231 (Paris, 1970), 9 ILM 1031 (1970).
30. An excellent survey of the status of world space activities and applicable international and national laws is available at http://unoosa.org and sources identified there.
31. A helpful index of materials is presented at http://archeologyfieldwork.com, a Web site listing U.S. federal laws, regulations, standards and guidelines, presidential executive orders, and other related documents, including a list of national archaeological organizations with posted codes of ethics in anthropology and archaeology.
32. See Capelotti, P.J. Space: The Final [Archaeological] Frontier. *Archaeology* 57(6), 2004. At the Sixth World Archaeological Congress, Dublin, Ireland, June 29 to July 4, 2008, there was a session dedicated to space archaeology, titled Nostalgia for Infinity: Exploring the Archaeology of the Final Frontier. An interesting introductory survey of related information is in the Astrobiology selection at http://spacearchaeology.org/wiki.
33. At http://www.hightechscience.org/russian_space_artifacts.htm is a listing of "authentic Russian space artifacts," part of the collection of http://HighTechScience.org. Another purveyor is at http://collectspace.com, with a subset http://collectspace.com/buyspace/. See also http://hobbyspace.com.
34. In the United States, one repository for the Apollo Moon rocks is the Lunar Sample Building at the Johnson Space Center in Houston, Texas. A smaller collection is maintained at Brooks Air Force Base at San Antonio, Texas. The rocks are usually stored in nitrogen, free of moisture. They are handled indirectly using special tools. Papike, J., G. Ryder, and C. Shearer. Lunar Samples. *Reviews in Mineralogy and Geochemistry* 36: 5.1–5.234, 1998. Spudis, P.D. *The Once and Future Moon*. Smithsonian Institution Press, 1996. The Soviet Lunar Program involved numerous missions to the Moon, three of which returned samples [Luna 16 in 1970 returned 100 grams (.22 lb.), Luna 20 in 1972 returned 50 grams (.11 lb.), and Luna 24 in 1976 returned 170 grams (.37 lb)]. A recent report by scientists studying these materials at the Institute of Geology of Ore Deposits, Petrography, Mineralogy, and Geochemistry, of the Russian Academy of Sciences in Moscow, has been published. See Mokhov, A.V., P.M. Kartashov, O.A. Bogatikov, L.O. Magazina, N.A. Ashikhmina, and E.V. Koporulina. Association of Carbon-Rich Substance and Native Molybdenum in a Lunar Regolith from Mare Crisium. *Doklady Akademii Nauk* 415(5): 663–666, 2007.
35. See note 28.
36. See note 20 (1).
37. World Heritage Convention, 1037 U.N.T.S.151, 11 ILM 1358. As of June, 2003, 176 states are parties to the Convention. See also the *Convention on the Means of Prohibiting and Preventing the Illicit Import, Export and Transfer of Ownership of Cultural Property*, 823 U.N.T.S. 231 (Paris, 1970), 9 ILM 1031 (1970).
38. Ibid.
39. Ibid.

40. See as examples the U.S. Department of Energy policy designated DOE P 141.1, approved in 2001; the NASA procedural requirement designated NPR 4301.1, established March 16, 1999; the Department of Interior Organization Manual, Part 145: National Park Service, Chapter 5: Associate Director for Cultural Resources; also, the National Park Service has not republished "The Secretary of the Interior's Standards and Guidelines for Archeology and Historic Preservation" since 1983 (48 FR 44716), to be used by "Federal agency personnel responsible for cultural resource management pursuant to section 110 of the National Historic Preservation Act, as amended, in areas under Federal jurisdiction. A separate series of guidelines advising Federal agencies on their specific historic preservation activities under section 110 is in preparation." Other sources are listed at http://archeologyfieldwork.com.
41. Compare, as examples, Internal Revenue Code of 1986 as amended (Qualified Conservation Contributions) (26 U.S.C. 170(h)), and Internal Revenue Code of 1990 as amended (Rehabilitation Credit) (26 USC 47).
42. The four agreed-upon conventions were as follows: (1) T*he Convention on the Territorial Sea and Contiguous Zone*, entered into force on September 10, 1964; (2) *The Convention on the Continental Shelf*, entered into force on June 10, 1964; (3) *The Convention on the High Seas*, entered into force on September 30, 1962; and (4) *The Convention on Fishing and Conservation of Living Resources of the High Seas*, entered into force on March 20, 1966. See 1833 United Nations Treaty Services U.N.T.S. 3.
43. A useful summary overview of the three UNCLOS conferences and discussion of their results can be seen at the Wikipedia Web site on UNCLOS, from which some materials in the following paragraphs were condensed.
44. UN Convention on the law of the Sea, Part XI, Section 2, Article 49, at 555 U.N.T.S. 285.
45. Ibid.
46. Convention on the Protection of the Underwater Cultural Heritage, 2001, Article 303; see text at 21 International Law Materials 37 (2002).
47. 35 United States Code, Ch. 10 Sec 105.
48. See note 21 (1).
49. See source at footnote 21 (1) above.
50. See Capelotti 2004 (note 32), containing a discussion of a space archaeology working group organized by Beth O'Leary of New Mexico State University. See the World Archaeological Congress Space Heritage Task Force work, located at the task force Web site, http://www.worldarchaeologicalcongress.org/site/active_spac.php.

40 One Giant Leap: Preserving Cultural Resources on the Moon

Beth Laura O'Leary

CONTENTS

INTRODUCTION

For most people the idea that there are archaeological sites that humans have created on the Moon is a stretch of the imagination. The idea that this group of sites, the majority of which are less than 50 years old, is "historical" does not fit most people's idea of history. For anyone over 45, the Apollo lunar landings are part of their own experience of the past. The world watched and listened as *Apollo 11* astronaut Neil Armstrong set down his foot on the Moon on July 20, 1969, and the event forever changed the view that the Moon was inaccessible. Equally, that this site, the other Apollo lunar landing sites, and earlier robotic sites created by both the United States and the former Soviet Union on the surface of the Moon may be in need of recognition and preservation seems premature. No humans have gone back since 1972, so what is the rush? But the increase in the commercial use of space and plans to return to the Moon have accelerated in the last 10 years. This chapter discusses the cultural and legal ownership of the Moon and presents as a case study the Lunar Legacy Project, which tests the current U.S. federal preservation and international law by proposing that the first lunar landing site on the Moon should be considered eligible to be listed on the U.S. National Register of Historic Places and as a U.S. National Historic Landmark and become part of the World Heritage List. The chapter begins with the cultural relationship between the Moon and humans and moves into a legal discussion of the preservation of significant sites on the Moon, focusing on the first lunar landing site.

THE MOON AND HUMAN CULTURES

The moon has always been on our human minds. It has been important to us as long as we have been humans and maybe even before we became *Homo sapiens*. It also affects the lives of other animals: the neap tides brought on by the full moon provide the spawning beds for fishes, wolves howl on a full moon, and all primates, including ourselves, tend to give birth at night on a full moon. The moon has clocked human days as far back as the Paleolithic. Many cultures divide the year in lunations, or literally moons. A lunation begins with the first appearance of the new crescent. Each and every culture has appropriated the moon, from Australia to the Arctic. Each human population has a relationship with and rights to the moon through its culture: stories, art, architecture, calendars and subsistence practices. At a prehistoric Anasazi planetarium in Chaco Canyon in New Mexico, there is a pictograph of what is probably the crab nebula supernova next to a crescent moon that may date to AD 1054 (Figure 40.1). Prehistoric peoples tracked the seasons and analyzed the

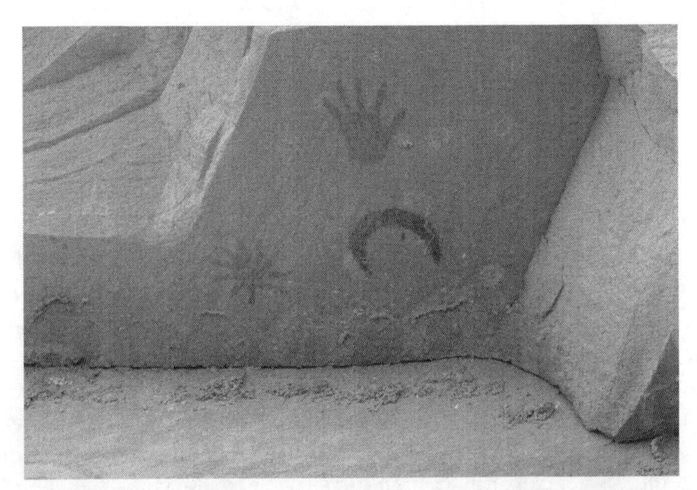

FIGURE 40.1 Chaco Planetarium. (From Chaco Canyon. Photo by Ron Lussier.)

phases of the moon. In Great Britain, 4,500 years ago enormous stones were placed in barrows that were oriented toward the moon's northernmost setting (midwinter moonset) and its southernmost moonrise (midsummer moonrise) (Figure 40.2).

Traditional indigenous narratives about the moon are almost countless and are unique to many different cultures. In Susan Milbrath's Chapter 10 of this volume, one can see a house post from the Tlingit Moon House in Yakutat, Alaska. It represents a boy who was blown up to the moon when he went out to fetch water during a storm. In the Tlingit culture, the moon has also been conceived as Raven's jealous uncle who controls the tides.[1] In the neighboring Southern Tutchone culture in the Yukon, Canada, the moon was thought to be owned and kept in a box by a big chief. The trickster god Crow wanted it. He turned himself into a pine needle and was swallowed by the big chief's daughter, to whom he was born as baby. When he cried for his grandfather's moon, he was given it and threw it up into the sky, and that is how it came to be there today.[2,3,4]

Some cultures have chosen the moon as a symbol and depicted it in both their art and language. In a detail on a vase from the eighth century AD, the Mayan moon goddess holds a rabbit. The Maya were great astronomers as well as storytellers. They have a narrative that was first recorded from the Aztecs in the sixteenth century that tells of how the sun and moon were created in the sky. They were created by the gods from pieces of a giant fire. Not happy with two equally bright disks, one of gods took a rabbit who was hopping nearby and slapped it onto one of the disks to make it

FIGURE 40.2 Stonehenge. (Author's image.)

less bright. The brightest object is the sun and the other is the moon, where the Maya still see the outlines of a rabbit.[5] The form of the glyph for the moon also depicts a body of water, as the moon is so closely associated with water and the tides.[6]

The sophisticated knowledge that prehistoric people had of the moon and the heavens is still being investigated, as more archaeological sites are discovered by archaeoastronomers to be observatories or structures that charted and recorded the movements and changes in the night sky. All oral traditions include nature in human affairs. Although Julie Cruikshank writes about landscape features such as glaciers, the moon also serves as a point of reference for community knowledge and is transformed by the traditional narratives into a place of significance.[7] It is a place in the sky where memories are stored. As Cruikshank also notes, these kinds of narratives provide empirical knowledge about geophysical changes, as well as social wisdom and lessons about the correct way to live life[8] (see Millbrath, Chapter 10 of this volume). The moon is rightly everyone's cultural property, but no human had physically been there before 1969. That year is a watershed for human involvement on the moon and for lunar archaeology.

LUNAR SITES

When *Apollo 11* astronaut Neil Armstrong first stepped onto the surface of the Moon on July 20, 1969, he said a now-famous line. It was a small (albeit extraordinary) step he took, and it remains one of the most fabulous leaps for humankind. But after he and fellow astronaut Buzz Aldrin returned to Earth, he also said, "Yeah, we left a few things up there." What Armstrong and Aldrin had done was to create the first historical archaeological site of actual human presence on another celestial body. By the end of 1972, the Apollo program had left 20 archaeological sites on the lunar surface, including pieces from the failed Apollo 13 mission. If all the other robotic sites from both the United States and the USSR are added to the first lunar landing site at Tranquility Base, there are currently more than eighty historic archaeological sites on the Moon. This represents an estimated 100 metric tons of cultural material and debris.[9] The Soviets created 15 Luna robotic sites, with *Luna 2* the first to successfully land on the far side of the Moon. The United States has five Ranger and seven Surveyor robotic sites and five Lunar Orbiters (Table 40.1). What some would term space "junk" represents the history of the Space Age. These material remains are the critical components of the cultural landscape between Earth and the Moon.

As Edward Staski (Chapter 2 of this volume) points out, archaeologists study the material record of human behavior. The popular image of an archaeologist is usually a man with a pith helmet and safari clothes digging up Egyptian tombs with dental picks. In the Hollywood *Indiana Jones* version, he is a handsome guy with his own exciting tools—a whip and a gun—discovering dangerous ancient sites by reading old maps. But in truth, archaeologists today of both genders, who study recent past properties in space or on other celestial bodies, have created a new kind of archaeology that has been variously called exoarchaeology, space archaeology, or space heritage (see O'Leary, Chapter 3 of this volume). What these archaeologists have come to realize is that the basic theories and methods of archaeology are the same whether one deals with historic periods on Earth or in space. Archaeology

TABLE 40.1
List of Sites on the Moon (1959–1976)

United States	Launch Date	USSR	Launch Date
Ranger 4	April 23, 1962	*Luna 2*	September 12, 1959
Ranger 6	January 30, 1964	*Luna 5*	May 9, 1969
Ranger 7	July 28, 1964	*Luna 7*	October 4, 1965
Ranger 8	February 17, 1965	*Luna 8*	December 3, 1965
Ranger 9	March 21, 1965	*Luna 9*	January 31, 1966
Surveyor 1	May 30, 1966	*Luna 10*	March 31, 1966
Surveyor 2	September 20, 1966	*Luna 13*	December 21, 1966
Surveyor 3	April 17, 1967	*Luna 15*	July 13, 1969
Surveyor 4	July 14, 1967	*Luna 16*	September 12, 1970
Surveyor 5	September 8, 1967	*Luna 17/Lunokhod 1*	November 10, 1970
Surveyor 6	November 7, 1967	*Luna 18*	September 2, 1971
Surveyor 7	January 7, 1968	*Luna 20*	February 14, 1972
Apollo 10	May 18, 1969	*Luna 21/Lunokhod 2*	January 8, 1973
Apollo 11	July 16, 1969	*Luna 23*	October 28, 1974
Apollo 12	November 14, 1969	*Luna 24*	August 9, 1976
Apollo 13 Saturn IVB	April 11, 1970		
Apollo 14	January 13, 1971		
Apollo 15	July 26, 1971		
Apollo 16	April 16, 1972		
Apollo 17	December 7, 1972		
Lunar Orbiter 1	August 10, 1966		
Lunar Orbiter 2	November 6, 1966		
Lunar Orbiter 3	February 6, 1967		
Lunar Orbiter 4	May 4, 1967		
Lunar Orbiter 5 (IMP-E)	July 19, 1967		
Explorer 49 (RAE-B)	June 10, 1973		

involves recording, describing, and analyzing material remains in order to answer questions about human behavior and what is called the life history of artifacts. Archaeologists can record the recent past as easily as the remote one. One critical function of archaeology and its applied subdiscipline of cultural resource management is to evaluate the significance of any site and determine how to best preserve it for future generations. This usually involves surveying and mapping a site, or sometimes undertaking an excavation according to a data recovery plan where the site is scientifically destroyed, but the records and analyses preserve all the information from the site. The site can also be surveyed and mapped and then simply avoided by current projects that could adversely affect it; it is left as it is in order to preserve it.

LUNAR LEGACY PROJECT

One of the earliest archaeological research in space began because an Anthropology graduate student, Ralph Gibson, at New Mexico State University (NMSU) in 1999, asked during my seminar in cultural resource management: "Does U.S. federal preservation law apply on the Moon?" It turned out to be a more complex question than anyone had anticipated. I, Gibson, John Versluis (another graduate student), and Co-PI, Dr. Jon Hunner (History Department at NMSU), sought funds from the New Mexico Space Grant Consortium (NMSGC), which is a member of the Space Grant College Fellowship Program administered by NASA, to promote educational programs in aeronautics, mathematics, technology, and space-related fields.[10] We created the Lunar Legacy Project, which continues to investigate the answer to this question.[11] Because there were many sites on the Moon, we chose to focus on the Apollo 11 Tranquility Base site as the best test case for the application of U.S. federal preservation law. It is the first time archaeologists had looked at a site on the Moon as a historical archaeological property and evaluated how it could be preserved. It was also the first time the NMSGC had given funds to NMSU's Anthropology Department.

The research has focused on three areas: first, a description of the location and the archaeological assemblage at the site; second, providing the historic context for determining its significance; and third, applying existing American federal preservation laws and regulations to the lunar site. Surprisingly, the description proved to be a challenge, while the second and third goals required beginning the legal process of determining the site's significance under U.S. federal law and gaining its recognition as such by various federal agencies.

The Apollo 11 Tranquility Base Site, the first lunar landing site, was created in approximately 21 hours. It was what archaeologists would call a single-component site, which means a site that has evidence of only one occupation, instead of multiple occupations. It is very rare for any historic site not to be reused or revisited. A nineteenth-century homestead in New Mexico, even if it was abandoned after several years, has an assemblage of artifacts, structures, and features such as barns, corrals, and so forth that were returned to and reused multiple times. Even a historic cattle camp for spring branding may have a short-term occupation (i.e., seasonally, when cattle were castrated, vaccinated, and branded), but the area, with its corrals, hearths for cooking, discard piles of typical trash, and so forth, would be returned to the next season, and similar activities would take place again. These multiple assemblages would mix with or overlay the earlier occupations. In prehistoric times, many large structures such as adobe pueblos created in the twelfth century AD were remodeled and reoccupied over hundreds of years.

The Tranquility Base site is what could be called an "extreme" single-component site, in that only two people actually created and occupied the site. Most archaeological sites are created by groups of people (e.g., families, tribes). If we include the infrastructure that helped the two astronauts create the site, there was a significant international community of scientists and engineers who were located 378,000 km away on another celestial body.

The site was also, to a large extent, self-documenting. The cameras and films, as well as written documents, made before, during, and after the site's occupation recorded a significant amount of information about the assemblage and the human behavior that happened at the site. The period from the earliest touchdown on the lunar surface to just before takeoff is documented perhaps more completely than any other historic event in the twentieth century. Surprisingly, many of the photos made

FIGURE 40.3 Armstrong seen in Aldrin's visor. (Courtesy of NASA.)

on the surface of the Moon capture only one astronaut's image—Buzz Aldrin's—as Neil Armstrong held the camera. The still camera usually does not reveal the photographer, just the subject. One picture of Armstrong on the Moon is a picture of his own reflection in Aldrin's visor (Figure 40.3). The site can also be considered extreme because it was created by two people doing fairly scripted behavior, and it was an event and behavior that had never been done before.

ARCHAEOLOGICAL ASSEMBLAGE AT APOLLO 11 TRANQUILITY BASE LUNAR LANDING SITE

What do we know about Tranquility Base as an archaeological site? The grant from NMSGC for the Lunar Legacy Project, unfortunately, did not cover the costs of a revisit to the site, so the inventory of archaeological assemblage was put together by sifting through several archives, including the Smithsonian National Air and Space Museum, the Lunar and Planetary Institute, and NASA's Johnson Space Center. A catalogue, which is probably not yet truly complete, of more than 106 artifacts and features was developed. It is a complex assemblage of parts and pieces of equipment, sentimental objects, and scientific experiments left in situ, some of which are still working. Table 40.2 shows the estimated number and kinds of artifacts and features at the site.[12]

Essentially an artifact is "any portable object used modified or made by humans."[13] Examples would include stone projectile points, copper pennies, and the Portable Life Support System left at Tranquility Base. A feature is best described as more permanent

TABLE 40.2
List of Artifacts on the Moon at Tranquility Base

1. Apollo 11 LM descent stage
2. U.S. flag
3. Laser ranging retroreflector
4. Passive seismic experiment
5. Neil Armstrong's PLSS
6. Neil Armstrong's Apollo space boots, model A7L (2)
7. Buzz Aldrin's PLSS
8. Buzz Aldrin's Apollo space boots, model A7L (2)
9. Empty food bags (2+)
10. Silicon disc carrying statements from presidents Eisenhower, Kennedy, Johnson, and Nixon, as well as statements from leaders of seventy-three other nations
11. Gold replica of an olive branch: traditional symbol of peace
12. Mission patch from Apollo 1 commemorating astronauts Gus Grissom, Ed White, and Roger Chaffee
13. Commemorative plaque attached to LM descent stage leg: "Here men from the planet Earth first set foot upon the Moon. July 1969, AD. We came in peace for all mankind." Signed by Apollo 11 astronauts and President Nixon
14. TV camera
15. Spring scales (2)
16. Tongs
17. Small scoop
18. Scongs
19. Bulk sample scoop
20. Trenching tool
21. Camera (Hasselblad El Data)
22. Armrests (4)
23. Mesa bracket
24. Solar wind composition staff
25. Handle of contingency lunar sample return container
26. Medals commemorating two dead cosmonauts (2)
27. Document sample box seal
28. Storage container
29. Hasselblad pack
30. Film magazines (2+)
31. Filter, polarizing
32. Remote control unit for PLSS packs (2)
33. Defecation collection device (4)
34. Overshoes, lunar (2)
35. Covers, PGA gas connectors (2)
36. Kit, electric waist tether
37. Bag assay, lunar equipment conveyor and waist tether
38. Conveyor assay, lunar equipment
39. Bag deployment lifeline
40. Bag deployment lunar equipment conveyor
41. Lifeline, light weight

TABLE 40.2 (*Continued*)
List of Artifacts on the Moon at Tranquility Base

42. Tether, waist, extravehicular activity (4)
43. Food assembly, LM (4 man-days)
44. TV subsystem, Lunar
45. Lens, TV wide angle
46. Lens, TV lunar day
47. Cable assembly, TV (100 ft.)
48. Adapter, SRC/OPS (2)
49. Canister, ECS LIOH (2)
50. Urine collection assembly, small (2)
51. Urine collection assembly, large (2)
52. Bag, emesis (4)
53. Container assembly, disposal
54. Filter, oxygen bacterial
55. Container, PLSS condensate
56. Antenna, S-band
57. Cable, S-band antenna
58. Bag, lunar equipment transfer
59. Pallet assembly #1
60. Central station
61. Pallet assembly #2
62. Primary structure assembly
63. Hammer
64. Gnomon (excludes mount)
65. Tripod
66. Handle/cable assembly (cord for TV camera)
67. York mesh packing material
68. SWC bag (extra)
69. Core tube bits (2)
70. SRC seal protectors (2)
71. Environmental sample container O rings (2+)
72. Apollo lunar surface close-up camera
73. Lunar equipment conveyor
74. ECS canister
75. ECS bracket
76. OPS brackets (2+)
77. Left hand side stowage compartment
78. Extension handle
79. Stainless steel cover for plaque on LM leg
80. Plastic covering for flag
81. 8-ft. aluminum tube
82. Retaining pins for flag and staff storage (2+)
83. Insulating blanket

Sources: NASA Apollo 11 Press Release Kit; NASA Apollo Query Book; Apollo 11: The NASA Mission Reports, vols. 1 and 2; NASM Archives; and NASA Johnson Space Center Archives.

LM, lunar module; PLSS, Portable Life Support System.

or "a non portable artifact."[14] Examples would be hearths, soil stains, or human footprints on the lunar surface. The most critical components for Apollo 11 were the tools and structures that were used to get to the Moon and those used to collect data about the Moon. The artifacts at the site are examples of what was then cutting-edge technology. For example, the scong was invented for the mission. It is a combination scoop and tong that was used to collect samples of the regolith and rocks on the Moon. These samples provided dates from the geological sediments in the Sea of Tranquility to between 3.6 and 3.9 million years.[15] A silicon disk was left at the site carrying statements from U.S. presidents Eisenhower, Kennedy, Johnson, and Nixon, and leaders from 73 nations, although the Soviet Union was not among them. There was also a small, colored, woven mission patch from Apollo 1, with the names Grissom, White, and Chaffee, left at the site to honor the astronauts who had died earlier in a terrible fire during a test. The lunar laser ranging retroreflector was placed there to allow scientists to measure, for the first time, the exact distance from the Earth to the Moon. It is still providing data today, more than 39 years later. Its location is at lunar coordinates 0.67266° N latitude, 23.47298° E longitude.[16] It acts as an archaeological datum and is one of the defined coordinates on the lunar surface[17] (see Stooke, Chapter 24 of this volume).

Of course, some of the most important features at the site are the footprints and trails made at the site. The first footprint when Armstrong stepped off the ladder onto the lunar surface was probably walked over by both astronauts. However, because of the lack of any wind, atmosphere, or major erosion or later impacts, the collection of footprints left there in 1969 should be as crisp as the day they were made. The trail of the first human presence on the Moon is still there. The importance of these human tracks rivals that of the 3.6-million-year-old footprint tracks of our hominid ancestor preserved in geological sediments in Laetoli, Tanzania.[18] Features such as tracks are rarely preserved in either the fossil or archaeological record on Earth.

ARCHAEOLOGICAL MAPS

Good archaeology demands a good map that shows the exact locations, or the *provenience,* of all the different artifacts and features. Provenience is the term used by archaeologists to denote the exact location of artifacts and features within a matrix at a site. It refers to the exact location of a find by giving its horizontal and vertical coordinates. The provenience, along with its association among the other finds, is crucial to understanding its context.[19] Even with all the photos and documentation at the time of the event, this is still a challenge. The mapping of Tranquility Base has been attempted after the event by several researchers using various photo images. It has been mapped using camera angles, and the actual area has been compared to the size of a baseball diamond. Most maps have been based on an interpretation of the documentation available at the time of the event. Many panoramas exist of photos taken at the time of the visit. These kinds of maps have never been field checked by a site revisit, even a remote one. Although the technology exists, no remote imaging at a scale useful to archaeologists has been completed during or after the actual landing. Future missions such as the Lunar Reconnaissance Orbiter, using remote imaging techniques, may be able to provide data that would allow archaeologists to see details at the site that would enable them to make a complete archaeological site map. While the locations of placed scientific equipments such as the laser ranging retroreflector are known, the *Apollo 11* ascent stage was abandoned in orbit and deorbited; its location on the lunar surface is unknown[20,21] (see Stooke, Chapter 24 of this volume).

Archaeologists study a site at a very fine scale and map the exact locations of the different artifacts and features, but exactly where some of the artifacts from the mission are located is still a mystery. Before taking off from the surface of the Moon, astronauts Armstrong and Aldrin jettisoned objects and refuse from the hatch of the lunar module *Eagle*.[22] For approximately 8 minutes, they stood and tossed out stuff. The artifacts were thrown into an environment with 1/6th the Earth's gravity. The pattern of material they created is similar to what archaeologist Lewis Binford has referred to as a "toss zone."[23] Although he was referring to an area of discarded material culture for the Nunamiut Eskimos in hunting camps in Alaska, it can be applied to the astronauts' behavior. Within this lunar toss zone area are objects such as Armstrong's portable life support system pack, a trenching tool, armrest, film magazines, and a hammer. More prosaic, like true trash, but equally demonstrative of human presence on the lunar surface, are four empty food bags, urine collection devices, and the plastic covering for the flag. Because no photographs exist of the artifacts after they were tossed, only their approximate position can be surmised on a site map. The Lunar Legacy Project revised a site map based on the United States Geological Survey map created shortly after the event to include a toss zone for the artifacts (Figure 40.4). Without subsequent investigations at the site, the provenience and current condition of these artifacts are unknown.

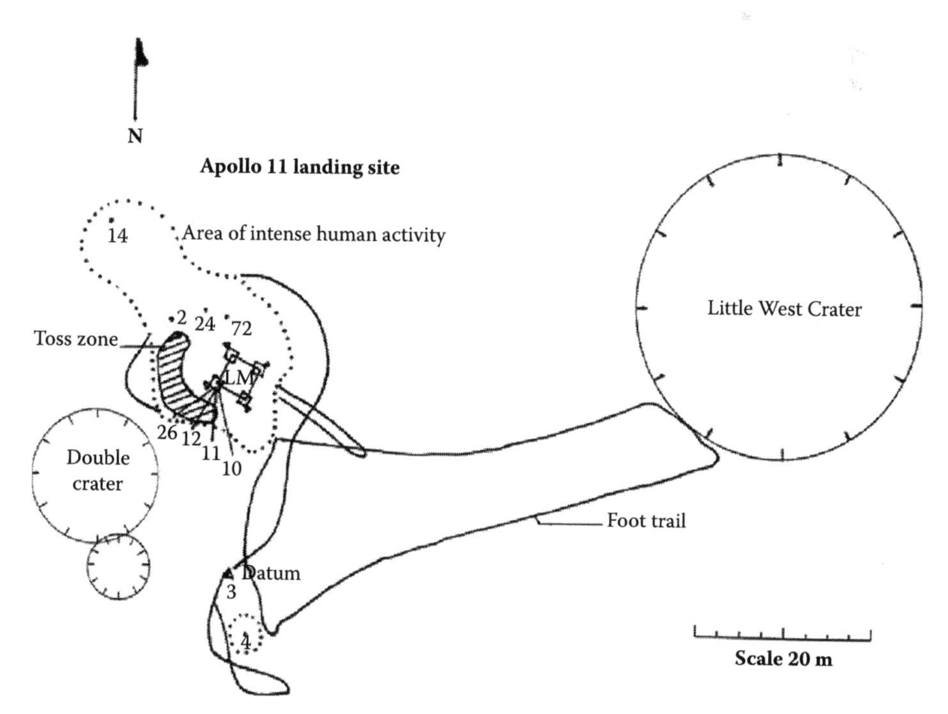

FIGURE 40.4 Revised U.S. Geological Survey map—Lunar Legacy Project.

LEGAL STATUS UNDER FEDERAL PRESERVATION LAW OF THE FIRST LUNAR LANDING SITE

How does this site fare under American federal preservation law? The majority of the laws and regulations evolved because important historic properties (also included in this category are sites dating to prehistoric times) were being destroyed by looting and modern development. The most important federal preservation law is the National Historic Preservation Act (NHPA), which expresses the general policy that preservation should be encouraged for future generations by directing federal agencies to assume responsibility for considering significant cultural resources in their activities. Section 110 of NHPA requires each responsible federal agency to have a program to identify, evaluate, and nominate historic properties to the National Register of Historic Places. The most important part of NHPA is Section 106, which reads:

> The head of any Federal agency having direct or indirect jurisdiction over a proposed Federal undertaking or federally assisted undertaking in any State and the head of any Federal department or independent agency having authority to license any undertaking shall, prior to the approval of the expenditure of any Federal funds on the undertaking or prior to the issuance of any license, as the case may be, take into account the effect of the undertaking on any district, site, building, structure, or object that is included in or eligible for inclusion in the National Register. The Head of any federal agency shall afford the Advisory Council on Historic Preservation ... a reasonable opportunity to comment with regard to such undertaking.[24]

The Section 106 procedures are codified in 36 CFR Part 800, NHPA's implementing regulations. What the Section 106 (NHPA) process requires is to seek a resolution for the federal agencies to take into account their actions on significant cultural resources. The implementing regulation "seeks to accommodate historic preservation concerns with needs of federal undertaking through consultation among agency official and other parties with interests in the effects of the undertaking on historic properties."[25] The definition of an undertaking is the first to be put into play. An undertaking

> means any project, activity, or program funded in whole or in part under the direct or indirect jurisdiction of a Federal agency, including
>
> (A) those carried out by or on behalf of the agency;
> (B) those carried out with Federal financial assistance;
> (C) those requiring a Federal permit, license, or approval; and
> (D) those subject to State or local regulation administered pursuant to a delegation or approval by a Federal agency.[26]

King writes that the key question to ask about an action in terms of the statute is whether there is federal involvement in an action.[27] The majority of actions occur on lands that are owned by federal agencies, but it includes those actions that are carried out with federal assistance or needing permits, licensing, or approval. The statutes do not specifically address outer space or the Moon, although the Apollo 11 mission (as all NASA missions) was carried out with federal monies, licensing, and approval.

The Section 106 process should start early if it is going to be effective; frequently it starts later.

Definitions of the categories of historic properties are more fully described in National Register Bulletin 15.[28] The federal agency is thus the responsible party. Also involved in the process created under NHPA are the State Historic Preservation Officer (SHPO) appointed by the governor of each state and U.S. territory, and the Advisory Council on Historic Preservation, which recommends and mediates solutions for compliance and preservation.

The National Register of Historic Places (also known as the National Register) was created and is maintained by the National Park Service and the Keeper of the Register as a list of significant historic properties. Significant properties covered under law include those that are actually listed on the National Register and those that are eligible for the National Register. There are also significant properties that are considered National Historic Landmarks. The historic properties are described in the laws and regulations as districts, sites, buildings, structures, or objects and must be determined to be significant at the national, state, and/or local level.[29] The criteria that must be met for National Register singly or in multiples (criteria a–d) are as follows:

> The quality of significance in American history, architecture, archaeology, engineering, and culture is present in districts, sites, buildings, structures, and objects, that possess integrity of location, design, setting, materials, workmanship, feeling, and association and
>
> (a) that are associated with events that have made a significant contribution to the broad patterns of our history; or
> (b) that are associated with the lives of persons significant in our past; or
> (c) that embody the distinctive characteristics of a type, period, or method of construction, or that represent the work of a master, or that possess high artistic values, or that represent a significant and distinguishable entity whose components may lack individual distinction; or
> (d) that have yielded or may be likely to yield information important in prehistory or history.[30]

The key idea embodied in these laws is "integrity." A site must possess integrity in the form of location, design, setting, materials, workmanship, feeling, and association. This class of properties needs to retain the ability to convey its significance. As King points out, that integrity is very much in the eye of the beholder and the property must be important for those who value it.[31] King also notes that when considering integrity, if a person from the property's period of significance would recognize it then it would have integrity.[32] But history and culture are also dynamic, and a property can represent the character of its times and also have a separate or additional role in the current culture. There is one caveat about historic properties that have achieved significance within the past 50 years—they are not considered eligible for the National Register unless they are "of exceptional importance."[33,34] It is a sliding rule, however, as many properties from the Cold War era are less than 50 years old and have been nominated or are candidates in the near future to be considered eligible for the National Register.

The most important part of determining significance is to evaluate a property within its historic context; only then can its significance be judged and explained. Although historians, archaeologists, and anthropologists may describe the phenomena of trends, patterns, themes, or cultural affiliation differently, a historic context is the patterns or trends in history by which a historic property is understood and its meaning is clear. In order to assess whether a property is significant within its historic context, the National Register Bulletin 15 defines four criteria: (1) which part of history in the locality, state, or nation does it represent; (2) whether that facet of history is significant; (3) whether and how the type of property illustrates the historic context; and (4) whether it possesses the physical features necessary to convey that aspect of history with which it is associated.[35] If it both represents an important aspect of history and has integrity then it can qualify for the National Register. Given the constraints of the law as explained briefly above, the Apollo 11 Tranquility Base site needs to be evaluated in its historic context, which is the Cold War era.

HISTORIC CONTEXT

The history of the exploration of space and the Moon in particular is a Cold War phenomenon. At the end of World War II, the United States and the USSR engaged in a race to acquire both German rockets and German rocket scientists. Former German scientist Wernher von Braun, who had created the V-2 rocket, and more than one hundred of his working colleagues at Peenemünde decided to come to Allied lines to surrender just prior to the end of the war.[36] The V-2 was further developed at White Sands Proving Ground (later White Sands Missile Range) in New Mexico with the work of the German expatriate teams in the late 1940s and became the basis for missile technology. A decade later descendents of the V-2 launched the first satellites and later propelled the first humans to the Moon. The Cold War was played out through military, political, and social maneuvers, not only on Earth but in space as well.[37] A cartoon by Herb Block of the Boston Globe in the early 1960s epitomizes the idea of a competition. The cartoonist has President Kennedy in a space suit running from a nearby U.S. rocket to a service attendant at a gas pump. He says, "Fill 'er up, I'm in a space race." The conflict between the Soviets and the Americans was accentuated when the International Geophysical Year announced a competition in 1957–1958 to see the launch of a satellite into Earth orbit. The Soviets were the first, in 1957, to inject *Sputnik 1*, an 83-kg satellite, into orbit. With some trepidation, the whole world watched and listened to the night sky in October for *Sputnik*'s faint radio signal and flash. A month later the Soviets dramatically launched *Sputnik 2*, carrying a small, unfortunate dog named Laika.

On January 31, 1958, *Explorer 1* became the first U.S. satellite in orbit. It was *Explorer 1* and *Explorer 2* that aided in the discovery of the Van Allen radiation belts.[38] Also developed by the United States was the Vanguard series, in which, after a number of failures, a satellite was successfully launched in March 1958. *Vanguard* is a 2-kg, grapefruit-sized satellite that provided information on the pear shape of the Earth and the bulge around the equator. Today *Vanguard* is still circling the Earth every 2 hours and is projected by NASA to stay aloft for another 600 to 2,000 years. *Vanguard* currently holds the title for the oldest human artifact in space.

Space, in the 1950s and 1960s, and perhaps even today, was thought of as an "ideological vacuum" and a wild frontier to be conquered. No humans have met, so far, any other indigenous inhabitants.[39] The quest of going into space and to the Moon represents the highest level of technological achievement within the Cold War context. The Apollo program consisted of eleven manned missions, nine of which went to the Moon and six of which landed there. Beginning with Neil Armstrong's jump onto its surface in July of 1969 to Gene Cernan's traverse of its rugged face in December of 1972, only twelve men have experienced life on the Moon. For only 3 ½ years, humans made the risky voyage to the Moon and returned.[40,41] Former *Apollo 8* astronaut Col. Frank Borman in an interview in 2001 said, "The Apollo program was not a voyage of exploration or … expertise in advancing technology, it was a battle in the Cold War."[42] It was astronauts Borman, James Lovell, and William Anders who became the first humans to leave Earth orbit and orbit the Moon. The flight schedule was pushed up, according to Borman:

> There was an enormous drive to accomplish this before the Russians. That's why *Apollo 8*'s mission was changed from an earth orbit to a lunar orbital mission because NASA had word that—from the CIA—that the Russians were going to go around the Moon before the end of '68. So they changed our mission.[43]

Borman, commander of the mission, saw the risk in this ambitious flight plan but also saw it as a necessity to get a man to the Moon before the Soviets. Even after *Apollo 8*'s success, the Soviets continued to plan manned missions to the Moon. One Soviet launch occurred only 13 days before the scheduled launch of *Apollo 11*, but it failed.[44]

The final stage in the quest for the Moon was standing on it. The first lunar landing in 1969 was an event that more than 600 million people on earth either watched or listened to as astronauts Neil Armstrong and Buzz Aldrin walked on the Moon, while Michael Collins circled above. It is surreal to watch the grainy black-and-white television images from space. A blurry, small figure in bulky white clothes steps, with a little bounce, off the ladder and onto the lunar surface. One of the first activities for the two astronauts was to plant an American flag on the Moon. There are many iconic images of that mission, and one of them is of Aldrin standing erect, saluting a short, wrinkled American flag with the dull, vast blackness of space behind him. Even though no one person or nation can own the Moon, by international treaty, this act of dramatically setting the flag was symbolic of claiming both territory and victory set by historic precedent. For the Soviets, who had been farther ahead in the Space Race, with many of the firsts—first satellite, first astronaut in orbit, first woman in space—the act of setting an American flag on the Moon signaled a defeat in space. Although the event was hailed by Armstrong as a "giant leap for mankind," it was an American who stepped onto the surface of the Moon. The names of the spacecraft involved were the *Columbia* and the *Eagle*, names that are metaphors for America. An archaeologist visiting the Tranquility Base lunar landing site a thousand years in the future would see the assemblage as American material culture, although it was global in scope even at that time. The Apollo program was another battlefront in the Cold War.[45] Once the United States had won that battle by first placing humans on the Moon and returning them safely to earth, the Cold War in space had been won.

NATIONAL REGISTER OF HISTORIC PLACES CRITERIA

The Lunar Legacy Project focused on the Apollo 11 Tranquility site as the best, most unambiguous test case for the application of U.S. federal preservation law. Putting aside the jurisdictional issue noted above and discussed below, the site meets the entire set of criteria for being eligible for the National Register. The archaeological assemblage on the Moon is the material correlate of one of the most important historic events in all of human history, not to mention one of *the* historic events in American history. It meets criterion *a*. Never before in the history of humankind had humans set foot on another celestial body. The technology necessary for this endeavor did not even exist when the United States committed itself to achieving this goal.[46] The United States, in a race with the Soviet Union, spent billions of dollars in research, engineering, and construction of the spacecraft, launch complexes, and support needed for the Apollo 11 event. Criterion *a* can be applied to all the artifacts and features based on the significance of the event.

Criterion *b* is the association with important persons in our past, and there are several obvious fits here. First, the *Apollo 11* astronauts, Neil Armstrong, Buzz Aldrin, and Michael Collins, became instantly recognized as the world's first lunar explorers. As astronauts they were American heroes. They are to the twentieth century what Columbus and Lewis and Clark were to their centuries.[47] The Moon landings are also associated with three U.S. presidents, John F. Kennedy, Lyndon Baines Johnson, and Richard M. Nixon. Kennedy had set the goal; Johnson kept the mission going, even assisting in the donation of land near Houston, Texas, for the mission control center; and Nixon was in office and a signatory to the plaque of goodwill on the lunar module descent stage that remains on the Moon.

Criterion *c* is perhaps the most detailed criterion; it deals with qualities of a historic property that embody a distinctive style or method of construction or work by a master (or that represent a significant and distinguishable entity even though the components may lack individual importance). This is the most relevant criterion to the Apollo 11 site. The archaeological assemblage on the Moon represents both the astronaut behavior and the science and engineering that went into, making it possible. The dedication and commitment of literally thousands of scientists (e.g., astrophysicists, physiologists) and multiple types of engineers (e.g., astronautical, electrical, mechanical) was enormous, as well as the careful and innovative coordination and planning to accomplish the goal. The skills required to accomplish the goals of the Apollo 11 program were cutting edge, never before attempted, and integrated by a complex interdisciplinary systems approach seldom done at this scale. The combined effort of many masters, experts, and journeymen in numerous fields of science, engineering, and technology was required to put humans on the Moon. Thousands of people theorized, conceived, designed, built, tested, and piloted the spacecraft and equipment that was used to send humans to the Moon.[48] Several have achieved recognition as masters, such as

Harrison Storms of North American Aviation and Tom Kelly of Grumman, but many are unsung and largely unknown engineers who designed and coordinated one of the most complex endeavors in the twentieth century. They rival the nineteenth- and twentieth-century engineers who built the bridges, skyscrapers, and dams as heroes of American ingenuity and technological advancement. The research and design of each object on the Moon represents the era of lunar exploration. Most, if not all, artifacts at Tranquility Base reflect a distinctive style and method of construction.

Criterion *d* is the one most commonly used by archaeologists. The historic property must have yielded or be likely to yield information important to history or prehistory. The property and the records have already yielded information on the physical nature of the Moon and the feasibility of landing on other celestial bodies. Future investigations would yield information on the long-term effects of exposure on electronic equipment, various metals, plastics, and fabrics on the lunar surface (see Donegan, Chapter 6; Clemens and Leary, Chapter 8; Darrin and Mehoke, Chapter 9; Mehoke, Chapter 29; Heaps, Chapter 30; Darrin, Chapter 32; and Biermann, Chapter 34, this volume). Any evidence of decay rates on materials would be important to future long-term missions to the Moon and elsewhere in space.

Nominating Tranquility Base as a National Historic Landmark requires the application of the four National Register criteria, as well as two additional criteria:

> (1) That [it] represents some great idea or ideal of the American people; or …
> (5) that are composed of integral parts of the environment not sufficiently significant by reason of historical association but collectively compose an entity of exceptional historical or artistic significance, or outstandingly commemorate or illustrate a way of life or culture.[49,50]

Tranquility Base represents one of the most significant technological achievements by humans. It meets all of the qualifying criteria for the National Register and as a National Historic Landmark. The human footprints, along with the assemblage of artifacts, objects, and structures at the site, meet the criteria and are in what is assumed to be an excellent state of preservation. Clearly there is little to argue about its importance to the history of the nation, if not the world. A discussion of what can be done in the future to help preserve this site and others is part of O'Leary's Chapter 43 (this volume) on the plan for the future preservation of space.

Several historic properties important in the development of space in the Cold War era are included on the National Register. Launch Complex 33, on White Sands Missile Range in New Mexico, is a National Historic Landmark. It was here that the V-2 rocket and its successors were further developed by a team of scientists led by Werhner von Braun in the late 1940s and early 1950s. The "Man in Space" theme study done by the National Park Service identified twenty-six historic properties worthy of preservation that were closely related to the Apollo program.[51] NASA, as a federal agency, falls under U.S. federal preservation law. It is the owner of many facilities from the Cold War that have been included in the National Register, separately or as part of the "Man in Space" theme. The Tranquility Base site on the Moon is a critical component of this theme and to all terrestrial sites associated with lunar exploration.

Because NASA falls under U.S. federal preservation law, it is on this agency that the Lunar Legacy Project focused its efforts. It is the owner of the Apollo 11 launch pad in Cape Canaveral, Florida, which is listed on the U.S. National Register of Historic Places. The only difference between the launch pad in Florida and the Apollo 11 Tranquility Base site is that the archaeological assemblage from the latter is on the Moon. The Apollo 11 lunar landing site is a critical component to the cultural landscape of space exploration. The location of the artifacts in situ on the Moon is critical to their significance. Without being on the Moon, without their locational integrity, those artifacts lose part of their extraordinary significance. Their place on the Moon is as important as the assemblage; their context is as critical as the components themselves. Because Tranquility Base clearly meets the federal preservation criteria, then what is its legal standing, and what is the future for this or any historic site on the Moon?

OUTER SPACE TREATY

In terms of the legal standing of the Apollo 11 sites, there are complexities that supersede the disposition of sites on Earth. The Lunar Legacy Project looked at both its national and international standing. The Treaty on the Principles Governing the Activities of States in the Exploration and Use of Outer Space, Including the Moon and Other Celestial Bodies was promulgated by the United Nations on January 27, 1967, in the spirit of openness, shared scientific information, and peaceful exploration.[52] It comprises seventeen articles that define how states should and should not explore outer space and other celestial bodies. Articles II and VIII are relevant to the nomination of Tranquility Base to the National Register or as a National Historic Landmark (NHL). Article II states:

> Outer space, including the Moon and other celestial bodies, is not subject to national appropriation by claim of sovereignty, by means of use or occupation, or by any other means.[53]

Essentially, the Moon cannot be claimed by any nation. No one nation or individual can own the Moon. It is governed by treaty.

Article VIII states:

> A State Party to the Treaty on whose registry an object launched into outer space is carried shall retain jurisdiction and control over such object, and over any personnel thereof, while in outer space or on a celestial body. Ownership of objects launched into outer space, including objects landed or constructed on a celestial body, and of their component parts, is not affected by their presence in outer space or on a celestial body or by their return to earth...[54]

Essentially, the objects launched into outer space or on another celestial body remain under the jurisdiction of those who put them there. In this case it is the United States under its federal agency, NASA. This right of ownership of artifacts on the Moon is strengthened by the minutes of the NASA Artifacts Committee Meeting in

1985, which addressed the issue of transferring title of all the objects that remain at Tranquility Base to the Smithsonian's National Air and Space Museum.[55] Although no such transfer occurred, the document states that NASA maintains title to the property and does not consider the lunar artifacts to be abandoned property. NASA is then the federal agency who owns this historic property. Also, in 1983 NASA actually did transfer title to the Smithsonian Institution the ownership of the *Viking Lander 1*. The *Viking Lander 1* is located on Mars.[56] As the responsible federal agency, NASA's actions fall under both Section 110 and Section 106 of NHPA, which requires a program of inventory of its cultural resources and a consideration of the effects their actions may have on properties eligible for the National Register.

LEGAL APPROACHES TO PRESERVATION

One of the clear problems with the Moon is that no one owns the surface; thus, it is the cultural materials on the lunar surface that are owned. The Lunar Legacy Project wanted to test U.S. federal preservation law to preserve the objects, structures, and features left at Tranquility Base. It approached both NASA and the Keeper of the National Register of Historic Places by proposing the site as a National Historic Landmark. One of the best outcomes was bringing the issue of lunar preservation directly to the relevant U.S. preservation authorities in 2000. This request seemed to be both surprising and a bit threatening. In the opinion of Robert Stephens, NASA's Deputy General Counsel, the listing the site as an NHL was contrary to the United Nations (UN) Outer Space Treaty of 1967 and would be thought of as a U.S. claim of sovereignty. He states:

> I must inform you that we cannot support your proposal to have Tranquility Base declared a National Historic Landmark (NHL). The Treaty declares that there can be no claims of sovereignty or territory by nations over locations in space … 'by means of use or occupation, or by any other means.' The listing of lunar areas as NHL's is likely to be perceived by the international community as a claim over the Moon.[57]

The Keeper of the National Register in 2000 was Carol Shull. She referenced the letter by Robert Stephens. Both were clearly a coordinated response. Ms Shull's response to the inquiry stated:

> It has been determined as a matter of policy that it would not be appropriate to designate National Historic Landmarks on the Moon. Even if, as a matter of policy, we did not consider it inappropriate to nominate Landmarks on the Moon, we do not consider that we have sufficient jurisdiction and authority over the land mass of the Moon to exercise our nominating authority over resources on the Moon.[58]

The responses of Keeper of the Register centered on the (recent) policy that the Moon could not have NHLs, nor was it appropriate, and even if it were, the National Park Service does not have sufficient jurisdiction. The U.S. astronauts visited the lunar surface, conducted research there, leaving artifacts in place, and returned to Earth. The UN Space Treaty recognizes the U.S. rights to the artifacts, the Smithsonian has title to NASA artifacts on another celestial body (i.e., *Viking Lander 1* on Mars), and federal preservation laws apply, yet there is, so far, a lack of commitment by the U.S. federal government to respond to preservation needs.

Tranquility Base is clearly an internationally significant site. It also qualifies under several criteria for UNESCO's World Heritage List, established for the preservation of sites of "outstanding universal value."[59] The nations who signed the World Heritage Convention, including the United States, pledge to preserve the listed landmarks, but before this can occur the United States, as a state party, must nominate the site. Who or what will preserve the Tranquility Base site in the future?

The remoteness of the Moon has kept any human impacts from disturbing the Apollo 11 site. This remoteness may be enough to keep it safe for the next 40 years. Also, if as archaeologists we can embrace new cutting-edge technology for recording and mapping the lunar archaeological sites, why do we need to worry about preservation? Future missions from several countries plan to map the Moon's surface. If archaeological data can be retrieved on these missions, this could obviate the need to ever go there again. There currently seems to be a lack of political commitment and responsibility for the preservation of this site. It could be argued that there may be no pressing need for preservation now if these untouchable sites have not yet been touched. The conditions on the Moon are, in many ways, benign, when it comes to the preservation of material culture, so the site is under no imminent natural threat of being destroyed.

One of the dangers of accepting the status quo for the preservation of this lunar site is that when there are threats to the site there will be no framework in place, and ad hoc solutions are not usually the best. Also, there are many other Apollo sites and earlier Soviet and U.S. robotic sites equally deserving of consideration in terms of preservation. Without a commitment to preservation and given the apparent lack of legal structures to deal with sites in space as cultural and historic resources, they are left vulnerable to impacts in the future by many varieties of space travel. It would be a waste for archaeologists to develop methods to survey and record sites in space without a preservation framework in place. Because of the nature of space, the heritage approach has to be international in scope.

The Outer Space Treaty did not specifically address historic preservation. In the early decades of the Space Race, not too many people thought about it. But the Outer Space Treaty did attempt to regulate how Earth's people treated this new area beyond its bounds. Sites on the Moon were not the kind of property that was envisioned by the National Historic Preservation Act or the World Heritage Convention. The properties in space and on other celestial bodies are recent past properties for which archaeologists do not have a large body of expertise. The artifacts are on another world, at a scale different from what archaeologists are used to dealing with, and they are not within the boundaries of any nation's territory; they are not even on Earth. Space exploration is still a functioning, ongoing, and increasingly complex system. NASA does not particularly want to be in the preservation business; it wants to continue space engineering. What has happened in the past, for the most part, is left behind in the dust. Space exploration is different today from what it was in the Cold War era.

NEW SPACE AGE

What was once inaccessible because of the great technological investment to even get to the Moon and the lack of desire to return has changed dramatically in the last few years. There is a new space race that has nationalistic interests, as in the Cold War, but there are more players now with commercial interests. This is always a scarier scenario for preservation. There are plans to go back to the Moon by many nations. Also, there is a new approach to space in the concept of personal space flight, or what Spennemann has called "extreme tourism."[60] It is possible to buy your way into space, as space tourists such as Dennis Tito and Anoush Ansari have done. Beginning in 2007, Google has had a competition that offers a $30M award for sending a robotic vehicle back to the Moon, with a $5M bonus for taking a picture of an Apollo artifact there.[61] To get to most of the artifacts on the Moon risks the obliteration of the original footprints with the vehicle tracks or damaging the physical integrity of any Apollo or other site. Without a commitment to preserve space sites, they are vulnerable to impacts in the future. Ironically, the landing of the first commercial robotic mission would itself create a significant archaeological site. But a plan to protect those historic sites that are deemed significant, such as the Apollo 11 site, should precede any revisits. Humans will go back to the Moon in the near or distant future, and the earlier sites will be affected.

Although the response of NASA and the Keeper of the Register was not the preferred answer, the political issues for both the involved parties can change. If the political will of the United States, in coordination with other countries, deems lunar preservation to be important, then a series of agreements and protocols can be created. Recent discussions on the heritage of the Antarctic have focused on the crucial importance of a legislative and management regime for heritage conservation.[62] Management issues there include the Antarctic Treaty, diplomacy, accelerating tourism, and climate change.[63]

CONCLUSION: SPACE HERITAGE

One outgrowth of the Lunar Legacy Project was to bring the concept of space heritage to the professional archaeological community. Actions at the Fifth World Archaeological Congress (WAC-5) in 2003 include a resolution to recognize the material culture and places associated with space exploration:

> As space industries and eventual space colonization develop in the twenty-first century, it is necessary to consider what and how elements of this cultural heritage should be preserved for the benefit of present and future generations.[64]

An international Space Heritage Task Force was created at the WAC-5 to draft language to protect space sites. It continued to work at the WAC-6 in Dublin in 2008. Its goals include identifying themes related to space exploration; investigating ways of assessing significance at the individual, local, national, and international levels; and identifying places that have exceptional cultural heritage value and whose preservation will benefit all humankind. It proposes to identify relevant international and national organizations; create a set of cultural, historical, social, and archaeological criteria; investigate avenues for preservation within existing structures; and explore options for preservation.[65]

The State of New Mexico Historic Preservation Division took a dramatic symbolic step in 2006 when its Laboratory of Anthropology Archaeological Resource Management Section, under the Direction of Dr. William Doleman and the SHPO Katherine Slick, recognized the Apollo 11 Tranquility Base site as a Laboratory of Anthropology site (LA) 2,000,000. LA 2,000,000 is part of the largest archaeological database in the United States, and it has more than 150,000 sites in New Mexico. The coordinates for the Apollo 11 site on the Moon are linked to the Universal Transverse Mercator (UTM) coordinates for the New Mexico Museum of Space History in Alamogordo, New Mexico, which serves as its host on Earth. It marks one of the first efforts by a state to preserve the cultural heritage of the Moon.

While the work needs to take place within the professional archaeological community, it also has to be brought to the attention of the general public. Only with recognition by broad coalitions of scientists, astronomers, engineers, space enthusiasts, and those starting careers in space exploration can the mandate to take into consideration ways of evaluating and preserving significant space sites be taken further. The U.S. government and the international community need to respond to space heritage preservation needs. Those involved in space heritage need to persuade the U.S. government to take action and cause appropriate international recognition to be given to the necessity of cultural resource management on the Moon (see O'Leary, Chapter 43 of this volume). Education of a new generation needs to include its members in preserving what happened before they were born. Members of the general public have to see the value of preserving a place that they will in all probability never see but that needs to be kept pristine because of its significance and iconic nature as the first place humans visited in the cosmos. These efforts are only the beginning of what needs to happen to preserve significant sites on the Moon. We otherwise stand to lose extraordinary sites on the celestial body where we first left our human mark.

REFERENCES

1. de Laguna, F. 1972. *Under Mount Saint Elias: The History and Culture of the Yakutat Tlingit.* Vol 7, Part Two. Washington, DC: Smithsonian Contributions to Anthropology.
2. McClellan, C. 1975. *My Old People Say: An Ethnographic Survey of the Southern Yukon Territory.* 2 vols. Mercury Series, Canadian Ethnology Service Paper 137. Ottawa, Canada: Canadian Museum of Civilization.
3. Cruikshank, J., in collaboration with A. Sydney, K. Smith, and A. Ned. 1990. *Life Lived Like a Story: Life Stories of Three Yukon Native Elders.* Lincoln, NE: University of Nebraska Press/Vancouver, Canada: University of British Columbia Press.
4. Jackson, M., with B. O'Leary. 2007. *My Country is Alive: A Southern Tutchone Life.* MS on file. Yukon Archives, Yukon College, Whitehorse, Yukon, Canada.
5. de Sahagún, B. 1950. *General History of Things in New Spain.* Florentine Codex. Translated from the Aztec into English with notes and illustrations by A.J.O. Anderson and C. Dibble. Salt Lake City, UT: University of Utah Press.
6. Lamb, Weldon. 2007. Personal communication.

7. Cruikshank, J. 2005. *Do Glaciers Listen? Local Knowledge, Colonial Encounters & Social Imagination*. Vancouver, Canada: University of British Columbia Press.
8. Ibid.
9. Johnson, N.L. 1999. *Man-Made Debris in and from Lunar Orbit*. American Institute of Aeronautics and Astronautics. IAA-99-IAA. 7.1.03
10. New Mexico Space Grant Consortium Web site. http://spacegrant.nmsu.edu/.
11. Lunar Legacy Web site. http://spacegrant.nmsu.edu/lunarlegacies/.
12. Gibson, R. 2001. Lunar Archaeology: The Application of Federal Historic Preservation Law to the Site Where Humans First Set Foot Upon the Moon. MA thesis, New Mexico State University, Las Cruces, NM. (On file at Rio Grande Archives.)
13. Renfrew, C., and P. Bahn. 2007. *Archaeology Essentials: Theories, Methods, and Practice*. London: Thames and Hudson (see specifically p. 290).
14. Ibid., p. 292.
15. Lunar and Planetary Institute Lunar Sample Overview Web site. http://www.lpi.usra.edu/lunar/missions/apollo/apollo_11/samples/.
16. Davies, M.E., and T.R. Colvin. 2000. Lunar Coordinates in the Regions of the Apollo Landers. *Journal of Geophysical Research* 105(E8): 20277–20280.
17. Stooke, P. 2007. *The International Atlas of Lunar Exploration*. Cambridge, United Kingdom: Cambridge University Press.
18. Nelson, H., R. Jurman, and L. Kilgore. 1992. *Essentials of Physical Anthropology*. St. Paul, MN: West Publishing Company.
19. Renfrew, C., and P. Bahn. 2007. *Archaeology Essentials: Theories, Methods, and Practice*. London: Thames and Hudson (see specifically p. 40).
20. O'Leary, B. 2007. Historic Preservation at the Edge: Archaeology on the Moon, in Space and on Other Celestial Bodies. Paper presented at the Extreme Heritage Conference ICOMOS-Australia, Cairns, Australia.
21. Stooke 2007 (note 17) (see specifically p. 212).
22. Godwin, R. 1999. *Apollo 11: The NASA Mission Reports*. Vol. 1. Toronto, Canada: Apogee Books.
23. Binford, Lewis S. 1982. *In Pursuit of the Past: Decoding the Archaeological Record.* Berkeley, CA: University of California Press (see specifically p. 156).
24. National Historic Preservation Act of 1966, (17 U.S.C. 470 et seq.) as amended.
25. 36 CFR Part 800. Protection of Historic Properties. 800.1(a).
26. National Historic Preservation Act of 1966 (17 U.S.C. 470 et seq.) as amended, Section 301(7).
27. King, T.F. 2004. *Cultural Resource Laws & Practice: An Introductory Guide.* Walnut Creek CA: Altamira Press (see specifically p. 89).
28. National Register Bulletin 15. How to Apply the National Register Criteria for Evaluation. Washington, DC: National Park Service.
29. National Historic Preservation Act of 1966, (17 U.S.C. 470 et seq.) as amended, Section 301(5)
30. 36 CFR Part 60.4 The National Register of Historic Places.
31. King 2004 (note 27) (see specifically p. 116).
32. Ibid., 117.
33. 36 CFR Part 60. The National Register of Historic Places.
34. Shefy, M., and W.R. Luce. 1998. National Register Bulletin 22: Guidelines for Evaluating and Nominating Properties That Have Achieved Significance within The Last Fifty Years. http://www.nps.gov/history/NR/publications/bulletins/nrb22/.
35. National Register (note 28) (see specifically p. 7).
36. Stuhlinger, E. 1963. *Astronautical Engineering and Science from Peenemunde to Planetary Science. Honoring the Fiftieth Birthday of Wernher von Braun.* New York: McGraw-Hill.

37. Gorman, A., and B. O'Leary. 2007. An Ideological Vacuum: The Cold War in Outer Space. In *A Fearsome Heritage: Diverse Legacies of the Cold War,* ed. J. Schofield and W. Cocroft. Walnut Creek, CA: Left Coast Press
38. Chapman, S. 1959. *IGY: Year of Discovery: The Story of the International Geophysical Year*. Ann Arbor, MI: University of Michigan Press (see specifically p. 85).
39. Gorman and O'Leary 2007 (note 37).
40. Chaikin, A. 1994. *A Man on the Moon*. Alexandria, VA: Time-Life Books.
41. Burrows, W. 1998. *This New Ocean.* New York: Random House.
42. Borman, F. 2001.Videotaped interview on January 23, 2001 with Beth O'Leary and Ralph Gibson, Las Cruces, NM. On file at the Rio Grande Archives, New Mexico State University, Las Cruces, NM.
43. Ibid.
44. Chaikin 1994 (note 40) (see specifically p. 358).
45. Gibson 2001 (note 12) (see specifically p. 56).
46. Chaikin 1994 (note 40).
47. Gibson 2001 (note 12) (see specifically p. 12).
48. Gray, M. 1992. *Angle of Attack: Harrison Storms and the Race to the Moon*. New York: W.W. Norton and Company.
49. 36 CFR Part 65. National Historic Landmarks Program.
50. National Register (note 28) (see specifically p. 15).
51. Butowsky, H.A. 1984. *Man in Space: A National Historic Landmark Theme Study*. Washington, DC: U.S. National Park Service, U.S. Department of Interior.
52. United Nations. 1967. The Treaty on the Principles Governing the Activities of States in the Exploration and Use of Outer Space, Including the Moon and Other Celestial Bodies. http://www1.umn.edu/humanrts/peace/docs/treatyouterspace.html.
53. Ibid.
54. Ibid.
55. Jones, A. 1985. Minutes From Alan Jones, Executive Secretary of NASA's Artifacts Committee, re: Property on the Lunar Surface, March 15, 1985. On file at the Smithsonian National Air and Space Museum.
56. Milestones of Flight—Viking Lander. http://www.hrw.com/science/si-science/physical/motion/milestones/viking.html.
57. Stephens, R.M. 2000. Letter dated August 18, 2000, from NASA Deputy General Counsel to the Lunar Legacy Projects regarding NHL designation of lunar artifacts at Tranquility Base. On file at the Rio Grande Archives, New Mexico State University, Las Cruces, NM.
58. Shull, C. 2000. Letter dated August 18, 2000, from Keeper of the National Register to the Lunar Legacy Project regarding the NHL designation of lunar artifacts at Tranquility Base. On file at the Rio Grande Archives, New Mexico State University, Las Cruces, NM.
59. World Heritage Convention. http://www.whc.unesco.org/.
60. Spennemann, D. 2007. Extreme Cultural Tourism from Antarctica to the Moon. *Annals of Tourism Research* 34(4): 898–918.
61. Google Lunar X Prize Web site. http://www.googlelunarxprize.org/.
62. Evans, S.-L. 2007. "Heritage at Risk": Cultural Heritage Management in the Antarctic. Paper presented at the ICOMOS Extreme Heritage Conference, Cairns, Australia.
63. Powell, S. 2007. *Beyond the Heroic Huts: Managing Australia's Antarctic Heritage*. Paper presented at the ICOMOS Extreme Heritage Conference, Cairns, Australia.
64. Fifth World Archaeological Congress, Space Heritage Task Force. http://www.worldarchaeologicalcongress.org/site/active_spac.php.
65. World Archaeological Congress International Space Heritage Task Force Web site. http://ehlt.flinders/au/wac/site/active_space.php

41 On the Nature of the Cultural Heritage Values of Spacecraft Crash Sites

Dirk H. R. Spennemann

CONTENTS

INTRODUCTION

In recent years, a small group of cultural heritage managers has begun to systematically consider not only the management of space heritage sites on Earth[1,2,3] but also the future heritage management of objects in Earth's orbit[4,5] and sites in existence on other celestial bodies—but currently out of reach of human interference—such as the Moon[6] and Mars.[7] For this chapter, I will only be concerned with the management of tangible, human-created sites on the surfaces of planetary bodies.

Many of the papers have highlighted the need for these tangible sites to be managed as a truly unique and highly significant part of humanity's heritage, especially the lunar landing sites of the Apollo program,[8] with Tranquility Base, the landing site of *Apollo 11*, possessing the most outstanding heritage values for human development as a whole.[9] Similar cases have been made for Mars.[10] Apart from the formal recognition as sites worthy of protection, a driving force was the appreciation that the lunar sites are truly unique and fragile. Unlike sites associated with any other key event in terrestrial history, the lunar sites are in an enviable state of preservation, retaining, *inter alia*, the actual footprints of the first humans to ever set foot on another celestial body. This makes these sites particularly fragile to any impact by future space tourists.[11]

Common to all cited work is that the heritage is embodied in the intentionally created landing sites and their assorted artifactual evidence. While this is fully understandable, and while such sites have arguably a very high level of cultural significance attached to them, there is a whole range of other sites that needs to be considered in this context: the sites of intentional and accidental crashes on interplanetary surfaces.

The aim of this chapter is to examine the conceptual differences between the types of crashes and to advance a theoretical framework that allows us to determine the cultural significance of the sites generated by the crashes.

CRASH LANDINGS ON THE MOON, MARS, AND VENUS

Given the intense political rivalry between the United States and the (then) USSR following World War II and the development of missile technology as a weapons platform and as a means of delivering objects into space, a Space Race soon ensued because space was rightfully seen as the means of surveillance, command, and control for an anticipated war event between the two powers. At the same, the enchantment of the developed world with technological inventions and achievements continued unabated. For both the United States and the USSR, much prestige on the international stage and also the national stage rode on being the first to deliver an object into Earth orbit (USSR, *Sputnik*, April 10, 1957) and to deliver into—and successfully retrieve from— Earth orbit an animal (USSR, August 19, 1960), a human (USSR, Yuri Gagarin, April 12, 1961), and the first woman (USSR, Valentina Tereshkova, June 16, 1963). A similar milestone in technological achievement and prestige was the exploration of other celestial bodies, such as the Moon (see discussion below), Venus (USSR, *Venera 3*, March 1, 1966, crashed), and Mars (USSR, *Mars 2*, May 19, 1971). Given the prestige that this engendered, it was paramount for both countries to be the first to reach the surface of the Moon, Earth's faithful companion in the night sky.

On September 14, 1959, the USSR succeeded in crashing the remote probe *Luna 2* onto the Lunar surface in the Palus Putredinus region at approximately 0°, 29.1° N.[12] Upon impact, the sphere shattered and scattered a collection of medals and ribbons, thus firmly and, importantly, in perpetuity documenting the Soviet achievement. On the same day, the third stage of the rocket that had propelled *Luna 2* on its way also crashed into the Moon. The *Luna 3* spacecraft followed and returned the first views ever of the far side of the Moon. The first image was taken at 03:30 Universal Time on October 7 at a distance of 63,500 km, after *Luna 3* had passed the Moon and looked back at the sunlit far side (Figure 41.1). Two and half years later, the United

FIGURE 41.1 This is the first image returned by *Luna 3*. Taken by the wide-angle lens, it showed that the far side of the Moon was very different from the near side, most noticeably in its lack of lunar maria (the dark areas). (Courtesy of NASA.)

States followed, on April 26, 1962, intentionally crashing *Ranger 4* into the Moon at 15.5° S, 130.7°W.[13] The United States sent nine more missions intentionally crashing into the Moon. Commencing with *Luna 5* (in May 1965), the Soviet Union attempted soft landings, most of which crashed accidentally. The first successful soft landing occurred on February 3, 1966, when *Luna 9* landed at 7.08° N, 73.37° W.[14] The United States followed 4 months later with *Surveyor 1*.[15]

In the early days of interplanetary exploration, soft landings on the Moon or Mars were complicated. As is evident from the compilation of crash landings on the lunar surface (Table 41.1), any crashes that occurred after soft landings had been developed were either accidental (because of equipment malfunction in the approach phase or because of telemetry errors), or they were probes, initially placed in lunar orbit, crashing uncontrolled onto the lunar surface as their orbits decayed.

After the astronauts of *Apollo 11* had set up a seismometer at Tranquility Base in July 1969,[16] it became customary for Apollo missions to set the ascent stages of the lunar modules, once jettisoned, on a trajectory for impact on the lunar surface.[17] In addition, the third stage of the Saturn IVB (SIVB), which had contained and protected the lunar module (LM) during the launch phase, was targeted for lunar impact (commencing with *Apollo 13*), preceding the landing of the LM by 4–5 days. The controlled crashes of both the LM and the SIVB formed part of seismic experiments established by astronauts during activities on the lunar surface. The crash of the SIVB is the only material culture on the lunar surface associated with the aborted mission Apollo 13.

TABLE 41.1
Crash Sites on the Lunar Surface

Mission	Event Date	Location	Item	Country	Nature of Crash
Luna 2	September 14, 1959	0 29.1 N	Remote probe	USSR	Intentional
Luna 2	September 14, 1959	Unknown	Third stage of Luna 2's rocket	USSR	Intentional
Ranger 4	April 26, 1962	15.5 S, 130.7 W	Remote probe	United States	Intentional
Ranger 6	February 2, 1964	Sea of Tranquility	Remote probe	United States	Intentional
Ranger 7	July 31, 1964	10.35 S, 20.58 W	Remote probe	United States	Intentional
Ranger 8	February 20, 1965	2.67 N, 24.65 E	Remote probe	United States	Intentional
Ranger 9	March 24, 1965	12.83 S, 2.37 W	Remote probe	United States	Intentional
Luna 5	May 1, 1965	Sea of Clouds	Remote probe	USSR	Accidental
Luna 7	October 1, 1965	Sea of Storms	Remote probe	USSR	Accidental
Luna 8	December 1, 1965	Sea of Storms	Remote probe	USSR	Accidental
Luna 10	> May 30, 1966	Unknown	Remote probe, orbiter	USSR	Orbital decay
Luna 11	≥ October 1, 1966	Unknown	Remote probe, orbiter	USSR	Orbital decay
Lunar Orbiter 1	October 29, 1966	7 N, 161 E	Photographic probe	United States	Intentional
Luna 12	≥ January 19, 1967	Unknown	Remote probe, orbiter	USSR	Orbital decay
Lunar Orbiter 3	October 9, 1967	14.3 N, 97.7 W	Photographic probe	United States	Orbital decay
Lunar Orbiter 2	October 11, 1967	3 N, 119.1 E	Photographic probe	United States	Intentional
Lunar Orbiter 4	< October 31, 1967	22–30 W	Photographic probe	United States	Orbital decay
Lunar Orbiter 5	January 31, 1968	2.79 S, 83 W	Photographic probe	United States	Intentional
Luna 14	> April 10, 1968	Unknown	Remote probe, orbiter	USSR	Orbital decay
Luna 15	July 21, 1969	Sea of Crises?	Remote probe, orbiter	USSR	Orbital decay
Apollo 12	November 20, 1969	3.94 S, 21.20 W	Lunar Module Ascent Stage	United States	Intentional
Apollo 11	1969/1970	Unknown	Lunar Module Ascent Stage	United States	Orbital decay
Apollo 13	April 14, 1970	2.75 S, 27.86 W	Saturn IVB stage	United States	Intentional
Apollo 14	February 4, 1971	8.09 S, 26.02 W	Saturn IVB stage	United States	Intentional
Apollo 14	February 8, 1971	3.42 S, 19.67 W	Lunar Module Ascent Stage	United States	Intentional

TABLE 41.1 *(Continued)*
Crash Sites on the Lunar Surface

Mission	Event Date	Location	Item	Country	Nature of Crash
Apollo 15	July 29, 1971	1.51 S, 11.81 W	Saturn IVB stage	United States	Intentional
Apollo 15	August 4, 1971	26.36 N, 0.25 E	Lunar Module Ascent Stage	United States	Intentional
Luna 18	September 11, 1971	3.57 N, 50.50 E	Remote probe, orbiter	USSR	Orbital decay
Luna 19	> October 3, 1971	3.34 N, 56.30 E	Remote probe, orbiter	USSR	Intentional
Apollo 16	April 19, 1972	~ 1.3 N, 23.8 W	Saturn IVB stage	United States	Intentional
Apollo 17	December 10, 1972	4.21 S, 12.31 W	Saturn IVB stage	United States	Intentional
Apollo 17	December 15, 1972	19.96 N, 30.50 E	Lunar Module Ascent Stage	United States	Intentional
Apollo 15	Late January 1973	Unknown	Subsatellite	United States	Orbital decay
Apollo 16	1973/1974	Unknown	Lunar Module Ascent Stage	United States	Orbital decay
Luna 22	May 27, 1905	Unknown	Remote probe, orbiter	USSR	Orbital decay
Hagoromo	February 21, 1991	Unknown	Remote probe, orbiter	Japan	Orbital decay
Hiten	April 10, 1993	34.0 S, 55.3 E	Remote probe	Japan	Intentional

Since the telemetric controls and technical requirements had been sufficiently developed by the time the first missions were sent to Mars and Venus, almost all crash landings on Mars (Table 41.2) and Venus (Table 41.3) were either accidental or caused by orbital decay. The more recent missions, where impact studies were carried out, are exceptions.

INVESTIGATING CRASH REMAINS

CRASH MECHANICS

Both intentional and accidental crash landings follow a projected landing trajectory that would normally see a steep descent through the outer reaches of the atmosphere. The (re)entry velocity is reduced due to atmospheric drag as the probe penetrates the denser conditions closer to the planetary surface, with the excess energy resulting in heat emissions. Controlled landings would deploy an air braking device, commonly a set of parachutes, once the probe has reached sufficiently dense atmospheric conditions. In the case of intentional crash landings, that final air braking is dispensed with, and the spacecraft will connect with the planetary surface at its terminal velocity. The same applies to situations where due to technical failure the air braking system does not deploy.

TABLE 41.2
Crash Sites on the Martian Surface

Mission	Event Date	Location	Item	Country	Nature of Crash
Mars 2	November 27, 1971	45° S, 313° W	Descent module, crash landed	USSR	Accidental
Mars 2	>November 27, 1971		Orbiter	USSR	Orbital decay
Mars 3	>December 2, 1971	45° S, 158° W	Orbiter	USSR	Orbital decay
Mars 5	>March 12, 1974		Orbiter	USSR	Orbital decay
Mars 6	March 12, 1974	23.90° S, 19.42° W	Descent module, crash landed	USSR	Accidental
Mars 6	>March 12, 1974		Orbiter	USSR	Orbital decay
Phobos 2	March 27, 1989		Lander 1 (to land on Phobos)	USSR	Accidental
Phobos 2	March 27, 1989		Lander 2 (to land on Phobos)	USSR	Accidental
Mars Climate Orbiter	September 23, 1997		Orbiter	United States	Accidental
Deep Space 1	December 3, 1999		Impact probe A	United States	Accidental
Deep Space 2	December 3, 1999		Impact probe B	United States	Accidental
Beagle 2	December 25, 2003		Descent module, Rover	European Space Agency	Accidental
Mars Polar Lander	December 3, 1999		Descent module	United States	Accidental

Orbital Decay

Crashes brought about by orbital decay, however, follow different mechanics. Depending on the nature of the orbit, low Earth orbit (200–1,200 km above the Earth's surface), medium Earth orbit (1,200–35,790 km), geosynchronous/geostationary Earth orbit (35,790 km), or high Earth orbit (above 35,790 km), the movement of the satellite is subject to a different velocity relative to Earth. Moreover, circular and elliptical orbits exhibit different characteristics. Decaying orbits, caused by the failure (or cessation) of a satellite's navigational rockets and concomitant effects of a planet's gravitational field, will lower a satellite's altitude to a level at which atmospheric drag will cause gradual braking, followed by a (re)entry. Common to all reentries due to orbital decay is the shallow angle of (re)entry. If that (re)entry occurs uncontrolled, the satellite/probe will be exposed to heating and will break up in denser atmospheres, even it has some level of (re)entry protection. While smaller objects will burn up in their entirety, debris of larger objects may reach the planet's surface, creating a spread-out debris field along the downrange path of (re)entry.

Examples of this on Earth are the breakup and debris dispersal patterns of major events. As of 2005, a Center for Orbital and Reentry Debris Studies compilation[18]

TABLE 41.3
Crash Sites on the Venusian Surface

Mission	Event Date	Item	Country	Nature of Crash
Venera 5	May 16, 1969	Atmospheric probe	USSR	Accidental
Venera 6	May 17, 1969	Atmospheric probe	USSR	Accidental
Venera 9	>October 22, 1975	Orbiter	USSR	Orbital decay
Venera 10	>October 25, 1975	Orbiter	USSR	Orbital decay
Venera 4	October 18, 1978	Atmospheric probe	USSR	Accidental
Pioneer Venus	December 9, 1978	Multiprobe (large)	United States	Accidental
Pioneer Venus	December 9, 1978	Multiprobe (small) night side probe	United States	Accidental
Pioneer Venus	December 9, 1978	Multiprobe (small) north probe	United States	Accidental
Pioneer Venus	December 9, 1978	Bus for multiprobes	United States	Accidental
Venera 12	>December 21, 1978	Orbiter	USSR	Orbital decay
Venera 11	>December 25, 1978	Orbiter	USSR	Orbital decay
Venera 13	>March 1, 1982	Orbiter	USSR	Orbital decay
Venera 14	>March 5, 1982	Orbiter	USSR	Orbital decay
Venera 15	1984	Orbiter	USSR	Orbital decay
Venera 16	1984	Orbiter	USSR	Orbital decay
Pioneer Venus	August 1993	Orbiter	United States	Orbital decay
Magellan	October 13, 1994	Orbiter mapping	United States	Orbital decay

had documented a total of 57 events, not counting the *Columbia* or the *Challenger*, where fallen space debris could be recovered on Earth. Among the more significant are the following:

- *Cosmos 954* in January 1978, which resulted in an area of approximately 124,000 km^2 covered with some level of radioactive debris over northern Canada[19]
- *Skylab* in July 1979, remains of which were scattered over an area of 6,500 km in length and 150 km wide, with Australia in the path[20]
- *Cosmos 1402* in February 1983, which fell into the Indian Ocean south of Diego Garcia, with the debris field sinking into the depths of the sea[21]
- A Delta II second stage in January 1997, spreading a debris field more than 750 km in length over Oklahoma and Texas[22]
- A Delta II second stage in April 2000, spreading a debris field near Cape Town in South Africa[23]
- *Salyut 7* in February 1991, breaking up over South America and leaving a debris field scattered over Argentina[24]
- *Mir*, being deorbited on March 23, 2001, breaking up over Fiji, with a debris field scattered over the South Pacific Ocean[25]
- The *Columbia* space shuttle accident that left a debris field across a number of U.S. states on February 1, 2003[26]

Some of the debris could be retrieved and subjected to a range of material sciences tests, such as an oxygen tank from *Skylab*,[27] beryllium cylinders recovered from *Cosmos 954*,[28] and a 255-kg fuel/oxidizer tank from the Delta II second stage both in the Oklahoma-Texas debris field[29] and the South African debris field.[30,31]

Orbital Decay on the Moon and Mars

The above cases demonstrate that Earth's atmosphere, which compared with that of the Moon or Mars is quite dense, is conducive to breaking the velocity of the deorbiting spacecraft, causing the heating and gradual disintegration of an object during (re) entry, leaving an elongated, lenticular field along the reentry path, with overall debris size a major factor in the lateral dispersal from the reentry axis. The reduced gravity of the Moon (about 1/6th of that of Earth) and the concomitant virtual absence of an atmosphere on the Moon would cause a crashing spacecraft neither to disintegrate nor to slow down during the descent. The spacecraft would connect with the lunar surface without any appreciable reduction of velocity. In all cases, the crash would have occurred with great velocity, thus in all likelihood smashing the probe/module/rocket stage into small fragments.

While there is an atmosphere on Mars, it only has an air pressure of 0.6 kPa, less than 1/100th that of the Earth.[32] Thus, the entry of orbiting and landing spacecraft into the Martian atmosphere will not be slowed down to the same degree as a (re) entry into Earth's atmosphere. While that can be taken into account in controlled landings, it plays a role in projecting the effects of accidental crashes. Unlike the situation on Earth, the resulting debris field can be assumed to be much smaller and very localized. Importantly, however, the lack of atmosphere implies that the spacecraft will not break up upon entry but will impact the surface largely intact. This will result in both an impact crater and a debris field circumscribed by the mass of the object, the spacecraft design, and the specific construction techniques employed.

SPACECRAFT INVESTIGATIONS ON EARTH AND OTHER PLANETARY SURFACES

While there are spacecraft crash sites on Earth, the Moon, Mars, and Venus, reasons of accessibility have confined any detailed analysis to sites on Earth. As intentional spacecraft crashes—as opposed to ballistic missile testing, such as at the Ronald Reagan Ballistic Missile Defense Test Site on Kwajalein Atoll, Marshall Islands—have not been carried out on Earth, they can be excluded in this review (but see discussion below). The examination of spacecraft crash sites on Earth has been limited to an investigation of the explosion of *Challenger* after takeoff on January 28, 1986,[33] and the breakup of *Columbia* upon reentry on February 1, 2003. The *Challenger* event is not really relevant to this discussion as it occurred soon after takeoff, well before the space shuttle had reached orbit. The subsequent investigation comprised the recovery of as many spacecraft parts as could be collected, and a forensic examination of these remains akin to an aircraft crash investigation.

The breakup of *Columbia* upon reentry and the associated descent of debris were caught on the weather radar of a number of stations in Texas and Louisiana, showing

FIGURE 41.2 Reconstruction of the disintegrated space shuttle *Columbia* from the recovered fragments. (Courtesy of NASA.)

the spread of the debris as it rained to Earth.[34] The debris field was 260 km long and 60 km wide, ranging from near Nacogdoches, Texas, to Alexandria, Louisiana. Almost 80,000 pieces of debris were collected, making up an estimated total of about 40% of the spacecraft,[35] with an aim of reassembling the spacecraft as far as possible, using standard aviation crash investigation techniques (Figure 41.2).

Even though people are officially required to hand in space debris derived from U.S. spacecraft (as the U.S. government does not abandon its ownership even when the spacecraft disintegrates), material was retained in private hands and has been auctioned off on eBay and other dedicated auction sites.[36,37]

While a number of space-flown satellites and space environment exposure platforms have been retrieved by several space shuttle flights,[38] only two spacecraft inspections have been carried out on other celestial bodies.

The first occurred on the lunar surface, when *Apollo 12* landed just 180 m away from *Surveyor 3*, which had touched down 31 months earlier.[39] Astronauts Pete Conrad and Alan Bean visited the spacecraft on their second moonwalk on November 20, 1969, examining and photographing *Surveyor 3* and its surroundings and then removing about 10 kg of parts from the spacecraft for later examination back on Earth. Items removed from the craft included the TV camera, electrical cables, the sample scoop, and pieces of aluminum tubing.[40]

The second instance occurred on Mars, when the rover *Opportunity* photographed its own heat shield, as well as the impact crater it had left upon landing.[41]

The survival of elements of material culture from crashed spacecraft depends on a number of factors, primarily the terminal velocity with which the spacecraft

TABLE 41.4
Scenarios of Mission Failure and Anticipated Associated Tangible Evidence

		Moon		Mars	
Point of Mission Failure	**Effect**	**Surface Evidence**	**Size**	**Surface Evidence**	**Size**
Descent angle too steep	Spacecraft descends faster than programmed, response devices (parachutes, etc.) do not deploy	Impact crater	Very small to small	Impact carter	Very small to large
	Heat shield fails, space craft overheats and disintegrates	n/a	n/a	Short debris field along entry path, impact crater	Very small to large
Descent angle too shallow	Spacecraft descends slower than programmed, heat shield may not be effective at that angle, space craft overheats and disintegrates	n/a	n/a	Long debris field along entry path (impact crater?)	Very small to large

n/a, not applicable.

connected with the planetary body. That in itself is determined by the nature of the mission, that is, intentional or accidental impact, and in case of the latter, the point in the mission cycle at which the mission failed. In principle, one can classify a range of broad failure scenarios and can approximate the expected tangible evidence (Table 41.4).

Role of Forensic Science

In terrestrial contexts, archaeological methodology has been employed in a forensic science context to document exhumations,[42] crime scenes,[43] and graves of war crimes.[44,45] In all such situations, a diligent and detailed documentation is made of the extant remains, and their relative position is recorded in three dimensions before the removal of the artifacts or human remains for analysis in a laboratory.

Another investigative paradigm has been espoused in the examination of air crashes. In order to understand the cause of the air crash, investigation teams diligently document, spatially record, and then collect all pieces of the aircraft wreckage. Brought to a central location, these fragments are then reassembled to their original location in the plane's fuselage in order to examine patterns of destruction and the like.

SPACECRAFT CRASHES AND CULTURAL HERITAGE

On one level, almost all spacecraft crashes possess some level of *historical* value. Unlike car crashes, which are all too ubiquitous, crashes of spacecraft are few in number. All space missions have large amounts of financial and intellectual investment riding on them. Intentional crashes provided the scientific data they were designed to collect and thus provided the expected advancement in science. These continue to occur with the LCROSS mission to crash into the Moon in 2009. Accidental crashes are representative of failed missions that did not deliver the expected scientific data. Given the costs of repeating a mission and the limited launch capacities, these crashes also possess *heritage* value: they altered the outcomes of the space programs because other missions were reprogrammed to fill the missing data and thus could do the tasks for which they had been earmarked, or the original mission was never replaced and the anticipated data were never collected.

Question of Integrity

Whether the crash sites have heritage value depends, *inter alia*, on whether the crash can be demonstrated to have effected some wide-ranging change in knowledge, social conditions, or policy. Whether the crash site has *heritage significance* depends on the condition of the site. One of the globally accepted key criteria in the assessment of the cultural significance of sites and objects is their integrity (viz. the various International Council on Monuments and Sites charters). While the integrity of a crashed spacecraft is obviously no longer given, the crash site itself has, immediately after impact, total integrity. Thereafter, this integrity can become diminished through actions of nature (e.g., material decay due to climatological, biological, or geological agents) and through the actions of people (from official investigators to souvenir hunters).

Natural decay agents are commonly well understood on Earth[46] or can be inferred on other planetary surfaces[47] and can be deemed manageable. The actions of people pose the most immediate and also most contentious threat.

One can safely assume that any investigation of spacecraft crash sites will follow the methodologies used for terrestrial aircraft crash investigations. Common to their approaches is that the need to understand the cause of the accident drives the total removal of artifactual evidence from the crash location to an investigatory laboratory setting. In terms of cultural heritage management, this means that all physical evidence of the crash is systematically removed from the site. All that remains is the historical record of the event and an admittedly very detailed documentation of the former site. Frequently, the event is memorialized at its location through a plaque or a structure. However, this removal of all material implies that the site itself has lost all heritage integrity. If and when the accident acquires cultural significance, which can be instant, as in the case of the *Challenger* disaster, or which may require the passage of time, there is no artifactual evidence left in situ to illustrate the significance. In the case of crash sites on interplanetary bodies, there are, in many instances, both the passage of time required to make an assessment and the unique opportunity presented by the fact that the sites have not yet been investigated. Thus, the debris field is still intact.

From a cultural heritage perspective, one needs to differentiate among intentional crashes, accidental crashes, and orbital decay. Not all of these are of the same cultural significance, and one has to be aware that even among the same type of crash the level of significance will vary.

ACCIDENTS

Overall, space accidents with human casualties, or events with the potential for casualties, are few in U.S. space history. When the crew of *Apollo 1* perished due to a fire in the command capsule during training, the United States resolve to be the first to land a human on the Moon was only spurred on even more. The near disaster of *Apollo 13* highlighted the dangers faced by the astronauts but did not dent U.S. interest. Both events occurred at a time when the Cold War was in full swing. With the collapse of the Soviet Union in 1989, the ground rules changed irrevocably.

There can be no doubt that the spectacular accidents of the space shuttles, both the explosion of *Challenger* after takeoff on January 28, 1986, and the breakup of *Columbia* upon reentry on February 1, 2003, shaped the attitude of the American public to space flight. While the aftermath to *Challenger* saw severe criticism of NASA's operations methods and overall culture of decision making,[48,49] its widely viewed live coverage (due the presence of teacher Christa McAuliffe) by a large proportion of American schoolchildren[50,51] had wide-reaching effects on space policy.

For example, NASA decided that nonprofessional astronauts would no longer be permitted, which had a direct bearing on the fledgling space tourism. Similarly, the very public breakup of *Columbia* upon reentry was also attributed to a breakdown of risk management and operations methods by NASA,[52] further undermining public confidence in the space program. The debris fields of both events, one scattered on the bottom of sea off Florida and occasionally washing ashore well after the event,[53] the other spread across a number of states in America,[54,55] thus form heritage sites associated with the modern manifestation of the post-Apollo space programs.

A Soviet equipment failure resulted in a significant crash site, the condition of which is totally unknown. *Venera 3*, launched 4 days after *Venera 2*, successfully left Earth orbit, reaching Venus on March 1, 1966. The spacecraft contained a radio communication system, a series of scientific instruments, and electrical power sources. Unfortunately, the communications systems had failed before planetary data could be returned. Despite this, *Venera 3* has the distinction of being the first human-made spacecraft to impact on the surface of another *planet*.[56]

On the Soviet side, we also have the crash of *Soyuz 1* with cosmonaut Vladimir Komarov due to parachute failure on April 24, 1967.[57] This extent of event was never published, as it occurred during the height of the Cold War, when such information was deemed politically sensitive; however, since the end of the Soviet Union, such information is now more easily accessible. The commercial Russian space industry has suffered a range of launch failures,[58] such as the crash of a Dnepr rocket soon after takeoff in 2006, which destroyed eighteen satellites.[59]

However, rocket crashes and associated debris fields are not confined to Russia and its predecessor, the USSR. They have occurred in the United States, with the European Space Program, and with the programs of Brazil, China, India, Israel, Japan, North Korea, and the Ukraine.[60]

The impact of the Dnepr rocket on July 26, 2006, was the greatest, as it destroyed Belka, an optical remote sensing satellite by the government of Belarus, Baumanets, an Optical remote-sensing cube owned by Bauman MVTU Technical University (Russia), *UniSat-4* research satellite owned by La Sapienza University, Rome (Italy), and a group of fifteen microsatellites from eleven different research institutions in the United States, Italy, Japan, Korea, and Norway.

ORBITAL DECAY

Debris fields associated with intentional orbital decay have, on the whole, little cultural heritage value. There are, in essence, sites associated with technological obsolescence, yet this depends on the nature of the deorbiting events. For example, the accumulation of debris of deorbiting spacecraft in the southeastern Pacific Ocean, an area that is colloquially called the spacecraft graveyard, has significance from the cumulative nature of the debris and the long practice by all powers to use that area as the "target zone for any chunks that may survive the reentry." Even though that debris is currently largely inaccessible at the bottom of the Pacific Ocean, it is not totally out of reach, as the investigations of deep sea shipwrecks such as the Titanic have shown.[61]

Likewise, it can be argued that the crash of the Soviet radar ocean reconnaissance satellite *Cosmos 954* on January 24, 1978, is of great cultural significance. It was the first time that a satellite with an onboard nuclear reactor had crashed back onto Earth, contaminating a wide area of northern Canada with radioactive material. An area estimated at 124,000 km^2 was covered with some level of radioactive debris,[62] with single pieces as heavy as 18 kg.[63] At the time, the events underlined the vulnerability of the environment to nuclear pollution caused by military activities. Notwithstanding that it was estimated later that the long-term health effects in the sparsely populated area would be small,[64] the crash provided both the antinuclear and the peace movements with ample political ammunition and shaped the attitudes of a generation of young people.

INTENTIONAL CRASHES

On the other hand, the cultural heritage value of intentional crashes into planetary surfaces can be more easily demonstrated if these crashes resulted in significant discoveries. There are only two crashes where the significance is most readily demonstrated: the most obvious, of course, is the site of the *Luna 2* crash in the Sea of Serenity on September 14, 1959, embodying the first human-made object to reach the lunar surface. The other exception is the crash site of the third stage of the Saturn IVB of *Apollo 13*. A total of seven Apollo landing missions had been planned (Apollo 11–18). The last mission was cancelled due to budgetary constraints and lack of public and government enthusiasm for the lunar program. All missions went according to plan with the exception of *Apollo 13*. Launched on April 11, 1970, the command module, then en route to the Moon, suffered an explosion of one of the oxygen tanks and subsequent venting of oxygen into space. In what has been

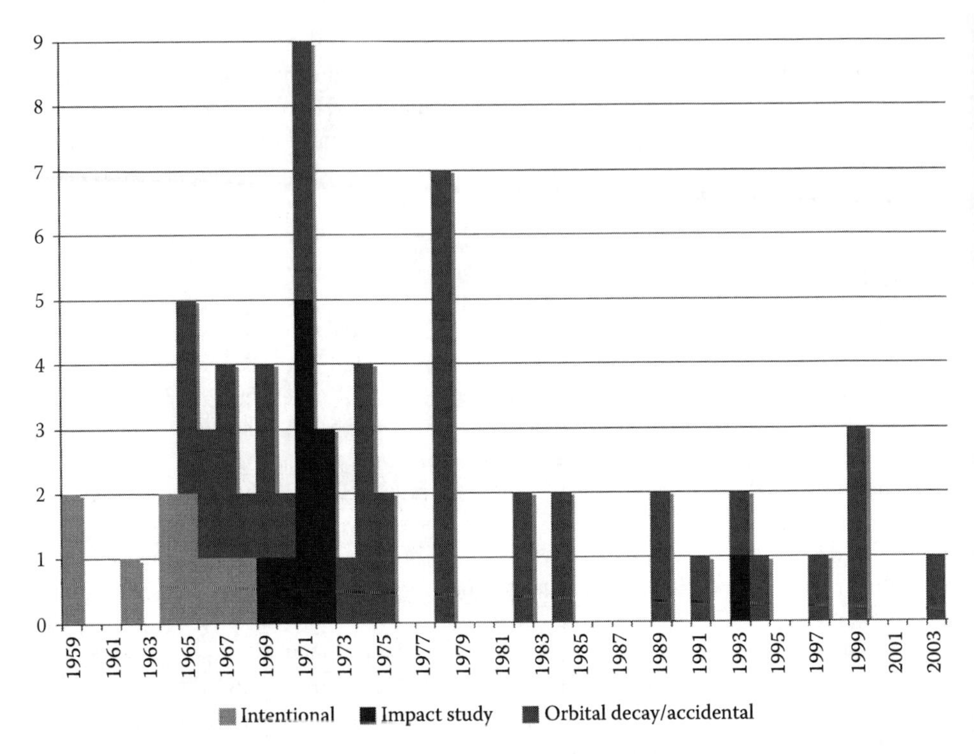

FIGURE 41.3 Chronology of intentional vs. accidental crashes.

described as NASA's finest hour, engineering ingenuity on the fly and commitment by both Apollo crew and ground control staff saw the safe return of the crew. The lunar landing, however, had to be abandoned. The third stage of the Saturn IVB of *Apollo 13*, which had been sent on a trajectory to the Moon, crashed on April 15, 1970, into the lunar surface at 2.75° S, 27.86° W with a terminal velocity of approximately 2.58 km/s.[65] The crash site is the only artifact on the lunar surface that is associated with the ill-fated mission of *Apollo 13* (Figure 41.3).

CLASSIFICATION OF CRASH SITES

The previous discussion has implications for the management of space heritage. Space crash sites are quite common on Earth, with another sixty-five on the Moon, Mars, and Venus. Table 41.5 synthesizes the information provided in Tables 41.1–41.3 and sets out the frequency of intentional and accidental crashes by celestial body. Clear trends in spacecraft development are evident. Intentional crashes are limited to the Moon. By the time Venus and Mars had become the focus of attention, soft landings were attempted on both planets as a matter of course. The fact that little was known about the atmospheric conditions on both is reflected by the high percentage of crash landings over orbital decay.

TABLE 41.5
Typology of Crashes

	Intentional	Impact Study	Accidental	Orbital Decay	*n*
Moon	27.0	29.7	8.1	35.1	37
Mars			66.7	33.3	12
Venus			41.2	58.8	17

In the light of this, an assessment of the cultural heritage significance of space crash sites cannot be carried out with a uniform paradigm but requires conditional responses. For the purposes of the remaining section of this paper we need to differentiate among four levels of cultural significance: (i) scientific, (ii) national, (iii) global (world), and (iv) universal.

Table 41.6 sets set out a matrix that provides guidance on the determining level of significance of sites associated with the various crash scenarios. Clearly, irrespective of the nature of the crash, a crash site will have *universal value for humankind* as a whole if that crash site represents the first human objects on that planetary surface. That level of significance will not change, irrespective of (re)evaluations of the cultural heritage significance that all other crash events may undergo by future generations. The first human objects will always remains the first. This, for example, applies to the crash sites of *Luna 2* on the Moon (intentional crash), of *Venera 3* on Venus (accidental crash), and of *Mars 2* on Mars (accidental crash). At a *national level*, the same applies, for example, in the case of the United States, to several probes of *Pioneer Venus*, which accidentally crashed in 1978.

Global heritage significance is much more subjective. The principles used for the World Heritage List may serve as a tool (although they are not applicable as

TABLE 41.6
Typology of Cultural Significance

	Universal Significance	Global Significance	National Significance	Scientific Significance
Orbital decay	If earliest human object to orbit planet	If connected with multinational (emotional) response	If earliest national object on planet	Limited
Accidental crashes	If earliest human object on planet	If connected with multinational (emotional) response	If earliest national object on planet; if connected with national (emotional) response	Yes
Intentional crashes	If earliest human object on planet	If connected with defining scientific experiment	If earliest national object on planet	Yes

a protective instrument, as a state entity must nominate, which assumes territorial control; see discussion below). It can be argued that the crash of *Cosmos 954* (orbital decay) exhibits that level of significance, as it provided a significant argument for the then fledgling antinuclear movement throughout the Western world. Likewise, crashes that are connected with defining scientific experiments can be argued to possess global value. An example of this would be the crash site of the third stage of the Saturn IVB of *Apollo 13* (intentional crash). It was intentionally directed into the lunar surface to create a physical impact that could be measured by the seismic station set up as part of the Apollo 11 and 12 missions.

CONCLUSION: THE FUTURE

It can be predicted that there will be considerable desire to access some of the crash sites in order to understand why a mission failed. In each case, the scientific benefit and value of the examination will have to be weighed against the loss of integrity of the site and its implications for cultural significance. It is posited by the author that the heritage sites of universal value should be off limits, irrespective of the scientific value put forward. Ultimately, this scientific value will only be short-lived in the greater scheme of humanity's future, yet these sites will remain in perpetuity to represent the first human objects on these planetary surfaces.

Finally, it is worth considering some of the realities of management. The country or agency that first visits a lander crash site may not be the same one that sent it. The data in the probe and the spatial distribution of spacecraft parts on the lunar/planetary surface are critical to understanding what happened. To ensure that these are not imperiled by ill-considered actions, regulations and protocols must be laid down as to how crash sites are treated. As has been addressed in another context,[66] there is no national jurisdiction on the lunar and other planetary surfaces. While the spacecraft, in complete or in crashed condition, remains the property (and responsibility) of the nation that launched it, the crash site cannot be protected through unilateral national protection. It is incumbent on us to develop international agreements that will protect our collective heritage for the future.

REFERENCES

1. Butowsky, H.A. 1984. *Man in Space: A National Historic Landmark Theme Study*. Washington, DC: National Park Service, U.S. Department of the Interior.
2. Spennemann, D.H.R. 2005. The Naval Heritage of Project Apollo: A Case of Losses. *Journal of Maritime Research*. http://www.jmr.nmm.ac.uk/spennemann.
3. Spennemann, D.H.R., and L. Kosmer. 2005. Heritage Sites of the US Space Program in Australia: Are We Managing Them Adequately? *QUEST—The History of Spaceflight Quarterly* 12(2): 52–64.
4. Barclay, R.L., and R. Brooks. 2002. In situ preservation of historic spacecraft. *Journal of the British Interplanetary Society* 55(5–6): 173–181.
5. Gorman, A.C. 2005. The Cultural Landscape of Interplanetary Space. *Journal of Social Archaeology* 5(1): 85–107. Gorman, A.C. 2005. The Archaeology of Orbital Space. In *Australian Space Science Conference 2005*, 338–357. Melbourne, Australia: RMIT University.

6. See, for example: (1) Fewer, G. 2002. Towards an LSMR and MSMR (Lunar and Martian Sites and Monuments Records): Recording Planetary Spacecraft Landing Sites as Archaeological Monuments of the future. In *Digging Holes in Popular Culture*, ed. M. Russell, 112–120. Oxford, United Kingdom: Oxbow. (2) Fewer, G. 2007. Conserving Space Heritage: The Case of Tranquillity Base. *Journal of the British Interplanetary Society* 60(1), 3–8. (3) Rogers, T.F. 2004. Safeguarding Tranquility Base: Why the Earth's Moon Base Should Become a World Heritage Site. *Space Policy* 20(1): 5–6. (4) O'Leary, B.L. 2006. The Cultural Heritage of Space, the Moon and Other Celestial Bodies. *Antiquity* 80(307): Project Gallery, http://antiquity.ac.uk/ProjGall/oleary/index.html. (5) O'Leary, B.L., R. Gibson, J. Versluis, and L. Brown. 2003. Lunar Archaeology: A View of Federal U.S. Historic Preservation Law on the Moon. World Archaeological Congress 5. Themes: The Heavens Above: Archaeoastronomy, Space Heritage and SETI. http://godot.unisa.edu.au/wac/paper.php?paper=1052. (6) Spennemann, D.H.R. 2004. The Ethics of Treading on Neil Armstrong's Footsteps. *Space Policy* 20(4): 279–290. (7) Spennemann, D.H.R. 2006. Out of This World: Issues of Managing Tourism and Humanity's Heritage on the Moon. *International Journal of Heritage Studies* 12(4): 356–371.
7. Fewer 2002 (note 6). Spennemann, Dirk H.R. and Murphy, Guy 2007. Technological Heritage on Mars: Towards a Future of Terrestrial Artifacts on the Martian Surface. *Journal of the British Interplanetary Society* 60(2): 42–53. Spennemann, Dirk H.R. Forthcoming. Human Heritage on Venus: Will There Be Sites Worth Managing?
8. Fewer 2002 (note 6). Rogers 2004 (note 6). O'Leary 2006 (note 6). Spennemann 2004, 2006 (note 6).
9. O'Leary 2006 (note 6). Spennemann 2004 (note 6). Spennemann, D.H.R. 2007. A Line in the Sand? Explorations of the Cultural Heritage Value of Hominid, Pongid and Robotid Artefacts. *International Journal Cultural Property* 14: 241–266.
10. Fewer 2002 (note 6). Spennemann and Murphy 2007 (note 7).
11. Spennemann 2004, 2006 (note 6). Spennemann, Dirk H.R. 2007. Extreme Cultural Tourism: From Antarctica to the Moon. *Annals of Tourism Research* 34: 898–918. doi:10.1016/j.annals.2007.04.003.
12. National Space Science Data Center. NSSDC Master Catalog: Spacecraft. Luna 2. NSSDC ID: 1959-014A. http://nssdc.gsfc.nasa.gov/nmc/spacecraftDisplay.do?id=1959-014A.
13. National Space Science Data Center. NSSDC Master Catalog: Spacecraft. Ranger 4. NSSDC ID: 1962-012A. http://nssdc.gsfc.nasa.gov/nmc/spacecraftDisplay.do?id=1962-012A.
14. National Space Science Data Center. NSSDC Master Catalog: Spacecraft. Luna 9. NSSDC ID: 1966-006A. http://nssdc.gsfc.nasa.gov/nmc/spacecraftDisplay.do?id=1966-006A.
15. National Space Science Data Center. NSSDC Master Catalog: Spacecraft. Surveyor 1. NSSDC ID: 1966-045A. http://nssdc.gsfc.nasa.gov/nmc/spacecraftDisplay.do?id=1966-045A.
16. National Space Science Data Center. NSSDC Master Catalog: Spacecraft. Apollo 11 Lunar Module/EASEP. NSSDC ID: 1969-059C. http://nssdc.gsfc.nasa.gov/nmc/spacecraftDisplay.do?id=1969-059C.
17. Lammlein, D.R. 1977. Lunar Seismicity, Structure, and Tectonics. *Philosophical Transactions of the Royal Society of London. Series A, Mathematical and Physical Sciences* 285(1327): 451–461.
18. Aerospace Corporation. 2005. Orbital Debris Summary of Recovered Reentry Debris. http://www.reentrynews.com/recovered.html.

19. Patera, R.P., and W.H. Ailor. 1998. The Realities of Reentry Disposal. *Advances in Astronautical Science* 99: 1059–1071. National Space Science Data Center. NSSDC Master Catalog: Spacecraft. Cosmos 945. NSSDC ID: 1977-079G. http://nssdc.gsfc.nasa.gov/nmc/spacecraftDisplay.do?id=1977-079G.
20. Kaplan, M.H., D.J. Cwynar, and S.G. Alexander. 1979. Simulation of Skylab Orbit Decay and Attitude Dynamics. *Journal of Guidance, Control, and Dynamics* 2(6): 511–516.
21. National Space Science Data Center. NSSDC Master Catalog: Spacecraft. Cosmos 1402. NSSDC ID: 1982-084A. http://nssdc.gsfc.nasa.gov/nmc/spacecraftDisplay.do?id=1982-084A.
22. Ailor, W.H. 2000. New Hazards for a New Age. *Crosslink: The Aerospace Corporation Magazine for Advances in Aerospace Technology* 1(1). http://www.aero.org/publications/crosslink/winter2000/04.html
23. Ibid.
24. National Space Science Data Center. NSSDC Master Catalog: Spacecraft. Salyut 7. NSSDC ID: 1982-033A. http://nssdc.gsfc.nasa.gov/nmc/spacecraftDisplay.do?id=1982-033A.
25. Aerospace Corp. 2001. The Final Days of Mir. Centre of Orbital and Reentry Debris Studies. http://www.reentrynews.com/Mir/sequence.html.
26. Carroll, C. 2004. Trail of Tragedy. High Tech Maps Aided Shuttle Columbia Recovery Effort. *National Geographic* 205(2): x–xii.
27. Gupta, A., and I.O. Jones. 1979. Skylab Debris: Analysis of Some Fragments. *Speculations in Science and Technology* 2: 499–504.
28. Patera, R.P., and W.H. Ailor. 1998. The Realities of Reentry Disposal. *Advances in Astronautical Science* 99: 1059–1071.
29. Ailor 2000 (note 22).
30. Anonymous. 2000. Delta II Second Stage Reentry. http://fuse.pha.jhu.edu/users/reentry.html.
31. Aerospace Corp. 2005. Orbital Debris Summary of Recovered Reentry Debris. http://www.reentrynews.com/recovered.html.
32. Lemmon, M.T., M.J. Wolff, M.D. Smith, R.T. Clancy, D. Banfield, G.A. Landis, et al. 2004. Atmospheric Imaging Results from the Mars Exploration Rovers: Spirit and Opportunity. *Science* 306(5702): 1753–1756.
33. O'Connor, E.A. 1986. Search, Recovery, and Reconstruction Task Force Team Report of the Challenger Mishap. May 14, 1986. Report of the Presidential Commission on the Space Shuttle Challenger Accident. Vol. 3. Appendix O. NASA Search, Recovery and Reconstruction Task Force Team Report.
34. GIS Lounge 2003. STS-107 Debris Field—Space Shuttle Columbia. http://gislounge.com/freisin/radar.shtml.
35. Federal Emergency Management Agency. 2003a. FEMA Closing Disaster Field Office in Lufkin on May 10. Press Release Number: 3171-69. Release Date: April 28, 2003. Federal Emergency Management Agency. http://www.fema.gov/news/newsrelease.fema?id=2774. FEMA 2003b. Environmental Protection Agency Sent 1,900 Responders To Columbia Shuttle Search; Made 48,400 Debris Collections. Press Release Number: 3172-70. Release Date: May 5, 2003. Federal Emergency Management Agency. http://www.fema.gov/news/newsrelease.fema?id=2809.
36. http://www.astro-auction.com/cgi-bin/auction/auction.pl?category=collectibles2&select=all.
37. The Register. 2003. Ebay Shuts Down Columbia Auctions. *The Register* February 3, 2003. http://www.theregister.co.uk/2003/02/03/ebay_shuts_down_columbia_auctions/.
38. Spennemann, D.H.R. Forthcoming. Managing Humanity's Moveable Cultural Heritage in Orbit—Practical and Ethical Considerations.

39. National Space Science Data Center. NSSDC Master Catalog: Spacecraft. Surveyor 3. NSSDC ID: 1967-035A. http://nssdc.gsfc.nasa.gov/nmc/spacecraftDisplay.do?id=1967-035A.
40. NASA. 2002. Experiments: Operations During Apollo EVA's. http://www.ares.jsc.nasa.gov/HumanExplore/Exploration/EXLibrary/docs/. NASA. 2003. Apollo XII Surveyor III Analysis. http://www./pi.usra.edu/expmoon/Apollo12/A12_Experiments_III.html.
41. David, L. 2004. Mars Rover Inspects Its Own Debris. http://www.space.com/missionlaunches/rover_debris_041227.html. ESA. 2004. ESA Mars Express: Facts about Mars. http://www.esa.int/SPECIALS/Mars_Express/SEM52E5V9ED_0.html.
42. Spennemann, D.H.R., and B. Franke. 1995. Archaeological Techniques for Exhumations: A Unique Data Source for Crime Scene Investigations. *Forensic Science International* 74(1): 5–15.
43. Boddington, A., A.N. Garland, and R.C. Janaway. 1988. *Death, Decay, and Reconstruction: Approaches to Archaeology and Forensic Science*. Manchester, United Kingdom: Manchester University Press.
44. Haglund, W.D. 2002. Recent Mass Graves: An Introduction. In *Advances in Forensic Taphonomy: Method, Theory and Archaeological Perspectives*, ed. W.D. Haglund and M.H. Sorg, 243–261. New York: CRC Press.
45. Skinner, M., D. Alempijevic, and M. Djuric-Srejic. 2003. Guidelines for International Forensic Bio-archaeology Monitors of Mass Grave Exhumations. *Forensic Science International* 134: 81–92.
46. Weaver, M.E. 1993. *Conserving Buildings: A Guide to Techniques and Materials.* New York: J. Wiley and Sons.
47. Spennemann and Murphy 2007 (note 7).
48. Hirokawa, R.Y., D.S. Gouran, and A.E. Martz. 1988. Understanding the Sources of Faulty Group Decision Making: A Lesson from the Challenger Disaster. *Small Group Research* 19(4): 411–433.
49. Heimann, C.F.L. 1993. Understanding the Challenger Disaster: Organizational Structure and the Design of Reliable Systems. *The American Political Science Review* 87(2): 421–435.
50. Wright, J.C., D.P. Kunkel, M. Pinon, and A.C. Huston. 1989. How Children Reacted to Televised Coverage of the Space Shuttle Disaster. *Journal of Communication* 39(2): 27–45.
51. Terr, L.C., D.A. Bloch, B.A. Michel, H. Shi, J.A. Reinhardt, and S. Metayer. 1999. Children's Symptoms in the Wake of Challenger: A Field Study of Distant-Traumatic Effects and an Outline of Related Conditions. *American Journal of Psychiatry* 156: 1536–1544.
52. Hall, J.L. 2003. Columbia and Challenger: Organizational Failure at NASA. *Space Policy* 19: 239–247.
53. CNN. 1996. Shuttle Challenger Debris Washes up on Shore. December 17, 1996. http://www.cnn.com/TECH/9612/17/challenger.debris/index.html.
54. GIS Lounge 2003 (note 34).
55. Carroll, C. 2004. Trail of Tragedy: High Tech Maps Aided Shuttle Columbia Recovery Effort. *National Geographic* 205(2): x–xii.
56. National Space Science Data Center. NSSDC Master Catalog: Spacecraft. Venera 3. NSSDC ID: 1965-092A. http://nssdc.gsfc.nasa.gov/nmc/spacecraftDisplay.do?id=1965-092A.
57. National Space Science Data Center. NSSDC Master Catalog: Spacecraft. Soyuz 1. NSSDC ID: 1967-037A. http://nssdc.gsfc.nasa.gov/nmc/spacecraftDisplay.do?id=1967-037A.

58. See compilation by Astronautix 2008. The Wrong Stuff: A Catalogue of Launch Vehicle Failures. http://www.astronautix.com/articles/thelures.htm.
59. BBC. 2006. Russian Satellite Rocket Crashes. July 27, 2006. http://news.bbc.co.uk/2/hi/europe/5219468.stm.
60. Astronautix 2008 (note 58).
61. BBC 2006 (note 59).
62. Patera, R.P., and W.H. Ailor. 1998. The Realities of Reentry Disposal. *Advances in Astronautical Science* 99: 1059–1071.
63. Gummer, W.K., F.R. Campbell, G.B. Knight, and J.L. Richard. 1980. *COSMOS 954: The Occurrence and Nature of Recovered Debris. INFO-0006*. Ottawa, Canada: Atomic Energy Control Board. http://adsabs.harvard.edu/abs/1980STIN.8225285G.
64. Tracy, B.L., F.A. Prantl, and J.M. Quinn, J.M. 1984. Health Impact of Radioactive Debris from the Satellite Cosmos 954. *Health Physics* 47(2): 225–233.
65. National Space Science Data Center. *NSSDC Master Catalog: Spacecraft*. Apollo 13 Command and Service Module (CSM). NSSDC ID: 1970-029A. http://nssdc.gsfc.nasa.gov/nmc/spacecraftDisplay.do?id=1970-029B.
66. Spennemann 2004 (note 6).

42 Mission Planning—Space Archaeology and Preservation Planning: System Engineering Perspective

Aaron Q. Rogers and Ann Garrison Darrin

CONTENTS

INTRODUCTION

The development of space systems is subject to numerous constraints and requirements regarding system performance, personnel safety, and system reliability. Germane to the discipline of space archaeology and heritage, these requirements also include general provisions for data archiving, program longevity, survivability, responsible stewardship of space, and planetary protection. The majority of these requirements are intended to ensure knowledge capture and data retention only to the point of mission termination or some small period of time thereafter. Only in the case of scientific data are provisions made for longer term retention. This chapter presents the source of these requirements, the extent of the requirements, and how these requirements influence the design and development of space systems. The chapter also explores how the products of these requirements could be used by future archaeological and heritage studies of space systems. Finally, recommendations for requirements that foster better historical records regarding the development of future space system will be offered for consideration. This chapter is written from the system engineers' perspective, as requirements would be directed at mission planning phases. The archivists, preservationist, historians, and curators can determine what is to be saved and the specific methodology. From the system engineers' perspective, the emphasis is that all considerations are to be planned and performed in a controlled, disciplined manner to avoid unexpected results. This chapter begins with several examples of just such outcomes and then the requirements of the respective mission development phases, before concluding with plans for future missions. A discussion of the last phase of a mission—end-of-life decommissioning—is included, as this plan directly affects the ultimate material artifacts and evidence of the mission.

LOST RECORDS: DATA AND MATERIALS

In July 1969, the world watched a grainy yet unforgettable television transmission of Neil Armstrong and his first tentative steps onto another world. There will never again be a first step on the Moon, yet we have lost the true and extremely clear images. The original magnetic tapes that recorded the first Moon walk—beamed to the world via three tracking stations—have gone missing at NASA's Goddard Space Center in Maryland. If not found, over the course of time these tapes may decompose to an irrevocably unusable state. The Apollo 11 television broadcast from the Moon was actually a high-quality transmission, far sharper than the blurry version viewed across the world. This was due to the fact that the format was not compatible with the prevailing commercial technology used by television networks. Instead, the recorded video was replayed on screens mounted in front of conventional television cameras.[1] The tapes were originally stored at Goddard, but in 1970 they were moved to the U.S. National Archives. No one knows why, but in 1984 about 700 boxes of space flight tapes there were returned to Goddard. Also among tapes feared missing are the original recordings of the other five Apollo Moon landings. This example reinforces several points:

1. If these tapes still exist, the fact that they are misplaced eliminates their value.
2. If these tapes have been inadvertently destroyed, we have permanently lost a significant artifact of truly historic events.
3. If these tapes still exist but are not recovered soon, aging may have or will permanently destroy the materials.
4. If the tapes are recovered, we would need special equipment only available at one NASA center in order to read out these tapes; copying them may not be possible.
5. The traceability of these tapes has been lost, as the people who might help solve this mystery may have retired or died.

With the hindsight famous to Monday morning quarterbacks, today we could have digitized, clear images of the first step, sharing the same sharpness and clarity of the first astronaut on the Moon. Museums and archives struggle continually with these kinds of preservation issues. Data and materials that are irreplaceable support much of our space exploration heritage in past, current, and future missions. The magnetic tapes of these first steps are like objects in destructive deorbit and may be lost forever. *Mir* (in Russian, МиР, which can mean both *peace* and *world*) was a Soviet (and later Russian) orbital station (Figure 42.1). *Mir* was humanity's first consistently inhabited long-term research station in space and the first third generation–evolved space station, constructed over a number of years with a modular design.

FIGURE 42.1 The space shuttle *Atlantis* docked to *Mir*. (Courtesy of NASA.)

The historic *Mir* was made internationally accessible to cosmonauts and astronauts of many countries. The most notable program making use of this access, the joint U.S. Space Transportation System shuttle-*Mir* program, saw American space shuttles visiting the station eleven times, bringing supplies and providing crew rotation. *Mir* was assembled in orbit by successively connecting several modules, each launched separately from 1986 to 1996. The station existed until March 23, 2001, when it was deliberately deorbited, breaking apart during atmospheric reentry over the South Pacific Ocean. Near the end of its life, several plans to save *Mir* were attempted, including private interests to purchase *Mir*, possibly for use as the first orbital television/movie studio. No viable method appeared, and *Mir* was deorbited when the 15-year-old space station entered Earth's atmosphere. Unburned fragments fell into the South Pacific Ocean around 6:00 Coordinated Universal Time.[2,3] The demise of *Mir* captured the attention of the world media, and as an odd aside, the owners of Taco Bell towed a large target out into the Pacific Ocean. If the target was hit by a falling piece of *Mir*, every person in the United States would be entitled to a free Taco Bell taco. The company bought a sizable insurance policy for this "gamble."[4] No piece of the station struck the target. Perhaps the best solution (certainly the best, from a cost perspective) was to deorbit the *Mir*. For future missions, consideration of alternative solutions is important. In this volume, Chapter 37 by Barclay and Brooks addresses the concept of in situ preservation. It is tragic that the *Mir* is gone forever.

Fortunately, we have built a significant hardware heritage record.[5] There is of course the argument that one cannot keep everything. That argument will be left for the archivists and preservationists. From the systems engineering perspective, again, the view is that all activities are to be planned. The implication is that through planned and defined activities, the optimized solution set will be available.

MISSION DESIGN

Globally, there is a widely accepted practice of stages to pragmatically developing a space mission. Missions may be initiated by a variety of methods driven by commercial, military, science, or sociopolitical objectives, but in general they follow the same milestone-driven programmatic approach, where the commonly employed terms of Phases A through F are often accepted. As a project matures, the effort typically advances through these associated phases:

- Pre-Phase A, conceptual study
- Phase A, preliminary analysis
- Phase B, definition
- Phase C/D, design and development
- Phase E, operations phase
- Phase F, systems decommissioning/disposal

During Phase F, the project implements the systems decommissioning/disposal plan developed in Phase E and performs analyses of the returned data and any returned samples. The NASA example would come from NPR 7131.1 and include the following:

- Complete analysis and archiving of mission and science data and curation of any returned samples, as well as archiving of project engineering and technical management data and documentation and lessons learned in accordance with agreements, the project plan and program plan, and center and agency policies
- Implementation of the systems decommissioning/disposal plan and safe disposition of project systems

Formal reviews are typically used as control gates at critical points in the full system life cycle to determine whether the system development process should continue from one phase to the next, or what modifications may be required. At the end of Phase A, there is normally a preliminary plan, and the mission truly gets "off the ground" with full funding commitments in Phase B. It is in Phase B, the definition stage, where mission planning fidelity will include full archiving and data retention requirements. The definition phase converts the preliminary plan into a baseline technical solution. Requirements are defined, schedules are determined, and specifications are prepared to initiate system design and development. Major reviews commonly conducted as part of the definition phase are system requirements review, system design review, and nonadvocate review.[6] Table 42.1 shows mission planning

TABLE 42.1
List of Mission Requirements by Phases

Stage/Phase	Document Title	Description	Due by	Update
Pre–Phase A formulation	Advanced studies	Advanced studies are performed to define needed engineering, technology, or commercial activities, and provide center resources to fund these studies and analyses.	Pre–Phase A	
Pre–Phase A formulation	Formulation plan	The document that defines the expected products, schedule, reporting, relationships, and required resources for formulation.	n/a	n/a
Pre–Phase A formulation	Operations concept document	The operations concept serves as a validation for the system design throughout the life cycle. Design choices are consistent with the operations concept. Later in the project cycle, following refinement during design and development, the concept serves as the basis for the Flight Operations Plan.	Phase B	Phase C/D and PDR
Phase A formulation	Risk management plan	Provides a description of how risks will be identified, assessed, tracked, mitigated, and documented.	SRR	PDR/NAR
Phase A formulation	Software management plan	Describes the work to be performed and the resources needed to accomplish the goals and objectives established in the customer agreement. The Software Management Plan includes the design planning information and the process management information.	MDR	PDR
Phase B formulation	Configuration management procedure	Defines how configuration identification, control, status accounting, and auditing will be performed for a program or project.	SRR	PDR/NAR
Phase B formulation	Contamination control	This document defines the contamination control requirements for fabrication, assembly, integration, test, and launch operations. It also specifies the verification approach to ensure that the requirements are met (author's note: risks versus level of preservation).	PDR	CDR
Phase B formulation	Data management procedure	This procedure details the mechanisms by which all documentation is identified, released, maintained, dispositioned, and stored in a project library.	SRR	NAR

TABLE 42.1 *(Continued)*
List of Mission Requirements by Phases

Stage/Phase	Document Title	Description	Due by	Update
Phase B formulation	Disposal, reentry, and decommission plans	Disposal planning (also known as postmission planning) involves how all mission processes and products are terminated. It addresses the disposition of hardware, software, facilities, processes, data, and personnel—the entire mission system. Disposal planning occurs throughout a project's life cycle, during both formulation and implementation.	CDR Mission Definition Review (MDR)	PDR, CDR and DR
Phase B formulation	Documentation/ drawing tree	Portrays the hierarchy of documents and/or drawings for a configuration item, system, or subsystem.	SRR	PDR, CDR
Phase B formulation	Electromagnetic compatibility (EMC) plan	This document describes the test objectives, requirements, and methods to verify the EMC of an observatory.	PDR	CDR
Phase B formulation	Environmental assessment	Document that ensures that environmental impacts have been considered in project planning and decision-making. See also "NHPA Guide" in the Guidance section.	SDR	n/a
Phase B formulation	Integrated independent review plan	This document establishes the plan for conducting a comprehensive set of integrated independent reviews	Confirmation Review	
Phase B formulation	Integration and test plan and verification documents	This document defines the requirements, tests, facilities, equipment, and actions necessary to integrate and functionally test an observatory.	PDR	CDR
Phase B formulation	Interface control document	Defines and controls detailed designs/specifications between systems and/or subsystems.	PDR	CDR
Phase B formulation	Level 1 requirements document	Mission requirements defined by headquarters.	SSR	PDR
Phase B formulation	Mission assurance guidelines	This Mission Assurance Guidelines document provides guidelines and recommendations for elements of quality assurance that are important to consider in any space flight assurance program. These guidelines are provided to assist potential developers.	PDR	CDR

(Continued)

TABLE 42.1 *(Continued)*
List of Mission Requirements by Phases

Stage/Phase	Document Title	Description	Due by	Update
Phase B formulation	Mission assurance requirements	The purpose of this document is to concisely present the safety and mission assurance requirements that may be necessary for projects.	PDR	CDR
Phase B formulation	Orbital debris	Debris assessments address thepotential for orbital debris generation that result from normal operations, malfunction conditions, and on-orbit collisions. The assessment also addresses provisions for post mission disposal.	PDR	CDR
Phase B formulation	Program commitment agreement	Agreement between the administrator and enterprise associate administrator that documents the agency's commitment to execute the program requirements within established constraints.	NAR	Annually validate
Phase B formulation	Program plan	Approach and plans for formulating, approving, implementing, and evaluating the project.	PDR/NAR	
Phase B formulation	Project plan	Product of project formulation; describes implementation of a project.	SDR; draft prior to IA, IAR, NAR	
Phase B formulation	Safety data packages (SDPs)	SDPs are developed to demonstrate a payloads compliance with launch range requirements. For GSFC projects, Code 302, the Systems Safety and Reliability Office will either prepare or review the SDPs.	Mission Definition Reviews (MDRs)	SDR, PDR, CDR, SAR, and FRR
Phase B formulation	Schedule baseline document	Establishes the project baseline schedule with major and control milestones and the I&T schedule.	PDR	CDR
Phase B formulation	Schedule management	Encompasses the project schedule development, management, and maintenance approach.	PDR	n/a
Phase B formulation	Software requirements document	This document forms the basis for software design.	PDR	Contract Award and PDR
Phase B formulation	Software test plan	This document lists the procedures used to test and validate software.	PDR	CDR
Phase B formulation	System safety plan	This safety plan defines the policy, procedures, and guidelines for achieving safety and mission success on a project.	SDR	PDR

TABLE 42.1 *(Continued)*
List of Mission Requirements by Phases

Stage/Phase	Document Title	Description	Due by	Update
Phase B formulation	Technology and commercialization plan	This plan describes the establishment of partnerships to transfer technologies, discoveries, and processes with potential for commercialization.	PDR/NAR	n/a
Phase D implementation	Contingency plan	Identifies specific risks and work-around solutions which are activated upon a triggering event. Establishes responsibilities for coping with and reporting contingencies.	MDR	SRR

documents and the points in a mission's life when they are first submitted for review. Those documents relative to data and material planning, deployment, and archiving are highlighted and discussed.

DECOMMISSIONING

At the end of Phase D, a decommissioning review is held; if new needs are developed the project life cycle starts over. Phases E and F encompass the problem of dealing with the system when it has completed its mission; the time at which this occurs depends on many factors. For a flight system with a short mission duration, such as a *Spacelab* payload, disposal may require little more than deintegration of the hardware and its return to its owner. On large flight projects of long duration, disposal may proceed according to long-established plans or may begin as a result of unplanned events, such as accidents. Alternatively, technological advances may make it uneconomical to continue operating the system in either its current configuration or an improved one. In addition to the uncertainty as to when this part of the phase begins, the activities associated with safely decommissioning and disposing of a system may be long and complex. Consequently, the costs and risks associated with different designs should be considered during the project's earlier phases.[7] As demonstrated by the earlier discussion on *Mir*, the decommissioning plan should be generated at the earliest possible time, perhaps Phase B. Having several potential options to the plan provides contingencies, while early planning may enable issues such as economic and political ones to not derail the long-term plan.

DEORBIT, IN SITU, AND THE GRAVEYARD ORBIT

There are several options for a decommissioning plan and numerous trade spaces. In situ preservation for spacecraft will not be viable for many orbits. In situ is practical for landers, probes, and rovers that reside on bodies. For lower orbits, in situ may present danger to space assets and the human presence in space.

Uncontrolled Deorbit

The deorbiting is initiated by one or more deceleration maneuvers, either with relatively short pulses of a high-thrust chemical propulsion system or with extended low-thrust maneuvers of electric propulsion systems. The deceleration maneuvers reduce the perigee altitude of the orbit, thus increasing the aerodynamic deceleration and therefore reducing the orbital lifetime. The satellite will be transferred into an orbit in which, using conservative projections for solar activity and atmospheric drag, the lifetime will be limited to no more than, for example, 25 years.

Direct Retrieval and Deorbiting (Controlled Deorbiting)

In cases in which the atmospheric destruction process is expected to be incomplete, an uncontrolled reentry is not permissible because of the risk to life on the ground. In these cases, the reentry maneuver has to be controlled, which means that the point of reentry—and thus also the point of impact on the ground—will have to be predetermined. The controlled reentry is initiated by one or several propulsive retroburn maneuvers of sufficient Δv, in order to lower the perigee altitude to a level that will result in an immediate atmospheric capture.

Controlled Deorbiting

This option might be attractive from an energetic point of view for missions at medium Earth orbit altitudes above 1,500 km, with a transfer to a disposal orbit of about 2,500 km altitude, similar to that of the geostationary Earth orbit (GEO) satellites' end-of-life graveyard.[8]

DESIGN FOR DEMISE: A CASE STUDY

The Global Precipitation Measurement (GPM) mission offers an example of *design for demise*. The GPM is designed for an uncontrolled deorbit such that none of its components survive the reentry process in significant shape to cause harm to life or property—the spacecraft can be allowed to simply fall to Earth under the influence of gravity alone (Figure 42.2). To this end, GPM has come up with a design-for-demise philosophy for the core spacecraft. The plan calls for the satellite to be designed in such a way that the need for a controlled reentry is eliminated. There are several advantages to this decision. First, it is inherently more reliable to design for an uncontrolled reentry; no spacecraft system has to perform an action to ensure safe disposal of the spacecraft. Also, it will enable maximum use of spacecraft resources for scientific objectives, since no unique hardware will be needed to accomplish reentry. In addition, GPM will avoid the costs (which can easily exceed $1M) associated with the controlled reentry process.

To ensure adequate destruction of spacecraft components upon reentry, vehicle and instrument assembly materials must be carefully selected. Standard designs for spacecraft propellant tanks and reaction wheels utilize titanium and stainless steel,

FIGURE 42.2 In 2001, this titanium fuel tank (from a successful Delta launch vehicle mission) survived the reentry process, landing in Saudi Arabia. (Courtesy of NASA.)

respectively. Titanium is also frequently used for satellite structural components; these materials tend to survive the reentry process. Alternative materials, such as composites that will not survive reentry, are often acceptable substitutions that are critical to the success of the design for demise.[9]

Beyond deorbiting, a viable decommissioning plan is to move into a higher orbit, often termed the graveyard orbit, where defunct missions present less danger to other space assets while remaining in a state of preservation. *Inmarsat 2F3's* successful decommissioning plan contained detailed provisions for moving to a higher orbit before being operationally retired. The Inmarsat satellites are located in GEO, at an altitude of 35,786 km above the equator. In October 2004, during the final stages of a regular east-west station-keeping maneuver, *Inmarsat 2F3* suffered an attitude control gyroscope anomaly, which caused it to point off the Earth and lose communications traffic. This eventually led to a shortened life period. The Inmarsat decommissioning involved attempting to raise the satellite's altitude more than 200 km above the GEO arc and then passivating any remaining onboard energy sources. Passivation, or preservation, consists of, once in the graveyard orbit, the reconfiguration of all satellite subsystems to minimize residual energy levels. This is achieved by venting all pressure vessels, discharging any stored electrical energy, and removing any sources of kinetic energy. Any subsequent collisions with human-made objects or micrometeorites should then cause less debris from the lower-energy secondary ejecta.[10]

DECOMMISSIONING AND INTERNATIONAL STANDARDS

International bodies have sought to provide decommissioning guidance. The Inter-Agency Space Debris Coordination Committee (IADC) provides guidance, including calculation of the minimum height above the GEO arc that a satellite should be placed in order to minimize the risk of collisions with other operational satellites. The most widely cited equation for performing this calculation is defined by the IADC[11] by

$$Hp(\text{km}) = 235 + 1{,}000 \times Cr \times A/M \qquad (42.1)$$

The figure of 235 km represents the sum of the minimum 200-km limit of the GEO protection region and the maximum 35-km descent that is possible because of luni-solar and geopotential perturbations. The remaining terms pertain to the long-term solar flux effects on eccentricity [A/M, effective satellite sun-facing area to-mass ratio (m^2/kg); *C*r, solar radiation pressure coefficient (–); *H*p, perigee height above GEO (km)]. International guidelines for the formulation of decommissioning plans include the European Code of Conduct for Space Debris Mitigation,[12] the ESA Space Debris Mitigation Handbook,[13] and the Federal Communications Commission Orbital Debris Mitigation Notice.[14] Of note is the International Standards Organization, which, together with other standards bodies, is producing a set of orbital debris mitigation standards, some of which are aimed at satellite manufacturers and operators. Lastly, the ESA has a tool called LEGO that allows generation of long-term GEO orbit propagations for periods up to 100 years and in perpetuity.

ARCHIVING

Many of the significant historical records created by NASA are now in the custody of the National Archives and Records Administration (NARA). NASA records are being stored by NARA in eight locations, with more than 26,000 cubic feet of them appraised as having permanent value. Unfortunately, decentralized, inconsistent archiving is often typical in the space community.

ARCHAEOLOGY AND HISTORIC PRESERVATION: THE SYSTEMS ENGINEERS APPROACH

With the earlier examples of the Apollo tapes and the *Mir* deorbit, it is evident that extensive planning is necessary for our off-world heritage. Predicting what will become historic is not an exact science. However, the first trip to Europa or the massive James Webb Space Telescope, one would speculate, has the potential. As early as Phase B, a full draft of the archaeology and historic preservation plan should supplement the traditionally required decommission plan. The fundamental principles that drive historic preservation are directly applicable:[15]

- Important historic properties cannot be replaced if they are destroyed. Preservation planning provides for conservative use of these properties,

preserving them in place and avoiding harm when possible and altering or destroying properties only when necessary.
- If planning for the preservation of historic properties is to have positive effects, it must begin before the identification of all significant properties has been completed. To make responsible decisions about historic properties, existing information must be used to the maximum extent, new information must be acquired as needed, and a set of criteria defining significance should be created.
- Preservation planning includes public participation. The planning process should provide a forum for open discussion of preservation issues. Public involvement is most meaningful when it is used to assist in defining values of properties and preservation planning issues, rather than when it is limited to review of decisions already made. Early and continuing public participation is essential to the broad acceptance of preservation planning decisions.

The very first step is to manage the planning process, which includes an implementation approach, a review and revision system, and a mechanism for resolving conflicts of the preservation plan with the scientific, military, or commercial mission goals. Space mission planning is a continuous process. This life cycle approach is seen in the phases, where preliminary and updated program elements are the norm. Preliminary planning steps should be completed by the preliminary design review. There is an organic element to the planning process, and the historic context will evolve. New information must be reviewed regularly and systematically and the plan revised accordingly. No one had the true foresight to understand the historic context of the Hubble Space Telescope or for *Vanguard 2* at the planning stage. *Sputnik* revolutionized our relationship with space, yet much of the historical material was not preserved. Novel in this expanded planning approach is the early involvement of the public. The planning process is directed first toward resolving conflicts in goals for historic preservation, and second toward resolving conflicts between historic preservation goals and other mission goals. Public participation assists in addressing sociopolitical aspects and may involve historians, architectural historians, archaeologists, scientific organizations and communities, folklorists, and people from related disciplines to define, review, and revise the historic contexts, goals, and priorities. Involving prospective users of the preservation plan in defining issues, goals, and priorities, a good plan will coordinate with other mission planning efforts and create mechanisms for identifying and resolving conflicts about historic preservation issues. The development of historic contexts, for example, should be based on the professional input of all disciplines involved in preservation and not limited to a single discipline. As various parts of the planning process are reviewed and revised to reflect current information, related documents must also be updated. Planning documents should be created in a form that can be easily revised or updated as goals change. It is also recommended that the format language and organization of any documents or other materials (visual aids, etc.) containing preservation planning information meet the needs of prospective users. The preservation planning is of course scalable.

FUTURE ARCHAEOLOGICAL MISSIONS

The archaeology and preservation planning step is fundamental for any mission of potential historical significance. This same planning process is fundamental for future archaeological missions. Significant archaeological missions will be available in the future. It has been suggested that *Vanguard 1* or capturing *Deep Space 1* in 3002 or preserving the Apollo 11 landing site are all viable space archaeology campaigns.

- *Vanguard 1* has the distinction of being the oldest artificial satellite orbiting the Earth. Its predecessors, *Sputnik 1* and *Sputnik 2* and *Explorer 1*, have since fallen out of orbit. *Vanguard 1* is 6 in. in diameter and weighs about 3 lbs. Its small size, compared with the Soviet's 200-lb. *Sputnik 1*, caused then-Soviet Premier Nikita Khrushchev to dub it "the grapefruit satellite." By 2000, Vanguard was in an orbit with an apogee height of 2,073 nautical miles above the Earth's surface, a perigee of 352 nautical miles, and an orbit period of 132.8 minutes. It may stay in orbit for up to 2,000 years and may remain preserved in situ preservation or be a candidate for retrieval.[16]
- *Deep Space 1* is predicted to return to Earth orbit around 3002. As part of its exploratory mission, there was a CD made of school children with 800 drawings of their vision of the world in 1,000 years. Retrieval of *Deep Space 1* would provide a fascinating time capsule for future researchers.[17] One hopes there will have been archived a CD player and powering conventions to read the CD.
- With an anticipated return of human presence to the moon in the next decade, it will be essential that steps be taken to preserve Tranquility Base in the form it was left, as an Apollo 11 heritage site.

Retrieving *Vanguard* as a mission would provide insights not possible otherwise regarding the long-term environmental effects of space and space debris on a spacecraft orbiting much higher than the International Space Station and other vehicles accessible to astronauts. Precedent was set in 1984, when the space shuttle *Discovery* brought home two communications satellites. Both were later launched again, and shuttles have visited other satellites to repair them and have brought back replaced hardware from them. Shuttles have also deployed and then retrieved science probes on the same mission. But *Vanguard 1* would be an entirely new kind of quarry, using entirely new kinds of tools.[18]

Planning these missions would involve a project design and the curation plan. Curation elements to be considered at the planning stage include the identification of a repository or repositories to curate the collections, appropriate budgeting for the short-term and long-term care of the collection, and a well-designed collection strategy. Traditional assessment criteria relative to end-of-life decisions are factored on the following items.

- Cost
  - Hardware- and software-related costs

- Increased launch cost due to increased mass and complexity of spacecraft
- Ground control station operation cost during deorbit maneuver
- Reliability
- Power demand
- Mass/volume
  - Volume constraints with respect to satellite and launcher
- Maneuver time duration
- Debris casualty area
- Complexity
- Operational capability

In addition to this set of criteria, the requirements for considering the space object in terms of our cultural heritage should be considered. As has been clearly demonstrated, these decisions must be included at the earliest possible point in the design process.

CONCLUSION

This chapter advocates for a archaeological and historic preservation planning process to enhance current mission planning when the potential (albeit not always known) for historic significance exists. Currently, the United States is faced with a decision on the Hubble Space Telescope's future. All elements are involved, including science, economics, politics, orbital mechanics, and risk avoidance. If left alone, Hubble will reenter and break up in an uncontrolled manner as it careens through Earth's atmosphere, with several elements likely to survive. Uncontrolled reentry is not allowed under the official policy of NASA's Office of Space Science. That office currently has provisions, with allocated funding, for one more Hubble servicing mission in 2009 to upgrade critical sensors and components. That would ultimately be followed by a final mission approximately five years later to "decommission" the telescope safely: in other words, to ensure the safety of the surface population from reentry harm. While the current baseline approach includes a shuttle flight to support the decommission, numerous variables must be evaluated, including the safety of an aging shuttle fleet, complexity and costs of robotic servicing missions, and trade spaces between the resultant science gaps until James Webb Space Telescope comes on line.[19] The likely outcome will be a controlled destruction of what has been a true "off-world" landmark. For future missions, the decommissioning plan initiated in Phase B under the auspices of an orbital debris plan is not adequate, and a broader perspective, with the inclusion of preservation planning for items of historical or cultural heritage significance, may be the prudent approach.

REFERENCES

1. Macey, R. 2006. *The Sydney Morning Herald*, August 5, 2006.
2. Mir Space Station Reentry Page. *Space Online*. http://www.ik1sld.org/mirreentry_page.htm.
3. The Aerospace Corporation. *The Final Days of Mir*. http://www.reentrynews.com/Mir/sequence.html.
4. Taco Bell Press Release. 2001. http://www.spaceref.com/news/viewpr.html?pid=4152.

5. Portee, D.S.F. 1995. *MIR Hardware Heritage.* NASA RE 1357. Houston, TX: Johnson Space Center.
6. NASA. 2007. *NASA Systems Engineering Handbook.* NASA/SP-2007-6105Rev1, December 2007. http://ntrs.nasa.gov/archive/nasa/casi.ntrs.nasa.gov/20080008301_2008008500.pdf.
7. NASA. 2007. *NASA Procedural Requirements. NPR 7120.5D. Effective Date: March 06, 2007. Expiration Date: March 06, 2012.* NASA Space Flight Program and Project Management. Washington, DC: NASA.
8. Burkhardt, H., M. Sippel, G. Krülle, R. Janovsky, M. Kassebom, H. Lübberstedt, O. Romberg, and B. Fritsche. 2002. *Evaluation of Propulsion Systems for Satellite End-of-Life De-Orbing AIAA.* AIAA-2002-4208. Reston, VA: AIAA.
9. http://gpm.gsfc.nasa.gov/Newsletter/june02/spacecraft.htm#top#top.
10. Hope, D. R. 2007. Special Issue Paper 933. *Decommissioning of the Inmarsat 2F3 Satellite.* London: Inmarsat Ltd. doi: 10.1243/09544100JAERO188.
11. Inter-Agency Space Debris Coordination Committee. 2002. *IADC Space Debris Mitigation Guidelines.* IADC-02-01, sec. 5.3.1, p. 5.
12. European Space Agency. 2005. *European Code of Conduct for Space Debris Mitigation.* ESA/IRC.
13. European Space Agency. 1999. *ESA Space Debris Mitigation Handbook.* Release 1.0. ESA Publications.
14. U.S. Federal Communications Commission. *FCC Orbital Debris Mitigation Notice.* FCC 04–130, 2004.
15. Archaeology and Historic Preservation: Secretary of the Interior's Standards and Guidelines [As Amended and Annotated]. http://www.nps.gov/history/local-law/arch_stnds_7.htmUS NPSlaw/arch_stnds_7.htm.
16. Tanner, R.L. Vanguard I Celebrates 50 Years in Space. http://www.nrl.navy.mil/vanguard50/own_words.php.
17. http://science.nasa.gov/headlines/y2002/20dec_ds1.html.
18. Oberg, J. 2008. Satellite Turns 50 Years Old. In Orbit! MSNBC. http://www.msnbc.msn.com/id/23639980/.
19. David, L. 2003. Bringing Down the House: How to Decommission Hubble Safely. August 4, 2003. http://www.space.com/businesstechnology/technology/hubble_demise_030804.

Section IX

The Future and Space Archaeology

Space Retirements:
The Baby Boomers New Frontier.

43 Plan for the Future Preservation of Space

Beth Laura O'Leary

CONTENTS

"Well here's the problem.
Specimen #AB5 is a nuclear Missile.
Specimen #AB6 is an ink cartridge."

INTRODUCTION

Space is a place few humans have ever been. It is the most recent frontier, both physical and technological, of the latter half of the twentieth and beginning of the twenty-first centuries. In the public's view, because space sites are often perceived as recent and technological, their material culture may not be regarded as worthy of preservation or considered as heritage. The past is usually thought of as ancient. In fact, for more than 99% of human history (about 3 million years), material culture is the only source of information that archaeologists can study. Looking at the key technological developments of human history, the first humans to leave the planet and step onto the Moon should be in the sequence that includes the creation of complex art (ca. 35,000 years ago), the control of fire (ca. 1 million years ago), and the first stone tools (ca. 2.5 million years ago). A common misconception is that all that is important to know about modern culture and technology can be conveyed by written or photographic records and documents. In a world of accelerating change, technologies are quickly superseded, and without recognition of their heritage value, knowledge about them can be quickly lost.

Surprisingly, a significant space object such as the *Vanguard* satellite is not completely documented (see Szczepanowska, Chapter 35 of this volume). When NASA recently (2008) wanted to design lunar digging equipment and needed to know the measurements of the *Surveyor 3* scoop recovered by Apollo 12 astronauts in 1969 on the Moon (*Surveyor 7* had a similar scoop), they found out that the all the blueprints had been lost. Only the actual scoop itself, which was preserved in the Kansas State Cosmosphere, could provide those critical measurements. Thus, the curation of space material served to provide essential data for future lunar exploration.[1]

Material culture is the tangible, physical remains of culture; it ranges from stone tools to command modules that can be used as sources of data, regardless of their age, to answer questions about cultural change and human behavior. Space exploration is still a living system, and most space agencies are rarely deeply invested in the historic preservation business; they want to continue space engineering. Moreover, commercial confidentiality and military secrecy can prevent the study of the documentary record. The Cold War era, when much of the early exploration took place, was cloaked in secrecy by both the United States and USSR. Already the archaeological record of the Space Age is at risk from neglect, deliberate destruction, and vandalism.

The sheer quantity of space heritage components, including sites, features, structures, landscapes, and objects, is daunting. Add to this space debris or parts and pieces of equipment floating in space or on other celestial bodies, and the volume of the material culture space heritage is overwhelming (see Gorman, Chapter 19 of this volume). Even in the best conditions on Earth, not all pieces of heritage can or should be preserved or treated as significant. Unlike much of the archaeological record on Earth, which can be assembled, described, and preserved in some kind of facility such as a museum or left in situ at a site that was not completely excavated, a critical portion of the record of space exploration is not in place but is moving above Earth all the time and, for the most part, will eventually deorbit and be destroyed as it does so. Most museums know where their heritage objects are stored, but some space debris is being preserved only by its constant movement within a path in space.

As future missions are planned to the Moon or other celestial bodies, the preservation of significant historic sites still in existence must be considered. Existing national and international historic preservation laws do not specifically address sites off Earth. For example, the lunar sites created by U.S. Apollo astronauts from 1969 to 1972 were not the kind of property that was envisioned when the U.S. National Register of Historic Places was created in 1966. Currently there are no off-Earth historic properties listed on the National Register. The U.S. National Historic Preservation Act of 1966[2] was created to establish a program for the preservation of historic properties in recognition that many such properties significant to the nation's heritage were being lost or altered with increasing frequency. Historic properties significant to the nation's heritage are considered "irreplaceable," and it is in the public interest that they be maintained for "future generations of Americans."[3] So far, that heritage is imperfectly maintained in the United States, but there is a legal framework for its preservation.

The United Nations Educational, Scientific and Cultural Organization's (UNESCO's) World Heritage Convention attempts to "identify, protect and preserve cultural or natural areas that provide outstanding value to humanity,"[4] but as of yet no areas on the World Heritage List are located in outer space or on other celestial bodies. UNESCO's international effort also recognizes the abilities and responsibilities of many nations (especially the wealthier ones) to preserve outstanding sites and to assist in the preservation of important sites in nations without the resources to do so.

The material culture of space heritage comprises recent past property for which archaeologists do not a large body of expertise. They need to think outside the box when it comes to space. Archaeology can be done anywhere there is a relationship between material culture and human behavior (see Staski, Chapter 2 of this volume). Archaeologists necessarily have to be dependent on the knowledge of and interaction with aerospace engineers, physicists, and astronomers. They must also have a robust knowledge of the history of science and scientific thought, as well as an appreciation of the recent history and the context of the Space Age within the last 50 years. This period encompasses the lifespan of many historians and archaeologists working today. Writing about the Space Age does not allow for a depth perspective of thousands of years that archaeologists have today when they look at ancient Greece or Mesopotamia. The recent past is more difficult to reflect on.

The avoidance of sites, which protects them from direct or indirect impacts, is a time-honored traditional method of archaeological preservation and can be put to good use for many Space Age sites. When the costs of returning to space sites are extraordinarily high because of their remoteness and a lack of desire to return, then avoidance is relatively easy. But to those scientists interested in studying the success or failure of past space endeavors, or trying to avoid damage by earlier space objects to current ones launched into space, their rights to investigation, even to the extent that cultural resources may be altered or destroyed, must also be considered in preservation management plans for the future.

Given the ongoing and new space initiatives, including space tourism, in the twenty-first century, there will be more diverse national and commercial issues concerning space. Significant space sites need to be claimable and treatable under amended or new laws, agreements, treaties, and protocols that are international in scope. Early space history was, more or less, an endeavor between the superpowers, the United States and USSR, and their allies. Today, more nations are involved, and the contemporary use of space is already heavily commercialized. While most space heritage sites are currently inaccessible, this will probably not be the case over the next 50–100 years. It is necessary now to think about what elements of space heritage should be preserved, and how, for the future generations of humankind, as the space industry further develops in the twenty-first century. This chapter examines the kinds of space material culture, ways of recognizing and evaluating its values, threats to its preservation, requirements and options for preserving it, and the existing and proposed legal instruments for dealing with space heritage.

KINDS OF MATERIAL CULTURE

Many authors have categorized the kinds of material culture in space.[5,6,7,8] The material culture of space shows great variability: the launch complex for the early Apollo missions in Cape Canaveral, Florida; the Apollo 11 astronauts' footprints at the first lunar landing; and the *Vanguard 1* satellite, launched on March 17, 1958, which is still circling the Earth every 2 hours and is expected to continue this orbit another 600 years, until its orbit degrades enough and it falls to Earth.[9] The terrestrial space sites include rocket ranges on a global scale, such as White Sands Missile Range in the United States and the Woomera Rocket Range in Australia. The material culture in orbit includes satellites and the Hubble Space Telescope. In deep space, *Voyager 1* keeps moving away from our solar system. There are also sites on other planets that were created by robotic rovers, such as those on the surface of Mars. There is even evidence of human material culture on asteroids.[10] An archaeologist should even consider the radiation released in space from the Starfish Prime exoatmospheric nuclear test in 1962. This created an area of radiation that increased the radioactivity in the Van Allen belts. This area could be considered an archaeological feature, albeit a dangerous one, of the Space Age.[11]

Space debris (also known as "space junk") is perhaps the most varied of these features, and much is unstable. For example, on June 23, 2007, astronauts on the International Space Station threw out an obsolete, refrigerator-sized ammonia reservoir called the Early Ammonia Servicer (EAS).[12] The 1,400-lb artifact has been circling Earth, and its orbit has decayed to the extent that it can be seen by the naked eye from Earth as it grows brighter in its descent. It is expected to burn up in Earth's atmosphere in late 2008 or early 2009.[13] The problem of preservation for that particular artifact is complex; perhaps it was meant to be discarded, although with its short existence in space, its importance can still be evaluated even if the artifact is gone forever. Complete documentation, models, and prototypes of the EAS should exist now, but they stand the chance of being lost in the future. It should also be noted that NASA has guidelines for limiting orbital debris and recommends that an object should not remain in its mission orbit for more than 25 years.[14] For an in-depth discussion of mission planning see Rogers and Darrin, Chapter 42 of this volume. Shorter amounts of time or brief life history of an artifact certainly limit the time that preservationists would have to evaluate an artifact's significance. Another historic example of a space object no longer in space is *Sputnik 1*, which degraded in orbit and burnt up 4 months after its launch on October 4, 1957. Although the actual first Sputnik is gone, no one would contest its significance as the first human satellite in Earth orbit; its launch into space is a historic event with multiple meanings to both the United States and the USSR, and its place is critical to the geopolitics in the Cold War era. Prototypes and replicas of Sputnik occur in many museums world wide, and it stands as a cultural icon for space exploration.

Barclay and Brooks (Chapter 37 of this volume) mourn the loss of the space station *Mir*, which became space junk when it was purposely brought down into the Pacific in 2001 and physically today exists mostly as inaccessible debris at the bottom of the ocean. Other space debris consists of droplets of urine and feces in space and on satellites and parts of rockets and satellites. There are also paint chips, which are plentiful and redundant in space. In totality they may be of questionable significance, but a representative sample may be worthy of preservation.

Gorman (Chapter 16 of this volume) views the shift in the discipline of cultural heritage or resource management away from the assessment of sites as discrete locations toward looking at places, objects, and values as being embedded in a cultural landscape. She terms this landscape a *spacescape* of the material culture of space.[15] The World Heritage Convention recognizes a cultural landscape as "the combined works of man and nature."[16] Although space can be seen as a true wilderness with no indigenous inhabitants discovered so far, in 1957 it began to collect the material culture of humans. It can also be argued that even as a *terra nullius*, space was and is a powerful associative landscape that is important to many different cultures, as well as to scientific inquiry and the history of technology.[17,18]

WAYS OF RECOGNIZING AND EVALUATING THE SIGNIFICANCE OF SPACE MATERIAL CULTURE

Archaeology as a discipline investigates the past using a body of theory and methods. But beyond reasons of scientific enquiry, what meaning does the past have? Are its meanings the same for all people?

Reaching the Moon was a great achievement for humankind. As early as 1984, the U.S. National Park Service commissioned a Man in Space Landmark Theme Study, which inventoried and assessed sites (including those associated with the Apollo program) for inclusion into a protective regime. One of the earliest attempts, in 1999, which recognized the significance of the *Apollo 11* lunar landing site was undertaken by the Lunar Legacy Project.[19] By using existing criteria already in place for the U.S. National Register, the project evaluated the small landing site at Tranquility Base on the Moon. This 1-day history of the landing was created by two astronauts' behavior. It is a unique moment in time, much like another rare archaeological moment: the one frozen in time at Pompeii.[20] The first lunar landing site meets all four criteria for eligibility to the National Register, which are applied to sites in the United States. The site is evidence of an event important to American history, it is associated with important persons in history, the technological components are the work of many experts or masters in their craft, and it contains information important to history.[21] It also meets the criteria for being of national significance, thus qualifying as a National Historic Landmark,[22] and although less than 50 years old, it is of exceptionally significance.[23] The only problem is that the first lunar landing site is not within the boundaries of the United States, although according to the 1967 Outer Space Treaty, the artifact assemblage is under U.S. control. It was a federal project (NASA) that was funded and permitted by the federal government, which is a condition for sites that can qualify for consideration under the National Historic Preservation Act. The process of determining significance involves a nomination to the National Register by the federal agency that is responsible for it and a determination of eligibility by the Keeper of the National Register. The outcome of the initial nomination process by the Lunar Legacy Project was not embraced by either federal authority and is currently unresolved (O'Leary, Chapter 40 of this volume).

LUNAR ARCHAEOLOGICAL SITES

Other archaeologists and geographers have noted the exceptional importance of the first lunar landing site. Fewer calls it one of the "most important and spectacular events in the history of space exploration."[24] It has been described as "a unique experience in the development and history of the human race."[25] Spennemann has referred to the first lunar landing as when "the world changed irrevocably."[26] All stress the idea that this is an "ultimate" heritage site that also still has great physical and locational integrity because of the qualities of the Moon environment. Spennemann[27] writes that not only is there a record of all the movements of the astronauts in the lunar dust, but no heritage site on Earth has the entire interaction at the site preserved. The atmosphere on Earth, along with erosion, changes the physical characteristics of a terrestrial site over time, and revisits and reuse affect the condition and patterning of the archaeological assemblage. With the exception of the *Apollo 12* visit to the earlier *Surveyor 3*, there have been no revisits to the currently existing lunar sites.

In a recent paper, Stooke proposed that at a minimum, until guidelines and formal arrangements are made, all sites from the first lunar exploration era (1959–1976) be considered significant and be protected.[28] The first era of lunar exploration contains the origins of all future exploration with nations and space agencies being the main players. Stooke has documented and noted the problems with defining the exact locations of several of the first lunar sites in his extensive atlas of lunar exploration.[29] He also finds significant the "first sites associated with new Moon-faring agencies, regardless of age."[30] The new era of lunar exploration, which he terms the "Moon 2.0 Era," is within the current era of proposed lunar exploration, beginning with companies such as Lunacorp, Transorbital, and the current (2008) advent of the Google Lunar X Prize. Ironically, the Google Lunar X Prize has the potential to be both a threat to previous lunar sites and a "first" in creating lunar sites that are commercial in nature. Spennemann[31] has looked at the heritage of the entire Apollo program and divided the heritage components into five units: (1) physical sites associated with the development and execution of the program, (2) artifacts associated with the program that remained on Earth, (3) artifacts that went into space and returned to Earth, (4) artifacts still in space, and (5) lunar samples collected on the Moon (i.e., regolith and rocks) and brought back to Earth. Although the borrow areas where the samples were collected are archaeological features at the lunar sites, the actual material collected is more geological than archaeological in nature.

Although there is general agreement on the first lunar landing site being significant, how should all the other sites, objects, and features of the Space Age be recognized and evaluated? How space artifacts become worthy of preservation can be viewed by looking at the accretive technical and social processes as Barclay and Brooks (Chapter 37 of this volume) have done.[32] They present a strong argument for preserving representative examples of all technologies. Their discussion of the *Mir* space station and the Hubble Space Telescope focuses on key concepts of objects being transient (or what is obsolete, expedient, and disposable) or those that are durable (or historically significant and therefore worthy of protection). *Mir* in this sense was thought of as disposable and purposefully deorbited into the Pacific Ocean. Brooks and Barclay (Chapter 37 of this volume) note its loss because it was not only the first international space station but contained a vivid social history of danger and triumph. Though *Mir* was battered, it kept flying for 15 years, and its operational accidents were offset by its remarkable duration and the skill and bravery of its different astronauts. It was a site in space where Cold War enemies worked together. It is argued that it could easily have been thought of as durable and historic and worthy of preservation.[33] To some, it was an orbiting disaster waiting to happen; to others it was as symbol of a period of space exploration and triumph of the human spirit. These two ideas are frequently considered when looking at modern material, and the decision one makes about a space or any historic artifact being transient or durable has definite consequences for preservation decisions. Although philosophically the cost and the consequences of preserving an object should not enter into whether it is significant or not, in reality they often do. Today, the Hubble Space Telescope in space sits on the edge between being transient and being durable.

Gorman[34] has proposed a three-tiered vertical landscape for evaluating places associated with space exploration: designed space landscapes, such as tracking stations on Earth; organically evolved landscapes in orbit and on the surfaces of celestial bodies (e.g., satellites and landers); and those places beyond the solar system. Gorman questions for whom is this spacescape significant and why. One example of a space heritage site is the former Nazi site of Peenemünde, Germany, which is referred to as the cradle of spaceflight; it is also a site where much of its significance was achieved through the forced labor of captives during World War II. Obviously, this is an area of conflicting heritage, a heritage of both spectacular technology and the death of concentration camp laborers. Gorman (Chapter 16 of this volume) argues strongly for evaluating the social significance of laborers' sacrifice at Peenemünde, as well as the space technology. She also argues for the use of the Burra Charter, whose guidelines were adopted by the Australian National Committee of the International Council on Monuments and Sites. The charter especially promotes the idea that cultural values can coexist and all should be recognized, even in cases where they conflict. The charter outlines four categories of significance:

> aesthetic: consideration of form, scale, color, texture, materials, smells, sounds and setting; historic: association with historic figures, events, phases or activities; scientific: importance of the data in terms of rarity, quality, representativeness and the degree to which a place can contribute further substantial information and social: the qualities of which a place has become a focus of spiritual, political, national or other cultural sentiment to a majority or minority group.[35]

When the Space Heritage Task Force was created at the Fifth World Archaeological Congress in Washington, D.C., in 2003, that international body passed a resolution to recognize that the material culture and places associated with space exploration are significant and that significance must be determined at individual, local, organizational, national, and international levels.[36] The resolution stated that it is necessary to consider what elements should be preserved and how. Most of these ideas discussed above are complementary to each other in terms of assessing the significance of space exploration places and artifacts. It is clear that space archaeology and heritage is a unique area of the discipline of archaeology and preservation still under development.

REQUIREMENTS AND OPTIONS FOR PRESERVATION

Usually when one considers the management of archaeological sites, the first step is an inventory or survey of the all the sites in an area in order to understand the nature and extent of the total populations of sites. For example, within an area on Earth, usually a check of the literature and state database of all the previously recorded prehistoric and historic sites is done. In a case (which happens frequently) where the total area has not been completely archaeologically surveyed, a sample survey may be done to help predict the occurrence and frequency of sites and prehistoric and historic land use of the area. In the case of space heritage, there are two major problems: the basic survey has not yet been done either by individual nations or by an international group, and some of the locations of significant sites are unknown. For example, a location is not known for the *Apollo 11* lunar module ascent stage, which was unloaded when it docked with the command module, deorbited, and about 4 months later crashed at an unknown location somewhere near the lunar equator.[37] It is not clear exactly how many sites are on the Moon, where they are located, and what exists in the artifactual assemblage at some of the sites. Gibson[38] compiled a preliminary inventory of artifacts left behind at Tranquility Base by Armstrong and Aldrin on *Apollo 11*, but other Apollo site assemblages have not been equally documented. Spennemann[39] has presented a list of spacecraft or major spacecraft components on the Moon, Mars, and Venus. A basic inventory is a necessary first step in any preservation plan.

The two Apollo 11 astronauts thought if they could return to the Moon 1 million years in the future, because of the Moon's physical characteristics they would basically see the same scene.[40] Both the physical isolation and costs of return have kept the lunar sites from experiencing human impacts. The Moon is large, even though most of the sites are congregated near the lunar equator. Avoidance of direct or indirect impacts to sites has kept many archaeological sites intact on Earth and continues to do so on the Moon. The Burra Charter advocates a similar tenet: to do as much as is necessary and as little as possible in order to maintain the cultural significance of place (Gorman, Chapter 16 of this volume).[41]

Brooks and Barclay (Chapter 37 of this volume) discuss the option of the in situ preservation of space vehicles currently in space with reference to safety, monitoring, potential future public access, and long-term costs. Using *Mir* as an example, financial constraints were one of the primary reasons it was disposed of, and the decision was made by Russia in the absence of an international structure of preservation and support for continuing to keep it in space. The Hubble Space Telescope, launched in 1990, has its last scheduled servicing mission in 2008 and planned "retirement" in 2013.[42] It is almost impossible for it to be returned to Earth and put into a terrestrial museum, and it may have a fate similar to that of *Mir*. NASA has designed the James Webb Space Telescope to be positioned 1.5 million kilometers from Earth, at the second Lagrange point (L_2). Space objects can be placed in a graveyard orbit where they are potentially retrievable for further study or they may be left in situ, and with the right safeguards, in the future, they could serve as space tourist destinations. The Lagrange points could become places for durable objects. There may be other places in space for space heritage objects to be curated. Today, lower orbit slots are assigned appropriate use by international committees; some of these may be also considered for use as heritage locations in the future. Using the existing preservation laws and regulations of nations who have space-related sites within their boundaries and linking those sites to sites in space or on other celestial bodies would at least put some sites within a preservation framework.

LEGAL DIRECTIONS FOR PRESERVATION

Who owns and is responsible the management and preservation of the material culture of space? Article 8 of the Outer Space Treaty of 1967 indicates that what is in space or on other celestial bodies remains the property of the state that put it there, but it does not specify that the title means it should be preserved in any way for posterity.[43] For the Moon, Stooke's ad hoc suggestion that the lunar sites dating from first exploration era be considered significant and be protected from future impacts is a good stopgap measure. The question is how could it be implemented? He argues for protection for artifacts themselves and all features (e.g., foot trails, rover tracks) and that future visits to observe lunar sites should be allowed only if "nothing is disturbed."[44] Using the Apollo 17 site as an example, Stooke argues it could be approached from new landing areas a few kilometers distant and even closer on the north or south side.[45] As such, he sees the Google Lunar X prize as not incompatible with protection strategies, given planning. The main problems are who will be doing the planning and what sanctions are in place if the plans are not adequate or there are, through accident or design, adverse impacts to a lunar site? In this scenario, only those competing to get to the Moon are responsible for deciding the boundaries of a revisit.

Making lunar historic parks or restricted areas has been proposed by many authors (Capelotti, Chapter 21 of this volume).[46,47,48] Like the prehistoric ruins in the Chaco Culture National Historic Park, where a larger area is set up around them, a series of lunar historical parks could encompass the landing or impact sites. Stooke[49] suggests that the boundaries should be out to the horizon visible from each site, a kind of viewshed approach that is proposed for sites on Earth such as historic trails (e.g., the Camino Real in New Mexico). Changes to the viewshed, such as development in close proximity and looting or vandalizing within a national park or monument, can be seen as adverse effects and are covered under law and regulations in the United States and for many other nations.

Several authors have argued for the intervention of the United Nations in the form of UNESCO's World Heritage Convention to be used to address space heritage. (O'Leary, Chapter 40 of this volume; Brooks and Barclay, Chapter 37).[50,51,52] It seems clear that the heritage of space at this juncture is the responsibility of the international community, since all its members are, or have the potential to be, players in the new space era. Organizations such as the European Space Agency did not even exist at the time of the first lunar landing. Japan has a historic site on the lunar surface. It seems that those nations with space sites or objects should participate in an international agreement through the United Nations (UN).

Fewer has argued that until recently the greatest threat to Tranquility Base was likely to be from a meteor impact, although with the advent of spacecraft and humans returning to the Moon this will change.[53] He reported a plea in 1997 for an internationally binding agreement that would criminalize theft and behavior that would damage any part of the artifactual assemblage on the Moon.[54] If the lunar sites are the heritage of humankind and not just the nations who created them, then attaining World Heritage status would be important and could be applied to all sites on the other celestial bodies. One of the ways in which a site attains World Heritage status currently is that it is proposed to be placed on the list by the country that owns or has jurisdiction over the site and has considered it significant by their own legislation. The earlier attempt by the Lunar Legacy Project to start this process reached an impasse in 2001, when the both relevant U.S. federal agencies considered that action to be a claim of sovereignty over the Moon, not under their jurisdiction, or inappropriate.[55] One suggestion has been for the United States to request the World Heritage Site list to include the solar system and off-Earth sites, such as the site of the first lunar landing, and further for the United States to transfer ownership of the assemblage there to the World Heritage as the first off-Earth site under its new, expanded program.[56,57] Spennemann[58] has described the problems with World Heritage status in enforcing protection of a site, and as an example uses the destruction by the Taliban regime in Afghanistan of the giant Buddha in 2001. There is no legal authority for the international body to intervene should the host country be negligent or active in the destruction of properties in its own country. What would the situation be in space or on other celestial bodies in the face of the complex, developing space industry?

As noted by Rogers,[59] there are a large number of countries (about 200) that are now members of the UN, with 170 as World Heritage members, which means that a great number of interests need to be served before a change in the World Heritage policies and procedures could be agreed upon for preserving space heritage. Equally, the Outer Space Treaty of 1967 states that space activities should forward humankind's use of space for peaceful purposes, but historic preservation is not a specified peaceful activity. Also, as noted by Rogers,[60] initial and ongoing costs are a factor in any preservation strategy, and to look at some alternatives for space involves expensive operating and maintenance costs. As Barclay and Brooks (Chapter 37 of this volume) note, because the costs and collaborative effort required to preserve space objects and sites are high, there is a great need for a clearly developed decision-making process. They argue persuasively for an evaluation strategy that is research-based and guides a collection strategy, whether on Earth or in space. UNESCO has been involved in a similar endeavor concerning the preservation of marine or underwater sites.[61] There has been no international legal instrument that adequately protects the underwater cultural heritage, and it has suffered much more damage and loss from pillage and natural damage than other classes of resources. Shipwrecks, like space crash or landing sites, are a kind of time capsule, capable of providing data that can be found nowhere else.

Doyle (Chapter 39 of this volume), as a lawyer, provides an excellent overview of the international space community and the law that has been established for its management, and of the application of this body of law to historic preservation of objects in space and on other celestial bodies. He also focuses on UNESCO Convention on Underwater Cultural Heritage and thus argues for the existence of a body of law in the United States and international precedents that could extend such laws regarding archaeology to outer space. Parallels for managing space heritage have also been drawn to managing the historic resources associated with the exploration of Antarctica. With the recent boom in tourism to the only continent managed by treaty, there have been impacts to the natural and historic resources with tourist visits doubling to more than 24,000 visitors in 2003–2004.[62] No country owns all or has legal possession of portions of Antarctica, but each nation has been allowed to manage the resources in its national interests, which are subject to treaty and signed protocols on the environment by signatory nations. Spennemann[63] argues for extending this concept to at least the lunar surface and by logical extension to space and other celestial bodies. No one owns space or any other celestial body, and the analogy with both marine heritage and that on Antarctica is inexact, but it does provide ways of looking at how to legally manage places and objects as space tourism and national, scientific, and commercial use of space becomes more frequent and threatens to destroy this heritage in the next century and beyond.

CONCLUSION: THREATS TO PRESERVATION

The market for Space Age collectibles has always been strong. There have been auctions at Sotheby's for Russian space artifacts from the former Soviet Union, and a check of online auction houses showed that such items are regularly offered (O'Leary, Chapter 40 of this volume).[64] The return to actual sites of the exploration of space could provide a treasure trove for collectors, as well as an increased likelihood of damage to a site(s) without clear and formal arrangements prior to a revisit.[65]

Astronaut Buzz Aldrin, as part of the Apollo 11 mission on the Moon, deployed a solar powered seismometer near the spacecraft and was told not to walk in front of the solar panels or he would cause some damage. He remarked he did walk there and one could see some of his footprints on a revisit: "I guess I wouldn't mind too much if somebody went up there to brush those away."[66] Brushing them away would, in archaeological terms, destroy part of the site. Will Pomerantz, X Prize Foundation director of space projects, believes there is a generation gap in public perceptions of the issue of lunar preservation. Essentially, those who were alive to witness the Apollo missions say the sites "should effectively be put under glass," while those not alive in that period say the opposite about revisiting those sites—"it's a good thing."[67] Pomerantz argues that the X Prize Foundation "should be the body that makes sure this is done respectfully, rather than the body defining what 'respectfully' is."[68] This approach has overtones of the fox guarding the hen house. While national space agencies (and those signatories to the 1967 Outer Space Treaty) in the future should respect the sites and artifacts as a matter of diplomacy and international agreement, private entities are not as restrained, and some of the prize money from the Lunar X Prize is specifically intended for viewing and returning to space heritage sites.

The earlier proposed Transorbital mission required approval from the Federal Communications Commission and State Department and a permit from the National Oceanic and Atmospheric Administration to fulfill the U.S. treaty obligations and to ensure that the proposed disposal of the Transorbital orbiter at the end of the mission crashed it far away from the historic lunar sites.[69] However, there has been no clear attempt at compliance with or specific linkage to the National Historic Preservation Act or any federal preservation law. In the case of the Google Lunar X prize, many of the contenders for the prize are not U.S. entities subject to U.S. sanctions.

What is the next step? In the absence of nations to assume leadership in space heritage, it is likely that little will happen soon on the protection of space heritage. Doyle (Chapter 39 of this volume) argues that the UN and UNESCO are not self-motivating but require that, in the case of the United States, the State Department bring the matter before the appropriate bodies of the UN; he recommends a series of "appropriate events." He advocates the use of proposals to manage space heritage crafted by professional archaeologists and historic preservationists to be presented to the U.S. State Department. A working group or task force needs to be assembled on both national and international levels. Some of these steps have already been initiated on several fronts. The Lunar Legacy Project has, at least, brought the issue into the public domain and to the attention of federal preservation authorities.[70] The New Mexico Office of Cultural Affairs Laboratory of Anthropology has included the first lunar landing site as LA 2,000,000 in its archaeological database for the state. This is the first state to officially recognize a space heritage site off Earth (O'Leary, Chapter 40 of this volume). The World Archaeological Congress in 2003, as an international body, signed a resolution that recognizes the importance of the material culture associated with space and created an ongoing Space Heritage Task Force (SHTF).[71] This international working group has met three times since its initial inception and continues to pursue a discussion of space heritage. If the SHTF and other workshops address the UN as an appropriate body, then productive approaches to that source need to be undertaken, with national representatives bringing up the matter to the UN. Finally there is a need to identify expert advisors to meet with officials and provide proposals for national governments to forward to relevant international officials. The treaty on outer space was a huge international effort that attempted to recognize and regulate activities in outer space and on other celestial bodies that enhanced the use of space for humankind. An international agreement on how to best preserve the heritage of outer space is the next step for humankind.

If, as McNutt sees (Chapter 44 of this volume)[72] in his excellent reflection on space history in the last 50 years, future space travel in the next 100 years is expensive and the provenance of government-sponsored missions for both manned and unmanned missions, then space travel may be stalled until resources are committed by international entities. McNutt finds that while commercialization of space may lead to lower costs, the markets for space are not yet clear and defined. His predictions for future space exploration, which will undoubtedly add to scientific knowledge in many areas, are in the more distant future than those of many others. Robotic missions will be selected first because of costs, and then humans will return to the Moon and/or Mars in 2030.[73] If these are fairly accurate predictions about a return to these celestial bodies and the trajectory of space exploration, there may be more time to consider feasible and prudent alternatives to preserving space heritage, but not much.

REFERENCES

1. Bell, T.E. 2008. Apollo Relic Reveals its Secrets. http://science.nasa.gov/headlines/y2008/20jun_apollorelic.htm?list.
2. *National Historic Preservation Act of 1966.* 17 U.S.C. 470 et seq., as amended.
3. Ibid., section 1(b)(4).
4. World Heritage Convention. http://www.whc.unesco.org/.
5. Spennemann, D.H.R. 2006. Out of This World: Issues of Managing Tourism and Humanity's Heritage on the Moon. *International Journal of Heritage Studies* 12: 356–371.

6. Spennemann, D.H.R. 2004. The Ethics of Treading on Neil Armstrong's Footprints. *Space Policy* 20(4): 279–290.
7. Gorman, A. 2005. The Cultural Landscape of Interplanetary Space. *Journal of Social Archaeology* 5: 85–107.
8. O'Leary, B. 2006. The Cultural Heritage of Space, the Moon and Other Celestial Bodies. *Antiquity* 80: 307. http://www.antiquity.ac.uk/projgall/oleary/index.html.
9. Gorman, A., and B. O'Leary. 2007. An Ideological Vacuum: The Cold War in Outer Space. In *A Fearsome Heritage: Diverse Legacies of the Cold War*, ed. J. Schofield and W. Cocroft. Walnut Creek, CA: Left Coast Press.
10. Report on NEAR and its landing on the asteroid Eros. http://near.jhuapl.edu/.
11. U.S. Army. 1962. A Quick Look at the Technical Results of Starfish Prime. Document released under the Freedom of Information Act. GetTRdoc.pdf
12. Space Weather News. 2008. Descending Space Junk. SpaceWeather.com, July 22, 2008. http://spaceweather.com.
13. Ibid.
14. NASA Management Instructions 1700.8: Policy for Limiting Orbital Debris Generation. NASA Safety Standards 1740.14: Guidelines and Assessment Procedures for Limiting Orbital Debris.
15. Gorman 2005 (note 7) (see specifically pp. 85, 89).
16. Operational Guidelines for the Implementation of the World Heritage Convention. 1998. Sections 35–42. http://whc.unesco.org/nwhc/pages/doc/main.htm.
17. Gorman 2005 (note 7).
18. Gorman and O'Leary 2007 (note 9).
19. Lunar Legacy Web site. http://spacegrant.nmsu.edu/lunarlegacies/.
20. Shanks, M., D. Platt, and W.L. Rathje. The Perfume of Garbage: Modernity and the Archaeological. *Moderism/Modernity* 11(1): 61–83.
21. National Register of Historic Places Criteria of Eligibility. 36 CFR Part 60.4
22. National Historic Landmark Criteria 36 CFR 65
23. Shafy, M. and W.R. Luce. 1998. *Guidelines for Evaluating and Nominating Properties That Have Achieved Significance within the Past Fifty Years.* National Register Bulletin 22. http://www/nps.gov/history/NR/publications/bulletin/nrb22.
24. Fewer, G. 2007. Conserving Space Heritage: The Case of Tranquility Base. *Journal of the British Interplanetary Society* 60: 3–8.
25. Rogers, T.F. 2004. Safeguarding Tranquility Base: Why the Earth's Moon Base Should Become a World Heritage Site. *Space Policy* 20(1): 5–6.
26. Spennemann 2004 (note 6).
27. Spennemann 2004 (note 6) (see specifically p. 283).
28. Stooke, P.J. 2008. Preserving Exploration Heritage in the Moon 2.0 Era. Poster presented at the NASA Lunar Science Conference, July 20–23, 2008, NASA Research Center, Moffat Field, CA.
29. Stooke, P.J. 2007. *The International Atlas of Lunar Exploration*. Cambridge, United Kingdom: Cambridge University Press.
30. Stooke 2008 (note 28).
31. Spennemann 2004 (note 6) (see specifically p. 281).
32. Barclay, R., and R. Brooks. 2002. In Situ Preservation of Historic Spacecraft. *Journal of the British Interplanetary Society* 55: 173–181.
33. Ibid.
34. Gorman 2005 (note 7).
35. ICOMOS Australia. 1999. Burra Charter. http://www.icomos.org/australia/.
36. Fifth World Archaeological Congress, Space Heritage Task Force. http://www.worldarchaeologicalcongress.org/site/active_spac.php.
37. Stooke 2007 (note 29) (see specifically p. 212).

38. Gibson, R. 2001. Lunar Archaeology: The Application of Federal Historic Preservation Law to the Site Where Humans First Set Foot upon the Moon. MA thesis, Department of Sociology and Anthropology, New Mexico State University, Las Cruces, NM.
39. Spennemann 2006 (note 5).
40. Chaikin, A. 1994. *A Man on the Moon*. Alexandria, VA: Time-Life Books.
41. Gorman 2005 (note 7).
42. Barclay and Brooks 2002 (note 32).
43. O'Leary 2006 (note 8).
44. Stooke 2008 (note 28).
45. Stooke 2008 (note 28).
46. Stooke 2008 (note 28).
47. O'Leary 2006 (note 8).
48. Capelotti, P.J. 2004. Space: The Final Archaeological Frontier. *Archaeology Magazine* 57(Nov–Dec): 48–55.
49. Stooke 2008 (note 28).
50. Fewer 2007 (note 24).
51. Rogers 2004 (note 25).
52. Spennemann 2006 (note 5).
53. Fewer 2007 (note 24).
54. Fewer 2007 (note 24).
55. O'Leary 2006 (note 8).
56. Rogers 2004 (note 25).
57. Fewer 2007 (note 24).
58. Spennemann 2004 (note 6) (see specifically p. 281).
59. Rogers 2004 (note 25).
60. Ibid.
61. UNESCO. Culture and UNESCO/Legal Protection and Heritage/Underwater Cultural Heritage. www.unesco.org/culture/legalprotection/water/html_eng/index_en.shtml.
62. Spennemann, D. 2007. Extreme Cultural Tourism From Antarctica to the Moon. *Annals of Tourism Research* 34(4): 898–918.
63. Ibid.
64. Fewer 2007 (note 24).
65. David, L. 2008. The Moon: An Archaeological Treasure Trove. http://www.spacecoalition.com/blog/1/2008/06/THE-MOON-AN-ARCHAEOLOGICAL-TREASURE-TROVE.html.
66. Billings, L. 2008. Preserving Tranquility. *Seed Magazine* August: 16.
67. Ibid.
68. Ibid.
69. Stooke 2008 (note 28).
70. Lunar Legacy Web site. http://spacegrant.nmsu.edu/lunarlegacies/.
71. Fifth World Archaeological Congress, Space Heritage Task Force. http://www.worldarchaeologicalcongress.org/site/active_spac.php.
72. McNutt, R.L., Jr. 2006. Solar System Exploration: A Vision for the Next 100 Years. *Johns Hopkins APL Technical Digest* 27(2): 306–319.
73. Ibid.

44 Space Exploration: The Next 100 Years

Ralph L. McNutt, Jr.

CONTENTS

INTRODUCTION: THE PERILS OF PREDICTION

> Alice laughed. "There's no use trying," she said: "one *can't* believe impossible things." "I daresay you haven't had much practice," said the Queen. "When I was your age, I always did it for half-an-hour a day. Why, sometimes I've believed as many as six impossible things before breakfast."[1]

Prognostication means (and conjures up) the ideas of forecasting and prophesying. When combined with technological subjects, such as the future of space exploration, the expositor is usually expected to pronounce the former—but experience shows the result to be closer to the latter. Projecting the future of technology was certainly a pursuit of Jules Verne, although his masterwork in this regard was not printed until well after his death.[2] His work, and that of others (Edward Bellamy, H.G. Wells, and Robert A. Heinlein,[3] in particular), was focused largely on the human element for which technology—and occasionally space travel—provided a backdrop. Hugo Gernsback had less interest in people and more in the things of the future, a shift that presaged "classic" science fiction.[4]

The Cold War led to the collaboration of artist Chesley Bonestell, Wernher von Braun, and Willey Ley on a variety of projects. The most notable ones led to a series of articles in *Collier's* magazine, as well as short programs on space travel made in conjunction with Walt Disney.[5] These popular outlets soon led to expectations of men in space, as well as connections between space and more down-to-earth products such as cigarette-brand advertising (Figure 44.1). The approach of illustration, prediction, and description expanded upon even more in the early 1960s although the purported date of 2015 for some predictions was optimistic even for the time.[6]

Von Braun made the first solid technical calculations for a human mission to Mars.[7] Based on the type of expedition that the U.S. Navy had conducted to Antarctica under Operation Highjump and transport issues during the Berlin Airlift,[8,9] Von Braun envisioned an undertaking the size of a "small war" that would take 70 men in seven ships to Mars and back (with three supply ships making a one-way voyage).

FIGURE 44.1 *Men into Space* was a weekly 30-minute television show. The show was sponsored by Lucky Strike cigarettes.

This was a scenario that would shrink as fiscal realities of such an undertaking sunk in over the years.[5]

The then-biochemist, science fiction, and general writer Isaac Asimov took up the specific task of predicting space developments early in the manned U.S. space program.[10] He noted that "an unmanned probe to Jupiter and beyond" was possible and that "by the year 2000, we might well have launched one or more probes to every one of the planets in the solar system. The results of these probes will not, however, be all known by then, for trips to the outer reaches of the solar system take a great deal of time. *Mariner IV* took more than eight months to reach the vicinity of Mars. If it were traveling to Pluto, many years would be required for the flight." (*New Horizons* launched on January 19, 2006, and will have its closest approach to Pluto on July 14, 2015.) Asimov also noted that robotic probes to other star systems were not probable: "We do not have the ability to explore far beyond the solar system [with a vehicle]. We may not have that ability for many centuries, if ever."

Turning to human flight, he noted that the effects of weightlessness and particle radiation—both high-energy particles from the Sun and galactic cosmic rays—were the real questions. Flights to the Moon would be possible, and he noted that a permanent lunar base might exist "by 1980 or 1985" (such a base had been discussed by Von Braun several years earlier in a now-declassified study for the U.S. Army called "Project Horizon"[11]). He notes that missions of months to Mars, Venus, and Mercury by humans should be feasible. He discusses implanting a robotic base on the asteroid Icarus as a means of getting close to the Sun. ("Any closer approach to the sun by man than Icarus would seem unlikely. Spaceships, manned or unmanned, could be made to skim about the sun at closer distances, but the heat and radiation would very probably be fatal not only to men, but also to instruments, unless they were particularly well-protected.")

Asimov notes that voyages that are years long will be required to reach the outer solar system, but that we may have landed humans on the largest asteroid, Ceres, "perhaps as early as 2000." Other asteroids could be stepping-stones. "Astronauts could tackle the outer planets one by one, establishing themselves firmly on one, then progressing to the next one." But without a "new kind of rocket … it may well be that man will never pass beyond the asteroids."

Extrapolating further, Asimov notes: "A generation later, say by 2025, we may well have landed on one or another of Jupiter's satellites. A century from now [i.e., 2066] a [human] landing may have been made within Saturn's satellite system, with plans in the making for reaching the satellites of Uranus and Neptune. By 2100, perhaps, man will stand on Pluto, at the very limits of the solar system." Little could Asimov have known—or suspected—that by 2008, a debate would be in progress about whether Pluto is a planet or a dwarf planet, like Ceres. Neither could he have known about our "new solar system," which includes a set of icy Kuiper-Belt Objects extending at least as far as the aphelion of the trans-Neptunian object 90377 (also known as Sedna), some 975 astronomical units from the Sun,[12] well into local interstellar space.

Asimov concludes with a discussion of "The fourth stage of space exploration—voyages lasting centuries" that would take humans, physically, to the stars. "Why bother? Well, nowhere in our solar system is there another planet on which man could live comfortably …. Nowhere else in the [solar] system, outside the earth, can there be anything more than primitive life-forms. Out there among the stars, however, there are sure to be other earthlike planets, which may very likely bear life. Some of them might even bear intelligent life. Unfortunately, we cannot detect them until spaceships can get fairly close to the stars that the planets circle." But here is a likely additional error made more than 40 years ago. With the possibility of space interferometry and large-scale items such as the NASA's *Terrestrial Planet Finder*, such observations will not require a crewed starship, although building such an observatory itself, say at the Sun-Earth L_2 point (see McNutt, Chapter 15 of this volume), is not a trivial undertaking. Asimov notes that even if we have reached Pluto with humans by 2100, an expedition to the stars by then is doubtful; however, travel to the stars may still, someday, be possible. Travel close to light speed to "slow" aging by time dilation, use of suspended animation, and "world-ships" that land the tenth generation of descendants are all noted as possibilities—and have all been discussed by a variety of writers of both science and science fiction. While the human exploration of the system could well be completed by 2100, Asimov concludes that "Those space feats which mankind will not have accomplished by 2100 (a landing on the giant planets, a very close approach to the sun, a voyage to the stars) may not actually be impossible, but they are so difficult that mankind may not even attempt them for many centuries after 2100" (Figure 44.2). Coming from a major science fiction writer, these words were sobering in 1966, and, given our lack of accomplishing "simple" things by 2008, such as a permanently staffed lunar base, they can only be more so today.

Similar conclusions regarding the need for world-ships to accomplish interstellar travel in the future have been reached by a variety of authors.[6,13,14,15] The eventual movement of humanity from a dying solar system has also been a favorite science

FIGURE 44.2 Isaac Asimov. (Courtesy of NASA.)

fiction topic, but it was also considered by Robert H. Goddard, the father of American rocketry, to be an ultimate requirement.[16]

Near-term programs were discussed by many of the contributors to hearings on future space programs held in 1975. Ehricke[17] outlined a plan of space industrialization through the rest of the twentieth century. Glaser[18] discussed establishment of a system of solar-powered satellites before the century was out. The most far-reaching proposal was that of Robert Forward for government funding of a program of manned interstellar travel and exploration,[19] commencing with definition studies in 1975 and launching the first crewed expedition in 2025. His program called for the launch of an Alpha Centauri probe in 1995 (first data return in 2015), followed by probes to the systems of Barnard's Star, Sirius, and Lalande 21185 in 2000 (first data return in 2020).[20] These hearings were held at the same time as the completion of NASA's *Outlook for Space* study[21] for the programs to conduct through the year 2000 (see Chapter 15 of this volume).

Funding requirements have always been a central issue for space exploration. For example, while the solar-powered satellite concept promised a return on investment, concepts such as a space colony that would eventually capitalize on such a venture were estimated as requiring $190B [fiscal year (FY) 1975 dollars] over a 22-year program. In addition to multiple new launch vehicle developments, the concept required a lunar material supply infrastructure and the transport of people and material to L_5 on more than 2,000 augmented-space-shuttle flights over a 22-year program.[22]

Predictions of possibilities a few decades away have typically been overly optimistic. In the fictional realm, Arthur C. Clarke's vision, in *2001: A Space Odyssey*, of space stations and Moon bases and trips to the Saturn system (or that of Jupiter in Stanley Kubrick's film adaptation) is well known.[23] The prediction by the Orion Project that Saturn's moon Enceladus could be reached by a human crew in 1975[19, 24, 25] is now seen as overly optimistic at best, but even more technically oriented predictions have fallen wide of the mark. In looking from 1982 to 2002 to 2022[26] at "what could happen" solar-powered satellites and explosive growth in communications channels led the list. Missions such as Mars rovers ("by mid-1990s") and landers on Titan ("mid-to-late 1990s") have come in about a decade late, while others such as a Mercury lander ("about 1993") fall wide of the mark. Space stations, orbital infrastructure, and next-generation space shuttles are all in the mix, as are single-stage-to-orbit reusable vehicles and heavy-lift launch vehicles. As with many such projections, these extrapolations were technology concept-limited and were not limited by capital investments or cash flow. "Futures that never were" ("paleo-futures") even now have their place on the Internet[27]

On the other hand, predictions that do not go so far into the future typically fall short of the mark. The ubiquitous use of personal computers, cell phones, and the great communications enabler, the Internet, were hinted at but not really predicted per se. Once again, Arthur C. Clarke, the "inventor" of the geosynchronous satellite concept, comes to mind with his vision of a not-so-future world linked via satellite in instant communication, including electronic newspapers, mail, and the Electronic Blackboard (which sounds like a primitive web browser).[28] The article describing this vision appeared in a magazine issue on the building of the Saturn V and the race to the Moon, with an artist's conception of a launch from Pad-39C, a launch pad never built (Figure 44.3).

FIGURE 44.3 Plans for going to the Moon. The Saturn V has just launched from pad 39C, which was never built. Launch pads 39B and 39A are south (toward the left) and were used for Apollo missions. They were then converted to use for space shuttle launches. (Courtesy of NASA.)

TECHNOLOGY VERSUS ECONOMY

With a fairly mature notion of what can be done in space—and at what cost (Chapter 15 of this volume)—we can make some predictions and, indeed, lay out a vision for the next century in space. The four ingredients are as follows[8, 9, 20]:

1. National policy/science: The case to go
2. Technology: The means to go
3. Strategy: The agreement to go
4. Programmatics: The funds to go

A well-thought-out approach with each key element is required to promote and accomplish a successful exploration plan. *National policy* includes the gamut of concepts, but national, including economic, advantage is always part of the mix. This was certainly the case for the Panama Canal, the Manhattan Project, and Project Apollo. Part of the project includes a means toward the needed technology, that is, the technology development comes along as part of the project. The case is perhaps more obvious for the Apollo and Manhattan efforts, but it is worth remembering that without understanding how to suppress yellow fever and malaria by controlling mosquitoes, the canal effort by the United States might have ended as unhappily as did that of the French before them. *Agreement* means that all of the potential stakeholders—real and imagined—must be sufficiently motivated not to stop the project even if they do not actively support it. That latter action maps onto *programmatics*, which means not only up-front monies, but also the tactical spending of those monies in order to meet the project requirements, including the timescale for accomplishment. Even the Manhattan Project did not have a totally blank check for expenditures, as the rest of the war effort also required funding, but it certainly is true that projects on a short financial leash will not fare as well as those projects that are not on such a leash. However, at the same time, there is nothing like a fixed budget to force program managers to focus on the task at hand. As with any human endeavor, the mix of requirements, resources, practicality, and motivation end up dictating the results, and efforts on the cutting edge of possibility do not come with a guarantee of success.

The use of low Earth orbit (LEO) for a variety of tasks—civil, military, and commercial, including the use of the International Space Station (ISS) and suborbital flights for tourism per se or rapid international transport—has been discussed for some time. While the advances in materials and manufacturing required to make these items economically viable ventures will help with other space initiatives, we do not concentrate on those items here; deep space (beyond Earth's magnetosphere—about 10 Earth radii in the sunward direction) have a very different set of economic and technological issues. Dealing with the latter are required to open a true space frontier.

The *Vision for Space Exploration*[30] has four items:

1. Implement a sustained and affordable human and robotic program to explore the solar system and beyond
2. Extend human presence across the solar system, starting with a human return to the Moon by the year 2020, in preparation for human exploration of Mars and other destinations
3. Develop the innovative technologies, knowledge, and infrastructures both to explore and to support decisions about the destinations for human exploration
4. Promote international and commercial participation in exploration to further U.S. scientific, security, and economic interests

The points are worded very carefully to be all-encompassing (strategy) and yet not dictate implementation (tactics). The key point is in 1: "a sustained and affordable human and robotic program." The one exception is the specification that a human return to the Moon is to occur by 2020. Cost is not otherwise mentioned, although "economic interests" are in point 4.

"Big Science" and Costs

Marburger has noted that human space flight and high-energy particle physics fall into the category of big science, which brings into play a dynamics all its own.[31] The current multibillion-dollar estimated costs for the ITER tokamak fusion experiment, as well as for NASA and NASA/European Space Agency outer-planet, robotic-mission projects puts these programs into that category as well.

In space, only a relatively small number of missions have gone beyond Earth orbit. Of these, only some of the Apollo missions have carried humans (McNutt, Chapter 15 of this volume). The use of humans "on the ground" is clear, the most obvious example being the selection of Moon rocks, as well as the mass of Moon rocks, delivered back to Earth by Apollo. There were 381 kg of Moon rocks divided into 2,196 samples returned as the result of twelve astronauts who spent a total of 81.1 hours outside of their lunar modules. The current Apollo cost (in FY2009 dollars) can be estimated as ~$110B (McNutt, Chapter 15 of this volume), or about $288K a gram or about $58,000 per carat, attributing the entire program cost to the return of the Moon rocks. For comparison, world mine production of gemstone-grade diamond for 2007 is estimated as 104 million carats (20,800 kg). If this is taken as the approximate world demand, then the U.S. demand was ~35% or 7,280 kg for $17.7B,[32] or ~$500 per carat. Larger diamonds have higher prices; a diamond solitaire in excess of 10 carats would start to approach the average effective cost of the Moon rocks. Put differently, the Apollo program cost about as much as 500 Hope Diamonds, or ~6 times the current yearly U.S. gemstone diamond market. Without the returns, one could estimate the time spent on the surface as $\$110B/(81.1 \times 2) = \$678M$ per person-hour on the lunar surface, amortized over the entire program.

One of the NASA New Frontiers missions[33] is suggested for returning lunar samples from the South Pole's Aitken Basin for a cost cap of ~$750M. The return of 1 kg of material would effectively cost $750K per gram or $150,000 per carat. Hence, the unit cost from Apollo is lower by a factor of ~2.6, but Apollo as a program cost more by a factor of ~147. Apollo also allowed for on-site selection and context characterization, significant issues for automated sample returns, a subject debated especially surrounding the return of a (very costly) sample from Mars. As an example, if $6B were required to design and implement a mission to bring back 3 kg of material from Mars, the implied cost would be $2M per gram or $400K per carat.

On a per-mass basis, human missions potentially provide a better "deal." For example, even a $1.2T human mission to Mars that brought back 1,200 kg of Mars rock samples would have an effective cost of $200K per carat. A return of 4,150 kg of Mars rocks would bring the sample cost per unit mass down to that of the Moon rocks.

A different item would be to keep humans close but "on site" in order to operate robots for doing sample gathering and investigations. This situation would be equivalent to human operators on Earth running lunar robots with a ~1.5-s time delay. Technology cannot overcome the delay in closed-loop operations that is enforced by the finite speed of light. Even the Earth-Moon delay is noticeable (it took ground controllers several tries—several missions—to track the ascent of the lunar module with the television camera on the lunar rovers that were parked nearby). This limitation has been felt even more closely by those operating remote equipment on Mars: the *Spirit* and *Opportunity* rovers and, more recently, the *Phoenix* sample-handling system. A human base on Deimos (close to synchronous orbit) could allow for human-in-the-loop operation without adding an excursion to the surface from an already hazardous mission to the Mars system. Such a philosophy has perhaps even more validity for telerobotic operations in outer planet systems,[34] although the lack of a landing would likely be frustrating for the crew, who might spend literally years in transit. A significant question for the next century is not whether detailed in situ exploration of the farthest reaches of the solar system make sense, along with the return of samples for detailed research, but whether sufficiently autonomous robots can be built such that multibillion-dollar space missions can—and will—be done "in the blind." The alternative is to send humans—or at least send them to be close enough for telerobotic operation. The typical prejudice is to send humans in any event, at least until there is a close look at the implied costs, as uncertain as they are.[35]

Stages in Exploration

People talk a lot about space exploration, but the underlying assumptions are worth making explicit, as different people (and sometimes the same people) mean different things. A scheme for integrating across these notions is shown in Figure 44.4. The items listed are representative but not inclusive. America (denoting the New World in the context of Europeans and not Native Americans) is included, as well as the continent of Antarctica. Planets (here including Pluto/Charon as well), moons (our Moon and Titan), and small bodies (asteroids and comets as classes of bodies rather than specific targets) are also called out.

Every item has been "discovered" (by definition), and an initial reconnaissance (by sailing ship for America and Antarctica) has been carried out for all—except the Pluto system, which will be visited by the *New Horizons* spacecraft in 2015. *MESSENGER* is on its way to orbit Mercury (2011), and *Rosetta* is on its way to orbit Comet 67P/Churyumov-Gerasimenko (2014). Properly speaking, no stage of exploration of either the Americas or Antarctica (or any other place on Earth) incorporated

Mercury	Venus	America	Antarctica	Moon	Asteroids	Comets	Mars	Jupiter	Saturn	Titan	Uranus	Neptune	Pluto/charon	Stage	Exploration is in the eye of the beholder Motives are many, but at some level tie to national objectives Stages range from discovery to new worlds
X	X	X	X	X	X	X	X	X	X	X	X	X	X	0	**Discovery**
X	X	X	X	X	X	X	X	X	X	X	X	X	~	1	**Reconnaissance - robotic flyby**
~	X	X	X	X	X	~	X	X	X	X				2	**Exploration - robotic orbiter**
		X	X	X	X	~	X			X				3	**Investigation - robotic lander**
		X	X	X			X							4	**Detailed investigation - robotic rover**
		X	X	X										5	**Human exploration - crewed orbiter**
		X	X	X										6	**Human investigation - crewed lander**
		X	X											7	**Human outpost with telepresence**
		X	X											8	**Human permanent base**
		X												9	**Human colony - dependent w/trade**
		X												10	**New world - self-sufficient w/trade**

FIGURE 44.4 The various stages of exploration. X, done; ~, in progress.

"robots," but there were very specialized initial human explorers. By the time one gets to mobile robots in the figure, only Mars (*Spirit* and *Opportunity*) and the Moon (the Soviet Lunokhods) qualify. Even Antarctica, with a permanent human presence only since 1957, does not engage in trade of any sort—unless knowledge qualifies—and is totally dependent upon expendables (e.g., food and fuel for power) being shipped in.

Robotic Missions

Robotic exploration across the solar system is something that NASA knows how to do; however, frustration at transit times to targets beyond the asteroid belt continues. The Prometheus/Jupiter Icy Moons Orbiter exercise, while expensive (about the cost of a Discovery mission) taught a valuable lesson about scalability issues with currently conceived approaches to nuclear electric propulsion (McNutt, Chapter 15 of this volume).

Cost overruns have been, and continue to be, an issue.[36,37,38] Typically, the only remedies are to absorb unanticipated costs by slipping other initiatives out into the future. An approach to dealing with this has been to provide detailed mission studies that are then independently reviewed in the same fashion as the completed Discovery, Mars Scouts, Explorer, and New Frontiers missions. This approach has been applied recently to four outer planet missions.[39]

Current robotic planning by NASA for solar system exploration[40] shows a decision to launch a Europa orbiter to be made in 2008 with a launch in 2015 followed by a new mission to Titan in 2020, a rover to Venus in 2025, and then either a lander to Europa or a large orbital mission to the Neptune system, with a focus on Neptune's large moon, Triton, to launch by 2030. (In early 2009, a decision was made to plan for a Europa orbiter mission to launch in 2020.) While technically feasible, such a launch rate of multibillion-dollar robotic missions would require a higher level of funding than has been present in the immediate past.[41] This type of approach has diverted from the older NASA science strategy of (1) reconnaissance (i.e., robotic flyby), (2) exploration (i.e., robotic orbiter), and (3) intensive study (i.e., robotic multiple probes, landers, rovers).[42]

For the longer goals of the next century, autonomous and/or teleoperated probes and rovers will be just the start if science continues to be a driver. Answering ultimate questions of the suitability of *any* off-Earth environment in the solar system to support life in the past, present, or future will require the return of appropriate samples to Earth. The less ambitious goal of establishing an accurate, absolute chronology of solar system beginnings and evolution to the present will also require well-chosen samples from a variety of solar system locations.

Human Missions to Deep Space

Human missions to the Moon are already difficult. Missions to Mars have been studied and debated for more than 50 years,[5] and, by most accounts, will cost multiples of what Apollo cost (Von Braun initially characterized the cost as that of "a small war"[7]). While human flyby missions of Venus were considered as part of the return from Mars in the Early Manned Planetary-Interplanetary Roundtrip Expedition (EMPIRE) study of the 1960s,[43, 44] human landings on Venus have not even been mentioned since the initial findings on the surface by *Mariner II* in 1962. A human mission to Mercury has been mentioned, but only in passing,[45] and before it was realized that Mercury is not tidally locked to synchronous rotation.

Human missions to orbit Venus and Mercury are more difficult than to Mars and Jupiter, respectively, because of the thermal environment and, in the case of Mercury, the higher speeds associated with the location in the Sun's gravity field. The thermal problem can be solved with mass and power and the rendezvous change in velocity (or "delta-vee," Δv) with a sufficiently advanced propulsion system. With the idea that the ultimate goals of space exploration are outward, it makes sense to inquire what is required to send humans to the Jovian system and beyond.

An integrated strategy of robotic sample-return/assay missions followed by human crews would be required for the intensive study of the solar system. An ARchitecture for Going to the Outer solar SYstem (ARGOSY) would use an incremental approach to reach throughout the solar system by the end of the twenty-first century (Table 44.1).[34, 46]

The possibility of a Mars sample return as early as 2015 is already out of date. While there has been recent discussion within NASA of a Mars sample return mission as early as the beginning of the 2020s, current funding levels for the required technology development may also render this time frame overly optimistic. A human return to the Moon by 2020 remains an operative part of the vision. Reaching the Sun-Earth second Lagrange point (L_2) and an appropriate near-Earth object for

TABLE 44.1
A Timetable for Human Exploration of the Outer Solar System

Year	Goal
2015	Mars sample return
2020	Human launch to L2: servicing mission; human launch to a near-Earth object
2025	Human return to the Moon
2030	Permanently staffed lunar base; human mission to Mars
2035	Commence robotic sample—return missions to the outer solar system: ARGOSY-R
2045	Sample return from Jupiter system
2050	Human mission to Callisto: ARGOSY I
2055	Sample return from Saturn system
2065	Sample return from Uranus system
2070	Sample return from Neptune system
2075	Human mission to Enceladus: ARGOSY II
2080	Permanently staffed Mars base
2085	Humans to Miranda (Uranus system): ARGOSY III
2090	Humans to Triton (Neptune system): ARGOSY IV
2095	Sample return from Pluto/Charon system
2110	Human mission reaches Pluto before its aphelion: ARGOSY V
2110+	Permanent human bases in outer solar system: post-ARGOSY
Twenty-second century	Systemwide commerce

maintaining telescope infrastructure and as a dress rehearsal for Mars, respectively, are also possible.[47]

Certainly, the technologies required for reaching Mars and returning to Earth, as well as operating a permanently staffed base on the Moon, will be prerequisites for considering a farther journey. Whether these events occur by 2030 or later in that decade[48] do not matter as much as it matters that they happen. Given the issues with the testing of nuclear thermal rocket (NTR) engines[49] and the likely need for their inclusion in missions to Mars and beyond,[50] the best reason for using the Moon as a stepping stone to Mars is likely as a test and development site for fully qualified interplanetary NTR engines.

Requirements

> There is almost universal agreement among astronautical engineers that planetary bases must be preceded in time by at least three other developments: (1) nuclear propulsion or other equally sophisticated propulsion systems for spacecraft; (2) a vast increase in basic knowledge of the planets obtained by unmanned probes and by observations from manned spacecraft and the lunar surface; and (3) fully reliable and efficient closed ecological systems capable of supporting life during long space voyages and in the planetary environments.[51]

While we have made progress on (2) since this was written 47 years ago, and we have learned to what detailed level such knowledge is needed, the other two points have eluded us, and the most vexing issues, sustained weightlessness, shelter from solar energetic particles, and sustained exposure to galactic cosmic rays (GCRs) were not appreciated at all. While schemes for "artificial gravity" by spinning parts of a large interplanetary transport vehicle, and the use of "storm cellars" for solar energetic particles (assuming sufficient warning) can provide mitigation, the GCR problem likely provides the ultimate limit (but not far behind the problems of expendables and isolation) to deep-space travel. Recycling of water and oxygen on the ISS is standard, but the current implementation is a long way from a closed system. A closed ecosystem is still a technology awaiting development. While short trips to the Moon or even month-long trips to L_2 do not require such systems, maintaining a human crew to Mars and beyond likely will. While foodstuffs could be stocked for voyages lasting 3–5 years, shelf life for that amount of time for a sufficient assortment of food may well be problematic.[52]

Ultimately, the initial missions will be limited by the GCR background to about 5 years, including the round-trip transit time and whatever time is spent at the target or in the target system.[35] While most requirements have been looked at from the perspective of human missions to Mars, missions even farther afield will have the same requirements for oxygen, potable water, and food. The "requirements" vary a lot—between ~3 and 7 kg person^{-1} day^{-1} (kppd) (with the high side being close to that of the ISS).[35] An in-depth study concluded that 4.98 kppd is the correct number[53] and is certainly within that range. To supply a crew of six for 5 years would take 54.5 metric tons (mt; 1 mt = 1,000 kg) of supplies, and it would take 90.9 mt for a crew of 10.

Similar types of zeroth-order considerations for a minimal crew of six with minimal shielding of ~20 g cm^{-2} yield a living section ~110 mt and a need of ~50 mt of expendable supplies for a 5-year mission. As an example, an interplanetary transport including all other subsystems as well (tankage, engines, power system, etc.) could conceivably be about the final mass of the ISS, or ~420 mt. A mass ratio of ~4 (initial propellant mass fraction of 75%), a nuclear system with a specific mass of the power system of ~0.1 kg/kWe, and a continuous low-thrust system with a specific impulse of ~20,000 s would enable a 2-year crossing to Neptune. If the ~1,300 mt of propellant could not be acquired in the Neptune system, or pre-emplaced, then that mass would have to be transported outbound as well. This is by far the most difficult mission (getting to Pluto is similar), but it does set the scale—and technology challenges—for what would be required.[35]

Launch Vehicles and Launch Sites

Assuming that the Neptune example is similar to other, closer excursions in the outer solar system (same flight time with larger power plant specific masses at lower technology levels), the requirements for the initial mass in low Earth orbit (IMLEO) are set. For initial human voyages, it would be unwise to rely on "fueling up" at the destination for the return. The ship plus propellant for return then imply leaving Earth obit with an initial mass ~6,800 mt (= 4 × 1,700 mt). This mass could be decreased to ~2,000 mt total if the specific impulse could be doubled, an illustration of the trades between difficult technological advances and mass on orbit. Such large masses on orbit require new heavy lift launch vehicles or extremely heavy lift launch vehicles that are in the previous class of Nova vehicles or well beyond.[34] It is generally agreed that assembly in orbit should be kept to a minimum if costs are to be contained,[54] and this has been made even more explicit by the assembly of the ISS.[55]

Capabilities for payloads of up to 1,000 mt have been studied in some detail. Most studies focused on the Nova vehicle that could take "a million pounds"—that is, some 450 mt—to LEO. Needs for new engines, dealing with liquid hydrogen boil-off during fueling, manufacturing challenges, and reusability were all studied.[45, 56, 57] The need for even larger vehicles—"Supernovas"—was noted early on,[58] but "the main problem is not the lack of appropriate technology or financial resources, but the lack of a program deemed socially and politically desirable!"[54] NASA planning documents from the early 1960s had left room at Cape Canaveral for three Saturn V, two "advanced Saturn," and three (some references have four) Nova launch sites, as well as possibilities for an additional vehicle assembly building with facilities for assembling rockets with nuclear upper stages[34] Figure 44.5).

COSTS AND PROFITS

Costs

No human interplanetary mission will be cheap. A round trip to Mars will take over 3 years if a conjunction-class mission is chosen and almost 2 years with an opposition-class mission. Both trajectories are ballistic, but the latter requires far more

FIGURE 44.5 The Kennedy Space Center as it currently appears (left) and with the addition of other launch pads and complexes contemplated in the early 1960s, including pad 39C, two launch pads for advanced Saturn vehicles and Nova vehicles (right). Physical spacing was determined by the likely area of destruction in a worst-case explosion scenario. (From Google Earth. With permission.)

propellant, especially if NTR engines are not used.[59] Predicted costs were large: the Paine report presented a \$700B plan to the Reagan administration; the Space Exploration Initiative was costed at over \$500B in the early 1990s.[5] Neither was implemented because of the large associated costs. Early cost estimates[60] all assumed that per-mass costs would decrease dramatically with time. However, access-to-space costs have remained high, with reported values ~\$20,000 per kg. One "sanity check" is that the completed ISS will have a mass of ~420,000 kg, which would suggest a cost of \$8.4B, low by an order of magnitude for the assembled station. The 6,000 mt on orbit for the ARGOSY expeditions at \$200,000 per kilogram would suggest an assembled cost ~\$1.2T ($= \$1.2 \times 10^{12}$).

Were we to build a vehicle to travel to Mars and back—and be reusable—the total Δv required would be on the order of 10 km/s. An all-chemical system with liquid oxygen and liquid hydrogen has a specific impulse ~450 s, and that for an NTR system using liquid hydrogen (a system that does not exist but which we know how to build) is ~900 s. The IMLEO for a 420 mt Mars ship would then be ~4,000 mt for chemical and ~1,300 mt, respectively, or ~\$800B and ~\$260B for chemical and nuclear propulsion—and these are for the low-energy, conjunction-class missions. Both examples assume that we can solve the problem of long-term storage of massive amounts of liquid hydrogen, a nontrivial task. This illustrates the problem of cost. With respect to mass, at 20 mt of payload a flight, it would take 65 space shuttle flights or about 12 Saturn V flights to deliver the mass needed on orbit using the nuclear approach.

From the beginning of the Space Age through its first 50 years, we have delivered about 29,000 mt of material to Earth orbit (or beyond). At an effective cost of

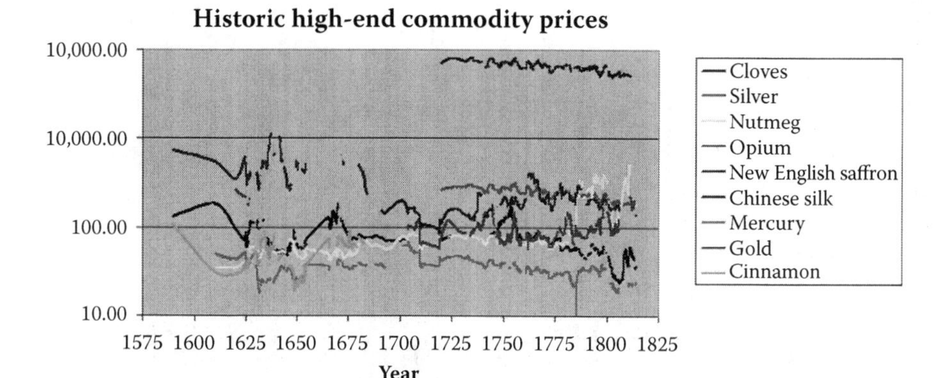

FIGURE 44.6 Historic commodity prices in 2002 euros from the late sixteenth century to the early nineteenth century, from a variety of sources. (From Google Earth. With permission.)

~$200,000 per kilogram, this amounts to an overall worldwide investment of some $5.8T in space, ~$120B per year. Could we have done this more cheaply? The likely answer is no. Could we have dropped the price to orbit? Here the answer (to the correct question) is likely yes—but only to the extent that the expenditures had taken place over a shorter period of time.

Profits

Proponents of space exploration like to call attention to the Age of Exploration as an example of a human imperative to explore. However, study of that age and its imperatives reveals that profits above all else, including loss of crew and sailing stock, were what drove Western European trade and exploration.[8, 9] The analogy of returned samples, in cost, to gemstones has already been mentioned. While spices were precious in the Roman period, the "Dark Ages," and the Middle Ages, they were never that precious. Tracing prices is difficult because of the absence of much quantitative data in the earlier historic record.

First-century Roman gold traded pound-for-pound for Chinese silk, and Pliny the Elder recorded and commented on spice prices, for example, 15 Roman denarii for 12 oz. of long pepper (about $26 an ounce, or just over a day's wages for a laborer[61]).[62] The preciousness of spices was well known in later years, but quantitative prices, with the exception of the Edict of Diocletian in AD 301, are lacking until about the thirteenth century,[63] and only later is there a fairly continuous record (Figure 44.6).

This situation can be summed up succinctly as "the cloves problem": that is, we have yet to identify any item that can be returned from deep space that has sufficient value on a commercial market to pay for the trouble of retrieving it.[35*]

* The only ship of Magellan's expedition around the world that returned to Spain was the *Victoria*, an 85-ton ship carrying 27,300 kg of cloves in 381 bags. The 18 survivors under the command of Juan Sebastían de Elcarno brought back sufficient cargo to make up for all of the losses, including four of the five ships, a pension for the pilot, advances to the crew, and back pay for the crew themselves for the 3-year expedition with a remainder to return a net profit.[64]

FIGURE 44.7 McMurdo Station near the Antarctic coast. (From Google Earth. With permission.)

INTERNATIONAL PLANETARY YEAR

As important as the International Geophysical Year (IGY) was for initiating the Space Age, it was also the central event that led to the permanent human occupation of Antarctica and the South Pole.[65] The Antarctic analogy for stations on the Moon and Mars is well known.[35, 50, 66, 67] What is less appreciated is the effort that went into setting up first the logistics base at McMurdo, and then the Amundsen-Scott base at the South Pole, two of the many bases that not only supported the IGY but have also remained permanently inhabited to this day.

The logistics are not trivial. McMurdo, still the primary staging area for the continent (Figure 44.7) was established during the austral summer of 1955/1956 with 1,800 men and ~23,000 m^3 of cargo. The following year, 64 aircraft sorties dropped 730 tons of supplies that were used to build the first South Pole station. Eighteen men wintered over the first year; it estimated that the cost in FY 2006 dollars was about $7.2M per man. The Scott expedition of 1911 spent about one-tenth this amount—with no survivors.

On the other hand, all of the targets for (initial) human landings in the outer solar system are smaller than Earth, with "interesting areas" on the order of the size of Antarctica or less (Figure 44.8). From 1980 through the present (currently available database), the Antarctic budget has run ~1.5% of the NASA budget or ~0.07% of the Department of Defense budget. The NASA budget is now ~$50 per person per year, and the corresponding expenditure is ~90¢ per person per year for the Antarctic program.

CONCLUSION

There is a great deal of knowledge to be gleaned from in situ exploration of our solar system. The history of our beginnings is certainly there. Whether "cousins" or others did, do, or may eventually exist off Earth will require even closer looks. In the same way that the IGY opened up the last doors on Earth in 1957, an International Planetary Year in 2107, with human visits to all of the outer planets, would open the

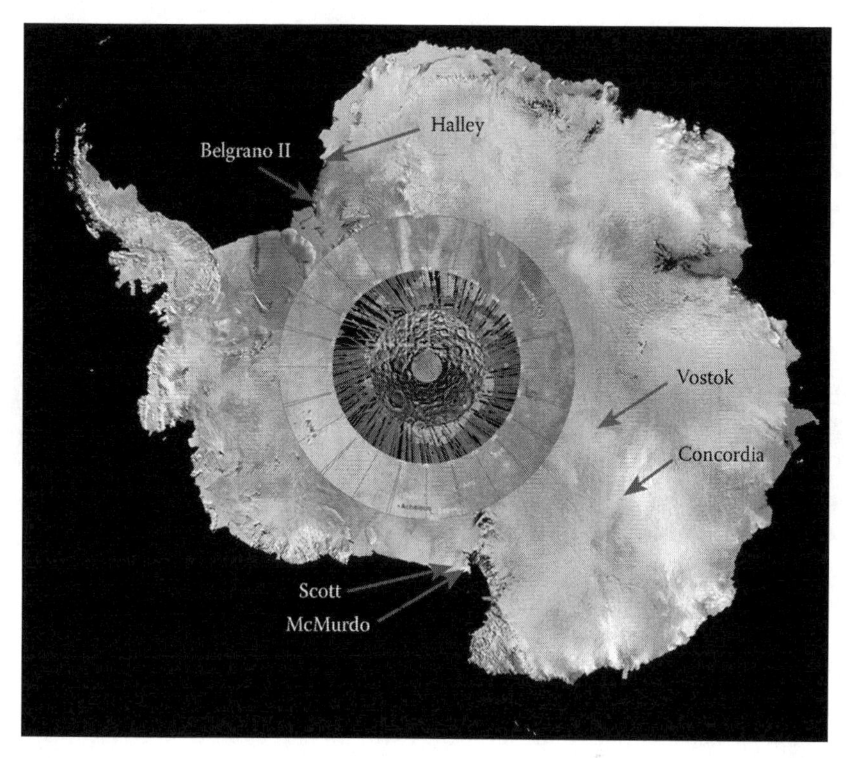

FIGURE 44.8 A composite photograph of Antarctica showing various past and present bases. Centered on the south pole are polar projections of Ganymede from 90° N to 60° N, Mars down to 75° N, the Moon from 90° S to 70° S, and Enceladus from 90° S to 60° S; Miranda (not shown) and Enceladus are about the same size. (http://commons.wikimedia.org/wiki/Image:Antarctica 6400px from Blue Marble.jpg).

last doors of our own planetary system. Such an undertaking would be nontrivial and, as with the IGY, would require a truly international effort. The price for success would be significant, certainly a large fraction of the current world gross domestic product. At the same time, the price for failure could be larger: what the world missed from a continuance of sailings of the treasure fleets under the early Ming dynasty will never be known.[68]

While the establishment of trade per se, including traffic in profitable commodities, across the solar system cannot conclusively be ruled out, its use as a motivator for solar system exploration is unlikely. Access-to-space costs are, and will likely remain, high. A decrease in the cost of transport to orbit will occur because the net tonnage increases, rather than there being some technological breakthrough. At the same time, such a breakthrough in propulsion will be required for missions that are sufficiently rapid for human excursions to the far reaches of the solar system in as little as 5 years' round trip.

Projected power densities are simply too small; a paradigm shift similar to that from steam-powered to gasoline-powered airplanes will be necessary. The massive

FIGURE 44.9 An undistorted view of the Venus landscape. (From http://www.mentallandscape.com. With permission from Don Mitchell.)

requirements for shielding against GCRs to some level approaching that on Earth will likely never be implemented on human craft; a round-trip time of more than ~5 years begins to have a dubious value in any case.

Humans can reach Mars (and return), but it will not be a cheap undertaking. Human excursions farther away, in reality, will await the outcome of that first trip and a frank assessment of its cost and difficulty. Whether humans will actually stand on Venus (or Mercury) in the twenty-first century (Figure 44.9) seems doubtful. Telepresence from orbiting stations may make sense, but more reliance on more autonomous robotic craft will require a great deal of the human/robotic mix in space exploration in any event.

REFERENCES

1. Carroll, L. 2003. *Alice's Adventures in Wonderland and Through the Looking-Glass and What Alice Found There*. New York: Barnes and Noble Classics (cited quotation can be found on page 207). *Through the Looking-Glass and What Alice Found There* was first published in 1871 by Macmillan and Co.
2. Verne, J. 1997. *Paris au XXe siècle*. [Paris in the Twentieth Century], Richard Howard, transl. First English ed. New York: Ballantine Books (222 pp.).
3. Heinlein, R.A. 2004. *For Us, The Living: A Comedy of Customs*. New York: Scribner (263 pp.).
4. Gernsback, H. 1958. *Ralph 124C41+*. 2nd ed. New York: Fawcett World Library (see 142 pp.).
5. Portree, D.S.F. 2001. *Humans to Mars: Fifty Years of Mission Planning, 1950–2000*. Monographs in Aerospace History #21. NASA SP-2001-4521. Washington, DC: NASA History Division (152 pp.).
6. Cole, D.M. 1965. *Beyond Tomorrow: The Next 50 Years in Space*. 1st ed. Amherst, WI: Amherst Press (168 pp.).
7. Von Braun, W. 1991. *The Mars Project*. 1991, Urbana, IL: University of Illinois Press (91 pp.).

8. McNutt, R.L., Jr. 2004. Solar System Exploration: A Vision for the Next Hundred Years. In *55th International Astronautical Congress*, paper IAC-04-IAA.3.8.1.02, October 4–8, 2004, Vancouver, Canada.
9. McNutt, R.L., Jr. 2006. Solar System Exploration: A Vision for the Next 100 Years. *Johns Hopkins APL Technical Digest* 27(2): 168–181.
10. Asimov, I. 1966. How Far Will We Go in Space? In *The 1966 World Book Year Book*, 148–163. Chicago, IL: Field Enterprises Educational Corporation.
11. U.S. Army. 1959. Project Horizon Report: A U.S. Army Study for the Establishment of a Lunar Outpost. In *Encyclopedia Astronautica*.
12. Sedna, W. 2008. 90377 Sedna. Wikipedia. http://en.wikipedia.org/wiki/90377_Sedna.
13. Bernal, J.D. 1969. *The World, the Flesh, and the Devil: An Enquiry into the Future of the Three Enemies of the Soul*. 2nd ed. Bloomington, IN: Indiana University Press.
14. Heinlein, R.A., *Orphans of the Sky*. 1965, New York: Signet Books.
15. Aldiss, B. 1960. *Starship*. New York: Signet Books.
16. Goddard, R.H. 1983. The Ultimate Migration. *Journal of the British Interplanetary Society* 36: 552–554.
17. Ehricke, K.A. 1975. Space Industrial Productivity: New Options for the Future. In *Subcommittee on Space Science and Applications of the Committee on Science and Technology, U.S. House of Representatives*, Vol. II: 59–245. Washington DC: U.S. Government Printing Office.
18. Glaser, P.E. 1975. The Satellite Solar Power Station—A Focus for Future Space Activities. In *Subcommittee on Space Science and Applications of the Committee on Science and Technology, U.S. House of Representatives*, Vol. II: 431–462. Washington DC: U.S. Government Printing Office.
19. Forward, R.L. 1975. A National Space Program for Interstellar Exploration. In *Subcommittee on Space Science and Applications of the Committee on Science and Technology, U.S. House of Representatives*, Vol. II: 279–387. Washington DC: U.S. Government Printing Office.
20. McNutt, R.L., Jr. 2006. Space Exploration in the 21st Century. In *10th International Workshop on Combustion and Propulsion: In-Space Propulsion*. 30-1 = 30-15, de Luca, L. T., Sackheim, R. L., and Palaszewski, B. A., editors. Lerici, La Spezia, Italy: grafiche g. s. s., Arzago d'Adda (BG) Italy.
21. Allen, J.P., J.L. Baker, T.C. Bannister, J. Billingham, A.B. Chambers, P.B. Culbertson, W.D. Erickson, J. Hamel, D.P Hearth, J.O. Kerwin, et al. 1976. Outlook for Space: Report to the NASA Administrator by the Outlook for Space Study Group. In *NASA SP-386,* ed. D.P. Heath. Washington, DC: Technical Information Office, NASA.
22. Johnson, R.D., and C. Holbrow, 1977. *Space Settlements: A Design Study*. NASA SP-413. NASA Special Publications. Washington, DC: Scientific and Technical Information Office, NASA.
23. Clarke, A.C. 1968. *2001: A Space Odyssey*. New York: New American Library, Inc. (222 pp.).
24. Dyson, F.J. 1965. Death of a Project. *Science* 149: 141–144.
25. Dyson, G. 2002. *Project Orion: The True Story of the Atomic Spaceship*. New York: Henry Holt and Company.
26. Brodsky, R.F., and B.G. Morais. 1982. Space 2020: The Technology, the Missions Likely 20–40 Years from Now. *Astronautics & Aeronautics* 20(4): 54–73.
27. Novak, M. 2008. *Paleo-Future: A Look into the Future That Never Was*. http://www.paleofuture.com.
28. Clarke, A.C. 1964. Everybody in Instant Touch. *Life* 118–131.
29. Brooks, C.G., J.M. Grimwood, and L.S. Swenson. 1979. Apollo 6: Saturn V's Shaky Dress Rehearsal. In *Chariots for Apollo: A History of Manned Lunar Spacecraft*. NASA SP-4205. Washington, DC: U.S. Government Printing Office.

30. NASA. 2004. *The Vision for Space Exploration*. Washington, DC: NASA (32 pp.).
31. Marburger, J. H., III. 2006. Science and Government. *Physics Today* 38–42.
32. U.S. Geological Survey. 2008. Gemstones. http://minerals.usgs.gov/minerals/pubs/commodity/gemstones.
33. National Research Council. 2008. *Opening New Frontiers in Space: Choices for the Next New Frontiers Announcement of Opportunity*. Washington, DC: National Academies Press (see specifically p. 58).
34. McNutt, R.L., Jr. 2007. ARGOSY: ARchitecture for Going to the Outer solar SYstem. *Johns Hopkins APL Technical Digest* 27(3): 261–273.
35. McNutt, R.L., Jr., J. Horsewood, and D.I. Fiehler. 2007. Human Missions throughout the Solar System: Requirements and Implementations. In *58th International Astronautical Congress* Paper IAC-07-D3.1.05, Hyderabad, India.
36. National Research Council. 2006. *Principal-Investigator-Led Missions in the Space Sciences*. Washington, DC: The National Academies Press (see specifically p. 120).
37. United States General Accounting Office. 2004. *NASA: Lack of Disciplined Cost-Estimating Processes Hinders Effective Program Management*. United States General Accounting Office (71 pp.).
38. Congressional Budget Office. 2004. *A Budgetary Analysis of NASA's New Vision for Space Exploration* (xvii+38 pp.). http://www.cbo.gov/ftpdocs/57xx/doc5772/09-02-NASA.pdf.
39. Jet Propulsion Laboratory. *Outer Planet Flagship Mission*, NASA Final and Joint Summary Reports/ESA Assessment Studies, http://opfm.jpl.nasa.gov.
40. National Aeronautics and Space Administration. 2006. *Solar System Exploration*. JPL 400-1292. Pasadena, CA: Jet Propulsion Laboratory (vi+142 pp).
41. National Research Council. 2008. *Grading NASA's Solar System Exploration Program: A Midterm Review*. Washington, DC: Space Studies Board, Commission on Physical Sciences, Mathematics, and Applications, National Research Council of the National Academies (xii+88 pp.).
42. National Aeronautics and Space Administration. 1991. *Solar System Exploration Division Strategic Plan*, 6 Vols., Washington, DC.
43. Ordway, F.I., M.R. Sharpe, and R.C. Wakeford. 1993. EMPIRE: Early Manned Planetary-Interplanetary Roundtrip Expeditions, Part I: Aeroneutronic and General Dynamics Studies. *Journal of the British Interplanetary Society* 46: 179–190.
44. Ordway, F.I., M.R. Sharpe, and R.C. Wakeford. 1994. EMPIRE: Early Manned Planetary-Interplanetary Roundtrip Expeditions, Part II: Lockheed Missiles and Space Studies. *Journal of the British Interplanetary Society* 47: 181–190.
45. Scala, K.J., and G.E. Swanson. 1992. They Might Be … Giants: A History of Project NOVA 1959–1964, Part I. *Quest* 1(3): 12–27.
46. McNutt, R.L., Jr. 2006. ARGOSY—ARchitecture for Going to the Outer solar SYstem. In *57th International Astronautical Congress*, Paper IAC-06-D3.1.5, Valencia, Spain.
47. Farquhar, R.W., D.W. Dunham, Y. Guo, and J.V. McAdams. 2004. Utilization of Libration Points for Human Exploration in the Sun–Earth–Moon System and Beyond. *Acta Astronautica* 55(3–9): August–November, 687–700. New Opportunities for Space. Selected Proceedings of the 54th International Astronautical Federation Congress.
48. Cockrell, C.S., ed. 2006. *Project Boreas: A Station for the Martian Geographic North Pole*. London: The British Interplanetary Society (192pp.).
49. Gunn, S. 1992. The Case for Nuclear Propulsion. *Threshold* 9: 2–11.
50. McNutt, R.L., Jr., S.V. Gunn, R.L. Sackheim, V.V. Siniavskiy, M. Mulas, and B.A. Palaszewski. 2006. Propulsion for Manned Mars Missions: Roundtable 3, Paper 33. In *10th International Workshop on Combustion and Propulsion: In-Space Propulsion* L.T. de Luca, R.L. Sackheim, and B.A. Palaszewski, eds. Lerici, La Spezia, Italy: grafiche g. s. s., Arzago d'Adda (BG) Italy.

51. Lowe, H.N., Jr. 1961. §28.5 Construction of Extraterrestrial Bases. In *Handbook of Astronautical Engineering*, ed. H.H. Koelle. New York: McGraw-Hill Book Company, Inc. (pp. 28.128–28.134).
52. Stilwell, D., R. Boutros, and J.H. Connolly. 1999. Crew Accommodations. In *Human Spaceflight: Mission Analysis and Design*, ed. W.J. Larson and L.K. Pranke, 575–606. New York: The McGraw-Hill Companies, Inc.
53. Reed, R.D., and G.R. Coulter. 1999. Physiology of Spaceflight. In Larson and Pranke 1999 (note 52), 103–132.
54. Koelle, H.H. 2001. *Nova and Beyond: A Review of Heavy Lift Launch Vehicle Concepts in the Post-Saturn Class*, Report ILR Mitt. 352. Berlin, Germany: Technical University Berlin, Aerospace Institute.
55. Griffin, M.D. 2008. The Constellation Architecture: Remarks to the Space Transportation Association, January 22, 2008. Published remarks released by NASA Headquarters. http://www.spaceref.com/news/viewsr.html?pid=26756.
56. Scala, K.J., and G.E. Swanson. 1993. They Might Be ... Giants: A History of Project NOVA 1959—1964, Part II. *Quest* 2(1): 16–21, 24–36.
57. Scala, K.J., and G.E. Swanson. 1993. They Might Be ... Giants: A History of Project NOVA 1959—1964, Part III. *Quest* 2(2): 4–20.
58. Von Braun, W. 1962. *Appendix B: Manned Lunar Landing Program Mode Comparison*. In NASA Technical Memorandum TM-74929, July 30, 1962. Washington, DC: NASA.
59. Wickenheiser, T.J., K.S. Gessner, and S.W. Alexander. 1991. *Performance Impact on NTR Propulsion of Piloted Mars Missions with Short Transit Times*. Paper AIAA 91-3401.
60. Koelle, H.H. and W.G. Huber. 1961. §1.9 Economy of Space Flight. In *Handbook of Astronautical Engineering*, ed. H.H. Koelle, pp. 1-50–1-82. New York: McGraw-Hill Book Company, Inc.
61. Denarius. Wikipedia. http://en.wikipedia.org/wiki/Denarius.
62. Miller, J.I. 1998. *The Spice Trade of the Roman Empire 29 B.C. to A.D. 641*. Oxford: The Clarendon Press (294 pp.).
63. Rogers, J.E.T. 1866, 1882, 1887, 1902. *A History of Agriculture and Prices in England: From the Year after the Oxford Parliament (1259) to the Commencement of the Continental War (1793)*. In 7 Vols. Oxford, United Kingdom: The Clarendon Press.
64. Turner, J. 2004. *Spice: The History of a Temptation*. New York: Alfred A. Knopf (352 pp.).
65. Belanger, D.O. 2006. *Deep Freeze: The United States, the International Geophysical Year, and the Origins of Antarctica's Age of Science*. Boulder, CO: University Press of Colorado. (476 pp.).
66. Ardanuy, P.E., C. Otto, J. Head, N. Powell, B.K. Grant, and T. Howard. 2005. Telepresence Enabling Human and Robotic Space Exploration and Discovery: Antarctic Lessons Learned. Paper AIAA2005-6756 in *Space 2005*, August 30–September 1. Long Beach, CA: AIAA.
67. Boehne, R.T., J. Wright, S. Dunbar, T. Howard, and P. Ardanuy. 2005. Logistics Vision for Space Exploration: Spiral Development Processes Applied to Antarctic Field Sciences and Traverses. Paper 2005-6755 in *Space 2005*, August 30–September 1. Long Beach, CA: AIAA.
68. Ming Dynasty. Wikipedia. http://en.wikipedia.org/wiki/Ming_Dynasty.

45 Surveying Fermi's Paradox, Mapping Dyson's Sphere: Approaches to Archaeological Field Research in Space

P. J. Capelotti

CONTENTS

INTRODUCTION

Aerospace archaeology and cultural resource management conducted on the Moon (see Chapter 21) will form the methodological and theoretical basis for the survey and stabilization of similar sites in the solar system, as well as the search for potential signatures of intelligent life throughout the universe. In this chapter, we examine larger related points that derive directly from the earlier consideration of the study and preservation of planetary sites from the history of aerospace exploration. These larger considerations encompass the potential for intelligent life elsewhere in the universe, the possible archaeological signatures of such life, and the parameters by which such signatures might be recognized and archaeological surveys conducted on them on several levels: galactic, planetary, area, and site.

SOUL SEARCHING: INTELLIGENT LIFE IN THE GALAXY AND ITS ARCHAEOLOGY

"Where is everybody?" asked Enrico Fermi over lunch at Los Alamos Laboratory in the summer of 1950.[1] Fermi's three colleagues at lunch—Herbert York, Emil Konopinksi, and Edward Teller—understood right away what he was referring to. If popular estimates of thousands of extraterrestrial civilizations are accurate, why don't we see evidence of such alien civilizations in the form of communications, spacecraft, and other alien artifacts, or even visits to Earth?

A decade after Fermi formed what has since become known as his universal paradox, astronomer Frank Drake[2] created an equation to estimate the number of technological civilizations that reside in our galaxy. The functional aspects of such a construct are no doubt sound. If it holds a conceit, it is perhaps the assumption that if intelligent life has developed something similar to *Homo sapiens'* civilization, such a civilization exists now, in real time, somewhere in the galaxy, in a mode capable of contacting Earth.

It was Isaac Asimov who first explored this notion as it might relate directly to aerospace archaeology.[3] Asimov used a more reasonable approximation of the lengths of the stages of technological development in human civilization to calculate the probability that similar such civilizations exist in our own galaxy.

As in all such constructs, one makes a few dramatic assumptions. For Asimov, writing in the Cold War era of the 1970s, it was that a civilization such as that of *Homo sapiens* destroys itself within, at most, several generations of developing nuclear power. Asimov further assumed that every habitable planet with a life-bearing span of 12 billion years developed an intelligent species after 4.6 billion years, which then developed an increasingly sophisticated and lethal civilization over the ensuing 600,000 years.

He continued:

> Since 600,000 is 1/20,000 of 12 billion, we can divide the 650 million habitable planets in our Galaxy by 20,000 and find that only 32,500 of them would be in that 600,000-year period in which a species the intellectual equivalent of *Homo sapiens* is expanding in power.
>
> Judging by the length of time human beings have spent at different stages in their development and taking that as an average, we could suppose that 540 habitable planets bear an intelligent species that, at least in the more advanced parts of the planet, are practicing agriculture and living in cities.
>
> In 270 planets in our Galaxy, intelligent species have developed writing; in 20 planets modern science has developed; in 10 the equivalent of the industrial revolution has taken place; and in 2 nuclear energy has been developed, and those 2 civilizations are, of course, near extinction.
>
> Since our 600,000 years of humanity occur near the middle of the sun's lifetime, and since we are taking the human experience as average, then all but 1/20,000 of the habitable planets fall outside that period, half earlier and half later. That means that on about 325 million such planets no intelligent species has as yet appeared, and on 325 million planets there are signs of civilizations in ruins. And nowhere is there a planet with a civilization not only alive but substantially farther advanced than we are.
>
> If all this is so, then even though … hundreds of millions of civilizations [may have arisen] in our Galaxy … it is no wonder that we haven't heard from them."

Asimov's analysis may be dispiriting to exobiologists hoping to make contact with an intelligent form of extraterrestrial life. For archaeologists, however, the calculation that "on 325 million planets there are signs of civilizations in ruins" is a notion to stagger the imagination. Not even the National Historic Preservation Act of 1969 could have envisioned a requirement for 325 million cultural resource professionals, much less the need to equip each of them with survey and transport technologies the equivalent of the Apollo program.

To this potential planetary database we need to add two additional considerations. The first is obvious: on Earth there is evidence not just of one civilization in ruins but several. So while we might speak of 325 million planets with signs of civilization in ruins, such a reality would translate to actual ruins numbering in the *billions*.

Second, we need to consider the possible forms such ruined civilizations might take if they managed to achieve, prior to their destruction, a level of technological development far beyond anything contemplated in the near-term cultural evolution of *Homo sapiens*. Freeman Dyson[4] suggested the possibility that potentially massive engineering works might exist in the universe when he proposed a galactic search for sources of infrared radiation as a necessary corollary to the search for radio communications. Dyson assumed that, given the enormous scales of time and distance in the universe, any technological civilization observable from Earth would have been in existence for many millions of years longer than comparable civilizations on Earth. [Dyson was (again) assuming a living civilization, but the argument holds perhaps even more closely for a civilization that survives only as an archaeological entity.] Such an advanced civilization—if our own are any kind of galactic model—would have long ago outstripped its planetary resources. In response, it conceivably might have developed solar system-scale technological structures to provide for its post-Malthusian energy requirements and, as Dyson wrote, its *lebensraum*.

Dyson proposed such an artifact in the form of a shell or sphere that would surround a solar system's star, effectively capturing all of its radiant energy and enabling the material resources of all the planetary and asteroid bodies in the system to be mined. Using our solar system as a model, Dyson conceived of an industrial operation that, over the course of 800 years, would disassemble the planet Jupiter and reassemble its mass as a 2–3-m-thick shell at a distance of twice the distance of the Earth from the sun. People occupying the inside surface of this sphere would have access to the entire output of the sun's energy. Presumably, given this enormous energy source, this entire inside surface would resemble a tropical rain forest.

In terms of a galactic archaeological survey, such a notion would require archaeologists to search not just for those places most visible to radio telescopes, but those dark areas where the light of an entire solar system is being harnessed for occupants living on the inside edge of a Dyson sphere. In terms of cultural anthropology, the magnitude of such an effort, both in terms of technology and time (800 years by Dyson's lights), would require the concentrated efforts of an entire planet over the course of forty generations or more.

There are no international Earth corollaries to such an effort, although national(ist) structures such as the Egyptian pyramids or the Great Wall of China perhaps come closest to predicting what a planetary-scale effort would require. The former, a response to a spiritual requirement, and the latter, a response to security threats, would suggest the difficulty of predicting the precise rationale for an undertaking of this magnitude.

In the end, the implication of Asimov's calculations and, to a lesser extent, Dyson's conjecture, is evident. SETI stations are hearing only static through their radio telescopes because they are, in effect, listening for a message from the interplanetary equivalent of the Mayans, or the Sumerians, or any of dozens of dead civilizations who can speak to us now only through their archaeology.

ORBITAL ARCHAEOLOGY: IS THERE INTELLIGENT LIFE DOWN THERE?

Given the time spans and likely cultural resistance (and/or résistance) that such an effort as a Dyson sphere would generate, it is more likely that a long-term project in extraterrestrial cultural survival would result in a loose federation of hundreds of smaller outposts. And this, of course, is assuming that such an advanced technological society, upon completing such a structure, would then retreat to a band-level tropical rain forest existence that its forebearers presumably walked away from millions of years earlier. Each of these scattered outposts, given the time and distances involved even in solar system-level travel, would within several generations have developed a variety of distinctive cultural adaptations and the different dialects or distinct languages that evolve with them.

An Earth corollary to such a planetary scenario would be the Norse colonization of the North Atlantic.[5] A mother culture for a variety of reasons spins perhaps 10% of its population away from the fjords of Norway, with some of the voyagers settling Iceland. Once this area became too crowded, 10% of this Icelandic population eventually settles even further to the west, in Greenland. Explorers from this population push on to Vínland, but are as quickly pushed back when Native Americans oppose the new colonization. A combination of factors then besets the Greenland colony over the course of several hundred years, until it is finally cut off from its originating cultures of Norway and then Iceland. Eventually, the Greenland colony vanishes, leaving the barest trace of written records, none of which testify to their fate. With no or indecipherable written records, as in the Norse Greenland colony or the civilization of Mohenjo Daro, the archaeological record then becomes the primary means of studying the vanished civilization.

Like a sliver of space junk that becomes impossibly heavy the faster it goes, it seems the closer archaeologists get to the question of potentially non-human artifacts of intelligence, the more they live in fear of embracing the subject too closely. Yet if one accepts the idea of archaeological research on sites from the history of human exploration in space, which as this volume testifies is increasingly a subject of mainstream discussion in archaeology, it is hardly a giant leap to consider the potential for archaeological fieldwork on the evidence of extraterrestrial civilizations.

Such material records of past civilization have been proposed even within our own solar system. Highly speculative landforms such as the unexplained linear markings on the Tharsis plateau of Mars,[6] or "The Face" and the pyramidal shapes in the Cydonia region of Mars,[7] which eventually were identified as ancient buttes worn down by sandstorms, have in the past been advanced as sites to be examined as potential artifacts of intelligence.[8]

The "scientific politics" of discussing such notions is evident from an article by a professor of physics, describing the *Mars Pathfinder* probe as it was being readied to explore for fossil life on Mars.[9] Faculty and graduate students in the Department of Geophysical Sciences at the University of Chicago gathered to discuss the more speculative aspects of the mission. Like Enrico Fermi, they felt comfortable bringing the subject up only outside the formal parameters of a scheduled seminar. Questions of life on Mars were brought into the open only after the scientists had consumed large quantities of beer—and this even though their multimillion-dollar mission was already an accomplished fact.

Such fears are increasingly unfounded. The subfield of exobiology has been a commonplace for half a century and consistently generates news from its mostly hypothetical undertakings. The icy Jovian moon of Europa, with an apparent frozen ocean similar to Earth's Arctic Ocean, is now under study by such biologists who, with intensive planning by spacecraft engineers, seek to burrow an oceanographic-style Icepick probe through Europa's pack ice in search of microbial life (Figures 45.1 and 45.2).[10]

Carl Sagan[11] suggested that the search for evidence of intelligent life in the universe had to begin by first verifying the criteria for intelligent life with the corollary of

FIGURE 45.1 Artist's conception of the so-called Icepick mission to Europa. (Courtesy of NASA–Jet Propulsion Laboratory.)

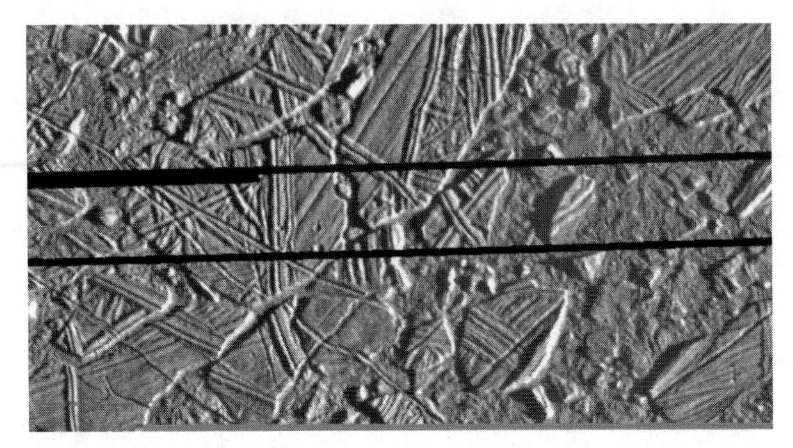

FIGURE 45.2 The apparent pack ice of Europa, taken from the *Galileo* spacecraft in 1998. Resolution is 70 m. Note the road-like lines, which in this case do not act as an indicator of intelligent life but do suggest an Arctic Ocean–like environment covering the surface of this Jovian moon. (Courtesy of NASA–Jet Propulsion Laboratory.)

Earth. Sagan who, unlike Asimov, believed there were more than a million currently active galactic civilizations, expressed his optimism in a chapter called "Is There Intelligent Life on Earth?" He argued that with 1970s-era satellite imagery of 1-mile resolution, one could stare for hours at the entire eastern seaboard of the United States and see no sign of life, intelligent or otherwise.

Without finer-resolution imagery, one would fall back on basic requirements for life in general: water and oxygen. To this, add carbon dioxide to warm the planet and the presence of methane as an indicator of life. Chlorophyll or some other mechanism for absorbing solar radiation and transmuting it into a form necessary to both sustain life and produce oxygen would also be present.

The search for intelligent life on Earth begins with the search for radio wave transmissions that are sequential and irregular, and travel along predictable frequencies. With space-borne imagery of less than 100-m resolution, one would begin to see a planetary penchant for geometric shapes and linear settlements and pathways. Perhaps one would mistake modes and means of transportation as intelligent life rather than its by-product.

For archaeologists, this raises the question that if structures similar to the Jonglei Canal (Figure 45.3) or the Aswan High Dam exist on Alpha Centauri, would we be able to recognize them as artifacts of intelligence? If in fact what we will find in space is not life but the traces of past life, not civilization but the traces of past civilization, it makes sense for us to start discussing what forms such life and such remains might take based on existing terrestrial analogues. As noted, this process is well along in the biological sciences but has been all but ignored in archaeology.

At a minimum, existing satellite imagery of Earth could be used to create analogues for the kinds of artifacts and structures that might be encountered on future exploratory space missions, analogues that could begin by drawing upon examples of similar work already accomplished.[12] Constructing such a database of structural

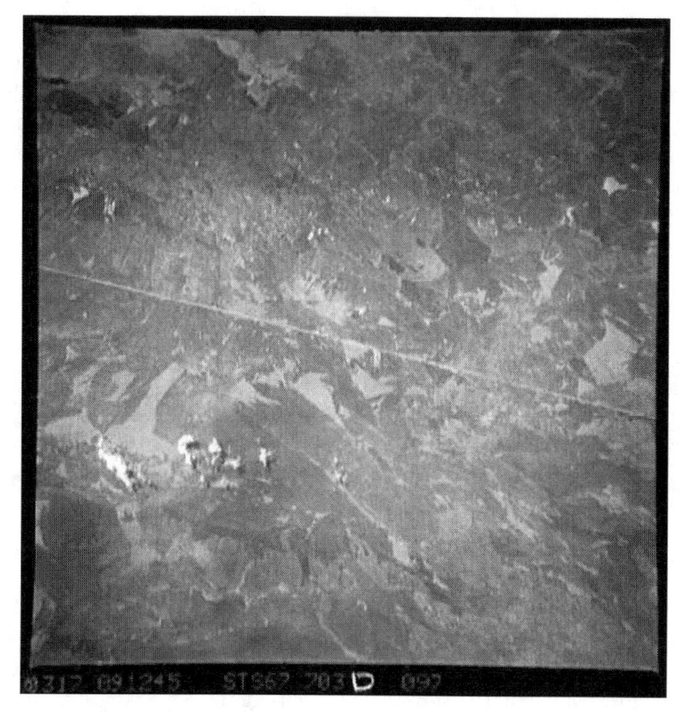

FIGURE 45.3 Jonglei Canal, seen from space. The canal is the straight line. (Courtesy of NASA.)

analogues for potential signatures of advanced civilizations is well within the province of aerospace archaeology.

It seems appropriate—not least given its unique place amid discussions of extraterrestrial visits to Earth—to apply here a cautionary note from Easter Island (Rapa Nui). During his 1955–1956 archaeological expedition to the island, the Norwegian explorer Thor Heyerdahl[13] discovered a subculture of Rapa Nui culture that had gone underground, living a cave-like existence. A society under the stress of resource depletion, external security threats, climate change, overpopulation, pollution, or any combination of the above might well radically reduce its planetary archaeological footprint to a problematic level for external space-based sensors. In such a case, the absence of a planetary surface archaeological signature does not rule out a cultural adaptation that exists (or existed) on the outer edges of our theoretical constructs and is all but invisible to much of the current methodology and technology of our remote sensing.

CONCENTRATION ON CONCENTRATIONS: REMOTE SENSING OF AREAS

As noted above, given the limitations on direct human exploration of the universe, archaeological expeditions to search, survey and study interplanetary or intergalactic sites will of necessity be conducted through the techniques and technologies of remote sensing. Given the requirements of water and oceans as essential to life, and current models, such as the planning of the mission to Europa, these remote sensing techniques will owe as much to the history of underwater archaeology as to aerial archaeology.

The shift toward remote sensing of archaeological sites began in the 1970s as one method to elevate archaeology from its reliance on "the limited observational capacity of the human senses."[14] It took on greater momentum in the latter part of the twentieth century not only from the accelerating refinement of survey technology[15] but also because of encroaching political considerations that threaten to overwhelm traditional archaeology.

As early as 1975, Americanist archaeologists began to acknowledge the potential for political limits on their global historical explorations stemming from foreign nationalism.[16] Since such an admission implied that excavations would be limited in both their scope and content—if they were allowed at all—it was suggested that Americanist archaeologists, "who tend to be intensive excavators rather than observers of surface remains,"[17] begin to systemically adopt two underemployed methods, one well-established and one relatively new and emerging.

The first was the British technique of noninvasive "field archaeology" espoused by O.G.S. Crawford. In the years before World War I, Crawford "longed to see [archaeological sites] not obliquely but in plan, as would be possible in an aeroplane" or balloon,[18] and after the war he went on to invent and refine the techniques of aerial photography of archaeological sites. The second method, one that relied on modern high-technology inventions, was to adopt, adapt, and develop sophisticated scientific technologies, similar to those already employed by investigators in other "hard" sciences, for archaeological investigations.

In addition to Crawford's classic methods of field archaeology and aerial photogrammetry, these techniques included proton magnetometry to measure magnetic anomalies, soil resistivity to measure electrical resistance, and ground-penetrating radar (see also Doleman, Chapter 38 of this volume). At the high end of the technological spectrum, advocated methods included thermal and infrared remote sensing by aircraft and satellites.

For underwater sites, exploration with submersibles such as Cousteau's *Soucoupe*[19] or Bass' *Asherah*, a submersible developed specifically for archaeological research in 1964,[20] could be combined with manned undersea field stations such as the Conshelf Two station envisioned and constructed in the Red Sea by Cousteau[21] and, since the 1960s, with unmanned remotely operated undersea research vehicles (AUVs) currently under development. Remotely operated instruments have been placed around underwater artifact concentrations in northern Europe to monitor currents, salinity, and other processes affecting wreck sites.[22]

These "alternative[s] to the traditional approach to archaeological exploration, discovery and investigation … emphasize the acquisition and sophisticated analysis of a variety of remotely sensed imagery and data as the *primary* tools of exploration, discovery and recording."[23] With the biosphere, geosphere, and archaeosphere thus delineated, traditional field surveys and excavations would automatically revert to a status as secondary methods for verifying, or *ground-truthing,* the data collected with the primary remote sensors.

The dependence on massive technological infrastructure such as submarines and the bases and personnel they require will obtain until the perfection of independent undersea/space archaeology robotic probes. The array of such an "archaeoprobe," equipped with real-time telemetry capability and a miniature terrain rover, could be modified, like its oceanographic cousins, to meet virtually any exigency in the field. More, such a probe could be sent to a planetary site and operated remotely by a single explorer, independent of a massive science and technology bureaucracy.

Such technological development will proceed alongside a persistent discussion in oceanographic technology circles over the course of subsea exploration[24] whether with submarines, submersibles, remotely operated vehicles (ROVs), or AUVs, or some combination of all of them.

What is clear is that there is a core of scientific explorers who do not want to leave exploration solely to robotic sensors. Willard Bascom, one of the most famous of American oceanographers, acknowledges this same human yearning when he writes in his autobiography of using satellites to study the oceans, remarking at one point that

> in recent years my interest [in satellite oceanography] has waned again, probably because this is not a very adventurous form of oceanography. Although the scientific findings can be intellectually exciting, it takes a special sort of person to sit at a computer all day long sorting the data from a big dish antenna pointed at the sky or to spend years figuring out algorithms for converting the millions of data bits into useful information about what the ocean is doing.[25]

The equation for archaeologists is simple: either develop postdoctoral programs at oceanographic and space research centers to get us up to speed on the latest survey technology and simultaneously develop new technologies for our own research designs, or see archaeological exploration descend into arbitrary deconstructionism, "a kind of literary criticism, in which equally stimulating and internally consistent interpretations abound, but where no basis exists for deciding which one best approximates the historical reality of the past."[26] New and effective combinations of already available technologies, combined with the development of autonomous vehicles, employed in a research design that frees the archaeologist from reliance on the technological debates of oceanographers, will provide the necessary data for the application of the scientific method by archaeological explorers.

More than 25 years ago, the archaeologist Bass entered the technological end of this controversy when the paradigm was manned versus unmanned exploration of space, and he seemed to endorse the conception of technological archaeology, in this case a sonar/ROV-equipped long-range lock-out submarine for archaeology, one that could also serve either as a supply vessel for an underwater habitat for archaeologists or as the habitat itself. Bass wrote:

> Submarines will allow archaeologists to map the visible remains with stereophotography and to clear away the sand with portable, neutrally buoyant air lifts directed by remotely controlled manipulators attached to the submarines.
>
> Only one thing is missing, and that is the sure touch of the archaeologist's hand on the site. The present scientific controversy over whether manned or unmanned vehicles are more practical in the exploration of outer space is easily answered for the archaeological exploration of inner space: only a vast array of the most delicate

manipulators imaginable could clean and raise the fragile and fragmentary pieces of wood which are so easily and gently handled by human divers.[27]

It has now been more than 40 years since Bass commissioned *Asherah*, the first (and only) submarine constructed specifically for archaeology. Since then, Robert Ballard, in his Black Sea research on board the nuclear submarine *NR-1*, has demonstrated that it is in fact possible to retrieve fragile artifacts without harm. And while some might see a submarine for archaeology as an expensive luxury, Bass notes that more has been paid by museums for a single work of classical art than was spent on his submarine, which held the potential for revealing entire shiploads of such art (again, as Ballard subsequently demonstrated).

SITE LEVEL: THE BASE CAMP OF THE IMAGINATION

The twentieth century stands as the first in which the human species rose above the Earth both to study its habitat as well as its capability to visit and potentially inhabit other worlds. Questions of where, how, and why humankind first sought to use air and space technology for scientific and geographic exploration have a direct bearing on behavioral questions of why we have become perhaps the "most inquisitive, exploring animal."[28] Defining the archaeological signatures by which this cultural transformation took place will provide a comparative model for similar cultural responses by extraterrestrial civilizations.

If remote survey technologies can be combined with techniques for identifying such surface remains in extreme environments, they will provide the basis for site evaluation without the requirements of surface survey. The current avenue for such remote evaluation of extreme environment sites is satellite photography on the order of high-altitude (150 miles perigee) advanced KH-11 satellite imagery.[29] The fine 1–2 ft. spatial resolution of KH-11 imagery and its use of metric markings for use as maps, combined with more recent Vega satellites that employ imaging radar for penetrating cloud cover, would make it possible to draw plans for virtually any small-scale site.

The current pathway of such research is clear. Part of the mission profile of the *Mars Reconnaissance Orbiter* was to search for archaeological traces of the lost *Mars Global Surveyor*, the *Mars Polar Lander*, and the British *Beagle 2* (see also Spenneman, Chapter 23 of this volume).[30] The expedition has thus far succeeded in capturing not only high-resolution images of the *Spirit* and *Opportunity* rovers that have been exploring the surface of Mars since 2004 (Figure 45.4), but also of the two Viking probes that reached Mars in the 1970s.[31]

This very human desire to search for archaeological traces of former expeditions becomes, in space, an almost spiritual requirement to relocate familiar human landmarks along expansionary pathways. It means that future space missions will devote significant energies not just toward pioneering new routes through space but to rediscovering the techniques and technologies of earlier explorers.

With the emerging capabilities of archaeologists to combine digital photography and photogrammetry in a regional GIS context in a Google Earth environment,[32] it

FIGURE 45.4 An image taken by the Mars Reconnaissance Orbiter shows the small crater where the airbag-cushioned lander of the Mars *Opportunity* probe came to rest. (Courtesy of NASA–Jet Propulsion Laboratory/University of Arizona.)

will be possible to delineate the exploratory archaeosphere, both topographically and, eventually, bathymetrically, in such a way as to enable provenance maps and charts of artifact clusters, sites, and site environments, as well as site and regional archaeological histories.

The requirement of life for water, along with the proposed mission to the potential frozen ocean of Europa, highlights the necessity to consider undersea sensing systems for archaeological research in a space context. Corollaries to human space missions, such as those that have established model Mars bases in the Arctic,[33] should be developed to test the requirements for similar archaeological missions. Simultaneously, archaeologists can prepare theoretically for eventual field work in space by using the extant volume of historical, technological, and planetary data to develop a combined pattern-recognition system and catalog of structural signatures of intelligent life.

NASA alone has eight separate Web sites devoted exclusively to data just on its current missions, and each of these sites is further broken down into distinct descriptive pages on human aspects of exploration, emerging hardware and other technologies, spaceport and research facilities, and mission goals and accomplishments. Then there are the sites for the eleven separate federal space launch, control, or research centers. Nearly all of this information is directly relevant to human evolution and cultural diversity. All of it is readily accessible over the Internet.

Even in terms of direct access, there are the earthbound challenges of completing archaeological surveys of abandoned rocket gantries in Florida or Kazakhstan. Such surveys, undertaken in part as corollaries to projected similar missions to extraterrestrial archaeological structures, could well provide the methodological and theoretical basis for future surveys of similar structures in space.

CONCLUSION

Given the vastness of space, the chances of real-time communication with other civilizations are small; on the other hand, the chances of excavating or otherwise remote-sensing traces of other civilizations seems, by comparison, rather high. The role of archaeology in space exploration in the near term, then, is twofold.

The first is to shape the currently available raw material of historical, technological, and planetary data into a catalog of analogues for defining the presence of extraterrestrial civilizations. This catalog should be both methodological and theoretical and should center on notions of *Homo sapiens* as an exploratory, migratory species.

Secondly, a department of archaeological engineering, colocated perhaps with departments with similar mission sets, such as those for ocean engineering, should develop a model of a remote probe to be employed throughout the galaxy to explore for traces of these signatures of civilization. With this mixture of anthropological theory and technological experience, an archaeoprobe can be designed and developed even though it might not see its first deployment for fifteen generations or more. Operating in a sense as medieval monks, we can at least prepare a sort of illuminated manuscript for the edification of researchers not yet born, and within it a blueprint for the machine required to explore for the traces of galactic civilization defined by that manuscript.

NOTES AND REFERENCES

1. Jones, E.M. 1985. *"Where is Everybody?" An Account of Fermi's Question*. Los Alamos, New Mexico: Los Alamos National Laboratory Report LA-10311-MS.
2. Drake, F.D. 1965. The Radio Search for Intelligent Extraterrestrial Life. In *Current Aspects of Exobiology*, ed. G. Mamikunian and M.H. Briggs, 323–346. Jet Propulsion Laboratory Technical Report.
3. Asimov, Isaac. 1979. *Extraterrestrial Civilizations*. New York: Crown. Associated Press. 2006. NASA Enlists Mars Probe to Search for Missing Craft, p. 193. http://www.space.com/news/061118_mro_update.html.
4. Dyson, F.J. 1960. Search for Artificial Stellar Sources of Infrared Radiation. *Science* 131: 1667–1668.
5. Seaver, K.A. 1996. *The Frozen Echo: Greenland and the Exploration of North America, ca. A.D. 1000–1500*. Stanford, CA: Stanford University Press.
6. Sagan, C. 1980. *Cosmos*. New York: Random House (see specifically p. 129).
7. Carlotto, M.J. 1991. *The Martian Enigmas*. Berkeley, CA: North Atlantic.
8. Sagan, Carl. 1980. *Cosmos*. New York: Random House (see specifically p. 130).
9. Trefil, J. 1995. Ah, but There May Have *Been* Life on Mars. *Smithsonian* 26(5).
10. See, for example, Isbell, D., and M.B. Murrill. 1996. *Jupiter's Europa Harbors Possible "Warm Ice" or Liquid Water*. NASA Release: 96–164, August 13, 1996 (see specifically Figures 1 and 2) http://nssdc.gsfc.nasa.gov/planetary/text/gal_eur_water.txt.
11. Sagan, C. 1994. *Pale Blue Dot: A Vision of the Human Future in Space*. New York: Random House (see specifically pp. 59–70).
12. Blom, R., R. Crippen, C. Elachi, J. Zarins, N. Clapp, and G. Hedges. 1992. *Space Technology and the Discovery of the Lost City of Ubar*. Pasadena, CA: JPL. http://trs-new.jpl.nasa.gov/dspace/bitstream/2014/27582/1/96-1629.pdf.
13. Heyerdahl, Thor. 1958. *Aku-Aku: The Secret of Easter Island*. London: George Allen and Unwin.

14. Lyons, T.R., and D.H. Scovill. 1978. Non-destructive Archaeology and Remote Sensing: A Conceptual and Methodological Stance. In *Remote Sensing and Non-destructive Archaeology,* no. 36, eds. T.R. Lyons and J.I. Ebert, 3–19. Albuquerue, NM: National Park Service and University of New Mexico.
15. e.g., Ebert J.I. 1984. Remote Sensing Applications in Archaeology. *Adv. Arch. Methods & Theory* (7): 293–362.
16. Hester, T.R., R.F. Heizer, and J.A. Graham. 1975. *Field Methods in Archaeology.* Mountain View, CA: Mayfield Publishing Company (see specifically p. 309).
17. Ibid., 309.
18. Crawford, O.G.S. 1953. *Archaeology in the Field.* London: Phoenix House (see specifically p. 46).
19. Cousteau, J.-Y. 1964. At Home in the Sea. *National Geographic Magazine* 125(4): 465–507.
20. Bass, G. 1970. *Archaeology Under Water.* Baltimore: Penguin Books.
21. Cousteau, J.-Y., with J. Dugan. 1965. *World Without Sun.* New York: Harper & Row. Miller, J.W., and I.G. Koblick. 1984. *Living and Working in the Sea.* New York: Van Nostrand Reinhold.
22. Whitfield, J. 2002. Shipwreck Network Launched: Three-Year European Project Aims to Safeguard Shipwrecks. *Nature news*, published on-line, January 23, 2002. http://www.nature.com/news/2002/020121/full/news020121-5.html.
23. Lyons and Scovill 1978 (note 14).
24. Travis, J. 1993. ABE and Odyssey: AUVs Test the Waters. *Science* 259: 1534–1535.
25. Bascom, W. 1988. *The Crest of the Wave: Adventures in Oceanography.* New York: Harper & Row (see specifically p. 314).
26. Gould, R.A. 1990. *Recovering the Past.* Albuquerque, NM: University of New Mexico Press (see specifically p. 239).
27. Bass 1970 (note 20) (see specifically p. 152).
28 Finney, B.R. 1992. *From Sea to Space.* Palmerston North, New Zealand: Massey University (see specifically p. 105).
29. Richelson, J.T. 1999. U.S. Satellite Imagery, 1960–1999. National Security Archive Electronic Briefing Book No. 13. http://www.gwu.edu/~nsarchiv/NSAEBB/NSAEBB13/.
30. David, L. 2006. Orbiter to Look for Lost-to-Mars Probes. Space.com. November 1, 2006. http://www.space.com/businesstechnology/061101_mro_search.html. Associated Press. 2006. NASA Enlists Mars Probe to Search for Missing Craft. http://www.space.com/news/061118_mro_update.html.
31. Goudarzi, S. 2006. Mars Orbiter Photographs Three Old Spacecraft. http://www.space.com/news/061205_mars_reconnaissance.html.
32. See, for example, Handwerk, B. 2006. Google Earth, Satellite Maps Boost Armchair Archaeology. *National Geographic News*, November 7, 2006. http://news.nationalgeographic.com/news/2006/11/061107-archaeology.html.
33. Fox, W.L. 2006. *Driving to Mars: In the Arctic with NASA on the Human Journey to the Red Planet.* Emeryville, CA: Shoemaker & Hoard.

Section X

The Mind and the Cosmos

46 Developing Exoarchaeology in the Solar System and Beyond

John B. Campbell

CONTENTS

INTRODUCTION

For human space heritage, arguing that we should protect off-world sites and landscapes as at Tranquility Base, *Apollo 11*'s landing site on the Moon, is reasonably straightforward, though the procedures for actually doing this can be complicated (see also O'Leary, Chapters 3, 40, and 43; Gorman, Chapters 16 and 19; Capelotti, Chapter 21). However, for all its potential problems and various levels of significance, our own space heritage is only one aspect of the universe's potential range of space heritages. That is, what should we do about archaeological sites or objects that might have been created by other intelligent species in our galaxy, the Milky Way? With continuing refinement of methods for detecting and analyzing exoplanets (planets in orbit round other stars), of which more than 340 are now known, predominantly in our galactic neighborhood, would it be premature for astronomy, astrophysics, space engineering, and archaeology to work together to develop more refined techniques for detecting possible evidence of the technologies of other clever species? As we ourselves expand further into space, what would we do if we were to encounter probes parked in the solar system that we did not create? Some astronomers have seriously suggested that these could exist, particularly at some of the Lagrange points, and have been searching for them on and off since 1979. Would the human species treat such probes with respect, or would we feel threatened by them? In addition to developing protocols for human space heritage, it is recommended that we develop more widely applicable protocols for nonhuman space heritage and that we participate more actively in the development and application of astrobiology and "xenoscience."

BACKGROUND

The Principle of Mediocrity argues that there is nothing special about the status or position in the universe of our planet, Earth. This was first espoused by Nicolaus Copernicus. He argued on scientific grounds that the Earth was not the center of the universe, and this argument was published posthumously in 1543. Had he been alive when it was published, he would have been persecuted or put under house arrest by the church for his heretical hypothesis, as was Galileo Galilei some years later when he found evidence for other celestial bodies not orbiting the Earth such as the moons of Jupiter and stars of the Milky Way, among other telescopic observations. Their work became known as the Copernican Revolution, which contradicted the established dogma of Aristotle's and Ptolemy's positions. The latter had argued that everything in the universe was fixed, with Earth at the center, and even species of organisms were seen as fixed or immutable. The very term "revolution" came from our newly hypothesized revolutions around the Sun, but now with political and philosophical meanings. By extension, the Principle of Mediocrity also argues for life being abundant in the universe, though whether it would automatically support intelligent life being abundant is more debatable.[1,2,3,4] By current estimates, the known universe is 13.7 billion years (13.7 Gyr) old; our galaxy, the Milky Way, is about 11.8 Gyr old; and the Solar System is about 4.8 Gyr old. Life has been evolving on Earth for close to 4 Gyr, but complex life is well under 1 Gyr old, and intelligent life is only a few million years old at most (or less than 100,000 years if one takes a very narrow and conservative view on the evolution of human language abilities). The latter, of course, is not counting what might have happened with the further evolution of some clever, possibly warm-blooded reptiles if the Earth had not been hit by an asteroid about 65 million years (65 Myr) ago, giving mammals a chance to reset or rather exploit a new evolutionary clock.[5,6] On a galactic scale, Gillman and Erenler[7] argue that the passage of the solar system through the spiral arms of the Milky Way correlates with major terrestrial extinction events. They calculate that the total time to pass through the combined arms is 703.8 Myr. Complex life on Earth evolved successfully during this time, despite six major extinction events of both cosmic and terrestrial origin.

The search for extraterrestrial intelligence (SETI), as represented by radio signals from somewhere else in the Milky Way, effectively began in 1961 with Frank Drake's Project Ozma (on development of SETI research, see Cocconi and Morrison, Tarter, Ekers et al., Weinberger and Hartl, Edmondson and Stevens, Shostak, and Dick[8,9,10,11,12,13,14]). By 1979–1980, searches had also begun for interstellar probes parked in stable orbits in the solar system, as well as for infrared excesses from astroengineering projects such as Dyson spheres, designed to capture the energy output of Sun-like stars.[15] While the Cold War was still under way in the 1980s, partly influenced by its mindset, some scientists in the Soviet Union and the United States began looking for evidence of stellar nuclear waste dumps and nuclear wars elsewhere in the Milky Way.[16] Though archaeologists are experts in the detection of physical evidence of a wide range of technologies, we were not involved in this work, even though the term "archaeology" was actually used at times.[17] I would propose that the primary reason was that in the 1970s and 1980s, we archaeologists were being "stung" by the impact of Erich Von Daniken's popular ideas on the alien origins of human technologies and civilizations on Earth. This pseudoscience nonsense put us right off any interest in the possible technologies of other intelligent species, other than the simpler ones of our own hominid ancestors and relatives on Earth, such as *Homo erectus* and *Homo habilis*. Spennemann[18] has recently addressed some of the issues of working on the cultural heritage of other primates and of human-designed robots on Earth and off-world (i.e., including remotely controlled space probes, landers, and rovers; see also Spennemann and Murphy, Chapter 23 of this volume).

The future development of new space telescopes such as the European Space Agency's (ESA's) Darwin and NASA's Terrestrial Planet Finder (TPF) for detection and direct imaging of Earth-like exoplanets in our galactic neighborhood represents an opportunity to develop and test new techniques for detecting signatures of technological activities in other star systems. Using nulling interferometry to block out the light of the parent star, both Darwin and TPF will be able to provide spectroscopic data on the chemistry and biochemistry of the atmospheres of Earth-like exoplanets and thus to detect some of the signs of life. Evidence of technological activities such as nuclear accidents, atmospheric pollution, rapid climate change, and wars should be detectable, among other signs of technology. Infrared signatures of technologies established off exoplanets, such as asteroid mining, might also be detectable. Direct visual imaging of any massive extraterrestrial structures will probably not be feasible until we have even more powerful telescopes or actually send long-term interstellar probes of our own. Full Dyson spheres would block our view of exoplanets within their shell, but if oriented at an appropriate angle in relation to our line of sight, we should be able to detect combinations of Dyson rings and exoplanets, unless the inner planets have been completely mined away (see also Capelotti, Chapter 45 of this volume).

ARCHAEOLOGY, EXOARCHAEOLOGY, AND SETI

Archaeology is the scientific study of the physical evidence of human and protohuman behavior and evolution from the split between our ancestors (members of the genera *Sahelanthropus*, *Ardipithecus*, *Australopithecus*, *Homo*, etc.) and those of our closest living relatives, chimpanzees and bonobos (members of genus *Pan*), some 6 to 7 Myr ago in Africa, right up through time to modern forensic archaeology carried out anywhere on Earth or, in the near future, off-Earth (see also Spennemann and Murphy, Chapter 23 of this volume). Archaeologists study sites, artifacts, cultural development, evolution of intelligence, cultural landscapes, seascapes, and spacescapes.[19,20] Exoarchaeology is the study of the evolution, technology, society, and impact of any intelligent species in the universe.[21,22] It is limited at any given time by what is feasible.

Space exploration is comparatively new, and conducting the archaeology of it is even newer. The World Archaeological Congress established a Task Force on Space Heritage in 2003, which is cochaired by Alice Gorman (Flinders University, Adelaide, South Australia, Australia) and the author. The Task Force also includes as principal members Beth O'Leary (New Mexico State University, Las Cruces, NM), Leslie Brown (Connecticut College, New London, CT), Kathryn Denning (York University, Toronto, Ontario, Canada), Robert Barclay (Conservation Consultant, Ottawa, Ontario, Canada), and Randall Brooks (Canada Science and Technology, Ottawa, Ontario, Canada). It is charged with developing draft protocols for the management and protection of the more significant places and objects associated with space exploration, as well as considering the ethical implications of the development of the space industry and our own work on space heritage. Some of the initial sites on Earth are protected at least in part either by continued use or by declaration as places of significant heritage, but sites on the Moon and Mars have no real protection other than by their current isolation. With new developments being planned by a widening range of spacefaring nations and organizations, lunar and Martian sites will come under direct threat of interference and looting. Sites and objects further afield will also eventually be threatened. We have the opportunity now to propose, draft, ratify, and implement protocols before serious impact occurs, unlike the varying treatment of World Heritage places on Earth before they were listed and protected under the World Heritage Convention. But what will we do about any objects encountered that might have been created by other intelligent species?

Cocconi and Morrison's[23] proposals for SETI and the first radioastronomical project, Project Ozma, run by Frank Drake in 1961,[24,25,26] led to the start of SETI research, some of which also involved at times a search for other signs of extraterrestrial technology, sometimes referred to as the search for extraterrestrial artifacts (SETA). This earlier work was carried out by astronomers and other space scientists. Having been negatively influenced by the efforts of popular authors such as Erich Von Daniken, mentioned above, to discredit Earth-based ingenuity and to claim that all of Earth's ancient civilizations owe their origins to one form or another of extraterrestrial visitor, many professional archaeologists have avoided widening their associations with the space sciences, other than satellite imaging of Earth and studies in archaeoastronomy, namely, studies of ancient or traditional naked-eye observations of the sky as recorded in texts; represented in monuments; represented in rock art; or recorded by archaeologists, ethnographers, or astronomers working with living indigenous peoples[27] (see also Milbrath, Chapter 10 of this volume). Unfortunately, the term SETA has been adopted by Von Daniken and his colleagues and supporters, so I do not recommend employing it in professional exoarchaeology.

For radioastronomers working in SETI, the intelligence of a technologically capable species elsewhere in the Milky Way or beyond would be indicated by the ability to control radio waves and/or to fire pulsed laser beams that outshine their host stars. This may seem a limited definition of intelligence, but defining it in a broader way is also problematic. The behavioral and social sciences that look at intelligence and culture in humans have never come to a universal agreement on how to define intelligence (and even "culture" has various definitions). Furthermore, as scientists in other disciplines we cannot agree on whether certain primates (e.g., bonobos), certain cetaceans (e.g., bottlenose dolphins), some birds (e.g., New Caledonian crows), and some mollusks (particularly cephalopods, such as octopuses) are intelligent, even when they clearly exhibit behaviors such as complex vocal learning and self-recognition.[28,29,30] The relevance of detecting and understanding intelligence in other Earth-bound species is enormous. If we cannot manage it here, how will we be able to recognize and understand a signal from beyond the solar system, other than saying we have found a carrier beam? The issues of communication and understanding in interstellar message construction are being addressed by workshops and conferences organized by Doug Vakoch of the SETI Institute.[31,32,33]

EXOPLANETS, OTHER SUNS, AND DYSON STRUCTURES

The idea of looking for scientific evidence of planets in orbit around other stars is not new,[34] but it was not till the 1990s that technology and knowledge had reached a point where it became actually possible to do it and have the evidence confirmed by others.[35,36] Now that more than 340 exoplanets have been detected, even if they are primarily gas giants like Jupiter, and the technology is being developed to detect and image smaller Earth-like exoplanets (ESA's Darwin, NASA's TPF, etc.), is it not time for archaeology to reconsider its inactive roles in astronomy and astrobiology? For a detailed review of the comparatively new but very broad multidisciplinary field known variously as astrobiology, bioastronomy, exobiology, and so forth, see Chyba and Hand,[37] and for a quick synopsis of the challenges it faces by a Nobel Laureate and Foundation Director of the NASA Astrobiology Institute, see Blumberg.[38]

Thus far, the search for astroengineering structures such as Dyson spheres and rings round other suns or broadly Sun-like stars (represented by infrared excesses; these structures are named after the English astrophysicist Freeman Dyson, at the Princeton Institute for Advanced Study) has been unsuccessful, or at best not successfully confirmed. Equally, the search for interstellar probes positioned at certain Lagrange gravitational equilibrium points in the solar system, placed in other special orbits, or left on planetary surfaces has been unsuccessful.[39,40,41] The possible exceptions are objects 1991 VG and 2000 SG344, which are or appear to be in heliocentric orbits near Earth, with calculated orbits of 379.6 and 354 days, respectively. They are possibly artificial and yet not easily attributable to any known or recorded rockets, probes, or debris from Earth, except possibly a Saturn IV-B stage from the Apollo program. The main candidate would be Apollo 12's IV-B stage. Whether either one or both of these objects represents nonhuman technology and is "passing through" on a survey of a number of star systems, or examining the solar system in detail, one cannot yet know.[42,43] The expected return date for object 1991 VG is October 26, 2016, and that for 2000 SG344 later, with the remote possibility of impact with Earth in 2071.[44] A mission to 1991 VG could be undertaken, as it will probably be only slightly farther away than the Moon. For the moment, the simplest hypothesis is that these objects are oblong asteroids that seem to be behaving like spent rocket stages. We will not know for certain until we examine them more closely.

FERMI PARADOX

Does all or most of this then automatically mean that the answer to Enrico Fermi's famous but only ever informally stated paradox, "Don't you ever wonder where everybody is?" is either zero or one intelligent species (i.e., in a supposedly crowded universe; see Horowitz for something close to the real history of this momentous comment in 1950 during work on the hydrogen bomb[45]). That is, $n = 1 + 0$, with the "1" being us, the only known and confirmed intelligent species in the Milky Way, at least from our biased human point of view (i.e., not counting other intelligent species on Earth, as they do not have radio!). Ward and Brownlee[46] argue that although simple life could be common in the universe, complex life is very rare, and only an Earth-Moon system exactly like ours could protect metazoa (animals or the equivalent) over the eons and allow the evolution of intelligent life. Morris[47] develops a similar argument for his "Lonely Universe," as does Watson,[48] in part, in a current assessment. Their arguments are both anthropocentric and geocentric (Earth-biased), though see also Waltham[49] on the large-moon hypothesis. Although the vast majority of known exoplanetary systems are very different from the solar system, the frequency of what is known so far is partly an artifact of the exoplanet detection techniques employed to date.[50]

In fact, the so-called Fermi paradox is more of a puzzle than a paradox. It is also particularly anthropocentric. Webb[51] presents 50 solutions to the paradox. Many of his solutions are minor variations of each other. For critiques of the reasoning behind the Fermi paradox, as well as those behind the more dominant kinds of astrobiology, which are generally Earth-biased, see the irreverent but well argued xenoscience views of Cohen and Stewart.[52,53] Dyson[54] presents some other alternative views on how to look for unexpected life off planets in vacuum-based environments. I would argue that our instruments and search strategies have been inadequate, and that they have not allowed for a sufficiently wide range of options or possibilities. We already have tremendous trouble understanding and appreciating fully the highly varying worldviews of different human cultures on Earth (for variations simply among the world's many kinds of surviving gatherer-hunters, our so-called oldest way of life, see Lee and Daly[55]). Furthermore, many of the earlier searches for extrasolar technology were done during the paranoia of the Cold War and before the newer cosmologies of the universe had been proposed (cf. Davies).[56,57] The various sorts of interference and outright blockage of line of sight detection that are now imaginable are much greater than 20 or 30 years ago. Have there in fact been no signs of extrasolar nuclear accidents, nuclear wars, or stellar dumping of nuclear waste in the Milky Way, as it would seem, or have we in fact missed the evidence? I think one can argue for the latter, that is, the evidence could exist but so far we have been unsuccessful in detecting it. It might also still be impossible, or at best highly unlikely, for us to be able to detect or recognize an extremely advanced civilization, say a Kardashev Type III (by Nikolai Kardashev's classification, we are not even Type I, as we cannot control our planet; Type II can control its star, and Type III can control the energy and fate of stars in an entire galaxy).[58] Freitas[59] espoused this view a quarter of a century ago. There is also the so-called Interdict Hypothesis,[60] which could "explain" in part the Fermi paradox. Although the idea has not been particularly popular, it is even conceivable that the greatest intelligence in the galaxy is not only far above us but entirely artificial. Among others, Dick[61] and Tipler[62] proposed this and even suggested that the apparent shortfall in silicon in the galaxy could be an indirect trace of the passage of electronic "life" across the Milky Way. If the imagined entity is complex, self-replicating, self-aware, and extremely, highly intelligent, then it would be a form of higher level "life," at least as broadly defined by Cohen and Stewart.[63,64] It might even consider us not worthy of investigation. Alternatively, we might occupy part of a "wilderness reserve" or "protected zoo,"[65] or we could conceivably be under some other galactic directive, say, similar to the Prime Directive of the *Star Trek* science fiction sagas. We might even be considered dangerous and best left to our own demise.

If the Fermi paradox has a logical basis beyond anthropocentric views of the universe, I would propose five possible solutions, though of course there could be others (see also Capelotti, Chapter 45 of this volume):

1. Evolution of intelligence is usually slow, and other intelligent species in the Milky Way are still evolving ["we are first" view; if Earth is removed in the simulations of the galactic habitable zone (GHZ) by Lineweaver et al.,[66] then only 30% of the stars in the GHZ that could host life are older than the Earth-Sun system].
2. Complex life in general is in fact very rare, and we are effectively alone as the only intelligent species in this galaxy ("rare Earth" view, primarily as developed by Ward and Brownlee,[67] supported to some extent by Morris,[68] though I stress again that this is very geocentric or Earth-biased; see also Cohen and Stewart[69,70]).
3. $n = 2$, and the other intelligent species is on the opposite side of the galaxy and at a similar stage of evolution ("galactic shadow" view; I suggest this partly with tongue in cheek, but it is conceivable and one never knows whether something like Murphy's Law might not be universal; certainly chaos and randomness have roles to play).
4. >10,000 (or >10M) civilizations in the galaxy, but we have not yet hit the right radio beam, or detected the right pulsed laser beam, or had sufficient time ("Sagan/Drake" view; basically an optimistic run of the Drake Equation with no confirmed results so far; the Allen Array and other new or proposed radiotelescope configurations on Earth and in space, such as on the far side of the Moon, might make a difference; see Oliver[71] and Ekers et al.[72]; see also Campbell[73] for a tweaked version of the Drake Equation that is more optimistic than Ward and Brownlee[74] but less optimistic than the usual SETI calculations have been; cf. Burchell[75]); Bounama et al.[76] estimate that the distance to the closest planet with primitive life is about 30 light-years (ly), and to the closest planet with complex life, about 130–200 ly.
5. Sentience varies enormously and uses very different media to communicate, including self-replicating electronic "machines" that either use parts

of the electromagnetic spectrum we consider very unlikely or are able to use quantum physics for instantaneous communication in ways we can only just begin to imagine ("quantum artificial intelligence" view; this steps up the ideas proposed by Dick[77] and Tipler[78]).

DETECTING EARTH-LIKE EXOPLANETS

The situation could change with the development of ESA's Darwin, NASA's TPF, and other space-based multiple telescope systems that are being designed to be launched and parked at Lagrange point 2 (L_2) in the Earth-Sun system. If the chemistry and biochemistry of Earth-like exoplanets within some 30 parsecs (97.8 ly, or roughly 100 ly; 1 parsec is 3.26 ly) can be detected and analyzed, then in theory we could detect some possible signs of technologies (pollution, greenhouse effects, infrared excesses, nuclear accidents, wars, etc.; though one would hope to find more optimistic evidence than that for conflict or bad planet management). On Venus, Earth, Mars, and Jupiter's moon Io, volcanic activity causes natural pollution. On Earth, we can distinguish between natural and artificial pollution, though there is an overlap in the characteristics. What we assume is a major, natural, runaway greenhouse scenario has long been established on Venus. On Earth, the natural climate has varied considerably over time between the extremes of "snowball" Earth (with ice and snow to the equator) and the Jurassic "hotbed" (with no polar ice and high air and sea temperatures). This, of course, is not to mention the initial Hadean phase of Earth history, when any toehold that life had established would have been regularly extinguished by further bombardment. Kaltenegger et al.[79,80] have modeled Earth's long-term global environmental changes with a view to extrapolating their data to Earth-like exoplanets, which, when successfully detected, could be at any stage in their development. In a nutshell, "worlds around other stars are likely to be very different from Earth and further modeling has to be done."[81]

Our own influence on a new "greenhouse" event is becoming more clearly established, though there are ongoing debates on what is evidence for a natural climate cycle and what is evidence for human interference with nature in the last 300 years. The warming of the Arctic Ocean and changes to the Antarctic ice shelves have probably reached their tipping points (pack ice reflects >90% of solar radiation, whereas the exposed dark or black polar seas do exactly the opposite). The mass rate of extinctions is going up so much (and is continuing to climb) that future paleontologists looking back on the present time will possibly interpret it as equal to the K/T boundary event (the Cretaceous/Tertiary extinction of the dinosaurs and many other forms of life at about 65M years ago, after a major asteroid impact). There might eventually be an argument for shifting the boundary of the Holocene/Pleistocene to the present, as the total extinctions will vastly outnumber the extinctions that occurred at the end of the Pleistocene as currently defined (11,000 years ago).

CANDIDATE STARS AND HABITABLE ZONES

The list of candidate target stars for Earth-like exoplanets that Kaltenegger et al.[82,83] have developed is very instructive. Furthermore, a closer parallel to the solar system has finally been detected. It has a slightly smaller star and a proportionally smaller exo-Jupiter and exo-Saturn in orbits relatively comparable to those of the gas giants of our system. It is expected that rocky planets will be detected, including in the habitable zone. Computer modeling of sufficiently wide binary, or double, star systems, and in particular our nearest neighbor, Alpha Centauri at 4.3 ly, suggests they could also have rocky planets.[84,85] The two stars Alpha Centauri A and B are never closer to each other than 11 astronomical units (AU), which is a little more than twice the distance of Jupiter from the Sun at 5.2 AU (the Earth is at 1 AU, the average distance between the Sun and the Earth). In other words, these Sun-like stars act as gas giants in relation to each other and their hypothetical Earth-like planets in their respective stellar habitable zones, cleaning up their systems of debris, asteroids, and comets, which would otherwise frequently threaten the rocky worlds. Direct imaging of the Alpha Centauri system is definitely feasible in the future, and using optical interferometric technology, the currently hypothetical rocky planets could be observed. Indeed, this system is so close that it would be feasible in the not too distant future to send probes and eventually multigenerational spacecraft.[86,87] Some other authors have more pessimistic views on the percentage of nearby stars that might be "astrobiologically interesting" (e.g., Porto de Mello et al.[88] estimate that within 10 parsecs of us, only about 7% of the stars have relevant research potential, but then even that is better than 0%!). There is also the possibility that some of the hypothetical Earth-like exoplanets are actually Earth-sized moons in orbit around some of the enormous gas giants that have been found in what could be habitable zones.[89]

The limits of what we might be able to achieve are being increased every year. At the moment, only Earth is known to have life. Nevertheless, the biochemical building blocks of life are known to be distributed in a fairly universal way, with clusters of the requisite molecules in various parts of the solar system and more widely in the Milky Way and other galaxies.[90,91] Whether intelligence in the universe has already evolved beyond a biochemical basis to fully reproducing artificial intelligence is not known to us, but it is certainly being speculated about. Some see this as the "natural" and "inevitable" product of cosmic evolution, and indeed as something ultimately required for the much longer term survival of the universe.[92,93,94,95,96] Would this be an aspect of the "Strong Anthropic Principle," or should one say the "Strong Artificial Intelligence Principle"?

CONCLUSION

Although firm evidence of extraterrestrial intelligence, or what I would prefer to label other intelligent species, has yet to be obtained, with the continued development and refinement of techniques in astronomy, astrophysics, and astrobiology, it is clear to me that it would be remiss of archaeology not to be involved in the searches. As professionals, archaeologists are best equipped to study material culture and the evolution of technology. It does not matter whether all of archaeology approves; all of astronomy is not in full agreement on the value of SETI research, but enough professionals, as well as supporters such as Paul Allen and The Planetary Society, are committed to making it happen. The continued discovery and analysis of large exoplanets and the "neonatal" phase of work on searching for Earth-like exoplanets make it all the more appropriate that archaeology should be involved. Calling this branch of archaeology *exoarchaeology* both indicates what it is and allows those professional archaeologists who might disapprove to maintain their distance or even complete indifference. Not to be involved and committed to this would be both naïve and arrogant, in my view at least. In broadening and strengthening some of the wider links that archaeology has, or could have, I would also advocate keeping in mind some of the concepts of xenoscience. Many disciplines currently participate in proposals and the development of research in astrobiology and the direct imaging of exoplanets, and archaeology is now among them, in however small a way.[97,98]

The popular question, "Are we alone?" may yet be answered in a more positive way than what Ward and Brownlee[99] and Morris,[100] among others, expect. If interstellar probes are eventually discovered in the solar system, either at Lagrange points or on planetary or lunar surfaces, we should have appropriate protocols in place for determining how to treat them, just as we must develop protocols for dealing with our own space heritage off-world and on Earth (see also Gorman, Chapters 16 and 19 of this volume; O'Leary, Chapters 3, 40, and 43; Spennemann, Chapter 41).

ACKNOWLEDGMENTS

This research was supported in part by JCU SSP research funding. Parts of this chapter were presented as oral papers at the 2007 Australia ICOMOS conference in Cairns and 6th World Archaeological Congress (WAC-6) in Dublin in 2008. I thank Susan McIntyre-Tamwoy, Lisa Kaltenegger, Beth Laura O'Leary, Ann Garrison Darrin, and Leslie Brown for their comments and advice. If further errors are detected, the blame lies with the author.

REFERENCES

1. Shostak, S. 1998. *Sharing the Universe: The Quest for Extraterrestrial Life*. Sydney: Lansdowne.
2. Davies, P. 2006. *The Goldilocks Enigma: Why Is the Universe Just Right for Life?* London: Allen Lane/Penguin Group.
3. Cirkovic, M.M. 2007. Evolutionary Catastrophies and the Goldilocks Problem. *International Journal of Astrobiology* 6: 325–329.
4. Franck, S., von Bloh, W., and Bounama, C. 2007. Maximum Number of Habitable Planets at the Time of Earth's Origin: New Hints for Panspermia and the Mediocrity Principle. *International Journal of Astrobiology* 6: 153–157.
5. Foote, M. 2003. Origination and Extinction through the Phanerozoic: A New Approach. *Journal of Geology* 111: 125–148.

6. Pellegrino, C. 2004. *Ghosts of Vesuvius: A New Look at the Last Days of Pompeii, How Towers Fall, and Other Strange Connections*. New York: Harper Perennial.
7. Gillman, M., and Erenler, H. 2008. The Galactic Cycle of Extinction. *International Journal of Astrobiology* 7: 17–26.
8. Cocconi, G., and Morrison, P. 1959. Search for Interstellar Communication. *Nature* 184: 844–846.
9. Tarter, J. 2001. The Search for Extraterrestrial Intelligence (SETI). *Annual Review of Astronomy and Astrophysics* 39: 511–548.
10. Ekers, R., Cullers, D.K., Billingham, J., and Scheffer, L.K., eds. 2002. SETI 2020: *A Roadmap for the Search for Extraterrestrial Intelligence.* Mountain View, CA: SETI Press.
11. Weinberger, R., and Hartl, H. 2002. A Search for 'Frozen Optical Messages' from Extraterrestrial Civilizations. *International Journal of Astrobiology* 1: 71–73.
12. Edmondson, W.H., and Stevens, I.R. 2003. The Utilization of Pulsars as SETI Beacons. *International Journal of Astrobiology* 2: 231–271.
13. Shostak, S. 2003. Searching for Sentience: SETI Today. *International Journal of Astrobiology* 2: 111–114.
14. Dick, S.J. 2006. Anthropology and the Search for Extraterrestrial Intelligence: An Historical View. *Anthropology Today* 22 (2): 3–7.
15. Tarter, J. 2002. Archive of SETI Searches. In Ekers et al. 2002 (note 10), 381–425.
16. Ibid.
17. Freeman, J., and Lampton, M. 1975. Interstellar Archaeology and the Prevalence of Intelligence. *Icarus* 25: 368–369.
18. Spennemann, D.H.R. 2007. A Line in the Sand? Explorations of the Cultural Heritage Value of Hominid, Pongid and Robotid Artifacts. *International Journal of Cultural Property* 14: 241–266.
19. Gorman, A. 2005. The Cultural Landscape of Interplanetary Space. *Journal of Social Archaeology* 5: 85–107.
20. Gorman, A. 2007. La Terre et l'Espace: Rockets, Prisons, Protests and Heritage in Australia and French Guiana. *Archaeologies: Journal of the World Archaeological Congress* 3: 153–168.
21. Campbell, J.B. 2004. The Potential for Archaeology within and beyond the Habitable Zones of the Milky Way. In *Bioastronomy 2002: Life among the Stars,* ed. R. Norris and F. Stootman, 505–510. San Francisco: Astronomical Society of the Pacific (International Astronomical Union Symposium 213).
22. Campbell, J.B. 2006. Archaeology and Direct Imaging of Exoplanets. In *Direct Imaging of Exoplanets: Science and Techniques,* ed. C. Aime, and F. Vakili, 247–250. Cambridge, United Kingdom: Cambridge University Press (International Astronomical Union Colloquium 200).
23. Cocconi and Morrison 1959 (note 8).
24. Tarter 2001 (note 9).
25. Tarter 2002 (note 15).
26. Dick 2006 (note 14).
27. Ruggles, C.L.N. 2007. Ancient Astronomy: *An Encyclopedia of Cosmologies and Myth.* Boulder, CO: University Press of Colorado.
28. Fitch, W.T. 2000. The Evolution of Speech: A Comparative Review. *Trends in Cognitive Science* 4: 258–267.
29. Fisher, S.E., and Marcus, G.F. 2006. The Eloquent Ape: Genes, Brains and the Evolution of Language. *Nature Reviews: Genetics* 7: 9–20.
30. Reiss, D., and Marino, L. 2001. Mirror Self-Recognition in the Bottlenose Dolphin: A Case of Cognitive Convergence. *Proceedings of the National Academy of Sciences* 98: 5937–5942.

31. Vakoch, D.A. 1998. Constructing Messages to Extraterrestrials: An Exosemiotic Perspective. *Acta Astronautica* 42: 697–704.
32. Vakoch, D.A. 1998. Signs of Life beyond Earth: A Semiotic Analysis of Interstellar Messages. *Leonardo* 31: 313–319.
33. Vakoch, D.A. 2008. Representing Culture in Interstellar Messages. *Acta Astronautica* 63: 657–664.
34. Bracewell, R.N., and McPhie, R.H. 1979. Searching for Nonsolar Planets. *Icarus* 38: 136–147.
35. Kaltenegger, L. 2004. Search for Extra-terrestrial Planets: The Darwin Mission, Target Stars and Array Architectures. PhD diss., Institut für Geophysik, Astrophysik und Meteorologie, Naturwissenschaftlichen Universität der Karl-Franzens Universität Graz, Graz, Austria (original thesis in English, used with permission).
36. Papaloizou, J.C.B. 2008. Planetary System Formation. *Science* 321: 777–778.
37. Chyba, C.F., and Hand, K.P. 2005. Astrobiology: The Study of the Living Universe. *Annual Review of Astronomy and Astrophysics* 43: 31–74.
38. Blumberg, B.S. 2004. Astrobiology: Process and Discovery. In Norris and Stootman 2004 (note 21), 3–7.
39. Carlotto, M.J., and Stein, M.C. 1990. A Method for Searching for Artificial Objects on Planetary Surfaces. *Journal of the British Interplanetary Society* 43: 209–216.
40. Shostak 1998 (note 1).
41. Tarter 2002 (note 15).
42. Steel, D. 1995. SETA and 1991 VG. *The Observatory* 115: 78–83.
43. Steel, D. 1998. The Fermi Paradox and 1991 VG. *The Observatory* 118: 226–229.
44. Phillips, T. 2000. Much Ado about 2000 SG344. http://science.nasa.gov/headlines/y2000/ast06nov_2.htm.
45. Horowitz, P. 2002. Appendix J: The Fermi Paradox. In Ekers et al. 2002 (note 10), 373–374.
46. Ward, P.D., and Brownlee, D. 2000. *Rare Earth: Why Complex Life Is Uncommon in the Universe.* New York: Copernicus.
47. Morris, S.C. 2003. *Life's Solution: Inevitable Humans in a Lonely Universe.* Cambridge, United Kingdom: Cambridge University Press.
48. Watson, A.J. 2008. Implications of an Anthropic Model of Evolution for Emergence of Complex Life and Intelligence. *Astrobiology* 8: 175–185.
49. Waltham, D. 2006. The Large-Moon Hypothesis: Can It Be Tested? *International Journal of Astrobiology* 5: 327–331.
50. Thommes, E.W., Matsumura, S., and Rasio, F.A. 2008. Gas Disks to Gas Giants: Simulating the Birth of Planetary Systems. Science 321: 814–817.
51. Webb, S. 2002. *If the Universe Is Teeming with Aliens ... Where Is Everybody? Fifty Solutions to the Fermi Paradox and the Problem of Extraterrestrial Life.* New York: Praxis/Copernicus.
52. Cohen, J. and Stewart, I. 2002. *Evolving the Alien: the Science of Extraterrestrial Life.* London: Ebury Press.
53. Cohen, J., and Stewart, I. 2004. *What Does a Martian Look Like? The Science of Extraterrestrial Life.* New York: Random.
54. Dyson, F.J. 2003. Looking for Life in Unlikely Places: Reasons Why Planets May Not Be the Best Places to Look for Life. *International Journal of Astrobiology* 2: 103–110.
55. Lee, R., and Daly, B., eds. 2004. *The Cambridge Encyclopedia of Hunters and Gatherers.* 1st paperback edition, revised. Cambridge, United Kingdom: Cambridge University Press.
56. Davies 2006 (note 2).
57. Tarter 2001 (note 9).
58. Tarter 2002 (note 15).

59. Freitas, R.A. 1983. Extraterrestrial Intelligence in the Solar System: Resolving the Fermi Paradox. *Journal of the British Interplanetary Society* 36: 496–500.
60. Fogg, M.J. 1987. Temporal Aspects of the Interaction among the First Galactic Civilizations: The "Interdict Hypothesis." *Icarus* 69: 370–384.
61. Dick, S.J. 2003. Cultural Evolution, the Postbiological Universe and SETI. *International Journal of Astrobiology* 2: 65–74.
62. Tipler, F.J. 2003. Intelligent Life in Cosmology. *International Journal of Astrobiology* 2: 141–148.
63. Cohen and Stewart 2004 (note 53).
64. Cohen and Stewart 2002 (note 52).
65. Ball, J.A. 1973. The Zoo Hypothesis. *Icarus* 19: 347–349.
66. Lineweaver, C.H., Fenner, Y., and Gibson, B.K. 2004. The galactic habitable zone and the age distribution of complex life in the Milky Way. *Science* 303: 59–62.
67. Ward and Brownlee 2000 (note 46).
68. Morris 2003 (note 47).
69. Cohen and Stewart 2002 (note 52).
70. Cohen and Stewart 2004 (note 53).
71. Oliver, B.M. 1975. Proximity of Galactic Civilizations. *Icarus* 25: 360–367.
72. Ekers et al. 2002 (note 10).
73. Campbell 2004 (note 21).
74. Ward and Brownlee 2000 (note 46).
75. Burchell, M.J. 2006. W(h)ither the Drake Equation? *International Journal of Astrobiology* 5: 243–250.
76. Bounama, C., von Bloh, W., and Franck, S. 2007. How Rare is Complex Life in the Milky Way? *Astrobiology* 7: 745–755.
77. Dick 2003 (note 61).
78. Tipler 2003 (note 62).
79. Kaltenegger, L., Jucks, K., and Traub, W. 2006. Atmospheric Biomarkers and Their Evolution over Geological Timescales. In Aime and Vakili 2006 (note 22), 259–264.
80. Kaltenegger, L., Traub, W.A., and Jucks, K.W. 2007. Spectral Evolution of an Earth-Like Planet. *The Astrophysical Journal* 658: 598–616.
81. Kaltenegger 2004 (note 35) (see specifically p. 87).
82. Kaltenegger 2004 (note 35).
83. Kaltenegger, L., Eiroa, C., Stankov, A., and Fridlund, M. 2006. Target Star Catalogue for Darwin: Nearby Habitable Star Systems. In Aime and Vakili 2006 (note 22), 89–92.
84. Quintana, E.V., Adams, F.C., Lissauer, J.J., and Chambers, J.E. 2007. Terrestrial Planet Formation around Individual Stars within Binary Star Systems. *The Astrophysical Journal* 660: 807–822.
85. Guedes, J.M., Rivera, E.J., Davis, E., Laughlin, G., Quintana, E.V., and Fischer, D.A. 2008. Formation and Detectability of Terrestrial Planets around Alpha Centauri B. *The Astrophysical Journal* 679: 1582–1587.
86. Kondo, Y., Bruhweiler, F.C., Moore, J., and Sheffield, C., eds. 2003. *Interstellar Travel and Multi-Generation Space Ships*. Burlington, Ontario, Canada: Apogee Books.
87. Bjørk, R. 2007. Exploring the Galaxy Using Space Probes. *International Journal of Astrobiology* 6: 89–93.
88. Porto de Mello, G., Fernandex del Peloso, E., and Ghezzi, L. 2006. Astrobiologically Interesting Stars within 10 Parsecs of the Sun. *Astrobiology* 6: 308–331.
89. Williams, D.M., and Knacke, R.F. 2004. Looking for Planetary Moons in the Spectra of Distant Jupiters. *Astrobiology* 4: 400–403.
90. Chela-Flores, J. 2007. Testing the Universality of Biology: A Review. *International Journal of Astrobiology* 6: 241–248.
91. Napier, W.M., Wickramasinghe, J.T., and Wickramasinghe, N.C. 2007. The Origin of Life in Comets. *International Journal of Astrobiology* 6: 321–323.

92. Barrow, J.B., and Tipler, F.J. 1986. *The Anthropic Cosmological Principle.* Oxford, United Kingdom: Oxford University Press.
93. Dick 2003 (note 61).
94. Tipler 2003 (note 62).
95. Davies 2006 (note 2).
96. Watson 2008 (note 48).
97. Campbell 2004 (note 21).
98. Campbell 2006 (note 22).
99. Ward and Brownlee 2000 (note 46).
100. Morris 2003 (note 47).

47 Technology and Material Culture in Science Fiction

Thomas S. Mehoke

CONTENTS

INTRODUCTION

With the advent of the first era in which there are old and unused spacecraft in orbit around Earth, combined with burgeoning capitalistic visions of a legitimate market for space tourism, we are for the first time faced with the historical significance that these objects may have for future humans. We have long had the idea that the spacecraft we put into space for specific missions may one day be found and looked at for historical value by other life, should it exist, as has been evident ever since the inclusion of the Golden Record in the Voyager space probes, meant to teach anyone that comes across it of the location and life on our planet. However, although we are only now beginning to envision these significant objects as things we could one day go see as space tourists, many science fiction authors have already capitalized on this notion (whether human discovery or alien discovery) and written of the significance of such archaeological finds. In L. Ron Hubbard's *Battlefield Earth*, the alien race that discovers one of the Voyager Golden Records proceeds to locate and conquer Earth. While we may not be concerned with alien species coming across our spacecraft, these stories provide an interesting perspective as to what our own society will do when "discovering" these "time capsules" that we put into space many years earlier. Much of the archaeological interest of science fiction works has been devoted to this discovering of relics of either terrestrial or extraterrestrial origin. This chapter will focus on science fiction concerning relics of terrestrial origin, as well as the portrayal of the general science behind the condition of space relics and how they are dealt with (e.g., are they brought aboard another ship to be studied, are they studied in situ, etc.) and the reasoning behind their examination (e.g., historical/scientific curiosity, nostalgia, etc.).

SUSPENSION OF BELIEF AND SPURIOUS CLAIMS

> It is a fallacy to argue (or assume) that since there is no certainty, there is no reason to believe anything. The most common form of this fallacy is the argument (or assumption) that since no descriptive claim about the world is justified with certainty, then all descriptive claims are alike in the sense that any one is as justified as any other. ... If all ideas are equal, then why not believe mine? But not all ideas are equal. There is more reason to believe some than others, even if none is perfectly justified to the point of certainty. ... Evidence can be found to support almost any claim; what counts is how much strain the interpretation of that evidence puts on the coherence of the rest of our network of beliefs.[1]

This quote by Peter Kosso demonstrates how many spurious claims can be made regarding archaeological artifacts, citing one piece of "evidence" as support. However, when the full breadth of information surrounding that artifact and the culture that produced the artifact is taken into account, most, if not all, of the spurious claims can be immediately rejected because of the strain they place on this epistemology. This relationship between unauthentic claims and potentially true claims can also be seen in terms of science fiction stories, via our framework of beliefs surrounding life on other planets. Part of the allure and fun of reading science fiction stories is their potential to be true. Maybe *Star Trek* actually shows what life might be like in 300 years if our civilization develops the ability to travel through space at faster than the speed of light. If the premise of a story is completely impossible, such as a story of an advanced civilization living on the surface of the Moon, it could not be envisioned by the readers because we have been to the Moon and know for certain that there is not such a civilization in existence there. Thus, science fiction writers will often write stories that fit into our framework of beliefs concerning space (at least to the average reader) such that their stories can be at least somewhat plausible, though this is still not always the case. Additionally, and unfortunately, science fiction makes no distinction between factual, evidence-supported knowledge and random proposed theories such as those put forward by pseudoarchaeologists. Erich Von Däniken's theories in particular have proven to be quite popular with Hollywood (e.g., *Stargate*,[2] *Alien vs. Predator*[3]). However, despite these completely false assumptions, many sci-fi visions of the future do bring up interesting points as to what our future life in space may be like. Will *Futurama* prove prophetic in portraying the *Apollo 11* landing site as a long-forgotten landmark on the Moon once a lunar theme park becomes the main attraction?[4] Will old space probes be merely used for Klingon target practice?[5] Or perhaps our descendants will be more interested in the history and examination of such historic objects and sites. Only time will tell for sure, but the effort we put toward establishing such landmarks as truly historic landmarks today may shape the course this future takes.

HISTORY OF SPACE TRAVEL IN SCIENCE FICTION

When the first descriptions of travel through space were written 2,000 years ago, not only was the nature of space was uncertain, but the concept of how humans could get there was beyond comprehension. In AD 170, Lucian of Samosata's *True History*[6] depicts a giant waterspout accidentally lifting a ship sailing on the Mediterranean to the Moon, while the Indian epic *Rāmāyana*,[7] written between 500 and 100 BC, had its own vimāna flying ship, the *Pushpaka*, which was described as flying by the force of its master's will (Válmíki 3.32.55–56; 3.48.20–21). Science fiction works concerning space travel were thus on the one hand limited to the current societal understanding of space and flight in general, but on the other hand constrained by the limits of the human imagination. It comes as no surprise then that the next real entries into the genre of science fiction would arrive shortly after the Copernican theories and around the same time as the creation of Galileo's first telescope in the early 1600s. With the telescope's ability to first see in more detail the surface of the Moon came theories of structures on the Moon, the lifeforms that created them, and where else in space life might be,[8,9,10] and these theories would be met with varying degrees of acceptance (Bruno was burned at the stake in 1600 by the Roman Inquisition). However, even at this point, the means of traveling to the Moon was still uncertain; Kepler would later use occult forces to transport characters to the Moon.[11] Many proposed the existence of water on other celestial bodies, as it was established that water would be the main requirement for extraterrestrial life.[12,13] This hope for the existence of water shows the existence of a desire to find life outside of our planet, despite the lack of an ability at this time to test this hypothesis.

A working concept for space travel in science fiction was first truly proposed by Jules Verne with his publication of *From the Earth to the Moon* in 1865.[14] This story was soon followed in 1869 by Edward Everett Hale's story *The Brick Moon*, with the first depiction of a human-made satellite and space station.[15] By the end of the 1800s, H.G. Wells had continued Jules Verne's idea with his own *The First Men in the Moon*,[16] which also contributed substantially to the plot of the first science fiction film.[17] In these last two stories of lunar travel, an indigenous species (the Selenites) was discovered living on the surface of the Moon, and these stories would help to firmly plant the idea of extraterrestrial species into popular culture. The idea of Mars as possibly containing life also came into public awareness at the end of the nineteenth century through the works of Percival Lowell and a mistranslation of the Italian word *canali* from Giovanni Schiaparelli's detailed maps of Mars (translated as *canals*, when in reality it means *channels*).[18,19] Lowell's ideas would place Mars at the focal point of humanity's interest in extraterrestrial life for the next 50 years. The idea of Martian life was evident through all genres of science fiction, from H.G. Wells' tale of Martian invaders, *The War of the Worlds*, in 1898 and Orson Welles' all too believable radio adaptation in 1938 to Hollywood films, such as *Flash Gordon's Trip to Mars*[21] and *Rocketship X-M*.[22] The idea was so widespread that in 1921, and again in 1924, the U.S. government ordering all U.S. radio stations silent so that Guglielmo Marconi might possibly pick up a Martian radio signal.[23] Early films, such as *Flash Gordon*, were highly fictionalized and fantastical; science fiction films would take a much more realistic and serious turn with the Cold War concerns of the 1950s. Films such as *Destination Moon*,[24] *Rocketship X-M*, and *The Day the Earth Stood Still*[25] not only seriously dealt with the prospect of space travel and the problems, and technology required for it but also shifted away from the more fantastical ideas of civilizations on Mars, introducing concepts such as archaeological ruins on planetary bodies and the idea that aliens might be as civilized as human beings, if not more so. This idea of noble aliens protecting the "galactic community" from human carelessness would also later be incorporated into the *Star Trek* story of human-alien first contact,[26] as well as the video game *Mass Effect*.[27]

The advent of actual human spaceflight in 1961 with Yuri Gagarin's orbit of the Earth, as well as unmanned probes to the Moon (1959), Venus (1962), and Mars (1964), only further increased our understanding of space travel and the planets with which we share our solar system. Accordingly, science fiction shifted to accommodate this new evidence with a new focus on what the actual experience of life in space might actually be like; rather than stories reminiscent of *Flash Gordon*, we have films such as *Total Recall*,[28] in which direct exposure of humans to the Martian atmosphere has disastrous consequences. The emphasis on the experience of life in space was nowhere more evident—and had nowhere more of a societal impact—than in Stanley Kubrick's 1968 film *2001: A Space Odyssey*.[29] With this film, the public's perception of space exploration would be forever changed. The film *2001* created a very real and believable futuristic portrayal of Earth that would go on to shape the general public's belief in what space travel would actually be like. From the idea of the giant rotating wheel of a space station to the unusual but accurate portrayal of the complete lack of sound propagation through space, the film made an indelible mark on the collective public understanding and belief in what space travel "should" be like.[30]

With the new information collected from the Moon and Mars no longer allowing the concept of a Lunar or Martian civilization to be plausible, science fiction writers adopted other conventions and places in order to retain the allure of human-alien encounters. The main plot devices that allowed this retention were the discovery of alien artifacts in our own solar system (already seen in *Rocketship X-M* and *2001*) or the idea of having aliens living in different solar systems. In order to achieve this interaction between multiple solar systems, one of our species would have to travel many light-years to reach the other's planet, and the only ways to traverse such a distance while keeping the main characters alive would be traveling faster than the speed of light (e.g., *Star Wars*, *Star Trek*) or going into stasis pods (e.g., *Alien*[31]). While stasis pods may be more realistic in terms of our current understanding of physics and relativity, light-speed travel is so much more interesting plot-wise, as it allows the characters to remain awake during the journey and also to visit multiple planets in a short amount of time (and we all know that any time the whole crew goes into stasis, something terrible is bound to happen[28,31]). It is the discovery of artifacts in space, however, that is the more interesting and applicable plot element given the scope of this chapter, and we will now examine it in more detail.

SCIENCE FICTION GENERAL CATEGORIES OF MATERIAL INVESTIGATIONS

For our purposes in this chapter, I will employ a simplistic definition of archaeology that was put forth by Garrett G. Fagan in *Diagnosing Pseudoarchaeology*. "Put simply, archaeology is the recovery, analysis, and interpretation of the physical remains of past human activity."[32] It should go without saying that within this chapter's discussion of science fiction, in particular when discussing the alien species that are so often portrayed in science fiction stories, the phrase "human activity" in the above definition can be assumed to include activity from these other sentient alien species as well. A science fiction story that includes the archaeological examination of alien artifacts may still present interesting archaeological insight, even though the premise of alien-created artifacts is clearly fictional (see Doleman, Chapter 38 of this volume). What is centrally important in this definition is the interaction with non-natural objects and how they are reached, identified, and interpreted. In terms of life in space, and particularly referenced in science fiction, this definition of archaeology can be thought of in three general categories: (1) alien artifacts on Earth, on other planets, or drifting in space discovered by humans; (2) human artifacts as discovered by alien species, and (3) the identification of items free-flying through space. This last category will prove to be the most relevant of the three to our actual world.

The first category is the discovery of an alien-built object unearthed on the Moon, Mars, or Earth or discovered drifting through space (e.g., the obelisk from *2001: A Space Odyssey*, the whale-communicating cylinder from *Star Trek: The Voyage Home*,[33] or the Allspark from *Transformers*[34]). The objects more often than not have a profound effect on the human race and/or Earth, yet they usually remain a mystery in terms of their origins or mechanics, and the plot often does not stay too close to the artifact. This typically ancient alien artifact is central to the storyline; however, the storyline usually does not contain real archaeological methods to analyze and interpret the artifact. The story usually just follows the effects the artifact has on the humans that discover it. This theme is particularly evident with the black monoliths in *2001: A Space Odyssey* and the astronauts' quest to first excavate the one buried on the Moon and later to rendezvous with the one orbiting Jupiter. However, it is never really explained or hypothesized what they actually are and where they came from, two crucial archaeological questions that go unanswered. Similar sci-fi settings incorporating artifacts from ancient, and usually long-dead, civilizations are pervasive throughout the science fiction genre, as can be seen through the 200,000-year-old Krell ruins in *Forbidden Planet*,[35] as well as the numerous encounters that *Star Trek* crews have had with ancient artifacts more than 100,000 years old.[36,37] Similarly, several recent video games have had storylines revolving around ancient alien ruins.[38,27] The basis for the stories for both *Halo* and *Mass Effect* are the archaeological ruins from ancient civilizations (the Forerunners from 100,000 years ago and the Protheans from 50,000 years ago, respectively). Interestingly, in addition to the use of ancient archaeological ruins for its setting, the latest installment of the *Halo* series, *Halo 3*, can be tackled "archaeologically" by the player as well. Hidden throughout the worlds of the game are seven computer terminals, unnecessary to the game's storyline, that if found reveal fragments of communications between ancient Forerunners and can be used to develop a picture of what happened 100,000 years ago to destroy their civilization. The information is incomplete and must be recorded by hand and read thoroughly; however, once all of the fragments are collected, a better idea of the mysterious individuals who built the Halos can be determined by those so inclined.

This initial category concerning the direct interaction and interpretation of alien artifacts in the traditional sense is an interesting category as it is most similar to the archaeology we have known on Earth. The material remains of past human ancestors encountered on Earth range from hominid stone tools that date to around

2 million years ago to ancient civilizations beginning around 10,000 years before the present.

The second category is something of a corollary to the first and centers on what would happen if an alien species recovered an object that we have sent into space. Interestingly, this idea has actually been considered in real life when sending spacecraft into space, ever since the inclusion of the Golden Record in the Voyager satellites. The Golden Record not only contains the probe's name and the date it was launched but also cultural information about the human race in the form of spoken greetings in fifty-five different languages, music from many different cultures and eras, and sounds and pictures of various animals and weather here on Earth.[39] This inclusion of culturally relevant material in itself is a very interesting idea; how does one go about creating an archaeological cache such that when it is discovered by future archaeologists it can be interpreted correctly? Language is obviously a challenge in future communication; the Voyager team came up with a fairly ingenious means of achieving this communication, creating a kind of scientific "language" for the record based on a binary system in which the units are the fundamental transition states of the hydrogen atom. However, even without an understanding of this language, or if the engraving is too damaged or not understandable symbolically for another "culture" (if aliens can be said to have a culture), future archaeologists (whether they be humans from a future Earth or possibly from another planet) may be able to date the spacecraft by measuring the radioactive decay of an ultra-pure sample of uranium-237 that was included on the surface of the record, which, with a half-life of 4.51 billion years, could allow it to be accurately dated for quite some time.

This concept of future discovery of actual human spacecraft, in particular the Voyager Golden Record, is often incorporated into science fiction works, resulting in some very interesting plotlines. This connection from the sci-fi fantasy world to the true world we live in makes the science fiction story all the more plausible and even potentially realistic. In both *Battlefield Earth* and the animated Transformers series *Beast Wars*, the Voyager Golden Record is discovered by an alien race (the Psychlos and Predacons, respectively), deciphered, and then used to triangulate the location of Earth. In *Battlefield Earth*, this leads to the enslavement of the human race,[40] and in Beast Wars the "Golden Disk" brings the Predacons and Maximals to prehistoric Earth in the initial episode[41] and was an integral part of the storyline for the next 34 episodes. These two stories bring up the interesting and somewhat worrisome concept of what might happen if the "map" of Earth's location sent out on the Golden Record was discovered by a hostile alien race. This concept is also examined in Carl Sagan's novel *Contact*,[42] in which it is debated whether a transmission (seemingly from the Vega system) of blueprints hidden in a retransmission of Adolf Hitler's speech at the 1936 Berlin Olympics was in fact a hostile message. In the first *Star Trek* movie,[43] a twist on this concept arises when an alien race that encountered a fictional *Voyager 6* probe modified it and sent it back to us. This concept that what we send into space may be found by other civilizations at some point in the future, even if they just come from Earth and desire to learn about their history, is an attractive idea in much the same way that time capsules are an attractive idea. Such a time capsule sent into space, containing an incredible amount of information and material culture about our human cultures and our experiences of life here on Earth, could survive for potentially billions of years.

Along the same lines as this idea of where the spacecraft we launch may one day end up, questions have arisen as to what effect we are having on our planet's environment, as well as our local space environment. Similarly, attention is also being paid in the sci-fi realm as to what effect an encounter with our technology might have on an alien civilization that comes across it. The 1999 movie *Galaxy Quest* revolves around a species living 20 light-years from Earth that constructs actual interstellar spacecraft simply from watching 20-year-old episodes of a show similar to *Star Trek*, without having the experience to deal with the galactic community of which they then become a part.[44] A more environmental question is discussed in a 2001 *Star Trek: Voyager* episode[45] in which a futuristic "warp-capable" (able to travel faster than light speed) probe named *Friendship One* is launched by future humans as a gesture of friendship to whoever might come across it. However, the probe crashes on a primitive planet in the Delta Quadrant, and the antimatter technology that it contains is used by the citizens of the planet to create terrible weapons and wage a war that ravages the entire planet's surface. While the actual likelihood of something we send into interstellar space landing on the surface of another planet, let alone another habited planet, is quite low, the idea that what we send into space could at some future time have environmental repercussions is good for the plot and interesting to consider.

Once we move our archaeological attention to space, it is much more likely that the large majority of objects humans will encounter will not be directly from another planet but will rather be of our own creation, adrift in space, orbiting some body of mass or Lagrange point. Therefore, while the use of a kind of archaeological scenario of the discovery an alien artifact on Earth or in space is the easiest to understand in terms of our current understanding of archaeological methods (and certainly the most marketable from a Hollywood standpoint), it is not the most relevant archaeological scenario to our current and future explorations in space. In particular, given the scope of this book, our strongest archaeological need with respect to space will be our ability to locate, identify, and interact with known satellites and rovers that we have sent into space. We are lucky in that the science fiction genre, not being constrained by such trifles as archaeological theory and methods, funding, or even physics, has already devoted considerable time to dealing with this third archaeological category. The *Star Trek* franchise will be used for examples of some of the technologies that have been used to aid this interaction.

In order to interact with an object, first it must be located (see also Capelotti, Chapter 45 of this volume). *Star Trek* ships, as well as spaceships from most science fiction stories, have both long-range and short-range sensors (with no clear explanation given as to the range limitations of either). Usually, short-range sensors allow a ship to scan objects that it can directly interact with, either immediately or after a short trip (minutes), whereas long-range sensors usually function at a distance of hundreds of thousands of kilometers, which usually is a few hours away. Sensors provide information as to the identity of an object, usually by hull configuration, propulsion signature, or hull composition, and if close enough the tactical information of a ship (shields, weapons, and other internal configurations) as well as the "life signs" of any living crew on board will be transmitted back. The trace elements of hull fragments have even been used to identify the exact ship from which they came.[46] If scanning a planet, sensors can provide detailed information as to the terrain and buildings on the surface, as well as any mineral deposits or life signs. These sensors are similar to what can be obtained by using a handheld tricorder to scan an object. Tricorders, and other handheld scanners, can usually provide detailed atomic analysis of any nearby object (usually, the scanner does not have to touch the object) and can also be used to detect any illness or infection. These readings are all also virtually automatic in the sense that a member of the crew can initiate a scan at a console and almost immediately will be provided with a complete list of the pertinent information via a screen readout, quickly and fully analyzed (apparently) by the ship's computer. Sensors are also linked with the communication array such that short- and long-range communications with little to no time-lag due to distance traveled are possible. In the real world, the location of objects in space is limited to detailed orbit calculations, with communication transmissions updating the ground station with any changes in the predetermined trajectory. When communication with an object is lost, the only way to determine where it is now is to perform detailed calculations as to what orbit an object of that mass should follow if exposed to the various gravitational forces in space and then make a guess as to where the spacecraft should be. Clearly, a capability such as *Star Trek* sensors would provide real-life space archaeologists and aerospace engineers with an amazing opportunity to locate and catalogue where many of the "lost" spacecraft in orbit really are.

Back in the sci-fi realm, once sensors have been used to locate an object, a ship often then will travel to the location of that object, whether it is a friendly (or hostile) ship, a nearby planet or moon, or just an interesting piece of debris. Most science-fiction ships are equipped with engines that allow them to travel incredibly fast, whether going below the speed of light (impulse engines, up to ~0.5 the speed of light), or above it (warp drive), making it possible to basically go anywhere you want as fast as your engines can take you there. In the real world, space travel usually depends on waiting until planetary orbits are just right in order to send a spacecraft to a desired location without having to expend excess fuel. This notion is almost unheard-of in sci-fi stories (the action would be terribly slow), and so in the sci-fi realm, fuel is either very plentiful (and lightweight) or very efficient, allowing spaceships to rely solely on engines to travel between points A and B, seemingly without regard to anything else in between. A spaceship travels in basically a straight line when the engines are on and then is able to stop once it gets to its desired location, or if near a planet, the ship will be placed into a standard (or low or high) orbit. The mechanics would obviously be too boring to include in a television show or film; however, it is interesting to consider how one would actually go about piloting a spaceship and how much time would actually be required to put it into orbit around a planet. One thing is for certain: traveling through space in a sci-fi story is certainly much easier and much less technically demanding than it is in the real world.

While both sensors and propulsion systems in science fiction stories can be viewed as highly perfected versions of systems we can build currently, one of the most difficult aspects of real-life space archaeology will most likely not be locating or getting to an object of interest but rather interacting with that object once there. In science fiction, this need to interact with another object in space is almost always dealt with in one of two ways: a tractor beam or a transporter. A tractor beam allows a spaceship to lock onto another ship or object in space and hold it close, or even pull it into a docking bay. A tractor beam could possibly be simulated in the real world with a physical tether that could be used in a similar fashion to grab onto a nearby object; however, the forces that would result from such an action may in fact prove to be too unstable to be useful. In conjunction with, or often in lieu of, a tractor beam, science fiction ships often have a transporter system, which allows matter to be transported almost instantaneously from one location to another. Using this capability, a drifting damaged satellite could be transported into the ship's cargo bay, allowing for much easier inspection and analysis than would be possible in situ. Similarly, a crew can be transported over to an abandoned spaceship to try to determine what caused it to be abandoned. These two means of "investigating" objects in space are exactly the kind of thing that would be needed for real world archaeological analysis of old spacecraft. Current means of interacting with other objects in space usually require highly specialized docking ports if two ships are coming into contact, or they require an astronaut to put on an extravehicular activity suit and go on a potentially dangerous spacewalk. Theoretically, if a spacecraft such as the shuttle, with large hangar doors, could place itself in an orbit directly above or below a sufficiently small satellite and had thrusters that could push it up or down with respect to the satellite such that the satellite would end up inside the hangar, this could potentially be one means of collecting and interacting with an object in space. However, problems could arise when attempting to return the object back into a stable orbit.

These thrusters, sensors, and transporters are crucial in science fiction in order to interact with objects in space, and it is not immediately apparent how we might apply these ideas in the creation and development of real-world technologies. While the idea of transporting is clearly the most farfetched, being able to scan objects and planets is already possible to some extent (though certainly nowhere near *Star Trek* levels), and we already have a fairly good handle on propulsion systems, at least to a workable degree. By developing capabilities in these three areas, our ability to locate what we have sent up into space and then either research how the materials have held up, salvage parts, or perhaps even collect them and put them in a museum would be immensely benefited.

CONCLUSION

It is easy to argue that because science fiction is rooted in fantasy, no valuable comment on the scientific validity it possesses can be made. This argument is strengthened by works from people such as von Däniken, Hancock, and others that make arguments for a sci-fi world in which aliens have supposedly come to Earth in the past and created many of our current archaeological sites, deeds we know from archaeological evidence to not be even remotely plausible. A separation therefore must be made between obvious falsehoods such as these and the genuine speculation of other science fiction works of what a possible future might entail. Often science fiction relies on assumptions about what future life might be like in creating its premise, such as with a utopian/dystopian government (e.g., *Brazil*,[47] *Equilibrium*[48]) or with a civilization that has developed faster-than-light travel (*Star Trek*, *Star Wars*). However, along with this basic premise, science fiction is also firmly rooted in the available scientific knowledge of the time, as we saw from the early science fiction stories. Because of this dependence on current technology (whether in directly applying it or using it to extrapolate possible future technology), science fiction works often become dated very quickly once the futuristic state-of-the-art vision becomes replaced in the real world by something of better quality (e.g., communicators from the original series of *Star Trek* compared with cell phones of today). The various *Star Trek* series are particularly interesting to examine because the franchise has existed since 1966; over that time, the technology of the real-world Earth developed very rapidly and very extensively, including the birth of the Internet, flat-screen televisions, and cell phones, not to mention incredibly powerful personal computers. Despite this technological lag, as long as the time in which the science fiction work was written is taken into account, useful ideas about the future may yet be gained. Examples of potential future methods for archaeologists and scientists have been explored through many *Star Trek* episodes. Humans have been discovered cryogenically frozen for hundreds of years[49,50] or suspended midtransport for up to 80 years,[51] and DNA from around the galaxy has both revealed hidden messages[52] and linked species from opposite sides of the galaxy to a common ancestor.[53] Interestingly, holographic life forms buried for hundreds of years have been used in Star Trek as "living witnesses," correcting misinterpreted evidence from an event centuries ago.[54] While many of these ideas are completely ridiculous, some of them may prove to be as useful for developing future technologies as a discussion of the sensors and propulsion systems is now.

REFERENCES

1. Kosso, P. The Epistemology of Archaeology. In *Archaeological Fantasies*, ed. G. Fagan, 3–22. New York: Routledge, 2006.
2. *Stargate*. Directed by Roland Emmerich. Carolco Pictures, 1994.
3. *Alien vs. Predator.* Directed by Paul W.S. Anderson. Twentieth Century-Fox Film Corporation, 2004.
4. *Futurama*. The Series Has Landed. Fox Broadcasting Company. First broadcast April 4, 1999.
5. *Star Trek V: The Final Frontier*. Directed by William Shatner. Paramount Pictures, 1989.
6. Lucian of Samosata. *The True History*. AD 170. Translated by H.W. Fowler & F.G. Fowler. http://oddlots.digitalspace.net/guests/lucian_true_history.html.
7. Válmíki. *The Rámáyan of Válmíki, translated into English verse*. 500–100 BC. Translated by Ralph T.H. Griffith. http://www.gutenberg.org/etext/24869 or http://www.sacred-texts.com/hin/rama/index.htm.
8. Kepler, J. *Dissertatio cum Nuncio Siderio*, 1610. Translated by Edward Rosen. Reprint, Johnson Reprint Corporation, 1965.
9. Bruno, G. *De L'Infinito Universo et Mondi*, 1584. English translation retrieved from http://www.positiveatheism.org/hist/brunoiuw0.htm
10. Campanella, T. *Civitas Solis*, 1602. Translated by Daniel John Donno. University of California Press, 1981.
11. Kepler, J. *Somnium*, 1634. Translated by Edward Rosen. Courier Dover Publications, 2003.
12. Huygens, C. *Cosmotheoros*, English translation. London: Timothy Childe, 1698. http://www.phys.uu.nl/~huygens/cosmotheoros_en.htm.

13. Herschel, W. On the Remarkable Appearances at the Polar Regions of the Planet Mars, the Inclination of its Axis, the Position of its Poles, and its spheroidal Figure; with a few Hints relating to its real Diameter and Atmosphere. *Philosophical Transactions of the Royal Society of London*. 1784. 74: 233–273.
14. Verne, J. *From the Earth to the Moon*. Norm Wolcott, Gregory Margo, and PG Distributed Proofreaders, 1865. http://www.gutenberg.org/etext/12901.
15. Hale, E. *The Brick Moon*, 1869. http://etext.lib.virginia.edu/toc/modeng/public/HalBric.html.
16. Wells, H.G. *The First Men in the Moon*, 1901. http://www.gutenberg.org/etext/1013.
17. *La Voyage dans la Lune*. Directed by Georges Méliès. Star Film, 1902.
18. Lowell, P. *Mars*, 1895 Reprint, Forgotten Books, 2008. http://books.google.com/books?id=4nNtjCUGhQIC&printsec=frontcover&source=gbs_summary_r&cad=0.
19. Lowell, P. *Mars as the Abode of Life*. New York: Macmillan Company. 1908.
20. Wells, H.G. *War of the Worlds*, 1898. Plain Label Books. http://books.google.com/books?id=1HoBYmku9uQC&printsec=frontcover&source=gbs_summary_r&cad=0#PPA1,M1.
21. *Flash Gordon's Trip to Mars*. Directed by Ford Beebe and Robert Hill. King Features Production, 1938.
22. *Rocketship X-M*. Directed by Kurt Neumann. Lippert Pictures, 1950.
23. Wilford, John N. Mars Is Getting Close, and Maybe So Are Those Little Green Men. *The New York Times*, August 24, 2003.
24. *Destination Moon*. Directed by Irving Pichel. George Pal Productions, 1950.
25. *The Day the Earth Stood Still*. Directed by Robert Wise. Twentieth Century-Fox Film Corporation, 1951.
26. *Star Trek: First Contact*. Directed by Jonathan Frakes. Paramount Pictures, 1996.
27. *Mass Effect*. Bioware, 2007.
28. *Total Recall*. Directed by Paul Verhoeven. Carolco Pictures, 1990.
29. *2001: A Space Odyssey*. Directed by Stanley Kubrick. Metro-Goldwyn-Mayer, 1968.
30. McCurdy, H., and D. Darling. 2001: A Space Odyssey. http://www.daviddarling.info/encyclopedia/A/2001.html.
31. *Alien*. Directed by Ridley Scott. Brandywine Productions, 1979.
32. Fagan, G. Diagnosing Pseudoarchaeology. In Fagan 2006 (note 1) (see specially p. 24).
33. *Star Trek IV: The Voyage Home*. Directed by Leonard Nimoy. Paramount Pictures, 1986.
34. *Transformers*. Directed by Michael Bay. Dreamworks SKG, 2007.
35. *Forbidden Planet*. Directed by Fred M. Wilcox. Metro-Goldwyn-Mayer, 1965.
36. *Star Trek Voyager*. Thirty Days. CBS Paramount Television. First broadcast December 9, 1998.
37. *Star Trek Voyager*. Message in a Bottle. CBS Paramount Television. First broadcast January 21, 1998.
38. *Halo*. Bungie, 2001.
39. Golden Record. http://voyager.jpl.nasa.gov/spacecraft/goldenrec.html, 2008.
40. Hubbard, L.R. *Battlefield Earth*. New York: St. Martin's Press, 1982.
41. *Beast Wars*. Beast Wars: Part 1. Mainframe Entertainment. First broadcast September 16, 1996.
42. Sagan, C. *Contact*. New York: Simon & Schuster, 1985.
43. *Star Trek: The Motion Picture*. Directed by Robert Wise. Century Associates, 1979.
44. *Galaxy Quest*. Directed by Dean Parisot. Dreamworks SKG, 1999.
45. *Star Trek Voyager*. Friendship One. CBS Paramount Television. First broadcast April 25, 2001.
46. *Star Trek: The Next Generation*. Unification: Part 1. CBS Paramount Television. First broadcast November 4, 1991.
47. *Brazil*. Directed by Terry Gilliam. Embassy International Pictures, 1985.
48. *Equilibrium*. Directed by Kurt Wimmer. Dimension Films, 2002.

49. *Star Trek: The Next Generation.* The Neutral Zone. CBS Paramount Television. First broadcast May 16, 1988.
50. *Star Trek Voyager.* The 37's. CBS Paramount Television. First broadcast August 25, 1995.
51. *Star Trek: The Next Generation.* Relics. Directed by Alexander Singer. CBS Paramount Television. First broadcast October 12, 1992.
52. *Star Trek: The Next Generation.* The Chase. CBS Paramount Television. First broadcast April 26, 1993.
53. *Star Trek Voyager.* Distant Origin. CBS Paramount Television. First broadcast April 30, 1997.
54. *Star Trek Voyager.* Living Witness. CBS Paramount Television. First broadcast April 29, 1998.

48 Space Archaeology and Science Fiction

Larry J. Paxton

CONTENTS

INTRODUCTION

In this chapter, we consider some of the basic constraints on our ability to do space archaeology and what popular culture (especially science fiction) and legitimate scientific research have told us we should expect in our search for artifacts of an extraterrestrial civilization. Science fiction is a mirror in which we can see at least some of the fears and aspirations for our society. Science fiction also provides a viewpoint from which we can consider what we would hope to learn from the study of our own space artifacts or those of another world. The existence of science fiction is an expression of our need to understand our place in the world and our place in the cosmos; it is ultimately an affirmation of our humanity. Admittedly, much of science fiction, at least vis-à-vis alien civilizations, is little different from a cargo cult in which the deus ex machina of an alien technology is used to change the setting from Earth to some other poorly realized version of Earth: even so, there are some interesting ideas to be gleaned from the study of science fiction with regard to what it is that we hope to learn from the artifacts of another civilization.

The overlap between science fiction and archaeology is that both endeavors attempt to answer the same questions: Where did we come from? Why are things the way they are? Archaeology has been invoked in science fiction as a means of establishing context and, more often, as a means of establishing a means for the further adventures of the human race; mechanisms, technologies, constructs, and powers that are beyond the linear extrapolation of our current capabilities are often provided to the human race through "space archaeology" in science fiction.

Science fiction has been called the literature of ideas. It is also known as speculative fiction. For most of us, science fiction falls into the "know it when we see it" category.[1] The hallmark of science fiction is a concern with the extrapolation of ideas from the author's understanding of the forces that shape or have shaped the natural world.[2] Science fiction writers and its fans often cite its ability to predict what will happen in the future. In reality, science fiction's success is more akin to "the one hundred monkey" problem.[3] What science fiction has done, and done successfully, is hold up a mirror to our own fears and aspirations.

Science fiction is not entirely the product of twentieth-century America or even of the twentieth century, but much of what it is was refined during the last 50 years in America. The ideas that make up the texture of science fiction go back, at least, to the origins of recorded history.[4] The legends of the Greeks and Romans tell of voyages to the Moon, automatons, and other "high tech" concepts. Cyrano de Bergerac wrote of travels to the Moon and sun[5] nearly 400 years ago. There were many other imaginative writers, but it was H.G Wells who shaped the initial form of science fiction as a genre long on imagination and short on character. The H.G. Wells stories are part of our culture; *The War of the Worlds* and *The Time Machine* provided a template for contact with alien races that persists to this day. Science fiction began to establish a real identity through the publishing activities of Hugo Gernsback. Gernsback began publishing science fiction in 1911 and found that, in 1926, he could establish a magazine, *Amazing Stories*, devoted to science fiction. It was in the early years of science fiction that much of what we see in, now, popular culture was established as the basic content of the medium and much of the best work was published as short fiction. Two examples are Ray Bradbury's brilliant, elegiac stories about a civilization on Mars in *The Martian Chronicles*[6] and Isaac Asimov's *I, Robot*[7] stories. By the 1960s the space operas of the movie serials had been translated to television, with the most important series being *Star Trek. Star Trek* defined for many what science fiction was and is. While this is somewhat akin to equating comic books and graphic novels, the distinction is lost to all but the cognoscenti. *Star Trek* in its various incarnations had, and continues to have, an essentially optimistic view of the future.[8] *Star Trek* dealt with ancient and contemporary civilizations. As part of that mythmaking, contact with the artifacts of another, usually previous, civilization was an important vehicle for explicating our own experience. For example, in the *Star Trek* episode "City on the Edge of Forever,"[9] an alien artifact that provides a gateway to any time is all that remains of a civilization that died millennia before. The story dealt not with the civilization that created the gateway, but with the concepts of destiny and free will, thus putting the idea of a time portal that could be used for "archaeology" in a human context. The idea of traveling back through time to do "archaeology" is more fully developed by Connie Willis.[10] Her novel *Doomsday Book* describes the use of a time machine to study the past. The story hinges upon the humanity of archaeology—that is, discovering through the study of an alien culture truths about our own.

Archaeology is the systematic study of the physical remains of past cultures. In our world, one of the reasons we study archaeology is to construct a cultural chronology. Archaeology is, however, more than a chronology: a major part of the appeal of the science of archaeology is that it provides an insight into the way that the past inhabitants lived. The overarching theme of archaeology is that by learning about how past civilizations dealt with the world we can gain insight into the processes and imperatives that shape human behavior. The physical remains of a culture may be rich and varied or exist in fragments that are either dispersed or incomprehensible. Archaeology is a science: observations are collected, and hypotheses are postulated and either proved or disproved (if enough information exists). One of the most important effects of archaeology is that a unified timeline among disparate cultures can be established.

The intersection of archaeology and science fiction occurs when we consider the physical remains of a past, though not necessarily an extinct culture; or a culture that is either of this planet (e.g., the rediscovery of a space probe); or some other artifact in space or on another body in this solar system or elsewhere; or our own culture by means of some postulated scientific advance, such as the time portal in *Star Trek*'s "The City on the Edge of Forever." In this chapter, we will look at what science and science fiction can tell us about the artifacts of another civilization.

A NEEDLE IN A HAYSTACK

In 1960, in order to provide a motivation for conducting a radio survey in the search for extraterrestrial intelligence (SETI), Frank Drake presented a formulation of the number of intelligent civilizations in the galaxy that would be capable of communicating.[11]

The Drake equation is:

$$N = R^{*} f_p n_e f_l f_i f_c L \tag{48.1}$$

where N is the number of technological civilizations in our galaxy that we might be able to communicate with at this time, R^* is the average rate of star formation in the galaxy, f_p is the fraction of stars with planets, n_e is the average number if planets in the stellar system that could support life, f_l is the fraction that go on to develop life, f_c is the fraction of civilizations that develop a technology that emits signs of their existence, and L is the length of time that such a civilization can emit signals.

Following Drake, we can express this in terms of the number of stars in the galaxy by equating the number of stars, N^*, to the product of the rate of formation, R^*, and the age of the galaxy, T_g. Thus

$$N = N^{*} f_p n_e f_l f_i f_c \, {}^{L}\!/_{T_g} \tag{48.2}$$

Not everyone agrees on the definition of a habitable world or a world in which life as we would recognize it would exist. Generally science fiction deals with life that is carbon-based and occurs on a planet where liquid water is stable.[12] Currently, the *Phoenix Lander* is looking for evidence of frozen water and life on the surface of Mars.

The original value for the Drake equation was 10. Estimates range from of the order of 1 to of the order of 10,000.[13] While the upper range of the estimates seems like a very large number, this number or any other reasonable value actually implies a very small number of civilizations that could contact us by means that we are currently capable of interpreting. The value of the number of technological civilizations should, of course, be compared with the number of stars in the galaxy, which is about 100×10^9. Thus, even at the highest range of the estimates, there would be only one civilization per 10,000,000 stellar systems.

Science fiction has always dealt with alien civilizations. The Kardashev scale classifies civilizations by the amount of energy that they have at their disposal.[14] Each of these levels of civilization can then be expected to produce different kinds of artifacts—artifacts that we may or may not be able to interpret (Table 48.1) At Kardashev level IV, we enter the realm of godhood, but as Arthur C. Clarke said, "Any sufficiently advanced technology is indistinguishable from magic."[15] It is notable that as far as we can tell, all the observable galaxies have failed to develop a Type III or Type IV civilization. Given that, based on our current understanding, there are stars older than our own sun that should have, in all likelihood, had planets upon which intelligent life could develop, and that these planets would have formed billions of years before our own: where is everyone? We see no evidence of a Type III or IV civilization in any of the tens of thousands of galaxies imaged by

TABLE 48.1
The Kardashev Scale

Classification	Amount of Power Available	Examples from Science Fiction
Type I	Current energy usage of the Earth up to Type II	The Krell from the movie *Forbidden Planet;* the spacefaring societies of the *Star Trek* videos, movies, and books.
Type II	Energy output of a star	*Ringworld* series by Larry Niven, Dyson spheres,[1] Asimov's *Foundation, Star Wars, Star Trek*'s Borg civilization, the Gene Wolfe novels of the New Sun,[2] the "sender" race in *Contact,* the Star Wars Empire.
Type III	Energy output of a galaxy	The Forerunner race of the Halo universe,[3] the Protheans of *Mass Effect*.[4]
Type IV	Energy output greater than a galaxy	The Q continuum of *Star Trek*, the Time Lords of *Doctor Who*, the Ancients from the *Stargate* series.

Source: Adapted from http://en.wikipedia.org/wiki/Kardashev_scale.

[1] Dyson, F.J. 1998. *Imagined Worlds*. Cambridge, MA: Harvard University Press.

[2] Wolfe, G. *The Book of the New Sun* and *Urth of the New Sun*, 1997 Orb Books (in which the Earth's sun is restored to full luminosity).

[3] For example, Nylund, E. 2001. *Halo*: *The Fall of Reach*. Dietz, W.C. 2003. *Halo: The Flood*. Nylund, E. 2006. *Halo: Ghosts of Onyx*. See also *Halo, Combat Evolved*, 2006 Microsoft Games.

[4] Karpyshyn, D. 2007. *Mass Effect: Revelation*. Del Rey Books.

the Hubble Space telescope, and there are something like 500,000,000 galaxies in the universe.

The density of stars in the galaxy can be represented by the value $\rho(r,\theta,\phi)$, where this density is expressed in stars per light-year as a function of the location within our galaxy. The coordinates r, θ, and ϕ are the radial distance, galactic latitude, and galactic longitude, respectively. The number of stars within the neighborhood of our sun is given by

$$N' = \oiiint_s \rho(r,\theta,\phi)\,ds\,d\theta\,d\phi \qquad (48.3)$$

where the integral is over a volume centered about our sun of radius s. We introduce the variable s to indicate that this is a distance from our sun that defines the neighborhood that we have access to in some form. Where s is sufficiently large, this integral reduces to the total number of stars in the galaxy. Figure 48.1 shows a simple plot of the probability that a stellar system will be encountered with life as a function of distance in light-years. In this calculation, the simplifying assumptions that the galaxy is spherical and that the average distance to a star is fixed as 4 light-years have been made.

Figure 48.1 shows that the search volume must be very large in order to have a reasonable chance of a direct encounter with a technological civilization. The values of the total number of civilizations can be related to the average lifetime of a

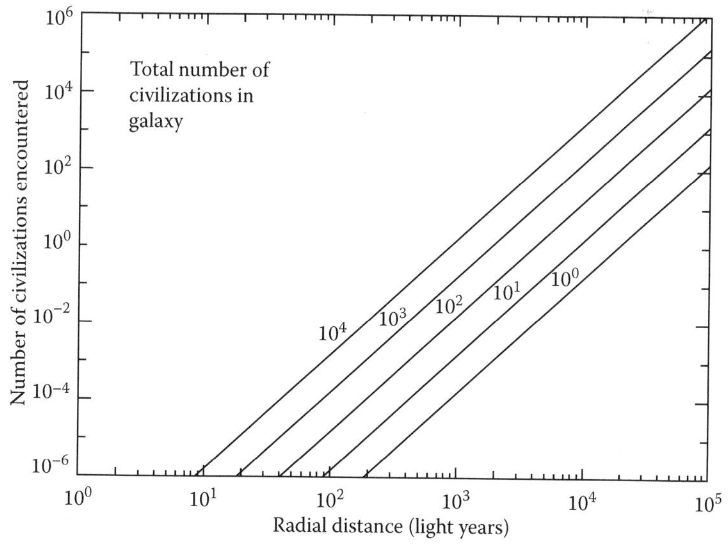

FIGURE 48.1 The number of civilizations as a function of radial distance from the Earth for five assumed values of the total number of civilizations in the galaxy that are capable of communicating with the Earth. Here, a high value for the average distance between stars of 4 light-years is assumed. A spherical distribution of uniformly spaced, suitable stars is assumed. This is arguably the best-case scenario.

technological civilization provided the other values are specified. If we assume values for the Drake equation as follows:

$$R^* = 10$$

$$f_p = 0.5$$

$$n_e = 2$$

$$f_l = 0.01$$

$$f_c = 0.01$$

then we can relate the number of civilizations in the galaxy to the lifetime of the civilization as in Figure 48.2 (see also Table 48.2).

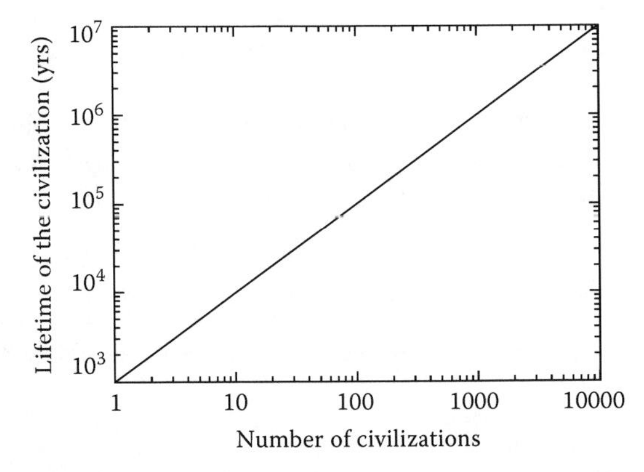

FIGURE 48.2 Relationship between the number of civilizations in the communicative stage and the lifetime of the civilization.

TABLE 48.2
The Odds of Finding a Civilization Still in Its Technological Stage as a Function of the Total Number of Such Civilizations and the Radial Distance to That Civilization

Total Number of Technological Civilizations in the Galaxy	Radial Distance Required to Encounter at Least One Civilization	Total Number of Stars within Volume
10^4	9×10^2	10^7
10^3	2×10^3	10^8
10^2	4×10^3	10^9
10	9×10^3	10^{10}
1	2×10^4	10^{11}

Let us consider this problem from a different perspective. How many stars will have planets around them that will retain the signature of life? To determine that, we can write the number as the product of the production rate of planets that have life of any kind and the lifetime of the fossil signature of that life, as follows:

$$N = R^{*} f_p n_e f_l L_f \tag{48.4}$$

where we have introduced a term L_f for the lifetime of a fossil signature of life.

To evaluate this, we take the following:

$$R^{*} = 6$$

$$f_p = 0.5$$

$$n_e = 0.001 \text{ to } 2$$

$$f_l = 0.1 \text{ to } 1$$

$$L_f = 1 \times 10^9 \text{ years.}$$

This expression yields a distribution of values illustrated in Figure 48.3. This leads to an astonishingly large number of possible sites for life of some kind that would have left behind a fossil remnant.

If we do reach another planet, will we find signs of life on the macroscopic level? This is, it turns out, no simple problem. As a case in point, consider the story of the Allan Hills meteorite ALH 84001.[16] This meteorite, believed to have originated on Mars, was, for a time, thought to exhibit the signature of nanobacteria. These nanobacteria are smaller than currently extant bacteria and have not been observed on Earth. This naturally leads to the question of whether we can detect the signature

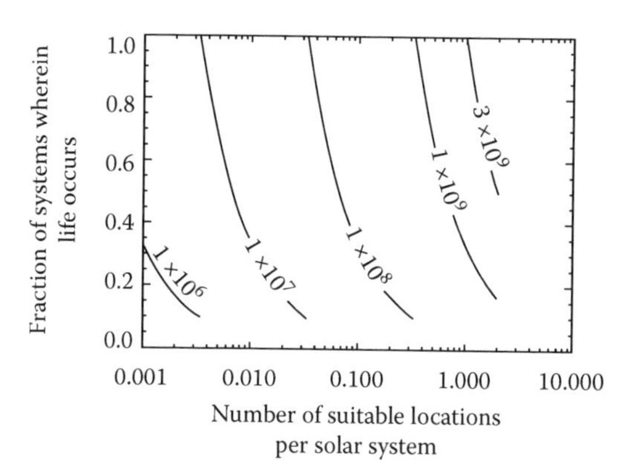

FIGURE 48.3 Number of locations in the galaxy that might harbor fossil evidence of life. The variation is due to the assumed value for the number of locations in each solar system that could produce life in a recognizable form. The lowest value considered is 0.0001; the greatest value is 1, with each curve representing a change of a factor of 10 in the number of objects in a solar system upon which life of some kind could evolve.

of carbon-based life unambiguously, but also whether, if there are other forms of life, we would recognize them.

While we may be interested in knowing whether life existed, archaeology is concerned with the lifetimes of artifacts. The picture becomes very different if we consider the lifetime of a technological artifact as evidence of the remains of an ancient technological civilization. The Drake equation can be written in a modified form as follows:

$$N = R^* f_p n_e f_l f_i f_c L_a \tag{48.5}$$

where L_a is the lifetime of an artifact.

What is the lifetime of an artifact? There has been a spate of interest recently in representing and projecting the state of our world after the demise of the human species.[17] To a first-order approximation, the lifetime of artifacts appears to be in inverse proportion to technological level, or information content, as illustrated in Table 48.3. Each of the various categories of information content requires the ability to command successively greater amounts of resources, as well as being imbued with greater information content. One might argue that technology actually serves to perpetuate knowledge by providing the means to produce many copies of the knowledge. There are two aspects to this issue: what is the lifetime of technological media, and what is the lifetime of the ability to read those media?

What is the lifetime of the information content of an artifact? Digital information storage media have a lifetime of much less than 100 years, short compared with the lifetime of stone, which has a lifetime, as a recording medium, of thousands of years under ideal conditions. The ability to read digital information may be much more fleeting than the media themselves. Consider the example of the Rosetta Stone.[18]

TABLE 48.3
Information Pyramid

Artifact	Lifetime (Years)
Cryptodata	?
Microfossils	10^9
Macrofossils	10^8
Hominid fossils	10^6
Tools and artifacts	10^4
Writing and cave paintings	10^4
Clay tablets	10^3
Papyrus	10^3
Paper	10^2
Digital media	10^1
Virtual information	10^0?

The lifetime of an artifact is, approximately, inversely proportional to its information content. The lifetime should be considered as an e-folding time for the information content.

It was to some degree by chance that this Egyptian artifact was found, valued, and studied rather than ending up repurposed. This stone contained the same inscription in two Egyptian scripts, including hieroglyphics, and classical Greek. The Greek text enabled scholar to translate hieroglyphics. Will there be a Rosetta Stone for digital records 10,000 years from now?

The major difficulty in interpreting information is that there may be no Rosetta Stone that enables one to make a connection between the object in hand and the information contained within the object. An example of this is the Inca knot messages or khipu. The khipu are a textile-based technique for recoding information for census and tax accounting purposes.[19] The word khipu comes from the Quechua word for "knot". Khipu are textile artifacts consisting of a primary cord with a number of subsidiary cords that hang from it. We know that khipu served a purpose other than decorative because colonial records indicated that these were used for record keeping and messages. These messages from the past are difficult to interpret but at least we have some common referents. In fact, the interpreted use of the artifact itself tells us about that civilization. It appears, by the way, that the khipu records are in base 10, as might be assumed on the basis of our common heritage.[20] What would have happened if they had been in another numbering scheme?

The longest-lived information is that which is either directly a product of the civilization, a tool, a piece of art, or the physical remains of an organism. Some tools, though, while recognizably the product of a civilization, will rapidly lose their identity in terms of function and their place in society once the interrelationship of the many objects that make up a culture is destroyed. We can still interpret a flint knife or an arrow head because we still have objects that serve those purposes. Contrast this with, for example, a beer bottle opener. Interpreting the ritual significance of this object presupposes a knowledge of the existence of beer, bottles, and a particular kind of sealing device, in order to be interpreted even at the first level of its significance in the hierarchy of tools within our culture. The cultural significance of the object is even more elusive. As an example of a technological implement, a slide rule is still recognizable to some of a certain generation who may, if pressed, be able to operate the device, but how long will that be the case? Evolving technology has erased its utility—though not the underlying principle. The more sophisticated the technology, the more likely it is to be both culturally specific and inaccessible. Basic information (e.g., the means and mechanisms of survival) will remain somewhat more readily interpretable for a longer period of time, but by their very nature, space travel and the artifacts of space exploration or artifacts encountered by virtue of space travel will be the most difficult to place in context.[21]

There are many books devoted to preserving our extant knowledge of the arts and crafts of earlier societies.[22] Those cultures existed for thousands or hundreds of years. Will the "mechanical culture" of the Industrial Revolution or "computer culture" of the Information Age be preserved? I would assert that it will not: there are too many skills required to preserve the ability to make a steam engine, for example, once we have transcended our current, physical culture. L. Sprague de Camp expressed it in his classic story of a modern man, the scholar of ancient Rome, Martin Padway, who is transported to sixth century AD Rome. Padway discovers that in order to prevent the fall of Rome, he has to invent the tools required to make the tools to prevent that fall, and to do that he must train the artisans in a new craft.[23] H.G. Wells recognized that even a late Victorian scientific genius would need knowledge outside of his own direct experience to rebuild a degenerate society: the question faced by the protagonist at the end of *The Time Machine*[24] is what knowledge is sufficient to rebuild a society? That story invites the question of whether having the abstract knowledge is sufficient to reconstruct the culture. This issue is handled in a largely perfunctory and self-aggrandizing fashion in the book *Lucifer's Hammer*,[25] which deals with the efforts to rebuild a disrupted but largely extant civilization after a large meteor impact. The decay of the knowledge and civilization and the need for a large population base skilled in that knowledge was dealt with half a century ago in *Earth Abides*,[26] with a reiteration of the idea that civilization is not a linear, inexorable process. This theme is also developed in the postapocalyptic novel *A Canticle for Leibowitz*.[27]

Archaeology uses several tools for developing an understanding. Archaeologists often reason by analogy: they begin with the known and then use that to form a basis for understanding. One of the most interesting ways of developing an understanding is by experimental archaeology: an attempt is made to replicate a technique such as flint knapping[28] or to replicate a lifestyle.[29] This would be particularly difficult, though, if there is no analogy readily at hand. That lack of reference could be due cultural or technological differences. Science fiction does, in fact, deal with other civilizations and their artifacts, customs, languages, and cultures.

The last entry in Table 48.3 describes "virtual information." This is the newest form of information in human society; it exists in an abstract form and largely relies on the connectedness of one piece of virtual information with another for both context and meaning. For example, little of Wikipedia exists in hard copy in the same form that it exists on the Web. How long would that information exist if there were some catastrophe that fragmented society to the degree that the Internet was no longer available? How much can be reconstructed from just one or even a thousand isolated computer systems? How could that information, which is stored in a variety of formats, be recovered and reconnected? How much value, in hyperlinked metadocuments, does one isolated instantiation have? What cultural assumptions or external information would be necessary in order to read the contents of a hard drive or to even understand what it was?

The first element, "cryptodata," is that information about the universe that is both embedded in the very structure of the universe and yet hidden from us. In fact, we do not even know where or how to look for it. Cryptodata, if it were discovered, would be the most profound artifact, as it could, potentially, change our own perspective on ourselves and the world. Science fiction has dealt with whether there is a hidden meaning in our world.

Archaeology in an alien environment must contend with the tendency of the human mind to look for patterns: in particular, the human mind strives to determine significance in vague or random phenomena. An example of this is that we tend to see faces in inanimate, random patterns. This phenomenon is termed pareidolia. Pareidolia enables humans to use only minimal details to recognize faces from a distance and in poor visibility, but it can also lead us to interpret random images or patterns of light and shade as being faces.[30] Simple examples include seeing the animals or faces in clouds or the moon. One example of pareidolia that impinged upon the scientific community is the "face on Mars." *Viking* imagery of Cydonia Mensae in photo #35A72 seemed to some to show a feature that looked like a human face.[31] Higher-resolution imagery and a changed viewing geometry show that it was the predilection for recognizing faces that led some observers to conclude that the object was artificial.

Science, in a sense, is the search for patterns and the codification of those patterns. The challenge in space archaeology will be to recognize the difference between real message and a message that arises from internal needs. For example, Oliver Morton has proposed that the place to search for alien messages is the human genome.[32]

Carl Sagan, in *Contact*, provides an interesting example of cryptodata. In that story, a radio telescope search detects a message from an extraterrestrial intelligence near Vega, a star 26 light-years away. That message contains an early television broadcast and, hidden within it, the blueprint for a machine that will enable five explorers to travel to Vega through a wormhole in space-time. It is there that the main character, Eleanor Arroway, meets a representative from an ancient race that tells her to look within the number π for traces from the creator of the universe. This is a message recorded in the fabric of the universe by the creator. The aliens who created the original message are just as ignorant of this message's meaning as Arroway, who eventually discovers the content if not the meaning.

In an earlier story, Stanislaw Lem wrote[33] about the efforts of a group of scientists to understand an alien transmission. The story, told in memoir form, contains within it discussions of the motives of extraterrestrial intelligence, as well as a consideration of the ethics of military-sponsored research. Interestingly enough, the scientists in *His Master's Voice* turn to science fiction for ideas but soon reject them. The protagonist lets us know early on that there are few real discoveries and in the end he is left, as are we, to wonder whether the message was not just random noise.

What would the human race make of the discovery of a message from an alien race? In *Contact*, some refuse to accept that reality. Some might embrace it, as in Robert J. Sawyer's *Factoring Humanity*.[34] Central to the story is the relationship between Heather, a psychology professor struggling to decipher extraterrestrial messages, and her estranged husband, Kyle. In this story, a very personal redemption comes about because the messages from Alpha Centauri that Heather has studied prove to be plan for a "psychospace" device. This device enables one to know anyone's deepest thoughts and have access to humanity's "overmind." In John Varley's *Ophiuchi Hotline*,[35] the human race has been displaced by an alien race called the Invaders, which has destroyed all human technology in 2050. The Invaders recognize themselves, whales, and dolphins as intelligent. Humanity, as a civilization, has moved off planet and has relied for 500 years on information embedded in a radio signal from the vicinity of 70 Ophiuchi. Interestingly enough, one of the central ideas of the story is that biological manipulation in all forms, including cloning, is readily available, including the technology to record the memories of a human and restore them to a cloned body. Ultimately, the price that the aliens demand is the ability to inhabit a human body. Other interesting takes on the SETI include Gregory Benford's *Dance to Strange Musics*, in which the first expedition to Alpha Centauri finds that a collective life form is sending out messages.[36] Cultural biases are also going to play a role in how we react to messages, whether they are delivered in person[37] or via radio.[38]

Science fiction offers amusing and informative takes on the problem of cultural biases. One of the best and most accessible is the book by David MacAuley,[39] *Motel of the Mysteries*, which is set in the year 4022. The ancient country of Usa has been covered by debris. A very amateur archaeologist, Howard Carson, stumbles upon a ceremonial burial chamber marked with a "Do Not Disturb" sign hanging from an archaic doorknob. From Carson's discoveries he is able to deduce (in a silly way) the entire fabric of that civilization. Some of the most important artifacts are a body lying on a ceremonial bed facing an altar containing the means of communicating with the gods (a television). Another body lies within a porcelain sarcophagus (a bathtub) in the "Inner Chamber" (the bathroom). Amusingly enough, Howard Carson makes reference to the fictional archaeologist Heinrich von Hooligan (a play on Heinrich Schliemann), who believes that the black and/or grey stripes of material found throughout Usa are alien landing strips.

One of the best examples of misreading cultural artifacts is Erich von Däniken's *Chariots of the Gods? Unsolved Mysteries of the Past*,[40] which purports to describe artifacts that were created with a technology inconsistent with that which the indigenous people were conversant in. Von Däniken, whose work was very popular when it was first published, described the Pyramids at Giza, Stonehenge, the statues at Easter Island, and other works, as alien artifacts. He went so far as to ascribe religion to the interpretation by primitive peoples of contact with extraterrestrials. Clifford Wilson in *Crash Go The Chariots*[41] and Ronald Story in *The Space Gods Revealed*[42] examine the evidence presented by von Däniken and refute his case.

The fact that we are broadcasting our presence via radio and television broadcasts is also the source of some humor in science fiction. While it is easy to recognize the humorous intent in these stories, without a cultural reference other than TV, what conclusions would actually be drawn from a study of these electromagnetic artifacts?

In the story "In Hoka Signo Vinces,"[43] the Hokas, inhabitants of the planet Toka, like to imitate the behavior they see in Earth entertainment and history. They have been watching the popular children's science fiction program *Tom Bracken of the Space Patrol* and decide to organize a local branch of the Space Patrol, taking that television program as truth. The Hokas commandeer the Earth ambassador Plenipotentiary Alexander Jones' spaceboat and rename it *Space Patrol Ship Number One.* Jones is worried about the Pornians, a militaristic race set to fall upon a demilitarized society. A might be expected, the Patrol manages to board a Pornian space dreadnaught, capture it, and force the Pornians to disarm and hold free elections.

Interestingly enough this plot device was revisited in the movie *Galaxy Quest.* Tim Allen plays Jason Nesmith, a parody of William Shatner as Captain Kirk of the original *Star Trek* television series. "Captain Nesmith" is taken from Earth by the Thermians to fight the Sarrians, a militaristic race. The Thermians have "reconstructed" the NSEA *Protector*, the mighty spaceship of the *Galaxy Quest* series, on the basis of the intercepted "historic documents" that they have received from the Earth. The Thermians do not realize that what they interpret as historic documents are fantasy.

The implications of a bona fide message from another civilization have been considered from a number of viewpoints. We sometimes forget that, as W.H. Auden observed of the human condition:

> About suffering they were never wrong,
> The Old Masters; how well, they understood
> Its human position; how it takes place
> While someone else is eating or opening a window or just walking dully along.[44]

In "Nothing Ever Happens in Rock City," the first radio SETI discovery is seen from the perspective of the owner of the liquor store closest to the observatory.[45] The liquor store owner fails to realize that the cause for the celebration is the discovery of the first evidence of an extraterrestrial intelligence. To him, it is of a piece with the rest of his life—lacking in a broader significance. The reader is left to wonder just how much significance such a discovery would have for vast majority of the people of Earth.

WHERE ARE WE GOING?

Archaeology is not usually concerned with our future but with our past. Science fiction and archaeology can intersect when they look at what will be left after our story, or that of some other civilization, ends.

Enrico Fermi is widely held to have first stated the question that if life is ubiquitous in the galaxy, why haven't we been contacted?[46] The logical extension of this question is, why aren't there artifacts of alien civilizations here? This is the Fermi paradox. There has been some significant work on this question in the scientific literature. The Fermi paradox can be addressed by a number of "solutions" (see also Capelotti, Chapter 45 of this volume). The first question is whether the galaxy would be colonized by an early emergent life form. Simulations show that this is certainly possible even with relatively slow colonization speeds.[47] Science fiction has explored the concept of the "generation ship"—a spaceship that travels at sublight speeds for hundreds of years to reach another star system.[48] Thus, if life did evolve elsewhere and it had a significant technological lifetime, the galaxy should be rich in life. Why haven't we been contacted then? One idea is that there is a generally benevolent galactic society that is preserving us from contact[49]—the Zoo Hypothesis,[50] the Prime Directive, or the Interdict Hypothesis. Now, one might argue that there could indeed by an interstellar "zoo" in our galaxy, but would it hold for all the observable galaxies? What does the lack of observable signatures tell us about the level of civilization that has actually been achieved elsewhere?

So what of the Fermi paradox? One of the answers to the Fermi paradox is that we have explored very little of the Universe or even our local environment as described above and that, as Freitas argued,[51] there really is no Fermi paradox. Now, one might argue based on "the Singularity" (see below) that life and the signs of life would be ubiquitous. Perhaps with more sensitive instruments or with better techniques we will discover evidence of an alien presence. Arthur C. Clarke explored the notion that an alien civilization could have left an artifact for us to discover, perhaps as a way to apply to leave the "cosmic zoo," in "The Sentinel"[52] and his novel *2001: A Space Odyssey*.[53] Our ability to reach the artifacts on the Moon and in orbit around Jupiter is a sign of our readiness to enter the broader community of the galaxy.

The other course, less optimistic, is to consider that we have not experienced an extraterrestrial intervention because the lifetime of a communicative technological civilization is very short. It is instructive then to consider the causes of the end of civilization, to which I have added three new members to the Horsemen of the Apocalypse. The traditional members are war, conquest, pestilence,[54] and death.[55] Science fiction and speculation or extrapolation based on current trends would indicate that death, or the elimination of death, can cause the end of a civilization as it transforms itself into a new modality. In addition, there are transformation, in which a civilization takes on an entirely different form that may not be outwardly directed; senescence, in which a civilization loses some or all of its desire or abilities to look outward; and obliteration, in which a civilization is destroyed by an astrophysical phenomenon.[56]

War has long been the subject of science fiction. Our technological society is complex enough that a war of sufficiently broad scale not only would destroy our technical capability but could bring about the end of the human race. Some of the more important stories that deal with the aftereffects of a nuclear war are *Farnham's Freehold* by Robert Heinlein, *Alas, Babylon* by Pat Frank, *On the Beach* by Nevil Shute, *Level 7* by Mordecai Roshwald, *Dr. Strangelove or: How I Learned to Stop Worrying and Love the Bomb* by Peter George, "I Have No Mouth and I Must Scream" by Harlan Ellison, *Beyond Armageddon* by Walter M. Miller and Martin H. Greenberg, "Fermi and Frost" by Fred Pohl, and *The Road* by Cormac McCarthy. Of these stories, "Fermi and Frost" deals particularly with the notion of the fragility of our society and, by extension, the question of why we have not been contacted by an electromagnetic signal from another civilization. Pohl's point is that of the many paths our civilization can follow, only one leads to a time when we are able to receive signals from an extraterrestrial intelligence.

Fear of conquest is a deep-rooted concern of the human race. Science fiction has considered both external and internal conquerors. Perhaps the best known scenario is that articulated by H.G. Wells in *War of the Worlds*. There is an interesting variant on this notion: conquest by intelligent machines, whether it created internally[57] or by an external source.[58]

This question of the role of intelligent machines has its roots in the earliest and most important science fiction stories, many of them written by Isaac Asimov more than a half century ago.[59] Asimov in "The Last Question"[60] follows humanity through six vignettes, starting in 2061, when the Multivac computer is asked whether there is a way for humanity to avoid the heat death of the universe. In the last scene of the story, which follows the evolution of computers through millions of years, the ultimate descendants of humanity merge with the Cosmic AC but ask the question one last time. The Cosmic AC exists in another dimension, outside of the space-time continuum that we are familiar with, and, consequently, is able to continue to work on the question. Eventually the Cosmic AC discovers the answer but there is no one left to report to. The Cosmic AC decides to demonstrate the answer—the story ends with AC's decree "Let there be light." The existence of the Prime Mover is then "cryptodata."

The Cosmic AC may seem to be the ultimate conceit, a product of the now quaintly optimistic and materialistic worldview of 1950s American culture, but that transformative view still holds power. Vernor Vinge, a science writer of some renown, uses the term *Singularity* to denote the point at which machines become able to recursively redesign themselves, with improvements.[61] Ray Kurzweil believes that mind and machine will merge and predicts that by 2020 a simple computer will have the processing power of the human brain.[62] He argues that the "Law of Accelerating Returns" ensures that progress will continue exponentially. In fact, by 2055, at the current rate of acceleration, a personal computer will have the processing power of all human brains combined. William Gibson's *Neuromancer*[63] deals with a world in which one artificial intelligence (AI), in a world in which AIs capable of improving their software are regulated by special police, strives to circumvent this control. How will humans remain in control? Will they want to? What happens if, as Kurzweil and others have suggested, human "mind files" can be created and AIs can have a human personality? As suggested by Anne McCaffrey, a human-AI interface may create spaceships that are capable of autonomous operation far from Earth.[64] Could that be the mechanism through which we explore other worlds? The question is that if this Singularity is an inevitable consequence of the development of a technological society, where is everyone? Perhaps they are there, on their way, heedless of our own world, in world ships such as that central to *Rendezvous with Rama.*[65] It is interesting to consider the dominance of the human race by machines in light of the Kurzweil Singularity. If machine intelligence is the endpoint of biological evolution, then there may be no interest in contacting biological entities. In the Kurzweillian future, it may be that machines hold the collective conscious in a virtual reality; there would, then, be no imperative to look beyond that internal reality.

Cultural senescence and transformation have been explored in science fiction in a number of stories.[66] These stories, of course, are strongly shaped by the social and political systems of the times. One of the most interesting is *Houston, Houston, Do You Read?* by James Tiptree, Jr. In this story, written in 1976 as the Apollo program faded into memory, the all-male crew of the spaceship *Sunbird*, from our near future, finds itself transported into a future that is technologically like the one it left but completely transformed socially. There, modern men find themselves in a world made up largely of female clones of the survivors of a global pandemic. That culture is matriarchal and much more inward-looking the one the male astronauts left behind, and it is no longer interested in technological advances purely for the sake of being the first. Ultimately, the crew members of the *Sunbird* and their patriarchal values represent a past so antithetical—and a presence that is so disruptive—to the current society that they have to be killed. One interpretation of the story is that the author is asking how much of the Space Race and exploration, in general, is driven by the perceived male need to dominate. In the context of finding other civilizations, this story encourages one to ask whether a culture aggressive enough to explore outside of their solar system is stable enough to have a long technological lifetime.

CONCLUSION

The greatest and most enduring questions of human society are, not coincidentally, the hardest to answer: Where did we come from? Where are we going? Are we alone? Science fiction views the cultural zeitgeist through many lenses, and much of science fiction is little more than the stories of today or the myths of our past transplanted to another setting. It does, however, capture and embody in an important, powerful, and immediate way the need of a segment of our society to explore the "big questions." Science fiction provides a way of asking questions that we have little hope of having answered within our lifetime.[67]

Science fiction embodies the principle that "wherever you go, there you are." We take with us our cultural assumptions and aspirations. If we hope to explore the cosmos and discover our place in the greater scheme of things, we must be aware of the challenges that face us as a civilization. As summarized in this chapter, many challenges face us. If we fail to surmount those challenges, the human race may be destined to tread the path trodden by uncounted civilizations elsewhere in the galaxy: to grow from the most primitive of origins, to reach a supreme crisis, and then to fall back from the stars. The only way for us to answer these questions is to continue our outward voyage, away from our nursery, to find what lies beyond "Childhood's End."

NOTES AND REFERENCES

1. Knight, D.F. 1967. *In Search of Wonder: Essays on Modern Science Fiction.* Advent Publishing, Inc. (see specifically p. xiii).
2. Wilson, D.H. 2007. *Where's My Jetpack? A Guide to the Amazing Science Fiction Future That Never Arrived.* Bloomsbury, USA.
3. See Wikipedia, http://en.wikipedia.org/wiki/Infinite_monkey_theorem, for a discussion of the idea that a random process can generate a sequence of characters that, when viewed by an external observer, can be mistakenly believed to have meaning. If one hundred monkeys were to be seated before typewriters and allowed to type at random for a long enough period of time they would eventually generate a Shakespearean sonnet.
4. "Mute Inglorious Tam" depicts what it would be like to be a person with an organic need to tell an imaginative, science fiction, story in a culture that denies that: in this case a serf in medieval England. Pohl, F., and Kornbluth, C.M. 1974. Mute Inglorious Tam. *The Magazine of Fantasy & Science Fiction*, October 1974: 111–120. Also published in Del Rey, L. 1974. *Best Science Fiction Stories of the Year (1974).* Dutton. 1987. *Our Best: The Best of Frederik Pohl and C.M. Kornbluth*, ed. F. Pohl. Baen. Discussions of the origin of science fiction are to be found in (1) Roberts, A. 2007. *The History of Science Fiction.* Palgrave McMillan. (2) Clute, J., and P. Nicholls. 1995. *The Encyclopedia of Science Fiction.* St. Martin's Press. (3) James, E., and F. Mendelsohn. 2003. *The Cambridge Companion to Science Fiction.* Cambridge University Press.
5. http://en.wikipedia.org/wiki/Cyrano_de_Bergerac. Cyrano appears in the science fiction series *Riverworld* by Phillip Jose Farmer, by the way.
6. Bradbury, R. 1950. *The Martian Chronicles.* New York: Doubleday.
7. Asimov, I. 1950. *I, Robot.* Gnome Press. Asimov, I. 1954. *Caves of Steel.* Fawcett.
8. Five *Star Trek* series have appeared on television: *Star Trek*, *Star Trek: The Next Generation*, *Star Trek: Deep Space Nine*, *Star Trek: Voyager*, and *Star Trek: Enterprise*, as well as an animated series. In addition, a Web "series" is also available http://www.startreknewvoyages.com/.
9. *Star Trek.* Episode 28, City on the Edge of Forever. First broadcast April 6, 1967. See http://en.wikipedia.org/wiki/The_City_on_the_Edge_of_Forever.
10. Willis, C. 1992. *Doomsday Book.* Bantam. Willis, C. 1982. Fire Watch. *Isaac Asimov's Science Fiction Magazine*, February 1982. Connie Willis has won ten Hugo awards, with an additional thirteen nominations, as well as six Nebula awards, with an additional eight nominations.

11. http://www.msnbc.com/modules/drake/default.asp. Note that inclusion on the Web is neither a necessary nor a sufficient demonstration that an idea is profound, as evidenced by the Web itself, but there are few other examples of an equation that has such appeal to people, as evidenced by these message boards. It is worth noting that the message boards indicate surprise at how the low the number is. Is there something in us that refuses to believe that we are alone?
12. See Kasting, J.F., and D. Catling. 2003. Evolution of a Habitable Planet. *Annual Reviews of Astronomy and Astrophysics* 41: 429–463, and the references cited therein.
13. See http://en.wikipedia.org/wiki/Drake_equation and the discussion therein for more details on the range of parameters.
14. See http://en.wikipedia.org/wiki/Kardashev_scale for a more complete discussion, including additional references from science fiction.
15. Clarke, A.C. 1962. *Profiles of The Future*. Harper and Row.
16. See http://en.wikipedia.org/wiki/ALH84001 and the references cited therein, in particular, the article that appeared in the journal *Science*: McKay, D.S., E.K. Gibson, K.L. Thomas Keprta, H. Vali, C.S. Romanek, and S.J. Clemett et al. 1996. Search for Past Life on Mars: Possible Relic Biogenic Activity in Martian Meteorite ALH84001. *Science* 273(5277): 924–930.
17. See, e.g., (1) *Aftermath: Population Zero*. 2008. National Geographic Television. http://channel.nationalgeographic.com/channel/aftermath/. (2) *Life After People*. 2008. History Channel Video. (3) Weisman, A. 2008. *The World Without Us*. Thomas Dunne Books.
18. See, e.g., Ray, J. 2007. *The Rosetta Stone and the Rebirth of Ancient Egypt*. Harvard University Press.
19. http://khipukamayuq.fas.harvard.edu/.
20. In the early 1900s, Leland Locke was the first to recognize the numerical significance of the knots. See, e.g., Urton, G. 2003. *Signs of the Inka Khipu: Binary Coding in the Andean Knotted-String Records*. Austin: University of Texas Press.
21. See also Lem, S. 1983. *His Master's Voice*. San Diego, CA: Harcourt Trade Publishers. Also published as Lem, S. 1999. *His Master's Voice*. Northwest University Press.
22. There are, of course, many books on the subject. The Bishop Museum in Honolulu, Hawai'i, preserves many artifacts from what was a stone age culture (see http://www.bishopmuseum.org/). The use of natural materials to create tools, ritual objects, and objects of adornment and status is described in, e.g., (1) Buck, P.H. 2003. Arts and Crafts of Hawai'i. Honolulu, HI: Bishop Museum Press. (2) Summers, C.S. 1999. *Material Culture: The J.S. Emerson Collection of Hawai'ian Artifacts*. Bishop Museum Press, 1999.
23. de Camp, L.S. 1941. *Lest Darkness Fall*. New York: Henry Holt Publishing. (A short story version appeared in the pulp *Unknown*, December 1939.) Poul Anderson, notably, dealt with the difficulties of a "civilized" being in dealing with a more primitive culture in *The High Crusade* (New York: Doubleday, 1960) and "The Man Who Came Early," which appeared in the June 1956 issue of *The Magazine of Fantasy and Science Fiction* and was reprinted in *The Best from Fantasy and Science Fiction* (New York: Doubleday and Co., 1957).
24. Wells, H.G. 2005. *The Time Machine*. Penguin Classics.
25. Niven, L., and Pournelle, J. 1985. *Lucifer's Hammer*. Del Rey Books.
26. Stewart, G.R. 2006. *Earth Abides*. Del Rey Books.
27. Miller, W.M., Jr. 1984. *A Canticle for Leibowitz*. Bantam.
28. It is worthwhile and entertaining to search the Web for references to flintknapping. There are more than 30,000 references to the terms "flintknapping how to." There are even books, such as Whittaker, J.C. 1994. *Flintknapping: Making and Understanding Stone Tools*. University of Texas Press.

29. Reenactors of various kinds, degrees of authenticity, and sophistication keep alive or at least interpret the many skills, ranging from those of primitive hunter-gatherers to those of Iron Age civilizations.
30. Sagan, C. 1995. *The Demon-Haunted World—Science as a Candle in the Dark.* New York: Random House. The Web site http://www.flickr.com, for example, has thousands of examples of faces or other significant images revealed in everyday objects: enter "pareidolia" as a search term.
31. See JPL Public Release at http://www.msss.com/education/facepage/pio.html, http://www.msss.com/education/facepage/face.html, and http://science.nasa.gov/headlines/y2001/ast24may_1.htm.
32. Morton, O. 2006. The Albian Message. In *Year's Best SF 11*, ed. D. Hartwell, and K. Cramer. Eos.
33. Lem, S. 1968. *His Master's Voice.* Note that this is undoubtedly a reference to the RCA advertising campaign from the turn of the century that depicted, based on the 1889 painting "Dog Looking at and Listening to a Phonograph" by Francis Barraud, a dog puzzled by a voice coming from a phonograph.
34. Sawyer, R.J. 1998. *Factoring Humanity.* Torr Books.
35. Varley, J. 1977. *Ophiuchi Hotline.* Dial Press.
36. Benford, G. 1999. Dance to Strange Musics. In *Year's Best Science Fiction 4,* ed. D. Hartwell. Eos/HarperCollins.
37. Clarke, A.C. 1987. *Childhood's End.* Reprint, Del Rey Books. Knight, D. 1950. To Serve Man. *Galaxy Science Fiction* 1(2).
38. Spinrad, N. 1991. The Helping Hand. *Full Spectrum 3.* Broadway.
39. MacAuley, D. 1979. *Motel of the Mysteries.* Houghton Mifflin.
40. Erich von Dänikin. 1969. *Chariots of the Gods? Unsolved Mysteries of the Past.* G.P. Putnam's Sons.
41. Wilson, C. 1972. *Crash Go The Chariots.* New York: Lancer Books.
42. Story, R. 1976. *The Space Gods Revealed: A Close Look at the Theories of Enrich von Daniken.* New York: Harper and Row.
43. See, e.g., Anderson, P. and G.R. Dickson. 1998. *Hoka! Hoka! Hoka!* Baen Books. First published in 1953 in the pulp *Other Worlds.*
44. Auden, W.H. 1966. Musée des Beaux Arts. *Collected Shorter Poems 1927–1957.* New York: Random House (pp.123–124).
45. McDevitt, J. 2004. Nothing Ever Happens in Rock City. In *Nebula Awards Showcase 2004*, ed. V. McIntyre. ROC/Penguin.
46. Jones, E. 1985. *"Where Is Everybody?": An Account of Fermi's Question.* Los Alamos Technical Report LA-10311-MS.
47. See Hart, M.H. 1975. An Explanation for the Absence of Extraterrestrial Life on Earth. *Quarterly Journal of the Royal Astronomical Society.* 16: 128–135, for a list of some of the reasons that, even if life were ubiquitous, we might not have been contacted. See also Frietas, R.A. 1985. There Is No Fermi Paradox. *Icarus* 62: 518–520, for a discussion of the lack of mathematical rigor in the statement of the Fermi Paradox.
48. For example, Harrison, H. 1969. *Captive Universe.* New York: Ace Books; Heinlein, R. 1951, republished 1989. *Orphans of the Sky.* Baen.
49. Gregory Benford speaks eloquently and elegiacally of this civilization in *Against Infinity*, 1983, p. 246, London: New English Library.
50. Ball, J.A. 1973. The Zoo Hypothesis. *Icarus* 19: 347–349. Danger Human (Dickinson, G.R. 1973. In *The Book of Gordon R. Dickinson.* DAW Publishing) provides another, interesting view of the "zoo," in which contact with humans is avoided because they are dangerous. For a further, serious, study of the idea that the Prime Directive, popularized in the *Star Trek* series, might actually be in force, refer to Fogg, M.J. 1987.

Temporal Aspects of the Interaction among the First Galactic Civilizations: The Interdict Hypothesis. *Icarus* 69: 370–384. The Prime Directive is widely used in science fiction and is discussed at http://www.wikipedia.com/prime_directive.

51. Frietas, R.A. 1985. There is no Fermi Paradox. *Icarus* 62: 518–520.
52. Clarke, A.C. 1951. The Sentinel. In *The Avon Science Fiction and Fantasy Reader*. Avon Periodicals.
53. Clarke, A.C. 1968. *2001: A Space Odyssey*, and the film by Stanley Kubrick.
54. Pestilence plays an important role in speculative, allegorical, and science fiction. The pestilence may be induced by our own actions, as in *The Andromeda Strain* (Michael Crichton, Knopf, 1969). Global pandemics can, of course, arise as a consequence of population density and spread quickly. George R. Stewart considered this in *Earth Abides* (Random House, 1949). Global pandemic might be brought about for another reason, as in James Tiptree's "The Screwfly Solution" (*Analog Science Fiction/Science Fact*, 1977).
55. See Ian Macdonald, *Terminal Café* (Bantam, 1995), for a consideration of the impact of immortality on the evolution of human society.
56. A few of the earliest and more significant stories about cosmic phenomena are Nightfall (Isaac Asimov), *When Worlds Collide* (Philip Wylie), "Pebble in the Sky" (Isaac Asimov), and *Lucifer's Hammer* (Larry Niven and Jerry Pournelle, 1985). See also Annis, J. 1999. An Astrophysical Explanation for the Great Silence. *Journal of the British Interplanetary Society* 52(19), in which Gamma Ray Bursters are proposed as a mechanism by which planets could be sterilized. The European Space Agency sponsors ESLAB 2008 Cosmic Cataclysms and Life, a scientific conference (http://www.congrex.nl/08c16/). See also Steel, D. 2008. Planetary Science: Tunguska at 100. *Nature* 453: 1157–1159.
57. Shelley, M. 1818. *Frankenstein, or The Modern Prometheus*. See also http://en.wikipedia.org/wiki/The_Terminator for more on the Terminator stories. Arthur C. Clarke explored the power of computers in human destiny in several stories, including "Dial "F" for Frankenstein" (*Playboy*, January 1964, collected in *The Collected Stories of Arthur C. Clarke*, Orb Books, 2002) and the character of the HAL9000 unit in *2001: A Space Odyssey*. See also Jones, D.F. 1966. *Colossus: The Forbin Project*. Berkeley Medallion.
58. For example, the Borg machine-humanoid construct of the *Star Trek* stories and the Berserker stories of Fred Saberhagen embody the concept of an external threat whose goal is the annihilation of our society. For more on the Borg, see http://en.wikipedia.org/wiki/Borg_%28Star_Trek%29. There are several Berserker novels: the first short story collection is *Berserker Stories*, 1967.
59. Capek, K. 1970. *R.U.R.* Pocket Books. This book put the term *robot* into common usage. Isaac Asimov's "robot stories" include the short story collection *I, Robot* (Gnome Press, 1950) and "The Bicentennial Man," which won the Hugo Award in 1977 for Best Novellete. This story and many others are collected in *The Complete Robot* (Voyager Books, 1983).
60. Asimov, I. 1956. The Last Question. *Science Fiction Quarterly*. Also in a number of collections including Asimov, I. 1986. *Robot Dreams*. Rutherford, NJ: Berkeley Trade. See also http://www.multivax.com/last_question.html for the full text.
61. Vinge, V. The Coming Technological Singularity. Vision-21: Interdisciplinary Science & Engineering in the Era of CyberSpace, proceedings of a symposium held at NASA Lewis Research Center March 30–31, 1993. NASA Conference Publication CP-10129, referenced by http://rohan.sdsu.edu/faculty/vinge/misc/singularity.html. See also his transformative writing *True Names*, 1981; *The Peace War*, 1984; and *Marooned in Realtime*, 1989.
62. Kurzweil, R. 2008. The Coming Merging of Mind and Machine. *Scientific American Reports*. See also *The Age of Intelligent Machines*, 1990.

63. Gibson, W. 2004. *Neuromancer*. New York: Ace.
64. McCaffrey, A. 1969. *The Ship Who Sang*. Del Rey Books.
65. Clarke, A.C. 1972. *Rendezvous with Rama*. See also O'Neill, G.K. 1977. *The High Frontier: Human Colonies in Space*. Morrow.
66. See, for example, Wells, H. G., *The Time Machine*; Forester, E.M. 1910. *The Machine Stops*; Bradbury, R. *The Martian Chronicles*; Zelazny, R. 1963. A Rose for Ecclesiastes. *The Magazine of Fantasy and Science Fiction*. Also available in many collections, including Silverburg, R. ed. 1970. *The Science Fiction Hall of Fame*. New York: Doubleday; Atwood, M. *The Handmaid's Tale*; James, P.D., *Children of Men*; Pohl, F. 1966. Day Million. *Impulse S.F.* 1(3). Collected in 1970, *Day Million*, Ballentine Books; and Bear, G. 1985. *Blood Music*. Arbor House.
67. The person who is skeptical of the value of science fiction would be wise to consider the words of the great science fiction writer Theodore Sturgeon, "Ninety percent of science fiction is crap but then ninety percent of everything is crap," before totally devaluing science fiction.

49 Space Archaeology and the Historiography of Science Fiction

Bradley G. Boone

CONTENTS

"Eat the grass in perfect circles.
It drives them crazy."

INTRODUCTION

> Far
> and unreachably back in geological time
> Lie footsteps in the sand traced by man.
> But in future's unborn path, in the wind,
> Our footsteps will fade
> into the stars.[1]

The juxtaposition of two words, *space* and *archaeology*, seems almost arbitrary if not contradictory, but by fusing these two seemingly opposite fields, we find, as so often is the case, ideas and ways of looking at the world that are totally unique and fresh. The landscape of archaeology is not just confined to the past—it also encompasses what will become the past and what is contemporaneous with our storied endeavors as individuals, who happen to be a small but perhaps influential part of a race of beings on this third world from the sun, an average main sequence type G2 star in the Perseus arm of the Milky Way galaxy in the Virgo cluster of galaxies, one filament in the warp and weft of space-time left by the Big Bang, the latest, perhaps, of innumerable cycles of creation that some theorize may be the underpinning of all that we can see or imagine. The records that we have left or will leave of our presence (in at least the solar system) will tell a story that may yet be read by some far future historian or archaeologist. This is the subject of many previous writers of science fiction, who in many cases have been prophets of the Space Age. A recent and notable work by Gregory Benford, a professional astrophysicist at the University of California, Irvine, titled *Deep Time: How Humanity Communicates Across Millennia* (1999),[2] describes how what we do today and leave behind for our far future posterity becomes the artifacts of their archaeology and perhaps, more importantly, communicates our legacy to them.

This notion of looking back at ourselves can be extended to what we will leave for our descendants to decipher: not just how we built the Saturn V rocket or the *Voyager* spacecraft (which is now meandering out beyond the solar system, faintly communicating, perhaps, until its radioisotope power supply dies around 2020), but to understand what we thought or dreamed about in the broader sense. Thus, science fiction plays and has played a role in stimulating the imaginations and passions of youth, who have grown up to become inspired leaders in the space business. Probably the single most notable proponent of this view whom most of the educated public is familiar, is Arthur C. Clarke (Figure 49.1). But there are many predecessors and workaday individuals who pursued a line that eventually led to Apollo and other historic space missions. After all, Glen Curtiss started out tinkering with motorcycles before he got into airplanes. At an age when some of us were in grade school, Chesley Bonestell was creating planetary landscapes that stimulated our imaginations and ambitions to get involved in space exploration. The interior artifacts of our mental life are thus the subject of retrospection and understanding. They are the cultural monuments and imaginative artifacts that have inspired the present and the future, but equally important, they too are the accessible artifacts of space archaeology.

BEGINNINGS

More than 30,000 years ago, light left the center of our galaxy on its trip to human eyes. Up until that time, Neanderthal humans dominated northwest Europe (as well as other parts of the Near East and western Asia), and well before that light even started on its way, modern humans (traditionally deemed Cro-Magnon, as well as related modern humans) emerged from Africa to conquer the world. They soon overran Europe, likely driving the Neanderthal westward to extinction (or possibly assimilation) by the time that light was on its way to us. One of the key reasons for this had far less to do with modern humans' tolerance of the cold climate, because they were decidedly gracile as compared with Neanderthal's robust frame and hence less suited to cold climates. It was likely their greater ability to communicate, cooperate, and cogitate, which enabled them to succeed in spite of the ice and other adversities following the last ice age (although scientists are still unsure how *Homo sapiens* came to be the dominant hominid species and why *Homo neanderthalensis* disappeared). The conquest of the next plane of human

FIGURE 49.1 Sir Arthur C. Clarke, who died recently (March 18, 2008) was one of the most enduring and visible writers of science fiction and speculative science. He was a visionary and prophet of space exploration and the future prospect of humans spreading amongst the planets of our solar system and perhaps among the stars in our galaxy. Many of his stories and novels focused on the prospect of a human-alien encounter and what its ramifications would be for human culture. He also envisioned the far future of human evolution and was hopeful about our eventual transformation. He admitted that he was influenced by the earlier works of H.G. Wells, as well as the lesser-known Olaf Stapledon, an early- to mid-twentieth-century English philosopher whose works included *The Last and First Men*[3] and *The Star Maker*.[4] Clarke's many books, stories, and technical articles are listed online at http://www.clarkefoundation.org. (Courtesy of the Estate of Arthur C. Clarke.)

existence, space, is physically a far more daunting prospect. It is true that there are important, if not vital, parallel cultural attributes that we should recognize.

Many science fiction movie themes have stressed this, one of the more popular ones being *The Day the Earth Stood Still*, which derives from a work by an obscure author (Harry Bates) titled "Farewell to the Master,"[5] featured in a 1940 issue of *Astounding Science Fiction* magazine. This movie featured Michael Rennie as an emissary named Klaatu, accompanied by his sidekick robot and protector, Gort, who brings a warning message from all extraterrestrial civilizations amongst the stars in our galaxy: that Earth cannot export its aggression to the stars, an apt moral for a movie couched in the postwar era, when ordinary citizens lived under the fear of nuclear war and the threat of communist expansion exaggerated by a rampant and growing McCarthyism. Those ballistic missiles that were soon to be developed for delivery of nuclear weapons, which were perfected in World War II, would become the means of entering space. Fortunately, the moral (and closing) message of that film is a quality that we have struggled mightily to exercise and preserve as we have proceeded through history and especially the twentieth century: *go in peace*.

This and the radical environmental challenges in our future portend perhaps not just a mental or moral transformation but a physical metamorphosis as well. If recent progress in exploring the human genome and developing nanotechnology is any indication, the twenty-first century promises to not only enable us to suppress diseases but also give us the power to enhance human resistance to the vicissitudes of the space environment, particularly ionizing radiation. This is not all: as many science fiction authors have anticipated for a long time, as well as serious scientists and science writers in more recent literature, "Robo-sapiens"[6] may be the next step beyond genetically engineered humans, to wit, "cyborgs" (cybernetic organisms). This theme, of course, was stimulated by Norbert Weiner's seminal works in cybernetics[7,8] in the 1940s, as well as the revolution in microelectronics since the 1950s. Another prodigy of the 1940s, John von Neumann, was a polymath who worked over many disciplines such as quantum mechanics, nuclear weapons, and computer architectures, who invented his famous von Neumann computer architecture,[9] which was subsequently extrapolated as the basis of potential alien or future human means of exponentially expanding among the stars in our galaxy.[10] In science fiction, however, the idea of self-replicating machines dates back to a 1920s play by Karel Čapek, titled *R.U.R.*[11] Well-known science fiction authors such as John W. Campbell introduced the subject of nano-scale organisms that replicate humans in his story *Who Goes There?*[12] which was made into a movie titled *The Thing from Another World* (twice, once in the 1950s and once in the 1980s). NASA's Advanced Automation for Space Missions[13] study addressed such machines and stimulated more science fiction on this subject. Von Neumann began studying the replication of automata probably because of his interest in computing stimulated by the Manhattan Project and the huge volume of calculations required for bomb design. The notion evolved that a von Neumann machine could travel the relatively short distance between adjacent stars in our galaxy, and whenever it arrives in a planetary system with sufficient resources, it would have the capacity to replicate itself and send copies on to other stars or planetary systems. In the long span of galactic evolution, where the Milky Way completes one revolution in approximately 250 million years, it would take a relatively short time (in multiples or submultiples of such a period) to proliferate within the galaxy. This idea also rests upon the assumption that stars, especially with planetary systems with attractive real-estate to justify the replication of sentient entities (whether human or cybernetic), are not too few and far between.

Recent serious discussion on this latter issue has been stimulated by works describing the Rare Earth Hypothesis[14] and the underpinnings of the Drake equation,[15] that is, whether early estimates of the probability of intelligent civilizations within the galaxy are now regarded as overly optimistic (see the boxed text inset below, "Some Basic Definitions"). Perhaps this argument carries more weight merely because of Fermi's paradox,[16] which was attributed to Enrico Fermi, who posed the question, if the galaxy is so laden with intelligent life, why aren't they already here? One possible answer to this conundrum is that they *are* here, and they are us! (See Capelotti, Chapter 45 of this volume.) Of course, this bit of space folklore only perpetuates speculation and counterspeculation, but fortunately this has been refined and made more cogent by scientific discoveries about the origins of the solar system and life, established by objective scientific inquiry. These more informed speculations have been fueled by new discoveries ranging from remote sensing of extrasolar planets to a better understanding of the gravitational effects of large gas giants on the solar system, and our large moon's gravitational stabilization of Earth's precession axis, to geological evidence of heavier asteroid and cometary bombardment in Earth's early history.

Because of the global and planetary reach of our space science, human knowledge of the world has achieved something approaching (in some quarters) a consilience[17] (in E.O. Wilson's spirit) unprecedented in human history. This same perspective is often fostered in young readers of science fiction. In this context, Einstein's observation "Imagination is more important than knowledge" is compelling.

Much of science fiction literature shares a common perspective that imbues archaeology: comprehending the long stretches of geologic and cosmic time and changes wrought over those times that determine our state of affairs today. The relative lack of this insight among the masses has been an obstacle to human understanding of evolution and to gaining insight into true human nature. This form of enlightenment is woefully lacking even in our modern Western civilization, not only among the less educated public but even among our national leaders, who make decisions on policy affecting populations far from our shore and investments in causes that could bring greater enlightenment and benefits rather than fostering the aggressive reactionary and destructive consequences of global ideological conflict.

Science fiction, or more generally speculative literature, has had a long and amicable relationship with science and those who practice it, whether professional or amateur (see the boxed text inset below, "Some Basic Definitions"). Many of the scientists involved in the war effort, such as those at Los Alamos, read pulp science fiction. In fact, I once met an old friend and neighbor of my uncle, who was a facilities support worker at Los Alamos during the war and who had a large chest full of *Astounding* he had collected since the 1940s (see Figure 49.2 for a couple of examples). Many future scientists encountered scientific ideas juxtaposed with imaginative thinking in science fiction, which often occurs at a tender adolescent age when they are most impressionable to new ideas. John Pierce, who taught physics and electrical engineering at Cal Tech, wrote science fiction for *Astounding Fiction Magazine* under the pen name of J.J. Coupling. More notable writers include Isaac Asimov (biochemistry), Arthur C. Clarke (electrical engineering), Poul Anderson (physics), Gregory Benford (astrophysics), and Stephen Baxter (aerospace engineering). These writers are also noteworthy for their style of writing: hard science fiction or space opera.

FIGURE 49.2 Examples of *Astounding Science Fiction*, the longest-running and most exemplary pulp science fiction magazine (in continuous circulation since 1930), which extolled the virtues of technological progress. These two covers feature cover art by Chesley Bonestell, an influential space artist of the period, who stimulated many future space enthusiasts. (Copyright © 1948 by Street & Smith Publications, Inc. Reprinted by permission of Dell Magazines, a division of Crosstown Publications.)

SOME BASIC DEFINITIONS

Science Fiction

Isaac Asimov has written that science fiction is that branch of literature that is concerned with the impact of scientific advance upon human beings. Science fiction is, of course, storytelling, usually imaginative as distinct from realistic fiction, that poses the effects of current or extrapolated scientific discoveries, or a single discovery, on the behavior of individuals of society. Robert A. Heinlein wrote[18] the following description of science fiction:

> Science Fiction is speculative fiction in which the author takes as his first postulate the real world as we know it, including all established facts and natural laws. The result can be extremely fantastic in content, but it is not fantasy; it is

> legitimate—and often very tightly reasoned—speculation about the possibilities of the real world. This category excludes rocket ships that make U-turns, serpent men of Neptune that lust after human maidens, and stories by authors who flunked their Boy Scout merit badge tests in descriptive astronomy.

A less well-known but influential editor and science fiction fan, Sam Moskowitz,[19] said,

> Science fiction is a branch of fantasy identifiable by the fact that it eases the "willing suspension of disbelief" on the part of its readers by utilizing an atmosphere of scientific credibility for its imaginative speculations in physical science, space, time, social science, and philosophy.

Historiography of Science Fiction

The historiography of science fiction is the study of the writing and methods of science fiction, drawing upon elements such as authorship, editorial influence, interpretation, and style (such as hard science fiction), as well as readership. In science fiction, the subject is usually the future or possible futures extrapolated usually from a scientifically plausible premise. It also refers to a body of work, such as, for instance, pulp magazines, as developed in the United States in the 1930s and 1940s onward, as well as many other sources. Thus, the term historiography is somewhat context dependent and has several possible meanings. Science fiction has been described in numerous ways, in part because it has numerous subgenres, such as space opera, time travel, the scientific romance, and utopian novels, but is generally distinct from fantasy in that science fiction has a strong basis in scientific realism.

Drake Equation and Fermi Paradox

The Drake equation, devised by Dr. Frank Drake in 1960, is a famous result in the field of exobiology and the search for extraterrestrial intelligence (SETI), which is fundamentally based on a number of conditional probabilities that, when multiplied together, give an overall estimate the number of possible extraterrestrial civilizations in our galaxy (N) with which we might come into contact. It is closely related to the Fermi paradox in that, given a large number of extraterrestrial civilizations predicted, the lack of evidence of such civilizations (the Fermi paradox) suggests that technological civilizations tend to destroy themselves rather quickly. It is given by the following equation:

$$N = R \times f_P \times n_e \times f_l \times f_i \times f_c \times L \tag{49.1}$$

and consists of nominally seven factors: R, the average rate of star formation in our galaxy; f_p, the fraction of those stars that have planets; n_e, the average number of planets that can potentially support life per star that has planets; f_l, the fraction of the above that actually go on to develop life at some point; f_i, the fraction of the above that actually go on to develop intelligent life; f_c, the fraction of civilizations that develop technology that enables their detection; and L, the length of time such civilizations release detectable signals into space.

Considerable disagreement on the values of most of these parameters exists; many of them are still guesses, but the values initially used by Drake and others in 1961 yielded a value of $N = 10$. Much debate, subsequent theorizing, and even some respectable tangential evidence[14,15,16] have been brought to bear to revise the numbers and assumptions that go with the component values, yielding values of 1 (us!) to greater than a few thousand. The upshot of this and other arguments suggests that intelligent life may be rare, despite the fact that bacterial life may be very abundant in the cosmos.

Hard science fiction is a subgenre that focuses on telling a story laced with scientific reasoning and/or lengthy descriptions of technological artifacts, such as new discoveries or encounters by humans with extraterrestrial life on other worlds, perhaps around other stars, and describing the means of getting there, whether within the laws of physics or extrapolating beyond the laws of physics. It is closely associated with another subgenre, space opera, which in its earlier forms was rather unsophisticated and decidedly unrefined as literature. Appealing to the adolescent mind, particularly teenage boys, it often contained descriptions of devices that could enable its inventors to break a sacred law of physics, such as the speed of light. Often this was merely an artifice to enable the story action to proceed at the usual fast pace so attractive to young male readers. In a sense, it was an American brand of the pulp fiction so popular from the early twentieth century. Stories and reprints of works of more notable authors, such as Jules Verne or H.G. Wells, began appearing in various magazines after World War I, including *Amazing Stories* and *Astounding Stories*. The first was largely the brainchild of Hugo Gernsback, a Luxembourg immigrant with a penchant for gadgets and electronics, who started several pulp magazines, including those related to science and invention (Figure 49.3).

This was the heroic age of lone inventors such as Thomas Edison (electrical gadgets of all kinds), Lee de Forest (radio), Philo T. Farnsworth (raster television), and Nikola Tesla (generators, radio, and the infamous Tesla coil), and many others. This was the time of individualism in American and European culture. Almost coincident with the revolution in electrical science was the rise of rocketry, expressed in the work of Konstantin Tsiolkovsky (left-hand side of Figure 49.4), Robert H. Goddard (right-hand side of Figure 49.4), Hermann Oberth, and Wernher von Braun (pictured on the left in Figure 49.5), the latter culminating in the development of the V-2 ballistic missile (on the right side of Figure 49.5), launched from Peenemünde on London and Antwerp, but later from White Sands and Kapustin Yar. But as World War II came to a close, punctuated with the birth of the atomic age over Hiroshima, the large scientific project represented by these inventions and its vast sums of money and administration emerged as the cauldron of new ideas and inventions. At this point,

FIGURE 49.3 Hugo Gernsback (left), a European immigrant to the United States who launched a number of magazines, both of an electrical experimental and radio amateur orientation (as shown on the right) and others, including both science fact and fiction. (Courtesy of the Hugo Gernsback Collection at Syracuse University.)

FIGURE 49.4 Konstantin Tsiolkovsky (left) and Robert H. Goddard (right), Russian and American, respectively, were visionary thinkers and doers in the infant era of space exploration and rocketry in their respective countries. In their early careers, they were revered—or in some cases ignored— by their governments, which really were just not ready for their visions of the future until after World War II, which played such a pivotal role in the twentieth century, especially in political and technological changes that led to the Space Age. (Soviet stamp and NASA.)

FIGURE 49.5 Wernher von Braun (left photograph: second from right) and Hermann Oberth (left photograph: center) were leaders in Germany in rocketry and space travel prior to World War II. Von Braun, of course, led the team at Peenemünde, which developed and launched the V-2 rocket, the first ballistic missile used as an instrument of war. Many of these missiles were captured, along with their support personnel, by American (and Soviet) forces near the end of hostilities in Europe, but it was the Americans who got the prize leaders, including von Braun. The V-2 (on the right) was subsequently launched at White Sands Missile Range not long after the war, not too far from the Trinity Site, location of the first atomic explosion, which ultimately led to a final cessation of hostilities with Japan after two bombs were dropped over Hiroshima and Nagasaki. These two aspects of the period (rockets and nuclear weapons) would play a pivotal role in the start of the Space Age, as well as becoming artifacts of the age. (Courtesy of NASA.)

in the late 1940s, science fiction, through the encouragement of editors such as John W. Campbell at *Astounding,* tutored a whole new generation of writers and readers, helping to digest the portents of the technical and scientific fruits of the wartime research and development in atomic energy, rocketry, radar, and electronics. These were assimilated by the genre, indeed some were anticipated: certainly the rocket as a spaceship and the spaceship as a means of global communications with Arthur C. Clarke's vision of geosynchronous satellites.[20] (Even before Clarke's vision there was a series of short stories published in *Astounding Science Fiction* between 1942 and 1945 by George O. Smith, titled the *Venus Equilateral*[21] series, which concerned the "Venus Equilateral Relay Station," an interplanetary communications hub located at the L_4 Lagrange point of the Sun-Venus system, which would be considered rather quaint today in that the imagined hardware using vacuum tubes!) Moreover, atomic energy was discussed in the pages of *Astounding* in 1944 in such detail[22] that it drew national security attention and a visit by the FBI to the editorial offices of the magazine. Even as far back as the writing of Hugo Gernsback's novel[23] titled *Ralph 124C 41+* (part of it read as "one to foresee for one") in 1911, was the prediction of radar.

It was in the late 1940s that the art of Chesley Bonestell began to appear in books published by Viking, the first titled *The Conquest of Space*[24] by Willy Ley

FIGURE 49.6 Lunar Lander. Recreation of Chesley Bonestell artwork for the cover of *Collier's* magazine, October 18, 1952, by Ron Miller 1991 (left). *Touchdown of the Moon.* NASA Constellation concept (right). [Courtesy of the Chesley Bonestell Estate (left) and NASA (right).]

(1949). This was followed every few years with a whole series of books, most notably *Across the Space Frontier*[25] (1952), *Conquest of the Moon*[26] (1953), *The Exploration of Mars*[27] (1956), and *Beyond the Solar System*[28] (1964). (An inspiring example is shown in Figure 49.6) These books inspired budding space enthusiasts; the general public; certainly science fiction readers, writers, publishers; and a new crop of space artists who eventually found jobs or commissions with NASA. As Ron Miller, a noteworthy contemporary space artist, pointed out in a later publication[29] about Bonestell's space art,

> What makes astronautics uniquely special among the sciences is that its roots lay so deeply buried in imagination and fantasy. For centuries it existed only in the works of such torchbearers as Edgar Allan Poe, Jules Verne and H.G. Wells, and those scientists—professional and amateur—who believed in such a patently unlikely proposition as the possibility of flying into space and visiting other planets. Given the very real influence of literature on the history and development of astronautics—where would it be today had Hermann Oberth, Robert Goddard and Konstantin Tsiolkovksy not read science fiction?—it's not the least bit surprising that an artist could also have had a major impact on that history. That artist was Chesley Bonestell.

Today, a crater on Mars and the asteroid 3129 Bonestell are named after him.

Although there were many B-grade movies produced in the 1940s and 1950s about space, aliens, and invasions of the Earth, there were more serious and credible attempts by film makers to produce visionary films. One in particular, *Destination Moon*, produced by George Pal (and for which Bonestell was an adviser) was based on a novel by Robert Heinlein titled *Rocketship Galileo*,[30] the first in a series of juvenile stories written by Heinlein up until the early 1960s. Another one, also produced by George Pal, was *The Conquest of Space*, based on Willy Ley's book by that name.[24] Almost concurrently, *Collier's* magazine joined the party with a series of articles[31] from 1952 through 1954 on the details of space exploration, treating in colorful detail the construction of Earth orbiting space stations, multistage rockets, shuttle-like upper-stage space planes (as shown on the left-hand side of Figure 49.6), lunar landers, and manned Martian missions. These were the brainchild of Wernher von Braun and were compiled in some of the books mentioned earlier.[24,25,26,27,28] In Figure 49.6 on the left-hand side is an illustration of a possible lunar lander circa 1952, as envisioned by Wernher von Braun and illustrated by Chesley Bonestell, and on the right-hand side is the next-generation lunar lander for the contemporary U.S. Constellation manned lunar program.[26]

Analog Science Fact and Science Fiction is the modern-day equivalent of *Astounding*, the longest-running science fiction magazine still in circulation. No other magazine has had content that was so consistently, optimistically biased toward technology, the future, and the exploration of space. It has reflected its readership's interest in space travel and optimism for the future and a streak of rational and humanistic concern for the human welfare by its editors over the decades since its inaugural issue in January 1930; the magazine is now more than 78 years old and still going strong. Its pages have been graced with good science and occasionally some really bad science or pseudoscientific concepts, such as inertia-free space drives, so-called "psionic powers," and a once-respected editor's forgettable obsession with the cultish aberration of Dianetics. In recent years, very respectable physicists such as John Cramer have written a physics-oriented column[32] for the magazine. In the average "Joe Public's" opinion it would be labeled "geeky," but in reality it is a rather well-informed, socially conscious forum for scientifically minded writers and readers of contemporary, usually hard science fiction. It continues to nurture new writers who continue to capture and maintain interested readers of space travel stories. It is the undercurrent that reflects the fact that space exploration is a cogent and tangible force within our culture that still has the power to inspire people's aspirations for the future. At its Web site, there are links to the Artemis Society[33] (for establishing a permanent human presence on the Moon) and the Planetary Society[34] (the world's largest space interest group).

SUMMARY

Usually, in the short time horizon of our present information-laden world (which is the subject of our weekly news magazines and popular press, with their focus on politics, social inanities, and warfare), it is easy to overlook the long-term prospects of the human race. Many times they seem, of course, rather dim prospects, given current events. Science fiction and archaeology both offer us the long view. Look at our knowledge of human nature garnered in part from archaeological studies, as well as scientifically disciplined inferences about the human mind and its influence on prehistorical culture over the long periods of time between the last ice age and the rise of agriculture, writing, and recorded history. This has been the subject of a recent spate of books on "Big History."[35,36,37,38] You will see that this knowledge is similar to what is conveyed by our fictional extrapolations of the future in science fiction. Despite earlier works that predicted a near future of prolific exploration *and* exploitation of space, it is likely that our future in space will develop over equally long stretches of time characteristic of recent prehistory, not mere historical epochs. In retrospect, humans have evolved from prehuman species over hundreds of thousands of years, from small, almost irrelevant numbers to virtually a "geologic force."[39] The future transformations required of human culture, technology, and indeed human physiology, which may be critical to our continued expansion into the universe, will be far more radical than those that have occurred over the 50,000 years since the last diaspora out of Africa (consistent with evidence supporting the "Out of Africa" hypothesis).[40] They may be, in fact, more than most people have imagined. We can only hope that the artifacts that we leave and the stories that we record will inform our descendants in ways that truly define our common thread of humanity. Perhaps Arthur C. Clarke had it right when he said:

> They will have time enough, in those endless aeons, to attempt all things, and to gather all knowledge. They will be like gods, because no gods imagined by our minds have ever possessed the powers they will command. But for all that, they may envy us, basking in the bright afterglow of creation; for we knew the universe when it was young.[41]

FURTHER READING

The breadth and depth of this subject cannot be adequately addressed in an essay or chapter such as this, so the serious researcher should consult the following sources for an in-depth coverage of both the historiography and artistic roots of science fiction and its relation to space exploration.

Ash, B., ed. 1977. *The Visual Encyclopedia of Science Fiction.* New York: Harmony Books.

Clute, J. 1995. *Science Fiction: The Illustrated Encyclopedia.* London: Dorling Kindersley.

DiFate, V.1997. *Infinite Worlds: The Fantastic Visions of Science Fiction Art.* New York: The Wonderland Press (Penguin).

Gilster, P. 2004. *Centauri Dreams: Imagining and Planning Interstellar Exploration.* New York: Springer Science (Copernicus Books).

Gunn, J.E. 1975. *Alternate worlds: The Illustrated History of Science Fiction.* Englewood Cliffs, NJ: Prentice-Hall.

Hardy, D.A. 1989. *Visions of Space.* Limpsfield Surrey, United Kingdom: Dragon's World.

Harrison, H., and M. Edwards. 1979. *Spacecraft in Fact and Fiction.* New York: Exeter Books.

Hartmann, W.K., A. Soklov, R. Miller, et al. 1990. *In the Stream of Stars: The Soviet/American Space Art Book.* New York: Workman Publishing Company.

Hartmann, W.K., and R. Miller. 2005. *The Grand Tour: A Traveler's Guide to the Solar System.* New York: Workman Publishing Company.

Kyle, D. 1977. *The Illustrated Book of Science Fiction Ideas and Dreams.* London: Hamlyn Publishing.

Launius, R.D., and E. McCurdy. 2001. *Imagining Space: Achievements. Predictions. Possibilities.* San Francisco: Chronicle Books.

Mallove, E.F., and G.L. Matloff. *The Starflight Handbook: A Pioneer's Guide to Interstellar Travel.* New York: John Wiley and Sons.

McCall, R. 1992. *The Art of Robert McCall.* New York: Bantam Books.

Miller, R. 1993. *The Dream Machines: An Illustrated History of the Spaceship in Art, Science and Literature.* Malabar, FL: Krieger Publishing Company.

Nichols, P. 1982. *The Science in Science Fiction.* London: The Roxby and Lindsey Press.

O'Neill, G.K. 2000. *The High Frontier: Human Colonies in Space.* Burlington, Canada: Apogee Books (Collector's Guide Publishing Inc).

Ordway, F.I., III, and R. Liebermann, eds. 1992. *Blueprint for Space, Science Fiction to Science Fact.* Washington, DC: Smithsonian Institution Press.

Ordway, F.L., III. 2001. *Visions of Spaceflight: Images from the Ordway Collection.* Berkeley, CA: Publishers Group West.

Robinson, F.M. 1999. *Science Fiction of the 20th Century: An Illustrated History.* Portland: Collectors Press.

Sagan, C. 1980. *Cosmos.* New York: Random House.

Sagan, C. 1994. *Pale Blue Dot.* New York: Random House.

Stine, G.H. 1998. *Halfway to Anywhere: Achieving America's Destiny in Space*, New York: M. Evans and Company, Inc.

Villard, R., and L.R. Cook. 2005. *Infinite Worlds: An Illustrated Voyage to Planets beyond Our Sun.* Berkeley, CA: University of California Press.

Von Braun, W., and H.J. White. 1962. *The Mars Project.* Urbana, IL: University of Illinois Press.

Webb, S. 2002. *If the Universe Is Teeming with Aliens ... Where Is Everybody? Fifty Solutions to Fermi's Paradox and the Problem of Extraterrestrial Life.* Springer.

Westfahl, G., ed. 2000. *Space and Beyond: The Frontier Theme in Science Fiction.* Westport, CT: Greenwood Press.

Zubrin, R. 1999. *Entering Space: Creating a Spacefaring Civilization.* New York: Jeremy Tarcher (Penguin Putnam).

Zubrin, Z., R. Wagner, and A.C. Clarke. 1996. *The Case for Mars: The Plan to Settle the Red Planet and Why We Must.* New York: Touchstone Press (Simon & Schuster).

NOTES AND REFERENCES

1. Boone, B.G. 2006. *The Twinkled Light of Far Distant Suns.* Booksurge LLC.
2. Benford, G. 1999. *Deep Time: How Humanity Communicates Across Millennia.*
3. Stapledon, O. 1931. *The Last and First Men.* London: Methuen.
4. Stapledon, O. 1937. *The Star Maker.* London: Methuen.
5. Bates, H. 1940. *Farewell to the Master.* Astounding Science Fiction, October.
6. Menzel, P., and F. D'Aliusio. 2000. *Robo Sapiens: Evolution of a New Species.* Cambridge, MA: MIT Press.
7. Weiner, N. 1948. *Cybernetics: Or the Control and Communication in the Animal and the Machine.* Cambridge, MA: MIT Press.
8. Weiner, N. 1950. *The Human Use of Human Beings.* Da Capo Press.
9. Von Neumann, J. 1966. *The Theory of Self-Reproducing Automata.* Urbana, IL: University of Illinois Press.
10. Bracewell, R.N. 1960 Communications from Superior Galactic Communities. *Nature* 186: 670–671.
11. Čapek, K. 2004. RUR (*Rossum's Universal Robots*). Penguin Classics.
12. Campbell, J.W., Jr. 1948. *Who Goes There?* Shasta Publishers.
13. Freitas, Jr., R.A., and W.P. Gilbreath. 1980. Advanced Automation for Space Missions. Proceedings of the 1980 NASA/ASEE Summer Study, Santa Clara, California, June 23-August 29, 1980. NASA Conference Publication 2255.
14. Ward, P., and D. Brownlee, 2000. *Rare Earth: Why Complex Life is Uncommon in the Universe.* Copernicus Books.
15. Drake, F., and D. Sobel. 1992. *Is Anyone Out There? The Scientific Search for Extraterrestrial Intelligence,* New York: Delacorte Press. (Dr. Arroway, played by Jodie Foster, paraphrases the Drake equation several times in the movie *Contact*, using it to justify the SETI program. The movie was based on the novel with the same title by Carl Sagan.)

16. Jones, E. 1985. "Where is everybody?" An Account of Fermi's Question. Los Alamos Technical report LA-10311-MS, March.
17. Wilson, E.O. 1999. *Consilience: The Unity of Knowledge*. New York: Vintage.
18. Heinlein, R.A. 1980. Ray Guns and Spaceships. In *Expanded Universe*. New York: Ace Books.
19. Moskowitz, S. 1974. *Explorers of the Infinite, Shapers of Science Fiction*. Hyperion Press.
20. Clarke, A.C. 1945. Extraterrestrial Relays. *Wireless World*, October: 305–308.
21. Smith, G.O. 1947. *Venus Equilateral*. Philadelphia: Prime Press.
22. Cartmill, C. 1944. Deadline. *Astounding Science Fiction*, March.
23. Gernsback, H. 1925. *Ralph 124C 41+*. Stratford Co.
24. Ley, W. 1949. *The Conquest of Space*. New York: Viking.
25. Ryan, C. ed. 1952. *Across the Space Frontier*. New York: Viking Press.
26. Ryan, C. ed. 1953. *Conquest of the Moon*. New York: Viking Press.
27. Von Braun, W., and W. Ley. 1956. *Exploration of Mars*. New York: Viking Press.
28. Ley, W., and C. Bonestell. 1964. *Beyond the Solar System*. New York: Viking Press.
29. Miller, R., and F.C. Durant III. 2001. *The Art of Chesley Bonestell*, London: Paper Tiger.
30. Heinlein, R. 1947. *Rocketship Galileo*. New York: Scribner's.
31. *Collier's* magazine: March 22, 1952, Man Will Conquer Space Soon. October 18, 1952, Man on the Moon. October 25, 1952, More about Man on the Moon. February 28, 1953, World's First Space Suit. March 7, 1953, More about Man's Survival in Space. March 14, 1953, How Man Will Meet Emergency in Space Travel. June 27, 1953, The Baby Space Station: First Step in the Conquest of Space. April 30, 1954, Can We Get to Mars?/Is There Life on Mars?
32. Cramer, J. The Alternate View. *Analog Science Fiction and Science Fact*.
33. The Artemis Project. http://www.asi.org/.
34. The Planetary Society, http://www.planetary.org/home/.
35. Brown, C.S. 2007. *Big History: From the Big Bang to the Present*. New Press.
36. Christian, D., and W.H. McNeill. 2005. *Maps of Time: An Introduction to Big History*. University of California Press.
37. McNeill, R., and W.H. McNeill. 2003. *The Human Web: A Bird's-Eye View of World History*. W.W. Norton.
38. Mithen, S. 2006. *After the Ice: A Global Human History 20,000–5000 BC*. Harvard University Press.
39. Klein, R.G. 2002. *The Dawn of Human Culture*. New York: John Wiley and Sons.
40. Wells, S. 2002. *Journey of Man: A Genetic Odyssey*. Princeton University Press and Random House.
41. Clarke, A.C. 1962. *Profiles of the Future*. New York: Harper and Row.

Appendix A

Toward an Understanding of Terminology in Space Engineering, Archaeology, and Heritage

Ann Garrison Darrin and Beth Laura O'Leary

Reviewing the terminology and history of words and phrases used in this text yields many interesting juxtapositions. Clearly, some of these derive from the dynamic nature of space versus the static nature of ground-based material studies. Others come from the very different applications of the same word. Understanding terms as used by engineers and archaeologists is critical to the preservation of space material culture in the future.

Space research and development has added a dimension to archaeological work in providing satellites that transmit locational data to handheld global positioning system (GPS) units. The GPS units compute their location by analyzing the microwave signals from a constellation of medium Earth orbit satellites. The exact position of an archaeological site is important to understanding the context of a site within the topography of a landscape (e.g., is it located on an alluvial fan or next to a playa?) and in relation to other sites dating to the same or other temporal periods. Likewise, when data recovery or excavation of buried remains is carried out, locational data in three dimensions are critical to interpreting the results. A grid system usually overlays an excavation tied to a datum point or reference point. The location of artifacts within an excavated site is known in three dimensions. The context of each piece of data is critical to interpretation. The context of an artifact refers to its surrounding matrix (the soil or material around it), its horizontal and vertical position, and its association with other artifacts or material culture. Because a site is systematically destroyed by archaeological excavation, all information must be recorded as accurately as possible. The spatial and temporal patterning among the site's soil layers, artifacts, features, ecological evidence, and components is critical to understanding past human activities and social processes. Removal by looting or vandalism destroys the context of the material remains.

An analysis of the term *in situ* illustrates the different perspectives held by engineers and archaeologists. For the terrestrial-based archaeologist, it refers to the location of biological, physical, or material culture objects in their original physical and cultural context, that is, in the natural or original position or place. The term is usually used for features that exist within a site that are nonportable (e.g., post molds and building foundations). For example, within a cave site with good preservation, there may be multiple or a series of hearths or charcoals stains that overlay each other in different strata. These are said to be in situ. For the aerospace engineer, *in situ* is a localized state or condition. In situ sensing refers to data gathering at the local environs of the source or collector where there is normally direct physical contact with the surrounding environment. Remote sensing would be small- or large-scale acquisition of information of an object or phenomenon by the use of either recording or real-time sensing devices that are not in physical or intimate contact with the object.

Another interesting example is the use of the law of superposition for terrestrially based debris investigations. The law of superposition (or the principle of superposition) is a foundational principle of sedimentary stratigraphy in geology; sedimentary layers are deposited in a time sequence, with the oldest on the bottom and the youngest on the top. Superposition as used by archaeologist during excavation is different because past or present human-made intrusions and forms of bioturbation in the archaeological record can affect the chronological sequence. Superposition in archaeology is complex and requires a great degree of interpretation to accurately identify chronological sequences, and in this sense superposition in archaeology is more multidimensional. Archaeologists have become well aware of what is called formation processes, which are the ways in which archaeological materials came to be buried and what has happened to them subsequently. Formation processes include the deliberate or inadvertent activities of humans mentioned above, as well as natural processes (e.g., erosion) that govern the visibility and survival of the material culture. Superposition applies to sedimentary layers in a fairly static way compared with orbital mechanics for space debris, that is, the positioning of unrelated materials in orbit does not reflect age or common origins. For the purposes of understanding space debris, we can refer to this as the principle of dynamic positioning. Dynamic positioning refers to the site of materials in orbit where constant change in location, relative to an observer, is an accepted premise.

For understanding very different uses of the same term, one can look at the term *artifact*. Traditionally, archaeologists have defined *artifact* as an object fashioned or altered by humans. Engineers use the same term to define a product of artificial character (as in a scientific test) due usually to extraneous (such as human) agency.

In discussing our fields, the editors often used the same word with numerous meanings, or different words have been used in the same manner. In some cases, two definitions are included, mostly in the social sciences, when there are contested meanings. The pronounced and subtle differences lend themselves to a truly intriguing study.

GLOSSARY KEY

[SS] Term as used in the social sciences.

[PS] Term as used in the physical sciences.

Reference numbers precede physical or social science usage indicator. When the definition originates from the current volume, the chapter number follows endnote, preceded by a hyphen (e.g.,[11-2]).

TERMS

Ablation: The erosion of a solid body by a high-temperature gas stream moving with high velocity, for example, a reentry vehicle's heat shield that melts or chars under the effects of air friction.[1] [PS]

Absolute zero: The temperature at which all heat action ceases, –273.16°C (–459.69°F).[1] [PS]

Absolute dating: The determination of age with reference to an absolute or specific date, such as a calendar or other units of absolute time. Also referred to as chronometric dating.[18] [SS]

Absorption: A process whereby a material extracts one or more substances present in an atmosphere or mixture of gases or liquids, accompanied by the material's physical and/or chemical changes.[2] [PS]

Acceleration: A change in velocity, including changes of direction and decreases, as well as increases, in speed.[1] [PS]

Accelerometer: A device that senses changes in speed along its axis.[1] [PS]

Acculturation: The process by which a culture absorbs the traits or customs of another culture with which it is in direct contact[12] [SS]; large cultural changes that people are forced to make as a consequence of intensive firsthand contact between their own group and another, more powerful society.[19] [SS]

ACS: Attitude control system.[1] [PS]

Active heating: The use of resistive electric heaters or radioisotope heaters to keep spacecraft components above their minimum allowable temperatures.[1] [PS]

Active sun: The sun during times of frequent solar activity, such as sunspots, flares, and associated phenomena.[1] [PS]

Actuator: A device that transforms an electric signal into a measured motion using hydraulic, pneumatic or pyrotechnic (explosive) action.[1] [PS]

Aerobraking: The process of decelerating by converting velocity into heat through friction with a planetary atmosphere.[1] [PS]

Aerodynamic heating: The heating of a body due to the passage of air or other gases over the body; caused by friction and compression processes.[1] [PS]

Aerozine 50: A storable liquid fuel: 50% hydrazine, 50% unsymmetrical dimethylhydrazine (UDMH).[1] [PS]

Albedo: Reflectivity; the ratio of reflected light to incident light. The fraction of the sunlight that is reflected off a planet.[1] [PS]

Altimeter: A device that measures altitude above the surface of a planet or moon. Spacecraft altimeters work by timing the round trip of radio signals bounced off the surface.[1] [PS]

Ambient: Environmental conditions, such as pressure or temperature.[1] [PS]

Analog computer: A computing machine that works on the principle of measuring, as distinct from counting, in which the measurements obtained (as voltages, resistances, etc.) are translated into desired data.[1] [PS]

Angstrom: A unit for the measurement of wavelength. Equals one hundred millionth of a centimeter (0.003937 millionth of an inch).[1] [PS]

Angular momentum: A quantity obtained by multiplying the mass of an orbiting body by its velocity and the radius of its orbit. According to the conservation laws of physics, the angular momentum of any orbiting body must remain constant at all points in the orbit, that is, it cannot be created or destroyed. If the orbit is elliptical, the radius will vary. Since the mass is constant, the velocity changes. Thus, planets in elliptical orbits travel faster at perihelion and more slowly at aphelion. A spinning body also possesses spin angular momentum.[3] [PS]

Annular: Pertaining to or having the form of a ring.[1] [PS]

Anomaly: The angular distance between the position of a planet and its last perihelion or between that of a satellite and its last perigee.[1] [PS] Deviation from the common rule; something anomalous: something different, abnormal, peculiar, or not easily classified.[3] [PS]

Anthropology: The scientific and humanistic study of humankind's present and past biological, linguistic, social, and cultural variations. Its major subfields are archaeology, physical/biological anthropology, cultural anthropology, and anthropological linguistics.[12] [SS]

Aphelion: That point in a solar orbit that is farthest from the sun.[1] [PS]

Apoapsis: That point in an orbit that is farthest from the primary.[1] [PS]

Apogee: That point in a terrestrial orbit that is farthest from the Earth.[1] [PS]

Apolune: That point in a lunar orbit that is farthest from the Moon.[1] [PS]

Archaeoastronomy: A focus on what can be learned about ancient astronomy through the study of archaeological sites and material culture.[11-10] [SS]

Archaeological context: The physical setting, location, and cultural association of artifacts and features within the archaeological record.[12] [SS]

Archaeological formation processes: Those processes affecting the way in which archaeological materials came to be buried and their subsequent history afterward. Cultural formation processes include the deliberate or accidental activities of humans; natural processes are natural or environmental events that govern the burial or survival of the archaeological record.[18] [SS]

Archaeological survey: On-ground or remote inspection of the archaeological record[11] [SS]; the systematic observation of the ground surface, or the testing of the subsurface with the recording of relevant landscape features and material culture.[11–2] [SS]

Archaeology (also spelled archeology): The scientific study of the physical evidence of past human societies. Archaeologists not only attempt to discover and describe past cultures but also to formulate explanations for the development of cultures[12] [SS]; that subdiscipline of anthropology that studies the relationships between material culture and human behavior in all times and places where human material culture exists.[11–2] [SS]

Argument of periapsis: In an orbit, the angular distance between the point of periapsis and the ascending node.[1] [PS]

Argument: Angular distance.[1] [PS]

Artifact: 1. Any portable object used, modified, or made by humans, for example, stone tools, pottery, metal weapons.[18] [SS] 2. A product of artificial character (as in a scientific test) due usually to extraneous (such as human) agency.[4] [PS]

Artificial gravity: Use of centrifugal force to simulate weight reaction in a condition of free fall. May be achieved by spinning the vehicle to make the centrifugal force of the outer periphery or bodies within the vehicle to replace the weight reaction experienced at Earth's surface.[1] [PS]

ASAT: Antisatellite.[1] [PS]

Ascending node: The point at which an orbiting object or spacecraft, traveling from south to north, crosses the plane of the equator.[1] [PS]

Ascent module: That part of a spacecraft that ascends from the surface of a planet or moon to rendezvous and dock with an orbiting spacecraft.[1] [PS]

Assemblage: A group of artifacts related to each other based upon recovery from a common archaeological context. Assemblage examples are artifacts from a site and/or features.[12] [SS] Assemblages represent the sum of human activities.[18] [SS]

Association: The co-occurrence of an artifact with other archaeological remains.[18] [SS]

Associative cultural landscape: Cultural landscape that may be valued because of the religious, artistic, or cultural associations to the natural element. May not contain material culture.[11-16] [SS]

Asteroid belt: A ½ astronomical unit wide region between the orbits of Mars and Jupiter where most asteroids are found.[1] [PS]

Asteroid: A small, usually irregularly shaped body orbiting the sun, most often at least partially between the orbits of Mars and Jupiter.[1] [PS]

Astronomical unit: The mean distance of Earth from the sun, that is, 92,955,807 miles (149,597,870 km). Referred to in multiples if greater than one.[1] [PS]

Atmospheric balloon: An instrumented package suspended from a buoyant gas bag, deployed in a planet's atmosphere to study wind circulation patterns.[1] [PS]

Atmospheric pressure: The weight of air on surfaces within Earth's atmosphere, about 14.7 psi (101 kPa) at sea level. Such pressure is also supplied artificially in spacecraft and spacesuits.[1] [PS]

Atmospheric probe: A small, instrumented craft that separates from the main spacecraft prior to its closest approach to a planet to study the gaseous atmosphere of the body as it drops through it.[1] [PS]

Attenuation: The decrease of a propagating physical quantity, such as a radio signal, with increasing distance from the source, or from some obstruction.[1] [PS]

Attitude and articulation control subsystem: The onboard computer that manages the tasks involved in spacecraft stabilization via its interface equipment. For attitude reference, star trackers, star scanners, solar trackers, sun sensors, and planetary limb trackers are used.[1] [PS]

Attitude control: The system that turns and maintains a spacecraft in the required direction as indicated by its sensors.[1] [PS]

Attitude: Orientation of a space vehicle as determined by the relationship between its axes and some reference plane, such as the horizon.[1] [PS]

AU: Astronomical unit.[1] [PS]

Autonomy: The quality or state of being self-governing.[4] [PS]

Azimuth: The angular position of an object measured in the observer's horizontal plane, usually from north through east. Bearing or direction in the horizontal plane. As one of the coordinates expressing celestial location, it is sometimes used in tracking spacecraft.[1] [PS]

Backscattering: Reflecting light back in the direction of the source.[1] [PS]

Backup: An item kept available to replace an item that fails to perform satisfactorily.[1] [PS]

Ballute: An aerodynamic braking device that is both balloon and parachute.[1] [PS]

Basin: A large, >200 km, circular depression from the explosive impact of an asteroid or similar sized body on a planet surface, usually rimmed by mountains.[1] [PS]

Bipropellant: A rocket propellant consisting of two unmixed or uncombined chemicals (fuel and oxidizer) fed separately into the combustion chamber.[1] [PS]

Black powder: A mixture of saltpeter (potassium nitrate), sulfur, and charcoal, used in explosives and as an early propellant for rockets.[1] [PS]

Boilerplate: A metal replica of the flight model (e.g., of a spacecraft) but usually heavier and cruder, for test purposes.[1] [PS]

BOL: Beginning of life.[1] [PS]

Bolide: A large meteor. Fireball; especially, one that explodes.[4] [PS]

Boost: The extra power given to a rocket or space vehicle during liftoff, climb, or flight, as with a booster rocket.[1] [PS]

Booster: The first stage of a missile or rocket.[1] [PS]

Bow shock wave: The compressed wave that forms in front of a spacecraft or satellite as it moves rapidly through Earth's atmosphere; more generally, any such wave that forms between an object and a fluid medium.[1] [PS]

Burn: Combustion action in rockets. Propulsion in space is achieved through a sequence of burns.[1] [PS]

Bus: A major part of the structural subsystem of a spacecraft that provides a place to attach components internally and externally and to house delicate modules requiring a measure of thermal and mechanical stability. The bus also establishes the basic geometry of the spacecraft.[1] [PS]

Calibration: Setting a measuring instrument before measuring for accurate results.[1] [PS]

Catalytic decomposition engine: A monopropellant engine in which a liquid fuel decomposes into hot gas in the presence of a catalyst. The fuel is most commonly hydrazine.[1] [PS]

Celestial sphere: The apparent sphere of sky that surrounds the Earth; used as a convention for specifying the location of a celestial object.[1] [PS]

Celsius temperature scale: A thermometric scale in which the freezing point of water is called 0°C and its boiling point 100°C at normal atmospheric pressure.[2] [PS]

Centrifugal force: A force that is directed away from the center of rotation.[1] [PS]

Centripetal force: A force that is directed toward the center of rotation.[1] [PS]

CEO: Close Earth orbit.[1] [PS]

Chaff: Metallic foil ejected by a reentry module to enhance its radar image.[1] [PS]

Charged coupled device: An imaging device consisting of a large-scale integrated circuit that has a two-dimensional array of hundreds of thousands of charge-isolated wells, each representing a pixel.[1] [PS]

Cislunar: Relating to the space between the Earth and the orbit of the Moon.[1] [PS]

Coherence: The property of being coherent, for example, waves having similar direction, amplitude, and phase that are capable of exhibiting interference.[1] [PS]

Command and data subsystem: The onboard computer responsible for overall management of a spacecraft's activity.[1] [PS]

Commodification: A process that is conferred upon an object as a result of its manufacture to suit a specific need.[11–37] [SS]

Composites: Structural materials of metal alloys or plastics with built-in strengthening agents, for example, carbon fibers.[1] [PS]

Conduction: Transfer of heat by direct contact.[2] [PS]

Conservation archaeology: A subfield of archaeology that focuses on the preservation of archaeological resources. This position encourages the stabilization and preservation of archaeological sites as opposed to their immediate excavation. Also referred to as historic preservation or heritage preservation.[12] [SS]

Constellation: A group of stars that make a shape, often named and included in social narratives by human cultures after mythological characters, people, animals, and things.[11–10] [SS]. In the physical sciences, an assemblage, collection, or group of usually related persons, qualities, or things, such as a constellation of communication satellites.[1][PS]

Context (archaeological): An artifact's context usually consists of its immediate matrix or the material around it, its provenience (horizontal and vertical position in the matrix), and its association with other archaeological remains, usually within the same matrix.[18][SS]

Continuing landscape: Landscape that retains an active social role in contemporary society closely associated with the traditional way of life and in which the evolutionary process is still in progress.[11–16] [SS]

Convection: The movement of a mass of fluid (liquid or gas) caused by differences in density in different parts of the fluid; the differences in density are caused by differences in temperature. As the fluid moves, it carries with it its contained heat energy, which is then transferred from one part of the fluid to another and from the fluid to the surroundings.[2] [PS]

Coolant: A medium, usually a fluid, that transfers heat from an object.[1] [PS]

Corona: The sun's outer layer. The corona's changing appearance reflects changing solar activity.[1] [PS]

Coronal mass ejection: A huge cloud of hot plasma, expelled sometimes from the sun. It may accelerate ions and electrons and may travel through interplanetary space beyond the Earth's orbit, often preceded by a shock front. When the shock reaches Earth, a magnetic storm may result.[1] [PS]

Cosmic background radiation: In every direction, there is a very low energy and very uniform radiation that we see filling the universe. This is called the *3*

Kelvin background radiation, the *cosmic background radiation*, or the *microwave background*. These names come about because this radiation is essentially a black body with temperature slightly less than 3 Kelvin (about 2.76 K), which peaks in the microwave portion of the spectrum.[13] [PS]

Cosmic ray: An extremely energetic (relativistic) charged particle.[1] [PS]

Cosmic year: The time it takes the sun to revolve around the center of the galaxy, approximately 225 million years.[1] [PS]

Cosmology: The study of the origin and evolution of the universe as a whole.[11–10] [PS, SS]

Crater: A round impression left in a planet or satellite from a meteoroid.[1] [PS]

Crust: The outer layer of Earth and other terrestrial planets.[1] [PS]

Cryogenic: A rocket fuel or oxidizer that is liquid only at very low temperatures, for example, liquid hydrogen, which has a boiling point of −217.2°C (−423°F).[1] [PS]

C-stoff: A rocket fuel used by Germany in World War II[1] [PS]: 30% hydrazine hydrate, 57% methanol, 13% water, with traces of potassium cuprocyanate. Used in conjunction with T-stoff oxidizer[1] [PS]: 80% hydrogen peroxide, with 1%–2% oxiquinoline as a stabilizer.[1] [PS]

Cultural anthropology: A subdiscipline of anthropology concerned with the nonbiological, behavioral aspects of society, that is, the social, linguistic, and technological components underlying human behavior.[18] [SS]

Cultural heritage management: The process of making conscious choices about what happens to cultural heritage so as to ensure the cultural significance is retained.[16] [SS]

Cultural landscape: The combined works of nature and humans (World Heritage Convention).[11-3] [SS]

Cultural resource management: A branch of archaeology that is concerned with developing policies and action in regard to the preservation and use of cultural resources.[12] [SS] Generally within a framework of law and regulation: the management both of the cultural resources and of the effects on them that may result from land use and other activities in the contemporary world[20] [SS]; a field of applied archaeology providing the majority of jobs in archaeology in the United States.

Cultural resources: Sites, structures, landscapes, and objects of some importance to a culture or community for scientific, traditional, religious, or other reasons.[12] [SS] Broadly, any resource that is cultural in character. Examples include social, institutions, natural environments, historic places, artifacts, and documents.[20] [SS]

Debitage: Byproducts or waste materials left over from the manufacture of stone tools (French). A term referring to all the pieces of shatter and flakes produced and not used when stone tools are made; also called waste.[23] [SS] The physical sciences corollary might be rocket slag.

Debris field: Any area, locale, space, or contour, that contains the debris of wreckage, impact, or other material that once constituted a complete object.[11-3] [SS]

Decay: The action of air drag upon an artificial satellite causing it to spiral back into the atmosphere, eventually to disintegrate or burn up.[1] [PS]

Deceleration: Negative acceleration, slowing.[1] [PS]

Declination: One of the coordinates, measured in degrees, used to designate the location of an object on the celestial sphere. Declination is a north-south value similar to latitude on Earth.[1] [PS]

Delta v **(Δv):** Difference or change in velocity.[1] [PS]

Descending node: The point at which an orbiting object or spacecraft, moving from north to south, crosses the plane of the equator.[1] [PS]

Descent engine: The rocket used to power a spacecraft as it makes a controlled landing on the surface of a planet or moon.[1] [PS]

Descent module: That part of a spacecraft that descends from orbit to the surface of a planet or moon.[1] [PS]

Digital computer: An electronic device for solving numerically a variety of problems.[1] [PS]

Dipole: A compact source of magnetic force, with two magnetic poles. A bar magnet, coil, or current loop, if its size is small, creates a dipole field. The Earth's field, as a first approximation, also resembles that of a dipole.[1] [PS]

Direct sensing: Instruments that interact with phenomena in their immediate vicinity and register characteristics of them.[1] [PS]

Dish: A reflector for radio waves, usually a paraboloid.[1] [PS]

Doppler effect: A phenomenon in which waves appear to compress as their source approaches the observer or stretch out as the source recedes from the observer.[1] [PS]

Dose equivalent: A dose normally applied to biological effects and including scaling factors to account for the more severe effects of certain kinds of radiation.[1] [PS]

Dose: A quantity of radiation delivered at a position. In the context of space energetic particle radiation effects, it usually refers to the energy absorbed locally per unit mass as a result of radiation exposure.[1] [PS]

Downlink: The radio signal transmitted from a spacecraft to Earth.[1] [PS]

Drag: The resistance offered by a gas or liquid to a body moving through it.[1] [PS]

Drogue: A small parachute used to slow and stabilize a spacecraft returning to the atmosphere, usually preceding deployment of a main landing parachute.[1] [PS]

DSN: Deep Space Network.[1] [PS]

Dust: Particulates that have a direct relation to a specific solar system body and that are usually found close to the surface of this body (e.g., lunar, Martian, or cometary dust).[1] [PS]

Dyne: A unit of force equal to the force required to accelerate a 1-g mass 1 cm per second squared.[1] [PS]

Eccentric: Noncircular, elliptical (applied to an orbit).[1] [PS]

Eccentricity: The amount of separation between the two foci of an ellipse and, hence, the degree to which an elliptical orbit deviates from a circular shape.[1] [PS]

EELV: Evolved expendable launch vehicle.[1] [PS]

Electric propulsion: A form of rocket propulsion that depends on some form of electric acceleration of propellant to achieve low thrust over long periods of time (e.g., an ion or magnetohydrodynamic engine).[1] [PS]

Electromagnetic waves: A wave propagated through space by simultaneous periodic variation in the intensity of electric and magnetic field at right angles to each other and to the direction of propagation. The electromagnetic spectrum includes radio waves, microwaves, infrared, visible and ultraviolet radiation, x-rays, gamma rays, and cosmic rays.[1] [PS]

Electromagnetic: Relating to the interplay between electric and magnetic fields.[1] [PS]

Elevation: The angular measure of the height of an object above the horizon; with azimuth, one of the coordinates defining celestial location and sometimes used in tracking spacecraft.[1] In terrestrial mapping, the vertical distance above sea level; at an archaeological site excavation, the vertical distance from a datum.[22] [PS]

ELV: Expendable launch vehicle.[1] [PS]

Endoatmospheric: Taking place inside the atmosphere

Energetic particle: Particles that can penetrate outer surfaces of spacecraft. For electrons, this is typically above 100 keV, while for protons and other ions this is above 1 MeV. Neutrons, gamma rays, and x-rays are also considered energetic particles in this context.[1] [PS]

EOL: End of life.[1] [PS]

Ephemeris time: A measurement of time defined by orbital motions. Equates to mean solar time corrected for irregularities in Earth's motions.[1] [PS]

Ephemeris: Table of predicted positions of bodies in the solar system.[1] [PS]

Epoch: An instant in time that is arbitrarily selected as a point of reference, for example, for a set of orbital elements.[1] [PS]

Equatorial orbit: An orbit in the plane of the equator.[1] [PS]

Escape velocity: The precise velocity necessary to escape from a given point in a gravitational field. A body in a parabolic orbit has escape velocity at any point in that orbit. The velocity necessary to escape from the Earth's surface is 6.95 miles/second (11.2 km/second).[1] [PS]

Ethnoarchaeology: The approach to archaeology that makes use of observations of the "archaeological" remains of living peoples in an attempt to gain understanding of the nature of the material evidence associated with specific activities.[14] [SS]

Ethnoastronomy: The collection of integrated data on traditional astronomical beliefs and practices among indigenous cultures.[11-10] [SS]

Ethnography: The descriptive study of living human cultures.[14] [SS]

Ethnology: The comparison of cultures using ethnographic evidence.[18] [SS]

EVA: Extravehicular activity.[1] [PS]

Excavation: The controlled removal of physical remains of culture from beneath the ground surface, designed to ensure the recording of spatial and formal information before it is destroyed.[11-2] [SS]

Exhaust velocity: The velocity of the exhaust leaving the nozzle of a rocket.[1] [PS]

Exoatmospheric: Taking place outside the atmosphere. [PS]

Exosphere: The part of the Earth atmosphere above the thermosphere that extends into space. Hydrogen and helium atoms can attain escape velocities at the outer rim of the exosphere.[1] [PS]

Experimental archaeology: Scientific studies designed to discover processes that produced and/or modified artifacts and sites.[12] [SS]

Fairing: A structure whose main function is to streamline and smooth the surface of an aircraft or space vehicle.[1] [PS]

Fault protection: Algorithms, which reside in a spacecraft's subsystems, that insure the ability of the spacecraft both to prevent a mishap and to reestablish contact with Earth if a mishap occurs and contact is interrupted.[1] [PS]

Fault: A crack or break in the crust of a planet along which slippage or movement can take place.[1] [PS]

Faunalturbation: Disturbance of the archaeological record resulting from insect and animal burrowing.[11-26] [SS]

Feature: A nonportable artifact, for example, hearths, architectural elements, or soil stains.[18] [SS]

Ferret: Satellite using electromagnetic surveillance techniques.[1] [PS]

Fluorescence: The phenomenon of emitting light upon absorbing radiation of an invisible wavelength.[1] [PS]

Flux: The amount of radiation crossing a surface per unit of time, often expressed in "integral form" as particles per unit area per unit time.[1] [PS]

Flyby spacecraft: A spacecraft that follows a continuous trajectory past a target object, never to be captured into an orbit. It must carry instruments that are capable of observing passing targets by compensating for the target's apparent motion.[1] [PS]

Flyby: Space flight past a heavenly body without orbiting.[1] [PS]

Force: A vector quantity that tends to produce an acceleration of a body in the direction of its application.[1] [PS]

Formational archaeology: The study of environmental and cultural forces upon the life history of human artifacts on earth or in space.[11-3,11-26] [SS]

Forward scattering: Reflecting light approximately away from the source.[1] [PS]

FOV: Field of view.[1] [PS]

Free fall: The motion of any unpowered body moving in a gravitational field.[1] [PS]

Free-return trajectory: Path of a spacecraft that provides for a return to Earth.[1] [PS]

Frequency: The number of oscillations per second of an electromagnetic (or other) wave.[1] [PS]

Fuel cell: A cell in which chemical reaction is used directly to produce electricity.[1] [PS]

Furrow: A gouge in the surface of a planetary body caused by the impact of a crash landing. [PS]

***g*:** The symbol for the acceleration of a freely moving body due to gravity at the surface of the Earth. Alternatively, 1 *g*.[1] [PS]

Gamma rays: Very short, highly penetrative electromagnetic radiation with a shorter wavelength than x-rays; produced in general by emission from atomic nuclei.[1] [PS]

Garbology: The anthropological study of (usually) a recent society by studying its garbage and discard practices, coined by archaeologist William Rathje.[11-3] [SS]

Gauss: Unit of magnetic induction in the centimeter-gram-second system (after the German mathematician Karl F. Gauss).[1] [PS]

GEO: Geostationary orbit. Also abbreviated GO.[1] [PS]

Geo-: Prefix referring to the Earth.[1] [PS]

Geocentric: Earth-centered.[1] [PS]

Geomagnetic storm: A worldwide disturbance of the Earth's magnetic field, distinct from regular diurnal variations.[1] [PS]

Geospace: Also called the solar-terrestrial environment, geospace is the domain of Sun-Earth interactions. It consists of the particles, fields, and radiation environment from the Sun to Earth's space plasma environment and upper atmosphere. Geospace is considered to be the fourth physical geosphere (after solid earth, oceans, and atmosphere).[1] [PS]

Geostationary orbit: A geosynchronous orbit with an inclination of zero degrees. A spacecraft in such an orbit appears to remain fixed above one particular point on the Earth's equator.[1] [PS]

Geostationary transfer orbit: An elliptical orbit used to transfer a space vehicle from low Earth orbit to geostationary orbit.[1] [PS]

Geosynchronous orbit: A prograde, circular, low-inclination orbit about Earth having a period of 23 hours 56 minutes 4 seconds. A spacecraft in such an orbit appears to remain above Earth at a constant longitude, although it may seem to wander north and south.[1] [PS]

***g*-Force:** A force caused by acceleration expressed in *g*'s.[1] [PS]

GH_2: Gaseous hydrogen.[1] [PS]

GHz: Gigahertz, equal to 1 billion hertz.[1] [PS]

Gimbal: A mechanical frame for a gyroscope or power unit, usually with two perpendicular axes of rotation.[1] [PS]

GMT: Greenwich Mean Time.[1] [PS]

GN&C: Guidance, navigation, and control.[1] [PS]

GO: Geostationary orbit. Also abbreviated GEO.[1] [PS]

GOX: Gaseous oxygen.[1] [PS]

GPS: Global positioning system. [PS]

Graveyard orbit: Also called a supersynchronous orbit, junk orbit, or disposal orbit; an orbit significantly above synchronous orbit where spacecraft are intentionally placed at the end of their operational life. It is a measure taken in order to lower the probability of collisions with operational spacecraft and generation of additional space debris. However, recent findings have shown that satellites left in a graveyard orbit will slowly break apart as micrometeorites hit them, and the smaller fragments will filter back down to lower altitudes. Thus, satellites boosted to higher disposal orbits will eventually endanger operational satellites. It is used when the Δv required to perform a deorbit maneuver would be too high. Deorbiting a geostationary satellite would require a Δv of about 1,500 m/s, while reorbiting it to a graveyard orbit would require about 11 m/s.[5] [PS]

Gravity assist trajectory: A trajectory in which angular momentum is transferred from an orbiting planet to a spacecraft approaching from behind. The result is an increase in the spacecraft's velocity.[1] [PS]

Gravity: The force responsible for the mutual attraction of separate masses.

Grid system: A three-dimensional Cartesian coordinate system consisting of x and y axes as well as the depth below datum, in archaeology used to map and excavate a site. [SS]

GSO: Geosynchronous orbit.[1] [PS]

GTO: Geostationary transfer orbit.[1] [PS]

Guillotine: A device equipped with explosive blades used to cut cables, water lines, wires, and so forth during separation of spacecraft modules.[1] [PS]

Gyration: The circular motion of ions and electrons around magnetic field lines.[1] [PS]

Gyroscope: A spinning, wheel-like device that resists any force that tries to tilt its axis. Gyroscopes are used for stabilizing the attitude of rockets and spacecraft in motion.[1] [PS]

Heat shield: A device that protects people or equipment from heat, such as a shield in front of a reentry capsule.[1] [PS]

Helio-: Referring to the sun.[1] [PS]

Heliocentric: Centered on the sun.[1] [PS]

HEO: Highly elliptical orbit.[1] [PS]

Heritage: Those places, objects, cultures, and indigenous languages that have aesthetic, historic, scientific, or social significance or other special value for future generations, as well as for the community today[16] [SS]; also used as a synonym for cultural resources.

Hertz: A unit of frequency equal to one cycle per second, named after Heinrich Hertz.[1] [PS]

Heterosphere: The Earth atmosphere above 105 km altitude where species-wise concentration profiles establish because of diffusive equilibrium, with N_2 dominance below 200 km, O dominance from 200 to 600 km, and He dominance as of 600 km altitude.[1] [PS]

Historic property: Any district, site, building, structure, or object included in or eligible for inclusion in the U.S. National Register of Historic Places.[20] [SS]

High-gain antenna: A dish-shaped spacecraft antenna principally used for high rate communication with Earth. This type of antenna is highly directionally and must be pointed to within a fraction of a degree of Earth.[1] [PS]

Hohmann transfer orbit: An interplanetary trajectory in which a spacecraft is launched into an elliptical solar orbit whose perihelion (inner planet) or aphelion (outer planet) reaches the orbit of the target planet on the opposite side of the sun. Uses least propellant.[1] [PS]

Homosphere: The Earth atmosphere below 105 km altitude, where complete vertical mixing yields a near-homogeneous composition of about 78.1% N_2, 20.9% O_2, 0.9% Ar, and 0.1% CO_2 and trace constituents. The homopause (or turbopause) marks the ceiling of the homosphere. The homosphere can be broadly divided into three distinct regimes: the troposphere (0–12 km), the stratosphere (12–50 km), and the mesosphere (50–90 km).[1] [PS]

Horizon: The line marking the apparent junction of Earth and sky.[1] [PS]

HTPB: Hydroxy-terminator polybutadiene. A polymeric fuel binder.[1] [PS]

Hydrazine: A rocket fuel that burns spontaneously with nitric acid or nitrogen tetroxide. Can also be used as a monopropellant: used in small thrusters for orbit modification and attitude control of spacecraft. Also see monomethyl hydrazine (MMH) and unsymmetrical dimethylhydrazine (UDMH).[1] [PS]

Hyperbolic: A trajectory path to a planet shaped like a hyperbola.[1] [PS]

ICBM: Intercontinental ballistic missile (range >5,500 km).[1] [PS]

ICO: Intermediate circular orbit.[1] [PS]

IGY: International Geophysical Year (1957–1958). [PS]

Impulse: The product of the average force acting on a body and the interval of time during which it acts, being a vector quantity equal to the change of momentum of the body during the same time interval.[1] [PS]

IMU: Inertial measurement unit. [PS]

In situ: 1. Location of biological, physical, or material culture objects in their original physical and cultural context[16] [SS]; the location in which an object is found; in the natural or original position or place.[3] [SS] 2. In a localized state or condition. In situ sensing would be at the satellite proper as opposed to remote. [PS]

Inclination: The angular distance between a satellite's orbital plane and the equator of its primary.[1] [PS]

Inertial guidance: An onboard system for launch vehicles and spacecraft where gyroscopes, accelerometers, and other devices satisfy guidance requirements.[1] [PS]

Inertial measurement unit: An onboard instrument system that measures the attitude of a spacecraft. It includes accelerometers and gyroscopes.[1] [PS]

Inferior planets: Planets whose orbits are closer to the sun than Earth's, that is, Mercury and Venus. Also called inner planets.[1] [PS]

Infrared radiometer: A telescope-based instrument that measures the intensity of infrared energy radiated by the targets.[1] [PS]

Infrared: Electromagnetic radiation of wavelengths between 7,500 A, the limit of the visible light spectrum at the red end, and centimetric radio waves.[1] [PS]

Injection angle: The angle at which a spacecraft's return trajectory intersects the Earth's atmosphere.[1] [PS]

Interferometer: Any of several optical, acoustic, or radio frequency instruments that use interference phenomena between a reference wave and an experimental wave or between two parts of an experimental wave to determine wavelengths and wave velocities, measure very small distances and thicknesses, and measure indices of refraction.[1] [PS]

Interplanetary magnetic field: The weak magnetic field filling interplanetary space, with field lines usually connected to the sun. The interplanetary magnetic field is kept out of the Earth's magnetosphere, but the interaction of the two plays a major role in the flow of energy from the solar wind to the Earth's environment.[1] [PS]

Interplanetary probe: Unmanned instrumented spacecraft capable of reaching the planets.[1] [PS]

Inverse-square law: The mathematical description of how the strength of some forces, including gravity, changes in inverse proportion to the square of the distance from the source.[1] [PS]

Ion engine: A rocket engine, the thrust of which is obtained by the electrostatic acceleration of ionized particles.[1] [PS]

Ion: An atom that has lost or acquired one or more electrons.[1] [PS]

Ionization: Formation of electrically charged particles. Can be produced by high-energy radiation, such as light or ultraviolet rays, or by collision of particles in thermal agitation. [PS]

Ionosphere: An atmospheric layer dominated by charged, or ionized, atoms that extend from about 38 to 400 miles above the Earth's surface.[1] [PS]

IR: Infrared.[1] [PS]

Isp: Specific impulse. Also abbreviated SI.[1] [PS]

ITO: Indium tin oxide; its main feature is the combination of electrical conductivity and optical transparency. [PS]

Jovian planet: Any of the four biggest planets: Jupiter, Saturn, Uranus, and Neptune.[1] [PS]

Kachina: Literally means "life-bringer" in Hopi and can be anything that exists in the natural world or cosmos; often used in Hopi and Pueblo cosmology and religious practices.[11-10] [SS]

Karman line: A line that lies at an altitude of 100 km (62.1 miles) above Earth's sea level and is commonly used to define the boundary between the Earth's atmosphere and outer space.[7] [PS]

Kelvin: Scale of temperature named after the English physicist Lord Kelvin, based on the average kinetic energy per molecule of a perfect gas. Absolute zero is equivalent to −273.16°C (−459.69°F).[1] [PS]

Kepler's first law: A planet orbits the sun in an ellipse with the sun at one focus. [PS]

Kepler's second law: A line directed from the sun to a planet sweeps out equal areas in equal times as the planet orbits the sun. [PS]

Kepler's third law: The square of the period of a planet's orbit is proportional to the cube of that planet's semimajor axis; the constant of proportionality is the same for all planets.[3] [PS]

Kinetic energy: The ability of an object to do work by virtue of its motion. (Water moving in a pipe, e.g., has kinetic energy.) The energy terms that are usually used to describe the operation of a pump are *pressure* and *head*. In classical mechanics, equal to one half of the body's mass times the square of its speed.[2] [PS]

Kiva: Structure used by prehistoric and modern Puebloans for religious rituals, many of which are associated with the kachina belief system.[11-10] [SS]

Lander spacecraft: A spacecraft designed to reach the surface of a planet or moon and survive long enough to telemeter data back to Earth.[1] [PS]

Landscape archaeology: The methodological and theoretical specialization that studies how people interact within their social and natural environments.[11-2] [SS]

Lagrange point (Langrangian point): In a system dominated by two attracting bodies (such as sun and Earth), a point at which a third, much smaller body (such as a satellite) keeps the same position relative to the other two. Theoretically, the Sun-Earth system has five Lagrange points, but only two are important: L_1, on the sunward side of Earth, about 4 times the distance of the Moon, and L_2, at approximately the same distance on the midnight side.[1] [PS]

Law of superposition (principle of superposition): 1. Superposition is a foundational principle of sedimentary stratigraphy and so of other geology-dependent natural sciences: Sedimentary layers are deposited in a time sequence, with the oldest on the bottom and the youngest on the top.[15] [PS (Geography)] 2. Superposition in archaeology and especially in stratification identified during excavation is slightly different as the human-made intrusions and activity in the archaeological record need not form chronologically from top to bottom or be deformed from the horizontal as natural strata are by equivalent processes. Superposition in archaeology requires a degree of complex interpretation to accurately identify chronological sequences, and in this sense superposition in archaeology is more multidimensional.[15] [SS] 3. Superposition applies to sedimentary layers: it is a foundation principle for archaeology; it is replaced by orbital mechanics for space debris. That is, the positioning of unrelated materials in orbit does not reflect age or common origins. For the purposes of understanding space debris, we can refer to this as the principle of dynamic positioning.

LEM: Lunar Exploration Module, used in the Apollo program [PS]

LEO: Low Earth orbit.[1] [PS]

LGA: Low-gain antenna.[1] [PS]

LH_2: Liquid hydrogen.[1] [PS]

Light speed: 299,792,458 m/s ± 1.2 m/s (186,282.39 miles/s). U.S. National Bureau of Standards, 1971. [PS]

Light time: The amount of time it takes light or radio signals to travel a certain distance at light speed.[1] [PS]

Light year: The distance light travels in 1 year, approximately 9.46 trillion km (5.88 trillion miles).[1] [PS]

Light: Electromagnetic radiation that is visible to the eye, in the neighborhood of 1 nanometer wavelength.[1] [PS]

LiOH: Lithium hydroxide.[1] [PS]

Liquid hydrogen: A cryogenic rocket fuel that becomes liquid at –423°F.[1] [PS]

Liquid oxygen: A cryogenic oxidizer that becomes liquid at –279°F.[1] [PS]

Lissajous orbit: In orbital mechanics, a Lissajous orbit is a quasiperiodic orbital trajectory that an object can follow around a collinear libration point (Lagrange point) of a three-body system without requiring any propulsion. Lyapunov orbits around a libration point are curved paths that lie *entirely* in the plane of the two primary bodies.[8] [PS]

Lithosphere: The crust of a planet.[1] [PS]

LM: Lunar module.[1] [PS]

LO_2: Liquid oxygen. Also abbreviated LOX.[1] [PS]

LOI: Lunar orbit insertion.[1] [PS]

Longitude of ascending node: In an orbit, the celestial longitude of the ascending node.[1] [PS]

Longitudinal axis: The fore-and-aft line through the center of a space vehicle.[1] [PS]

Low Earth orbit: An orbit in the region of space extending from the Earth's surface to an altitude of 2,000 km. Given the rapid orbital decay of objects close to Earth, the commonly accepted definition is 160–2,000 km above the Earth's surface.[1] [PS]

Low-gain antenna: An omnidirectional spacecraft antenna that provides relatively low data rates at close range, several astronomical units, for example.[1] [PS]

LOX: Liquid oxygen. Also abbreviated LO_2.[1] [PS]

LRBM: Long-range ballistic missile.[1] [PS]

Lunar: Of or pertaining to the Moon.[1] [PS]

LV: Launch vehicle.[1] [PS]

Magnetic field line: Lines everywhere pointing in the direction of the magnetic force, used as a device to help visualize magnetic fields. In a plasma, magnetic field lines also guide the motion of ions and electrons and direct the flow of some electric currents.[1] [PS]

Magnetic field: A region of space near a magnetized body where magnetic forces can be detected.[1] [PS]

Magnetopause: The boundary of the magnetosphere, lying inside the bow shock. The location in space where Earth's magnetic field balances the pressure of the solar wind. It is located about 63,000 km from Earth in the direction of the sun.[1] [PS]

Magnetosphere: That region of space surrounding the Earth that is dominated by the magnetic field.[1] [PS]

Maria: Dark areas on the Moon, actually lava plains, once believed to be seas.[1] [PS]

Mass: The quantity of matter in a body. It can be determined by measuring the force of gravity acting on it (weight) and dividing this by the gravitational acceleration at that point. Thus, the mass of a given body remains the same everywhere, while its weight changes with the gravitational attraction.[1] [PS]

Material culture: Objects made, manufactured, or in other ways manipulated by humans and natural objects of cultural significance.[16] [SS]

Max Q: Maximum dynamic pressure; the point during launch when the vehicle is subjected to its greatest aerodynamic stress.[1] [PS]

Medium Earth orbit: An orbit in the region of space above low Earth orbit (2,000 km) and below geosynchronous orbit (35,786 km). Sometimes called intermediate circular orbit.[1] [PS]

Medium-gain antenna: A spacecraft antenna that provides greater data rates than a low-gain antenna, with wider angles of coverage than a high gain antenna, about 20–30°.[1] [PS]

Megalithic astronomy: The study of astronomical interpretations of megalithic sites, such as Stonehenge.[11-10] [SS]

MEO: Medium Earth orbit.[1] [PS]

Mesosphere: A division of the Earth's atmosphere extending from altitudes ranging from 18–30 miles to 48–55 miles.[1] [PS]

Meteor: The luminous phenomenon seen when a meteoroid enters the atmosphere, commonly known as a shooting star.[1] [PS]

Meteorite: A part of a meteoroid that survives through the Earth's atmosphere.[1] [PS]

Meteoroid: A solid body, moving in space, that is smaller than an asteroid and at least as large as a speck of dust. Nearly all meteoroids originate from asteroids or comets.[1] [PS]

MeV: One million electron volts.[1] [PS]

MHz: Megahertz, equal to 1 million hertz.[1] [PS]

Microgravity: An environment of very weak gravitational forces, such as those within an orbiting spacecraft. Microgravity conditions in space stations may allow experiments or manufacturing processes that are not possible on Earth.[1] [PS]

Micrometeoroid protection: Shielding used to protect spacecraft components from micrometeoroid impacts. Interplanetary spacecraft typically use tough blankets of Kevlar or other strong fabrics to absorb the energy from high-velocity particles.[1] [PS]

Micrometeoroid: Meteoroid less than 1/250th of an inch in diameter.[1] [PS]

Microwaves: Radio waves having wavelengths of less than 20 cm.[1] [PS]

MMH: Monomethyl hydrazine, CH_3NHNH_2. A liquid hypergolic fuel.[1] [PS]

Mock-up: A full-size replica or dummy of a vehicle (e.g., a spacecraft), often made of some substitute material, such as wood, to assess design features.[1] [PS]

Molniya Orbit: A type of highly elliptical orbit with an inclination of 63.4° and an orbital period of about 12 hours. Molniya orbits are named after a series of Soviet/Russian Molniya (Russian: "Lightning") communications satellites that have been using this type of orbit since the mid-1960s.[10] [PS]

Momentum: The product of the mass of a body and its velocity.[1] [PS]

Monopropellant: A rocket propellant consisting of a single substance, especially a liquid containing both fuel and oxidizer, either combined or mixed together. [PS]

Moon: A small natural body that orbits a larger one. A natural satellite.[1] [PS]

Motor: In spacecraft, a rocket that burns solid propellants.[1] [PS]

Multilayer insulation (MLI): Thermal insulation composed of multiple layers of thin sheets. It is mainly intended to reduce losses by thermal radiation. In its most common form, it does not appreciably insulate against other thermal losses, such as heat conduction or convection. It is therefore most commonly used on satellites and other applications in vacuum where conduction and convection are not important and radiation dominates. MLI gives many satellites and other space probes the appearance of being covered with gold foil.[9] [PS]

N_2O_4: Nitrogen tetroxide. Also abbreviated NTO.[1] [PS]

Nadir: The direction from a spacecraft directly down toward the center of a planet. Opposite the zenith.[1] [PS]

National Historic Landmark: A highly significant property eligible for the National Register of Historic Places, as well as meeting the criteria in US 36 CFR Part 65. [SS]

National Historic Preservation Act (NHPA) of 1966 as amended; 16 U.S.C. 470: Best-known U.S. authority for dealing with historic properties, including archaeological resources. Section 106 of the act requires agencies to consider the effects of their actions on historic properties.[20] [SS]

National Register of Historic Places: 36 CFR Part 60, a list maintained by the U.S. National Park Service of districts, sites, buildings, structures, and objects, each determined by National Park Service to be of historic, cultural, architectural, archaeological, or engineering significance at the national, state, or local level. [SS]

Neutron: Atomic particles having approximately the same mass as a hydrogen atom; very penetrating.[1] [PS]

Newton: That force that gives a mass of 1 kg an acceleration of 1 m per second per second; equal to 100,000 dynes.[1] [PS]

NiCd: Nickel cadmium.[1] [PS]

Nose shroud: A cover on the nose of a rocket or spacecraft that jettisons before insertion into orbit.[1] [PS]

NTO: Nitrogen tetroxide. Also abbreviated N_2O_4.[1] [PS]

Orbit insertion: The placing of a spacecraft into orbit around a planet or moon.[1] [PS]

Orbit trim maneuver: The firing of control rockets to refine a spacecraft's speed and trajectory.[1] [PS]

Orbit: The path of a body acted upon by the force of gravity. Under the influence of a single attracting body, all orbital paths trace out simple conic sections. Although all ballistic or free-fall trajectories follow an orbital path, the word *orbit* is more usually associated with the continuous path of a body that does not impact with its primary.[1] [PS]

Orbital elements: Six quantities used to mathematically describe an orbit; that is, semimajor axis, eccentricity, inclination, argument of periapsis, time of periapsis passage, and longitude of ascending node.[1] [PS]

Orbital period: The time taken by an orbiting body to complete one orbit.[1] [PS]

Orbital velocity: The velocity necessary to overcome the gravitational attraction of the Earth and so keep a satellite in orbit, about 17,450 mi/h (28,080 km/h) close to the Earth.[1] [PS]

Orbiter spacecraft: A spacecraft designed to travel to a distant planet or moon and enter orbit. It must carry a substantial propulsive capability to decelerate it at the right moment to achieve orbit insertion.[1] [PS]

O-stage: Rocket boosters that operate during part of the burning time of the first stage of a launch vehicle to provide additional thrust.[1] [PS]

Oxidizer: An agent that releases oxygen for combination with another substance, creating combustion and gas for propulsion.[1] [PS]

Parking orbit: Orbit in which a space vehicle awaits the next phase of its planned mission.[1] [PS]

Parsec: Measure of distance; 1 parsec = approximately 3.26 light years.[1] [PS]

Pascal: A unit of pressure equal to 1 newton per square meter.[1] [PS]

Passive cooling: The use of painting, shading, lectors, and other techniques to cool a spacecraft.[1] [PS]

Payload: Revenue-producing or useful cargo carried by a spacecraft; also, anything carried in a rocket or spacecraft that is not part of the structure, propellant, or guidance systems.[1] [PS]

Periapsis: That point in an orbit that is nearest the primary.[1] [PS]

Perigee: That point in a terrestrial orbit that is nearest the Earth.[1] [PS]

Perihelion: That point in a solar orbit that is nearest the sun.[1] [PS]

Perilune: That point in a lunar orbit that is nearest the Moon.[1] [PS]

Period of revolution: Time of one complete cycle in orbital motion; referred to as a year when applied to Earth.[1] [PS]

Period of rotation: Time of one complete cycle; referred to as a day when applied to Earth.[1] [PS]

Perturbation: Modifications to simple conic section orbits caused by such disturbances as air drag, nonuniformity of the Earth, and gravitational fields of more distant bodies, such as the Moon.[1] [PS]

Phase: 1. The particular appearance of a body's state of illumination, such as the full phase of the moon. [PS] 2. As applied to electromagnetic waves, the relative measurement of the alignment of two waveforms of similar frequency.[1] [PS] 3. A distinguishable part in a course, development, or cycle.[3] [PS] 4. An archaeological complex that is sufficiently distinctive so as to be distinguishable from adjacent contemporary complexes and from those that precede and succeed it. A phase may be viewed as a complex that is bounded in time as well as space.[14] A phase is generally the basic classificatory unit of archaeological "cultures." [SS]

Photon: A quantum of radiant energy.[1] [PS]

Photovoltaic cells: Crystalline wafers called solar cells that convert sunlight directly into electricity without moving parts.[1] [PS]

Pitch: The rotation of a vehicle about its lateral (y) axis, that is, movement in elevation.[1] [PS]

Planet: A nonluminous celestial body larger than an asteroid or a comet, illuminated by light from a star, such as the Sun, around which it revolves. The only known planets are those of the Sun, but others have been detected on physical (nonobservational) grounds around some of the nearer stars.[1] [PS]

Planetoid: An asteroid.[1] [PS]

Plasma engine: A rocket engine in which thrust is obtained from the acceleration of a plasma with crossed electrical and magnetic fields.[1] [PS]

Plasma: A gas-like association of ionized particles that responds collectively to electric and magnetic fields.[1] [PS]

Plasmasphere: The region of the atmosphere consisting of cold dense plasma originating in the ionosphere and trapped by the Earth's magnetic field.[1] [PS]

PO: Polar orbit.[1] [PS]

Polar orbit: An orbit that passes over the poles.[1] [PS]

Polysemy: The capacity for a sign to have multiple meanings.[11-37] [SS]

Potential energy: The energy of a body due to its position in a field.[1] [PS]

ppm: Parts per million.[1] [PS]

Precession: A change in the direction of the axis of spin of a rotating body.[1] [PS]

Preservation: Maintaining the material of places or objects in their existing state and retarding physical deterioration[16] [SS]; includes the identification, evaluation, recordation, documentation, curation, acquisition, protection, management, rehabilitation, restoration, stabilization, maintenance, research, interpretation, conservation, and education and training regarding the foregoing activities or any combination thereof.[20] [SS]

Pressurized: Containing air or other gas at a pressure higher than the pressure outside the chamber.[1] [PS]

Primary: The body around which a satellite orbits.[1] [PS]

Principle of dynamic positioning: The positioning of unrelated materials in orbit does not reflect age or common origins.[11-9] [PS]

Probe: An unmanned instrumented vehicle sent into space to gather information.[1] [PS]

Prograde: Orbital motion in the same direction as the primary's rotation.[1] [PS]

Propellant: A chemical or chemical mixture burned to create the thrust for a rocket or spacecraft.[1] [PS]

Propulsion: The process of driving or propelling.[1] [PS]

Putative: Purported, commonly put forth or accepted as true on inconclusive grounds.[15] [SS]

Pyrotechnics: The use of electrically initiated explosive devices to operate valves, ignite solid rocket motors, and explode bolts to separate from or jettison hardware, or to deploy appendages.[1] [PS]

Radiation belt: The region of high-energy particles trapped in the Earth's magnetic field, also known as the Van Allen belts.[1] [PS]

Radiation: Energy in the form of electromagnetic waves or particles.[1] [PS]

Radioisotope thermoelectric generator: A device that converts the heat produced by the radioactive decay of plutonium-238 into electricity by an array of thermocouples made of silicon-germanium junctions. The plutonium-238 is contained within a crash-resistant housing.[1] [PS]

Radioisotopes: Atomic particles that decay by natural radioactivity.[1] [PS]

RCS: Reaction control system.[1] [PS]

RE: Unit of distance equal to the radius of the Earth, 6,371.2 km.[1] [PS]

Reaction control system: System of thrusters used to control spacecraft attitude.[1] [PS]

Reaction wheels: Electrically powered wheels mounted in three orthogonal axes aboard a spacecraft. To rotate the vehicle in one direction, the proper wheel is spun up in the opposite direction. To rotate the vehicle back, the wheel is slowed down.[1] [PS]

Record (archaeological): All archaeological evidence that includes the physical remains of past human activities.[11-38] [SS]

Record (depositional): Archaeological evidence that has been discovered, recorded, and categorized by stratigraphy.[11-38] [SS]

Redundancy: The duplication of certain critical components in a space vehicle.[1] [PS]

Reentry interface: An altitude of 400,000 ft.; the point at which reentering spacecraft are considered to enter the Earth's atmosphere.[1] [PS]

Reentry: The descent into Earth's atmosphere from space.[1] [PS]

Reflection law: For a wavefront intersecting a reflecting surface, the angle of incidence is equal to the angle of reflection, in the plane defined by the ray of incidence and the normal.[3] [PS]

Regenerative cooling: Circulation of a propellant through a jacket around the combustion chamber in order to cool the chamber wall, the propellant subsequently being injected into the combustion chamber.[1] [PS]

Relativity principle: The principle, employed by Einstein's relativity theories, that the laws of physics are the same, at least locally, in all coordinate frames. This principle and that of the constancy of the speed of light are the founding principles of special relativity.[3] [PS]

Relativity, theory of: Theory of motion developed by Albert Einstein, for which he is justifiably famous. Relativity more accurately describes the motions of bodies in strong gravitational fields or at near the speed of light than Newtonian mechanics. All experiments done to date agree with relativity's predictions to a high degree of accuracy. (Curiously, Einstein received the Nobel prize in 1921 not for relativity but rather for his 1905 work on the photoelectric effect.)[3] [PS]

Relict (or fossil) landscape: Landscape in which an evolutionary process came to an end at some time in the past, either abruptly or over a period.[11-16] Its significant distinguishing features are, however, still visible in material form. [SS]

Rem: Roentgen equivalent man. A measure of nuclear radiation causing biological damage.[1] [PS]

Remote sensing: Instruments and techniques that record characteristics of objects at a distance, sometimes forming an image by gathering, focusing, and recording reflected light from the sun or reflected radio waves emitted by the spacecraft.[1] [PS] Used in archaeology to locate sites or features.

Rendezvous: A place of meeting at a given time, for example, a spaceship with a space station.[1] [PS]

Resolution: Ability to distinguish visual detail, usually expressed in terms of the size (in kilometers) of the smallest features that can be distinguished.[1] [PS]

Resonance: A relationship in which the orbital period of one body is related to that of another by a simple integer fraction, such as 1/2, 2/3, or 3/5.[1] [PS]

Retrograde: Orbital motion in the direction opposite to the primary's rotation.

Retrorocket: A rocket fired to reduce the speed of a spacecraft. [PS]

Revolution: Orbital motion about a primary.[1] [PS]

RF: Radio frequency.[1] [PS]

Right ascension: With declination, one of the coordinates used to designate the location of an object on the celestial sphere. Right ascension is measured in hours, minutes, and seconds and is similar to longitude on Earth.[1] [PS]

Rille (also spelled rill): Any of several long narrow valleys on the Moon's surface.[4] [PS]

Ring current: A very spread-out electric current circling around the Earth, carried by trapped ions and electrons.[1] [PS]

RJ-1: A hydrocarbon rocket fuel (a refined kerosene).[1] [PS]

RLV: Reusable launch vehicle.[1] [PS]

Rocket: A missile or vehicle propelled by the combustion of a fuel and a contained oxygen supply. The forward thrust of a rocket results when exhaust products are ejected from the tail.[1] [PS]

Roll: The rotational movement of a vehicle about a longitudinal (x) axis.[1] [PS]

Rotation: Rotary motion about an axis.[1] [PS]

Round-trip light time: The elapsed time it takes for light, or a radio signal, to travel from Earth, be received and immediately transmitted or reflected, and return to the starting point.[1] [PS]

RTG: Radioisotope thermoelectric generator.[1] [PS]

Salvage archaeology: Archaeology conducted primarily because a site or area is in imminent danger of destruction by natural forces or by construction or development.

The British equivalent to this term—rescue archaeology—is self-explanatory.[12] Archaic term in United States; supplanted by cultural resource management that is driven by compliance with federal and state law. [SS]

Satellite: Any body, natural or artificial, in orbit around a planet. The term is used most often to describe moons and spacecraft.[1] [PS]

Semimajor axis: Half the major axis of an ellipse. The mean distance of a planet or satellite from its primary.[1] [PS]

Semiology: The study of sign processes, or signification and communication, signs and symbols, both individually and grouped into sign systems. It includes the study of how meaning is constructed and understood.[11-37] [SS]

Sensor: An electronic device for measuring or indicating a direction or movement.[1] [PS]

Service module: That part of a spacecraft that usually carries a maneuvering engine, thrusters, electrical supply, oxygen, and other consumables external to the descent module. Discarded prior to reentry.[1] [PS]

SETI: Search for Extraterrestrial Intelligence.[1] [PS]

Sidereal time: Time relative to the stars other than the Sun.[1] [PS]

Significance: Under the U.S. National Historic Preservation Act, the historical, cultural, archaeological, architectural, or engineering importance of a prehistoric or historic property.[20] [SS]

Simulator: A device that mimics the operational conditions of equipment or vehicles.[1] [PS]

Singularization: The process of turning a disposable commodity into an object of unique value[11-37] [SS]; also, rendering a commodity no longer exchangeable as its value becomes too great. [SS]

Solar array: See solar panel.[1] [PS]

Solar cell: A cell that converts sunlight into electrical energy. The light falling on certain substances (e.g., a silicon cell) causes an electric current to flow.[1] [PS]

Solar constant: The electromagnetic radiation from the sun that falls on a unit area of surface normal to the line from the sun, per unit of time, outside the atmosphere, at 1 astronomical unit.[1] [PS]

Solar flare: A sudden brightening in some part of the sun, followed by the emission of jets of gas and a flood of ultraviolet radiation. The gale of protons that accompanies a flare can be very dangerous to astronauts.[1] [PS]

Solar panel: An array of light-sensitive cells attached to a spacecraft and used to generate electrical power for the vehicle in space. Also called solar array.[1] [PS]

Solar sensors: Light-sensitive diodes that indicate the direction of the sun. [PS]

Solar wind: A current of charged particles that streams outward from the sun.[1] [PS]

Solar: Of or pertaining to the sun.[1] [PS]

Solid propellant: A rocket propellant in solid form, usually consisting of a mixture of fuel and oxidizer.[1] [PS]

Solid rocket booster: A rocket, powered by solid propellants, used to launch spacecraft into orbit.[1] [PS]

Sounding rocket: A research rocket used to obtain data from the upper atmosphere.[1] [PS]

Space: The universe beyond Earth's atmosphere. The boundary at which the atmosphere ends and space begins is not sharp, but it starts at approximately 100 miles above Earth's surface.[1] [PS]

Space (spatial) archaeology: One of the three dimensions used by archaeologists to conceptualize and order patterns of material culture and patterns of behavior (the other two dimensions are time and form).[11-2] [SS]

Space archaeology/heritage: The archaeological study of material culture found in outer space relevant to the exploration of space, that is, exoatmospheric material that is clearly the result of human behavior and the evaluations of its significance in terms of preservation for the future.[11-2] [SS]

Space debris: Human-made objects or parts thereof in space that do not serve any useful purpose[1] [PS]; also, the material culture described above that may be evaluated for its significance and contribution to understanding human behavior.[11-43] [SS]

Space station: An orbiting spacecraft designed to support human activity for an extended time.[1] [PS]

Space weather: The popular name for energy-releasing phenomena in the magnetosphere, associated with magnetic storms, substorms, and shocks. [PS]

Spacecraft clock: A counter maintained by the command and data subsystem. It meters the passing of time during the life of the spacecraft and regulates nearly all activity within the spacecraft systems.[1] [PS]

Spacecraft: A piloted or unpiloted vehicle designed for travel in space.[1] [PS]

Specific impulse: Parameter for rating the performance of a rocket engine. Indicates how many pounds or kilograms of thrust are obtained by consumption of a pound or kilogram of propellant in 1 second.[1] [PS]

Spectrometer: An optical instrument that splits the light received from an object into its component wavelengths by means of a diffraction grating and then measures the amplitudes of the individual wavelengths.[1] [PS]

Spectroscopy: The study of the production, measurement, and interpretation of electromagnetic spectra.[1] [PS]

Spectrum: A particular distribution of wavelengths and frequencies. [PS]

Speed of light.[1] The speed of light in a vacuum is an important physical constant usually denoted by the symbol c (or less frequently by c_0). The meter is defined such that the speed of light in free space is exactly 299,792,458 meters per second (m/s). [PS]

Spin stabilization: Spacecraft stabilization accomplished by rotating the spacecraft mass, thus using gyroscopic action as the stabilizing mechanism.[1] [PS]

SRB propellant: Composite propellant used in the space shuttle solid rocket boosters. The final product is a rubbery material not unlike a typewriter eraser.[1] [PS]

SRB: Solid rocket booster.[1] [PS]

SRBM: Short-range ballistic missile (range <800 km).[1] [PS]

SSO: Sun-synchronous orbit.[1] [PS]

SSPO: Sun-synchronous polar orbit.[1] [PS]

Stage: An independently powered section of a rocket or spacecraft, often combined with others to form multistage vehicles.[1] [PS]

State Historic Preservation Officer (SHPO): The state official designated by the governor, who carries out the historic preservation functions described in the National Historic Preservation Act. [SS]

Static firing: The firing of a rocket on a special test stand to measure thrust and so forth.[1] [PS]

Stratigraphy: Geological deposits being arranged in a series of layers. According to the law of superposition, when one deposit overlies another, the higher must have been laid down more recently. As a consequence, any artifacts found in the upper layer must be younger than those from the lower layer[14] [SS]; also requires interpretation in an archaeological site because of complexities or changes to the layers not covered by the law. [SS]

Stratosphere: A division of the Earth's atmosphere extending from altitudes ranging from 5–10 miles to 18–30 miles.[1] [PS]

Structure: A complex entity constructed of many parts[15] [SS]; used to distinguish from buildings, those functional constructions made usually for purposes other than human shelter.[21] [SS]

Suborbital: Not attaining orbit, that is, a ballistic space shot.[1] [PS]

Subsatellite: A secondary object released from a parent satellite in orbit, for example, an electronic "ferret" released by a reconnaissance satellite.[1] [PS]

Sun-synchronous orbit: A walking orbit whose orbital plan precesses with the same period as the planet's solar orbital period. In such an orbit, a satellite crosses periapsis at about the same local time every orbit.[1] [PS]

Sunspot cycle: The recurring, 11-year rise and fall in the number of sunspots.[1] [PS]

Sunspots: Dark regions on the sun that are the centers of large vortices and possess powerful magnetic fields. Maximum sunspot activity occurs in cycles with a period of about 11 years.[1] [PS]

Superior planets: Planets whose orbits are farther from the sun than Earth's, that is, Mars, Jupiter, Saturn, Uranus, Neptune, and, until recently, Pluto. Also called outer planets.[1] [PS]

Surface penetrator: A probe designed to penetrate the surface of a body, surviving an impact of hundreds of *g*'s, measuring and telemetering the properties of the penetrated surface.[1] [PS]

Surface rover: A semiautonomous roving vehicle deployed on the surface of a planet or other body, taking images and soil analyses for telemetering back to Earth.[1] [PS]

Sustainer engine: An engine that maintains propulsion of a launch vehicle once it has discarded its boosters.[1] [PS]

Telemetry: The system for radioing information, including instrument readings and recordings, from a space vehicle to the ground.[1] [PS]

Terrestrial: Of or pertaining to the Earth.[1] [PS]

Terrestrial planet: Any of the four planets closest to the sun: Mercury, Venus, Earth, and Mars.[1] [PS]

Thermal energy: Energy in the form of heat.[1] [PS]

Thermal tile: Silica fiber insulation used to protect 70% of the exterior of the space shuttle orbiter against reentry temperatures of up to 1,430°C. Surface heat dissipates so rapidly that an uncoated tile can be held by its edges with the bare hand while its interior glows red hot.[1] [PS]

Thermosphere: The Earth atmosphere between 120 and 250–400 km (depending on the solar and geomagnetic activity levels), where temperature has an exponential increase up to a limiting value Temperature-exo-atmospheric Texo at the thermopause. The temperature Texo is called the exospheric temperature.[1] [PS]

Three-axis stabilization: Stabilization accomplished by nudging a spacecraft back and forth within a deadband of allowed attitude error, using small thrusters or reaction wheels.[1] [PS]

Throttle: To decrease the supply of propellant to an engine, reducing thrust. Liquid propellant rocket engines can be throttled; solid rocket motors cannot.[1] [PS]

Thrust vector control: Control of the thrust vector direction to steer a rocket or spacecraft during powered flight. Thrust vector control is most often achieved by hydraulically gimbaled engines.[1] [PS]

Thrust: The force that propels a rocket or spacecraft measured in pounds, kilograms, or newtons. Thrust is generated by a high-speed jet of gases discharging through a nozzle.[1] [PS]

Thruster: Rocket engines used for maneuvering spacecraft in space.[1] [PS]

Time of periapsis passage: The time in which a planet or satellite moves through its point of periapsis.[1] [PS]

Traditional archaeology: Studying the past in order to better understand the past.[11-2] [SS]

Traditional cultural property: A district, site, building, structure, or object that is valued by a human community for the role it plays in sustaining the community's cultural integrity; may be eligible for the National Register of Historic Places.[20] [SS]

Trajectory: The flight path of a projectile, missile, rocket, or satellite.[1] [PS]

Transducer: Device for changing one kind of energy into another, typically from heat, position, or pressure into a varying electrical voltage or vice versa, such as a microphone or speaker.[1] [PS]

Transmitter: An electronic device that generates and amplifies a carrier wave, modulates it with a meaningful signal, and radiates the resulting signal from an antenna.[1] [PS]

Transponder: A device that transmits a response signal automatically when activated by an incoming signal.[1] [PS]

Trojan relay system: A method of ensuring uninterrupted radio contact with the surface of any planet in the solar system at any time, first proposed by James Strong in 1967. Two radio satellites, keeping station along the Earth orbit, 60° ahead and 60° behind the Earth, would transmit/receive signals from a similar pair of relay satellites at the Trojan equilaterals of another planet. Radio communications via these satellite links, from surface to surface, would then become possible day and night, despite planetary rotation or orbital displacement. It could be used, for example, in steering a remotely controlled vehicle on the surface of Mars.[1] [PS]

Tropopause: The level separating the troposphere and the stratosphere, occurring at an altitude of 5–10 miles.[1] [PS]

Troposphere: A division of the Earth's atmosphere extending from ground level to altitudes ranging from 5 to 10 miles.[1] [PS]

True anomaly: The angular distance of a point in an orbit past the point of periapsis, measured in degrees.[1] [PS]

TT&C: Tracking, telemetry, and command.[1] [PS]

Ultraviolet: A band of electromagnetic radiation with a higher frequency and shorter wavelength than visible blue light. Ultraviolet astronomy is generally performed in space, since Earth's atmosphere absorbs most ultraviolet radiation.[1] [PS]

Umbilical: A cable conveying power to a rocket or spacecraft before liftoff. Also a tethering or supply line for an astronaut outside a spacecraft.[1] [PS]

Underwater archaeology: The process of excavating or otherwise recovering archaeological material covered by fresh or seawater.[14] Also called Marine Archaeology. [SS]

Universal time coordinated: The worldwide scientific standard of timekeeping; based upon carefully maintained atomic clocks and accurate to within microseconds. The addition or subtraction of leap seconds, as necessary, keeps it in step with Earth's rotation. Its reference point is Greenwich, United Kingdom; when it is midnight there, it is midnight UTC.[1] [PS]

Universal time: The mean solar time of the meridian of Greenwich, United Kingdom. Formerly called Greenwich mean time.[1] [PS]

UV: Ultraviolet. [PS]

UMDH: Unsymmetrical dimethylhydrazine.

Van Allen radiation belts: Two doughnut-shaped zones of radiation about the Earth, concentrated at altitudes of 3,000 and 10,000 miles; named after James A. Van Allen, who instrumented the satellite *Explorer 1*. The belts contain charged particles generated by solar flares and trapped by the Earth's magnetic field.[1] [PS]

Vector: A quantity that is specified by magnitude, direction, and sense.[1] [PS]

Velocity vector: Magnitude of speed plus direction.[1] [PS]

Velocity: In mechanics, the time rate of change of position of a body in a specified direction.[6] [PS]

Vernier: Rocket engine of small thrust used for fine adjustments in velocity and trajectory.[1] [PS]

Visible: Electromagnetic radiation at wavelengths that the human eye can see. We perceive this radiation as colors ranging from red (longer wavelengths; ~ 700 nanometers) to violet (shorter wavelengths; ~400 nanometers).[3] [PS]

Walking orbit: An orbit in which gravitational influences are used to induce a precession in a satellite's orbital plane.[1] [PS]

Weight: The force acting on a body in a gravitational field, equal to the product of its mass and the acceleration of the body produced by the field. [PS]

Weightlessness: A state experienced in a ballistic trajectory (i.e., in orbit or free fall) when, because the gravitational attraction is opposed by equal and opposite inertial forces, a body experiences no mechanical stress. [PS]

***x*-Axis:** See roll.[1] [PS]

X-rays: A band of electromagnetic radiation intermediate in wavelength between ultraviolet radiation and gamma rays. Because x-rays are absorbed by the atmosphere, x-ray astronomy is performed in space.[1] See table following. [PS]

Yaw: The rotation of a vehicle about its vertical (z) axis, that is, movement in azimuth.[1] [PS]

***y*-Axis:** See pitch.[1] [PS]

***z*-Axis:** See yaw.[1] [PS]

Zenith: The point on the celestial sphere directly above the observer. Opposite the nadir.[1] [PS]

Zero gravity: A condition in which gravity appears to be absent. Zero gravity occurs when gravitational forces are balanced by the acceleration of a body in orbit or free fall.[1] [PS]

Zero lift trajectory: A trajectory in which the control system acts to maintain a condition of no aerodynamic lift on the rocket.[1] [PS]

TABLE A1
Electromagnetic Spectrum

	Wavelength (m)	Frequency (Hz)	Energy (J)
Radio	$>1 \times 10^{-1}$	$<3 \times 10^{9}$	$<2 \times 10^{-24}$
Microwave	1×10^{-3} to 1×10^{-1}	3×10^{9} to 3×10^{11}	2×10^{-24} to 2×10^{-22}
Infrared	7×10^{-7} to 1×10^{-3}	3×10^{11} to 4×10^{14}	2×10^{-22} to 3×10^{-19}
Optical	4×10^{-7} to 7×10^{-7}	4×10^{14} to 7.5×10^{14}	3×10^{-19} to 5×10^{-19}
Ultraviolet	1×10^{-8} to 4×10^{-7}	7.5×10^{14} to 3×10^{16}	5×10^{-19} to 2×10^{-17}
X-ray	1×10^{-11} to 1×10^{-8}	3×10^{16} to 3×10^{19}	2×10^{-17} to 2×10^{-14}
Gamma ray	$<1 \times 10^{-11}$	$>3 \times 10^{19}$	$>2 \times 10^{-14}$

REFERENCES

1. Glossary of Space Technology. Compiled and edited by Robert A. Braeunig, 1996, 1998, 2005, 2006. http://www.braeunig.us/space/glossary.htm.
2. Power Engineering Dictionary. http://www.massengineers.com/Power%20Plant%20Dictionary/PowerEngineeringDictionaryTable.htm.
3. NASA. Imagine the Universe Dictionary. http://imagine.gsfc.nasa.gov/docs/dictionary.html.
4. Merriam-Webster Online Dictionary. 2008. Merriam-Webster Online. http://www.merriam-webster.com/dictionary.
5. http://en.wikipedia.org/wiki/Graveyard_orbit.
6. Dictionary.com Unabridged (v 1.1). Based on the Random House Unabridged Dictionary, Random House, Inc. 2006.
7. http://en.wikipedia.org/wiki/K%C3%A1rm%C3%A1n_line.
8. http://en.wikipedia.org/wiki/Lissajous_orbit.
9. http://en.wikipedia.org/wiki/Multi-layer_insulation.
10. http://en.wikipedia.org/wiki/Molniya_orbit.
11. Handbook of Space Engineering, Archaeology, and Heritage (this volume). All citations include the chapter number.
12. http://smu.edu/anthro/collections/glossary2.html.
13. http://csep10.phys.utk.edu/astr162/lect/cosmology/cbr.html.
14. http://www.archaeolink.com/glossary-p.htm.
15. Princeton WordNet. http://wordnet.princeton.edu/perl/webwn.
16. Australian Natural and Cultural Heritage Theme Report. http://www.environment.gov.au/soe/2001/publications/theme-reports/heritage/glossary.html.

17. http://en.wikipedia.org/wiki/Law_of_superposition.
18. Renfrew, C., and P. Bahn. 2007. *Archaeology Essentials: Theories, Methods, and Practice*. London: Thames & Hudson.
19. Haviland, W.A., H.E.L. Prins, D. Walrath, and B. McBride. 2005. *Anthropology: The Human Challenge*. 11th ed. Belmont, CA: Thomson Wadworth.
20. King, T.P. 2004. *Cultural Resource Laws & Practice: An Introductory Guide*. 2nd ed. Walnut Creek, CA: Altamira Press.
21. U.S. Park Service. 1995. National Register Bulletin 15: How to Apply National Register Criteria for Evaluation. Washington, DC: U.S. Park Service.
22. Brinker, R., and P.L. Wolf. 1977. *Elementary Surveying, 6th Edition*. NY: Harper & Row.
23. Price, T.D. 2007. *Principles of Archaeology*. Boston: McGraw Hill.

Appendix B

Space Programs and Organizations

Cheryl L. B. Reed

Fifty years ago, identifying and defining the spacefaring nations was a straightforward task. There were two prominent nations, the USSR, first to orbit with *Sputnik 1* on October 4, 1957, and then the United States, with *Explorer 1* on February 1, 1958. Western Europeans followed thereafter on November 26, 1965, with *Asterix*. The organizations and capabilities behind these early accomplishments were close-held derivatives of each nation's department of defense and intelligence agencies. For the United States, a full-scale crisis resulted on October 4, 1957 when the Soviets launched *Sputnik 1*, the world's first artificial satellite as its International Geophysical Year (IGY) entry. This had a "Pearl Harbor" effect on American public opinion, creating an illusion of a technological gap and provided the impetus for increased spending for aerospace endeavors, technical and scientific programs, and the chartering of new federal agencies to manage air and space research and development.[1]

A swift U.S. government organizational response and structured changes quickly followed. By 2008, more than fifty nation-states and organizations had satellites in orbit, including many nations that the West still considers "third world." This ever-increasing number of nations entering into the world of space evolved from the fact that space or space services become a "commodity" of the information age. Any nation or private or public entity can now both "purchase" space services and even begin creating its own space capability if its funding accommodates such transactions. This possibility has come about largely through the expansion of space science and engineering academic programs and a growing commercial space market. There is money to be made, and the commercial satellite market, predominantly communications-based, has well exceeded any government agency's investment made some 20 years ago. Both the investments of the private sector and the United States' increasingly severe export control policies increased the space market. Presidential directives of the 1990s, whose creation was the result of the large U.S. space corporations' maneuvers to protect their own launch vehicle market share, directly served to grow the international space market. Although these actions served to proliferate space capabilities in terms of knowledge and know-how; they were in fact trying to protect or isolate their own markets as a space power. Participation in international space is thriving, with leaders such as China and India who are now even leaving low Earth orbit (LEO) to explore the Moon and beyond. Other countries, such as Algeria and Malaysia, to name a few, not only have satellites in orbit but have legally purchased knowledge and technology transfer programs from commercial organizations to develop indigenous space capabilities. In this appendix, the global history and evolution of space agencies and organizations and their policies will be described and summarized.

DESCRIPTION OF SPACE AGENCIES IN THE WORLD

The brief description below of each international space agency is based on publicly available information. Each space agency in the world is in alphabetical order by country[2] (Soviet materials are under Russia) and by name.

National Commission for Space (CNIE)/Comision Nacional de Actividades Espaciales (CONAE) (Argentina)

The Comisión Nacional de Actividades Espaciales (National Space Activities Commission) is the civilian agency of the government in charge of the national space program. CONAE is the forerunner of the CNIE founded in the 1960s. The new commission was created on May 28, 1991, during the government of President Carlos Menem, after the cancellation of the military Condor missile program in an attempt to move all the commission efforts to civilian purposes. It received the Air Force aerospace facilities in Córdoba and Buenos Aires of the former CNIE, as well as some of the civil personnel involved in the canceled project. Since the 1990s, the new commission has signed agreements with NASA and European agencies and developed a number of earth observation satellites, including SAC-A, the failed mission SAC-B, and the SAC-C, launched in 2000 and still operating. CONAE's Córdoba Ground Station has been in operation since 1997.[3,4]

Commonwealth Scientific and Industrial Research Organization (CSIRO) (Australia)

Australia's space programs are generally executed by one quasigovernment organization, CSIRO, which is Australia's premier research organization for scientific solutions for problems in industry, governments, and communities around the world. CSIRO is Australia's national science agency and one of the largest and most diverse scientific research organizations in the world.[5]

Austrian Space Agency (ASA) (Austria)

The Austrian Space Agency was an organization whose original purpose has been to coordinate Austrian space exploration-related activities. It was established in 1972 by authorities in Vienna. In 1987, Austria became a member state of the European Space Agency.[6]

Space Research and Remote Sensing Organization (SPARRSO) (Bangladesh)

Space Research and Remote Sensing Organization is a government-controlled astronomy and space research organization of Bangladesh. Established in 1980 as an autonomous, multisectioned research and development organization of the government of the People's Republic of Bangladesh, it acts as the center of excellence and national focal point for the peaceful applications of space science, Remote Sensing and Geographic Information System (GIS) in Bangladesh. Space technology applications in Bangladesh started in 1968 through the establishment of a ground station in the then Atomic Energy Center. It is an autonomous organization under the Ministry of Defense.[7, 8]

Belgian Institute for Space Aeronomy (Belgium)

The Belgian Institute for Space Aeronomy is one of the youngest Belgian scientific institutions. The Belgian Institute for Space Aeronomy [Belgisch Instituut voor Ruimte-Aëronomie (BIRA), Institut d'Aéronomie Spatiale de Belgique (IASB)] is a scientific institute providing public service and performing research in the field of the space aeronomy. The institute provides Belgium with access to advanced knowledge resulting from the application of space investigation methods. Created in 1964, its main tasks are public service and research in the field of the space aeronomy, or tasks that require data knowledge, gathered using ground-based, balloon, rocket, and satellite observations within the framework of physics and chemistry of the atmosphere and outer space.[9]

Brazilian Space Agency (AEB) (Brazil)

The Brazilian Space Agency (Agência Espacial Brasileira) is the civilian authority that is in charge of the country's burgeoning space program. It operates a spaceport at Alcântara and a rocket launch site at Barreira do Inferno. The agency has given Brazil a leading role in space in the Latin American region and has made Brazil a valuable and dependable partner for cooperation in the International Space Station. The Brazilian Space Agency is the heir to its space program that previously had been under the control of the Brazilian military. The program was transferred into civilian control on February 10, 1994. It suffered a major setback in 2003, when a rocket explosion killed 21 technicians. Brazil successfully launched its first rocket into space on October 23, 2004, from the Alcântara Launch Center; it was a VSB-30 launched on a suborbital mission. Several other successful launches have followed.[10]

Britain: See United Kingdom

Bulgarian Aerospace Agency (Bulgaria)

The Bulgarian Aerospace Agency, known as the Space Research Institute (SRI) (Институт за Космически изследвания, ИКИ), founded in 1969, is among world's oldest space agencies. It is among the leading sections in the Bulgarian Academy of Sciences.[11]

Canadian Space Agency (CSA) (Canada)

The CSA [l'Agence spatiale canadienne (ASC)] is the government space agency responsible for Canada's space program. It was established in March 1989 by the Canadian Space Agency Act and sanctioned in December 1990.[12]

China National Space Administration (CNSA) (People's Republic of China)

The CNSA (Guó Jiā Háng Tiān Jú) is the national space agency of the People's Republic of China, responsible for the national space program. The agency was created in 1993 when the Ministry of Aerospace Industry was split into CNSA and the China Aerospace Corporation (CASC). The former was to be responsible for policy, while the latter was to be responsible for execution. This arrangement proved unsatisfactory, as these two agencies were, in effect, one large agency, sharing both personnel and management. As part of a massive restructuring, in 1998, CASC was split into a number of smaller state-owned companies. The intention appeared to have been to create a system similar to that characteristic of Western defense procurement in which entities that are government agencies, setting operational policy, then contract out their operational requirements to entities that are government-owned but not government-managed.[13]

Colombian Space Commission (CCE) (Colombia)

The general mission of the CEC is to optimize the space contribution of sciences and technologies to the social, economic, and cultural development of Colombia by means of applying its knowledge to the solution of national problems; the fortification of state, academic, and productive sectors; sustainable development; and the competitiveness of the country. Sectorial and specific objectives are telecommunications; satellite navigation; earth observation; astronautics, astronomy, and aerospace medicine; and political and legal management of knowledge and investigation.

Czech Space Office (Czech Republic)

The Czech Space Office (Česká kosmická kancelář) is not the national space agency of the Czech Republic. It is a nonprofit company contracted on an annual basis by the Ministry of Education to administer its space interests. It has no executive powers. It was created in November 2003 by several managers/owners of companies with space business interests and the academy of sciences with its own interests.[14]

National Space Center (Denmark)

The Danish National Space Center [DNSC; Danmarks Rumcenter (DRC)] is a Danish sector research institute and a part of the Danish Ministry of Science, Technology and Innovation. It came about as a result of combining the Danish Space Research Institute with the geodesy part of the National Survey and Cadaster of Denmark on January 1, 2005. The center conducts research in astrophysics, solar system physics, geodesy, and space technology. To conduct the research, the center collaborates with the Niels Bohr Institute for Astronomy, Geophysics and Physics.[15]

Danish Space Research Institute (DSRI) (Denmark)

The DSRI [Dansk Rumforskningsinstitut (DRI)] was a research institute under the Danish Ministry of Science Technology and Innovation. Its primary areas of research were astrophysics and solar system physics. A great deal of the research was concentrated on x-rays coming from astronomical objects. DRI has x-ray equipment onboard the Russian satellite *Granat* and the European *EURECA* satellite. The Danish Small Satellite Program was headed by DRI. On January 1, 2005, DRI merged with the geodesy part of Kort og Matrikelstyrelsen to form the Danish National Space Center.[16]

European Space Agency (ESA) (Seventeen Member States)

The ESA, established in 1974, is an international, intergovernmental organization dedicated to the exploration of space, currently with seventeen member states. By coordinating the financial and intellectual resources of its members, it can undertake programs and activities far beyond the scope of any single European country. Its member states are Austria, Belgium, Denmark, Finland, France, Germany, Greece, Ireland, Italy, Luxembourg, the Netherlands, Norway, Portugal, Spain, Sweden, Switzerland, and the United Kingdom. Canada takes part in some projects under a cooperative agreement. Hungary, Poland, Romania and the Czech Republic are European Cooperating States.

ESA also has centers in a number of European countries, each of which has different responsibilities: the European Astronauts Center (EAC) in Cologne, Germany; the European Space Astronomy Center (ESAC) in Villafranca del Castillo, Madrid, Spain; the European Space Operations Center (ESOC) in Darmstadt, Germany; the ESA Center for Earth Observation (ESRIN) in Frascati, near Rome, Italy; and the European Space Research and Technology Center (ESTEC) Noordwijk, the Netherlands. ESA's main spaceport is the Guiana Space Center in Kourou, a site made available by France. It is close to the equator; hence, commercially important orbits are easier to access. ESA became the market leader in commercial space launches in the 1990s. In recent years, ESA has also established itself as a major player in space exploration. ESA's science missions are based at ESTEC in Noordwijk, the Netherlands; Earth Observation missions at ESRIN in Frascati, Italy; ESA Mission Control (ESOC) is in Darmstadt, Germany, and the EAC, which trains astronauts for future missions, is situated in Cologne, Germany.

Chronology of the European Space Agency

After World War II, many European scientists had left Western Europe to work in either the United States or the Soviet Union. Although Western European countries could still invest in research and space-related activities, European scientists realized that solely national projects would be unable to compete with the major superpowers. The following paragraphs summarize important events by year:

1958: Pierre Auger (France) and Edoardo Amaldi (Italy), two prominent members of the Western European scientific community, recommend that European governments set up a "purely scientific" joint organization for space research, taking CERN as a model.

1960: Scientists from ten European countries established the Groupe d'études europeen pour la collaboration dans le domaine des recherches spatiales (GEERS).

1961: The Commission préparatoire européenne de recherches spatiales (COPERS) has a scientific program, an 8-year budget, and an administrative structure for the envisaged European Space Research Organization (ESRO).

1964: European nations decide to have two different agencies, one, the European Launch Development Organization (ELDO), to develop a launch system and the other, the European Space Research Organization (ESRO), to develop spacecraft.

1966: ESRIN is set up as part of ESRO in Frascati, near Rome, Italy, and begins acquiring data from environmental satellites in the 1970s.

1967: European Space Operations Center (ESOC) set up in Darmstadt, Germany.

1968: ESTEC moves to its present site in Noordwijk, the Netherlands.

1972: Although ESRO is establishing itself as a leader in space exploration, ELDO deals with technological problems, cost overruns, and political dispute. The idea of a new single European space organization is first discussed.

1973: ESRO and NASA agree to build *Spacelab*, a modular science package for use on space shuttle flights.

1975: ESA is created in its current form, merging ELDO with ESRO. There are 10 founding members: Belgium, Germany, Denmark, France, United Kingdom, Italy, the Netherlands, Sweden, Switzerland, Spain, and Ireland.

1978: ESA joins NASA and the United Kingdom in launching International Ultraviolet Explorer, the first high-orbit telescope.

1979: First Ariane launched.

1980: A French company, Arianespace, is formed to produce, operate, and market the Ariane 5 rocket as part of ESA's Ariane program.

1990s: SOHO, Ulysses, and the Hubble Space Telescope all jointly carried out with NASA.

1997: The first flight of *Ariane 5* ends in failure, but later flights establish Ariane in the highly competitive commercial space launch market.

2003: *Mars Express* orbiter and its lander, *Beagle 2*, launched.

2005: The ESA *Huygens* probe lands on the surface of Titan, Saturn's largest moon—the first ever to land on a world in the outer solar system.

2007: The European Space Policy was signed on May 22, 2007, unifying the approach of the ESA with those of the individual European Union member states.

2008: ESA's Columbus laboratory is launched on the space shuttle *Atlantis* to the International Space Station. ESA now becomes a fully responsible partner in the operations and utilization of the International Space Station (ISS) and is thus entitled to fly its own astronauts for long-duration missions as members of the resident ISS crew. Automated transfer vehicle (ATV) *Jules Verne*, ESA's first ATV, is also launched to take vital supplies to the ISS.[17,18]

National Center of Space Research [Center National d' Études Spatiales (CNES)] (France)

CNES is the French government space agency or, administratively, a "public establishment of industrial and commercial character." Its headquarters are located in central Paris. It operates out of the Center Spatial Guyanais but also has payloads launched from other space centers operated by other countries. CNES was formerly responsible for the training of French astronauts, but the last of them were transferred to the European Space Agency in 2001. Founded in 1961, CNES is the government agency responsible for shaping and implementing France's space policy in Europe.[19,20]

German Aerospace Center (Germany)

The German Aerospace Center [Deutsches Zentrum für Luft- und Raumfahrt e.V. (DLR)] is the national research center for aviation and space flight of the Federal Republic of Germany and the German Space Agency. DLR is a national research center for aeronautics and space. Its extensive research and development work in aeronautics, space, transportation, and energy is integrated into national and international cooperative ventures. The federal government has given DLR, as Germany's space agency, responsibility for the forward planning and implementation of its space program, as well as international representation of its interests.[21,22]

Institute for Space Applications and Remotes Sensing (ISARS) (Greece)

ISARS is one of the five research Institutes of the National Observatory of Athens (NOA). ISARS was founded in 1955, under the name Ionospheric Institute. In 1990, the institute was renamed to Institute of Ionospheric and Space Research and in 1999 took its current title in order to reflect its expanded activities, which cover a variety of aspects of space research and applications.[23]

Hungarian Space Office (HSO) (Hungary)

The Hungarian Space Office is the national space agency of the Republic of Hungary and was established by the Hungarian government in 1992. The Hungarian Space Board (HSB), headed by the secretary of state, helps the work of the minister responsible for space affairs. The members of the HSB are delegated by Ministries and Government Offices interested in space activities, complemented by outstanding experts from different fields of space research. The HSO manages, coordinates, and represents the Hungarian space activities.[24,25]

Indian Space Research Organization (ISRO) (India)

Dr. Vikram Sarabhai was the founding father of the Indian space program and is considered a scientific visionary by many, as well as a national hero. After the launch of *Sputnik* in 1957, he recognized the potential that satellites provided. India's first prime minister, Jawaharlal Nehru, who saw scientific development as an essential part of India's future, placed space research under the jurisdiction of the Department of Atomic Energy (DAE) in 1961. The DAE director Homi Bhabha, the father of India's atomic program, then established the Indian National Committee for Space Research (INCOSPAR), with Dr. Sarabhai as chairman, in 1962. The Indian Rohini program continued to launch sounding rockets of greater size and complexity, and the space program was expanded and eventually given its own government department, separate from the DAE. On August 15, 1969, ISRO was created from the INCOSPAR program under the DAE and continued under the Space Commission and finally the Department of Space, created in June of 1972. ISRO's hardware and services are available commercially through Antrix Corporation.

ISRO has entered the lucrative market of launching payloads of other nations. Prominent among them are the launches of Israel Space Agency's TecSAR spy satellite and the Israeli Tauvex-II satellite module.[26,27]

National Institute of Aeronautics and Space (LAPAN) (Indonesia)

National Institute of Aeronautics and Space (Lembaga Penerbangan dan Antariksa Nasional) is the Indonesian government space agency. It was established on November 27, 1964, by Indonesian President Suharto after 1 year of informal space agency organization. For more than two decades, LAPAN has launched many satellites and most are telecommunication satellites, including the Palapa project satellites. On May 31, 1962, Indonesia started its aeronautics exploration when the Aeronautics Committee was established. For more than 20 years, LAPAN has created and launched several telecommunication and other types of satellites. Recently, LAPAN launched LAPSAT-1 and LAPSAT-2, which were equipped with black and white charge-coupled device cameras. Known as the Palapa constellation, the satellites provide the Indonesia domestic communication system. Palapa, which means *fruits of labor* in Indonesian, was established in 1976 to provide a communication system for about 6,000 inhabited islands in Indonesia.[28,29]

Iranian Space Agency (ISA) (Iran)

ISA is the governmental space agency. The president of the Iranian Space Agency is one of the deputies of the Ministry of Communication and Information Technology.

ISA was established to conduct research in the field of space and technology. Iran developed a satellite launch vehicle of the Shahab family, quite similar to the DPRK's *Taepodong 2*, named *IRIS* and called *Shahab SLV*. After 2000, Iran had acquired the necessary skills to begin initial production of the Shahab-3 missile. This was followed by indigenous Iranian modifications and improvements, leading to the test firing of an improved version (*Shahab SLV*) in late 2004, which would be used to launch a completely indigenous Safir Iran satellite. This would be followed by the Mesbah, developed in collaboration with Italy in May 2005, with the Mesbah-2 satellite to follow. In January 2005, the Zohreh geosynchronous satellite project was approved, with a contract signed with Russia. *Sinah-1* was the first Iranian satellite, built and launched by the Russians on October 28, 2005, on a Kosmos-3 booster rocket from the Plesetsk Cosmodrome, making Iran the forty-third country to possess its own satellite. The next Iranian satellite, *Mesbah*, should be built by Iran with Italian assistance and launched on a domestically made rocket. The Shahab-4 rocket, still in development, with an estimated range of 2,000 miles (3,200 km), is said to be able to launch satellites in space.[30, 31]

Israeli Space Agency (ISA) (Israel)

The history of Israel in space is short but remarkable. It started in 1988 with the launch of *Ofeq 1* by the Shavit launcher, affiliating Israel to the very exclusive club of seven countries who launched a self-developed satellite with a domestically made launcher. Geographical constraints, as well as safety considerations, caused the Israeli space program to focus on very small satellites, loaded with sophisticated payloads. Among the leading projects of ISA is the renewal of the Israeli projects of the ultraviolet (UV) telescope, the TAUVEX (Tel Aviv University Ultra Violet Experiment). A UV telescope for astronomical observations that had been developed in the 1990s can be accommodated on the Indian geosynchronous satellite G Sat-4 and is jointly operated and utilized by Indian and Israeli scientists. The partners for this project are the Israeli Ministry of Science and Technology and ISA; The Indian Space Research Organization (ISRO); Tel Aviv University; the Electrio-Optical Israeli Industry EL-OP, Elbit Ltd; and the Indian Institute for Astronomy in Mumbai.

Recently, a new international collaboration created a new project. ISA and CNES of France reached an agreement to jointly work on a design and construction project of an innovative microsatellite and a ground station for scientific purposes. The high cost of space research and the development and execution of space projects has created the necessity for international cooperation. Such collaboration is very common in the field of space research and development in many countries all over the world, mostly of same technological capabilities.

ISA has signed cooperation agreements with the following bodies: CNES (France), NASA (United States), CSA (Canada), ISRO (India), DLR (Germany), NSAU (Ukraine), RKA (Russia), and NLR (Holland). Agreements with the following bodies are yet to be signed: ACE (Chile), AER (Brazil), and KARI (Korea). Usually such agreements concern research cooperation, yet sometimes the cooperation is manifested in collaborated planning and building of space systems.[32, 33]

Italian Space Agency (ASI) (Italy)

The Italian Space Agency (Agenzia Spaziale Italiana; ASI) was founded in 1988 to promote, coordinate, and conduct space activities in Italy that had begun in the 1960s. Operating under the Ministry of the Universities and Scientific and Technological Research, the agency cooperates with numerous international and Italian entities that are active in space technology and with the Italian President of the Council of Ministers. Internationally, the ASI provides Italy's delegation to the Council of the European Space Agency and to its subordinate bodies. Within 20 years' time, ASI became one of the most significant players in the world for space science, satellite technologies, and the development of mobile systems for exploring the universe.[34,35]

Japan Aerospace Exploration Agency (JAXA) (Japan)

The Japan Aerospace Exploration Agency (宇宙航空研究開発機構, Uchū-Kōkū-Kenkyū-Kaihatsu-Kikō) is Japan's national aerospace agency. JAXA was formed on October 1, 2003, through the merger of three previously independent organizations. JAXA is responsible for the research, development, and launch of satellites into orbit and is fundamentally involved in many missions, such as asteroid exploration and a possible human mission to the Moon. Under a corporate message, "Reaching for the skies, exploring space," JAXA is pursuing its possibilities in various aerospace fields and strives to succeed with various research and development missions in order to contribute to the peace and happiness of humankind. On October 1, 2003, the Institute of Space and Astronautical Science (ISAS), the National Aerospace Laboratory of Japan (NAL), and the National Space Development Agency of Japan (NASDA) were merged into one independent administrative institution called the Japan Aerospace Exploration Agency. Before the merger, ISAS was responsible for space and planetary research, whereas NAL was focused on aviation research. NASDA, which was founded on October 1, 1969, had developed rockets and satellites and built the Japanese Experiment Module for the International Space Station. NASDA also trained Japanese astronauts, who flew with the U.S. space shuttles. The history of ISAS began in 1955 with the PENCIL rocket launch experiment at the University of Tokyo. In 1964, the Institute of Space and Aeronautical Science was founded in the University of Tokyo. Japan's first artificial satellite, OHSUMI, was launched and put into orbit by an L-4S rocket using solid propellant. Since then, ISAS has cultivated its unique climate where its missions are achieved based on the concurrent and synergetic efforts of two groups of people: space science staff, who research the mysteries of space, and engineering research and development staff, who work to comply with the needs of space science. In 1981, ISAS was born as a joint research organization among Japanese universities.

NAL was established as the National Aeronautical Laboratory in July 1955, assuming its present name with the addition of the Aerospace Division in 1963. Since its establishment, it has pursued research on aircraft, rockets, and other aeronautical transportation systems, as well as peripheral technology. It has endeavored to develop and enhance large-scale test facilities and make them available for use by related organizations, with the aim of improving test technology in these facilities. NASDA was established on October 1, 1969, under the National Space Development Agency Law, to act as the nucleus for the development of space and promote the peaceful use of space. Prior to the establishment of JAXA, ISAS had been most successful in its space program in the field of x-ray astronomy during the 1980s and 1990s. Another successful area for Japan has been Very Long Baseline Interferometry (VLBI), with the HALCA mission. Additional success was achieved with solar observation and research on the magnetosphere, among other areas.[36, 37]

North Korean Space Agency (People's Democratic Republic of Korea)

North Korea launched the first medium-range Taepo Dong 1 ballistic missile from the northeastern part of North Korea shortly after noon on August 31, 1998. It plunged into the Pacific Ocean off the Sanriku coast of Japan after flying over the Japanese island of Honshu. North Korea's test evoked swift condemnation in the region and the world. Many analysts speculated on the country's possible motives for conducting its provocative launch at that time. North Korea may have been intent on demonstrating a show of force in advance of the 50th anniversary of its founding on September 9, 1948, and the expected installation of Kim Jong-Il as "paramount leader" of the nation. The launch probably had multiple purposes: to serve as an advertisement for the country's missile technology and as a bargaining chip to win concessions in negotiation with the United States.

On September 9, 1998, the Korean Central News Agency broadcast a report claiming the successful launch of the first North Korean artificial satellite, *Kwangmyongsong-1*. The U.S. Space Command has not been able to confirm North Korean assertions. It has not observed any object orbiting the Earth that correlates to the orbital data the North Koreans have provided in their public statements.[38]

Korea (South) Aerospace Research Institute (KARI) (South Korea, Republic of Korea)

KARI is the aeronautics and space agency of South Korea. Its main laboratories are located in the Daedeok Science Town. Since its establishment in 1989, KARI has been a driving force behind research and development in Korea's aerospace science and has made every possible effort to function as a leading national research and development institute in this field. South Korea first gained experience with missiles provided by the United States to counter North Korea. KARI began in 1990 to develop its own rockets, the KSR-I and KSR-II, one and two-stage rockets. In December 1997, it began development of a liquid oxygen /kerosene rocket engine. KARI wished to develop satellite launch capability. Past projects include the 1999 Arirang-1 satellite. A test launch of the KSR-III took place in 2002. An agreement was made on October 24, 2005, to send a Korean astronaut into space aboard a Russian Soyuz spaceflight. Through the Korean Astronaut Program, Russia has trained two South Koreans and sent one to the International Space Station in 2008. KARI is also involved in the development of Korea Multi Purpose Satellite

(KOMPSAT) 3; KOMPSAT 5, equipped with synthetic aperture radar; and the Communication, Ocean and Meteorological Satellite (COMS).[39, 40,41]

Malaysian National Space Agency (MNSA) (Malaysia)

MNSA [Agensi Angkasa Negara (ANGKASA)] is the Malaysian space agency; it was established in 2002. It is responsible for providing leadership in Malaysian space program, space education, and research, as well as assisting the Malaysian government in formulating national policies on space.[42]

Mexican Space Agency (Mexico)

The Mexican Space Agency [Agencia Espacial Mexicana (AEXA)] is a proposed space agency awaiting its approval in the Senate of Mexico after receiving a significant vote of confidence in the Chamber of Deputies on April 26, 2006. The project, lobbied by many Mexican scientists, but particularly by former astronaut Rodolfo Neri Vela and former NASA contractor Fernando de la Peña, intends to promote the development of space-related technologies, increase competitiveness among Mexican companies, and resume research performed by the former National Commission for Outer Space [Comisión Nacional del Espacio Exterior (CONEE)], which existed between 1962 and 1977. [43]

National Space Research and Development Agency (NASRDA) (Nigeria)

NASRDA is the national space agency of Nigeria. It is a part of the Federal Ministry of Science and Technology, and it is overseen by the National Council on Space Science Technology. NASRDA was established in 1998 by the government with a primary objective of establishing a "fundamental policy for the development of space science and technology," with an initial budget of $93 million.[44,45]

Netherlands Institute for Space Research (SRON) (Netherlands)

SRON is under the Dutch Organization for Scientific Research (NWO) and is responsible for promoting, coordinating, and supporting Dutch activities in space research. SRON develops satellite instruments for astrophysics and earth sciences. SRON was founded in 1983 under the name Stichting Ruimteonderzoek Nederland/Space Research Organization Netherlands. Its stakeholders include major space agencies, such as ESA and NASA.[46, 47]

Norwegian Space Center (NSC) (Norway)

The Norwegian Space Center (Norsk Romsenter) is a government agency that promotes space exploration. The NSC is a government agency under the Ministry of Trade and Industry. NSC promotes the development, coordination, and evaluation of national space activities, as well as supporting Norwegian interests in the ESA. The high northern latitude of the country, particularly the Svalbard archipelago, is an asset. Andøya is an ideal place to launch rockets that help study the aurora borealis and a ground station on Svalbard can access data from all satellites in polar orbit. NSC aims to strengthen these activities.[48, 49]

Pakistan Space and Upper Atmosphere Research Commission (SUPARCO) (Pakistan)

SUPARCO is the government space agency responsible for Pakistan's space program. It was formed in September 1961 by the order of President Ayub Khan on the advice of Professor Dr. Abdus Salam, a Nobel laureate, who was also made its founding director. SUPARCO is an autonomous research and development organization under the government of Pakistan. In May 2007, China (as a strategic partner) signed an accord with Pakistan to enhance cooperation in the areas of space science and technology.[50,51]

Peru Space Agency (CONIDA) (Peru)

CONIDA is the governing body of the aerospace activities in Peru and the space agency of Peru. Every year the institution renews its commitment with the country to dedicate all capacities for the development of aerospace technologies to the benefit of the national development.[52]

Space Research Center (Poland)

The Space Research Center (Centrum Badań Kosmicznych) is Poland's space agency. It is a scientific institute of the Polish Academy of Sciences. It was established in 1976. From 1977 to 2001, the SRC staff developed, constructed, and prepared for launch forty-four instruments, of which nineteen were used in suborbital rocket flights and twenty-five in satellite missions. The first experiments of the institute were launched chiefly on Soviet VERTICAL rockets within the framework of INTERCOSMOS cooperative program, but the last rocket experiment was launched in 1992 onboard a NASA Terrier Black Brent rocket.[53, 54]

Gabinete de Relacoes Internacionais da Ciencia (GRICES) (Portugal)

Portugal became the 15th member state to join ESA in 2000 and interfaces through GRICES, the department of Ministry for Science and Technology and Higher Education.

National Space Organization (NSPO) (Taiwan)

NSPO, formerly known as the National Space Program Office, is the civilian space agency of Taiwan, under the auspices of the Executive Yuan's National Science Council. NSPO originated in October 1991 as the National Space Program Office for carrying out the first stage plan of the 15-year Space Technology Long

Term Developmental Program. NSPO is currently involved in both the development of space- and satellite-related technologies and infrastructure (including the FORMOSAT series of Earth observation satellites) and related research in aerospace engineering, remote sensing, astrophysics, atmospheric science, and information science. FORMOSAT missions include FORMOSAT-1 (communications and ionospheric research satellite, launched in January 1999), FORMOSAT-2 (ionospheric research and surface mapping satellite, launched in May 2004), and FORMOSAT-3/COSMIC (constellation of six microsatellites to perform GPS occultation studies of the upper atmosphere). A collaborative project with U.S. agencies, including NASA and the National Center for Atmospheric Research, was launched in April 2006. Other efforts include YamSat, which is a series of picosatellites (volume 10 cm^3, weight roughly 850 g) designed to carry out simple, short-duration spectroscopy missions. YamSat had been planned to launch in 2003 by a Russian launch vehicle, but the launch was canceled because of political pressure from the PRC.[55,56]

Romanian Space Agency (ROSA) (Romania)

The national coordinating body of the space activities is ROSA, established in 1991 and reorganized by government decision in 1995 as an independent public institution under the auspices of the Ministry of Education and Research. ROSA is the national representative in the cooperative agreements with international organizations, such as ESA and the Committee on Space Research (COSPAR), as well as bilateral governmental agreements. Together with the Ministry of Foreign Affairs, ROSA represents Romania in the sessions of the United Nations Committee on the Peaceful Use of Outer Space (COPUOS) and its subcommittees.[57]

USSR and Russia Today (Russian Federation)

Russian Federal Space Agency (RKA/RSA)

RKA, or RSA [Федеральное космическое агентство России (Federal'noe kosmicheskoe agentstvo Rossii), commonly known as Roskosmos], formerly the Russian Aviation and Space Agency [Российское авиационно-космическое агентство (Rossiyskoe aviatsionno-kosmicheskoe agentsvo), commonly known as Rosaviakosmos], is the government agency responsible for Russia's space science program and general aerospace research. RKA was formed after the breakup of the former Soviet Union and the dissolution of the Soviet space program. RKA uses the technology and launch sites that belonged to the former Soviet space program. The 1990s saw a decreased cash flow, which encouraged Roskosmos to improvise and seek other ways to keep space programs running. This resulted in Roskosmos' leading role in commercial satellite launches and space tourism. While scientific missions, such as interplanetary probes or astronomy missions during the 1990s, played a very small role, Roskosmos managed to operate the space station *Mir* well past its lifetime, contribute to the International Space Station, and continue to fly additional Soyuz and Progress missions.

The Russian Space Agency is one of the partners in the International Space Station (ISS) program; it contributed the core space modules *Zarya* and *Zvezda,* which were both launched by Proton rockets and later were joined by NASA's Unity Module. RKA also provides space tourism for fare-paying passengers to ISS through the Space Adventures company. As of 2007, five space tourists have contracted with Roskosmos and have flown into space, each for an estimated fee of at least US$20 million. Currently, rocket development encompasses both the new rocket system, Angara, and enhancements of the Soyuz rocket, Soyuz-2 and Soyuz-2–3. Two modifications of the Soyuz, the Soyuz-2.1a and Soyuz-2.1b, have already been successfully tested, enhancing the launch capacity to 8.5 tons to LEO. RKA manages by far the most commercial launches per year; in 2005, it performed nearly 50% of all commercial satellite launches into space.[58, 59]

Russian Space Research Institute (IKI) (Russia)

The Russian Space Research Institute [Институт космических исследований Российской Академии Наук (Space Research Institute of the Russian Academy of Sciences); Russian abbreviation, ИКИ РАН (IKI RAN)] is the leading organization of the Russian Academy of Sciences for space exploration to benefit fundamental science. Originally it was founded May 15, 1965, in Soviet Union by Council of Ministers decree as the Space Research Institute of the USSR Academy of Sciences [Russian abbreviation, ИКИ АН СССР (IKI AN SSSR)]. In 1992 it received its present name. It is located in Moscow and has staff of 290 scientists.[60]

Soviet Space Program (USSR)

The Soviet space program consisted of a number of projects within the Union of Soviet Socialist Republics by competing design groups. Being primarily a military program, it was classified. Unlike the U.S. program, which was mostly open and in public view, announcements of the outcomes of missions were delayed until success was certain and failures were sometimes kept secret. Ultimately, as a result President Gorbachev's policy of *glasnost*, many facts about the space program (which was heavily connected to the military) were declassified. The Soviet space program was dissolved with the fall of the Soviet Union, with Russia and Ukraine becoming its immediate heirs. Russia continued its program by creating the Russian Aviation and Space Agency, now known as the Russian Federal Space Agency (RKA), while Ukraine created the National Space Agency of Ukraine (NSAU).

In its earliest phase, January 1956, plans were approved for Earth-orbiting satellites to gain knowledge of space (*Sputnik*) and four unmanned military reconnaissance satellites (*Zenit*). Further planned developments called for a manned Earth orbit flight by 1964 and an unmanned lunar mission at an earlier date. After the first *Sputnik* proved to be a successful propaganda coup, Sergei Korolev was charged to accelerate the manned program, the design of which was combined with the Zenit program to produce the *Vostok* spacecraft. Two days after the United States announced its intention to launch an artificial satellite, on July 31, 1956, the Soviet Union announced its intention to do the same. *Sputnik 1* was the first satellite launched, on October 4, 1957, beating the United States and stunning people all over the world. The Soviet space program pioneered many aspects of space exploration.

Instituto Nacional de Tecnica Aeroespacial [The Spanish Space Agency (INTA)] (Spain)

INTA (National Institute of Aerospace Technology) is Spain's space agency. Since its creation, INTA has developed intense activity in the aeronautical field and in space. The effort of generations of INTA scientists and technicians has formed the backbone of Spanish aerospace activities and has contributed to the strengthening of the industrial fabric of the country. INTA is the public research organization specializing in aerospace research and technology development.[61,62]

Swedish National Space Board (SNSB)/Swedish Space Corporation (SSC) (Sweden)

SNSB (Rymdstyrelsen) is a government agency operating under the Ministry of Industry, Employment and Communications. It receives funding for research from Ministry of Education and Science. The board distributes government grants for research and development, initiates research and development in space and remote sensing, and is Sweden's contact in international cooperation. Examples range from large-scale projects such as the Swedish part in the International Space Station to the Odin and SMART-1 satellites. The Swedish space program is carried out through extensive international cooperation, in particular through Sweden's membership in the European Space Agency.[63, 64]

SSC (Rymdbolaget) is a government-owned company that was established in 1972 to develop and implement space projects primarily on behalf of the Swedish National Space Board and European Space Agency. It operates Esrange, a rocket launch facility and research center in Kiruna, Sweden.[65,66]

Swiss Space Office (SSO) (Switzerland)

The State Secretariat for Education and Research (SER) is responsible for creating and implementing policy in the area of science, research, university, and space policy in Switzerland. It coordinates related activities within the Federal Administration and ensures cooperation with the cantons. Switzerland's involvement as a full-fledged member in ESA programs and core activities enables the country to develop its own technological capacities, allowing it to make use of scientific discoveries and build a competitive, Europe-wide industry.[67]

Geo-Informatics and Space Technology Development Agency (GISTDA) (Thailand)

GISTDA, also known as the Thai Ministry of Science and Technology's Space Agency, is a space research organization. It is responsible for remote sensing and technology development satellites. GISTDA owns the THEOS satellite, which was launched by a Dnepr rocket from the Dombarovskiy Cosmodrome in Russia in 2008.[68]

National Space Agency of Ukraine (NSAU) (Ukraine)

NSAU [Національне космічне агентство України (Natsional'ne kosmichne ahentstvo Ukrayiny), НКАУ (NKAU)] is the government agency responsible for space policy and programs. NSAU is a civil body in charge of coordinating the efforts of government installations, research, and industrial companies (mostly state-owned). Several space-related institutes and industries are directly subordinated to NSAU. However, it is not a united and centralized system participating in all stages and details of space programs (like NASA in the United States). A special space force in the military of Ukraine is also absent. The agency oversees launch vehicle and satellite programs, cooperative programs with the Russian Aviation and Space Agency, the European Space Agency, NASA, and commercial ventures and is an international participant with Sea Launch. Launches are conducted at Kazakhstan's Baikonur, at Russia's Plesetsk Cosmodromes, and on Sea Launch's floating platform. NSAU has ground control and tracking facilities in Eupatoria, Crimea. Spacecraft include the Sich and Okean Earth observation satellites and the Coronas solar observatory with Russia. NSAU has engineered, constructed, and launched a total of six satellites since 1992. The latest successful Ukrainian satellite, Sich-1M, was launched in 2004. In 2007, Sich-2 was launched with a goal of scientific research and space exploration. NSAU is currently working on further Sich series satellites: Sich-3, Sich-3-O, and Sich-3-P. Prior to Ukraine's independence, Ukrainian cosmonauts flew under the Soviet flag. The first NSAU cosmonaut enter space under the Ukrainian flag was Leonid K. Kadenyuk on May 13, 1997. He was a payload specialist for the NASA STS-87 mission. His was an international spaceflight mission, involving cosmonauts of NASA (United States), NSAU (Ukraine), and NASDA (Japan). Kadeniuk holds the rank of Air Force Major General. Sea Launch is a joint venture space transportation company, partially owned by companies in Ukraine that handle operations for the National Space Agency. Sea Launch offers a mobile sea platform, used for spacecraft launches of commercial payloads on specialized Ukrainian Zenit 3SL rockets. The main advantage of floating cosmodrome is its location at the directly at the equator. It allows the use of the greatest effect of Earth's rotation to deliver payloads into orbit at low expense. Ukrainian companies Yangel Yuzhnoye State Design Office and Makarov Yuzhny Machine-Building Plant (Yuzhmash) have engineered and produced seven types of launch vehicles. Adding strapon boosters to launch vehicles may expand the family of Mayak, which is the latest launch vehicle developed.[69,70]

United Nations Office for Outer Space Affairs (UNOOSA) (United Nations)

The United Nations Office for Outer Space Affairs (OOSA) is an organization of the General Assembly charged with implementing the Assembly's space-related policies. It is located in the United Nations Office in Vienna. The office implements the Program on Space Applications and maintains the Register of Objects Launched into Outer Space. Through the United Nations Program on Space Applications, UNOOSA conducts international workshops, training courses, and pilot projects on topics that include remote sensing, satellite navigation, satellite meteorology, tele-education, and basic space sciences for the benefit of developing nations.[71,72]

British National Space Center (BNSC) (United Kingdom)

BNSC is a government body that coordinates civil space activities. It operates as a voluntary partnership of ten British government departments and agencies and research councils. The civilian portion of the British space program focuses on space science, Earth observation, satellite telecommunications, and global navigation (for example, GPS and Galileo.).[73]

United States of America

National Aeronautics and Space Administration (NASA)

On July 29, 1958, President Dwight D. Eisenhower signed the National Aeronautics and Space Act, establishing NASA. When it began operations on October 1, 1958, NASA consisted mainly of the four laboratories and some eighty employees of the government's 46-year-old research agency, the National Advisory Committee for Aeronautics (NACA). A significant contribution to NASA's entry into the Space Race was the technology from the German rocket program, directed by Wernher von Braun, who became a naturalized citizen of the United States after World War II. He is today regarded as the father of the United States space program. Elements of the Army Ballistic Missile Agency (of which von Braun's team was a part) and the Naval Research Laboratory were incorporated into NASA. NASA's earliest programs involved research into human spaceflight and were conducted under the pressure of the competition between the U.S. and the USSR in the form of a Space Race that existed during the Cold War. Project Mercury, initiated in 1958, started NASA down the path of human space exploration with missions designed to discover simply whether humans could survive in space.

On May 5, 1961, astronaut Alan Shepard became the first American in space when he piloted *Freedom 7* on a 15-minute suborbital flight. John Glenn became the first American to orbit the Earth on February 20, 1962, during the 5¼-hour flight of *Friendship 7*. Once the Mercury project proved that human spaceflight was possible, Project Gemini was launched to conduct experiments and work out issues relating to a Moon mission. The first Gemini flight with astronauts on board, *Gemini 3*, was flown by Gus Grissom and John Young on March 23, 1965.

The Apollo program was designed to land humans on the Moon and bring them safely back to Earth. *Apollo 1* ended tragically when all the astronauts died because of fire inside the command module during an experimental simulation. Because of this incident, there were a few unmanned tests before men boarded the spacecraft. *Apollo 8*, *9*, and *Apollo 10* tested various components while orbiting the Moon and returned photography. On July 20, 1969, *Apollo 11* landed the first humans on the Moon, Neil Armstrong and Buzz Aldrin. *Apollo 13* did not land on the Moon because of a malfunction, but it did return photographs.

Skylab was the first space station the United States launched into orbit. The 75-ton station was in Earth orbit from 1973 to 1979 and was visited by crews three times, in 1973 and 1974. This was followed by the Apollo-Soyuz Test Project (ASTP), which was the first joint flight of the U.S. and Soviet space programs. The mission took place in July 1975. The space shuttle became the major focus of NASA in the late 1970s and the 1980s. Planned to be a frequently launchable and mostly reusable vehicle, four space shuttles were built by 1985. The first to launch, *Columbia*, did so on April 12, 1981. The 1986 *Challenger* disaster highlighted the risks of human space flight. Nonetheless, the shuttle has been used to launch milestone projects such as the Hubble Space Telescope (HST). The HST was created with a relatively small budget of $2 billion, but it has continued operation since 1990 and has delighted both scientists and the public. Some of the images it has returned have become near-legendary, such as the groundbreaking Hubble Deep Field images. The HST is a joint project between NASA and ESA, and its success has paved the way for greater collaboration between the agencies.

In 1995, Russian-American interaction would again be achieved as the shuttle-*Mir* missions began, and again an American vehicle docked with a Russian craft (this time a full-fledged space station). This cooperation continues to the present day, with Russia and America the two biggest partners in the largest space station ever built, the International Space Station (ISS). The strength of their cooperation on this project was even more evident when NASA began relying on Russian launch vehicles to service the ISS following the 2003 *Columbia* disaster, which grounded the shuttle fleet for well over 2 years.

During much of the 1990s, NASA was faced with shrinking annual budgets because of congressional belt-tightening in Washington, D.C. In response, NASA's ninth administrator, Daniel Goldin, pioneered the "faster, better, cheaper" approach that enabled NASA to cut costs while still delivering a wide variety of aerospace programs (Discovery Program). NASA's ongoing investigations include in-depth surveys of Mars and Saturn and studies of the Earth and the sun. Other NASA spacecraft are presently en route to Mercury and Pluto. With missions to Jupiter in planning stages, NASA's current itinerary covers more than half the solar system.

On January 14, 2004, 10 days after the landing of the Mars Exploration Rover *Spirit*, President George W. Bush announced a new plan for NASA's future, dubbed the Vision for Space Exploration. According to this plan, humankind will return to the Moon by 2018 and set up outposts as a test bed and potential resource for future missions. The space shuttle will be retired in 2010, and Orion will replace it by 2014. Orion is capable of both docking with the ISS and leaving the Earth's orbit. The future of the ISS is somewhat uncertain, and its mission beyond construction is less clear. Although the plan initially met with skepticism from Congress, in late 2004 Congress agreed to provide start-up funds for the first year's worth of the new space vision. Hoping to spur innovation from the private sector, NASA established a series of Centennial Challenges and technology prizes for nongovernment teams, in 2004. This includes tasks useful for implementing the Vision for Space Exploration, such as building more efficient astronaut gloves. NASA's 2002 mission statement, used in budget and planning documents, stated that on December 4, 2006, it was planning to build a permanent Moon base. NASA Associate Administrator Scott Horowitz said the goal was to start building the Moon base by 2020, and by 2024 to have a fully functional base, that would allow for crew rotations like the International Space Station. Additionally, NASA plans to collaborate and partner with other nations for this project. On September 28, 2007, NASA administrator Michael D. Griffin stated that NASA aims to put a man on Mars by 2037, and in 2057, "We should be celebrating 20 years of man on Mars."[74]

Missile Defense Agency (MDA), U.S. Air Force, National Reconnaissance Office (NRO), and Other U.S. Agencies

The United States has numerous organizations—military, private, public, and academic—all with a vested interest in space exploration or exploitation. A brief description of these stakeholders follows showing the diversity of the "interested parties" in space. This list then is expanded by scientific communities of interest and hobbyists. In North America, there is U.S. Northern Command (NORTHCOM) and North American Aerospace Defense Command (NORAD). At the U.S. national level, the White House National Security Council, in conjunction with the Office of Science and Technology Policy, established the Space Policy Coordinating Committee (Space PCC) in 2001 for the purpose of coordinating national space policy matters affecting multiple agencies of the federal government. Other resource planning entities include the Department of Commerce (representing the interests of both the commercial space sector and National Oceanic and Atmospheric Administration's civilian space program); the Office of Space Commercialization; Space.gov; Strategic Planning & Architectures Collaborative Environment, an essential resource for collaborative space planning for users across the Department of Defense and Intelligence Community enterprise; National Security Space Office (NSSO); and NRO. NRO designs, builds and operates the reconnaissance satellites of the United States government. It also coordinates collection and analysis of information from both airplane and satellite reconnaissance by the military services and the Central Intelligence Agency. It is funded through the National Reconnaissance Program, which is part of the National Foreign Intelligence Program. NRO is part of the Department of Defense and was established on August 25, 1960, as a classified agency to develop the nation's revolutionary satellite reconnaissance systems. Missions of the NRO subsequent to 1972 are still classified, and portions of many earlier programs remain unavailable to the public.

The U.S. Department of Defense (DoD) has joint forces, including Air Force Space Command (AFSPC) and National Security Space Institute (NSSI), a joint training facility complementing existing space education programs at Air University, the Naval Postgraduate School and the Air Force Institute of Technology. Other joint DoD elements include the Joint Space Operations Center; U.S. Strategic Command (USSTRATCOM);[75] Military Space Forces; U.S. Northern Command (NORTHCOM); MDA, formerly known as Ballistic Missile Defense Organization (BMDO); Defense Intelligence Agency (DIA); National Geospatial-Intelligence

Agency (NGA), formerly the National Imagery and Mapping Agency (NIMA); Unified Command Plan (UCP); Defense Advanced Research Projects Agency (DARPA); ACQWeb, which is under the Secretary of Defense for Acquisition and Technology; the DoD GPS Support Center; Joint Modeling and Simulations; and USJFCOM. The U.S. Air Force organizations are Space Air Force (USSPACEAF); numerous space wings; Space and Missile Systems Center (SMC); U.S. Air Force Space Innovation and Development Center (SIDC), formerly Space Warfare Center (SWC); and Space Battlelab. The U.S. Army, Navy, Marine Corps, and Coast Guard also have space support elements. Civilian agencies include the National Oceanic and Atmospheric Administration, U.S. Geological Survey; Associate Administrator for Commercial Space Transportation (AST); Federal Aviation Administration (FAA); Federal Communications Commission (FCC); Department of Transportation (DoT); Department of Commerce (DoC); and the often-referenced NASA.[76]

NOTES AND REFERENCES

1. NASA Office of External Relations. A Brief History of NASA. NASA History Division Factsheet (online archive of NASA history and historical data). http://www.hq.nasa.gov/office/pao/History/factsheet.htm.
2. This list states the country names (official short names in English) in alphabetical order as given in ISO 3166-1.
3. http://en.wikipedia.org/wiki/CNIE.
4. http://www.conae.gov.ar/eng/sobre/sobreconae.html.
5. http://www.csiro.au/org/AboutCSIROOverview.html.
6. http://en.wikipedia.org/wiki/Austrian_Space_Agency.
7. http://en.wikipedia.org/wiki/Space_Research_and_Remote_Sensing_Organization.
8. http://sparrso.gov.bd/.
9. http://www.aeronomie.be/en/contact/whoarewe.htm.
10. http://en.wikipedia.org/wiki/Brazilian_Space_Agency.
11. http://en.wikipedia.org/wiki/Bulgarian_Aerospace_Agency.
12. http://en.wikipedia.org/wiki/Canadian_Space_Agency.
13. http://en.wikipedia.org/wiki/China_National_Space_Administration.
14. http://en.wikipedia.org/wiki/Czech_Space_Office.
15. http://en.wikipedia.org/wiki/Danish_National_Space_Center.
16. http://en.wikipedia.org/wiki/Danish_Space_Research_Institute.
17. http://en.wikipedia.org/wiki/European_Space_Agency.
18. http://www.esa.int/SPECIALS/About_ESA/SEM7VFEVL2F_0.html.
19. http://en.wikipedia.org/wiki/CNES.
20. http://www.cnes.fr/web/455-cnes-en.php.
21. http://www.dlr.de/.
22. http://en.wikipedia.org/wiki/German_Aerospace_Center.
23. http://www.space.noa.gr/isars/.
24. http://www.hso.hu/cgi-bin/page.php?page=215.
25. http://en.wikipedia.org/wiki/Hungarian_Space_Office.
26. http://en.wikipedia.org/wiki/ISRO.
27. http://www.isro.org/about_isro.htm.
28. http://en.wikipedia.org/wiki/LAPAN.
29. http://www.lapan.go.id/.

30. http://en.wikipedia.org/wiki/Iranian_Space_Agency.
31. http://www.isa.ir/en/.
32. http://www.most.gov.il/English/Units/Science/Israel+Space+Agency/.
33. http://en.wikipedia.org/wiki/Israeli_Space_Agency.
34. http://en.wikipedia.org/wiki/Italian_Space_Agency.
35. http://www.asi.it/SiteEN/Default.aspx.
36. http://en.wikipedia.org/wiki/Japan_Aerospace_Exploration_Agency.
37. http://www.jaxa.jp/index_e.html.
38. http://www.fas.org/spp/guide/dprk/index.html.
39. http://en.wikipedia.org/wiki/KARI.
40. http://www.kari.re.kr/.
41. http://www.angkasa.gov.my/welcome/about_us/introduction.php.
42. http://en.wikipedia.org/wiki/Malaysian_National_Space_Agency.
43. http://en.wikipedia.org/wiki/Mexican_Space_Agency.
44. http://en.wikipedia.org/wiki/NASRDA.
45. http://www.nasrda.org/.
46. http://en.wikipedia.org/wiki/SRON.
47. http://www.sron.nl/.
48. http://en.wikipedia.org/wiki/Norwegian_Space_Center.
49. http://www.spacecenter.no/english/.
50 http://en.wikipedia.org/wiki/SUPARCO.
51. http://www.suparco.gov.pk/.
52. http://www.conida.gob.pe/#.
53. http://en.wikipedia.org/wiki/Space_Research_Center.
77. http://en.wikipedia.org/wiki/UNOOSA.
54. http://www2.cbk.waw.pl/index.php?id=80.
55. http://en.wikipedia.org/wiki/National_Space_Organization.
56. http://www.nspo.org.tw/2005e/.
57. http://portal.rosa.ro/index.php?l_id=2.
58. http://en.wikipedia.org/wiki/Russian_space_agency.
59. http://www.roscosmos.ru/index.asp?Lang=ENG.
60. http://en.wikipedia.org/wiki/Russian_Space_Research_Institute.
61. http://www.inta.es/index.asp.
62. http://en.wikipedia.org/wiki/Instituto_Nacional_de_Tecnica_Aeroespacial.
63. http://www.rymdstyrelsen.se/dyn_aktuellt.asp?languageId=2.
64. http://en.wikipedia.org/wiki/Swedish_National_Space_Board.
65. http://en.wikipedia.org/wiki/Swedish_Space_Corporation.
66. http://www.ssc.se/.
67. http://www.sbf.admin.ch/htm/themen/weltraum_en.html.
68. http://en.wikipedia.org/wiki/GISTDA.
69. http://en.wikipedia.org/wiki/NSAU.
70. http://www.nkau.gov.ua/nsau/nkau.nsf/indexE.
71. http://www.unoosa.org/oosa/index.html.
72. http://www.unoosa.org/oosa/index.html.
73. http://en.wikipedia.org/wiki/British_National_Space_Center.
74. http://en.wikipedia.org/wiki/NASA.
75. *USSTRATCOM Space Components.* U.S. Army Space and Missile Defense Command, Arlington, VA , Naval Network Warfare Command, Dahlgren, VA, and Air Force Space Command, Peterson AFB, CO.
76. http://space.au.af.mil/orgs.htm.

Index

D

E

F

G

H

I

J

K

L

M

N

S

U

V

W

X

Y

Z